PERIODIC OPERATION OF REACTORS

PERIODIC OPERATION OF REACTORS

Edited by

P. L. Silveston
Waterloo, Ontario, Canada

R. R. Hudgins
Waterloo, Ontario, Canada

ELSEVIER

AMSTERDAM • BOSTON • HEIDELBERG • LONDON • NEW YORK • OXFORD
PARIS • SAN DIEGO • SAN FRANCISCO • SINGAPORE • SYDNEY • TOKYO

Butterworth-Heinemann is an Imprint of Elsevier

BH

Butterworth-Heinemann is an imprint of Elsevier
The Boulevard, Langford Lane, Kidlington, Oxford OX5 1GB, UK
225 Wyman Street, Waltham, MA 02451, USA

First edition 2013

British Library Cataloguing in Publication Data
A catalogue record for this book is available from the British Library

Library of Congress Cataloging-in-Publication Data
A catalog record for this book is availabe from the Library of Congress

ISBN: 978-0-12-391854-3

For information on all Elsevier publications
visit our web site at books.elsevier.com

Printed in the United States of America

Transferred to Digital Printing, 2012

Contents

Preface ix
About the Authors xi

1. Introduction 1

ROBERT ROSS HUDGINS, PETER LEWIS SILVESTON, ALBERT RENKEN, YURII SH. MATROS

1.1. Periodic Operation 1
1.2. Origins of Periodic Operation 3
1.3. Variables in Periodic Operation 4
1.4. Cycle Structure in Periodic Operation 6
1.5. Measuring Improvement 8
1.6. Inherently Periodic Processes 9
1.7. Objectives of Periodic Operation 10
1.8. Strategies in Periodic Operation 10
1.9. Equipment for Periodic Operation 13
1.10. Reaction Systems Examined 16
1.11. New Directions 17
1.12. A Brief History of the Study of Periodic Operation 19

2. Hydrogenation Processes 23

ROBERT ROSS HUDGINS, PETER LEWIS SILVESTON

2.1. Ammonia Synthesis 23
2.2. NO_x Reduction 33
2.3. Methanation 34
2.4. Methanol Synthesis 39
2.5. Ethylene Hydrogenation 43
2.6. Aromatics Hydrogenation 45
2.7. Oscillatory Behavior 47

3. Catalytic Oxidation and Reduction of Gases 49

ROBERT ROSS HUDGINS, PETER LEWIS SILVESTON, ALBERT RENKEN

3.1. Introduction 49
3.2. CO Oxidation 50
3.3. Sulfur Dioxide Oxidation 66
3.4. Reduction of SO_3 by CO Over Platinum 70
3.5. Reduction of Nitrogen Oxides 70
3.6. Traveling Waves in Packed Beds 76

4. Partial Oxidation and Dehydrogenation of Hydrocarbons 79

ROBERT ROSS HUDGINS, PETER LEWIS SILVESTON, CHENG-YUE LI, ADESOJI A. ADESINA

4.1. Introduction 79
4.2. Partial Oxidation and Reforming of Methane to Syngas 80
4.3. Oxidative Coupling of Methane 88
4.4. Epoxidation 92
4.5. Propene and Butene Partial Oxidation and Ammoxidation 95
4.6. Catalytic Dehydrogenation of Propane, Butane and Higher Hydrocarbons 102
4.7. Maleic Anhydride from Butane 108
4.8. Anhydrides and Aldehydes from Aromatic Hydrocarbons 119
4.9. Aromatic Nitriles 121

5. Combustion Systems 123

PETER LEWIS SILVESTON, ROBERT ROSS HUDGINS

5.1. Non-Catalytic Combustion Reactions 124
5.2. Catalytic Combustion 127
5.3. Looping Combustion 134
5.4. Simulated Loop Reactors 134

6. Automotive Exhaust Catalysis 141

PETER LEWIS SILVESTON, WILLIAM S. EPLING

6.1. Internal Combustion Engines 142
6.2. Modulation of Detoxification Reactions 146

6.3. Modeling Studies 155
6.4. Studies on Modulating Automotive Exhaust 158
6.5. Effect of A/F Modulation on Poisoning and Sintering 160
6.6. Effects of Irregular A/F Variation 161
6.7. Lean Burn Spark-Ignited Engines 162
6.8. Application of NSR to Diesel Exhausts 168
6.9. Does A/F Modulation Improve Converter Performance? 169

7. Polymerization Under Modulation 171
GREGORIO R. MEIRA, PETER L. SILVESTON

7.1. Introduction 171
7.2. Simulation of Polymerization Under Input Modulation 177
7.3. Experiments on Polymerization Under Input Modulation 188
7.4. Spontaneous Oscillations 194
7.5. Saturation of Polymers 200
7.6. Assessment 201

8. Catalytic Gas-Solid Reactions 205
PETER LEWIS SILVESTON

8.1. Partial Oxidation and Oxidative Dehydrogenation of Hydrocarbons 206
8.2. Methane Cracking 209
8.3. Non-Catalytic Gas-Solid Reactions 209
8.4. Catalytic Gasification Under Modulation 216
8.5. Gasification Employing a Circulating Solid Oxygen Carrier 223
8.6. Combustion in Circulating Fluidized Beds 228
8.7. Periodic Reaction Switching 233

9. Electrochemical Processes 235
MARK PRITZKER, PETER LEWIS SILVESTON

9.1. Introduction 235
9.2. Electroplating 246
9.3. Electroforming 262
9.4. Anodization 263
9.5. Electrochemical Machining and Polishing 264
9.6. Electrowinning and Electrorefining 265
9.7. Galvanic Cells 266
9.8. Electrolytic Production of Chemicals 267
9.9. Applicability of Principles or Practices to Non-Electrochemical Reactions 268

10. Modulation of Biological Processes 273
JENO M. SCHARER

10.1. Introduction 273
10.2. Theoretical Considerations 275
10.3. Substrate and Flow Rate Modulation 280
10.4. Dissolved Oxygen Modulation 283
10.5. Culture Medium Tuning 287
10.6. Survival in Mixed Cultures 289
10.7. Stabilization of Recombinant Cell Cultures 292
10.8. Applications to Immobilized Cells or Enzymes 298
10.9. Fed-Batch Operations 303
10.10. Overview 312

11. Miscellaneous Reactions 317
ROBERT ROSS HUDGINS, PETER LEWIS SILVESTON, ALBERT RENKEN

11.1. Ethyl Acetate from Ethylene and Acetic Acid 317
11.2. Claus Reaction 324
11.3. Dehydrogenation of Methanol 325
11.4. Deamination and Alcohol Dehydration Reactions 326
11.5. Photocatalytic Degradation of AZO Dyes 332
11.6. The Minimal Bromate Reaction 332
11.7. Propanol Dehydrogenation 334
11.8. Glucose Oxidation 336
11.9. Overview 337

12. Modulation of Multiple Reactions 341
ROBERT ROSS HUDGINS, PETER LEWIS SILVESTON, ALBERT RENKEN

12.1. Introduction 341
12.2. Homogeneous Reactions 349
12.3. Solids Catalyzed Reactions 353
12.4. Competitive Reactions 359
12.5. Methane Homologation 361
12.6. Oligomerization of Ethene 365
12.7. Modulation of Multiple Inputs 366

13. Use of Modulation in Mechanistic Studies 369
ALBERT RENKEN, ROBERT ROSS HUDGINS, PETER LEWIS SILVESTON

13.1. Introduction 369
13.2. Qualitative Applications 370
13.3. Quantitative Applications 371
13.4. Modulation of Light Intensity 374
13.5. Application of Modulation to the Testing of Rival Models 375
13.6. Overview 385

14. Evaluation of Periodic Processes 387
MENKA PETKOVSKA, ANDREAS SEIDEL-MORGENSTERN

14.1. Introduction 387
14.2. Nonlinear Frequency Response and Higher Order Frequency Response Functions 389
14.3. Estimation of the Time Average Performance of Periodic Processes Using Nonlinear Frequency Response Analysis 393
14.4. Application of Nonlinear Frequency Response Analysis for the Estimation of the Periodic Steady States of Cyclic Processes 405
14.5. Summary 412

15. Pressure Modulation 415
PETER LEWIS SILVESTON, ROBERT ROSS HUDGINS

15.1. Introduction 415
15.2. Acceleration of Mass Transfer 416
15.3. Sonocatalysis 427
15.4. Periodic Pressure Reduction 430
15.5. Combined Compression and Reaction 431
15.6. Application to Rate and Equilibrium Measurements 432
15.7. Assessment and Research Opportunities 433

16. Temperature Modulation 435
JÜRGEN J. BRANDNER, PETER LEWIS SILVESTON, ROBERT ROSS HUDGINS

16.1. Introduction 435
16.2. Theoretical Studies 436
16.3. Simulation Studies 437
16.4. Experimental Studies with Conventional Laboratory Equipment 439
16.5. Temperature Modulation of Trickle Beds 444
16.6. Experimental Studies with Microreactors 444
16.7. Overview and Comments 462

17. Flow Interruption in Trickle Beds 463
PATRICIA HAURE, JIRI HANIKA, PETER LEWIS SILVESTON

17.1. Introduction 463
17.2. Steady-State Operation of a Trickle Bed Reactor 464
17.3. Periodic Operation of Trickle Bed Reactors 474
17.4. Liquid Flow Modulation with Multiple Reactions 481
17.5. Hydrodynamics Under Liquid Flow Modulation 486
17.6. Modeling of the Periodic Operation of Trickle Bed Reactors 487
17.7. Summary 493

18. Periodic Flow Reversal 495
GRIGORII BUNIMOVICH, HRISTO SAPOUNDJIEV

18.1. The Heat-Trapping Concept 496
18.2. Theoretical Aspects 500
18.3. Oxidation of Volatile Organic Compounds 511
18.4. Other Applications of Reverse Flow Reactors 521
18.5. Thermal Reactors 523
18.6. Endothermic Processes 528
18.7. Mass Trapping Reactors 534
18.8. Biofilters 537
18.9. Miscellaneous Applications 537
18.10. Commercial Applications 538

19. Control of Periodically Operated Reactors 543
HECTOR BUDMAN, PETER LEWIS SILVESTON

19.1. Formulation of an Optimal Control Problem for a Periodically Operated Reactor 545
19.2. Chattering Controls 546
19.3. Controls for Stirred Slurry and Fluidized Bed Reactors 548
19.4. Controls for Packed Bed Reactors 553

19.5. Control of Packed Bed Reactors with Flow-Direction Switching 553
19.6. Control of Simulated Moving Bed Chromatographic Reactors 562
19.7. Other Control Schemes for Periodically Operated Reactors 564
19.8. Comments and Research Needs 565

20. Chromatographic Reactors 569

PETER LEWIS SILVESTON, KENJI HASHIMOTO, MOTOAKI KAWASE

20.1. Introduction 569
20.2. Concept and Types 570
20.3. General Models 573
20.4. Cyclic Steady State 579
20.5. Pulse Chromatographic Reactor 580
20.6. Countercurrent Moving Bed Chromatographic Reactor 587
20.7. Continuous Rotating Annular Chromatographic Reactor 590
20.8. Stepwise, Countercurrent Multi-Stage Fluidized Bed Chromatographic Reactor 591
20.9. Fixed Bed Chromatographic Reactor With Flow Direction Switching 592
20.10. Extractive Reactor Systems 593
20.11. Centrifugal Partition Chromatographic Reactor 593

21. Simulated Moving Bed Chromatographic Reactors 597

KENJI HASHIMOTO, MOTOAKI KAWASE, PETER LEWIS SILVESTON

21.1. Operation and Application 598
21.2. Modeling and Simulation 599
21.3. Experimental Studies 620
21.4. Other Reactor Applications of Simulated Moving Beds 632

22. Pressure and Temperature Swing Reactors 637

ALIRIO RODRIGUES, PETER LEWIS SILVESTON

22.1. Concepts and Types of Pressure Swing Reactors 638
22.2. Models for Swing Reactors 640
22.3. Computational Considerations 642
22.4. Simulations of Pressure Swing Systems 644
22.5. Experimental Studies 652
22.6. Temperature Swing Reactors 657
22.7. Simulation of Temperature Swing Systems 660
22.8. Temperature Swing Reactor Networks 663
22.9. Experimental 663
22.10. Combined Pressure and Temperature Swing Reactors 671
22.11. Overview and Assessment 674

23. New Directions—Research and Development Challenges 679

KENJI HASHIMOTO, MOTOAKI KAWASE, MENKA PETSKOVSKA, JIRI HANIKA, CHENG-YUE LI, ADESOJI ADESINA, HRISTO SAPOUNDJIEV, JENO SCHARER, PETER LEWIS SILVESTON

23.1. Challenges 680
23.2. New Directions 686

References 691
Index 749

Preface

Interest in the periodic operation of reactors of all types, chemical, catalytic and biochemical, has expanded rapidly in the last several decades. The time has come to undertake an overview of not only the research activity, but its translation into engineering application. Such an overview, perhaps a summary, is the intent of "Periodic Operation of Reactors". It is our hope that this book will offer an introduction to the large literature, as well as a platform for evaluating both new and past contributions. As well, it may serve as a resource for any engineers considering utilizing non-steady-state operation in reaction processes they now operate. As authors, editors, and researchers, we trust that this book may provide direction for planned or on-going research. For this reason, several chapters and the final one in particular identify research needs and challenges.

Although steady-state operation appears to be an industry paradigm, it is one with many exceptions. A great many metallurgical processes take place under unsteady conditions. Steady-state operation is hardly the rule in biochemical systems, as will be evident in Chapter 14. Indeed, electrochemical processes have taken advantage of periodic operation (apart from just the use of alternating current) for more than 60 years, as can be seen in Chapter 9. Yet whether the standard is steady or transient operation is not relevant to the intent of this book. What we strive to demonstrate is that periodic operation can often enhance process performance.

Another important aim of this book is to recognize the pioneers, the innovators, and the synthesizers of process periodic operation. As authors we have sought to identify the important contributions to the development of periodic operations. Undoubtedly, some, regrettably, have been overlooked. In the years to come, advances will surely follow. It is our hope that in a decade or two, this book will be superseded by a superior overview.

This book has been organized to make it useful. Your benefit from our effort depends, of course, on what you are looking for. If you seek to build your general knowledge about the periodic operation of reactors, you should begin with Chapter 1 which presents an overview of the subject. If your interest is in a specific reaction, electrochemistry, a specific modulation process or in the combination of reaction and separation, only a few chapters need be consulted. If your interest is limited to a specific reaction or a group of reactions, look over chapters 2 to 12 or consult the Table of Contents or the Index. The latter is quite broad and many specific reactions are mentioned. If you want to examine the contributions of a specific researcher, see the bibliography. It gives the chapters in which the reference appears. Chapters 13 and 14 deal with theory, while Chapters 15 to 18 examine modulation of a specific operating condition. Chapter 19 discusses control while Chapters 20 to 22 consider combined reaction and separation. As a further convenience, the literature introduced in a chapter is summarized in a table early on. A good starting point for any reader would be to begin with such tables.

As editors, it is with pleasure and gratitude that we acknowledge the effort and scholarship the authors of the individual chapters have contributed to this book. They have also volunteered comments, argued interpretations or drawn attention to omissions, along with their written contributions. Both authors and editors wish to thank our institutional librarians for their help locating publications. In addition, we acknowledge the support of the secretarial staff at our various institutions. The editorial staff at Elsevier, particularly Fiona Geraghty and Pauline Wilkinson, have provided useful advice, services, and constant prodding towards deadlines. We are grateful for their engagement. Finally, we acknowledge the patience of spouses who have graciously accepted the constraints borne by those who would write books.

Peter Silveston, Robert Hudgins

About the Authors

Adesoji Adesina

Adesina obtained a PhD (1986) from the University of Waterloo, Canada. He taught at the University of Port-Harcourt, Nigeria, before joining the University of New South Wales, Sydney, Australia, in 1991 where he is presently a full professor leading a major research group in Catalysis and Reactor Engineering. His activities are primarily focused on energy and environmental themes. He is credited with over 350 technical papers and a book in metal carbide catalysts for Fischer-Tropsch synthesis. He is on the editorial board of Catalysis Communications and member of the International Advisory Board for Environmental Catalysis. He enjoys playing soccer.

Juergen Brandner

Currently heads the Thermal Processes Department and Heat Exchange Group of the Institute for Micro Process Engineering at the Karlsruhe Institute of Technology (KIT), Germany. He is also a lecturer for Micro Process Engineering at the Dresden Technical University. Brandner is author or co-author of some 260 technical articles and 7 technical books. After studies in Chemistry he obtained a Diploma in Electrical Engineering and an Engineering Doctorate in Mechanical Engineering, both from KIT, as well as a habilitation degree from the Dresden Technical University. Living close to Heidelberg, Germany, he also teaches several martial arts courses as recreation.

Hector Budman

Hector is currently a Professor of Chemical Engineering at the University of Waterloo, came to Canada from Argentina via Israel. He has authored or co-authored over 150 journal publications and conference papers in the areas of process modeling and control with applications to biotechnology and reaction systems. During his career he has undertaken collaborative work in the pharmaceuticals, water treatment, mineral processing and pulp and paper industries. He holds a Mechanical Engineering doctorate from the Israel Institute of Technology (Technion).

Grigori Bunimovich

Grigori is a key engineering specialist at Matros Technologies, Inc. St. Louis, Missouri. After graduating from the Technical University of Tomsk, he obtained a Ph. D. in Chemical Engineering from and worked at the Boreskov' Institute of Catalysis (Novosibirsk, Russia). In 1993 he joined Matros Technologies Inc. where he serves as the Director of Catalyst Applications. Grigori co-authored more than 50 technical publications concerning various applications of periodic reverse-flow reactors and catalytic processes for air pollution control.

William Epling

Bill Epling is an Associate Professor in the Department of Chemical and Biomolecular Engineering at the University of Houston where his research focuses on Environmental

Catalysis. He has authored over 80 peer-reviewed papers in archival journals and given over 100 presentations or seminars. He won the Early Researcher Award from the Province of Ontario in 2007. Bill graduated with a B.S. in Chemical Engineering from Virginia Tech and a PhD in Chemical Engineering from the University of Florida. Outside of work, Bill fosters stray dogs, pretending to find them homes although few seem to ever leave.

Jiri Hanika

Currently a "Director Emeritus" of the Institute of Chemical Process Fundamentals, Czech Academy of Sciences and a Professor of the Institute of Chemical Technology Prague (ICT), Jiří retired in 2012 after a 40 years carrier. However he still continue teaching, research and scholarly writing. He has written a number of technical and/or text books and co-authored several chapters in monographs as well as more than 150 technical papers. He is a graduate of ICT and holds an Engineering Doctorate and "Doctor of Sciences" degree from ICT as well. He has resided nearly his whole life in Prague where he regularly attends cultural and/or musical events.

Kenji Hashimoto

Currently a Professor Emeritus of Kyoto University, Kenji retired from the University in 1999, but he has continued teaching and research at Fukui University of Technology. He has published some 210 technical papers on chemical reaction engineering and chromatographic separation and written and /or edited 7 technical books. Kenji has served as the president of the Japanese Society of Chemical Engineering and received the highest Society award for his academic achievements. He is a graduate of Kyoto University and holds an Engineering Doctorate from that University. He now resides in Kobe, where he enjoys golfing and watercolor painting.

Patricia Haure

Patricia is currently a Professor of Chemical Engineering at the Universidad Nacional of Mar del Plata in Argentina. She graduated in 1981 from that University and in 1989 obtained a PhD from the University of Waterloo, Canada, where she pioneered research on the periodic operation of trickle bed reactors. Haure continues exploring the effect of liquid flow modulation on reactor performances and has expanded her research to detoxification of waste waters. Her many publications are on those subjects. She is a pet lover involved as a volunteer in rescue actions. She also enjoys painting, gardening and outside activities.

Robert Hudgins

A graduate of University of Toronto and Princeton University, Bob met Pete Silveston, his co-editor of this book, at University of Waterloo where they were faculty members in its fledgling Department of Chemical Engineering. A common interest in teaching and research on reaction engineering led to a close collaboration on several topics, primary among them an examination of catalytic reactors under periodic operation lasting over three-decades. Currently, Bob is a professor emeritus, living in Kitchener-Waterloo, Canada, where he also plays piano and rings bells, helps spouse Virginia in the garden, and (too seldom) visits his Vancouver-based grandchildren.

Motoaki Kawase

Motoaki is currently a Professor in the Department of Chemical Engineering, Kyoto University, where he supervises the Laboratory of Chemical Reaction Engineering. He graduated from that department in 1987 and later obtained an Engineering Doctorate from the University. He is an early pioneer of simulated moving bed chromatographic reactors. He has authored or co-authored about 100 technical publications or conference papers in the field of Chemical Reaction Engineering. He has lived in Kyoto since birth.

Cheng-Yue Li

Li recently retired from the Chemical Engineering Department of the Beijing University of Chemical Technology where he served as a Professor for some twenty years. He was also a member of the State Key Laboratory of Catalysis. With over 200 authored or co-authored journal publications, he is a Chinese pioneer of the study of reactor transient operation. Cheng-Yue and his wife reside in Beijing.

Yurii Matros

An engineer, a scientist and an entrepreneur, Yurii originates from Odessa in the Ukraine. He is currently president of Matros Technologies Inc. located in St. Louis, Missouri, a design and engineering company he founded in 1992. He completed an engineering degree at the Technical University in Odessa. After a few years in the Russian chemical industry, he joined the Boreskov Institute of Catalysis (Novosibirsk, Russia) where he spent 28 years and won a doctorate degree. He is a pioneer of the application of periodic operation, particularly periodic flow reversal, to chemical processes. He has written and/or edited 6 technical books, and authored or co-authored more than 300 technical papers.

Gregorio Meira

Greg Meira is a Chemical and Petroleum Engineer (University of Buenos Aires), and holds M.Sc. and Ph.D. degrees in Control Engineering (University of Bradford, UK). Since 1981, he has been a Full Professor at the Universidad del Litoral (Santa Fe, Argentina); and Director of the Polymer Group at the "Institute for the Technological Development of the Chemical Industry" (INTEC, CONICET), also in Santa Fe. He is an author of over 100 scientific publications and book chapters on the mathematical modeling and control of polymerization reactors, and on the characterization of polymers by size exclusion chromatography and other analytical techniques.

Menka Petkovska

Currently a full professor at the Department of Chemical Engineering of the Faculty of Technology and Metallurgy at the University of Belgrade, in Belgrade, Serbia. She received her Diploma, Master and Ph.D. degrees from the same department. During her 28-years long academic career, she was a visiting researcher in France, Germany and Australia. She teaches courses in the area of process dynamics and control and process intensification. She is an author or co-author of two textbooks and about 50 scientific and technical papers. Her main areas of scientific interest are: nonlinear dynamics and non-stationary processes, adsorption and process intensification.

Mark Pritzker

From Montreal, Mark is a McGill University graduate and holds a Ph.D. from Virginia Polytechnic Institute (1984). After a few years at the University of Arizona, he joined the University of Waterloo where he is now a Full Professor directing research on a wide variety of electrochemical systems. Currently he works on high frequency pulse electroplating, modeling and experimental studies of PEM and SOFC fuel cells, use of electrochemical methods for nano materials fabrication and on process modeling of leaching and adsorption, and strand board manufacturing processes. Mark is the author/co-author of more than 100 technical publications.

Albert Renken

Albert obtained his academic degrees at the Technical University of Hanover, Germany. In 1977 he became Professor of Chemical Reaction Engineering at the Swiss Federal Institute of

Technology, Lausanne. He is currently an Honorary Professor at that institution. He is author of over 450 scientific publications, numerous patents and author / editor of 5 technical books. From 1992 to 2000, he was a science councilor of the Swiss National Science Foundation, and from 1996 to 2002 he served as chairman of the European Federation of Chemical Reaction Engineering Working Party. He is also member of the Swiss Academy of Engineering Sciences. Albert is widely recognized as a pioneer of the application of periodic operations to chemical processes.

Alirio Rodrigues

Currently a Full Professor at the Chemical Engineering Department of the University of Porto in Portugal, Alírio is the Director of the Laboratory of Separation and Reaction Engineering (LSRE). Over a 40+ year career, he has supervised over 50 PhD students and edited and/or written 7 technical books and authored or co-authored some 450 technical papers and 5 patents. He is a graduate of University of Porto and holds an Engineering Doctorate from the University of Nancy (France). His research focuses on cyclic adsorption/reaction processes, perfume engineering and microencapsulation, CO_2 capture and utilization and lignin valorization.

Hristo Sapoundjiev

Hristo is a senior researcher at CanmetENERGY, the Natural Resources Canada Research Centre in Varennes, Quebec, and adjunct professor at the University of Alberta, Canada. He has authored or co-authored some 75 technical papers and he holds about 15 patents on development of environmentally friendly and energy saving technologies. He is a graduate of Sofia Technological University (Bulgaria) and holds an engineering doctorate from the Boreskov Institute of Catalysis, Novosibirsk, Russia. Hristo received the prestigious Canadian Public Service Award for Excellence in 2002 and Natural Resources Canada Departmental Merit Award in 2001.

Jeno Scharer

Jeno (BSc, PhD, Chemical Engineering, University of Pennsylvania) is a Professor Emeritus in the Department of Chemical Engineering, University of Waterloo, Canada. Although retired in 2004, he continues research that combines engineering , biological and biochemical concepts to the study of biological processes ranging from mammalian cell culture, to recombinant microbial fermentations for producing therapeutic agents and to diagnostic tools for the microbial oxidation of sulfide minerals or acid mine drainage. He has co-authored a large number of research papers as well as two books. For leisure, he trains championship dogs.

Andreas Seidel-Morgenstern

Andreas studied Process Engineering at the Technische Hochscule Leuna, graduating in 1982; his doctorate in science is from the German Academy of Science in Berlin (1987) with a later Habilitation in 1994. Since 1995, he has been a Professor of Chemical Process Engineering at the University of Magdeburg. In addition, he heads a large research team on the physical and chemical foundations of process engineering for the Max Planck Institute in Magdeburg. Andreas is now Managing Director of the Max Planck Institute for Dynamics of Complex Systems. He is the author/coauthor of over 350 journal papers and editor/coauthor of 6 technical books. His special interests are in system dynamics, chromatography and reaction + separation.

Peter Silveston

Currently a "Distinguished Professor Emeritus" of the Chemical Engineering Department of the University of Waterloo in Canada, Pete retired in 1997 after 35 years, but he has continued teaching, research and scholarly writing. Over a 50+ year career, he has written and/or edited 4 technical books and authored or co-authored some 280 technical papers. He is a graduate of M.I.T. and holds an Engineering Doctorate from the Technical University of Munich (Germany). He currently resides on the Carolina coast in the U.S. where he indulges in hiking, bicycling and water sports.

CHAPTER

1

Introduction

Robert Ross Hudgins, Peter Lewis Silveston*, Albert Renken†, Yurii Sh. Matros***

*Waterloo, Ontario, Canada, †Lausanne, Switzerland and **Springfield, Missouri, U.S.A

OUTLINE

1.1 Periodic Operation 1

1.2 Origins of Periodic Operation 3

1.3 Variables in Periodic Operation 4

1.4 Cycle Structure in Periodic Operation 6

1.5 Measuring Improvement 8

1.6 Inherently Periodic Processes 9

1.7 Objectives of Periodic Operation 10

1.8 Strategies in Periodic Operation 10

1.9 Equipment for Periodic Operation 13
- *1.9.1 Laboratory Reactors* 13
- *1.9.2 Industrial-Scale Equipment* 14

1.10 Reaction Systems Examined 16

1.11 New Directions 17
- *1.11.1 Applications to Systems with Multiple Objectives* 17
- *1.11.2 Control of Carbon Deposition* 17
- *1.11.3 Modulation of Multiple Reactor Inputs* 18
- *1.11.4 Improvement of Photocatalysis* 18

1.12 A Brief History of the Study of Periodic Operation 19
- *1.12.1 Pioneers* 19
- *1.12.2 Research Groups* 20
- *1.12.3 Industrial Applications* 21

The objective of this monograph is to summarize the large and growing literature on the periodic operation of chemical reactors. Research on such reactors is worldwide and the subject continues to attract new investigators, even though penetration of the technology into industry has been limited primarily to flow direction switching, simulated moving bed chromatographic reactors and swing reactors.

1.1 PERIODIC OPERATION

Periodic operation is a technique for enhancing the performance of a catalytic reactor, be it higher reactant conversion, better selectivity to a desired product, longer catalyst life or lower energy consumption. It involves periodically manipulating reactor inputs or the reactor environment. For example, composition or temperature of the feed to the reactor might be switched between two levels, as illustrated in **Figure 1-1**. This type of operation seems best suited to heterogeneous systems in which the catalyst is placed in a bed with reactants percolating through it, but examples for homogeneous, noncatalytic reactors are known (*Renken, 1972*). Simple cycles have been used in most research studies so far with just two composition or temperature levels employed. Some applications have combined separation and reaction so that more complicated cycles are involved.

Figure 1-1 introduces the term "modulation," the main term that will be used to describe simple periodic operation. However, the term cycling and occasionally forcing or periodic forcing will also be used. In all cases, this means one or more reactant concentration and/or temperature entering the reactor vary with time but in such a way that each "state" of an input is revisited after a time τ or τ_{cycle} corresponding to the period. A result of the manipulation shown in the figure is that most reactor outputs are modulated as well. Usually, it is not the instantaneous values of the outputs but rather their mean or time-averaged values that are of interest. It has been widely observed that over a range of modulation frequencies these means differ from outputs obtained under steady-state operation even when the reactor functions at the respective means of the inputs.

"Modulation" is not used for flow direction switching, a technique for energy saving or reactant recovery that is necessarily periodic. Indeed, work on this technology at the Boreskov Laboratory for Catalysis (Russia) beginning in the late 1960s (*Boreskov et al., 1977a; Boreskov and Matros, 1983a; Matros and Bunimovich, 1995*) initiated the present interest in periodic operations. Periodic flow direction switching is also referred to as periodic flow reversal.

Also considered in this monograph are other processes that are inherently periodic, such as adsorption or catalyst regeneration. In adsorption, a solid becomes fully or partially saturated after an interval of contact with a fluid phase and the adsorbate must be recovered or removed before a subsequent cycle commences. In catalytic processes, fouling residues or poisons must be stripped from the catalyst to maintain operation. Combined reaction-separation processes, considered in **Chapters 20 to 22**, are necessarily periodic. However, strictly adsorption processes, many of which are also periodic, will not be discussed. There is a large literature on those processes and current science and technology has been ably summarized (*Ruthven, 1984; Suzuki, 1990; Crittendon and Thomas, 1998*).

Although catalyst regeneration is indeed a periodic process, limits must be placed as catalyst regeneration is an enormous topic. Somewhat arbitrarily then, examination of periodic

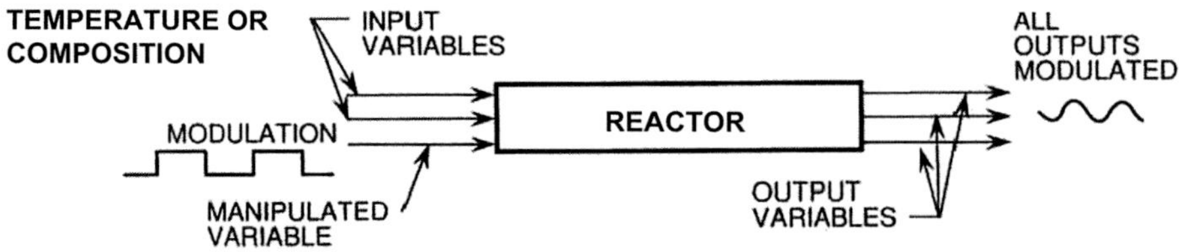

FIGURE 1-1 **Schematic of periodic reactor operation.** *(Figure reproduced from* Silveston (1998) *with permission.)*

systems will be restricted to those whose maximum cycle time or period is of the order of an hour. Catalytic cracking will be bypassed, even though it falls within the time definition of a periodic operation, because this subject has been so thoroughly treated in the literature (see, e.g., *Magee and Mitchell, 1993; Occelli and O'Connor, 1998; Sadeghbeigi, 2000; Occelli, 2004, 2007*). The other extreme of milli or microsecond cycling will also be left out. This is the realm of acoustical stimulation or sono-catalysis. A large literature also exists in this field (e.g. *Mason, 1997; Mason and Peters, 2002; Mason and Lorimer, 2002; Nowak, 2010*).

The intent in this chapter is to present the concept of periodic operation. This includes objectives, potential variables, measurements of the operation, equipment used, cycle structure and strategies. For background, the historical development of this reactor application will be outlined. A few examples of applications will be given.

1.2 ORIGINS OF PERIODIC OPERATION

Periodic operation is not a new technique in Reactor Engineering. Repeated operation of a batch reactor over a duration of time is essentially a periodic operation. Batchwise operation, of course, goes back to the beginning of the discipline; indeed, to the origins of Chemical Engineering.

Any reactor in which the catalyst is regenerated in situ operates periodically. Even those reactors whose catalyst charge is removed for regeneration or replaced periodically share many similarities with periodic operation. Normally, time periods encountered for systems involving catalyst regeneration are measured in days, in months or even in years. In such cases, the reactor operates at steady state or quasi-steady state on the scale set by the residence time of reactants in the system. By quasi-steady state is meant a system that changes so slowly that the change is negligible on the time scale being used. Not all catalytic systems employing regeneration operate with long time periods. Catalytic cracking of gas oils, perhaps the largest regenerative systems in terms of production volume, is carried out using periods of the order of minutes. Of course, carrying out cracking in a fluidized bed becomes a steady-state operation for reactants and products. Only the catalyst particles experience a periodic environment as they move between reactor and regenerator.

In what follows, reaction systems are considered that are transient on the scale of residence times. The purpose of periodic operation discussed in this monograph differs from periodic catalyst regeneration in that it is intended to achieve performance that is either extremely expensive or impossible to attain under steady-state operation, whereas the intent of regeneration is simply to restore performance. With reference to separating reactors discussed in the last chapters, periodic operation enables the separation function.

Periodic operation is also encountered in other fields of Process Engineering. Heat recuperators have been used in steel making as well as other processes for well over a century. In steel making, these systems employ stacks of bricks to recover heat from stack gases, heat that is used to raise the temperature of air entering a furnace.

In the last two decades, periodically operated separating reactors (also called combined reaction-separation systems) have appeared in industry. Such reactors fall into two broad groups: 1) Chromatographic reactors in which a fluid stream desorbs products and unconverted reactant from solid surfaces and 2) swing reactors, in which pressure or temperature change is employed for desorption. More recently, systems operating with both pressure and temperature changes have been explored.

1.3 VARIABLES IN PERIODIC OPERATION

Composition modulation has been and remains a widely studied type of periodic operation. **Figure 1-2** illustrates the operation. The figure also introduces the manipulated variables that define composition modulation: Period (τ or τ_{cycle})—the time between repetitions of a change in an input condition, split (s)—the duration of a partial cycle relative to the period, amplitude (A)—the change in the value of an input condition from its mean, and the mean composition. The split, sometimes referred to as the duty fraction, measures the symmetry of a cycle. A value of $s = 0.5$ indicates a symmetrical cycle with each part of equal duration. For symmetrical cycling, these parts are commonly referred to as half-cycles. The split must be defined relative to one of the reactants and the convention is that it measures the relative duration of the partial cycle in which that reactant is at its highest concentration. Amplitude has just a single value for symmetrical forcing, but, if $s \neq 0.5$, two amplitudes must be given, one for each partial cycle. Mean composition is also a variable for composition forcing, just as it is when a reactor operates at steady state. A further variable possible in composition modulation, but not illustrated in **Figure 1-2**, is phase lag. The composition changes shown in the figure are 180° or π radians out of phase. Other phase lags could be used. In a pulsed operation, phase lag is not a variable.

Two partial cycles shown in the figure are possible when just one reactant is modulated or, if two reactants are modulated, only when the phase lag is 0 or 180°. Other phase lags, multiple amplitudes, or variation of an additional feed component lead to three or more

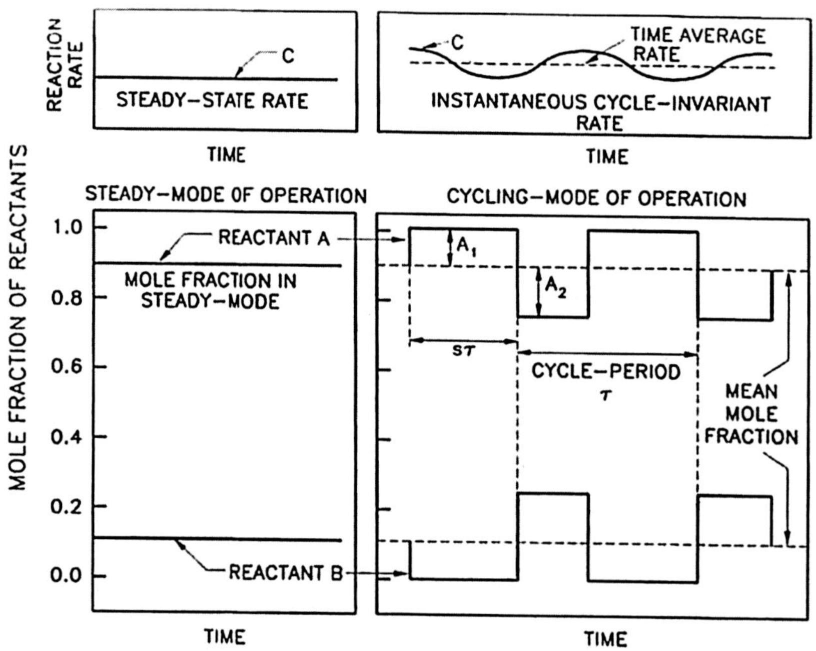

FIGURE 1-2 **Comparison of steady-state (left side) and periodic (right side) operation showing definition of the cycling variables: cycle period (frequency),** τ**, cycle split (duty fraction),** s**, amplitudes,** A_1**,** A_2**.** *(Figure reproduced from* Silveston (1998) *with permission of the author.)*

partial cycles. Most research to date has dealt with two-part cycles and no phase lag. In such cases for those shown in the above figure, flow rate or residence time in the reactor does not change within a cycle. If just one reactant is modulated or the phase lag differs from 0 or 180°, another process variable must change. Usually, this variable will be the flow rate.

Cycle period or its inverse, frequency, is a manipulated variable in almost all cycling applications, but this cannot be said of the other variables shown in **Figure 1-2**. Even when some particular variable can be used, another variable might be a better choice. For example, split is not the appropriate variable in pulsing. By this term, a change of short duration but large magnitude is meant. The variable for pulsing should be pulse length or duration. Similarly, amplitude should be replaced by pulse magnitude: For example, a flow rate, or the reactant concentration in the pulse. Phase lag becomes a variable only when more than one input is changing with time. Time-average feed composition is a variable only in a periodic mode; it is usually not a variable for an isolated pulse. Even cycle period becomes a restricted variable in the extreme modes of periodic operation, i.e., in the relaxed and quasi-steady-state modes. In the former, frequency is so high that most reactor outlet variables, including the product composition, are invariant, while in the quasi-steady state, the cycle period is so much longer than the relaxation time after a variable change that steady state is maintained through most of the half-cycle. In such cases, choice of period is important only to insure that pulse or quasi-steady-state mode of operation is established. In both extremes of periodic operation, performance is independent of period.

When there are two components and the phase lag is not held at 0 or 180°, a cycle will have three or more parts. Stated differently, composition forcing can have two cycle periods and two cycle splits, one for each component. Even when the phase lag is 180° out of phase, cycling must be described by two amplitudes if $s \neq 0.50$. When the number of components becomes greater than two, multiple partial cycles become the general rule and the number of variables that must be considered increases rapidly. There have been no simulations or experimental investigations of systems having more than three time-changing inputs.

Other choices also exist in reactor input modulation, such as the choice of input(s) to be modulated. Here a distinction can be made between strategy and operating variables. Strategy refers to the choice of input(s) or mode of variation or even cycle structure. Inputs could be inlet concentration, system pressure, feed or coolant temperature, or flow rate to name the obvious ones. Mode of variation could be pulse, square wave or sinusoidal. Cycle structure could be symmetrical or asymmetrical. Two amplitudes or multiple amplitudes of a variable could be used. Once the strategy is specified, operating variables become cycle period, split, phase lag and amplitude.

Most of what will be discussed in the early chapters deals with concentration modulation or forcing. Another strategy that may be used is flow modulation. An example of such a strategy is "stop flow" in which a reactant is periodically removed from the feed. This mode of operation has been employed for reactions strongly inhibited by one reactant (*Thullie and Renken, 1991*). Modulating just the flow of one reactant, however, forces flow rate and residence time to change, unless flow rate of another reactant or a diluent is also modulated for compensation.

Modulation of the total pressure is also possible for gaseous reactants. This strategy has been applied to increase transport rates in porous catalysts by inducing convective flow in macropore structures (*Silveston and Hudgins, 2004a*). This operation causes simultaneous changes in flow rate and residence time within the reactor. Temperature modulation has also been examined (*Silveston and Hudgins, 2004b*).

Changing the temperature of the feed is not effective because of the very large difference in the heat capacity of a gas flowing through a bed of catalyst and that of the catalyst. Even with liquids, for which heat capacities are much higher, it is difficult to achieve abrupt changes in the catalyst temperature. Thus, temperature modulation has been investigated by using heating coils placed within the catalyst bed (*Abdul-Kareem et al., 1980a*) and by using microstructured catalytic reactors (*Luther et al., 2008a, b*). Again, abrupt changes in temperature cannot be realized because of transport processes. Results using periodic temperature forcing will be discussed in **Chapter 16**. As with other variables, temperature variations initiate variations in concentration and residence time for gaseous reactants.

Periodic flow reversal, unlike other modes mentioned so far, has seen significant industrial exploitation (e.g., *Bunimovich et al., 1990a*). Flow reversal with an exothermic reaction traps the heat evolved in a bed of solid catalyst and permits this heat to be used to raise the reactant temperature to a point at which ignition can take place. An important application has been to the combustion of dilute volatile organic compounds (VOCs) for pollution mitigation. Many hundreds of installations are now running worldwide.

Other process variables can be manipulated in special processes. A few catalysts, such as TiO_2, can be activated by illumination at a specific wavelength. It might be useful to investigate whether intermittent illumination is better than continuous or whether variation of wavelength or intensity stimulates catalyst activity. In some processes, catalyst promoters are continuously added to the reactor feed. Could intermittent feeding achieve a higher rate or selectivity or could it reduce the quantity of promoter needed? Velocity variation in packed beds holds little promise, but flow has a large effect on fluidization through bubble and mixing phenomena. Perhaps, flow modulation in fluidized beds should be explored. This manipulation, referred to as base-peak modulation, has been utilized for catalytic trickle bed reactors (e.g., *Silveston and Hanika, 2004*). Much more research has been done on periodic liquid flow interruption, also termed on-off modulation (e.g., *Haure et al., 1989*). This topic is the subject of **Chapter 17**. These other manipulated variables, however, will not be discussed in what follows for purposes of brevity.

With separating reactors, variables are step duration, choice of carrier fluid, and flow rate of that fluid for moving-bed chromatographic reactors, whereas for swing reactors, additional variables must be considered. These are the number of steps in the cycle, pressure differences at the reactor entrance and exit ports, and the use of different temperatures in the flushing steps. **Chapters 20 to 22** deal with these types of operation.

1.4 CYCLE STRUCTURE IN PERIODIC OPERATION

Different cycle structures may be used in periodic operations. The simplest types are characterized by the split variable, *s*. There is a uniform division of a cycle as illustrated in **Figure 1-2** in which duration of the partial cycles are the same ($s = 0.5$). This cycle, identified as a symmetric modulation, has been well studied. Other two-part cycles are asymmetric. The extreme asymmetric cycle is periodic pulsing, which will be referred to as a pulse mode of operation.

Feed conditions may remain constant during a partial cycle or one or more feed variables, such as reactant concentration or space velocity, could change with time. The periodic operations literature has considered just the first option up to now. Similarly, only two-part cycles have been reported, although, in principle, multi-part cycles are conceivable. For example, a cycle could be built up from three parts, each with a different feed composition or a different reactor temperature. Such a complex cycle would be interesting, perhaps, with catalysts whose activity varies strongly with feed composition.

A distinction between types of periodic operation may also be made with respect to the characteristic response or relaxation time of the catalytic reaction or reactor that is being periodically forced, where relaxation time, t_c, is the time required to closely approach steady state after a disturbance. One extreme would be cycles with long periods with respect to the relaxation time, that is, $\tau >> t_c$. At this extreme, the reactor remains essentially at steady state. The transient portion of a cycle, roughly 2 t_c, is a negligible portion of the cycle period. Referring to **Figure 1-2**, $\bar{r}$ becomes the time average of the steady-state rates for the two feed concentrations shown in the figure. Even though the operation is periodic, the time-average reaction rate, $\bar{r}$, is uniquely determined by the steady-state behavior of the reaction or reactor. The term "quasi-steady state" (qss) is commonly used for this mode of operation.

At the other extreme of very short cycle periods, two situations can arise. If mixing occurs, as it does in all real situations, and the characteristic time of mixing is shorter than the cycle period, time variations in an input, such as a reactant concentration, will be smoothed out. In this extreme, the reaction or reactor behaves as though it is at steady state at a condition represented by the time average of the forced input. This situation arises with recycle reactors or reactors in which large-scale backmixing takes place.

If mixing is negligible, i.e., for systems that approach the plug flow limit, and $t_c >> \tau$, the "relaxed steady state" (rss) mode arises. The catalyst and transport to the catalyst surface are no longer able to keep pace with the passing composition variation. Catalyst surface and product concentrations then become time-invariant, but the state is often different from that seen with a continuous steady feed. The relaxed steady state is used in analytical treatments of composition modulation [see, for example, *Bailey (1973, 1977)*], but is reported only rarely in experimental studies. Reaction rate at the relaxed steady-state limit will be represented by $(r)_{rss}$. Both $(r)_{qss}$ and $(r)_{rss}$ are asymptotic conditions for periodic forcing.

Between the relatively fast and slow cycling extremes lies the region where $\tau \approx t_c$. It is on this region that most research focuses. Even for this cycling mode, several different types of operations can be distinguished. Consider the common situation in catalytic reactors where two vastly different characteristic response times exist—one associated with response of the surface (adsorbate concentrations, surface state, temperature) to changes in reactant concentration or reactor temperature and the other associated with catalyst deactivation. Response times of the first type are typically measured in seconds or at most in minutes, whereas those associated with deactivation may be measured in weeks or months of time-on-stream. One of the operations in the $\tau \approx t_c$ region is a process with significant decay of catalyst activity. Once fouling or surface poisoning has brought catalyst activity down to a low fraction of its initial level, the catalyst is regenerated or replaced and the cycle begins again. Clearly, the period, that is, the time between regeneration or replacement, is about equal to the characteristic response time for deactivation. *Bailey (1973)* has labeled such cases "process life cycle" operations. With respect to the first characteristic response time, the process life cycle is much longer than this response time so that at any instant, if temperature or feed composition is undisturbed, the reaction system can be treated as though it operates at steady state. The steady-state assumption is standard for models of catalytic reactors. Process life-cycle systems are not considered in this monograph.

The two modes of periodic operation that arise in the $\tau \approx t_c$ region have been introduced already but in a somewhat different manner. The first of these is shown in **Figure 1-2**. In this mode, the changes in reactant concentration (or volumetric flow, if **Figure 1-1** is considered) are 180° out of phase. While reactant "A" increases in concentration, reactant "B" decreases. This is a "standard" cyclic mode. Phase lags other than 180° are possible as has been mentioned. Several investigators have used a mode in which the concentration or flow changes have a phase angle of zero. Usually the duration of reactant flow to the reactor is short compared with the cycle period and this is referred to as a pulse mode or as periodic pulsing. If the parts of the cycle are about equal, it is termed a stop-flow mode of operation. **Figure 1-3** illustrates feed to the reactor in stop-flow for a reaction system: **A + B → C**. For some systems, if a reactant is just weakly adsorbed, only the flow of the strongly adsorbed reactant will be interrupted.

The pulse mode has attracted considerable research attention because it offers mechanistic insight — see, for example, *Hattori and Murakami (1968)*. The mode is also interesting because chromatographic effects can arise leading to significant yield and/or selectivity improvement when equilibrium limitations exist.

Cycle structure with separating reactors is in all cases more complicated than those illustrated above. A bed of mixed catalyst and adsorbent in simulated moving-bed chromatographic reactors (SMBCRs) undergoes a three- to four-step sequence in a cycle because the ports at the entrance and end of the bed take on different functions. In a typical SMBCR, there is a step in which a reactant is added. This step will be followed by one in which the raffinate (the strongly adsorbed component) is removed at the exit of the bed and then by a step in which desorbent or eluant is added at the front of the bed and finally by the removal of an extract (the weakly adsorbed component) from the end of the bed. The duration of each of these steps is the same. With swing reactors, the situation described above arises with the exception that the duration of each step can be different.

1.5 MEASURING IMPROVEMENT

Comparison of periodic and steady-state operation depends, of course, on the objective

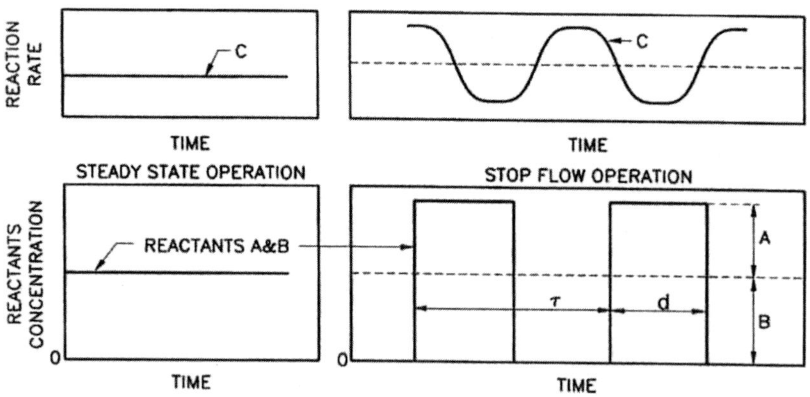

FIGURE 1-3 **Comparison of stop-flow and steady-state operation. In the latter, during the portion τ - d of a cycle, only diluent flows through the reactor. The upper part of the figure shows the rate of formation of product C. Variation of reactant A and reactant B concentrations appear below. Concentrations are taken to be the same.** *(Figure reproduced from Silveston (1998) with permission.)*

of periodic operation. If the objective is enhanced reactor performance, Ψ may be used to represent enhancement under periodic operation:

$$\Psi = \bar{r}/r_{ss} \qquad (1\text{-}1)$$

where $\bar{r}$ is the time-average rate of a modulated reaction and r_{ss} is the rate at steady state for a reactant composition corresponding to the time-average composition under cycling. Enhancement can also be defined in the same way for selectivity, for product concentration or for yield.

Enhancement with respect to the steady-state rate at the time-average feed composition may not be the most significant comparison. It can be argued that if the feed composition can be changed (as it is in modulation), it should be possible to find a composition that will improve reactor performance through steady-state operation. The meaningful comparison for composition modulation, then, is the maximum rate under steady-state operation when all other operating conditions (temperature, pressure, space velocity, etc.) are held constant. This comparison will be referred to as the global enhancement and defined as:

$$\Psi' = \bar{r}/(r_{ss})_{\max} \qquad (1\text{-}2)$$

where the denominator is the maximum rate achievable at steady state at any composition with space velocity, temperature and pressure of the system held constant.

Another result of periodic operation might be an extension of useful catalyst life. An enhancement factor can be defined for such an objective in the same way as the factor for rate of reaction. Such is not always the case, however. Modulation might reduce energy demand. In this case, as well as in those just mentioned, a "relative" improvement measure can be used:

$$\psi = \frac{\bar{r} - r_{ss.}}{r_{ss}} \qquad (1\text{-}3)$$

Measurements mentioned above may not be appropriate for separating reactors; indeed, the operations are so different that simple ratio comparisons are difficult to interpret. Often, a comparison between use of a separating reactor and of a more conventional reactor plus a separations chain must be made using economic criteria such as operating or cost per unit of product.

1.6 INHERENTLY PERIODIC PROCESSES

A distinction should be made between reactions that oscillate naturally and those combined with a separation step in which the separating agent, such as an adsorbant or extractant, must be periodically replaced or regenerated for the process to continue. With respect to the former, *Nicolis and Prigogine (1977)* predicted in their book the possibility of spontaneous oscillation for systems far removed from thermodynamic equilibrium. A number of examples of such oscillations exist, the most famous of which is the Belousov-Zhabotinsky (BZ) reaction discovered in 1959 by *Belousov* and confirmed in 1964 by *Zhabotinsky (1964)*. In this reaction, transition metal ions such as Ce^{3+} and Ce^{4+} catalyze the oxidation of organic substances, e.g., citric and malonic acids, using a bromate ion that is reduced to bromide. Reduction reverses as organic matter is consumed, changing the Ce oxidation state and the solution color. The color changes during this process make observation of the oscillations visible in both time and space. Another example of such self-oscillating systems is the Briggs-Rauscher reaction (*Furrow, 1995*) involving an iodate ion and hydrogen peroxide. The catalyst in this case is a manganese ion. Both of these reactions have attracted a great deal of attention because they are easily performed and have visual appeal. The BZ reaction yields chemical waves that interact resulting in a variety of remarkable spatial patterns. The mechanisms of both of these reactions are complex and beyond the scope of this review. Suffice it to say that a negative feedback loop is involved that includes a time delay. Details are given in various sources—*Field and Noyes (1974)* among many others. There appears to be no commercial applications of these spontaneous oscillations.

Separating reactors are inherently periodic. **Chapters 20 to 22** are devoted to this class of reactors.

1.7 OBJECTIVES OF PERIODIC OPERATION

Why consider periodic operation when steady state would appear to be a much simpler way of running a catalytic reactor?

A simple answer is that periodic operation may sufficiently improve reactor performance to offset any higher capital and operating costs. Enormous possibilities are suggested by laboratory results. For example, in ammonia synthesis, a 1000-fold increase in the synthesis rate has been demonstrated for an Ru catalyst by rapid cycling between H_2 and N_2 (*Rambeau and Amariglio, 1981b*). The yield of acrolein in the partial oxidation of propene over a Sb-SnO catalyst can be doubled by switching between air and propene mixtures of different composition (*Silveston and Forissier, 1985*). Catalyst life can be extended by multi-step composition and temperature cycles (*Ruettinger et al., 2004*). Periodic flow direction switching permits inexpensive disposal of highly diluted hydrocarbons through combustion (*Matros and Bunimovich, 1995*). Moreover, periodic operation—specifically, composition modulation—allows investigation of the kinetics of a catalytic process and permits separation of model parameters that are lumped when determined from steady-state measurements (*Renken, 1990a*).

TABLE 1-1 Objectives in Studies of Periodic Operation

1. Greater Conversion of a Reactant per Unit Weight of Catalyst
2. Better Product Recovery
3. Higher Yield of a Desired Product
4. Higher Rate of Reaction
5. Improved Selectivity
6. Increased Heat Recovery
7. Higher Reactor Throughput
8. Longer Catalyst Life
9. Lower Reactor Temperature
10. Maintenance of Catalyst Activity
11. Prevention of Reactor Thermal Runaway or Reaction Extinction
12. Determination of Parameters in Reactor Models
13. Measurement of Mixing in a Reactor
14. Estimation of Reaction Kinetics

Benefits of periodic operation or periodically operated separating reactors often cited by researchers in their publications are summarized in **Table 1-1**.

1.8 STRATEGIES IN PERIODIC OPERATION

There are many ways of operating a reactor periodically. It is this richness of opportunity that makes it possible to find at least one way that achieves an objective listed above. On the other hand, this very richness presents a challenge: How to efficiently discover which of the various modes should be used. The terms "strategy" and "variables" were introduced earlier. By "strategy" the method or means of periodic forcing is meant, while "variables" are those variations that can be made in the operation without changing the method or means.

INTRODUCTION

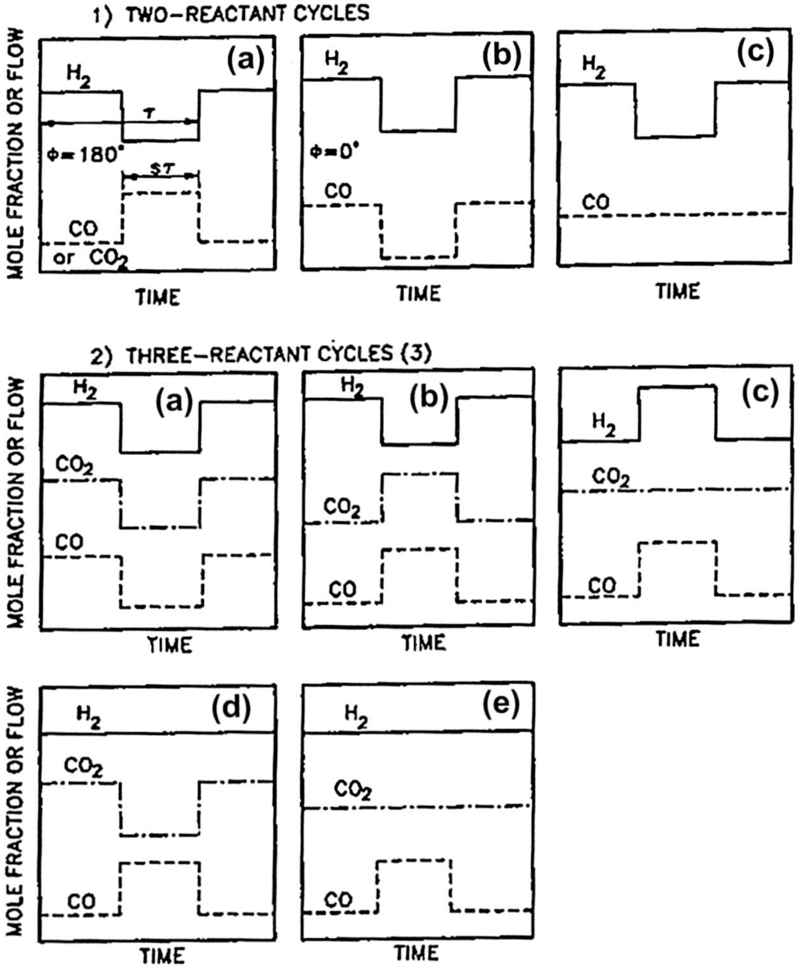

FIGURE 1-4 **Different possible composition modulation operations with two and three reactants for methanol synthesis.** *(Figure reproduced from* Chanchlani et al. (1994) *with the authors' permission.)*

Most reactor inputs are manipulated variables that can be forced periodically. The choice of input is part of the strategy. Reactants, promoters, or diluents are candidates. Usually, more than one input will be modulated. Often these will be the concentrations of reactants. In this situation, there are two strategy options. **Figure 1-4**, reproduced from a study of the application of composition modulation to methanol synthesis (*Chanchlani et al., 1994*), shows these in 1) "Two Reactant Cycles." They: (a) Vary both reactants, in this case y_{H2} and y_{CO}, simultaneously but out of phase, (b) vary both simultaneously but in phase, or (c) hold the concentration of one reactant constant, such as CO, and vary the other reactant concentration periodically, in this case y_{H2}.

If the reaction system has three reactants, some 13 different periodic operations are possible, reflecting three manipulation strategies. In 2) "Three Reactant Cycles," operations (a) and (b) represent modulation of all three reactants in phase or with one reactant modulated out of phase. Operation (b) shows only one of three possible combinations. Two reactants are forced and one is held constant in the operations shown by (c) and (d). There are six variants, depending on the reactant held constant and the two permissible phase lags of 0 and 180°. The third manipulation strategy is to force the concentration of one reactant and hold the remaining two reactant concentrations constant—see (e). There are three variants of this strategy.

In several of the operations in **Figure 1-4**, for example (b) in "Two Reactant Cycles" and (b) or (e) in "Three Reactant Cycles," both flow rate through the reactor and feed composition must vary together. If flow rate variation must be avoided, a further component, an inert, must be introduced and must vary periodically with time. Adding the inert component expands the strategies available, although holding the flow rate through the reactor constant limits those possible operations considerably.

Further strategies can arise when more than two components are modulated. Although only two-part cycles are shown in **Figure 1-4**, multi-part cycles are possible. Sometimes, a multi-part cycle is useful—for example, higher selectivity might be obtained by flushing the catalyst surface to desorb reactant or product before exposing the surface to a second reactant. *Park et al. (1983)* and *Lang et al. (1989b, 1991)* employed multi-part cycles in their investigations of the partial oxidation of C_4 hydrocarbons. **Figure 1-5**, reproduced from *Lang et al. (1991)*, illustrates 3-part cycles in (a) and (b), while (c) and (d) show 4-part cycles. The location of the N_2 flush differs in (a) and (b). Diluent flows to the reactor after each exposure to a reactant in (c) and (d). The latter illustrates a pulse mode in which a pulse of a reactant is followed by a long duration of diluent flow before the next reactant pulse.

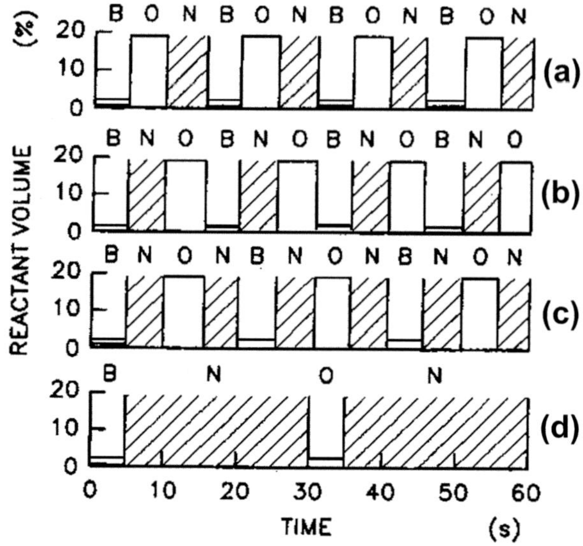

FIGURE 1-5 **Modulation strategies with 3 and 4-part cycles for partial oxidation of butadiene to maleic anhydride with and without diluent flushing: O = air, B = butadiene, N = diluent (N_2).** *(Figure adapted from* Lang et al. (1991) *with the authors' permission.)*

The manipulation strategies just discussed can be viewed as modes of operation, just like the relaxed and quasi-steady-state modes introduced earlier. **Figure 1-4** represents switching between reactant mixtures. Alternating between single reactants and a diluent is shown in **Figure 1-5**. All of these examples, except (d), represent a periodic mode. The scheme shown in (d) would be a pulse mode.

Selecting a modulation strategy requires choosing the variables to be manipulated and the mode of operation. It is just a point of view in differentiating between strategy and operating variables. Are relaxed, periodic and quasi-steady state really strategies? The difference is only the choice of cycle period. Similarly, the difference in switching between mixtures and a single reactant is the amplitude, which again is a variable. We use "strategy" in this context because the behavior of the reaction system may differ dramatically when the type of modulation is changed.

With periodically operated separating reactors, strategy enters into the design of the cycle as well as cycle operation. Important now are: choice of carrier fluid, ratio of catalyst to adsorbent, arrangement of catalyst and adsorbent within a single bed (it is not necessary to have the solids phase well mixed), and duration of the separate steps in a cycle.

1.9 EQUIPMENT FOR PERIODIC OPERATION

Periodically altering flow rates is quite straightforward and needs no comment. For temperature or reactant composition in a reactor, the alternatives are to change reactant feed rates or temperature or to use two reactors each having continuous feeds of different composition or temperature and continuously transporting catalyst from one reactor to the other. Composition or temperature switching using a single reactor is simpler to build and operate so such systems are widely used in laboratory investigations. A two-reactor design with a circulating catalyst, however, may be favored on an industrial scale. Equipment for periodically operated separating reactors is too complicated for this introduction and will be deferred to **Chapters 20** to **22**.

1.9.1 Laboratory Reactors

Studies of composition modulation reported in the literature invariably use valving to periodically change the composition of the feed flowing to a reactor. Valve networks are easily fabricated and inexpensive. **Figure 1-6** shows a lab-scale reactor employed to investigate composition modulation by a University of Waterloo team. This reactor was made from 1 cm (o.d.) steel tubing. The catalyst was supported on a steel screen held in place by a fitting

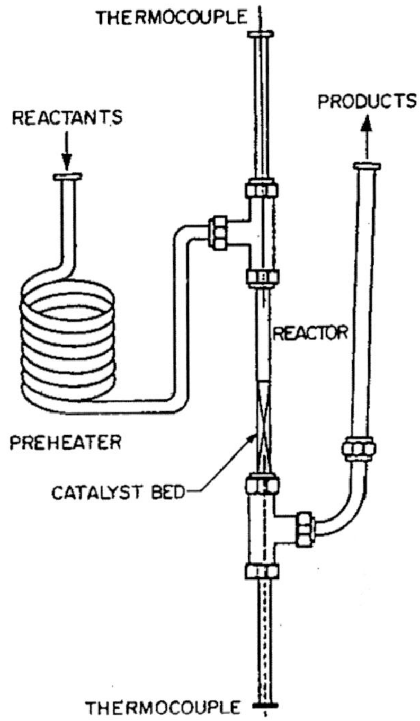

FIGURE 1-6 Schematic of a laboratory reactor. *(Reproduced from* Chanchlani et al. (1992) *with the authors' permission. © 1992, Elsevier Science Publishers, Ltd.)*

at the bottom of the tube. A screen was press-fitted on top of the bed to prevent disturbance of the catalyst when the reactor was moved. Prior to pressing the top screen in place, the tube was rapped to settle the catalyst bed to its maximum random-packed density. The figure shows a separate coil for preheating reactants to within 3 to 5 K of the reaction temperature. Extending the reactor tube and/or filling the tube above the catalyst with ceramic or steel beads can also be used. However, this procedure can result in reactor-preheater lengths greater than 15 cm that may exceed the depth of many commercial heating baths or furnaces. Generally, a coil preheater provides a more compact design as the coil can be wound around the reactor. Two thermocouples are shown in **Figure 1-6**. The upper is fixed above the bed and measures the temperature reached in the preheater. The second thermocouple penetrates the bed from the bottom and passes through the supporting screen. It measures the bed temperature. Catalyst bed depths in these tube reactors range from 1 to 6 cm. The depth depends on catalyst activity at the reaction temperature. *Silveston (1998)* discusses various types of laboratory reactors used in studies of periodic operation.

Since the 1990s, technical advances have made miniaturization of laboratory reactors possible. These "microreactors" have made temperature modulation of a catalyst bed feasible by sharply reducing the mass of the reactor enclosure. They also have the potential to dramatically increase the ease of performing composition and flow modulation experiments. **Figure 1-7** illustrates the construction of a microreactor. This reactor is so small that it fits inside a laboratory T-junction. Many other designs exist and are discussed in **Chapter 16**.

1.9.2 Industrial-Scale Equipment

A single reactor system for periodic operation is not well-suited for large-scale use because of problems associated with rapidly opening and then tightly shutting large valves. On the downstream side of the reactor, a hold-up vessel would usually be needed to smooth out composition variation forced on the reactor. Two reactors simplify the design. Feed streams could then be periodically switched between the reactors so problems of sudden starting and stopping of flow are avoided.

Circulating catalyst between the two reactors each using a different feed composition and, if desirable, a different flow rate, provides composition modulation for the catalyst, but allows segregating of the discharge streams; a possible advantage for product recovery. Consider partial oxidation using this system. One reactor would be fed a hydrocarbon stream with little or no air, while the second reactor would be fed just air

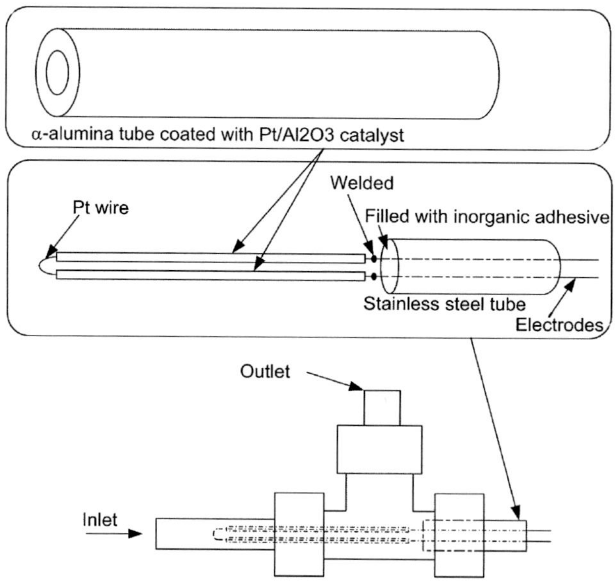

FIGURE 1-7 Tee-junction microreactor: catalyst is impregnated onto the surface of the ceramic tubes that are heated by a continuous Pt wire. *(Figure adapted from* Sotowa et al. (2008) *with permission. © 2008 by Elsevier Inc.)*

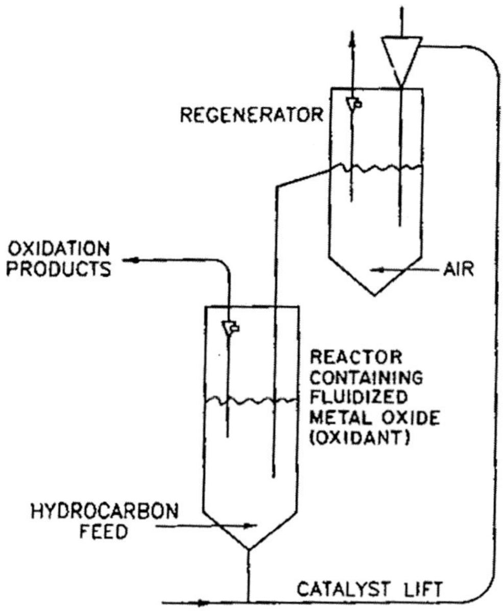

FIGURE 1-8 Conceptual system for composition forcing of partial oxidation reactions using fluidized beds with catalyst transfer between beds. *(Figure reproduced from* Lewis et al. (1949) *with permission. © 1949, the American Chemical Society.)*

for catalyst regeneration. Only the hydrocarbon stream contains product and so it alone passes to the product recovery train. The product is at higher concentration than the comparable discharge stream from a continuous, steady-state operation and separable from the discharge at lower cost. Separation advantages were recognized many years ago, well before periodic operation was discussed in the literature. **Figure 1-8** shows a system of two fluidized beds and catalyst transfer via a pneumatic lift proposed by *Lewis et al. (1949)*. The arrangement resembles early versions of catalytic cracking. One of the beds is fluidized by a hydrocarbon that is selectively oxidized while the catalyst is reduced. The second bed is fluidized with air that re-oxidizes the catalyst. The catalyst, thus, functions as an O_2 carrier.

The advantages of this system are quite obvious. Molecular oxygen is absent during oxidation of the hydrocarbon, reducing the extent of total oxidation. Higher hydrocarbon concentrations are possible in the reactor because a detonation hazard no longer exists. Higher concentrations also reduce the cost of downstream separation. Air can be used as the oxidant in place of oxygen because the oxidant is separated from hydrocarbon so that N_2 does not become a load on downstream separation. Furthermore, catalyst oxidation and reduction steps can be carried out at different temperatures to minimize contact time because each step occurs in a different reactor. A less apparent advantage is that heat released by the highly exothermic partial oxidation reaction is split between two vessels, thereby reducing the cost of cooling.

The two-fluidized bed circulating-catalyst system was studied experimentally in the late 1960s (*Callahan et al., 1970*). It was not adopted for commercial use because, according to the authors, the advantages just mentioned are offset by the cost of circulating large amounts of catalyst. Because redox catalysts are only seldom reducible reversibly to the metal, a large mass of catalyst must be circulated for every unit mass of product created. Furthermore, subjecting a catalyst to transport, often at high speed and alternating environments, leads to attrition and catalyst loss.

For a chemicals application, the balance between advantages and disadvantages of circulating solids can be improved if composition modulation provides higher selectivities. Such is the case for the partial oxidation of butane to maleic anhydride as long as short contact times between butane and the catalyst are used. To achieve the short contact time, a riser reactor or a fast fluidized bed replaces one of the conventional fluidized beds shown in **Figure 1-8**. The Du Pont Company operated a two-bed, circulating catalyst scheme for several years before problems with catalyst attrition forced its abandonment. **Figure 1-9** shows schematically the riser-fluidized bed system used for the partial

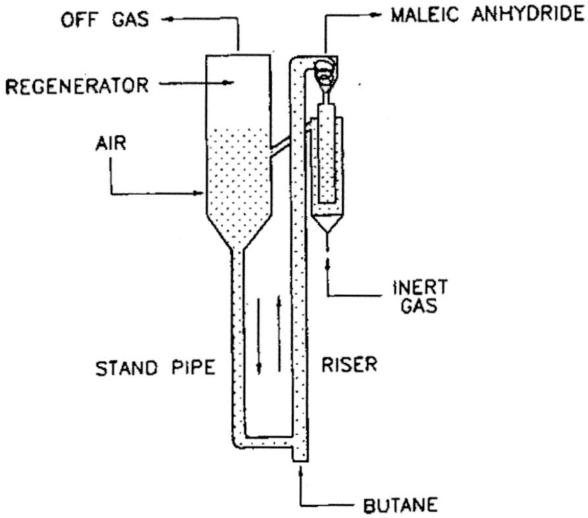

FIGURE 1-9 Schematic of a recirculating-solids riser reactor/fluidized bed for the partial oxidation of butane. *(Figure reproduced from* Contractor et al. (1987) *with permission. © 1987, Elsevier Science Publishers.)*

oxidation of butane to maleic anhydride (*Contractor et al., 1987*). Re-oxidation of the catalyst is not a critical step for high maleic anhydride yield so this step was performed in a fluidized bed. In partial oxidation, multi-part cycles arise when the catalyst is stripped with a gaseous inert between contact with the reactants. With recirculating solids in fast fluidized beds, catalyst separation and transport steps provide the opportunity for stripping so a further vessel is not needed. Such an operation can be seen in **Figure 1-9**. Stripping of the hydrocarbons and oxygenates from the catalyst is accomplished in the accumulator following the riser cyclone and in the line transferring the catalyst to the fluidized bed regenerator.

Du Pont was not the first company to commercialize a two-bed, recirculating-catalyst system. The technology was available some 10 years before (*Sze and Gelbein, 1975, 1976*) and, apparently, a commercial unit existed (*Schwendeman et al., 1983*).

It is also possible to circulate a catalyst between zones in a single bed. A design for hydrocarbon partial oxidation in a single bed was proposed by *Park and Gau (1986)* but not built. A spouted bed is another means of circulating a catalyst while using different parts of the bed for the two parts of a cycle.

Circulating solids reactors have become a subject of intense activity during the last decade for the combustion of carbon fuels because they appear to offer an economic separation of a CO_2 rich gas for sequestration from the gaseous products of combustion. Well over 100 research papers were published up to 2011 in scientific and engineering journals. Several reviews are available (*Scott et al., 2006; Anthony, 2008; Fan and Li, 2010*) and a monograph on looping combustion has appeared (*Fan, 2010*). Looping combustion will be examined briefly in **Chapter 5** and in much more detail in a discussion of circulating catalyst systems in **Chapter 8**.

1.10 REACTION SYSTEMS EXAMINED

Periodic operation has been applied already to a large number of reaction systems as may be gleaned from the chapter index for this monograph. For automotive systems considered in **Chapter 6**, periodic modulation of the combustion reactions is unintentional as it originates through engine operation. As will be seen, modulation in such systems has only a small effect on the reactions occurring. On the other hand in large-scale combustion or gasification systems examined in **Chapter 5** and **Chapter 8**, the circulating solid systems, a type of periodic operation, is essential for the purpose of separating a high purity CO_2 or synthesis gas stream from combustion products. Similarly, periodic operation is unavoidable with the separating reactors, discussed in **Chapters 20** to **22**, in order to carry out the separation function. This is also the case for periodic flow reversal employed for heat and mass trapping considered in **Chapter 18**.

Apart from the specialized considerations in **Chapter 13, 14** and **19**, dealing with discovering reaction mechanisms, performance predictions or operational controls, and perhaps **Chapter 9**, electrochemical systems, other chapters in this monograph are concerned with the use of modulation to enhance conversion and selectivity or yield of a catalytic reaction network. Buried in some of those chapters, particularly **Chapters 3** and **10**, the interaction of external modulation with spontaneous or autonomous oscillation is fleetingly examined. This field of research has become a part of surface science and has been dominated by physicists and theoretical physical chemists. It has a huge and still growing literature. Zhdanov (2004) provides a cogent review and a guide to the literature.

1.11 NEW DIRECTIONS

The final chapter in this book is devoted to this subject. Nevertheless, some of the most compelling of these research opportunities deserve mention in this introduction.

1.11.1 Applications to Systems with Multiple Objectives

Virtually all of the systems considered in the literature on periodic operations have dealt with improving a single performance measure such as selectivity to a desired product or accelerating the rate of reaction. In an industrial environment, optimization problems often involve multiple objectives. This subject has drawn a good deal of attention for steady-state systems and often leads to the condition of a Pareto Set, i.e., situations in which not all of the objectives can be optimized for the same choice of operating conditions. A significant literature exists on this topic. Recently, researchers at Petrochina Company Ltd. working together with Tsinghua University (*Gao et al., 2008a*) have investigated the production of lower olefins from naphtha pyrolysis. The problem specifically was to optimize the production of either ethylene or propylene. Pyrolysis of naphtha becomes a periodic process on a timescale of days because the cracking furnace must be periodically shut down for coke removal. In an earlier paper, *Gao et al. (2005)* considered the periodic burn-off of coke in attempting to optimize the yield of ethylene. Consideration of yields of both ethylene and propylene generates the Pareto Set because yields of these products cannot be increased at the same time.

The researchers developed a model for the naphtha process and attempted to use conventional multi-objective algorithms to determine a solution. These algorithms turned out to be extremely slow, so the authors modified one of them to produce what they termed a "parallel hybrid algorithm." Details of this algorithm and multi-objective optimization are beyond the scope of our discussion. The paper by *Gao et al. (2008a)* contains a well-written discussion. The system considered by *Gao et al. (2005, 2008a)*, and earlier by *Xu et al. (2001)*, introduced a type of periodic operation rarely examined in the literature. Each part of the cycle considered has a duration of days so that most investigators including *Gao et al.* replace the process dynamics in their models with a pseudo-steady-state assumption. Nevertheless, future problems with multiple objectives will most likely arise in a context of systems described by dynamic models, such as those considered in this monograph.

1.11.2 Control of Carbon Deposition

Despite claims in the literature that types of periodic operation extend catalyst life, there has been little study of this application. Catalyst deactivation through carbon deposition was tackled by *Jaeger et al. (1992)* in an examination of CO methanation on a Ni-faujasite catalyst. The authors used a vibrationally fluidized bed in which the reactants CO and H_2 could be added separately or combined in a single feed. In the separate-feed operation, H_2- and CO-rich gas regions exist in the fluidized bed. Catalyst particles circulate through these regions. Hydrogen reacts with active carbon deposited by the Boudouard reaction to form methane, thereby removing carbon and regenerating the Ni catalyst. The authors demonstrate that the catalyst activity can be maintained indefinitely in the separate feed situation. The result is an enhanced methanation rate for long on-stream times. Even so, this interesting experimental study has not been pursued.

1.11.3 Modulation of Multiple Reactor Inputs

The periodic operation literature up to just recently has considered just modulation of a single input such as concentration, reactor temperature or reactor flow rate. *Parulekar (2003)* appears to be the first to examine modulating more than one reactor input simultaneously. He provides, in this important paper, the conditions that must be fulfilled for an improvement to occur in reactor performance using the π-criterion. This criterion is discussed briefly in **Chapter 14**. *Parulekar's* analysis shows that the interaction between the modulated inputs dominates the magnitude of the improvement achievable.

Two examples are given in his 2003 paper to illustrate the improvement possible with multiple input modulation. The first of these deals with the competition between recombinant microorganisms and nonrecombinant ones. This system was dealt with by *Lee and Parulekar (1996)* among others and is discussed further in **Chapter 11**. The authors examined the use of antibiotics in a continuously operated stirred tank reactor to control the nonrecombinant cells and demonstrated through modeling that by modulating different inputs to the reactor they could maintain a higher concentration of desirable recombinant organisms. Improvements were predicted through modulating substrate concentration and dilution rate. Also, improvements occurred when, in addition to these, the antibiotic concentration was also modulated. *Parulekar* considered two different types of antibiotics in his analysis.

A second example utilized a series-parallel reaction in which the intermediate product was valuable. This system had been analyzed under periodic operation by *Watanabe et al. (1981)*. Modulation of reactant flow rates was examined. The authors observed that the frequency region in which periodic operation improved performance was extended significantly when both of the feed rates were modulated. *Parulekar's* analysis applied to small variations in the forced variables that allow linearization of the descriptive equations. Experimental verification of these findings would be useful and suggests a new research direction.

1.11.4 Improvement of Photocatalysis

Photocatalytic processes using a TiO_2 catalyst have attracted attention for environmental applications to decompose organic contaminants into simple inorganic species. *Wang and Ku (2006)* examined the application of periodic illumination using the ultraviolet light from a light-emitting diode as the UV light source for the photocatalytic decomposition of a Reactive Red 22 dye. The dye was assumed to represent a contaminant of a water source. The UV-LED was used with TiO_2 in a rectangular planar fixed-film reactor operated with recirculation of a low pH water. Results indicated that the photonic efficiency with periodic illumination was much higher than that with continuous illumination. However, photonic efficiencies for these two arrangements became nearly identical at higher light intensities.

1.12 A BRIEF HISTORY OF THE STUDY OF PERIODIC OPERATION

1.12.1 Pioneers

Interest in the use of periodic operation with chemical reactions developed independently in several countries during the 1960s. In Russia, this took the shape of an interest in heat recovery by the Academician *Boreskov* working with his protégé, *Yu. Sh. Matros*. These investigators built on the use of heat recuperators in the metallurgical industry and applied the concept to exothermic reactions. In steelmaking, the recuperators are beds of firebrick. Exhaust from an open hearth furnace is directed to one bed of brick until it reaches a desired temperature. The flow through that bed is then switched to fresh air that is preheated before entering the furnace. This practice dates from the nineteenth century. *Boreskov* and *Matros* combined the inert recuperator beds with a catalytic reactor and demonstrated that the system could be used to react feed mixtures entering a reactor below their ignition point (*Boreskov et al., 1977a*). Investigations by the team directed by *Matros* extended to other types of modulation such as composition and velocity changes (*Boreskov and Matros, 1983a*). Early research, both experimental and theoretical, was publish in Russian journals and escaped the attention of other workers. These pioneering contributions preceded experimental investigations by *Hudgins* and *Silveston* (*Unni et al., 1973*) and by *Schügerl* and *Renken* (*Helmrich et al., 1974*).

In North America, Princeton's *R.H. Wilhelm* became interested in the modulation of different reactor variables in the early 1960s. One aspect of this interest was in what he termed "parametric pumping." *Wilhelm's* interest in modulation was carried on by his co-workers, one of whom (*Schügerl*) returned to Germany, and another of whom (*Hudgins*) returned to Canada. In Hannover, *Renken*, working under the supervision of *Schügerl*, explored the use of composition modulation on catalytic reactions in the early 1970s; while at the University of Waterloo, *Hudgins* initiated experimentation that, like the work of *Renken*, extended over more than three decades.

Theoretical studies began at about the same time by different researchers, notably, *F.J.M. Horn*, who demonstrated through work on optimization that composition modulation could improve reactor yield for certain reaction kinetics (*Horn and Lin, 1967*). *Douglas and Rippin (1966)* undertook numerical simulations of concentration modulation in stirred tank reactors. Shortly thereafter, *Renken (1972, 1974a)* demonstrated that the yield and selectivity of the intermediate products for consecutive-competing reactions in stirred tanks can exceed those obtainable from steady-state operation. *Horn's* work was extended by *Bailey* and his students and this yielded the first review of the application of periodic operation to reactors (*Bailey, 1977*).

Although composition modulation in the guise of in situ catalyst regeneration or the gasification of coke in cracking tubes has been practiced for 60 years, application to chemical reactors was first considered in the middle 1960s through numerical simulation by *Douglas (1967)* and later through experimentation by others. The latter were published almost simultaneously in Germany and Canada (*Wandrey and Renken, 1973; Helmrich et al., 1974; Unni et al., 1973*). This early experimental work was discussed by *Renken (1982, 1984)*. *Renken's* review addressed the question of the proper comparison of periodic and steady-state operations, a subject discussed earlier in this chapter. *Renken* also noted that models that predicted improved performance under modulation were not supported by experimental data, whereas experimental systems for which improvement had been found had either inadequate models or none at all.

Since this early work, several journal reviews have appeared (*Meira, 1981a; Boreskov and Matros, 1983a; Renken, 1982, 1984; Silveston, 1987; Stankiewicz and Kuczynski, 1995*). A monograph has also been published (*Silveston, 1998*).

Interest in periodically operated separating reactors developed almost a decade after the start of research on periodic operation. The concept of using a simulated moving-bed as a separating reactor followed surprisingly quickly after the application of this concept to adsorptive separations. A US patent by *Zabransky and Anderson (1977)* discussed using a simulated moving bed as a separating reactor. *Hashimoto* and his co-workers built such a system soon thereafter and published the first experimental results early in the next decade (*Hashimoto et al., 1983*). At about the same time, *Villermaux* and his co-workers began developing the theoretical framework for chromatographic reactors (*Schweich and Villermaux, 1978, 1982a*). The theory was extended to countercurrent moving-bed chromatographic reactors by *Aris* and *Carr* (*Viswanathan and Aris, 1974; Cho et al., 1982; Fish et al., 1986*) as well as *Takeuchi and Uraguchi (1976a, b)*. Simulated moving-bed chromatographic reactors were analyzed by *Ray et al. (1990)*. Mention should also be made of *Barker* and his co-workers who successfully applied separating reactors to biochemical systems (*Barker and Ganetsos, 1988*).

Pressure swing adsorption has been an industrial process since the late 1970s. By the middle 1980s, *Kadlec* and co-workers discussed building and operating a pressure swing reactor (*Vaporciyan and Kadlec, 1987*). Experimental results were published shortly thereafter (*Vaporciyan and Kadlec, 1989*). Temperature swing reactors were mentioned in the literature later (*Han and Harrison, 1994*). Combined pressure and temperature swing systems followed 10 years thereafter (*Xiu et al., 2003a*).

1.12.2 Research Groups

Long-term experimental studies of composition modulation of important industrial catalytic reactions were undertaken by *Renken* working at EPFL Lausanne and by *Hudgins* and *Silveston* at the University of Waterloo. These were the first two groups to experiment significantly with the modulation of chemical reactors. Both teams were active for about 30 years. In the 1980s, a study of modulating CO oxidation began at the University of Michigan under *Gulari's* leadership that lasted for almost 10 years. At the CNRS laboratory in Nancy, France, *Rambeau* and *Amariglio* investigated ammonia synthesis. Their work with several research students formed another important group. Noble metal catalytic converters for automotive exhausts were studied by the automotive industry and by several university teams during the 1980s and 1990s. Since then, efforts on composition modulation of chemical reactors have been initiated in China by *C.-Y. Li* in Beijing and *W.-K. Yuan* in Shanghai. In Sydney, Australia, *Adesina* continues applying modulation to partial oxidation. Research groups have been established at all of these locations.

The analytical literature has concentrated on three questions: 1) Can modulation improve the performance of a chemical reaction system? 2) Can performance under modulation exceed the best possible performance under steady state? 3) What properties of a catalytic system result in improved performance under modulation? However, no long-term activity developed, with the possible exception of *Hofmann's* team at the University of Erlangen in the 1980s. Instead, studies have been spread over a number of institutions. The first question was extensively discussed in the 1970s where the issue was the criteria for determining optimal modulation conditions. This discussion continued into the next decade (*Watanabe et al., 1981; Farhadpour and Gibilaro, 1981; Sterman and Ydstie, 1990a*). Several contributions by *Nowobilski and Takoudis*

(1986), and by *Thullie et al. (1986)* have dealt with the second question, while *Schädlich et al. (1983)* and *Grabmueller et al. (1985)*, among others, consider both of the final questions. *Suman (2004)* discussed the first two questions.

Work on flow direction switching, started at the Boreskov Institute of Catalysis in Novosibirsk, is now widespread. A leader of that work, *Yu.Sh. Matros*, moved into private industry in about 1990 and continues to work on applications of this type of periodic operation. Most academic research has also focused on applications, e.g. *Nieken et al. (1994a)* or *Marín et al. (2008)*. Research centers have arisen in Stuttgart, Houston and Orviedo.

Application of flow modulation to trickle beds has led to the development of research groups in Argentina (Universities of Mar del Plata and Buenos Aires) and the Czech Republic. A large, four-year, multi-university–industry European project called "CYCLOP" was also created. In that effort, now ended, a team at Eindhoven studied the hydrodynamics of flow modulation in trickle beds (*Boelhouwer et al., 2002a*). Other groups, supported through CYCLOP, dealt with flow modulation in trickle beds and were located at the Technical Universities of Delft, Turin, Prague, Dortmund, Nancy (ENSIC) and in Saloniki (*Giakoumakis et al., 2005; Borremans et al., 2007; Brkljac et al., 2007*).

Important centers of research on temperature modulation using microreactors have arisen at EPFL Lausanne and at the Karlsruhe Research Center as well as in Lynby, Denmark (*Brandner et al., 2004; Jensen et al., 2008a*).

Both modeling and experimental work on the application of composition modulation to polymerization reactions was undertaken by *Meira* in Argentina (*Meira, 1981a*), but after a decade this effort ended. Most research on this topic has been through single studies at different institutions, so that research groups have not developed.

Research on periodically operated separating reactors has developed along two lines: Rotary or simulating moving bed chromatographic reactors and pressure- or temperature-swing reactors. Vigorous research groups have worked on each of these systems. One of these is at Kyoto University, organized by *Hashimoto* and now led by *Kawase* (*Pilgrim et al., 2006*), deals with simulated moving-bed chromatographic reactors. Another active group for the two last decades has operated in England at the University of Ashton (*Barker and Ganetsos, 1988*). A group at the University of Minnesota that worked on moving-bed chromatographic reactors lasted for about 10 years, but now seems inactive. There is considerable activity at the ETH Zürich on the theory of simulated moving-bed chromatographic reactors by a group headed by *Morbidelli* (*Lode et al., 2001*) and by *Ray's* team at the University of Western Ontario (*Zhang et al., 2007a*).

Several research groups dedicated to swing reactors have developed since the end of the 1980s. Important groups led by *Alpay* at Imperial College, and *Harrison* at Louisiana State University are no longer active. However, *Rodriguez* at the University of Porto (Portugal) and *Agar* at the University of Dortmund continue to explore various applications of swing reactors (*Pereira et al., 2009; Lawrence et al., 2005*).

1.12.3 Industrial Applications

The greatest penetration of periodic operation in industry has been with periodic flow reversal. There exist, at present, several hundred units operating worldwide for the decontamination of low levels of organic pollutants in waste gas streams. These systems are discussed in **Chapter 18**. In addition, simulated moving-bed chromatographic reactors have found application in biochemical systems. The number of such reactors and their application are difficult to ascertain because companies using the technology have not published descriptions of their applications. Pressure swing reactors have just advanced out of the development stage. Otherwise, despite the voluminous research described in the following pages, there has been negligible industrial adoption.

Stankiewicz and Kuczynski (1995) discussed the state of industrial adoption. They ascribed the limited use of periodic operation to uncertainty, namely a lack of basic understanding of effects, that impair a reliable basis for scale-up. They further point out that integrating a periodic operation into a plant that operates mainly under steady state poses difficulties.

Nomenclature

A	=	amplitude of modulation (various units)
A_n	=	amplitude of the nth part cycle
A,B	=	reactant species
BZ	=	Belousov-Zhabotinsky (reaction)
C	=	product species
d	=	halfcycle duration (s)
o.d.	=	outer diameter (cm)
r	=	rate of reaction (mol/cm^{3}s)
$\bar{r}$	=	time-average rate of reaction
r_{ss}	=	steady-state rate of reaction
rss	=	relaxed steady state
SMBCR	=	simulated moving-bed chromatographic reactor
s	=	cycle split (-)
t_c	=	characteristic (relaxation) time (s)
VOC	=	volatile organic compound
y_i	=	mole fraction of the ith species

Greek

ϕ	=	phase lag (°)
Ψ	=	enhancement (-)
Ψ'	=	global enhancement
τ, τ_{cycle}	=	cycle period (s, min.)

Subscripts

c	=	characteristic
max	=	maximum
qss	=	quasisteady state
rss	=	relaxed steady state
ss	=	steady-state

CHAPTER

2

Hydrogenation Processes

Robert Ross Hudgins, Peter Lewis Silveston
Waterloo, Ontario, Canada

OUTLINE

2.1 Ammonia Synthesis 23
2.1.1 Iron Catalysts 24
2.1.1.1 Mass Transfer Interference 28
2.1.1.2 Near-Adiabatic Operation 29
2.1.2 Ruthenium Catalyst 30
2.1.3 Osmium Catalyst 30
2.1.4 Interpretation 30
2.1.4.1 Relaxed Steady State 30
2.1.4.2 Reactant Inhibition 31
2.1.4.3 Reactant Storage 31
2.1.4.4 Surface Activation/Restructuring 33
2.1.5 Comments on the Ammonia Synthesis Application 33

2.2 NO_x Reduction 33
2.3 Methanation 34
2.4 Methanol Synthesis 39
2.4.1 Copper-Zinc-Alumina Catalysts 39
2.4.2 Copper Zinc Catalyst 41
2.4.3 Commercial Methanol Catalysts 43
2.5 Ethylene Hydrogenation 43
2.6 Aromatics Hydrogenation 45
2.7 Oscillatory Behavior 47

In this chapter, the hydrogenation of small molecules, such as nitrogen, nitrogen oxides, carbon monoxide, ethylene and the simple aromatics are examined under periodic operation. under periodic operation are examined. There is no mention in the literature of the application of modulation to the hydrogenation of complex molecules such as found in hydrocracking, hydrodesulfurization, hydrodenitrogenation or in the hydrotreating of polymers, though they certainly warrant attention. The Fischer-Tropsch synthesis may be considered a hydrogenation reaction, but it is also a polymerization one, so it will be discussed in **Chapter 7**.

2.1 AMMONIA SYNTHESIS

In industrial hydrogenation processes, ammonia synthesis is among the most important. Between 1980 and 2000, some 15 studies of the periodic operation of this synthesis were published. Employing a triply promoted iron catalyst, it is one of the largest commercial applications of catalysis.

Thermodynamics suggest that low temperatures should be used, but rates are far too slow for practical purposes. At higher temperatures where rates are much higher, conversion is equilibrium-limited, usually to less than 5%. This limitation is overcome by taking advantage of the stoichiometry and conducting the synthesis at pressures as high as 10 MPa, but high pressure and low conversion-per-pass makes reactor operation expensive.

Motivation for research on periodic operation of the synthesis lies in its kinetics which suggests the reaction is reactant-inhibited over most of the range of the thermodynamically feasible operating conditions. Controlling the exposure of the surface to reactants might be a potential means to reduce inhibition. Research has been summarized up to 1997 by *Silveston (1998)*. No further publications have appeared since then.

As a convenience, experimental studies of ammonia synthesis under periodic operation are summarized in **Table 2-1**. Most observations mentioned in the table are discussed further and are interpreted in the text that follows.

2.1.1 Iron Catalysts

In an early unsteady state study of the synthesis attacking the apparent problem of reactant inhibition, *Unger and Rinker (1976)* employed a chromatographic reactor and performed experiments at temperatures normal for NH_3 synthesis but at about one-tenth of the commercial operating pressure. The operation represented a form of composition modulation by alternating between the reaction mixture and a diluent. Succeeding researchers (*Wilson and Rinker 1982; Chiao et al., 1987; Jain et al., 1982a, 1983*) employed a short reactor bed and composition modulation in place of on-off reactant flow. They used the same triply promoted iron catalyst as *Unger and Rinker.*

A significant improvement, a rate increase of 30%, was observed by *Jain et al. (1983)* over a fairly broad range of cycling conditions as shown in the two parts of **Figure 2-1**. Steady-state rates seen in **Figure 2-1(a)** were measured at the mean composition of the cycling mixture, while in **Figure 2-1(b)** the envelopes show the maximum time-average and steady-state rates observed for different cycle splits and cycle periods. The cycling measurements showing improvement in **Figure 2-1(b)** were obtained over just a restricted feed composition range as indicated by the cross-hatched area in the figure. Step-change experiments (*Jain et al., 1982a*) showed transients lasting up to 90 min after N_2 was removed from the feed and almost 10 min when H_2 was removed. An abrupt peak in NH_3 production occurred when the synthesis gas mixture was replaced by H_2. This initial peak did not occur on a switch to N_2, suggesting the catalyst surface was largely covered by adsorbed N_2 or partially hydrogenated intermediates.

Using the same catalyst, *Wilson and Rinker (1982)* and *Chiao et al. (1987)* generally confirmed the *Jain* observations (**Figure 2-2**). The figure shows enhancement approaching 50% at a 1 min cycle period. *Wilson and Rinker* cycled between reactants in a diluent (bang-bang cycling). They observed a strong influence of cycle split on enhancement. Limited measurements by *Jain et al. (1982a)* also suggested cycle split is an important variable. Examination of **Figure 2-2** shows enhancement decreasing as total pressure and temperature increase. The highest pressure used in this figure is 4.14 MPa, less than half the operating pressures of commercial reactors and it results in no enhancement. Further experiments by *Chiao et al. (1987)* observed that cycling at 4.14 MPa enhances the rate of NH_3 production just slightly when switching occurs between H_2

TABLE 2-1 Summary of Experimental Studies of Ammonia Synthesis under Periodic Operation

Authors	Objective	Reaction	Modulated Variable	Reaction Conditions	Observations	Comments
Unger and Rinker (1976)	Surmounting steady state equilibrium limits through composition modulation	$N_2 + 3H_2 = 2NH_3$ over a triply promoted iron oxide synthesis catalyst	N_2 with steady flow of H_2	$P = 928$ kPa; $728 < T < 752°C$; plug-flow reactor packed with mixture of 5A, 13X molecular sieves and Girdler G-82 catalyst	Pulsing N_2 once every 2 min increased conversion to 57% from 41% under steady-state operation	Explanation for the improvement was not explored
Amariglio and Rambeau (1977); Rambeau and Amariglio (1978a, b)	As above	$N_2 + 3H_2 = 2NH_3$ over a triply promoted iron oxide synthesis (KMI) catalyst	Details not given	Details not given	No rate improvement through composition modulation was observed	
Jain (1981a); Jain et al. (1982)	Investigate effects of composition modulation under near-adiabatic operation	$N_2 + 3H_2 = 2NH_3$ over a triply promoted iron oxide catalyst (Girdler G-82)	Both N_2 and H_2 concentrations; $0.2 < s < 0.85$; $1 < \tau < 60$ min	$P = 2.38$ MPa; $T = 673$ K; packed-bed reactor	20 to 30% increase in the time-average synthesis rate was observed. Maximum temperatures in the front 20 to 40% of the packed bed were about 1° higher than the steady state maximum	Composition modulation flattened the temperature profile and moved the maximum temperature deeper into the bed. Temperature fluctuations increased with increasing period

(Continued)

TABLE 2-1 Summary of Experimental Studies of Ammonia Synthesis under Periodic Operation (*cont'd*)

Authors	Objective	Reaction	Modulated Variable	Reaction Conditions	Observations	Comments
Rambeau and Amariglio (1981a, b)	Study of the concentration modulation of NH_3 synthesis over an Ru catalyst to circumvent strong inhibition by H_2	$N_2 + 3H_2 = 2NH_3$ over an unsupported Ru catalyst	Bang-bang cycling between H_2 and N_2	$373 < T < 773$ K; 1 atm pressure; 10 g catalyst sample; used optimal periods (5 to 10 s) for each temperature tested.	Modulation significantly enhanced the synthesis rate by reducing H_2 inhibition. Improvement explained by steady-state rates of N_2 adsorption on Ru that are 1000-fold greater than on Fe of similar specific surface area	Authors claim high inherent activity of Ru catalyst is lost because of the strong inhibition by H_2.
Rambeau et al. (1982)	As above but using an osmium catalyst	As above, over an unsupported Os catalyst	Bang-bang cycling between H_2 and N_2; also used H_2 pulsing	$574 < T < 671$ K; 1 atm pressure	As above. N_2 adsorption rates on Os are more than 100-fold greater than on Fe of similar specific surface area.	Results show that H_2 inhibits NH_3 formation on Os, but to a lesser extent than on Ru.
Jain et al. (1982a, 1983)	Study of the modulation of H_2 and N_2 concentrations to surmount steady-state equilibrium limits	$N_2 + 3H_2 = 2NH_3$ over a triply promoted iron oxide catalyst (Girdler G-82)	Both N_2 and H_2 concentrations with $0.4 < s < 0.6$ and from $1 < \tau < 60$ min	$P = 2.38$ MPa; $T = 673$ K; using a Berty-type gradientless reactor; $SV = 19\,000\ h^{-1}$; $d_p = 0.3$ mm	Maximum time-average rates were up to 30% higher than corresponding steady-state ones.	Step-change experiments showed that H_2 and especially N_2 were retained in the catalyst: N_2 was detected for up to 90 min after removal from the feed. Gradientless reactor is poorly suited for studying modulation.

Wilson and Rinker (1982); Chiao et al. (1987)	Investigation of H_2 and N_2 concentration modulation on NH_3 synthesis	As above	Cycling between pure H_2 and N_2, and between pure H_2 and 50-50 mixture of H_2 and N_2	$P = 0.13$, 0.57 and 1.13 MPa; $598 < T < 698$ K; $\tau = 1$ min; $0.32 < s < 0.75$	Rate enhancement reached 1.6 × at 598 K and 1.13 MPa	Enhancement increased at lower temperatures and pressures
Jain et al. (1985)	Study of the effect of mass transfer interference on enhancement by composition modulation	As above	Both N_2 and H_2 concentrations with $0.2 < s < 0.85$ and $6 < \tau < 20$ min	$P = 2.38$ MPa; $T = 673$ K; in a Berty-type gradientless reactor; $SV = 19\,000\ h^{-1}$; $d_p = 0.3$ and 5 mm	Enhancements with 5 mm particles were substantially lower than with 0.3 mm particles. Resonance in the time-average rate of NH_3 synthesis seen at $\tau = 10$ min	Steady-state effectiveness factors provided satisfactory estimates of the effectiveness factors under modulation
Krylova and Abdeligani (1987); Krylova (1990)	Study of the pulsing of synthesis gas into streams of pure H_2 or N_2	$N_2 + 3H_2 = 2NH_3$ over a Russian catalyst (no details given)	Pulses of synthesis gas (stoichiometric H_2 in N_2) added to pure H_2 or N_2; also pulses of H_2 in N_2 added to stoichiometric H_2 in N_2	$10 < T < 500$°C; $0.1 < P < 5$ MPa; pulses of synthesis gas 0.125 to 20 mL; gas flows 10 to 222 mL/min; $s = 0.5$	Synthesis strongly increased ($\psi = 1.5$) in terms of H_2 conversion by pulsing H_2 into a stoichiometric synthesis gas mixture at $\tau = 0.5$ min. Even greater improvement occurs for the opposite strategy ($\psi = 2.4$)	Higher conversions were explained in terms of a catalyst surface assumed to be covered by adsorbed N_2 or partially hydrogenated N atoms that are swept from the surface in the pulse thereby improving H_2 access to catalyst sites

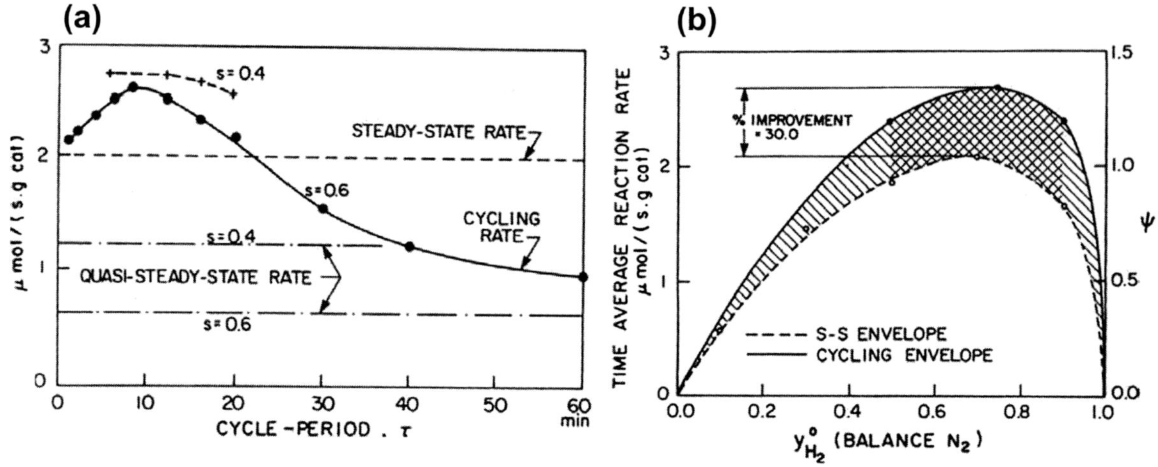

FIGURE 2-1 Comparison of steady state and time-average cycling rates of NH_3 formation as a function in (a) of cycle structure and period and as a function of mean feed composition in (b). Experiments performed at 400°C, 2.38 MPa and $SV = 19{,}000\ h^{-1}$ using United Catalyst C73-1-01. *(Figure reproduced from* Jain et al. (1983) *with the authors' permission.)*

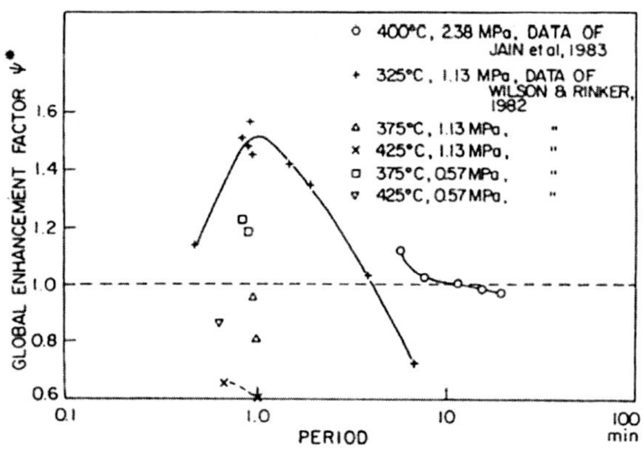

FIGURE 2-2 Normalized rates of NH_3 synthesis over a triply promoted iron catalyst (United Catalyst C73-1-01). The *Wilson and Rinker* data are for bang-bang cycling between reactants. *(Figure reproduced from* Silveston (1987) *with the author's permission.)*

and N_2 mixtures. However, cycle periods in the range of seconds must be used.

Early experiments by *Amariglio and Rambeau (1977)* and *Rambeau and Amariglio (1978a, b)* used a triply promoted iron oxide catalyst but their experiments showed no rate enhancement through composition modulation. A Russian synthesis catalyst was studied by *Krylova and Abdeligani (1987)* and is summarized by *Krylova (1990)*. These investigators introduced pulses of synthesis gas into a flow of pure H_2 or N_2. *Krylova* observed a large enhancement of the synthesis ($\psi = 1.5$) in terms of H_2 conversion for pulsing H_2 into a stoichiometric synthesis gas mixture at $\tau = 0.5$ min. Even greater improvement occurred for the opposite strategy ($\psi = 2.4$). *Krylova* explained the higher conversions in terms of coverage of the catalyst surface by adsorbed N_2 or partially hydrogenated N atoms. These are swept from the surface during a pulse thereby improving H_2 access to a catalyst site. A similar argument applied to N_2 pulsing. *Krylova* also observed that enhancement under cycling decreased as the total pressure was raised.

2.1.1.1 Mass Transfer Interference

Little attention has been paid in the literature to the effect of intraparticle diffusion on enhancement through composition modulation. Concentration gradients arise in porous catalysts under both steady state and periodic operation through diffusion of reactants towards the particle center and result in reaction rates that are lower than rates that would be predicted from concentrations prevailing at the external surface of the particle. Only *Jain et al. (1985)* have investigated the effect of intraparticle diffusion on enhancement through composition modulation using NH_3 synthesis over a triply promoted iron oxide catalyst. They observed that enhancement by composition forcing is reduced but does not disappear. To evaluate diffusional interference, *Jain et al.* compared the performance under modulation as well as steady state for fine particles with that of a particle 16 times greater in diameter. They observed that the ratios of rates for the large and fine particles were virtually the same. This result suggests that an effectiveness factor calculated for steady-state can be used as a first approximation for the effectiveness factor under periodic operation.

2.1.1.2 Near-Adiabatic Operation

Temperature profiles in full-scale ammonia synthesis reactors reflect the exothermicity of the synthesis, the existence of a significant reverse reaction at conversion greater than ca. 10%, and heat exchange with fresh feed. There is a sharp rise in temperature just after the bed entrance to a maximum and then a slow drop with increasing axial depth in the bed. *Jain et al. (1982a)* utilized a near-adiabatic fixed bed, integral reactor to examine composition forcing under the conditions of a commercial operation, except that the reactor pressure was roughly a tenth of typical industrial levels. Questions addressed by their study were:

1) Can composition modulation raise the time-average synthesis rate?
2) How does the improvement compare with that found under an isothermal, differential operation?
3) How does cycling affect the temperature profile in the reactor?

Jain et al. observed that the absolute increase in rate due to composition modulation is greater for the near-adiabatic fixed bed reactor than for the differential reactor. With respect to periodic versus steady-state operation, there was an enhancement of 27% at the maximum rate observed in the latter operation. This result was about the same as that observed for the differential reactor (**Figure 2-1(b)**), so these experiments suggest that modulation observations made in a differential reactor represent what will be observed at full-scale. Periodic operation raises and flattens the temperature profile and moves the temperature maximum deeper into the bed as **Figure 2-3** shows. The changes depend on cycle period: As the period increases axial temperature variations become greater. Thus periodic operation can mitigate hot spots in near-adiabatic synthesis reactors. It appears that the temperature profile is leveled by both thermal inertia and the axial movement of temperature waves in the bed initiated by changing rates of ammonia formation following a switch in composition. The largest fluctuation occurs near

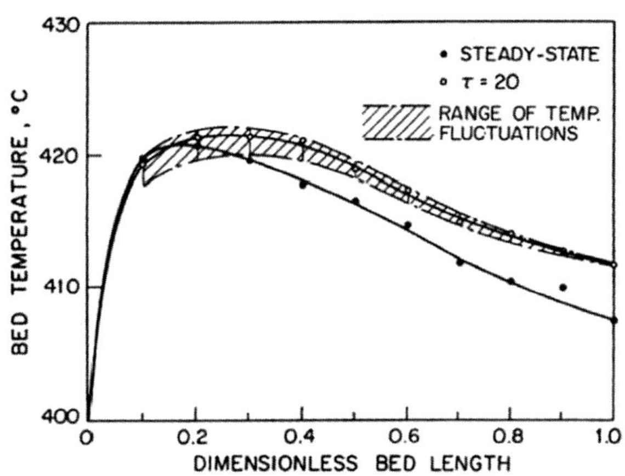

FIGURE 2-3 **Temperature profiles and range of fluctuations in a near-adiabatic packed bed of triply promoted iron catalyst under periodic and steady-state operation. Cycle period = 20 min and cycle split = 0.2 with an equimolar N_2-H_2 feed mixture.** *(Figure reproduced from* Jain et al. (1982a) *with permission. © 1982 American Chemical Society.)*

the maximum bed temperature. Fluctuations vanish at the inlet (where feed temperature determines bed temperature) and at the outlet (where conversion is almost complete).

2.1.2 Ruthenium Catalyst

Several commercial synthesis reactors apparently now use ruthenium despite its much higher cost. *Rambeau and Amariglio (1981a, b)* observed that the activity of an unsupported ruthenium catalyst is about 500 times greater than those of conventional iron catalysts at 400°C and 2000 times greater at 250°C. However, the synthesis reaction is inhibited by H_2 much more strongly than with iron catalysts. Indeed, at any temperature, the rate over an Ru catalyst increases as the H_2 partial pressure decreases. This observation led the French researchers to explore composition modulation. Earlier experiments (*Rambeau and Amariglio, 1981a, b*) found that the synthesis is limited by the rate of N_2 adsorption. The maximum synthesis rate, then, is close to the rate of N_2 adsorption on a clean Ru surface.

Periodically forcing the synthesis by bang-bang switching between pure reactants significantly raised the rate obtained using a ruthenium catalyst in a powder form. Except for at temperatures above 600 K, enhancement through composition modulation is 2- to 5-fold greater than under steady-state operation. Under similar conditions, enhancements were only about 10% greater for the triply promoted iron catalyst.

Modulation of the reactor feed composition appears to be attractive with a ruthenium catalyst. It is surprising that there are no reports in the process literature on pilot plant experiments.

2.1.3 Osmium Catalyst

Osmium, like ruthenium, is an active catalyst for NH_3 synthesis. *Rambeau et al. (1982)* undertook composition modulation studies on this catalyst using bang-bang cycling between H_2 and N_2. They observed time-average rates of synthesis significantly higher than those over Fe catalysts but lower than those seen for ruthenium. For time-average NH_3 synthesis, a bang-bang cycling strategy resulted in the highest rates. The time-average synthesis rates are reduced for cycling between H_2 and a stream with N_2 and 0.2 vol% H_2 and further reduced when the mixed stream contains four times the amount of H_2. The presence of H_2 in the N_2 half-cycle suppressed the rate of reaction at low temperatures but appeared to have little effect at the highest temperatures used where equilibrium was approached. Just as for the Ru catalyst, adsorption of N_2 on the osmium surface is rate-controlling. The synthesis rate is inhibited by H_2 but to a lesser extent than it was for the Ru catalyst.

2.1.4 Interpretation

Many investigators of modulation have invoked mechanistic or kinetic models of ammonia synthesis to interpret their results and to project what might be expected at higher pressures typical of commercial reactors. These models will now be discussed and their predictions compared with the limited experimental measurements available.

2.1.4.1 Relaxed Steady State

Wilson and Rinker (1982) contend that the relatively high cycling frequency used in their experiments forced their single-pass, packed bed reactor to operate in the relaxed steady state. In this mode of operation the time-average rate becomes independent of cycling frequency. To use relaxed steady-state projections, *Wilson and Rinker* assumed their experiments were undertaken in an isothermal plug-flow reactor and that steady-state models adequately describe the kinetics of NH_3 synthesis in the relaxed steady state. They tested a number of steady-state kinetic models—the Temkin, OTB (for Ozaki-Taylor-Boudart), Brill, and Tamaru models—against their composition modulation data. These models are discussed by *Wilson and Rinker (1982)* and by *Silveston (1998)*. From their tests, *Wilson and Rinker* concluded that the Tamaru model best described their cycling data. In a further development (*Thullie et al., 1986, 1987*) an Eley-Rideal mechanism was considered involving the hydrogenation of adsorbed N_2 by gas-phase H_2. The *Thullie* work indicated that enhancements greater than one were possible and the maximum values would be attained in the relaxed steady state. Unsymmetrical cycling predicted rates higher than those attained at $s = 0.5$.

The assumption of operating in the relaxed steady state was extended to a non-isothermal NH_3 synthesis system by *Chiao and Rinker (1989)*. A pseudo-homogeneous, plug flow model was used with kinetic expressions for the synthesis reaction that had been successfully tested against both steady-state and dynamic data. The authors demonstrated that the model correctly predicted temperature and concentration profiles as well as multiple steady states that have been observed when more sophisticated models are used. Their simulation showed that large enhancements of the synthesis rate are achievable by high-frequency composition forcing. Enhancement was influenced by split and amplitude.

The relaxed steady-state approach used by *Rinker* and co-workers appears to yield useful predictions. Their analysis predicts $\psi > 1$ for the NH_3 synthesis models in an isothermal reactor and that $s > 0.5$ should maximize ψ. The kinetic models used, however, resulted in low values of ψ at low reaction pressures and high values at 4.14 MPa, just opposite to what was observed experimentally. An interesting but untested prediction of the analysis and kinetic models is that composition modulation should result in higher NH_3 production when rate in the reactor is equilibrium-limited under steady-state operation. A critical evaluation of the work by *Rinker* and co-workers is given by *Silveston (1998)*.

2.1.4.2 Reactant Inhibition

This model was developed by *Rambeau and Amariglio (1984)*. It assumes that NH_3 synthesis is rate-limited by the dissociative adsorption of N_2. Adsorbed nitrogen is then rapidly hydrogenated stepwise to adsorbed NH_3 (through NH, NH_2). These steps are so fast that they can be considered to occur under equilibrium conditions. Atomic hydrogen dominates the surface even at very low H_2 partial pressures. The abundance of surface hydrogen, forces N-s towards NH_3-s. Nevertheless, according to *Rambeau and Amariglio (1984)*, N-s is the most abundant nitrogen species on the surface.

For the surface environment envisioned by *Rambeau and Amariglio* for cycling between pure N_2 and H_2 at the limit of high frequencies:

$$r = kP_{N_2} \frac{t_{N_2}}{t_{N_2} + t_{H_2}} \tag{2-1}$$

where t represents the duration of pure gas exposure in a partial cycle. They further assumed that the partial period, t_{H_2}, is long enough to hydrogenate all the N_2 adsorbed during the previous cycle. In addition, $t_{N_2}$is short enough that the metal surface does not become saturated with N_2. Thus, the rate of adsorption of N_2 onto a clean surface is:

$$r_{ad} = k_{ad} P_{N_2}. \tag{2-2}$$

Because the synthesis rate is controlled by N_2 adsorption, **Eq. (2-2)** gives the synthesis rate. *Rambeau and Amariglio* showed that these various assumptions are applicable to their experiments on the Ru and Os catalysts. By separating the reactants competing for the same adsorption site, cycling between pure N_2 and H_2 reduces inhibition, raising the rate of synthesis.

2.1.4.3 Reactant Storage

Storage of a reactant by the catalyst offers another explanation of enhancement through cycling. Thus, *Li et al. (1985b)* and *Nam et al. (1990)* developed a model based on the apparent storage of N_2 in the catalyst bulk. *Li et al.* suggested that there could be separate adsorption sites for nitrogen and hydrogen whereby only the sites for nitrogen provide access to the catalyst bulk. As **Figure 2-4** suggests, a separate-sites model matches experimental data for a step change from pure H_2 to a mixture of H_2 and N_2 at 400°C much better than a competitive adsorption model.

Reactant inhibition cannot occur if reactants adsorb on separate sites. *Rambeau and Amariglio (1978a, b)* did not report reactant inhibition for their promoted iron catalyst and the widely-used Temkin Pyzhev model does not indicate reactant inhibition. However, application of the separate adsorption sites model does not predict experimental observations. Storage of nitrogen by the catalyst appears to be necessary. Such considerations led *Nam et al. (1990)* to systematically examine several storage models. The catalyst surface was divided into catalytically active or inactive. N_2 adsorption, however, could occur on either type and diffusion could proceed between them. Catalysts below the surface could store N_2 and, in one model, exchange it with the gas phase. The sub-surface thus was a source of N_2 when the gas phase was predominantly H_2. Details of the models are given by *Nam et al. (1990)* and summarized by *Silveston (1998)*.

Parameters for the models were obtained by fitting the step-change, steady-state and composition-forcing data of *Jain et al. (1982a, 1983)* and the models were tested against experimental data. Only surface storage and bulk-dissolution models predicted resonance with an enhancement greater than one. **Figure 2-5** shows the agreement between experiments and the bulk dissolution models. This model allows for exchange of N_2 between the bulk catalyst and the gas phase. The single-site version of this model assumes that N_2 and H_2 adsorb on the

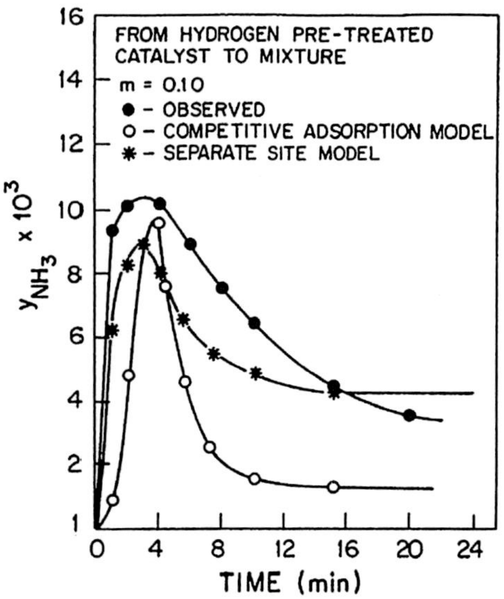

FIGURE 2-4 **Comparison of predicted response and experimental data for a step-change from H_2 to a synthesis gas mixture at 400°C and 2.38 MPa in a gradientless reactor.** *(Figure drawn from* Li et al. (1985a) *with permission of the authors.)*

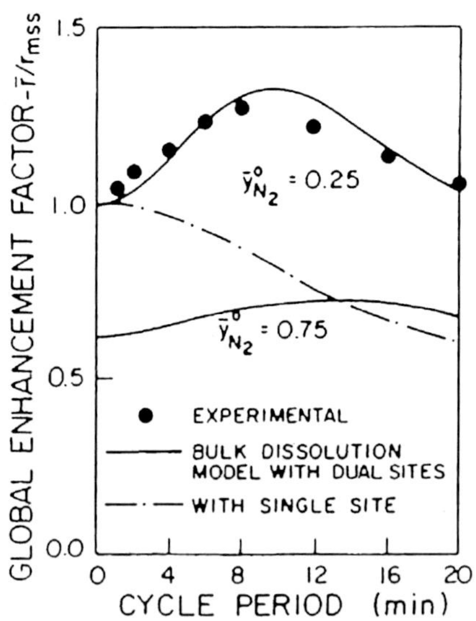

FIGURE 2-5 **Prediction of the bulk dissolution model as a function of the cycle period.** *(Figure reproduced from* Nam et al. (1990) *with the authors' permission.)*

same site. That assumption does not properly reproduce the experimental results. Details are given by *Nam et al. (1990)*.

2.1.4.4 *Surface Activation/Restructuring*

Restructuring of the Fe catalyst surface was proposed by *Chiao et al. (1987)* to explain the observed activation/deactivation behavior after a switch from steady-state to periodic operation. Restructuring resulted in a super-activated catalyst and reflects the observations of *Amariglio and Rambeau (1977)* and *Rambeau and Amariglio (1978c)* that the presence of NH_3 increased catalyst activity by permitting the formation of an iron nitride, whereas exposure to the pure reactants appeared to deactivate the catalyst. According to *Chiao et al. (1987)*, the nitride can be derived only from ammonia. *Chiao and Rinker (1989)* assumed competitive adsorption of the reactants on a single activated site. Parameters for this model were obtained from step-change data and the result was tested against data that the authors acquired from an integral reactor operating at a steadystate. Their model adequately represented reactor performance.

2.1.5 Comments on the Ammonia Synthesis Application

With the promoted iron catalysts universally used, conversion per pass in commercial synthesis reactors is equilibrium-limited and low, under 10%. Reactant recycle is essential. However, recycle with composition switching would likely require separation of the two reactants. Separation of non-condensable gases is an expensive operation. Thus, in order for switching between reactants to be practical, conversion per pass must be high. This requirement seems to be unobtainable with the iron catalysts. Without economic incentive, interest in periodic operation cannot be sustained so it is not surprising that there have been no further studies of modulating ammonia synthesis since *Silveston's (1998)* review.

Many controversies introduced by the work reviewed in this chapter remain unresolved. It is clear, though, that switching between reactants or their mixtures can significantly increase the rate of the NH_3 synthesis reaction. The rate controlling step under conditions of periodic operation appears to be nitrogen transport to active sites of the catalyst or, possibly, nitrogen storage on or in the catalyst. Models based on different rate controlling steps or different surface compositions and structure have been advanced and the most successful of these predict experimental measurements.

Although there is little justification besides scientific curiosity to pursuing cycling studies on iron catalysts, such a statement may not be true for the ruthenium catalyst that can be up to 2000 times more active than the promoted iron catalyst at low temperatures where equilibrium is more favorable. What is needed is an economic study to see if a single pass reactor, without reactant recycle, is feasible. A single pass operation would allow composition modulation.

A variety of hydrogenation reactions under composition modulation have been investigated. Modulated hydrogenation reactions are interesting because with just several exceptions reactant mixtures are reducing so performance improvement will usually arise from the manipulation of adsorbate or adatom concentrations and/or their distribution on the catalyst surface.

2.2 NO_X REDUCTION

NO_x is a group of nitrogen oxides that arise mainly in combustion and include NO. Reduction of NO_x is an important mitigation process for industrial as well as automotive emissions. There are distinct differences between the noble metals-catalyzed reduction of NO_x with H_2 and with CO, so observations made on the more heavily studied CO reaction cannot be transferred to reduction with H_2. Noble metals catalyze the reduction and Pt exhibits the highest NO conversion below 573 K. At a space velocity of 30,000 h^{-1}, conversion goes almost to completion above 670 K using H_2 as the reductant and the activities of Pt, Pd and Rh catalysts appear to be equal.

Table 2-2 is intended to summarize the quite small literature on NO_x reduction under modulation. There is a much larger literature on treatment of vehicle exhausts of which NO_x is a component. This literature is examined in **Chapter 6**.

Modulation experiments undertaken by *Shinjoh et al. (1987)* used a microreactor packed with a noble metal catalyst. These researchers used Rh with symmetrical cycling between an He feed containing 0.3 vol% NO and one with 0.3 vol% H_2 (the stoichiometric ratio). They observed that cycle period and reactor temperature had a large effect on NO conversion under modulation at low temperatures at least. Above 523 K, a strong dependence on frequency was evident. Also, the frequency that maximized conversion increased with temperature so that at 573 K, the period that maximized conversion decreased to 0.2 s from about 0.5 s at 523 K. However, the enhancement of conversion achieved through composition modulation decreased as the catalyst temperatures rose.

Shinjoh et al. mention that the Rh surface appears to be oxidized by NO and this suppressed further NO adsorption. Their kinetic measurements indicate a negative order with respect to NO, a result that suggests NO inhibition of reduction. The order with respect to H_2 is large and positive; as a result, hydrogen probably reduces the Rh surface while functioning as a reactant. Thus, cycling seems to maintain the catalyst surface in a reduced state. NO dissociation is rate controlling and this seems to proceed faster on the rhodium metal. Cycling may also equalize the mean reactant adatom concentrations on rhodium, increasing the surface reaction rate in this way.

NO has been studied under UHVac conditions by surface scientists over many years. *Schütz et al. (1998)*, however, seem to be the only investigation that considered modulation. For these researchers it served as a tool for exploring the fundamental surface processes that proceed in catalysis.

2.3 METHANATION

Table 2-3 summarizes the modulated methanation literature. *Stuchlý and Klusáček (1990)* found that composition modulation increased the time-average rate of CO hydrogenation by up to 25% for an optimal choice of the cycling frequency. The objective of their study was to determine if time-average rates under modulation can exceed the maximum rate attainable under steady-state operation. A feature of this reaction was the slow rise in both reactants after a composition switch: 1 to 2 s were needed. Thereafter, the slope of the CO trace was positive during flow of the mixture over the catalyst, while the slope of H_2 was negative. When only H_2 passed over the catalyst, the H_2 partial pressure rose slowly during the half-cycle and there was a significant CO tail extending 3 to 4 s. The behavior of the two product partial pressures was also different. The CH_4 variation was large and its partial pressure highest when only H_2 passed through the reactor. Water showed just small variation between the half-cycles.

Stuchlý and Klusáček concluded from their observations that methane is formed by hydrogenation of a carbon residue created by the dissociative adsorption of CO. They suggested that the hydrogenation step is rate controlling over this Ni catalyst at the temperatures used. When CO is present in the feed, even below 10 vol%, it floods the catalyst surface, inhibiting the adsorption of hydrogen. Adsorbed hydrogen is apparently a prerequisite for hydrogenation. These observations would

TABLE 2-2 NO_x Reduction

Authors	Objective	Reaction	Modulated Variable	Reaction Conditions	Observations	Comments
Shinjoh et al. (1987)	Catalyst development for automotive emission catalysis.	NO reduction with H_2 and CO	Symmetrical cycling between 0.3 vol% NO and 0.3 vol% H_2.	Rh catalyst; 523 K; $SV = 30\,000\ s^{-1}$; τ up to 2 s.	Temperatures of 300 to 400°C and periods less than 0.25 s were needed in order to have NO conversions above 80%.	NO may oxidize Rh surface. NO adsorption appears to inhibit reaction; H_2 may reduce the surface as well as act as a reactant. NO dissociation is rate-limiting.
Schütz et al. (1998)	Investigation of reactions at domain interfaces and observation of surface reaction waves	NO reduction and NO + CO pulsing on microscopic Pt(100) surface implanted on Rh or TiO_2	H_2 partial pressure in the 10^{-3} mm bar range	$T = 563, 588$ K, modulation frequency from 5 to 500 cycles/s	Large effect of modulation on rate and on selectivity of the hydrogenation reaction. Domain size of surface implants affected modulation behavior.	Surface science experiments were used to elucidate the effect of catalyst microstructure on rate and selectivity under modulation. Very rapid diffusion of H adatoms was observed.

explain the large amount of CH_4 formed after CO is removed from the feed. Water is strongly adsorbed. The large increase in the methanation rate under periodic flushing of the catalyst with H_2 is because flushing increases the hydrogen adatom concentration on the surface and that carbon residues, after exposure to a CO-containing stream, greatly exceed monolayer coverage.

Resonance with respect to cycle period was evident in the authors' results. Between cycle periods of 30 to 60 s, the time-average rate of methanation is 75% greater than the steady-state rate at the corresponding time-average composition of the reactor feed. At a space velocity of 200 min^{-1}, $\tau = 30$ s and a catalyst temperature of 478 K, the rate increased by 125% over the steady-state rate. The largest enhancement in catalyst activity under composition forcing occurred under conditions that are optimal for the steady-state process. This is rarely observed.

The investigators did not model their observations. A dynamic model of a methanation reactor, however, has been proposed by *Van Doesburg and De Jong (1978)* for an alumina supported nickel catalyst and tested under typical industrial operating conditions (0.6 to 2.5 vol% CO or CO_2 in a H_2 stream, $453 \leq T \leq 523$ K). The pseudo-homogeneous, plug flow model was able to closely represent both steady state and step-changes in feed conditions. *Stuchlý and Klusáček* cautioned that a model capable of rendering steady-state behavior would not be expected to reproduce behavior under cycling if rates are at their maxima for the same time-average feed condition and there is global enhancement as well.

Fluidized bed hydrogenation of CO over a nickel faujasite catalyst was described by *Jaeger et al. (1990)*. Nickel faujasite is a zeolite containing strong acid sites that provoke coke formation. The catalyst deactivates rapidly with feed compositions used for methanation, but coking and deactivation can be suppressed by increasing the H_2 partial pressure. While the experiments of *Jaeger et al.* are not periodic composition forcing in the sense used in this chapter, they are interesting for revealing performance that might be expected with composition modulation on a commercial scale. As discussed in **Chapter 1**, a likely method would be to transport catalyst between beds or zones. *Jaeger et al.* constructed a microbed of nickel faujasite powder that could be fluidized by the reactant mixture or by one of the reactants, in their case H_2. The other reactant (CO) was injected at a point within the fluidized bed. The injected reactant created a zone relatively rich in that reactant while elsewhere the bed contained some CO, because of mixing, but the gas phase was rich in the fluidizing reactant, H_2. Fluidization ensures the catalyst passes continuously through the two zones of this bed. The measurements made by *Jaeger et al.* show a large difference between fluidizing the catalyst powder with H_2 and fluidizing it with the premixed reactants. In a following paper, *Jaeger et al. (1992)* explained this observation by noting that, on contact with CO, decomposition or disproportionation deposits carbon on the catalyst surface occurred. Methane forms by hydrogenating these deposits. However, hydrogenation proceeds only through a hydrogen adatom. Excessive carbon deposition prevents adatom formation. Circulating the catalyst through H_2-rich zones in the fluidized bed precludes carbon build-up and allows H_2 adsorption. A further discussion is given by *Silveston (1998)*.

Carbon dioxide can be hydrogenated to methane with virtually 100% selectivity on Ru/TiO_2. *Marwood et al. (1994)* employed modulation of the CO_2 flow rate to investigate the mechanism of this hydrogenation over a 2 wt% Ru on TiO_2 (see **Table 2-3**). The only variable considered was cycle period. Adsorbed formate or carbonate and linearly bonded CO were observed by IR spectroscopy and the concentration of these species varied with the modulation of the CO_2 partial pressure in the feed. Lags of

TABLE 2-3 Methanation

Authors	Objective	Reaction	Modulated Variable	Reaction Conditions	Observations	Comments
Van Doesburg and De Jong (1978)	Investigation of models for CO methanation under unsteady-state operation.	CO or CO_2 methanation in an H_2 stream; Ni/alumina catalyst	0.6 to 2.5 vol% CO or CO_2 in an H_2 stream under for steady-state and for step-change experiments.	Temperatures between 180 and 250°C	Authors chose a pseudo-homogeneous, plug flow model that represented both steady-state and step-change results closely.	
Stuchlý and Klusàček (1990)	Investigation of enhancement under composition modulation.	CO hydrogenation over 29 wt% Ni/SiO_2	Symmetrical modulation between a CO-H_2 mixture and H_2 alone.	Temperatures of 478 and 488 K; quadrupole mass spectrometer monitored reactant consumption and methane formation during a cycle; $SV = 200\ min^{-1}$; $\tau = 20$ s	Variation in methane concentration was greatest when H_2 passed over the catalyst. Water showed just small variation between half-cycles. Resonance in time-average methanation rate was observed at periods between 30 and 60 s.	CH_4 was apparently formed by hydrogenation of a carbon residue created by dissociative adsorption of CO. Hydrogenation step was assumed rate-controlling. Greatest enhancement occurred under conditions optimal for steady state.
Jaeger et al. (1990)	Study of CO hydrogenation in a fluidized bed.	CO hydrogenation over Ni/faujasite catalyst.	Modulation was achieved when fluidized catalyst particles pass between CO-rich and H_2-rich zones of the bed, zones created by feeding the two reactants at different levels in the bed.		Deactivation occurred because of coking on the zeolite catalyst, but was relieved by raising the partial pressure of H_2. Deactivation was suppressed by separating the feeds.	Large difference was observed between fluidizing the catalyst with H_2 and fluidizing it with premixed CO and H_2.

(Continued)

TABLE 2-3 Methanation (*cont'd*)

Authors	Objective	Reaction	Modulated Variable	Reaction Conditions	Observations	Comments
Jaeger et al. (1992)	Study of a fluidized bed with separate reactant feed locations	As above	As above, fluidized bed was mechanically stirred	$573 < T < 673$ K, $8000 <$ GHSV $<$ 10,000 h^{-1}, $d_p \approx 0.35$ mm	Conversion held under 3% to maintain isothermality. Rate kinetically controlled. Separate introduction allowed operation at a constant rate and selectivity.	Separate introduction of H_2 and CO created zones within bed. Circulation of the catalyst between zones allowed for regeneration. Thus catalyst functioned in a periodic regime.
Marwood et al. (1994)	Examined behavior of surface species during catalytic CO_2 hydrogenation to methane under modulation.	CO_2 hydrogenation over 2 wt% Ru/TiO_2 catalyst	Modulated CO_2 concentration; $7 < \tau < 93$ min; symmetrical cycles.	383 K; packed bed reactor with an external recycle loop; FTIR tracked changes in surface species; CO_2 mol fraction varied from 0 to 0.2; H_2:He mol ratio was 4:5.	Reaction occurred with almost 100% selectivity in steady state. Modulating CO_2 concentration enhanced rate of reaction over the comparable steady-state rate.	Surface tracking showed variation in adsorbate concentrations with modulation of CO_2 in the feed.

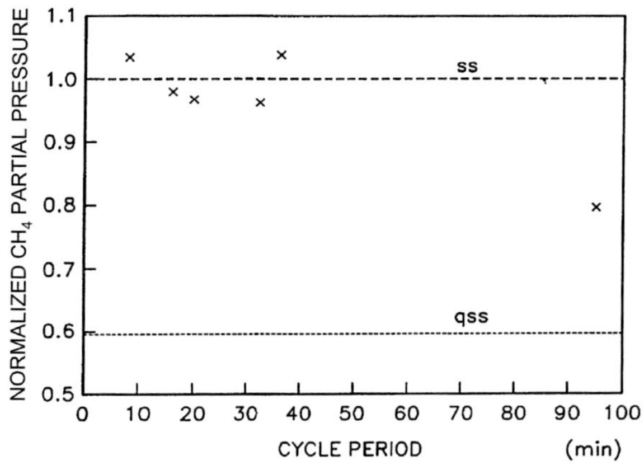

FIGURE 2-6 **Exiting methane partial pressure under modulation normalized by steady state partial pressure obtained operating at the time average of the modulated feed composition for CO2 hydrogenation at 383 K and 103 kPa over a 2 wt.% Ru/TiO2 catalyst.** *(Figure redrawn from Marwood et al. (1994) with permission.)*

signal peaks for each of these adsorbates and CO_2 and CH_4 indicated that formate or carbonate arises on chemisorption of CO_2. These surface species decompose to CO, which is subsequently hydrogenated to CH_4.

Sufficient data were collected to show that modulating the CO_2 concentration enhances the methane partial pressure at the reactor outlet for $\tau < 10$ min; thus, the rate is enhanced as may be seen by a linear fit of the normalized data points in **Figure 2-6**. The research of *Marwood et al.* will be examined further in **Chapter 13**.

2.4 METHANOL SYNTHESIS

Composition forcing of this high-pressure industrial reaction has been examined by two research teams with generally complementary results (**Table 2-4**). Both teams found enhancement under particular modulation strategies and experimental conditions. Objectives shared by the teams were: 1) Behavior of methanol synthesis under composition modulation and 2) identification of those cycling parameters that exert the largest effect on the time-average synthesis rate. *McNeil and Rinker (1994)* focused on whether the global enhancement factor, Ψ^*, exceeded unity. They were also interested to see if modulation provided mechanistic insight. Different catalysts were used by each team.

Methanol synthesis has been also investigated utilizing flow direction switching and simulated moving bed chromatographic reactors. These investigations will be discussed in **Chapters 18** and **21**.

2.4.1 Copper-Zinc-Alumina Catalysts

Nappi et al. (1985) describe an exploratory study using a CuO:ZnO:Al_2O_3 catalyst (see **Table 2-4**). Synthesis gas for methanol usually contains CO_2. Both carbon oxides apparently can be hydrogenated to MeOH; thus several different cycling strategies are possible. *Nappi et al.* tested two that are listed in **Table 2-4**. Variables considered were cycle split, s, (defined in terms of the duration of the high CO exposure), cycle period, τ, and the time-average CO concentration in the reactor feed. **Table 2-5** shows the enhancement, Ψ, for the CO:H_2 modulation experiments. Evidently only bang-bang switching between CO and H_2 increased the synthesis rate. However, the increase was small and within the bounds of measurement error. The enhancement factor, Ψ, increased uniformly as the cycle period was reduced so shorter cycle periods should have been used. Unfortunately, mixing in the gradientless reactor prevented the use of periods shorter than six minutes. Some methane was formed, probably due to iron impurities in the Al_2O_3, but the amounts were small and interpretation was clouded by the presence of CH_4 in the CO source.

Periodic dosing of a constant CO:H_2 feed mixture with 2 vol% CO_2 appeared to be a more effective means of increasing the synthesis rate. At this level, up to a 27% increase in the synthesis rate was projected based on estimated values of the steady-state synthesis rates.

TABLE 2-4 Methanol Synthesis

Authors	Objective	Reaction	Modulated Variable	Reaction Conditions	Observations	Comments
Nappi et al. (1985)	Exploration of methanol synthesis under concentration modulation.	H_2 and CO reactants with varied CO_2 content; catalyst was CuO/ZnO/Al_2O_3 = 27:44:29 (wt%)	1) Modulation of CO/H_2 ratio at constant CO_2; 2) modulation of CO_2 concentration from 0 to 2 vol% at a constant CO/H_2 ratio.	$P = 2.51$ MPa; $T = 250°C$; τ from 6 to 60 min; CSTR (Berty type)	For strategy 1), steady-state conversion exceeded that under modulation, except at the shortest cycle period. Strategy 2) enhanced performance by pulsing CO_2 in a 12 min cycle period.	Small variation in rates of methanol formation suggested its strong adsorption on the catalyst.
Chanchlani et al. (1992, 1994)	Study of methanol synthesis using different modulation strategies and catalysts.	As with *Nappi et al.* using two catalysts a) 30:70 CuO/ZnO catalyst and b) 60:30:10 CuO/ZnO/Al_2O_3 catalyst.	As above but 1) maintaining one reactant constant and modulating the other two; 2) pulsing one reactant while holding the other two constant.	Variations made in τ, s and CO_2 content; $P = 2.86$ MPa; $T = 498$ K, $SV = 37$ mL(STP)/g_{cat}•s in an isothermal, fixed bed reactor.	Symmetrical modulation gave yields up to 4-fold higher than under steady-state. Methanol synthesis rates doubled under modulation of H_2/CO reactor feed for CuO/ZnO catalyst; increased up to 35% for CuO/ZnO/Al_2O_3 catalyst.	Methanol synthesis rates showed no enhancement through modulation in the absence of CO_2 in the feed.
McNeil and Rinker (1994)	Study of composition modulation of a high-pressure industrial reactor.	H_2 and CO reactants with addition of 2 or 3 vol% CO_2; ICI 51-2 catalyst; BASF S 2-85 catalyst	Bang-bang cycling of syngas reactants.	Isothermal, differential reactor with cycle period τ as short as 12 s; $P = 2.86$ MPa; $T = 513$ K.	Results were similar to those of *Chanchlani et al. (1992)*: 25% rate enhancement with BASF catalyst; 15% enhancement with ICI catalyst.	Unlike *Chanchlani et al. (1992)*, these authors found a cycle period dependence for the BASF catalyst, though not for the ICI catalyst.

TABLE 2-5 Influence of Modulation Variables on Enhancement of Methanol Synthesis Rate over a $CuO/ZnO/Al_2O_3$ Catalyst at 2.51 MPa and 523 K (*Nappi et al., 1985*)

τ (min)	s (-)	$(y_{CO})_1/(y_{CO})_2$[#]	Ψ (-)
6	0.333	∞	1.05
9	0.333	∞	1.05
12	0.333	∞	1.03
30	0.325	3	0.87
60	0.325	3	0.86
18	0.583	4.7	0.87
30	0.583	4.7	0.87
60	0.583	4.7	0.86

[#] *ratio of the CO mol fraction in two parts of the cycle*

Variations of methanol and carbon monoxide leaving the reactor with time were measured for two of the CO-H_2 modulation experiments. Variations of CO with time in the cycle for the bang-bang experiment were remarkably small. This lack of variation was due in part to physical mixing in the reactor used, but the data nonetheless suggested strong adsorption of CO on the catalyst surface. Methanol variation was small as well and lagged the introduction of CO by four to five minutes, indicating CO inhibition and probably rate control by a surface reaction. The small methanol variation also suggested that methanol was strongly adsorbed by the catalyst.

2.4.2 Copper Zinc Catalyst

Considering that CO_2 might be directly hydrogenated over CuO/ZnO catalysts, *Chanchlani et al. (1992, 1994)* re-examined composition modulation of methanol synthesis because it offered the two parallel reaction paths to the same product and three reactants for possible manipulation. They employed two-component (CuO:ZnO = 30:70) and three-component ($CuO:ZnO:Al_2O_3$ = 60:30:10) catalyst formulations. These oxide ratios were chosen because each exhibited the highest methanol activity in its respective two- or three-component class (*Herman et al., 1979*) and each had been previously characterized by other investigators. Composition modulation of $H_2:CO_2$ mixtures retarded the rate of deactivation by about 50%, compared with deactivation in steady-state operation. When an H_2:CO mixture was used, modulation of the H_2:CO ratio did not affect deactivation. Deactivation was much slower when CO_2 was present in the reactor feed, suggesting that catalyst reduction may cause deactivation.

Strategies tested by *Chanchlani et al. (1994)* are given in **Table 2-4**. Not all of the strategies tested enhanced the synthesis rate. Those that provided enhancement were cycling between H_2-CO_2 mixtures, H_2-CO mixtures at high CO_2 partial pressures, and between CO-CO_2 mixtures in the presence of H_2 but with low fractions of CO_2 in the mixtures. Most strategies were tested on the two-component catalyst.

Chanchlani et al. (1994) carried out further experiments to measure the influence of the forcing parameters on rate and yield enhancement under modulation of the $H_2:CO_2$ ratio in the absence of CO. Two groups of experiments were performed. In the first, the time-average $H_2:CO_2$ ratio in the feed corresponded to the ratio that maximized the methanol synthesis under steady-state operation, while in the second, the time-average ratio was appreciably larger than the optimum steady-state ratio. In both experiments, rate enhancement was observed and was as much as 30% greater than the highest synthesis rate measured under steady-state operation at the same pressure, temperature and space velocity. Yield is defined as the fraction of the carbon oxide converted that is recovered as methanol. Results indicated a remarkable increase in yield through symmetrical modulation. Furthermore, in the range studied, cycle period and amplitude had just a small and uncertain influence on yield. The rate data also showed a small but consistent effect on the enhancement. Yield results can be explained by suppression of the reverse water gas shift reaction for the two-component (CuO/ZnO) catalyst. Some methanation occurred on the CuO/ZnO/Al_2O_3 catalyst but this too was suppressed by feed composition modulation.

Cycle periods were long enough to measure composition and rate variation during a cycle after a reproducible cycling state was attained. About two to three minutes were needed for the CO_2 in the reactor effluent to reach the feed concentration. Methanol formation reached a maximum in the CO_2-rich half-cycle, well before the CO_2 in the outlet became constant. Both methanol and water rose immediately after the switch to an H_2-rich mixture and then declined. These observations suggest competitive adsorption of CO_2 with at least one other species, probably H_2O and/or H_2. The slow rise in methanol formation after the switch to the hydrogen-rich feed indicated that surface hydrogenation is rate controlling. Methane is attributed to iron impurities in the alumina, but impurity levels were not tested. Observations made using a 60 min cycle period showed that the CH_3OH concentration in the reactor outlet was still changing in the half-cycle. This observation probably accounts for the surprisingly small influence of period on rate and yield.

Figure 2-7 sums up the *Chanchlani* study. The data points are for the cycling experiments discussed above dealing with the modulation of the H_2:CO_2 ratio in the reactor feed at 2.86 MPa and 225°C. The dashed line joining the extreme points indicates the maximum methanol synthesis rates achievable under composition modulation.

Figure 2-7(a) for the CuO/ZnO catalyst shows composition modulation of the H_2:CO_2 reactor feed is capable of doubling the methanol synthesis rate at the conditions considered by *Chanchlani et al.* The synthesis rate increased

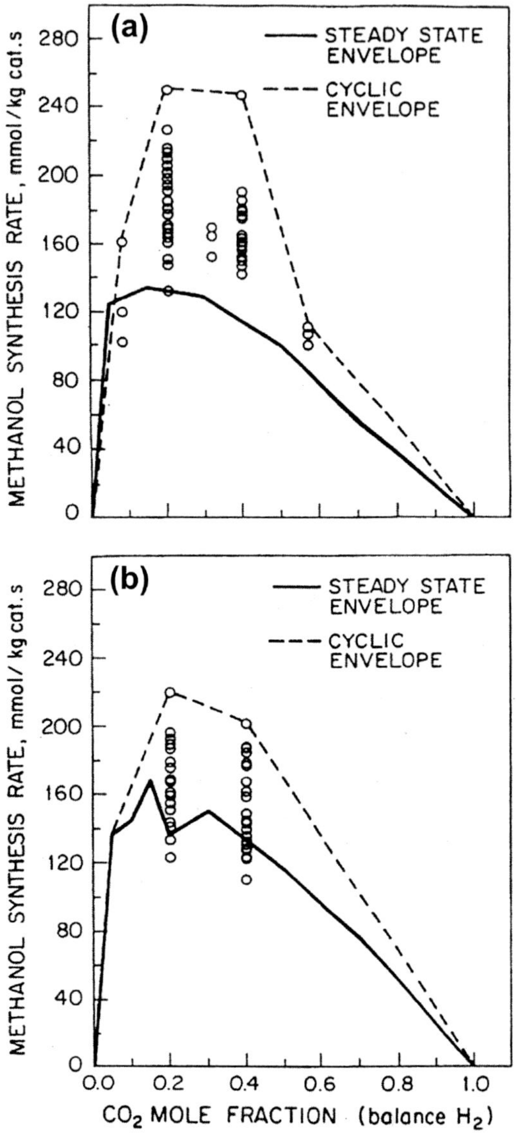

FIGURE 2-7 **Comparison of modulation and steady-state rates of methanol synthesis as a function of feed composition at 2.86 MPa and 225°C: (a) Results for the CuO/ZnO catalyst, (b) results for the CuO/ZnO/Al_2O_3 catalyst.** *(Figure reproduced from* Chanchlani et al. (1994) *with the authors' permission.)*

by up to 35% for the $CuO/ZnO/Al_2O_3$ catalyst as **Figure 2-7(b)** shows. Contrary to the rate results, the yield enhancement was much greater for the three component catalyst than for the two component example.

Chanchlani et al. (1994) did not intend to investigate the synthesis mechanism in their study. They remarked, nevertheless, that the large rate and yield enhancements they observed for $H_2:CO_2$ modulation were caused by an alteration of the catalyst surface that provided a more optimal distribution of carbon oxide and H atom species than that possible in steady-state operation.

2.4.3 Commercial Methanol Catalysts

McNeil and Rinker (1994) obtained results similar to **Figure 2-7(b)** employing two commercial catalysts (ICI 51-2 and BASF S 2-85) and cycling between H_2 and CO in the presence of 2 or 3 vol% CO_2. These two catalysts have compositions close to the $CuO/ZnO/Al_2O_3$ catalyst used by *Chanchlani et al. McNeil and Rinker* built an isothermal, plug flow microreactor for differential conversion to operate with cycle periods as short as 12 s. They observed that synthesis rates were the highest at short periods. With the BASF catalyst, this bang-bang modulation gave a global enhancement of up to 25% at the conditions studied. For the ICI catalyst, the global enhancement was 15% at best.

Unlike the data of *Chanchlani et al.* discussed above, *McNeil and Rinker* found that the rate enhancement depended on cycle period for the BASF catalyst, but not for the ICI catalyst. The effect of cycle split on the enhancement was small or negligible. For similar experiments, there is some agreement between the bang-bang cycling results of *Chanchlani et al.* and those of *McNeil and Rinker.* Both show enhancement, but *McNeil and Rinker* used just a low CO_2 mol fraction (0.02 to 0.03). *Chanchlani* observed that enhancement occurs only if the mol fraction of CO_2 is high enough (> 0.1). *McNeil and Rinker,* like *Chanchlani et al.*, experimented with cycling between H_2-CO mixtures in the presence of 2 vol% CO_2, but they did not observe enhancement. *Chanchlani et al.*, however, did see rate enhancement.

Experiments were run by *McNeil and Rinker* on periodically switching between CO and CO_2 while holding H_2 at 61% by volume. Their ICI catalyst was used. Global enhancement was not found for the 12 and 24 s period and $0.25 \leq s \leq 0.8$ range examined, although these authors observed enhancement over the steady-state rate measured at the time-average CO_2 mol fraction in the feed. Low cycle splits with respect to CO_2 suppressed the synthesis rate, but cycle period did not affect the enhancement. Results of *Chanchlani et al.* for this strategy, showed some rate enhancement, but, as with *McNeil and Rinker,* it was not deemed promising.

2.5 ETHYLENE HYDROGENATION

Helmrich et al. (1974) used this reaction to demonstrate experimentally the activation of a heterogeneous catalytic reaction by composition modulation. The reaction was also used by *Baiker and Richarz (1976)* to examine the effect of modulating other reactor operating conditions. *Prairie and Bailey (1987)* modulated ethylene hydrogenation to test dynamic models for the reaction (see **Table 2-6**).

Helmrich et al. (1974) employed a $Pt/\gamma\text{-}Al_2O_3$ catalyst but at low temperature so that only double bond hydrogenation occurred. Their experimental system employed a laboratory scale tubular reactor operated non-differentially. Their results showed a substantial conversion improvement through modulation. The maximum increase in conversion exceeded 50% and appeared to be independent of space velocity. At 75°C, the percentage increase in the conversion is much smaller, but conversion might have been greater if a shorter cycle period had been chosen. The time-average conversion under modulation was independent of the vol%

TABLE 2-6 Ethylene Hydrogenation

Authors	Objective	Reaction	Modulated Variable	Reaction conditions	Observations	Comments
Helmrich et al. (1974); Baiker and Richarz (1976)	Investigation of composition modulation	C_2H_4 hydrogenation over Pt/γ-alumina catalyst	H_2 concentration	$0.075 < y_{ethylene} < 0.30$; 64°C < T < 75°C; 0.106 < SV < .303 s^{-1}; 0.6 < s < 0.95; 2 < τ < 80 s	Conversions of C_2H_4 under modulation were greater than under steady-state operation over a wide range of conditions.	Rate of reaction seemed to be inhibited by C_2H_4. A model was developed that closely predicted the modulation results.
Prairie and Bailey (1986)	Use of feedback to cause bifurcation as a means of model identification.	As above	H_2 and C_2H_4 flow rates were augmented based on reactor output to induce bifurcation.	CSTR with SV = 0.1 min^{-1}; P = 1 atm; T = 30°C	Hopf bifurcations were observed but difficulty occurred as described in the paper.	Observations indicated a large reservoir for surface hydrogen, suggesting that hydrogen spillover occurred between Pt and support.
Prairie and Bailey (1987)	Examination of dynamic methods of estimating parameters for kinetic models	As above	H_2 and C_2H_4 concentrations were cycled symmetrically. Feedback-induced Hopf bifurcation experiments were also performed.	P = 1 atm; T = 80°C; SV = 0.25 s^{-1}	Composition modulation rates were higher than steady-state rates for similar time-average compositions for all conditions studied. The authors' dynamic model failed to reproduce their results.	Authors concluded that a comparison with cycling data is likely the most rigorous test for the validity of a model.
Pinkerton and Luss (2007)	Investigation of oscillating hot spots in catalytic hydrogenation of acetylene and CO oxidation	Hydrogenation of acetylene in ethylene-rich mixtures in a shallow bed of Pd catalyst pellets	99.99% hydrogen, 99.99% ethylene, 4% acetylene in methane were fed at constant flow rates	Cylindrical catalyst pellets (4 mm o.d. and 4 mm long) containing 0.02% Pd placed on a stainless steel mesh to form a shallow bed (one pellet high ≈ 4 mm). Observations used an IR camera.	Moving hot zones in the shallow packed bed exhibited periodic antiphase behavior in most experiments. Periods were 35 to 50 min. Temperatures oscillated from 3 to 30° for maximum catalyst temperatures of 78 to 161°C and ethylene feeds ≥ 22 vol%.	Results appear to confirm theoretical predictions that hot zones are likely to form when the catalytic reaction rate oscillates under constant ambient conditions.

ethylene in the feed for the narrow range of concentrations used. The maximum conversion under H_2 pulsing for the three experiments performed varied between 75 and 83% and were well above the 47% obtained at steady state.

Helmrich et al. measured the steady-state kinetics of the Pt catalyzed hydrogenation and demonstrated that it is reactant inhibited in the temperature range studied. They worked out a simple kinetic model that they applied to their modulation measurements, but this model under-predicted the experimental conversions. They interpreted this failure to mean that the rate controlling step is the surface hydrogenation of adsorbed ethylene, as implied by their asymmetrical cycling results.

Experimental bifurcation analysis was employed by *Prairie and Bailey (1986)* to discriminate between rival models for ethylene hydrogenation. These early experiments concluded that H_2 spillover was important for this system.

The objective of *Prairie and Bailey (1987)* was primarily to test dynamic methods of determining parameters for kinetic models. Composition modulation was used to test the models obtained. Switching periodically from a mixture to H_2 or to the diluent alone significantly raised the ethylene conversion for some cycle frequencies. At the optimal period, there was about a 30% increase in rate. The steady-state rate reached a maximum at 10 vol% ethylene, but the hydrogenation rate was less than that for the modulated reaction at the same mean feed condition. Consequently, the time-averaged rate represented a true global enhancement. Indeed, modulation increased the hydrogenation rate over that at steady state for all concentrations of ethylene tested.

From their cycling experiments, *Prairie and Bailey* observed a slow desorption of C_2H_4 after introduction of H_2. They also found a weak deactivation of the catalyst at long cycle periods that they attributed to ethylene poisoning. Results for a third set of experiments, switching between a reactant mixture and diluent, confirmed time-average hydrogenation rates were higher than the corresponding steady-state rates. This finding implies that deactivation can be reversed by removing ethylene from the reactor feed. Finally, *Prairie and Bailey (1987)* observed that their dynamic model could not reproduce the results discussed above.

2.6 AROMATICS HYDROGENATION

Toluene hydrogenation was employed by *Baiker and Bergougnan (1985a, b)* in their ambitious study of the use of fixed bed reactor dynamics to estimate rate and transport parameters through measurements obtained under modulation. Their modulation consisted of: 1) A ramp up followed immediately by a ramp down, and 2) imposition of a sinusoidal variation. These forcing disturbances were applied simultaneously to each of several variables. Results presented by these researchers appeared to demonstrate the efficiency of their estimation technique. However, to obtain reliable results, the forcing frequencies had to be chosen to reflect the relaxation dynamics of the reactor-reaction system. *Silveston (1998)* provides a summary of this difficult study.

Benzene hydrogenation was studied by *Rao et al. (1992)* through some experiments summarized in **Table 2-7**. Three sets were performed: 1) Cycling of only the benzene flow, 2) cycling of only the H_2 flow and 3) steady-state operation with the ratio of benzene:hydrogen feed rates maintained at 1:4. In 2) the time-average benzene:hydrogen ratio was the same as in 3). Forcing was symmetrical and cycle periods were 20 and 60 minutes. Variation of benzene conversion and mean catalyst pellet temperature between successive cycles as well as the limited experimentation made interpretation of the result uncertain. Nevertheless, time-average benzene conversions at $\tau = 20$ min appeared to be about 20 to 25% greater than under steady state,

TABLE 2-7 Hydrogenation of Aromatics

Authors	Objective	Reaction	Modulated Variable	Reaction conditions	Observations	Comments
Baiker and Bergougnan (1985a, b)	Derivation of rate parameters from composition modulation.	Toluene hydrogenation	Sinusoidal and ramp (up-down) simultaneous forcing of toluene concentration, total pressure, jacket temperature, and gas flow rate were carried out employing different periods.	Reaction carried out in a 0.4 m jacketed tube.	Kinetic and transport numbers were obtained from radial and axial profiles by matching first derivatives of the model at collocation points to those calculated from the data.	Forcing frequencies were chosen to reflect relaxation dynamics of the reactor-reaction system.
Rao et al. (1992)	Investigation of composition modulation	Hydrogenation of benzene	1) Modulation of benzene flow rate with a constant H_2 flow rate; 2) modulation of H_2 flow rate with a constant benzene flow rate.	$T = 172°C$; $P = 2$ MPa; $\tau = 20$, 60 min	20 to 25% enhancements occurred in benzene conversion for 20 min period. For 60 min period, conversions were below the comparable steady-state values.	

whereas at $\tau = 60$ min, conversions were less than at steady state. *Rao et al.* attributed the increased conversion to alteration of heat storage and heat removal from the catalyst pellet through the periodic change in composition but did not provide support for their views.

Visser et al. (1994) examined modulating the gas phase composition in a 3-phase reactor running the hydrogenation of phenylacetylene. Although the investigators were concerned with phenylacetylene conversion, hydrogenation is a consecutive reaction so selectivity becomes an important issue. The *Visser et al.* paper will be discussed in **Chapter 17**.

Reshetnikov et al. (2003), working in cooperation with Renken's laboratory developed a model for the hydrogenation of benzene and the hydrogenolysis of thiophene that allowed for their mutual inhibition effects. Later, *Reshetnikov et al. (2007)* used step-change experiments to evaluate parameters for this model. The authors then used their model to predict that modulation between a benzene-in-H_2 stream and a stream that also contained thiophene (in the ratio of Bz:Th = 9:1) could substantially increase the hydrogenation of benzene to cyclohexane. According to the model, the mean conversion could reach 24%. This conversion would represent a large improvement over a steady-state operation using either benzene or the Bz-Th mixture in the feed.

2.7 OSCILLATORY BEHAVIOR

Experiments by *Pinkerton and Luss (2007)* demonstrated the presence of periodic-like temperature variations in space and time across a shallow catalyst packed bed during the hydrogenation of mixtures of ethylene and acetylene. Two different temperature patterns were observed: 1) A hot spot that moved across the catalyst bed but eventually returned to its original location and 2) a hot spot that remained stationary. In both cases, the temperature distribution was nonuniform under steady-state operation and effluent compositions remained constant. However, temperature inhomogeneities under steady state fall outside of the focus in this chapter.

Nomenclature

Bz	=	benzene
C	=	concentration (mol/m^3 usually)
C$_i$	=	concentration of species i (mol/m^3)
CSTR	=	continuous stirred tank reactor
c	=	concentration in the catalyst phase (mol/cm^3)
IR	=	infra red
k	=	reaction rate constant (various units)
k$_{ad}$	=	reaction rate constant for adsorption (various units)
k$_p$	=	reaction rate constant in terms of partial pressure (various units)
m	=	reactant ratio in cycling mixtures (-)
P	=	total pressure (kPa)
P$_i$	=	partial pressure of component i
qss	=	quasi-steady state
r	=	rate of change (usually reaction) (mol/m^3·s)
r$_{ad}$	=	rate of adsorption (various units)
SV	=	space velocity (s^{-1})
s	=	cycle split (-)
ss	=	steady-state
T	=	temperature (K)
t	=	time (s)
t_{N_2} t_{H_2}	=	exposure durations for N_2, H_2
y	=	mol fraction (usually in the gas phase) (-)
y$_i$	=	mol fraction of the ith component (-)
y°	=	mol fraction at time 0 or in the reactor feed (-)
$\bar{y}$	=	time-average mol fraction

Greek

ψ	=	enhancement (-)
ψ*	=	global enhancement (-)
τ	=	cycle period (s)

Subscripts

ad	=	adsorption
i	=	reactant species
p	=	pressure

Superscripts

0	=	initial or entering
*	=	optimal or maximum

CHAPTER

3

Catalytic Oxidation and Reduction of Gases

Robert Ross Hudgins, Peter Lewis Silveston*, Albert Renken†*

*Waterloo, Ontario, Canada and †Lausanne, Switzerland

OUTLINE

3.1 Introduction 49

3.2 CO Oxidation 50

3.2.1 Noble Metal Catalysts 50

3.2.1.1 Modulation in the Presence of Autonomous Oscillations 50

3.2.2 Single Crystal Catalysts 58

3.2.3 Studies at Temperatures above 250°C 61

3.2.4 Miscellaneous Catalysts 61

3.2.5 Modeling 63

3.3 Sulfur Dioxide Oxidation 66

3.3.1 Modeling 69

3.4 Reduction of SO_3 by CO Over Platinum 70

3.5 Reduction of Nitrogen Oxides 70

3.5.1 NO Reduction over Supported Catalyst 71

3.5.1.1 Pt Single Crystals 75

3.5.2 N_2O Reduction 76

3.6 Traveling Waves in Packed Beds 76

3.1 INTRODUCTION

Much of the research on the periodic operation of chemical reactors has focused on oxidation or reduction reactions. Periodically operated combustors, discussed in **Chapter 5**, are commercial devices and early studies of such systems may have inspired the exploration of the processes dealt with in part of this chapter.

Much of the periodic operations research on oxidation has dealt with CO oxidation. It is a reaction that is easily studied: It proceeds at temperatures below 100°C on many catalysts; the only side-reaction, CO disproportionation, does not take place in an oxidative environment;

and the reverse reaction, $CO_2 \rightarrow CO + 1/2\ O_2$, is not important at low temperatures. The single oxidation product, CO_2, can be accurately measured by various methods. The reaction is particularly interesting to explore because multiplicity and autonomous oscillations can arise. Furthermore, carbon monoxide is an important industrial air and vehicular pollutant. CO also functions as a reductant for the nitrogen oxides, important pollutants in many industrial off-gases.

3.2 CO OXIDATION

3.2.1 Noble Metal Catalysts

Much of the research on CO oxidation has been conducted using supported noble metal catalysts such as Pt or Pt alloyed with other metals such as Rh, Pd and Zn. The support generally has been Al_2O_3. Considerable effort has been devoted to single crystals of these noble metals. Experiments have also been conducted on oxide catalysts: NiO/Al_2O_3, K_2O-V_2O_5, CuO/Al_2O_3, as well as hopcalite (a mixed oxide of Mn, Cu and Co). Much of the experimental literature on CO oxidation under modulation is summarized in **Table 3-1**.

One of the earliest studies (*Cutlip, 1979*) used a Pt catalyst and established that the surface coverage changed drastically in each partial cycle. Reasoning that the reforming of surface was a source of higher catalyst activity, *Barshad and Gulari (1985a)* undertook a set of experiments which showed that composition modulation improved the distribution of adsorbed CO and oxygen on the surface. **Figure 3-1** shows CO conversion as a function of the period and split for composition modulation over supported Pt. The maximum concentration of CO_2 measured by *Barshad and Gulari* is almost ten times that under steady-state operation. The interpretation made by these authors is shown in **Figure 3-2**, namely, periodic operation improves distribution on the surface through a finer division of regions occupied by adsorbed oxygen and CO. In addition, regions of mixed adsorption of these reactants are observed. It is these mixed regions and the larger regions of contact between domains of pure reactant adsorption that account for the enhancement, as suggested by the figure.

A thorough examination of composition modulation on 0.5 wt% Pt/alumina was reported by *Lynch* and co-workers and is summarized in the table. They explained the observed behavior through varying coverage of O-atoms and CO adsorbed on the surface. They used this concept in their model to explain, for example, the effect of frequency on enhancement of the oxidation rate. *Graham and Lynch (1984, 1987, 1988)* also observed an effect of phase lag on enhancement. Strong enhancement required introduction of CO after that of O_2. Further details of the *Lynch* papers—indeed of all the pre-1998 papers mentioned in **Table 3-1**—are given by *Silveston (1998)*.

3.2.1.1 Modulation in the Presence of Autonomous Oscillations

A few chemical systems exhibit regions of autonomous oscillations as the gas phase composition varies or system temperature (or pressure) changes. CO oxidation over Pt surfaces is perhaps the most widely studied of such systems. *Kapička and Marek (1989a, b)* described the transition from steady state to chaotic oscillations for CO oxidation over an industrial, alumina-supported Pt catalyst either in a packed bed or as an individual pellet. The authors attributed the oscillations to synchronization between many individual active centers on the catalyst. This behavior is remarkably similar to those seen on single crystals as discussed in **Section 3.2.2**.

Interaction of composition modulation and autonomous oscillations over a supported Pt catalyst was explored by *Keil and Wicke (1980)* under conditions indicated in **Table 3-1**. Behavior depended on the levels of CO coverage

TABLE 3-1 Modulation Studies of the Catalytic Oxidation of Carbon Monoxide

Reference	Study Objective	Catalyst and Reaction Conditions	Modulated Variable	Observations	Comments
Cutlip (1979)	Mechanism of rate enhancement by composition modulation	0.5 wt% Pt/γ-alumina catalyst, 60°C, 1 atm	CO/O_2, cycle split	Sharp resonance in oxidation rate at $s = 0.67$, $\tau = 1$ min; rate falls to zero at $\tau = 3$ min.	Surface saturated by CO lowered time-average rate of CO oxidation.
Casanova et al. (1979); Baiker et al. (1980)	Effect of periodic operation on catalyst and reactor behavior for CO oxidation	Hopcalite catalyst (Mn, Cu, Co oxides); bed was well-insulated to approximate an adiabatic condition	CO switching. Feed mole fractions of CO were within the range 0.012 to 0.028	Maximum temperature was 470°C under composition modulation; steady-state temperature at the same time-average CO concentration was 500°C.	Composition modulation reduced the maximum temperature within the bed.
Keil and Wicke (1980)	Composition forcing in the presence of multiplicity and autonomous oscillation	CO oxidation; 0.3 wt% Pt/γ-alumina; 3 mm × 3 mm cyl. particles, 275°C	CO and O_2 concentrations in symmetrical cycles for $\tau = 8$ s; $s = 0.5$	Three regions were identified: I ignition II limit cycle and III quenched state.	Cycling between I and II did not increase conversion. Cycling between II and III more than doubled conversion over steady state.
Abdul-Kareem et al. (1980a); Silveston and Hudgins (1981); Silveston et al. (1983); Chakrabarty et al. (1982)	Influence of the catalyst in periodic forcing	CO oxidation; 0.1 wt% Pt/γ-alumina catalyst	CO and O_2 concentrations with time-average vol. ratios of 0.36 and 0.86; $T = 247$°C	Enhancements over steady-state CO reaction rates were found. Large rate increases occurred at $\tau < 2$ min.	For periodic operation at time-average $CO/O_2 = 0.36$, enhancement explained by system ignition (multiplicity). At 0.86, however, multiplicity cannot account for enhancement.
Oh and Hegedus (1982)	Effect of rapid concentration forcing	CO oxidation; 1 wt% Pt/γ-alumina wafer; Rh also used. Fast response IR spectroscopy for adsorption on wafer mounted in transmission cell at 450°C	CO concentration	From CO concentration step-changes monitored at 2070 cm^{-1}, at 1 Hz, adsorbed CO closely tracks the gas phase modulation.	At 450°C, adsorption and desorption were rapid enough that equilibrium can be assumed, even though diffusion slowed response within the wafer.

(Continued)

TABLE 3-1 Modulation Studies of the Catalytic Oxidation of Carbon Monoxide (*cont'd*)

Reference	Study Objective	Catalyst and Reaction Conditions	Modulated Variable	Observations	Comments
Cho (1983); Cho and West (1986)	Role of intraparticle diffusion on concentration modulation	0.1 wt% Pt/γ-alumina catalyst; simulated and experimental results. $200 < T < 550°C$; shell and uniform impregnated catalysts were used; $\omega = 0.5$ Hz	CO and O_2 concentrations; symmetrical cycles	Simulation suggested that cycling effects are confined to outer layers of a catalyst pellet. Major benefits of concentration modulation occur < 300°C.	Modulation phenomena arise from interactions amongst stoichiometric number, temperature, diffusion and reaction rates.
Graham and Lynch (1984, 1987, 1988, 1990)	Effect of modulation for a Pt catalyst with introduction of phase-lead between two feed concentrations	CO oxidation; 0.5 wt% Pt/γ-alumina catalyst, $T = 90$ to 110°C, $P = 1$ atm, $\tau < 60$ s; packed bed reactor with recycle so as to approach CSTR behavior	Phase-lead between forced CO and O_2 was studied, 3rd gas component was N_2	Optimal phase-lead was between 45 and 90°.	Optimal phase-lead varied only slowly with forcing frequency, ω.
Marek (1985)	Conditions for autonomous oscillations	CO oxidation; Pt/γ-alumina, switching between 7 vol% CO and air at $T = 428$ K, $\tau = 400$ and 600 s; $s = 0.5$	Highly asymmetrical cycles	Autonomous oscillations observed for 5 vol% CO in air at a natural period of 1410 s.	Entrainment of the autonomous oscillations was observed on the 3/1 and 2/1 horns.
Barshad and Gulari (1985a)	Investigate the effect of different cycling variables	0.5 wt% Pt/γ-alumina catalyst, $90 < T < 120°C$, $P = 1$ atm	CO/O_2 ratio, cycle split under fast cycling	Periods were as short as $\tau = 2$ s; split was also important; rates of CO oxidation varied 19-fold.	
Barshad and Gulari (1985b); Barshad et al. (1985); Zhou and Gulari (1986a, b, c)	Identify the nature of CO bonding on Pt	0.5 wt% Pd /γ-alumina catalyst, $T < 413$ K, $P = 1$ atm; use of FTIR and time resolution of 0.6 s in transient experiments.	CO/O_2, cycle split studied, used fast cycling	Different forms of CO-Pt bonding identified.	Island adsorption model was invoked to explain differences between steady state and periodic operation.

Muraki et al. (1985)	Modulation of components of automotive exhaust	CO oxidation; 0.006 wt% of five different noble metals on α -Al_2O_3 catalyst, time-average concentrations of 1.33 vol% CO and 0.665 vol% O_2, $SV = 30\,000\ h^{-1}$; $200 < T < 450°C$	CO and O_2 concentrations, $0.5 < \omega < 5$ Hz; $s = 0.5$	Steady-state and periodic operation compared at same time-average concentrations. Enhancements were greatest at the lowest temperatures; optimum cycle period decreased with increased temperature.	Lowering the stoichiometric number had similar effect to lowering temperature. CO inhibition important in kinetics of CO oxidation over noble metals. Oxidation is controlled by reaction between adsorbed species.
Hugo (1986); Stanitsas (1986)	Effect of periodic operation on catalyst performance	CO oxidation; 10 to 12 wt% NiO/γ-alumina (Girdler (United Catalysts) G66 and T-310 catalysts), $150 < T < 250°C$	CO and O_2 concentrations; symmetrical cycles	Greater time-average CO oxidation rates were inconsistently observed.	Conflicting results likely caused by oxidation state of the catalyst. Step-change experiments revealed transients in catalyst behavior of between 1000 and 3000 s after a concentration step-change.
Onken and Wicke (1986)	Exploration of spontaneous oscillations in a flow reactor	CO oxidation; 0.3 wt% Pt/γ-alumina; adiabatic, 90-mm-deep packed bed reactor, $T_{feed} = 440$ K; feed was 1 vol% CO	CO and O_2 concentrations; symmetrical cycles	Spontaneous oscillations amplified as they moved through the bed.	Power spectral analysis showed conditions for entrainment as well as multiple period response.
Capsaskis and Kenney (1986)	Composition forcing in the presence of autonomous oscillations	0.5 wt% Pt/Al_2O_3, 3 mm particles, well mixed, micro-reactor, $135 < T < 150°C$, Argon diluent	CO and O_2 concentrations, CO stream contained C_3H_6, $20 < \tau < 50$ s	Presence of C_3H_6 resulted in spontaneous oscillations which depended also on gas concentrations and temperature. Behavior was affected by composition modulation.	Responses were predicted from simulation. Entrainment, period doubling and chaos were observed.

(*Continued*)

TABLE 3-1 Modulation Studies of the Catalytic Oxidation of Carbon Monoxide (*cont'd*)

Reference	Study Objective	Catalyst and Reaction Conditions	Modulated Variable	Observations	Comments
Prokopowicz et al. (1987)	Effect of composition modulation with copper catalysts	CO oxidation; 10 wt% CuO/γ-alumina catalyst; 10 wt% CuO/silica catalyst, $T = 220°C$; transmission FTIR used to follow CO_2 produced	CO flow rate	CO_2 wrong-way behavior was identified with alumina catalyst but not with silica catalyst.	Similar observations were made by Barshad et al. (1985a).
Schwankner et al. (1987)	Composition forcing on Pt single crystals	Pt(100) face, $525 < T < 530$ K, high vacuum (10^{-3} Pa)	O_2 partial pressure (CO pressure held constant) in some experiments using temperature modulation	Quasi-periodic behavior occurred under steady operation. Forcing of O_2 partial pressure caused strong entrainment. Period doubling and quasi-periodic response was seen as ω varied.	LEED measurements suggested surface reconstruction occurred under composition forcing.
Shanks and Bailey (1987, 1988)	C_2H_4 epoxidation under periodic operation	CO oxidation; 1 wt% Ag/α-alumina catalyst, $T = 147°C$	CO and O_2 concentrations; symmetrical and asymmetrical cycles, $30 < \tau < 3500$ s	Time-average oxidation rates were between 20% higher and 42% lower than comparable steady-state rates depending on τ and s.	Authors' model assumed O_2 was adsorbed by an Eley-Rideal mechanism and stored on the catalyst.
Vaporciyan et al. (1988)	Periodic operation over bimetallic catalyst	Pt/SnO_2 catalyst; monolith support, $T = 390$ K, $P = 1$ atm	CO/O_2 cycle split, cycle period studied	Autocorrelation analysis revealed quasi-periodic phenomena: once every 12 cycles, another every 175 cycles.	Quasi-periodic phenomena were linked to storage of potentially reactive intermediates on the catalyst surface; only a 13-cycle periodicity was identified for CO oxidation over 0.5 wt% Pt/γ-alumina.
Svensson et al. (1988); Jaeger et al. (1990)	Effect of sinusoidal and sawtooth modulation of CO partial pressure	CO oxidation; Pd/zeolite, $T = 483$ K	CO in N_2 and air; used power spectral analysis	System showed almost complete entrainment under certain CO amplitudes and frequencies; entrainment was lost under other conditions.	Multiple resonances were suggested by the results.

Eiswirth and Ertl (1988)	Composition forcing on Pt single crystals	CO oxidation on a Pt 110 face, $525 < T < 530$ K, high vacuum (10^{-3} Pa)	O_2 partial pressure (CO pressure held constant)	System exhibited autonomous oscillations. Modulation of O_2 partial pressure caused entrainment, period doubling and subharmonic response.	Subharmonic response was quasi-periodic and may have increased amplitude. Behavior was successfully modeled.
Eiswirth et al. (1989)	Composition forcing on Pt single crystals	CO oxidation on a Pt 110 face, $525 < T < 530$ K, high vacuum (10^{-3} Pa)	O_2 partial pressure (CO pressure held constant)	Period doubling was observed at the $t/t_o =$ 1/2 horn, while at the 2 horn, a low amplitude subharmonic appeared. At higher forcing periods, response was quasi-periodic.	Subharmonic response of varying amplitude suggested lower frequency oscillations arise when low forcing frequencies are used.
Luu et al. (1992)	Composition modulation with a vanadia catalyst	CO oxidation; unpromoted vanadia catalyst and K-promoted vanadia catalyst similar to that of *Abdul-Kareem et al. (1980a, b)*	CO and O_2 concentrations; symmetrical cycles	Authors searched for regions of multiplicity and autonomous behavior with unpromoted vanadia catalyst.	No evidence of multiple resonance peaks with either catalyst. Low-frequency oscillation showed low amplitude, quasi-periodic response. Unable to show that the oscillations arose from the catalyst system.
Qin and Wolf (1995)	Stabilization of spontaneous oscillations	Rh/SiO_2 catalyst wafer, $T = 390$ K, $P = 1$ atm	CO/O_2 ratio, cycle period and amplitude	Quenching and variation of the CO_2 product occur with large forcing amplitude and $\tau < 50$ s; enhancements to 30% were measured.	Modulation reduced the turning point temperatures for reaction extinction and ignition on the catalyst.

(*Continued*)

TABLE 3-1 Modulation Studies of the Catalytic Oxidation of Carbon Monoxide (*cont'd*)

Reference	Study Objective	Catalyst and Reaction Conditions	Modulated Variable	Observations	Comments
Nibbelke et al. (1998)	Modulation experiments for model development	CO oxidation over a $Pt/Rh/CeO_2/\gamma\text{-}Al_2O_3$ catalyst for $393 < T < 433$ K	Symmetrical, bang-bang cycles switching between 0.5 vol% CO and 0.5 vol% O_2, both in He with $0.1 < \omega < 0.25$ Hz	Models developed from experiments assumed 3 oxidation pathways: 1) CO and NO adsorbed on separate sites followed by a surface reaction, 2) CO adsorbed on an adsorbed NO molecule and 3) Ceria also contributed oxygen to oxidation.	Authors' model was not tested against experimental data.
Jaree et al. (2001a)	Temperature response to reactant concentration perturbations where multiple steady states are observed	CO oxidation over Pt/A_2O_3 in a packed-bed tubular reactor of 2.5 cm i.d. containing 50 cm depth of catalyst; temperature profile was monitored axially.	CO concentrations varying from 0.3 to 0.8 vol% in an O_2/N_2 mixture were fed at 130°C to the reactor.	For concentration changes within the multiplicity region, temperature fluctuations were within the adiabatic ΔT. Adiabatic ΔT could be surpassed if one concentration extreme lay within the multiplicity region and the other outside it.	Temperature and concentration profiles in the packed-bed reactor at steady state depended on the dynamic history of the reactor prior to the steady-state condition. The results imply serious difficulties for reactor control in the presence of multiplicity.
Sedmak et al. (2004a, b)	Modulation of $CO:O_2$ over a catalyst with O_2 storage	CO oxidation over a $Cu_{0.1}\,Ce_{0.9}\,O_{2\text{-}y}$ catalyst, isothermal operation at $50 < T < 250$°C	Symmetrical bang-bang switching between 1 vol% CO and 1 vol% O_2 both in He, $30 < \tau < 250$ s	CO oxidation proceeds at a high rate in both O_2 lean and O_2 rich parts of a cycle. Catalyst capacity is 0.03 mol/kg. cat.	Catalyst reoxidation in < 15 s at 200°C.
Marek et al. (2006)	Investigation of modulation imposed on autonomous oscillations	CO oxidation on noble metal catalysts and automotive TWCs	Numerical study and simulation of a reaction network of not only CO oxidation, but also oxidation of light hydrocarbons and NO reduction	Lumped parameter model used that demonstrated entrainment of autonomous oscillations leading to period doubling and eventually chaos as amplitude increases.	Model and analysis extended to systems containing light hydrocarbons and to NO reduction by CO.

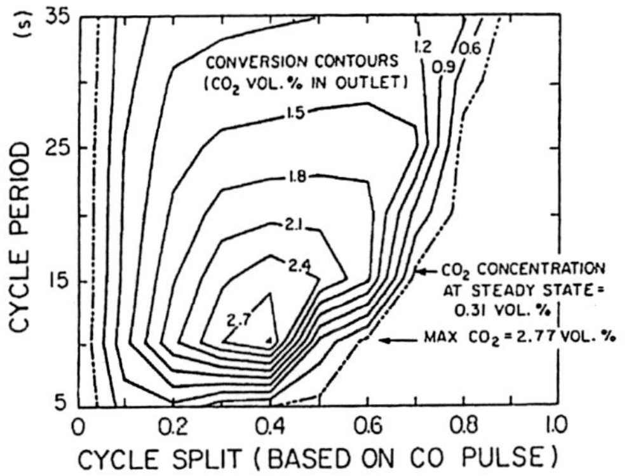

FIGURE 3-1 Contours of equal CO conversion for bang-bang composition modulation as a function of τ and s using a supported Pt catalyst at 1 bar and 120°C. *(Figure adapted from* Barshad and Gulari (1985a) *with permission. © 1985 by the AIChE.)*

of the surface, characterized as an extinguished state, an ignited state, and a state of low coverage. Autonomous oscillations were observed only in the ignited state. Modulation increased the reaction rate in the ignited and extinguished states but not where CO coverage was low.

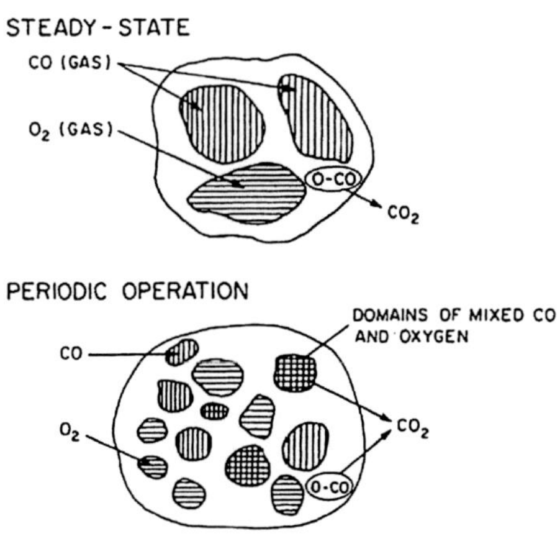

FIGURE 3-2 Schematic of the distribution of adsorbate islands on a Pd catalyst surface under steady state and composition modulation. *(Figure adapted from* Zhou et al. (1986c) *with permission. ©1986 by Elsevier Inc.)*

Greater than a doubling of the oxidation rate was observed in cycling between the ignited and extinguished states. The experiments of *Marek (1985)*, also with a Pt catalyst, showed entrainment of autonomous oscillations by concentration forcing using a highly asymmetric modulation cycle.

Experiments by *Svensson et al. (1988)* and *Jaeger et al. (1990)* continued the examination of the interaction of modulation and autonomous oscillation; however, these investigators employed a Pd catalyst supported on a zeolite. They observed rate enhancement and suggested that the autonomous oscillations resulted from a transition between a catalytically active partially oxidized Pd surface and an inactive oxidized surface. The presence of CO reverses the transition and leads to a burst of CO_2 production. Modulation of the CO feed prevents the transition and leads to higher rates of CO_2 production. A study of the oscillating response indicated entrainment of autonomous oscillations, a phenomenon that disappears when the modulation frequency approaches the autonomous oscillation frequency. Essentially the same observations were made by *Liauw et al. (1995)*. They found that a triangular sawtooth waveform changed the response and enhanced the rate. The greatest oxidation rates under modulation were observed for a pulse-mode of cycling.

Onken and Wicke (1986), working with a supported Pt catalyst in a packed bed, found that autonomous oscillations were amplified as they moved through the bed. Modulation of the CO feed at low amplitude showed that the frequency of the autonomous oscillations dominated the response; however, as amplitude increased, entrainment of those oscillations occurred. Period doubling was also seen. Quasi-periodic behavior disappeared, however.

A quasi-periodic response appears under some conditions for CO oxidation. Studies by *Vaporciyan et al. (1988)* on a Pt/SnO_2 catalyst found low-frequency variations, once every

12 cycles and also once every 175 cycles. They attributed this behavior to the storage of reactive intermediates on the catalyst surface. Simulations available in the literature (*Kevrekidis et al.,1984; McKarnin et al.,1988*) have shown quasi-periodic behavior in the concentration region associated with autonomous oscillations.

Qin and Wolf (1995) specifically examined composition modulation in the presence of autonomous oscillations. They observed that oscillations can be suppressed. Their experiments suggested that autonomous oscillations resulted from a hot-spot moving across the catalyst surface.

Numerical studies of the effect of composition modulation on autonomous oscillation show a variety of responses including entrainment of the oscillations at the forcing frequency, a descent into chaotic behavior and an entrainment involving doubling of the cycling period. In entrainment, the spontaneous oscillation is shifted in frequency and changed in amplitude. *Cutlip et al. (1983)* and later *Capsaskis and Kenney (1986)* undertook experimental studies of forcing in the presence of spontaneous oscillations with CO oxidation but using propene in the CO feed to assist in establishing such oscillations. They observed entrainment of the autonomous oscillations by composition modulation as well as period doubling and chaotic response. Behavior of oscillatory systems under modulation is described by *Marek et al. (2006)* employing a complex network for CO oxidation involving transient reaction intermediates.

3.2.2 Single Crystal Catalysts

Single-crystal studies of the interaction of forcing and autonomous oscillations under high vacuum have been undertaken by a number of researchers and are discussed in two reviews *(Ertl, 1990; Zhdanov, 2004)* as well as in a large number of research papers, e.g., *Bär et al., 1992; Rose et al., 1996; von Oertzen et al., 1998, 2000; Krefting et al. (2009)*. These studies used constant CO and modulated oxygen flow on Pt single crystals as well as vice versa and are both experimental and simulated based on a surface-reaction model. Experimental results were formulated as excitation diagrams by plotting response amplitude against the response period normalized by the period of forcing. These excitation diagrams show entrainment of the autonomous oscillations, as well as a quasi-periodic behavior characterized by sub-harmonic oscillations in the rate of CO_2 formation. Many of these experiments show additional peaks of low amplitude at frequencies greater than the forcing frequency. These are referred to as subharmonic behavior. All of these experiments were carried out in ultra high vacuum on carefully cleaned surfaces. Their purpose was to understand the fundamental processes occurring on catalyst surfaces. Perhaps at some future date, such information will become useful in the design of commercial catalytic systems and even in the application of composition modulation. The research is fascinating; however, for the purposes of this monograph only a sketch can be given of the voluminous literature that has developed.

The type of data collected in these ultra-high vacuum experiments is shown in **Figure 3-3**, which is a plot of the detector response (in mV) to O_2 partial pressure modulation (in Pa) shown above the response. The uppermost traces in (a) of the figure show period doubling. The second group (b) illustrates a subharmonic response at twice the forcing frequency as well as a phase shift. An increase in forcing frequency in the third frame from the top (c) leads to new subharmonics. In the bottom frame (d), a further change in frequency as well as an amplitude change results in quasi-periodic behavior.

Bodega et al. (2007), Marts et al. (2007), and *Bodega et al. (2009)* demonstrated experimentally that unforced oscillations (not shown in the figure) on the Pt(110) surface can be suppressed by modulating the CO concentration in the gas

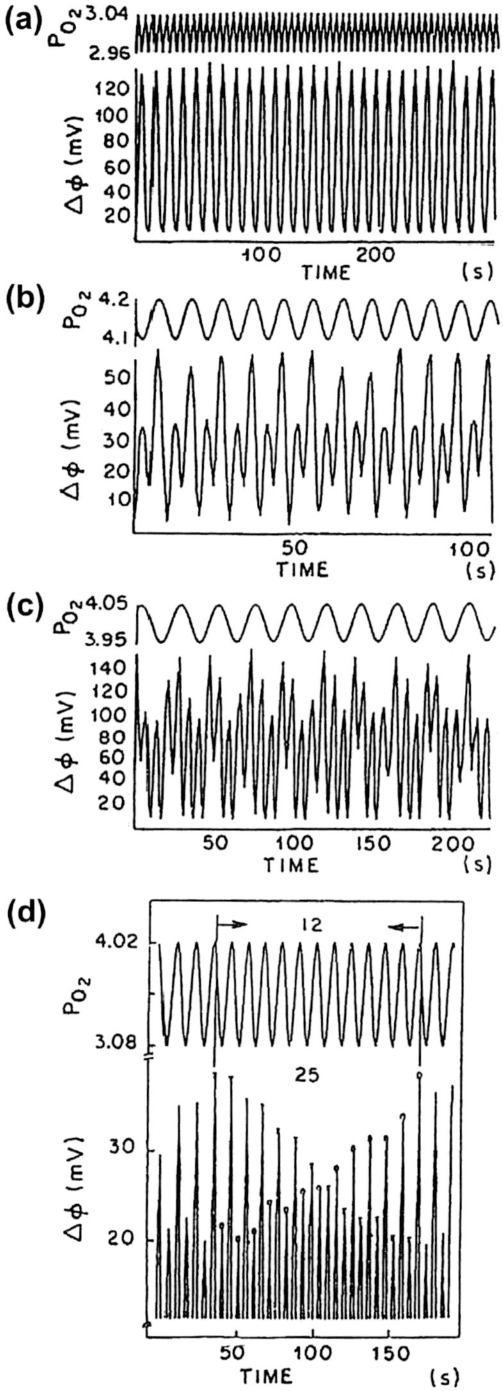

FIGURE 3-3 **Detector response in millivolts to O_2 partial pressure (in Pa) forcing of CO oxidation on a Pt(110) single crystal face: (a) At $\tau/\tau_0 = ½$, (b) near the 2/1 horn at $\tau/\tau_0 = 2.275$, (c) in the quasi-periodic region at $\tau/\tau_0 = 2.135$, (d) magnification of the quasi-periodic response.** *(Figure adapted from* Eiswirth et al. (1989) *with permission. © 1989 by Surface Science.)*

phase. Simulation by *Kiss et al. (2008)* supported this finding. Also, *Bodega et al. (2009)* have shown that turbulent surface oscillations are damped by random fluctuations introduced into the CO stream passing over the surface. Boundaries of the turbulent region are also altered. Suppression of spontaneous oscillation by modulating a reactant has been demonstrated as well for the liquid-phase Belousov-Zhabotinsky reaction *(Marts et al., 2006)*.

Forcing experiments by *Schwankner et al. (1987)* on the Pt(100) face are summarized in **Table 3-1**. These destroyed a time-varying response under steady-state operation, locking the response onto that of the forcing frequency. In addition to the entrainment and quasi-periodic behavior just mentioned, *Schwankner et al.* found that distorting the forcing sinusoid affected the response. They also undertook temperature modulation and observed responses similar to those for composition modulation.

Simulation and experimental work by *Krefting et al. (2009)* on the Pt (110) face have extended the range of behavior observed on single crystals. Simulation used sinusoidal forcing of the Krischer-Eiswirth-Ertl model that allows for phase transitions on the Pt plane and surface diffusion of CO as well as adsorption and reaction kinetics. The authors assumed sinusoidal modulation of the gas-phase CO concentration at about twice the natural frequency of the oscillatory system (ca. 0.59 Hz) and increased the amplitude stepwise in their simulation. At sufficiently high amplitudes, the natural oscillations in CO surface coverage were entrained by the sinusoidal forcing of the gas-phase CO concentration. This represented period doubling.

A further increase in amplitude of the forcing doubled that period again. This continued to eight times the CO concentration modulation period; thereafter, the oscillations in CO coverage, roughly corresponding to CO_2 production, became chaotic, as revealed in the frequency analysis by the absence of subharmonics of the forcing frequency.

The *Krefting* experiments employed photoemission electron microscopy whose signals may be interpreted as surface coverage by CO or the rate of CO_2 production. Ultrahigh vacuum was used and measurements were undertaken on an argon-sputtered Pt surface at 546 K. The Pt sample was dosed with O_2 at 1.5×10^{-4} mbar. The mean CO partial pressure feeding the experimental chamber was 6.22×10^{-5} mbar. Results of the experiment are shown in **Figure 3-4**. This figure plots the Fast Fourier transforming of the number of CO adsorbate sites versus the forcing frequency normalized by the dominant frequency of the spontaneous oscillation of the unforced reaction. In (a), the unforced system shows a peak at a normalized frequency of about ½. Introducing CO modulation at low amplitude focuses the peak at ½ and sub-harmonics appear at 1, 1.5, and 2. Increasing the amplitude results in period doubling and stronger sub-harmonics. Further increases led to (e) in the figure, in which the natural oscillation at ½ has almost disappeared along with the sub-harmonics. This result is interpreted as the onset of chaotic behavior.

These experiments demonstrate that spontaneous oscillatory behavior can, indeed, be controlled by the modulation of a reactant

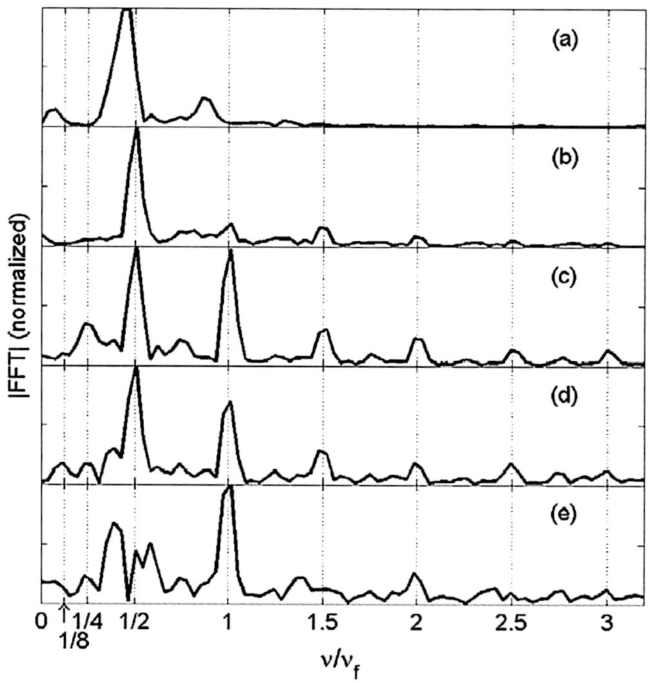

FIGURE 3-4 **Fast Fourier transforms of surface coverage of an initially clean Pt(110) face by CO under modulation of the CO gas phase concentration plotted versus the modulation frequency relative to the dominant frequency of the random oscillation on the surface under ultra-high vacuum and 546 K.** *(Figure reproduced from* Krefting et al. (2009) *with permission. © 2009 by American Physical Society.)*

concentration. However, they also show that amplitude of the modulation is important. If the amplitude is too large, random oscillations set in once again.

Ertl and co-workers in the last two decades have discovered the existence of both stationary and traveling waves for CO oxidation on the Pt(110) face (e.g., *Bär et al., 1992; Rose et al., 1996; von Oertzen et al., 1998, 2000*). Standing waves are an oscillatory pattern in which a periodic, spatial array of lines on the surface represents the same adsorbate. During a finite period these lines move toward each other, merge into a single line, and then split into two lines that move away from each other. The patterns formed extend over the crystal surface indicating temporal synchronization. This synchronization occurs, it seems, through the gas phase. Experiments as well as simulation suggest that surface reconstruction on adsorption/desorption of CO along with the formation of oxygen reservoirs below the outermost layers of Pt atoms are associated with these traveling and standing waves. Wave patterns are reinforced by external modulation of the CO pressure. *Von Oertzen et al. (2000)* show splendid images of the pattern formed.

3.2.3 Studies at Temperatures above 250°C

Many of the composition modulation studies of CO oxidation above 250°C over the noble metal catalysts have been undertaken because of its importance in automotive exhaust catalysis. **Table 3-1** summarizes several studies of composition modulation. *Abdul-Kareem et al. (1981)* and *Silveston et al. (1983)* examined rather low-frequency modulation and found up to a four-fold increase in oxidation rate at the shortest periods considered. They explained their results by ignition considerations in which cycling periodically ignites the oxidation reaction and ignition spills over into the composition range in which the oxidation reaction is inhibited by CO coverage of the surface.

Oh and Hegedus (1982), Muraki et al. (1985a) and *Prairie et al. (1988)* report experiments undertaken as part of a research program on automotive exhaust catalysis. These experiments used very low metal loadings with symmetrical cycling at frequencies of 0.5 to 5 Hz. The effect of the stoichiometric number (SN) was studied. Its value is one for a stoichiometric CO-air mixture. Enhancement is greater for stoichiometric numbers below one. These investigators observed an induction time that depended on temperature. They explained their results by reaction inhibition through the CO coverage of the surface. *Oh and Hegedus* used IR spectroscopy to show that adsorbed CO closely tracks modulation of the gas phase composition. Experiments by *Prairie et al. (1988)* supported the *Oh and Hegedus* observations. Results of *Muraki et al. (1985a)* are seen in **Figure 3-5**.

The effect of intraparticle diffusion was examined in the work of *Cho (1983)* and *Cho and West (1986)*. These studies of both shell and uniformly impregnated catalysts indicated that intraparticle diffusion has a large influence on enhancement through modulation. The authors found that the interaction of diffusion and composition modulation resulted in storage of reacting species within a partial cycle. The stored reactants were then available for reaction in the succeeding half-cycle. The studies are summarized by *Silveston (1998)*, but the original papers should be consulted for details.

3.2.4 Miscellaneous Catalysts

A number of other catalysts have been studied, mainly by the research group at the University of Waterloo. Experiments by *Shanks and Bailey (1987, 1988)* used a supported silver catalyst. Only the O_2 concentration was modulated and a strong dependence of the oxidation rate on cycle period and split was observed. The experiments became reproducible after only about 36 h, suggesting that the Ag surface was probably converted to an oxidized form with time-on-stream.

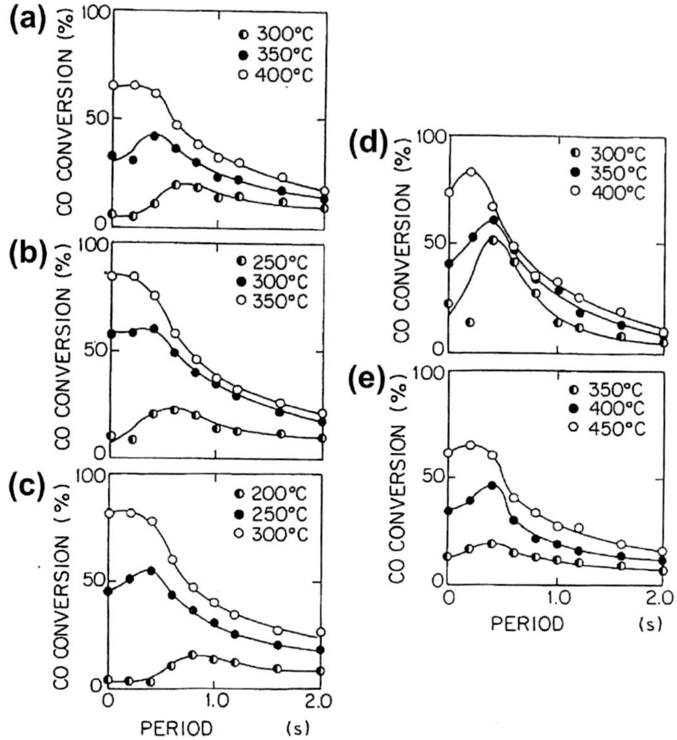

FIGURE 3-5 Effect of τ and $T_{reactor}$ on time average CO conversion in a micro-reactor packed with 3 mm catalyst particles. Catalysts were supported on α-Al_2O_3 at about 0.006 wt% metal: (a) Pt, (b) Pd, (c) Rh, (d) Ru, (e) Ir with cyclic around the stoichiometric ratio. *(Reproduced from* Muraki et al. (1985a) *with permission. © 1985 Nippon Kagaku Kaishi.)*

Experiments by *Hugo et al. (1986)* on a commercial Ni catalyst showed that modulation enhanced the CO oxidation rates at high frequencies provided a sufficiently large amplitude was used. Extremely long transients were observed before the cycling system became reproducible, suggesting a slow conversion of the Ni surface to a form that responds to concentration changes. Apparently an adjustment of the Ni oxidation state between the surface and the bulk is involved.

Experiments using a vanadia catalyst undertaken by *Abdul-Kareem et al. (1980a)* found that amplitude was much less important than frequency in enhancing the oxidation rate. A striking discovery was the existence of a resonance phenomenon as shown in **Figure 3-6.**

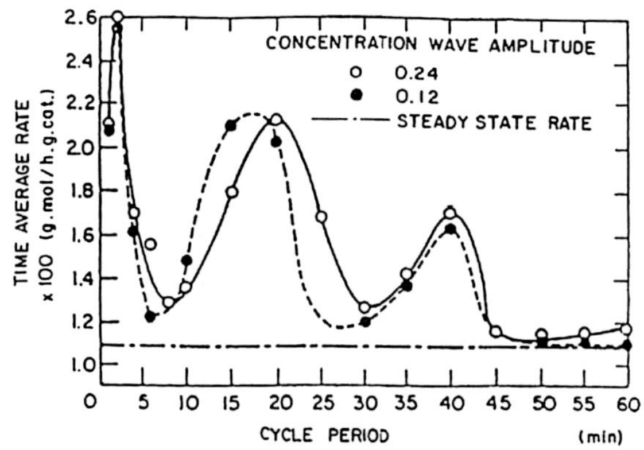

FIGURE 3-6 Multiple resonance for CO oxidation over a vanadia catalyst at 440°C. *(Figure adapted from* Abdul-Kareem et al. (1980a) *with permission. © 1983 Elsevier Science Publ.)*

The largest enhancement of reaction rate was observed at a cycle period of less than five minutes. A twofold enhancement was observed at a cycle period between 10 and 15 min. A third enhancement peak of about 50% above the steady-state rate appeared at a cycle period of about 40 min. Cycle periods greater than 60 min were not used so it is possible that there is a further rate enhancement at a lower frequency. The results suggest a harmonic behavior; however, there are no other reports of such behavior in the literature. Autonomous oscillations were not observed for this catalyst so entrainment does not seem to be an explanation. *Luu et al. (1992)* searched for regions of multiplicity in autonomous oscillation with this catalyst without success. This research is discussed further by *Silveston (1998)*.

The Al_2O_3-supported CuO catalyst used by *Prokopowicz et al. (1988)* exhibited a curious wrong-way behavior that was picked up by FTIR analysis. When CO increased in a cycle, the rate of CO_2 production dropped for about 30 s before increasing. The opposite occurred after a decrease. Previous investigations had only observed a lag in response, not wrong-way behavior. This behavior disappeared when a silica-supported catalyst was used.

Casanova et al. (1979) and *Baiker et al. (1980)* examined hopcalite, a mixed oxide combustion catalyst. Composition modulation experiments were performed around the ignition temperature for this catalyst. They suggested that modulation might be an operating strategy in situations where parametric sensitivity is encountered or a maximum temperature limit exists. In their experiments, modulation reduced the maximum temperature in the bed.

3.2.5 Modeling

Reactor response under composition modulation has been modeled by various investigators (*Hegedus et al., 1977; Oh et al., 1978, 1980; Goodman et al., 1982; Oh and Hegedus, 1982; Cutlip et al., 1983; Jain et al., 1981b; Cho, 1983; Lynch, 1983, 1984; Mukesh et al., 1983; Graham and Lynch, 1984, 1987, 1990; Lynch et al., 1986; Shanks and Bailey, 1987; McKarnin et al., 1988; Eiswirth et al., 1989; Ertl, 1990; Krischer et al., 1991; Nibbelke et al., 1998*). These models generally assume that oxidation proceeds on the catalyst surface by the following steps:

$$\begin{array}{ll} \text{(a)} & CO(g) + s \rightleftarrows CO\cdot s \\ \text{(b)} & O_2(g) + 2\,s \rightarrow 2\,O\cdot s \\ \text{(c)} & CO\cdot s + O\cdot s \rightarrow CO_2(g) + 2\,s \\ \text{(d)} & CO_2(g) + S \rightleftarrows CO_2\cdot S \end{array} \quad (3\text{-}1)$$

where s is a surface site on a noble metal capable of bonding CO or O_2, while an S site, assumed to be on the support, can adsorb CO_2 only. Some investigators have simplified the sequence by neglecting adsorption of CO_2, neglecting CO adsorption and assuming a reaction between gaseous CO and adsorbed oxygen. More recent versions of the sequence allow for a change in surface activity as CO adsorption changes, a change attributed to surface restructuring. Some researchers (Graham and Lynch, 1990) dispense with the Langmuir adsorption assumption. The mechanism shown above is incorporated into a material balance such as one presented in the Graham and Lynch (1984) model below. This balance applies to a single crystal, a differential reactor or a fully backmixed reactor. Almost all researchers have assumed isothermal operation. **Figure 3-7** shows that this model predicts the resonance behavior that can be seen in **Figures 3-1, 3-5 and 3-6**. For the choice of parameters used by the investigators, a 10-fold enhancement was predicted. The quasi-steady-state line in **Figure 3-7** is for an operation with a cycle period long enough to attain steady state in each half-cycle.

The Graham and Lynch (1984) model is shown in **Eqs (3-2) to (3-9)** below (*Lynch, 1983, 1984; Lynch et al., 1986; Graham and Lynch, 1984, 1987, 1990*). These are material balance equations for the oxidation of CO(g) in a differential or fully backmixed bed of catalyst.

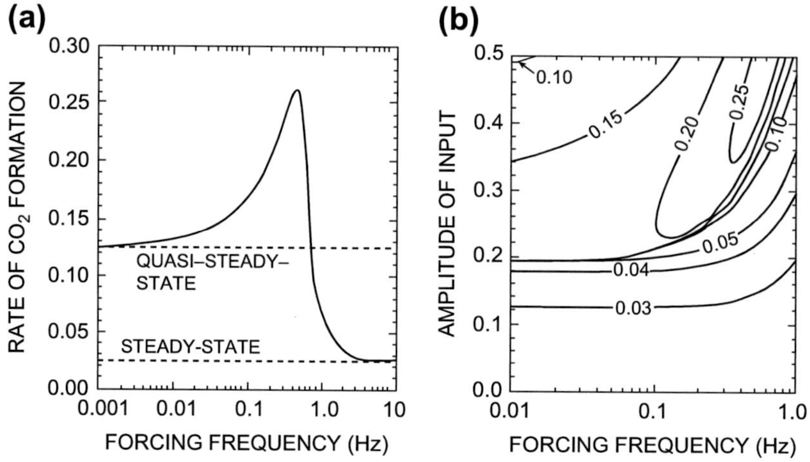

FIGURE 3-7 Predicted rates of CO oxidation as a function of cycle frequency and amplitude from *Lynch's* model using arbitrarily chosen parameters: (a) Predictions for a time average equimolar feed mixture of CO and O_2 and switching between $y_{co} = 0.1$ and 0.9; (b) lines of constant oxidation rate for the same time average feed mixture. *(Figure adapted from* Lynch (1983) *with permission of the Can. J. Chem. Eng.)*

Gas Phase:

$$V\frac{dC_i}{dt} = Q_0(C_i)_0 - qQ_0C_i - \left(\frac{S_pm}{\rho}\right)_{cat} r_i - \left(\frac{S_pm}{\rho}\right)_{sup} r_i^S \quad i = 1, 2, 3. \tag{3-2}$$

Catalyst Surface:

$$\frac{d\theta_i}{dt} = (\nu_{ad})_i r_i - r_3 \quad i = 1, 2. \tag{3-3}$$

Support Surface:

$$\frac{d\theta_i^S}{dt} = (\nu_{ad})_i^S r_i^S \quad i = 3. \tag{3-4}$$

Gas Shrinkage:

$$q = 1 - y_1r_1 - y_2r_2 + y_1r_3 - y_1r_3^S. \tag{3-5}$$

Initial Condition:

$$t = 0: \quad C_1 = 1, \quad C_2 = C_3 = 0, \quad \theta_1 = 1, \quad \theta_2 = \theta_3^S = 0. \tag{3-6}$$

where 1 = CO, 2 = O_2, 3 = CO_2 and r_1, r_2, r_3, r_1^S, r_2^S, r_3^S depend on the kinetics assumed. *Lynch* and co-workers assumed Langmuir-Hinshelwood kinetics, namely:

$$\text{CO:} \quad r_1 = S_t(k_a)_1C_1(1-\theta_1-\theta_2) - S_t(k_d)_1\theta_1 \quad r_1^S = 0. \tag{3-7}$$

$$O_2\text{:} \quad r_2 = S_t^2(k_a)_2C_2(1-\theta_1-\theta_2)^2 - S_t(k_d)_2\theta_2 \quad r_2^S = 0. \tag{3-8}$$

$$CO_2\text{:} \quad r_3 = S_t^2k\theta_1\theta_2 \qquad r_3^S = (k_a^S)_3C_3(1-\theta_3^S) - (k_d^S)_3\theta_3^S. \tag{3-9}$$

Critical evaluation of this model was undertaken by *Graham and Lynch (1984)* for a supported Pt catalyst by switching between 2.0 mol% CO and 1.0 mol% O_2 diluted in N_2. Observed enhancement was 10 to 23 times the rate at steady state at a resonant frequency. The lower value is quite close to the predicted resonance maximum shown in **Figure 3-7**. *Graham and Lynch (1984)* observed that the experiments were undertaken in a parameter space outside the region of multiplicity determined experimentally. However, the parameters used in the model suggest that the simulation shown in **Figure 3-7** encompassed a region in which multiple steady states may arise. This region is in the upper right corner of the figure where lines of constant rate come together.

Successful simulation of composition modulation with a porous catalyst was carried out by *Oh et al. (1978)*. The model used required the addition of transport steps and boundary conditions. Details are given by *Oh et al.* and are summarized by *Silveston (1998)*.

Table 3-2 summarizes the discussion of the last few pages. A variety of effects and phenomena are apparent. Some are catalyst specific as indicated by the bracketed comment for many of the table entries. Several of these, such as 1) to 4) are found in other reaction and catalyst systems. Phenomena such as 5) to 8) are probably general as well, but the necessary experiments to observe them have not been carried out.

Long relaxation times over nickel oxides were found when the catalyst was reduced prior to use. They were not seen when the catalyst was oxidized. A stationary cycling condition is reached after just several cycles for most noble metal catalysts (*Graham and Lynch, 1984; Barshad and Gulari, 1985a; Zhou and Gulari, 1986a*). The term, *excitation*, introduced in the table, means only that the CO oxidation rate depends on cycle period. It does not mean rates are higher than those at a comparable steady state. *Enhancement* is used to indicate that case.

Extended relaxation times imply interaction via diffusion between the bulk of the catalyst and the surface. This condition is likely when the catalyst contains an oxygen storage material,

TABLE 3-2 Summary of Behavior under Composition Modulation of Catalytic CO Oxidation

1) Enhancement (increase of oxidation rate above the steady-state rate at the time-average feed composition).
2) Rate resonance in which the oxidation is stimulated within a relatively narrow range of cycling frequencies (noble metal catalysts, supported V_2O_5).
3) Global enhancement where rates exceed the highest rates attainable at steady state for a specified temperature and pressure (Pd/Al_2O_3, $Pt\text{-}SnO_2/Al_2O_3$, V_2O_5/SiO_2).
4) Excitation where the cycle split or duration of exposure to one reactant is as important as frequency or period (noble metal catalysts).
5) Excitation or enhancement dependence on intraparticle diffusion and/or the distribution of the catalytically active material in the porous catalyst (Pt/Al_2O_3).
6) Excitation dependence on the oxidizing/reducing (stoichiometric number) character of the reactants (Pt/Al_2O_3, Pd/Al_2O_3).
7) Decrease in temperature gradients in non-isothermal packed beds (Hopcalite).
8) Suppression of the natural frequency, period doubling, quasi-periodic response when changing compositions cross regions of multiplicity or autonomous oscillations ($Pt\text{-}SnO_2/Al_2O_3$, Pt(110), Pd/zeolite).
9) Multiple rate resonance with changing frequency with possible harmonic excitation of rate (V_2O_5/SiO_2).
10) Wrong way behavior under isothermal conditions (CuO/Al_2O_3).
11) Long relaxation times to reach a stationary, reproducible cycling condition (NiO/Al_2O_3).
12) Induction periods as large as 5 to 10 s after a composition switch has occurred (noble metal catalysts).

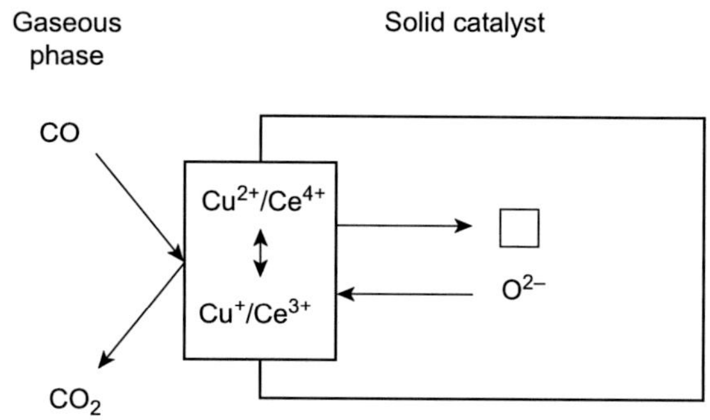

FIGURE 3-8 Schematic of CO oxidation over a dual catalyst in which one component exhibits an oxygen storage capacity. Arrows represent diffusion pathways between catalyst bulk and surface. *(Figure reproduced from* Sedmak et al. (2004a) *with permission. © 2004 by Elsevier Inc.)*

such as CeO_2. **Figure 3-8** illustrates the situation that arises. Oxygen storage requires the addition of mass balances for the second catalyst component on the catalyst surface and the subsurface. Additional kinetic models must be devised and new initial conditions added. This has been undertaken by *Sedmak et al. (2004a)* resulting in a model much more complex than the Graham and Lynch (1984) model given previously.

Several investigators have questioned whether global enhancement of reaction rate is possible through composition modulation (*Nowobilski and Takoudis, 1986; Yadav and Rinker, 1989*). Research summarized in **Table 3-1** demonstrates that true global enhancement is attainable, although it may not be a general rule. Global increases in the CO oxidation rate were found for the Pd/Al_2O_3, $Pt\text{-}SnO_2/Al_2O_3$ and the V_2O_5/SiO_2 systems, but rates measured for the Pt/Al_2O_3 system appear to be below rates in the high rate branch in the region of multiple steady states.

Significant differences are evident between the noble metal and the oxide catalysts: For example, phenomena 4), 8) and 12) in **Table 3-2** are seen only for noble metals, while responses 9) to 11) were observed only for oxide catalysts. Furthermore, rate increases between 10- and 40-fold have been measured for the noble metal catalysts, while for the oxide catalyst these seldom exceed 2-fold. These results suggest differences in rate-controlling steps and/or mechanism.

Explanations of rate enhancement with the noble metal catalysts involve reactant storage that increases the extent of adsorbed oxygen on the catalyst surface thereby reducing CO coverage of that surface. As a result, adsorbed CO and oxygen concentrations are better balanced. Mathematical models incorporating storage and competitive adsorption (*Oh et al., 1978, 1980; Cho, 1983; Lynch, 1983; Graham and Lynch, 1984*) agree well with experiments. A further factor suggested by *Gulari* and coworkers (*Zhou and Gulari, 1986b*) is that the spatial distribution of adsorbed CO and oxygen as islands is increased by composition modulation. Improved spatial distribution may also arise with the Pt/Al_2O_3 catalysts and this could account for the quantitative failure of the *Graham and Lynch (1984)* model with this catalyst.

Oxygen storage is probably part of the explanation of enhanced CO oxidation rates with the oxide catalysts, but understanding of these catalysts is much poorer than for the noble metals. *Shanks and Bailey (1987)* were able to find good agreement with experimental data using a mathematical model that assumes the presence of an inactive form of oxygen on the catalyst surface or as part of the surface. *Jain et al. (1981b)* proposed oxygen diffusion into the bulk of the catalyst to explain their observation of multiple resonances with a vanadia catalyst. Certainly, the long times needed to obtain reproducible cycles for the NiO catalyst suggest that changes in the catalyst phase are important with this system.

3.3 SULFUR DIOXIDE OXIDATION

SO_2 oxidation was one of the first catalytic reactions subjected to composition modulation

(*Unni et al., 1973*). This study followed the discovery that a rise in the feed SO_2 concentration induced an increase in SO_3 concentration followed by a gradual decline, lasting as long as 6 h (*Yeramian et al., 1970*). The limited research on SO_2 oxidation is summarized in **Table 3-3**.

Unni et al. (1973) chose a commercial catalyst and experiments were run at 405°C and 1 atm. **Figure 3-9** shows the experimental results at a mean SO_2:O_2 mol ratio of 0.6. Feed compositions were cycled symmetrically between an SO_2/O_2 ratio of 0.3 and 0.9. The lower curve shows results at a mean ratio of 0.9, but in this case, the amplitude in terms of the SO_2/O_2 ratio was 0.1. A lower ratio was also examined but almost no enhancement was observed. Regardless of these mean ratios, the largest excitation of the oxidation rate occurred at long cycling periods lasting between four and six hours. The figure also compares two different strategies—the highest enhancement, shown by the upper curve, occurs when concentrations of both SO_2

TABLE 3-3 Modulation Studies of the Reactions of Sulfur Dioxide

Reference	Objective	Catalyst and Reaction Conditions	Modulated Variable	Observations	Comments
Unni et al. (1973)	Effect of composition modulation on conversion	30/40 mesh K and Fe promoted V_2O_5 supported on diatomaceous earth, $T = 405°C$, $P = 1$ bar	SO_2:O_2 concentration ratios	Slow modulation enhanced SO_2 oxidation rates by up to 20%. Mean reactant ratio and the amplitude of the ratio change affected enhancement.	Enhancement at low frequency is consistent with a reaction that occurs within a melt phase.
Briggs et al. (1977, 1978)	Application of modulation to commercial reactors	As in *Unni et al. (1973)* but with 12/20 mesh particle size, $T = 405°C$, $P = 1$ bar	Air and SO_2-SO_3 mixtures	Modulation with $\tau = 26$ min increased overall SO_2 conversion to > 98%. Also periodic heated air flushing of catalyst bed with unsymmetrical cycles at $14 < \tau < 29$ min. also raised conversion	Air flow reactivates the catalyst by stripping out SO_3. Periodic blowing of the final stage of an SO_2 converter with hot air could raise SO_2 conversion to close to its equilibrium limit.
Olsson and Schöön (1985)	Reduction of SO_3 in automotive catalytic converters.	Pt automotive muffler catalyst, $T = 180°C$, very low reactant partial pressures	Switching between SO_3-O_2 and CO-O_2 mixtures, cycle period = 8 min	Modulation eliminated SO_3 in the off-gas in both half-cycles.	Observations suggest strong adsorption of reactants on the Pt surface and CO reduction of SO_3 formed.
Olsson and Schöön (1986)	As above	Pt/Rh automotive muffler catalyst, $T = 212$ and 242°C,	Asymmetric modulation of O_2 partial pressure for constant CO and SO_2 at 1Pa and 2.4 Pa respectively, $3 < \tau < 5$ min	Modulation suppressed SO_2 oxidation and increased CO oxidation.	Aperiodic responses at the higher temperature indicated autonomous oscillations.

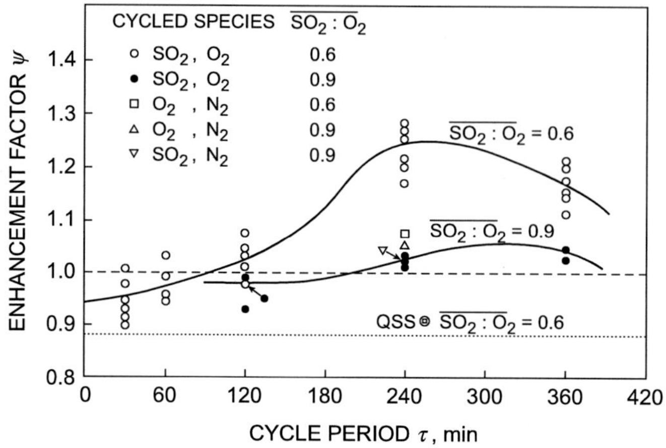

FIGURE 3-9 Rate enhancement for SO_2 oxidation over a commercial, promoted vanadia catalyst as a function of τ, time-average feed SO_2:O_2 ratio, and forcing strategy (reactant cycled) for $P = 1$ bar, $T = 405°C$, $s = 0.5$. *(Figure reproduced from* Unni et al. (1973) *with authors' permission.)*

and O_2 were varied. The lower curve includes data points to a period of 4 h at which the concentration of just one component is varied even though the mean ratio continued to be 0.6.

The influence of cycle amplitude with symmetrical forcing was examined at a time average of SO_2:$O_2 = 0.6$ in the feed and $\tau = 240$ min. **Figure 3-10** indicates that a threshold amplitude between 0.1 and 0.2 as a SO_2:O_2 ratio exists for the system. The largest amplitude shown represents switching between a feed containing just one of the two reactants and a mixture that approaches the stoichiometric ratio. The absence of one reactant appears to be detrimental.

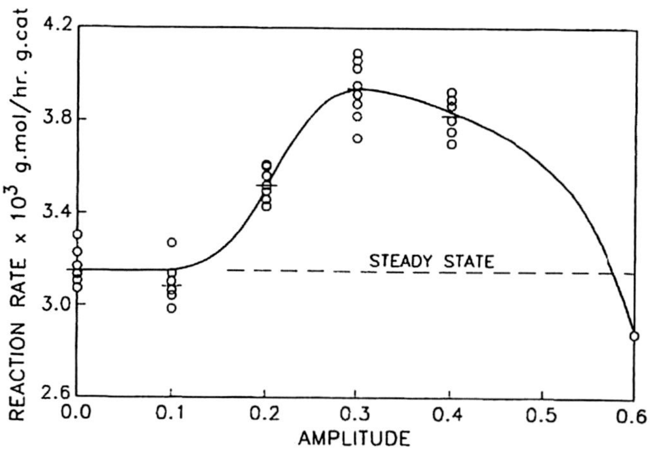

FIGURE 3-10 Dependence of rate enhancement on cycle amplitude as the SO_2:O_2 ratio for $\tau = 240$min, $s = 0.5$ and a time average SO_2:O_2 ratio = 0.6 in the feed. Catalyst and other conditions are given in Figure 3-9. *(Figure adapted from* Unni et al.,1973.*)*

Measurements within a half-cycle indicate that at the highest mol ratio, the rate of oxidation is approximately equal to the steady-state rate at that ratio.

However, in the half-cycle at a ratio equal to 0.3, the rate of SO_3 production is three times the steady-state rate. A slow change in the rate of formation was observed in each half-cycle and seems to explain the long cycle period needed for maximum enhancement. The observations can be explained only by assuming that SO_3 and/or SO_2 dissolve in a highly viscous liquid phase (see Silveston (1998) for further discussion).

SO_2 oxidation becomes equilibrium limited at high SO_2 conversions. *Briggs et al. (1977)* explored the application of composition forcing to the final stage of a multi-stage SO_2 converter using the commercial catalyst employed by *Unni et al. (1973)*. Isothermal experiments at 405°C were undertaken cycling the feed to a packed bed of catalyst between air and an SO_2:SO_3 mixture approximating the concentrations leaving the first stage of a commercial SO_2 converter. Presumably air would strip adsorbed SO_3 from the surface as well as re-oxidize the catalyst causing a much higher rate of reaction in the second half of the cycle. This was indeed the case. The result was a 98.8% conversion of SO_2, compared with 95.8% under steady-state operation. Some step-change observations indicated even higher time-average conversions could be attained by shortening the half-cycle. In these experiments, only traces of SO_2 were detected after switching back to air. Thus, SO_2 was not desorbed from the catalyst during air flushing. Segregating the flows from the reactor as the feed is switched would lead to an operation representing almost 99% conversion of the SO_2 feed. Replacing air by either N_2 or O_2 did not affect the conversion obtained in the experiment just described. Consequently, the higher conversion must be due solely to stripping SO_3 from the catalyst.

In a further set of experiments, *Briggs et al. (1978)* used two packed bed reactors each holding about 30 g of vanadia catalyst. This system was operated with two feed streams, one typical of a feed from a sulfur-burning plant and the other typical of a smelter effluent. With the former, the gas exiting the cyclically operated final bed had a time-varying SO_2 concentration whose lowest value was only about 8% of that of steady-state operation. With the smelter effluent gas, the lowest SO_2 emission was just 13% of the steady-state value. Cycle periods of four to five minutes appeared to be optimum. The beds exhibited breakthrough behavior. Temperature variations in the beds showed temperature waves moving through the bed after a change in feed composition. The experiments are complicated; details can be found in the literature (*Briggs et al., 1977, 1978*). Results of the *Briggs* studies have been interpreted by *Silveston and Hudgins (1981)* and are summarized by *Silveston (1998)*.

3.3.1 Modeling

Russian researchers (*Balzhinimaev et al., 1989*) have developed a detailed mechanism for SO_2 oxidation over K-promoted vanadia catalysts. It has been known for decades that commercial potassium oxide promoted vanadia catalysts function in a melt phase under reaction conditions and that the oxidation proceeds through a redox mechanism. According to the *Balzhinimaev* mechanism, K_2O is converted to the pyrosulphate in the presence of SO_3 and this compound acts as a fluxing agent for vanadia. Potassium promoted vanadia catalysts oxidize SO_2 through a cycle in which three different binuclear vanadium complexes participate. The redox mechanism involves the transformation of these complexes among one another through oxidation, SO_2 absorption and the oxidation of SO_2 to SO_3. The bulk of the SO_3 produced comes from the transfer between the V^{5+} complexes and does not involve a change in the coordination of vanadium. SO_3 is released from the melt by decomposition of the pyrosulphate. Oxygen enters the catalyst through absorption in the melt where it oxidizes a vanadium species. Its absorption may be rate-controlling.

A complicated kinetic model (*Ivanov and Balzhinimaev, 1987; Balzhinimaev et al., 1989*) assumes that the steps in the mechanism are elementary and that the active and inactive forms of the vanadia complexes are in equilibrium. Employing this model for a reactor is difficult and many assumptions are required. An effectiveness factor was used to eliminate intraparticle dynamics. Heat balances are required for the gas and catalyst phases, and material balances are needed for each reactive species. Details are to be found in a paper by *Silveston et al. (1990)* and are summarized by *Silveston (1998)*.

Strots et al. (1992) applied the *Balzhinimaev* model to the simplest reactor situation: A back-mixed reactor that operated both isobarically and isothermally. Equilibrium between the phases was assumed and heat and mass transport were neglected. Results predicted rate enhancement under cycling and suggested multiple resonance such as seen in **Figure 3-6**. It is interesting to note that the enhancements predicted by *Strots et al.* and the cycle periods for which these enhancements should occur are in order-of-magnitude agreement with the *Briggs* results discussed earlier. The model also predicted that enhancement disappears when SO_2 is absent from the feed in a half-cycle, as was observed in one of *Unni's* experimental studies. *Silveston (1998)* provides a summary of the modeling studies.

3.4 REDUCTION OF SO_3 BY CO OVER PLATINUM

Sulfuric acid mists were sometimes observed in automotive exhausts, which has led to some work on reduction of SO_3 by CO. Because automotive emission catalysts operate in a periodic mode, reduction was studied in this mode by *Olsson and Schöön (1985, 1986)*. Experiments were performed using a commercial Pt/Rh monolith catalyst mounted in a temperature-controlled furnace. They also used Pt impregnated on a monolith support. Two different experiments were performed: (a) Switching between SO_2 and CO streams each containing the same partial pressure of O_2 and (b) forcing O_2 partial pressure while maintaining constant flows of the other reactants. Aperiodic behavior was observed for experiment (a) and traced to autonomous oscillations under steady-state operation. These oscillations were attributed to deactivation of Pt surface sites where oxidation by adsorbed SO_3 occurred. Abrupt SO_2 peaks were seen in the off-gas in the (b) experiment when the O_2 partial pressure dropped.

3.5 REDUCTION OF NITROGEN OXIDES

Nitrogen oxides (NO_x) attract interest because they are implicated in smog formation and contribute to acid rain. As pollutants, they are removed from waste gases by reduction with H_2, CO, hydrocarbons or ammonia. Composition modulation has been applied to those reduction reactions in a number of studies.

Nitrous oxide (N_2O), nitric oxide (NO) and nitrogen dioxide (NO_2 or N_2O_4) constitute the nitrogen species referred to as NO_x. In the past much of the NO_x pollution came from automotive exhausts. Over several decades, this pollution burden has been greatly reduced by a three-way catalyst (TWC) bed placed between the engine and the vehicle exhaust (see **Chapter 6**). TWC beds are subjected to a quasi-periodic, low amplitude disturbance at about 0.5 to 2 Hz. In this bed, the dominant reaction is the reduction of the oxides by CO. It is hardly surprising then that most of the modulation studies have been concerned with this reaction. The noble metals are the chief components of three-way catalysts, so these have been the target catalysts.

3.5.1 NO Reduction over Supported Catalyst

The overall reaction in NO_X reduction with CO is:

$$2\,CO + 2\,NO \rightarrow 2\,CO_2 + N_2 \qquad (3\text{-}10)$$

where NO represents the mixture of oxides present in stack gas or automotive exhaust. The reaction is believed to proceeds stepwise from NO as:

$$CO + 2\,NO \rightarrow CO_2 + N_2O \qquad (3\text{-}11)$$

and

$$CO + N_2O \rightarrow CO_2 + N_2. \qquad (3\text{-}12)$$

Experimental investigations on modulating the reduction are summarized in **Table 3-4**. As might be expected, results show dependence on cycle period, cycle split and the stoichiometric ratio (SN). The stoichiometric ratio, SN, is defined for NO reduction by CO as the volumetric NO:CO ratio. Therefore, if the formation of N_2O can be neglected, SN = 1 for a stoichiometric mixture. If the nitrous oxide reaction proceeds, SN = 2. Work by the Toyota Research Laboratory (*Muraki and Fujitani, 1986*) led to **Figure 3-11** that shows the behavior under composition modulation at 400°C for the five noble metals usually employed in TWC formulations. As can be seen, Rh is the most effective catalyst for NO reduction; however, because of its high cost, it is usually replaced by Pt. In order for Pt to approach the efficacy of Rh, a lower cycling frequency must be used. *Muraki and Fujitani* reported that modulation shifts the reduction from N_2O to N_2. It also sharply reduced the light-off temperature for the TWC.

Muraki and Fujitani (1986) and *Muraki et al. (1986)* interpreted their observations to suggest that NO adsorbs dissociatively on weakly bonding sites, whereas CO adsorbs on both strong and weak sites. The enhancement through modulation appears to be due to the scavenging of adsorbed CO from the catalyst surface by NO during NO exposure, a process that increased NO adsorption. Scavenging of adsorbed atomic oxygen by CO during the CO portion of a cycle resulted in CO_2 formation. Further details are given in the *Muraki* papers and by *Silveston (1998)*. Cycling measurements by *Cho et al. (1989)* employing a supported Rh catalyst (see **Table 3-4**) supported the conclusions of *Muraki and Fujitani (1986)*.

Sadhankar and Lynch (1996a) modulated NO reduction over a 0.5 wt% Pt/γ-alumina catalyst and observed a remarkable enhancement that may be due to steady-state multiplicity in the composition range studied. These authors further investigated the effect of phase lag when the concentrations of both reactants are cycled. They did not observe autonomous oscillations; however, these have been seen under high vacuum at about 100°C when NO reduction is undertaken on Pt(100) single crystal surfaces (*Schwartz and Schmidt, 1987, 1988; Fink et al.,1991; Dath et al., 1992*).

Sadhankar and Lynch (1996a) studied a supported Pt catalyst using both bang-bang cycling and two-part cycling with either an in-phase (0°) or out-of-phase (180°) lag and found a pronounced frequency resonance for both NO and CO conversion. This result is nicely illustrated in **Figure 3-12**. Despite the strong frequency dependence, CO conversion is less than the almost 100% conversion obtained in the high-rate branch for multiple steady states observed for this system. Similarly, NO conversion to N_2O is well below that in the high-rate branch. **Figure 3-12** shows a large increase in NO conversion so that, together with the small conversion to N_2O, it seems that bang-bang modulation substantially increased the rate of conversion to elemental N_2. It should be mentioned that the absence of improvement at $\omega > 6$ mHz in the figure appears to be due to the presence of mixing, which forced the system to operate in the low-rate branch of the multiplicity region. Multiplicity in the NO-CO reaction was explored further by *Sadhankar and Lynch (1997)*.

TABLE 3-4 Modulation Studies of NO_X Reduction

Authors	Objective	Catalyst and Reaction Conditions	Modulated Variable	Observations	Comments
Muraki and Fujitani (1986)	Effect of composition modulation on NO reduction	Single metals (Pt, Pd, Rh, Ru, Ir) supported on α-Al_2O_3 at 0.006 wt% except for Pd, $T = 400°C$, $SV = 30{,}000\ h^{-1}$	Composition (0.3 vol% NO and 0.3 vol% CO) in terms of SN; $0.2 < \tau < 5$ s; $s = 0.5$	Large reduction in NO was observed for modulation with the Pt catalyst. For Ru and Rh catalysts, which achieved 100% conversion under steady-state operation, modulation reduced conversion.	Modulation of SN makes much cheaper Pt as effective as Rh or Ru.
Muraki et al. (1986)	As above	Pt supported on α-Al_2O_3, $350 < T < 500°C$, $SV = 30{,}000\ h^{-1}$	As above	Under cycling, conversion increased with temperature and increasing SN. Optimum cycle period depended on temperature. Fast cycling reduced selectivity to N_2O.	Light-off temperature of catalyst decreased under composition modulation. Results offered mechanistic insights.
Cho et al. (1989)	Effect of composition modulation on NO reduction.	Rh on 0.09 wt% on γ-Al_2O_3, or 0.04 wt% on CeO_2, $150 < T < 500°C$, $SV = 80{,}000\ h^{-1}$	Bang-bang cycling between 0.008 vol% NO and 0.008 vol% CO	Light-off temperature was about 200°C. NO conversion under cyclic operation was less than under steady state.	Cerium oxide was a better support than alumina.
Dath et al. (1992)	Effect of modulating NO reduction on a single crystal	Pt (100) face, $T = 20°C$, high vacuum	Temperature or gas composition.	Modulation induced oscillations in CO_2 partial pressure. These showed period doubling and subharmonic behavior.	The reaction system on a uniform catalyst face seemed to behave as a linear oscillator.
Sadhankar and Lynch (1994)	Modulation of N_2O reduction	0.5 wt% Pt / γ-Al_2O_3, $T = 226°C$, $P = 1$ bar	Switching between 2.4 vol% N_2O and 2.4 vol% CO. Use of phase-lead	N_2O reduction increased with cycling frequency to a maximum. A further increase led to a collapse to the relaxed steady state.	Phase-lead improved performance. A 90° phase-lead and a cycling frequency of 5 mHz yielded > 90% N_2O conversion.

Sadhankar and Lynch (1995)	Investigation of reproducibility for catalytic NO reduction under modulation	0.5 wt% Pt on γ-Al_2O_3, $T = 212°C$, well mixed reactor, phase-lead operation	Bang-bang switching between various CO and NO concentrations while maintaining a time-average 1.2 vol% of each gas in the feed.	Catalyst activity abruptly dropped to near zero after hundreds of modulation cycles. Time needed to achieve this failure depended on composition and flow rate.	Abrupt loss in activity may be corrected by a periodic exposure to oxygen.
Sadhankar and Lynch (1996a)	Low temperature modulation of NO reduction by CO	As above	As above	Conversion of NO increased with rising cycling frequency until a critical value where it dropped to almost zero. Conversion to N_2O followed the same pattern. Phase-lead of the CO pulse affected conversion.	Mixing in reactor eliminated concentration variation with increasing cycling frequency and reactor operated at steady state. Phase-lead resulted in four-part cycles. These improved the performance.
Sadhankar and Lynch (1996b)	Investigation of the time needed to reach a cycle invariant state	0.5 wt% Pt on γ-Al_2O_3, $465 < T < 505$ K, $1 < \omega < 10$ mHz, well mixed reactor, various NO:CO ratios, but most at 0.7 vol% NO, 0.7 vol% CO	Bang-bang switching between various CO and NO concentrations but varying the time-average vol% of the reactants	For a pre-oxidized catalyst at 0.7 vol% NO and 0.7 vol% CO, high conversion of NO (ca. 60%) maintained over 1750 cycles (55 h), then slowly declined to very low conversion. Duration of high conversion branch depended on vol% of reactants and T. Pre-reduced catalyst operated in low conversion branch of multiple steady states.	Behavior explained by the ratio of the contact time and concentration of CO to the duration of a half cycle indicating that in the high conversion branch the rate of CO replacement on the surface roughly matched CO consumption by reaction. In low conversion branch, CO dominates the surface.

(Continued)

TABLE 3-4 Modulation Studies of NO_X Reduction (*cont'd*)

Authors	Objective	Catalyst and Reaction Conditions	Modulated Variable	Observations	Comments
Aida et al. (1997, 1999)	Effect of porous catalysts on modulation of the NO-CO reaction	Singly and mixed impregnated Pt/Al_2O_3 and Rh/Al_2O_3 catalysts, metal loading at 1 wt%, $T = 150$ and 300°C,	Symmetrical, bang-bang switching between 4.46×10^{-2} mol/m^3 NO and 0.223 mol/m^3 CO for $0 < \tau < 240$ s	Type of impregnation affected conversion and the optimal cycling frequency. Conversion changed by internal distribution of metals as well as by ordering different catalysts in layers in a reactor.	Intraparticle diffusion has a strong effect on performance. Mixed catalysts and layering of different catalyst within a catalyst pellet can significantly improve performance.
Harmsen et al. (2001c)	Mechanism of NO-CO reaction over TWC automotive catalysts	$Pt\text{-}Rh\text{-}CeO_2/\gamma\text{-}Al_2O_3$ catalyst and meta components on $\gamma\text{-}Al_2O_3$ at 548 K, $P = 110$ kPa	Symmetrical, bang-bang switching between CO at 0.55 kPa and NO at 0.1 kPa with $0.35 < O_2 <$ 1.05 kPa	In the absence of CeO_2, conversion of NO and oxidation of CO occurs only on switching gas composition.	Interpretation of cycling measurements indicates CO does not react with NO, but only with products of NO dissociation. Ceria does not dissociate NO, but rather stores O^{2-} dissociated on noble metals.

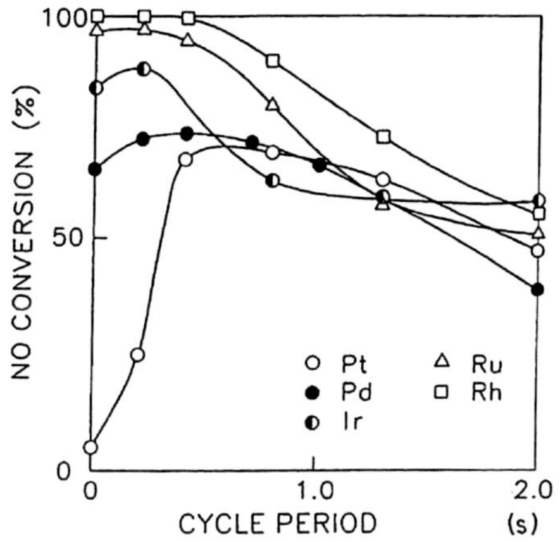

FIGURE 3-11 Comparison of NO conversion on noble metal catalysts at 400°C, $SN = 1$, and $SV = 30{,}000\ h^{-1}$ for symmetrical modulation between 0.3 vol% NO and 0.3 vol% CO. *(Figure adapted from* Muraki and Fujitani (1986) *with permission. © 1986 American Chemical Society.)*

Sadhankar and Lynch (1996a) examined the effect of phase-lead or lag of modulated CO or NO on conversion. With a lag or lead other than 0 or 180°, a cycle will have four parts. These authors found that phase is important at the lower frequencies studied. As frequencies approached 15 mHz, the effect of phase-lead, arising from mixing in the reactor, disappeared.

A second paper by *Sadhankar and Lynch (1996b)* examined the slow attainment of a stationary cyclic state at high forcing frequencies. They observed that the duration of the transient to a stationary cycling state depended on the space time, the flow through the reactor, and an adsorption time, that is, the ratio of adsorption capacity for the adsorbate divided by the mass flow rate of that adsorbate into the reactor. The authors suggest that the long transients offer means of exciting the reaction through the use of superimposed cycling frequencies. The original literature should be consulted for a full discussion; an abbreviated one is offered by *Silveston (1998)*.

3.5.1.1 Pt Single Crystals

Modulation of NO reduction by CO has been examined only on the Pt(100) face (*Dath et al., 1992*). NO reduction on this face in the absence of forcing has been intensively studied (*Schwartz and Schmidt, 1987, 1988; Fink et al., 1991*) because oscillations are observed when the temperature of the single crystal is disturbed; however, these oscillations are damped and eventually disappear if the temperature is held constant. Damped oscillations could not be initiated by varying the NO or the CO partial pressures.

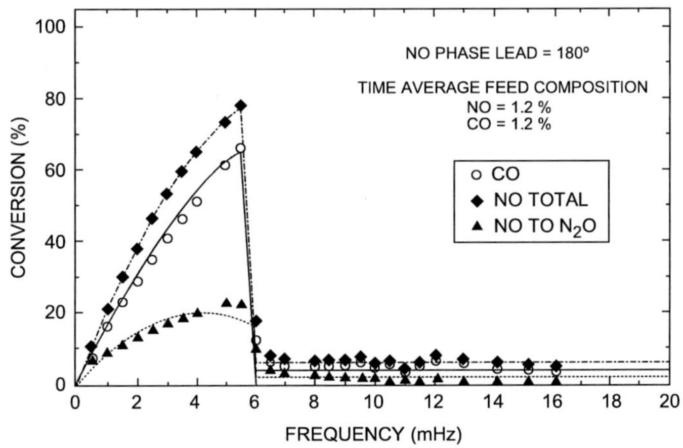

FIGURE 3-12 Conversion as a function of cycling frequency for bang-bang (φ = 180°) switching between feed mixtures containing 1.2 vol% CO and 1.2 vol% NO at 212°C and 103 kPa over a 0.5 wt% Pt/Al_2O_3 catalyst. *(Figure adapted from* Sadhankar and Lynch (1996a) *with authors' permission.)*

However, a sinusoidal oscillation of the CO partial pressure caused sustained harmonic oscillations once these were initiated by a 5°C temperature disturbance. The amplitude of the harmonic response (CO_2 production) was a function of the cycling amplitude and its frequency. The authors suggested the NO+CO reaction on the crystal face behaves as a linear oscillator.

3.5.2 N_2O Reduction

Sadhankar and Lynch (1994) extended their investigation of NO_X reduction by CO to the reduction of N_2O, an intermediate in that reaction. Experiments were carried out in the neighborhood of the region of isothermal multiplicity for the $N_2O + CO$ system. Modulation essentially extended that region, therefore CO conversions of about 90% were achieved at mean compositions in which the conversion is only about 30% at steady state. Just as for the NO + CO system, high frequencies led to a relaxed steady state while very low frequencies resulted in zero conversion as bang-bang cycling was used between N_2O and CO. The authors were able to closely model the behavior under composition modulation by assuming surface exclusion in which the presence of an adsorbed CO molecule prevented the adsorption of another CO molecule on an adjacent site.

The importance of the formation of N_2O during NO_X reduction by CO was explored by *Cho (1994)* using isotopes of C and N. *Cho* concluded that the reaction proceeds through the N_2O intermediate but this species is rapidly reduced by CO and thus appears only as a minor product in NO_X reduction.

3.6 TRAVELING WAVES IN PACKED BEDS

A series of simulation studies by *Yakhnin et al. (1995b)* and by *Yakhnin and Menzinger (1999)* of catalyzed CO oxidation examined temperature excursions in packed beds arising from abrupt changes in inlet temperatures or concentrations. These authors observed that their models predicted large temperature excursions for the parameters used in their model. Their explanation for these excursions invoked the interaction of temperature and concentration waves in the reactor. This interaction has been known for many years and results in the phenomenon known as wrong-way behavior.

The Yakhnin-Menzinger simulations were examined experimentally by *Jaree et al. (2003a, b)* in a thin-walled, vacuum-jacketed tubular reactor using CO oxidation over a Pt/Al_2O_3 catalyst and by *Na-Ranong et al. (2003)* who substituted a supported copper oxide catalyst for Pt. Using both step-changes and square-wave concentration modulation, these investigators observed a large variation in temperature response to small amplitude disturbances in the

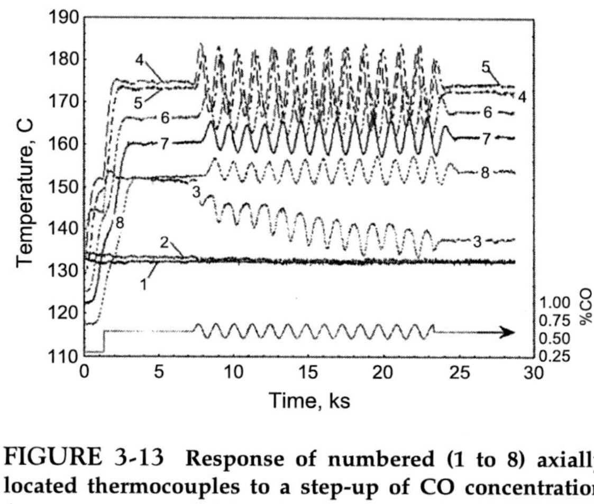

FIGURE 3-13 Response of numbered (1 to 8) axially located thermocouples to a step-up of CO concentration from 0.25 to 0.5 vol% in a feed mixture of 19.5 vol% O_2 in an N_2 carrier gas flowing at 0.373 mol/min. Pressure was atmospheric and entering temperature was 117°C. Thermocouple (TC) 1 located 6.3 cm above a catalyst bed packed with 0.2 wt% Pt on Al_2O_3 pellets. TC 2 located at the bed interface and remaining TCs located 7.6 cm apart. A concentration oscillation with a period of 20 min and amplitude of 0.125 vol% CO was introduced at 7.5 ks and discontinued at 23.5 ks. *(Figure reproduced from* Jaree et al. (2003b) *with the authors' permission.)*

CO feed concentration just as *Yakhnin and Menzinger* predicted. **Figure 3-13** shows these observations. The top of the figure shows the temperature response to a step-change in feed concentration at about one minute after start up, followed by a small sinusoidal variation that began at 7.5 and ended at 23.5 ks. These concentration variations appear at the bottom of the figure. It should be noted that the temperature fluctuations increase to about midway in the reactor (profile 4). Thereafter, the amplitude decreases because of heat loss and conduction along the reactor wall. The effect of conduction along the wall of the reactor was tested using a reactor with a heavier wall and a much smaller amplification was observed.

Poisoning, when it occurs in packed bed reactors, begins at start-up at the bed entrance and moves progressively deeper into the bed with time, creating an axial variation in catalyst activity. *Jaree et al. (2004)* used simulation to explore the effect of this variation on disturbance amplification. They chose to exaggerate the variation effect by examining a bed with alternating layers of catalyst and inert support. The layer depth was a variable. They observed that the bed structure had a strong effect on resonance. Shallow layers had little effect on maximum gain. However, as depth increased, the gain grew larger and went through a maximum.

Notation

a	=	volume specific interfacial area (m^{-1})
C	=	concentration (mol/m^3)
C_i	=	concentration of species i
C^s	=	concentration on surface (mol or molec./cm^2)
CSTR	=	continuous stirred tank reactor
FFT	=	fast fourier transform
FTIR	=	fourier transform infra-red spectroscope
g	=	gas
IR	=	infra red
i	=	species indicator
k	=	reaction rate constant (various units)
k_a,	=	adsorption rate constants (mol/s cm^2)
k_d	=	desorption rate constants (mol/s cm^2)
LEED	=	low-energy electron diffraction
M	=	particle mass
P	=	pressure (kPa, MPa)
Q	=	volumetric flow rate (m^3/min)
Q_0	=	entering volumetric flow rate
q	=	volume shrinkage
r_i	=	rate of reaction of species I (mol/sgm catalyst)
r_k	=	rate of the kth reaction where $k = 1, 2, \ldots$
r_i^S	=	rate of reaction on catalyst support
S	=	surface site on support
s	=	surface site (adsorption or reaction)
	=	cycle split (-)
S_p	=	particle or pellet surface area (cm^2 or cm^2/g)
S_t	=	total surface area
SN	=	stoichiometric number
SS	=	steady state
STP	=	standard temperature and pressure
SV	=	space velocity (s^{-1})
s	=	cycle split (-)
T	=	temperature (K)
T_{feed}	=	entering temperature
TWC	=	three-way catalyst
t	=	time (s)
V	=	reactor volume (L)
y_i	=	mol fraction of species i

Greek

ν_I	=	stoichiometric coefficient for species i
θ_I	=	fractional occupancy of species i
ρ	=	density (g/cm^3)
τ	=	cycle period (s)
τ_o	=	period of the forcing flow or composition
ω	=	frequency (Hz)
ψ	=	enhancement (-)

Subscripts

a, ad	=	adsorption
cat	=	catalyst
d, de	=	desorption
i	=	species
o,0	=	initial or entrance
sup	=	support
t	=	total

Superscripts

S	=	support

CHAPTER

4

Partial Oxidation and Dehydrogenation of Hydrocarbons

Robert Ross Hudgins, Peter Lewis Silveston*, Cheng-Yue Li†, Adesoji A. Adesina***

*Waterloo, Ontario, Canada, †Beijing, China and **Sydney, New South Wales, Australia

OUTLINE

4.1 Introduction 79

4.2 Partial Oxidation and Reforming of Methane to Syngas 80
- *4.2.1 Noble Metal Catalysts Supported On Ce-Containing Oxides* 80
- *4.2.2 Perovskite Oxide Catalysts* 85
- *4.2.3 Methane Reforming* 86

4.3 Oxidative Coupling of Methane 88

4.4 Epoxidation 92

4.5 Propene and Butene Partial Oxidation and Ammoxidation 95

4.6 Catalytic Dehydrogenation of Propane, Butane and Higher Hydrocarbons 102

4.7 Maleic Anhydride from Butane 108
- *4.7.1 Du Pont Research* 108
- *4.7.2 Chinese Research* 113
- *4.7.3 European Research* 117

4.8 Anhydrides and Aldehydes from Aromatic Hydrocarbons 119
- *4.8.1 Maleic Anhydride* 119
- *4.8.2 Phthalic Anhydride from o-Xylene* 120

4.9 Aromatic Nitriles 121

4.1 INTRODUCTION

These reactions, many of which are commercially important, appear to be attractive candidates for composition modulation. In them, oxygen (O_2), is reproduced up by the catalyst. Consequently, the catalyst functions as an oxygen carrier allowing the dehydrogenation or partial oxidation reaction to be split into separate steps—catalyst oxidation by exposure to an

O_2-containing gas in one step and reduction by a hydrocarbon in the second step. In this way, combustible mixtures can be avoided. Even with separation into two steps, selectivity remains an important consideration because over-oxidation resulting in CO_2 may occur.

4.2 PARTIAL OXIDATION AND REFORMING OF METHANE TO SYNGAS

The conventional production of syngas by steam reforming is energy-wasteful: About 60% of the cost of syngas is due to the steam reforming step (*Haggin, 1992*). Catalytic partial oxidation of CH_4 to syngas has several advantages over steam reforming, such as providing a desirable molar ratio, H_2:CO, of 2, better energy-efficiency and higher selectivities to CO and H_2 (*Li et al., 2000*). Efforts to avoid the problems such as a high temperature gradient in the reactor and the risk of explosion with premixed CH_4/O_2 have focused on either autothermal reforming of CH_4 using a mixture of O_2 and steam or partial oxidation of CH_4 to syngas using lattice oxygen in oxygen-storage materials in place of gaseous O_2. The latter tactic results in a cyclic process.

4.2.1 Noble Metal Catalysts Supported On Ce-Containing Oxides

With certain catalysts, direct partial oxidation of CH_4 to syngas is possible (*Fathi et al., 2000a*). *Otsuka et al. (1998, 1999)* proposed periodically switching between CH_4 and CO_2 or H_2O feeds to a packed bed of cerium oxides as shown in **Figure 4-1**. CH_4 reacts with CeO_2 in a first step, producing H_2 and CO with a molar ratio (H_2:CO) of 2, the reduced cerium oxide, CeO_{2-x} is then reoxidized to CeO_2 by CO_2 or H_2O in a second step, at the same time forming either CO or H_2. *Otsuka's* preliminarily investigation (see **Table 4-1**) showed this periodic process

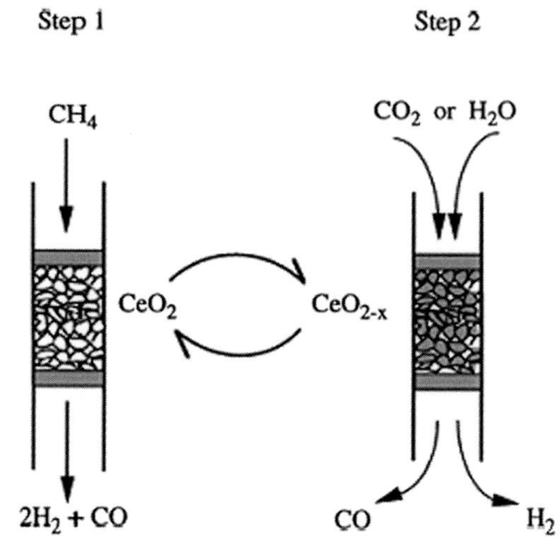

FIGURE 4-1 Periodic composition modulation for direct conversion of CH_4 to syngas over a CeO_2 catalyst. *(Figure reproduced from* Otsuka et al. (1998) *with permission. © 1998 by Elsevier Inc.)*

was feasible. About 95% of the O_2 consumed in the first step originated in the catalyst lattice. Unfortunately time-average rates of product formation were very low at the conditions used.

The effects of additives on the formation rates of H_2 and CO were the subject of *Otsuka et al. (1998)*. They observed that the addition of Pt or Pd black significantly accelerated the formation of these products. Other additives (BeO, MgO, CaO) offered just minor increases in rate, or none at all (BaO, Li_2O, K_2O). With Pd black, the molar ratio of H_2 to CO was about 6.0, suggesting the deposition of carbon on the Pd modified CeO_2. With Pt black, on the other hand, the molar ratio was 2.05, implying that no carbon deposited on the oxide.

Fathi et al. (2000a, b), using repetitive pulsing of CH_4, confirmed the well-known function of CeO_2 as an oxygen carrier. The noble metal additives improved the activity of the cerium oxide for partial oxidation of CH_4 to syngas and carbon deposition on the oxide was reduced. On pulsing the reduced catalyst repetitively with O_2 or CO_2, CO was formed in the early pulses, confirming

TABLE 4-1 Methane Partial Oxidation and Reforming Experiments

Reference	Objective	Reaction and Reaction Conditions	Modulated Variable	Observations	Comments
Otsuka et al. (1998, 1999)	Investigation of the partial oxidation of CH_4 to synthesis gas using a solid oxygen carrier	$CH_4 + CeO_2 \rightarrow CO + 2\,H_2 + CeO$, $O_2 + 2CeO \rightarrow 2CeO_2$, $P = 1$ atm; $723 < T < 1073$ K	CH_4 pulses used to reduce CeO_2.	Reaction between CH_4 and CeO_2 produced a synthesis gas with H_2/CO ratio = 2. Adding Pt black or Pd black promoter accelerated this process, decreasing the temperature needed.	H_2 exchange between CH_4 and CD_4 suggested that the cleavage of the C−H bond of CH_4 was not rate-limiting, but rather recombination or desorption of H_2. It was suggested that Pt accelerated this step via a reverse spillover mechanism.
Zhang and Amiridis (1998); Aiello et al. (2000)	Investigation of catalytic cracking of CH_4 to produce pure H_2 followed by steam regeneration of the catalyst	$CH_4 \rightarrow C + 2H_2$; a fixed-bed quartz lab reactor, CH_4 was cracked over the catalyst at 923 K at GHSV = 30 000 h^{-1} for 3 h with N_2 for 30 min followed by steam regeneration	Cycles consisted of two independent reactions—cracking followed by steam regeneration	Catalyst was fully regenerated at 923 K with steam for up to 10 successive cracking/regeneration cycles without any significant loss of catalytic activity.	Filamentous carbon deactivated the 15 wt% Ni/SiO_2 catalyst.
Opoku-Gyamfi and Adesina (1999); Opoku-Gyamfi et al. (2000)	Demonstration of a periodic autothermal process for syngas production	$CH_4 + H_2O + O_2 \rightarrow CO + H_2 + CO_2$ (not balanced), $T = 773$ to 873 K; P = 101.3 kPa; co-precipitated Ni/Co/alumina	Cycle period and split for different compositions in the partial cycles	Periodic operation increased H_2 production and the H_2/CO ratio significantly. Shorter exposure to the O_2–containing mixture gave highest H_2 concentration. Cycling between equimolar mixtures in the partial cycles is autothermal for $s \geq 3$.	Authors note that periodic operation provides flexibility to meet composition requirements of downstream processes.

(*Continued*)

TABLE 4-1 Methane Partial Oxidation and Reforming Experiments (*cont'd*)

Reference	Objective	Reaction and Reaction Conditions	Modulated Variable	Observations	Comments
Fathi et al. (2000a)	Exploration of the partial oxidation of CH_4 to synthesis gas using Pt and Rh on $CeO_2/\gamma\text{-}Al_2O_3$ as catalysts	$CH_4 + CeO_2 \rightarrow CO + 2\,H_2 + CeO$, $X + 2\,CeO \rightarrow 2\,CeO_2$ where $X = O_2$ or CO_2, $T = 973$ K, $P = 180$ kPa	Feed composition, switching between pulses of CH_4 and oxidants	Modulation was interpreted as CeO_2 acting as an intermediate oxygen carrier – being reduced by CH_4 and reoxidized by oxidant pulses. The redox cycle was strongly promoted by Pt and Rh.	No direct comparison was made between steady operation and modulation by pulsing, except to note that modulation required a temperature of ~700°C vs. ~900°C for conventional syngas production at steady state.
Fathi et al. (2000b)	Exploration of O_2 species active in the partial oxidation of CH_4 to synthesis gas using a Pt gauze	$CH_4 + O{\cdot}Pt \rightarrow CO + H_2 + Pt$, $O_2 + Pt \rightarrow O{\cdot}Pt$, (equations not balanced), $T = 1073$ to 1373 K, $P \approx 100$ kPa	Pulsing CH_4 and/or O_2 experiments using a TAP reactor	Low concentration or mobility of adsorbed O species was observed to be important for syngas formation Thus, periodic operation is necessary. Addition of Rh reduces mobility of adsorbed O species.	Authors observed subsurface oxidation of Pt gauze at higher temperatures.
Monnerat et al. (2001)	Development of an H_2 reduction process	Cracking of CH_4 and sequential gasification of carbon deposition. Tubular reactor packed with Ni gauze treated to create a Raney Ni surface, $P = 150$ kPa;	Gas-phase composition with switching between CH_4 and O_2. $0.1 < s < 0.9$; $2 < \tau_{cycle} < 20$ min	Ni gauze was active at $723 < T < 773$ K. CH_4 conversion was highest at $\tau = 2$ to 4 min and $s \approx 0.1$; highest H_2 productivity at $\tau = 4$ to 6 min, with $s = 0.5$; carbon deposited as filaments.	Gauze structure offers low pressure drop but low surface area for reaction. Treatment of the surface increased area by 27-fold.
Shen et al. (2003)	Effect of perovskite composition on catalyst performance	$CH_4 + MO_3 \rightarrow CO + 2\,H_2 + MO_2$, $O_2 + 2\,MO_2 \rightarrow 2\,MO_3$, $T = 1173$ K	Asymmetrical composition modulation switching between 11 mol% O_2 in Ar and 11 mol% CH_4 in He, $\tau_{cycle} = 36$ s	Additional TPSR experiments indicated two oxygen species available in the perovskite. The high temperature species provided partial oxidation.	

Galvita and Sundmacher (2005)	Production of ultrahigh purity H_2	2-step process using CH_4 to reduce Fe_3O_4 and steam to reoxidize Fe to Fe_3O_4. Two layers of catalyst ($Pt\text{-}CeO_2\text{-}ZrO_2$ and $Fe_3O_4\text{-} CeO_2\text{-}ZrO_2$). Quartz tube lab reactor (1 cm i.d.); $873 < T < 1073$ K	Bang-bang switching between CH_4 and steam	Authors found that using two layers of different catalyst increased the rate of reduction of Fe_3O_4. O consumption from steam reached 79% at 1073 K but was accompanied by the presence of ppm levels of CO in the H_2 product.	Authors recommended keeping O_2 consumption below 60% in order to drop CO concentration in the H_2 product below 40 ppm.
Dai et al. (2006a, 2006b)	Effect of perovskite composition on catalyst performance	$CH_4 + MO_3 \rightarrow CO + 2\ H_2 + MO_2$, $O_2 + 2\ MO_2 \rightarrow 2\ MO_3$, $T = 1173$ K; $P = 1$ atm; fixed bed of $LaFeO_3$ perovskite oxide catalyst	Switching between 11 mol% O_2/Ar for 10 s and 11 mol% CH_4/He for 30 s; successive pulsing of CH_4 until CH_4 conversion ends	Results suggested depleted oxygen species on lanthanum-based catalysts could be regenerated and CH_4 oxidized to syngas by lattice oxygen with high selectivity.	Authors observed that the La-bearing perovskite were superior catalysts in terms of stability, safety, ease of operation, and cost efficiency for the partial oxidation of CH_4
Promaros et al. (2007)	Investigation of H_2 production using CO_2 for CH_4 reforming	$CH_4 \rightarrow C + H_2$, $C + CO_2 \rightarrow CO$ (not balanced) over $Ni/SiO_2{\cdot}MgO$ catalyst, 1.1-cm i.d. × 50 cm quartz tube reactor; $T = 923$ K, 1023 K	Composition modulation between CH_4 and CO_2, $10 < \tau_{cycle} < 40$ min	Authors observed a decrease in CH_4 as well as CO_2 conversion with time-on-stream in periodic operation. At 1023 K, H_2 yield was below yield at steady state but matched this yield at 923 K. Cycle split was not important.	Carbon accumulation occurred from cycle to cycle in periodic operation.
Dai et al. (2008)	Investigation of continuous composition modulation	$CH_4 + MO_3 \rightarrow CO + 2\ H_2 + MO_2$, $O_2 + 2\ MO_2 \rightarrow 2\ MO_3$, $T = 1173$ K; $P = 1$ atm; fixed bed of La-bearing oxide catalysts	Switching between 11 mol% O_2/Ar for 10 s and 11 mol% CH_4/He for 30 s	As in the authors' previous publication.	Authors suggested that their fixed bed catalyst could be applied to continuous fluidized bed reactors.

(*Continued*)

TABLE 4-1 Methane Partial Oxidation and Reforming Experiments (*cont'd*)

Reference	Objective	Reaction and Reaction Conditions	Modulated Variable	Observations	Comments
Pholjaroen et al. (2009)	Investigation of CO_2 reforming of CH_4 to produce pure H_2.	Thermal cracking of CH_4 followed by C gasification with CO_2. Tubular reactor packed with mixture of 55 wt% Ni/SiO_2–MgO and SiO_2; $823 < T < 1023$ K	Gas composition modulation: CH_4 and CO_2 using bang-bang switching. $\tau_{cycle} = 20$ min; $s = 0.5$	Catalyst lost activity at $T > 923$ K in periodic operation due to C encapsulation; steady-state operation showed much higher CH_4 conversion than composition modulation.	Authors noted that cracking and coke gasification provided autothermal operation. Carbon formation at 923 K is filamentous and can be readily gasified by CO_2.
Assabumrungrat et al. (2009)	Improvement of CH_4 cracking process.	Thermal cracking of CH_4 followed by C gasification with CO_2. $T = 923$, 1023 K; $CO_2:O_2 = 7:2$ in regeneration partial cycle	Gas composition: CH_4 and $CO_2 + O_2$, using bang-bang switching, $s = 0.5$; $\tau = 20$ min	Addition of O_2 to regeneration partial cycle stabilized catalyst by gasifying all the coke deposited. Highest H_2 yield obtained for regeneration partial cycle $CO_2:O_2 = 8:2$ at 923 K and $= 9:1$ at 1023 K. Steady-state operation with O_2 addition to the feed gave higher H_2 production than the periodic operation investigated.	Successful results indicate O_2 is needed to gasify coke deposits at higher temperatures.

the presence of carbon on the catalyst surface. The *Fathi* results suggest a highly selective catalytic route from CH_4 to syngas by a periodic redox cycle involving CH_4 and O_2 employing Pt/cerium oxide/γ-Al_2O_3. Also, 1,000 K is a significant reduction in temperature from conventional operation.

4.2.2 Perovskite Oxide Catalysts

The perovskite-type oxides with excellent redox properties are attractive catalysts for partial oxidation. These oxides have the general formula ABO_3, in which A and B are respectively rare earth and transition metal cations. They exhibit high O_2 mobility and a large oxygen storage capacity. Although perovskites have been widely studied as catalysts for oxidation and combustion in the past (*Ciambelli et al., 2002*), they have not been applied to the partial oxidation of CH_4 to syngas until recently (*Shen et al., 2003; Dai et al., 2006a, b, 2008).*

Shen et al. (2003) prepared $La_{1-x}Sr_xFeO_3$ (x = 0, 01, 02, 05) oxides whose XRD patterns revealed a single crystalline phase with a perovskite structure. The oxygen species in or on the catalysts and their reactions with CH_4 were studied by switching asymmetrically between O_2/Ar and CH_4/He. Results indicated that two different oxygen species exist. The one reacting at a lower temperature was strongly oxidizing and resulted in complete oxidation, while the other, reacting at over 968 K, was moderately oxidizing and resulted mostly in the selective oxidation to CO and H_2, with just minor amounts of CO_2 formed. The fraction of the strongly oxidizing species in the crystals decreased with a lower Sr content in the $La_{1-x}Sr_xFeO_3$ catalysts. $La_{0.8}Sr_{0.2}FeO_3$, $La_{0.9}Sr_{0.1}FeO_3$ and $LaFeO_3$ catalysts selectively oxidized CH_4 to CO and H_2 through lattice oxygen.

Partial oxidation of CH_4 to syngas with asymmetrical forcing over the $La_{1-x}Sr_xFeO_3$ catalyst was carried out at 1173 K by switching between feeds of 11 mol% O_2 in Ar and 11 mol% CH_4 in

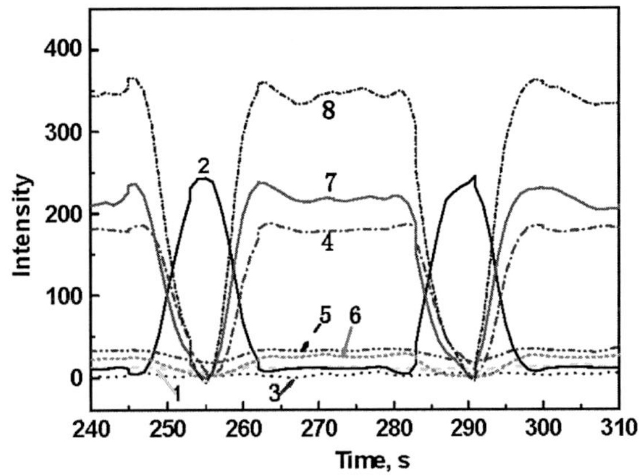

FIGURE 4-2 **Transient response to reactant switching between 11 mol% O_2 in Ar for 11 s and 11 mol% CH_4 in He for 25 s over a $La_{0.8}Sr_{0.2}FeO_3$ catalyst at 1173 K. In the figure: 1 = CO_2, 2 = Ar/4, 3 = O_2, 4 = CO, 5 = H_2O, 6 = CH_4, 7 = He/4, 8 = H_2.** *(Figure reproduced from* Shen et al. (2003) *with permission. © 2003 by Chinese Journal of Chemical Engineering.)*

He. **Figure 4-2** shows the measurements for one catalyst. No significant amounts of CO and CO_2 were detected after switching from CH_4/He to O_2/Ar. After the reverse switch, large amounts of CO and H_2 were formed, while the CO_2 intensity remained at a background level. Conversion of CH_4 was 90%; selectivities to H_2 and CO exceeded 93%.

Dai et al. (2006a, b, 2008) undertook periodic pulsing of CH_4 for its partial oxidation over $AFeO_3$ (where A = La, Nd or Eu) catalysts. XRD and XPS spectra were used to identify phase structure and surface oxygen species for pre- and post-reaction $AFeO_3$. TPSR/MS measurements indicated that the $AFeO_3$ oxides provide two types of surface oxygen species associated with CO_2 and syngas formation respectively. These different types were confirmed by XPS spectra.

The effect of reaction temperature with periodic CH_4 pulsing was studied over the $LaFeO_3$ catalyst, as shown in **Figure 4-3**. It can be seen that CH_4 conversion and CO selectivity are strongly affected by the reaction temperature, presumably because the partial oxidation rate is dominated by lattice oxygen diffusion from the catalyst bulk.

The $LaFeO_3$ catalyst maintained relatively high catalytic activity at 1173 K for the 50 redox cycles used experimentally (see **Figure 4-4**). Structural stability was confirmed by XRD characterization. Based on their results, *Dai* and coworkers proposed producing H_2 from natural gas by employing a circulating fluidized perovskite catalyst bed followed by successive water-gas shift reactors (*Dai et al., 2006b*). Their proposal was later demonstrated on a laboratory scale (*Dai et al., 2008*).

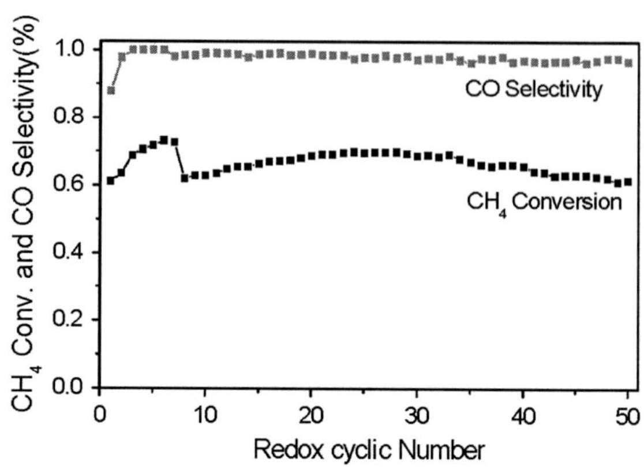

FIGURE 4-4 CH_4 conversion and CO selectivity as functions of the cycle number for composition modulation between 11 mol% O_2 in Ar for 10 s and 11 mol% CH_4 in He for 30 s at 1173 K for an $LaFeO_3$ catalyst. *(Figure reproduced from* Dai et al. (2006a) *with permission of the authors.)*

4.2.3 Methane Reforming

Steam reforming is the conventional route to synthesis gas and high-purity H_2. As early as 1993, however, an iron redox cycle was proposed by *Fukase and Suzuka (1993)*. Since then, this cycle has been frequently studied (e.g., *Hacker et al., 2000; Otsuka et al., 2003*). *Galvita and Sundmacher (2005)* found that adding CeO_2 increased the amount of oxygen that could be removed from or added to the solid phase, while ZrO_2 in the catalyst raised the storage capacity of CeO_2. Higher rates of O_2 transfer meant that reforming could be carried out at lower temperatures. These investigators noted

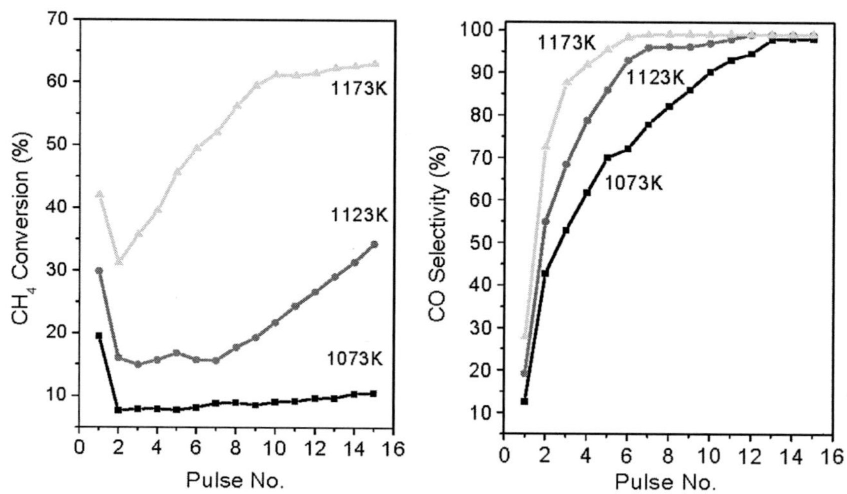

FIGURE 4-3 CH_4 conversion and CO selectivity vs. pulse number in successive pulsing at different temperatures over an $LaFeO_3$ oxide. *(Figure reproduced from* Dai et al. (2006a) *with permission of the authors.)*

that the most active formulation for CH_4 conversion was Fe_3O_4-$Ce_{0.5}Zr_{0.5}O_2$.

Reforming has also been explored as a means of consuming some of the vast quantities of CO_2 produced through combustion and other industrial processes. Periodic operation has been explored for this application by *Pholjaroen et al. (2009)* using an industrial steam reforming catalyst, Ni/SiO_2·MgO. The authors' process produced a high-purity H_2 stream through cycling between CH_4 and CO_2. The first part of the cycle cracks CH_4 to produce H_2 and lay down carbon on the catalyst; carbon that is then gasified by CO_2 in the following half-cycle. The authors' experiments (**Table 4-1**) showed that a high-purity H_2 was obtained via this cycle. SEM studies indicated that at low temperatures CH_4 cracking deposits a filamentous carbon but that carbon becomes refractive at temperatures above 923 K and encapsulates the catalyst surface.

Assabumrungrat et al. (2009) added O_2 to the CO_2 gasification half-cycle and demonstrated stable periodic operation at temperatures up to 1023 K for over 12 cycles. **Figure 4-5** shows that CH_4 and CO_2 conversion were independent of the CO_2:O_2 ratio in the gasification half-cycle. In the absence of O_2, however, conversion dropped rapidly with time-on-stream. Periodic operation failed to increase CH_4 conversion over steady state at the time-average feed composition under identical residence times and temperatures.

Opoku-Gyamfi and Adesina (1999) suggested that CH_4 cracking and carbon gasification, an autothermal route to syngas, could be made more flexible if it were operated periodically instead of at steady state. They proposed employing a fluidized bed and switching the feed to the bed between CH_4:H_2O and CH_4:O_2. Alternating these feeds would allow the exothermic gasification reaction to furnish heat for the endothermic reforming reaction. Operating at steady state in the presence of both

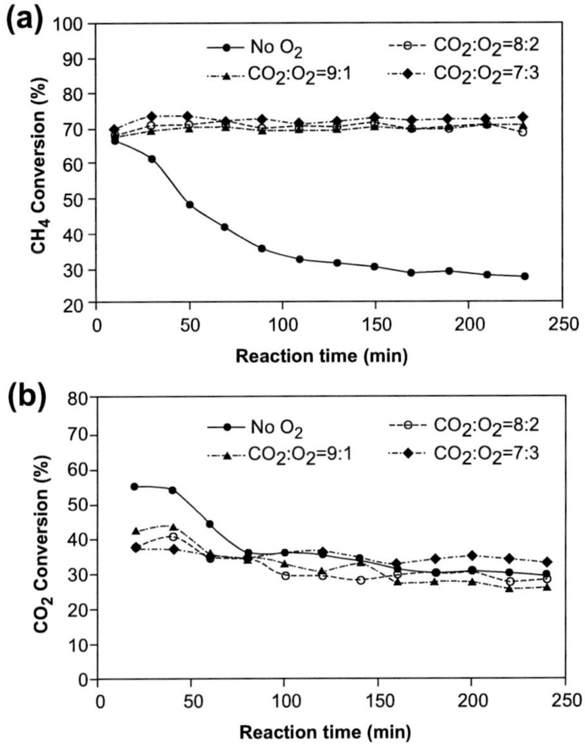

FIGURE 4-5 **Results of symmetric modulation between CH_4 and CO_2/O_2 mixtures at $\tau_{cycle} = 20$ min., $T = 750°C$ over 12 consecutive cycles employing a Ni/SiO_2·MgO catalyst diluted with SiC in a 1:3.5 ratio. (a) shows CH_4 conversion while (b) gives CO_2 conversion.** *(Figure reproduced from* Assabumrungrat et al. (2009) *with the authors' permission.)*

steam and O_2, oxygen consumes H_2 as well as CH_4, reducing H_2 yields.

Experiments were undertaken in a fluidized bed using a co-precipitated Ni/Co catalyst supported on alumina. Co-precipitation increased catalyst activity above that of each metal used alone. Modulation consisted of switching between equimolar mixtures of CH_4 and steam and CH_4 and O_2. The authors reported autothermal operation for $s \geq 0.3$. Furthermore, they found a 25% increase in hydrogen yield and an 80% higher H_2:CO molar ratio for composition modulation. Their encouraging results led to a further study of the modulation variables using the same catalyst and fluidized bed reactor (*Opoku-Gyamfi et al., 2000*). The authors reported obtaining an H_2:CO molar ratio of 7 through cycling with $s = 0.1$; this ratio fell to 2.5 at $s = 0.9$. At steady state, the ratio was 5.2. Cycle period had almost no effect on this ratio at $s > 0.3$. Also, cycling increased H_2 production by up to 15%.

Opoku-Gyamfi et al. observed that the peak in CO formation trailed the introduction of the CH_4-steam mixture, whereas the other products were in-phase, indicating that CO is formed during the CH_4:O_2 partial cycle and H_2 arises primarily in the CH_4:H_2O portion of the cycle. They interpreted this to mean that dissociative adsorption of CH_4 and steam occurs and CO and H_2O are formed by oxidation of these adsorption fragments.

Further work on the cracking/gasification cycle was undertaken by *Monnerat et al. (2001)* employing a novel reactor consisting of Ni gauze packed in a tube. Such a treatment increased the specific surface 27-fold. **Table 4-1** summarizes the experiments carried out. Using cycle periods of 10 to 20 min, the authors found both period and cycle split to be important. Pure H_2 could be obtained only in the latter half of the methane feed partial cycle, as can be seen in **Figure 4-6**.

Promaros et al. (2007) revisited the production of H_2 from CH_4 using CO_2 reforming. They worked at high temperatures employing an Ni catalyst and observed that, although periodic operation does not offer the H_2 yield attained under steady state, it has the advantage of producing an H_2 stream almost free of carbon oxides. These investigators used rather long cycles, up to 40 min. They observed a continual drop in the CH_4 cracking rate during the CH_4 half-cycle. Cycle period and cycle split had just a small effect on performance for the conditions studied. Two temperatures, 923 and 1023 K, were considered. H_2 yield was better in steady state than in periodic operation at the upper temperature.

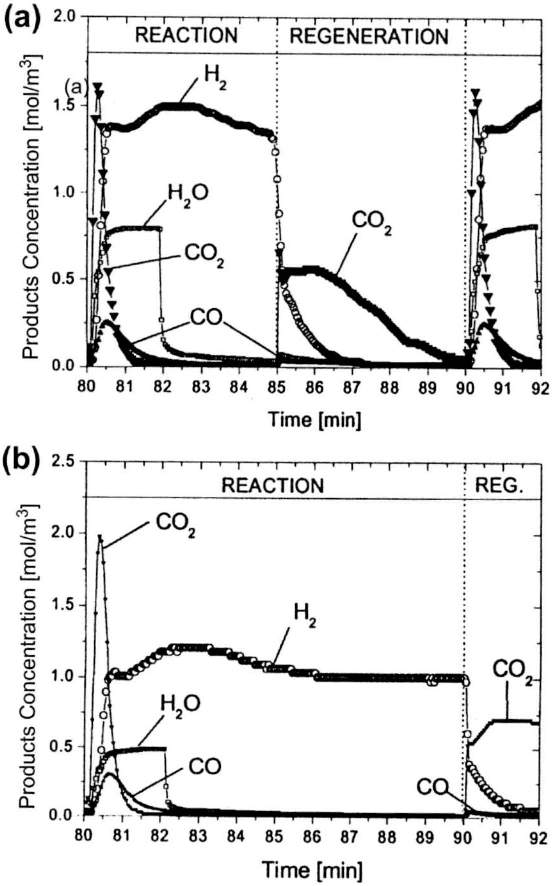

FIGURE 4-6 **Reactant and product transients during symmetric cycling between CH_4 at 4.67 mol/m^3 and O_2 at 1.87 mol/m^3 in an Ar diluent at 773 K. In (a), $\tau_{cycle} = 10$ min, while in (b), $\tau_{cycle} = 20$ min. (The full cycle is not shown in (b).)** *(Figure adapted from* Monnerat et al. (2001) *with permission. © 2001 Pergamon Science Publisher.)*

4.3 OXIDATIVE COUPLING OF METHANE

During the 1990s, the conversion of CH_4 to C_2H_4 and higher hydrocarbons was intensively explored by research teams around the world. Oxidative coupling to C_2H_4 and partial oxidation to CH_3OH or formaldehyde attracted much attention. Both operations have severe selectivity problems because their products are further oxidized under the reaction conditions used. For oxidative coupling, the catalyst's primary function is to break a C-H bond to create a methyl radical. Selectivity drops rapidly as conversion rises above 10% resulting in what has been called a yield barrier (*Amenomiya et al., 1990*). Several U.S. patents recognized an opportunity for composition modulation, but *Keller and Bhasin (1982)* first published results that employed modulation between CH_4 and air. Similar experiments were reported by *Sofranko et al. (1987)*. **Table 4-2** summarizes experimental studies on modulating oxidative coupling.

Significant amounts of higher hydrocarbons were found under composition modulation. Selectivities to C_3 reached 7% for the Mn catalysts, while those to C_4 and benzene were about 4%. *Sofranko et al. (1987)* commented that product distributions by carbon number were similar regardless of conversion and the catalyst used. *Jones et al. (1987a)* concluded that the C_2's and higher hydrocarbons are formed by a common mechanism involving just gas species. Thus, the primary role of the catalyst was to generate methyl radicals. Over-oxidation of the catalyst had to be avoided to maintain high selectivity. At contact times under 0.1 s, when conversion is very small, the main product was C_2H_6. Increasing contact time led to the formation of C_2H_4, higher hydrocarbons, coke and total oxidation products. *Jones et al. (1987a)* reasoned that the primary coupling product is C_2H_6, which is then dehydrogenated. The higher hydrocarbons originated from methyl radical-olefin reactions and olefin dimerization.

Mortazavi et al. (1992, 1996) attempted to improve performance under modulation through the choice of a forcing strategy. They also explored the influence of cycle variables on the C_2 selectivity and CH_4 conversion. Two catalysts were employed: The Li-doped magnesium oxide studied intensively by Lunsford and his co-workers (*Ito et al., 1985*) and a mixed cerium oxide/Li-doped magnesium oxide investigated by *Bi et al. (1988)* and by *Bartsch et al. (1989)*. One of *Mortazavi's* strategies was methane-cycling, the other oxygen-cycling. The bang-bang operation used by the earlier investigators mentioned above was the third strategy investigated. All experiments are summarized in **Table 4-2**.

Mortazavi et al. (1996) observed that stationary, reproducible cycles were reached after one hour on stream. The Li/MgO catalyst enhanced CH_4 conversion and C_2 formation provided that cycle periods were less than 10 s. Amplitude appeared to have little effect on selectivity to C_2 products and on conversion. Cycle split, defined as the ratio of the duration of mixture flow to the cycle period, influenced CH_4 conversion but not selectivity to C_2. A quasi-steady state seemed to be reached for cycle periods exceeding 50 s. In CH_4 cycling, C_2 was observed only during exposure of the catalyst to the CH_4-O_2 mixture. There was little selectivity enhancement and it was largely independent of cycle period. C_2H_4 yield increased and showed a dependence on the CH_4:O_2 ratio. Amplitude affected selectivity to $C_2^{=}$ and CH_4 conversion.

Bang-bang cycling gave similar results to those for O_2-cycling for the Li/MgO catalyst. Results for that catalyst and the Ce promoted catalyst confirmed the moderate enhancement seen by *Sofranco et al. (1987)* and *Jones et al. (1987b)*.

Mortazavi et al. (1996) also examined flushing with a 3-part cycle: 10 vol% CH_4 in He, 10 vol% O_2 in He followed by He alone as well as a 3-part cycle with overlapping CH_4 and O_2 exposures so that the catalyst sees a CH_4-O_2 mixture for part of a cycle. Overlapping raised the C_2 selectivity to 56%, while the $C_2^{=}$ selectivity went to 15% so it was an effective choice. However, none of the strategies explored by *Mortazavi et al.* were able to reach the 25% C_2+ yield cited in the 1980s literature as the target for oxidative coupling. Highest C_2+ yields achieved were obtained with CH_4 cycling and were only 10%. Symmetrical bang-bang cycling

TABLE 4-2 Oxidative Coupling Experiments

Reference	Objective	Reaction and Reaction Conditions	Modulated Variable	Observations	Comments
Keller and Bhasin (1982)	Effect of modulation on oxidative coupling with various catalysts for process and catalyst development	$CH_4 + MO \rightarrow 0.5\ C_2H_4 + H_2O + M$, $O_2 + 2M \rightarrow 2MO$, $SV = 1900$ (STP) h^{-1}; $773 < T < 1273$ K. Employed 25 individual metal/metal oxide catalysts on α-Al_2O_3	4 part symmetrical cycle for CH_4 and air (20 s each) with intermediate 5 s N_2 flushes	At 873 K, only Mg and Sr showed activity; at 1073 K, the most active metal/metal oxide systems were Mn, Cd, Tl and Pb; at 1273 K, Li, Tl, Mg, Sn, Pb and Sb showed good activity. C_2 yields under composition modulation were 10-fold greater than at steady state. CO formation was also noted.	Segregating the CH_4 and air feeds by small N_2 flushes reduces the cost of separating C_2H_4 in the product. O_2 removed during the cycling experiments exceeded monolayer coverage for most catalysts by more than 10 x. Thus, the catalysts function by a redox mechanism.
Sofranko et al. (1987); Jones et al. (1987a, b)	Investigation of feed modulation on conversion and selectivity to C_2+ with a manganese oxide catalyst supported on silica	$CH_4 + O_2 \rightarrow C_2^{=} + H_2O$ over 15 wt% Mn/5 wt% $Na_4P_2O_7/SiO_2$, also Mn, Pb, Sn alone with SiO_2 support, SV for methane = 600 to 800 (STP) h^{-1}; $973 < T < 1073$ K	Asymmetrical cycles for CH_4 and air, some with an N_2 flush between reactants	High CH_4 conversion at 1073 K at C_2 selectivity of ca. 10%. NaOH or pyrophosphate promoters or CH_3Cl addition improved selectivity. C_2 yields were much less than 25%. Under feed modulation, C_3+ selectivity reached 7% for Mn catalysts. Selectivities to C_3 and benzene were about 4%. Product yields up to C_8 were measured.	Selectivity was only slightly improved over steady state, contrary to Keller and Bhasin (1982). Carbon number selectivity vs. carbon number is independent of catalyst used. Authors posited that the catalyst's role is to generate methyl radicals. CO_2 in the air partial cycle suggests coke-like deposits on the surface that forms from olefins or condensation residues, rather than oxidation of CH_4 or CH_3 radicals.

Mortazavi et al. (1992, 1996)	Use of feed modulation to overcome the C_2 "yield barrier" using catalysts: 1) Li-doped MgO and 2) mixed Ce oxide/Li-doped MgO	$CH_4 + O_2 \rightarrow C_2^{=} + H_2O$ using as catalysts Li doped MgO and a mixed Ce-Li doped MgO. $SV = 1$ L/g catalyst· min; 1 bar pressure; $T = 1023$ K; He comprised 75 to 90% of the time-average reactor feed; all experiments began with a fully oxidized catalyst	(i) CH_4 cycling: switching between CH_4 and CH_4:O_2 molar mixtures (2:1 to 4:1) representing operating over a reduced catalyst; (ii) O_2 cycling: switching between O_2 and CH_4:O_2 molar mixtures (2:1 to 4:1) representing operating over an oxidized catalyst, (iii) switching between CH_4:He and O_2:He mixtures, a bang-bang strategy	Both CH_4 and O_2 cycling using a CH_4:$O_2 = 3$:1 mixture globally enhanced C_2 formation for $\tau_{cycle} < 10$ s for both catalysts; C_2H_6 is the dominant product. QSS attained in about 20 s for both modes. Selectivity to $C_2 >$ 60% at long τ_{cycle} for CH_4 cycling, but only at short τ_{cycle} for O_2 cycling; QSS was reached for $\tau_{cycle} > 50$ s. A 3-part cycle with an inert flush improved C_2 selectivity. Overlapping CH_4 and O_2 exposures gave the highest C_2 selectivity (56%).	CH_4 and bang-bang cycling are the most attractive strategies. This implies that the best performance results from short exposure of the catalyst to the CH_4 stream. Bang-bang experiments imply C_2H_6 dehydrogenation occurs and that O_2 adsorption causes total oxidation. The 25% C_2 yield barrier was not exceeded. The best yields found were only 14%.
Ramos et al. (1996); Herguido et al. (2005)	Investigation of a 2-zone fluidized bed reactor for oxidative coupling	Oxidative CH_4 coupling but with measurements of C_2+ products. Fluidized bed reactor with 3 cm i.d.; $P = 1$ atm; $T = 1073$ K; reactor bed is fluidized with O_2; catalyst was Mn/P/SiO_2	2-zone fluidized bed as an alternative to composition modulation. Reactant was introduced through a distributor in mid-bed	Demonstrated stable operation with an O_2 concentration below 5% in the reaction zone. No improvement was found in selectivity to C_2 over single fluidized bed operation.	Operating variables were not exhaustively studied. The Mn/P/SiO_2 catalyst was not satisfactory.

gave 9.5%. Introducing flushing or overlapping could increase the yield to perhaps 13 or 14%. O_2 cycling was not attractive. Coupling research is discussed in more detail by *Silveston (1998)*.

Ramos et al. (1996) applied a two-zone oxidizing-reducing fluidized bed to the coupling reaction. Oxygen with a carrier gas was introduced only at the bottom of the bed, while a dilute CH_4 stream was injected at the midpoint of the 3 cm i.d. reactor. They demonstrated long-term stable operation but found the yields of C_2 and higher hydrocarbons were small. *Herguido et al. (2005)* summarizes the *Ramos* work and other two-zone fluidized bed applications.

The claims made by *Keller and Bhasin (1982)* in their pioneering study of oxidative coupling have not been borne out in subsequent work. It now appears that yield increases through composition modulation are not sufficient to overcome the yield barrier discussed by *Amenomiya et al. (1990)*. A possible solution might be to superimpose periodic flow reversal on composition modulation, a solution that appears practicable because cycle periods for flow reversal are in hours, while those for composition modulation are in seconds.

Neural network modeling is often useful for exploring optimal conditions for producing a desired product. *Abdolahi et al. (2005)* applied this type of modeling to a transient reaction system. A description of their approach is beyond the scope of our treatment. It suffices to say that the authors showed that the neural network model they developed successfully predicted experimental measurements.

A second route through methanol to the higher hydrocarbons has not been as thoroughly studied as oxidative coupling at least with respect to composition forcing, nor has modulation been applied to the selective oxidation of CH_4 to MeOH. This application could be attractive because composition modulation can reduce the total oxidation side-reaction, a key problem in CH_4 to MeOH technology.

4.4 EPOXIDATION

Side reactions and high activation energies make epoxidation a difficult reaction to control. Consequently epoxidation was one of the first industrially important reactions to be considered for periodic operation (see **Table 4-3**). In 1976, *Renken et al.* investigated both the selectivity–conversion behavior for ethene epoxidation under composition modulation and its effect on the temperature profile in an integral reactor. Higher selectivity to the epoxide relative to total oxidation reduces the heat generated in a reactor. Thus, selectivity has a marked effect on the temperature profile. Both epoxidation and combustion exhibited parametric sensitivity so *Renken et al. (1976)* compared the temperature profiles for steady state and modulated operation as the reactor jacket temperature increased. They observed that $C_2^=$ modulation accomplished a remarkable leveling of the temperature in the bed. Increasing the cycle period suppressed sensitivity of the temperature profile and $C_2^=$ conversion to variation of the jacket temperature. Composition forcing caused temperatures in the catalyst bed to change with time. Up to $\tau_{cycle} = 40$ s, these temporal variations did not exceed 3°C. Increasing τ_{cycle} to 120 s forced the maximum temperature to oscillate axially.

Further experiments were undertaken by *Renken's* team to elucidate the epoxidation mechanism and develop a model for this reaction (*Renken et al., 1989*). Their step-change experiments suggested that the epoxidation proceeds through reversible adsorption of O_2 on the silver oxide catalyst. A model successfully reproduced their experiments. *Silveston (1998)* devotes several pages to the *Renken* work.

Park et al. (1983) used 4- and 8-part cycles to investigate epoxidation and total oxidation mechanisms. The purpose of these complex cycles, particularly the 8-part one, was to investigate the role of reaction residues on C_2H_4 conversion and selectivity to C_2H_4O. Their

TABLE 4-3 Epoxidation Reactions

Reference	Objective	Reaction and Reaction Conditions	Modulated Variable	Observations	Comments
Renken et al. (1976)	Effect of composition modulation on selectivity and conversion	$C_2H_4 + 0.5\ O_2 \rightarrow C_2H_4O$ with an Ag catalyst. Fixed bed reactor held at 473 K in a constant temperature bath	C_2H_4 cycled between 0 and 12 mol% at periods between 2 and 40 s; O_2 remained constant at 6 mol%; balance was N_2 to keep space velocity constant. Symmetric cycles ($s = 0.5$)	Under identical conditions, C_2H_4 modulation resultsed in lower conversions than steady state; however, yield increased because of higher selectivity. Selectivity to C_2H_4O exhibited resonance at $\tau_{cycle} = 30$ s. Steady-state and periodic operation displayed parametric sensitivity. Composition modulation leveled temperatures in the reactor.	Observations were explained by inhibition through C_2H_4, C_2H_4O and decomposition products that reduce the adsorption of O_2 on the surface. Authors suggested that adsorbed C_2H_4 and adsorbed O_2 react in epoxidation. Composition modulation removed other adsorbates and increased the concentration of adsorbed O_2.
Park et al. (1983); Park and Gau (1986)	Effect of modulation and cycle structure on selectivity and conversion	$C_2^{=} + O_2 \rightarrow C_2H_4O$ using Ag catalyst with α-Al_2O_3 or SiO_2 supports and a dichloroethane promoter. $T = 453$ K; fixed bed reactor	(i) 4-part cycles: C_2H_4:$O_2 = 38$:12 (or just C_2H_4 alone); N_2 flush; O_2; N_2 flush. Duration of the N_2 flushes were usually equal but those of C_2H_4 and/or O_2 varied widely. (ii) 8-part cycles: retained the N_2 flushes after each reactant but added a pulse with H_2 and another with O_2	Dichloroethane promoter increased activity by 10- to 30-fold over a commercial catalyst. 92 and 98% selectivities to C_2H_4O were measured in the 4-part cycles compared with ca. 50% selectivity in commercial reactors.	A circulating fluidized bed was simulated by a fixed bed operated periodically. A silica gel adsorbent added to the bed trapped C_2H_4. If the duration of the N_2 flush was long enough to recover adsorbed C_2H_4O, modulation increased the selectivity to C_2H_4O to 70.4%; conversion of $C_2^{=}$ increased by 27% for $\tau_{cycle} = 50$ s.

(*Continued*)

TABLE 4-3 Epoxidation Reactions (*cont'd*)

Reference	Objective	Reaction and Reaction Conditions	Modulated Variable	Observations	Comments
Balzhinimaev et al. (1984)	Effect of cycling strategy on selectivity and conversion	$C_3^= + O_2 \rightarrow C_3H_6O$ using Ag catalyst with α-Al_2O_3 and Ba or Cd and Se promoters. Ba-promoted catalyst was used at 686 K; the second catalyst at 708 K was used in the 4-part cycles and at 666 K in the 8-part ones	Cycling strategies were: (1) C_3H_6:$O_2/N_2/O_2/N_2$; (2) as in (1) but with ppm amounts of $C_2H_8Cl_2$ in the C_3H_8 feed; (3) as in (2) but with the O_2 pulse at 20°C $> T$; (4) C_3H_6:$O_2/N_2/H_2/N_2$; (5) C_3H_6:$O_2/N_2/O_2/N_2/H_2/N_2$ with ppm quantities of $C_2H_8Cl_2$ added and with the O_2 pulse at $> T$. Variables were pulse duration for O_2 and feed vol% C_3H_6	Selectivities reached 29% in 8-part cycles using 96% C_3H_6 in the reactive mixture pulse. The selectivity to the epoxide in that pulse exceeded 80%. Increasing the vol% O_2 in the pulse reduced selectivity. $C_2H_8Cl_2$ in the feed improved selectivity, but reduced propylene conversion and thus the epoxide yield.	Molecularly adsorbed oxygen participated in epoxide formation. $C_2H_8Cl_2$ inhibited O_2 dissociation on adsorption, but dissociation increased as the O_2 partial pressure rose and led to a competing dehydrogenation side reaction. H_2 pulses removed surface carbonates but also adsorbed oxygen and possibly reduced the oxide. Raising the temperature in the O_2 pulse increased residue removal from the surface.
Li et al. (1992b)	Investigation of the modulation of ethylene epoxidation	$C_2^= + O_2 \rightarrow C_2H_4O$ over a commercial Ag/γ-Al_2O_3 catalyst. Packed-bed microreactor at about 100 kPa and 523 K	Modulation of $C_2^=$: O_2 ratios with τ and s as variables.	Time-average selectivity increased from 65% at the optimal steady state operation to 70% under modulation; maximum increase observed at a period of 60 s.	Selectivity to the epoxide is improved by composition modulation, but improvement depends on modulation frequency. Predictions using a model based on step-change experiments did not agree with experimental observations.

work supplemented a modeling study by *Ghazali et al. (1983)* which suggested that different hydrocarbon adsorbates as well as three different forms of adsorbed O_2 are present on the catalyst surface. Atomic oxygen was held responsible for total oxidation through a reaction with the epoxide or its $C_2^{=}$ precursor. Adsorbed molecular O_2, on the other hand, reacted with $C_2^{=}$ to form the epoxide.

Based on their research, a multitubular circulating fluidized bed for epoxidation, using a high-surface-area, supported Ag catalyst, was proposed by *Park and Gau (1986)* but not built. Instead a fixed bed was used. These researchers mixed their Ag catalyst with a silica gel that adsorbed ethylene oxide (Et_2O) as it formed so their cycle then alternated between a 50:50 mixture of C_2H_4 and O_2 and an N_2 flush to desorb product. Experimental variables were the durations of the mixture and the N_2 exposures, and the N_2 flow rate during the flush. **Table 4-3** summarizes these experiments. More details are given by *Silveston (1998)*.

Balzhinimaev et al. (1984) studied cycling strategies for propene epoxidation using an Ag catalyst promoted with barium or with a mixture of cadmium and selenium. Large increases in selectivity were achieved with these complex cycles relative to the steady-state selectivity at the time-average feed composition. Nevertheless, both selectivities and yields appear to be almost an order of magnitude lower than those attained with ethene epoxidation even in steady-state operation. The four-part cycle with an O_2 pulse gave 4% selectivity, compared with an almost zero yield of the epoxide under steady-state operation.

A noticeable lag occurred before products appeared and reached 20 s when H_2 was used to clean the catalyst surface. The quantities of epoxide and CO_2 formed in this cycle were much less than in cycles using an O_2 pulse. Selectivity to the epoxide is much higher during the reactant mixture pulse than for the overall cycle for both strategies. Highest selectivities observed by *Balzhinimaev et al.* reached 29% and were attained in the eight-part cycles using 9% C_3H_6 in the reactant mixture pulse. The selectivity to the epoxide in that pulse exceeded 80%. Using only C_3H_6 in place of the mixture increased the epoxide selectivity, but substantially suppressed the yield. These observations led *Balzhinimaev et al.* to conclude that molecularly adsorbed oxygen participated in epoxide formation.

C.-Y. Li and co-workers took up the composition forcing of ethene epoxidation over an Ag catalyst a decade later. Their experiments employed cycling between different C_2H_4:O_2 ratios (*Li et al., 1992b*) and examined the influence of cycle period and split on selectivity to C_2H_4O. For symmetrical modulation, selectivity increased from 65% under optimal steady-state operation to about 70%. Resonance was also observed and maximum selectivity occurred at $\tau_{cycle} = 60$ s. Decreasing the cycle split brought the selectivity to about 72% at $s = 0.3$.

A model was prepared assuming a redox mechanism and two classes of reactive sites on the surface: AgO and Ag_2O_2. The former was assumed to be active for C_2H_4O, while the Ag_2O_2 site strongly bound C_2H_4 and led to total oxidation. The CSTR model with kinetic parameters estimated from step-change data (*Li et al., 1992a*) as well as from cycling experiments qualitatively reproduced the experimental observations, but the predicted resonance and maximum in the selectivity vs. cycle period was at a τ_{cycle} well below the measured value.

4.5 PROPENE AND BUTENE PARTIAL OXIDATION AND AMMOXIDATION

Attempts to incorporate periodic operation into propene partial oxidation began almost 50 years ago. The concept used the lattice oxygen of a readily reducible metal oxide in place of air. This was realized by shuttling a metal oxide between a reducing bed and an oxidizing one. The early effort identified catalysts that could be used in this way and established that complex oxides of molybdenum or phosphomolybdic acids on silica were effective when promoted with copper, antimony, tin or bismuth (*Callahan et al., 1970*). Acrolein yields of up to 60% were obtained by passing propene through a fluidized bed of bismuth phosphomolybdate. However, cyclic operation was rejected for industrial use at that time because of the solids circulation requirement. *Callahan et al.* estimated that up to 300 kg oxidant/kg of C_2H_4 would have to circulate between fluidized beds if the catalyst was to maintain its most active oxidation state.

In another early composition modulation study, *Niwa and Murakami (1972a)* investigated the partial oxidation of C_2H_4 to acrolein over a variety of metal oxides (see **Table 4-4**). The symmetric cycling employed increased neither the selectivity nor the yield. Acrolein production rate, however, was sensitive to τ_{cycle}.

Formation of acrylonitrile from the partial oxidation of a propene-ammonia mixture was explored by *Brazdil et al. (1980)* for several bismuth molybdates using a sequence of pulses of the mixture alternating with pulses of O_2-He. Their intent was to separate the reduction and reoxidation steps in a redox system and to investigate mechanisms occurring on the catalyst surface. They observed that oxygen drawn from the catalyst either partially oxidized the $C_3^=$-NH_3 mixture to acrylonitrile or caused total oxidation. Ammoxidation sites were those with partial oxygen coordination of the metal ions in the solid so that reduction of the catalyst transformed combustion sites into ammoxidation ones prior to a total loss of activity. Catalyst reoxidation was first order in O_2 partial pressure and was also the slow step in the redox cycle. *Silveston (1998)* offers a thorough discussion of this important work.

Silveston (1980) and *Silveston and Forissier (1985)* investigated the composition modulation of C_2H_4 partial oxidation on the widely studied $SbSnO_2$ and α-$Bi_2Mo_3O_{12}$ catalysts. Their objective was to test composition modulation on redox catalysts with different semiconductor characteristics: Sb-Sn catalyst appears to a p-type semiconductor whereas the bismuth molybdate catalyst is an n-type. Rather long cycle periods were used on the basis of step-change experiments that showed acrolein formation still evident 90 min after the change from an oxidizing feed to a reducing one. Despite some data scatter, low frequency, symmetric forcing increased selectivity to acrolein by about 20% for the Bi-Mo catalyst and by 80% for the Sb-Sn catalyst. There appeared to be an optimal cycle period between two and four hours, a surprisingly long τ_{cycle}. More details of this study are given by *Silveston (1998)*.

A brief study of periodically operating ammoxidation was published by *Magagula and van Steen (1999)*. These authors observed that a periodic operation was effective for an Fe-Sb catalyst but not for a Bi_2MoO_6 catalyst. The difference appeared to be due to the extraction of lattice oxygen. Only the surface layer in the Fe-Sb catalyst seems to participate in hydrocarbon oxidation, whereas for Bi_2MoO_6 deeper extraction occurred. Results for that catalyst suggested much weaker adsorption of C_3H_6. Periodic exposure of the surface to higher O_2 concentrations apparently increased the rate of reoxidation, an observation that seemed to account for the higher rates of product formation under modulation.

Partial oxidation of propene under transient conditions was studied on a copper molybdate, $Cu_2Mo_3O_{10}$, by *Labastida-Bardales et al. (1989)*. This catalyst was chosen because it undergoes a reversible change from Cu(I) to Cu(II) on alternating exposure to air and to $C_3^=$. Re-crystallization also takes place separating out MoO_3 and restructuring the catalyst surface (*Haber, 1980; Haber et al., 1980a, b*). *Haber's* observations indicated that the acrolein yield could be substantially increased by periodically

TABLE 4-4 Organic Acids, Aldehydes and Ketones from Olefin Partial Oxidation

Reference	Objective	Reaction and Reaction Conditions	Modulated Variable	Observations	Comments
Lewis et al. (1949); Callahan et al. (1970)	Exploration of circulating catalyst to separate reduction and reoxidation steps	$C_3H_6 + O_2 \rightarrow C_4H_2O_3$ (not balanced). One fluidized bed reactor oxidized propene while the other reoxidized the catalyst	Modulation was achieved by continuously circulating catalyst between a bed for catalyst reduction and one for reoxidizing the catalyst	No fluidized bed results were given. Early effort (*Callahan et al.*) was to identify catalysts useful for a two-bed process. Complex oxides of molybdenum or phosphomolybdic acids on silica were effective when promoted with Cu, Sb, Sn or Bi.	The two-bed concept of *Callahan et al.* was rejected because of the burden of solids circulation.
Niwa and Murakami (1972a)	Application of composition modulation to partial oxidation of propene to acrolein.	$C_3^{=} + O_2 \rightarrow C_3H_6O_2$ over binary Bi, Mo, Sb, Sn, W, P unsupported catalysts, also Bi-Mo/SiO_2, $P = 1$ atm; $T = 659$ K	Symmetric cycling between 25 vol% $C_3^{=}$ in N_2 and air for $\tau = 30$ s	Very little acrolein formed during O_2 exposure; except for Sb-Sn and Sb-Mo and MoO_3 catalysts, more than 70% of CO_x originated in this part of the cycle. Only the Sb-Mo binary oxide catalyst enhanced selectivity to acrolein under modulation.	Authors suggested strong C_3H_6 adsorption on catalysts during the reducing part-cycle. Combustion of residues occurred in the O_2 portion. O_2 adsorbed on the Sb-Sn catalyst resulted in higher CO_X production in the reducing part-cycle. Bi increased the reoxidation rate of Mo and W and catalysts through higher O_2 mobility.
Stamicarbon (1974)	Composition modulation of olefin concentration for ketone production.	$C_3H_6 + O_2 \rightarrow C_3H_4O$ and other ketones. Fluidized bed, but process details were not provided. Sn-Mo oxide catalysts	Catayst is alternately exposed to the hydrocarbon and the O_2-containing streams	Using $C_3^{=}$, selectivity to acetone reached 98% at yields of up to 91%. Similar results were cited for a butylene where the product is a methylethyl ketone.	Little detail in patent.

(*Continued*)

TABLE 4-4 Organic Acids, Aldehydes and Ketones from Olefin Partial Oxidation (*cont'd*)

Reference	Objective	Reaction and Reaction Conditions	Modulated Variable	Observations	Comments
Brazdil et al. (1980)	Exploration of alternate reactant - O_2 pulses on ammoxidation	$C_3^= + NH_3 + O_2 \rightarrow C_3H_3N$ (acrylonitrile) over α-$Bi_2Mo_3O_{12}$, β-$Bi_2Mo_2O_9$, γ-Bi_2MoO_6 and Bi-Mo co-precipitated with other metal oxides catalysts. Catalysts calcined at 703 K	Second-long pulses of $C_3^=$: NH_3 mixture alternating with O_2: He pulses	O extracted from the catalyst can either partially oxidize $C_3^=$ to acrylonitrile or cause total oxidation. For the bismuth molybdates, strength of the M-O bond at surface sites controlled ammoxidation selectivity. Reduction of the catalyst destroys reaction sites, but increases M-O bond strength.	Separating the reactants leads to higher conversion rates because the slow reoxidation step can be carried out under optimal conditions. For the active catalysts, reduction is necessary to achieve high selectivity.
Silveston (1980); Silveston and Forissier (1985)	Composition modulation of propylene partial oxidation using redox catalysts with p-type and n-type semiconducters.	$C_3^= + O_2 \rightarrow$ acrolein over $SbSnO_2$ (10:90, a p-type semiconductor) and α-$Bi_2Mo_3O_{12}$ (an n-type semiconductor). Differential reactor operating at 1 bar, 673 K	Symmetrical cycling between 6.6 vol% $C_3^=$ and 6.6 vol% O_2 with an N_2 diluent for $0.5 < \tau < 6.5$ h	Strong enhancement of both selectivity to and production rate of acrolein observed under modulation for both catalysts. Enhancement decreased as cycling amplitude increased. With Sb-Sn catalyst, vol% acrolein in the product gas stayed almost constant through a cycle. The Bi-Mo catalyst provided higher production rates under forcing than under steady state.	Bang-bang switching between $C_3^=$ and O_2 less attractive than cycling between $C_3^=$ and O_2 mixtures. For Sb-Sn catalysts, only small changes in acrolein concentration under cycling due to strong adsorption of C_3H_6 or acrolein precursors. For Bi-Mo catalysts, higher rates due to much weaker adsorption of C_3H_6.
Greger et al. (1984); Labastida-Bardales et al. (1989); Labastida-Bardales (1991).	Investigation of the effect of surface restructuring during modulation on partial oxidation	$C_3^= + O_2 \rightarrow C_3H_4O_2$ over a $Cu_2Mo_3O_{10}$ catalyst at 623 K and on an O_2 pretreated Cu_2MO_{10}	Cycling between $C_3^=$: O_2 mixtures of 6:1 and 3-1; 1:1 and 1:3 for $2 < \tau < 60$ min. with $0.5 < s < 0.9$	Time-average rates of acrolein and acetaldehyde formation increased by 170% over the highest steady-state rates measured. Selectivity to acrolein unchanged by modulation. Rates declined with increasing τ and amplitude. Acrolein formation rates, measured under decreasing $C_3^=$ partial pressure, were less than rates measured under increasing pressure.	A steady-state hysteresis observed for partial oxidation on a 1.2 wt% Cu/Al_2O_3 was attributed to stabilization of a copper oxidation state, ordinarily reducible in the absence of reaction, by the C_3^+ oxidation reaction.

Saleh-Alhamed et al. (1992, 1993)	Investigation of a 1-step process to make acrylic acid	$C_3^= + O_2 \rightarrow C_3H_4O_2$ (acrylic acid) using a proprietary Sb/Sn/V (2:1:1) oxide and a $Bi_2Mo_3O_{12}$ catalyst. Differential fixed bed of catalyst at 613 K for Sb/Sn/V oxide and at 633 K for Bi-Mo oxide. $SV = 0.25$ L/ min·g cat., $P = 1$ atm	Three strategies were tested: 1) modulation of the steam content of the feed, 2) modulation of the C_3H_6:O_2 ratio in the feed, and 3) periodic switching between feeds containing C_3H_6 in air and just air. 30 s $\leq \tau_{cycle} \leq$ 50 min with symmetrical cycles	Modulation of water content in the feed decreased selectivity and the rate of acrylic acid production. Cycling of the $C_3^=$ concentration in the feed increased rate to acrylic acid by 50% and to acrolein by 85% over rates at steady-state for $1 < \tau < 2$ min. Cycle split is a primary variable: oxygenate production and selectivity increased with an increase in split.	Cycling strategy was found to greatly affect selectivity and product rate of formation.
Chen et al. (1996)	Ammoxidation of propene in a circulating fluidized bed	$C_3/C_3^= + NH_3 + O_2 \rightarrow$ Acrylonitrile + H_2O (not balanced). Experimental details not given	Catalyst circulated between an oxidizing riser and a reducing fluidized bed	Circulating fluidized bed increased selectivity to acrylonitrile slightly.	Paper discusses comparison of model with small-scale pilot reactors.
Magagula and van Steen (1999)	Investigation of ammoxidation under periodic operation	$C_3/C_3^= + NH_3 + O_2 \rightarrow$ Acrylonitrile + H_2O. Tubular plug flow reactor; 1 cm i.d. packed with Fe-Sb oxide or Bi_2MoO_6 catalyst and sand. Ammoxidation run at 653 K for $C_3^=$ and 743 K for C_3	Composition modulation between O_2 and hydrocarbon + O_2 mixture	Different behavior observed for Fe-Sb oxide and Bi_2MoO_6 catalysts. Yield with the former is greatly increased under periodic operation. No improvement was seen for the Bi_2MoO_6 catalyst.	

(Continued)

TABLE 4-4 Organic Acids, Aldehydes and Ketones from Olefin Partial Oxidation (*cont'd*)

Reference	Objective	Reaction and Reaction Conditions	Modulated Variable	Observations	Comments
Song et al. (2005)	Application of periodic operation to partial oxidation	Tubular reactor, 4.4-mm i.d., containing various $BiMoO_x$ catalysts diluted with SiO_2; $250 < d_p < 300$ μm; $633 < T < 663$ K	Bang-bang switching between $C_4^{=}$/Ar-He and O_2 /He	With $Bi_2Mo_2O_9$, authors observed that initial methacrolein production in periodic operation exceeded that for steady-state operation. CO production and C deposition disappeared with time-on-stream. With promoted catalyst, deposition was observed.	Promoted catalysts gave higher yields of methacrolein under steady state than under periodic operation.
Hu et al. (2007)	Investigation of a circulating fluidized bed for ammoxidation	$C_3^{=} + O_2 \rightarrow$ acrylonitrile with a Bi-Mo-O/α-Al_2O_3 catalysts; $660 < T < 730$ K; $25 < t_{contact} < 230$ g·h/mol $C_3^{=}$	Circulating fluidized bed with $C_3^{=}$ and NH_3 injected into the riser and air injected into the fluidized bed regenerator	Acrylonitrile yields were recorded from 75 to 83% at 716 to 724 K at 99% $C_3^{=}$ conversion with the circulating fluidized bed reactor, representing a 3 to 7% increase over the yield for a conventional fixed bed reactor.	Authors observed a gas throughput four times that of a fixed bed for the same $C_3^{=}$ conversion.

introducing O_2 pulses into the reactor feed. The objective of the *Labastida-Bardales* study was to determine the performance, under composition modulation, for a catalyst that responded to the gas environment with changes in the metal ion coordination, phase separation and surface restructuring. Composition modulation experiments (**Table 4-4**) employed cycle split, *s*, defined by the duration of the $C_3^=$ rich mixture and cycle period. Time-average rates of both acrolein and acetaldehyde were as much as 170% greater than the maximum steady-state rates, measured at $C_3^=:O_2 = 6:1$. Selectivity to acrolein, however, was not changed by modulation. The cycling variables, split and period, did not exert a significant effect on the time-average rate, except at the extremes of the $C_3^=:O_2$ ratios. Increasing amplitude suppressed the time-average rates of partial oxidation.

Acrylic acid, a commercially important monomer, is now produced from propene in a two-step catalytic partial oxidation process. *Saleh-Alhamed et al. (1992)* explored a single-step process to this acid using composition modulation. Three cycling strategies and two catalysts were used to examine the interaction of catalyst and forcing strategy. The first of the two, an Sb/Sn/V oxide in the atomic ratios 2:1:1, is an industrial catalyst, while the second, $Bi_2Mo_3O_{12}$, is a frequently used partial oxidation catalyst. The first cycling strategy modulated the water content of the reactor feed. Although water cycling affected rates of formation and selectivities, no global enhancement was observed. In a companion study, *Saleh-Alhamed et al. (1993)* examined the role of water in propene partial oxidation and why the water cycling strategy was unsuccessful.

Modulating the concentration of $C_3^=$ in the air stream passing over the catalyst, the second strategy, had a remarkable effect on rates of acrylic acid and acrolein formation. For $1 < \tau_{cycle} < 2$ min, there was a 50 and 85% enhancement of the acrolein and acrylic acid rates of formation with respect to the steady-state rates at the time-average feed composition, while for CO_2 the enhancement was negligible. Selectivity ratio is defined as the ratio of the rate of acrylic acid or acrolein formation to the rate of CO_2 formation. Consequently, the best selectivity ratios for acrolein and acrylic acid were 40 and 75% greater than the steady-state ratios. *Saleh-Alhamed et al. (1992)* explained the enhancement through the reduced flooding of the catalyst surface by $C_3^=$ and oxidation products. Access of O_2 to the surface in the next part cycle was greater and this increased the coordination of surface and near surface metal ions.

The third strategy investigated corresponded to switching between an air feed and a 1:1 $C_3^=:O_2$ feed mixture. This strategy was advantageous for acrolein formation but not for acrylic acid. A more extensive discussion of the *Saleh-Alhamed* work is provided by *Silveston (1998)*.

Song et al. (2005) were able to achieve almost 100% selectivity to methacrolein, using bang-bang modulation between isobutene in Ar-He, and O_2 in He at 833 K. Catalysts for their experiments were bismuth molybdate with various Bi/Mo ratios and the molybdate promoted by Co and Fe. Selectivity to methacrolein increased to almost 100% with time-on-stream for the unpromoted catalysts; however, conversion dropped. Promoting the catalyst with Co or Fe decreased selectivity by a small amount but raised the i-$C_4^=$ conversion. *Song et al.* concluded that this gain in selectivity was outweighed by the complexity of periodic operation on an industrial scale.

Ammoxidation experiments in a circulating fluidized bed with an internal baffle to decrease backmixing are summarized by *Chen et al. (1996)* and compared with the authors' model for a three-zone looping reactor in which the bismuth molybdate catalyst is reoxidized in air in an up-flow riser and the $C_3^=$ - O_2 - NH_3 reactions takes place in a baffled fluidized bed. *Chen et al.* observed just a small increase in selectivity to acrlonitrile over that in a conventionable fluidized bed.

A paper by *Hu et al. (2007)* continued exploration of circulating fluidized beds (CFB) for ammoxidation. CFB reactors generally offer high throughput and relatively low backmixing. This is important for $C_3^=$ ammoxidation because the products, acrylonitrile and acrolein, are easily oxidized to carbon oxides, NO_x and water. The *Chen* paper examined the effect of contact time, reactor temperature, air-to-propene ratio, and NH_3:$C_3^=$ ratio on the yield of acrylonitrile and total oxidation products. Temperature and contact time dominated conversion and product yield. In a further experiment, the authors staged the introduction of O_2 in the riser reactor and observed a significant increase in the acrylonitrile yield.

4.6 CATALYTIC DEHYDROGENATION OF PROPANE, BUTANE AND HIGHER HYDROCARBONS

Olefins and diolefins, particularly butenes and butadiene, are currently produced on a commercial scale through catalytic dehydrogenation. Selectivity, endothermicity and moderately high temperatures are problems with this technology. Despite the high steam demand, catalyst coking remains a difficulty, requiring occasional shutdowns for regeneration. These considerations led to investigations of oxidative dehydrogenation under modulation with alternate streams containing paraffins and O_2 (see **Table 4-5**) to separate the reactants. The oxidized catalyst delivered the oxygen required for partial oxidation.

During the 1990s, the Dow Chemical Company explored making butadiene by oxidative dehydrogenation employing a circulating fluidized bed (*Murchison et al., 1993; Khazai et al., 1995; Vrieland and Murchison, 1995; Vrieland et al., 1995*). The Dow work was motivated by several publications from the Ukraine describing an MoO_3/MgO catalyst (*Doroshenko et al., 1986; Shapovalova et al., 1987; Luk'yanenko et al., 1987*). Some of the Dow research used a cyclically operated packed bed. The Ukrainian development, however, employed composition modulation of a fluidized bed. Dow workers employed a four-part cycle involving exposure to C_4 and O_2-containing streams separated by 60 s He flushes. The selectivities measured by the different researchers were similar, even though conversions in the Ukrainian studies were sometimes double those of *Murchison et al. (1993). Vrieland and Murchison (1995)* and *Vrieland et al. (1995)* showed high conversions reduced butadiene selectivity. For a further discussion of this research see *Silveston (1998)*.

With these results in mind, *Del Rosso et al. (1998)* examined the role of oxygen within a β-$NiMoO_4$ catalyst during the oxidative dehydrogenation of propane to propene (**Table 4-5**). At temperatures up to 788 K, they observed selectivities greater than 89% for $\tau_{cycle} = 40$ s. However, at 838 K and a 90 s period, selectivity to $C_3^=$ gradually fell from 82% after one hour to zero after five hours. At the same time, Ni deposits and carbon filaments formed on the catalyst surface, deactivating the catalyst. These researchers concluded that at the higher temperature, lattice oxygen had been depleted. This was confirmed by catalyst weight loss measurements.

Creaser et al. (1999a, b, c) undertook bang-bang modulation of propane and O_2 diluted in He over an MgO-$Mg_3(VO_4)_2$ catalyst packed in a small fixed bed. Varying cycle period and split, they observed a 50% increase in the time-average selectivity to the olefin relative to steady-state operation (**Figure 4-7**). Selectivity changed little as τ_{cycle} became larger as long as the C_3 part cycle remained at 50 s. With symmetrical cycling, conversion decreased as τ_{cycle} was extended, while selectivity increased. This conversion-selectivity trade-off resulted in a sharp maximum in the propene yield. At this maximum, the yield was 30% higher than at steady state. Cycle split was studied for symmetrical cycles, but it had

TABLE 4-5 Oxidative Dehydrogenation of Paraffins and Olefins

Reference	Objective	Reaction and Reaction Conditions	Modulated Variable	Observations	Comments
Doroshenko et al. (1986), Luk'yanenko et al. (1987), Shapovalova et al. (1987)	Exploration of the use of composition modulation in partial oxidation	$C_4 + O_2 \rightarrow C_4^{2=} + H_2O$ over MgO/MoO_3 and V_2O_5 promoted catalysts. $843 < T < 873$ K	Symmetrical cycling between C_4 or $C_4 + O_2$ mixtures and O_2 or air using Ar or steam as diluents for $\tau = 60$ s	A 70 to 75% selectivity to $C_4^{2=} + C_4^{=}$ at 60 to 72% C_4 conversion was achieved using the vanadia-promoted MgO/MoO_3 catalyst.	
Murchison et al. (1993)	Simulation of a circulating fluidized bed for the oxidative dehydrogenation of butane	$C_4H_{10} + 0.5\ O_2 \rightarrow C_4H_8 + H_2O$, 10 mL packed bed of MoO_3/MgO catalyst at $833 < T < 853$ K	A 4-part cycle was used: a 5 to 30 s catalyst exposure to C_4–He, a 60 s flush with He, a 60 s exposure to O_2-He mixture, then a 60 s He flush	The 4-part cycle using He flushes between the C_4H_{10} and O_2 feeds approximated the single-pass riser system using He to transport the catalyst in the riser and fluidize the downcomer leg.	Researchers assumed that the performance measured in a periodically operated packed bed represented the performance of a recirculating solids reactor.
Khazai et al. (1995); Vrieland et al. (1995); Vrieland and Murchison (1995)	Application of composition modulation to oxidative dehydrogenation	$C_4H_{10} + 0.5\ O_2 \rightarrow C_4H_8 + H_2O$. Catalyst: MoO_3/MgO with various promoters. $T = 833–853$ K, $P = 1$ atm	Complex 4-part cycles between butane and air with He flushing between exposure to reactants	High selectivities were observed to butadiene and $C_3^{=}$. Butadiene selectivity decreased with increasing conversion.	Alkali promoters improved selectivities but decreased catalyst strength. Spinels increased strength but reduced $C_4^{2=}$ selectivity.
Del Rosso et al. (1998)	Investigation of the role of lattice oxygen in oxidative dehydrogenation	Propane oxidative dehydrogenation over β-$NiMoO_4$ catalyst. Four parallel tubular reactors each received the flow sequence: C_3H_8, He, O_2, He for equal durations. $733 < T < 838$ K	$1 < \tau_{cycle} < 180$ s	At low T and short τ_{cycle}, conversion was low and $C_3^{=}$ selectivity high. At high T and long τ_{cycle}, lattice oxygen was depleted; complete oxidation of C_3 was observed with metallic Ni and carbon filaments on catalyst.	Apparently, the catalyst is permanently deactivated by metallic Ni deposits and depletion of lattice oxygen.

(*Continued*)

TABLE 4-5 Oxidative Dehydrogenation of Paraffins and Olefins (*cont'd*)

Reference	Objective	Reaction and Reaction Conditions	Modulated Variable	Observations	Comments
Creaser et al. (1999a, b, c)	Application of composition modulation to partial oxidation and analysis of the reaction mechanism	$C_3 + O_2 \rightarrow C_3^{=} + H_2O$ over a $MgO/Mg_3(VO_4)_2$ catalyst. $P = 1$ bar; $T = 783$ K; mass spectrometer was used to follow product concentrations within a cycle	Symmetrical cycling between 6 vol% C_3 and 6 vol% O_2 in an He diluent for $10 < \tau_{cycle} < 400$ s. Some runs used cycling between C_3 and an O_2-C_3 mixture	Symmetrical cycling enhances selectivity by 60% at $\tau = 400$ s but suppresses conversion; $C_3^{=}$ yield maximized at $\tau = 60$ s. Researchers showed $C_3^{=}$ formed only through O extracted from ortho-vanadate; some extracted O and adsorbed O_2 combusts $C_3^{=}$. Cycle split has a small effect on selectivity and yield.	The oxygen source for reaction is the catalyst; the rate of dehydrogenation is diffusion-limited. Some adsorbed hydrocarbon is burned off by gas phase or adsorbed O_2. Catalyst reoxidation is rapid.
Genser and Pietrzyk (1999); Grabowski et al. (2002)	Mechanism investigation and modeling of a catalyst under modulation	$C_3 + O_2 \rightarrow C_3^{=} + H_2O$ over a multilayer V_2O_5/TiO_2 catalyst, also with a K-doped catalyst. Packed bed, $P = 1$ bar, $T = 650$ K	4-part asymmetric cycle: 20 vol% C_3 for 2 min; 30 s flush; 20 vol% O_2 for 2 min; 30 s flush. He was diluent and flushing medium	Selectivity to $C_3^{=}$ rose to 70% as C_3 conversion dropped from 10% to 1%. Selectivity reached 85% with K doped sample. Steady state selectivity was nil.	High selectivity results from O extraction from V_2O_5. Coke residues on catalyst were gasified during reoxidation.
Grasselli et al. (1999); Tsikoyiannis et al. (1999)	Investigation of a dehydrogenation + H_2 oxidation catalyst to overcome equilibrium limitation	$C_3 + O_2 \rightarrow C_3^{=} + H_2O$ with a 50:50 mixture of 0.7 wt% Pt-Sn/ZSM-5, and 42 wt% Bi_2O_3/SiO_2. Additional experiments tested different H_2 oxidation catalysts: Bi_2O_3, $Bi_2Mo_3O_{12}$ and In, La, Al. Cr, and Fe replacing Bi in the molybdate. Packed bed used with $723 < T < 823$ K	4-part cycle employing 15 vol% $C_3^{=}$ + 15 vol % H_2 in He for 5 min, then reoxidation in air with an He purge	Bi_2O_3 catalyst gave 99% selectivity to H_2 oxidation and highest rate, while $Bi_2Mo_3O_{12}$ had same selectivity but a much lower rate. Other molybdates showed just poor oxidation rates.	Authors demonstrated that the use of a selective H oxidation catalyst significantly raised $C_3^{=}$ selectivity and yield.

Soler et al. (1999); Herguido et al. (2005)	Investigation of a 2-zone fluidized bed reactor for oxidative dehydrogenation	Fluidized bed reactor 3 cm i.d. × 30 cm; $P = 1$ atm; $T = 823$ K; 4 vol% hydrocarbon in feed; reactor bed fluidized with O_2	2-zone fluidized bed as an alternative to modulation. Reactant introduced through a mid-bed distributor	Selectivity to dehydrogenation products approached 54% at a 40% n-butane conversion.	Selectivity in the 2-zone system was improved over a fixed-bed only for greater than 75% conversion. However, higher $C_4^=$ concentrations were obtained by solids circulation.
Kiwi-Minsker et al. (2002)	Non-oxidative dehydrogenation of propane using an H_2 permeable membrane and fiber catalysts	Microreactor consisting of two bundles of alumina coated fibers impregnated with a Pt/Sn catalyst with bundles separated by a Pd membrane; operated at 823 K and 0.14 MPa	Flow cycled between fiber bundles, switching between C_3 to air at $\tau_{cycle} = 85$ min and $s = 0.5$	Removing H_2 through a semi-permeable membrane raised C_3 conversion to about 30 mol% with a $C_3^=$ selectivity of 96% vs. initial steady state conversion of 22% (equilibrium limit) and a selectivity of 88%. In extended periodic operation, C_3 conversion is about 6%.	Use of impregnated fiber catalyst eliminated mass transfer interference. The problem with the novel reactor is the fragility of the Pd membrane.
Lobera et al. (2009)	As for the *Herguido et al. (2005)* study.	As for the *Herguido et al. (2005)* study: $743 < T < 823$ K; 10 vol% C_4; reactor bed fluidized with $O_2 + Ar$	As for the *Herguido et al. (2005)* study	Stable operation attained provided O_2 in the fluidizing gas < 5%. Highest selectivities to $C_4^=$ and $C_4^{2=}$ were obtained for $O_2 < 2\%$. Fluidized bed operated with coke deposits. Selectivity to $C_4^=$ and $C_4^{2=}$ depended slightly on temperature but strongly on conversion.	The detonation region for the C_4/O_2 mixture was avoided by maintaining a low O_2 content in the fluidizing gas.

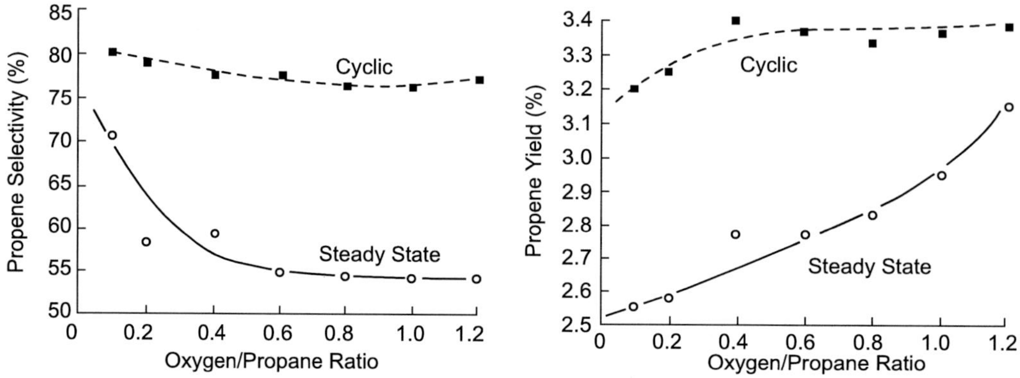

FIGURE 4-7 Comparison of bang-bang cycling between 6 vol% propane in He and 6 vol% oxygen He at 783 K over a V-Mg-O catalyst at varying τ_{cycle} but with a constant C_3 partial cycle of 50 s. *(Figure reproduced from* Creaser et al. (1999a) *with the authors permission. © 1999 by Elsevier Science Ltd.)*

just a small effect on selectivity and yield. At 783 K, reoxidation of the catalyst took just 20 s but suggested that the bulk of the catalyst was the source of most of the O_2 consumed.

In a further investigation, *Creaser et al. (1999d)* observed that high selectivity could be obtained for steady-state operation at the same degree of C_3 conversion by drastically reducing the O_2 partial pressure in the mixture. However, to do so required a lower C_3 feed concentration. This result suggested another pathway must exist to carbon oxides besides further oxidation of $C_3^=$. A later modeling study (*Creaser et al., 2000*) showed that models that assumed different sources of oxygen for oxidation of C_3 or $C_3^=$ provided a much better fit of the experimental data than those assuming a single source.

Spanish researchers mentioned in **Section 4.3** applied their two-zone fluidized bed to oxidative dehydrogenation using a conventional V/MgO catalyst (*Soler et al., 1999; Herguido et al., 2005*). An increase in product yield was attained as selectivity increased above those measured in fixed beds. By repeating their experiments in larger beds with better fluidization, even greater product yields were found. Further improvement resulted through promoting internal circulation in the bed (*Rubio et al., 2004*).

Butane dehydrogenation was re-examined by *Lobera et al. (2009)* using a two-zone fluidized bed reactor with a supported Pt-Sn catalyst (**Table 4-5**). These authors were concerned with possible excursions into the detonation region for the C_4/O_2 mixture and were thus able to study reactor performance with different concentrations of O_2 in the fluidizing gas. Transport of hydrocarbons or O_2 occurs by bubbling in aggregative fluidization so that mixing of the gases cannot be avoided. Their experimental results are summarized in **Figure 4-8** which shows the evolution of conversion and product concentrations as a function of time after start-up. Stable operation is evident for O_2 concentrations greater than 1%. The amount of coke circulating depended on temperature and on the concentration of O_2 in the fluidizing gas.

A two-zone system, different from that of the Spanish researchers, was used by *Kiwi-Minsker et al. (2002)* for the dehydrogenation of propane. That system employed a microreactor consisting of a bundle of parallel, alumina coated glass fibers placed in each of the zones (see **Figure 4-9**) and separated by a thin Pd membrane. The coated fibers were impregnated with a Pt and Sn solution to create a catalytically active bundle. Hydrocarbon was fed to one zone while air

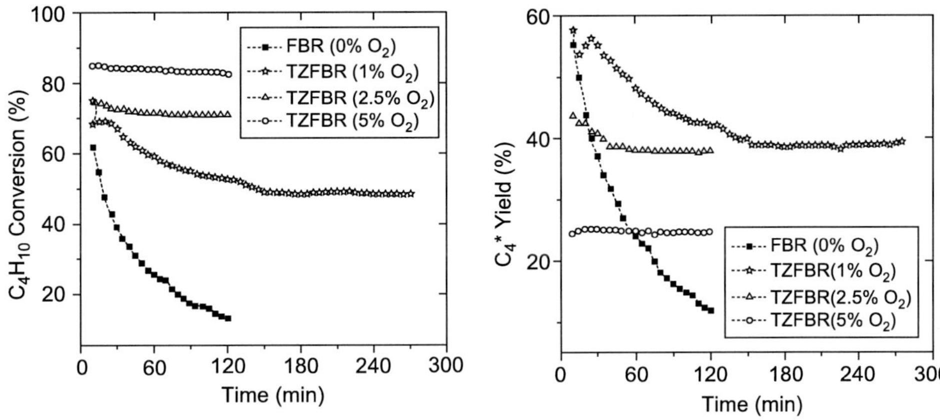

FIGURE 4-8 Evolution of butane conversion and product concentrations as a function of time-on-stream and different vol% O_2 in the fluidizing gas. *FBR = Fluidized Bed Reactor, TZFBR = Two-Zone Fluidized Bed Reactor TZFBR refers to a two-zone fluidized bed. (Figure adapted from* Lobera et al. (2009) *with permission. © 2009 by the American Chemical Society.)*

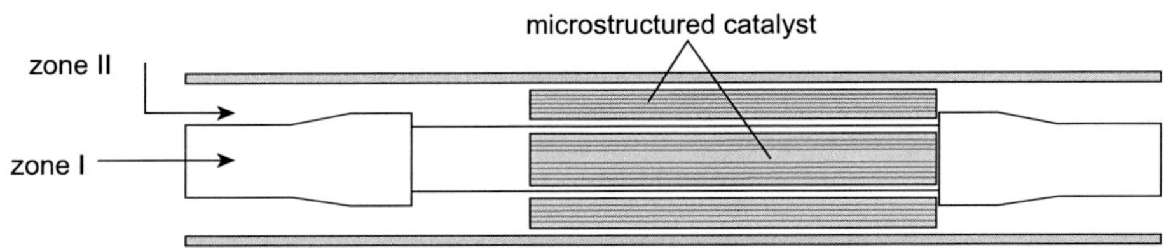

FIGURE 4-9 Concentric catalytically active glass fiber bundles contained in two zones and separated by a thin Pd membrane permeable to just H_2. Zones are fed with either a hydrocarbon or air and flows are periodically switched. *(Figure reproduced from* Kiwi-Minsker et al. (2002) *with permission. © 2002 by Elsevier Inc.)*

went to the other. The catalyst stripped H_2 from the C_3 as atomic hydrogen and created a gradient between the two bundles that drew hydrogen through the separating membrane into the air-fed bundle where it was oxidized to water. The combustion heat offset the endothermic dehydrogenation reaction. Cracking occurred over the fibers at their 823 K operating temperature, eventually deactivating them for dehydrogenation. At an appropriate time, flows to the bundles were switched and O_2 burned off the carbon. Thus, periodic operation served to maintain catalyst activity. The Swiss team demonstrated that their zoned reactor increased conversion appreciably beyond that at equilibrium and had a propane selectivity of almost 97%.

Zagoruiko (2008a) offered a novel proposal for an autothermal operation of propane dehydrogenation: Two fixed catalytic beds operating in parallel with the first bed fed propane from the top and the second bed fed air from the bottom, both feeds at ambient temperature. In the first bed, dehydrogenation occurs, drawing oxygen from the catalyst, while in the second, the catalyst is reoxidized. Reaction fronts move slowly through each bed, but in opposite directions. Thus, from the first bed, a propene stream emerges while from the second bed carbon oxides flow. After an appropriate interval, the feed streams are switched.

This system was simulated adopting a quasi-homogeneous model that neglected dispersion and assumed the kinetics for oxidative dehydrogenation, minor combustion of C_3's and catalyst reoxidation proposed by *Grabowski et al. (2002)*. By simulation, *Zagoruiko* demonstrated that, through careful choice of air and C_3H_8 feed rates, autothermal operation is possible once the system is brought to operating temperature at start-up. Advantages of the two-bed arrangement, *Zagoruiko (2008a)* points out, are that the temperature in the moving dehydrogenation zone decreases towards the outlet, thus significantly reducing the oxidation of $C_3^{=}$, there is no contact between C_3 and O_2 (eliminating the explosion hazard), and diluent is not present, simplifying recovery of propene. However, stable operation limited the yield of propene to 4%.

4.7 MALEIC ANHYDRIDE FROM BUTANE

Butane has largely replaced aromatics as the feedstock for maleic anhydride (MA) production through partial oxidation. Composition modulation, particularly in a recirculating solids system, may be attractive because the catalyst can act as the oxygen carrier, reducing the cost of product recovery and recycle of unused reactant. Heat generation is distributed between butane partial oxidation and catalyst reoxidation. The heat of reaction can thus be removed at two locations rather than one.

A tellurium-promoted vanadium molybdate supported on TiO_2 was used by *Lang et al. (1989b, 1991)* in a study of the oxidative dehydrogenation of butadiene to maleic anhydride (MA). Strategies considered were concentration modulation of the 1) C_4H_6, 2) O_2, 3) on-off modulation of both O_2 and $C_4^{2=}$, and 4) on-off reactant modulation with inert flushes between reactant pulses. Step-change experiments indicated a rapid response occurring over a minute's duration and a slower concentration adjustment needing about 20 to 40 min (*Lang et al., 1989a*). $C_4^{2=}$ modulation suppressed rates of formation but increased the selectivity to the furan intermediate at $\tau_{cycle} = 1$ min. Modulating the O_2 concentration resulted in a small increase in the formation of MA. Variations of the $C_4^{2=}$ and product volume% during a cycle for the $C_4^{2=}$ cycling showed an overshoot-undershoot pattern attributed to adsorption competition between $C_4^{2=}$ and the oxygenates. Bang-bang modulation of both hydrocarbon and oxygen was a more effective means of exciting this reaction system, but only for production of the furan intermediate. A 10% increase in the rate of furan formation was found for $\tau_{cycle} = 12$ s. The rapid rise in oxygenate formation at the beginning of the $C_4^{2=}$ half-cycle indicated the O_2 half-cycle depleted the store of adsorbates on the surface and recharged the oxygen held by the catalyst. The relatively slow rise of $C_4^{2=}$ points to adsorption and conversion. Fully oxidized surface sites form the anhydride or split butadiene to form acrolein, while partially oxidized sites yield furan. When O_2 is re-introduced, adsorbed butadiene was oxidized to maleic anhydride, accounting for its continuous appearance in the product stream. For a further discussion see *Silveston (1998)*.

4.7.1 Du Pont Research

Experiments on the partial oxidation of butane to maleic anhydride (see **Table 4-6**) undertaken by Du Pont are described in a patent (*Contractor, 1987*), an overview paper (*Contractor et al., 1987*) and several contributions dealing with specific aspects of the process (*Contractor et al., 1988; Contractor and Sleight, 1988; Contractor et al., 1990*). Use of a recirculating solids reactor was motivated by speculation that adsorbed oxygen is responsible for combustion so that segregating the reactants could raise selectivity to maleic anhydride. The catalyst developed for the Du Pont process was a vanadium phosphate ($(VO)_2P_2O_7$) promoted by silica and small amounts of indium, antimony, or

TABLE 4-6 Partial Oxidation of Paraffins and Olefins to Anhydrides

Reference	Objective	Reaction and Reaction Conditions	Modulated Variable	Observations	Comments
Contractor (1987); Contractor et al. (1987, 1988, 1990); Contractor and Sleight (1988)	Process development: investigation of a recirculating solids reactor to partially oxidize C_4 to MA	$C_4H_{10} + O_2 \rightarrow MA + H_2O$; Catalyst was $VO_2P_2O_7/K_2PO_4$, promoted by Si, Sb, In,Ta, encased in a SiO_2 shell. $623 < T_{Riser} < 636K$, $623 < T_{Fluidized\ Bed} < 633$ K in the two sections of the recirculating solids reactor	Cyclic experiments used a circulating solids reactor with duration of anaerobic C_4 oxidation = 3 to 7 s; mean duration of reoxidation = 4 s	Although laboratory experiments gave > 85% selectivity to MA at n-C_4 conversion of 30 to 48%, pilot plant experiments at $T = 633$ K showed an MA selectivity of ~75% at n-C_4 conversion of 30 to 45%. The flushing experiments demonstrated that stripping O_2 from the catalyst was important.	Oxidation-reduction experiments on VPO catalysts indicated that O_2 drawn from the catalyst lattice is the primary source of partial oxidation through oxidative dehydrogenation and O_2 insertion. Lattice O_2 also participated in combustion.
Lang et al. (1989a, b; 1991)	Application of composition modulation to butadiene partial oxidation to maleic anhydride	$C_4^{2=} + O_2 \rightarrow MA + H_2O$. Catalyst = $TeO_2/V_2O_5/MoO_3$ catalyst supported on TiO_2. $T = 823$ K, $P = 1$ atm. Feed contained 0.4 vol% C_4H_6, 10 vol% O_2 both in N_2	Cyclic variation of either C_4^{2+} or air with $1 < \tau < 40$ min and symmetrical 2, 3 and 4 part cycles using $C_4^{2=}$, air, and N_2 (flushing) for $6 < \tau < 60$ s	Modulation of individual reactants ineffective, bang-bang modulation increased furan yield by about 75%. Production rate and selectivity to MA decreased under modulation.	Partial oxidation is associated with O^- surface sites while $O^=$ (peroxide) sites cause combustion. CO_2 formed from combustion of adsorbed reaction intermediates. Researchers used the *Fiolitakis et al. (1983)* catalyst, suggesting reaction mechanism is important under modulation.
Huang et al. (2001)	Optimization of composition and combined composition and temperature modulation	$C_4H_{10} + 3.5\ O_2 \rightarrow MA + 4\ H_2O$. $(VO)_2P_2O_7/VOPO_4$ catalyst. $T = 689$, 703 K, $P = 1$ atm. Micro and bench-scale reactors used	**1)** Modulation between C_4-O_2 in He and O_2 in He. **2)** Modulation of both composition and temperature. Temperature cycling between 689 and 833 K; $\tau_{cycle} = 4$ min	1) An 8% enhancement was observed in selectivity, but yield and conversion were suppressed. 2) Modulation of both operating variables increased conversion, yield and selectivity significantly.	Orthogonal design of experiments was used to search for optimal operating conditions.

(*Continued*)

TABLE 4-6 Partial Oxidation of Paraffins and Olefins to Anhydrides (*cont'd*)

Reference	Objective	Reaction and Reaction Conditions	Modulated Variable	Observations	Comments
Hess et al. (2001a, b)	Investigation of C_4 partial oxidation in a 2-bed riser reactor -regenerator	Air oxidation of C_4H_{10} using a VPO catalyst. Tube reactor used stainless 60 mm × 4.6 mm i.d. electrically heated tube operating at 573 to 923 K	2 to 8 vol% butane in 20% O_2 in Ne was used for both steady-state and pulsed operation. Pulses were as short as 150 ms	Furan selectivity reached 10%. Comparison of the authors' model with experimental results showed good agreement.	Pulse operation was used to simulate the behavior of the riser portion of a fluidized bed. Authors' model suggests that the catalyst provides an oxygen reservoir for the reaction.
Shen et al. (2002)	Development of a VPO catalyst with greater O_2 availability	$C_4H_{10} + 3.5\ O_2 \rightarrow MA + 4\ H_2O$. Catalyst: $(VO)_2P_2O_7/VOPO_4$ promoted by Ce-Fe oxides. $T = 673$ K; $P = 0.12$ MPa	Pulses of a few mL of 20% n- C_4H_{10} in He or Ar were introduced into the reactor	Ce–Fe promoted VPO catalysts had more available lattice oxygen and gave higher conversion and selectivity than the unpromoted catalyst.	Catalytic role of Ce–Fe complex oxides remains to be clarified, but significantly increased selectivity, conversion of C_4 and MA yield.
Huang et al. (2002a)	Exploration of composition modulation and transient behavior	$C_4H_{10} + 3.5\ O_2 \rightarrow MA + 4\ H_2O$ on a $(VO)_2P_2O_7/VOPO_4$ catalyst at $T = 688$ K	Composition modulation between 2.78% C_4 -15.5% O_2 in He and 20% O_2 in He, $\tau_{cycle} = 6$ min; $s = 0.5$	Authors observed stable cyclic operation and measured transient responses.	The model developed for a somewhat simplified reaction network reproduced experimental results closely.
Huang et al. (2002b)	Investigation of reoxidation kinetics for a VPO catalyst reduced by C_4H_{10}	$(VO)_2P_2O_7 + 1/2\ O_2 \rightarrow 2\ VOPO_4$	Combined temperature and composition modulation between 4% C_4 -20% O_2 in He at 689 K and 21% O_2 in He at 853 K	A large increase in yield of MA observed when the regeneration temperature was increased from 689 to 853 K. Active catalyst was a mixture of $(VO)_2P_2O_7/VOPO_4$.	Performance improved under modulation for a two-bed process if the catalyst regeneration step was carried out at a higher temperature.

Ye et al. (2005)	Exploration of catalyst reoxidation in the partial oxidation of n-C_4	$C_4H_{10} + O_2 \rightarrow MA + H_2O$ (not balanced) in a VPO-catalyzed electrochemical membrane reactor using yttria-stabilized zirconia as an electrolyte membrane. $723 < T < 784$ K. Constant current was imposed between the two electrodes	O_2 required for anodic C_4 oxidation was generated by electrical oxygen pumping. n-C_4 diluted in N_2 was fed to the anode at $P = 1$ atm; air was fed to the cathode	Periodic redox experiments confirmed that reduced activity of VPO catalyst in C_4 oxidation could be regenerated by oxygen ion pumping. Membrane reactor gave MA yield of 10% with selectivity up to 53% at 753 K.	Reoxidation mechanism involved either direct electrochemical oxidation of the surface layer of the VPO catalyst or indirect oxidation of the subsurface layer.
Rubio et al. (2002); Herguido et al. (2005)	Investigation of a 2-zone fluidized bed reactor for oxidative dehydrogenation	Rubio et al. discussed C_4 partial oxidation. *Herguido* examined several different examples. Fluidized bed reactor with 3 cm i.d.; $P = 1$ atm; $T = 823$ K; reactor fluidized with O_2; catalyst was V/PO	2-zone fluidized bed was used as an alternative to composition modulation. Hydrocarbon reactant was introduced through a mid-bed distributor	Selectivity to MA reached 57% at a 60% n-C_4 conversion; selectivity was lower than those found in circulating fluidized beds or in pulse reactors.	*Herguido et al.* discussed the successful application of a 2-zone fluidized bed to several partial oxidation and oxidative dehydrogenation reactions.

(*Continued*)

TABLE 4-6 Partial Oxidation of Paraffins and Olefins to Anhydrides (*cont'd*)

Reference	Objective	Reaction and Reaction Conditions	Modulated Variable	Observations	Comments
Song et al. (2006)	Investigation of the partial oxidation of n-C_4 to MA using vanadia catalysts	Partial oxidation of n-C_4H_{10} to MA, with and without gaseous O_2. Reactor was a 3 mm i.d. tube with 0.25 g VPO or VPD catalyst; $P = 1$ bar; $T = 623$ K	Tubular flow reactor was pulsed with C_4/air or C_4/Ar followed by O_2 to reoxidize the catalyst surface: 250 μL pulses of C_4 from 0.9 to 2.9% in air or Ar carrier	Bang-bang periodic operation enhanced MA yields slightly at low conversion but vanished under commercial operating conditions. Similar experiments for VPD catalyst gave lower C_4 conversion and MA yields. Significant furan production occurred under periodic operation.	One of the few experiments that showed periodic operation is less attractive than steady operation.
Chalakov et al. (2007)	Partial and total oxidation of n-C_4 in an electrochemical cell with power generation	Electrochemical membrane reactor with ytteria stabilized ZrO_2 as electrolyte and a VPO catalyst deposited on an Au anode was fed n-C_4 and air separately at $823 < T < 873$ K	Butane feed switched with air in an asymmetric cycle at $\tau_{cycle} = 50$ min and $s = 0.2$ (based on the air flow duration)	Membrane reactor partially oxidized n-C_4 to maleic anhydride and generated an electrical potential.	Researchers explored electrochemical cells for chemical reactions and power production.

tantalum. The Du Pont design used a riser or transport reactor for C_4 oxidation and a fluidized bed for reoxidation of the catalyst.

Selectivity to maleic anhydride using the riser reactor ranged from about 83 to 93% at C_4 conversions from about 5 to 46% with the promoted $(VO)_2P_2O_7$ catalyst. Employing the same catalyst in a fluidized bed gave significantly lower selectivity to maleic anhydride but achieved much higher conversions under almost identical operating conditions. C_4 vol% in the riser feed ranged between 1.2 and 3%, while the C_4:O_2 ratios were above 1:15 for the highest conversions. Both conversion and selectivity depended on the C_4:O_2 ratio (*Contractor, 1987*).

Catalyst attrition was a major concern. Attractive catalysts are often mechanically weak and require a strong support for reactor use. The spray-dried promoted and unpromoted vanadium phosphates studied by Du Pont had low crushing strength so attrition losses were high in the Du Pont riser reactor. Strength was significantly increased by coprecipitating the vanadium phosphate with colloidal silica, but this decreased selectivity substantially (*Contractor et al., 1987*).

Experiments indicated that conversion and MA selectivity are highest at 633 K, the lowest temperature employed in pilot plant experiments. Selectivity remains constant up to about 40% conversion and then decreases slowly with rising conversion. At conversions approaching 90%, C_4 must be less than 6 mol% to avoid over-reduction of the catalyst. Adding a little additional O_2 to the C_4 feed maintains selectivity whilst reducing the catalyst recirculation rate and flushing the catalyst with He between the C_4 and air portions of the catalyst cycle improves anhydride yield (*Contractor, 1987*).

Oxidation-reduction studies of vanadium phosphate catalysts indicated oxygen drawn from the catalyst lattice was involved in the oxidative dehydrogenation of C_4 as well as in the oxygen insertion steps (*Contractor et al., 1990*). This mechanism is supported by the experiments of various researchers (*Arnold and Sundaresan, 1989; Emig et al., 1994; Abon et al., 1997*). Lattice oxygen also participated in the parallel combustion reaction.

Durability of the polysilicate reinforced catalyst was thoroughly investigated (*Contractor et al., 1990*). Satisfactory results led Du Pont to construct a full-scale plant that operated for several years (*Contractor, 1987, 1999; Contractor et al., 1994*). Problems with catalyst loss through attrition, however, caused a premature shutdown. *Emig et al. (1994)* discuss potential limitations of processes using a circulating fluidized bed. *Silveston (1998)* devotes nine pages to the Du Pont undertaking.

4.7.2 Chinese Research

Modulation to enhance selectivity for the partial oxidation of butane to maleic anhydride was undertaken by a Chinese research team using a domestic VPO catalyst (*Qiao, 1998; Huang, 1999; Huang et al., 2001, 2002a, 2002b; Chen et al., 2002; Liang et al., 2003a, 2003b, 2004*). A commercial laboratory reactor consisting of a 9.5 mm o.d. × 250 mm stainless steel tube packed with about 0.6 g of 40/60 mesh catalyst and usually diluted with quartz sand was used by the above authors. Reactor effluent flowed to an absorber to trap MA and then to a gas chromatograph. A mass spectrometer was employed for transient measurements. A commercial VPO catalyst was used that had been characterized by XRD, XPS and FTIR measurements. These showed that fresh catalyst was primarily $(VO)_2P_2O_7$ with some orthophosphate, $VOPO_4$.

Results of composition modulation experiments, summarized in **Table 4-7**, are shown in **Figure 4-10**. These suggest that the optimal hydrocarbon/oxygen molar ratio is 1:5, a value between the stoichiometric ratios for selective oxidation and total oxidation. The transient response of MA is shown in **Figure 4-11**.

TABLE 4-7 Anhydrides and Aldehydes from Aromatics

References	Objective	Reaction and Reaction Conditions	Modulated Variable	Observations	Comments
Wainwright and Hoffman (1974).	Investigation of the oxidation of o-xylene under composition modulation	$C_8H_{10} + O_2 \rightarrow C_8H_4O_3$ with catalysts (1) a K_2SO_4 promoted V_2O_5 supported on SiO_2; (2) a K promoted vanadium antimonate supported on titania. Reactions were carried out at 603 to 663 K using 1 to 2 vol% o-xylene in streams containing either 10 or 30 vol% O_2 in an N_2 carrier	Comparisons were made between only start-up and steady-state activities. Authors used a packed bed as well as a lab-scale riser reactor	Vanadia catalyst lost about 90% of start-up activity after 60 to 120 s. Steady-state activity was 30 to 40 times lower than the start-up activity. Short contact times in a riser reactor permitted operation at lower temperatures with higher selectivity to oxygenates.	Authors found greater than a 10-fold increase in conversions for a riser reactor over those for packed beds, even at lower operating temperatures.
Cordova and Gau (1983)	Investigation of the production of maleic anhydride from benzene under composition modulation	$C_6H_6 + O_2 \rightarrow MA + H_2O$, catalyst: coprecipitate of ammonium vanadate + ammonium molybdate + other salts on a silica support. $T = 573$ K, $P = 1$ bar, $SV = 3$ L/h g.cat	4 part symmetrical cycle at $\tau = 4$ min. with equal durations of exposure to 0.27 vol% Bz in air or 0.47 vol% Bz in N_2 and air, separated by flushing with N_2 for an equal duration	12 to 24 h were needed to attain a stationary condition after a change in composition or temperature. Selectivity to MA was 5% higher than the steady-state selectivity during the reducing exposure, but for the full cycle, it dropped slightly. Fully segregating the reactants is not a suitable forcing strategy.	Oxygen incorporated in the catalyst bulk participated in both partial oxidation and combustion, but its mobility was low. Authors proposed that a circulating fluidized bed with a distributed feed of air and Bz would be attractive for MA production. A second fluidized bed would reoxidize the catalyst.

Fiolitakis et al. (1983).	Investigation of mechanism and catalyst changes on switching between oxidizing and reducing atmospheres	$C_6H_6 + O_2 \rightarrow MA + H_2O$ using a $TeO_2/V_2O_5/MoO_3$ catalyst supported on TiO_2; $T = 693$ K, but varying in some experiments, $P = 1$ bar, $SV = 1.14\ s^{-1}$	Symmetrical cycling between 5 vol% Bz in air and air. Also cycling between mixtures with O_2:Bz volumetric ratios of 2.4 and 4.6 for $10 < \tau < 360$ s	After start up, catalyst activity and selectivity to MA increased for several cycles to reach 70 to 100%. With extended use, activity and selectivity declined; the latter to as low as 20 to 35%. Exposure to air in the absence of Bz restored lost selectivity. Cycling between O_2:Bz ratios doubled the selectivity, but at a 20% drop in Bz conversion.	Wave-front analysis was used on the 5th Bz-rich pulse. Results suggest that selectivity of the catalyst could be improved by composition modulation using feeds with different Bz:air ratios. See *Lang et al. (1989a).*
Matros (1985)	Uncertain	o-Xylene + $O_2 \rightarrow$ Phthalic Anhydride. Vanadia catalyst	Down flow, raining solids bed used for anaerobic oxidation of o-xylene, catalyst reoxidized in a riser reactor	Selectivity to the anhydride reached 75% at an o-xylene conversion of 80% but dropped slowly at higher conversions.	Results are given in a book with little detail.
Bulushev et al. (2000a, b; 2004; 2005)	Benzaldehyde formation mechanism and reaction modeling	Toluene + $O_2 \rightarrow$ BA + H_2O over VO_x/TiO_2, K-doped V_2O_5/TiO_2 catalysts. Experiments performed at 523 to 673 K	Repeated pulses of toluene, toluene-O_2 mixtures, or O_2 in N_2 were used	Nucleophilic oxygen associated with the surface V-O species appeared to be responsible for partial oxidation of toluene. Electrophilic oxygen arising from reoxidation of V^{4+} caused toluene combustion.	

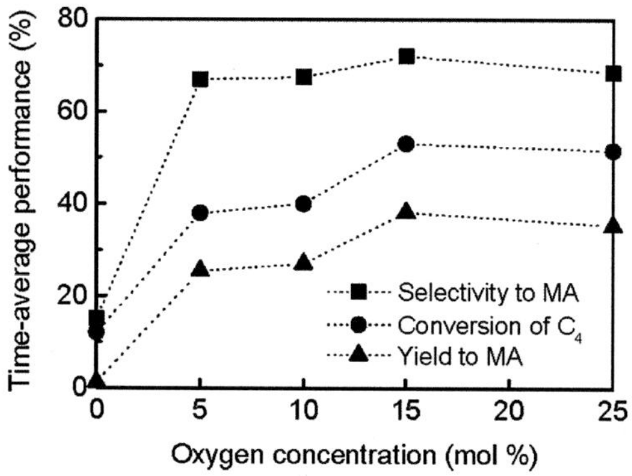

FIGURE 4-10 **Effect of O_2 concentration in the C_4-containing feed on reactor performance at $T = 703$ K, $s = 0.5$, $\tau_{cycle} = 4$ min for modulating the feed between 20 mol% O_2 in He and a mixture of 3 mol% n-C_4, 0-25 mol% O_2 in He.** *(Figure reproduced from* Huang et al. (2001) *with permission of the authors.)*

A sharp overshoot in MA concentration just after switching can be seen, but the concentration dropped to a low level in **(a)** of the figure after several seconds owing to the limited rate of oxygen supply from the bulk phase of the catalyst. A performance improvement may be achieved by using a mixture of butane and O_2, provided a suitable cycle period is used as may be seen in **(b)** of **Figure 4-11**.

Indeed at 416°C, $\tau_{cycle} = 4$ min and $s = 0.5$ for modulation between air and a feed with 3% n-C_4, 15% O_2 was optimal. These experiments demonstrated about a 5% higher selectivity to MA under modulation but about a 6% reduction in yield with respect to steady-state operation at the time-average composition in the cycle.

Temperature programmed oxidation experiments (*Huang et al., 2001*) suggested that simultaneous modulation of both composition and temperature would improve time-average performance by increasing the rate of reoxidation of the catalyst. Cyclic experiments were performed with optimal feed conditions given by *Huang et al. (2001)*. Modulating just composition resulted in a C_4 conversion of 78% and an MA yield of 50%. However, modulating both composition and temperature increased conversion to 90% and raised the yield to 60%. Such a remarkable improvement suggests further investigation of modulating two reactor inputs. Perhaps this difference arises from more storage of lattice-oxygen in the near-surface layer of the catalyst. FTIR spectra of the catalyst indicated a partial transformation from $(VO)_2P_2O_7$ to β-$VOPO_4$ *(Huang, 1999)*.

Further examination of simultaneous modulation of two inputs was carried out in a bench-scale fixed-bed reactor held at a constant wall temperature and packed with a 30 g catalyst. The VPO catalyst was diluted with ceramic beads and packed in the middle section of the tube. With this reactor, steady-state experiments, carried out at 683 K and SV = 2000 h^{-1}, showed a selectivity of 62 at 83% conversion, while two-input modulation increased the selectivity up to 73%, depending on the cycle period. Conversion dropped, however, so the yield was somewhat lower. Experiments were carried out according to an orthogonal experimental design described by *Huang et al. (2001)*. Using the yield of MA as the reference, modulation always improved performance over steady state.

The study of the kinetics of C_4 oxidation to MA over the VPO catalyst under transient conditions was advanced by *Huang et al. (2002a)*, *Chen et al. (2002)* and *Liang et al. (2003a, 2004)*. These researchers proposed a simplified reaction network and combined this with their observations on the roles of lattice and adsorbed oxygen to develop a transient kinetic model allowing for oxygen storage and diffusion in the bulk-phase of the catalyst. This model was applied to their composition modulation experiments (*Huang et al., 2002a*). Good agreement was found. The model was also applied to a circulating fluidized bed reactor for selective oxidation of n-butane (*Liu et al., 2001*) and used to explore temperatures in the riser and the fluidized-bed.

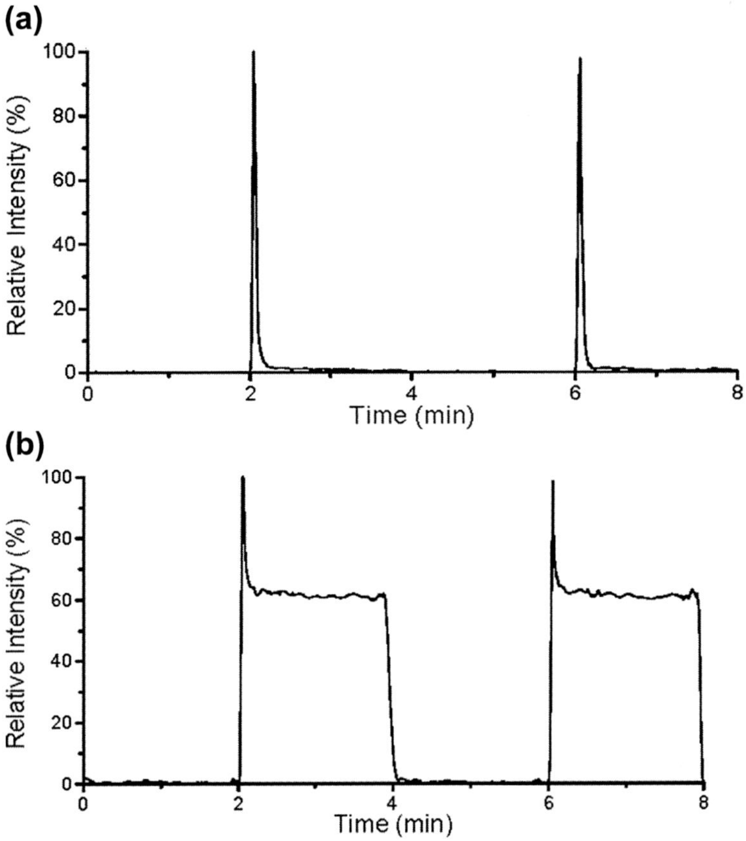

FIGURE 4-11 Transient responses of MA for composition modulation between 20 mol% O_2 in He and a feed with (a) 3 mol% C_4 in He and (b) 3 mol% C_4 and 15 mol% O_2 in He at $T = 689$ K, s = 0.5, $\tau_{cycle} = 4$ min. He flushing used between the two half-cycles. *(Figures reproduced from* Huang et al. (2001) *with permission of the authors.)*

Shen et al. (2002) investigated an improved VPO catalyst through adding Ce-Fe complex oxides to increase the available lattice oxygen. A comparative study showed the complex oxides promoted catalysts increased selectivity to MA while the conversion of C_4 decreased. The atomic ratio of Ce to Fe was important and the improvement was related to the amount of mobile oxygen species on or near the catalyst surface. *Shen et al.* used an $^{18}O_2$ exchanged catalyst and observed that the amount of labeled oxygen appearing in the MA product was roughly double the amount from the VPO catalyst when a 4:1 Ce-Fe oxide promoter was incorporated. They speculated that this may be due to the co-existence of γ-$VOPO_4$ and $(VO)_2P_2O_7$ when the complex oxide is present.

4.7.3 European Research

A two-zone fluidized bed was also examined for the production of MA from C_4 by *Rubio et al. (2002)* and *Mallada et al. (2002)*. These authors reported using higher C_4 concentrations than was possible with fixed bed reactors; however, the reducing atmosphere that resulted diminished the activity of the commercial VPO catalyst used. Thus, they chose to use much lower C_4 concentrations in their experiments. Selectivities to MA approaching 60% were measured.

A performance comparison of two VPO catalysts under periodic and steady-state operation has been published by *Song et al. (2006)* who showed that periodic operation improved selectivity to MA at conversions below 3% for the standard VPO catalyst. However, at higher conversions, selectivity and yield were much greater for steady-state operation. Periodic operation, however, generated significant amounts of furan. Experiments were also undertaken using complex promoted catalyst representative of those used industrially for MA production. No enhanced formation of acrolein was seen, but the generation of CO_2 was much lower under periodic operation. The authors concluded that periodic operation is beneficial for the unpromoted Bi-Mo catalyst but not for catalysts used in commercial operations.

A novel cyclic system for investigating mechanisms in C_4 partial oxidation used by *Ye et al. (2005)* and later by *Munder et al. (2007)* is illustrated in **Figure 4-12**. As discussed above, it is generally accepted that partial oxidation over the VPO catalyst occurs with oxygen extracted from the catalyst. In turn, gas-phase O_2 reoxidizes the catalyst. Periodic operation experiments in which electrical pumping is turned on for 120 min and then switched off demonstrated that this operation reoxidizes the VPO catalyst. However, electrical pumping produces electrophilic oxygen species (O^{2-}, O_2^{2-}, O^-) that are more active than O or O_2 for partial oxidation but also result in total oxidation. *Ye et al. (2006)* verified this in steady-state experiments and found that electrical pumping shifted selectivity from MA to CO_2. Significant increases in reaction rates were observed.

In a second paper, *Munder et al. (2007)* proposed a model with which they analyzed the experimental results of *Ye et al. (2005)*. Using this model, they explored optimal operation and found that the selectivity to MA under steady-state operation could not exceed about 65%.

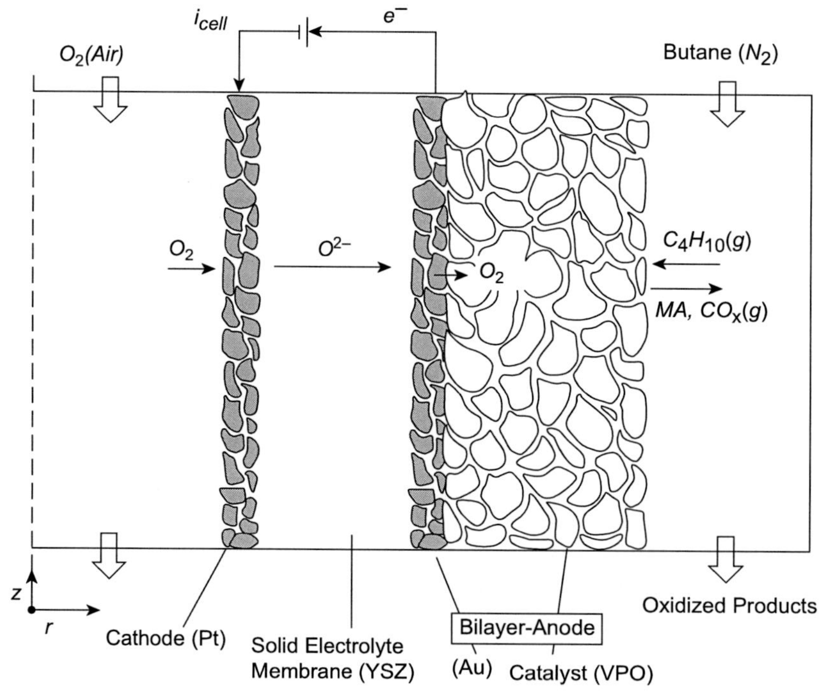

FIGURE 4-12 Schematic cross-section of the Sundmacher-Munder cell for the voltaic assisted partial oxidation of butane. *(Figure adapted from* Munder et al. (2007) *with permission. © 2007 by Elsevier Science Publishers.)*

They went on to consider a periodic operation in which butane was introduced in successive pulses interrupted by electrical pumping to reoxidize the VPO catalyst. Between these different pulses, the VPO catalyst was flushed with N_2. In this system, the ratio of lattice to surface oxygen concentration varied cyclically with sharp changes during pulsing as is shown in **Figure 4-13**. Conditions assumed for the figure are $\tau_{cycle} < 5$ s, C_4 pulse duration $\cong$ N_2 flush duration $\cong$ 2 $\times$ duration of current input, 1 vol% C_4 in feed, $N_{Da} = 2$, Thiele modulus $= 0.01$. Operating above the diagonal yields selectivities under pulsing that exceed steady-state operation. Indeed, for the oxygen cycle shown in the figure, the simulation by *Munder et al.* showed a large enhancement in the selectivity to MA provided cycle periods were held to a few seconds. Indeed, time-average yields of 92% were possible, but high oxygen/butane feed ratios were required.

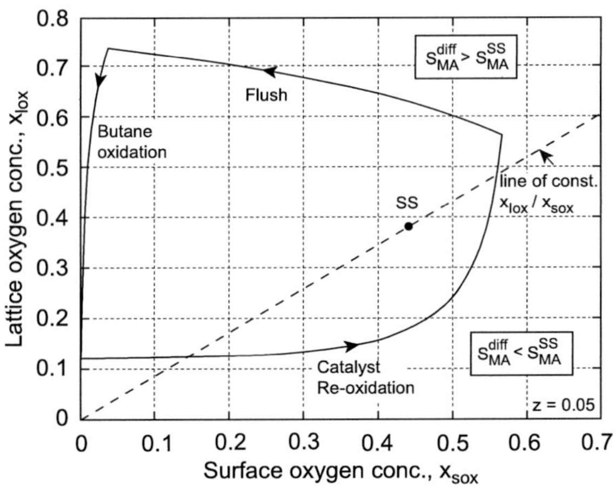

FIGURE 4-13 Simulation of the time variation of available surface oxygen to lattice oxygen in the bulk of VPO catalyst under periodic electrochemical O_2 pumping and C_4 pulses separated by inert flow for a tubular membrane reactor with structure shown in Figure 4-12. Steady-state operation given by the diagonal marked "ss," z is axial position. *(Figure adapted from* Munder et al. (2007) *with permission. © 2007 by Elsevier Science Publishers.)*

Operating a fuel cell with partial oxidation to produce a commercial product rather than total oxidation has led to several experimental programs employing methane to yield synthesis gas (*Ishihara et al., 1999; Sobyanin and Belyaev., 2000; Kiatkittipong et al., 2004*). Ethane and butane have also been used *(York et al., 1996)*. None of these programs employed periodic operation. However, *Chalakov et al. (2007)*, using the partial oxidation of butane to maleic anhydride, observed a gradual decrease in the cell current with time. They attributed this to carbon deposition on the VPO catalyst. These investigators found that periodically switching asymmetrically between C_4 and air with $\tau_{cycle} = 50$ min and $s = 0.2$ maintained a constant current output and rate of maleic anhydride formation.

4.8 ANHYDRIDES AND ALDEHYDES FROM AROMATIC HYDROCARBONS

4.8.1 Maleic Anhydride

Partial oxidation of benzene (Bz) to maleic anhydride (MA) over vanadium molybdate catalysts was studied by composition modulation in two contributions with the objective of probing mechanisms (**Table 4-7**). *Cordova and Gau (1983)* used a co-precipitate of ammonium vanadate and molybdate with various other metals onto a spherosil (silica) carrier that was then calcined. Their modulation experiments were modeled on those of *Niwa and Murakami (1972a, b)* discussed earlier. In a four-part cycle, Bz was mixed with air to begin the cycle; this was followed by an N_2 flush, then air and a second N_2 flush. Selectivity to MA increased by 5% compared with steady state for the part of a cycle when Bz was present in the feed. However, time-average selectivity was slightly lower for the entire cycle. Air was absent in the benzene partial cycle for the second experiment and selectivity to MA was substantially reduced. Thus, segregating reactants did not improve selectivity or yield. Reducing the duration of the Bz exposure greatly improved the selectivity but drastically reduced conversion. *Cordova and Gau (1983)* concluded from their experiments that lattice oxygen is responsible for both partial oxidation and combustion, but its mobility is low so only a small fraction of the oxygen contained in the catalyst participated during the reducing part of a cycle. These researchers claimed strong adsorption of MA and Bz on the catalyst surface.

A tellurium-promoted vanadium molybdate supported on TiO_2 was used by *Fiolitakis et al. (1983)* to investigate reaction mechanism and catalyst changes on exposure to different reacting mixtures. A two-part cycle with switching between air and 5% Bz in air was used in several experiments (see **Table 4-7**). The first 12–15 cycles were examined. Bz in the first pulse of hydrocarbon was almost completely consumed but without a large draw down of O_2. Products amounts were much smaller than in subsequent pulses. Most of the Bz appeared to be irreversibly adsorbed onto the catalyst. The temperature dropped, even though adsorption is exothermic, so changes occurred in the catalyst as well. MA, H_2O and CO_2 increased in the next two or three pulses and then declined. By the eighth pulse, the combustion activity of the catalyst had greatly decreased, indicating catalyst reduction. In subsequent pulses, the primary reaction was MA formation and the pulses became reproducible despite increasing temperature. Low initial selectivity followed by a large increase was found to be typical of fresh catalysts. Selectivity to MA climbed to 70–100%. With extended use, activity and selectivity declined. Exposure to air restored much of the lost selectivity and activity.

Fiolitakis et al. (1983) applied a novel wave front analysis (*Fiolitakis and Hofmann, 1983*) to the fifth Bz-rich pulse to show that the aldehyde product is strongly adsorbed and oxidized further to the anhydride. The authors speculated that there was a build-up of a peroxide on the surface that totally oxidized adsorbed MA and Bz. After the eighth pulse, combustion disappeared and selectivity to MA increased sharply.

Results of the *Fiolitakis et al.* analysis suggested that selectivity could be improved by composition modulation using different Bz-air ratios and a short τ_{cycle}. This was tested by *Fiolitakis et al.* at 693 K. Feed to a fixed bed of the tellurium-promoted catalyst was symmetrically modulated between O_2:Bz volumetric ratios of 2.4 and 4.6 (thus between reducing and stoichiometric conditions) at periods ranging from about 10 to 360 s. Results showed a 100% increase in selectivity to MA, but a 17% decrease in conversion. Highest selectivities occurred for $30 < \tau < 120$ s. *Silveston (1998)* gives further details.

The catalyst used by *Fiolitakis et al.* was also used by *Lang et al. (1989b, 1991)* for the partial oxidation of butadiene to MA as discussed earlier. The intention was to compare composition modulation when the catalyst and product were the same but the chemical route was altered. *Lang et al.* observed no improvement in $C_4^{2=}$ selectivity to MA through modulation.

Toluene partial oxidation over vanadia catalysts, some potassium promoted, supported on Ti/O_2, have been studied over several years by a Russian-Swiss team (*Bulushev et al., 2000a, b; 2004, 2005*). Using successive pulsing, they observed that nucleophilic O_2 provided selective partial oxidation of toluene to benzaldehyde. However, electrophilic O_2 oxidized benzaldehyde all the way to CO_2. **Chapter 8** discusses this work further.

4.8.2 Phthalic Anhydride from o-Xylene

One of the early publications on composition modulation dealt with the partial oxidation of o-xylene in a simulated circulating fluidized bed (*Wainwright and Hoffman, 1974*). Two potassium-promoted vanadia catalysts on different supports were used (**Table 4-7**). The catalyst was not regenerated so only one part of a catalyst exposure cycle was examined. *Wainwright and Hoffman* observed that the vanadia catalyst lost about 90% of its start-up activity after 60 to 120 s use. Steady-state activity was 30 to 40 times lower than at start up. Transient operation using a riser reactor showed an order of magnitude increase in conversion. The vanadia catalyst used produced o-tolualdehyde rather than phthalic anhydride in the 506 to 615 K temperature range. Selectivity to this aldehyde was about 95 at 10% xylene conversion. Industrial reactors attain at best 70% selectivity to this aldehyde.

Step-change experiments undertaken with a vanadium antimonate catalyst showed that 73 to 80% selectivity to phthalic anhydride could be obtained with short catalyst exposures at xylene conversions; about twice those obtained in steady-state operation. Russian researchers quoted by *Matros (1985)* worked with vanadia catalyst using a looping system in which o-xylene was oxidized countercurrently in a downflow reactor and catalyst was reoxidized in a riser. Segregation of reactants in this way substantially increased selectivity to phthalic anhydride for high o-xylene conversions.

4.9 AROMATIC NITRILES

In the 1960s, ABB Lummus Crest Inc. developed a process for producing terephthalic acid from p-xylene via hydrolysis of terephthalonitrile. Severe catalyst deactivation was encountered and traced to the reduction of the metal oxides. However, the initial activity could be restored by air oxidation of the catalyst at just above the reaction temperature. These observations led the Lummus researchers to use metal oxide as the oxygen carrier. Thus, their nitrile process shuttled their oxide catalysts between two fluidized beds: An ammoxidation reactor fed with xylene, ammonia and molecular oxygen, and a regenerator blown with air. The pilot plant stage was reached in the development and was described in a paper presented at a technical meeting. Experiments with other feedstocks were undertaken by Lummus during their development effort and it was observed that their vanadium-based catalyst produced different aromatic nitriles in good yields depending on the feedstock used. Unfortunately, only meagre descriptions of the Lummus technology exist in the literature (*Sze and Gelbein, 1975, 1976; Schwendeman et al., 1983*) and experimental data are not available. The Lummus work is summarized by *Silveston (1998)*.

Nomenclature

Bz	=	benzene
C	=	concentration (mol/m^3)
CFB	=	circulating fluidized bed
d_p	=	particle diameter (mm)
Et_2O	=	ethylene oxide
FTIR	=	fourier transform infra red
GHSV	=	gas hourly space velocity (h^{-1})
i.d.	=	inner diameter (cm)
MA	=	maleic anhydride
MS	=	mass spectrometer
MeOH	=	methanol
N_{Da}	=	Damköhler Number
o.d.	=	outer diameter (cm)
P	=	pressure (kPa, bar)
S	=	selectivity
SS	=	steady state
STP	=	standard temperature and pressure
SV	=	space velocity (s^{-1})
s	=	cycle split (-)
T	=	temperature (K)
TPSR	=	temperature programmed surface reaction
t	=	time (s, min.)
$t_{contact}$	=	contact time
VPO	=	vanadate phosphate
XRD	=	X-ray diffraction
XPS	=	X-ray photoelectron spectroscopy
x	=	mol fraction or concentration in the solid phase
YSZ	=	ytteria stabilized zirconia
z	=	axial position

Greek

τ = cycle period (s)
τ_{cycle} = cycle period (s)

Subscripts

lox = lattice oxygen
MA = maleic anhydride
sox = surface oxygen

Superscripts

ss = steady state

CHAPTER

5

Combustion Systems

Peter Lewis Silveston, Robert Ross Hudgins
Waterloo, Ontario, Canada

OUTLINE

5.1 Non-Catalytic Combustion Reactions *124*
5.1.1 Pulsed Combustion **124**
5.1.2 Modulated Combustion **124**
5.2 Catalytic Combustion *127*
5.2.1 Single Component Combustion **127**
5.2.2 Fuels Combustion **130**
5.3 Looping Combustion *134*
5.4 Simulated Loop Reactors *134*

Combustion is a technically familiar process undertaken to generate heat or power, to propel a vehicle, or to dispose of a noxious gas. When the fuel is a gas, technical considerations are the rate of mixing with the oxidant, ignition and stabilization to prevent blowout. With liquid or solid fuels further considerations arise, such as liquid droplet or particle size and mass transfer between phases. Periodic operation of combustion systems offers a means of accelerating mixing, greater size reduction and increasing mass transfer rates. Combustion may be carried out homogeneously in just a single phase, or heterogeneously with gaseous, liquid or solid phases. Heterogeneous combustion may employ a catalyst, usually as a separate solid phase. In this chapter, both catalytic and non-catalytic combustion will be discussed.

One type of periodic operation, cycling a solid oxidant offers a means of utilizing air, an abundant low cost oxidant, in combustion without diluting the stack gas with nitrogen. Dilution makes CO_2 sequestration expensive. This operation, employing circulating fluidized or moving beds of a reducible oxide, removes oxygen from air and transports that oxidant into a combustor. Oxidation of a fuel proceeds there in the absence of N_2 generating a hot gas for use in power generation or for heat supply consisting primarily of CO_2 and H_2O. The hot solids circulate back to a regenerator where they are re-oxidized and returned to the combustor. This cycling operation is now widely referred to as chemical looping.

Chemical looping has become an immensely popular R & D topic over the last decade. Most of the development activity has focused on combustion of carbonaceous fuels as looping has the potential of substantially reducing the cost of CO_2 capture and sequestration. Looping, however, has other potential applications, such as gasification and partial oxidation. However, the quintessential reaction is a gas-solid one. Thus, we reserve discussion of chemical looping for **Chapter 8**.

A few of the large number of published combustion studies, both catalytic and non-catalytic, are summarized in **Table 5-1**. Most of these are discussed in the following sections.

5.1 NON-CATALYTIC COMBUSTION REACTIONS

5.1.1 Pulsed Combustion

Modulation of combustion through pulse injection of fuel has been practiced on a commercial scale since W.W.II. It appears to have begun with the V-1 rocket systems which employed the periodic injection of a fuel into a combustor to provide a gas stream that propelled the vehicle. It should be mentioned too that the internal combustion engine, which goes back to the nineteenth century, is in principle a periodic pulse operation because the fuel is injected into the engine cylinder and then ignited. Rockets and combustion engines, however, are beyond the scope of this monograph.

Another form of pulse combustion continues to be popular for domestic and commercial gas-fired furnaces and water heaters. These units use either mechanical valving or aerovalves, whose operation is controlled by combustion. The latter operate at high frequencies between 15 and 230 Hz. Their frequencies depend on fuel type, feed rate, and air/fuel ratio and are ultimately governed by valve design. The frequencies of these systems can be as much as an order of magnitude greater than those employed in other modulation schemes discussed in this monograph. The advantage of pulse combustors is that they produce oscillatory flow as well as pressure waves that provide rapid gas-fuel mixing in the combustor and high rates of heat transfer to surfaces. Rapid mixing leads to higher combustion efficiencies and lower NOx emissions than those observed in continuously operated combustors or furnaces. Higher rates of heat transfer are achieved so that heat exchanger size can be significantly reduced.

Putnam et al. (1986) provide a comprehensive survey of pulse combustors and some performance data. The literature on pulsed combustion is very large. Recent references from active researchers are *Hermanson et al. (2004a, b)* and *Fregeau and Hermanson (2009)*; however, the major research effort took place in several decades prior to 2000. State of the art has been summarized by *Putnam et al. (1986)* and by *Zinn (1992)*. *Reuter et al. (1986)* provide a condensed review of the development.

5.1.2 Modulated Combustion

Research on pulsed combustion in the 1950s appears to have spun off an interest in modulated combustion. All of the early work was done in Russia. It is discussed by *Putnam et al. (1986)* in their survey of pulse combustion and in reviews published by *Severyanin (1971, 1982)*. In modulated combustion, the fuel supply to the burner or combustion chamber is switched on and off so it represents bang-bang forcing. Motor driven rotary valves and fluidic devices are used so that frequencies between 1 and 20 Hz are readily achieved. These are well below the natural or acoustic frequencies at which pulse combustors operate. Just as for pulse combustors, the advantage of combustion

TABLE 5-1 Modulation Studies of Catalytic and Non-catalytic Combustion

Authors	Study Objective	Catalyst and Reactor System	Modulated Variables	Reaction Conditions	Observations	Comments
Wandrey and Renken (1977)	Stabilization of a saddle point	Pt gauze	$C_3H_6 - O_2$ - He mixture and He alone	Gas preheat temperatures from 70 to 250°C, $12\ s < \tau_{cycle} < 20\ s$	Higher frequency stabilized the catalyst temperature in the saddle point region. Lower frequency caused oscillations around the excited state.	Experiments demonstrated that operation at an unstable saddle point in the variable space can be stabilized by modulation.
Shinjoh et al. (1987, 1989)	Investigation of the effects of forced cycling behavior on catalysts' performance	Pt, Pd, Rh on $\alpha-Al_2O_3$	Air-Fuel (C_3H_6) ratio as 0.01 to 0.4 vol% hydrocarbon	$180 < T < 550°C$, $0.2 < \tau_{cycle} < 2\ s$, $SV = 30{,}000\ h^{-1}$	Modulation of the air-fuel ratio increased C_3H_6 conversion for the Pt catalyst only at $T < 250°C$; for the Rh catalyst, conversion increased below 500°C.	Optimal cycling frequency depended on temperature and catalyst. Inhibition by C_3H_6 when modulation improved performance.
Shinjoh et al. (1989)	As above	Pt on $\alpha-Al_2O_3$	Air-propane ratio	$180 < T < 500°C$, $1 < \tau_{cycle} < 20\ s$, $SV = 30{,}000\ h^{-1}$, modulation of air-fuel ratio as well as bang-bang cycling	Modulation of the air-fuel ratio increased C_3H_6 conversion for the Pt catalyst at the temperatures used.	Lag in CO_2 formation in the O_2 half cycle and no lag in the C_3H_8 half cycle indicate dissociative O_2 adsorption on the Pt surface.
Charon et al. (1991)	Reduction of NO_x formation	Non catalytic	Methane-oxygen ratio, oxygen and methane flow	Maximum temperature in furnace # 1 = 1350°C, in furnace # 2 = 1800°C, symmetrical cycles	Modulation decreased NO_x production. Largest reduction at low frequency. CO formation increased with frequency.	Concurrent modulation of both fuel and oxygen more effective than modulating individual flows.
Carlsson et al. (2007)	Comparison of catalyst behavior for low temperature CH_4 combustion	5 wt% Pt/γ-Al_2O_3 and 5 wt.% Pd/γ-Al_2O_3	Feed composition with periodic addition of 1250 ppmv O_2 to flow of 500 ppmv of CH_4 in Ar	$T = 623\ K$, $GHSV = 7000\ h^{-1}$, pulse addition of O_2, with a 5 min pulse duration	Enhancement of combustion through O_2 pulsing much greater for the Pd catalyst.	Authors suggested the lower combustion activity for the Pt catalyst caused by formation of a surface oxide that weakly adsorbs CH_4.

(*Continued*)

TABLE 5-1 Modulation Studies of Catalytic and Non-catalytic Combustion (*cont'd*)

Authors	Study Objective	Catalyst and Reactor System	Modulated Variables	Reaction Conditions	Observations	Comments
Becker et al. (2009)	As with Carlsson et al. (2007) above	1 wt% Pt/γ-Al_2O_3 and 1 wt% Pt/CeO_2 impregnated onto a cordierite support	As with Carlsson et al. (2007) above	$T = 623$ K, $GHSV = 13000\ h^{-1}$, pulse addition of O_2, with a 5 min. pulse duration	Modulation ineffective with the Al_2O_3 supported catalyst, but significantly increased CH_4 combustion on the CeO_2 supported catalyst.	Authors attribute failure to enhance Pt/Al_2O_3 to O_2 poisoning of the catalyst surface.
Noorman et al. (2010a, b)	Evaluation of anaerobic CH_4 combustion in a fixed bed of metal oxide	1.1 mm particles of 12.5 wt% CuO/ Al_2O_3 in a 30 mm i.d. × 0.4 m bed	Cycling between CH_4 oxidation with CuO reduction and re-oxidation of Cu_xO_y with air	$T_0 = 923$ K, Feed was 4 vol% CH_4 in N_2. Bed re-oxidized with air	High degree of CuO reduction observed but just a small amount of CH_4 cracking was measured.	Authors' model matched measurements closely.
Noorman et al. (2011)	As above with a limited parametric study of operating variables	As above, but also with Ti stabilized $CaMnO_3$ with a perovskite structure, $2 < d_p < 4$ mm	Cycling between CH_4 oxidation with catalyst reduction and re-oxidation of the catalyst with O_2 in N_2	$723 < T_0 < 1073$ K, Feed was 4 to 10% CH_4 in N_2 and catalyst re-oxidation used 1 to 25 vol% O_2 in N_2	Raising T_0 of feed to the CuO catalyst increased C deposition but H_2 was oxidized. Addition of steam to feed suppressed cracking but CH_4 reforming occured.	Use of the $CaMnO_3$ catalyst avoided the cracking side reaction, but catalyst activity is low and mechanical strength was poor.

modulation is that flow and pressure oscillations increase mixing and rates of heat transfer from hot combustion gases to heat exchanger surfaces. Mixing significantly lowers NO_x formation. *Corliss et al. (1984)* review this subject.

Recent activity on modulated combustion has focused on NO_x generation in high-temperature industrial furnaces. An example of this research is given in a paper by *Charon et al. (1991)* describing experiments on modulating the combustion of natural gas in oxygen. The application is to furnaces in which temperatures in excess of 1200°C are needed. NO_x levels can reach 20 vol% in the stack gas from such furnaces if natural gas is burnt with O_2-enriched air. Modulating this air and fuel mixture reduces NO_x production significantly. This observation was made in the lower portion of the frequency range studied by *Charon et al.* It is evident from **Figure 5-1** which plots the ratio of the time-average NO_x emerging from the furnace under modulation to that emerging at steady state against the modulation frequency. Amplitude was 0.2 SN, where SN = stoichiometric number defined as the moles of oxygen per mole of methane fed, divided by the moles of oxygen needed to completely oxidize methane. The figure shows that periodically lowering the air-fuel ratio increases the formation of CO, although *Charon et al.* claim that if the emerging gas was completely mixed, CO would be oxidized to CO_2 at the flue gas temperature of 1300°C. The *Charon* experiments were undertaken in a model furnace. Further experiments in a much larger furnace gave similar results.

Figure 5-1(b) and **(c)** show the effect of modulating the natural gas flow rate and the oxygen flow rate separately. The amplitude of modulation is half that in (a) and just a 45 to 60% reduction in the NO_x emission was realized. CO formation was also lower. The effect of frequency on NO_x formation was the same in all three modulation experiments.

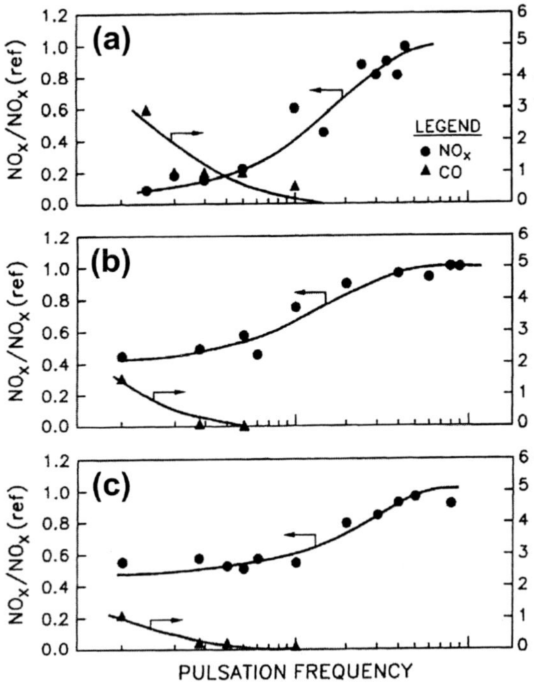

FIGURE 5-1 Percent CO in the furnace off-gas and normalized NO_x emission versus combustion modulation frequency: (a) Modulation of both natural gas and oxygen, (b) modulation of natural gas alone, (c) oxygen alone. Amplitude of the modulation in (b) and (c) is 1/2 that in (a). *(Figure adapted from* Charon et al. (1991) *with permission of the authors.)*

5.2 CATALYTIC COMBUSTION

5.2.1 Single Component Combustion

Wandrey and Renken (1977) studied the combustion of propylene in air over a Pt gauze to demonstrate that modulation could be used to operate a reaction system in the region of an unstable steady state. For their demonstration, *Wandrey and Renken* switched between helium (He) and a feed containing the combustion mixture (1.8 vol% C_3H_6 and 8.1 vol% O_2 in He). Using a symmetrical modulation with cycle periods between 12 and 20 s, these researchers investigated the influence of inlet feed temperature and cycle period on the time-average conversion of $C_3^{=}$ and on the temperature of the Pt screen.

The region of multiplicity for the mean mixture at an inlet temperature of 200°C lies between ignition/extinction boundaries at 230° and 370°C. Alternating between a diluent and a reaction mixture allows the system to operate at a time-average temperature in the multiplicity region. Choice of period determined the point of operation. The platinum gauze fluctuated between ignition when a reactant mixture flows across it and extinction when it sees only the He diluent. The maximum gauze temperature increased with cycle period and caused an increase in the time average temperature. The period effect arises from a gauze temperature lag with respect to ignition caused by the thermal inertia of the gauze and the reactor walls. Thus, at a period of 12 s, modulation caused a small oscillation around either the ignition or extinction stable states depending on which steady state occurred prior to the initiation of cycling. A small increase of the period to 13.6 s brought the system into the multiplicity region. Above 20 s, the system moves out of the unstable region and oscillates with a large amplitude around the high-temperature stable state. Using steady-state kinetics of propylene combustion and assuming the gauze behaves as a CSTR, *Wandrey and Renken* simulated this behavior. *Silveston (1998)* provides more detail.

Propylene and propane combustion under modulation of the air/fuel ratio were studied by the Toyota Central Research and Development Laboratories as part of a program to understand the effect of air/fuel ratio modulation on automotive emission catalysis (*Shinjoh et al. 1987, 1989*). Measurements were made for each of the noble metal catalysts used in commercial three-way catalytic (TWC) mufflers. Catalyst support was α-Al_2O_3, a low surface area alumina. This support was chosen to minimize support-catalyst interactions so the catalysts were not typical of those employed commercially. The metal loading on the support was approximately the total metal loading used in TWCs. Temperature and cycle period were the independent variables. Feed compositions typified those in rich and lean fuel mixtures. Both hydrocarbon and O_2 were varied in a symmetrical cycle ($s = 0.5$). Results showed that composition modulation increased conversion of the fuel for Pt and Rh but not for the Pd catalyst. The cycle period for maximum conversion depended on temperature as can be seen in **Figure 5-2**, a plot of time-average conversion vs. cycle period for different temperatures. Steady-state data for the time-average feed composition are given by the point at a cycle period of 0. Consequently, at 330°C, the cycle period would have to be less than 1 s for an improvement to be seen.

Measurements of propane oxidation for the Pt catalyst showed that conversions were one-half to one-third lower than for propylene. It seems likely that oxidative dehydrogenation to propylene is the first step in propane oxidation over Pt. Also, C_3H_8 is probably less strongly adsorbed than C_3H_6 so that at lower

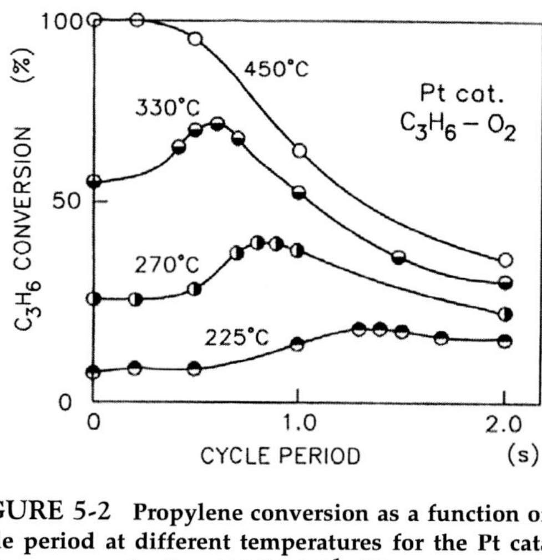

FIGURE 5-2 **Propylene conversion as a function of the cycle period at different temperatures for the Pt catalyst with a space velocity of 30,000 h^{-1}.** *(Figure reproduced from* Shinjoh et al. (1989) *with permission. © 1989, Elsevier Science Publishers B.V.)*

temperatures adsorbed oxygen dominated the surface.

Transient concentrations of reactant and product species after a composition switch are given in **Figure 5-3**. This figure shows a short drop followed by a relatively slow rise in O_2 leaving the catalyst bed after a change from a $C_3^=$/He to an O_2/He feed, whereas O_2 disappears from the outlet instantaneously after the $C_3^=$/He feed is re-introduced. The CO_2 maximum appears about 2 s after O_2 is injected into the He carrier, roughly at the point where O_2 shows a minimum. The CO_2 peak in the gas leaving the catalyst bed appears almost instantaneously after C_3H_6 is re-introduced, reaches a maximum and vanishes after about 3 s. Steady-state kinetic measurements indicated

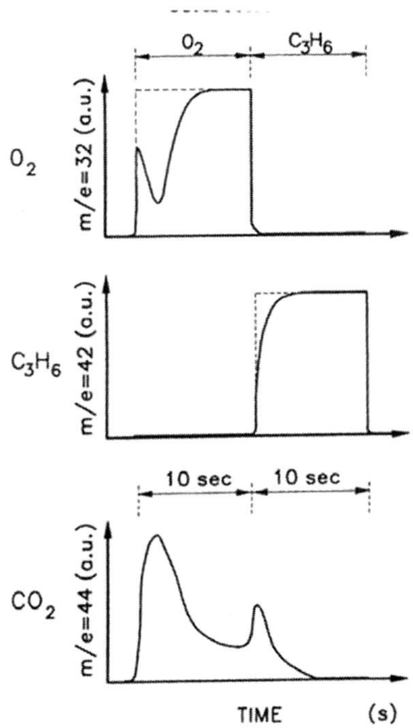

FIGURE 5-3 **Transients for O_2, C_3H_6 and CO_2 under bang-bang switching between O_2 and C_3H_6 diluted by He streams flowing over a Pt/α-Al_2O_3 catalyst at 160°C.** *(Figure reproduced from* Shinjoh et al. (1989) *with permission. © 1989, Elsevier Science Publishers B.V.)*

a negative order of reaction for propylene on Pt, while the order is positive for oxygen. These observations mean that the hydrocarbon or its partially oxygenated intermediates are more strongly adsorbed on the metal than O_2. The complex and slow rise in O_2 partial pressure shown in the figure indicates oxidation of adsorbed C_3H_6 or hydrocarbon fragments on the surface. The smaller CO_2 peak on C_3H_6 injection may reflect either stoichiometry or that the propylene must adsorb first before reacting with adsorbed oxygen. The optimal cycle periods indicated in **Figure 5-2** may result from the time needed for O_2 to scavenge adsorbed hydrocarbons or their fragments from the catalyst surface. *Shinjoh et al. (1989)* observed that under steady-state operation the noble metal catalysts are reactant-inhibited, suggesting that the mechanism for the improvement through periodic operation is to balance the adsorbate concentrations on the surface.

Low temperature combustion of methane has attracted attention because of CH_4's importance as a greenhouse gas and its presence at low concentrations in industrial emissions, flushing gas discharge from fire prevention systems and, of course, from natural sources (swamp gas). These streams usually have quite low calorific value so it is costly to heat the streams to temperatures required for ignition. Noble metals are often chosen for low temperature combustion of methane. *Carlsson et al. (2007)* compared the performance of Pd and Pt catalysts using the same support and weight of metal under oxygen pulsing in a differential reactor at 623 K (**Table 5-1**). They observed that O_2 pulsing enhances the activity of both catalysts, but the effect was greatest for the Pd catalyst. *Carlsson's* interpretation is that bulk oxides form with Pd, but only surface oxides arise with Pt. These latter oxides may partially deactivate the Pt surface.

Becker et al. (2009) report on the effect of oxygen pulsing on CH_4 combustion for two common catalysts at 623 K. For the alumina supported catalyst, cracking predominated prior to and during the O_2 pulse as **Figure 5-4** shows. H_2 produced was burnt by O_2 in the pulse and some of the deposited carbon was gasified. Almost no CH_4 combusted. On the other hand, carbon burn-off, H_2 and CH_4 combustion all took place for the ceria supported catalyst. Cracking occurred when O_2 was absent from the feed. Under identical operating conditions, Pt/CeO_2 was a far superior CH_4 combustion catalyst, as the comparison of the two catalysts in **Figure 5-4** demonstrates.

5.2.2 Fuels Combustion

The explosion of interest in "chemical looping," has overlooked another periodic alternative employing a fixed bed with two or more partial cycles of, usually, different durations. Use of a fixed bed substantially reduces the problem of attrition and the associated difficulties of fines separation and continual catalyst replacement. Gas mixing is reduced so there would be a closer approach to plug flow implying a much shorter reactor. Less purge gas would be required, if purging between oxidative and reducing steps is necessary. Perhaps of most importance, the power cost of circulating fine solids is avoided. Of course, the use of a fixed bed demands the use of a robust, durable oxygen carrier.

The periodically operated packed bed is inexpensive and quickly constructed so it could be used to screen oxygen carrying solids or

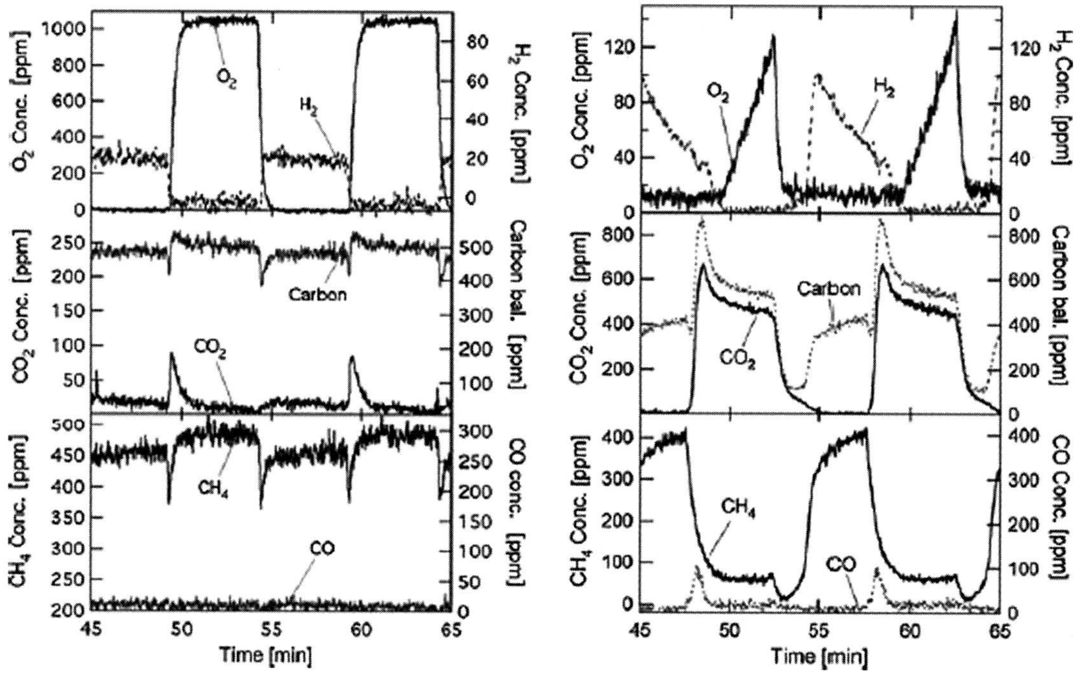

FIGURE 5-4 **Comparison of effluent composition with time under periodic 5 min pulsing of 1250 ppm (vol.) O_2 in Ar into a stream of 500 ppm (vol.) CH_4 in Ar at 623 K over a 1 wt% Pt/Al_2O_3 on the left and a 1 wt% Pt/CeO_2 on the right.** *(Figure reproduced from* Becker et al. (2009) *with permission. © 2009 by Springer Science Media LLC.)*

catalysts for looping experiments. *Noorman et al. (2007)* seemed to have a looping alternative as well as a carrier screening system in mind when they proposed and built such a system. Laboratory-scale experiments demonstrating performance are given in further publications (*Noorman et al., 2010a, b, 2011*) and are summarized in **Table 5-1** for a CuO and for a perovskite oxygen carrier. Experiments were undertaken in a 30 mm i.d. × 1.5 m steel tube. Reactor layout is shown in **Figure 5-5**. Closely spaced thermocouples measured the temporal evolution of the temperature distribution. Thermogravimetric experiments showed that CuO could be reduced by CH_4 to CuO_{1-y} and then fully reoxidized (*Noorman et al., 2010b*). Some CH_4 decomposition accompanied the reduction of the oxide.

Experiments were conducted for a gas entering temperature of 923 K. Heat was generated wholly in the carrier re-oxidation portion of the two-part cycle. Temperature fronts travelled in the flow direction across the bed. These are shown in **Figure 5-6** for CuO re-oxidation. Early in re-oxidation the front of the bed heats up, but after the reaction front has passed through, cooling of that part of the CuO bed sets in as a moving front as shown in parts **(a)** and **(b)** of the figure. Temperature rise during re-oxidation reached almost 200 K. Velocities were calculated as 12×10^{-3} m/s for reaction and as 1.2×10^{-3} m/s for the much slower moving temperature front. These values agreed with model prediction when carbon deposition was reproduced into consideration. If the bed is restored to its initial 923 K temperature, the oxide reduction step with combustion of CH_4 is also modestly exothermic and raises the bed temperature by about 50 K. A further paper (*Noorman et al., 2011*) examined another catalyst, a Ti stabilized $CaMnO_3$, and operational parameters, such as feed concentration, O_2 content in regeneration, feed and reoxidation temperatures and the addition of steam to suppress cracking. Steam addition was effective, strongly retarding the appearance of CH_4 in the gas leaving the oxide bed and causing the formation of CO_2 to continue throughout the reduction cycle as can be seen in **Figure 5-7(a)**. H_2 production climbs rapidly in the latter part of the step as the reduction of CuO proceeds towards completion. Cu_2O or possibly copper metal now catalyzes the steam reforming of CH_4. Effluent composition during re-oxidation of the catalyst with air appears in **(b)** of the figure with and without the addition of steam. The absence of carbon oxides when steam was added to the feed demonstrates that CH_4 decomposition was totally suppressed.

The perovskite catalyst, also considered by *Leion et al. (2009a)* as a carrier for chemical looping, was explored by *Noorman et al. (2011)* using the reactor shown in **Figure 5-5** and in most of the experiments undertaken for the CuO

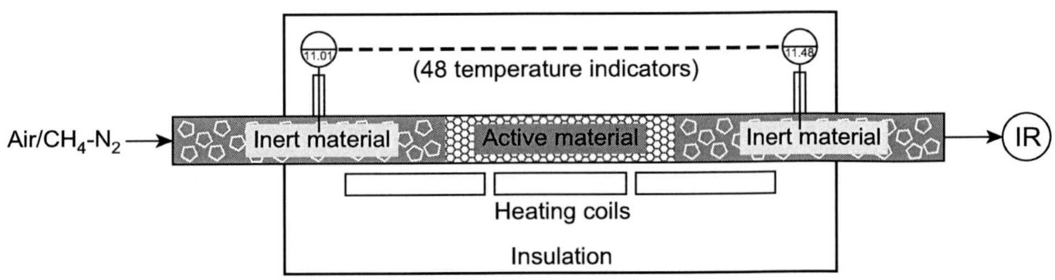

FIGURE 5-5 **Schematic of the experimental, periodically operated, packed bed reactor used by *Noorman* and co-workers to study methane combustion with metal oxide oxygen carriers.** *(Figure reproduced from* Noorman et al. (2010a) *with permission. © 2010 by Elsevier Ltd.)*

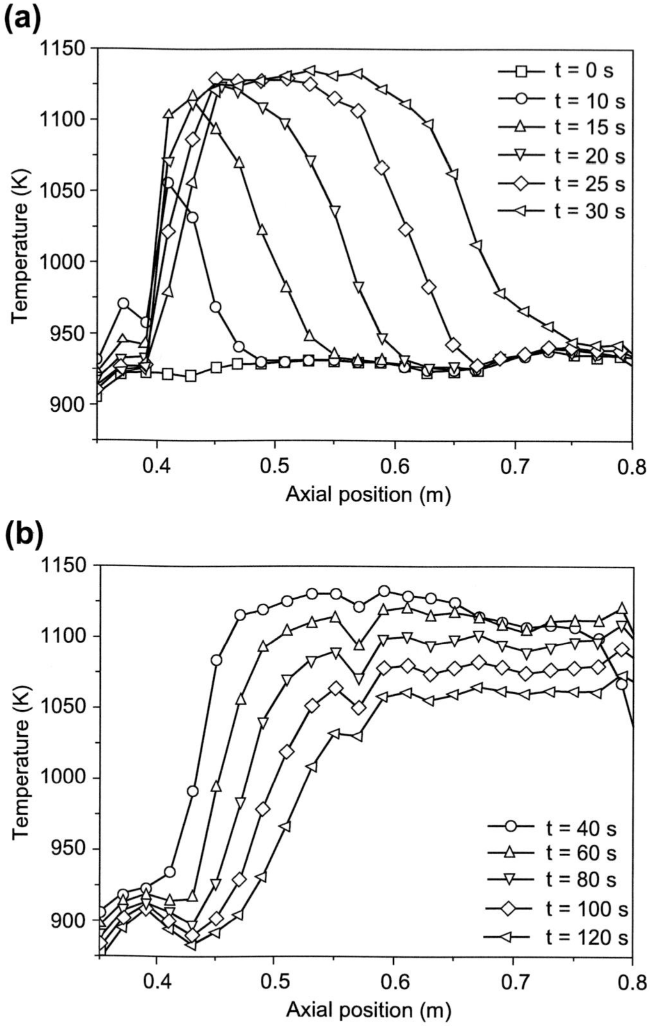

FIGURE 5-6 Spatiotemporal temperature distribution in an experimental periodically operated fixed bed packed with a CuO/Al_2O_3 oxygen carrier during the re-oxidation partial cycle using air entering at 923 K. In (a) re-oxidation of the carrier has begun, by (b) re-oxidation is largely complete. *(Figure reproduced from* Noorman et al. (2010a) *with permission. © 2010 by Elsevier Ltd.)*

material. Although the Ti stabilized $CaMnO_3$ exhibited a much greater oxygen capacity per unit weight than CuO, the oxide had poor stability and showed low as well as erratic reactivity at the temperatures used.

Noorman et al. concluded from their study that neither of the two catalysts tested were attractive: The CuO catalyst because of the low melting point of the oxide that limits the possible reactor temperatures and the perovskite oxide because of stability and reactivity considerations. These investigators found that, in principle, periodic operation of a fixed bed of a metal oxide for anaerobic combustion

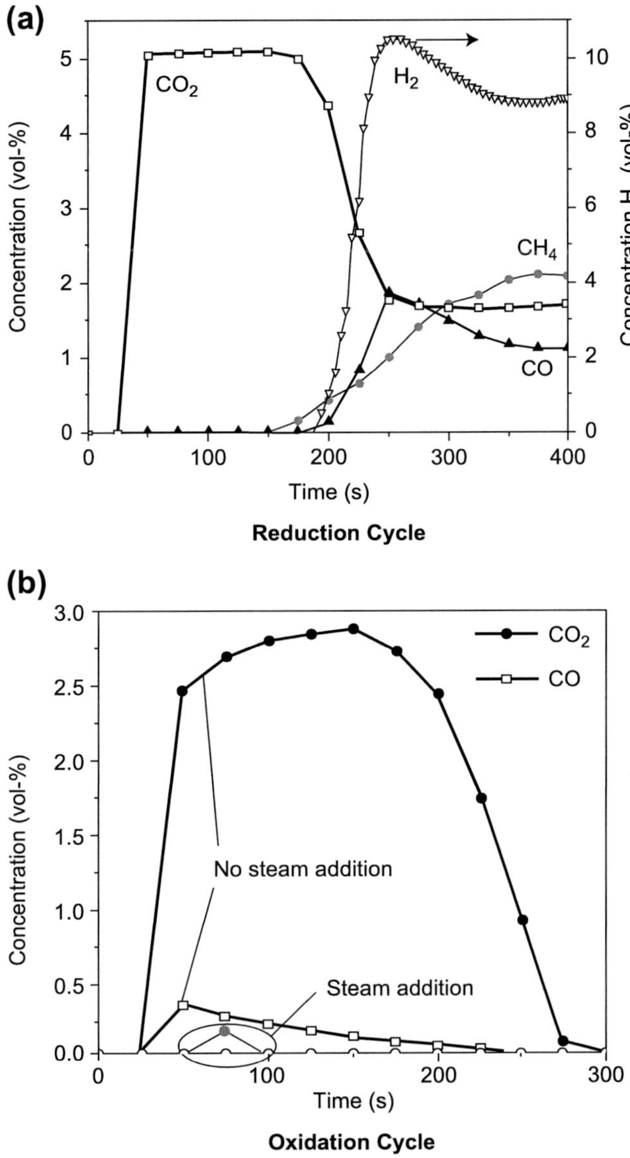

FIGURE 5-7 **Comparison of the reactor effluents for the anaerobic combustion of 4 vol% CH_4 mixed with N_2 in a fixed bed packed with CuO/Al_2O_3 particles in the presence and absence of steam in the feed at CH_4:Steam = 1:4. Bed is fully oxidized and feed enters at 923 K. Temperature rise in the reactor is ca. 40 K. The upper figure, (a), shows effluents in the reduction half cycle, while the lower figure, (b), shows these in the oxidation half cycle.** *(Figure reproduced from* Noorman et al. (2011) *with permission. © 2011 by the American Chemical Society.)*

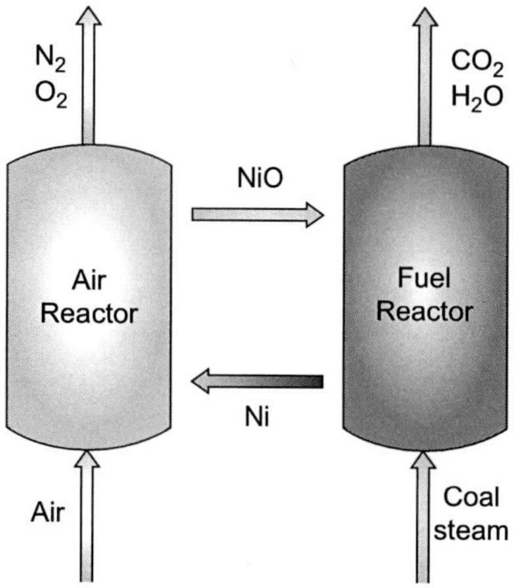

FIGURE 5-8 Schematic of looping combustion or chemical looping using an oxide (NiO) as an oxygen carrier. Both reactors are run at high temperature and pressure so that the off-gas streams can be used in turbines for power generation. *(Figure reproduced from* Gao et al. (2008b) *with permission. © 2008 by the American Chemical Society.)*

with generation of an off-gas of high CO_2 content was a feasible concept.

5.3 LOOPING COMBUSTION

Looping combustion is now rather loosely applied to two quite different reactor systems: 1) Those in which the catalyst moves as either a fluidized or falling bed between oxidizing and reducing environments, as discussed in the preceding section, and 2) a system of multiple fixed beds in which a reaction zone rotates through the beds forced by periodically changing the entrance and take off locations. As mentioned in the introduction to this chapter, the first system, usually referred to as chemical looping, is undergoing study internationally. It is illustrated by **Figure 5-8**. We defer discussion of chemical looping until **Chapter 8**. Development of the second system, the loop reactor, has not advanced beyond conceptual studies utilizing numerical simulation.

5.4 SIMULATED LOOP REACTORS

As discussed in previous chapters, catalyst attrition, fine particulates generation and the operational problems that arise from particle movement in beds and pipes can be avoided by sectioning the catalyst bed with multiple inlet and withdrawal valving and, periodically as well as sequentially, switching inlet and takeoff locations to create a simulated moving bed. Since the 1980s, this approach has been pursued for combustion applications (See **Table 5-2**). Systems containing loops of just three reactors have been considered for the most part. **Figure 5-9** shows the operation of the three-bed system. Injection of cold reactants at the inlet to a bed of hot solids, say "1" in the figure, creates a reaction front by heating the reactants up to reaction temperature through heat transfer, cooling thereby the hot catalytic solids. Ignition takes place and the high temperature front created moves through the bed forced by the continued addition of fuel and oxidant. Behind the front, catalyst is cooled by the incoming cold reactants. Once the hot front reaches reactor "2," inlet and outlet shift one reactor forward and so on. The result of the periodic shift in inlet and outlet is at least one hot reaction zone that rotates through the loop. Exhaust leaves the loop after passing through "cold" packing and thus takes rather little heat from the reactor.

The loop reactor arose from the early research on reverse flow reactors (*Matros, 1985, 1989*). These reactors are discussed in **Chapter 18**. Just as for periodic flow reversal, the attraction of loop reactors is for detoxification of gas streams of low calorific value. *Haynes and Caram (1994)* carried out a numerical simulation of a two-bed loop, showing that the performance closely approached that of a packed bed operated under periodic reversal of the flow direction through the bed. Both reversible and irreversible reactions were considered. They observed that there is just a narrow window of

TABLE 5-2 Loop Combustion Systems

Authors	Study Objective	System Considered	Observations	Comments
Matros (1985, 1989)	Exploration of alternatives to periodic reversal of flow direction	Array of reactors connected serially in a closed loop with T junctions to allow feed or product withdrawal	Avoids the discharge of unreacted material (washout) at a reactor entrance on flow direction reversal.	
Haynes and Caram (1994)	Investigation of the operation and performance of a loop reactor	2-reactor loop with irreversible and reversible, exothermic reactions	Performance is very similar to that of a periodic flow reversal reactor. Range of stable operating periods was very limited.	Numerical simulation.
Brinkmann et al. (1999)	Investigation of control and performance of a loop reactor	3-reactor loop with each reactor having inert and active packing. Different switching strategies were considered.	A two-point strategy for switching worked best. After start-up, the second point controls switching.	Autothermal operation achieved for the model parameters considered.
Fissore and Barresi (2002)	Comparison of a loop reactor with a periodic flow reversal reactor	2- and 3-reactor loops, with reactors packed partially with inert solids	Demonstrated a very narrow range of τ_{switch} for stable operation in comparison with a periodic flow reversal reactor.	Range of τ_{switch} for stable operation increases with the number of reactors in the loop and with $\Delta T_{adiabatic}$ of the fuel mixture.
Russo et al. (2002)	Investigation of the stability of 2 and 3 bed loop reactors	Exothermic and irreversible reactions, primarily for 3-reactor loops	Spatiotemporal symmetry dominates bifurcation behavior. Reactor model is not important.	
Sheintuch and Nekhamkina (2004)	Comparison of a loop reactor with a periodic flow reversal reactor and with a steady-state internal recirculation reactor	4- to 8-reactor loops packed with inert solids in front of and after the bed of catalyst with start-up at uniform temperature	Slow switching, when the switching velocity roughly equals the front velocity, provides high conversion and bed temperature, but τ_{switch} for stable operation is very limited.	Fast switching alternative greatly enlarges range of τ_{switch} for stable operation; maximum bed temperatures depend on kinetic parameters and bed length.

(*Continued*)

TABLE 5-2 Loop Combustion Systems (*cont'd*)

Authors	Study Objective	System Considered	Observations	Comments
Sheintuch and Nekhamkina (2005)	Operational analysis of loop reactors using asymptotic pseudo-homogeneous models	N-bed loop reactor for VOC combustion	N-bed asymptotic models closely represent combustion for $N > 4$ beds. Properties of slow switching and fast switching examined and the effect of loop operating variables on performance and stability were observed.	Intermediate region of switching velocity exhibits some stable operation but results in complex spatiotemporal behavior, often with multiple moving fronts in a reactor bed.
Altimari et al. (2006)	Bifurcation analysis of a 3-bed loop reactor considered τ_{switch}, feed temperature and heat loss	3-bed reactor loops for reactors with inert sections	Identified the existence of asymmetric regimes in which reactors in the loop display different spatiotemporal patterns; feed temperature and heat loss strongly affect stability.	Continuation method used in place of an asymptotic model. Authors comment that a pseudo-homogeneous reactor model and large loops do not show the complex dynamics possible with 2- and 3-reactor loops.
Continillo et al. (2006)	Birfucation analysis of a 2-bed loop reactor considered τ_{switch} and catalyst heat capacity	2-reactor loop without inert sections for VOC combustion	Region of stable operating states found for slow cycling bounded by reaction front blowout and the N_{Lewis}. Extent of τ_{switch} stable region affected by catalyst heat capacity.	Applied pseudo-arc length continuation method to reduce model dimension and applied parallel computing for integration of system model.
Mancusi et al. (2007)	Examination of the effect of switching strategy on stability	3-reactor loop with an irreversible exothermic reaction. Non-adiabatic systems considered	Switching inlet to previous reactor instead of next reactor affects spatiotemporal temperature patterns in the loop as well as stability.	Strategy of switching to the previous reactor doubles the range of τ_{switch} for stable operation, but strategy has little effect on the extent of conversion and maximum bed temperature.

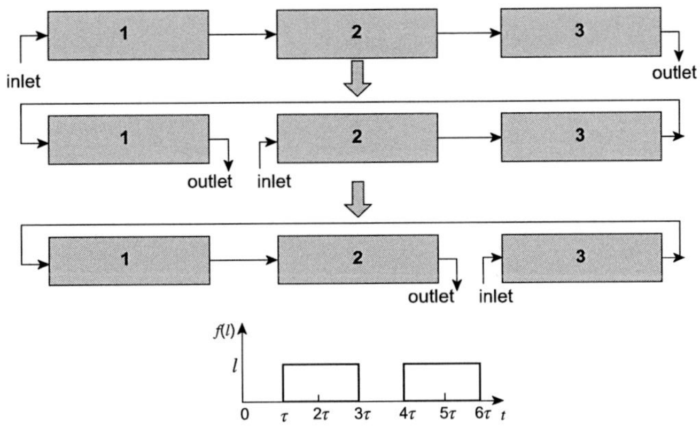

FIGURE 5-9 Three bed simulated loop reactor showing the 3-reactant inlet-product outlet connections for each portion of a cycle. *(Figure reproduced from* Altimari et al. (2006) *with permission. © 2006 by the American Institute of Chemical Engineers.)*

stable periodic operation (τ_{switch}) that depends on the thermodynamics of combustion, feed rates and feed properties, thermal properties of the catalyst and the inlet conditions. Like reactors operating with periodic flow reversal, the loop reactor was quite stable with respect to brief disturbances as *Matros (1989)* had shown for flow reversal.

Brinkmann et al. (1999) provided the first analysis of a three-bed system. A reactor bed consisting of an inert packing followed by a catalytically active packing was investigated. The researchers assumed that all three fixed beds were brought to a constant temperature sufficient to ignite the fuel-oxidant mixture through external heating. After start-up, to maintain the operation, temperature set points controlled switching of the feed and draw-off locations. The first of the two control points was located at the point where reactants enter the first bed (see **Figure 5-9**), and the second was placed at the entrance to the second bed. Switching of the feed and draw-off location moves forward in the direction of flow when the temperature at the first point drops below the first set point (the bed start-up temperature) but exceeds the second set point which depends on the adiabatic temperature rise ($\Delta T_{adiabatic}$) for the combustion mixture. The switching thus initiated determines τ_{switch} for the 3-reactor loop. Changes in $\Delta T_{adiabatic}$ or a loss of catalyst activity instigates an adjustment of τ_{switch}. *Brinkmann et al.* show through simulation that the loop operated autothermally when disturbances were absent.

A set switching time was considered by *Fissore and Barresi (2002)* in their comparison of loop reactors with a periodic flow reversal reactor for combustion of VOC gases. They observed that performance with respect to VOC conversion and the pseudo-steady state temperature profile in a reactor bed were similar and the effect of variation of bed length, fraction of inert packing in a two-part bed, flow rate of the combustion mixture and fuel concentration in the latter was about the same for both modes of operation. The largest difference between the modes was in the range of τ_{switch} for a stable operation. This range was very much narrower for the loop reactor.

Limited experimental work was reported by *Sheintuch's* laboratory (*Ben Tullilah et al., 2003a*) on periodic flow reversal and steady-state internal recirculation reactors, alternatives to loop reactors, but not on loop reactors themselves. In a follow-up, numerical study, *Sheintuch and Nekhamkina (2004)* examined loop reactors with four to eight separate beds. These researchers observed that such a number of beds represented an N bed loop. Two operating regimes were identified: A slow cycling one in which the switching velocity, defined as the bed length divided by τ_{switch}, roughly matched the velocity of the moving temperature front, and a fast cycling regime. The slow cycling regime offered essentially complete combustion and a maximum bed temperature close to that for periodic flow reversal.

In a second contribution, Sheintuch and Nekhamkina (2005) derived an N-reactor loop, asymptotic model that can be used for bifurcation analysis. They showed through simulation that their asymptotic model closely

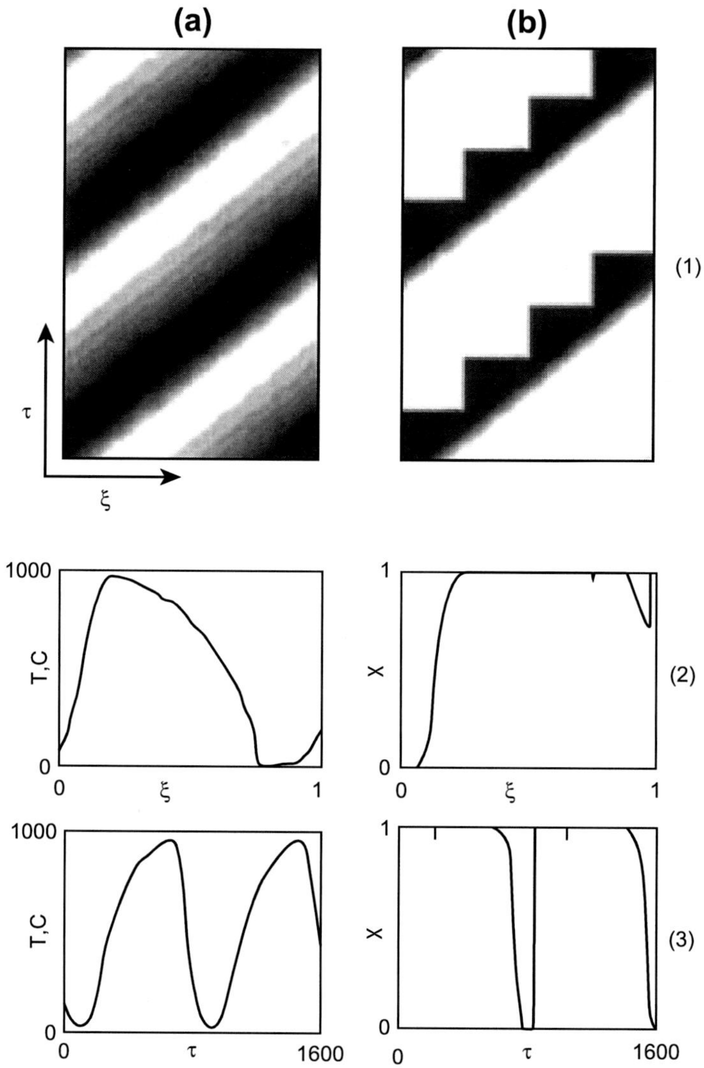

FIGURE 5-10 Stable spatiotemporal patterns of temperature and fuel conversion in a bed of an N-bed loop reactor for VOC combustion. In the upper row (1), clear is start-up bed temperature while black indicates maximum temperature, ≅ 1000°C; likewise clear = 0 conversion and black = complete conversion. Row (2) shows the bed temperature (°C) and fuel conversion as a function of dimensionless position, while row (3) shows snapshots of temperature and conversion at a point within the bed as time evolves. *(Figure reproduced from* Sheintuch and Nekhamkina (2005) *with permission. © 2005 by the American Institute of Chemical Engineers.)*

represented loop performance, fuel conversion and temporal temperature patterns in a loop for $N > 4$ and provided a useful approximation for $N > 2$. The asymptotic model predicted a stable rotating temperature in the loop for a slow switching velocity, v_{switch}, greater than the velocity of the reaction front in a fixed bed and less than the velocity of the thermal front. This stability requirement can be expressed as:

$$v_{switch} << Le^{-1} << 1 \qquad (5\text{-}1)$$

where *Le* is the Lewis Number. **Figure 5-10** shows the progression of a high temperature wave through a four-bed loop. The top and bottom figures in the second column show virtually complete conversion as the fuel streams through the bed. The figure assumes fuel entering at 566 K and a concentration of 1.21×10^{-4} kmol/m^3 with a velocity of 1 m/s. $\Delta T_{adiabatic}$ of the feed was just 50 K.

Stable rotating temperature waves arise for fast switching but only for sets of model parameters that are within the neighborhood of a Hopf bifurcation of the asymptotic model. *Sheintuch and Nekhamkina* found that unstable and complex rotating temperature waves will appear in loop reactors with switching times in between the fast and slow ranges.

The study of loop reactors has been dominated during the last decade by a research team led by *Crescitelli* of the University of Naples (Italy) and a large and growing literature has been created. Switching strategy was addressed by *Mancusi et al. (2007)*. Bifurcation analyses were provided by *Russo et al. (2002, 2006)*, *Altimari et al. (2006)* and *Continillo et al. (2006)*. The last-named study used a heterogeneous model and examined operating states possible as τ_{switch}, T_0 and heat loss from the reactor network are varied. Generally, a rotating temperature and conversion wave travels through the network with identical spatiotemporal behavior in each reactor. This behavior was lost when the switching velocity equals the front velocity at a large τ_{switch} and also at a small τ_{switch} when the switching velocity is fast. At low switching times, multi-period regimes occur, that is, two or more thermal waves travel through a bed between switches of the inlet location. At a very slow switching velocity, an unstable operation exists with a single temperature-conversion wave that moves through a bed (*Altimari et al., 2006*). *Continillo et al. (2006)*, treating a two-bed loop, observed a similar behavior. They found also that the region of stable operation depends on the thermal properties of the catalytic solids, primarily the heat capacity.

A literature has also developed around the application of loop reactors to reactions other than combustion. See, for example, *Vanden Busche and Froment (1996)* and *Velardi and Barresi (2002)* who applied the reactor to methanol synthesis. The latter team suggested the loop reactor is worth examining for many equilibrium-limited catalytic reactions.

Nomenclature

CSTR	=	completely stirred tank reactor or fully back mixed reactor
d_p	=	paricle diameter (mm)
GHSV	=	gas hourly space velocity (h^{-1})
Hz	=	Herz, (measure of frequency)
i.d.	=	inner diameter (cm)
Le	=	Lewis number
N	=	number of beds or reactors connected in a loop
$\mathbf{N_{Lewis}}$	=	Lewis number
SN	=	Stoichiometric number (ratio of the moles of reactant fed or used to the moles needed to complete the reaction)
SV	=	space velocity (s^{-1})
s	=	cycle split (= ratio of the duration of a part of a cycle to the cycle period)
T	=	temperature (K)
T_0	=	initial or entrance temperature
TWC	=	three-way catalyst
t	=	time (s, h)
VOC	=	volatile organic carbon (pollutant)
v	=	velocity (m/s)
v_{switch}	=	switching velocity

Greek

$\Delta T_{adiabatic}$ = adiabatic temperature rise (°C)
τ, τ_{cycle} = cycle period (s, h)
τ_{switch} = switching or cycle period
ξ = normalized axial position (-)

Subscripts

p = particle or pellet
0 = inlet or initial

CHAPTER

6

Automotive Exhaust Catalysis

Peter Lewis Silveston, William S. Epling
Waterloo, Ontario, Canada

OUTLINE

6.1 Internal Combustion Engines 142
6.1.1 Engine Behavior, Exhaust Condition and the Catalytic Converter System 142
6.1.2 Experimental Systems 146

6.2 Modulation of Detoxification Reactions 146

6.3 Modeling Studies 155

6.4 Studies on Modulating Automotive Exhaust 158

6.5 Effect of A/F Modulation on Poisoning and Sintering 160

6.6 Effects of Irregular A/F Variation 161

6.7 Lean Burn Spark-Ignited Engines 162

6.8 Application of NSR to Diesel Exhausts 168

6.9 Does A/F Modulation Improve Converter Performance? 169

Automotive engines generally operate on a four-step cycle, and the rapid adjustment of the air-fuel ratios imposed upon engine cylinders results in relatively high frequency composition variations in the engine exhaust. The task of catalytic devices now mounted on almost all vehicles is to convert the noxious components in the engine exhaust to less harmful gases. As a consequence of the enormous numbers of vehicles operating, automotive catalytic converters form by far the largest group of periodically operating catalytic systems.

Because of its environmental and economic importance, a very large literature has built up on automotive catalytic converters. A part of this deals with converters exposed to a periodically varying feed composition. *Silveston (1998)* devoted a chapter to the subject in "*Composition Modulation of Catalytic Reactors*" that summarized the source and nature of automotive exhaust composition variation and reviewed the development and performance of three-way catalysts (TWC). Much information can also be found in books or reviews (*Wei, 1975; Cooper et al., 1987; Cybulski and Moulijn, 1994*) and symposia collections (*Crucq and Frennet, 1987; Crucq, 1991*) or in collections of technical papers published by the Society of Automotive

Engineers (SAE). Recent review papers (*Twigg, 2007; Eigenberger et al., 2007*) discuss the current development of emission control systems. Special attention is given to diesel or lean burn systems, where NO_x control is a key concern. Reviews by *Epling et al. (2004)* and *Roy and Baiker (2009)* discuss this subject.

Since the 1990s, research on automotive exhaust catalysis has moved from the development of TWCs, their support, and the feedback control systems to oxygen storage and release, control of sintering of platinum group materials, NO_x storage and reduction, and more recently to the refinement of the nitrogen storage and reduction catalysts and their accompanying control systems. *Matsumoto and Shinjoh*, important pioneers of TWC research, have published a concise summary of reduction catalysts (*Matsumoto* and *Shinjoh, 2007*). Detoxification of diesel exhaust has become a dominant R & D focus during the last two decades.

In addition to the discussion in this chapter, individual reactions taking place in catalytic converters are discussed in earlier chapters: Hydrocarbon oxidation in **Chapters 2** and **4**, CO oxidation and NO reduction in **Chapter 3**. Discussion in this chapter is restricted to catalysts employing noble metals. There has been a continuing effort to explore catalysts free of these metals, but such catalysts (see, for example, *Stegenga et al., 1991*) have not yet advanced out of the laboratory, even though their response to *A/F* modulation has been studied.

6.1 INTERNAL COMBUSTION ENGINES

Several types of combustion engines are currently in use. The most common of these is the Otto engine which uses a four-stroke cycle and injects fuel into the engine cylinder that vaporizes and is ignited by a spark. The engine operates at close to a stoichiometric mixture of fuel and air. A two-stroke version of this engine is used in motorcycles and in boating. For several years, a lean burn version of the Otto engine has been available to improve fuel economy. Such engines operate with excess combustion air.

Diesel engines also have four- and two-stroke versions with the former universally used for vehicles. These introduce a mostly non-volatile fuel through an atomizing nozzle. The engine operates with excess air. The fuel composition and combustion conditions differ significantly in the two systems. The conventional Otto engine will be discussed first in what follows.

6.1.1 Engine Behavior, Exhaust Condition and the Catalytic Converter System

Reactions occurring during combustion of a fuel depend on the air fuel ratio (*A/F*), temperature and fuel composition. Once start-up is complete, the *A/F* governs vehicle emissions. **Figure 6-1** shows how the percentage or ppm levels of the major exhaust components change with *A/F*. It is clear from this figure that *A/F* control alone is insufficient for the detoxification of vehicle exhaust. An $A/F > 16$ brings CO to well below 1% by volume, but a ratio < 12 is needed to bring NO_x to that level.

Moreover, the *A/F* varies with vehicle acceleration or deceleration. Of course, emissions depend on engine temperature and change during acceleration, idling or start-up. As a consequence, a catalyst is necessary to meet the vehicle emissions restrictions now enforced in almost all industrialized nations. A large R & D effort by the automotive industry during the 1970s resulted in three-way catalysts (TWC) that were introduced in Europe and the US along with exhaust gas recirculation in 1977. Design of that catalytic system had to meet requirements of low pressure drop, highest possible surface area, ability to withstand high and time-varying temperatures and

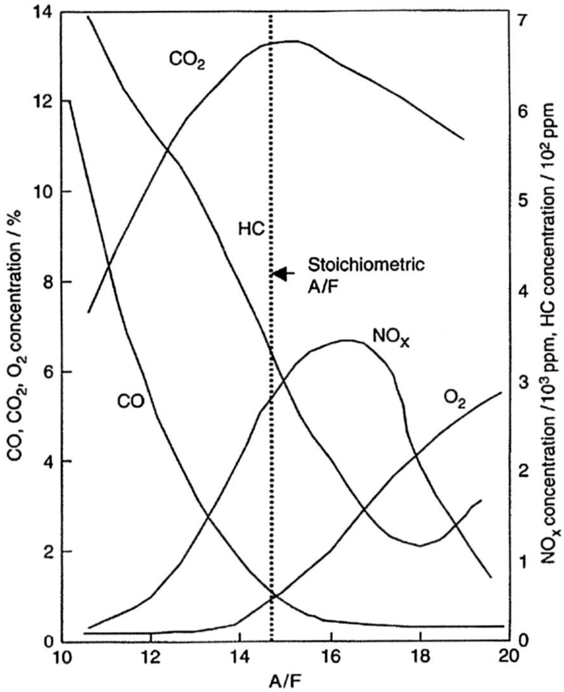

FIGURE 6-1 Variation of CO, CO_2, hydrocarbon, O_2 and NO_x concentrations in an automotive exhaust with the *A/F* ratio. *(Figure adapted from* Matsumoto and Shinjoh (2007) *with permission. © 2007 by Elsevier, Inc.)*

a low content of noble metals. TWC catalysts are now based on Pt or Pd with significant amounts of Rh. They are promoted by cerium and/or barium oxide and alkaline metal oxides such as K_2O. These active components are supported on an alumina wash-coated onto cordierite ($2MgO \cdot 2Al_2O_3 \cdot 5SiO_2$) or, in a few cases, a metal sheet. The active metals and promoters are generally impregnated as dilute metal salts onto a high surface area alumina, γ-Al_2O_3. The alumina is wash-coated, some hundred microns thick, onto a mechanically strong Cordierite ($2MgO_2 \cdot Al_2O_3 \cdot 5SiO_2$) substrate. **Figure 6-2** shows a schematic of a typical TWC structure for vehicle use.

Roles of the noble metals are well established: Pt catalyses CO and HC oxidation. It is required to oxidize paraffins; Pd is a more active CO, olefin and oxygenate catalyst than Pt, but it has poor tolerance for poisons, while Rh improves the NO reducing ability of the catalyst. Rh is also an active catalyst for other important reactions, including the water gas shift reaction. These metals are most active in a reduced state. Cerium oxide or barium oxide is incorporated into the catalyst to provide oxygen storage, while base metals are added to increase stability.

Vehicle emissions are set through an *A/F* ratio controller that is generally operated at $A/F = 14.6$ to 14.7 (as weight), which is about the range of the stoichiometric ratio for most automotive fuels. Flow rate increases sharply on acceleration and somewhat less with deceleration. Both the *A/F* ratio and the space velocity changes force excursions in the CO emissions when the $A/F < 14.5$ and for NO_x when $A/F > 16$. It can be seen from **Figure 6-1** that the catalytic converter operates on just 0.5 to 2% of the exhaust gas. Typical gas conditions entering the converter

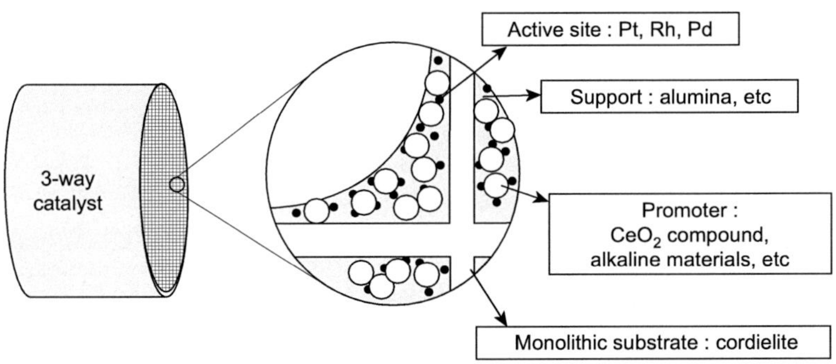

FIGURE 6-2 Schematic of a TWC monolith element in a vehicular catalyst module showing the composition and the coating of the active catalyst phase on the monolith support. *(Figure reproduced from* Matsumoto and Shinjoh (2007) *with permission. © 2007 by Elsevier, Inc.)*

are 750 K and 103 kPa. Flow rates depend on engine displacement, but average about 33,000 cm^3/s (STP) which for most converters works out to about an SV = 28,000 h^{-1} (*Taylor, 1984*). Temperature of the gas entering the catalyst bed depends on the *A/F* ratio increasing as the ratio decreases. Temperatures approach 1300 K, and SVs reach over 100,000 h^{-1} when a vehicle operates at full throttle, that is, with a rich mixture. Cyclic discharge of the engine cylinders means that both flow and temperature oscillate at a high frequency.

Three-way catalysts function efficiently only in a narrow range of exhaust gas compositions, that is the *A/F* ratio, as shown in **(c)** of **Figure 6-3**. The role of a controller is to maintain the ratio within this essential range. A schematic of the system appears in **Figure 6-3(a)**. A sensing device measures the oxygen partial pressure in the exhaust. The active element is a solid electrolyte, a doped zirconium oxide. The sensor signal is periodically monitored by the controller which initiates a change in the *A/F* ratio depending on the signal magnitude. This system maintains the *A/F* ratio close to its stoichiometric value under constant speed, cruising operation, and stays quite close to this value even when abrupt changes in motor rpm occur (see **(b)** in **Figure 6-3**). *Herz and Shinouskis (1985)* present data showing that the ratio is within 0.1 of the stoichiometric value for about 70% of the time for typical throttle variation once the motor-sensor-converter system has warmed up.

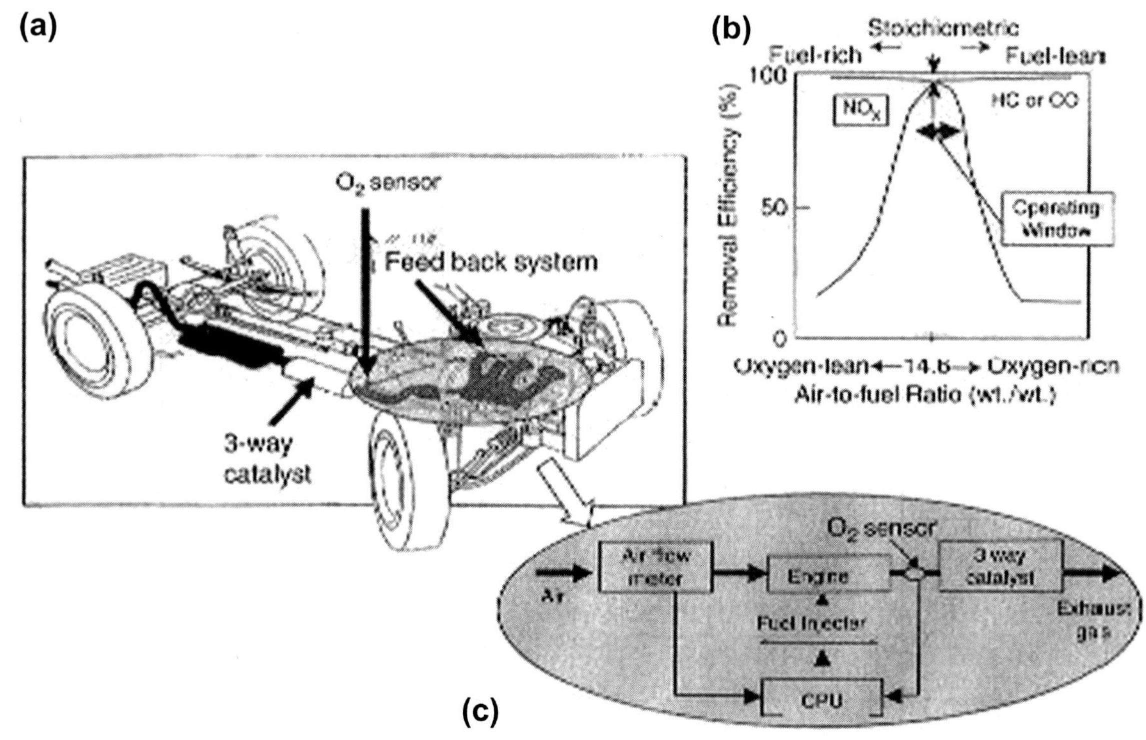

FIGURE 6-3 **Schematic of a catalytic converter, oxygen sensor and the *A/F* ratio feedback control loop system showing (a) component location, (b) schematic of signal flows, (c) controller operating range and toxic component removal efficiency in terms of the *A/F* ratio.** *(Adapted from* Matsumoto and Shinjoh (2007) *with permission. © 2007, Elsevier Inc.)*

Sensor hysteresis and measurement lags combined with lags due to the flow path, which change with acceleration/deceleration, result in a low amplitude oscillation of the *A/F* ratio even under constant motor speed. The frequency varies with the exhaust gas flow rate from 0.5 to 2 Hz, but typically holds at about 1 Hz.

The fluctuating *A/F* ratio results in time-varying concentrations at the converter inlet. Variations in the CO concentrations measured in situ by *Herz et al. (1983)* using IR spectrophotometry are shown in **Figure 6-4(a)**. The amplitude is about 0.5 for a time average concentration of 0.75 vol% but depends on the *A/F* ratio and motor operating conditions. Other data (*Herz, 1982, 1987; Herz et al., 1983*) show amplitudes as low as 0.05 vol%.

In addition to the regular ca. 1 Hz oscillations and the irregular oscillations at 0.01 to 0.1 Hz

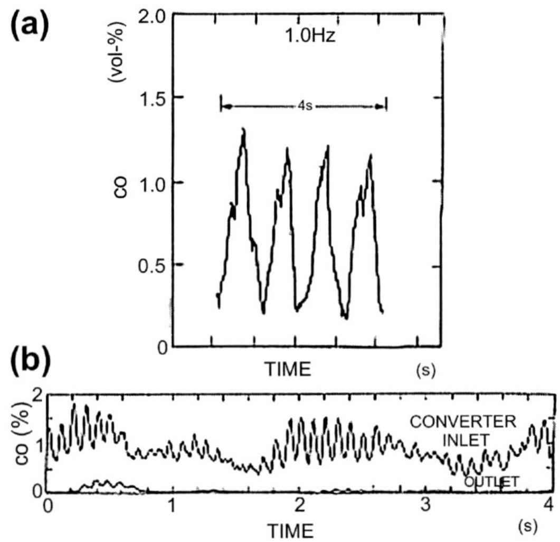

FIGURE 6-4 Time variation of CO concentrations in exhaust systems: (a) Modulation of CO in the exhaust of a dynamometer-mounted automotive engine by intentionally cycling the *A/F* ratio at 1 Hz to represent forcing by the feedback control loop, (b) time record of catalytic converter inlet and outlet CO concentrations for an exhaust from a dynamometer-mounted automotive engine under *A/F* ratio feedback control. *(Figures adapted from* Herz et al. (1983) *with permission. © 1983 by the American Chemical Society.)*

caused by variation in motor speed and throttle control, *Herz (1982)* mentions a higher frequency oscillation at about 10 Hz caused by fuel maldistribution and modulation in the fuel injection pump. This quite regular disturbance is shown in **Figure 6-4(b)** along with a ca. 0.7 Hz oscillation resulting from the *A/F* ratio, feedback control loop. Variation from the cyclic discharge of the engine cylinders appears as low amplitude noise in the upper trace. The significance of this figure is that the 10 Hz variation is damped out in the outlet CO concentration by the catalytic system; the 0.7 Hz signal is damped too, but it is still present in the outlet. Irregular disturbances in composition, flow rate and temperature occur at 0.01 to 0.1 Hz, but their amplitude is low so that their effect on converter performance can be neglected as can be seen along the bottom axis in **Figure 6-4(b)**. The figure also suggests that the regular 10 Hz modulation does not affect the converter performance so it too will be disregarded.

An early question posed in vehicular emissions control was whether or not fluctuations in emission composition and/or flow rates enhance or depress pollutant conversion. Later questions have dealt with conditions under which enhancement is observed, its magnitude and sources of improved performance. In recent years, the question has become how this unavoidable modulation affects the choice of catalyst or converter design.

Answers to the above questions have been sought through studies of: 1) Single component oxidation and reduction reactions, 2) groups of reactions, 3) simulated exhaust under carefully controlled conditions, and 4) the exhaust of test engines. A shortcoming of 2) is the extrapolation of results to an exhaust under actual operating conditions, while for 3) and 4) it is that these approaches often make observations difficult to explain. In this chapter, consideration begins with individual reactions, then to simulated and finally to actual exhausts.

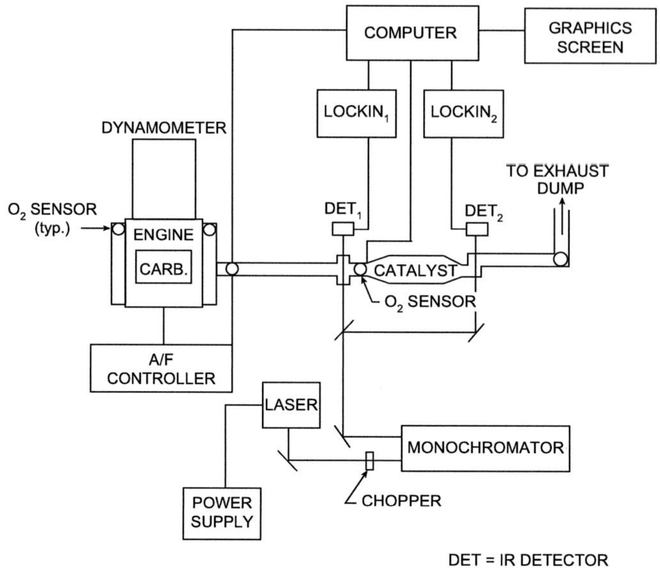

FIGURE 6-5 **Schematic of a dynamometer-mounted engine under either direct computer control or operated under a feedback control loop employed to provide an exhaust gas feed to a catalytic converter. The converter is equipped with sapphire or BaF_2 windows at the inlet and outlet for use in IR measurements with a tunable laser diode as the IR source. The IR system is shown schematically.** *(Figure from* Herz (1982) *with permission. © 1982, American Chemical Society.)*

6.1.2 Experimental Systems

Equipment and experimental procedures described in earlier chapters are limited to modulation frequencies well below 1 Hz. Flow modulation through an ordinary solenoid valve at 1 Hz produces a sinusoid rather than a square wave with a diminished amplitude (*Muraki et al., 1985c*). General Motors (GM) researchers (*Schlatter et al., 1983*) employed special fast-acting valves which allowed them to reach 4 Hz before square waves were damped into sinusoids. Feeds for both parts of a cycle were blended from as many as seven pure gases to represent automotive exhaust at different *A/F* ratios. Actual exhausts have been supplied from an engine attached to a dynamometer so that the behavior of vehicles under diverse driving conditions could be simulated. The set-up used by GM is illustrated in **Figure 6-5**. Modulation of the *A/F* ratio could be forced by the feedback control loop or through a computer as shown in the figure (*Herz, 1982*). The system employed by Toyota is similar (*Matsunaga et al., 1987*).

With time-varying inputs, continuous on-line analysis is needed, such as IR spectroscopy. **Figure 6-5** shows the application of a laser diode IR spectrophotometer to CO measurements. This system was used for the data shown in **Figure 6-4(b)**. Other researchers (*Yokota et al., 1985*; *Matsunaga et al., 1987*) leaked converter effluent into a quadrupole mass spectrometer.

6.2 MODULATION OF DETOXIFICATION REACTIONS

Some of the research on this topic was examined in **Chapters 2** and **3** or by *Silveston (1998)*, and this discussion will not be repeated. **Table 6-1** summarizes the modulation literature on automotive exhaust components. This literature is quite large so the table contains

TABLE 6-1 Modulation Studies on Automotive Exhausts and Exhaust Components

Reference	Study Objective	Fuel/Component, Catalyst, Operating Conditions	Observations	Comments
SINGLE REACTANT (CO)				
Schlatter and Mitchell (1980)	Assess the benefit of CeO in TWC catalysts	CO with $14.2 < \text{Air/CO}$ ratios < 15.2, 773 K, $SV = 104{,}000\ h^{-1}$, Pt/Rh catalyst	*A/F* modulation suppressed CO conversion in the 0.25 to 2 Hz range, but addition of CeO enhanced conversion from 20 to 40% for Rh catalyst.	With HC components, enhancement of the water gas shift reaction (WGS) appeared to be important in the presence of CeO.
Herz (1981; 1982)	Investigate catalyst oxidation state under modulation for different catalyst formulations and deposition patterns	$0.045 < \text{wt\% Pt} < 0.087$; $0.024 < \text{wt\% Pd} < 0.032$; $0.002 < \text{wt\% Rh} < 0.006$; $1.0 < \text{wt\% Ce} < 2.6$ with γ-Al_2O_3 support; egg-shell deposition of metals, $723 < T < 773$ K; $SV \cong 50{,}000\ h^{-1}$	A wide range of experiments was performed. Oxygen storage was identified through oxidation/reduction of Ce oxides. The presence of Ce improved CO conversion under *A/F* modulation.	Although earlier investigators reported a similar observation, the *Herz* papers supported the observation with wide-ranging data.
Herz et al. (1983)	Assess the effects of modulation on catalyst performance	CO oxidation under *A/F* modulation at 1 Hz and under forcing from an O_2 sensor. Various catalysts tested under conditions as above	*A/F* modulation frequency affected enhancement under modulation. Reduction and re-oxidation of Ce oxides was rapid enough to account for the enhancement.	
Herz and Sell (1985)	Role of water gas shift reaction with CeO promoted catalysts	Catalyst composition and experimental conditions were within range given above	Only Pt/Rh/Ce catalyst formulation when used with *A/F* modulation could match the steady-state performance at $A/F = 14.7$.	
Muraki et al. (1985a)	Assess modulation effects on CO oxidation on Pt, Pd or Rh/α-Al_2O_3 at a low loading	Modulation of either CO or O_2 in a feed with a mean composition 1.33 vol% CO, 0.77 vol% O_2, 0.0233 vol% C_3; $553 < T < 673$ K; $SV = 30{,}000\ h^{-1}$ (STP)	Modulation enhanced conversion for Pt and Pd catalysts over steady-state operation.	

(*Continued*)

TABLE 6-1 Modulation Studies on Automotive Exhausts and Exhaust Components (*cont'd*)

Reference	Study Objective	Fuel/Component, Catalyst, Operating Conditions	Observations	Comments
Aida et al. (1997b)	Investigate the modulation of an NO-HC mixture	Symmetrical cycling between 400 ppm C_3H_8 + 1000 ppm NO and 2 vol% O_2 + 1000 ppm NO over an Al_2O_3 catalyst at 775 to 975 K and 100 cm^3/min	Modulation at $T > 875$ K enhanced NO conversion to N_2. Optimal conversion, seen at $\tau = 10$ s and $T = 950$ K, was about 55% compared with about 40% under steady state at 875 K.	Reaction proceeds by oxidation of NO to NO_2 which adsorbs on alumina. C_3 reduces NO_2 and products N_2, CO_2 and H_2O desorb.
Nibbelke et al. (1998)	TWC model development for cold start conditions	Pt/Rh/CeO_2/γ-Al_2O_3 with symmetrical cycling between 0.5 vol% CO in He and 0.5 vol% O_2 in He at $0.1 < \omega < 0.25$ Hz, $393 < T < 433$ K and 110 kPa	Model was based on 10 adsorption/desorption and surface reactions. Three reaction pathways were proposed, two involving just metal sites and one a reaction between adsorbed CO and CeO_2.	
Harmsen et al. (2001a)	Mechanism of acetylene inhibition in automotive exhaust detoxification for a Pt/Rh/CeO_2/γ-Al_2O_3 catalyst	Symmetrical cycling between $C_2^{\equiv}$ at 0.15 kPa and O_2 at 0.0.55 kPa for $503 < T < 543$ K and 110 kPa. Some experiments with CO in rich feed	CO oxidation is inhibited by strongly adsorbed C_2H_2, but CO does not affect rate of acetylene oxidation. CO is an intermediate in acetylene oxidation. Authors' model was not validated.	Conditions used are those encountered in vehicle cold start.
Harmsen et al. (2001b)	Mechanism of NO reduction by CO in automotive exhaust detoxification for Pt/Rh/CeO_2/γ-Al_2O_3 and individual metal components on alumina	Symmetrical cycling between NO or NO + O_2 and CO (0. kPa and 1.05 kPa, and 0.55 kPa) at $0.05 < \omega < 0.1$ and $523 < T < 548$ K	Oxygen inhibition of NO adsorption was observed. N_2 formation proceeds through N_2O. CO does not react with adsorbed NO, but rather forces desorption.	See above.
Jahn et al. (2001)	Investigate development of the temperature distribution in a catalytic monolith under composition modulation	Pt/Rh/Al_2O_3 (5:1) supported in rectangular cell cordierite monolith or Pt/Al_2O_3 on cordierite and metal monoliths with cycling of hot CO and cold air at $\tau = 44$ s and with cycling of hot C_3 /NO and cold air at $\tau = 40$, 50 s	Light-off behavior occurred after a switch to the hot reactant mixture feeding the monolith. Temperatures were initially uneven but gradually evened out throughout the monolith.	Most experiments used uniform flow across a monolith face. Some experiments were done with parabolic flow distributed around the monolith axis.

Carlsson et al. (2004)	Low temperature modulation of O_2 for C_3 oxidation over Pt catalysts with different metal dispersions	Asymmetrical cycling by switching vol% O_2 in the feed between $SN = 1.1$ and 0.5 (lean/rich) or between $SN = 0.5$ and 4.5 at 523 K so that $(SN)_{mean} = 1$	For lean/rich cycling (HC pulse), conversion increased with Pt loading, while for rich/lean cycling (O_2 pulse), conversion became independent of loading. Oxygen consumed for C_3 oxidation comes mainly from adsorbed oxygen. Some Pt was oxidized, showing that a redox cycle is present.	Catalyst performance depended on access to the Pt surface: for $SN > 1$, adsorbed O dominates, while for $SN < 1$, C_3 dominates. Diffusion is not rate-limiting.
MULTIPLE REACTANTS				
Hegedus et al. (1980)	Investigate the role of a postulated isocyanate intermediate in CO-NO-O_2 reactions	IR transmission cell at 783 K and packed bed experiments at 878 K and $SV = 104{,}000\ h^{-1}$ using 1 wt% Pt/Al_2O_3	CO and NO conversions approached completeness at 1 Hz in a packed bed experiment. Surface isocyanate fluctuated with modulation.	Isocyanate was present on catalyst surface during CO-NO-O_2 reaction, although not part of the reaction sequence.
Muraki et al. (1986a)	Assess modulation effects on oxidation reactions with a CO + C_3 fuel and for the CO + NO + O_2 system on Pt, Pd or $Rh/\alpha\text{-}Al_2O_3$ at a low metal loading to simulate an aged catalyst	As above, but with $T = 673$ K, except for Pd catalyst which was also run at 553 K	Modulation greatly enhanced conversion for C_3 at 2 Hz, but increase in conversion for CO was much less. Frequency was important. Asymmetrical modulation with longer O_2 exposure increased conversion. For the CO + NO + O_2 system with Pt catalyst, modulation enhanced the rate by 10 to 15% at 3 to 4 Hz.	Authors speculated that the modulation effect caused by inhibition was due to stronger CO and/or C_3 adsorption on catalyst that excludes oxygen from the surface. Rapid response of surface species to gas-phase composition changes suggested that surface processes control the reaction rate.
Muraki et al. (1985b)	Determine portioning of oxidation function for CO-NO-O_2 system	As in above Muraki et al. with $T = 673$ K. Modulation based on stoichiometric number (normalized *A/F* ratio)	Partitioning of oxidation depends on stoichiometric number (*SN*): O_2 is the dominant source at low *SN*, but modulation shifts the source to NO as frequency increases.	Presence of O_2 in feed does not decrease NO reduction significantly, but CO oxidation rate increases.

(Continued)

TABLE 6-1 Modulation Studies on Automotive Exhausts and Exhaust Components (*cont'd*)

Reference	Study Objective	Fuel/Component, Catalyst, Operating Conditions	Observations	Comments
Harmsen et al. (2000)	Develop model for an exhaust oxidation catalyst	Commercial $Pt/Rh/CeO_2/\gamma\text{-}Al_2O_3$ catalyst used with $0.1 < \omega < 0.25$, $0.1 < s < 0.9$ of an $C_2^{=}$ - O_2 mixture at $393 < T < 443$ K, 110 kPa	Using parameters from experimental data, authors' model closely represented treated exhaust compositions in both parts of a cycle.	CO inhibits $C_2^{=}$ oxidation so the latter reaction starts after most CO of the CO is oxidized.
SIMULATED VEHICLE EXHAUST				
Hegedus et al. (1979)	Catalyst screening	Performance under modulation studied for 7 Pt-Rh-Pd-Ce formulations with different alumina supports	*A/F* modulation depressed oxidation/reduction reaction rates except for hydrocarbon oxidation.	
Schlatter et al. (1983)	Investigate the effect of modulation on catalyst performance	*A/F* modulation for Pt/Rh on alumina at 823 K and $SV = 52{,}000$ h^{-1}, fresh and aged catalyst tested	Modulation increased CO oxidation for $(A/F)_{mean} < 14.5$ and for NO_x conversion for > 14.7. Modulation frequency and amplitude affected the improvement. Conversions decreased as catalyst aged.	
Weibel et al. (1991)	Role of CeO in Rh catalysts ($Rh/\gamma\text{-}Al_2O_3$)	Authors used 1.02 wt% Ce; measurements made at 723 K and $SV = 100{,}000$ h^{-1} with modulation at 1 Hz and amplitude $= 0.25$ $(A/F)_{mean}$	Ce promoter substantially increased CO conversion under modulation at rich *A/F*, but had little effect on NO conversion.	Authors used low Rh loading (0.003 wt%). They suggested CeO increases WGS rates.
Muraki et al. (1985c)	Explore frequency and amplitude effects as well as light-off in *SN* modulation	Effects on HC, CO and NO conversion and light-off near $SN = 1$ for Pt, Pd or $Rh/\alpha\text{-}Al_2O_3$ catalysts at 673 K	Modulation enhanced conversion over the Pt catalyst, but only for $\omega \geq 2$ Hz for the Pd catalyst. Generally, enhancement decreased with greater amplitude. Modulation decreased light-off temperatures for Pt and Pd catalysts with NO reduction and C_3 oxidation, but increased the light-off temperature for CO oxidation and for C_3 oxidation on Rh catalyst.	*Muraki et al.* papers offer a comprehensive picture of modulation effects on the Pt, Pd and Rh catalysts. Light-off temperature was defined as the temperature needed for 50% conversion.

Taylor and Sinkevitch (1983)	Explore asymmetric modulation ($s \neq 0.5$) on component conversion and light-off	Pt/Pd/Rh TWC promoted with 2.7 wt.% Ce on γ-Al_2O_3. Experiments conducted at 823 K, $SV = 20{,}000\ h^{-1}$ with cycling at 1 Hz and $A = 0.15$ $(A/F)_{mean}$	Asymmetric cycling increased CO conversion at $s = 0.8$ and NO conversion at $s = 0.8$ in contrast to modulation at $s = 0.5$. Modulation decreased light-off temperature for C_3 oxidation, but increased light-off temperature for NO reduction. Light-off temperature was unaffected for CO oxidation.	
Padeste and Baiker (1994)	Exploration of short term breaks in engine operation on TWC performance	Effect of 12 wt% CeO on1 wt% Pt, 0.2 wt% Rh/Al_2O_3 catalyst at 533 and 583 K, $SV = 50{,}000\ h^{-1}$ (STP) with SN modulation around $SN = 1$. Also recovery of TWC performance after flow interruption	Modulation enhanced C_3 oxidation and NO reduction, but also suppressed N_2O formation at 1 Hz with amplitude $= 0.02\ (SN)_{mean}$ for unpromoted catalyst. Modulation greatly enhanced recovery of NO, C_3 and N_2O conversion after interruption by zero gas flow or an air pulse.	These investigators considered operation of hybrid vehicles in which the engine stops or the exhaust is flooded with air.
Yamazaki et al. (2004)	Investigate the addition of Fe_2O_3 to a TWC	Pt/BaO/CeO_2/Al_2O_3 catalyst with various amounts of Fe_2O_3 added under strongly asymmetric cycling between lean and rich feeds with constant levels of NO, CO_2 and H_2O at $\tau = 63$ s and $s = 0.05$ based on a rich mixture	Adding Fe_2O_3 to a TWC did not alter the performance for the fresh catalyst, but Fe_2O_3 significantly lowered the light-off temperature for the aged catalyst and increased reduction of NO_x by CO. The iron oxide provides oxygen for reaction and also mitigates CO inhibition in oxidation by O_2 and in NO_x reduction.	Addition of Fe_2O_3 improved the performance of NO_x storage and reduction catalysts.
TEST ENGINE EXHAUST				
Muraki et al. (1989), Muraki (1991)	Effect of A/F (SN) modulation on TWC performance	Rh/α-Al_2O_3, Pd/α-Al_2O_3, Pt/α-Al_2O_3, $613 < T_0 < 633$ K; also used simulated exhaust at 623 K for a Pd/α-Al_2O_3 catalyst but at high metal loading	Modulation frequency affected conversion, but for Pd and Pt catalysts conversions were above steady state at their optimal frequencies. Without CeO promoter, NO_x conversion was very low for Pd catalyst cycling above $(A/F)_{mean} > 14.7$.	

(*Continued*)

TABLE 6-1 Modulation Studies on Automotive Exhausts and Exhaust Components *(cont'd)*

Reference	Study Objective	Fuel/Component, Catalyst, Operating Conditions	Observations	Comments
Yokota et al. (1985)	Determine if Pt or Pd catalysts could achieve performance of an Rh based TWC using a test engine as the exhaust source	Catalysts were those used by *Muraki (1991)*, but aged by extended use. Modulation at 1 Hz with amplitudes from 0.3 to 1.2 $(A/F)_{mean}$, $613 < T < 633$ K	Symmetric modulation around $SN = 1$ enhanced component conversions only for Pt catalyst. For $SN < 1$, modulation enhanced conversions for Pt and Pd catalysts, but only for C_3 oxidation for the Rh catalyst. At $SN > 1$, modulation slightly increased conversion over Pt and Pd catalysts for CO oxidation.	
Matsunaga et al. (1987)	Investigate the effect of *A/F* modulation on a fresh and aged commercial monolithic TWC	$Pt/Rh/CeO_2/Al_2O_3$ catalyst under modulation for different sets of dual conditions: $13.5 < A/F < 15.5$, $2\% < A < 10\%$; $320 < T_0 < 670$ K, $0.5 < \omega < 5$ Hz holding remaining two conditions constant at $\omega = 1$ Hz or $T_0 = 660$ K or $(A/F)_{mean} = 14.5$ or $(A)_{mean} = 6\%$ of $(A/F)_{mean}$	Maximum conversion for the TWC was found at low amplitude and $A/F \approx 14.5$ for NO and CO, but at the highest amplitude for HC. Light-off with fresh catalyst depended strongly on amplitude at $\omega = 1$ or on ω at constant amplitude and $(A/F)_{mean} = 14.5$ for HC, but was independent of both frequency and amplitude for NO. CO oxidation was not ignited at the highest temperature used. Catalyst ageing suppressed light-off for HC and NO.	Modulation effect on light-off for a HC feed was much larger than that observed by *Muraki et al. (1986b)*, but was similar for NO and CO. (See above.)

SULFUR POISONING AND SINTERING				
Monroe et al. (1991)	Investigate the effect of SO_2 on the performance of TWCs under *A/F* modulation	Pelleted alumina supported Pt/Pd/Rh/CeO_2 catalyst and a Pt/Rh/CeO_2/LaO catalyst wash-coated onto a cordierite monolith. Experiments used a simulated exhaust at $\omega = 0.5$ Hz around $(A/F)_{mean} = 14.5$ with 120 min exposure to feed containing 5 or 20 ppm SO_2	SO_2 in the simulated exhaust suppressed light-off for single component streams as well as for simulated waste and the increase in light-off temperature was proportional to ppm of SO_2 in the range studied. Recovery of light-off behavior after SO_2 was removed from the exhaust was virtually complete for HC and NO_2 exhausts, but not for CO.	
Lööf et al. (1991)	As *Monroe et al. (1991)* above with attention to the role of CeO_2	Used 0.8 wt% Pt/γ-Al_2O_3 with and without 30 wt% CeO_2 for an experiment with $\omega = 0.5$ Hz for a simulated exhaust with 20 ppm SO_2 added	Under modulation the addition of CeO_2 enhanced conversion of HC, CO and NO. Addition of 20 ppm SO_2 reduced conversion of all waste components. Sulfur poisoning was greatest at low *A/F*, but was still evident in lean mixtures.	
Hirata et al. (2011)	Investigation of precious metal sintering on TWCs	Used 2 wt% Pt or 0.5 wt% Rh on alumina and mixed oxide (CeO_2, ZrO_2, Y_2O_3) supports and on oxides alone with symmetrical cycling between 3% H_2 and 20% O_2 both in He at $\tau = 120$ s at 873 K	Mixed oxide support eliminated Pt sintering which was pronounced for Pt/Al_2O_3. Loss of Rh activity on Al_2O_3 appeared to be due to the gradual incorporation of Rh into the alumina lattice during lean/rich cycling.	Lean/rich cycling was observed to re-disperse Pt on the mixed oxide support surface. Mixed oxides, namely ZrO_2 are better supports for Rh.

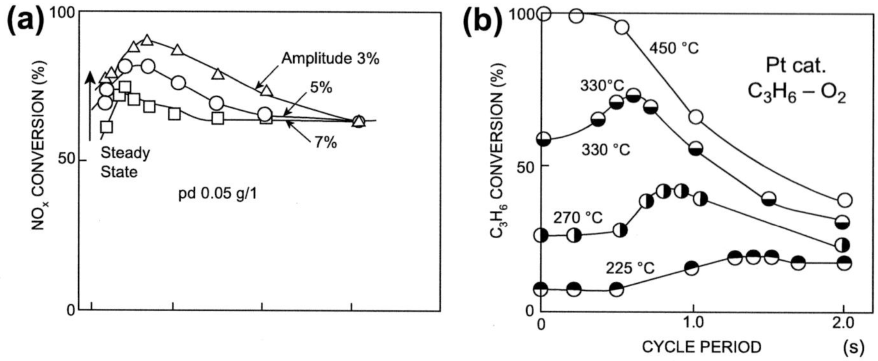

FIGURE 6-6 ***A/F*** **Modulation effects on automotive exhaust component conversion: (a) NOx reduction for an actual engine exhaust over a 0.05 g/L Pd/Al_2O_3 catalyst (2 L engine operating at 1600 rpm and 440 Torr), (b) oxidation of propene using O_2 over a Pt/γ-Al_2O_3 catalyst. Measurements made by modulating the C_3H_6-O_2 mixture.** *(Figure (a) modified from* Yokota et al. (1985) *with permission. © 1985 by the Society of Automotive Engineers; Figure (b) reproduced from* Shinjoh et al. (1987) *with permission. © 1987 by Elsevier Science Publishers.)*

representative publications only. Entries are organized by the fuel system studied.

Catalyst composition, namely the metals used and the quantity of promoters such as CeO_2, modulation of the *A/F* and temperature are the dominant factors in detoxification of exhaust components. Examples of the effect of modulation measured for different conditions and exhaust sources are shown in **Figure 6-6** with **(a)** representing NO_x reduction by CO and hydrocarbons using an actual engine exhaust and **(b)** showing propene oxidation in air measurements. The figure shows that conversion of these exhaust components depends on modulation frequency, but also on amplitude and the temperature of the catalyst. Amplitude and temperature affect conversion, while the latter influences the optimal frequency of the modulation. Note that neither of the figures reproduces the performance of modern TWC exhaust catalysts. NO_x reduction was also studied by *Aida et al. (1997b)* who reported a temperature effect on the reduction under modulation.

Problems arising in automotive engine cold start have been explored by *Harmsen et al. (2001a, b)* who observed in their cycling experiments between 520 and 548 K that O_2 inhibits NO adsorption on a TWC. In the rich portion of a symmetrical cycle, they reported that CO forces desorption of NO and that the reduction reaction proceeds through N_2O. Acetylene is observed in cold start-up and *Harmsen et al. (2001b)*, examining $C_2^{\equiv}/O_2$ cycling, found $C_2^{\equiv}$ inhibited CO oxidation, but that CO_2 did not interfere with CO oxidation.

Feedback through *A/F* control results in near symmetrical modulation at the TWC inlet. Asymmetric modulation, however, is possible through the onboard computer (**Figure 6-5**) and this modification was investigated by *Carlsson et al. (2004)* for C_3—O_2 cycling between $SN = 0.5$ and 11 or 4.5 with adjusting duration to maintain $(SN)_{mean} = 1$. These researchers found that rich pulses brought C_3 conversion quite close to those at a stoichiometric C_3—O_2 ratio. However, asymmetric operation brought no advantage over the usual symmetric modulation.

Jahn et al. (2001) explored the development of temperature distribution in a monolith TWC converter after cold start-up or a flow interruption using the cycling of CO/air or C_3/NO. They noted that the flow distribution at the face of a monolith is non-uniform and would be expected to affect light-off. Higher velocities at the centerline lead to light-off at that location which then propagates radially through the monolith. Initiation can occur in the center, back or front of the monolith. Additional experiments using switching between rich hot exhaust and cold air showed that the exposure duration could be adjusted to insure high pollutant conversion in a TWC, particularly for NO_x destruction.

Systems such as CO-C_3-O_2 and NO-CO-O_2 were heavily studied in the 1980s by *Muraki et al. (1985b, 1986b)* and by *Hegedus et al. (1980)*. The two systems form a simple reaction set for TWCs. The stoichiometric number:

$$SN = (2(O_2) + NO)/CO \qquad (6\text{-}1)$$

may be used to represent this oxidation-reduction system where $SN > 1$ signifies an oxidizing feed.

Muraki et al. (1986b) noticed that the optimal frequency to maximize the conversion decreased when changing from a single fuel to multiple ones. They observed that asymmetrical forcing showed higher conversions only when the oxidant exposure duration was extended and concluded that oxidation rates over the noble metal catalysts are CO and hydrocarbon inhibited. Modulation then improves performance by increasing O_2 access to the catalyst surface. Because it is better access that leads to rate enhancement through modulation for both single and simultaneous reactions, there is just a small difference in enhancement as a function of frequency between CO oxidation proceeding alone or in the presence of hydrocarbon oxidation or vice versa. *Hegedus et al.* found that surface species responded rapidly to gas phase composition fluctuations so that the performance of catalytic converters is governed by surface processes. *Muraki et al. (1985b)* observed that partitioning of CO between oxidation by O_2 and NO reduction depends on the SN. The former is favored by low SN. The presence of O_2 in the reactor feed did not depress NO reduction significantly.

6.3 MODELING STUDIES

Modeling of CO oxidation under composition modulation on a TWC monolith was undertaken in the 1990s (*Lie et al., 1993; Hoebink et al., 1999, 2001*). These studies are not included in **Table 6-1**. Since 1990, considerable mechanistic and kinetic data has been collected, much of this by *Hoebink* and co-workers (*Nibbelke et al., 1998; Harmsen et al., 2000, 2001a, b*). Indeed, the *Harmsen* papers provide a detailed synopsis of kinetic research on TWC catalysts (see, in particular, *Harmsen et al., 2001a*). This progress along with the enormous growth in computing power has shifted development from the laboratory and bench-scale pilot plant to modeling studies. *Kočí et al. (2004a)* used literature information to formulate a comprehensive dynamic model of a monolith TWC employing a γ-alumina supported Pt-Rh/CeO catalyst. This model assumed a well-mixed gas phase and allows for mass transfer to the solid surface and diffusion through a wash coat. Material balances were constructed in terms of the occupancy of surface sites, degree of ceria oxidation as well as for oxygen storage on the alumina support. The kinetic scheme assumed 29 elementary reactions for the nine molecular species. Hydrocarbon residue in the exhaust was assumed to be just ethylene and acetylene. Rate models and rate coefficients were drawn from the *Hoebink* studies mentioned above. Effective diffusivities and mass transfer coefficients were estimated from literature data. In place of step-changes in the *A/F* ratio, sinusoidal oscillation of the inlet O_2 concentration was assumed to drive the composition variations.

The model equations employed by *Kočí et al. (2004a)* are not shown as they are given in simpler forms in **Chapters 2** and **3**. Integration

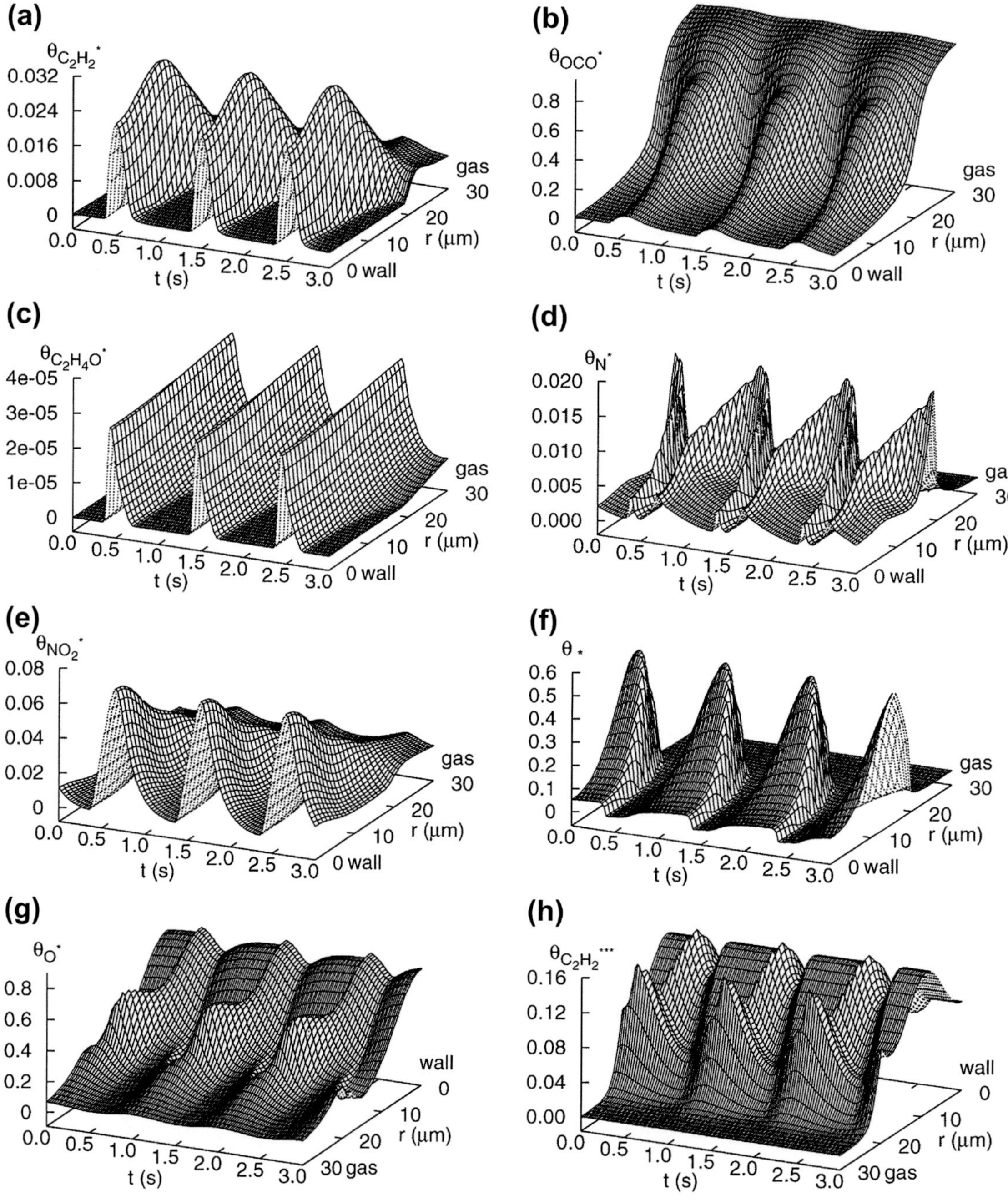

FIGURE 6-7 **Simulated depth and time profiles in a 30 μm Pt-Rh/CeO/γ-Al_2O_3 washcoat for detoxification of a CO-NO-HC exhaust under sinusoidal modulation of the inlet oxygen concentration around $y_{O2} = 0.72\%$ (stoichiometric value) and $\omega = 1$ Hz, $A = 10\%$ for a gas inlet temperature = 300°C with inlet CO = 1.2%, CO_2 = 12%, C_2H_2 = 200 ppm, C_2H_4 = 400 ppm and NO = 1000 ppm with the remainder N_2 and H_2O.** *(Figure reproduced from* Kočí et al. (2004a) *with permission. © 2004 by Elsevier B.V.)*

of the system of PDEs and ODEs (the latter group arising from a well-mixed gas phase) is a colossal undertaking. **Figure 6-7** represents the dynamic response to the inlet O_2 modulation in the porous washcoat predicted by the Kočí model. Profiles within the washcoat were determined at 473 and 573 K, shortly after light-off. Only the profiles at 573 K are given in the figure. As might be expected, adsorbates are concentrated at the gas-solid interface with the exception of relatively unreactive acetylene which penetrates to the wall. Adsorbed, dissociated oxygen in **(g)**, the oxidant, is concentrated in about the centerline of the washcoat and varies relatively little within a cycle. The differences between 473 and 573 K are the much higher adsorbate concentrations at the lower temperature and the steeper concentration gradients at the higher one. Indeed, the innermost 10 μm of the 30 μm washcoat does not participate in the detoxification reactions. *Kočí et al.* found that pollutant conversions under modulation exceeded 90% for CO and 80% for HC and NO. The primary purpose of the simulation was to explore light-off performance. Parameters considered were $1 < \omega < 2$ Hz and amplitudes $> \pm 10\%$ around $SN = 1$. An optimal cycling frequency was identified.

The *Kočí et al.* model was used several years later by *Marek et al. (2006)* to investigate the possibility of chaotic behavior in TWCs exposed to *A/F* forcing. The *Marek* team, indeed, found that the *Kočí et al.* model predicted regions of multiplicity, period doubling and even chaos when large forcing amplitudes were imposed. These were similar to the predictions of CO oxidation models they studied years earlier. However, unstable oscillatory response does not seem to have been reported in the exhaust detoxification literature.

Simulation has been used recently to measure the contribution of ceria to TWC converters. Of course, the use of ceria in TWCs goes back to early research studies (e.g., *Hegedus et al., 1979; Schlatter and Mitchell, 1980*). *Shamin (2005, 2008)* has attempted, through simulation, to evaluate ceria's contribution to TWC performance. A much simpler kinetic scheme was employed and diffusion in the washcoat was neglected. Shamin's model considered these reactions:

$$1)\ s + \frac{1}{2}O_2 \rightarrow Os;\ 2)\ Os + CO \rightarrow s + CO_2;\ 3)\ H_2O + CO \rightarrow H_2 + CO_2;\ 4)\ Os + H_2 \rightarrow s + H_2O;$$
$$5)\ CH_\alpha + H_2O \rightarrow CO + (1 + \frac{\alpha}{2})H_2;\ 6)\ \frac{3}{2}CH_\alpha + Os \rightarrow + \frac{1}{2}C + CO + \frac{3\alpha}{4}H_2;$$
$$7)\ C + O_2 \rightarrow CO_2;\ 8)\ s + NO \rightarrow Os + \frac{1}{2}N_2;\ 9)\ Os + \frac{2}{5}NH_3 \rightarrow s + \frac{2}{5}NO + \frac{3}{5}H_2O$$

and gas phase material balances for seven of the above species, surface material balance for two species, as well as gas phase and surface energy balances. These balances have been presented elsewhere. Details are given by *Shamin (2008)*. The model was used to compare the performance of catalysts formulated with and without ceria in lean and rich exhaust mixtures as well as an exhaust for $A/F = 14.7$ with A/F modulation at $\omega = 1$Hz and $A = 5\%$ of $(A/F)_{mean}$. Temperature modulation at 0.01 Hz was also tested. Independent CO, HC and NO modulation was considered at 1 Hz, but also at a 50% amplitude. The *Shamin* simulation showed that the presence of ceria improved pollutant conversion in *A/F* modulation (except for NO), but this improvement was small. It was somewhat larger for temperature modulation near the stoichiometric *A/F*. Much larger effects were found for CO, HC and NO modulation.

6.4 STUDIES ON MODULATING AUTOMOTIVE EXHAUST

These studies, representative ones summarized in **Table 6-1**, have been undertaken to develop catalysts and determine if the unavoidable low amplitude modulation of exhaust composition is beneficial. The search for effective and durable three-way catalysts have been from the outset a major focus in catalytic converter R & D. Besides catalytic activity at different SN, light-off temperature and stability with respect to poisons and temperature excursions, behavior under composition modulation has been a crucial concern. Consequently, SN modulation experiments are usually included in catalyst screening tests. The use of modulation in the context of catalyst development is nicely illustrated by the contributions of *Muraki (1991)* and *Muraki et al. (1989)*.

The composition of automotive exhausts as a function of vehicle operation is well understood providing confidence in the usefulness of performance research obtained from measurements with simulated automotive exhaust. Thus, many investigations have been undertaken with such feeds. As may be seen from **Table 6-1**, these studies showed that modulation, at a suitable frequency, can enhance pollutant destruction for less active catalysts and for all catalysts at lower temperatures. *Muraki et al. (1985c, 1986b)* observed that conversion of key pollutants was essentially complete under steady-state operation for the most active Rh/Al_2O_3 catalyst. Modulation decreased conversions noticeably. *Weibel et al. (1991)* also found modulation did not enhance conversions for a cerium oxide promoted Rh catalysts.

Muraki et al. (1985c, 1986b), *Weibel et al. (1991)* and earlier work on synthetic exhausts by *Kaneko et al. (1978)* and *Schlatter et al. (1983)* used symmetrical modulation in which durations of each part of a cycle are the same (cycle split, $s = 0.5$). Of course, this approximates the modulation forced on a catalytic converter by the *A/F* ratio control loop. Asymmetric forcing was examined by *Taylor and Sinkevitch (1983)* who found that for an aged, Ce promoted, noble metal catalyst operating at constant temperature, asymmetric forcing at $s = 0.8$ increased CO conversion for mixtures which were rich on average. Cycle split here is defined as the ratio of the duration of the lean mixture exposure to the cycle period, τ. On the other hand, cycling at $s = 0.2$ improved NO conversion in rich mixtures. Because steady-state CO conversion drops in rich mixtures while NO conversion falls in lean mixtures, *A/F* modulation is beneficial in that it raises conversions on either side of the stoichiometric *A/F* ratio. This is referred to as widening of the conversion window. Generally, asymmetric modulation is more effective for increasing conversion at $SN < 1$, or for NO reduction at $SN > 1$, than symmetric forcing (*Schlatter et al., 1983*).

An important property of TWCs is the conversion achieved after engine start-up, but before the engine attains a time-average stationary temperature. This is referred to as the light-off or cold-start performance. Light-off temperature is defined as the temperature at which conversion reaches 50% and it is usually measured by ramping the catalyst temperature by 2°C/min from ambient temperature. *Taylor and Sinkevitch (1983)* investigated light-off under symmetrical as well asymmetrical *A/F* modulation, while *Muraki et al. (1985c)* used just symmetrical forcing. The light-off results of these investigators are different and indicate a dependence on the catalyst used. Both teams of researchers found that modulation affected light-off over a Pt or Pd catalyst for C_3 oxidation and NO reduction. For the Rh catalyst, only light-off for NO reduction depended on modulation frequency. The researchers differed on the effect of modulation on light-off for CO oxidation on Pt. Explanations for the different behaviors were not pursued.

Some automotive hybrid engines generate electricity which is used along with energy recovered from braking to power the vehicle. In such systems, the vehicle engine is switched on and off while in use. The effect of *A/F* modulation on pollutant conversion as well as the superimposed influence of periodic exposure of the TWC to air or even "no" flow was examined by *Padeste and Baiker (1994)*. They first determined TWC performance in modulated operation and then examined how short term exposure to air altered conversions. *Padeste and Baiker* found the addition of Ce effective. They observed an optimal frequency of about 1 Hz for a Pt/Rh formulation without Ce. Breakthrough of pollutants was observed at lower frequencies suggesting that at longer cycle periods CO "swamps" the metal surface towards the end of the rich half-cycle while high coverage of O adatoms occurs at the same point in the lean half-cycle. Periodic air exposure affected operation under modulation primarily for NO conversion with little change in the conversion of other exhaust components. It was not apparent from the *Padeste and Baiker* contribution whether hybrid drive systems will require modification of the *A/F* ratio feedback control loop. A more detailed discussion is to be found in *Silveston (1998)*.

Research on three-way converters continues. Based on studies of *Yamazaki et al. (2004)*, Toyota now adds Fe_2O_3 to the TWCs in their vehicles. These researchers showed that the addition of this oxide to the washcoat lowers the light-off temperature for a simulated exhaust under *A/F* modulation and increases NO_x reduction by CO.

The effect of *A/F* modulation on conversion of toxic gases in actual automotive exhausts was examined by several research teams. *Yokota et al. (1985)* and *Matsunaga et al. (1987)* employed an experimental system like that shown in **Figure 6-5**. Modulation was forced by the oxygen sensor and feedback control loop or by an external oscillator with the control loop open. The noble metal catalysts tested by *Yokota et al.* were those used by *Muraki et al. (1985c)* but were aged by an extended exposure to automotive exhaust. Converter inlet temperatures in their experiments, between 610 and 640 K, are those encountered leaving actual engines. *Muraki et al.* observed that *A/F* modulation at 1 Hz with large modulation amplitudes depressed conversions for Rh and Pd but elevated those for the Pt catalyst. These observations are consistent with those discussed in the previous section. They also agree with GM research which showed no conversion enhancement through modulation around the stoichiometric *A/F* ratio with TWCs containing Rh.

With lean mixtures, $SN > 1$, modulation raised NO conversion well above that possible for steady state operation. Larger amplitudes increased conversion. For the Pd and Pt catalysts, conversions were enhanced away from the stoichiometric even in rich mixtures. For hydrocarbon oxidation, modulation raised conversion substantially for the Rh catalyst. Modulation effects found by *Yokota et al.* for non-stoichiometric *A/F* ratios are considerably greater than those reported by *Schlatter et al. (1983)* or by *Taylor and Sinkevitch (1983)*, but the latter were made at a temperature about 100 K greater and employed a different catalyst.

The objective of *Yokota et al. (1985)* was to see if a noble metal mixture could match the performance of Rh catalysts. These investigators found that for $SN \neq 1$, catalysts formulated around Pd and Pt perform as well as an Rh catalyst, except for NO_x reduction, provided the exhaust was modulated at a suitable frequency. Even with NO_x, the performance of the Rh catalysts is approached. *Yokota et al.* also addressed the question of whether or not the steady-state performance at $SN = 1$ can be matched using periodic *A/F* forcing. They showed this was possible for NO_x reduction and HC oxidation for a Pd catalyst but falls well below this target for CO oxidation.

Matsunaga et al. (1987) examined the *A/F* modulation of an engine exhaust to quantify performance for a commercial 1.3-L Pt/Rh monolith catalyst. Their lower inlet temperature and SV reflected the then current engine design, which differed from designs used in the 1978 to 1984 period by GM researchers. *Matsunaga et al.* compared conversions for a fresh monolith catalyst and one aged to represent 60,000 km of driving. For the fresh catalyst, conversion enhancement rose with increasing amplitude of modulation at 1 Hz. However, an amplitude threshold seemed to exist. Comparing conversions for fresh and aged catalysts showed a small loss of activity for hydrocarbon oxidation, NO reduction and CO oxidation under *A/F* modulation around the stoichiometric ratio. At other ratios, the large improvement observed under modulation extended to aged catalysts so that modulation dramatically broadened the operating window for the TWC tested catalysts.

Modulation exerts a large effect on light-off temperatures for fresh and aged Pt/Rh catalysts. For the hydrocarbons and NO_x, forcing at 1 Hz at a typical *A/F* amplitude reduced light-off temperatures significantly for NO_x. However, neither ageing nor modulation affected light-off for CO oxidation. The modulation influence on light-off temperatures found by *Matsunaga et al. (1987)* was much greater than that observed by *Muraki et al. (1986b)* for C_3 but is comparable with what the earlier researchers found for NO_x. No investigators have found light-off improvement for CO oxidation through modulation.

6.5 EFFECT OF *A/F* MODULATION ON POISONING AND SINTERING

Although the preponderance of research has dealt with clean exhausts, attention has been paid to the effect of sulfur on converter performance. According to *Monroe et al. (1991)*, sulfur poisoning is a problem with rich or stoichiometric mixtures. An early study by *Joy et al. (1979)* found severe poisoning of the water gas shift and steam reforming reactions but saw little effect on converter performance because sulfur was stripped from the catalyst in the lean part of a cycle. Formation of H_2S in converters when operating with rich mixtures has been a problem. Prior to *Monroe et al. (1991)*, *A/F* modulation in the presence of SO_2 was explored by *Schlatter and Mitchell (1980)* and *Su and Rothschild (1986)*.

Two commercial catalysts, a pelleted, noble metal impregnated alumina and a monolith containing Pt, Rh, Ce and La were used as fresh and thermally aged samples by *Monroe et al. (1991)*. These researchers examined SO_2 poisoning and recovery after SO_2 is removed from the exhaust for warmed up conversions (at 770 K) and on light-off with a simulated automotive exhaust cycled around $SN = 1$ at 0.5 Hz. Light-off temperatures were severely affected with both catalysts and were sensitive to the SO_2 concentration in the exhaust. Once removed from the feed, the light-off temperature decreased and performance slowly recovered for NO_x reduction and HC oxidation with the monolith catalyst, but not for the pelleted catalyst. *Monroe et al. (1991)* also undertook experiments on an aged monolith catalyst with actual exhausts for fuels of different sulfur contents. Their data clearly showed a reduction in conversion as the sulfur content of the fuel increased with the largest effect on HC oxidation, while NO reduction was the least impacted. The authors found a mitigating influence of *A/F* modulation on sulfur poisoning. This had been recognized much earlier by *Joy et al. (1979)*.

As part of an investigation of the role of CeO_2 in TWCs, *Lööf et al. (1991)* measured the effect of *A/F* modulation on CO, HC and NO conversion in the presence and absence of 20 ppmv SO_2 in a simulated exhaust for a $Pt/\gamma\text{-}Al_2O_3$ catalyst and for the same catalyst containing CeO_2. They found that SO_2 in the exhaust decreased conversions of all target components. This decrease was largest at the lowest *A/F* ratio, but was still seen in lean mixtures. *Lööf et al.* concluded that the formation of a sulfate destroyed the ability of cerium to store O_2 and donate O_2 in the oxidative reactions.

In addition to poisoning, loss of TWC activity through catalyst ageing, primarily sintering, has drawn study. Sintering depends on the mobility of the noble metals on the catalyst support. It was observed early on that Pt mobility on alumina was high. Mobility and the sintering rate dropped substantially when a mixed CeO_2-ZrO_2 oxide replaced alumina. Indeed, CO adsorption experiments indicated no change in Pt cluster size on the mixed oxide after exposure to a vehicle exhaust (*Nagai et al., 2006*). Apparently formation of a Pt-O-Ce bond anchors the noble metal.

Matsumoto and Shinjoh (2007) quote Japanese researchers who observed Pd mobility on a Perovskite support as well as redispersion of Pt cluster on a mixed CeO_2-ZrO_2 support under modulation between reducing and oxidizing atmospheres. **Figure 6-8** illustrates the reported observations. These intriguing results suggest *A/F* modulation may contribute to catalyst stability if a suitable support material is used.

6.6 EFFECTS OF IRREGULAR *A/F* VARIATION

The Pt based, Ce modified TWC promoted with small amounts of Pd and Rh function very well under the low amplitude *A/F*

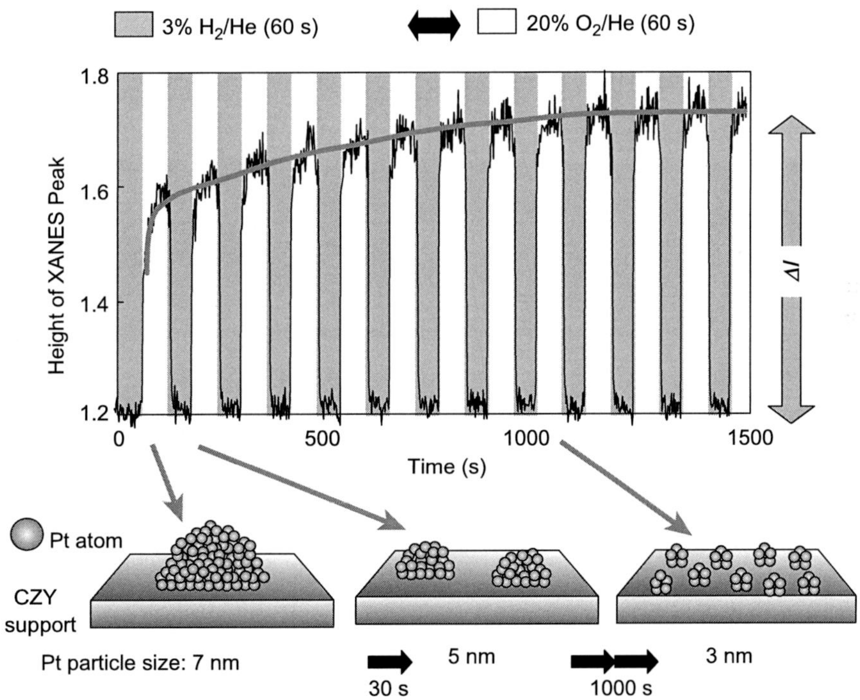

FIGURE 6-8 Time-resolved heights of XANES measurements for a sintered Pt supported on a mixed CeO_2-ZrO_2 with a highly ordered structure showing the reduction in cluster size under modulation of the gas phase between reducing and oxidizing atmospheres. Initial size of the Pt cluster was 7 nm. *(Figure reproduced from* Hirata et al. (2011) *with permission. © 2011 by Elsevier Inc.)*

modulation experienced at constant throttle. The exhaust contains the right balance of reducing components (CO, H_2, HCs) to react with NO_x and O_2, which in turn oxidizes those gases. Sudden acceleration or deceleration, however, forces the *A/F* away from its 14.5 stoichiometric ratio. During acceleration, sharp peaks of NO_x occur in vehicle emissions. Cerium oxide largely moderates the magnitude of these peaks by releasing O_2. As a TWC ages, Ce loses storage capacity allowing these peaks to pass through the converter. Eliminating NO_x release from these irregular *A/F* variations has been an important challenge in the last 20 years. Addition of lanthium to form mixed La-Ce oxides enhances the O atom mobility in the oxide lattice, while the addition of zirconium to yield mixed Ce-Zr oxides greatly decreases the sintering of CeO_2, which appears to be mainly responsible for the loss in O_2 capacity. Sintering can also be reduced by adding Al_2O_3 to the Ce-Zr mixed oxide (*Matsumoto and Shinjoh, 2007*). The alumina appears to build a barrier between the CeO_2 crystallites and reduces crystallite growth. The presence of Zr in a mixed crystal enlarges the lattice thereby increasing O_2 diffusion. Improved stability of the cerium oxides and higher O_2 mobility through the addition of La and Zr have substantially improved the performance of modern TWCs in mitigating NO_x breakthrough during vehicle acceleration.

6.7 LEAN BURN SPARK-IGNITED ENGINES

Vehicles employing lean burn, spark ignited engines continue to increase their market share over those operating near $SN = 1$. The TWCs discussed up to this point, however, are ineffective for lean exhausts from such engines. Indeed, these lean burn exhausts are so similar to diesel exhausts that converters for diesels can be used for those spark ignition engines, assuming the temperature windows can be made similar enough. The last twenty years has witnessed a burst of research activity on catalysts for lean burn exhausts (*Collins and Twigg, 2007*). **Table 6-2** summarizes representative experimental contributions on converters for lean burn spark ignition and diesel engines, with emphasis on papers that deal with lean/rich modulation.

These spark ignition engines discharge just ppm levels of unburnt hydrocarbons and CO but with more substantial amounts of NO_x. The CO and hydrocarbons are easily oxidized in the converter system, which typically consists of an oxidizing bed upstream of other components. Several technologies for NO_x have been explored. Among these is selective catalytic reduction, which uses urea, an ammonia precursor. Urea decomposes and selectively reacts with NO_X rather than O_2. Another technology is selective periodic catalytic reduction (NSR) using hydrogen, carbon monoxide or short chain hydrocarbons, all derived from the fuel through partial combustion. They operate with steady periodic pulses at low *A/F* that are superimposed on variations arising from acceleration or deceleration. The combustion motor and control system along with the periodic bursts of a rich mixture are shown in **Figure 6-9**. The essential function of the NO_X trap converter is trapping NO or NO_2 that arise in fuel combustion, and then during the periodic swing to low *A/F* ratios, the trapped NO_X species are reduced to N_2 or NH_3.

Most NSR catalysts are mixtures of Pt and an alkali earth oxide, often barium oxide (BaO), impregnated onto a silica or alumina support. Other noble metals may be added. Rhenium is effective, but extremely expensive. Potassium oxide is a suitable trap for NO_x and seems to be more stable than Ba at higher temperatures. Nevertheless, applications favor barium oxide. On oxidization in the converter, NO forms NO_2 on Pt or Pt-Rh and is captured by the alkaline or alkaline earth oxide as a nitrate (**Figure 6-10**). A brief pulse (1 to 10 s) of a reducing mixture

TABLE 6-2 Modulation Studies on Diesel and Lean Burn Exhausts

Reference	Study Objective	Fuel/Component, Catalyst, Operating Conditions	Observations	Comments
Nakatsuji et al. (1999)	Development of an NO_x trapping and reduction catalyst	Highly asymmetric cycling between a rich mixture (500 ppm NO, 40 ppm SO_2, 2000 ppm C_3, 2% H_2, 0.4% O_2, 9% H_2O, 2% CO) and a lean mixture prepared by injecting O_2 to reach 9% in the mixture for $0.5 < \tau < 120$ s, with s based on the rich mixture from 0.05 to 1, but mostly at $s = 0.1$, $470 < T < 770$ K	Authors observed that a $Pt/Rh/CeO_2/BaCO_3/K_2O/\gamma\text{-}Al_2O_3$ catalyst was not deactivated in the presence of 40 ppm SO_2 in the exhaust under the asymmetric lean-rich modulation.	
Dementhon et al. (2003)	Investigate the effect of low sulfur levels on NO_x trapping and conversion with periodic catalyst regeneration using engine exhaust for simulated driving conditions	$Pt/Pd/Rh/\gamma\text{-}Al_2O_3$ catalyst (10:10:1) on exhaust from light and heavy duty diesel engine burning cetane with varying sulfur content using an asymmetric, periodic pulsed, rich burn regeneration	With the presence of sulfur in a fuel, catalyst regeneration frequency needed to be increased. Periodic rich-burn pulse only partially removes sulfate from the catalyst. Significant increase in rich-burn temperature was needed to gasify sulfate trapped by the catalyst.	
Amberntsson et al. (2003)	The effect of SO_2 on NO_x storage catalysts and their regeneration	Several catalysts prepared with 1 to 4 wt.% Pt, 0.5 to 1.5 wt% Rh, impregnated into a $\gamma\text{-}Al_2O_3$ wash-coated on a cordierite monolith. Experiments used highly asymmetric cycles of rich (400 ppm NO, 500 ppm C_3) and lean (rich feed with 8 vol% O_2) with $\tau = 10$ min and $s = 0.5$	Experiments used regeneration with H_2 for 30 min followed by oxidation with O_2 for 30 min. This regeneration completely restored catalyst storage capacity and NO_x reduction activity after contact with SO_2.	Both Pt and Rh are required for efficient NO_x storage and reduction in a rich-mixture pulse.
Güthenke et al. (2007)	Model development and testing of a converter for NO_x storage and reduction	$Pt/Rh/BaO/CeO_2/\gamma\text{-}Al_2O_3$ catalyst held in a cordierite monolith subjected to a highly asymmetric lean-rich modulation (rich mixture pulsed) at $420 < T < 775$ K for $30{,}000 < SV < 90{,}000$ h^{-1}, also engine exhaust from simulated driving cycles	Input and output exhaust data were used to evaluate parameters of a converter model that encompassed 22 reactions. Model was validated using data from a test engine. Time variation of the mean treated exhaust was well represented.	Model parameters incorporated diffusion in washcoat, but must be re-evaluated for different cell designs and converter operations.

(*Continued*)

TABLE 6-2 Modulation Studies on Diesel and Lean Burn Exhausts *(cont'd)*

Reference	Study Objective	Fuel/Component, Catalyst, Operating Conditions	Observations	Comments
Sharma et al. (2007)	Investigation of multiplicity in NO_x trapping and reduction	Commercial $Pt/BaO/\gamma\text{-}Al_2O_3$ wash-coated onto a cordierite monolith with asymmetric modulation between a lean mixture (500 ppm NO + 5% O_2) and a rich mixture with just 1.4% C_3 added) for $t_{lean} = 60$ s and $10 < t_{rich} < 24$ s and $440 < T_0 < 520$ K	Region of multiplicity extended over a T_0 range of 25 K with a NO_x conversion of 60 to 75% in the upper branch and 15 to 25% in the lower branch. Extent of the multiplicity region was controlled by the duration of the rich-feed pulse.	
Schmeisser et al. (2007; 2009)	Development of an NO_x trapping and reduction model for a commercial monolith catalyst	$Pt/Rh/CeO_2/BaCO_3$ / Al_2O_3 catalyst on a cordierite monolith for $t_{lean} = 60$ s, $2 < t_{rich} < 5$ s, also $t_{lean} = 300$ s, $t_{rich} = 15$ s, $420 < T < 720$ K	With parameters reproduced from step-change data, a model containing 11 elementary reactions fitted experimental cycling data well after the rate parameter for NO oxidation in the lean portion of a cycle was substantially increased.	
Al-Harbi and Epling (2009)	Exploration of the effect of reductant composition on the performance of NO_x trapping and destruction	Commercial $Pt/Rh/CeO_2/BaO/Al_2O_3$ catalyst wash-coated onto a monolith support and an experimental $Pt/BaO/Al_2O_3$ catalyst. Experiments used $20 < t_{lean} < 150$ s and $t_{rich} = 5$ s at $470 < T < 770$ K	Above 570 K, performance of both the experimental and commercial catalyst is about the same for an H_2 or CO reductant or a mixture of the two in the rich pulse. At 470 K, CO inhibits NO_x trapping and reduction.	
Mandur et al. (2011)	Application of a shrinking core model to the operation of an NO_x storage and reduction catalyst	Catalysts in *Al-Harbi and Epling (2009)* above. Cycled gas compositions were 5 vol% CO_2, 5 vol% H_2O with lean gas also having 330 ppm NO and 10 vol% O_2 and the rich gas having 0 to 1.5 vol% CO, $\tau = 150$ s with 3 to 5 s duration of a rich pulse	Shrinking core model encompassing six gas-solid reactions with a depth-dependent diffusivity gave a reasonable representation of an independent set of cycling data obtained for validation purposes.	Authors suggest that the variable diffusivity relation allows the application of the model to different monolith geometry and catalyst formulations.

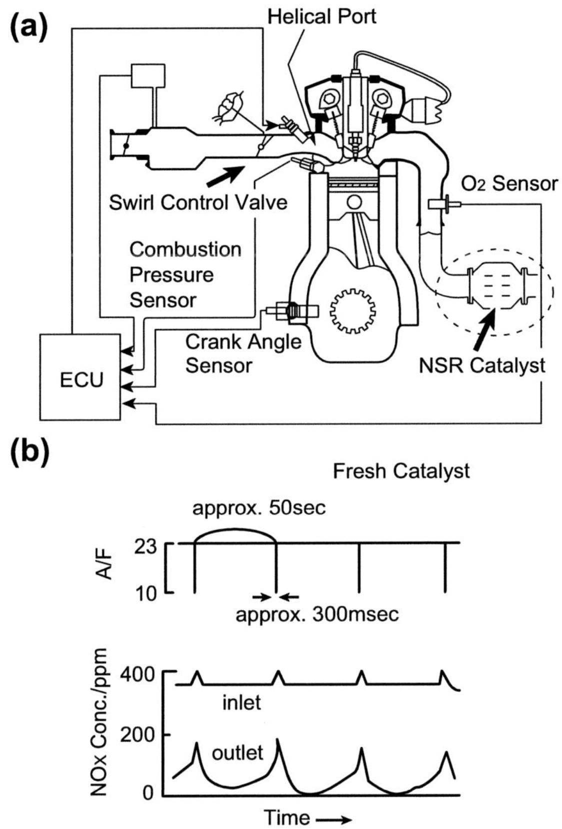

FIGURE 6-9 Schematic in (a) of a lean burn engine with sensors, swirl valve, helical port, control unit and NSR catalyst tank. In (b), the periodic rich mixture millisecond pulses (with *A/F* ≈ 10) are shown. NO_x trapping by the alkaline earth oxide at *A/F* ≈ 23 drops NO_x from almost 400 ppm to 30 to 50 ppm. *(Figure from* Matsumoto and Shinjoh (2007) *with the authors' permission.)*

decomposes the nitrate over the noble metal (Pt) involved either to N_2 or NH_3. NH_3 is sometimes preferred if a downstream SCR is used because it can lead to higher ultimate conversions. Although the description above applies to the overall behavior, there are many complications (*Prinetto et al., 2001; Nova et al., 2002; Forzatti et al., 2009*): Oxides and carbonates are formed and react with nitrates or with barium oxide. Mixed hydroxides and carbonates also occur. Nitrites can be present and are also trapped by the solid oxides. Not all the solid oxide is active in the reduction pulse. Adjacent or nearby BaO and Pt seem to be needed for effective reduction. Spillover has been shown to occur.

Sulfur poisoning remains a problem despite decreasing sulfur content in gasoline. SO_2 present in the exhaust stream reacts with BaO and other oxides to form stable sulfates thereby reducing capacity for NO_x capture in the oxidizing step. Sulfate decomposition and reduction can occur, but at higher temperatures and under reducing conditions. Promoting steam reforming of CO or HCs into H_2 in the rich pulse has been used. When available, H_2 reacts with the sulfate to remove it as S_x or

SHINICHI MATSUMOTO AND HIROFUMI SHINJOH

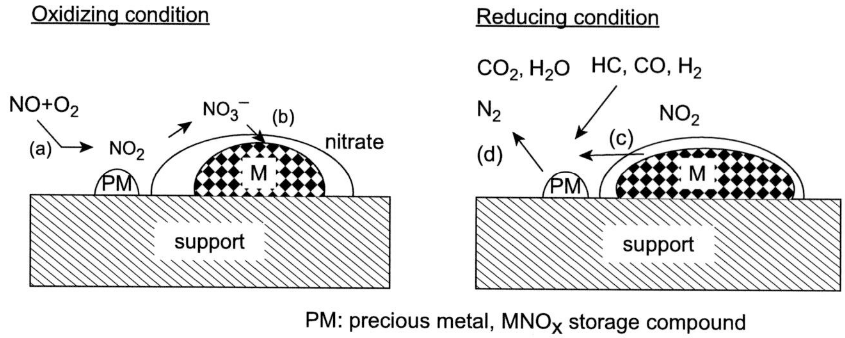

FIGURE 6-10 Schematic of an NSR system showing conversion to NO_2 and trapping by an alkaline earth oxide in an oxidative environment and decomposition of nitrates in a reducing pulse together with reduction to nitrogen. *(Figure reproduced from* Matsumoto and Shinjoh (2007) *with permission of the authors.)*

H_2S. Adding Rh/ZrO_2 to the Pt-BaO-γ-Al_2O_3 system and using a monolith support with a hexagonal cell (to give a shallower and more uniform washcoat) has been shown by *Matsumoto et al. (2000)* to improve NSR performance in the presence of SO_2. *Yamazaki et al. (2004)* have found that Fe affects SO_2 poisoning by catalyzing decomposition of $BaSO_4$ during the reducing pulse. Fe oxides seem to decrease thermal deterioration and depress CO poisoning as well. The addition of TiO_2 seems to disperse the sulfates and promote desorption in the rich pulse. Indeed, *Takahashi et al. (2009)* suggested that TiO_2-ZrO_2 is an effective K_2O storage support and helps convert the sulfate in the reducing pulse. Stability of this support with respect to sintering is increased by adding Al_2O_3 to an apparent solid solution of the oxides.

Kočí et al. (2004b) explored the problem of varying exhaust temperatures encountered as driving conditions change by simulating the use of low and high temperature beds for NO_x trapping and destruction. A metallic monolith investigated by *Maunula et al. (2001)*, as shown in **Figure 6-11**, was considered. The corrugated portion was wash-coated with a γ-$Al_2O_3/CeO_2/K_2O$ material and impregnated with a Pt/Rh catalyst, active at higher temperatures, while the flat portion was wash-coated with γ-$Al_2O_3/CeO_2/BaO$ and impregnated with just Pt to form a catalyst active at lower temperatures. *Kočí et al.* applied the model introduced earlier to the system using experimentally evaluated rate constants obtained from cyclic experiments. Their simulation demonstrated that the dual bed NO_x trap enhanced NO_x removal significantly.

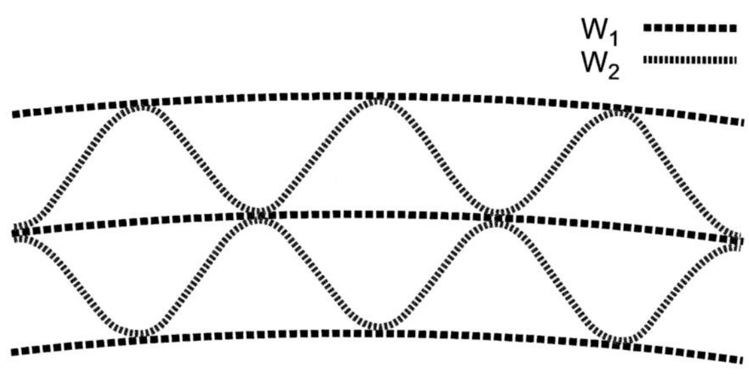

FIGURE 6-11 Two component (flat and corrugated) metal gauze packing supporting differentiated wash-coated catalysts for NO_x storage and destruction tested by *Maunula et al. (2001)* and simulated by *Kočí et al. (2004b)*. *(Figure reproduced from* Kočí et al. (2004b) *with permission. © 2003 by Elsevier B.V.)*

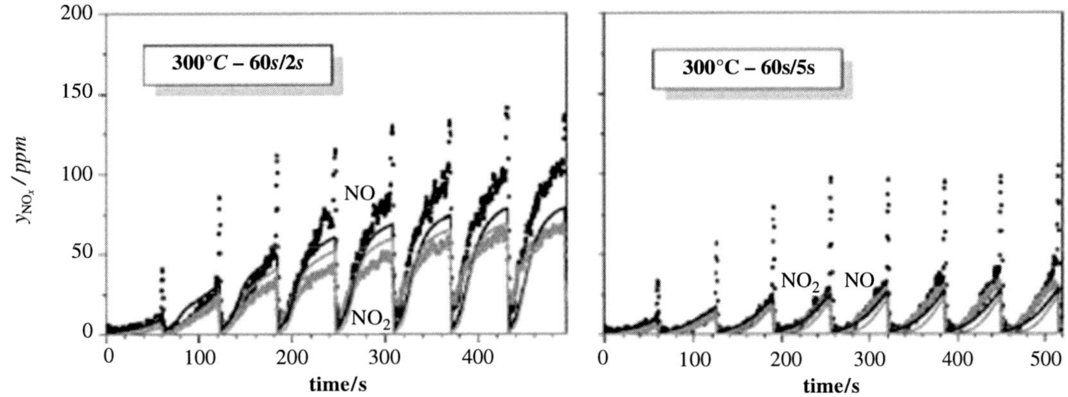

FIGURE 6-12 **Comparison of simulated and measured off gas NO and NO_2 concentration leaving an isothermal laboratory catalytic converter at 573 K under modulation between a lean mixture (10 H_2, 0.04 CO, 7 CO_2, 12% O_2, 200 ppm NO, 40 ppm NO_2, 60 ppm C_3H_6) and a rich mixture in which CO increased from 0.04 to 2.1% and O_2 dropped from 12 to 0.9%. Dots show measurements; lines the simulation.** *(Figure reproduced from* Schmeisser et al. (2009) *with permission. © 2004 by Springer Verlag.)*

Studies directed at improving NSR performance have been reported by *Olsson et al. (2006), Scholz et al. (2007) and Schmeisser et al. (2007, 2009)*. The latter two studies focused on the BaO component, the NO_x reactions and catalyst regeneration. *Schmeisser et al.* conjectured that once on stream, BaO forms a nitrate shell on the particle surface in the lean phase of the cycle which expands inward with time retarding NO_2 diffusion to the carbonate interface toward the particle center (see **Figure 6-10**). Later in the cycle, when the gas phase is briefly reducing, the barium nitrate shell disintegrates releasing NO_2, which is reduced to NO, and barium carbonate forms. Thus modulation switches the surface layer from nitrate to carbonate and back again. *Schmeisser et al.* modeled the NSR system in just one dimension, neglecting diffusion in the washcoat and using a much smaller set of reactions than *Kočí et al.* Oxygen storage on CeO was included. They could not reproduce the disintegration of the nitrate shell on a BaO partricle, but the model did capture much of the NO/NO_2 behavior. **Figure 6-12** compares simulation and measurement for NO and NO_2 at two regeneration durations.

A combination of separate catalysts was investigated and led to prevention of sulfur poisoning . The combination led to additional NO_x reduction in the high *A/F* environment. The system is illustrated by **Figure 6-13** using

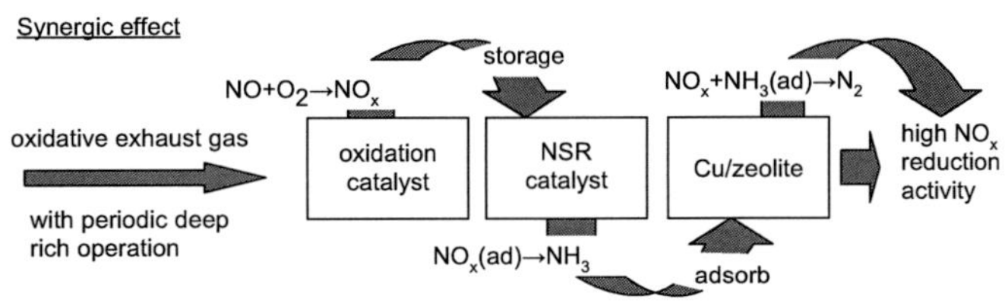

FIGURE 6-13 **Schematic of a sequence of different catalyst beds to improve nitrogen oxide removal.** *(Figure from* Matsumoto and Shinjoh (2007) *with permission. © 2007 by Elsevier, Inc.)*

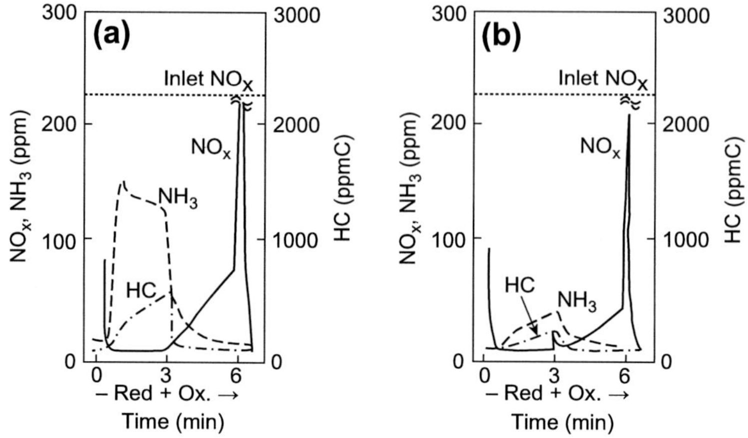

FIGURE 6-14 Improvement in pollutant removal through staging catalyst beds with Pd/Al_2O_3 followed by an NSR bed. Comparison of parts (a) and (b) shows a substantial reduction of NO_x by adding the second Cu/zeolite bed. *(Figure from* Matsumoto and Shinjoh (2007) *with permission. © 2007 by Elsevier, Inc.)*

Pd/Al_2O_3 upstream from the NSR unit to oxidize NO compounds to NO_2.The NSR catalyst further traps NO_2 on BaO coated onto the support and converts these oxides to N_2 and NH_3 during the short reducing pulse. By adding a copper/zeolite SCR catalyst downstream of the NSR catalyst, the NH_3 formed on the trap can adsorb onto the SCR catalyst and then react with NO escaping from the NSR catalyst to form N_2. **Figure 6-14** compares NO_x profiles when a separate zeolite bed is added. It can be seen that HC and NH_3 are lowered as well the NO_x escaping. Thus the sequence takes NO_x to N_2.

Harold and co-workers have discovered that multiple states can exist under periodic operation in a single bed NSR system (*Sharma et al., 2005; Sharma et al., 2007; Muncrief et al., 2004*). These states are observed as fluctuating bed temperatures and exit composition with time. This multiplicity is confined to a range of feed temperatures extending from about 450 to 520 K. The upper branch of that region represents the highest NO_x conversion. Indeed, the highest conversion is at the extinction point.

Fundamental studies continue on the mechanisms occurring in NO_x reduction. *Okumura et al. (2009)* studied NO reduction by hydrogen in pulsed operation using a wide range of catalysts and metal oxide supports and promoters. They observed that the activity of the Pt and Pd catalysts depended dramatically on the support used and resulted from the ability of the supported metal to dissociate NO. Transient formation of NH_3 was also observed for these catalysts.

6.8 APPLICATION OF NSR TO DIESEL EXHAUSTS

Over the last decade, simple catalytic mufflers have been installed in diesel passenger vehicles. These oxidize CO and unburnt hydrocarbons and operate continuously under lean conditions of $A/F > 14.7$. Commercial diesel vehicles, mainly heavy trucks, and many passenger cars within the European Union now must employ soot traps (*Twigg, 2007*). These can be regenerated through oxidation of the accumulated soot by NO_2 at temperatures typically above 300°C, or by periodically injecting a pulse of fuel into the exhaust prior to the oxidative regenerator, which induces a large temperature rise at the soot filter through combustion of this added fuel. At these higher temperatures (> 550°C), soot is gasified by O_2 in the stream. Although low amplitude

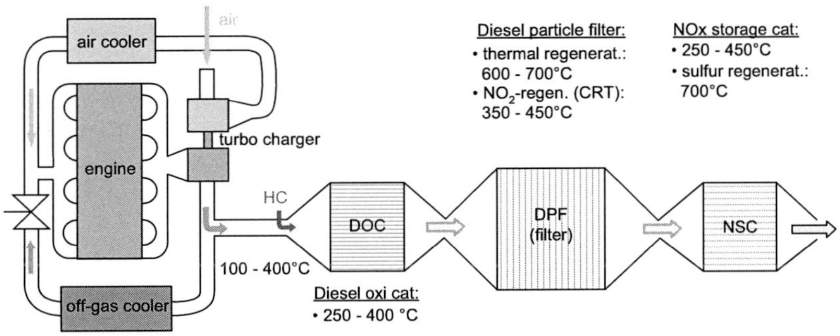

FIGURE 6-15 **Diesel exhaust detoxification train showing the temperature range for each process step. The system uses periodic injection of fuel upstream from the catalytic oxidation unit to generate a hot, lean gas to combust soot trapped on the filter. Operating temperature range shown for each process unit.** *(Figure adapted from* Eigenberger et al. (2007) *with permission. © 2007 Elsevier Ltd.)*

A/F modulation in diesel exhaust does not seem to play a role in detoxification, modulation through fuel injection is required to economically treat the exhausts.

Regulations requiring NO_x removal from diesel exhaust exist in both the European Union (*Eigenberger et al., 2007; Twigg, 2007*) and the US. Different architectures have been proposed, due to challenging heat integration issues. For example, placing either an ammonia-based SCR catalyst or a periodically regenerated NO_x trap after the particulate (soot) filter requires cooling the gas during soot filter regeneration, unless during this time NO_x slip is allowed. Furthermore, during soot filter regeneration, outlet temperatures will be higher because of the exothermic soot combustion reaction, which can lead to thermal degradation of the downstream components. With the NO_x catalysts downstream, there is also a longer warm-up period as a result of the thermal mass of the upstream soot filter. **Figure 6-15** illustrates the dilemma showing the temperature ranges of diesel engine exhaust and the three after-treatment components. The alternate is having the SCR or NSR catalysts upstream of the soot filter, but this leads to difficulties in getting the soot filter to temperatures high enough for regeneration with O_2, as little to no NO_X should reach the filter; thus, higher regeneration temperatures are required.

Possible solutions to what is essentially a heat integration problem have been proposed (*Eigenberger et al., 2007; Schejbal et al., 2010*). Exhaust control systems on heavy duty trucks can use a concentric arrangement in which the SCR or NSC unit surrounds the oxidation bed and the particulate filter (*Walker et al., 2003*). Heat loss from the outer NO_x unit cools the gas to close to its optimal temperature.

Research on decontamination of diesel exhaust has increased dramatically since the 1990s. **Table 6-2** summarizes some representative contributions.

6.9 DOES *A/F* MODULATION IMPROVE CONVERTER PERFORMANCE?

Chapter 3 described large increases (10- to 40-fold) in the rates of CO oxidation under composition modulation on supported Pt or Pd catalysts at temperatures under 200°C. Modulation substantially increased C_3 oxidation rates on supported Pt (**Chapters 4** and **5**). In **Chapter 3**, 10-fold enhancements of the NO reduction rate over Pt/Al_2O_3 through modulation of the CO and NO concentrations are described. On the other hand, in this chapter, *A/F* modulation was found to diminish NO, HC or CO conversions on TWCs containing these same noble metals. Why has the promise of the low temperature CO oxidation or NO reduction modulation studies on the noble metal catalysts not been realized as better performance by catalytic converters under *A/F* modulation? There appear to be several reasons. The first of these is that increasing reaction temperature reduces the optimal cycle period at which conversion is at its highest. With a Pt catalyst, τ_{opt} is 10 to 20 s at 370 to 420 K, but drops to 1 to 2 s in the 550 to 670 K range studied by the Toyota workers and appears to be under 1 s at 820 K (*Silveston, 1995*). The frequency of the *A/F* modulation caused by the feedback control loop ranges between 0.5 and 2 Hz so in the normal temperature range of converters, 620 to 880 K, the *A/F* modulation at the entrance to the converter is suboptimal. The explanation of this temperature effect is that converter reactions involve strong adsorption of a reactant which inhibits oxidation or reduction. For CO oxidation over Pt or Pd, oxidation is inhibited by either CO or O_2. The role of *A/F* modulation is to provide surface access for the less strongly adsorbed reactant, thus diminishing inhibition. Other contributing factors are discussed by *Silveston (1998)*.

Nomenclature

A = amplitude (various units)
$(A)_{mean}$ = time average amplitude in a cycle
A/F = air- fuel ratio (-)
$(A/F)_{mean}$ = time average air fuel ratio in a cycle
DOC = diesel oxidation catalyst
DPF = diesel particle filter
HC = hydrocarbon
IR = infra red
NSC = nitrogen (NO_x) storage catalyst
NSR = nitrogen (NO_x) storage and reduction catalyst
ODE = ordinary differential equation
PDE = partial differential equation
r = depth on surface (μm)
SCR = selective catalytic reduction
SN = stoichiometric number (see **Eq. (6-1)**)
$(SN)_{mean}$ = time average stoichiometric number in a cycle
STP = standard temperature and pressure
SV = space velocity (s^{-1})
s = cycle split (-)
s = catalytic site
T = temperature (°C, K)
T_0 = inlet or initial temperature
TWC = three-way catalyst
t = time (s, min.)
t_{lean} = duration of lean feed part of a cycle
t_{rich} = duration of rich part of a cycle
y_i = mol fraction of species i
WGS = water gas shift
w_i = gauze sheet

Greek

α = coefficient
θ = fractional occupation of catalyst sites
τ = cycle period (s, min)
ω = frequency (s^{-1})

Subscripts

α = stoichiometric coefficient
lean = oxygen deficient for combustion (SN < 1)
mean = time average
opt = optimum
rich = excess oxygen for combustion (SN > 1)

CHAPTER

7

Polymerization Under Modulation*

Gregorio R. Meira, Peter L. Silveston*[†]

[*]Santa Fe, Argentina and [†]Waterloo, Ontario, Canada

OUTLINE

7.1 Introduction 171

7.2 Simulation of Polymerization Under Input Modulation 177

7.2.1 Coordination Polymerization 179

7.2.2 Free Radical Polymerization 188

7.3 Experiments on Polymerization Under Input Modulation 188

7.3.1 Emulsion Polymerization 188

7.3.2 Anionic Polymerization 192

7.3.3 Free Radical Polymerization 193

7.3.4 Ziegler-Natta Catalysts 193

7.3.5 Deposition Polymerization 193

7.4 Spontaneous Oscillations 194

7.5 Saturation of Polymers 200

7.6 Assessment 201

7.1 INTRODUCTION

Polymerization was an early focus for periodic operation. *Ray (1968)* and *Laurence and Vasudevan (1968)* observed that the molecular weight distribution of a polymer may be modified by periodically varying the concentrations of monomer, initiator, or transfer agent fed to a polymerization reactor. Since those original studies, a large literature on this application of periodic operations has developed.

Concepts and terminology differ considerably from applications considered previously, so a short introduction to polymerization begins this chapter.

Polymers are high molar mass materials characterized by repeating units of one or several monomers. Homopolymers contain a single type of repeating unit, copolymers contain two or more. The total number of repeating units in a polymer is its chain length or degree of polymerization. Many synthetic polymers are linear molecules made up by

* The editors wish to acknowledge the help of professor J.B.P. Soares (university of waterloo) with the preparation of this chapter.

repeating units of dual functionality. Because of their high molar mass, between 100,000 and 300,000 g/mol, and because of chain entanglement, polymers are solids at room temperature but may become viscous liquids between 100 and 300°C. Their molar masses are a compromise between mechanical properties (such as elastic modulus and tensile strength) and processability in the molten state. If the polymer is crosslinked, the molar mass is essentially infinite and the polymer will not flow upon increasing temperatures.

There are two main polymerization mechanisms. In chain polymerization, the linear polymer grows by reaction between a monomer molecule and a reactive site on the chain end, with the reactive site being regenerated after each propagation step. Chain building may be *free-radical*, *anionic*, or *cationic*. Massively exploited free-radical polymerizations enable the production of polyethylene, polypropylene, polyvinyl chloride and polystyrene. In step-polymerization, all molecules have reactive chain ends so chains grow by reaction between molecules of different degrees of polymerization. Most step-polymerizations are polycondensations. Two important ones are Nylon 6,6 and polyethylene terephtalate or PET.

Most polymerizations are highly exothermic and employ liquid monomers. In bulk polymerization, the monomer is a liquid but the polymer may or may not be soluble in the liquid. Thus, polymers can be either homogeneous or heterogeneous according to polymer solubility. Liquid viscosity increases dramatically with conversion, so conversion is often limited to ensure adequate mixing and product extraction. Heat removal and the viscosity problem can be solved if a polymerization can be carried out in a dispersed phase, often in an aqueous medium. Thus, free-radical polymerization of nonpolar monomers is done in either a suspension or an emulsion.

Synthetic polymers exhibit a molar mass distribution (MMD), also referred to as a molecular weight distribution (MWD), which characterizes their molecular macrostructure. These distributions can be represented numerically by a mean molecular weight on a chain number basis, $\overline{M_n}$; or on a weight basis, (either $\overline{M_w}$).
The first and second moments are also used to represent distributions. Many polymers are branched and therefore have distributions of the number of long-chain branches per molecule, while copolymers also exhibit distributions of their chemical composition.

Homopolymer microstructure describes the orientation of the different isomers along the chain. For copolymers, microstructure shows the sequence distributions for the co-monomers. A homopolymer is atactic if there is no systematic or regular configuration in their repeating units. Atactic polymers are amorphous and have a glass transition temperature (T_g). Polymers with chain regularity are semi-crystalline when cooled slowly. Semi-crystalline polymers generally have improved mechanical properties compared with amorphous polymers. Copolymers widen the range of feasible physical properties with respect to the base homopolymers. According to their sequence distribution, copolymers are classified into: (a) statistical, (b) alternate, (c) linear block and (d) branched graft copolymers.

Although molar mass and number chain length distributions or moments of these distributions are calculated directly in most simulations and describe the distribution adequately, they are not widely used in practice. Instead, the number average chain length, $\overline{\mu}_n$, or mean molar masses (molecular weights), and the dispersity, D, are normally used for characterization. The latter, defined as $D \equiv \overline{M}_w/\overline{M}_n$, and is given by

$$D = \left(\frac{\sigma_n}{\overline{M}_n}\right)^2 + 1 \tag{7-1}$$

where σ_n is the standard deviation of the number molar mass distribution. Values of D between 1.05

and 1.2 are indicative of the very narrow MMDs that are typical of living anionic polymerization. Dispersities between 1.5 and 3.0 are typical of conventional free-radical polymerizations; dispersities of 20 or higher are typical of polyolefins obtained *via* Ziegler-Natta catalysts.

Table 7-1 presents the three basic polymerization mechanisms for the synthesis of linear homopolymers. When these reactions are carried out in a batch reactor, the evolution of the number-average chain length with conversion is given by **Figure 7-1**. Note that side-reactions and reactions with the accumulated

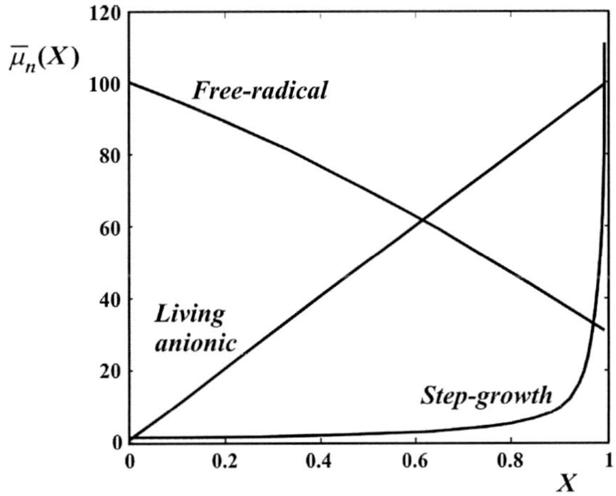

FIGURE 7-1 **Typical evolution of the number-average chain-length $\bar{\mu}_n$ with conversion X, for three ideal polymerization mechanisms: conventional free-radical, living anionic, and step-growth.**

TABLE 7-1 Basic Polymerization Mechanisms for the Synthesis of Linear Homopolymers

Chain Growth (Conventional Free-Radical)	**Chain Growth (Living Anionic)**	**Step Growth**
Initiation	Initiation	
$I \xrightarrow{k_1} 2\,R_c^*$	$I + M \xrightarrow{k_1} I\,P_1^*$	
$R_c^* + M \xrightarrow{k_2} R_1^*$		
Propagation	Propagation	Propagation
$R_1^* + M \xrightarrow{k_p} R_2^*$	$I\,P_1^* + M \xrightarrow{k_p} P_2^*$	$P_n^* + P_m^* \leftrightarrow P_{n+m}^* + C$
$R_n^* + M \xrightarrow{k_p} R_{n+1}^*$	$P_n^* + M \xrightarrow{k_p} P_{n+1}^*$	
Termination	Termination	
$R_n^* + R_m^* \xrightarrow{k_{td}} P_n + P_m$	$P_n^* + K \xrightarrow{k_t} P_n$	
$R_n^* + R_m^* \xrightarrow{k_t} P_n$		
Transfer		
$R_n^* + M \xrightarrow{k_{fm}} P_n + R_1^*$		
$R_n^* + S \xrightarrow{k_{fs}} P_n + S^*$		
$R_n^* + T \xrightarrow{k_{ft}} P_n + T^*$		
$S^* + M \xrightarrow{k_{sm}} R_1^* + S$		

I = initiator; M = Monomer; R_c^* = primary radical; R_1^* = activated monomer or free-radical of unit chain length; R_n^*, R_m^* = free-radicals of chain lengths n, m; P_n = dead polymer of chain length n; P_n^* = reactive anion or reactive polymer molecule of chain length n; S, S^* = solvent and solvent radical; T, T^* = transfer agent and transfer agent radical; K = deactivating (or "killing") agent; C = condensation product (e.g., water and methanol). For the living anionic mechanism the molecules are linear and have reacting groups at each chain end; P_1^* represents a special monomer.

dead polymer have been neglected in the table. Reactions are assumed to be simple and irreversible. In reality, many involve complex mechanisms. Examples of these are the propagation steps in the synthesis of polypropylene via a gas-solid Ziegler-Natta coordination polymerization and in the ring-opening polymerization of ε-caprolactam to produce Nylon 6.

In conventional free-radical polymerization, there is a slow, continuous generation of primary initiator radicals, R_c^*, via homolytic cleavage of an initiator (typically a peroxide). Not all primary initiator radicals initiate a polymer chain so an efficiency factor is introduced into the kinetics. The final initiator conversion is kept low to ensure an essentially constant generation of free-radicals during the reaction, and the concentration of free-radicals is always very low (in the order of 10^{-7} M). A balance exists between rates of generation and deactivation. Termination by disproportionation results in two dead polymer molecules. However, termination by recombination yields a single dead polymer molecule. Transfer reactions can occur to the monomer, to the solvent, or to a chain transfer agent. All terminate a growing chain and therefore lower the average molar mass but do not reduce the total concentration of free radicals. The lifetime of a growing radical is around 1 s and is negligible compared with the total reaction time. The final chain length results from a competition between the rates of propagation and termination. When consumption of monomer is faster than that of the initiator, the average molar mass decreases with conversion. However, this is subject to a "gel" effect at high viscosities that slows the effective rate of termination with respect to propagation and therefore increases the average molar mass. Branching reactions are not included in the above table, but they can occur when the polymer contains terminal double bonds and/or by chain transfer to the polymer. In the absence of reactions with the accumulated polymer, the instantaneous molar mass distribution exhibits a Schulz-Flory distribution ($D > 2$) for the fraction of free-radicals that terminate by disproportionation or chain transfer and a Schulz-Zimm distribution with ($D > 1.5$) for those that terminate by recombination.

In living anionic polymerization, bimolecular termination between living ends is avoided because anions repel each other. These polymerizations require high-purity conditions to avoid termination reactions by impurities and must be carried out in organic solvents. Ideally, the number of growing anions is constant and equal to the number of initiator molecules. Consequently, the number-average chain length is directly proportional to monomer conversion (see **Figure 7-1**) and depends on the ratio between moles of reacted monomer and initiator. Furthermore, if the rate of initiation is $\geq$ the rate of propagation, the living chains grow almost simultaneously and this yields the narrowest possible molar mass distribution: the Poisson distribution. If the rate of initiation $\leq$ the rate of propagation, the Gold distribution results. Apart from enabling the production of narrow homopolymers of controlled molar mass, living anionic polymerization also enables the synthesis of "tailor-made" molecular and morphological structures.

At a basic level, step-polymerization is represented by just a propagation reaction. For linear polymers, bifunctional monomers are required and intramolecular reactions must be avoided. Often, the monomer contains two different reacting groups, e.g., an amino acid. Any growing polymer molecule will exhibit the same types of reactive group at their chain ends. Alternatively, 50:50 mixtures of two co-monomer types with identical reacting groups can be employed. Unlike chain reactions, in step-polymerization, the monomer is rapidly consumed. For example, 1% of the monomer remains unreacted when the average chain length is only 10. Step-growth reactions are similar to living anionic polymerizations in that all the molecules remain potentially reactive while inside the reactor and the average molar masses increase monotonically with conversion. However, the increase of molar mass is highly nonlinear in step-polymerizations. Conversions very close to unity are required for high molar masses (see **Figure 7-1**). Ideal step-polymerization of bifunctional monomers produces linear polymers with Schulz-Flory distributions. Chain length distributions were mentioned previously. The most important of these are the Schulz-Flory and Poisson distributions — **Figure 7-2** shows both. For linear polymers, these chain length distributions along with mean molar masses (M_w, M_n) determine their mechanical properties. Contacting of monomers in a reactor as well as the polymerization kinetics shape the chain length or molar mass distributions and mean molar masses as indicated in **Table 7-2**.

Industry employs a variety of batch, semicontinuous, and continuous polymerization reactors. Among the last-named group, the plug flow reactor provides the narrowest residence time distributions (RTDs). Polymerization performance, in principle, under isothermal or adiabatic operation is equivalent to that of a batch stirred-tank reactor. The

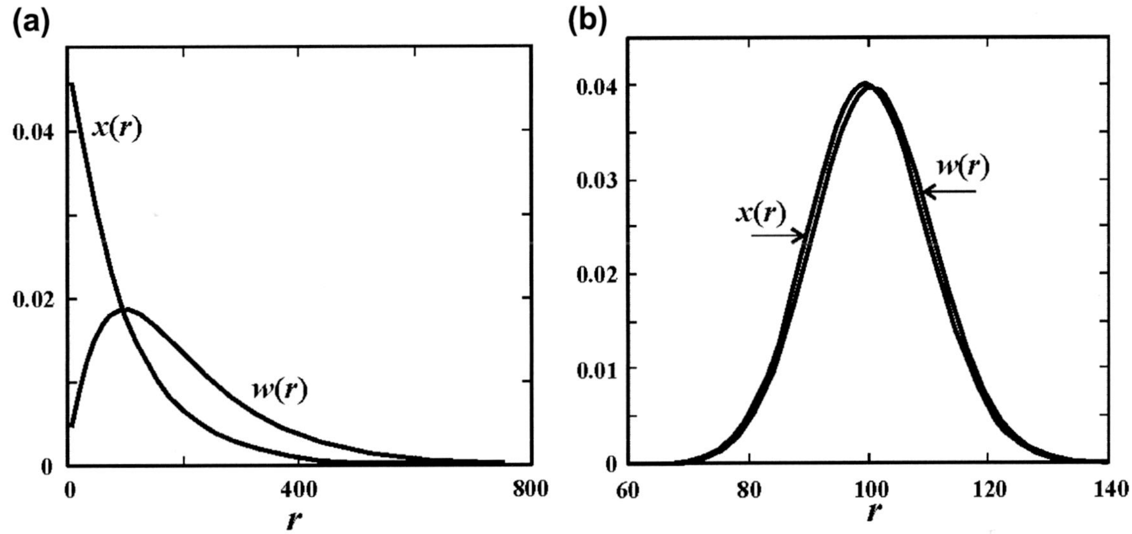

FIGURE 7-2 **Number- and weight-chain-length distributions of a number-average chain length $\overline{\mu_n} = 100$ for: (a) a Schulz-Flory distribution of $D \cong 2$; and (b) a Poisson distribution of $D \cong 1.01$. The ordinates are either molar fraction x or weight fraction w.**

broadest RTD is obtained from a micro-mixed CSTR. A cascade of two or three CSTRs or a macro-mixed CSTR results in an intermediate RTD.

Table 7-2 presents the theoretical molar mass distributions obtained when the ideal mechanisms of **Table 7-1** are combined with the three types of ideal reactors just mentioned. In living anionic or step-growth polymerizations all chains maintain their growing capacity while inside the reactor. Thus, when these processes are carried out in a micro-mixed CSTRs, the distributions are considerably broader than those for batch reactors. With ideal anionic mechanisms, distributions vary from Poisson- or Gold-type in batch reactors to Schulz-Flory

TABLE 7-2 Molar Mass Distributions of Linear Polymers for Different Combinations of Ideal Polymerization Mechanisms and Reactor Types

	Ideal Polymerization Mechanism		
Ideal Reactor	**Free-Radical**	**Living Anionic**	**Step Growth**
Batch or Plug Flow	With increasing conversion, molar mass distribution becomes broader than Schulz-Flory.	Poisson distribution for $k_i \geq k_p$ and Gold distribution for $k_i < k_p$.	Schulz-Flory distribution.
CSTR (micromixed)	Schulz-Flory distribution with $1.5 < D < 2$, according to type of termination.	Schulz-Flory distribution.	Much broader than Schulz-Flory distribution.
CSTR (macromixed)	With increasing conversion, molar mass distribution becomes broader than Schulz-Flory.	Schulz-Flory distribution for $p = 0$ and Poisson distribution for $p = 1$.	Molar mass distribution in between those for Free-Radical and Living Anionic.

in micro-mixed CSTRs as the table shows. Phase separation can occur in the later stages of the polymerization turning the reactor into a segregated flow system and switching CSTR operation from micro- to macro-mixed.

The brief summary in the previous paragraphs is intended to assist readers unfamiliar with polymerization concepts and terminology. Polymerization mechanisms and kinetics, effects of temperature, solvents, initiators, transfer agents, and deactivators are dealt with in several monographs and reviews. More detailed background will be found in books and reviews on polymerization reaction engineering, such as *Reichert and Moritz (1989a, b), Hamielec and Tobita (1992, 1998), Ray et al. (2004), Meyer and Keurentjes* (2005) and *Asua (2007).*

Polymerization processes consist of three stages: preparation, polymerization and separation. In the *first*, monomers and other reagents are purified. These are vital operations since small impurities can drastically affect the polymer produced. In the *polymerization* stage, heat exchange and fluid viscosity crucially affect polymerization and the degree of mixing. Polymer purity and state depends on the third stage that often involves thermal and mechanical operations, possibly including recycle of solvent and unreacted monomer. The final mechanical, physical and chemical properties of a polymer are acquired mainly at the reaction stage but may be modified during separation and post-reaction processing.

Polymerization reactors are multivariate dynamic systems with interrelationships between their inputs and outputs often only scarcely known. Close monitoring and control of a polymerization reactor is necessary. **Figure 7-3** illustrates the various feed-back loops associated with control of a continuous polymerization reactor for the production of a thermoplastic material. The inner loops are automatic and control base process variables (temperature, pressure, level and conversion). The second level is discontinuous and involves off-line measurements of mechanical properties.

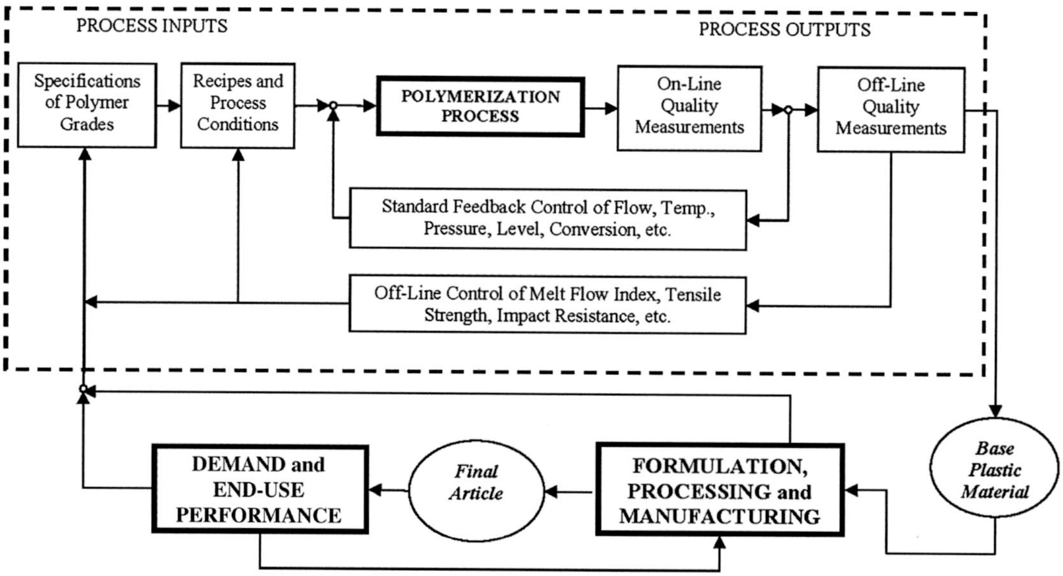

FIGURE 7-3 Feed-back control loops for a typical thermoplastic process. *(Figure adapted from* Meira et al. (2007) *with permission. © 2007 by Wiley-VCH.)*

In addition, because the manufacturing industries produce the final products after formulation and processing of the base plastic material, and the outer feedback loop suggests modifications of the base material specifications provided by end users.

Reviews of the monitoring and control of polymerization reactors are given *MacGregor et al. (1984)*, *Eliçabe and Meira (1988)* and *Richards and Congalidis (2006)*.

Periodic operation has been considered for polymerization since the earliest simulations mentioned at the beginning of this chapter. Although light modulation has long been used to investigate the mechanism and kinetics of free radical polymerization, this application of modulation will not be examined here. It is considered briefly in **Chapter 13**.

7.2 SIMULATION OF POLYMERIZATION UNDER INPUT MODULATION

Table 7-6 summarizes many of the published contributions to the simulation literature. It can be seen that a wide variety of processes, mechanisms, reactors types, modulated variables and waveforms have been considered. The long-chain material formed in polymerization can be described in several ways mentioned above and most have been used in the simulations as **Table 7-6** indicates.

Simulations use part or all of the set of elementary reactions shown in **Table 7-1** for the three major classes of polymerization mechanisms. Complications arise mainly in formulating the rate terms (see **Table 7-3**). Heterogeneously catalyzed polymerization can follow either a "living" chain or free radical mechanism. Like emulsion polymerization, they involve a second phase so that transport steps must be added to a reactor model.

In all mechanisms, polymers with different chain length will be present in a reactor. Material balances and, if temperature is manipulated, a heat balance lead to a large set of simultaneous, ordinary differential equations as **Table 7-3** illustrates for a free radical mechanism (neglecting the use of chain transfer agents).

The calculational problem of solving a large number of equations may be avoided by introducing population balances which replace polymer concentrations of different chain lengths by the moments of the polymer number chain length distribution. Moment generating functions or the z-transform (**Table 7-4**) can be used to provide the equations for these moments. **Table 7-5** gives these equations for the simplified free radical model given in **Table 7-1**. Details of the transform technique are given by *Ray (1967, 1972)*.

The number average chain length and the polydispersity may be calculated directly from the moments as:

$$\overline{\mu}_n = \left(\sigma_1/\sigma_o\right) \tag{7-2}$$

$$D = \left(\sigma_o\sigma_2/\sigma_1^2\right). \tag{7-3}$$

Simulation studies listed in **Table 7-6** have been concerned with the influence of periodic operation on the molar mass distribution or the number average chain length $\overline{\mu}_n$, the monomer conversion X_m and the dispersity, D. Chain length and monomer conversion, as in any chemical reaction, depend on residence time, operating conditions, contacting pattern in the reactor and, of course, on the polymerization mechanism. Increasing residence time, raising monomer concentration, introducing plug flow characteristics, for example, all increase $\overline{\mu}_n$ and X_m. These parameters of polymerization reactions are limited only by the viscosity of the monomer-polymer solution. Increasing viscosity as polymerization proceeds leads to a switching of rate control from kinetic to diffusional processes and, ultimately, gel or cage effects appear. With dispersity, however,

TABLE 7-3 Material and Heat Balances for a Simplified Free Radical Polymerization Mechanism (*Konopnicki and Kuester, 1974*)

Component mass balances:*[t]

$$\left({}^{dI}\!/_{dt}\right) = (1/\theta)(I^0 - I) - k_i I \tag{7-4}$$

$$\left({}^{dM}\!/_{dt}\right) = (1/\theta)(M^0 - M) - 2k_i fI - (k_p + k_{fm})MR_c \tag{7-5}$$

$$\left({}^{dR_1}\!/_{dt}\right) = 2k_i fI - (R_n/\theta) - (k_p + k_{fm})MR_1 + k_{fs}S\,(R_c - R_1) + R_c(k_{fm}M - k_t R_1) \tag{7-6}$$

$$\left({}^{dR_n}\!/_{dt}\right) = -(R_n/\theta) + k_p M(R_{n-1} - R_n) - R_n(k_t R_c + k_{fs}S + k_{fm}M) \tag{7-7}$$

$$\left({}^{dP_n}\!/_{dt}\right) = -(P_n/\theta) + (½)k_t \sum_{m=1}^{n-1} R_{n-m} + R_n(k_{fs}S + k_{fm}M) \tag{7-8}$$

where n, m = 1, 2, 3, …, ∞.

Energy balance:

$$v_T\left(S^o(C_p)_S + M^o(C_p)_M\right)\left(\frac{dT}{dt}\right) = v_T\left(S^o(C_p)_S + M^o(C_p)_M\right)(T^0 - T) - UA(T - T_i) - V(\Delta H_R)_p k_p M \sum_{n=1}^{\infty} R_n. \tag{7-9}$$

* *Assumes chain transfer agent not present.*
[t] *CSTR assumed; other assumptions are standard.*

limiting values may be observed even at low $\overline{M_n}$ where viscosity and gel phenomena are absent. For example, with "living" anionic polymerization in batch or plug flow reactors, *D* will be unity. A dispersity of two, indicating a Schultz-Flory distribution of polymer chain lengths, arises with step-growth polymerization in a plug flow or batch reactor, with "living" anionic polymerization in a fully back mixed reactor, or with free radical polymerization in the latter reactor when termination occurs only through disproportionation (and in the absence of high viscosity or gel formation). However, *D* becomes 1.5 if recombination is the only termination mechanism. In all these cases, steady-state operation is assumed.

With any given co- or homopolymer, physical properties and thus value of the polymer depends on $\overline{\mu}_n$ and *D*. The monomer conversion, X_m, of course, affects process economics. Perhaps because of the economic significance as well as the limits on *D* just discussed, investigators of periodic operation have focused on the dispersity.

Table 7-6 demonstrates that a considerable effort has gone into the investigation of the periodic operation of polymerization reactors through simulation. Much of this earlier work has been examined by *Meira (1981b)* in a very

TABLE 7-4 Transformations Converting Polymer Mass Balances to Moments (*Konopnicki and Kuester, 1974*)

Z – Transform :	$\overline{R(z,t)} = \sum_{n=1}^{\infty} R_n(t)z^{-n}.$	(7-10)
Moment Generation Theorem :	$\lambda_k = [-zd/dz]^k R(z,t)\|_{z=1}.$	(7-11)
Shift Theorem :	$z[R(n-k)] = z^{-k}R(z).$	(7-12)
Real Convolution Theorem :	$z\left[\sum_{n=1}^{k} R_1(n)R_2(n-k)\right] = R_1(z)R_2(z).$	(7-13)

thorough review. It is dealt with briefly in a review on control of polymerization reactors by *Eliçabe and Meira (1988)* and in somewhat more detail by *Silveston (1998)* in "*Composition Modulation of Catalytic Reactors.*" From these reviews, it is evident that imposed modulation significantly expands control over the MMD of a polymerization reactor. Modulation is also a means of reactor control as *Eliçabe et al. (1986)*, *Eliçabe and Meira (1989)*, *Vega et al. (1991)* and particularly *Chen et al. (1980)* demonstrate. This earlier work also showed that modulation of the monomer or initiator concentrations around a steady-state operating point has just a small effect on *D* and other MMD properties for free radical reactions. Modulation of the chain transfer agent concentration had a much greater effect on the MMD of living anionic polymerization processes. Indeed, in the absence of termination reactions, periodic operation can produce any MMD in principle. Of course, variations of the manipulated variables will no longer be sinusoids or square waves. Given the availability of the earlier reviews mentioned above, contributions to the literature after 1998 will be discussed primarily in the remainder of this section.

7.2.1 Coordination Polymerization

Nele and Pinto (2000) took up the question of broadening of the molecular weight distribution (MWD) for polyolefins produced using metallocene Ziegler-Natta catalysts in just a single CSTR. Changes in the process variables with time were not restricted to sinusoids or square waves, so following the lead of *Meira* and coworkers, (*Frontini et al., 1986* or *Vega et al., 1991*), *Nele and Pinto* undertook a search for a time variation in reactor temperature that would achieve the target MWD. Reactor temperature could be controlled by coolant flow rate through heat exchange coils in the reactor and/or by the temperature of the gaseous feed stream. As an alternative to temperature modulation, a chain transfer agent could be varied. Hydrogen is used in olefin polymerization as such an agent so H_2 partial pressure could be modulated for control. Another alternative was to control monomer partial pressure. *Nele and Pinto* showed that time variation within a cycle for either reactor temperature or monomer partial pressure required an increase in the first part of the cycle, followed by a decrease in the latter part to achieve the desired D. This variation was reversed for H_2 partial

TABLE 7-5 Moment Equations from Simplified Free Radical Polymerization Model given in Table 7-3 (*Konopnicki and Kuester, 1974*)

Live Polymer Moments:

$$\left(d\lambda_o/_{dt}\right) = -(\lambda_o/\theta) + 2k_i fI - k_t\lambda_o^2 \tag{7-14}$$

$$\left(d\lambda_1/_{dt}\right) = (\lambda_1/\theta) + 2k_i fI + \lambda_o(k_pM - k_t\lambda_1) - k_{fs}S(\lambda_1 - \lambda_o) - k_{fm}M(\lambda_1 - \lambda_o) \tag{7-15}$$

$$\left(d\lambda_2/_{dt}\right) = (1/\theta)(\lambda_1 - \lambda_2) + 2k_i fI + k_pM(2\lambda_1 + \lambda_o) + k_t\lambda_o(\lambda_1 - \lambda_2) + k_{fs}S(\lambda_o + \lambda_1 - \lambda_2) + k_{fm}M(\lambda_o + \lambda_1 - \lambda_2). \tag{7-16}$$

Dead moments:

$$\left(d\mu_o/_{dt}\right) = -(\mu_o/\theta) + \lambda_o\left[\tfrac{1}{2}k_t\lambda_o + k_{fs}S + k_{fm}M\right]. \tag{7-17}$$

$$\left(d\mu_1/_{dt}\right) = -(\mu_1/\theta) + \lambda_1\left[k_t\lambda_o + k_{fs}S + k_{fm}M\right]. \tag{7-18}$$

$$\left(d\mu_2/_{dt}\right) = (1/\theta)(\mu_1 - \mu_2) - \lambda_1\left[k_t\lambda_o - k_t\lambda_1 + k_{fs}S + k_{fm}M\right] + \lambda_2\left[k_t\lambda_o + k_{fs}S + k_{fm}M\right]. \tag{7-19}$$

Conversion of Heat Balance:

$$\sum_{n=1}^{\infty} R_n = \lambda_o. \tag{7-20}$$

Total moments:

$$\sigma_o = \lambda_o + \mu_o \tag{7-21}$$

$$\sigma_1 = \lambda_1 + \mu_1 \tag{7-22}$$

$$\sigma_2 = \lambda_2 + \mu_2. \tag{7-23}$$

Pseudo steady state assumption:

$$\left(d\lambda_o/_{dt}\right) = \left(d\lambda_1/_{dt}\right) = \left(d\lambda_2/_{dt}\right) = 0 \tag{7-24}$$

TABLE 7-6 Simulations of Polymerization Under Periodic Operation

Reference	Polymerization	Reactor	Manipulated Variable and Waveform	Observation	Comments
Ray (1968)	Free radical and step-growth (condensation)	Isothermal CSTR	Monomer concentration; sinusoid	Molar mass distribution for free radical mechanism was broadened by modulation; dispersity, *D*, maximized by slow cycling; conversion, chain length, μ_n and *D* decrease with cycle frequency for a step-growth mechanism.	Early use of moments and z-transform.
Laurence and Vasudevan (1968)	Caprolactam (free radical and living anionic)	Isothermal CSTR	Initiator, monomer concentration; square wave and sinusoid	Modulation increased dispersity, *D*; maximum *D* was found at the quasi-steady-state limit.	
Bandermann (1971)	Isoprene with n-butyllithium initiator (living anionic)	Isothermal CSTR	Initiator also total "living" ends; sinusoid	Either broadening or narrowing of the molar mass distribution was observed, depending on mean monomer concentration and other properties.	Author observed a narrowing of the molecular weight distribution can occur.
Lee and Bailey (1974)	Olefin polymerization with Ziegler-Natta catalyst (free radical, also step-growth)	Isothermal CSTR; also a two vessel cascade representing non-ideal mixing	Monomer and transfer agent; pulse inputs for the relaxed steady-state limit and square wave for the quasi-steady-state limit	Ideal mixing maximized *D* at the quasi-steady-state limit; imperfect mixing increased *D* at a constant mean chain length. *D* increases by almost 50% for slow cycling. The increase is greater at relaxed steady-state limit where $D = 13$.	Simulations examined polymerization at limits of fast cycling (relaxed steady state) and slow cycling (quasi-steady state).
Konopnicki and Kuester (1974)	Styrene polymerization with AIBN initiator (free radical)	Non-isothermal CSTR	Initiator concentration, cooling jacket temperature separately or simultaneously; square waves, sinusoids	Sinusoidal modulation of cooling jacket temperature and initiator concentration had a large effect at low frequency, but effects vanished at high frequency; only $D_{cycling} > D_{steady\ state}$ was observed; conversion and μ_n can be greater or less, however. Phase lag in simultaneous modulation had only a small effect; waveform had also a minor effect.	Authors allowed for non-isothermal operation and the effect of viscosity. They examined a wide range of perturbations. Simultaneous variation of variables caused the largest effect.

(*Continued*)

TABLE 7-6 Simulations of Polymerization Under Periodic Operation (*cont'd*)

Reference	Polymerization	Reactor	Manipulated Variable and Waveform	Observation	Comments
Spitz et al. (1976)	Styrene polymerization (free radical)	Isothermal CSTR	Initiator; pulses	Large amplitudes required for significant variation in polymer D or $\overline{M_n}$.	Limit on the ratio $\bar{t}/\tau$ imposed by requirement of constant polymer properties.
Hashimoto et al. (1976a)	Methyl methacrylate polymerization (free radical)	Non isothermal CSTR	Radiation dosage; pulses	Step-change variation of radiation broadened polymer chain length distribution.	Authors employed X-rays to excite initiator. They also induced temperature variation.
Langner and Bandermann (1978)	Isoprene polymerization (living anionic)	Isothermal CSTR	Inlet/outlet flow rates; on/off mode	Flow modulation decreased D from 2 to 1.375.	Modulation of reactor withdrawal rate changed reactor volume.
Klahn and Bandermann (1979)	Isoprene polymerization (living anionic)	Isothermal CSTR	Initiator, monomer flow rate	Symmetric modulation of feed flow rate with pulse withdrawal of liquid from reactor can produce $1 < D < \infty$; molecular weight distributions are usually multi-modal.	Modulation changed reactor volume.
Marconi and Pogni (1979)	Styrene polymerization (step-growth)	Isothermal CSTR; CSTR in series; non-ideal CSTR	Initiator, monomer; sinusoid oscillation	D increased with $\tau_{cycle}/\bar{t}$ and monomer or initiator concentration, reaching a maximum at the quasi-steady-state limit. D also increased as the plug flow limit was approached.	Authors examined operations where $0 < \tau_{cycle}/\bar{t} < 2$.
Meira and Johnson (1981)	Styrene polymerization (living anionic)	Isothermal tubular reactor (PFR)	Monomer; ramp, ramp + step-change	Monomer flow rate and duration were modulated to give a target molecular weight distribution; mean value of modulated flow will yield at steady state the desired molecular weight distribution.	Authors assumed very fast reactant mixing and initiation rates; termination reactions were neglected.
Gordon and Weidner (1981)	(emulsion).	Batch reactor	On-off emulsifier flow based on temperature sensor	Limiting excess emulsifier in reactor narrows the aggregate size distribution by reducing micelle formation.	Objective was the control of the polymer aggregate size distribution.

Couso and Meira (1984)	Styrene polymerization (living anionic)	Isothermal CSTR	Initiator, monomer, temperature; square waves, sinusoids	Modulation yields $D > 10$ at quasi-steady state with square wave modulation, but D depends on rate constants. High frequency modulation does not change D, $\overline{M_n}$, but increases monomer conversion.	Simultaneous variation of initiator and monomer flow rates in-phase and 180° out-of-phase; considered $60 < \tau_{cycle}/\bar{t} < 0.05$.
Thiele (1984)	Co-polymerization of styrene-acrylonitrile (step-growth)	Non-isothermal CSTR	Cooling jacket temperature	Up to a 10% increase in D was dependent on cycling amplitude.	Simulation indicated a smaller effect of temperature cycling than was observed experimentally.
Gupta et al. (1985)	Multi function monomer (step-growth)	CSTR (micro mixed, isothermal)	Monomer concentration; sinusoid	Bifurcation was observed when the number of functional groups exceeded 2, with one branching resulting in gelation. $D_{cycling}$ increases by 35% over $D_{steady\ state}$ at the quasi-steady-state limit.	Authors considered monomer with three and four functional groups. They observed gelation that prevented achieving a stationary cycling state.
Gugliotta and Meira (1986)	Styrene polymerization with potassium perchlorate initiator (emulsion)	Isothermal CSTR	Total flow rate; step-change plus ramp	A desired particle size distribution can be obtained by cycle split and ramp slope in a two-part cycle.	Because a steady state model was used, results are restricted to $\tau_{cycle} >> \bar{t}$.
Frontini et al. (1986, 1987)	Isoprene polymerization (living anionic)	CSTR (micro mixed, isothermal)	Initiator, monomer concentration or flow rates; irregular input waveforms and cycle periods	$1.58 < D < 4$ at constant monomer conversion and $\overline{M_n}$ through concentration and flow rate modulation. In most cases initiator and monomer concentrations were π radians out-of-phase.	Modulation variable was determined through a search for optimal control. The starting cycle period and perturbation influenced the search.
Eliçabe et al. (1986), Eliçabe and Meira (1989)	Isoprene polymerization (living anionic)	CSTR (micro mixed, isothermal)	Initiator, monomer concentrations, flow rate; sinusoid	180° out-of-phase sinusoidal modulation of initiator and monomer raised D. Adaptive cascade control was capable of maintaining D with input changes.	Simulation specified the periodic trajectories for the first two moments of the chain length distribution.

(Continued)

TABLE 7-6 Simulations of Polymerization Under Periodic Operation (*cont'd*)

Reference	Polymerization	Reactor	Manipulated Variable and Waveform	Observation	Comments
Gugliotta and Meira (1988)	Styrene-butadiene co-polymerization (emulsion)	Isothermal CSTR	Transfer agent, flow rate, UV radiation; pulses	Modulation of transfer agent concentration broadens emulsion particle size distribution, increasing D from 2 to 8 at the quasi-steady-state limit; solvent flow rate modulation decreased conversion.	Steady-state model was used so the results are useful for quasi-steady-state operation.
Vega et al. (1991)	Isoprene polymerization (living anionic)	CSTR (micro mixed, isothermal)	Initiator, monomer concentrations	Optimal modulation of monomer flow raised D from 2 at steady state to 4. The presence of impurities severely decreased the attainable D and resulted in an oscillatory output.	Authors employed optimal control strategy for systems where impurities cause "dead" polymer.
Hungenberg (1992)	Propene polymerization with Ziegler-Natta catalyst (coordination)	Stirred fluidized bed	Transfer agent (H_2) partial pressure; on-off, saw-tooth	On-off variation in H_2 partial pressure caused $D > 10$. Author observed large variations of the number average molar mass leaving reactor. With saw-tooth input, only $D = 8$ was achieved.	A steady state model was used to examine the effect of H_2 partial pressure on polymer chain length; the large $\overline{M_n}$ variation within a cycle will require an external mixing vessel.
Gosden et al. (1995)	Styrene polymerization (living anionic)	CSTR (micro mixed, isothermal)	Initiator, monomer concentrations; sinusoid	Asymmetrical modulation of monomer flow rate (at $s = 0.1$) increases D from 2 to 2.4; modulation caused multiple sharp peaks in M_w or μ_n distribution.	Constant flow was maintained by adjusting solvent flow.
Nele and Pinto (2000)	Propene polymerization with a metallocene catalyst (coordination)	Non-isothermal CSTR	Monomer or chain transfer agent concentrations, reactor temperature	Authors' optimization routine found a policy of reactor temperature, transfer agent or monomer partial pressure variation that realized their target parameters: $D = 5$ and $\overline{M_n} = 90{,}000$.	Modulation profile for a single variable was determined by an optimization routine which divided a cycle into finite periods.

Vicente et al. (2001)	Emulsion co-polymerization of styrene and n-butyl acrylate	Near-isothermal CSTR; system started up from a water filled reactor	Monomer and chain transfer agent flow rates	Authors employed a closed loop control strategy to set monomer and chain transfer agent feed rates using a nonlinear, model-based controller.	Controller operation produced low amplitude oscillations in manipulated and process variables. It is uncertain whether these perturbations affected performance.
Dünnebier et al. (2005)	Polymer system not disclosed	Commercial reactor with evaporative cooling	Manipulated variables were flow rates of fresh monomer, recycle and catalyst, $T_{Reactor}$	Linear, time-varying model predictive controller was used with a multivariate Kalman filter to maintain MWD for step-changes in load. MWD was measured by system viscosity.	As in *Vicente et al. (2001)* above. The *Dünnebier* paper gives simulation results and plant measurements.
Atasoy et al. (2007)	Acrylonitrile polymerization with addition of vinyl acetate	CSTR	Free radical polymerization mechanism; control variable was bisulfite radical concentration	Neural network operated with model predictive or nonlinear auto-regressive moving-average control increased polymer fiber molecular weight.	Controller operation induced irregular step-changes in manipulated variables.
M.A-H. Ali et al. (2008)	Propene polymerization using a multi-site Ziegler-Natta catalyst	Jacketed CSTR with gas and liquid feed, equipped with adjustable flow controllers	H_2 or catalyst feed rate; square wave	Minimizing the displacement of the realized MWD from the target MWD through feed rate control succeeded in forcing the MWD produced to closely approximate the target distribution.	The search for cycle period and amplitude was treated as an optimization problem with the objective function defined as departure from a reference *D*. MATLAB SPQ served as the optimization algorithm.
M.A-H. Ali et al. (2010)	Ethene polymerization using a multi-site Ziegler-Natta catalyst	Ideally back-mixed, non-isothermal fluidized bed equipped with adjustable flow controllers	Monomer and/or H_2 flow rate; square wave. Variations led to a ramp shaped modulation of the H_2/monomer ratio	Variation of cycle split and magnitude of the H_2 step changed the ratio of H_2 to monomer in the fluidized bed and thereby allowed low-and high-number molar masses. Simulation of the reactor and controller demonstrated that a narrow MWD under steady-state operation can be broadened and closely approach a target distribution.	Use of a nonlinear model-based predictive control incorporating a Kalman filter resulted in a reactor operating policy that allowed for unprogrammed disturbances.

pressure. Indeed, the required variation was a non-symmetric square wave modulation. Polymer MWD leaving the stirred reactor changes with time so a mixing vessel up to four times the reactor size would be needed to produce a polymer with minimal variation with time. *Nele and Pinto* set $\tau_{cycle} = \bar{t}$ so as to limit the size of the post reactor mixer. The object of their study was to achieve a dispersity, D, of three to four instead of the two normally observed for a steady-state operation using metallocene catalysts. Simulations by these researchers demonstrated the target D's or MWD's could be achieved using widely accepted kinetics for Ziegler-Natta catalyzed olefin polymerization with a common metallocene catalyst and rate constants derived experimentally.

The fluidized bed production of polyethylene, known commercially as the UNIPOL process, can exhibit spontaneous oscillations. In this process the $C_2^=$ monomer, any co-monomer, H_2 and N_2 diluents, and recycle enter as gas at the bottom of the reactor and fluidize the emulsion phase, generally at a velocity close to the minimum fluidizing velocity. Gas also serves as coolant for the exothermic polymerization reaction. Ziegler-Natta catalyst is injected into the bed as a fine powder and forms the nucleus for the growing polyethylene sheath. Polymerization occurs in the dense phase bordering the gas bubbles rising through the emulsion phase.

Polymer particle growth depends on the balance between the catalyst introduction rate, the gas feed rate and the temperature. The interrelationships of these variables, expressed in models for the reaction system, cause bifurcations in the operating states and spontaneous oscillation in the $C_2^=$ conversion, polymer production rate, polymer/catalyst ratio in the solid phase, and temperature of the dense phase. These oscillations appear for a range of polymer aggregate sizes and depend importantly on the catalyst feed rate and the inlet gas temperature. *Ghasem (1999)* discussed the bifurcation behavior of the UNIPOL process and condition under which spontaneous oscillations can arise. *McAuley et al. (1995)* also discuss spontaneous oscillation for gas phase, Ziegler-Natta ethylene polymerization.

Simulation of propene polymerization over a dual site Ziegler-Natta catalyst in a liquid-phase, well-stirred reactor was examined again by *Ali et al. (2008)* in a manner similar to the *Nele-Pinto* work. *Ali et al.* considered an optimization problem whose objective was to find a two-part policy of H_2 and catalyst feed rates that maximize an objective function consisting of weighted contributions of the squared fractional increase in *D* over an assumed dispersity of seven at steady state and the squared conversion, again referenced to steady state. Actual optimization variables were τ_{cycle} and s_{cycle}. Amplitudes, as maximum feed rate, were set for each search. The problem was solved by *Ali et al. (2008)* using nonlinear optimization routines in MATLAB SQP. They found that maximizing *D* required withdrawing H_2 from the reactor by a suitable blower. Frames **(a)** and **(b)** in **Figure 7-4** show the policy obtained on condition of maintaining a more or less constant monomer conversion.

These simulations showed that the near-optimal modulations of the two flow rates are several degrees out-of-phase. The remaining frames in the figure show the consequences of the selected operating policy. The dispersity fluctuates somewhat with time, but the average is about twice that for a steady-state operation; conversion fluctuations are larger and, on average, they are about 80% of conversion at steady state. The number molar mass in **(f)** changes significantly within a cycle period as anticipated by the change in *D*. A sharp temperature spike of ca. 5°C arises when the separation between the step-changes in H_2 and catalyst feed rate is the largest. Residence times in the reactor are about the same as the cycle periods, namely two to three ours. Other simulations presented by *Ali et al. (2008)* showed that

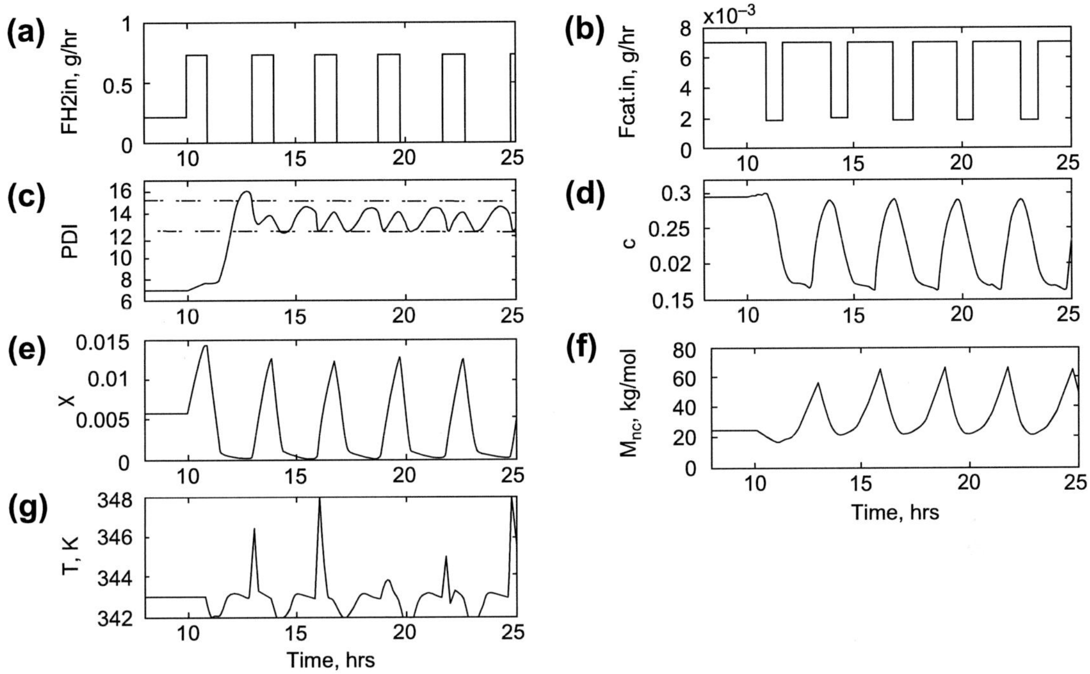

FIGURE 7-4 **Phase-shifted square-wave modulation of the flow rates of H_2, the transfer agent, in (a) and the Ziegler-Natta catalyst in (b) as policy for broadening the time-average molar mass distribution as represented by the dispersity, D. Effect of the policy on reactor operation shown in (c) as instantaneous dispersity, (d) monomer conversion, (e) ratio of H_2 and monomer feed rates, (f) instantaneous weight molar mass exiting the reactor, and (g) T_R, temperature of the well-mixed reactor.** *(Figure adapted from* Ali et al. (2008) *with permission. © 2008 by Wiley Interscience.)*

modulating the H_2 feed rate alone provided just a small broadening of the molar mass distribution.

A later paper deals with the modification, mainly broadening, of the MWD of polyethylene through using a multi-site Ziegler-Natta catalyst in an ideal, temperature-controlled fluidized bed (*Ali et al., 2010*). Their system used a nonlinear model-based predictive control algorithm on the controllers for the gaseous ethane monomer and the H_2 transfer agent feed rates to force the system stepwise to the desired MWD. Other control structures can be adopted to manage the MWD. Popular choices are nonlinear predictive control, control based on neural networks, and various combinations of proportional, integral, and derivative control. Indeed, *Ali et al. (2010)* mention several prior publications that have used various nonlinear process control algorithms to enhance polymer properties or increase production rates. Most of these process control systems generated a modulated reactor operation of low amplitude. It was unclear from these studies whether this small level of modulation favorably affected polymer properties, monomer conversion or polymer production rate.

Figure 7-5 shows the narrow, starting MWD, the target distribution and the one resulting from the modulation of the two feed flow rates. The target MWD is approached closely. The flow manipulations used in the simulation would cause changes in the temperature and pressure of the fluidized bed; however, separate

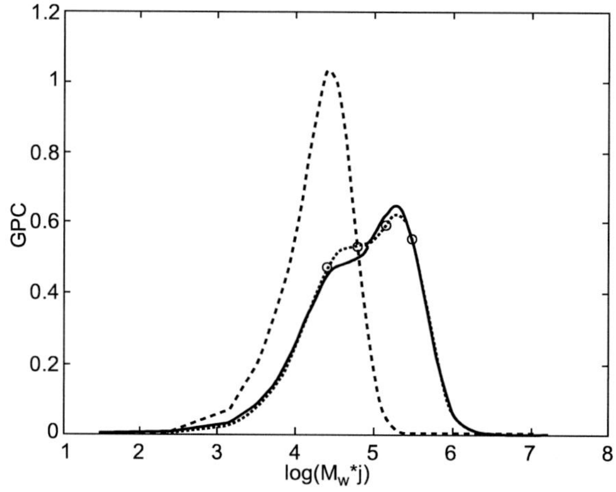

FIGURE 7-5 Simulation of the operation of a nonlinear model predictive controller operating on the monomer and hydrogen flow rates and using a Kalman filter to adjust model error to force the initial molecular weight distribution at steady state (dashed line) to a desired broader MWD (solid line) for ethene polymerization using a multi-site Ziegler-Natta catalyst in an ideally mixed, non-isothermal fluidized bed. Result of controller operation is shown by the dotted line. *(Figure reproduced from* Ali et al. (2010) *with permission. © 2010 by the Korean Journal of Chemical Engineering.)*

control of coolant, nitrogen flow and gas bleed rate were included in the simulation so that bed temperature and pressure remained constant.

7.2.2 Free Radical Polymerization

Polymethylmethacrylate or polystyrene are examples of industrially important polymerizations that proceed via radical mechanism.

Teymour and Ray (1989, 1992a, b) examined the dynamic behavior of a CSTR carrying out the free radical polymerization of vinyl acetate and observed bifurcation. One branch led to limit cycles or spontaneous oscillations in polymer properties as well as reactor temperature that converts a steady-state operation into a periodic one. In a subsequent paper, *Teymour (1997)* extended dynamic analysis to semi-batch reactors and demonstrated bifurcation into spontaneous oscillation when these reactors were operated with flow rate scheduling. Sequential filling and discharge of semi-batch reactors can result in oscillatory behavior with chaotic regimes possible.

7.3 EXPERIMENTS ON POLYMERIZATION UNDER INPUT MODULATION

The first experiments on periodically operating a polymerization reactor took place before 1969 and coincided approximately with the earliest theoretical considerations. **Table 7-7** summarizes many of the experimental reports published to date. It shows a steady effort over the years covering most types of polymerization reactions. For reasons mentioned above, most of the earlier work will be just summarized in **Table 7-7**.

7.3.1 Emulsion Polymerization

Self-sustained oscillations and the more general instability of emulsion polymerization reactors has been discussed in the literature for about 50 years (see discussion of these topics by *Kiparissides et al. (1979)* or *Rawlings and Ray (1987)*). *Ohmura et al. (1998)* suggested operating in the region of these oscillations because better

TABLE 7-7 Experiments on Polymerization Under Periodic Operation

Reference	Polymerization	Reactor	Manipulated Variables and Waveforms	Observation	Comments
Claybaugh (1969)	Propylene polymerization with a Ziegler-Natta catalyst; H_2 served as chain transfer agent (coordination)	Near CSTR	Chain transfer agent (H_2) flow rate switched on/off	Broad molecular weight distribution ($D = 10.8$) was obtained through modulation. $D = 8.0$ under steady-state operation.	Incomplete mixing in the reactor apparently prevented the expected bidispersed molecular weight distribution.
Spitz et al. (1976)	Styrene polymerization with AIBN initiator (free radical)	CSTR (isothermal, jacketed reactor)	Initiator concentration; on-off step-changes	D and chain length, μ_n, increased with increasing τ_{cycle}, also as $t_{½}/\tau_{cycle}$ decreased below 0.2, but monomer conversion and $\overline{M_n}$ decreased.	Important parameters were $\bar{t}/\tau_{cycle}$, a measure of damping of the concentration change and $t_{½}/\tau_{cycle}$, a measure of initiator consumption in a half-cycle.
Hashimoto et al. (1976b)	Methyl methacrylate polymerization (free radical)	Non-isothermal CSTR (50 mL batch reactor and a 180 mL flow reactor with coolant circulated through the reactor jacket)	γ-ray exposure; square wave	Dosage modulation (5.3 to 153 rad/s) broadens M_n distribution; degree of broadening depends on exposure duration and τ_{cycle}.	Dosage modulation was controlled by reactor temperature as radiation induced a change in reactor temperature.
Meira et al. (1979)	Methyl methacrylate polymerization with AIBN initiator (free radical)	CSTR (300 mL, steel jacketed reactor), feed valves computer controlled	Ratio of monomer/ initiator flow rates; near sinusoid or near square wave	Near sinusoid oscillations increased D from 1.77 to 2.0 and $\overline{M_n}$ by about 5%, conversion decreased; square waves gave about a 15% increase in $\overline{M_n}$ but a smaller change in D.	Experiments were performed at $\bar{t}/\tau_{cycle} = 0.44$. Amplitudes used were greater than those assumed in simulations. 4 to 5 cycles were needed to reach stationary state.
Crone and Renken (1979a, b)	Styrene Polymerization with AIBN initiator (free radical)	Isothermal CSTR in glass reactors from 250 to 1000 cm^3 volume. $T = 80°C$	Initiator or monomer concentrations; square wave (on-off); both (simultaneous) or just singly	At about the same D and $\overline{M_w}$, on-off modulation of monomer concentration gives significantly higher conversion than steady-state operation. D is larger for simultaneous modulation of initiator and monomer than for modulation of just one input.	Comparison with steady state requires constant or lower 1) total flow rate, 2) mean feed concentrations, and 3) reactor volume. Conditions for comparison affect the degree of improvement.

(Continued)

TABLE 7-7 Experiments on Polymerization Under Periodic Operation (*cont'd*)

Reference	Polymerization	Reactor	Manipulated Variables and Waveforms	Observation	Comments
Chen et al. (1980)	Styrene polymerization initiated by photo dissociation of BME (free radical)	Non-isothermal CSTR with radiation induced temperature change	UV radiation exposure, temperature; pulses	*D* decreased from 4.8 for operation at the low temperature stable state to 2.83 for operation at a higher temperature metastable state.	Operation at a metastable state was created by on/off UV irradiation.
Meira and Johnson (1981)	Styrene polymerization with sec-butyllithium initiator (living anionic)	Isothermal, 1.2 mm i.d. × 10 m tubular reactor (PFR)	Monomer flow rate; ramp and square wave	Ramp flow rate increased *D* to 2.57 from 1.25 under steady state. Square-wave input increased *D* to 4.34 from 1.63 under steady state.	Shape of the modulated input causes a change in the molecular weight distribution. Formation of dead polymer through impurities decreased the expected change.
Thiele (1984)	Styrene/acrylonitrile co-polymerization (free radical)	Water-jacketed CSTR (5 L)	Temperature of water coolant, $\tau_{cycle} = 60$ min, $A = 1$, 10°C	At 10°C amplitude, *D* increased by 5%.	Considered temperature modulation.
Hoedemakers and Thoenes (1990)	Styrene homo-polymerization (emulsion)	Down-flow packed bed (5 cm i.d. × 5 m)	Feed rate and pulsation rate		Prevention of plugging of a packed bed
Paquet and Ray (1994)	Methyl methacrylate homo-polymerization with Na-dodecyl sulfate surfactant and K-persulfate initiator (emulsion)	Coiled tubular reactor immersed in a cooling bath; also a coolant jacketed CSTR	Feed rate, pulsation rate	Tubular reactor exhibited frequent plugging at high monomer conversions. Flow pulsing prevented plugging. Tubular reactor performance equaled that of a batch reactor.	Authors used flow pulses to prevent reactor plugging and shutdown.
Gosden et al. (1997a, b)	Styrene polymerization with sec-butyllithium initiator (living anionic)	Isothermal CSTR (1 L, jacketed)	Initiator, monomer concentrations; square wave; *T* for initiator modulation = 50°C, for monomer modulation = 40°C	Modulation sharply raised and lowered instantaneous *D* from 2.0 at steady state and decreased instantaneous $\overline{M_n}$; however, mean *D* changes under modulation were small.	Authors examined instantaneous change on introduction of initiator or monomer pulse. Results were corrected for inadequate mixing.

Limb et al. (1998)	Fluorocarbon polymerization (CVD)	Plasma deposition reactor	On, off power; square wave	Target CF_2 composition of the coating was closely approached by control of on and off durations. Coating conditions, e.g., reactor pressure and temperature, were important.	Preparation of a polymer coating of a solid substrate used CVD under modulation of the power to the plasma generator.
Ohmura et al. (1998)	Vinyl acetate polymerization with Na-dodecyl sulfate emulsifier; NH_4^- peroxodisulfate initiator (emulsion)	Isothermal CSTR	Emulsifer feed rate, waveform not given	Better control of aggregate size distribution was achieved while operating in an oscillatory regime, but results were not reproducible.	Authors observed that operation in the region of spontaneous oscillation improved the aggregate size.
Ohmura et al. (2001)	Vinyl acetate polymerization with Na-dodecyl sulfate emulsifier; NH_4^- peroxodisulfate initiator (emulsion)	Isothermal CSTR	Flow rate; square wave	Flow rate modulation stabilized operation in the region of spontaneous oscillations, but induced a bimodal MWD.	Flow rate modulation was employed to stabilize reactor operation.
Ohmura et al. (2005)	As above	Two isothermal CSTRs in series	Steady-state operation with control of $\bar{t}$, and impeller rpm	Authors observed oscillations of mean molar mass under steady-state reactor operation. Impeller rpm affected amplitude, while $\bar{t}$ controlled period.	Emulsifier concentration was held below critical micelle concentration.
Brétagnol et al. (2006)	Ethylene oxide polymerization (CVD)	Plasma deposition reactor	Input power to a capacitatively coupled radio frequency vaporization system	Periodically varying the power input to the radio frequency generator produced a deposition film with desired properties.	Diethylene glycol dimethyl ether is decomposed in the glow discharge field producing the polymer deposited on a substrate.

control of polymer aggregate size distributions might be possible. They found that the period of these spontaneous oscillations was very sensitive to reactor operating conditions so that the particle size distribution was not reproducible from one experiment to the next.

It is well-known from oscillating chemical reactions that spontaneous oscillations can be entrained by externally imposed forcing. *Ohmura et al. (2001)* examined whetheror not such a strategy could be used to operate in a region of process variables space where oscillations occur. Their experiments were carried out for vinyl acetate polymerization (see **Table 7-7**) at an emulsifier concentration below the critical micelle concentration for the sodium dodecyl sulfate emulsifier used. Under conditions studied, vinyl acetate conversion jumps to about 60% from close to 0 when liquid flow rate to the authors' CSTR was decreased so that the mean residence time, $\bar{t}$ exceeded 22 min. Modulating the liquid flow rate by switching $\bar{t}$ between 27.2 and 19.7 min maintained the conversion at the low rate part of the cycle and the particle size distribution (PSD) leaving the reactor changed little between the two cycle parts. On the other hand, setting the flow rate to give $\bar{t} = 13.6$ min instead of 19.7, thus causing the average $\bar{t}$ to drop to 20.4 min, resulted in conversion falling gradually to about 20% for the high flow rate portion of the cycle. A bimodal PSD resulted from that experiment. Conversion changes in the half-cycles were mirrored in changes of the reactor temperature. Flow rate modulation thus significantly increased reactor throughput and stabilized reactor operation. Increasing the amplitude of the flow rate change offers the possibility of obtaining a bimodal in place of a unimodal PSD.

An extreme form of periodic operation is periodic pulsing. This type of periodic operation has been proposed as a means of avoiding plugging in the packed bed and tubular reactors employed for emulsion polymerization. *Hoedemakers and Thoenes (1990)* found that pulsing was effective for styrene polymerization in a laboratory-scale packed bed. Pulsing the emulsion flowing to a tubular reactor was studied by *Paquet and Ray (1994)* who attached a piston pump to the reactor feed line. They studied the effect of strokes/min for different feed rates on reactor behavior. These researchers reported that periodic pulsing prevented plugging, but they did not discover a critical frequency or amplitude for avoiding plugging. They did show that flow pulsing did not change the plug flow character of their tubular reactor. Thus, conversion, dispersity, and number molar mass closely approached values for a batch reactor operating under the same conditions as the tubular reactor.

7.3.2 Anionic Polymerization

Examples of anionic, living chain polymerization are the synthetic rubbers: styrene-butadiene copolymers, and polyisoprene. Simulations discussed in **Section 7.2** for this polymerization mechanism revealed that feed or initiator concentration modulation can increase product dispersity, while flow modulation could decrease this property. Several experiments have independently tested these predictions as discussed by *Silveston (1998)*. *Meira and Johnson (1981)* were the earliest experimental investigators. They employed periodic operation with irregular concentration patterns to obtain a specified $\bar{\mu}_n$ and D; their experiments used styrene polymerization in tetrahydrofuran (THF). A single cycling period of 15 min was used with different mean residence times. Experimental problems prevented achieving their target $\bar{\mu}_n$ and D. Later, *Gosden et al. (1997a, b)* used the same polymerization but with a different solvent and initiator. Initiator flow was pulsed in one experiment while monomer flow was cycled in the second. For an initiator pulse, the polydispersity index rose abruptly from 2 to about 10, while the number average molar mass dropped from 40,000 to 5000 g/mol. These changes reversed when the pulse ends but are repeated on a successive pulse. Time-averaging these opposite changes showed that modulation increased D and reduced M_n.

7.3.3 Free Radical Polymerization

Silveston (1998) also discusses this type of polymerization. An early contribution was the free radical polymerization of styrene using an on-off variation of the azobisisobutyronitrile (AIBN) initiator (*Spitz et al., 1976*). Ratios of mean residence time ($\bar{t}$) to period (τ) and initiator half-life ($t_{½}$) were found to be important. The former measured the damping occurring in the reactor, while the latter indicated the variation of initiator concentration during a cycle. Undesirable oscillating monomer conversions and number average chain lengths, μ_n, were observed only if $\bar{t}/\tau < 2.5$. Broadening of the MWD occurred when $\tau/t_{½} > 5$ as indicated by the dispersity, D. Large increases in D were seen when $\tau/t_{½}$ exceeded 4.3. Monomer or initiator modulation can have a strong effect on the instantaneous D leaving a CSTR depending on the $\tau_{cycle}/\bar{t}$ ratio as might be expected for a range of this ratio from ca. 0.25 to 2. However, the effect of modulation on the time-average value of D or the number molar mass was found to be much smaller. The explanation for the effect of τ on D, the sensitivity to damping and the importance of the cycle period to initiator half-life ratio is that the molecular weight of the polymer is controlled by the termination reactions and is sensitive to initiator concentration.

Crone and Renken (1979a, b) observed that the space-time yield of polymer could be increased by up to 44% above that obtained at steady state through periodic operation. D and μ_n were unchanged, but this required doubling the mean residence time. *Meira et al. (1979)*, however, found that concentration modulation led to higher dispersities and number average molar masses.

Hashimoto et al. (1976b) examining the X-ray induced free radical homo-polymerization of methyl methacrylate found broadening of the MWD for square-wave modulation of irradiation. Broadening depended on periodicity and pulse length relative to cycle period. *Chen et al. (1980)* worked with styrene polymerization initiated by the photo-dissociation of benzoin methyl ether. On-off control of the light enabled this research group to operate their reactor at a higher temperature, metastable state which narrowed the MWD significantly and raised μ_n and X_M by about six-fold.

7.3.4 Ziegler-Natta Catalysts

With Ziegler-Natta catalysts, polymerization can follow either a "living" chain or free radical mechanism. Experimental results reported in the *Claybaugh et al. (1969)* patent are limited and summarized in **Table 7-7**. However, that reference appears to be the only one for the Ziegler-Natta catalysts in the open literature.

7.3.5 Deposition Polymerization

Plasmas produced in the vapor phase by radiation or radio frequency variation from an organic source polymerize on deposition to form surface films with often unique properties. Research on this coating technique has been pursued over the last three decades. Modulation has been introduced into the plasma generation step and found to influence the composition and thus the properties of the films formed.

Fluorocarbon coatings were the subject of *Limb et al. (1998)* experiments who found that on-off modulation of the power fed to a radio frequency generator changed the F/C ratio in the deposited film. Through choice of τ_{cycle} and s, the cycle split, *Limb* and co-workers found that they could make the F/C = 2 atom ratio they wanted in the coating deposited from the plasma. *Limb et al. (1998)* compared composition of the deposited film for on/off and continuous RF power as well as the influence of process condition on composition. For their study, they used a stainless steel vessel with Al electrodes, spaced at 2.5 cm from each other. Substrates for coating silicon wafers were placed on the grounded electrode. Source gas for the plasma was hexafluoropropylene oxide, reactor pressure ranged from 400 to 1200 mtorr, temperature from −38 to 123°C and the radio frequency source was 13.65 MHz with peak power either 140 or 280 W. Under modulation, "on" times ranged from 2 to 150 ms and "off" times from 10 to 2000 ms. Teflon is a CF_2 polymer and the target of the *Limb et al.* study was to prepare coatings with this C/F ratio. About 70% CF_2 in the chemical vapor decomposition (CVD) coating was obtained through modulation. High rate of film growth per cycle was found to be important. High growth rates needed a low temperature, short "on" times and long "off" times.

Modulating the power input to a radio frequency discharge allowed *Brétagnol et al. (2006)* to lay down a polyethylene oxide coating on a substrate which had the properties of substrate adhesion and bio-interaction with cell matter and other proteins that the researchers wished to have. A square-wave modulation was used with different durations of each half-cycle. No details on the effect of cycle period or split on the coating properties are given in the publication.

7.4 SPONTANEOUS OSCILLATIONS

Within the spectrum of operating states of polymerization systems, regions of spontaneous oscillations have been observed experimentally and predicted through modeling. These oscillations exhibit properties and exert effects on time-average polymer properties that are quite similar to those of the forced oscillations examined previously in this chapter. A large literature has developed on spontaneous oscillations that cannot be reasonably summarized in this chapter. Furthermore, the existence of spontaneous oscillations is closely associated with the stability of polymerization reactors. Stability analysis is a subject, although of great import and interest, that is beyond the intended scope of this monograph. Several studies have attempted to improve reactor performance by operating in the region of spontaneous oscillations. It is these contributions that are the focus of the remainder of this chapter. They are summarized in **Table 7-8**.

Stability studies of polymerization reactors were pioneered by *W.H. Ray* and co-workers at the University of Wisconsin and somewhat later by *G.R. Meira* at the Universidad Nacional del Litoral in Argentina. *Hamer et al. (1981)* studying parametric sensitivity and multiplicity in polymerization reactors commented that oscillatory behavior was anticipated in the free radical polymerization of methyl methacrylate at high conversion. Later, *Schmidt et al. (1984)* explored the dynamic behavior of this polymerization system in a CSTR and observed transitions to instability where oscillatory behavior was expected. *Zacca and Ray (1993)* undertook a simulation of Ziegler-Natta catalyzed olefin polymerization in a loop reactor. As part of that study, they examined the effect of the recycle ratio on reactor performance. Starting at the lowest ratio they considered, $R_{Rec} = 15$, their simulation showed a large initial temperature excursion that would have boiled off reactants of a real system, followed by a transition to sustained oscillations with an amplitude of about 40°C and a period about equal to the mean residence time. In a continuation of their study, simulation showed a small temperature perturbation after which a new steady state established itself at a higher temperature at $R_{Rec} = 150$. At $R_{Rec} = 19$, however, an initial jump of 25°C was followed by a strongly damped oscillation to a lower constant temperature. The number mean molar mass also evidenced damped oscillation.

A general dynamic flow sheet package with the capability of analyzing the stability of

TABLE 7-8 Spontaneous Oscillations and Multiplicity Analysis of Polymerization Reactors

Reference	Polymerization System	Reactor	System Variables Examined	Observations	Comments
Hamer et al. (1981)	Simulation and experimental polymerization of methyl methacrylate, vinyl acetate	Non-isothermal CSTR	Effect of residence time on conversion and reactor temperature	Research was limited to exploring multiplicity. Good agreement was found between model and experiment.	This work was one of the earliest studies of multiplicity in polymerization reactors.
Schmidt et al. (1984)	Simulation and experimental polymerization of methyl methacrylate, vinyl acetate	Non-isothermal CSTR	Effect of residence time on conversion and reactor temperature	Research was limited to exploring multiplicity. Isolas were observed that offer higher conversion at lower reactor temperatures. Good agreement was found between model and experiment.	
Choi and Ray (1985)	Simulation of ethylene polymerization	Fluidized bed reactor		Authors identified regions of instability.	
Kim and Choi (1991)	Simulation of styrene polymerization	Cascade of two CSTRs		Mixing of initiators introduced chaotic oscillations.	
Teymour and Ray (1992a)	Simulation and experimental study of vinyl acetate polymerization	Non-isothermal CSTR		Spontaneous oscillations were associated with Hopf bifurcations. Isolas exhibiting sustained oscillations were observed.	Authors analyzed bifurcation behavior. Experiments were done in a laboratory CSTR.
Teymour and Ray (1992b)	Simulation of vinyl acetate polymerization	Non-isothermal CSTR		Authors predicted spontaneous oscillations and chaotic behavior in a small region of parameter space.	Authors considered a commercial-scale stirred reactor.
Zacca and Ray (1993)	Simulation of olefin homo-polymerization and co-polymerization with Ziegler-Natta catalysts	Non-isothermal loop reactor	Recycle ratio	Simulation predicted sustained and damped oscillation when the recycle ratio =15. CSTR behavior was found at a recycle ratio = 150.	This was only a limited examination of instability.

(Continued)

TABLE 7-8 Spontaneous Oscillations and Multiplicity Analysis of Polymerization Reactors (*cont'd*)

Reference	Polymerization System	Reactor	System Variables Examined	Observations	Comments
Pinto and Ray (1995a,b, 1996)	Free radical co-polymerization of methyl methacrylate and vinyl acetate	Non-isothermal CSTR	Residence time, initiator concentration in feed	Spontaneous oscillations arise with co-polymerization under steady-state conditions.	
Hyanek et al. (1995)	Ethylene polymerization with H_2 addition for control; propylene polymerization	Fluidized bed; loop reactor	Catalyst feed rate and H_2 partial pressure	Authors located Hopf points in bifurcation diagram and mapped oscillations for both process examples.	Authors provided examples of the application of their stability analysis package.
Melo et al. (2001a)	Generic free radical, solution polymerization	Non-isothermal CSTRs and loop reactors with CSTRs characteristics	Residence time, fraction monomer in feed stream, initiator concentration, exothermicity	They observed that regions of parameter space exist in which spontaneous oscillations occur for all free radical reaction systems regardless of reactor type. Oscillations can be in a stable limit cycle or may be random. Gel effects restrict oscillatory behavior.	Authors undertook mapping of Hopf bifurcation points using continuation methods.
Melo et al. (2003)	Homopolymerization of vinyl acetate in a MeOH solvent with a AIBN initiator	Non-isothermal loop reactor	Residence time, recycle ratio, initiator concentration in feed, $(N_{Pe})_{mass}$, $(N_{Pe})_{heat}$, $U_{overall}$ of heat exchange surface in loop	Thermal capacitance in recirculation pump and heat exchangers stabilizes the loop reactor so that spontaneous oscillations are unlikely in commercial units. Spontaneous oscillations may occur during loop reactor start-up.	Bifurcation diagrams were obtained for different continuation variables.

non-isothermal polymerization reactors has been published by *Hyanek et al. (1995)*. This package has the ability to identify parameter space where sustained, spontaneous oscillation may occur through a subprogram that carries out continuation analysis on a process variable. Based on the AUTO subroutine (*Doedel et al., 1997*), steady-state branches can be followed; branching points, limit points and Hopf bifurcation points can be identified. *Hyanek et al.* demonstrated the use of the package for two cases given in **Table 7-8**.

The *Pinto's* research team at the Universidade Federal do Rio de Janeiro has mounted a significant effort on the dynamics of loop reactors recently. This reactor greatly increases the surface area for heat exchange by using several externally cooled tubular reactors and a recycle pump in place of a CSTR. **Figure 7-6** shows a schematic of the reactor. Reactors based on the schematic are currently used for the commercial production of high density polyethylene, isotactic polypropylene and their co-polymers. *Melo et al. (2001b)* observed that the thermal capacity of the recycle pump and the tubes shown in the figure contribute significantly to the total thermal capacity of the loop system. Their experimental measurements found that the thermal capacitance for the recycle pump and heat exchange surfaces were more than 40 times the factor for flow in the reactor so that the pump and surface contributions to dynamic behavior cannot be neglected.

Simulation studies mentioned above and experiments of *Teymour and Ray (1992a)* suggest that spontaneous oscillations should be expected within the parameter space occupied by laboratory as well as commercial reactors undertaking free radical polymerizations. *Melo et al. (2001a)* embarked on a search to identify those oscillatory regions assuming a generic free radical polymerization consisting of the steps shown in the first column of **Table 7-1** with kinetic parameters drawn from the literature. For this search they employed a continuation algorithm based on the AUTO software package *(Doedel et al., 1997)* to construct continuation or bifurcation diagrams plotting the reactor temperature, T_R, against the dimensionless continuation variable, $d = \ln A_d \bar{t}$, where A_d is the pre-exponential factor in the Arrhenius expression for the initiator decomposition rate and $\bar{t} = V/Q_0$ is the mean residence time in the reactor. The d parameter appears in an exponential term in the normalized mass and energy balance for the reactor. To illustrate the search, **Figure 7-7a** shows the variation of T_R with d as a function of the volume fraction of monomer in the feed to the reactor, α. Two steady states are possible for $25 < d < 38$ and there is an unstable state as well. Hopf bifurcation points are the small, black squares. They represent a transition from a continuous operation to an oscillatory one. It can be seen from the figure that spontaneous oscillations arise for $30 < d < 41.5$ for a feed containing 30% monomer. A two continuation variable analysis is needed to determine if the oscillations are stable. **Figure 7-7b** reproduces a two continuation variable map for d and α reproduced from *Melo et al. (2001a)*. The region between the small, black

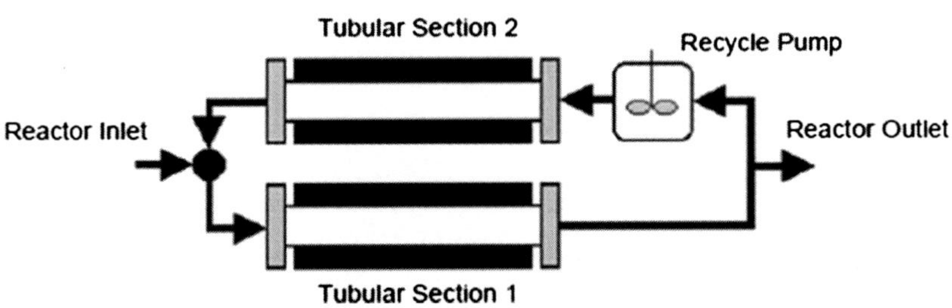

FIGURE 7-6 Schematic of a loop reactor. *(Figure reproduced from* Melo et al. (2003) *with permission. © 2003 by Elsevier Ltd.)*

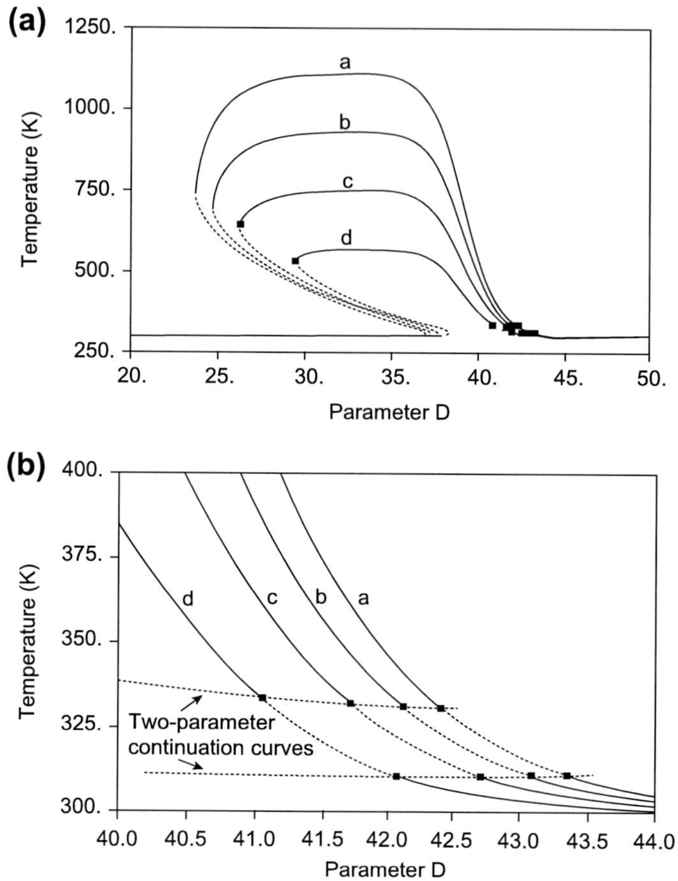

FIGURE 7-7 Parameter space in which spontaneous oscillations can arise based on (a) a single continuation variable, *d* representing mean residence time, and (b) a second continuation variable, α representing the volume fraction of monomer in the feed to a reactor. Curves a to d correspond to $\alpha = 0.9$, 0.7, 0.5 and 0.3 respectively. Small black squares are the Hopf bifurcation locations. *(Figure reproduced from* Melo et al. (2001b) *with permission. © 2001 by Elsevier Ltd.)*

squares is where the oscillatory behavior will be unstable by varying in period and amplitude. Other variables affect the parameter space permitting oscillation. These are held constant in **Figure 7-7**.

Additional continuation parameters discussed by *Melo et al.* are normalized measures of the initiator concentration in the reactor feed, heat transfer capacity of the reactor, exothermicity of the polymerization reaction, polymer kinetics (represented by activation energies) and initiator sensitivity to temperature variations. Dynamic behavior of free radical polymerization is quite sensitive to initiator concentration in the feed. Bifurcation diagrams exhibit strange branches when the initiator concentration is reproduced as a continuation variable and moves into regions of high reactor temperature, at least with respect to oscillations possible in **Figure 7-7**. The vinyl acetate polymerization study of *Teymour and Ray (1992a)* and that of *Pinto and Ray (1995a, b)* for the co-polymerization of methyl methacrylate and vinyl acetate demonstrated, several years earlier, the sensitivity of Hopf bifurcations points and the oscillatory branches from these points to the initiator concentration.

Melo et al. (2001a) concluded that spontaneous oscillations at high D, corresponding to low throughput, should be expected for free radical polymerization in well-mixed systems such as a CSTR or a loop reactor with a recycle ratio exceeding 0.5. These oscillations will be found at relatively low reactor temperatures, namely those encountered in commercial production units. However, when low initiator concentrations are fed to the reactor, isolas arise at higher temperatures and these are associated with sustained oscillations of reactor temperature. Isolas for steady-state operation also arise.

Examination of the bifurcation diagram for a loop reactor was undertaken by *Melo et al. (2003)* who introduced the recirculation ratio as the first continuation variable in place of the residence time. Polymerization kinetics were assumed to be the same as those for vinyl acetate using an AIBN initiator. Allowance was made for the recycle pump which was modeled as a CSTR and the heat balance incorporated the heat capacity of the reactor tubes, valves, the pump housing and the pump internals. These researchers employed a continuation algorithm along with the AUTO software to locate bifurcation points as well as steady state and oscillatory branches. Using the thermal Peclet number as a variable, the bifurcation diagram for mean residence time as the continuation variable resembles **Figure 7-7a** for a high recycle ratio. Simulation of the oscillatory behavior in the low temperature range of operation showed an amplitude of 60°C. **Figure 7-8** shows the bifurcation diagram for the recycle ratio as the continuation variable for different initiator concentrations in the feed stream. Hopf bifurcation points, bounding oscillatory ranges, appear in the figure. *Melo et al. (2003)* point out that the figure indicates that start-up of a loop reactor will involve increasing the recycle ratio during that operation; as a result, start-up will transit a region of spontaneous oscillations.

The heat transfer coefficient is also an important variable. When its normalized form is used as a second continuation variable, Hopf bifurcations disappear at low and high values so that oscillations will not be found. Low values of the coefficient indicate an approach to adiabatic

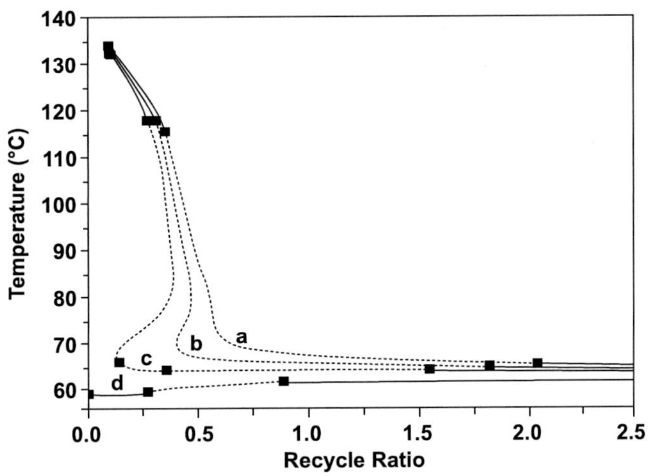

FIGURE 7-8 **Parameter space for spontaneous oscillations as a function of the recycle ratio and the initiator concentration where a to d correspond to initiator concentrations of 4, 3, 3.5 and 2 × 10^{-5} gmol/cm^3 for $\bar{t}$ = 5000 s. Dotted lines represent sustained oscillations.** *(Figure reproduced from* Melo et al. (2003) *with permission. © 2003 by Elsevier Ltd.)*

operation, while fast heat transfer at high values of the coefficient stabilizes reactor operation. Coolant temperature acts oppositely to the heat transfer coefficient. Low temperatures stabilize operation. The importance of thermal capacity of reactor externals, such as the heat transfer surfaces and the housing of the reactor was mentioned earlier. When thermal capacitance values typical of commercial operation were introduced into the system model, *Melo et al.* show that that Hopf points disappear from the T_R vs. $\bar{t}$ bifurcation diagram. This consideration and reactor operation in industry led the authors to comment that spontaneous oscillations are unlikely to occur except in start-up or a possible intentional change in the operating state.

In experiments undertaken some ten years earlier, *Belkhiria et al. (1994)* observed oscillations in a loop reactor running co-polymerization of styrene and maleic anhydride when the recycle ratio was significantly decreased. This reduced the rate of heat transfer from the reactor and raised temperatures in the loop. The *Melo et al.* analysis suggests that under such conditions oscillation could set in. *Fleury et al. (1992)* mention instability in their experimental study of the homo-polymerization of methyl methacrylate in a loop reactor, but failed to describe what they observed. These researchers did find multiplicity which they referred to as a hysteresis loop.

7.5 SATURATION OF POLYMERS

Modulation has been applied recently to the chemical modification of polymers. This post-polymerization processing is commonly used with polybutadiene, particularly for hydrogenation of the polymer. *Bussard and Dooley (2008)* explored on-off modulation of H_2 flow to a screw extruder processing polystyrene. Screw extruders find application to processing of molten polymer when they operate gas-starved as gas-filled voids are continuously created and destroyed by the moving screws. Thus, extruder operation periodically exposes the polystyrene melt to H_2.

A schematic of the *Bussard and Dooley* experimental extruder appears in **Figure 7-9**. A ceramic monolith wash-coated with a Pd/γ-Al_2O_3 slurry or a packed bed of impregnated alumina pellets

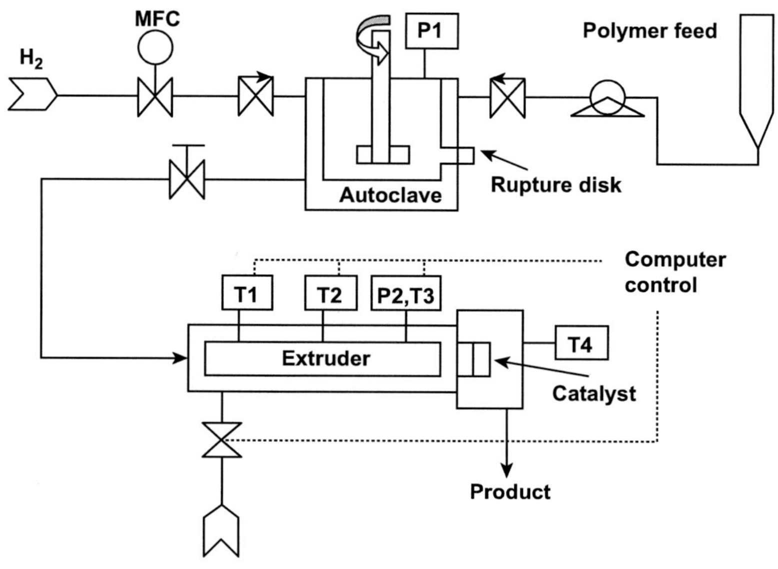

FIGURE 7-9 Experimental screw extruder for hydrogenating polystyrene. The catalyst bed is placed at the extruder outlet. *(Figure reproduced from* Bussard and Dooley (2008) *with permission. © 2008 by the American Institute of Chemical Engineers.)*

was fitted to the end of the extruder, replacing the die normally located there. Polystyrene was slurried in a THF/cyclohexane liquid phase, mixed with gaseous H_2 in a stirred autoclave and fed to the constant speed, bladed extruder whose rpm at slurry feed rate indicated it was liquid-starved. Thus, pulses of H_2 and the slurried polymer were forced through the monolith or packed bed of catalyst. Slurries containing 2 to 10 wt% polymer were used. H_2 pressures ranged mostly from 2 to 3.5 MPa; gas-liquid volumetric ratios were 0.5 to 1.2. In some experiments, H_2 gas was forced into the extruder inlet (see figure). With or without the additional H_2 flow, irregular pulses of gas flow, at a frequency of s^{-1} were observed. Results were compared with hydrogenation in a CSTR.

Measurements showed that with 2 wt% polymer slurry, hydrogenation rates in the extruder were comparable to those observed for a CSTR. The reaction, of course, is mass transfer controlled. Increasing the polymer concentration in the slurry to 10 wt% raised the liquid viscosity dramatically and substantially reduced the hydrogenation rate compared with the rate in a CSTR for the shear rate controlled slurry viscosity. The advantage of the *Bussard-Dooley* system is a much lower power input for roughly the same rate of reaction. For 10 wt% slurry, the power input/volume is just 5% of that needed for a CSTR. Pulsing additional H_2 into the extruder improved performance. *Bussard and Dooley (2008)* concluded that either the naturally pulsed or the externally pulsed (through additional H_2 feed) extruder is an attractive option for chemically modifying polymers.

7.6 ASSESSMENT

Simulation results reviewed in this chapter dealt with various polymerization systems but all assumed ideal operating conditions such as well- defined mixing (CSTR or PFR) and the absence of impurities or initiator side reactions which can cause chain terminations. These simulations predicted large increases in dispersity, particularly for anionic living chain polymerization. For Ziegler-Natta catalyzed polymerization, a better than three-fold increase in D through on-off transfer agent cycling was possible. Simulation predicted a two to eight fold increase in D for emulsion polymerization. With well-studied free radical polymerization, modulation of monomer and initiator concentrations in-phase or forcing of the CSTR jacket temperature increased D up to two fold. Indeed, complex variations of total flow, thus changing reactor volume, or of initiator, monomer and transfer agent/solvent flows at constant reactor volume can, in principle, either raise or lower D depending on the cycle. More recent work suggests that the τ_{cycle}, $\bar{t}$, and $t_{1/2}$ ratios are important because they control polymer property variations in the reactor discharge and thus the size of a post-reactor homogenization mixer. Modulation of reactor temperature for control of the MWD has also been demonstrated.

Recent work has discussed modification of molecular weight distributions using various optimal control strategies that often induce modulation. Simulations using these strategies either broaden or narrow polymer MWDs as well as drive them toward target distributions. This approach appears to be promising, but laboratory demonstration is needed. Periodic manipulation of two operating variables has been found to offer wider control of the MWD.

Is induced modulation important? The experimental evidence, unfortunately, is limited. Only free radical polymerization has been examined to any extent. Just two experimental studies on living chain polymerization appear in the literature; for Ziegler-Natta systems, a patent gives only incomplete experimental data. Modulation remains to be tested for industrially important emulsion polymerization. With anionic, "living" chain polymerization, periodic flow manipulation increased the dispersity by a factor of two, but this was much less than anticipated. Some control of the MWD and $\overline{\mu}_n$ through flow manipulation was achieved, but again it was much less than the models predicted. On-off H_2 addition provided only a 20% increase in *D*, rather than the two- to three-fold change expected. For free radical polymerization, just small increases in dispersity, ranging from 15 to 20%, were observed with composition modulation. Modulation of the reactor jacket temperature resulted in almost no change in *D*. Furthermore, in many of the experiments, reproducibility problems were indicated, so the results are suspect.

Certainly, the current challenge for periodic operation of polymerization reactors is to demonstrate experimentally the control of dispersity predicted by simulations. This situation may have contributed to the assessment of *Hungenberg (1992)* that periodic operation is not an attractive alternative for broadening the MWD of polypropylene polymers. *Hungenberg* dealt with the situation introduced by the new generation of Ziegler-Natta catalysts which produce polymer with $D = 4$ to 5, whereas several polymer markets require $D > 10$. A further problem with modulation is that the largest effects on dispersity are observed at the quasi-steady-state limit or at least under low frequency cycling. For this condition, the exiting polymer properties will vary with time. Such variation cannot be tolerated commercially; thus, a blender would have to follow the reactor to produce a product of constant properties. This blender could be as great as four times the reactor volume so that it is questionable whether a low frequency periodic polymerization process could ever be economic.

These considerations suggest that there is a region, $0.2 \leq \tau_{cycle}/\bar{t} \leq 2$ in which periodic forcing may be attractive. This operational window needs to be looked at in simulations and even more pressingly through experimentation. The literature examined in this chapter suggests that any experimental effort undertaken should focus on emulsion, step-growth, and Ziegler-Natta polymerization.

Several of the recent theoretical studies of polymerization indicate that the dynamics are "rich" in the sense that the parameter space contains multiple bifurcation points, some of which are Hopf bifurcations marking a boundary between steady and oscillatory operation. One study observed Hopf bifurcation points at cycle periods arising in reactor start-up suggesting transit through oscillatory operation might occur. These simulations need to be followed up by experimentation.

Nomenclature

A	=	Heat transfer surface area in CSTR (m^2) Amplitude
$\mathbf{A_d}$	=	Pre-exponential factor in the Arrhenius expression for initiator decomposition
C	=	Condensation product (e.g., water and methanol)
$\mathbf{C_P}$	=	Heat capacity (kJ/mol°C)
CSTR	=	Continuous stirred tank reactor
CVD	=	Chemical vapor deposition
D	=	Dispersity or polydispersity
d	=	Continuation variable (mean residence time)
$\mathbf{F_n}$	=	Feed rate of reactant "n"
F/C	=	Fluorine-carbon atomic ratio
f	=	Cycle frequency (s^{-1}) Number of functional groups on monomer
$\mathbf{\Delta H_{rxn}}$, $\mathbf{\Delta H_R}$	=	Heat of reaction (kJ/mol)
I	=	Initiator species or initiator concentration (mol/L)
K	=	Impurity, Impurity concentration, deactivating (or "killing") agent
k	=	Rate constant (different units), shift variable.
$\mathbf{k_i}$	=	Initiation rate constant
$\mathbf{k_p}$	=	Propagation rate constant
$\mathbf{k_t}$	=	Termination rate constant
M	=	Monomer species or monomer concentration (gmol/L)
$\overline{M_n}$	=	Number average molar mass or molecular weight (g/mol)
$\overline{M_w}$	=	Weight average molar mass or molecular weight (g/mol)
MMD	=	Molar mass distribution
MWD	=	Molecular weight distribution
m	=	Number of monomer units in chain, chain length index
$\mathbf{(N_{Pe})_{heat}}$	=	Peclet number for heat (-)
$\mathbf{(N_{Pe})_{mass}}$	=	Peclet number for mass (-)

n	=	Number of monomer units in chain, chain length index
	=	integer
PFR	=	Plug flow reactor
PSD	=	Particle size distribution
$\mathbf{P_n}$	=	"Dead" polymer species or living chain polymer species of chain length n, Chain length or degree of polymerization or dead polymer concentration (gmol/L)
$\mathbf{P}_n^*$	=	Reactive anion or reactive polymer molecule of chain length *n*
$\mathbf{P}_1^*$	=	Special monomer
RTD	=	Residence time distribution
$\mathbf{R_c}$	=	Free radical initiator, concentration of free radical initiator (gmol/L)
RF	=	radio frequency
$\mathbf{R_{Rec}}$	=	Recycle to fresh feed ratio
$\mathbf{R_n}$	=	"Live" polymer species of chain length n (free radical mechanism), "Live" polymer concentration (gmol/L)
$\overline{R}$**(z,t)**	=	z transform of function R
$\mathbf{R}_c^*$	=	Primary radical
$\mathbf{R}_1^*$	=	Activated monomer or free-radical of unit chain length
$\mathbf{R}_n^*, \mathbf{R}_m^*$	=	Free-radicals of chain lengths n and m
$\mathbf{Q_0}$	=	Volumetric feed rate (m^3/h)
S	=	Solvent species or solvent concentration (mol/L)
S*	=	Solvent radical
T	=	Temperature (°C, K)
$\mathbf{T_g}$	=	Glass transition temperature
$\mathbf{T_j}$	=	Jacket temperature
$\mathbf{T_R}$	=	Reactor temperature
T	=	Transfer agent species or transfer agent concentration (mol/L)
T*	=	Transfer agent radical
THF	=	Tetrahydrofuran
t	=	Time, age (s, min)
$\bar{t}$	=	Space time (s)
$t_{1/2}$	=	Initiator half-life (s)
U, $\mathbf{U_{overall}}$	=	Mean, overall heat transfer coefficient in CSTR (kJ/m^2·K·s)
V	=	Reactor volume (m^3)
$\mathbf{v_T}$	=	Volumetric flow rate (L/s)
w	=	Weight fraction
X	=	Conversion (-)
$\mathbf{X_M}, \mathbf{X_m}$	=	Monomer conversion (-)
x	=	Mole fraction
z	=	Transform variable (-)

Greek

α	=	Volume fraction of monomer in feed
λ_i	=	i^{th} moment of the chain number (or molecular weight) distribution of the living polymer
μ_i	=	i^{th} moment of the chain number (or molecular weight) distribution of the dead polymer
$\overline{\mu}_n$	=	Chain length as number of repeating units
σ_i	=	i^{th} moment of the chain number (or molecular weight) distribution of the total polymer
θ	=	Space time (s), Dimensionless time
τ, τ_{cycle}	=	Cycle period (s, h)

Subscripts

cat	=	Catalyst
fm	=	Chain transfer to monomer
fs	=	Chain transfer to solvent
fT	=	Chain transfer to transfer agent
I	=	Initiator
i	=	Initiator, initiation
j	=	Jacket
k	=	Moment index
M	=	Monomer
m	=	Monomer, chain number
n	=	Chain number, number variable
p	=	Propagation, constant pressure
R, rxn	=	Reaction
sm	=	Solvent monomer chain transfer
S, s	=	solvent
t	=	Termination by combination
w	=	Weight variable
0, 1, 2	=	Moments of chain number/molecular weight distribution
1, 2, 3,	=	Chain species, steps in initiation sequence

Superscripts

o, 0	=	Feed, initial
_	=	Time-average, mean or average
*****	=	Radical species

CHAPTER

8

Catalytic Gas-Solid Reactions

Peter Lewis Silveston

Waterloo, Ontario, Canada

OUTLINE

8.1 Partial Oxidation and Oxidative Dehydrogenation of Hydrocarbons 206

8.2 Methane Cracking 209

8.3 Non-Catalytic Gas-Solid Reactions 209

8.3.1 Lime Recovery from Waste Gypsum 209

8.3.2 Regeneration of Calcium Oxide 211

8.3.3 Chemical Heat Pumps 212

8.4 Catalytic Gasification Under Modulation 216

8.5 Gasification Employing a Circulating Solid Oxygen Carrier 223

8.6 Combustion in Circulating Fluidized Beds 228

8.6.1 Methane and Syngas 228

8.6.2 CO_2 Capture 230

8.7 Periodic Reaction Switching 233

Many of the modulated systems discussed heretofore have dealt with a solid phase functioning as a catalyst. Now attention turns to systems in which the solid phase is a reactant. On one hand, it is a reactant that is consumed or converted into another substance, but on the other, it is a reactant that provides oxygen to a gaseous species and is not otherwise consumed. Indeed, that reactant is regenerated in a second reaction step. Oxidation drawing O_2 from a solid phase is commonly referred to as anaerobic oxidation. There are several partial oxidation or dehydrogenation examples of anaerobic processes: O-xylene to maleic anhydride or phthalic anhydride, toluene to benzaldehyde, butane to butadiene, the formation of ethylene or propylene oxides from hydrocarbons and even syngas production from natural gas. In all of these systems, the solid phase is reduced by the hydrocarbon and that phase is reoxidized in a separate step by air. The solid phase functions acts primarily as an oxygen carrier. Hydrocarbon partial oxidation and oxidative dehydrogenation have been discussed in **Chapter 4**. They will be revisited just briefly in this chapter.

Composition modulation has been applied to gas-solid reaction systems also, such as the catalytic gasification of coal, the apparently non-catalytic decomposition of gypsum and anhydrite, and the regeneration of sulfided adsorbents. Interest in coal gasification has developed quite rapidly during the last two decades as a process for producing more "environmentally friendly" fuels. Oxygen carrier systems appear to offer a means of economically concentrating carbon dioxide, a necessity for sequestering this ubiquitous greenhouse gas.

Research and, recently, development have focused on the design of circulating fluidized beds, discovery of natural or synthetic materials that offer high O_2 capacity per weight of carrier and mechanical strength under time-varying environments. Separation of the carriers from ash in the solid fuels used may also pose a problem to be solved. This chapter is concerned primarily with circulating fluidized beds. The term chemical looping or looping reactor is often used to describe such systems. Gas-solid reactions under modulation have been discussed in earlier chapters with reference to the oxidation of SO_2 to SO_3 in **Chapter 3**, methane conversion over metals, partial oxidation, epoxidation and oxidative dehydrogenation in **Chapter 4**, automotive exhaust catalysis in **Chapter 6**, and, of course, combustion in **Chapter 5**.

8.1 PARTIAL OXIDATION AND OXIDATIVE DEHYDROGENATION OF HYDROCARBONS

Over the last several decades, surface science advances have established that surface oxides or adsorbed oxygen species can react with hydrocarbons to yield different products. In most systems some sort of periodic operation will be necessary to exploit these surface reactants. As a consequence, studies of periodically operated gas-solid reactions have multiplied. Summaries of these studies have been given in **Table 4-1** and **Tables 4-3** to **4-7** in **Chapter 4**.

Chapter 4 dealt with the DuPont process for the partial oxidation of butane to maleic anhydride using a vanadium phosphate oxygen carrier/catalyst which cycled between oxygen-rich and hydrocarbon environments (*Contractor et al., 1988, 1990*). More recent work on oxygen carriers by a Russian-Swiss team has dealt with the mechanism of toluene partial oxidation to benzaldehyde over various vanadia catalysts (*Bulushev et al., 2000c, 2004, 2005*). They observed that nucleophilic O_2 associated with a sub-monolayer in a K-doped V_2O_5 phase on a TiO_2 support provided selective partial oxidation of toluene to the aldehyde through a Mars-Van Krevelen reaction scheme. Excess surface vanadia as a pentoxide or a KVO_3 phase, however, resulted in electrophilic O_2 that oxidized benzaldehyde all the way to CO_2. A kinetic model devised by the team, based on the nucleophilic and electrophilic oxygen species, reproduced their pulse measurements closely. Electrophilic oxygen, apparently strongly adsorbed on the VO_2 surface, is probably responsible for the fission of the Bz ring. Adsorbed nucleophilic and electrophilic oxygen species identified on the vanadium surface appeared to yield benzaldehyde and CO_2, the latter apparently resulting from further oxidation of the aldehyde through benzoic acid. Control of the oxygen species on the catalyst surface through separating surface reduction and hydrocarbon partial oxidation from surface re-oxidation offered the prospect of significantly improving selectivity (*Pyatnitsky and Ilchenko, 1996*). To do this requires some sort of periodic operation if a single bed is used, or cycling of the oxygen carrier between reactors if multiple beds are chosen.

Of course, highly selective partial oxidation employing oxygen extracted from the catalyst lattice has been studied under a composition modulation for over 30 years for producing oxides, such as ethylene oxide, from the lower molecular weight olefins (*Renken et al., 1976; Park et al., 1983; Balzhinimaev et al., 1984: Li et al., 1992a, b*). Producing acrolein or acrylic acid from propene has been explored in the laboratory using oxygen extracted from antimony-tin and bismuth molybdate catalysts (*Silveston and Forissier, 1985; Labastida-Bardales et al., 1989; Saleh-Alhamed et al., 1992, 1993*). *Lang et al. (1989b, 1991)* applied periodic operation to the production of maleic anhydride from C_4's. Several investigators (*Fiolitakis et al., 1983; Fiolitakis and Hofmann, 1983; Cordova and Gau, 1983*) substituted benzene for butane. These modulated partial oxidation processes are discussed in more detail in **Chapter 4**. A much more exhaustive description as well as analysis and interpretation are given by *Silveston (1998)*.

Limited industrial utilization of circulating fluidized beds appears to have taken place at the end of the 1970s. In several publications in the technical press, the Lummus Company, now ABB Lummus Crest Inc., published sketchy descriptions of a process for producing isophthalonitrile, a herbicide-insecticide intermediate, from the ammoxidation of m-xylene (*Sze and Gelbein, 1975, 1976*). The processes are mentioned in **Chapter 4**. More details are given by *Silveston (1998)*.

The application of modulation to propane oxidative dehydrogenation has been explored worldwide (*Creaser et al., 1999a; Genser and Pietrzyk, 1999; Grasselli et al., 1999; Grabowski et al., 2002; Herguido et al., 2005*). Using a magnesium vanadate (MgO/V_2O_5) catalyst and a bang-bang periodic operation (switching between 6 vol% C_3H_8 and 6 vol% O_2 in a He carrier), *Creaser et al. (1999a, b, c)* measured a 50% increase in the time-average selectivity to propene relative to a steady state operation for a cycle duration of about 400 s, decreasing the period reduced selectivity. The selectivity enhancement was accompanied by a lower rate of $C_3^=$ production so that the net effect was a large increase in propene yield at a cycle period of about 60 s as illustrated in **Figure 8-1**. The *Creaser* experiments are discussed also in **Chapter 4**.

An industry-university team (*Tsikoyiannis et al., 1999; Grasselli et al., 1999, 2000*) explored a novel oxidative dehydrogenation process employing a mixed catalyst made up of a 50:50 mixture of 0.7 wt% Pt-Sn/ZSM-5, a dehydrogenation catalyst, and 42 wt% Bi_2O_3/SiO_2, a selective catalyst for H_2 combustion. They observed that at 813 K the time-average performance depended strongly on the duration of C_3 contact with the catalyst. Initially, selectivity to $C_3^=$ was 98%, but then dropped. Re-oxidation of the catalyst was rapid and could be carried out at 350°C. Oxygen concentration was ramped from 2 to 21 vol% during the partial cycle to avoid overheating and agglomeration of Pt in the catalyst mixture. The results of a second set of asymmetric experiments are shown in **Figure 8-2** in which the C_3 duration dropped to 2.8 min from 130 min while the O_2 exposure remained at 10 min. Prior to a feed switch, 5 min inert flushes were used.

Addition of the selective Bi_2O_3 catalyst for H_2 combustion greatly increased conversion and also improved selectivity with respect to the Pt-Sn catalyst alone under the modulation employed. However, performance fell with successive cycles. The researchers attributed this to structural changes and loss of active phase dispersion to be caused by deep reduction of the catalyst during hydrocarbon exposure. They concluded that the adoption of mixed reducing-oxidizing catalysts depends on developing a stable composition and finding suitable operating conditions.

In a different approach to finding a commercial process, *Zagoruiko (2007)* observed that with just a modest degree of propane total oxidation, the oxidative dehydrogenation of that paraffin to propylene would be autothermal. He proceeded from this observation to explore, through simulation, an autothermal dehydrogenation process employing a homogeneous, adiabatic plug flow model. The proposed process utilized periodic flow reversal with the hot

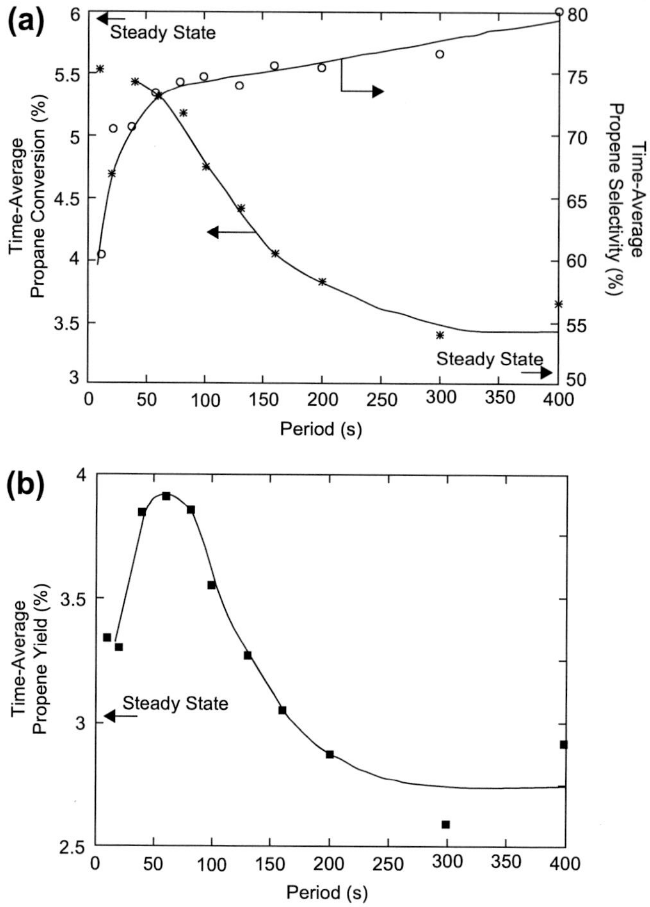

FIGURE 8-1 **Performance of bang-bang cycling between 6 vol% C_3H_8 in He and 6 vol% O_2 in He at 1 bar and 783 K over a packed bed of $MgO/Mg_3(VO_4)_2$ catalyst. Steady-state data for an equal mixture of the two streams is given on the ordinate axis.** *(Figure from* Creaser et al. (1999a), *with permission of the authors.)*

reduced Ti-supported vanadia catalyst re-oxidized by cold air flowing in one direction through the bed in one half-cycle and with cold propane flowing through the hot catalyst in the opposite direction. Dehydrogenation and catalyst regeneration proceed at markedly different rates at the catalyst temperature. *Zagoruiko* handled this problem by varying the gas velocity in the bed. Simulation indicated that for the conditions chosen, 85% selectivity to propylene could be achieved, but at a conversion of just 2% of the propane fed. This low conversion needs to be compared with conversion under a steady-state operation. Simulation of such an operation suggests that at a maximum feasible temperature of 750°C, conversion would be just 5% at a propane selectivity of 50%. *Zagoruiko's* interesting autothermal process requires heating the catalyst bed to 400°C on start-up. Advantages, in addition to autothermal operation, are the absence of an inert diluent in the feed and the separation of hydrocarbon and oxidant. **Chapter 4** also deals with the *Zagoruiko* simulation.

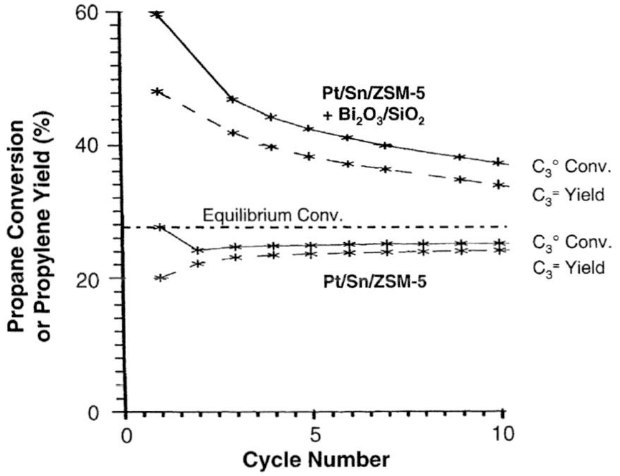

FIGURE 8-2 **Comparison of the performance of a mixed dehydrogenation-oxidation catalyst with that of a dehydrogenation catalyst alone for the oxidative dehydrogenation of propane under asymmetric modulation at 1 atm, 540°C and SV = 2 h^{-1}.** *(Figure adapted from* Grasselli et al. (1999) *with permission. © 1999 Elsevier Science B.V.)*

8.2 METHANE CRACKING

Production of hydrogen on a commercial scale employs a sequence of reaction and separation steps beginning with steam reforming of natural gas, successive water gas shift reactions separated by scrubbing with amine solutions and usually ending with methanation. Partial oxidation or autothermal reforming of hydrocarbons or alcohols has also been considered, but, as with steam reforming, such a process requires further conversion and separation steps. A periodic process based on cracking methane over metals avoids these further steps and has been investigated by *Zhang and Amiridis (1998), Otsuka et al. (2000, 2001)* and later by *Monnerat et al. (2001)*. These representative studies are summarized in **Table 8-1**.

The *Monnerat* study employed a woven Ni gauze which was treated to form a Raney nickel outer surface that increased the specific surface area from 1 to 26.7 m^2/g. Gauze stripes, tightly fitted into a quartz tube placed inside a temperature controlled furnace were used. In the *Monnerat* experiments, partial cycle feed pressures were 30 kPa for CH_4 and 12 kPa for the carbon gasifying O_2 stream. Gauze temperatures varied, but most experiments were made at 773 K. Effluent composition for a symmetrical cycle at $\tau_{cycle} = 4$ min are shown in **Figure 8-3**. Such a cycle gave the maximum time average H_2 flow rate. Small amounts of carbon oxides appeared in the product gas during the first minutes on stream, but then vanished. Flushing the gauze with O_2 gave only CO and CO_2; however, the fact that the gauze was partially oxidized resulted in carbon oxide production on switching to methane.

In **Figure 8-4**, effluent from a 20 min symmetrical cycle is shown. After 3 min into a composition switch, the effluent contains just ppm levels of water and carbon oxides. Recovering just a portion of the effluent after 3 min produces a satisfactory fuel cell feed.

Cracking over the Ni catalyst resulted in carbon whisker formation on the metal surface. The phenomenon is widely observed over supported metal catalysts and is well-studied (*Trimm, 1977; Bartholomew, 1982; Snoeck and Froment, 1997a, b; Otsuka et al., 2000; Ogihara et al., 2006*). It is accompanied by the slow loss of metal through a nano-particle at the whisker tip which is lost when the whisker is gasified in the regeneration step of the cycle.

8.3 NON-CATALYTIC GAS-SOLID REACTIONS

8.3.1 Lime Recovery from Waste Gypsum

Large quantities of low grade gypsum ($CaSO_4 \cdot 2H_2O$) are generated in the production of industrial acids, the treating of acidic wastewater and in some stack gas scrubbing systems. A phosphogypsum is generated also in vast amounts as a by-product in the production of

TABLE 8-1 Methane Cracking

Authors	Objective	Reaction	Modulated Variable	Reaction Conditions	Observations	Comments
Zhang and Amiridis (1998);	Process Development	$CH_4 \rightarrow C + 2\ H_2$; $C + O_2 \rightarrow CO_2$	Asymmetric cycling between 20 vol% CH_4 in He and steam or air over a 16.4 wt% Ni/SiO_2 catalyst	1 bar and 823 K in a packed bed of catalyst	Methane partial-cycles up to 200 min at $GHSV$ =15000 L/h were possible. Catalyst retained activity over > 10 cycles with steam regeneration, but O_2 regeneration destroyed catalyst.	Carbon whisker formation was identified: with steam regeneration, additional H_2 was produced, but process was no longer autothermal.
Otsuka et al. (2000); Ogihara et al. (2006)	Development of a transport - storage process for CH_4	$CH_4 \rightarrow C + 2\ H_2$; $C + 2\ H_2 \rightarrow CH_4$ over Ni/SiO_2; $CaNi_5$; and alloy catalysts, e.g., Pd/Al_2O_3; $Pd{:}Co/Al_2O_3$	2 part cycle switching between CH_4 at 2.7 to 67 kPa and H_2 at 60 kPa	CH_4 decomposition was run at 523 < 823 K, but most experiments were at 773 K; alloy experiments mainly at 973 K	Ni catalysts accumulated > 200 C/Ni, but 8 C/Ni were recovered on hydrogenation. $Ni/SO_2{:}CaNi_5$ mixture showed the best performance; Pd/Al_2O_3 accumulated > 450 C/Pd, but Pd-Ni/Al_2O_3 > 10,000 C/metal.	Oxidation/ hydrogenation to restore metal or alloy surface was not investigated.
Monnerat et al. (2001)	Process optimization	$CH_4 \rightarrow C + 2\ H_2$; $C + O_2 \rightarrow CO_2$ over an Ni gauze with a Raney nickel surface	Symmetrical and asymmetrical bang-bang switching between CH_4 at 30 kPa and O_2 at 12 kPa	Quartz vessel + packing with surface area = 26.7 m^2/g, $T = 773$ K, but $683 < T < 823$ K was investigated	Optimal formation of H_2 for symmetrical cycling occurred at $\tau_{cycle} = 4$ min. Carbon combustion in the O_2 half-cycle was temperature dependent.	Complete removal of coke was not possible even at the highest temperatures used. Carbon was deposited as whiskers with a nano-amount of Ni at tip.

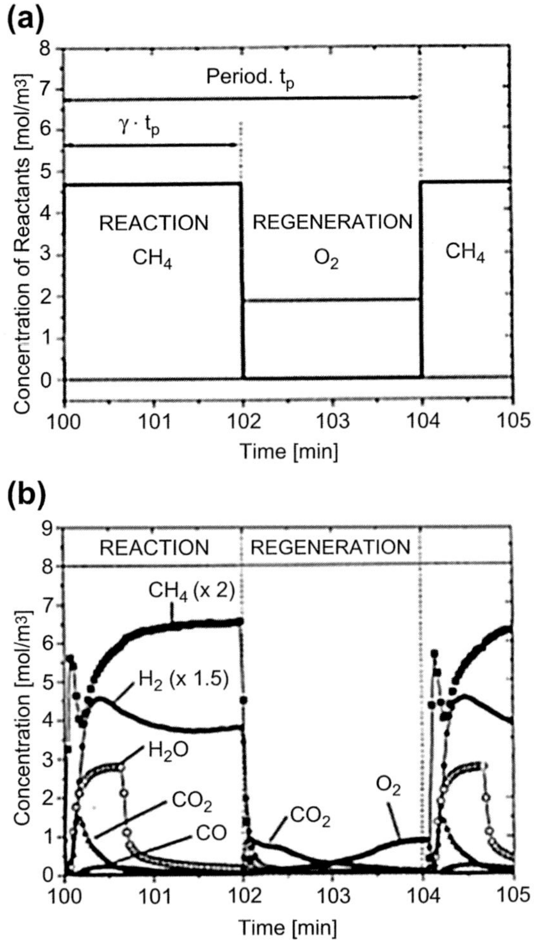

FIGURE 8-3 **Hydrogen production from CH_4 cracking showing in (a) the imposed feed concentration cycle and in (b) the off-gas composition in each partial cycle for continuous cyclic operation at $T_{Reactor}$ = 773 K, Q = 75 mL (NTP)/min, $m_{Catalyst}$ = 207 mg, C_{CH4} = 4.67 mol/m^3, C_{O2} = 1.87 mol/m^3.** *(Figure reproduced from* Monnerat et al. (2001) *with permission. © 2001 by Elsevier Science Ltd.)*

phosphate fertilizers. Most of this waste ends up as landfill because processing to recover building grade gypsum or Portland cement, or phosphates in the case of the phosphogypsum, cannot be justified economically. Perhaps, because of the rising cost of landfill, there has been an interest in reclaiming quicklime and sulfuric acid (via SO_2) from gypsum wastes. Recovery is possible by reductive decomposition at temperatures above 1270 K.

The primary reaction in reductive decomposition is the formation of calcium oxide and CO_2 or H_2O or both depending on the reductant used. Sulfur is recovered as SO_2. The reduction proceeds through various mineral phases with sulfide as an intermediate. Strongly reducing conditions and temperatures below 1370 K favor the sulfide rather than the oxide. Sintering limits reduction temperatures to 1470 K or below. Endothermicity and the rather narrow permissible temperature range dictate the use of a fluidized bed for reductive decomposition. Nevertheless, calcium sulfide persists as an impurity in the lime produced and affects the economics for some applications.

Wheelock and co-workers demonstrated that the sulfide impurity can be eliminated by using fuel injection to create oxidizing and reducing zones in the fluidized bed, or by periodically switching the air-fuel ratio in the fluidizing gas so that the entire bed becomes alternately oxidizing and then reducing. The modulated fluidized bed, realized through switching the A/F ratio, was investigated on a bench scale by *Wheelock and Riel (1991)*. An asymmetric cycle was used. *Composition Modulation of Catalytic Reactors* (*Silveston, 1998*) discusses this interesting research in some detail.

Table 8-2 summarizes this and the modulation of other non-catalytic processes.

8.3.2 Regeneration of Calcium Oxide

Jagtap and Wheelock (1996) applied the research results from Wheelock's gypsum experiments to the regeneration of calcium oxide, a potential sorbent in hot gas cleanup (removal of H_2S from reducing gas streams). They demonstrated that under modulation calcium oxide can be regenerated at 1225 to 1375 K rather than at 1675 K used in a continuous process. Below the latter temperature, the CaS formed in hot gas cleanup reacts with O_2

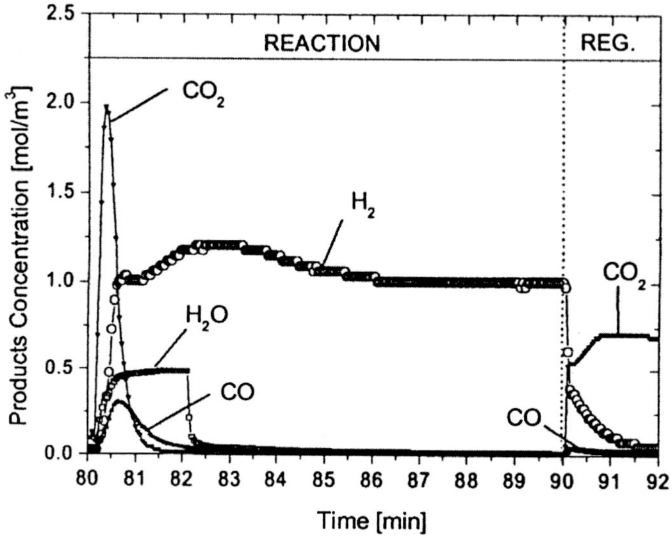

FIGURE 8-4 **Off-gas composition from a symmetrical cycle with τ_{cycle} = 20 min. Operating conditions are as given in Figure 8-3. Only the first two minutes of the regeneration half-cycle are shown.** *(Figure reproduced from* Monnerat et al. (2001) *with permission. © 2001 by Elsevier Science Ltd.)*

to form the sulfate. The sulfate is less dense and interferes with regeneration. *Jagtap and Wheelock* reasoned that if exposure to O_2 or air is followed by exposure to a reducing gas, some of the sulfate formed could be reduced to the oxide or back to the sulfide, thus opening the pore structure. To demonstrate their concept, they carried out modulation experiments and observed that essentially complete conversion of CaS to CaO was possible even at 1225 K. Further experiments used repeated sulfidation and regeneration cycles and verified that the lime produced did not lose its capacity for sulfur removal in successive cycles. *Silveston (1998)* provides further details of these studies.

Separating a gas-solid reaction into steps and carrying each out in a near optimal environment is essentially what Wheelock and his co-workers have done for gypsum decomposition. They were able in this way to avoid sintering and still minimize the impurities in their solid product. However, a better solution may be to use a two-part circulating fluidized bed with each part operated at steady state but with different gas compositions and temperatures. It seems unlikely that the recovery of lime and SO_2 from gypsum or spent lime regeneration are the only gas-solid reactions that can be separated into steps. Other industrial solid conversion reactions need to be examined.

8.3.3 Chemical Heat Pumps

Periodically changing solid compositions are encountered in chemical heat pumps. Indeed, the familiar refrigeration cycle is a "looping system" in which the refrigerant circulates between an evaporator, where heat is withdrawn, and a condenser where heat is discharged. Industrial systems often employ an ammonia cycle in which absorption and desorption replace refrigerant condensation and vaporization. Adsorption and desorption

TABLE 8-2 Non-Catalytic Fluid-Solid Reactions

Authors	Objective	Reaction	Modulated Variable	Reaction Conditions	Observations	Comments
FLUID-SOLID PROCESSES						
Swift and Wheelock (1975); Wheelock and Morris (1986); Morris et al. (1987)	Regeneration of lime adsorbants	$CaS(s) + O_2 \rightarrow CaO(s) + SO_2$ with an undesirable side reaction: $CaS(s) + O_2 \rightarrow CaSO_4(s)$	Employed solids circulating between oxidizing and reducing zones in a fluidized bed	1370 K to 1480 K in a 12-cm (i.d.) bed fluidized with either CH_4 or CO + air and with air injection at bed midpoint, also a 25-cm (i.d.) fluidized bed and larger i.d. pilot plant vessel were used.	Sulfide content (0.4 to 2.6%) of the fine solids entrained with the fluidizing gas was strongly affected by A/F ratio of the fluidizing gas. Sulfide content of the larger particles in the bed overflow was nil or very small.	SO_2 in the reducing partial cycle arises from reduction of $CaSO_4$ by CO or CH_4. Source in the oxidizing half-cycle is the oxidation of CaS.
Wheelock et al. (1988)	As above	As above using gypsum, waste gypsum, anhydrite, phosphogypsum	Employed solids circulating between oxidizing and reducing zones in a fluidized bed	$1370 < T < 1480$ with pilot plant bed fluidized with an air: CH_4 mixture of 9.5 to 12.6. Air injected at bed midpoint	0 to 0.9 wt% CaS in larger particles in bed overflow. Wt% in fines < 2.5 wt%	
Wheelock and Riel (1991)	As above	As above, but using a phosphogypsum	Single zone fluidized bed with switching of the fluidizing gas between air and a CH_4: air mixture	1370 K with a molar CH_4: $CaSO_4$ ratio = 1.54; A/F = 9.1	CaO yields ranged from 86 to 91%, increasing with a 50 K rise in temperature.	
Jagtap and Wheelock (1996)	Regeneration of CaS from a hot gas cleanup process	$CaS(s) + O_2 \rightarrow CaO(s) + SO_2$ with side reaction $CaS(s) + O_2 \rightarrow CaSO_4(s)$	Modulated environment in a laboratory TGA between air and 5 vol% CH_4 in air	$1230 < T < 1370$ K for $\tau_{cycle} = 2$ min. Besides CH_4, CO and C_3 were used as reductants	At 1370 K, 98% conversion of CaS to CaO required 40 min. At 1230 K, 90% conversion required 110 min.	H_2S trapping capacity was maintained through multiple cycles.

(Continued)

TABLE 8-2 Non-Catalytic Fluid-Solid Reactions (*cont'd*)

Authors	Objective	Reaction	Modulated Variable	Reaction Conditions	Observations	Comments
CHEMICAL HEAT PUMPS						
Ogura et al. (2003)	Performance testing	Low temperature cycle: $CaO + H_2O \leftrightarrow Ca(OH)_2$; high temperature cycle: $H_2O(v) \leftrightarrow H_2O(l)$	Operation testing	Performance tested for $273 < T_L < 426$ K; $683 < T_H < 773$ K. Coefficient of performance measured = 1.4	Authors found the CaO-based system outperformed other chemical heat pumps and could deliver heat at higher temperatures than mechanical heat pumps.	
Huang et al. (2004)	Model development	$SrCl_2 \cdot NH_3 + 7\ NH_3 \leftrightarrow SrCl_2 \cdot 8NH_3$	Investigation of cycle	Salt formation/decomposition. Formation: $3 < P < 5$ bar, $303 < T < 313$ K, decomposition: $1\ 2 < P < 15$ bar, $393 < T < 423$ K	Heat pump model satisfactorily represented experimental performance.	
Wang et al. (2008)	Improved ammonia traps for chemical heat pumps	Low temperature cycle: $CaCl_2 \cdot 2NH_3 + 2\ NH_3 \leftrightarrow CaCl_2 \cdot 4NH_3$; high temperature cycle: $MnCl_2 \cdot 2NH_3 + 4\ NH_3 \leftrightarrow MnCl_2 \cdot 6NH_3$	Temperature cycled between T_{M1} and T_L for low temperature cycle and between T_{M2} and T_H for high temperature cycle	For heat input T_M =120°C and discharge T_L = 30°C, heat was pumped over $\Delta T = 22$°C	Another salt, $CaCl_2 \cdot 8NH_3$ is possible by control of T_L and P. The ΔT achieved depends on the low temperature salt at T_L.	
Li et al. (2009b); Oliveira and Wang (2007, 2008)	Performance testing and system simulation	Low temperature cycle: $MCl_2 \cdot 2NH_3 + 2\ NH_3 \leftrightarrow MCl_2 \cdot 4NH_3$; where M = Ca, Ba, Mn; high temperature cycle: $NH_3(v) \leftrightarrow NH_3(l)$	$BaCl_2$ and $MnCl_2$ system operated between 283 K and 453 K. $CaCl_2$ system operated for ice-making with $253 < T_L < 263$K; $293 < T_H < 303$ K	$CaCl_2$ ice-making system achieved a specific cooling power > 1250 W/kg salt at $T_L = 258$ K and $T_H = 298$ K	Specific cooling power depended on the thickness of the graphite-salt layer in the reactor.	Performance results for ice-making application were obtained from system simulation.

or chemical formation and decomposition are also exploited. The latter involve gas-solid reactions. The use of chemical reactions in heat pumps and the technology involved encompass a growing literature. Fortunately, a useful review (*Wongsuwan et al., 2001*) and a good explanation of system function (*Goertz et al., 1993*) exist.

The principle of a chemical heat pump is illustrated by the reversible dehydrogenation of 2-propanol to acetone: $(CH_3)_2CHOH \leftrightarrow (CH_3)_2CO + H_2$. Heat is supplied at a low temperature for endothermic dehydrogenation, the gas products are pressurized by a compressor and heat is released to a high temperature sink by the exothermic hydrogenation that then takes place. Expansion of 2-propanol cools the gas and completes the cycle. Cycles are asymmetric and the period may vary depending on the heat source and sink temperatures.

Most of the current development effort on chemical heat pumps focuses on gas-solid reactions that do not involve a catalyst. Typical of such systems are $CaO/H_2O/Ca(OH)_2$ or $CaCl_2/NH_3/CaCl_2{\cdot}6NH_3$ (*Wang et al., 2008; Oliveira and Wang, 2008*). The presence of a solid complicates the operation of the chemical heat pump because of the problem of transporting a solid reactant. This is avoided using a periodically operated reaction couple such as the $CaCl_2/NH_3/CaCl_2{\cdot}4NH_3//MnCl_2/NH_3/MnCl_2{\cdot}6NH_3$ system or a salt formation/decomposition combined with NH_3 vaporization and condensation. In the former, heat supplied to $CaCl_2{\cdot}4NH_3$ causes decomposition and release of NH_3. The gas passes to another reactor where it reacts with $MnCl_2$. The removal of low temperature waste heat leads to the formation of $MnCl_2{\cdot}6NH_3$. Now at a higher pressure and temperature, that salt decomposes releasing heat at a higher temperature to a target sink. This is the heat pump step. The NH_3 released in the high temperature decomposition returns back to the $CaCl_2$ system in the first reactor and initiates a new cycle. The operation is illustrated in **Figure 8-5**. Details are given by *Wang et al. (2008)*. Changes to the cycle to improve the coefficient of performance have been proposed by *Li et al. (2009b)*. *Goertz et al. (1993)* discuss the choice of salt pairs to match heat source temperature and the temperature lift desired.

Cyclic ammoniate formation and decomposition take place in the same porous solid bed. Various designs are used, but the simplest, shown in **Figure 8-6**, illustrates the principle. The bed is heated in the decomposition step by a heat transfer fluid circulating on the periphery, while gas is withdrawn through the central tube. In the formation step, gas is fed under pressure through the tube and the peripheral heat transfer fluid cools the bed. Graphite is intimately mixed with the ammoniate salts to increase the heat transfer rate (*Oliveira and Wang, 2007, 2008*).

A heat pump can be operated as well with ammonia condensation and vaporization replacing the salt formation/decomposition reactions. This operation has been studied and modeled by *Huang et al. (2004)* for the formation/decomposition of $SrCl_2{\cdot}8NH_3$ and by *Oliveira and Wang (2008)* for $CaCl_2{\cdot}4NH3$. The diagram in **Figure 8-5** applies but the part 2 formation/decomposition is replaced by NH_3 vaporization and condensation.

Another version of the two reaction system uses just formation and decomposition of $CaCl_2{\cdot}8NH_3 \rightarrow CaCl_2{\cdot}4NH_3 + 4NH_3$ and $CaCl_2{\cdot}4NH_3 \rightarrow CaCl_2{\cdot}2NH_3 + 2NH_3$ (*Oliveira and Wang, 2007, 2008*). These contributions dealt with the problem of improving the rate of heat transfer through the solid packing by mixing the calcium salt with graphite powder. Transient behavior of the solid decomposition and gas-solid synthesis reaction and their influence on heat transfer through the surface of the packed bed and the thermal sink are discussed by *Aidoun and Ternan (2004)*. These investigators cautioned that the often used assumption of

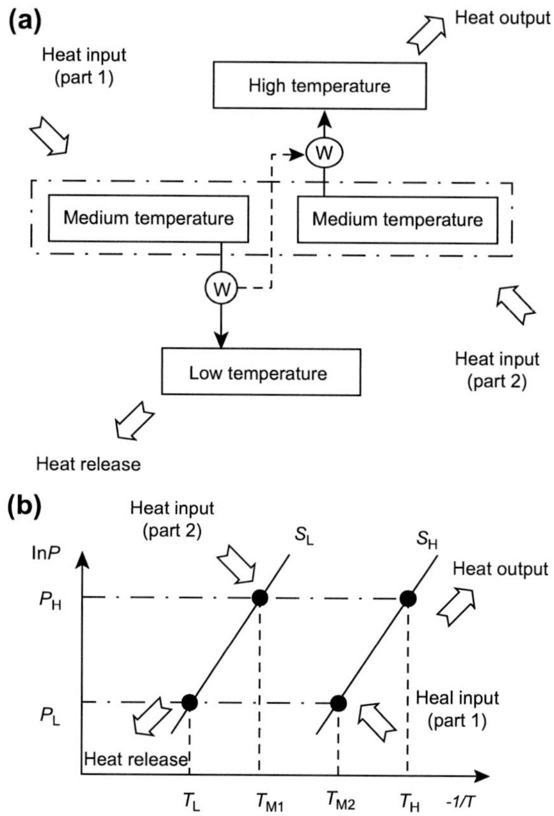

FIGURE 8-5 Operation of a chemical heat pump using salt transformations through temperature and pressure change: $CaCl_2 \cdot xNH_3 \rightarrow CaCl_2 \cdot yNH_3$ and $MnCl_2 \cdot wNH_3 \rightarrow MnCl_2 \cdot zNH_3$: (a) process diagram, (b) representation on the lnP vs. T plane. *(Figure reproduced from* Wang et al. (2008) *with permission. © 2008 AIChE)*

constant heat transfer coefficients during an operating cycle is not justified. Dynamic models for the gas-solid systems mentioned above have been developed by *Mbaye et al. (1998)* and *Huang et al. (2004)*.

A Beijing based research team (*Lai et al., 1992, 1993*) explored a methanolation-demethanolation cycle based on $CaCl_2$ that employed $CaCl_2 \cdot MeOH$ and MeOH condensation and vaporization. Using a $CaCl_2$ methanolation-demethanolation cycle, *Lai and Li (1996)* and *Lai (1999)* proposed employing periodic flow direction reversal in heating or cooling of the reactor block by their heat transfer fluid. Their simulation demonstrated higher COPs for both heating and cooling applications.

8.4 CATALYTIC GASIFICATION UNDER MODULATION

Gasification is a primary source of synthesis gas for industrial processes and a key step in integrated combined cycle and carbon sequestering systems now being developed to raise power generating efficiency and reduce CO_2 and SO_2 discharge to the atmosphere. Modulation of the gasification process will be

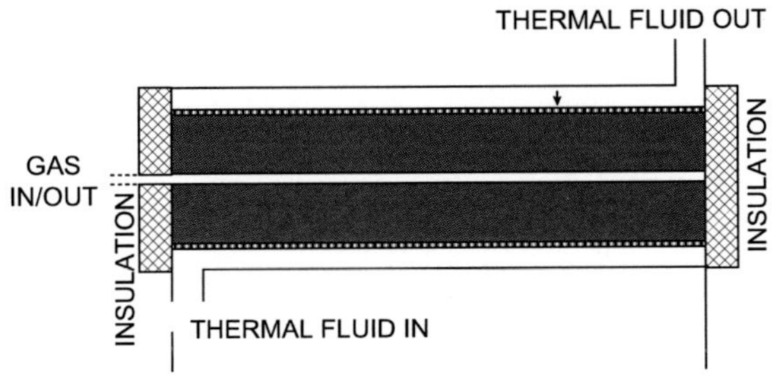

FIGURE 8-6 **Reactor block used for ammoniate formation and decomposition in chemical heat pumps. Blackened region is the reactor bed. Central tube is open to the solids bed.** *(Figure modified from* Mbaye et al. (1998) *and used with permission. © 1998 by Elsevier Science Ltd.)*

considered first on the scale of a coal particle and later on a process scale.

Investigations of gasification under modulation on the particle scale go back to the 1980s. All coals contain mineral matter, usually iron or alkali oxides that catalyze gasification. Early research dealt (see **Table 8-3**) with the mechanism of gasification in the presence of these materials. Indeed, adding oxide catalysts to a coal charge was considered seriously as a means of reducing combustion temperature and thereby decreasing NO_x generation and coal ash slagging. A number of common, reducible metal oxides, such as Fe_2O_3, NiO, CaO, take part in carbon gasification through the mobility of oxygen on the oxide surface. The overall reaction cycle can be represented for iron oxide as:

$$Fe_nO_{m+1}(s) + C \rightarrow Fe_nO_m(s) + CO(g). \quad (8.1)$$

$$2\ Fe_nO_m(s) + O_2(g) \rightarrow 2\ Fe_nO_{m+1}(s). \quad (8.2)$$

CO_2 and water exchange oxygen with the oxides to yield CO and H_2. Redox cycles of other metal-metal oxide couples also catalyze gasification. Of course, the gasification activity will differ depending on the oxides, as *Ohtsuka et al. (1987)* have demonstrated.

Early modulation studies used chars rather than coal because of complication introduced through volatilization during heating of the samples. Thus, *Suzuki et al. (1988, 1989)* used a char prepared from Yallourn coal (an Australian brown coal) impregnated with an iron nitrate solution. Symmetrical modulation was between argon (Ar) and a CO_2/Ar mixture. The operating temperature was 1070 K. Results indicated a 60 s cycle maximized the gasification rate. Up to 40% conversion, the gasification rate for this cycle period is about three times the steady-state rate. Further experiments by *Suzuki et al. (1990)* used chars prepared from iron impregnated Morwell and Loy Yang coals. These authors found about a 50% increase in gasification rate over rates under steady state.

Zhang et al. (1994a, 1994b, 1995) extended the *Suzuki* studies, but used chars made from a Forestburg coal, a Canadian sub-bituminous coal. Modulation employed symmetrical cycles, switching between CO_2 and N_2. Improvements in char burn-off, about 15 to 20%, were found under composition modulation at 1070 K, but they were much less than the rate enhancements reported above by *Suzuki et al.* These studies are discussed in more detail by *Silveston (1998)*.

The reactions proceeding in gasification are suggested by the variation of the CO content of the off-gas with time in a cycle. These are shown in **Figure 8-7**. In this figure, successive pulses of CO_2 increase in duration and are separated by N_2 exposures of constant duration (bottom portion of the figure). A burst of CO evolution, seen as a sharp peak, occurs 5 to 10 s after CO_2 is switched on. Height and area of the peak are determined by the duration of CO_2 exposure only for durations under 30 s. For durations above 30 s, peak height and width remain constant, but after 30 s a plateau appears. The height of this plateau is greater than the height of the CO signal under steady state operation at the same degree of char conversion. The plateau height and the steady-state signal are strong functions of temperature.

TABLE 8-3 Gasification of Carbonaceous Materials

Authors	Objective	Reaction	Modulated Variable	Reaction Conditions	Observations	Comments
GASIFICATION ON A PARTICLE SCALE						
Suzuki et al. (1988, 1989)	Investigation of CO_2 gasification of coal char under composition modulation	$C_xH_yO_z + CO_2 \rightarrow CO + H_2O$ catalyzed by Fe_2O_3. (Yallourn brown coal)	Symmetrical cycling between CO_2/Ar and Ar, $30 < \tau_{cycle} < 90$ s	$T = 1075$ K, $P = 1$ bar; catalyst/coal = 0.6 mmol FeO/g dry char	Up to 40% conversion, modulation at $\tau_{cycle} = 60$ s increased gasification rate 3 x, but enhancement dropped after 40% conversion.	Damping of gas composition due to mixing depressed rate enhancement through modulation for $\tau_{cycle} < 60$ s.
Suzuki et al., 1990)	As above	$C_xH_yO_z + CO_2 \rightarrow CO + H_2O$ catalyzed by Fe_2O_3, NaOH and mixtures (Morwell and Loy Yang sub-bituminous)	As above	$T = 1075$ K, $P = 1$ bar; various levels of char impregnation	Up to a 50% enhancement of gasification was observed for rates under modulation for $60 < \tau_{cycle} < 90$ s.	Gasification rates for Loy Yang coal were 10 x rates for Yallourn coal, while for Morwell coal, rates were 50 x.
Zhang et al. (1994a, 1994b, 1995)	As above	$C_xH_yO_z + CO_2 \rightarrow CO + H_2O$ catalyzed by Fe_2O_3; also experiments without Fe_2O_3 and with demineralized coal (Forestburg sub-bituminous, high ash coal)	Symmetrical cycling between CO_2 and N_2	$1075 < T < 1175$ K, $P = 1$ bar; 3 wt% Fe/C in char (some with 0 wt% Fe/C and demineralized char)	15 to 20% yield or conversion increase was observed at $\tau_{cycle} = 60$ s and 1075 K. Optimal τ_{cycle} depended on temperature.	Evolution of CO gas with time during modulation confirms a redox reaction and that oxide reduction is the slow reaction. Optimal cycling is asymmetric.
GASIFICATION/COMBUSTION USING A CIRCULATING SOLIDS OR A SIMULATED CIRCULATING SOLIDS OXYGEN CARRIER						
Jin and Ishida (2002)	Testing of O_2 carriers for circulating fluidized bed combustion of methane	$CH_4 + NiO/Ni/Al_2O_4 \rightarrow CO_2 + H_2O + Ni/Ni/Al_2O_4$; $Ni/Ni/Al_2O_4 + O_2 \rightarrow NiO/Ni/Al_2O_4$	Step change experiments with solids regeneration in air	$873 < T < 973$ K, $1 < P < 3$ atm.; experiments extended over 5 cycles	Conversions were not given. $NiO/NiAl_2O_4$ gave reproducible results and maintained activity.	Carbon deposition was observed.

Iyer et al. (2004)	Effect of SO_2 on CO_2 removal with CaO	$CaO + CO_2 \leftrightarrow CaCO_3$; $CaCO_3 + SO_2 \leftrightarrow CaSO_4 + CO_2$	Symmetrical cycling between 10 vol% CO_2, 4 vol% O_2 and 3000 ppm SO_2 in N_2 and N_2 for $\tau_{cycle} = 60$ min, 600 min	$T = 973$ K, $P = 1$ bar; Experiments performed in a TGA	Best result with CaO was 45 wt% CO_2 capture after 40 cycles; this result dropped to 36 wt% after 100 cycles.	Sulfated adsorbent can be regenerated at $T > 1270$ K.
Scott et al. (2006)	Development of circulating solids combustor for solid fuels	$C_xH_yO_z + CO_2 \rightarrow CO + H_2 + H_2O$; $CO + Fe_2O_3(s) \rightarrow CO_2 + Fe_3O_4 + FeO$; Experiments used Hambach lignite or char	Asymmetrical cycling between 27.5 vol% CO_2 in N_2 and air with lignite added in the reducing step	$T \approx 1173$ K, $P = 1$ bar	Authors observed 92% conversion of lignite char.	Observed that gasification of lignite or lignite char is the rate limiting step.
Siriwardane et al. (2007)	Investigation of O_2 carrier performance with syngas combustion	$NiO + CO + CO_2 + H_2 \rightarrow Ni + CO_2 + H_2O$: NiO supported on bentonite	4 part cycle: 1) reduction with 36 vol% CO, 27 vol% H_2, 12 vol% CO_2 in N_2 for 10 min, 2) & 4) N_2 or Ar flush for 5 min, 3) re-oxidation in air for 60 min	Experiments were conducted over 10 cycles in a TGA and in a fixed bed of supported NiO at $973 < T < 1173$ K and 0.67 MPa	100% of H_2 was consumed in the reduction step. Particle size and pressure had little effect on rates. Particle morphology was unchanged over 10 cycles. A small loss of surface area was seen at 1173 K.	Rate of reduction for NiO was significantly greater than the re-oxidation rate.
Pfeifer et al. (2007)	Testing of pilot plant (100 kW) performance using biomass	$C_xH_yO_z + H_2O + CaO \rightarrow CaCO_3 + H_2$; $CaCO_3 + O_2 \rightarrow CaO + CO_2$; Calcite oxygen carrier	Circulating solids consisting of a bubbling fluid. bed + riser regenerator and bubbling fluid. bed gasifier	$825 < T_{gasifier} < 975$ K; $1075 < T_{regenerator} < 1175$ K; experiments in a 100 kW (fuel) pilot plant	H_2 content of the product varied from 66 to 75 vol%, and was dependent on $T_{gasifier}$.	Off-gas contained tar. Tar production depended on $T_{gasifier}$, the ratio of steam/biomass, and the carrier circulation rate.

(*Continued*)

TABLE 8-3 Gasification of Carbonaceous Materials (*cont'd*)

Authors	Objective	Reaction	Modulated Variable	Reaction Conditions	Observations	Comments
Leion et al. (2008, 2009b)	Testing of the gasification of different fuels with different Fe and Ni oxygen carriers	$C_xH_yO_z + H_2O \rightarrow CO + CO_2 + H_2$ catalyzed by ilmenite, Fe_2O_3/$MgAl_2O_4$, $NiO/NiAl_2O_4$; fuel: bituminous or sub-bitum. coals or petroleum coke	4 part cycle of 1) fuel fed with H_2O in N_2 fluidizing gas, for 20 min, 2) N_2 flush for 3 min, 3) re-oxidation with 5 vol% O_2 in N_2, 4) N_2 flush for 3 min	$1123 < T < 1273$ K, but most experiments done at 1223 K, $P = 1$ bar; conical fluidized bed promoted mixing. In some experiments CH_4 or syngas replaced steam	Experiments were done over 2 to 6 cycles. Gasification rate depended on % steam in the fluidizing gas. No agglomeration or ash-oxygen carrier interaction was observed.	When sand replaced the O_2 carrier, gasification rate was reduced by ca. 50%. Ilmenite ore performed as well as Fe_2O_3/$MgAl_2O_4$, but NiO was poisoned by sulfur in fuel.
Berguerand and Lyngfelt (2008)	Performance testing of a circulating solids coal gasification pilot plant	$C_xH_yO_z + H_2O \rightarrow CO + CO_2 + H_2$ catalyzed by ilmenite ($FeTiO_2$). Coal was S. African sub-bituminous	Ilmenite circulated between a fluidized bed fed by $H_2O + N_2$ and a fluidized riser fed with coal and air	$T_{fluidized} = 1223$ K, gas velocity > 0.1 m/s. $T_{riiser} = 1273$ K, $P = 1$ bar; coal flow was 500 g/h	Pilot plant was run for 22 h with 12 h of stable operation. CO_2 capture (= fuel carbon converted to CO_2) ranged from 82 to 96%.	
Song et al. (2008)	Evaluation of $CaSO_4$ as an oxygen carrier for a circulating solids combustor for methane	$CaSO_4 + CH_4 \rightarrow CaS + CO_2 + H_2O$; $CaS + O_2 \rightarrow CaSO_4$	4-part cycle: 1) reduction with CH_4 for 600 min, 2) & 4) N_2 flush for 40 min, 3) re-oxidation for 340 min	Reduction step at 1223 K, re-oxidation at 1123 K, P = 1 bar	Reduction step was slowest at 1123 K. At 1273 K, SO_2 in effluent indicated the formation of CaO. Maximum CH_4 combustion was < 80%.	
Shen et al. (2009a, b)	Performance testing of a bench-scale (10 kW) circulating solids gasification plant	$C_xH_yO_z + CO_2 \rightarrow CO + H_2 + H_2O$; $CO + NiO(s) \rightarrow CO_2 + Ni$; experiments used Shenhua bitumen. coal	32.7 wt% NiO/$NiAl_2O_4$ oxygen carrier circulated between a spouted bed fuel reactor and a turbulent fluidized bed oxidation reactor	Fuel reactor: $1215 < T < 1235$ K; re-oxidizer: $1245 < T < 1310$ K	Reactor achieved 92.8% carbon conversion with 80% CO_2 capture efficiency. Efficiency was lowered by the carryover of CO/CO_2 from gasifier to re-oxidizer and elutriation of fine coal particles.	Slow activity loss of NiO oxygen carrier was related to sulfation of the oxide from sulfur in the coal.

Acharya et al. (2009)	Process development for H_2 production from biomass	$C_xH_yO_z + H_2O + CaO \rightarrow H_2 + CaCO_3$; $CaCO_3 \rightarrow CaO + CO_2$ with heat supplied by burning H_2: fuel or sawdust	Not cyclic; production followed by lime regeneration with continuous biomass and $CaCO_3$ feed	$T_{gasifier} = 850$ K; $T_{Regenerator} = 1070$ K, $P = 1$ bar	Gasifier achieved 71% H_2 in the off-gas.	Gasification step and lime regeneration step operated separately.
Mattisson et al. (2009a, b)	Development of a gasification/ combustion process for solid fuels	$C_xH_yO_z + CuO/Al_2O_3 \rightarrow CO + CO_2 + H_2O + Cu/Al_2O_3$	2 part cycle: 1) air oxidation for 70 s, 2) N_2 for 30 s	$T = 1258$ K, $\frac{u_{fluidization}}{u_{min.\ fluidiz.}} = 1.8$	Gasification rate with CuO was ca. 50 x rate obtained with a Fe_2O_3 carrier.	Single cycle was used. Experiments showed higher gasification rates are possible using CuO as the O_2 carrier.
Kolbitsch et al. (2009, 2010)	Measurement of the performance of a 120 kW pilot plant	Reduction: $H_2 + FeTiO_2 \rightarrow H_2O + FeTiO_{2-x}$; re-oxidation: $FeTiO_{2-x} + O_2 \rightarrow FeTiO_2$; also used CH_4 as fuel instead of H_2; artificial NIO carrier also used	Ilmenite circulation was 10 to 88 kg/m^2s for $0.82 < A/F < 1.10$. Fuel reactor was a fast fluidized bed (riser reactor) while re-oxidation employed a turbulent fluidized bed.	Fuel reactor operated $\approx$ 1173 K and 1 bar. With NiO carrier, 100% utilization of fuel was observed for $A/F \geq 1$ and a sufficient solids loading	Pilot plant functioned satisfactorily, but was not optimized. Carbon deposition was not observed with CO or CH_4 fuels.	Oxygen carrier operated at 10 to 18% of oxygen capacity in fuel reactor and at 20 to 25% in re-oxidation reactor.
Dennis et al. (2010), Dennis and Scott (2010)	Investigation of CO_2 gasification of coal in the presence of a CuO oxygen carrier at temperatures above the Cu melting point	$C_xH_yO_z + CO_2 \rightarrow CO + H_2O$ catalyzed by 21 wt% CuO/Al_2O_3 active phase supported on θ-Al_2O_3 (Taldinskaya, Illinois #5 biuminous coal, Hambach lignite)	3 part cycle consisting of 1) coal feed with CO_2 or H_2O as fluidizing gas, 2) fluidized gasification of charge, 3) re-oxidation of CuO with air as fluidizing gas	$T = 1203$ K, $P = 1$ bar. 78 mm (i.d.) fluidized bed; $\tau_{cycle} = 1650$ to 1700 s for bituminous coals with $\tau_1 =$ 570-600 s, $\tau_2 =$ 700-800s, $\tau_3 =$ 300-350s; for lignite, $\tau_{cycle} = 1190$ s	Lignite was gasified completely in cycle at 1203 K. Gasification of the bituminous coals was incomplete for τ_{cycle} and T used, but the presence of a CuO oxygen carrier enhanced gasification rates.	Reactive lignite fuels were successfully gasified in the cycling system. Oxygen carrier cycled between Cu and Cu_2O in experiments.

(Continued)

TABLE 8-3 Gasification of Carbonaceous Materials (*cont'd*)

Authors	Objective	Reaction	Modulated Variable	Reaction Conditions	Observations	Comments
Iggland et al. (2010)	Examination of the effect of particle size on char gasification rates	$C_xH_yO_z + H_2O \rightarrow CO + CO_2 + H_2$ catalyzed by ilmenite ($FeTiO_2$). Coal was a Colombian bituminous	5 part cycle: 1) coal fed to bed, 2) 92 vol% H_2O in N_2, $3 < \tau_2 < 30$ min, 3) and 5) N_2 purge $\tau_3 = 3$ min, 4) 10 vol% O_2 in N_2	$968 < T < 980$ K; conical fluidized bed	Coal particle size did not influence the rate of coal gasification.	
Xiao et al. (2010a, b)	Pressurized steam gasification of coal with a low-cost iron ore oxygen carrier	$C_xH_yO_z + H_2O \rightarrow CO + CO_2 + H_2$ catalyzed by CVRD iron ore. Experiments used Xuzhou bituminous coal	5 part cycle: 1) coal fed to bed, 2) 87 vol% H_2O in N_2, $\tau_2 = 75$ min, 3) N_2 purge, $\tau_3 = 15$ min, 4) 5 vol% O_2 in N_2, $\tau_4 = 40$ min, 5) N_2 purge, $\tau_5 = 5$ min	Downflow through packed coal bed; $973 < T < 1243$ K, $0.1 < P < 0.5$ MPa; iron ore = Fe_3O_4 Operation for 20 cycles	CO_2 in off-gas during reducing phase increased from ca. 80 vol% in 1st cycle to 96% in 20th cycle. Carbon gasification at 0.1 MPa was 76% and 85% at 0.5 MPa.	No agglomeration of ore particles and no coal ash-iron ore fusion were observed.
Li et al. (2009a, 2010a)	Development of a process for H_2 production from syngas + testing of Fe based oxygen carriers	Only countercurrent moving bed reducer studied: $Fe_2O_3 + CO \rightarrow FeO$ or $Fe + CO_2$; oxygen carrier = 60 wt% Fe_2O_3 on inert support	Oxide in downflow at 12.9 g/min with syngas containing 43.8 vol% CO, 29.2 vol% H_2	$1123 < T < 1173$ K, $P = 1$ bar; unit operated for 13 h; tested Fe_2O_3 through 100 oxidation/reduction cycles	Reducer achieved 49.5% reduction from Fe_2O_3 to a mixture of FeO + Fe at 99.5% conversion of CO to CO_2.	Fe_2O_3 was shown to be a suitable oxygen carrier.
Zheng et al. (2010)	Investigation of $CaSO_4$ carrier for chemical looping	$C_xH_yO_z + H_2O + CO_2 + CaSO_4 \rightarrow CO_2 + H_2O + CaS$; Shenhua bituminous coal	Only the reducing step was investigated: Coal + $CaSO_4$ fluidized by CO_2 or H_2O or mixtures	$1123 < T_{Reactor} < 1248$ K; $\tau_{reduction} = 30$ min	High H_2O/CO_2 in fluidizing gas enhances the efficiency of CO_2 capture, but coal conversion was independent of this ratio.	

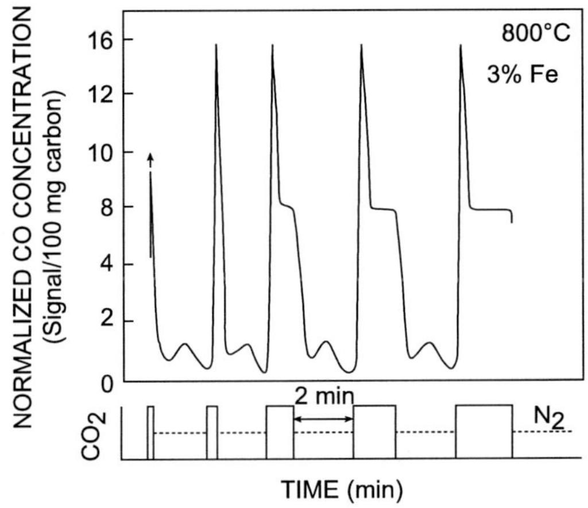

FIGURE 8-7 **Normalized IR signal for CO versus time in the initial exposure of an iron loaded char (3 wt% Fe/g carbon) from a Forestburg sub-bituminous coal for alternating exposure to CO_2 and N_2 at 1070 K and 1 bar. Duration of the time exposure increases from left to right, while the duration of the N_2 exposure is constant. Time scale is indicated in the schematic of the exposure sequence at the bottom of the figure.** *(Figure reproduced from* Fan, L.-S., Li, F., Ramkumar, S., 2008 *with permission. © 1994 Elsevier.)*

There is no initial CO burst when char from a demineralized coal is used. Only the step-up to a plateau and the step-down are seen. As might be expected, the CO concentration corresponding to the plateau at any temperature is largest for char from iron impregnated coal and smallest for the char from a demineralized coal. The difference is large. On the switch to N_2, a second burst of CO evolution is seen. The area under this peak is proportional to the area under the first peak. These peaks or bursts of CO evolution are evidence for a redox cycle. The first peak represents gasification by CO_2, while the second smaller peak arises from reduction of iron oxide by the char (**Eq. (8.1)**). The existence of these bursts and their interpretation have been known for several years (*Suzuki et al., 1988*).

A consequence of the above observations is that the cycle period is quite strongly temperature dependent. Below 1070−1120 K, where catalytic gasification would be commercially interesting, an asymmetric cycle must be used with the duration of the reducing partial cycle exceeding the oxidizing part. Asymmetric exposures are easily implemented using a circulating fluidized bed.

8.5 GASIFICATION EMPLOYING A CIRCULATING SOLID OXYGEN CARRIER

Interest in non-steady-state gasification has accompanied the recent development of circulating fluidized beds as chemical looping reactors. Indeed, it has become a popular R & D topic judging from the large number of publications that have appeared recently (see reviews by *Li and Fan (2008)*, *Fan et al. (2008)* and *Anthony (2008)* as well as a book, *Chemical Looping Systems for Fossil Energy Conversion*, by *Fan (2010)*). These reviews offer broad comparisons of competing technologies. They observe that it is possible, as well as feasible, to separate air for combustion into an O_2-rich stream that would permit sequestering CO_2 generated in combustion at a cost well below the cost for CO_2 capture through scrubbing. In their view, the most attractive option is to transfer oxygen for combusting fuels through a circulating metal/metal oxide material. This offers the opportunity of generating a nearly N_2-free gas without prior air separation.

Circulating fluidized beds of oxygen carriers are fed with a gaseous fuel or coal, coal char or biomass and a gasifying agent, such as steam or CO_2, in one part of a cycle and air in a second part. **Figure 8-8** shows a schematic of a double fluidized bed version of a looping reactor system. The circulating solid is a readily reducible oxide, such as NiO or Fe_3O_4 impregnated into an abrasion resistant support like γ-Al_2O_3.

In the solid fuels version, coal is fed into the first fluidized bed, operating at about 1200−1240 K and under a slight pressure, where it mixes with the hot circulating solid. Rapid

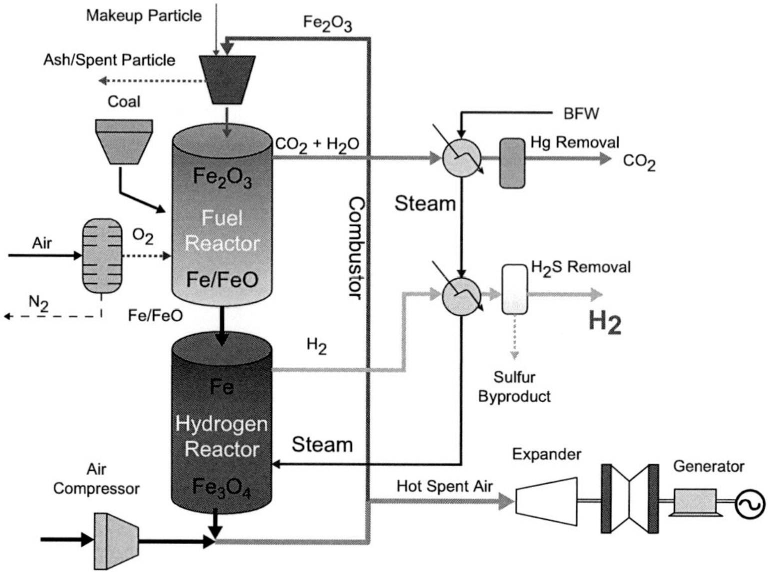

FIGURE 8-8 Simplified schematic diagram of a circulating fluidized bed system for coal gasification using separate fluidized beds for steam/ CO_2 gasification and air regeneration of the circulating solid oxygen carrier (Fe_2O_3).

heat exchange, characteristic of fluidized beds, pyrolyzes the coal converting it into a porous char. The bed of hot solid particles and powdered coal is fluidized by steam, carbon dioxide or a mixture of these gases. These oxidizing gases react to some extent exothermically with CH_4, higher hydrocarbons and tar driven out of the coal through pyrolysis, but primarily they gasify the char in an endothermic process. The heat demand of gasification is partially met by the oxidation of the reducing gases.

The remainder of the heat requirement is drawn from the hot solids that enter the gasifier at 1270−1340 K. The residual finely divided ash from the gasified coal is carried over with the hot gas leaving the gasifier and is removed in a gas cyclone. The ash-free hot gas at 1200−1240 K, consisting primarily of H_2O and CO_2 with some H_2 and CO, can be sent to a system of gas turbines, steam generators and steam turbines for power generation. The solids, now partially reduced, are circulated back to a fluidized bed regenerator where they are reoxidized with air and reheated. The hot gas leaving that regenerator is carbon-free and can be used to generate power in a system of turbines and steam boilers. Such a process is not suitable for all coals. Ash content and composition will be important because ash melting must be avoided. Sulfur in the coal or ash could contaminate the off-gas or react with the metal oxide carrier to form sulfates.

Investigation of chemical looping for coal or char combustion have generally used either a single fluidized or a packed bed operated periodically (*Scott et al., 2006*; *Leion et al., 2008, 2009b*; *Berguerand and Lyngfelt, 2008*; *Dennis and Scott, 2010*; *Iggland et al., 2010*; *Xiao et al., 2010a, b*). *Brown et al. (2010)* used such a system in their study of the gasification of a lignite char, prepared from a Hambach lignite, with CO_2 as the gasifying agent and iron oxide as the oxygen carrier. They observed that the presence of an oxygen carrier increased the rate of char gasification by oxidizing CO formed in gasification. CO appeared to inhibit gasification. *Dennis et al. (2010)* used a copper oxide and two bituminous coal chars from Taldinskaya (Russia) and Illinois # 5. These were tested along with a Hambach lignite at 1200 K, 1 bar using CO_2/N_2 as the gasifying agent. The 1500 s, three-part cycle

comprised gasification with continuous char addition, gasification of the remainder of the char with no further addition, and then a switch to air to reoxidize the copper oxygen carrier. **Figure 8-9** shows the variation of the CO_2 content during the three-part cycle with a lignite fuel. The fluctuating CO_2 flow during the feeding step reflects the pulsed feeding of coal char to the bed. The sharp peak when air is added results from O_2 removed from the entering gas to re-oxidize the Cu/Al_2O_3 carrier. The Dennis experiments also found that the presence of an oxygen carrier increased the rate of gasification. The mechanism for this appeared to be oxidation of the CO formed in gasification to CO_2. Temperature spikes when air is introduced cause dissociation of CuO to Cu_2O and O_2. Some of the O_2 released may have been available for gasification in the following partial cycle when coal char was fed.

Leion et al. (2008) also used a periodic operation consisting of a four-part cycle: Granular coal addition with steam fluidization of the metal oxide-coal mixture, N_2 fluidization for flushing (180 s), and air fluidization for re-oxidation of the metal. Experiments were conducted at 1 bar and 1223 K and employed several bituminous coals or petroleum coke, thus, devolatilization took place. These volatiles were rapidly oxidized by the metal oxide at the operating temperature. The residual char was gasified by hot steam. *Leion et al.* used a $MgAl_2O_3$ support impregnated with hematite (Fe_2O_3) or ilmenite ($FeTiO_3$), a natural mineral. Cycle period was varied to allow for the different reactivities of the coals studied.

Several investigators have used the $CaSO_4$–CaS cycle for O_2 transport (*Wang and Anthony, 2008; Song et al., 2008; Zheng et al., 2010*). The latter research team explored the kinetics of the sulfate reduction in the presence of CO_2 gasification and re-oxidation of CaS as well as the morphology changes of a particle during multiple cycles. Operating temperature was the major variable. Gasification rates and carbon conversion increased with temperature up to 1220 K, but sintering appearred at higher temperatures and suppressed the rate. As observed by many investigators, the gasification rate increases with an increasing H_2O/CO_2

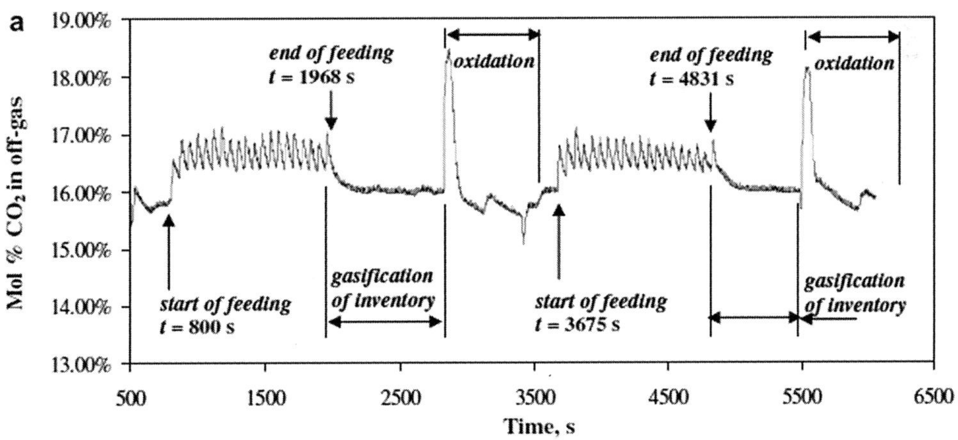

FIGURE 8-9 **CO_2 content of gas leaving the mixed fluidized bed of lignite char and CuO/Al_2O_3 particles as a function of time. $T_{Bed} = 1200$ K, the N_2-CO_2 fluidizing gas fed at 600 cm^3/s during all parts of the cycle contained 16 vol% CO_2 and the pulsed, char feed rate was 0.25 g/min. Air flow was added in the third part of the cycle at 530 cm^3/s to give a total flow of about 1100 cm^3/s.** *(Figure reproduced from* Berguerand and Lyngfelt (2008) *with permission. © 2008 by Elsevier Ltd.)*

ratio. Whisker and dendrite formation was seen for CO_2 gasification but not when H_2O was used. Small amounts of SO_2 and H_2S were found in the gases released from the gasification and re-oxidation steps.

A periodic operation was also used to examine the effect of coal particle size on gasification rates in the presence of oxygen carrier solids. *Iggland et al. (2010)* found no size effect in the range studied; devolatilization during heating, however, was affected by particle size.

Versions of the double fluidized bed system (**Figure 8-8**) have been tested in the laboratory by several research teams (*Cao et al., 2006; Berguerand and Lyngfelt, 2008; Shen et al., 2009a*). Small scale demonstration units have been built in China and Sweden with a gas production rate capable of generating 3 to 5 kW (*Leion et al., 2008; Shen et al., 2009a, b*). **Figure 8-10** gives a schematic of the nominal 10 kW Swedish system. Experiments with this system operating with ilmenite as the O_2 carrier were conducted at 1 bar using a South African sub-bituminous coal, fed at 500 g coal/h (*Berguerand and Lyngfelt, 2008*). The gasifier operated at about 1220 K while the ilmenite regenerator was held at 1270 K. Continuous operation over a 12-h period was achieved. Coal conversion ranged from 50 to 79% by weight, while carbon capture in terms of CO_2 produced per unit of carbon in the fuel charge varied from 82 to 96%. Design changes and polishing of the gasifier off-gas with a post combustion unit fed by oxygen would produce, as the authors suggest, a commercially acceptable process.

A Shenhua bituminous coal was used in experiments on the Chinese $10\,kW_{th}$ unit reported by *Shen et al. (2009a)*. This small scale pilot unit employed an up-flow fluidized bed such as that shown in **Figure 8-10**. However, in place of the segmented fluidized bed, a spouted bed served as the gasifier with powdered coal fed along with a $CO_2{-}H_2O$ mixture and/or recycled off-gas. The system operated at slightly above atmospheric pressure. Measurements in the gasifier indicated the gas temperature decreased with height from 1230 to 1210 K. In the upflow regenerator, the NiO/Al_2O_3 oxygen carrier entered at about 970°C and reached the cyclone at about 1035°C. Off-gas from the spouted bed was 94% CO_2 by volume. The CO_2 capture efficiency was 80%. This is the ratio of CO_2 leaving the gasifier to total carbon as CO_2, CO and carbon dust in the off-gas from both gasifier and regenerator. *Shen et al.* reported that the two largest detractors of capture efficiency were carbon in the ash carried over from that unit and gas entrained with the solids passing from the gasifier to the regenerator with the recirculating oxide.

Eliminating CO_2 in the gasifier fluidizing gas reduces the CO_2 content in the off-gas. Passing the off-gas separately through a gas turbine and steam boiler to cool the gas to ca. 370 K and recover further energy would provide a feed to a MEA/DEA scrubber so as to remove almost all CO_2 from this stream. After condensation, the gasifier off-gas would be 95 to 98% H_2 and suitable for process use. Indeed, depending on demand, such a process could be switched between power and H_2 generation. Either coal or biomass could fuel such a system.

A comprehensive review of gasification using chemical looping published by *Fan and Li (2010)* examines a variety of processes for chemical applications employing gasifier off-gas. Necessary properties for oxygen carriers are identified; perhaps of most interest is the authors' observation that moving bed contactors, rather than fluidized beds, would substantially reduce solids cycling by offering much greater oxidation and reduction of the oxygen carrier. Advantages of countercurrent moving-bed gasifiers have been tested by simulation using the Aspen Plus design package, and assuming thermodynamic

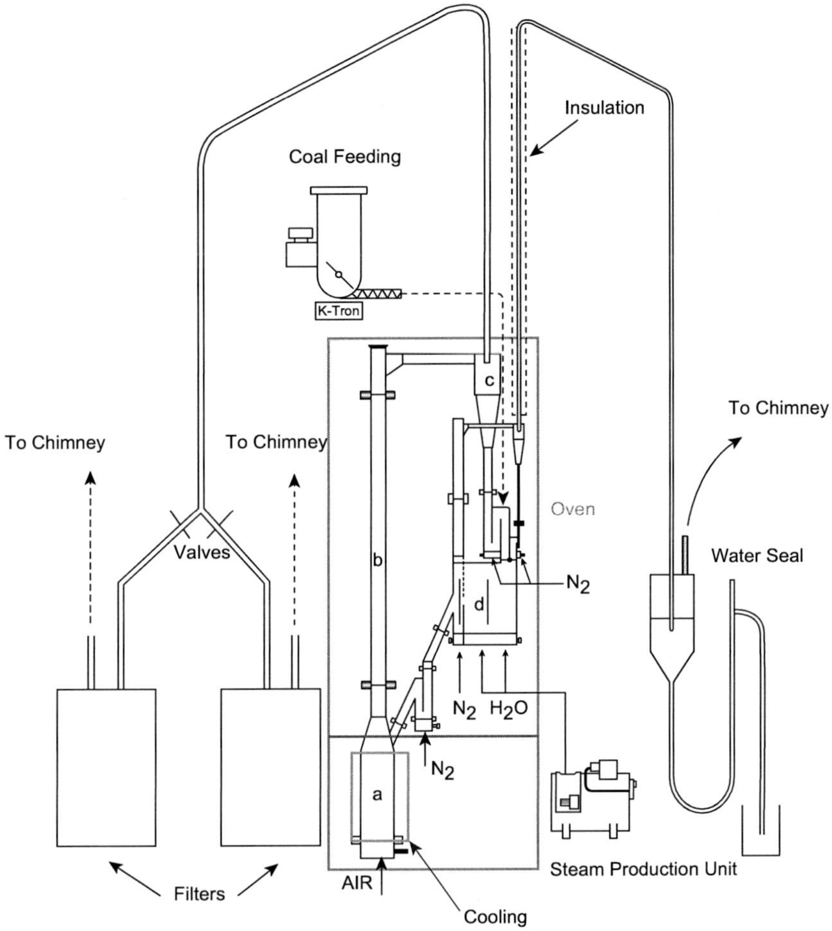

FIGURE 8-10 Schematic of the nominal 10 kW circulating fluidized bed coal combustion demonstration unit operating at the Chalmers University of Technology, Gothenburg, Sweden. In the schematic, "a" is the up flow fluidized bed regenerator, "b" is a riser reactor, "c" is a cyclone for capturing the ilmenite oxygen carrier and "d" is a compartmentalized gasifier. *(Figure reproduced from* Acharya et al. (2009) *with permission. © 2009 Amer. Chem. Soc.)*

equilibrium with further parameters from additional experiments (*Li et al., 2010a*). The program results show virtually complete conversion of the syn gas feed at almost 49% reduction of the Fe_2O_3 carrier at 30 atm and 1170 K. Synthesis gas contained 43.8 vol% CO, 29.2% H_2, 5.0% CO_2 with N_2 as the balance. Fe_2O_3 conversion under fluidized bed operation would be about 11% under those operating conditions.

In practice, char or coal gasification would be carried out under pressure, possibly up to 1–2 MPa, so that the hot gases leaving the gasifier and regeneration beds could be sent to gas turbines for power generation. Several investigators have studied the effect of pressure on reduction and/or re-oxidation of oxygen carriers (*Jin and Ishida, 2002; Garcia-Labiano, 2006* and *Siriwardane et al., 2007*). Generally, they observed that increasing pressure suppressed rates. *Xiao et al. (2010a, b)* examined a pressurized process using a periodically operated, 30 mm i.d. packed bed of a commercial iron ore. A powdered low rank, high ash bituminous coal (Xuzhou) was fed to the bed with a N_2 carrier gas. This was followed by 90 min exposure to 87% steam in N_2, then a 5 min purge and a switch to 5% O_2 in N_2 for 10 to 15 min. This cycle was tested at 1240 K and pressures up to 0.6 MPa. *Xiao et al.* observed a higher CO_2 concentration in the off-gas and increased conversion of the iron oxide as pressure increased; however, the devolatilization rate decreased. Porosity of the iron oxide carrier increased, without substantial loss through attrition, for three to five cycles.

Power or hydrogen production from biomass with CO_2 sequestration has also become a popular research topic in the last decade (*Hanaoka et al., 2005; Ni et al., 2006; Cao et al., 2006; Cao and Pan, 2006; Mahishi and Goswami, 2007; Pfeifer et al., 2007*). *Acharya et al. (2009)* proposed the biomass gasifier design shown in **Figure 8-11**, discussed the energy balance and reported separate batch performance tests on the gasifier and regenerator. They used CaO as a CO_2 trap to separate H_2 and CO_2 in the process off-gas. To balance the heat demand of the endothermic gasification step, some of the H_2 product was consumed to heat the lime regenerator. The cyclone following the regenerator captured and recycled the lime particles; fine dust from the small amount of ash in the biomass that was carried over with the CO_2 stream into the heat recovery and steam generation section. For a biomass feed rate of 1 kg/s at 100 K, 1.07 kg/s of 575 K steam entered the gasifier along with 2.41 kg/s of CaO at 1220 K. Endothermicity of gasification and heat loss dropped the temperature of the solids passing to the regenerator to 1070 K. This was also the temperature of the H_2 stream. The authors assumed, optimistically, that lime traps all of the CO_2 produced by biomass gasification, so 4.30 kg/s of $CaCO_3$ circulated to the regenerator. The process consumed 1.07 kg/s of water which was converted to steam using the hot off-gases. These inputs generated 0.10 kg/s of high purity H_2 and 1.89 kg/s of 100 K CO_2 for sequestering.

Acharya et al. (2009) reported that batch tests on a bubbling bed gasifier using sawdust as the biomass, supplemented by charcoal to generate extra heat, produced an off-gas containing about 70% H_2 at 580°C when CaO was fed to the bed. Most of the remaining gas was CH_4 and the CO_2 content was 1 to 2%. Separate batch experiments on the regenerator using an upflow fluidized bed heated externally to 1070 K showed about 40% of the $CaCO_3$ feed was converted to CaO within a 1-h contact time employing air as the fluidizing gas.

8.6 COMBUSTION IN CIRCULATING FLUIDIZED BEDS

8.6.1 Methane and Syngas

Combustion systems allow for sequestration of a CO_2 target, besides natural gas, refinery and chemical plant discharges containing hydrocarbons, coal mine ventilation emissions and gases generated in waste treatment. The basic process is identical to that shown in **Figure 8-9**, except that the powdered coal fed to the gasifier is replaced by a gaseous fuel. This input would be mixed with steam and/or CO_2 as extra gas is necessary for fluidization. Additional steam and/or CO_2 also reduces the amount of solid oxygen carrier that must be circulated to combust the fuel.

A number of laboratory studies, discussed by *Li and Fan (2008)* for example, have laid the foundation for small scale pilot units such as the 120 kW_{th} dual circulating fluidized bed combustor operated by the Vienna University of Technology on synthesis gas or on methane (*Kolbitsch et al., 2009, 2010*). A schematic of this

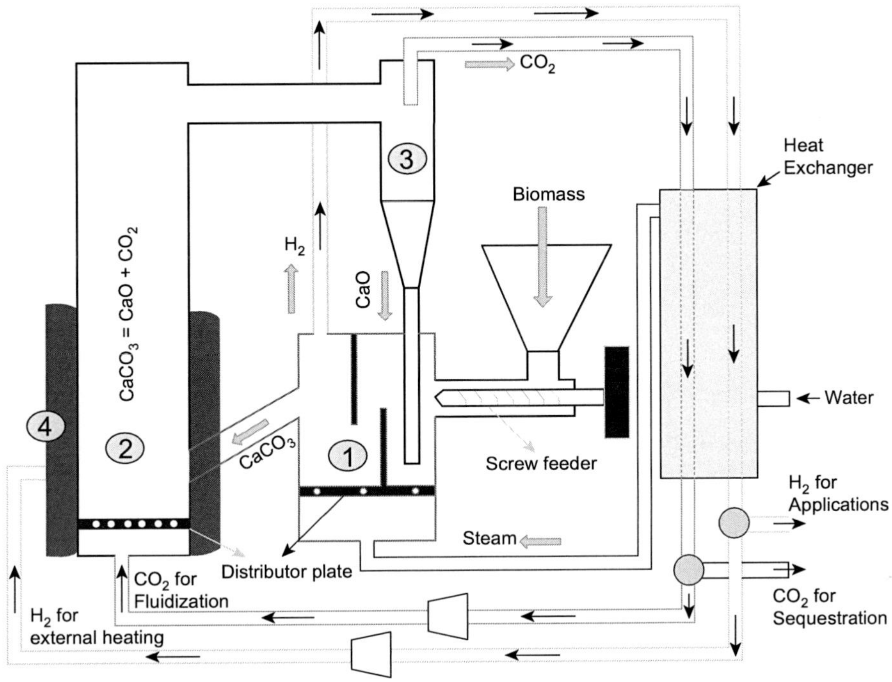

FIGURE 8-11 **Schematic of a double fluidized bed system for steam gasification of biomass with capture of CO_2 for H_2 production, or power generation or process application. "1" is the bubbling fluidized bed gasifier, "2" is a circulating fluidized bed, "3" is a cyclone to separate gas and ash from the solid oxygen carrier and "4" is an external gas-fired heater for the bottom of the circulating fluidized bed.** *(Figure reproduced from* Acharya et al. (2009) *with permission. © 2008 American Chemical Society.)*

unit is given in **Figure 8-12**. The regeneration reactor is usually operated as an up-flow fluidized bed, while the gas oxidation reactor operates as a bubbling fluidized bed, although it could be run as an up-flow fluidized bed also. Circulating solids carry oxygen from the regenerator to the oxidation reactor.

Several investigators have avoided the problem of the slow gasification of carbonaceous solids using circulating beds by employing conventional gasifiers using O_2 from an air separator and processing the $CO/CO_2/H_2/H_2O/CH_4$ gas mixture produced in the fuel portion of a circulating fluidized bed. Thus, *Li et al. (2009a, 2010a)* undertook experimental studies of their looping systems with such a synthetic gas. Their main interest was in the performance of the oxygen carrier in the combustor and the regenerator.

Choice of the oxygen carrier is important: Requirements such as oxygen capacity, rate of reduction and oxidation, thermal stability, mechanical strength and cost have been discussed by several research teams (*Adànez et al., 2004; Garcia-Labiano, 2005; Gupta et al., 2007*). Iron oxides seemed to be preferred. Reactivity and mechanical performance of the oxides in the $Fe_2O_3 \leftrightarrow Fe_3O_4 \leftrightarrow FeO$ cycle were investigated by *Li et al. (2009a)*. Iron oxide was coprecipitated with an alumina precursor. A TGA and quasi-periodic tests in a fixed bed were used to characterize the O_2 capacity and oxidation/reduction reactivity. A hydraulic press measured particle strength and an up

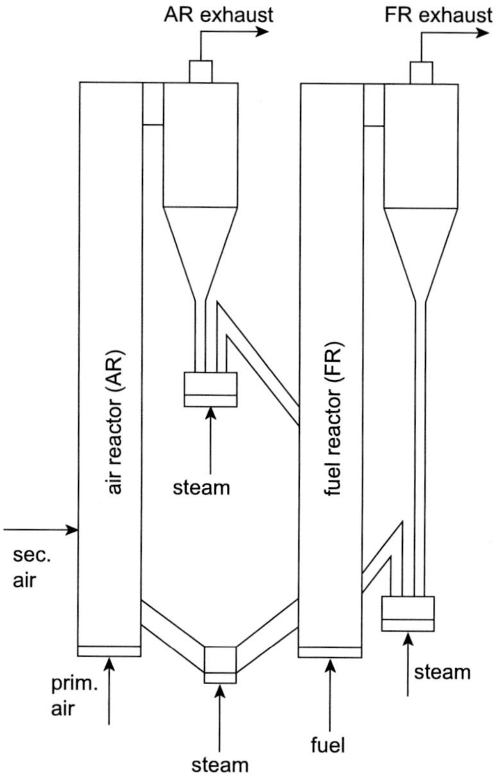

FIGURE 8-12 Schematic of the 120 kW$_{th}$ Chemical Looping Combustor used for burning either syngas, CO, H_2 or CH_4 with air to generate a high purity CO_2 stream after water condensation. Both exhaust streams can be used to generate power. *(Figure reproduced from* Kolbitsch et al. (2010) *with permission. © 2010 by Elsevier Ltd.)*

flow fluidized riser determined the entrainment rate. The *Li* experiments disclosed a strong increase in the oxidation rate with the number of cycles as may be seen in **Figure 8-13**. Reduction rate increased just slightly. Alumina supported iron oxide appeared to be an attractive oxygen carrier. Measured compressive strength was high and attrition rate at 0.57 wt%/cycle was low.

Song et al. (2008) have explored the operation of a methane combustor with separation of a N_2-rich, CO_2-free waste gas from a concentrated CO_2 stream using the $CaSO_4$/CaS oxygen carrier couple. Experiments were undertaken in a periodically operated fixed bed and focused on the effect of operating conditions (temperature, gas flow rates, $CaSO_4$ particle size, time on stream) on capacity and reactivity. The authors suggested a 1220 K operating temperature for the combustor, but that the regenerator temperature should not exceed 1270 K.

Under temperatures reached in the regenerator, some oxide partially decomposed releasing oxygen. Indeed, *Mattisson et al. (2009a)* have suggested that the oxygen released can be exploited to gasify chars. They identified CuO or Mn_2O_3 as potential sources. The former is in equilibrium with Cu_2O and gaseous O_2 at 1186 K. At that temperature, gaseous O_2 reacts exothermically with CO or a carbon char. This would raise bed temperature and further dissociate CuO. Cuprous oxide, however, melts at 1508 K, so temperature control would be necessary. Mn_2O_3 dissociates at 1048 K, but both manganese oxides have melting points above 1575 K so carrier agglomeration would not be a problem. *Mattisson's* experiments found that periodic exposure of a fluidized bed of CuO/Al_2O_3 at 120 K to air, N_2 and CH_4 showed almost complete conversion of CH_4 to CO_2 until the released O_2 was consumed (**Figure 8-14**). A similar experiment was undertaken at 1258 K with a petroleum coke contained in a fluidized bed of CuO/ZrO_2. Coke oxidation continued until the free O_2 was consumed whereupon diffusion became controlling. Details are given by *Mattisson et al. (2009a, b)*.

8.6.2 CO_2 Capture

Research on the application of chemical looping to gasification and combustion discussed above and summarized in **Table 8-3** has dealt with utilization of an oxide or a sulfate to introduce N_2-free O_2 into a reactor to produce a gas for further processing or for

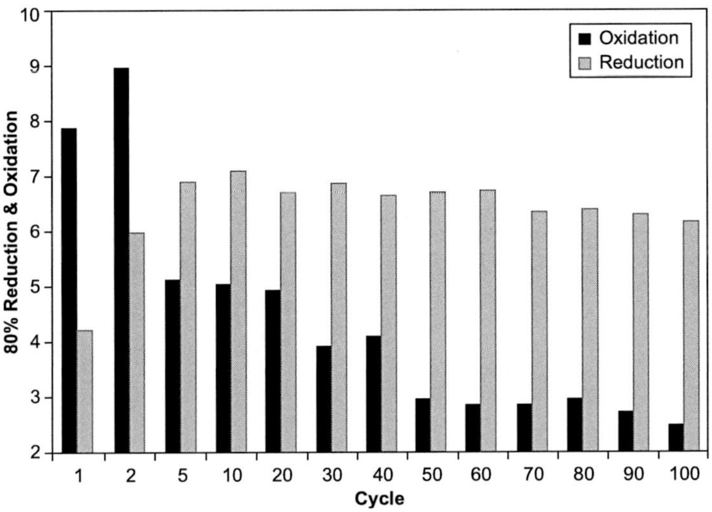

FIGURE 8-13 **Time in minutes for 80% oxidation or reduction of a single sample of Fe_2O_3/Al_2O_3, particle size: 850–1000 μm, at 830°C.** *(Figure reproduced from* Li et al. (2009a) *with permission. © 2009 by American Chemical Society).*

combustion that is not burdened by a high concentration of nitrogen. An alternative use of looping employs the carbonation of lime. This was discussed earlier in **Section 8.5** for H_2 production from biomass. Researchers at the Vienna University of Technology have also developed an H_2 from biomass process involving carbonation and carbonate decomposition that has advanced to pilot plant testing (*Pfeifer et al., 2007*).

Materials used for CO_2 capture are calcined limestone or, at lower temperatures, a potassium promoted hydrotalcite or a sodium promoted alumina. Operation of these gaseous chemical looping combusters is illustrated in **Figure 8-15**.

The calciner would require a moderately pure O_2 stream to prepare a CO_2 stream for sequestration. Decomposition of calcium carbonate is endothermic so additional fuel or a part of the producer gas must be combusted with O_2 to provide heat for calcination.

In the carbonation bed, CH_4 is oxidized and, with added steam, CO_2 adsorption forces the

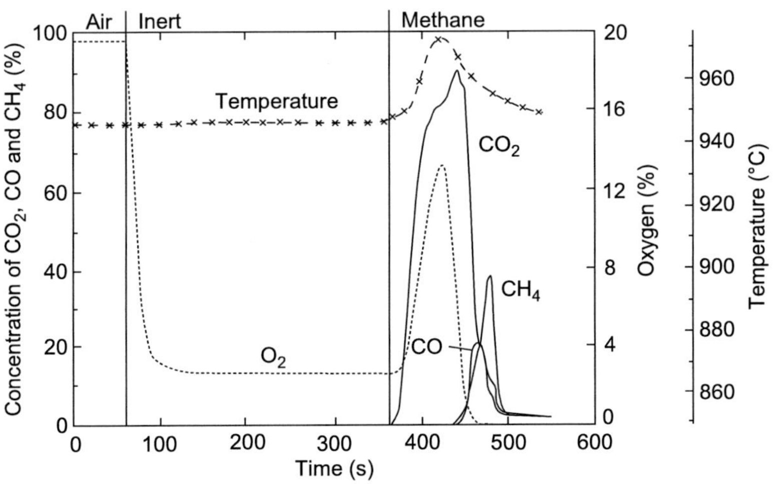

FIGURE 8-14 **Exit gas composition and bed temperature for the oxidation of methane by CuO/Al_2O_3 at 950°C and 1 bar in a periodically operated laboratory fluidized bed. Inert shown in the figure was N_2.** *(Figure reproduced from* Mattisson et al. (2009a) *with permission. © 2009 by Elsevier Ltd.)*

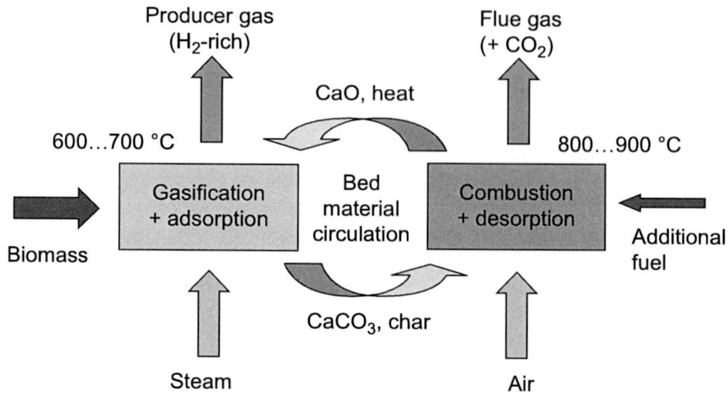

FIGURE 8-15 Schematic of a circulating adsorbent bed process for H_2 and CO_2 separation using calcined lime as the trapping media. *(Figure reproduced from* Pfeifer et al. (2007) *with permission. © 2007 The Berkeley Electronic Press.)*

water gas shift reaction to H_2. Use of an adsorber reduces substantially the steam demand (*Iyer et al., 2004*). Although the looping system can function at lower pressure, operating at 2 to 3 MPa would permit the capture of power through gas turbines. *Fan et al. (2008)* report a laboratory scale demonstration of H_2-CO_2 separation.

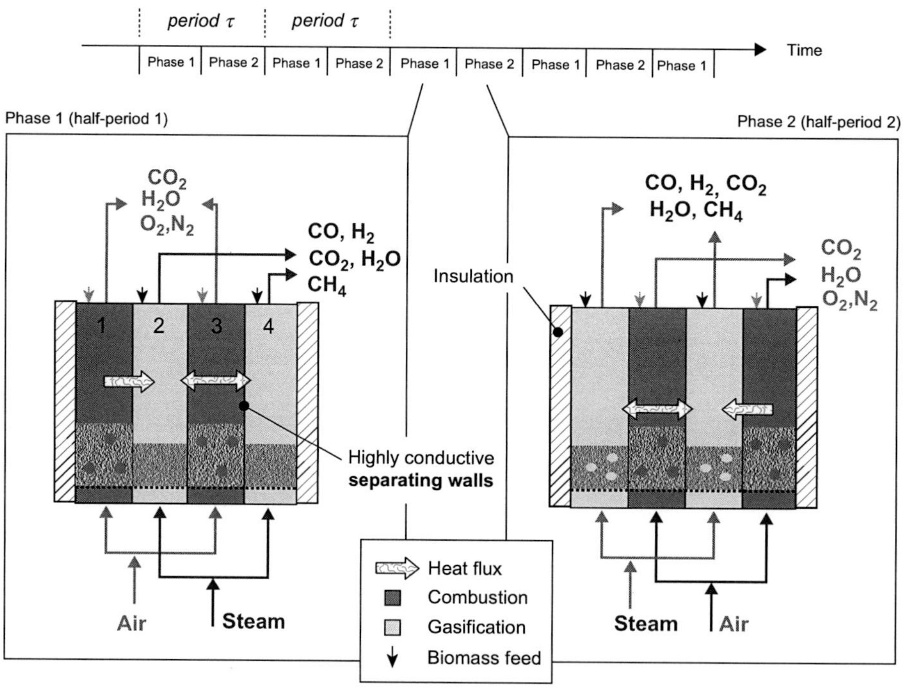

FIGURE 8-16 Representation of temporal and locational modulation of a solid fuel gasifier consisting of a honeycomb of hexagonal channels filled with a fluidized solid oxygen carrier and intermittently fed with solid fuels (biomass in the diagram) and undergoing periodic switching of the fluidizing gas. *(Figure adapted from* Iliuta et al. (2010) *with permission. © 2009 Elsevier Ltd.)*

8.7 PERIODIC REACTION SWITCHING

In chemical looping, the circulating solids encounter greatly different temperatures and environments making laboratory experiments difficult and costly. This has led to simulation of such systems using a stationary bed of an oxygen carrier and composition modulation of gases fed to the bed (see **Table 8-3**).

Larachi and co-workers (*Wang et al., 2009; Iliuta et al., 2010*) have cleverly applied composition modulation to perhaps solve problems of incomplete gasification, the mixing of different gas streams and heat loss encountered in chemical looping. They proposed undertaking gasification or combustion by arranging square or hexagonal bubbling fluidized bed cells containing the solids oxygen carrier, say Fe_2O_3/Al_2O_3, in a plane with continuous biomass feed, such as sawdust, to each cell. Fluidizing gas, also to each cell, would periodically switch between steam and air so that the cell would shift from a pyrolyzing and gasifying biomass to combusting the char formed. Off-gases would be collected separately. Contiguous combusting and gasifying cells would allow heat transfer from the hotter combustion cells to the cooler gasification ones. Their concept is illustrated in **Figure 8-16**.

Simulating the proposed operation assuming adjacent beds, *Iliuta et al. (2010)* found, using data and fluidization models from the literature, that an operation producing O_2- and N_2-free hydrogen could be sustained for a switching period, τ_{switch}, of about a minute, even with a 20% loss of heat. *Wang et al. (2009)* considered another version of this design in which the biomass particles are transported with the gas phase. They investigated just the hydrodynamics of the operation.

Notation

A/F	= air/fuel ratio (-)
Bz	= benzene
C_i	= concentration of species "I" (mol/m^3)
COP	= coefficient of performance
DEA	= diethanolamine
GHSV	= gas hourly space velocity (h^{-1})
i.d.	= inner diameter (cm)
l	= liquid
MEA	= monoethanolamine
m	= mass of catalyst (g)
P	= pressure (kPa, MPa)
P_H	= high pressure in heat pump cycle
P_L	= low pressure in heat pump cycle
P_M	= median pressure in heat pump cycle
Q	= volumetric flow rate (m^3/h)
S_H	= high temperature reaction couple in heat pump cycle
S_L	= low temperature reaction couple in heat pump cycle
SV	= space velocity (s^{-1}, h^{-1})
s	= cycle split (-)
T	= temperature (K)
T_H	= high temperature in heat pump cycle
T_L	= low temperature in heat pump cycle
T_M	= median temperature in heat pump cycle
TGA	= thermogravimetric scale or analysis
t	= time (s, min)
u	= velocity
v	= vapour
W	= work

Greek

ΔT	= temperature lift or difference in heat pump cycle
γ	= fraction or segment of a cycle (-)
τ	= cycle period, partial period, time within a period (s, min)
τ_{cycle}	= cycle period
τ_{switch}	= time between a switch of a manipulated variable
τ_1, τ_2, etc.	= duration of partial cycles (s, min)

Subscripts

H	= high or higher
L	= low or lower
M	= median
x, y, z	= stoichiometric coefficients
1, 2, 3, etc.	= segments or parts of a cycle

CHAPTER

9

Electrochemical Processes

Mark Pritzker, Peter Lewis Silveston
Waterloo, Ontario, Canada

OUTLINE

9.1 Introduction 235
9.1.1 Electrochemistry Background 236
9.1.2 Electrochemical Cells and Processes 243
9.1.3 Variables Affecting Performance of Electrochemical Systems 244
9.1.4 Dynamics in Electrolytic Cells 244

9.2 Electroplating 246
9.2.1 Electroplating of Single Metals 255
9.2.2 Electroplating of Alloys 256
9.2.3 Novel Electroplating Applications 259

9.3 Electroforming 262

9.4 Anodization 263

9.5 Electrochemical Machining and Polishing 264

9.6 Electrowinning and Electrorefining 265

9.7 Galvanic Cells 266

9.8 Electrolytic Production of Chemicals 267

9.9 Applicability of Principles or Practices to Non-Electrochemical Reactions 269

9.1 INTRODUCTION

Although both electrochemical and reaction engineering are concerned with chemical reactions, these disciplines have developed separately for the most part. Only a handful of engineers and scientists straddle the disciplines. There is, therefore, an opportunity for cross-fertilization.

Periodic operation, now under study for a variety of catalyzed gas and gas-liquid reactions (*Silveston, 1998*), has been known ever since the nineteenth century in electrochemistry. As alternating or pulsed current or potential, it has been applied commercially for more than a century. The operation has been applied primarily to electrolytic processes and to a lesser extent to galvanic ones. The objective of this chapter is to discuss periodic operation in electrochemical systems and thereby explore the potential for carryover of concepts or techniques into the engineering of non-electrochemical reaction systems. However, before doing so, let us consider some background on electrochemical processes and terminology.

9.1.1 Electrochemistry Background

Fundamentals of electrochemistry are presented by a number of excellent textbooks (e.g., *Newman and Thomas-Alyea, 2004; Scott, 1991; Bockris and Reddy, 1998a, b; Bard and Faulkner, 2001*). *Puippe and Leaman (1986)* provide a comprehensive overview of the principles and practice of pulse plating as of the mid-1980s.

An electrochemical process is unique in that it involves redox reactions which lead to the movement of electrons and ions from one part of the system to another. Consequently, an electrochemical cell requires a connection between electronic conductors (i.e., electrodes, electrical connections) to allow the transport of electrons and an ionic conductor (i.e., electrolyte) which provides the transport of ions. The changeover from electronic charge transfer to ionic transfer within a cell takes place at the electrode/electrolyte interface where half-cell reactions involving both electrons and ions occur. At a minimum, electrochemical cells or processes require the following components: Electronic conductors consisting of two electrodes (anode and cathode) with electrical connections, ionically conducting phase(s) (i.e., electrolyte) and charge carriers (electrons and ions). Two types of electrochemical cells operate. In the first, an external electrical power source is connected to the electrodes to drive processes that would otherwise not occur. In the second, electrochemical reactions occur spontaneously to generate current and/or desired chemical products. To simplify this overview, certain items have been excluded—cells involving semiconductors where charge carriers other than electrons and ions are possible.

Every electrochemical cell can be characterized by two quantities: cell current and voltage. Current is the flow rate of charge through the cell, whereas the voltage at any specific location measures the electrical energy at that position. At steady state, the total current remains constant across any cross-sections drawn between the two electrodes of an operating cell. Although both electronic and ionic charge carriers can, in principle, contribute to the total current at any position, usually one dominates within each cell component depending on whether the particular phase is an electronic or ionic conductor. The overall potential drop across a cell is the cell voltage. Cell current and voltage are important quantities in an electrochemical process since their product determines the required electrical power input (electrolytic process) or the generated electrical power output (galvanic process).

Each electrochemical reaction occurs at an electrode/electrolyte interface and involves the consumption or generation of electrons. For this reason, it is termed a half-cell reaction. In every cell, at least one half-cell reaction consuming electrons occurs at an electrode/electrolyte interface while at least one other half-cell reaction generates electrons at another or the same electrode/electrolyte interface. The overall rate of consumption of electrons must always balance the overall rate of generation of electrons in any cell. For the purposes of this discussion, consider a cell where the following half-cell reactions occur:

$$2\,H^+(aq) + 2\,e^- \rightarrow H_2(g), \tag{9-1}$$

$$H_2O(l) \rightarrow \frac{1}{2}\,O_2(g) + 2\,H^+(aq) + 2\,e^-. \tag{9-2}$$

The electrolyte in this case is an aqueous solution. **Eq. (9-1)** represents a reduction reaction that leads to the consumption of electrons, while **Eq. (9-2)** shows an oxidation reaction with the generation of electrons. Regardless of the type of cell, a reduction reaction occurs at the cathode/electrolyte interface and so is termed a cathodic reaction, whereas an oxidation reaction occurs at the anode/electrolyte interface and is termed an anodic reaction.

To facilitate electrochemical processes, cathodes and anodes must not only be good electronic conductors but also good electrocatalysts for their respective reactions. In some processes, electrodes themselves also directly participate in the electrode reactions and consequently are altered (e.g., electrodeposition, anodizing and corrosion), whereas in other cases, they remain inert as in conventional catalytic systems. Both reactions given by **Eqs (9-1)** and **(9-2)** fall into the second category.

When the above reactions occur simultaneously in a cell, the resulting overall reaction is the electrolysis of water, i.e.,

$$H_2O(l) \rightarrow H_2(g) + \frac{1}{2}\,O_2(g). \qquad (9\text{-}3)$$

Since this process is not spontaneous, the cell must be connected to an electric power source in order for the reactions of **Eqs (9-1)** and **(9-2)** to proceed as written above. A cell in which electric current drives half-cell reactions that would otherwise not occur is termed an electrolytic cell.

Consider now a cell in which the above reactions proceed in the opposite direction, i.e.,

$$H_2(g) \rightarrow 2\,H^+(aq) + 2\,e^-, \qquad (9\text{-}4)$$

$$\frac{1}{2}\,O_2(g) + 2\,H^+(aq) + 2\,e^- \rightarrow H_2O(l). \qquad (9\text{-}5)$$

Obviously, the reaction of **Eq. (9-4)** involves oxidation and occurs at the anode, while the reaction of **Eq. (9-5)** involves reduction and occurs at the cathode. These half-cell reactions will combine to produce the following overall reaction:

$$H_2(g) + \frac{1}{2}\,O_2(g) \rightarrow H_2O(l). \qquad (9\text{-}6)$$

Since this overall reaction proceeds spontaneously, connecting the anode and cathode to an external load enables electric current to flow through the load. These reactions form the basis for fuel cells that generate electrical energy. Such a cell is termed a galvanic cell. While fuel cells and batteries serve useful purposes, other galvanic processes, such as corrosion, are harmful and obviously undesirable. Galvanic processes are those that produce electrical energy.

It is not possible to independently control cell potential and current during the operation of an electrochemical cell. Consequently, a cell can be operated only in one of two possible modes—galvanostatic or potentiostatic. During galvanostatic operation, current is applied in the case of an electrolytic cell or generated in a galvanic cell according to some controlled manner and the resulting potential can be monitored. This mode is commonly used in industry, primarily because it enables direct control of the production rate. During potentiostatic operation, a potential is applied (or generated) in some controlled manner and the resulting current can be monitored. This mode is typically preferred in situations where galvanostatic operation leads to an inefficient process or poor deposit quality (in the case of plating). Such a situation often arises when more than one reaction occurs at an electrode surface and selectivity of the desired reaction is improved by careful control of potential.

When two or more electrochemical reactions occur simultaneously at an electrode, they share the total current I_{total} flowing through the electrode. The fraction of this total current used for a given reaction j is the current efficiency ε_j of the reaction,

$$\varepsilon_j = \frac{I_j}{I_{total}} \times 100\,\%. \qquad (9\text{-}7)$$

As with any other reaction, a half-cell reaction is affected by thermodynamics, kinetics and mass transfer. However, the variable that is most important to a half-cell reaction is the potential drop across the electrode/electrolyte interface, referred to simply as the electrode potential E. It determines the reactions that can occur at an electrode and the direction in which they proceed. This quantity is equivalent to the difference in the potential, V, on the metal side of an electrode/electrolyte interface and the potential, ϕ_0, on the solution side, i.e.,

$$E = V - \phi_0. \tag{9-8}$$

The factor determining the direction in which a given half-cell reaction *j* can proceed is the relation between its reversible electrode potential $E_{rev,j}$ and the electrode potential. $E_{rev,j}$ is a thermodynamic quantity determined using the Nernst equation. For the general half-cell reaction:

$$v_1\, A^{zA}(aq) \rightarrow v_2\, B^{zB}(aq) + n\, e^-. \tag{9-9}$$

$E_{rev,j}$ is given as:

$$E_{rev,j} = E^0_{rev,j} + \frac{RT}{nF} \ln \frac{a_B^{\upsilon_2}}{a_A^{\upsilon_1}}, \tag{9-10}$$

where $E^0{}_{rev,j}$ is the standard reversible electrode potential and a_A and a_B are the activities of $A^{zA}(aq)$ and $B^{zB}(aq)$, respectively. A half-cell reaction j is at thermodynamic equilibrium when $E = E_{rev,j}$, proceeds in the anodic direction when $E > E_{rev,j}$ and in the cathodic direction when $E < E_{rev,j}$. For a given half-cell reaction, $E_{rev,j}$, is only affected by the solution composition and not at all by the applied electrode potential or current. On the other hand, the electrode potential is an operating variable that can be adjusted either directly or indirectly by the cell operator and therefore can be used to control which reactions occur at an electrode.

An alternative way to view the electron transfer process is in terms of the electron energy levels in the metal electrode and the molecular orbitals of the chemical reactants on the solution side of the interface (**Figure 9-1**). A shift in the electrode potential in the negative direction raises the energy level of electrons in the electrode. When the electrode potential is shifted sufficiently that these electrons are raised to a higher energy than the lowest vacant or partially vacant molecular orbital of the chemical species on the solution side of the interface, electron transfer from the electrode to the molecular orbital occurs, causing the reduction of the chemical species. A positive shift in the electrode potential has the opposite effect by lowering the energy level of electrons in the electrode. When the energy level drops below that of the highest occupied or partially occupied molecular orbital of a chemical species on the solution side of the interface, electron transfer from the molecular orbital to the electrode takes place, causing the oxidation of the chemical species.

Of course, thermodynamics alone does not dictate whether an electrode reaction will actually be observed in a practical situation. As with non-electrochemical reactions, factors such as electrode kinetics and mass transfer govern the rate at which a thermodynamically feasible process actually occurs. Many of the sub-processes occurring at or near electrodes are similar to those involved in non-electrochemical heterogeneous reactions (**Figure 9-2**). As with other heterogeneous reactions, mass transfer and the surface reactions (i.e., charge transfer, chemical reactions, adsorption, crystallization, etc.) occur in sequence and the slowest of these steps controls the overall rate of the electrode reaction. Mass transfer in electrolytes differs from mass transfer in non-electrolytes in that charged species are subject to transport by migration in addition to transport by diffusion and convection. This mode of transport results from the response of a charged species to an electric field whereby cations move in the direction from higher electric potential to lower electric potential and anions move in the opposite direction. Thus, depending on the sign of the ionic charge and the direction of the electric field, migration can operate in the same direction as diffusion and convection or can oppose them. Of course, if a species is neutral, migration has no influence on transport; only diffusion and convection operate. The importance of migration to the overall transport of an ion depends on concentration.

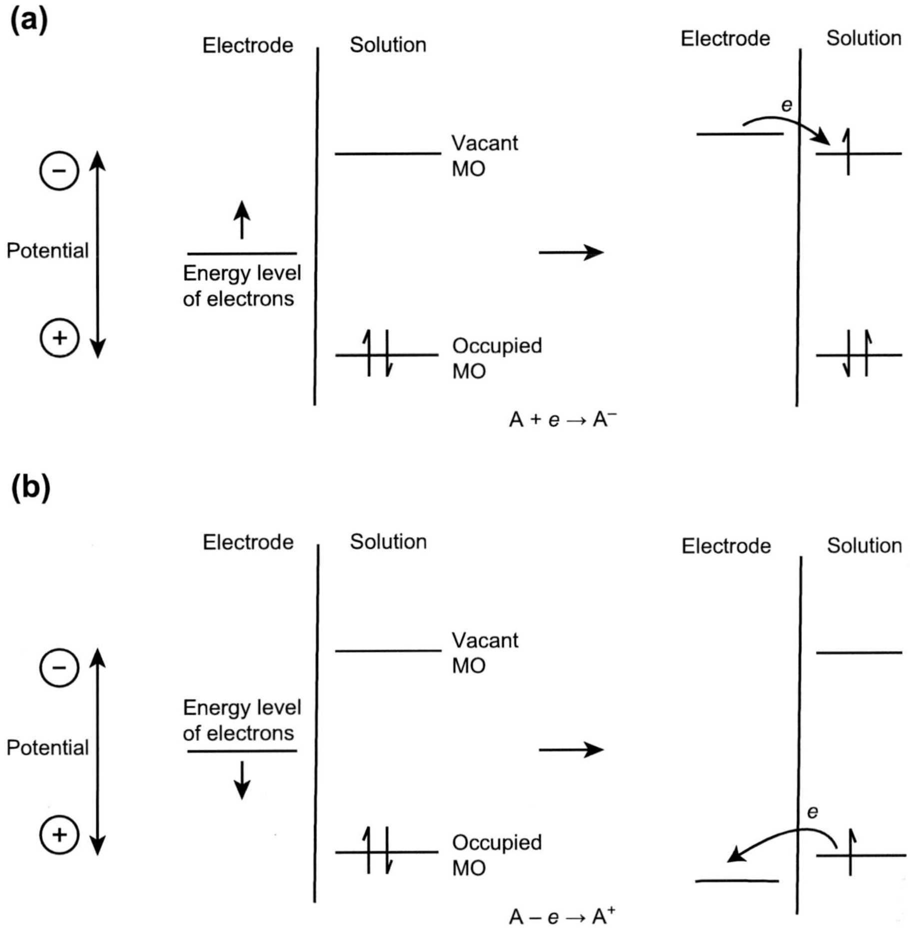

FIGURE 9-1 Charge transfer at the electrode surface showing the electron energy levels before and after transfer for an (a) cathode reaction and (b) anodic reaction. *(Figure reproduced from* Bard and Faulkner (1980) *with permission of the publisher. © 1980, John Wiley & Sons, New York.)*

Thus, if it is present in minor amounts relative to that of other species, the contribution of migration to the transport of that species becomes very small in comparison with that of diffusion and convection. Consequently, it is normal practice in experiments to maintain the concentration of reacting species very low in comparison with that of the supporting electrolyte when studying the dynamics of electrode reactions. Such a procedure allows migration to be ignored as a mode of transport for the reacting species and simplifies the interpretation and modeling of the experimental data.

The mechanisms for electrochemical reactions typically involve a complex series of elementary steps that necessarily include at least one electron charge-transfer step (i.e., an elementary half-cell reaction). However, most reaction mechanisms incorporate heterogeneous reactions such as mass transfer, adsorption and desorption. Whenever the electrode reactions involve phase change (e.g., gas evolution or electrodeposition), surface

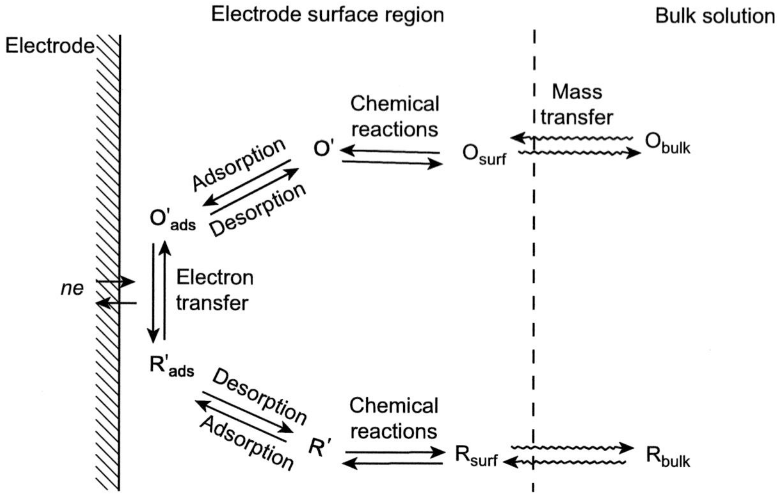

FIGURE 9-2 Processes at or near the electrode-electrolyte interface when there is no formation or dissolution of a non-ionic species. *(Figure reproduced from* Bard and Faulkner (1980) *with permission of the publisher. © 1980, John Wiley & Sons, New York)*

phenomena that are common to other systems become important. Surface phenomena such as nucleation of metal growth centres, surface diffusion and grain growth are the processes by which electrocrystallization of metals and alloys occur. The electrode potential strongly affects the nucleation of metal growth centers but does not usually directly affect surface diffusion or grain growth. As discussed later, electrode potential becomes a critical factor influencing deposit qualities such as smoothness, hardness, grain size, reflectance and surface resistance.

Perhaps one of the most important differences between electrochemical and non-electrochemical systems is that the operator has an additional variable available to influence reaction kinetics. In fact, since the electrode potential is so simple to adjust, it is by far and away the most widely used variable for controlling reaction kinetics. The adjustment of temperature and pressure, on the other hand, are much less commonly used. The relationship between electrode potential and reaction rate for a typical half-cell reaction is illustrated in **Figure 9-3** which presents a polarization curve (*Scott, 1991*). In such a curve, the current, i, or rate of an electrode reaction is plotted versus E or the driving force for the reaction. The example shown applies to a case where the bulk solution is being stirred so that a well-defined hydrodynamic boundary layer exists. When E is close to $E_{rev,j}$ for a particular half-cell reaction (i.e., small degree of polarization), the driving force for charge transfer due to this reaction is small so that electrode kinetics is slow compared with mass transfer from the well-stirred electrolyte and consequently becomes rate controlling.

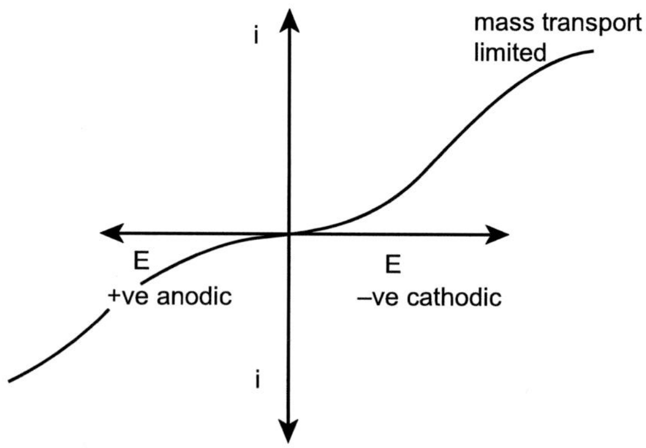

FIGURE 9-3 The effect of the applied potential E on current density i for single anodic or cathodic reactions. *(Figure reproduced from* Scott (1991) *with permission of the publisher. © 1991, Academic Press, London.)*

When E is adjusted to deviate further from $E_{rev,j}$, the driving force for the electrode reaction increases, and the electrode kinetics increase, whereas mass transfer remains unaffected. At some point, the process falls into a region of mixed kinetic-mass transfer control. With even further polarization, electrode kinetics increases sufficiently that mass transfer becomes rate limiting. In this region, the rate of the reaction depends only on factors affecting mass transfer (i.e., degree of agitation and bulk concentration of electroactive species) and becomes independent of E. Obviously, as long as the agitation and bulk concentration remain unchanged, no further increase in current is possible and so it levels off to a limiting plateau.

Consequently the controlling factor for a particular reaction is the deviation between E and $E_{rev,j}$. This deviation is called the overpotential, η_j, for reaction j,

$$\eta_j = E - E_{rev,j}. \tag{9-11}$$

This quantity represents the extra potential required to drive a reaction at a given rate and accounts for the barriers associated with the various sub-processes such as charge transfer activation, adsorption, crystallization and mass transfer. The overall overpotential is often split into two components, i.e.,

$$\eta_j = \eta_{s,j} + \eta_{conc,j}, \tag{9-12}$$

where the surface overpotential, $\eta_{s,j}$, includes all effects associated with the surface reaction, such as activation, crystallization, adsorption, and the concentration overpotential, $\eta_{conc,j}$, accounts for effects associated with mass transfer. In an electrolytic cell, extra electrical energy must be introduced to overcome these barriers and thus η_j represents the additional voltage (hence additional electrical energy) that must be applied to each electrode. In a galvanic cell, these barriers consume electrical energy so consequently η_j represents the loss in voltage from each electrode that would otherwise be delivered to the external load.

The surface overpotential $\eta_{s,j}$ and electrode kinetics are strongly affected by the presence of a structured region called the electrical double layer at the electrode-electrolyte interface (*Stojek, 2010*). It is important to note that such a region exists even when an electron transfer reaction is not occurring. Although different structures for this region have been proposed over the years, perhaps the most commonly accepted one is shown in **Figure 9-4**. Adsorbed water molecules are immediately adjacent to the electrode. In the diagram below, the potential of the electrode has been adjusted so that the metal surface contains an excess negative charge. This tends to orient the dipoles of the adsorbed water so that the positive hydrogens are directed toward the electrode surface, although not all of this water will be affected in exactly the same manner. Of course, if the electrode is positively charged, the oxygen end of the adsorbed water will be oriented toward the surface. Also, in direct contact with the electrode are neutral or charged species that have displaced water from the surface and so are considered to be adsorbed. Just outside this layer are the innermost solvated ions that do not displace any of the adsorbed water and so are not in direct contact with the surface. These solvated ions are not considered to be adsorbing.

The double layer structure is marked by two planes. The inner Helmholtz plane (IHP) passes through the centers of the specifically adsorbing species or the outside of the adsorbed water when no specifically adsorbing species are present. The outer Helmholtz plane (OHP) passes through the centers of the innermost solvated ions. Outside the OHP is the diffuse layer that is governed by long-range coulombic forces. Consequently, it contains a predominance of solvated counter-ions attracted to the oppositely charged surface over solvated co-ions that are repelled. Obviously, this electrostatic effect diminishes moving away from the electrode. The thickness of the double layer is

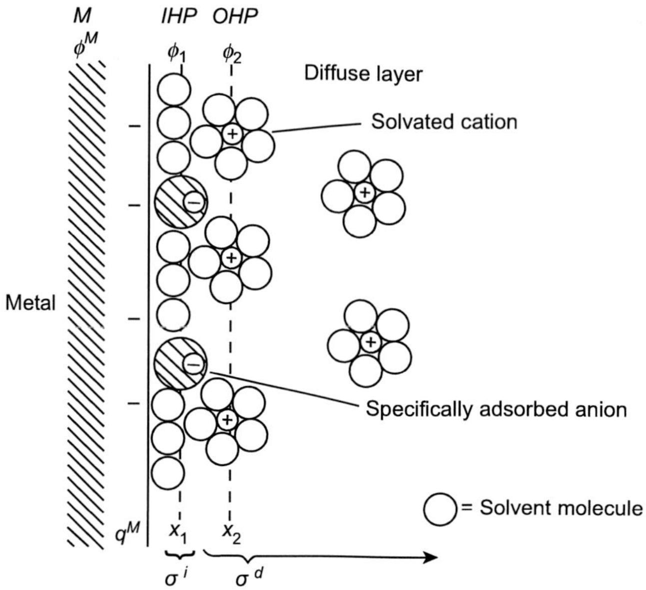

FIGURE 9-4 Schematic of the electrical double layer at the electrode-electrolyte interface. *(Figure reproduced from* Bard and Faulkner (1980) *with permission of the publisher. © 1980, John Wiley & Sons, New York.)*

strongly affected by the electrolyte composition and the electrode potential. An increase in either the ionic strength of the solution or the magnitude of the electrode potential has a similar effect: compression of the diffuse layer. Regardless of the conditions, the double layer is always much thinner than that of the hydrodynamic boundary layer and has a thickness much less than 1 μm under most practical conditions.

Electron transfer reactions can take place from both the IHP (termed inner sphere reactions) and the OHP (termed outer sphere reactions). Regardless of where electron transfer occurs, the electrical double layer has a profound effect. For example, inner sphere reactions require that adsorbed water be displaced or solvating water molecules be shed, phenomena that obviously tend to retard electrode kinetics. Some specifically adsorbing species at the IHP can enhance the rate of electrode reactions and so act as electrocatalysts or accelerators, whereas others can impede electron transfer and thus act as inhibitors. However, perhaps the most fundamental effect of the double layer on electron transfer is due to its role as a capacitor. Any change in the electrode potential causes the excess charge in the metal to vary which in turn causes a redistribution of species, particularly ions, within the double layer in response. Since the net movement of ions within the double layer constitutes a current, any change in the electrode potential is accompanied by a charging current due to this rearrangement of the double layer structure. This charging current density $i_{dc}(t)$ which can be likened to the charging or discharging of a capacitor flows only when the electrode potential is varied and can be described as follows:

$$i_{dc}(t) = C_{dl}\frac{dE}{dt}, \tag{9-13}$$

where C_{dl} is the double layer capacity. This flow of current occurs in parallel to the faradaic current that is responsible for the electrode reactions. Consequently, the total current density is split between the faradaic and charging current densities, i.e.,

$$i_{total}(t) = i_{farad}(t) + i_{dc}(t). \quad (9\text{-}14)$$

This relationship means that the charging current can affect a faradaic process when a transient method such as a periodic pulse operation is used to drive an electrochemical process. If the time constant for $i_{total}(t)$ becomes comparable to that for $i_{dc}(t)$, then $i_{farad}(t)$ becomes distorted by the charging current. Given that the relaxation time to charge or discharge the double layer is of the order of a ms, then distortion of the faradaic component of the electrode response begins to occur when the frequency of the input current or potential signal reaches approximately 1 kHz.

In summary, the objectives in electrochemical engineering are typically to obtain high current density to maximize production per unit area of an electrode surface, high current efficiency and uniform distribution of current density over the electrode surface with as low an overpotential as possible. Of course, some of the variables available are the same as for non-electrochemical systems. However, in many situations, the electrical variables unique to electrochemical systems enable a degree of reaction control not possible in non-electrochemical systems.

9.1.2 Electrochemical Cells and Processes

As discussed above, electrolysis cells consume electrical energy to produce chemical products, while reactions in galvanic cells occur spontaneously to produce either electrical energy or chemical products. In industry, electrolytic processes dominate. Examples are the production of Cl_2 and NaOH from brines or concentrated NaCl solutions in chloro-alkali cells, electrohydrodimerization of acrylonitrile to adiponitrile, epoxidation of propene and water electrolysis. **Figure 9-5** is a schematic diagram of a standard two-compartment chloro-alkali membrane cell used to produce Cl_2 and other chlorine-based products. The use of a cation-exchange membrane is key to the success of such a cell since it provides separation between the four streams discharged from the cell: Cl_2 at the anode, H_2 at the cathode, concentrated NaOH solution in the catholyte and spent NaCl solution in the anolyte. The mesh functions both as an electrode and as a medium for bubble formation. Spacing between the electrodes and membrane is minimized, consistent with a reasonable flow of electrolyte through the gap between the surfaces.

Electrolysis is also important in the primary production of metals (e.g., aluminum from Al_2O_3) and the purification of metals (e.g., copper electrorefining) during the metallurgical

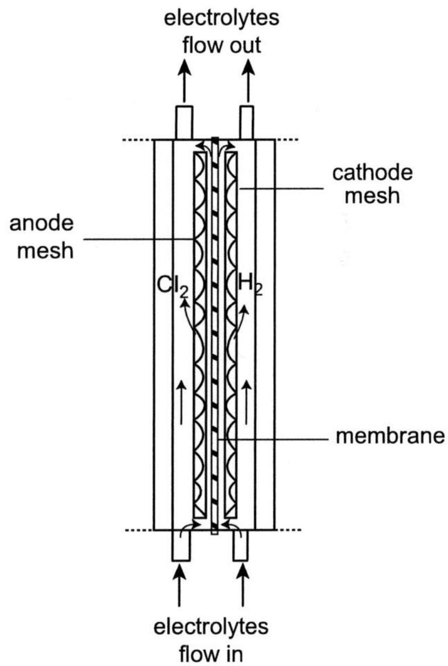

FIGURE 9-5 Schematic of a chloro-alkali cell using mesh electrodes and a gas impermeable membrane. *(Figure adapted from* Scott (1991) *with permission of the publisher. © 1991, Academic Press, London.)*

processing. It is a basic operation in electroforming processes whereby small metal parts are manufactured or in the fabrication of electronic components such as printed circuit boards. It also dominates the metal finishing industries (e.g., gold, nickel or chromium plating). Agitation of the solution over an electrode surface during electroplating is often carried out by simple stirring of the electrolyte, but in large operations pumping electrolyte across the part, surface moving on frames or belts may be used.

Cell designs vary according to the application. Often the electrodes are stacked as alternating anode and cathode vertical plates separated by a membrane. Three-dimensional electrodes are also used. One design uses a packed bed of metal particles serving as anodes surrounded by a cylindrical wall that functions as the cathode. Another design employs fluidized particles as electrodes. Such designs find use in organic electrosynthesis for the production of such species as sorbital and gluconic acid from glucose. In these designs, the product formed at the anode is swept out of the reactor by the flowing liquid. When a reactant is a gas, such as in the epoxidation of propene, the cell design must allow for its introduction. Along with providing sufficient hold-up to utilize the feed and to contain a potentially costly and hazardous electrolyte, cells must be designed to meet electrode surface requirements and mixing demands for the operating current density. Higher current densities can increase the production rate of the desired product and reduce surface area requirements, but may increase operating costs and could increase the overpotential to the point that an undesirable side reaction is initiated. A side reaction, in turn, will lower the current efficiency and/or can have a deleterious effect on the properties of the desired product. Higher flow rates and turbulence, however, can reduce the overpotentials at high current density, but they increase the capital and operating costs (*Bard and Faulkner, 2001*).

9.1.3 Variables Affecting Performance of Electrochemical Systems

Generally, the objectives in electrochemical system design are to achieve high current density, high current efficiency and uniform distribution of current density with as low an overpotential as possible. These tasks can be tackled by manipulating system variables as well as through proper cell design, as enumerated in **Figure 9-6**.

A unique feature of electrolytic cells is the fine control offered by the electrical inputs of applied current or potential. These can be varied over wide ranges and readily reversed or operated to vary with time. However, their effects on cell performance are complicated by the interactions possible between components of the system, both in the bath and on the electrode surface. Even a solvent can participate in an electrode reaction. Water is certainly the most common solvent used in electrochemical systems and commonly decomposes during electrochemical processes either intentionally (e.g., water electrolysis or electrolysis of NaCl to form Cl_2) or as undesired side reactions (e.g., H_2 evolution during metal electroplating). The interactions set limits on the range of the electrical variables used for control.

9.1.4 Dynamics in Electrolytic Cells

Cell design, applications and dynamic behavior are described in considerable detail in a number of textbooks (*Fahidy, 1985; Scott, 1991; Goodridge and Scott, 1995*). Electrode reaction dynamics are usually described by the temporal variation of potential and current density since these quantities are measurable but are also directly related to changes in the double layer and concentration gradients in the electrolyte that occur simultaneously near an electrode as well as the various processes that occur on the electrode surface. **Figure 9.7(b)** below shows the response to

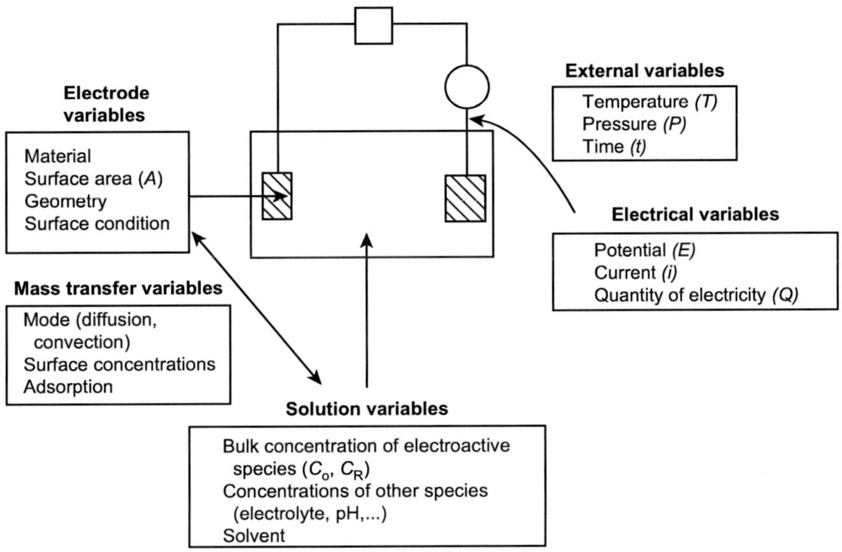

FIGURE 9-6 Variables influencing the performance of electrolytic or galvanic cells. *(Figure reproduced from* Bard and Faulkner (1980) *with permission. © 1980, John Wiley & Sons, New York.)*

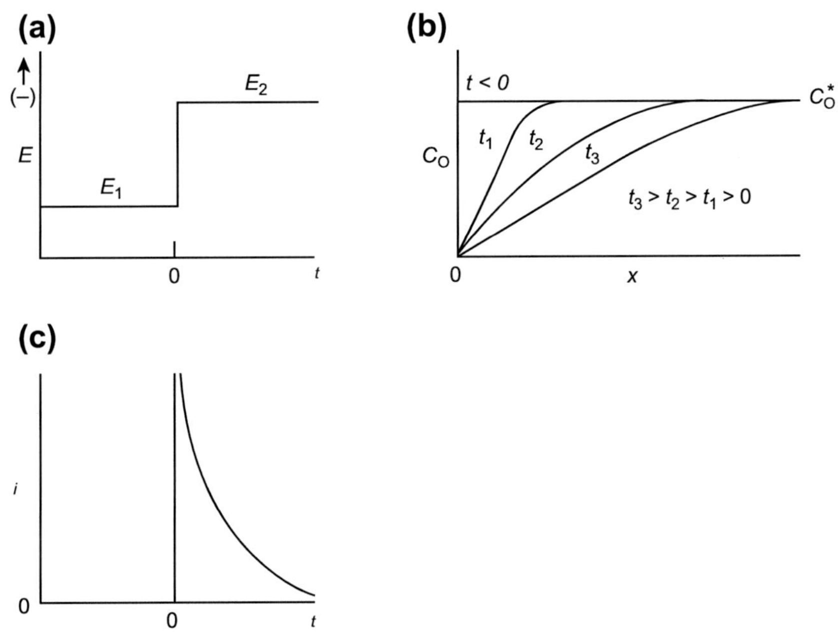

FIGURE 9-7 Response of ion concentration near the electrode-electrolyte interface and of current through the cathode when a step change in potential initiates a reduction reaction. *(Figure adapted from* Bard and Faulkner (1980) *with permission. © 1980, John Wiley & Sons, New York.)*

a step change in potential (**Figure 9.7(a)**) which initiates an electrode reaction under quiescent conditions. Before the step change, no current flows and consequently no reactant concentration gradient occurs in the vicinity of the electrode. Once the reaction begins at the electrode, a concentration gradient develops at the interface if mass transport of the reactant from the bulk cannot keep pace with the rate of consumption at the electrode. The decrease in the concentration of the reactant at the interface causes the current associated with the electrode reaction to diminish. Over time, the process becomes controlled completely by diffusion from the hydrodynamic boundary layer at the electrode surface leading to a further reduction in the current density j. Processes at the electrode such as surface diffusion or electrocrystallization that are important to the morphology of the deposit produced during plating, for example, are affected by the rate at which ions are discharged at the electrode. Consequently, electrode kinetics and mass transfer directly influence the quality and structure of the deposit that forms.

9.2 ELECTROPLATING

Since the 1950s, the superposition of pulses, potential reversal or alternating current on the basic direct current has become a widespread practice in the electrochemical industry. The technology is discussed in the most recent handbooks and is a component of most professional-level courses. Unlike the remainder of the chemical industry, periodic operation is now a mainstream approach to carrying out electrochemical processes.

The state of the art of periodic operation as applied to electroplating has been recently summarized by *Schlesinger and Paunovic (2010)*, although an earlier book by *Puippe and Leaman (1986)* is still widely consulted. At least one review paper provides a concise summary of its applications to Ni plating (*Vermeijlen, 2002*). Other authors (*Aroyo and Tzonev, 2002, 2003*) discuss the use of pulse reverse plating and compare it to unipolar pulsing. Advances are also summarized in the Biennial International Pulse Plating Symposia sponsored by the American Electroplaters & Surface Finishers Society. The discussion which follows is drawn extensively from these symposia.

Electroplating proceeds by a cathodic reaction that brings about the deposition of a metal. The purpose of this process is to produce coatings with the desired chemical, surface and mechanical properties; properties depend on the coating purity, morphology, crystal structure, grain size, extent of dislocations and the size and distribution of occlusions. The electroactive species in the plating bath can be a single- or a multi-charged bare metal ion or an ionic complex involving a metal and a ligand such as Cl^- or CN^-. In fact, bare metal ions do not exist in aqueous solution as they are solvated with water ligands. Other electroactive species can be present in the electrolyte and may be discharged by an undesired side reaction if the overpotential is high enough. A common example is the reduction of H^+ in acidic solutions and H_2O in neutral or alkaline solutions to evolve H_2. Additives which are added to plating baths to influence the reaction mechanism and/or affect the electrocrystallization process in order to modify the deposit structure often specifically adsorb on the electrode surface as well. The surface can thus be a complex environment with many different species present. On such a heterogeneous surface, metal adatoms formed by the cathodic reaction can form new nuclei that become centers for grain growth or diffuse to pre-existing nucleation centers and are incorporated into a growing crystal structure. These processes are depicted schematically in **Figure 9-8**. The morphology of the resulting deposit depends strongly on the rate controlling regime during deposition. Deposits formed when the metal

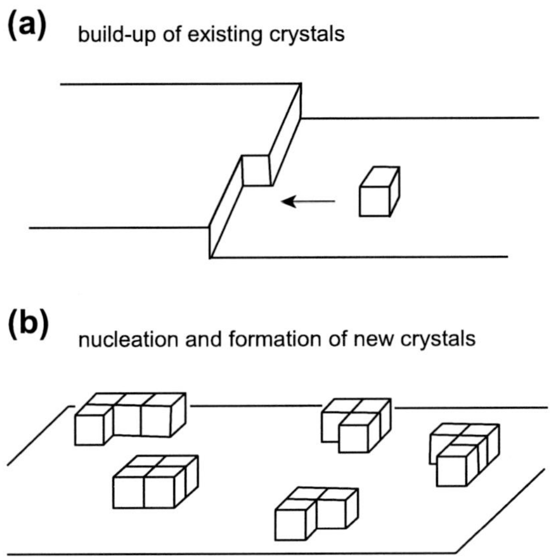

FIGURE 9-8 **Schematic diagram showing the role of metal adatoms during metal electroplating. a) crystal build up; b) nucleation.** *(Figure reproduced from* Puippe and Leaman (1986) *with permission. © 1986, Amer. Electroplater & Surface Finishers Society.)*

reduction reaction is under kinetic or mixed kinetics-mass transport control tend to be smoother and more fine-grained, whereas coatings tend to be very rough and coarse-grained and can exhibit dendrites in the extreme when electroplating is controlled primarily by mass transport. Since such coatings are typically undesirable, applied currents rarely exceed about 70% of the mass transfer limiting currents. One of the purposes of adding organic additives to plating baths is to promote the formation of smooth and fine-grained deposits and allow the process to operate at as high a current as possible without the deposit quality suffering. In some cases, the presence of these additives on the surface can lead to dislocations or to disordered granular deposits. In these situations, the additives, electrolyte or even H_2 can be trapped in the growing deposit on the electrode. The morphology and macroscopic smoothness of a coating or an electroformed object also depend on the current density distribution over the electrode surface.

Since it is difficult to meet all of the objectives simultaneously, operating conditions are typically chosen based on a compromise. The use of pulse plating is particularly advantageous over conventional DC plating in this regard since it offers additional control variables with which to obtain the desired deposit properties.

Figure 9-9 illustrates the square-wave pulse waveforms that have been used or explored by industry or used in research. Of these, simple unipolar on-off pulses and bipolar pulse reverse pulses are the most popular. Curiously, the more complex waveforms shown have been utilized in industry rather than in research studies. As with DC plating, either the electrode potential or current is the input. The input signals typically cover a wide range of frequencies, duty cycles and pulse amplitudes. In general, the duty cycle is defined as follows:

$$\text{duty cycle} = \frac{t_{\text{cathodic}}}{\tau} \times 100\,\%, \qquad (9\text{-}15)$$

where t_{cathodic} is the duration during each pulse cycle when cathodic current or potential is applied and τ is the duration of each cycle. Although square wave pulses are shown, other waveforms, such as a ramped current or potential, can also be used. AC operation, i.e., sinusoidal modulation of current over a wide range of frequencies, is also used, albeit less often. The frequencies that can be practically applied are restricted by the double layer charging and discharging effects described previously. The pulse on-time time, t_{on}, during unipolar pulse plating and $t_{on(c)}$ during bipolar pulse plating should be considerably longer than the time required to charge the double layer in response to the change in electrode potential during the cathodic portion of each cycle. Likewise, the discharge time of the double layer must be much shorter than the off-time, t_{off}, during unipolar pulsing and $t_{on(a)}$ during bipolar pulsing. Otherwise, a significant portion

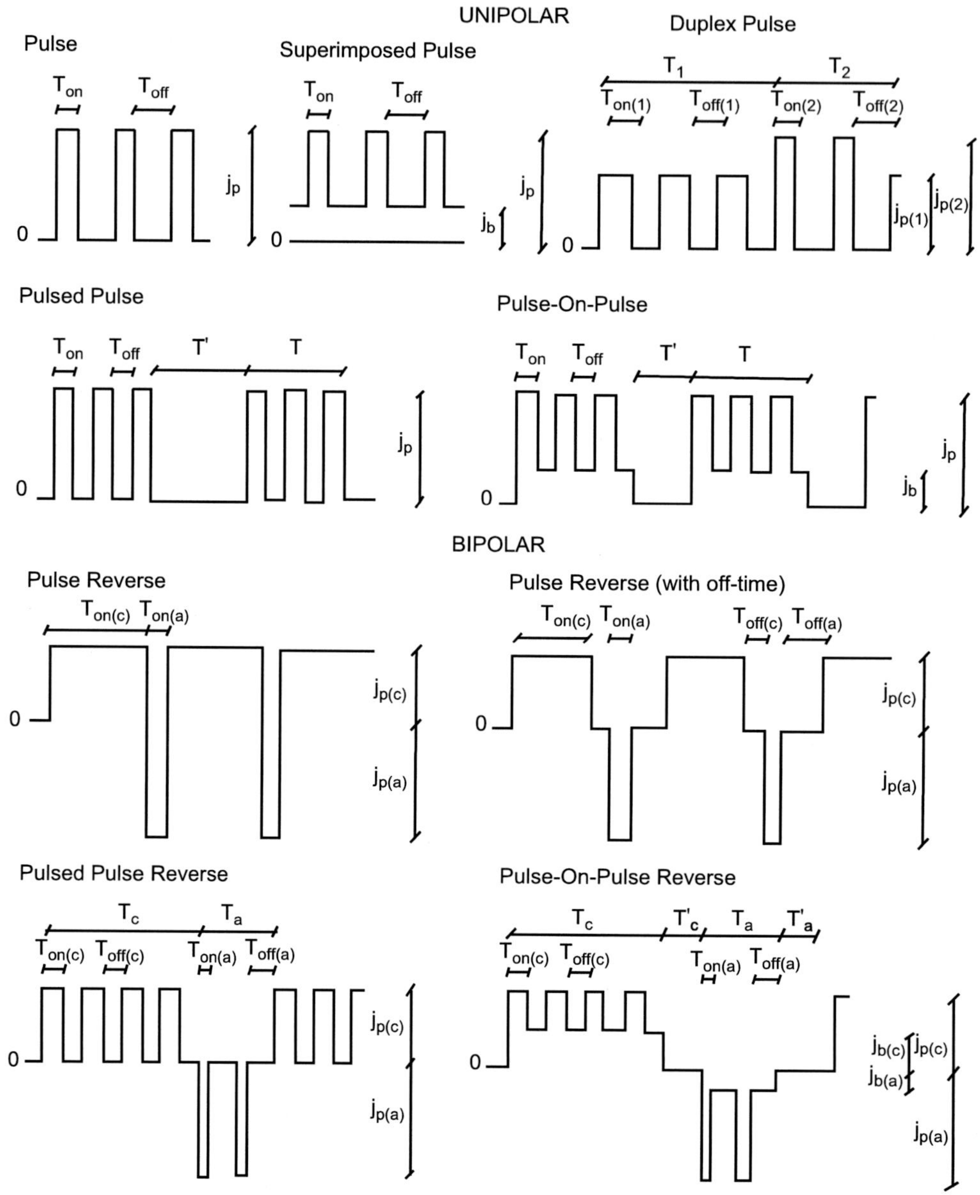

FIGURE 9-9 Square waves used in unipolar and bipolar pulse plating. Suggested nomenclature are given for each pulse pattern. *(Figure reproduced from* Puippe and Leaman (1986) *with permission. © 1986, Amer. Electroplater & Surface Finishers Society.)*

of the total current is diverted to double layer charging rather than to the intended faradaic processes (deposition). The double layer charging and discharging times depend on pulse amplitudes, but charging times in the order of 0.05 ms and discharge times of 0.35 ms are not unusual. Thus, frequencies much above 1 kHz are rarely applied. As with DC plating, the applied current density during a pulse is also limited to prevent mass transfer from having too large an effect, although one of the most important advantages of pulsed methods is that they enable mass transfer effects to be much more effectively offset than is possible using DC methods.

Unlike DC plating where mass transport depends on hydrodynamic conditions alone, mass transport during pulse plating is strongly affected by the waveform of the input pulse signal, i.e., pulse parameters, in addition to hydrodynamics. Pulsing sets up two boundary layers: A narrow pulsating boundary layer with thickness δ_p closest to the electrode that responds continually to changes in current or potential throughout each cycle and a thicker stationary layer with thickness δ_s that is essentially the same as the DC boundary layer at the time-average current density in the pulse operation. This structure is depicted in **Figure 9-10**. As metal ions adjacent to the electrode are consumed during the cathodic portion of a cycle, the concentration gradient in the pulsating layer becomes steeper. The profile subsequently relaxes during the off-time or anodic portion of the cycle as the metal ion concentration is replenished in this layer. Obviously, the higher the pulse current density, the larger is the metal ion depletion in the pulsating boundary layer, leading to a steeper concentration gradient. The use of small duty cycles or high frequencies allows the application of very high pulse current densities, perhaps as high as 1000 times the values during DC operation. The thickness of the pulsating boundary layer during simple unipolar on-off pulsing can be shown to be

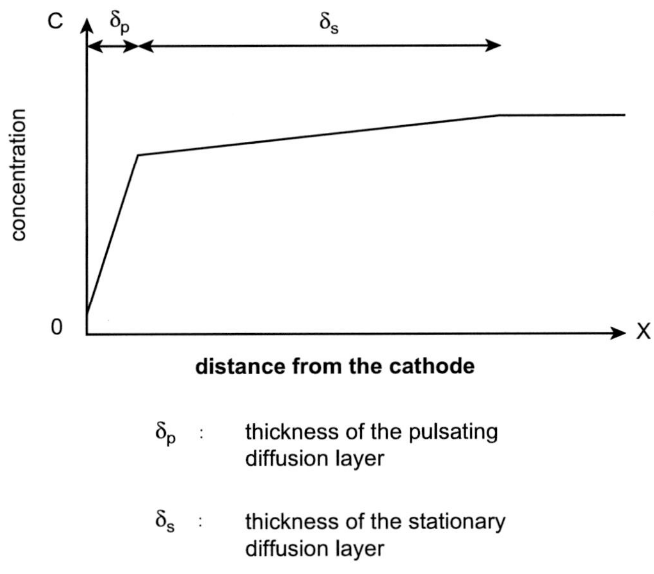

FIGURE 9-10 **Pulsating and stationary boundary layer at the electrode-electrolyte interface established by applying a pulsing current or potential to an electrochemical cell.** *(Figure reproduced from* Puippe (1986) *with permission. © 1986, Amer. Electroplater & Surface Finishers Society.)*

determined by the pulse parameters t_{on} and the duty cycle s expressed as a fraction as follows:

$$\delta_p = \sqrt{\frac{4}{\pi} D_{eff}\, t_{on}(1 - s)}, \qquad (9\text{-}16)$$

where D_{eff} is an effective diffusivity of metal ion accounting for transport by both diffusion and migration. On the other hand, the thickness of the stationary boundary layer is determined by cell hydrodynamics. Since the time-average limiting current density during pulsing is determined by this thickness, it cannot exceed the limiting current density attained during DC electrolysis. Consequently, pulse plating cannot be employed to exceed the DC limiting current density. However, relaxation of the pulsating boundary layer during each cycle does allow the cell to operate at higher time-average current densities without mass transfer adversely affecting deposit quality than is possible under DC conditions. Consequently, it effectively extends the range of current densities that can be used in practical applications. This is one of the main reasons why pulsed methods have become so popular in industrial plating operations.

During DC plating, the most important quantity characterizing mass transport during an electrolytic process when the solution is stirred externally is the limiting current density, j_l. It occurs when the concentration of the electroactive reactant at the electrode surface reaches zero and it corresponds to the maximum rate achievable by the reaction under the given hydrodynamic conditions. It is particularly important in electroplating since deposit quality often suffers dramatically when current densities approaching the limiting value are applied. The analogous quantity during pulse plating is the pulse limiting current density $(j_p)_{limiting}$. This limit corresponds to the amplitude at which the surface concentration of the reactant reaches zero just at the end of the cathodic portion of a cycle (*Viswanathan et al., 1978*). It has been shown that $(j_p)_{limiting}$ during unipolar pulsing can be related to the on-time, t_{on}, and the duty cycle, s, expressed as a fraction, namely:

$$(j_p)_{limiting} = \frac{j_l}{\sqrt{\frac{4D_{eff}t_{on}}{\pi\delta^2}(1 - s)^{3/2} + s}}, \qquad (9\text{-}17)$$

where δ is the stagnant boundary later thickness. An important advantage of pulsed methods is that smooth deposits can be obtained even if $(j_p)_{limiting}$ is reached as long as δ_p can be kept small. Diminishing mass transfer interference in this way allows the positive effects that high current densities have on deposit structure to be more effectively realized.

The relations above apply only to simple unipolar pulsing. With the bipolar pulse reversal, dissolution of the deposit occurs when the current direction is reversed. This enriches the pulsating boundary layer so that $(j_p)_{limiting}$ can be increased. The increase can be substantial but it depends on t_{off} and the ratio of $j_{p(a)}/j_{p(c)}$. On the other hand, because of current reversal, the time-average plating rate will always be less than DC and unipolar pulse plating rates at comparable conditions.

The grain size of coatings depends strongly on the electrode potential or current density during deposition. As mentioned above, electrocrystallization occurs primarily by a nucleation and growth process. Metal adatoms formed on the electrode by metal ion reduction can either form new nucleation centers or diffuse along the surface to existing centers and undergo incorporation into growing grains (see **Figure 9-8**). The relative rate of these two processes determines the final grain size of the deposit and many other properties. Of these two processes, only nucleation of new centers is affected directly by electrode potential or current density. The larger the cathodic overpotential or current density, the greater is the nucleation rate and the smaller the resulting grain size. Thus, pulsing with large cathodic pulse amplitudes is able to produce deposits with smaller grain size and thus greater tensile strength, hardness and often more ductility than is possible with DC plating. Many industrial plating applications exploit this property. The effect of current density on grain size in pulse plating is illustrated in **Figure 9-11** that shows SEM images of copper deposits plated from a bath containing 0.75 M $CuSO_4$ and 1 M H_2SO_4 at pulse current densities of 1000 and 4000 A/dm^2. In both cases, the on-time, t_{on}, was set to 0.03 ms and the time-average current density was 37.5 A/dm^2. As clearly evident, the grain size of the coating formed at $j_p = 4000$ A/dm^2 is much finer than the one formed at the lower pulse current density.

However, the situation is further complicated by events that can occur during the off-time of unipolar pulse plating in the case of some metals such as gold and copper. In the absence of current, these metals can undergo recrystallization whereby metal atoms migrate across grain boundaries based on the driving force to minimize their interfacial energies. Consequently, recrystallization leads to a growth of larger grains at the expense of smaller ones and overall coarsening of the grain structure. For these metals, the duration of the off-time is an important operating parameter, with the coatings becoming coarser-grained as the off-time is extended.

Pulse plating can also have a significant influence on the uniformity of deposit thickness across the cathode surface and produce smoother coatings than those obtained by DC methods under the appropriate conditions. This effect can be understood by making use of the concepts of a microprofile and macroprofile. A surface can be considered to be a microprofile when the thickness, δ, of the diffusion boundary layer is much thicker than the characteristic dimension, h, of the surface features (**Figure 9-12(a)**). When a surface reaction such as electroplating is strongly influenced by mass transfer, its rate will be fastest in the regions where the path length from the bulk to the surface is shortest. Under these conditions, electroplating at the peaks or crests of a microprofile will be greater than in the recesses and any differences in the heights of surface features will be amplified and grow over time. Such a trend, of course, leads to a rough coating. This behavior explains the well-known observation that deposits are usually very rough or powdery when formed near to or at the limiting current density. On the other hand, when the

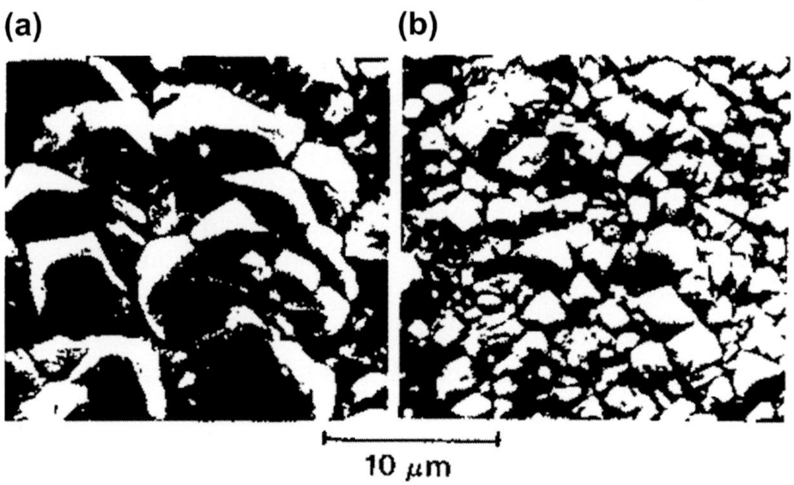

FIGURE 9-11 SEM micrographs of copper deposits produced by pulse plating at pulse current densities j_p of (a) 1000 and (b) 4000 A/dm^2. Other conditions are given in the text. *(Figure reproduced from* Puippe (1986) *with permission. © 1986, Amer. Electroplater & Surface Finishers Society.)*

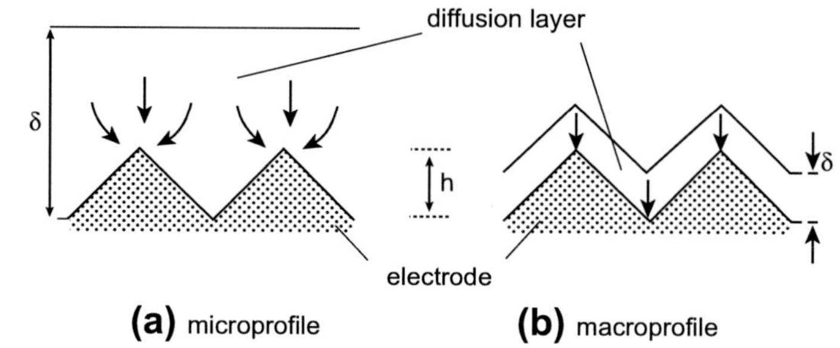

FIGURE 9-12 Effect of boundary layer thickness on current distribution over an electrode surface depending on whether it is an (a) microprofile or (b) macroprofile. *(Figure reproduced from* Dossenbach (1986) *with permission. © 1986, Amer. Electroplater & Surface Finishers Society)*

diffusion boundary layer thickness becomes comparable with or smaller than the characteristic dimension h of surface features, the surface is equivalent to a macroprofile (**Figure 9-12(b)**). All points on the surface are now equally accessible to species transported from the bulk solution. Thus, a rough or powdery deposit would not be expected when mass transfer controls electrodeposition.

During pulse plating, the thickness, δ_p, of the pulsating boundary layer is the important dimension when considering uniformity of mass transport to the electrode and determining whether a given surface can be considered a microprofile or macroprofile. As discussed above and evident from **Eq. (9-16)**, δ_p can be controlled by adjustment of the on-time during unipolar pulse plating. It therefore becomes possible by the application of pulses with short enough t_{on} to make δ_p small enough that a surface which would otherwise be a microprofile under DC conditions becomes a macroprofile during pulse plating. Consequently, deposits produced at high time-averaged current densities when mass transfer affects deposition are much smoother than is possible using DC plating at the comparable current densities.

Pulse reverse plating can be useful in controlling deposit thickness uniformity when the conditions at an electrode surface correspond to a microprofile. For the same reason that more metal is deposited at the crests in a microprofile than in the recesses during the cathodic portion of a cycle, the high points in the coating are preferentially dissolved during the anodic phase. To take full advantage of this approach, the anodic pulse current density must be selected independently of the cathodic pulse current density. Best results were obtained when short anodic pulses with high amplitudes were applied since this most effectively favors preferential dissolution of the protrusions on the surface over other portions. This effect has been exploited in bipolar pulse reversal plating of copper or nickel for example. A relatively long cathodic pulse of 16 ms at $j_{p(c)} \approx 1.3\, j_l$ followed by a short anodic pulse of 1 ms at $j_{p(a)} \approx -3\, j_l$ was found to substantially improve uniformity.

For the most part, pulse and pulse reversal plating have been limited to frequencies below 1000 Hz, resulting in pulse durations in the order of milliseconds. *Kristof and Pritzker (1998)* carried out pulse reverse plating of copper at frequencies as high as 200 kHz. Improved smoothness and microhardness compared with plating at frequencies in the 50 to 1000 Hz range were observed. A theoretical study of electrode dynamics and current distribution at very high frequency provided an explanation of these experimental results (*Pritzker and Hudson, 2000*). Examination of the electrolyte side of the electrode-electrolyte interface indicated that the charging and discharging of the double layer progressively dominated as cycle frequency increased. At low frequencies, 50 Hz, transport effects at the interface force the current density to the steady-state limiting density at the end of a pulse. However, this does not occur at 200 kHz. At such high frequencies, ion transport cannot keep pace with current changes at the electrode and the faradaic current density freezes. In the case of copper deposition from a $CuSO_4$ electrolyte, a negative cathodic current is maintained and copper continues to deposit on the cathode, even during the anodic portion of the cycle, since the current discharge from the cathodic double layer is greater than the current through the electrolyte. The constant cathodic faradaic current arising from an extremely thin pulsating boundary layer through the pulse reversal cycle resulted in deposit uniformity. Freezing of the current density at high frequencies also leads to small overpotentials and so may reduce the likelihood of undesirable side reactions during plating.

More recently, the effects of a high pulsing frequency on copper deposition from acidic sulphate solutions onto a rotating disc electrode were studied experimentally and also modeled by *Tantavichet and Pritzker (2002, 2003)*. They showed that double layer capacitance effects persisted at relatively low frequencies (50 Hz). On the other hand, at frequencies as high as 50 kHz, the electrode response did not become identical to that during DC operation, contrary to what was generally expected. Analysis of their model showed that double layer charging and discharging at these high frequencies did not dominate completely. Consequently, the maximum practical frequencies that could be used during pulse plating may not be as limited, due to capacitance effects, as had been previously assumed. To explain these observations and correctly fit their model to the measured electrode responses during pulse and pulse reverse plating, *Tantavichet and Pritzker* included a contribution from a pseudo-capacitance from the formation of an adsorbed intermediate during Cu^{2+} reduction. They also found that the double layer capacity, C_{dl}, did not remain constant when the pulse period was of the order of the relaxation time of a water molecule adsorbed on the electrode surface. Excellent agreement between their model and the experimental electrode responses was obtained when C_{dl} varied with pulse frequency, f, according to the relationship $C_{dl} = A/\sqrt{f}$, where A is an empirical parameter. This decrease in C_{dl} with an increase in pulse frequency explained why double layer effects are less dominant at frequencies as high as 50 kHz than was heretofore assumed.

A common way in which deposit uniformity is characterized is through its throwing power. Although this is defined differently depending on the object being plated, the throwing power measures the ability of a solution to plate into low current density areas with the same thickness as in the higher current density areas. In essence, this characterizes the ability of a plating bath to produce a uniform deposit on an irregularly shaped cathode. It is most commonly calculated as the ratio of the deviation between the thickness obtained in the portion of the surface deemed difficult to coat from the average plating thickness divided by the average thickness. Factors in DC plating related to the throwing power are combined together in a dimensionless Wagner number, $N_{Wa} = \frac{d\eta}{di}\frac{\kappa}{L}$, which gives a ratio of polarization resistance due to the electrode reactions to the electrolyte ohmic resistance. In this relation, κ is the electrolyte conductivity, L is the characteristic dimension of the part to be plated, $d\eta/di$ is the slope of the polarization curve where η is the overpotential and i is the current density. The larger the value of N_{Wa}, the more uniform is the current distribution and the better is the throwing power.

Particular challenges arise in plating (most commonly copper) of uniform thickness on printed circuit boards and other parts with irregular or difficult geometry. Examples are 0.08 mm lines or the edges of laser-drilled microvias (holes) as well as large surface areas termed pads. To obtain satisfactory coatings, plating operators must use additives and leveling reagents in their plating baths and carry out deposition at low current densities. Unfortunately, such operation leads to low production rates and tends to be expensive. However, significant improvements to these operations have recently been achieved by the use of bipolar pulse reverse plating that allows operation at much higher current densities with reasonable throwing power. Brighteners, additives and levellers are still required, but generally in lower amounts and in different combinations than is required by DC plating as discussed below.

Plating of printed circuit boards (PCB) has been the focus of a large number of studies concerned with the design of the most appropriate pulse plating cycle, use of additives and electrolyte composition as well as modeling studies of the plating of holes, vias and recesses. *Ward et al. (2000)* discussed solutions to these problems and summarized the state of the art. They presented a comprehensive study of the effect of cycle parameters such as $t_{on(a)}$, $t_{on(c)}$, i_a/i_c and i_c on the performance of plating onto actual PCBs in a pilot size plating line. The effect of agitation of the electrolyte bath with a liquid jet was compared with the more conventional agitation by compressed air injection. The electrolyte contained 0.3 M $CuSO_4$, 2.25 H_2SO_4, 50 ppm Cl^-, a proprietary additive-brightener and a proprietary carrier. Throwing power and plating properties, such as thickness, microhardness and morphology, were measured. *Ward et al.* observed that the pulse cycle parameters had a large effect on plating uniformity and throwing power, but coating appearance, morphology and microhardness did not vary greatly. Best uniformity was obtained by operating at $t_{on(c)} = 20$ ms and $t_{on(a)} = 1$ ms, $2.6 < i_a/i_c < 3$ and $3.3 < i_c < 3.5$ A/dm^2. These conditions are close to those used in the electroplating industry. Several experiments using DC plating at 3.3 and 4 A/dm^2 gave thinner deposits and poorer uniformity, but throwing power and plating properties were similar to those achieved with pulse reversal.

Three-dimensional printed circuit boards are now in wide use. To connect individual planar boards to each other, microvias are filled with copper using electroplating. Similarly, the connections between all the boards in the device are made by copper plated-through-holes (PTH). The challenge in both cases is to fill these holes without any voids since this reduces the conductivity of these conduits. This is particularly difficult in the case of microvias which are much smaller than PTHs and have diameters in the submicron range and aspect ratios of two to three. An approach based on bipolar pulse reverse plating and the use of particular chemical additives in the electrolyte has been successful for filling PTHs but less so in the case of microvias. *Taylor et al. (2000)* proposed a waveform consisting of two main segments for combined PTH/microvia plating that they termed charge-modulated plating. Its structure resembled the pulsed pulse reverse pattern in **Figure 9-9** but with anodic pulses in both parts of the cycle. The first portion consisted of a short cathodic pulse at a high current density followed by a much longer anodic pulse at low current density to enable conformal plating of the microvias. In the second segment of the cycle, a long cathodic pulse at a moderate current density was followed by a short anodic pulse at high current density to fill PTHs without significantly affecting the plating of the microvias. *Rasmussen and Kartchner (2000)* also studied PTH plating in an agitated bath using a multi-part cycle consisting of a sequence of on-off cathodic pulses with duty cycles above 70% interrupted at intervals by sharp anodic pulses, then followed by on-off cathodic pulses with duty cycles of 50%. Their study focused on the effects of agitation and hole diameter on the effectiveness of PTH plating. They showed that smaller diameter holes required lower anodic pulse amplitudes for satisfactory results.

Zone plates used in X-ray equipment require precise geometry and uniformity. They are composed of circular gratings made up of several hundred concentric zones that decrease in width as one moves outward in the radial direction. Plate performance depends on zone depth and the outermost zone width. Nickel is widely used as the metal of choice for these gratings and is electroplated through a mask onto a mold. *Lindblom et al. (2006)* described the use of pulse reverse plating to uniformly plate the zones and preserve the critical aspect ratio with the following pulse waveform: $t_{on} = 2$ ms, $t_{off} = 499$ ms and $j = 2.5$ A/cm^2 during the cathodic portion of the cycle; $t_{on} = 1$ ms, $t_{off} = 199$ ms and $j = 300$ mA/cm^2 superimposed on a DC current density of 12 mA/cm^2 during the anodic portion.

9.2.1 Electroplating of Single Metals

Gold is commonly used in the fabrication of PCBs and interconnects, although it is most widely used for corrosion protection and decorative purposes, such as custom jewelry. Because of cost, pulse plating is often employed for high tonnage processing of non-electronic products. The use of simple on-off cathodic pulses was studied heavily in the 1970s (*Avila and Brown, 1970; Lendvay and Raub, 1975; Rehrig, 1974; Bielinski et al., 1979*). Compared with DC plating, pulse operation enabled the application of higher plating densities that yielded smaller grain size, improved ductility and conductivity, eliminated inclusions (particularly polymer) and reduced plating porosity. Off-time was found to be particularly important for reducing foreign inclusions since gold undergoes recrystallization. The effect of time-averaged current density on the carbon content in gold coatings is shown in **Figure 9-13**.

Rhenium coatings formed by DC plating on a metal substrate such as tungsten often have a problem of forming cracks upon heat hardening (*Knödler, 1986; Puippe, 1986*). The use of pulsed plating to produce these films solves this problem, provided that j_p, j and t_p are controlled. Also, a higher current efficiency is possible using pulsing compared with the DC operation, although the coating quality is sensitive to j_p, t_{on} and τ reflecting the competing influences of the double layer capacity, mass transport through the layer and the adsorption of intermediates on the plating surface. The optimal pulsing parameters, such as t_{on}, depend on the time constants for ion mass transport or adsorption of the rhenium intermediate.

Corrosion protection of iron or steel by electroplating zinc or nickel coatings is common

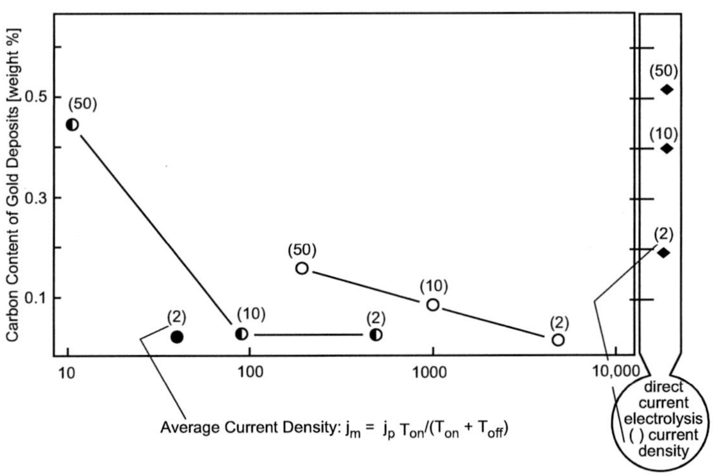

FIGURE 9-13 Dependence of carbon content in gold coatings produced by pulse plating on cathodic pulse current densities j_p of 1 (●), 10 (○) and 100 (○) A/dm^2 with t_{on} = 10 ms. *(Figure adapted from* Puippe (1986) *with permission of the publisher. © 1986, Amer. Electroplater & Surface Finishers Society.)*

practice. Pulse plating has been shown in many studies to produce a smoother, less nodular and a finer-grained coating with better mechanical properties and corrosion resistance than those obtained by DC plating. The influence of unipolar cycle parameters, such as pulse current density, duty cycle and period, on surface morphology, crystal size and crystal orientation for the electroplating of nickel using cycles of a millisecond's duration was reported by *El-Sherik et al. (1997)*. Surface morphology of electroplated Ni coatings was studied by *Krgulikov et al. (2000)* who showed that roughness can be controlled by a combination of pulse frequency and amplitude and the appropriate inclusion of leveling additives in the electrolyte bath. *Cheng et al. (1998a)* used unipolar pulse plating to produce nickel films on a Nd-Fe-B permanent magnet that had good adherence and excellent corrosion resistance. The effect of different waveforms in potentiostatic pulse plating was explored by *Wong et al. (1999, 2000, 2001)*. They showed that the use of ramped or triangular waveforms improved the surface finish, and they also presented a model to explain their observations. Research on the pulse plating of Ni continues. For example, *Krgulikov et al. (2008)* studied the effect of off-time on pulsed Ni deposition from an acetate-chloride bath and obtained 10-μm coatings with greater hardness and no porosity with current efficiencies as high as 90% by using pulse current densities up to 50 A/dm^2.

Many of the results reported for gold and nickel also apply to other metals such as cadmium, lead, palladium and particularly silver. A coating that contains voids that leave the base metal exposed is obviously undesirable and so coverage is an important aspect of deposit quality along with morphology. Pulse methods have been shown to improve coverage as well as plating quality for these other metals as well and so have been adopted to some extent. However, many DC plating problems can be solved with the use of chemical additives and/or by control of electrolyte composition, and so the use of pulse plating for surface property control or protection is not universal.

The application of chromium coatings to substrates to improve their hardness, corrosion resistance and lubricity is also common and usually carried out by DC plating from a Cr(VI) electrolyte. However, health, safety and environmental hazards associated with this electrolyte have prompted research into the alternative of plating from Cr(III) baths. DC plating often yields poor results with this electrolyte due to excessive H_2 evolution that leads to low current efficiencies and the formation of hydroxides on the deposit that degrade coating quality. The incorporation of rest intervals with zero current into a reverse pulse waveform was shown by *Renz et al. (2000)* to give high quality, thick Cr layers needed for functional applications. The cycle is essentially that shown in **(g)** of **Figure 9-9** except that no rest period was used prior to the anodic pulse. Parameters studied included τ_{cycle}, j, duty cycles and the ratio of the cathodic to the anodic current pulse amplitudes. Cr deposition and H_2 evolution occur during the cathodic pulse, while a species is oxidized in the anodic pulse to maintain a low pH and avoid hydroxide formation. Since the charge ratio in the two pulses is important, this waveform is sometimes referred to as charge modulation.

9.2.2 Electroplating of Alloys

The composition of an alloy coating is important for its proper function. For example, variations in composition of an alloy to be used as a solder can have a devastating effect if the solder is to be used in PCB fabrication since composition strongly influences a melting point. The composition of gold alloys used for decorative purposes affects the reflected color of the plated surface. It is often difficult to obtain the desired composition using DC methods and so typically complicated bath chemistry is required. Pulse plating of alloys is now widely practiced because it permits many more degrees of freedom through the design of the waveform to control deposit composition and morphology than does DC plating. Pulse plating has made it possible to achieve compositions that cannot be obtained by DC methods and to do so with cheaper and simpler baths and a lower reliance on chemical additives.

An example of an early approach to alloy pulse plating involved the unipolar deposition of brass from a Cu-Zn cyanide electrolyte using j_p close to the DC limit for Cu deposition and a t_{on} in the order of milliseconds (*Avila, 1986*). During the early portion of each cathodic pulse, Cu deposition dominated, but eventually the pulsed boundary layer became depleted of Cu and allowed Zn deposition to proceed. The boundary layer was allowed to replenish during the off-time. The values of t_{on} and t_{off} were found to determine the brass color which varied from green to yellow to red depending on the copper content. This, in turn, was found to depend on the electrolyte composition, pulse current density and duty cycle.

Silver-tin alloys containing each metal at roughly 50% are promising substitutes for pure silver coatings because of their tarnish resistance. Control of alloy composition is difficult and coatings tend to be heterogeneous with small regions of metallic silver or tin forming during DC plating; silver dendrites also form when DC plating is carried out near the limiting current density. *Leidheiser and Ghuman (1973)* found that pulsed plating resulted in a more uniform alloy composition and prevented nodule or dendrite formation. Coatings formed from a $AgCN/K_2SnO_3/KOH/KCN$ electrolyte were ductile and could be polished to a high luster. Alloy composition and the amount of elemental silver in the deposit were controlled through proper choice of j_p, τ_{cycle} and duty cycle. It is generally accepted that this alloy forms partially through a displacement reaction between metallic Sn on the electrode and Ag^{2+} at the electrode interface. As a result, it is not surprising that the pulse current density, t_{on} and t_{off} should significantly affect the alloy composition since they strongly influence the Ag^{2+} concentrations in the pulsating boundary layer. Another decorative substitution involving 18 karat gold in place of pure gold has also been examined using pulse plating to control composition and secure smoothness and brightness of the deposit. The cost of gold, greater hardness and superior wear properties justified the substitution of the pure metal with the alloy. Pulse plating significantly decreased the cladding porosity and provided better uniformity over complex surfaces. This enabled a reduction in plating thickness and further gold saving (*Puippe and Leaman, 1986*).

Opportunities for gold substitution in PCB electrical contacts have also been explored. Examples include the use of binary Au-Sb and ternary Au-Cu-Cd alloys to reduce the amount of gold required (*Mandich, 2000*). In these experiments, pulsed plating with a cycle represented by **(a)** in **Figure 9-9** was carried out in cyanide solutions. Comparisons were made between pulse plating at $\tau_{cycle} = 2$ ms and 50% duty cycle and DC plating at equal average current densities. The experiments confirmed that coating composition, color and brightness could be controlled by the pulse parameters. Pulse plating decreased the coating porosity so well that it could produce a deposit with the same corrosion resistance as in a DC deposit but with only half the thickness. Pulsing permitted use of a much higher average current density while still maintaining acceptable deposit brightness so that plating could be conducted at a higher electroplating rate.

The use of a simple unipolar waveform (in **Figure 9-9(a)**) has found wide commercial use in the electroplating of tin-lead solder on PCBs where the composition of the alloy and its distribution around holes where the PCB is anchored to the motherboard are both critical. Plant experience demonstrated that this waveform enabled good control of the Pb/Sn ratio in the solder and uniform distribution of the deposit over the edge of the holes. The improved distribution saved enough solder to significantly lower process costs (*Avila, 1986*). *Knödler et al. (1985)* discussed the influence of pulse current, cycle period and pulse duration on solder composition and throwing power.

Another important commercial application of pulse plating has been to form thin magnetic films by deposition of iron-group alloys (i.e., Ni-Fe, Co-Ni, Co-Fe). The magnetic properties of these films and their ageing both depend on alloy composition and grain size. Electroplating of these alloys is complicated by their well-known anomalous behavior whereby the deposition of the less noble component occurs preferentially over that of the more noble metal. Consequently, during DC deposition, a large excess of the more noble metal must be used in the plating bath to yield the desired alloy composition and the electrolyte composition has to be formulated carefully and checked repeatedly during operation. Furthermore, a stabilizing heat treatment is required at the completion of the plating step. Permalloy films containing 20% Fe and 80% Ni have a desirable high magnetic permeability and exhibit unusually high resistance to ageing. Consequently, the production of these alloys has been the subject of many electroplating studies, including those focused on pulsed methods. Some success was achieved using unipolar waveforms with long cycle periods, 40–55 s, and a high cathodic pulse over 15 s followed by a rest period. Good control of nickel-iron alloy composition and the magnetic properties of the permalloy was also achieved by superimposing an AC signal on a DC one. The frequency of the AC current was found to control alloy composition. For example, at 100 Hz, Ni and Fe deposited at the same rate from an electrolyte containing a Ni^{2+}:Fe^{2+} ratio of 4:1, whereas at frequencies above 100 Hz, the Fe deposition rate exceeded that of Ni. Pulse plating appears to be the best method of depositing soft magnetic films made of a nanocrystalline ternary CoNiFe alloy (*Phan et al., 1994; Natter et al., 1996*). This material has a high saturation magnetic flux density and a low coercivity making it attractive for magnetic recording heads. *Nakanishi et al. (2001)* demonstrated that the pulse waveform exerts an influence on the crystalline structure of the ternary alloy and consequently on alloy properties. They employed a cycle time of 100 ms at duty cycles of between 30 and 100%. The pulse current density had the largest effect on the crystalline structure. *Tang (2001)* was able to produce a ternary CoNiFe alloy employing the bipolar cycle (**Figure 9-9(f)**). Deposition of these alloys onto both silicon wafers and copper foils has been demonstrated (*Nakanishi et al.,* 2001).

Electroplating of alloys in place of a single metal for protective purposes was commercialized many years ago. *Eckler et al. (1980)* found that DC deposition of chromium-molybdenum on titanium could provide the wear, erosion and corrosion resistance that the substrate lacked. Later, further improvements in the cladding properties were achieved by using unipolar pulse plating. These experiments showed that surface morphology as well as the Mo content of the deposit was sensitive to pulse duration. **Figure 9-14** illustrates the observations reported by *Avila (1986)* on the effect of t_{on} on alloy deposit morphology obtained at a duty cycle of 50% and constant average current density of 46.5 A/dm^2 in an electrolyte containing a 4:1 weight ratio of chromic acid to ammonium molybdate.

Zinc-nickel alloys provide better mechanical properties and have superior corrosion resistance than do pure zinc coatings. Use of a simple unipolar cycle resulted in a more uniform alloy composition and improved corrosion resistance than that obtained by DC electrodeposition (*Pagotto, Jr. et al., 1999*). Bath temperature and pulse duration were found to be the most

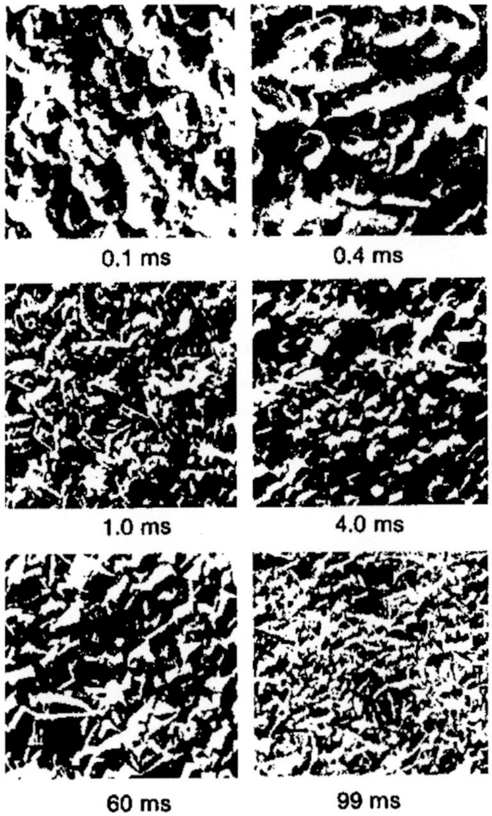

FIGURE 9-14 **Influence of t_{on} on deposit morphology of the electrodeposit for a simple unipolar on-off pulse with 50% duty cycle and constant average current density of 46.5 A/dm^2 (1000x magnification).** *(Figure reproduced from* Avila (1986) *with permission. © 1986, Amer. Electroplater & Surface Finishers Society.)*

influential operating variables. Nickel-chromium alloys have been used for the corrosion protection of iron and steel structural products. For these claddings, the use of unipolar pulse plating with cycles of millisecond duration have been shown to yield smoother deposits than those obtained by DC plating (*Lashmore et al., 1986*). The chromium content in the alloy obtained by plating from a chloride electrolyte at pH 4 was readily controlled by proper adjustment of j_p and the duty cycle. High chromium is usually desirable because it increases the microhardness and corrosion resistance of the coating. Examination of an etched cross-section obtained by pulse plating revealed a regular pattern of varying Cr content resulting from the temporal variation of the electrolyte composition in the pulsing boundary layer (*Lashmore et al., 1986*). On the other hand, an undesirable concentration gradient over the coating thickness was observed in the sample produced by DC plating. Unipolar pulse plating has also been investigated for producing nickel-copper alloys from a citrate electrolyte using a 100 Hz frequency, $j_p = 20$ A/dm^2 and average $j = 2$ A/dm^2 (*Ghosh et al., 2000*). They obtained coherent coatings containing about 36 wt% Ni with microhardness, internal stress and corrosion resistance significantly better than that possible with DC plating. At those pulsing conditions, the deposition had a nanocrystalline grain structure. The degree of bath agitation, electrolyte pH and temperature were found to influence alloy composition and properties in addition to the cycle parameters of j_p, τ_{cycle} and duty cycle.

Pulsed plating can also be performed in molten salt electrolytes. An example is the cladding of a graphite substrate with titanium diboride at 600°C using a K_2TiF_6:KBF_4 electrolyte in a 1:4 mass ratio and a fluoride mixed salt: LiF-NaF-KF (*Ett and Pessine, 1999*). Using simple pulsed cycles, as in **(a)** of **Figure 9-9**, at 5–100 Hz, duty cycles between 75 and 83% and $1.5 \leq j_p/j_m \leq 1.8$, these researchers obtained claddings that were more adherent to the substrate and contained fewer cracks than those produced by DC electrodeposition at comparable operating conditions.

9.2.3 Novel Electroplating Applications

The research literature during the last decade describes several novel applications of pulse plating. One of these is the formation of copper and gold single crystals with diameters in the range of 10^{-9} to 10^{-6} m and very high aspect ratios using ion tracks etched in a nonconducting polymer foil as a template *Dobrev et al. (1999, 2000)*. Bipolar reverse plating cycles were used (**Figure 9-9(g)**) in a potentiostatic mode. Best results were obtained when the electrolyte was agitated ultrasonically. Silver single crystals were also made using a glass capillary tube as a template via DC plating at a low current density.

Electrodeposition using pulsed potentials has been applied to catalyst preparation (*Popov and Pavlovic, 1993*), while several teams have investigated the use of pulsed current electrodeposition to produce catalysts (*Natter and Hempelmann, 1996; Zoval et al., 1996, 1998*). Expensive catalytic materials, such as palladium, must be extremely well distributed on support surfaces and, moreover, should be fixed on the support as fine crystallites to achieve high turnover rates per gram of metal. Cathodic electrodeposition distributes metals over flat surfaces far better than impregnation and subsequent reduction and calcining. However, DC plating fails when substrate surfaces are rough and high current densities are required to produce very fine crystallites result in low throwing power. Much more uniform metal distribution and smaller crystallite size can be achieved by unipolar or bipolar pulsing methods.

Since electrodeposition is not feasible on porous catalyst supports, it has been carried out on essentially non-porous graphite. The process involves first coating fine graphite particles with metal and then agglomerating them into larger sizes that are better suited for use in reactors. Agglomeration of fine particles results in a macroporous structure with relatively high effective diffusivities. For the plating step, *Bradley et al. (2001)* reported the use of pulse reverse plating of palladium onto fine graphite particles using a bipolar cell formed by immobilizing 1–2 μm graphite particles on high purity cellulose sheets that were then stacked between a graphite cathode and anode. Plating was carried out in an organic toluene-acetonitrile solution containing 1 mM $PdCl_2$ solution at potential fields up to 3000 V/cm and frequencies up to 20 kHz with symmetrical cycles. Surface areas as high as 200 m^2/g of metal were obtained at frequencies of 500 Hz and 20 kHz. Dispersion of the graphite particles with ultrasound was necessary to obtain high surface areas. An unmodulated potential gave surface areas just over 100 m^2/g of metal. Deposition occurred on one side of the particles, generally as nodules 20–25 nm in size, but nodules as small as 5 nm were observed.

As with metal catalysts, semiconductor crystals can also be formed on a substrate by electrodeposition. The use of modulated cycles has been found to produce a superior photoconverter than that achieved by DC operation. *Swaminathan et al. (2000)* measured the effect of a duty cycle on the properties of a CdSe thin film formed using a simple unipolar pulse cycle. The films so formed were polycrystalline with cubic and hexagonal structures. A conversion efficiency of light into electricity between 1.75 and 2.4% was found to depend on duty cycle and reflected the ratio of cubic to hexagonal structures in the film. The attractiveness of pulsed electrodeposition stems from the simple control of the crystal structure that it offers. The same research team explored the electrodeposition of other opto-electronic materials. *Murali et al. (2009)* reported on the effect of duty cycle on ZnSe films produced on Ti, graphite or conducting glass substrates by unipolar pulse deposition. The duty cycle was shown to affect the dislocation density, internal strain and grain size of the films so formed.

Periodic operation is also used for composition-modulated metal (CMM) or alloy (CMA) plating. These are multilayered coatings that often exhibit properties quite different from the individual metals or alloys making up the multilayers. Since layer thickness is important in some applications, coatings may be classified as micro or macro. In micro-CMMs, the properties arise from atomic-electronic interactions between the layers. With macro multilayers, it is the physical interaction between layers that brings about the different properties. Many metals can be electrodeposited alone or as alloys. *Leisner et al. (1996)* discussed those that can be introduced into multilayers. The technique now used is a type of periodic operation in which a substrate is immersed back and forth between two plating baths of different composition to deposit the separate layers. Double bath systems are hardly new. They have been used with Cu or Ni to improve the adhesion of gold or silver plated on iron or steel surfaces.

Another approach is to deposit the separate layers from a single bath containing the precursor compounds for each layer by using potentiostatic or galvanostatic pulse plating. The method is used to produce CMAs having nanometer layer thickness by employing millisecond cycle periods. Generally, one layer is made up of a relatively noble metal (e.g., Cu) that can be preferentially deposited at low currents (more positive potentials), while the other layer consists of an alloy rich in a baser metal (e.g., Ni) that can form at higher currents (more negative potentials). For example, when Ni and Cu are deposited onto a copper foil from a sulfate electrolyte containing Ni(II):Cu(II) at a mole ratio of about 30:1, pure Cu is deposited at low cathodic current density, while nearly pure Ni is deposited at high current density. Different types of behavior are possible. For example, when the more noble component is deposited at the beginning of a pulse, but its ionic concentration in the electrolyte is much lower than that of the other metal, it can become depleted in the boundary layer later in the pulse so that the less noble metal can begin to deposit. Another type of behavior termed displacement can occur during unipolar pulse plating. If the cathodic pulse current during the on-time is large, an alloy rich in the less noble metal can be produced. When the current is switched off, the more noble metal in solution can displace the less noble component in the alloy. Since this reaction occurs topochemically, the displacement progresses as a sharp front moving inward from the surface exposed to the electrolyte and results in one layer composed of the alloy rich in the less noble metal adjacent to a layer consisting essentially of only the more noble metal. The thickness of each layer can be independently controlled by the appropriate adjustment of t_{on} and t_{off}. A multi-layered structure of this type can be built up by applying many such pulse cycles. For the Ni-Cu system described above, Ni is deposited at high current density and replaced by Cu when the current is switched off. **Figure 9-15** shows examples of current switching waveforms used for CMA formation. Layer sharpness can be improved by varying the agitation of the electrolyte bath in phase with current changes.

Past research has focused on the effect of the pulse cycle on layer thickness and whether or not a CMA or a CMM is formed. Multilayer properties appear to differ depending on whether the structure is formed by potential or current pulsing. Agitation of the electrolyte bath as well as the pulse cycle parameters affect the magnetic properties of the Ni-Cu cladding and the surface roughness. *Roy et al. (2000)* and *Cavallotti et al. (2000)* have explored the use of a three-part pulsed cycle, such as given by a variation on **Figure 9-9(b)** where there is a zero current interval after the high i_p pulse, for the CMA deposition of Cu and Co-Ni-Cu alloy layers on various substrates from a sulfamate-Rochelle salt electrolyte containing the metal ions. *Cavallotti et al. (2000)* observed that the deposition of the copper layer is diffusion controlled whereas the formation of the Co-Ni-Cu alloy layer is kinetically controlled via adsorbed intermediates.

Corrosion protection afforded by zinc-containing CMAs has been examined over several years by Wilcox and co-workers (*Kalantary et al., 1995; Liao et al., 1998; Jensen et al., 1998; Nitipanyawong et al., 2000*). They found that corrosion of the cladding was driven by H_2 evolution and that the adsorption of H_{ads}

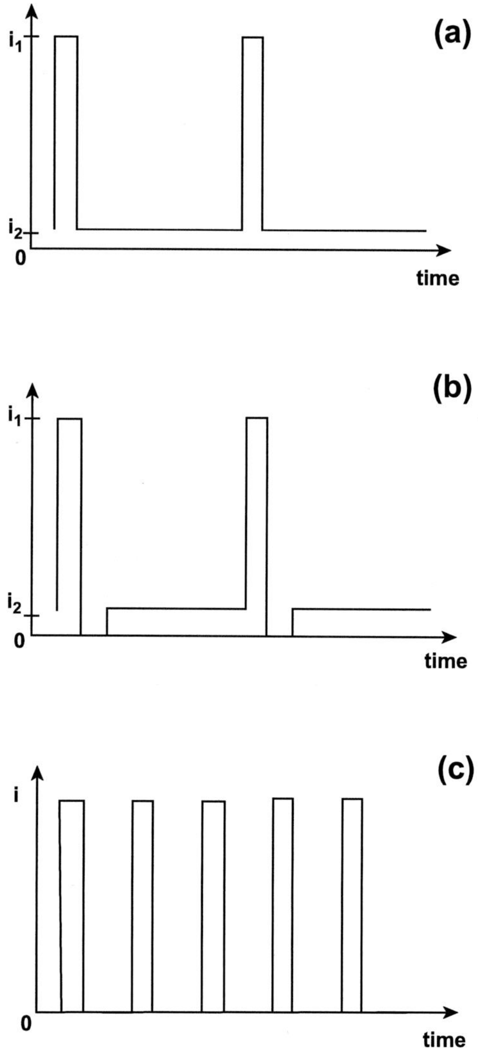

FIGURE 9-15 **Unipolar pulse plating waveforms used for CMA formation from a single electrolyte bath: (a) pulse-base, (b) interrupted pulse base, (c) successive pulses.** *(Figure adapted from* Leisner et al. (1996) *with permission of the publisher. © 1996, Elsevier Science S.A.)*

played a key role. This was overcome by the application of bipolar reverse pulse cycles that were able to remove the adsorbed hydrogen by oxidizing it to H^+. The application of such pulse cycles produced strong claddings with excellent corrosion protection.

9.3 ELECTROFORMING

The purpose of electroforming is to produce a complete small and high-precision part or component by electrodeposition. This is normally achieved in one of two ways. In the first approach, the cathode is a mandrel or mold with the same shape of the part to be formed and certain regions masked or insulated to prevent metal deposition. This process strongly resembles electroplating but differs mainly in the geometry of the deposit to be formed. In the second approach, an inert anode with the same shape of the part to be formed serves as a die for the cathode. Metal plating is carried out using a very small cathode-anode spacing so that the deposition rate varies inversely with the separation distance. This enables the geometry of the anode to be transferred to the deposit formed at the cathode. Hardness and/or ductility of the metal are important as well as surface properties such as smoothness, brightness or uniform thickness. All of these properties can be achieved through the appropriate choice of a pulse waveform (uni- or bipolar, with or without rest intervals) and pulse parameters. The principles governing electroplating apply equally to electroforming. Several examples of the preparation of metal forms and electroforming of CDs and DVDs, optical gratings and inserts for micro-fluidic devices are described by *Tang et al. (2004). Fauser and Zielonka (2004)* described electroforming experiments for preparing tiny medical inserts. The sequence for forming small mass-molded parts began with shaping the part from a soft material, such as a polymer, using masks and precision cutting tools (X-rays, lasers). A mold was prepared from the part, rendered conductive and used as a cathode. From such a mold, a negative gold alloy mandrel was made by electrodeposition using unipolar, low frequency pulses with rest periods. Gold height in the mold was 150 μm with a smooth surface and sharp edges. Electroforming by dissolution rather than deposition is also possible. *Dambrowsky et al. (2005)* described creating a mandrel for mass producing small, precise plastic parts by electro-machining a gold alloy plate using unipolar, pulse potentials with rest periods. Current densities were 0.2—0.4 A/dm^2.

9.4 ANODIZATION

Anodization is a widely used electrolytic process to form an oxide or oxyhydroxide film on metals such as aluminum and titanium to provide corrosion protection. Periodic operation based on AC or square-wave cycles is used to carry out this process. If the oxide film is non-conductive, the electrical resistance at the electrode-electrolyte interface increases and thereby slows and eventually stops the anodic reaction. If the oxide film formed is porous, oxide formation during anodic polarization slows down continually and eventually ceases altogether when the pores in the film are filled by the oxide and electrical contact between the metal and the electrolyte is lost. Disruption of electrical contact is the mechanism by which these oxide films provide corrosion resistance for the substrates when they are in use. Although, in some cases, the oxide films are not passivating but still protect the substrate because they promote anodic processes such as the oxidation of water to oxygen or chloride to chlorine rather than metal oxidation.

Aluminum is perhaps the most commonly anodized metal. The process provides desirable surface properties such as corrosion and wear resistance or electrical insulation. The oxide film can be colored, which is an important property in architectural applications. Two quite different types of oxide film can be laid down on aluminum depending on the electrolyte, pH and, to a lesser extent, the applied potential and current density. Most anodization operations form a thick porous film composed of an amorphous oxide penetrated by bundles of moderately uniform parallel pores that begin at the surface of the film and end at a thin barrier film on the aluminum surface, as shown below in the idealized structure in **Figure 9-16** (*Konno, 1986*). Many properties of anodized films depend on their pore structure and on the process by which they are formed.

The porous film grows inward at the dense, amorphous barrier film next to the metal surface. The barrier film thickness remains largely constant during anodization due to a balance between the oxidation of aluminum and the electric field-assisted dissolution of the oxide at the base of the pores. This dissolution feeds the growth of the porous film, as illustrated by the dashed lines in **Figure 9-16**. The pore walls tend to dissolve slowly during

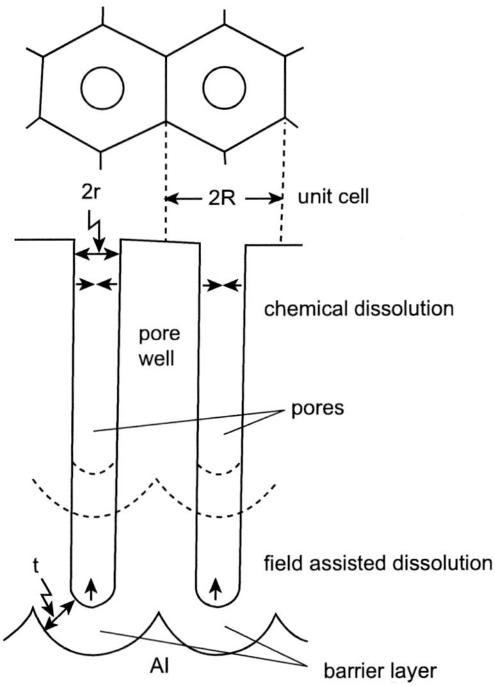

FIGURE 9-16 Structure of porous Al_2O_3 films formed on aluminum during anodizing using an acidic electrolyte. *(Figure reproduced from* Konno (1986) *with permission. © 1986, American Electroplaters and Surface Finishers Society.)*

anodization so that the pores remain narrow. Since the pores grow inward, the "oldest" walls are located at the film-electrolyte interface. Because of this growth pattern, the outer surface of the film exposed to the environment weakens with extended exposure and crumbles when abraded. During anodization, the surface is heated significantly due to the high voltages applied and the low electrical conductivity of the film. Since elevated temperature is detrimental to film hardness and some other properties, anodization is often carried out in refrigerated baths.

The thickness ε of the barrier film, pore diameter r and the dimension 2R of a unit cell in the film structure are largely controlled by the anode potential and tend to increase as the potential rises. The growth rate of the porous film also rises as the potential increases due to the concomitant increase in the current density. These effects lead to a decrease in the number of pores and the porosity in the resulting film as the potential becomes more positive. Electrolyte composition is an important factor as well, but temperature and electrolyte concentration are not. The factors controlling film structure are succinctly discussed by *Rasmussen (2001)* who also provides an introduction to pulse processes.

Pulse methods have largely replaced the older DC and AC anodization methods in the last decades. As in cathodic processes, the use of pulsed cycles with short, high potential anodic pulses and longer intervals of low anodic potential or open-circuit potential can maintain film uniformity, produce rapid growth with a suitable pore structure, while avoiding overheating. The success of this approach arises because more variables are available to control the various phenomena that occur during anodization. The use of pulse cycle **(b)** in **Figure 9-9**, with periods of several seconds up to several minutes was described by *Yokoyama et al. (1982)* and has been widely adopted by industry. The superposition of an AC signal on a low DC potential can also be used and is preferred to AC alone which generally leads to coarse but flexible coatings. The combination of AC and DC inputs has been shown to enable control of the film color through manipulation of the ratio of the AC to DC current amplitudes (*Konno, 1986*).

The use of a mixed sulfuric-oxalic acid electrolyte has become common; electrolyte composition and concentration, duration of the high and low anodic current or potential pulses affect film color, thickness and hardness. An advantage of pulse anodization is that it makes refrigeration of the anodizing bath unnecessary. The inclusion in the pulse cycle of an interval when little or no anodic current flows enables heat accumulated during the highly anodic portion of the pulse to dissipate as well as barrier film thickness across the surface to even out (*Yokoyama et al., 1982*). *Rasmussen (2000)* discussed the use of a pulse cycle with frequencies an order of magnitude greater than usually applied. He demonstrated that film hardness depends on t_{on}, t_{off}, j_b and j_p.

9.5 ELECTROCHEMICAL MACHINING AND POLISHING

Several machining or metal finishing operations are carried out on the anode rather than the cathode. Electrochemical machining or electromachining is usually carried out in a strong acid bath by anodic dissolution of the metal stock when precise and limited removal of material is required. With this method, parts are not mechanically stressed nor do they have defects or burrs imparted by mechanical machining. Electromachining operates in reverse to electroforming in that a small cathode serves as the machining tool or mandrel and is held at a narrow distance from the anode or working stock. During machining, the cathode moves while the anode is fixed in place and the current is confined to the region between the tool and the working stock. Since the local current varies inversely with the anode-cathode separation distance, the shape of the cathode mandrel is transferred to the anode. The choice of electrolyte depends on the material undergoing machining. For example, simple salt solutions may be used for iron and steel.

Electrochemical machining, however, has problems: Excessive heat generation, metal deposition on the cathode during operation that alters the mandrel shape and a rise in local metal ion concentration at the anode due to dissolution that can harm machining accuracy and precision. Working stock-tool gaps typically vary from 0.02 to 0.7 mm, while very high current densities ranging between 3000 and 20,000 A/dm^2 are applied at the tool surface. It is important to continually remove the dissolved metal and heat from the electrolyte between the anode and mandrel. In DC operation, fresh solution must be circulated through the gap at high flow rates. This difficulty can be overcome by the use of a simple unipolar cycle (*Datta and Landolt, 1981, 1982*) that permits dissolved metal and heat removal from the gap during the interval in each cycle when little or no current flows. A large flow rate is no longer required. A further advantage addresses metal stocks that are passivated during anodic polarization under DC conditions. In some cases, a high enough pulse current density can be applied during a unipolar cycle to move the potential beyond the passive zone into the transpassive region where the metal becomes re-activated and once again can be anodically dissolved.

Electropolishing is similar to electromachining, but is aimed at the removal of surface irregularities from castings or burrs from mechanically machined parts. Pulsed cycles are not used because surface faults such as burrs are typically large enough to protrude through the boundary layer and so can be selectively removed by DC current. Electrochemical etching, in contrast to polishing, roughens the surface or forms macroscopic patterns. Etching is normally carried out in neutral or slightly acidic electrolytes containing ionic salts such as NaCl or NH_4Cl and oxidizing agents. Although pulsing is not commonly used, etching can be carried out using AC or AC superimposed on DC.

9.6 ELECTROWINNING AND ELECTROREFINING

Two other processes that involve electrochemical deposition of a metal at a cathode are electrowinning and electrorefining; both are standard operations in extractive metallurgy. Although each is almost always used as a final step in the extraction of a metal, the objective of electrowinning is to recover dissolved metal from a solution, whereas electrorefining purifies an impure metal which is cast as the anode and dissolves during electrolysis. Both processes are carried out as continuous operations. Thus, the source of metal during electrowinning is the solution fed into the cell, whereas the metal source during electrorefining is the anode. Pulsed operation is not often used in either case, although an example of its potential has been demonstrated for the electrorefining of copper produced by the pyrometallurgical processing of sulphide ores. The smelting of these ores leads to the production of what is termed "blister" copper, which is approximately 99% pure, but contains metal impurities Fe, As, Zn, Pb, Ni, Co, Ag and Au depending on the ore mineralogy. Blister copper is cast into rectangular anodes for refining in electrolytic cells to purify the copper and to recover valuable metal impurities. During electrolysis, metals that are more basic than copper dissolve, whereas the more noble metals (e.g., Ag, Au) are not oxidized but slough off the anode as tiny particles and form what is called a "slime" as the electrode is consumed. During DC electrorefining, a problem that arises is the uneven consumption of the anode particularly along its edges, leading to deterioration and premature replacement. Use of a bipolar reverse waveform consisting of a short high current cathodic pulse followed by a long low current anodic interval has been shown to be effective at preventing early anode failure and extending its useful life.

9.7 GALVANIC CELLS

Although a rechargeable battery, a familiar galvanic cell, operates by switching between discharge and charge, such a mode is not periodic in the sense discussed in this chapter. Charging is carried out at a constant, fixed potential that is repeated from cycle to cycle, but the discharge, when the battery delivers electrical energy, changes between use due to the variable nature and/or requirements of the load. Indeed, batteries lose capacity over repeated charge/discharge cycles.

Fuel cells are also galvanic cells. During their operation, their efficiency is limited by the overpotential, particularly at the cathode, which decreases the voltage that can be delivered to the external load. In the case of a proton exchange membrane (PEM) fuel cell, the reduction of oxygen to water (i.e., **Eq. (9-5)**) is the cathodic reaction. Although periodic operation of a PEM fuel cell has not been considered as an alternative to the usual mode in which inlet and operating conditions are maintained fixed, some improvements may be possible by modulation of the operating pressure, temperature or inlet gas composition. For example, a periodic pulse of pure oxygen on the cathode side or modulation between methane and an alcohol (in case of a methanol fuel cell) on the anode side could increase the time-average energy efficiency. Large reductions of the overpotentials are probably unlikely, but exploratory experiments may be worthwhile.

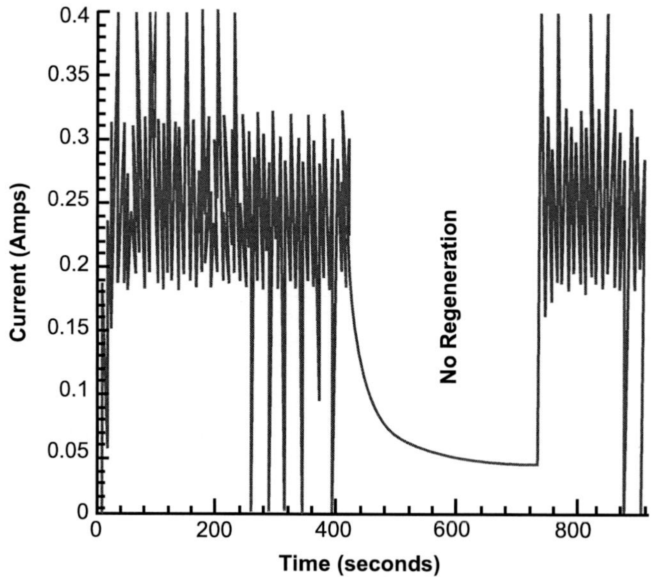

FIGURE 9-17 **Effect on the current by imposing a single 1.4 V, 300 ms pulse at 9 s intervals on a PEM methanol fuel cell operating at 0.55 V cell voltage. The pulsing was discontinued between 420 and 760 s and then re-imposed. The cell operated on 1.5 M MeOH using a Pt anode.** *(Figure reproduced from* Adams and Gardner (2005) *with permission. © 2005 The Electrochemical Society.)*

The PEM fuel cell utilizing H_2 as fuel appears to be the design of choice for small loads such as vehicles. It is compact, operates at low temperature and achieves a high power density. However, it has a number of operating problems associated with water management and temperature distribution in the polymer membrane, removal of impurities trapped in the cell and ageing of both membrane and electrodes at the current densities used. Periodic load (and thus current density) variation and flow direction switching in the fuel and air manifolds may mitigate these problems. *Adams and Gardner (2005)* imposed a periodic voltage pulse on a methanol PEM fuel cell to disperse ion buildup on the anode. An almost immediate and remarkable increase in cell current upon the initiation of the cell voltage pulses is clear in **Figure 9-17**.

9.8 ELECTROLYTIC PRODUCTION OF CHEMICALS

A wide variety of chemical reactions can be carried out in an electrolytic cell. The energy range needed to activate a reacting molecule ranges from just over 0 to 3.5 electron volts in electrolytic cells and is just a fraction of the energy required when thermal or photon sources are used. Furthermore, the potential applied over the cell strongly influences the reaction(s) that takes place. A widely cited example is the anodic reduction of nitrobenzene. Phenylhydroxylamine will be the primary product at lower overpotentials, whereas aniline is generated at higher overpotentials (*Fleischmann and Pletcher, 1969*). Radicals, carbonium ions or reactive adsorbates can be formed from alkyl halides, carboxylic acids or amines. Even hydrocarbons are anodically oxidized depending on the electrode, solvent and operating conditions. At the cathode, carbanions can be formed from alkyl halides, aromatic hydrocarbons, quinones or nitro-compounds. Consequently, a wide variety of products can be made. On the other hand, rates tend to be low because of activation limitations. Periodic operation can be used to reduce the polarization and together with the choice of electrode enhance selectivity and achieve higher yields. As an example, *Fleischmann et al. (1965a, b)* showed that periodic switching of the polarity of the two electrodes enabled the conversion of nitrobenzene into a range of nitrated aromatics depending on E, j_p and t_{on}.

Kolbe electrolysis is an anodic reaction in which a carboxylic acid adsorbs on the electrode and is either oxidized to an adsorbed intermediate or desorbs as a radical. The adsorbed intermediate or radical formed at the anode can then dimerize to yield stable compounds. Electrolysis is usually carried out on Pt anodes in a non-aqueous solvent such as methanol or acetic acid. If acetic acid or an acetate salt is used, ethane is a major product. Most salts, if present in the electrolyte, inhibit dimerization. DC electrolysis of carboxylic acids has been used to form symmetrical compounds (e.g., alkanes) or unsymmetrical compounds such as pheromones and fatty acids if two or more carboxylic acids are used (*Fleischmann et al., 1965a*). Yields of unsymmetrical compounds are poor, however, due to the preferential formation of symmetrical compounds. The reaction has been widely studied because the reaction product changes with time during DC electrolysis. Experiments using pulsed current have been carried out by a number of researchers (*Fleischmann et al., 1965b; Fleischmann and Goodridge, 1968; Feldhues and Schäfer, 1985*). The choice of the pulse frequency in a unipolar cycle is important. It must be high enough to generate sufficient radicals in the pulsating boundary layer, but the current efficiency drops if it becomes too high. For example, with decanoic acid, the current efficiency for oxidation of decanoic acid to octadecane is 60% at 0.05 Hz but drops to 27% at 5 Hz.

Fleischmann et al. (1965b) and *Fleischmann and Goodridge (1968)* used unipolar pulsing to determine the kinetics of Kolbe electrolysis in acetic acid. They observed that the initial reaction over a smooth Pt electrode involved water decomposition forming oxygen. A Pt-oxide layer quickly formed on the electrode and passivated the surface, halting production. *Feldhues and Schäfer (1985)* attempted to produce unsymmetrical products using a reverse pulse cycle with diacyl peroxide. The rationale behind this proposal was that during the cathodic pulse, the peroxide would be cleaved by reduction into two different carboxylates. Before the carboxylates diffuse from the pulsating boundary layer, they should be decarboxylated by the anodic pulse, thereby permitting coupling to occur. Experiments with octanoyl dodecanoyl peroxide dissolved in methanol in the presence of small amounts of sodium octanoate and octanoic acid favored the mixed dimer octadecane as expected, but the yields were low.

Lower molecular weight carboxylic acids in the C_1 to C_4 range are capable of complexing radioactive metal cations and have been proposed as reagents for reprocessing spent nuclear fuel. Once these complexes have formed, they can be electrochemically decomposed into CO_2 and water to recover the metals. With DC electrolysis, low current density must be applied to avoid inhibition due to oxidation and passivation of the electrode surface. An alternative is to use a pulse cycle. Past research has focused on studies with formic acid (*Beden et al. 1992*). *Le Naour et al. (1999)* carried out pulsed current decomposition experiments on mesoxalic acid which is hydrated to form dihydroxymalonic acid ($C_3H_4O_6$) when present in aqueous solutions. These investigators carried out the decomposition in nitric acid using a polycrystalline Pt electrode in a single-compartment cell. Under DC conditions, formation of an oxide layer on the electrode quickly inhibited the reaction. When a pulse reverse current or potential cycle was used, the platinum oxide film was at least partially removed during the cathodic pulse, which allowed the decomposition of dihydroxymalonic acid to take place during the anodic pulse; *Le Naour et al.* observed quantitative conversion of the acid to carbon dioxide and water. The order of the oxidation reaction on the anode was zero suggesting that the rate-limiting step is the oxidation of the adsorbed acid.

Although little prospect exists that ferrate (Na_2FeO_4) (Fe in VI oxidation state) will become an important industrial chemical, its preparation has been a scientific challenge for about 150 years. Its formation should occur by anodic oxidation of an iron electrode in a strongly basic ($pH = 14$) electrolyte. Cathodic pretreatment of the electrode is necessary to remove oxides. Even with such pretreatment, the current efficiency remained low. The problem is the formation of a passivating oxide layer on the iron surface. Almost a hundred years ago, *Grube and Gmelin (1920)* demonstrated that the current efficiency can be more than doubled by superimposing an AC current on DC with the AC amplitude about equal to the DC current. *Bouzek and Roušar (1993)* investigated the effect of AC frequency and amplitude on ferrate formation and attempted to find the reason for the current efficiency improvement. A 2-compartment electrolysis cell with a PVC membrane and separate anolyte and catholyte feed circuits was used. These authors confirmed the large increase in ferrate current efficiency when AC current was superimposed on a DC current. They found that the optimum AC current amplitude was 2.6 times the DC current and the optimum frequency was 5 Hz. An important aspect of their AC waveform was that the current was cathodic at the minimum in the input signal. *Bouzek and Roušar (1996)* suggested that the current efficiency is strongly affected by reduction during the cathodic pulses. The improvement in current efficiency was attributed to several possible reactions during that pulse. H_2 evolution in the interval helped break up the passive oxide layer formed in the anodic portion of the cycle. Secondly, direct reduction of the oxide layer could also occur. However, it is also possible that O_2 evolution, during the anodic portion, breaks up this layer. The dependence of the current efficiency on the frequency was linked to the growth of a passive layer on the metal and the reduction of Fe(VI) to Fe(III) that is catalyzed by fine grain carbon in the metal (*Bouzek and Roušar, 1996*).

9.9 APPLICABILITY OF PRINCIPLES OR PRACTICES TO NON-ELECTROCHEMICAL REACTIONS

The striking success of periodic operation for electrochemical processes described in this chapter can only invoke optimism with regard to the potential for the application of periodic operations to non-electrochemical reaction systems. Whether some of the concepts or practices used in electrochemical applications are transferrable to non-electrochemical reaction systems is doubtful. Periodic operation variables in electrochemical reaction systems are commonly applied current or potential. Current density and potential operate on the mass transfer steps in the electrolyte film adjacent to the electrode, charge transfer steps and diffusional transport on the electrode surface. They directly affect the rates of deposition, dissolution, crystal growth, crystal nucleation and non-electrochemical surface reactions if any occur. The electrode potential acts as a driving force for these processes and usually controls selectivity in electrochemical reaction systems.

Characteristics of the commonly employed pulse cycles are frequencies in the range of 1 to 10 kHz, use of interruption or a rest interval and current polarity reversal. Cycles are primarily based on square waves. Furthermore, many cycles in use such as waveforms **(c)**, **(d)**, **(h)** and **(i)** in **Figure 9-9** can be complex. A further characteristic is the relatively small amount of damping of the effects associated with manipulation of the potential or current. An explanation for the low level of damping is that potential and current density operate directly on the reaction surface at a microscopic level and so can be considered to act instantaneously. In contrast, changes in reactant concentration or temperature that are typical of the periodic operation of heterogeneous non-electrochemical reactions act simultaneously on both the surface and surrounding fluid. Although concentration and temperature changes are capable, in principle, of directly influencing molecular processes if they occur on timescales equivalent to the time constants of these processes, their effects are mediated by mass and heat transfer processes which are relatively slow in chemical systems. Thus, regardless of how rapidly the temperature or concentration are varied, their effects will not be transferred to the surface reactions at frequencies above a kHz. Thus, if changes on these time scales are required to directly influence molecular processes, they cannot be provided by manipulation of operating temperature or composition.

The successful application of periodic operation to electrochemical systems suggests that frequencies must be high enough that the manipulated variables can interact with molecular processes. They further suggest that complex cycling patterns can be beneficial and that simultaneous variation of process variables can be fruitful, although the electrochemical evidence for this is limited to a few examples in which stirring combined with current density modulation improved plating performance.

As for the application of periodic operation to non-electrochemical systems, only periodic reversal of flow direction has had significant commercial adoption. This operation, which captures the heat of reaction, employs cycle frequencies of 10^{-3} to 10^{-4} Hz. Composition cycling, which has been rather thoroughly investigated (*Silveston, 1998*), operates at frequencies of 0.01 to 1 Hz. Higher frequencies are damped out by mass transfer effects in most flow reactors. Some large effects have been seen, namely CO oxidation on noble metals. These seem to be explained by relatively slow adsorption/desorption processes which are probably several orders of magnitude slower than other surface reactions. Temperature cycling has not been thoroughly explored because the large thermal capacity of reactors limits the highest possible frequencies without damping to 0.005 to 0.01 Hz. Pressure cycling in gas-solid systems has been studied and effects on mass transfer controlled systems demonstrated. Besides the use of more complicated cycling patterns and simultaneous manipulation of more than one variable, most of the periodic practice in electrochemical systems cannot be imported into the operation of conventional gas or liquid phase reactors. The best prospects are for less conventional operating variables such as ultrasound or electromagnetic waves (e.g., laser or UV irradiation, magnetic fields or molecular beams) that can be manipulated at frequencies in the kHz range. Commercial applications will probably require a totally new reactor design.

A modest transfer of periodic technology may also be possible in the reverse direction. Variables other than current or potential have not been manipulated in electrochemical systems to much extent. Some benefit might be derived from the application of composition, pressure, temperature, and flow direction cycling now in use or under study in non-electrochemical systems.

Nomenclature

A	= area (m^2)
	= empirical parameter
AC	= alternating current
a_i	= activity of species or ion i (kmol/m^3)
aq	= aqueous
C_I	= concentration of species or ion i (kmol/m^3)
C_{dl}	= double layer capacity
CMA	= composition modulated alloy
CMM	= composition modulated metal
D_{eff}	= effective diffusivity (contributions of diffusion + migration)
DC	= direct current
E	= potential, half cell potential (V)
E_{rev}	= reversible half cell potential
E_0	= standard half cell potential at unit activity of oxidized and reduced species (V)
e^-	= electron
emf	= electromotive force
F	= Faraday constant
f	= pulse frequency
g	= gas or gas phase
h	= height of surface protruberence
I	= current
I_{total}	= sum of charging and faradaic currents
IHP	= inner Helmholtz plane
i	= current density, current (A/m^2, A)
i_d	= diffusion current density (A/m^2)
i_{dc}	= charging current density
$i_{faraday}$	= faradaic current density
i_j	= current density associated with ion j (A/m^2)
i_l	= limiting current density (A/m^2)
i_m	= migration current density (A/m^2)
J_i	= flux or ion flux of ion I (kmol/m^2·s)
$\bar{j}$	= mean current density
j	= current density, reaction marker
j_b	= base or sometimes steady state current density (A/m^2)
j_l	= limiting current density
j_p	= pulse current density
L	= critical or characteristic dimension (cm)
l	= liquid
M	= metal
MO	= molecular orbital
N	= pore density in aluminum anodization
N_{Wa}	= Wagner number
n	= number of charges per ion
OHP	= outer Helmholtz plane
P	= pressure (kPa)
PCB	= printed circuit board
PEM	= proton exchange membrane
PTH	= plated-through-hole
Q	= charge or quantity of electricity (coulombs)
q	= charge C
R	= universal gas constant (various units)
	= half dimension of a unit cell in aluminum anodization (Å)
	= surface roughness (μm)
r	= pore radius (Å)
SEM	= scanning electron microscope
s	= cycle split, duty cycle, usually t_{on}/τ
T	= temperature (°C, K)

t = time (s)
t_{on} = duration of current or voltage pulse (ms)
t_{off} = duration of base or zero current in a pulsing cycle (ms)
V = voltage (V)
x = position perpendicular to surface or interface (Å)

Greek

δ = boundary layer thickness (Å)
δ_p = pulsing boundary layer thickness
δ_s = stationary boundary layer thickness
ε_i = current efficiency for the ith reaction
ε = thickness of the barrier layer in aluminum anodization (μm)
ε_p = potential efficiency
η = over potential (V)
η_j = over potential for the jth reaction
η_{conc} = concentration over potential
η_s = surface over potential
φ = potential (V)
κ = electrolyte conductivity
σ = layer thickness (boundary or ionic) (Å)
τ = period, cycle duration
τ_{cycle} = period
ν_I = stoichiometric coefficient of species or ion i

Subscripts

A,B, C = species markers
a, (a) = anode, anodic
ads = adsorbed
b = base or sometimes steady state
c, (c) = cathode, cathodic
ct = charge transfer
d = diffusion
dc = direct current
dl = double layer
eff = effective
i = chemical species or ion i
j = ion j
m = migration
mt = mass transfer
O = oxidized state
on(a) = on time for the anode
on(c) = on time for the cathode
p = pulse, pulsating
p(a) = anodic pulse (in pulse reversal)
p(c) = cathodic pulse (in pulse reversal)
R = reduced state
rev = reversible
rxn = reaction
s = stationary
0 = initial, non-pulsating, usually steady state condition

Superscripts

i = inner
M = metal or electrode
z = charge
υ_I = stoichiometric coefficient
0 = standard
***** = equilibrium
′ = electrode region

CHAPTER

10

Modulation of Biological Processes

Jeno M. Scharer
Waterloo, Ontario, Canada

OUTLINE

10.1 Introduction 273
10.2 Theoretical Considerations 275
10.2.1 Application of the π Criterion 277
10.2.2 Application of Quasi-steady-state Analysis 280
10.3 Substrate and Flow Rate Modulation 280
10.4 Dissolved Oxygen Modulation 283
10.5 Culture Medium Tuning 287
10.6 Survival in Mixed Cultures 289
10.7 Stabilization of Recombinant Cell Cultures 292
10.8 Applications to Immobilized Cells or Enzymes 298
10.9 Fed-Batch Operations 303
10.10 Overview 312

10.1 INTRODUCTION

Bioreactors are vessels of many different types that employ living organisms to convert a reactant or substrate into one or more products. A wide variety of organisms are used in such vessels: bacteria, fungi, algae, plant or animal cells. The cells are either suspended or attached to inert surfaces. Chemical transformations undertaken by these organisms occur primarily through enzymes within the organisms or extracted from them as free enzymes. In the latter case, these enzymes are complex proteins and are no longer living entities. Because of their small size, enzymes and some microorganisms can be entrapped in a porous solid and thereby recovered and reused. Bioreactors are often vessels containing a slurry stirred mechanically by rising bubbles or by a flowing liquid. Conversely, they may be packed or structured beds of entrapped micro-organisms.

Cell growth, by nature a periodic event, is encountered in bioreactors. Cells, particularly eukaryotic cells exhibit periodic physiological phases between successive divisions. These events are collectively referred to as the cell cycle. A cell cycle of typical mammalian cells comprises four phases. These are the G_1 phase, the S (synthesis) phase, the G_2 phase and the M (mitosis) phase. The G_1 and G_2 phases, known as interphases, are when key enzymes are synthesized. Cells may enter also a reversible resting phase, the G_0 phase, with temporary stoppage of enzyme synthesis. In recombinant protein production by animal cell culture, one of the important characteristics determining specific productivity is the relationship of protein expression to the phases of the cell cycle. Expression patterns of various secreted recombinant proteins from CHO host cells have been evaluated from measurements, including external protein, internal protein, and messenger RNA of the protein, or inferred from plasmid construct (*Chevalot et al., 1994; Croughan and Wang, 1990; Fann et al., 2000; Gu et al., 1993; Hayter et al., 1993*). However, synthesis of a complete protein is regulated at many stages. Even after accurate translation of the relevant mRNA, many proteins produced in animal cells undergo a number of post-translational chemical modifications and processing steps before the mature protein appears (*Cartwright, 1994; Butler, 2007*). These processing steps, including secretory mechanisms, may be limiting in cells that produce high levels of heterologous mRNA (*Freedman et al., 1995*). In addition to being the rate-limiting step, secretion may itself be a cell cycle dependent process (*Gu et al., 1993*).

Non-linear behavior of the reacting system is necessary if reactor periodic operation (PO) is to improve yield of or selectivity to a desired product (*Silveston, 1998*). Even though bioreactors operate almost isothermally, they are attractive candidates for PO because the kinetics of biochemical reactions are non-linear. Enzyme mediated reactions often follow Michaelis-Menten kinetics:

$$r = kC_{enzyme}\frac{C_{substrate}}{K + C_{substrate}}. \qquad (10\text{-}1)$$

Modeling a bioreactor system by inserting this kinetic expression into a cell or substrate material balance results in a non-linear partial or ordinary differential equation.

Furthermore, living organisms exhibit rhythms or life cycles consisting of stages such as growth, maturity and wasting. The rates of substrate consumption and/or the generation of cell matter are different in each stage so that conditions for an optimal growth rate continually change. Existence of these rhythms insinuates that some pattern of periodic change is desirable. Indeed many biochemical reactions are conducted in batch reactors as an inherently periodic operation.

It is not surprising that theoretical and experimental studies of periodically operated bioreactors quickly followed work on chemical systems. In the 1970s, draw and fill operation of batch, activated sludge, waste treatment systems were first analyzed (*Irvine and Davis, 1971*), although this PO had been practiced previously. Staged continuous activated sludge systems were introduced earlier and it was soon realized that periodic shifts in aeration accomplished a similar result (*Chudoba et al., 1973*). At the same time, several investigators (*Zines and Rogers, 1971; Borzani et al., 1976; Welles and Blanch, 1976; Vairo et al., 1977*) began exploring flow modulation to improve chemostat performance.

More recently, the sequencing batch reactor (SBR) has attracted interest (*Irvine et al., 1997; Wilderer et al., 2001*). This system has characteristics of both batch and continuous operation. There is a sequence of batch steps, filling, reaction, settling, draw-down and idle, that are endlessly repeated. Simulation and experimental work predict higher volumetric efficiency than for an activated sludge CSTR operating under identical feed conditions. Explanation for the enhanced efficiency appears to be a reduction in the toxic or inhibiting intermediate products in the aqueous phase, specifically when denitrification is necessary.

There is a considerable literature on the PO of bioreactors for the production of chemicals or destruction of wastes. *Irvine et al. (1997)* have reviewed the earlier literature, but that on chemicals production has not been tackled. It is this that will be the focus of this chapter. Much is being written on fed-batch operations, a periodic reactor feeding operation similar to batch sequencing in waste treatment.

10.2 THEORETICAL CONSIDERATIONS

Any consideration of reactor PO poses questions: 1) Does PO offer an improvement over steady-state (SS)? 2) Is the limit cycle induced by PO asymptotically stable? 3) In what range of parameter and variable space does improvement appear? These questions can be answered by employing mathematical models of biological systems under steady-state and periodic operation. Solving these models gives their time behavior at a stationary cycling state from which yields, conversions, or selectivities, may be calculated. Comparison of PO and SS results answers the first and third performance questions posed above. Stability can be studied by introducing a perturbation into the input for the transient model and following the system with time. Such a simulation consumes much time and computing resources. However, quicker and less-resource-demanding methods have been developed. For the question of superiority and identification of spaces of interest, these methods make use of the maximum principle based on the Hamilton-Jacobi approach or a second variational approach, applied as the π criterion. Their use depends on frequencies and magnitude of the input change. At the extremes of the frequency range, quite simple methods, such as quasi-steady-state and relaxed steady-state analysis, can be used. For stability determinations, methods employing the Jacobian matrix of the model equations may be utilized. These methods are reviewed by *Bailey (1974)* and *Silveston (1998)*.

The generalized π criterion (*Sterman and Ydstie, 1990a, b*) employs a performance index, J, that is optimized for a process represented by $\frac{dx}{dt} = f(x,u)$ where x is a vector of state variables and u is a vector of control variables. A vector of products, $y = h(x)$. The criterion assumes that a stable steady-state operation, x^0, u^0 exists, which optimizes J as $(J^o)_{SS}$, and establishes the conditions and a frequency for an optimal, periodic operation, where $u(\tau) = u(t+\tau)$ *and* $x(\tau) = x(t+\tau)$, so that $(J^o)_P > (J^o)_{SS}$ where $J_P = \frac{1}{\tau}\int_0^\tau g(y,u)dt$. Allowance must be made for equality as well as inequality constraints, $\frac{1}{\tau}\int_0^\tau v(y,u)dt = 0$ *and* $\frac{1}{\tau}\int_0^\tau \omega(y,u)dt \leq 0$. These constraints are joined into the Hamiltonian for the optimization problem through Lagrangian multipliers. If all the functions are twice differentiable in x and u, the Hamiltonian is proper and the frequency domain π criterion is an n x n Hermitian matrix. Maximizing the periodic performance index, $(J^o)_p - (J^o)_{SS} + \lambda[f(x,u) + \nu(y,u) + \omega(y,u)]$ with respect to u is often difficult; so the original maximum problem is simplified by a second-order expansion of the performance index. Averaging the solution over the time period, τ, further simplifies the problem. The input is transformed from the time domain to the frequency domain using Fourier transforms, converting the time variables into frequency and/or phase variables. Maximization is done in the frequency domain on the Hermitian, π, the frequency domain matrix obtained from the transformation of the Hamiltonian. The search variable is the frequency or the phase shift if more than one control variable is used. In some applications, maximization may be avoided by testing the second partial derivative of the Hamiltonian H with respect to the vector u. If this second derivative is positive, the performance index function is convex so that an increase in conversion, yield or selectivity through periodic operation with respect to an optimal steady-state operation is possible.

Terms of the Hermitian, in various forms, are given by *Sterman and Ydstie (1990a, b)*, *Abulesz and Lyberatos (1987)* or *Parulekar (1998)*. If the Hermitian is negative definite, there is no PO that is better than SS. On the other hand, if a frequency ω exists that makes the π matrix positive, $J_P > (J^0)_{SS}$ at that frequency. For any set of model parameters or operating conditions, the region of operating variables for $J_P > (J^o)_{SS}$ can be determined by finding frequencies that make the Hermitian positive.

The starting point for any mathematical investigation must be a model of the biological system. Such a model must recognize that living organisms require a food source (substrate), an oxidant for energy generation (O_2 in aerobic systems), and nutrients. Living organisms grow and can subdivide. They also may excrete by-products (metabolites). Even with simplifying assumptions, several material balances are needed to describe behavior with time. With mixed cultures where one organism may feed on another, further balances must be introduced. **Table 10-1** gives a widely used set of model equations that assume a single culture, a sterile feed, a single limiting substrate and excess nutrients and oxidant. The reactor vessel is reproduced to be a CSTR.

In the expressions, X, S and P are the cell mass or concentration, limiting substrate concentration and the metabolite or product concentration, μ; σ and ε are specific growth, consumption and yield terms. Most terms are defined in the **Nomenclature**. Some simple models use $\alpha = \mu/Y_{X/S}$ where $Y_{X/S}$ is a cell mass yield constant. In any case μ, σ and ε are generally functions of the limiting substrate and may be also functions of a metabolite

TABLE 10-1 Simplified Model for a Cell Culture

$$\frac{dX}{dt} = \left[\mu - \frac{Q(t)}{V(t)}\right]X \tag{10-2}$$

$$\frac{dS}{dt} = \frac{Q(t)}{V(t)}[S_0 - S] - \sigma X \tag{10-3}$$

$$\frac{dP}{dt} = \varepsilon X - \frac{Q(t)}{V(t)}P \tag{10-4}$$

and/or a lag term Z. Also $\varepsilon = \upsilon\mu + \beta$ (Luedecking–Piret Equation) may be assumed. The rate constant υ is the growth associated, while β is the non-growth associated term. The Michaelis-Menten relation, **Eq. (10-1)** is one form of the specific growth rate. It is referred to as an unstructured model because balanced growth of all parts of the cell is assumed, that is, the growth rate of cell components (ribosomes, for example) is the same as the growth rate of structural cell proteins. The Michaelis-Menten relation and other forms are given in **Table 10-2**.

All of the constants in the above models, μ_{max}, K_S, K_i, are real positive numbers. K_S is called a saturation constant while K_i is an inhibition constant. In **Eq. (10-10)**, $\mu(S)$ could be any of the other models in the table. The specific consumption or yield models may take on similar forms.

Structural models assume that the different parts of a cell have different specific growth rates and yields. Simple models assume that the cell is made up of a synthetic portion consisting of ribosomes and metabolites and a structural portion containing cell structural proteins and DNA. Use of a structural model requires the addition of another material balance to the model. Of course, different biological systems need additional balances or, at least, modification of the terms in **Table 10-1**.

TABLE 10-2 Specific Growth Models

$$\mu = \frac{\mu_{\max} S}{K_S + S} \quad (10\text{-}5)$$

$$\mu = \frac{\mu_{\max} S}{K_S + S + S^2/K_i} \quad (10\text{-}6)$$

$$\mu = \mu_{\max}\left[1 - \exp\left(-\frac{S}{K_S}\right)\right] \quad (10\text{-}7)$$

$$\mu = \frac{\mu_{\max} S^y}{K_S + S^y} \quad (10\text{-}8)$$

$$\mu = \mu_{\max} S\exp(-\gamma S) \quad (10\text{-}9)$$

$$\mu = \left(1 - \frac{P}{P_m}\right)\mu\left(S\right) \quad (10\text{-}10)$$

10.2.1 Application of the π Criterion

Abulesz and Lyberatos (1987) were among the first researchers to apply the π criterion. They considered flow rate modulation of a process designed to produce cell protein so their model used material balances **Eqs. (10-2)** and **(10-3)** in **Table 10-1**. With just a single manipulated variable, π is a the scalar quantity making its evaluation facile. Assuming the specific growth rate is given by the Michaelis-Menten relation (**Eq. (10-5)** in **Table 10-2**), they found negative values of π for all ω. Thus, for systems governed by this growth rate expression, modulation of the flow or dilution rate is not effective.

These investigators suggested that their model was oversimplified by assuming instantaneous adaptation of the cell to the abiotic environment. They proposed a growth rate which is a function of not only the current substrate concentration but also of previous concentrations. They replaced $\mu(S)$ by $\mu(Z)$ where Z is given by $Z = \int_0^{\tau} S(t)f(t-\tau)dt$ and $f(t-\tau)$ is a delay or memory function. *Abulesz and Lyberatos* suggested using $f(t-\tau) = \varsigma\exp(-\varsigma\gamma)$ for the memory term function, where ζ is an adaptability constant. They replace the μ in **Eq. (10-2)** of **Table 10-1** by $\mu(Z)$ and introduced a balance for Z, namely $\frac{dZ}{dt} = \varsigma(S - Z)$. Of course, π remained a scalar which these investigators evaluated. **Figure 10-1** shows that flow rate forcing substantially increased cell production in a narrow low frequency region. The increase mirrors positive values of π.

More complex models may also predict a performance improvement through modulation. *Abulesz and Lyberatos* examined a structured model for the specific growth function. Even though they considered a cell model with just two different parts and growth rates, they showed that flow rate modulation improved performance at low frequency for some sets of model and economic parameters.

Ten years later, *Parulekar (1998)* undertook a comprehensive study of single input modulation employing the π criterion. He used the basic model given in **Table 10-1** and showed that low-frequency modulation of the dilution rate, *Q/V*, or the limiting substrate concentration, S_0, cannot increase metabolite production. Increased production occurred at higher frequencies for some values of the non-forced input, but that depended on the growth rate function and the relation of growth to consumption functions. He also observed that if cell washout is the only stable steady state, modulation of *Q/V* or S_0 will not produce a stable limit cycle.

Parulekar distinguished in his analysis three types of bioprocesses: 1) Specific growth, consumption and yield terms share the same model with μ, σ and ε all linearly related, such as $\varepsilon = \mu/Y_{P/S}$, 2) these terms are related by a linear equation ($c_1\mu + c_2\sigma + c_3\varepsilon = 0$), and 3) no linear relationship exists between the specific terms. *Parulekar* found that weak modulation of

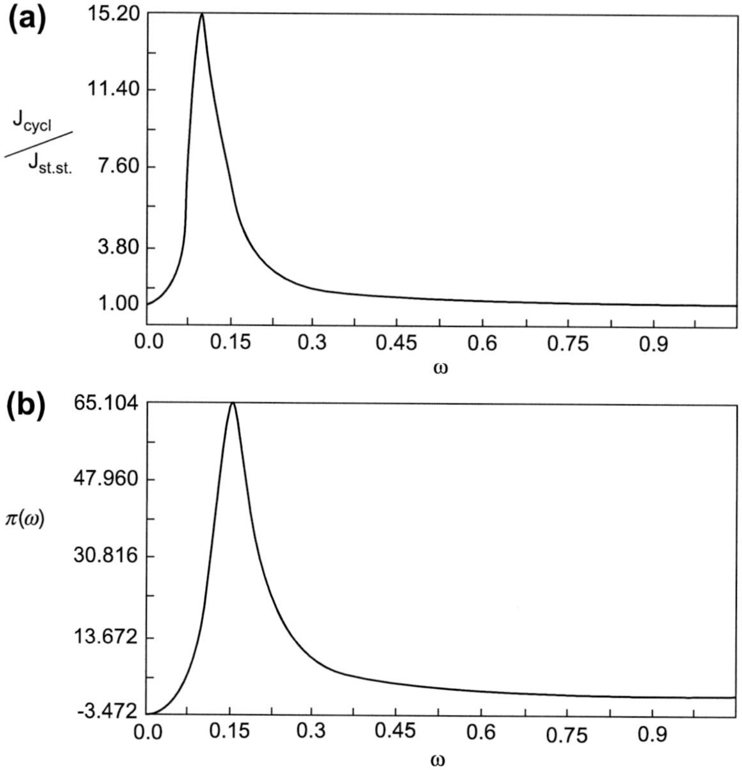

FIGURE 10-1 Modulation of the dilution rate around the optimal SS rate for a chemostat culture producing cell matter. Results are for a specific growth model incorporating lag and one set of model and economic parameters and substrate concentration in the feed: (a) Ratio of the periodic and optimal SS performance indices vs. cycling frequency, (b) value of the π criterion *(Figure reproduced from* Abulesz and Lyberatos (1987) *with permission. © 1987 by Wiley Periodicals Inc.)*

dilution rate for type 1 processes cannot improve the production rate. Fermentation of sugars or carbohydrates to form alcohol are type 1 processes. Dilution rate modulation, however, can increase production under restricted conditions for type 2 and 3 processes. On the other hand, modulation of the limiting substrate concentration, S_0, can increase production in limited regions of parameter space for type 1 and 2 processes if stable steady-state solutions exist. Numerical examples, using parameters obtained in experimental studies, suggest, however, that modulation of S_0 is not effective for type 1 processes, unless a time lag is introduced into the model. In that case, modulation of either Q/V or S_0 improves reactor performance for a suitable choice of frequency just as *Abulesz and Lyberatos* observed. *Parulekar* also examined a type 1 bioprocess in which μ was a function of X and P as well as S. Dilution rate forcing did not increase production, whereas substrate feed concentration forcing did offer an increase in some narrow regions of the Q/V parameter space. A type 2 process in which μ depends on P as well as S was also studied. Lactate fermentation and *Propionibacterium* fermentation to produce propionic acid represent such processes. *Parulekar* found that dilution rate modulation improved production over a triangular region of the $Q/V - S_0$

parameter space. Modulation of S_0 was successful, but over a smaller region. In a further type 2 example with μ and ε having a similar functional dependence on S and P, *Parulekar* observed that modulation of Q/V was ineffective, while modulation of S_0 improved bioreactor performance in a limited region of the $Q/V - S_0$ space. Two other examples of type 2 processes with different kinetic expressions gave similar results.

Simultaneous modulation of dilution rate and the limiting substrate concentration was treated in a second contribution using the generalized π criterion (*Parulekar, 2000*). The analysis suggested that two variable manipulations at the same frequency further improved the production rate over the manipulation of a single input. The former introduces two additional variables, phase shift and amplitude ratio, which may be used to increase production even more than by phase modulation at the same amplitudes. Examples considered in *Parulekar's* earlier paper were re-examined. In the first type 1 example considered, where μ depended only on S, modulation of Q/V and S_0 increased production across the range of practical frequencies and the $Q/V - S_0$ parameter space provided an asymptotically stable steady-state solution to the model equations in that space. For the type 1 example in which μ, σ and ε are functionally dependent on S, X and P, **Figure 10-2** illustrates Parulekar's findings. Modulation of S_0 improved bioreactor performance to the right of curve FIJ and below curve FGJ in part **(a)** of the figure. In-phase modulation of both Q/V and S_0 provided better performance vis-à-vis both the optimal SS performance and performance varying just S_0. While in part **(b)**, improvement of cell growth rates (curve 1) and of metabolite production appear as functions of limiting S_0 for a specified Q/V. The dashed curve shows the variation in the optimal phase shift between Q/V and S_0. Note that optimal modulations of Q/V and S_0 are out of phase at low values of the time-average S_0.

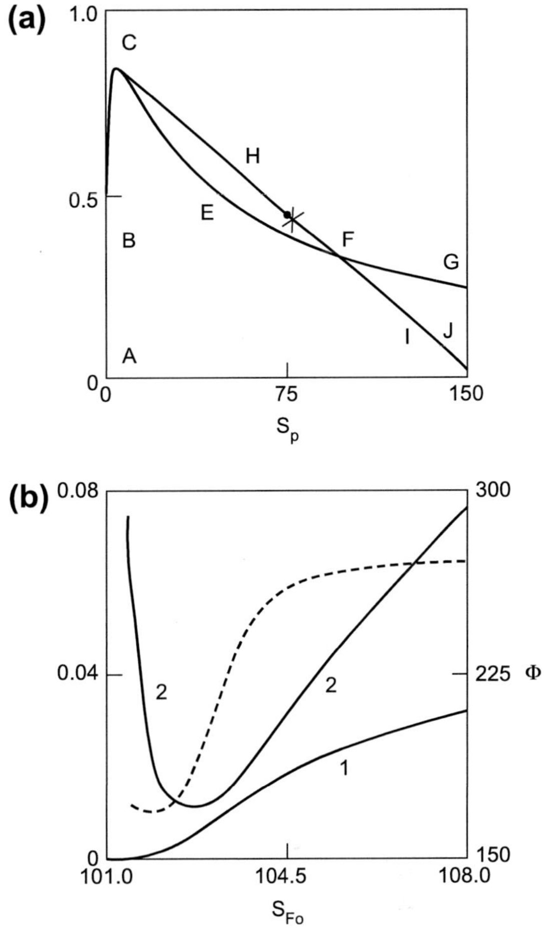

FIGURE 10-2 **Application of the generalized π criterion to a type 1 bioreactor with substrate consumption and metabolite formation dependent on X, S and P. a) Curve ABCEFG represents the condition $\mu(S_0) = Q/V$, admissible steady-state solutions exist only below this curve. There is one solution below the curve on segment CEF. There are two steady-state solutions in the envelope CEFHC. Conditions above curve ABCEFG represent steady-state cell washout; b) Plot of enhancement over steady state vs time-average S_0 for simultaneous modulation of Q/V and S_0: curve 1 is for enhancement of cell production, curve 2 is for enhancement of metabolite production. The dashed curve shows the optimal phase shift for enhancement. Curves are for one time-average, Q/V. ^{X} denotes optimum steady state.** *(Figure reproduced from* Parulekar (2000) *with permission. © 2000 by Elsevier Ltd.)*

The region of parameter space in which forcing is effective was significantly expanded when two inputs are modulated. For the type 2 examples considered, the system representing lactate fermentation was improved by cycling both Q/V and S_0. However, the region in $Q/V - S_0$ space in which modulation was effective could be expanded only by using impractically high frequencies.

Nguang and Chen (*1998*) applied the π criterion to evaluate the simultaneous modulation of the limiting substrate concentration and the space velocity. They considered a somewhat simpler formulation for high cycling frequencies. The model given in **Table 10-1** was used with $\sigma = \mu/Y_{X/S}$ and $\varepsilon = \alpha\mu + \beta$. For the specific growth rate, the researchers used expression of **Eq. (10-10)**, in **Table 10-2** with $\mu(S)$ given by expression of **Eq. (10-6)**. They found small increased metabolite production under modulation with respect to SS operation except at the optimum limiting S_0 for steady state. At that concentration, the two operation modes were equivalent. This may be a consequence of the second variational method, which assumes small perturbations around a steady state.

Further applications of the π criterion will be examined in sections devoted to specific types of modulation. The discussions in this section illustrate how the criterion may be applied.

10.2.2 Application of Quasi-steady-state Analysis

At very low frequency with bang-bang modulation of an input variable, a continuously operated bioreactor approaches steady-state behavior at each input level as long as there is no deactivation. The time-average the performance becomes then the average of the performance in each part of a cycle. *Roopsingh and Chidambaram (1999)* used this analysis to consider an enzyme catalyzed or glycolytic reaction, $A + B \rightarrow 2\ B$, $B \rightarrow C$, in which $r_1 = k_1 C_A C_B$ and $r_2 = k_2 C_B / (1 + K_i C_B)$, under cycling of either feed concentrations $(C_A)_0$ or $(C_B)_0$. They considered an isothermal, plug flow reactor, but their analysis applies to any reactor type in isothermal as well as non-isothermal operations. For the kinetics chosen, the yield of B goes through a maximum as the space time or flow rate through the reactor varies. Thus, at any yield less than the maximum, two space times will provide the same yield. They refer to this behavior as input multiplicity.

Roopsingh and Chidambaram evaluated the effect of input concentration modulation on this type of a system. In the quasi-steady-state limit, with modulation of just one input, the only cycling variable is the cycle split, s. They observed that for $(C_A)_0$ cycling, the yield goes through a maximum as the split decreases where the time-average yield exceeds the SS yields at the lower of the two space times for input multiplicity, whereas the opposite is found at the higher of the space times. If both $(C_A)_0$ and $(C_B)_0$ are cycled in phase, the yield of B increases with split, but the slopes are different at the two space times.

10.3 SUBSTRATE AND FLOW RATE MODULATION

Early application of periodic operations to bioprocesses dealt with the modulation of limiting substrate concentration or the space velocity. Substrate forcing effects cell growth and the production of metabolites. Reactor volume is invariably constant so dilution rate modulation implies periodically switching the volumetric liquid flow rate to the reactor. Washout of a cell population, of course, may occur at high space velocities.

In the 1970s cycling was used to study the behavior of waste systems or cell cultures under transient conditions or as a method of obtaining model parameters. Examples of the former use employed Saccharomyces cerevisiae for the anaerobic fermentation of molasses. Substrate was forced in a 24 h cycle using several different waveforms (*Borzani et al, 1976*). Later, the flow rate was cycled over the same period at a constant molasses concentration but again using different waveforms. Although steady-state measurements were also made, noisy data and the use of near-triangular waveforms made interpretation of the results uncertain. Several early investigators (*Gilley and Bungay, 1968; Zines and Rogers, 1971; Sunstrom et al., 1976*) used low-amplitude sinusoidal perturbation of substrate concentration and dilution rate. Their interest was in the dynamics of the culture and extracting model parameters from the response.

The studies examined in **Section 10.2** suggest that Q/V and S_0 modulation increases metabolite production in just a rather limited range of the Q/V-S_0 parameter space, whereas experimental investigations indicate that modulation of either of these variables is effective over a much wider range. Several investigators attribute this difference between simulation and experiment to the assumption that a change in a culture medium has an instantaneous effect on cell behavior. They have sought to introduce a more realistic representation by allowing the specific growth rate to depend on a previous cell state, Z. Thus, the RHS in the cell balance relations in **Table 10-1** which becomes $\mu Z - X(Q/V)$ and a new balance is added to reflect changing Z: $\frac{dZ}{dt} = \alpha(S - Z)$. In this relation, α is a delay constant. A larger value of α reduces the response lag.

Sastry et al. (*1994*) investigated dilution rate and substrate concentration cycling for the response lag assumption, but unlike other investigators they assumed cell growth rate kinetics given by **Eq. (10-6)** in **Table 10-2** and that, further, the specific growth rate can be inhibited by metabolite accumulation. Thus, they write the specific growth term as

$$\mu = \mu_{\max}\left(1 - \frac{P}{P_m}\right)\left(\frac{Z}{\left(K_S + Z + \left(Z^2/K_i\right)\right)}\right). \tag{10-11}$$

This form of growth kinetics leads to an input multiplicity in the limiting S_0 according to *Kumar et al.* (*1993*). An input multiplicity occurs when the yield of cell mass goes through one or more maxima with the same input to the bioreactor. Three values of S_0 were identified for steady-state operation, two of which were unstable. These investigators explored the effect of modulating S_0 in the feed or dilution rate. They observed that cycling of S_0 around the higher feed concentration corresponding to an unstable state increased metabolite production, whereas cycling S_0 around the lower feed concentration depressed production. The cycling of Q/V around dilution rates for both unstable states also gave a poorer performance.

On simulation of the *Kumar et al.* model, *Sastry et al. (1994)* found that including a lag in their culture model changed the observation of *Kumar et al.* and extended the region of Q/V-S_0 variable space in which PO increases metabolite production. Cycling S_0 around the upper steady state of input multiplicity increased production for any value of the delay constant, whereas cycling around the lower S_0 steady-state resulted in improvement only for large values of α. With respect to Q/V, the advantage of cycling depends also on the dilution rate. Improvement occurs through cycling at lower values for large values of α but not at higher dilution rates, regardless of the response lag.

In a further exploration of steady-state multiplicities, *Suresh and Chidambaram (1997)* examined the use of a more realistic CSTR model that allows for flow bypassing. They observed that three different values of S_0 gave the same metabolite production rate for the model given by **Eq. (10-6)** in **Table 10-2** and identified two of those input substrate concentrations as unstable. They assumed that modulation of S_0 would result in an unstable limit cycle so they examined the use of a relay feedback control to initiate oscillations in the substrate feed concentration. *Sterman and Ydstie* (*1991*) suggested the use of such a controller for chemical reactors.

Suresh and Chidambaram's numerical simulation showed that the application of relay feedback control for the two unstable substrate feed concentrations resulted in stable oscillations in S_0 and thus offered a method of input forcing.

Experiments on modulating the limiting S_0 were carried out by *Pickett et al.* (*1979, 1980*) employing *Escherichia coli* grown in a well-aerated 1 L chemostat at 37°C with a constant dilution rate of 0.144 h^{-1}. An optimal cycle period of ca. 2 h increased the protein production by about 15% and the *E. coli* cell yield by about 5%. Other results suggested an inverse relation between protein as a percentage of the dry weight and the cell yield. The investigators observed a response lag between *E. coli* and a step change in substrate concentration. The existence of such a lag has been assumed by several researchers in their theoretical studies, notably by *Lyberatos* and coworkers (*Stephens and Lyberatos, 1987, 1988; Abulesz and Lyberatos, 1987*). **Figure 10-3** shows the lag differs between step-up and step-down in concentration and depends on both the modulation frequency and the amplitude of the step change. Apparently, a stationary cyclic state was achieved after four to six cycles.

Later (*Pickett et al., 1980*), using the same system and examined the effect of amplitude at an apparently optimal period of 2 h. The lower glucose concentration was set at 0.2 g/L, sufficient to prevent washout, while the higher concentration was 5 g/L. Operating problems were encountered at that concentration along with a reduction in cell protein content. There was evidence, too, that the substrate was not growth-limiting. Experiments showed an increase in cell mass and protein production with increasing amplitude; however, this improvement probably resulted from a rising time-average S_0 rather than increasing amplitude. Steady-state experiments at different glucose levels in the feed were not carried out. Such experiments could have established the influence of amplitude.

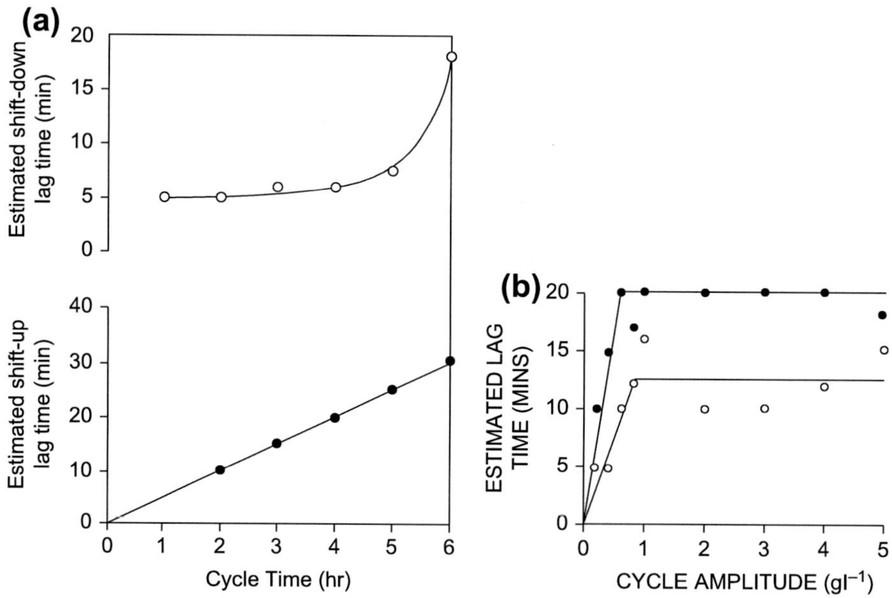

FIGURE 10-3 Lag in the response of *E. coli* to step changes in glucose feed concentration in a substrate modulation experiment: (a) cycle time effect; (b) amplitude effect. o = step down, • = step-up. *(Figure reproduced from* Pickett et al. *(1979) with permission. © 1979 by Wiley Periodicals Inc.)*

The effect of flow rate modulation on cell mass in a continuous chemostat was investigated by *Abulesz and Lyberatos* (*1989*) using the yeast Saccharomyces cerevisiae feeding on glucose. **Figure 10-4** is a schematic of the experimental unit employed by these investigators. The automation of control and monitoring is typical of equipment used for both SS and PO studies. *Abulesz and Lyberatos* examined the response of the yeast culture to step-up and step-downs of the glucose concentration in the feed. They observed that the Michaelis-Menten relation and the material balance given in **Table 10-1** failed to reproduce the observed response to a step change in either direction although that model adequately represented the experimental steady-state behavior. Assuming a lag in response corresponding to the expression for Z given in **Section 10.2**, they found that assuming a lag of about 3 h provided a reasonable representation for both step-up and step-down experiments in contrast to the *Pickett et al.* experiments just discussed.

Bang-bang flow rate cycling experiments were undertaken assuming that an optimal cycle period would be the same as the duration of the response lag. *Abulesz and Lyberatos* observed about a 25% increase in cell production over that obtained at steady state with the time-average flow rate for a 3 h cycle period. Cycle periods of 2 h or 4 h substantially decreased production.

10.4 DISSOLVED OXYGEN MODULATION

An organic acid bacteria, *Propionibacterium*, produces the commercially important vitamin,

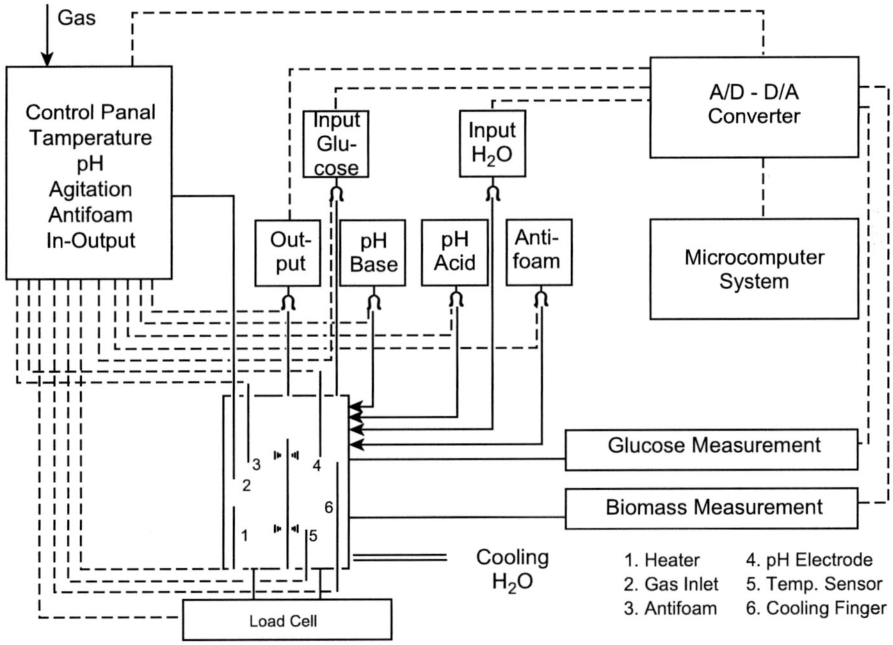

FIGURE 10-4 Schematic of an experimental unit used to study the dynamics of a yeast culture and the performance of the culture under flow rate modulation. *(Figure reproduced from* Abulesz and Lyberatos (1989) *with permission. ©1989 by Wiley Periodicals Inc.)*

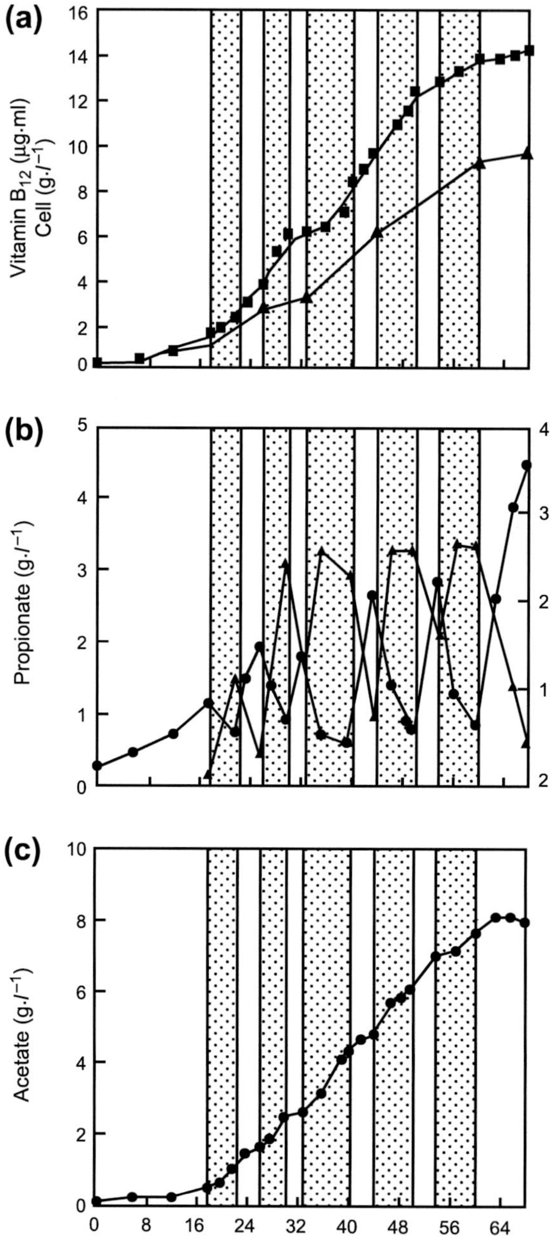

FIGURE 10-5 **Time profile, in hours, of cell density and product concentration in a periodically operated, mechanically stirred fermentor. Dotted columns show times of aerobic operation: (a) Cell Density = •, B_{12} conc. = ▲, (b) end product conc. – propionic acid = •, pyruvate = ▲, (c) acetate end product conc.** *(Figure reproduced from* Ye et al. (1996)*, with permission. © 1996 by Elsevier Science Ltd.)*

B_{12}, intracellularly from glucose while excreting propionic and acetic acids as its primary end products. Both end products inhibit cell growth and thus the production of the vitamin. As an alternative to separative fermentation processes, *Ye et al. (1996)* investigated a process employing switching between aerobic and anaerobic environments. For their experiments, they employed *Propionibacterium freudenreichii* (IFO 12424), an aerotolerant anaerobic species. Experiments were performed at 30°C in a well-stirred vessel, sparged alternately with air and nitrogen. Justification for the modulation strategy was that the aerobic environment decomposes propionic acid to a pyruvate, while it also inhibited cell growth and the bacterium's metabolic rate. The anaerobic environment increases cell growth until the increasing propionic acid concentration works to inhibit this growth.

Figure 10-5 shows a 60 h experiment. The narrower clear columns represent N_2 bubbling periods, while the dotted columns represent air bubbling. Switching was initiated by the slope of the observed cell density with time so that the period and the cycle split varied respectively from 8 to 10 h and 0.5 to 0.62 based on the aeration portion of the cycle. Steady build-up of cell mass and B_{12} concentration under periodic operation can be seen after the initial 18 h of steady anaerobic operation. Alternating peaks and valleys in propionic acid and pyruvate concentrations are evident in **Figure 10-5(b)**. A lag in the response to the environment shift is evident, but the sampling frequency makes the lag duration impossible to measure. Acetate build-up occurred throughout the experiments. Its final value, though, is well below the threshold for cell growth inhibition. The researchers explained the slower increase of B_{12} concentration with time than that for cell mass by a cessation of vitamin production in the presence of dissolved oxygen. When compared with continuous operation, the *Ye* experiments seem spectacular. **Table 10-3** compares the results. However, the advantages of PO remain uncertain. Neither PO nor SS operation was optimized.

Ye et al. (1996) undertook step-change experiments in DO levels in a chemostat to explain the improvement under PO. These clearly showed a shift in the chemical pathway that the bacterium uses to acquire energy from substrate level phosphorylation to oxidative phosphorylation when the cell environment becomes aerobic. They also showed that the duration of the aerobic portion should be held to 6 h or less while the anaerobic portion should not exceed 10 h.

Mixed cultures, often consisting of many different genuses as well as organisms of the

TABLE 10-3 Production and Productivity in Different Operating Strategies*

Operation	Cell Density g/L	Vitamin B_{12} Conc., μg/mL	Proprionate Conc., g/L	Acetate Conc., g/L	Vitamin B_{12} Productivity, mg/mL/h
Anaerobic	5.2	4.64	5.08	2.39	97
Aerobic[1]	6.29	4.81	0.48	2.88	100
Periodic[2]	15.02	9.54	2.78	8.52	140

* *after 50–60 h of operation at a* pH ~ *6.7 and 30°C*
[1] *at* DO ~ *5 ppm*
[2] *using* $\tau_{cycle} = 6$ *to 8 h,* s ~ *0.6 and maintaining* DO ~ *1 ppm during the aerobic portion*

same genus, are normal in biological waste treatment. They also are found intentionally in several biochemical processes, such as fermentation of xylose or of xylose-glucose mixtures. Of particular interest are systems in which one species produces a metabolite that is consumed by a second species to produce the desired chemical product. Examples are the production of nisin or acetic acid from glucose. *Tohyama et al.* (*2002*) investigated the use of *Lactobacillus delbrueckii* to convert glucose to a lactate that served as the substrate for *Ralstonia eutropha* which synthesized poly-β-hydroxybutyrate. This material is a biopolymer with several commercial applications. A single fermentor was employed with periodic forcing of dissolved oxygen in the broth. The operation was investigated experimentally and also simulated.

The model employed by *Tohyama et al.* consisted of the time-dependent material balances given in **Table 10-1**. These were applied to the first species, X_1. The final balance in the table must be modified by a term, - $\varepsilon_2 X_2$ for the consumption of the metabolite P by the second microorganism, X_2. For the species X_2, another set of three time-dependent balances must be written:

$$\frac{dX_2}{dt} = \left(\mu_2 - \frac{Q}{V}\right)X_2 \tag{10-12}$$

$$\frac{dN}{dt} = \sigma_2 X_2 - \frac{Q}{V}(N - N_0) \tag{10-13}$$

$$\frac{dP_2}{dt} = \varepsilon_3 X_2 - \frac{Q}{V}P_2. \tag{10-14}$$

N in the second equation represents ammonia, a critical nutrient for the second microorganism, *R. eutropha*. P_2 is the biopolymer. The specific growth functions used by *Tohyama et al.* were

$$\mu_1 = \frac{\mu_{\max_{11}} O_d S}{K_S + S}\left(1 - \frac{P}{P_{\max}}\right)^n \tag{10-15}$$

$$\varepsilon_1 = c_\alpha \mu_1 + \frac{c_{\beta_{\max}} O_d S}{K_S + S} \tag{10-16}$$

$$\varepsilon_2 = \frac{\mu_2}{Y_{X_2/P} O_d} \tag{10-17}$$

$$\sigma_1 = \frac{\mu_1}{Y_{X_1/S} O_d} + \frac{\varepsilon_1}{Y_{P/S} O_d} \tag{10-18}$$

$$\mu_2 = \left(\frac{\mu_{\max_2} O_d P}{K_P + P + \left[P^2/K_i\right]}\right)\left(\frac{N}{K_N + N}\right) \tag{10-19}$$

$$\sigma_2 = \frac{\mu_2}{Y_{X_2/N} O_d} \tag{10-20}$$

$$\varepsilon_3 = q_{\max} \frac{k_N}{k_N + N}. \tag{10-21}$$

The large numbers of parameters in this complicated model were evaluated from batch experiments with single microorganisms and steady-state runs with mixed cultures. Despite its complexity, the model was not able to predict accurately the experimental observation under cycling of S, O_d or N (ammonia). The researchers, however, used the model to determine a near-optimal operation of the mixed culture reactor.

The *Tohyama* experiments were carried out in an aerated chemostat with a 3 L working volume maintained at 37°C and a constant pH. Modulating N or S caused operating problems, so just the results for switching O_d between 3 and 0.5 ppm were reported. Best performance was achieved with $\tau = 2$ h and equal durations at the two dissolved oxygen concentrations. This achieved a productivity almost as high as the best observed at a constant O_d of 3 ppm, but it doubled the polymer yield per gram of glucose consumed. An attempt was made by the researchers to vary the durations of high and low O_d based on the lactate concentration in the chemostat. No improvement was obtained probably because of the lag in measuring the lactate concentration.

10.5 CULTURE MEDIUM TUNING

Often in bacterial plant or animal cell cultures used to produce chemicals, the culture medium conditions that impart high growth rates of cells do not yield high rates of metabolite formation. Alternatively, medium conditions that favor metabolite formation do not promote cell growth. It is possible to change the media and thereby alter the relative rates of cell growth and product (metabolite) formation by altering the nutrient mix, pH, temperature, hormone levels, etc. without altering the limiting substrate concentration or the space velocity. This is known as a perfusion operation, but it suffers from sterility problems. The staging of a reactor was used by *Yamada and Fujita* (*1983*) to enormously increase the production of shikonin pigment from a culture of Lithospermum erythrorhizon in a commercial operation. The result, however, could have been obtained in a single vessel by periodically switching the fluid medium. *Pedersen et al.* (*1987*) report other examples in which production from plant organisms have been enhanced by what may be called medium tuning: 1) Caffeine formation from *Coffea arabica* and 2) atropine production by *Atropa belladonna*.

Pedersen et al. (*1987*) and *Parulekar and Lee* (*1993*) have studied medium tuning theoretically for a relatively simple system by modifying the equations in **Table 10-1**. A tuning term, Δ_1, was added multiplying μX in the cell balance relation, the σX term in the substrate balance was replaced by $\frac{\Delta_1 \mu X}{Y_{X/S}}$ and $\frac{\Delta_2 \varepsilon X}{Y_{P/S}}$ and finally the second tuning term, Δ_2, was added to the product balance multiplying εX. *Pedersen et al.* simulated bang-bang tuning (with Δ_1 and Δ_2 switching between 0 and 1 on each half cycle). For a specific growth term given by **Eq. (10-1)**, a specific set of parameters

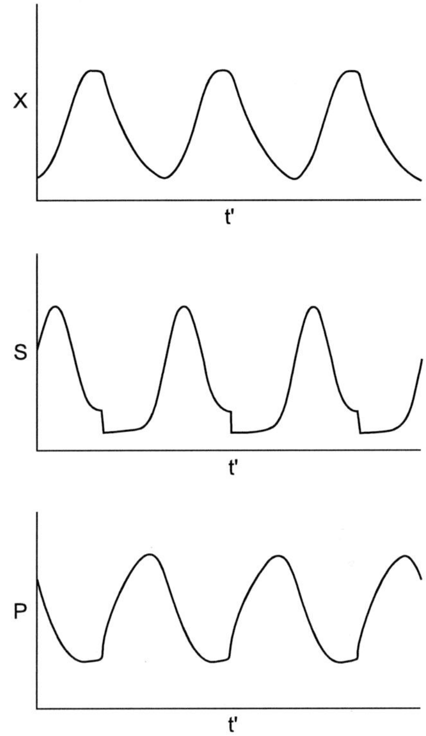

FIGURE 10-6 **Time behavior of cell mass, X, substrate, S, and product, P, concentrations for periodic bang-bang medium tuning of a cell culture where t = time.** *(Figure reproduced from* Parulekar and Lee (1993) *with permission.* *© 1993 by Elsevier Inc.)*

and operating variables, Q/V and S_0, the time behavior of cell mass, substrate and metabolite concentrations are given in **Figure 10-6** for a stationary cyclic state. The figure shows that periodic tuning of the medium causes the peaks in cell mass and metabolite product to move almost 180° out of phase. The researchers observed that the operation substantially increased the time-average metabolite production above what was attainable in SS operation using either one of the tuning media.

Parulekar and Lee (*1993*) undertook a stability analysis of periodic media tuning by examining the behavior of the Floquet multipliers for a choice of growth kinetics, model parameters and cycle variables as O/V and S_0 changed. As the multipliers crossed the unit circle in the complex plane allowing a washout solution to become unstable, three types of transitions were observed: l) a LIMIT cycle, 2) period doubling or 3) a Hopf bifurcation. The researchers examined the stability behavior for five cases of increasing kinetic complexity. As complexity rose, increasingly complex limit cycles were observed in the *X-P* plane, some of which were unstable. The *Parulekar and Lee* study suggests that periodic operation of bioreactors can result in strange stationary cycles and perhaps types of chaos. *Sinčić and Bailey (1977)* report bizarre cycles in their examination of the periodic forcing of chemical processes.

The simplest kinetics considered by *Parulekar and Lee* assumed specific growth and metabolite formation rates increased with increasing concentration of the limiting substrate, while formation rates could be reduced by metabolite accumulation. These were the kinetics treated by *Pedersen et al. (1987)*. **Figure 10-7(a)** shows the stable solutions to the system model for a set of kinetic parameters in the $Q/V - S_0$ operating space. Above the curve OAB only the cell washout solution is stable. In the region between the curves OAB and OCD, the washout solution is stable at high Q/V and low S_0; however, at any level of Q/V, there is a level of S_0 that results in a limit cycle bifurcation where a stable, nonwashout solution emerges. The curve OCD is the boundary for stable washout solutions, while below the curve, the washout solution is unstable and only stable PO solutions are possible. A nonwashout solution for SS operation using one of the tuning levels is possible whenever there is a nonwashout solution for periodic operation. Improvement through periodic tuning is not universal; it occurs in just a restricted range of the $Q/V - S_0$ variable space.

Phase portraits showing the approach to the stable operating states are given in **Figures 10-7(b)** and **(c)**. In **(b)**, the trajectories converge to zero, the washout situation. This is the behavior in region 1 and part of region 2 in **Figure 10-7(a)**, whereas in **(c)**, the trajectories converge to a limit cycle on the diagonal. The phase portrait represents the behavior in region 3, along the curve OCD and at low Q/V and high S_0 in region 2. A special case of medium tuning occurs when two nutrients are available for a microorganism. *Cail et al. (1979)* describe experiments with a continuous culture of a fodder yeast, *Candida utilis*, in which the glucose fed culture received periodic pulses of ethanol. These pulses substantially increased the expression of cell cytochromes even though the yeast can utilize either ethanol or glucose. Experiments were performed at constant space velocity, pH and temperature under conditions in which glucose was the rate-limiting substrate. Intervals between pulses were about 5 min, but a stationary periodic state was not

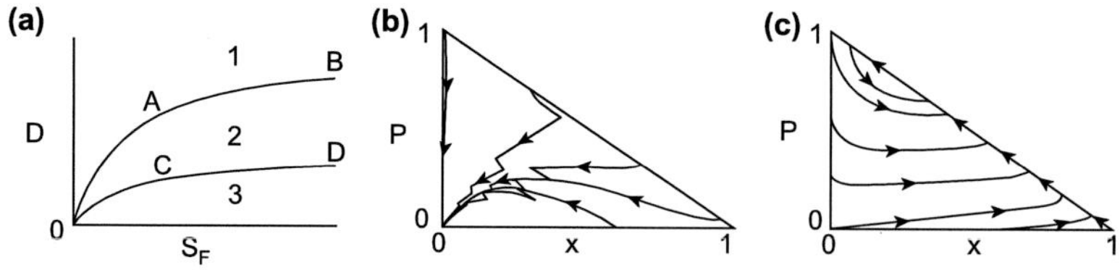

FIGURE 10-7 **Stable solutions for the model and phase trajectories for two solutions: (a) Regions of stable nonwashout solutions in the Q/V-S_0 variable space, (b) time trajectories in the P - X plane for a washout solution (Region 1 above curve OAB and part of Region 2) in A and C trajectories for a periodic (limit cycle) solution (Region 3 below curve OCD, on curve OCD and part of Region 2 in A).** *(Figure reproduced from* Parulekar and Lee (1993) *with permission. © 1993 by Elsevier Ltd.)*

achieved. Cytochrome levels were well above those for constant addition of ethanol and increased with time-on-stream after pulsing began, reached a maximum and then declined slowly. The researchers were interested primarily in cell respiratory mechanism, but suggested pulsing a second substrate as a method of stimulating chemical production in cultures. There appears to have been no follow-up of the *Cail et al.* discovery.

10.6 SURVIVAL IN MIXED CULTURES

Much attention has been given in the literature to the survival of a species in a mixed culture of microorganisms. Evidence suggests that if there is competition for a limited resource, either substrate, oxygen or nutrient, only one species will survive in a uniform environment with time-invariant inputs. Input modulation is a strategy that may be used to allow the survival of a desired species in face of invasion by a wild species or alternatively where the coexistence of two species is the objective.

Stephanopoulos et al. (*1979*) considered the situation in which the growth rates of two competing species cross over as the limiting substrate concentration increases, e.g., at low concentrations the growth of one type of microorganism is favored so that organism survives, whereas at a higher concentration growth of the second type is faster and it replaces the first type in the mixed culture. Using simulation and the unstructured model given in **Table 10-4** with the Monod relation for specific growth rate, these researchers demonstrated that modulation of either S_0 or Q/V ensures coexistence of both microorganisms provided an appropriate frequency and level of the unperturbed input are selected. The requirement was that one part of a cycle must favor one type of microorganism while the other part had to favor survival of the second type. Stability analysis of the cycles employing Floquet multipliers as well as simulation showed that cycles in which there was little difference in growth rates in one part led to instability. Coexistence was possible by singly varying an input in just narrow ranges of the chemostat operating variables. Behavior with time of the two cell types, X_1 and X_2, in a phase plane plot is shown in **Figure 10-8(a)** and **(b)** for dilution rate forcing. The first of these figures illustrates the requirement that each of the two halves of the cycle favor survival of one or the other species. *Stephanopoulos et al.* observed further that stable cycles were achieved by modulating Q/V or S_0 in just narrow ranges of the variable space.

TABLE 10-4 Mixed Culture Models with a Single Limiting Substrate

$$\frac{dX_1}{dt} = \left[\mu_1(S) - \frac{Q(t)}{V}\right]X_1 \qquad (10\text{-}22)$$

$$\frac{dX_2}{dt} = \left[\mu_2(S) - \frac{Q(t)}{V}\right]X_2 \qquad (10\text{-}23)$$

$$\frac{dS}{dt} = \frac{Q(t)}{V}\left[S_0(t) - S\right] - \frac{\mu_1(S)X_1}{Y_1} - \frac{\mu_2(S)X_2}{Y_2} \qquad (10\text{-}24)$$

Another policy, periodic harvesting of the biomass followed by the addition of a fresh mixed culture, also achieves coexistence of the different microorganisms. In this policy, Q/V and S_0 remain constant. Harvesting and the addition step are assumed to be instantaneous. **Figure 10-8(c)** shows the phase plane representation for harvesting 90% of the cell mass. Harvesting less than 80% resulted in unstable cycles for the operating variables examined. Simultaneous cycling of Q/V and S_0 permits coexistence over the widest range of operating variables. **Figure 10-9** gives the

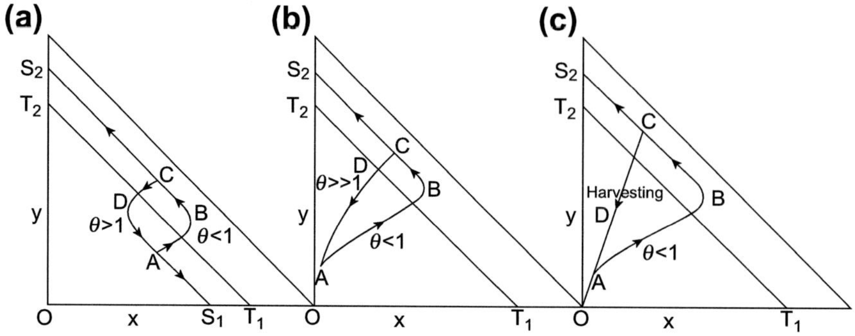

FIGURE 10-8 **Phase plane plots of competing microorganism species obeying Michaelis-Menten kinetics in a chemostat subject to input modulation or cell mass harvesting: (a) Modulation of *Q/V* between a low rate favoring survival of species 2 and a high rate favoring survival of species 1, (b) *Q/V* modulation between a low value as in (a) and a very high rate that ensures total cell washout, (c) periodic instantaneous harvesting of 90% of the cell mass and re-innoculation with a mixed culture with operation at a constant *Q/V* and S_0 favoring survival of species 2. Stable populations "x" and "y" co-exist along the line "T." "S_1 and S_2" denotes stable steady-state limits of population "x" and "y". "T" denotes boundary between steady-state population "x" and population "y." Along line "T" populations "x" and "y" coexist in indefinite proportions, "θ" denotes dimensionless dilution rate *(Q/V)* = $\theta\mu$.** *(Figure reproduced from* Stephanopoulos et al. (1979) *with permission. © 1979 by the AIChE.)*

phase plane representation of a stable cycle for in phase modulation of both operating variables. The area encompassed by the cycle depends on the cycling frequency. Note the small differences between the variables in the two half-cycles.

Stephens and Lyberatos (*1987*) considered a two culture system in which one of the growths rates was smaller than the other at all substrate concentrations. If unstructured specific growth models are used, coexistence cannot be attained by cycling either the limiting S_0 or *Q/V*, regardless of the frequency imposed. However, if the organisms of the two cultures respond at different speeds to environmental changes, *Stephens and Lyberatos* proposed exploiting this difference through modulating either S_0 or *Q/V*. They explored this possibility through simulation using the memory function model devised by *Abulesz and Lyberatos* and described in **Section 10-2**. This required replacement of the specific growth rate terms $\mu_1(S)$ in **Eq. (10-22)** and $\mu_2(S)$ in **Eq. (10-23)** in **Table 10-4** by $\mu_1(Z_1)$ and $\mu_2(Z_2)$ respectively and two additional material balances:

$$\frac{dZ_1}{dt} = \varsigma_1(S - Z_1) \quad and \tag{10-25}$$

$$\frac{dZ_2}{dt} = \varsigma_2(S - Z_2). \tag{10-26}$$

Michaelis-Menten kinetics (**(Eq. 10-5)** in **Table 10-2**) were assumed for μ_1 and μ_2. For these kinetics, the growth rate of X_1 always exceeds that of X_2, if $\frac{\mu_{1_{max}}}{\mu_{2_{max}}} > \frac{K_{S_1}}{K_{S_2}}$. For this condition, *Stephens and Lyberatos* found that coexistence under bang-bang cycling between $S_0 = 0$ and $(S_0)_{max}$ depended on cycling frequency and amplitude, the dilution rate and the ratio, $\eta = \varsigma_1/\varsigma_2$. Simulation results are summarized in **Figure 10-10** for a set of model parameters and $\eta = 2$. Bang-bang modulation of the dilution rate was also considered. Coexistence can be achieved with such modulation, but the investigators observed that impractically high levels of S_0 are needed. They suggested that

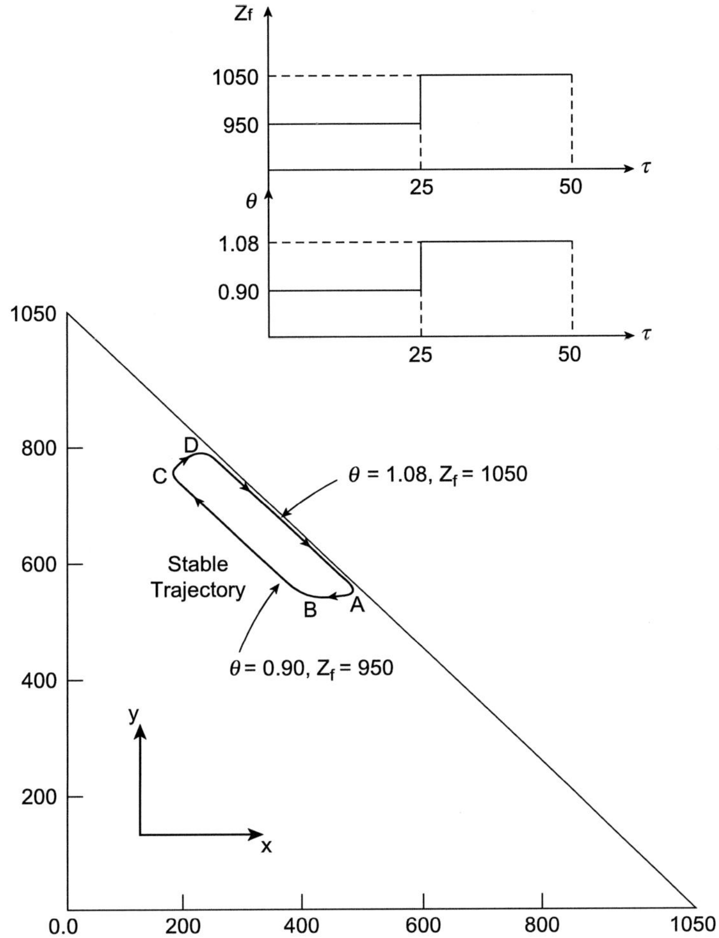

FIGURE 10-9 Phase plane for a stable cycle under in phase modulation of both dilution rate and limiting substrate concentration. Values of the variables in the half cycles are given on the figure. *(Figure reproduced from* Stephanopoulos et al. *(1979) with permission. © 1979 by the AIChE.)*

substrate modulation was a more feasible strategy.

Temperature, nutrients or pH are other inputs that might be manipulated to provide coexistence of competing organisms. Cycling of pH was explored experimentally by *Davison and Stephanopoulos (1986)*. A bacterium, *Escherichia coli*, and a yeast, *Saccharomyces cerevisiae*, were used. These organisms have different sizes and shapes so they could be readily distinguished using a Coulter Counter. Both consume glucose, which was the substrate employed. Glucose level was 1 g/L in a standard culture media for yeast that was buffered with phthalic acid. At pH = 5, the growth rate of *E. coli* exceeds that of *S. cerevisiae*, but this reverses when the pH drops below 4.5. *Davison and Stephanopoulos* observed that the fermentation products, ethanol and acetic acid, were not consumed by or effected the growth of either organism in the concentrations encountered in the experiments. Thus, both organisms

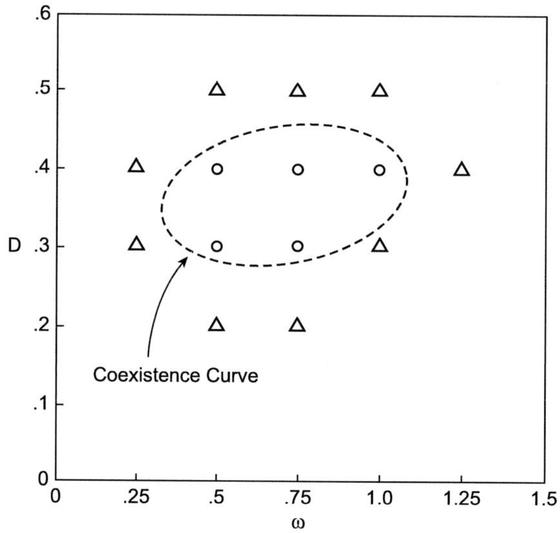

FIGURE 10-10 Band of coexistence in the dilution rate and cycling frequency operating space of two competing species with growth rate and adaptability differences: o = species 2 survival, Δ = species 1 survival. *(Figure reproduced from* Stephens and Lyberatos (1987) *with permission. © 1987 by Wiley Periodicals Inc.)*

competed for the limiting substrate, but did not otherwise interact.

Several cycles were tested with pH levels of 4.3 and 5 and periods ranging from 12 to 32 h. Data collected were quite noisy, rendering interpretation difficult, but they indicated stable coexistence of the two organisms in symmetrical 2-part ($s = 0.5$) cycles with $\tau_{cycle} =$ 32 h. Stable coexistence was also seen for an unsymmetrical cycle of 14 h at pH = 5 and 18 h at a pH = 4.3 ($s = 0.56$). However, when τ_{cycle} for symmetrical forcing was dropped from 32 h to 12 h, washout of *S. cerevisiae* took place. Simulation using the *Stephanopoulos et al. (1979)* model above, with parameters reproduced from batch and steady-state experiments to reflect pH change, qualitatively confirmed the experimental results. Stability analysis around the apparent limit cycle confirmed stability for symmetrical and unsymmetrical forcing for $\tau_{cycle} = 32\,h$; whereas simulation of the $\tau_{cycle} = 12\,h$ experiment found the limit cycle to be unstable. Indeed, for longer τ_{cycle}, a wide range of cycle splits produced stable coexistence of the microorganisms. Conditions permitting, pH cycling was found to be a suitable strategy for maintaining coexistence, or, in other circumstances, for eliminating an undesired organism from a mixed culture.

10.7 STABILIZATION OF RECOMBINANT CELL CULTURES

Recombinant human protein pharmaceuticals including cytokines (interleukins, interferons), hormones (insulin, human growth hormone, calcitonin, glucagons, somatropin), tissue plasminogen activator (TPA), and various vaccines (cholera B subunit) are often produced on a large scale by host microorganisms, usually the bacterium *Escherichia coli*, the yeast *Saccharomyces cerevisiae* or mammalian cell cultures such as hybridoma, Chinese hamster ovary (CHO) or human cell lines (*Ferrer-Miralles et al., 2009*). Gene transfer into bacterial or yeast host cells is usually accomplished by inserting the DNA code for the particular pharmaceutical (rDNA) into plasmid DNA and transforming the host cell with the genetically modified plasmid. Plasmids are circular extrachromosomal DNA molecules containing the genes of a few proteins. Since plasmids multiply independently of the nucleus, plasmid insertion allows for overproduction of the recombinant product. However, this overproduction imparts a metabolic burden on the plasmid-bearing host, hence the transformed cells grow slower than the wild type (plasmid-free) cells. A fraction of the plasmid bearing organisms reverts to the faster growing wild type by rejection of the plasmid. Consequently, if unchecked, the wild type cells rapidly outgrow the recombinant protein producers. Periodic manipulation of inputs to a bioreactor, a temperature controlled CSTR, is one of several methods that have been explored to stabilize a culture or, at a minimum, extend the production duration. Of course, this is a specific application of strategies that may be used to eliminate unwanted organisms in mixed cultures, examined in the previous section. Application of antibiotic selection pressure by introducing an antibiotic resistance gene along with the rDNA for the pharmaceutical product is a common means of overcoming the problem of outgrowth by wild type cells. The reverted cells lose their resistance to the antibiotic and either die or become seriously impaired by the presence of the antibiotic. However, continuous application of an antibiotic on a large scale is expensive and significantly raises the cost of the recombinant pharmaceutical. Periodic application of the antibiotic in a continuous culture may reduce that cost by improving productivity (*Lee and Parulekar, 1996*).

A numerical study of dilution rate modulation was undertaken by *Stephens and Lyberatos* (*1988*) to exploit differences in adaptation times to stabilize the plasmid-containing cells. They employed a version of a general unstructured model of a chemostat made up of material balances on two types of cells, X^+ and X^-, where the superscript indicates the presence or absence of plasmid. The models are given in **Table 10-5** as a set of ODEs. $S_0(t)$ and S (gL^{-1}) are the growth-limiting substrate concentrations in the feed and in the bioreactor, Y^+ and Y^- are the yield coefficients (g cells/g substrate) of the plasmid bearing and plasmid-free cells, P (gL^{-1}) is the recombinant product (biopharmaceutical) concentration and π^- is the specific productivity (g product/g cells·h) of the recombinant product. The specific growth rate of the plasmid containing cells (μ^+) and plasmid-free cells (μ^-) are defined as:

TABLE 10-5 Dynamic Model for a Mixed Recombinant and Host Cell Culture in a Chemostat

$$\frac{dX^+}{dt} = \left[\left(1-\alpha\right)\mu^+(S) - \frac{Q(t)}{V}\right]X^+ \tag{10-27}$$

$$\frac{dX^-}{dt} = \alpha\mu^+(S)X^+ + \left[\mu^-(S) - \frac{Q(t)}{V}\right]X^- \tag{10-28}$$

$$\frac{dS}{dt} = \frac{Q(t)}{V}\left[S_0(t) - S\right] - \frac{\mu^+(S)X^+}{Y^+} - \frac{\mu^-(S)X^-}{Y^-} \tag{10-29}$$

$$\frac{dP}{dt} = \pi^+(S) - \frac{Q(t)}{V}P \tag{10-30}$$

$$\mu^+ = \frac{\mu^+_{\max}S}{K^+_S + S} \quad \text{and} \quad \mu^- = \frac{\mu^-_{\max}S}{K^-_S + S}. \tag{10-31}$$

The parameters μ^+_{max} and μ^-_{max} (h^{-1}) are the maximum specific growth rates, K^+_S and K^-_S are the Monod half saturation constants of the plasmid containing and plasmid-free cells, respectively. In continuous SS operation, the derivatives in the set of equation are zero, while in stable PO, the condition in **Eq. (10-32)** arises:

$$C(t+\tau) = C(t) \quad for \quad C = X^+,\ X^-,\ S,\ A,\ P. \tag{10-32}$$

In their model, *Stephens and Lyberatos* neglected a balance for the metabolite. Using the specific growth rate model **Eq. (10-5)** in **Table 10-2**, *Stephens and Lyberatos* observed that dilution rate cycling could not prevent the washout of plasmid-bearing cells. They reasoned that if the plasmid-bearing cells had a slower growth rate, they might also have a slower adaptability to changes in the chemostat's substrate concentration. Thus, they introduced an adaptation model with a memory function as discussed in **Section 10.2.2**. They further assumed that the two types of cells had different adaptability constants, ζ^+ and ζ^-. Consequently, two more balances:

$$\frac{dZ^+}{dt} = \varsigma^+(S - Z^+) \quad and \qquad (10\text{-}33)$$

$$\frac{dZ^-}{dt} = \varsigma^-(S - Z^-) \qquad (10\text{-}34)$$

were introduced into the model. If $\zeta^+ < \zeta^-$, simulation showed that there were regions in the Q/V - ω space where plasmid-bearing and host cells co-exist.

Modulation of the dilution rate to provide stability was examined experimentally and by simulation. *Weber and San* (*1989*) employed *Escherichia coli* RR1 containing the plasmid pBR322. This plasmid expresses the β-lactamase enzyme. Experiments were performed in an aerated chemostat maintained at pH 7 and 37°C with a Luria broth and glucose as the substrate. Ampicillin, an antibiotic, was added to the broth. This gave a competitive advantage to the recombinant cells resistant to this antibiotic. Experiments were begun with a recombinant cell culture. Optical density measured cell mass. Plasmid stability, copy number and the β-lactamase activity were also monitored.

Weber and San found that Q/V square wave modulation at $\tau = 2$ h could not eliminate the eventual washout of the plasmid-bearing cells, but this strategy was able to extend by about $2\times$ the duration of a dominant, $> 90\%$, plasmid-bearing culture. The researchers speculated that the growth rate of the recombinant cells adapted better to the time-varying Q/V than the plasmid-free *E. coli* cells, but with time lose this adjustment ability so that washout eventually occurs.

Weber and San discovered that a model derived from **Table 10-5** by assuming a large excess of substrate so as to eliminate the **Eq. (10-29)** balance (c) and neglecting reversion of plasmid-containing cells back into host cells, represented their experimental data when model parameters are estimated from cycling runs. Parameters estimated from steady-state runs underestimated the stability of the plasmid-bearing cells.

Experiments similar to those of *Weber and San* were reported by *Stephens et al.* (*1992*). They also employed an *Escherichia coli* strain containing the plasmid pBR322. Their experiments employed, in a stirred, aerated chemostat maintained at pH 7 and 37°C, and fed with a minimum mineral medium containing glucose as the only substrate, two amino acids and a vitamin. Monitoring methods used by *Weber and San* were followed, except that the fraction of recombinant cells in the culture was followed by using their resistance to an antibiotic. SS and PO experiments were performed; both began with a mixture of plasmid-bearing and host cells. The objective of the *Stephens et al.* experiments was to determine if periodic forcing of Q/V could maintain coexistence of the two cell types. These researchers found that the fraction of recombinant cells in the mixed population decreased with time in SS operation, regardless of the operating variables Q/V and S_0. This behavior was not expected for a system obeying the Michaelis-Menten specific growth rate relation. *Stephens and Lyberatos* (*1988*) assumed this model in their simulations demonstrating that Q/V forcing stabilizes a mixed culture of recombinant and host cells.

Experiments on modulation of Q/V around the rate used in their SS experiments and at the same glucose concentration in the feed increased the fraction of recombinant cells in the culture. **Figure 10-11** shows their

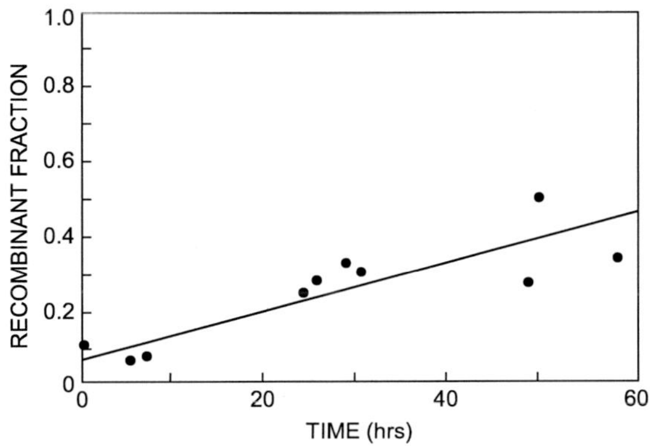

FIGURE 10-11 Variation of the fraction of plasmid-bearing cells with time on stream in a mixed *Escherichia coli* culture subject to bang-bang modulation of the flow rate to the chemostat at $\tau = 2$ h. *(Figure reproduced from* Stephens et al. (1992), *with permission. © 1992 by Wiley Periodicals Inc.)*

experimental results. With a possible reversal of a decrease in recombinant cells, a cycle frequency and amplitude is conceivable that would provide mixed culture stability.

Experimental evidence suggests that the *Stephens and Lyberato's* assumption of a constant α, fraction of recombinant cells created in cell division that revert to plasmid-free cells, is incorrect. Flow rate cycling was re-examined by *Patnaik (1994)* who allowed the reversion fraction, α, to depend on the substrate concentration, specific growth rate of the recombinant cells and the response time of the host cells, t^-. The relationship used was:

$$\alpha = c_1 \frac{\mu^+_{max} Z^+}{t^-(K_S^+ + Z^+)} - \frac{c_2\left[\frac{\mu^+_{max} Z^+}{c_3 t^-(K_S^+ + Z^+)}\right]^n}{1 + \left[\frac{\mu^+ Z^+}{c_3 t^-(K_S^+ + Z^+)}\right]^n} \tag{10-35}$$

where c_1, c_2 and c_3 are adjustable constants. *Patnaik* used values from the literature that were appropriate for *E. coli* ($t^- = 0.2$ h) and examined square-wave modulation of Q/V. The variables in the *Patnaik* study were cycling frequency and amplitude. Results obtained showed that flow rate modulation could not prevent the eventual washout of the recombinant cells, but that it appreciably delayed washout. Recombinant cell mass was almost double the mass under SS operation at the same time-on-stream provided optimal frequency and amplitude were used. Production of the desired metabolite, P, was also much higher. Amplitude affected performance more than frequency. *Patnaik's* simulation agreed well with the observations of *Weber and San (1989)*.

Observing that inserting a plasmid into a host cell may impart antibiotic resistance to the recombinant cell, *Lee and Parulekar (1996)* and later *Parulekar (2001)* simulated the periodic feeding of an antibiotic to stabilize a recombinant culture. They assumed a CSTR with continuous introduction of a limiting substrate and a control antibiotic into the culture medium. Their model is given in **Table 10-5**. Use of an antibiotic adds a material balance:

$$\frac{dA}{dt} = \frac{Q(t)}{V}[A_0(t) - A] - k_3 v(A) X^- - k_4 \omega(A) \tag{10-36}$$

and requires the addition of a term, $-f(A)X^-$, to the RHS of **Eq. (10-28)** in the table. In **Eq. (10-36),** $A_0(t)$ and A (UL^{-1}) are the feed and bioreactor antibiotic concentrations while k_3 and k_4 (h^{-1}) are kinetic constants relating to the cellular uptake and the stability of the antibiotic.

To compare periodic and SS operation, *Lee and Parulekar* used a performance index (PI) for PO:

$$PI = \frac{1}{\tau}\int_0^{\tau}\frac{Q}{V}\left[X^+ - c_A A_0(t) - c_S S_0(t)\right]dt. \tag{10-37}$$

Provided the SS substrate and antibiotic concentrations in the feed are the same as their time-average values in PO, the differences in *PI* measure the improvement through PO. Of course, process variables $A_0(t)$, $S_0(t)$ and $Q(t)$ (see **Table 10-5** and **Eq. (10-36)** are allowed to vary periodically, either singularly or in combination, at a cycling frequency, ω (h^{-1}).

Solutions to the set of equations lead to total washout, partial washout where only $X^+=0$ (reversion prevents $X^-=0$) or to coexistence where $X^+>0$ and $X^->0$. Using stability theory, through evaluating the eigenvalues of the monodromy matrix derived from the Jacobian of the equation set, **Eqns. (10-27)** to **(10-30)** in **Table 10-5** and **Eq. (10-36)** above, the local stability of each solution was evaluated. Not unexpectedly, the researchers showed that coexistence emerges from either washout or partial washout as model parameters change.

Further analysis of the stability of the periodic solution and a comparison with steady-state operation requires kinetic models: *Lee and Parulekar* assumed that specific growth rates followed the Michaelis Menton relation, **Eq. (10-1)**, and that $f(A)$, $v(A)$, $w(A)$ are linear functions of A. With these kinetics, they showed that multiplicity can exist but with only one or two coexisting stable stationary periodic states possible. Also they observed that coexistence in PO has a lower antibiotic demand than coexistence in SS operation.

Examples were discussed for two types of antibiotics: 1) Those destroying plasmid-free cells and 2) those inhibiting growth of such cells. These examined local stability and evaluated *PIs*. The examples considered bang-bang modulation of an antibiotic and in phase, bang-bang modulation of both the antibiotic and the limiting substrate. The maximum values of A_0 and S_0 depended on the cycle split, *s*. The stability analysis indicated that a stable coexistence solution to the model equations meant that a washout solution was unstable, although it did not deny the possible existence of a stable partial washout. With variation of an operating parameter $(A_0)_{SS}$, $(S_0)_{SS}$, Q/V, s or τ, the partial washout solution could bifurcate into two coexisting stationary states or transit into a coexistence solution.

Modulation of just the antibiotic alone failed to improve performance. However, in phase modulation of both antibiotic and the limiting substrate resulted in a significant increase in X^+, the concentration of the plasmid bearing cells, for low values of the time-average

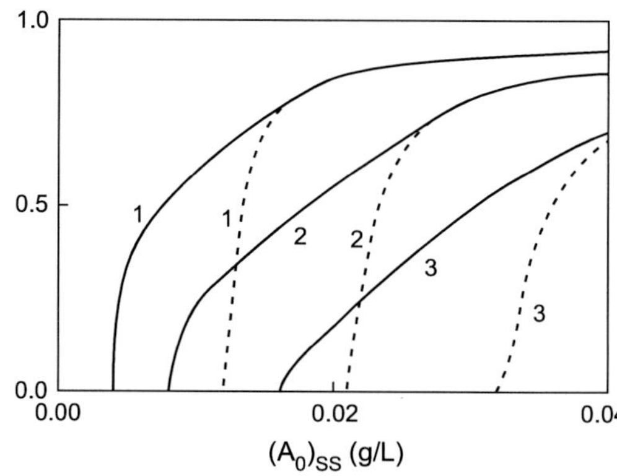

FIGURE 10-12 Variation of the normalized concentration of plasmid containing cells as a function of Q/V, **numbers, and time-average type 1) antibiotic concentration,** $(A_0)_{SS}$, **for specified values of the time-average limiting** $(S_0)_{SS}$ **and** τ_{cycle}. **Solid lines are for** $\tau_{cycle} = 40$ **h, the dashed lines are for SS operation. Numbers give results for three space velocity in increasing order with each higher by a factor of 2.** *(Figure reproduced from* Lee and Parulekar (1996) *with permission. © 1996 by Elsevier Ltd.)*

antibiotic concentration, $(A_0)_{SS}$. The $(A_0)_{SS}$ boundary for improvement depended on $\frac{Q}{V}$, and τ_{cycle}. The magnitude of the difference in *PIs* depended on the cycle split, *s*, and on $(S_0)_{SS}$, the time-average substrate concentration. **Figure 10-12** compares normalized concentrations of plasmid-bearing cells and thus the production of the metabolite versus the time-average feed concentration of a type 1) antibiotic for periodic and steady-state operations. It clearly shows the advantage of PO at the desirable low level of antibiotic use. There is a switch to total washout, however, if the antibiotic concentration is too low. The advantage for type 1) antibiotic was also demonstrated for type 2).

An extension of the above study compared periodic manipulation of the inputs with optimal SS operation (*Parulekar, 2001*). Conditions for improvement by PO of one or more input variables were obtained using the generalized π criterion discussed earlier. Use of the π criterion allowed identification of regions in the variable space that provided improvement through PO vis-à-vis optimal SS operation for the same set of time-averaged variables. Stable solutions to **Eqs (10-27)** to **(10-30)** in **Table 10-5** and to **Eq. (10-36)** above for both periodic and steady-state operation must be found to quantify the improvement. Modulation of space velocity, concentrations of antibiotic, A_0, and limiting substrate, S_0, alone or together but at the same frequency were considered.

Parulekar examined the optimal SS for five sets of parameters for the antibiotic effects (k_3 and k_4 in **Eq. (10-36)**) holding the maximum growth rate and Michaelis-Menten coefficients constant and showed through stability analysis that when two coexisting steady states were found that only one was asymptotically stable. The existence of such a state allowed the use of the generalized π criterion. Modulation of a single input, either S_0 or *Q/V*, improved the *PI*, but only for two of the five sets of k_3 and k_4 considered. Relatively high modulation frequencies were needed. But, at sufficiently high levels of S_0, modulation was unable to increase the index. As *Lee and Parulekar* observed in their first study, modulation of A_0 was not effective.

In-phase modulation of A_0 and S_0 improved the *PI* according to *Lee and Parulekar* (*1996*). *Parulekar* (*2001*) showed that in-phase modulation of A_0 and *Q/V* or S_0 and *Q/V* were also effective strategies. A phase plane plot, **Figure 10-13**, indicates the frequency and $(S_0)_{SS}$ ranges for which modulation of A_0 and *Q/V* elevate performance with respect to an optimal SS operation at $(A_0)_{SS}$ and $(Q/V)_{SS}$. The phase plane plots exhibit quite different shapes for simultaneous modulation of S_0 and *Q/V* or for A_0 and S_0. Antibiotic type, whether destroying cells or inhibiting growth, also exerts an effect on the shape and size of the regions for which in phase modulation of two inputs increased the *PI*. In phase modulation was not essential for improvement over the optimal SS. The π criterion can be used to obtain the optimal phase shift at any cycling frequency and value of $(S_0)_{SS}$. For the type 1) antibiotic, the phase shift is small at low values of modulating frequency ($\omega < 0.15\ h^{-1}$) for A_0 and S_0, but rises to about 90° at $\omega \sim 1\ h^{-1}$. For simultaneous modulation of A_0 and *Q/V*, in phase modulation is not optimal. At low frequencies, the optimal phase shift is about 140°.

Regions of improvement through modulation of all three input variables are difficult to represent compactly in two dimensions. Parulekar (2001) remarked that the region of variable space for improvement is substantially expanded when all the input variables are cycled compared with two variable modulation. They surmised that the *PI* also improved. According to the *PI* for Type 1 antibiotic response depicted in **Figure 10-13**, PO is superior to any SS operation for the range $0.807245 < \omega < 1.2895$. Performance can be improved further by adjusting the cycle split for each variable cycled, by altering the amplitude of the step change, and/

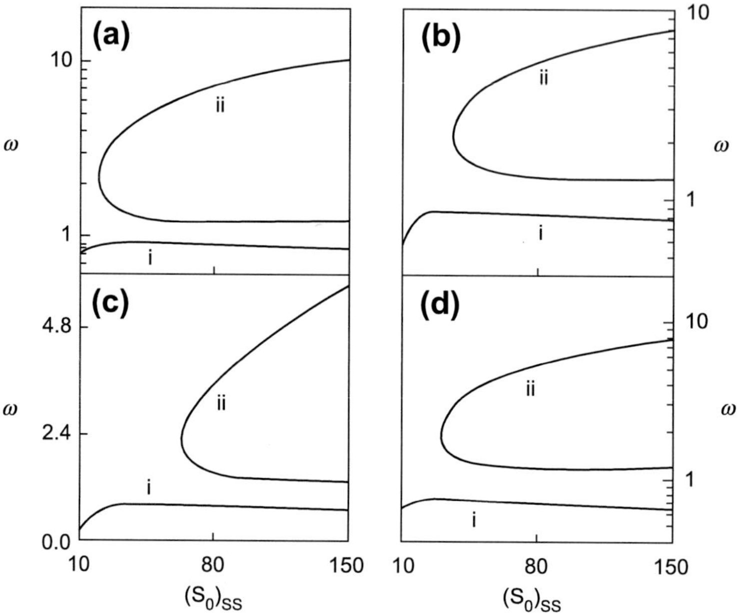

FIGURE 10-13 **Regions of *PI* improvement through in phase, simultaneous modulation of the type 1) antibiotic concentration and *Q/V.* (a) Improvement occurs in the region above curve i and to the left of curve ii. Individual figures (b), (c) and (d) show the variation for different combination of k_3 and k_4. Units of ω are cyles/h.** *(Figure reproduced from* Parulekar (2001) *with permission. © 2001 by Elsevier Ltd., this is Figure 5 from the paper.)*

or by introducing a phase shift. **Figure 10-14** shows optimal phase shift versus optimal frequency for a fixed $(S_0)_{SS}$. In determining the optimal phase shift, cycle splits for A_0 and *Q/V* have been allowed to vary. Results are for a type 2) antibiotic that inhibits cell growth. The *Parulekar* study suggests that there are a variety of cycling parameters that can be tweaked to improve biochemicals production from recombinant cell cultures.

10.8 APPLICATIONS TO IMMOBILIZED CELLS OR ENZYMES

Limits on continuous operation imposed by cell washout can be overcome by fixing living cells or enzymes onto a solid. This is particularly important for the production of antibiotics where the antibiotic is largely produced in the low growth stage of the organism's life cycle. Solids containing the microbes can be packed into a bed, fluidized by air in a bubble column, or by either air or liquid in a circulating fluidized bed. A variety of circulation methods are used. Several experimental and simulation studies of modulating such systems have been undertaken. *Sayles and Ollis* (*1989*) reported work by *Forberg, Enfors and Haggstrom* in which those researchers modulated the supply of a nitrogen-containing nutrient to an immobilized Cl. Acetobutylicum, fed by a constant flow of glucose, to promote the production of a butanol metabolite relative to biomass formation. A remarkably asymmetric 1:32 cycle was used, in which there was 0.25 h of nutrient feed in an 8 h period. This reduced the biomass leakage to butanol production by a factor of 25.

Arcuri et al. (*1983*), using an immobilized *S. cattleya* in a bubble column to produce an

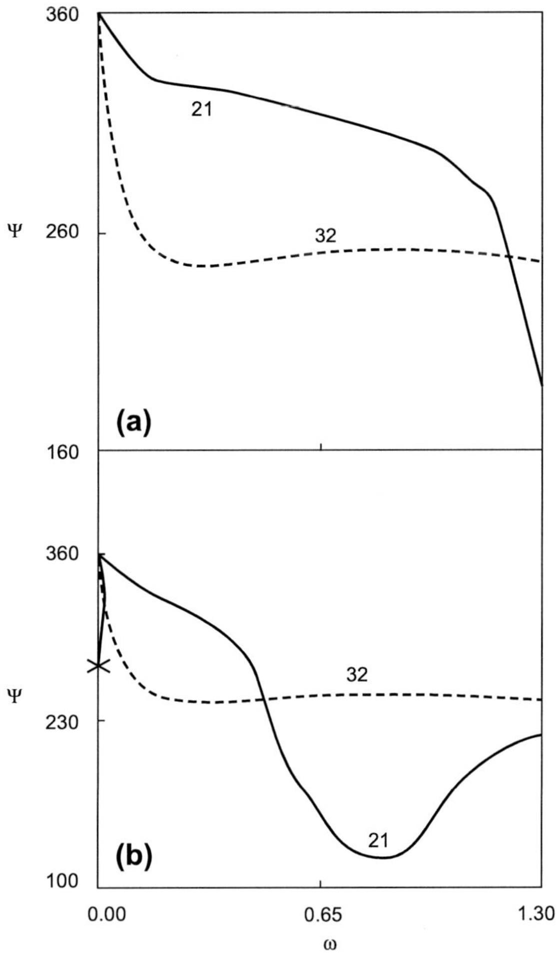

FIGURE 10-14 **Diagram of optimal phase shifts, ψ, and frequencies, ω, for modulation of three inputs (A_0, S_0, and Q/V) at a fixed value of $(S_0)_{SS}$ and $s = 0.5$ for S_0, but variable for A_0 and Q/V in a); with $s = 0.5$ for Q/V and variable for A_0 and S_0 in b). Numbers on the curves designate the shift: 21 is the phase shift of S_0 relative to A_0, 32 is the phase shift of Q/V relative to S_0. X is the phase shift 32 at $\omega = 0$.** *(Figure reproduced from* Parulekar (2001) *with permission. © 2001 by Elsevier Ltd.)*

antibiotic metabolite, thienamycin, demonstrated higher production of the antibiotic by periodically adding a maintenance media that restored organism growth to the continuous feed to a bubble column, a scheme that starved the organism and led to antibiotic expression. A dynamic model for this type of "starve-regenerate" periodic operation was advanced by *Sayles and Ollis (1989)*. It assumed uniform distribution of immobilized cells in a spherical, porous solid, however the total mass of cells could wane or wax up to a maximum value. A limiting nutrient was assumed that diffused into the porous solid where it was consumed by cells following a simple Monod model. Metabolite, formed within the solid, diffused outward. Diffusivity of nutrient and metabolite were different, but both depended on the ratio of cell mass to its maximum value, the void fraction in the porous solid and the ratios of the bulk diffusivities of nutrient and metabolite. Those assumptions resulted in a second-order partial differential equation. The solids containing the organisms were held in a small, well-mixed bath so that the nutrient concentration was effectively modulated as a square wave by a high dilution rate. Mass transfer at the surface of the particles was neglected, resulting in simple boundary conditions.

Assuming model parameters from experimental data on *Z. mobilis*, *Sayles and Ollis*, solved their model numerically for square-wave cycling of the substrate concentration fed to the immobilized cell reactor. An objective was to avoid cell leakage from the reactor. This was done by letting τ_{cycle} become a function of maximum and minimum cell mass for a non-leaking operation. *Sayles and Ollis* found that the advantage of substrate modulation depended critically on product formation kinetics. If this falls with increasing specific growth rate, the time-average metabolite formation rate can be many-fold the SS rate at the same time-average substrate concentration in the feed. Productivity, as product yield per unit reactor volume, also increased. On the other hand, if the specific formation rate increased linearly with μ or as a convex function of the specific growth rate, the time-average yield did not increase over that obtainable by SS operation. Productivity actually decreased substantially. *Sayles and Ollis* also observed a large radial cell density gradient and a waxing and waning of this gradient as the feed frequency of the substrate changed. The zero cell leakage solution, however, was unstable. Stable operation required some microorganism loss.

Constantinides and Mehta (*1991*) suggested a two-phase model that divides the reactor into a well-mixed fluid phase and an isotropic solid phase of immobilized cells. Their lumped parameter model replaced substrate and product diffusion into the supported cell mass by mass transfer terms. The model, given in **Table 10-6**, is much simpler than the *Sayles-Ollis* model just examined. It allows for cell leakage so that some reaction occurs in the fluid phase. Kinetics in the two phases were assumed to be the same and the Michaelis-Menten expression was used. *Constantinides and Mehta* applied their model to the production of candicidin, a non-growth metabolite and antibiotic, in the fermentation of glucose by *Streptomyces griseus*. Their simulations showed that the volumetric productivity of candicidin could be increased by nutrient modulation. The optimum cycle period depended on the saturation constant K_S in the specific growth rate model. Values of the order to 10^{-5} predicted τ_{opt} of less than 12 h. At this

TABLE 10-6 Two Phase Dynamic Model for a Fluidized Bed, Immobilized Live Cell Bioreactor

Fluid Phase

$$\frac{dX_f}{dt} = k_X\left(X_s - \varepsilon X_f\right) + \frac{\mu_{\max} S_f X_f}{K_S + S_f} - k_D X_f - (Q/V)X_f \quad (10\text{-}38)$$

$$\frac{dS_f}{dt} = k_S(S_s - \varepsilon S_f) - \frac{\mu_{\max} S_f X_f}{Y_X\left(K_S + S_f\right)} + \frac{Q}{V}(S_{f_0} - S_f) \quad (10\text{-}39)$$

$$\frac{dP_f}{dt} = k_P(P_s - \varepsilon P_f) + \frac{\sigma K_P X_f}{\left(K_P + S_f\right)} - \frac{QP_f}{V} \quad (10\text{-}40)$$

Immobilzed Cell Phase

$$\frac{dX_s}{dt} = -k_X(X_s - \varepsilon X_f) + \frac{\mu_{\max} S_s X_s}{K_S + S_s} - k_D X_s \quad (10\text{-}41)$$

$$\frac{dS_s}{dt} = -k_S(S_s - \varepsilon S_f) - \frac{\mu_{\max} S_s X_s}{Y_X(K_S + S_s)} \quad (10\text{-}42)$$

$$\frac{dP_s}{dt} = -k_P(P_s - \varepsilon P_f) + \frac{\sigma K_P X_s}{(K_P + S_s)} \quad (10\text{-}43)$$

relatively high cycling frequency the fluidized bioreactor appeared to function in a relaxed steady state. Decreasing K_S to about 10^{-8} shifted the optimal cycle period, τ_{opt}, to about 50 h and the simulation predicted sharp swings with time in the specific candicidin productivity.

Constantinides and Mehta tested their model experimentally; modulating a phosphate nutrient rather than glucose. This nutrient inhibits candicidin expression while simultaneously fostering mycelial growth. The *S. griseus* cells were immobilized in a κ carrageenan material as 2 mm beads. These were used in the liquid fluidized bed shown in **Figure 10-15**. Other vessels were an aerator used to saturate the liquid phase and a disengaging column. Operation was continuous with modulation performed by switching feed of the phosphate nutrient between two tanks as seen at the bottom of the figure.

Experimental results confirmed a simulation which predicted that forcing the phosphate nutrient flow significantly increases volumetric productivity of candicidin over SS operation. The largest increase was seen at a cycle period of 16 h. *Constantinides and Mehta* observed that at a cycle period of 12 h, the fluidized reactor operated in a relaxed steady state. Oscillations in candicidin concentration in the product stream were just perceptible. On the other hand, a 96 h cycle shifted the reactor system into a quasi-steady state.

A periodic operation consisting of aerobic and anoxic steps has been employed for waste treatment in sequencing batch reactors (SBRs). This has proven to be more flexible than continuous operation for small-scale waste treatment plants (*Irvine et al., 1997; Wilderer et al., 2001*). In continuous flow bioreactors, the specific biomass growth rate is determined by t Q/V. Dilution rates higher than the maximum specific growth rate result in biomass loss. An SBR operation is shown in **Figure 10-16**. It typically consists of a short feeding period, followed by a longer reaction period in anoxic and aerobic conditions, a settling period and a final period of wastewater withdrawal from the bioreactor. In SBRs, the duration of cyclic operations can be varied for the same sludge residence time. Unlike SS conditions in continuous bioreactors, the biomass grows under transient conditions in periodic processes. The specific growth rate is usually highest following the feeding cycle and gradually declines as the limiting substrate is exhausted.

Dionisi et al. (2001) used acetate as a substrate in their investigation of an activated sludge SBR. Variables included the loading rate, feeding cycle duration, sludge age, aerobic and anoxic reactor cycle times. They chose a multi-step model whose first step was the internal accumulation of substrate (X_{acc}) from the feed (S) during anoxic operation:

$$\frac{dX_{acc}}{dt} = q_{acc}\frac{S}{K_{acc}+S}X_h\left(1-\frac{X_{acc}/X_h}{f_{acc}}\right). \quad (10\text{-}44)$$

In this relation, X_h is the concentration of viable biomass. The term, q_{acc}, represents the maximum rate of internal acetate accumulation and f_{acc} is the maximum internal level of accumulated acetate. The latter is used either for growth or is converted to a storage compound. The authors claimed that the storage compound is mainly poly-hydroxybutyrate (PBH). The rate of internal PHB accumulation, X_{PHB}, is:

$$\frac{dX_{PHB}}{dt} = q_{PHB}\frac{X_{acc}/X_h}{K_{PHB}+X_{acc}/X_h}X_h\left(1-\frac{X_{PHB}/Xh}{f_{PHB}}\right). \quad (10\text{-}45)$$

In this relation, q_{PHB} is the maximum specific rate and f_{PHB} is the maximum internal content of PHB. Growth is assumed to be dependent on either the accumulated substrate (acetate) or the stored product (PHB), and the previous "famine" period during the previous reaction and withdrawal cycles:

$$\frac{dX_h}{dt} = \mu_{\max}(t)\frac{X_{acc}/X_h}{K_{acc}+X_{acc}/X_h}X_h \quad or \quad \mu_{\max}(t)\frac{X_{PHB}/X_h}{K_{PHB}+X_{PHB}/X_h}X_h. \quad (10\text{-}46)$$

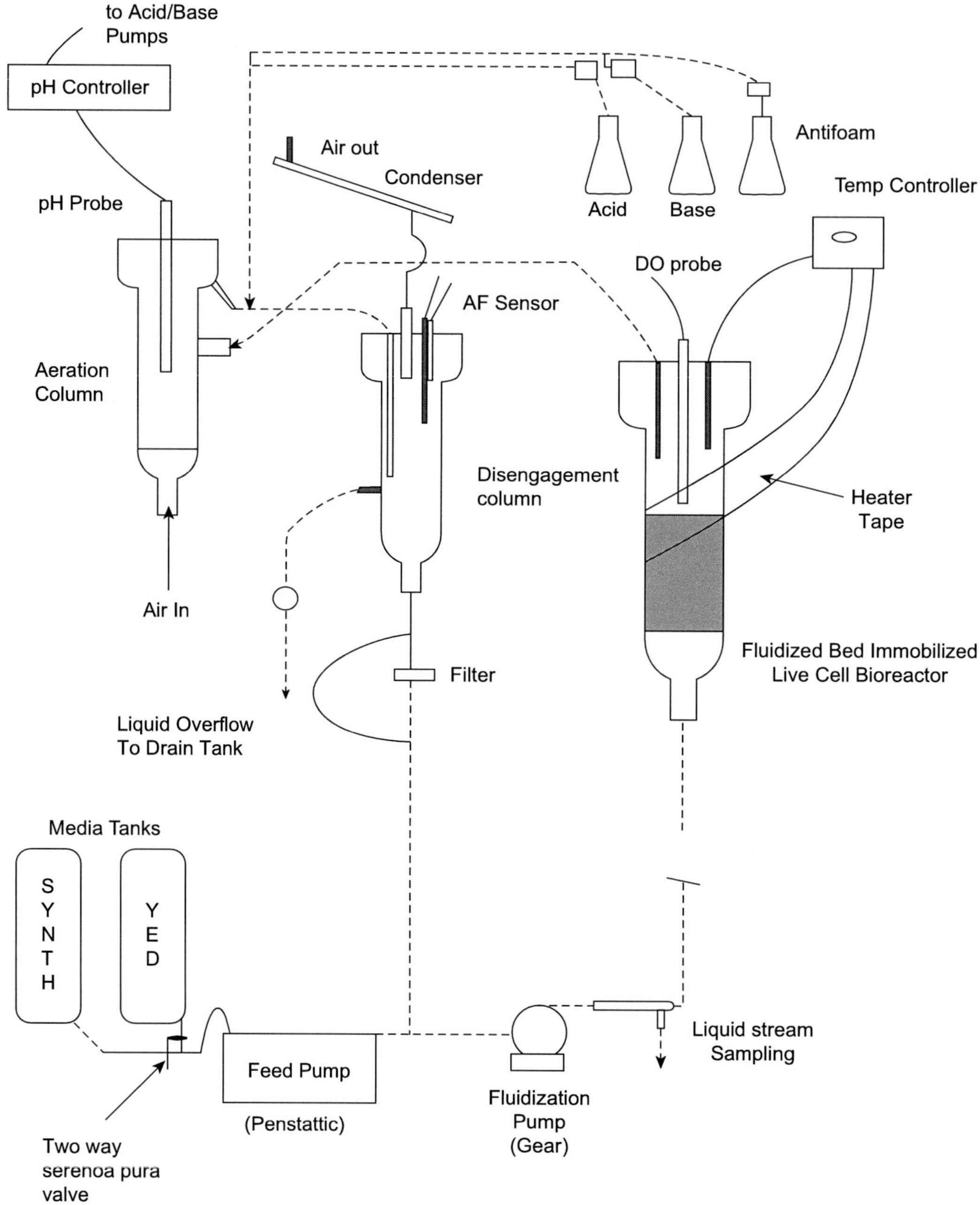

FIGURE 10-15 Liquid fluidized bed bioreactor used to study the modulation of a phosphate nutrient for aerobic glucose fermentation by *S. griseus*. *(Figure reproduced from* Constantinides and Mehta (1991) *with permission. © 1991 by Wiley Periodicals Inc.)*

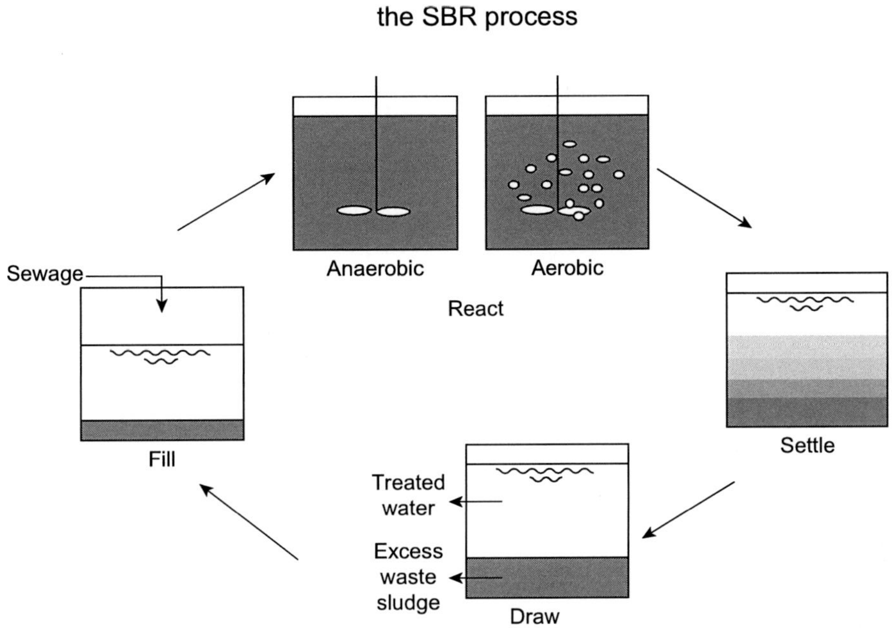

FIGURE 10-16 **Unit operations during Sequencing Batch Reactor (SBR) operation in a waste treatment system.**

The maximum specific growth rate term, $\mu_{\max}(t)$, is assumed to be time dependent. This dependence is given by:

$$\mu_{\max}(t) = \mu_{\max}(t_0) + \left[\mu_{\max}^{\infty} - \mu_{\max}(t_0)\right]\frac{t}{K_t + t}. \tag{10-47}$$

The *Dionisi et al.* (2001) model included a maintenance term for the non-growth associated consumption of the stored polymer. Its purpose was to explain the experimental results. However, the authors recognized its empirical nature and the necessity of parameter adjustment to fit different operating conditions. They recommended cautious application of the model to SBR design and operation, since the experimental SBR operated on a synthetic media. Nevertheless, the authors claimed that the SBR concept and their mathematical model can be adapted for treatment of industrial and domestic wastes.

10.9 FED-BATCH OPERATIONS

Fed-batch culture of various recombinant organisms, ranging from bacteria to mammalian cells, has become the biopharmaceutical industry's platform of choice for therapeutics production (*Dutton and Scharer, 2007*). The operation involves semi-continuous or periodic feeding of a variable amount of substrate to a growing culture. It begins as a batch culture until the late logarithmic phase of growth when fresh nutrient addition is initiated. The feed medium contains the growth-limiting substrate, but often secondary nutrients and growth factors are added. Stoichiometric feeding is often used for fed-batch culture of mammalian cells. This comprises the addition of essential nutrients in proportion to their uptake from the medium (*Xie and Wang, 1994a, b, c*). The controlled feeding strategy achieves high cell density without accumulating growth-limiting by-products.

Submerged fed-batch culture may operate at either constant or variable volume. Volume can be maintained essentially constant by supplying a highly concentrated substrate solution. This type of feeding is particularly advantageous if the desired cell product is growth associated, hence the product concentration is directly proportional to the cell concentration. In variable volume operation, the substrate concentration is more dilute; consequently, volume increases upon each scheduled feeding. A variable volume system is advantageous if the formation of the recombinant is not associated with growth; consequently, productivity depends on maintaining high viability for long time periods. In this case, non-replicating cells continue to synthesize the desired product. To avoid excessive dilution, the volume increase is usually less than double the starting volume. Less often in biopharmaceutical manufacture, some medium with in the product is withdrawn periodically after feeding, resulting in a cyclic fed-batch operation. This procedure is common in small-scale waste treatment systems, known as sequencing batch, or draw and fill operations (*Irvine et al., 1993; Wilderer et al., 2001*).

There are several advantages of fed-batch over other culture methods. Problems related to substrate inhibition and/or catabolite repression that often plague batch processes are minimized. Since product is harvested once, quality control is the same as in batch cultivation. Although productivity tends to be less than in continuous operation, the product titer at the end of a fed-batch operation is much higher, thus mitigating possible problems in product purification. The bioreactor configuration is the same for fed-batch and batch culture. Additional equipment and analytical tools needed are minimal. In general, fed-batch processes are easily automated and controlled.

Due to simplicity and straightforward implementation, dissolved oxygen-based control (DO-stat) is often employed in aerobic industrial fermentation to schedule nutrient feeding in fed-batch operations (*Yamane and Shimizu, 1984*). The objective is to maintain the DO concentration at a constant level above its critical value by manipulating the feed rate. The first step in applying DO-stat is to establish the critical oxygen concentration of the organism. This is defined as the minimum oxygen concentration needed to maintain a fully aerobic metabolism. It is a characteristic of the microorganism and ranges usually between 0.1 and 1 mg/L O_2. Then, the oxygen concentration set point can be established experimentally at some concentration above that critical concentration.

The key limiting substrate being fed is usually the energy source and its metabolism is directly linked to O_2 consumption. The rate equation for the DO concentration is:

$$\frac{dO_2}{dt} = k_{la}\left(O_2^* - O_2\right) - q_{OX}\,\mu(s)X. \quad (10\text{-}48)$$

In the above relation, k_{la} is the volumetric mass transfer coefficient, O_2^* the dissolved O_2 concentration at saturation and q_{OX} is the normalized oxygen uptake (mg O_2/g cells). Exhaustion of the limiting substrate causes the DO in the medium to rise. Thus, fresh substrate is pumped into the bioreactor when the DO rises above the set point. Since the O_2 concentration profile in the fermentor is often noisy, smoothing by a time series (moving average) filter may be required to avoid chatter of the feed pump. In principle, the DO-stat is an on-off controller. A drawback of the DO-stat, fed-batch system is that an inadvertent medium overfeed causes the O_2 concentration to fall below the critical value. Excess glucose feed in many yeast and microbial cultures causes a switch to an anaerobic metabolism even at relatively high O_2 concentrations (Pasteur Effect). This could result in sub-optimal operation in which the growth of the microorganism becomes linear rather than the more desirable exponential pattern (*Nor et al., 2001b*). Attempts to improve fed-batch performance include a hyperbaric oxygen supply (*Belo et al., 2003*) and the use of two different dissolved oxygen set points with optimized bandwidths to turn the feed pump on and off (*Lee et al., 2003*).

Based on the work of *Oh et al.* (*1998*), *Nor et al.* (*2001a*) developed a novel feed forward exponential feeding schedule based on the maximum substrate uptake rate (MSUR) employing the fungus, *Kluyveromyces fragilis*, as the test organism. This organism excretes the enzyme, lactase. Their on-line strategy utilized measurements of the DO concentration to estimate periodically the specific growth rate and consequently the substrate feed rate to an aerobic fed-batch culture. The estimate is based on evaluation of the MSUR by the use of a DO probe. The methodology is illustrated schematically in **Figure 10-17** which shows the accumulated lactose feed, the feed rate, and lactose and DO concentrations in the fermentor. Since the feed solution is highly concentrated, this is essentially a constant volume operation.

Lactose is fed continuously at a predetermined rate at fermentation time $t < t_1$. At time t_1, the flow of substrate is interrupted and the residual concentration declines toward zero. Since lactose metabolism is associated with oxidative phosphorylation, exhaustion of lactose in the medium results in a rapid rise of the DO concentration at time t_2. This rise is sensed and a given amount (G) of substrate is added rapidly to the system. As this substrate is utilized, the DO drops again. At time, t_3 the added amount of substrate is exhausted and the DO level rises again. Using this information, the "excess" substrate level (H_1) in the bioreactor at t_1 can be calculated:

$$H_1 = \frac{t_2 - t_1}{t_3 - t_2} G. \qquad (10\text{-}49)$$

Using the value for H_1, the "corrected" specific growth rate, μ_f, between fermentation time t_3 and the next MSUR test, t_f, can be predicted. The feed rate, F, in turn, is given by:

$$F = \frac{\mu_f}{Y_{X/S}\left(S_{in} - S\right)} V_3 X_3 \exp\left[\mu_f(t - t_3)\right]. \qquad (10\text{-}50)$$

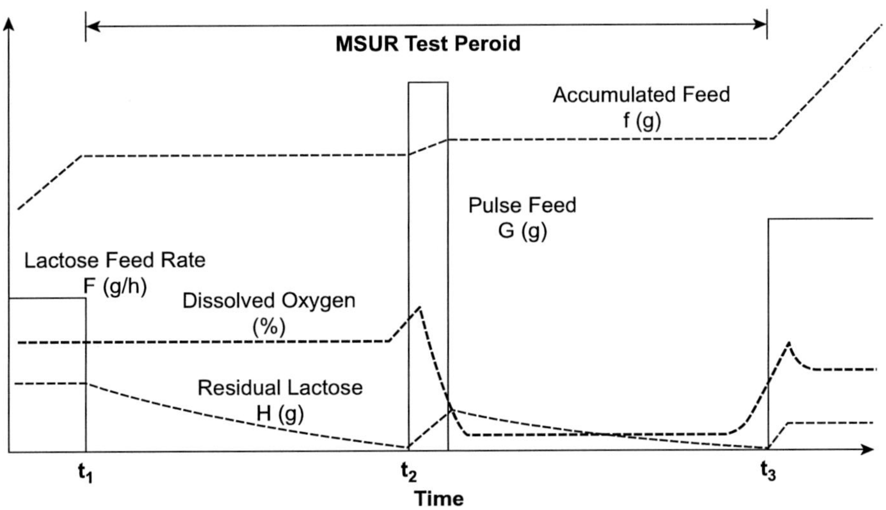

FIGURE 10-17 Time profile of the substrate (lactose) and dissolved oxygen (DO) during on-line measurement of the MSUR for estimating the fed-batch feed rate. *(Figure reproduced from* Nor et al. (2001a) *with permission. © 2001 by Elsevier Ltd.)*

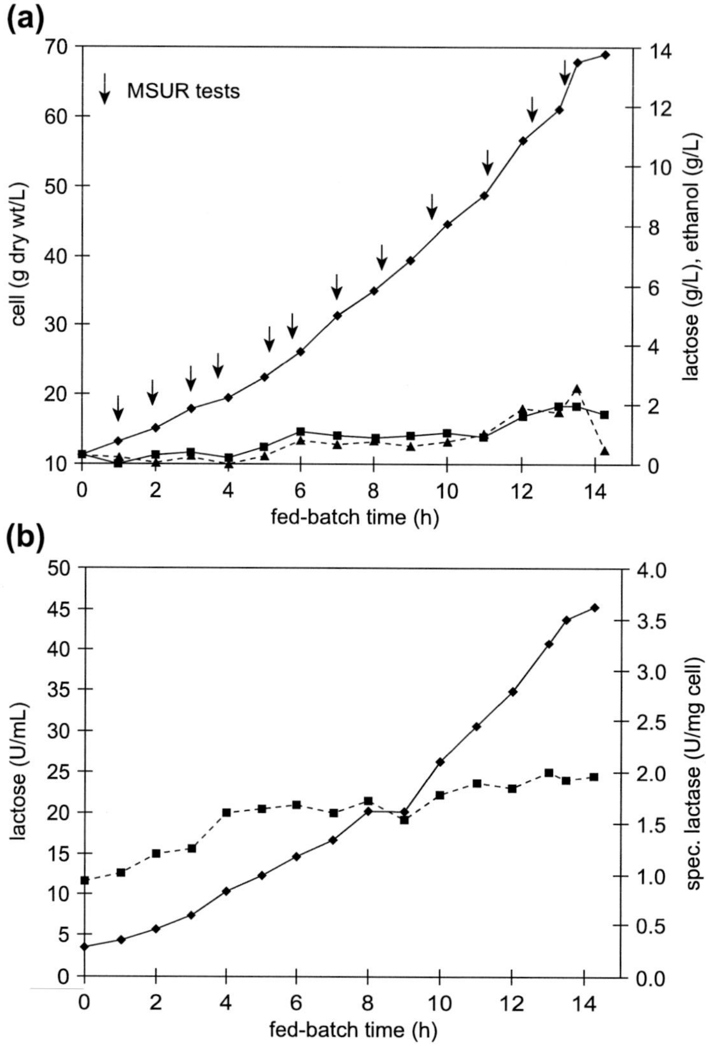

FIGURE 10-18 **Typical fed-batch culture of *Kluyveromyces fragilis* producing lactase with an adaptive, corrected feed-forward strategy: a) Biomass ◆, lactose ■, and ethanol ▲ concentrations, b) Lactase concentration ■ and specific activity ◆.** *(Figure reproduced from* Nor et al. (2001a) *with permission.* © *2001 by Elsevier Ltd.)*

$Y_{X/S}$ is the biomass yield for the substrate, S_{in} is the substrate concentration in the feed, S is the set-point concentration, V_3 and X_3 are the volume and the estimated biomass concentration at time t_3 in **Eq. (10-50)**.

Figure 10-18 shows a typical MSUR test schedule and the profiles of the biomass and lactase concentrations. Using a feed-forward feeding strategy, nearly exponential growth can be maintained throughout the fed-batch operation. A maximum biomass concentration of 69 g/L is significantly higher than reported previously for this organism. The increase of lactose concentration above the set point indicates a slight overfeeding, but without evident ill effect on growth rate or on biomass yield.

As shown in **Figure 10-18(b)**, the volumetric lactase concentration increased exponentially as the biomass increased indicating that lactase production is growth associated. The specific productivity rapidly approached a steady-state value of 1.75—2.00 U/mg cell.

Nutrient model based adaptive control of a fed-batch culture requires conventional mass balance models for the biomass, X, limiting substrate, S, desired product, P, and the liquid volume, V. Typical relations for fed-batch culture are given in **Table 10-7**.

Dependence of the specific growth rate (μ) on substrate concentration is often expressed by the Monod Equation (**Eq. (10-5)** in **Table 10-2**), but any of the relations in **Table 10-2** may be more appropriate. The flow rate (Q) is time-varying and one of the variables to be optimized. The medium may be added batch-wise rather than in a continuous manner. The rate of product formation includes both growth associated ($\alpha\mu$) and non-growth associated (β) terms.

Murine hybridoma 130-8F producing anti-F glycoprotein monoclonal antibody (MAb) was grown in batch and fed-batch modes with glutamine as the limiting substrate *(De Alwis at al., 2005; Dorka et al., 2009)*. The objective was to evaluate the enhancement of MAb productivity by fed-batch cultivation. MAb production was found to be partially growth associated. The fed-batch medium was analyzed off-line for glucose and glutamine and defined amounts of glutamine and glucose solutions were injected when the glutamine concentration fell below a setpoint level. The data were used by *Dorka et al. (2009)* to develop a mathematical model for optimizing a fed-batch operation; details will be found there.

TABLE 10-7 Dynamic Models for Fed-Batch Operation

$$\frac{dX}{dt} = \left[\mu(S) - \frac{Q(t)}{V}\right]X \qquad (10\text{-}51)$$

$$\frac{dS}{dt} = \frac{Q(t)}{V}\left[S_0(t) - S\right] - \frac{\mu(S)}{Y_{X/S}}X \qquad (10\text{-}52)$$

$$\frac{dP}{dt} = (\alpha\mu(S) + \beta)X - \frac{Q(t)}{V}P \qquad (10\text{-}53)$$

$$V = V_0 + \int Q(t)dt \qquad (10\text{-}54)$$

Usually, animal cell culture is employed for glycosylated biopharmaceutical and viral vaccine production. The expression of authentic glycosylated biopharmaceuticals in bacteria is a relatively recent development. Fed-batch culture has been proven to be advantageous for the production of recombinant glycoprotein vaccines by *E. coli* (*Ihssen et al., 2010*). Glycosylation enhances the efficacy of a vaccine by affecting long-term, T-lymphocyte dependent immunity. Normally bacteria do not naturally glycosylate, consequently the antigen and polysaccharides are combined chemically to enhance the vaccine's immunological properties. To eliminate tedious chemical procedures, separate antibiotic-selectable plasmids were used for the periplasmic expression of the toxoid (vaccine) and the polysaccharide polymerase and glycotransferase enzymes. Since correctly conjugated glycoprotein vaccines are synthesized later during the course of the culture, the induction and pulse feeding schedule greatly influenced glycosylation. By applying appropriate induction and fed-batch strategies, the *in vivo* synthesis of conjugated vaccines was increased 40-fold over batch processes according to the authors.

Optimization of fed-batch culture can be challenging, often involving resolution of high-order, nonlinear and multimodal systems. Several model-based strategies have been developed for optimizing fed-batch process parameters. *Mekarapiruk and Luus (2001)* applied iterative dynamic programming with unspecified initial conditions to optimize a feed-rate policy for producing penicillin. Considered were the feed-rate profile, initial volume and initial substrate concentration. Markov Chain-Monte Carlo (MCMC) statistical methods such as Gibbs parameter sampling and the Metropolis-Hasting algorithm also have been employed to estimate model parameters (*Skolpap et al., 2004*). Evolutionary algorithms, such as genetic algorithms mimicking the principles of natural biological evolution (*Chen and Nguang, 2002; Sarkar and Modak, 2003; Chen et al., 2004; Skolpap et al., 2008a*), or particle swarm optimization (*Jayaraman et al., 2001; Skolpap et al., 2008b*) have also been applied to solve for optimal feed-rate profiles. In addition to those population-based search techniques, many point-based search techniques have been applied to determine optimal feed-rate profiles. For instance, *Cuthrell and Biegler* (*1989*) employed orthogonal collocation-based sequential quadratic programming to optimize fed-batch culture for penicillin production. The model for the culture, which contains four state variables, was first studied by *Lim et al.* (*1986*).

TABLE 10-8 Performance Criterion for Fed-Batch Operation

Production	$J_T = C_P V$	(10-55)
Productivity	$J_p = \frac{J_T}{T_f}$	(10-56)
Product Yield	$J_Y = \frac{J_T}{\int S_0 \; Q(t)\, dt}$	(10-57)

Several performance criteria (*J*) can be applied for optimization (*Jadot et al., 1998*). Typical choices are given in **Table 10-8**.

The first criterion, J_T, is simply the harvested amount. The second is defined as the harvested amount per the total time of operation and includes batch and fed-batch operation, while the third is the ratio of the harvested product to the total amount of substrate added. *Jadot et al.* (*1998*) showed conflict between the criteria as may be seen in **Figure 10-19**, where the higher the product yield, the less is the productivity. To

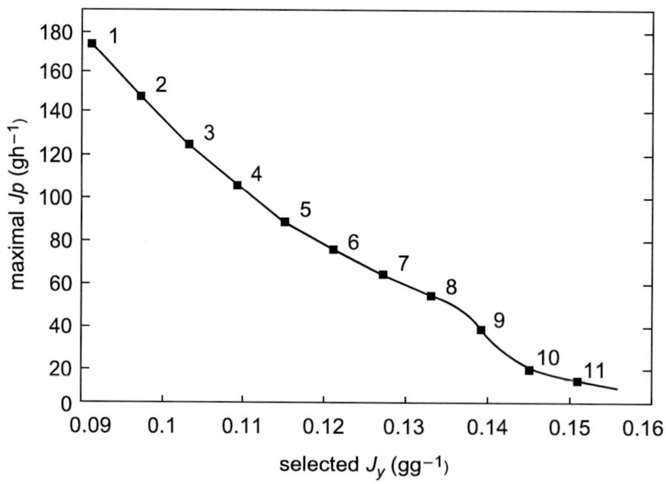

FIGURE 10-19 Maximal productivity in fed-batch operation at selected product yields. *(Figure from* Jadot et al. (1998) *reproduced with permission. © 1998 by Elsevier Science Ltd.)*

resolve this conflict, they proposed a parametric optimization algorithm which they tested on the model that generated the above figure. Fermentation was initiated as a batch process for 12 h followed by fed-batch operation until the initial volume of 250 L increased to 400 L. The algorithm of *Jadot et al.* allowed adaptive control of the limiting substrate feed. The design value of J_Y ranged from nearly 0 gg^{-1}, point 1 in **Figure 10-19**, to 0.16 gg^{-1}, point 11 in the figure. The optimum compromise between J_p and J_y was located at point 8 and the operation was halted then. There was an unexpected hysteresis between the substrate load and the J_y setpoint.

Several model-based strategies have been proposed to optimize feeding schedules in a fed-batch culture. MCMC methods (*Skolpap et al., 2004*) were compared with particle swarming and genetic algorithms by *Skolpap et al. (2008b)* for optimizing recombinant protein while minimizing unwanted protease production by *Bacillus subtilis*. The model considered cell mass, substrate uptake, target product and by-product formation, and channeling various intermediary metabolites (ethanol, acetate, lactate) into oxidative and overflow metabolic pathways. Model parameters for the differential equations were determined using batch fermentation. Fermentation started as a batch process while feeding began at some time, t_S. They were run under glucose limiting conditions. The substrate (glucose and salts) feed rate, $Q(t)$, was parameterized as follows:

$$Q(t) = Q_0 \; e^{\kappa(t-t_S)} \quad \text{if } t > t_S, \tag{10-58}$$

where t is fermentation time, t_S the start of the feeding time and κ is a specific flow parameter. Thus, the feed rate will be a piecewise exponential function. Indeed, *Nor et al. (2001a, b)* showed that the optimum feed rate for bacterial fermentations for growth-associated or partially growth-associated product formation is an exponential function. The parameters to be optimized for feed rate are t_S, u_0 and κ. In practice these parameters are subject to constraints, such as pump speed or fermentor volume.

Optimum values of those design parameters were estimated using the Metropolis-Hastings algorithm (*Gilks et al., 1998a*), an MCMC method. This method is a powerful tool, especially when there is prior knowledge about model parameters. It is applicable to any set of observations that conform to a Markov chain. A set of independent observations, $y_1, y_2, y_3 \ldots y_{n-1}, y_n$, form a Markov chain if the following conditional probability is obeyed, namely that the probability of a current observation (i.e., observation "*n*") is conditional on only the previous observation ("*n-1*"). Many data sets from natural processes such as growth, death, etc., are Markov chain observations. For further discussion of the Metropolis-Hastings algorithm, see *Gilks et al. (1998b)*.

To solve the fed-batch optimization problem using the genetic algorithm, the Genetic Algorithm Optimization Toolbox (*Houck et al., 1995*) was implemented in MATLAB Version 7.04 (MathWorks, Inc., Natick, MA). Because population size is an important parameter of GA optimization, an adequate population size for the fed-batch optimization problem was needed prior to implementation. The maximum numbers of generations as well as the crossover and mutation operations were set to be proportional to the population size. The particle swarm optimization (PSO), developed by *Eberhart and Kennedy (1995)*, a stochastic search technique, was also used for feed optimization.

The optimum biomass and α-amylase concentration profiles are shown in **Figure 10-20** where the data points correspond to the optimum trajectory predicted by the MCMC algorithm (*Skolpap et al., 2004*). Curves show the optimum trajectories predicted by the genetic algorithm (GA) and PSO algorithms. Optimal parameter values obtained by MCMC (*Skolpap et al., 2004*), GA and PSO (*Skolpap et al., 2008a, b*) were very similar.

Biological parameters, for example: The specific growth and death rate, product yield

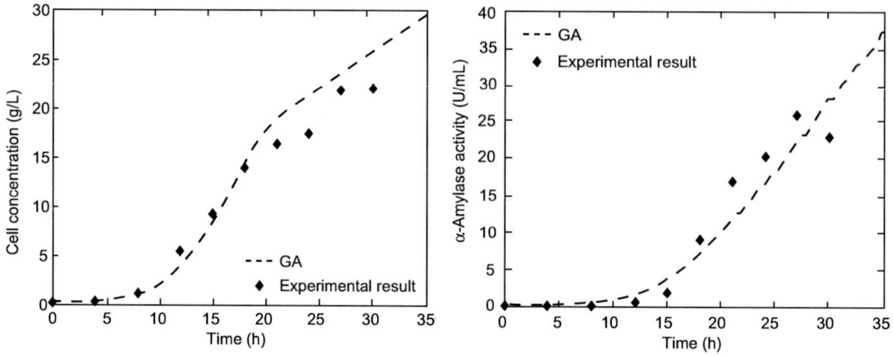

FIGURE 10-20 Optimized profiles of biomass and α-amylase concentration in fed-batch culture of recombinant *Bacillus subtilis*. *(Figure from* Skolpap et al. (2008b) *with permission. © 2008 by Elsevier Ltd.)*

and productivity, are often time variables or uncertain. For this reason adaptive control of fed-batch operations is often employed by industry. Adaptive control involves periodic modifications of the control relations to cope with uncertainties of the parameters.

Unlike robust control, adaptive control does not require a priori knowledge of the parameters. *Chen et al.* (*2004*) established the theoretical groundwork of a model-based adaptive controller, known as a Model Reference Adaptive Controller (MRAC), for the adaptive nonlinear regulation of fed-batch bioreactors. The authors applied such adaptive control to industrial scale bioethanol production using yeast. High glucose concentration tends to induce glycolysis which produces ethanol at the expense of biomass. On the other hand, low glucose concentration limits growth rate resulting in suboptimal biomass production. *Chen et al.*'s objective was a controller imposing a set point on the glucose concentration. The controlled parameter was the dilution rate (*Q/V*) for continuous yeast fermentation. The MRAC proposed was:

$$Q/V = \frac{(\lambda_1 + \lambda_2 X)\,(E^* - E) + \theta X - Q_E V^{-1}}{(yS_{in} - E)} \qquad (10\text{-}59)$$

where E^* and E represent the set-point and model-based estimates of the ethanol concentration respectively, X is the model-based biomass concentration, θ is the specific ethanol productivity, y is the ethanol yield from the substrate and Q_E refers to the ethanol outflow. The parameters λ_1 and λ_2 were defined as:

$$\frac{dE}{dt} = \lambda_1(E^* - E) \text{ and } \frac{dE}{ds} = \lambda_2(E^* - E) \text{ where } s(t) = \int_0^t X(t)dt. \qquad (10\text{-}60)$$

This algorithm was applied for about 60 h. Adaptive control was begun after 8 h of operation. To test controller robustness, control action was switched off for about 2 h after 48 h of operation, then on again. Glucose returned rapidly to the set point.

Adaptive control was also applied successfully to an animal cell culture of the HEK-293 cell line by *Siegwart et al. (1999)*. Glucose measurements were made on-line every 30 minutes during the culture period. It appeared to be beneficial to maintain low glucose concentration in the culture medium to prevent excess lactic formation by glycolysis, as lactic acid is a known inhibitor of animal cell growth. The control objective of 1.0 mM was achieved using a slightly modified version of **Eq. (10-59)**:

$$F_{t+1} = \left(Q/V\right)_{t+1} V = \frac{C(S^* - S_{t+1}) + \theta_{t+1} S_{t+1}}{S_{in} - S_{t+1}} V \quad (10\text{-}61)$$

where F = glucose feeding rate and S = measured glucose concentration (M). Other terms are defined in the **Nomenclature**. The control parameter, C, is proportional to the biomass concentration. However, C cannot not be evaluated precisely. The authors proposed estimating C as:

$$C = C_0 \left[1 + \frac{\theta}{\alpha \theta}\right]. \quad (10\text{-}62)$$

The value of C_0 was set at 0.4 and α typically varied between 3 and 5. The above algorithm gave a satisfactory performance at different phases of growth. Lactate production was significantly reduced compared with uncontrolled batch culture operating under the normal initial glucose concentration. Specific growth rate was higher in fed-batch operations, but the ultimate cell concentrations after 200 h of culture time were approximately the same.

Ronen et al. (2002) employed an evolutionary algorithm to optimize the fed-batch feeding profile of the yeast-like fungus, *Aureobasidium pullulans*, producing the polysaccharide, pollulan. Pollulan is used mainly to make edible films for oral hygiene products. Evolutionary algorithms mimic mechanisms in biology such as reproduction, evolution, mutation, recombination, and selection. These algorithms represent parameters of the fed-batch process as genes of individual chromosomes. The fitness value of the chromosome reflects the "goodness" of the given set of parameters with regard to optimum performance. A set of 20 chromosomes was established initially. The authors employed crossover mutation (roulette wheel or geometric selection functions) as the selection process. Either parameter selection function resulted in a rapid convergence to optimal conditions. The main advantage of this evolutionary approach is its applicability in situations where the model is complex and non-differentiable.

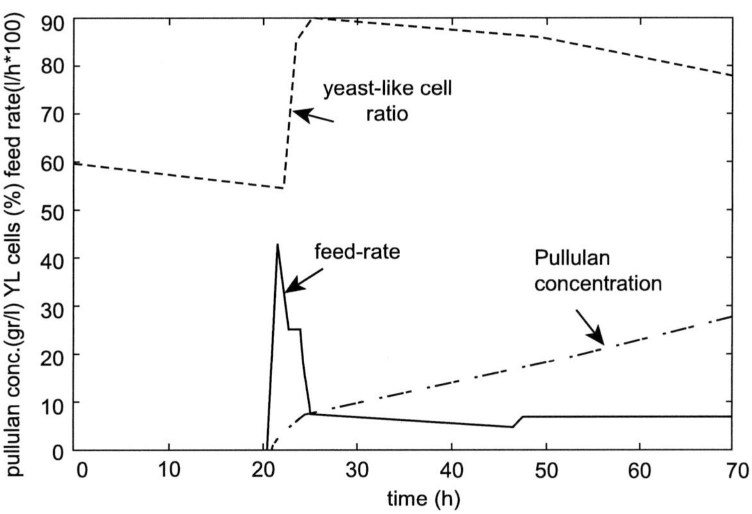

FIGURE 10-21 Optimum feeding profile, biomass ratio and product (pullulan) concentration obtained by off-line simulation using the evolutionary algorithm. *(Reproduced from* Ronen et al. (2002) *with permission. © 2002 by Elsevier Science Ltd.)*

The optimal feeding profile obtained off-line with geometric ranking is shown in **Figure 10-21**. The correlation between feed and biomass concentration is evident. The final product concentration is approximately 30% higher than predicted by an arbitrary feeding schedule. The evolutionary model has excellent learning capability so consequently the optimal feeding profile is rapidly found. This suggested a methodology adaptable for on-line optimization.

10.10 OVERVIEW

The diversity of organisms used in biological processes as well as the numerous metabolic pathways employed by these organisms to convert a substrate or a multitude of substrates into intra- and/or extra-cellular metabolites, cell matter or final oxidation products means that many modulation strategies are possible. Obvious ones exploit the different environments in which cells thrive or even just survive. Dissolved oxygen has been manipulated to provide an aerobic or anaerobic environment. This was accomplished by controlling air flow to a dispersion device. In waste treatment applications, it is done by manipulating flow of a liquid phase over a biofilm anchored on a packing. Foods for the organisms have been manipulated to periodically starve an organism and force it into a non-growth regime in which cell resources are utilized to produce a valuable secondary metabolite. In waste treatment, periodic starvation may be used to adjust ratios of food and grazing populations. In addition to substrate, organisms require nutrients so that these too have been manipulated. Most cells are sensitive to the pH of the medium. Hormones appear to affect cell growth. Others chemicals act as cell growth promoters or product excretion inducers. Plant cells respond to light intensity. All of these have or could be manipulated with time in a chemostat. Alternatively, the sensitivity of different populations to a poison such as an antibiotic has been employed to control the relative concentrations in a mixed culture.

In periodically operated batch bioreactors, different populations have been controlled by the mechanical mixing rate or by the periodic harvesting of biomass and its replacement by fresh culture (*Lenas et al., 1994; Baltzis et al., 1996*). Periodic harvesting is possible in continuous bioreactors. Consider **Figure 10-15**, the liquid fluidized bed has a disengager that continuously removes a product stream. It could be engineered to remove biomass and return the supernatant to the fluidized bed and then just operated to periodically separate out cell mass. Periodic addition of fresh biomass has been investigated for the elimination of polyaromatic hydrocarbons in a continuous activated sludge process (*Cardinal and Stenstrom, 1991*) and for the treatment of a cephalexin and volatile fatty acids containing waste in a batch-operated, anaerobic liquid fluidized bed (*Saravanane et al., 2001*). Both research teams used an enricher reactor operated for cell growth to supply a replacement mixed culture. Waste products in some experiments were added to the enricher feed to acclimatize the culture. Periodic harvesting and culture replacement should be applicable to the challenge of stabilizing a recombinant cell culture.

Another technique, periodic removal of an inhibitor, explored in a periodically operated batch reactor, could also be used in a continuously operated system. *Ghose et al.* (*1984*) examined the effect on the ethanol production rate of periodically removing ethanol from a bioreactor broth by flashing a portion of the broth under a vacuum high enough to prevent temperature from exceeding a 40°C limit imposed by fermentation. The application was to simultaneous saccharification and fermentation of lignocellulose (finely ground rice straw) to ethanol using Candida acidothermophilum, a cellulase enzyme obtained from Trichoderma reesei, and a β-glucosidase enzyme from *Aspergillus wentii*. Ethanol, it was found, inhibits the saccharification enzymes as well as the fermentation microorganism. Periodically circulating the fermentation broth to a flash vessel to strip out water and ethanol increased ethanol productivity by about 45%; cellulose utilization increased by about 95% and the ethanol yield went up by about 50%. If, in addition, the rice straw substrate was fed periodically to the fermentor, productivity increased 3-fold. The *Ghose et al.* study is an example of combining reaction and separation in a single process step. A combination of these two "unit operations" has been investigated during the past two decades to overcome equilibrium limitations, avoid inhibition and control selectivity in chemical reactions (*Aida and Silveston, 2005*). Familiar examples are catalytic distillation and pressure swing reactors.

Application of the π criterion and stability analysis employing matrix methods to the complex models that are needed to describe biochemical processes is impressive. Nevertheless, the papers examined by and large show that the unstructured models employed in these studies are unable to adequately reproduce the time response of the microorganism systems to concentration change (see, for example, *Pickett et al., 1979, 1980; Weber and San, 1989; Stephens et al., 1992; Ye et al., 1996*). Use of structured models or, at least, specific growth models that include all of the factors affecting growth and non-growth synthesis give a better representation of system dynamics. However, these models introduce such complexity that the determination of the π scalar or matrix and the Floquet terms becomes a staggering undertaking. There is therefore a need to search for simpler models.

Fed-batch culture platforms of prokaryotic and eukaryotic hosts for the production of recombinant biopharmaceuticals are an ever-expanding area in research and production. To prevent substrate inhibition, catabolite repression, or overflow production of by-products, fresh medium is added either batchwise, semi-continuously, or continuously to the viable culture in the bioreactor. The principal component of the feed medium is the limiting substrate; hence, the rate of addition controls the specific growth rate of the organisms. Depending on the extent of substrate enrichment in the feed, the fed-batch culture is said to be a constant or variable volume operation. Techniques to improve performance include adaptive or model based control measures. Fed-batch culture technology has been shown to provide significant improvements in either product concentration or productivity over conventional batch operations.

The preponderant number of theoretical studies or numerical simulations of the periodic operation of bioreactors relative to experimental investigations points to a need for experimental work. Indeed, weakness of the models discussed above reinforces the necessity for further experimentation. There is a need to see how robust modulation strategies, such as combined substrate and dilution rate cycling, perform when applied to different microbial systems.

Nomenclature

A	= antibiotic concentration (mg/L)
$\mathbf{A_0}$	= antibiotic concentration in the feed
A(t)	= time varying antibiotic concentration
A, B, C	= arbitrary chemical species
C	= concentration (mol/L or g/L)
	= control parameter
$\mathbf{C_P}$	= product concentration
C*	= concentration set point
CHO	= Chinese hamster ovary
CSTR	= continuous stirred tank reactor
c_i	= model constant where i = 1,2,....
c_α, c_β	= consumption model constants
c_k	= unit cost where k = material ($/g)
c_A	= antibiotic cost
c_S	= substrate cost
DO	= dissolved oxygen
E	= model based estimate of control variable
E*	= model based set-point of control variable

F = feed rate (mol/min)
f = function, usually of a substrate or a control variable, but also delay or memory
$\mathbf{f_{acc}}$ = maximum internal level of accumulated substrate
$\mathbf{f_{PHB}}$ = maximum internal level of PHB
G = amount of substrate added (g or mol)
GA = genetic algorithm for optimization
g = performance function, usually of the product and/or control variables
H = Hamiltonian
$\mathbf{H_1}$ = excess level of substrate 1
h = production function, usually of the state variables
J = performance index (criterion)
$\mathbf{J_T}$ = performance index based on total product
$\mathbf{J_P}$ = performance index based on productivity
$\mathbf{J_Y}$ = performance index based on yield
K = equilibrium constant
$\mathbf{K_S}$ = saturation constant
$\mathbf{K_i}$ = inhibition constant
$\mathbf{K_P}$ = product saturation constant
$\mathbf{K_N}$ = ammonia saturation constant
$\mathbf{K_t}$ = time constant in **Eq. (10-8)**
k = utilization rate constant
= consumption rate constant
= cellular uptake rate constant
= antibiotic stability rate constant
$\mathbf{k_{la}}$ = volumetric oxygen mass transfer coefficient (1/h)
$\mathbf{k_N}$ = consumption rate constant for ammonia
MAb = monoclonal antibody
MCMC = Markov chain Monte Carlo
MRAC = model reference adaptive controller
MSUR = maximum substrate uptake rate
mRNA = microcondrial RNA
N = ammonia or nitrogenous substance concentration (mg/L)
n = integer, exponent
= system dimension
$\mathbf{O_2}$ = oxygen concentration
$\mathbf{O_d}$ = dissolved oxygen concentration
P = product, metabolite concentration (mg/L)
PI = performance index (various units)
PHB = poly hydroxyl butyrate
PO = periodic operation
PSO = particle swarm optimization
Q = volumetric flow rate (L/min)
Q(t) = time varying volumetric flow rate
Q/V = dilution rate, space velocity (h^{-1})
$\mathbf{Q_E}$ = ethanol out flow rate
q = maximum rate of internal metabolite accumulation (g/g biomass *h)
$\mathbf{q_{OX}}$ = normalized oxygen uptake rate
r = rate (mol/min·L)
S = substrate concentration, reactant concentration (various units)
$\mathbf{S_0}$ = substrate concentration in the reactor feed (various units)
SBR = sequencing batch reactor
SS = steady state
s = cycle split, duty fraction
s(t) = biomass accumulation with time
T = response time (h)
$\mathbf{T_f}$ = fermentation time (h)
TPA = tissue plasminogen activator
t = time (h)
$\mathbf{t_0}$ = starting time
$\mathbf{t_s}$ = feeding time start
$\mathbf{t_3}$ = fermentation time
t- = response time of host cells
u = control variable or vector
$\mathbf{u_0}$ = initial control point
V = volume, reactor volume (L)
V(t) = time varying volume
v(), w() = linear function
Y, $\mathbf{Y_{X/S}}$ = cell mass to limiting substrate yield coefficient
$\mathbf{Y_{P/S}}$ = product mass to limiting substrate yield coefficient
$\mathbf{Y_{X/P}}$ = cell mass to product relation
$\mathbf{Y_{X/N}}$ = cell mass to limiting nitrogen yield coefficient
YL = yeast like cells
y = product variable or vector
= observation
= ethanol yield from substrate
= exponent on substrate concentration
X = cell, microorganism, enzyme concentration (mg/L)
$\mathbf{X_{acc}}$ = internal accumulation of substrate
$\mathbf{X_h}$ = concentration of viable biomass
$\mathbf{X^+}$ = concentration of plasmid bearing cells
$\mathbf{X^-}$ = concentration of plasmid-free or host cells
x = state variable or vector
Z = previous cell or organism concentration (mg/L)

Greek

α = reversion fraction
= delay constant
= tunable parameter
= growth associated rate constant
β = non-growth associated rate constant
Δ = tuning term
ε = specific yield (metabolite or product) rate
ϕ = phase difference
γ = delay (h)

η	=	ratio of adaptability constants
κ	=	specific flow parameter
λ	=	Lagrangian multiplier
μ	=	specific growth rate, cell or microorganism specific growth rate
μ_{max}	=	maximum growth rate
$\mu(S)$	=	substrate dependent specific growth rate
ν	=	consumption function
	=	equality constraint function
	=	growth associated rate constant
θ	=	system time constant
	=	specific ethanol productivity
	=	dimensionless dilution rate
π	=	specific consumption rate for metabolite formation
	=	Fourier transform of the expansion of the Hamiltonian of the performance index (π criterion), the Hermitian
$\rho(\omega)$	=	adjoint variable
σ	=	specific consumption rate
τ	=	period, cycle period (h)
τ_{cycle}	=	period, cycle period
τ_{opt}	=	optimum cycle period
ω	=	frequency
	=	consumption function
	=	inequality constraint function
ψ	=	phase shift
ζ	=	adaptability constant

Subscripts

A,B,C	=	reaction species
acc	=	internal accumulation
D	=	decay, death
E	=	ethanol
F	=	feed or flow
f	=	fluid, fluid, phase
	=	function
	=	test time
	=	fermentation
h	=	viable
i	=	inhibition
in	=	input
m	=	mass, maximum
max	=	maximum
opt	=	optimal
P	=	product, metabolite
	=	periodic
	=	productivity
PHB	=	poly hydroxyl butyrate
PO	=	periodic operation
P/S	=	yield of metabolite with respect to substrate
S	=	saturation
	=	substrate
SS	=	steady state
s	=	solid, solid phase
T	=	production
t	=	time
X	=	cell, microorganism, enzyme
X/N	=	cell mass with respect to ammonia or nitrogen source
X/P	=	cell mass – product ratio
X/S	=	yield of cell mass with respect to substrate
Y	=	product yield
0	=	feed or initial condition
1,2, 3..	=	cell species or microorganism index, sampling or test time

Superscripts

n	=	arbitrary exponent
o	=	optimal
y	=	exponent in specific growth model
0	=	initial or stationary
∞	=	limiting
*	=	saturation
+	=	plasmid bearing
-	=	host (plasmid free)

CHAPTER

11

Miscellaneous Reactions

Robert Ross Hudgins, Peter Lewis Silveston*, Albert Renken*[†]

*Waterloo, Ontario, Canada and [†]Lausanne, Switzerland

OUTLINE

11.1 Ethyl Acetate from Ethylene and Acetic Acid 317

11.2 Claus Reaction 324

11.3 Dehydrogenation of Methanol 325

11.4 Deamination and Alcohol Dehydration Reactions 326

11.4.1 Experimental 326

11.4.2 Modeling 330

11.5 Photocatalytic Degradation of AZO Dyes 332

11.6 The Minimal Bromate Reaction 332

11.7 Propanol Dehydrogenation 334

11.8 Glucose Oxidation 336

11.9 Overview 337

This chapter completes the discussion of the use of composition modulation to activate catalytic reactions. The examination of the literature is arranged more or less chronologically. **Table 11-1** summarizes the available literature.

11.1 ETHYL ACETATE FROM ETHYLENE AND ACETIC ACID

In one of the earliest explorations of composition modulation, Hoechst AG investigated composition modulation of this reaction as a possible replacement for traditional processes such as esterification or the Tishtshenko process. Their experiments were carried out at 6 bar and 140°C using sulfuric acid on silica as a supported liquid phase catalyst (SLPC) (*Leupold and Renken, 1977, 1978*; *Leupold et al., 1979*). Acetic acid (HAc) with 1 wt% H_2SO_4 added to maintain catalyst activity was brought to reaction temperature and fed to the reactor. Under the conditions used, acetic acid vaporized as it moved downward through the reactor. Ethene

TABLE 11-1 Miscellaneous Reactions

Author(s)	Objective	Reaction & Reaction Conditions	Modulated Variable	Observations	Comments
Leupold and Renken (1977, 1978); Leupold et al. (1979)	Production improvement	$C_2H_4 + CH_3COOH \rightarrow CH_3COOC_2H_5$ catalyzed by 25 wt% H_2SO_4 on a silica carrier. Time-average HAc mol fraction = 0.0825; $P = 6$ bar; $T = 140°C$; reaction carried out in a tubular packed bed 17.8 mm i.d. × 1000 mm, with 1 wt% H_2SO_4 added to the HAc to maintain catalyst activity.	Vaporized acetic acid flow was switched on and off, while ethene flow was continuous. $4 < \tau_{cycle} < 30$ min; $0.167 < s < 0.333$; $SV = 3030\ h^{-1}$.	Up to 170% increase in EtAc production was observed using HAc flow interruption at 140°C for $s = 0.33$. τ_{cycle} was less important than split provided τ_{cycle} exceeded 5 min. The ester mol fraction was 7 times the maximum under steady operation. Both change during a cycle and the large EtAc mol fraction enhancement indicated HAc inhibition of the reaction.	Composition modulation improved reactor performance by providing ethene access to the catalyst surface. On-off cycling of the HAc feed to the reactor prevented deactivation of the catalyst.
Koubek et al. (1980a; 1980b)	Investigation of the "stop-effect"	Deamination and dehydration of organic amines and alcohol over supported metal oxide (alumina, thoria and chromia) and decationized zeolites. Isothermal, shallow packed bed of catalyst at 1 bar and $460 < T < 580$ K.	Bang-bang (on-off) flow of the amine or alcohol over a bed of catalyst.	On interruption of alcohol or amine in the N_2 carrier gas flowing to the reactor, the rate of olefin production increased rapidly to as much as 50 times the rate measured with alcohol or amine in the feed. Time-average rates under on-off modulation can be 5 to 10 times greater than under steady-state operation.	Stop-effect was explained by reactant inhibition. Deamination or dehydration requires adjacent surface sites that can be blocked by reactant adsorption.
El Masry (1985); Hudgins et al. (1988)	Optimization of the Claus Reaction	$H_2S + SO_2 = S_x + H_2O$ (not balanced) over a commercial bauxite catalyst. Isothermal, differential packed bed was used; $P = 108$ kPa, $T = 483$ K.	$H_2S{:}SO_2$ ratio was varied in constant N_2 flow at constant space velocity. $2 < \tau_{cycle} < 12$ min; $0.2 < s$ (based on H_2S) < 0.4.	Maximum of time-average modulated rates appeared to coincide with the location of the steady-state rate. Enhancements of up to 80% were observed for H_2S-rich modulated feeds.	Optimal periodic performance required τ_{cycle} of less than 2 min. SO_2 sorption appeared to be a slow step in the reaction sequence.

Truffer and Renken (1986), Renken et al. (1984)	Enhancement of the ethyl acetate formation rate	$C_2H_4 + CH_3COOH \rightarrow CH_3COOC_2H_5$; catalyzed by H_2SO_4. Isothermal, differential packed bed, $P = 105$ bar, $T = 483$ K	HAc addition to an N_2 carrier was switched on and off; ethane feed was continuous. $\tau_{cycle} = 1500$ s, amplitude of HAc modulation = 10 mbar.	Enhancement was observed but was about 20% below levels predicted by reactor model under modulation when HAc was removed from the feed.	Researchers were able to closely simulate start-up of the modulated reactor.
Zaza et al. (1991)	Conversion and selectivity enhancement for the dehydrogenation of methanol to formaldehyde	$CH_3OH \rightarrow HCHO + H_2$ with Na_2C0_3 as a catalyst. Fluidized bed consisted of a quartz tube 550 mm long with 20 mm i.d., a preheater and a return leg. Circulating catalyst was 18 g Na_2C0_3, with $d_p = 50–80$ μm; $600 < T < 800$°C. Gas throughput was 20–60 cm^3 (STP)/s; $0.05 < y_{MeOH} < 0.25$ at bed entrance.	Composition modulation achieved by circulating the catalyst between regions of high and low methanol concentration.	Catalyst activity remained constant for about 10 h. Both selectivity and conversion in the circulating fluidized bed were significantly higher than in a fixed bed reactor. Source of these improvements was not identified.	In a related study, *Zaza et al. (1994)* used step-change experiments and deuterated MeOH to propose a mechanism for this reaction on an Na_2CO_3 catalyst.
Morávek (1992a)	Verification of stop-effect mechanism through pulse and step-change experiments	Dehydration of propanol-2 over an alumina catalyst. Experiments conducted in an isothermal, packed bed, differential reactor at 1 bar and 450 K. Outlet composition was followed by a mass spectrometer.	Pulse and step-change response examined. Cycling was not used.	A version of the Koubek II single site model (**Table 11-3**) fitted experimental measurements closely.	Stop-effect was attributed to adsorptive inhibition.
Sczechowski et al. (1993)	Investigation of photocatalytic decomposition	Decomposition of formate in a TiO_2 suspension using black light. Formate-TiO_2 suspension flowed through a rectangular cavity illuminated from above $T = 24$°C.	Pulse and continuous illumination compared	Researchers observed a significantly higher photonic efficiency under periodic illumination.	
Hohmann et al. (1997)	Investigation of the modulation of a "threshold" reaction system exhibiting Hopf bifurcation points	Minimal bromate reaction in a CSTR with Hopf bifurcation in a parameter or a turning point in a bistability region. Experiments undertaken in a small CSTR at 25°C using dilute concentrations of $Ce_2(SO_4)_3$, KBr and K_2BrO_3 in 1.5M H_2SO_4.	Step-pulse and sinusoidal variation of volumetric feed rate to a CSTR.	Square-wave modulation at $\tau_{cycle} = 100$ s gave a Ce^{4+} production rate that was 60% higher than highest steady-state rate. Non-symmetrical square-wave cycling of the flow rate gave larger production rate enhancement than sinusoidal cycling.	The minimal bromate reaction is the spontaneously oscillating interconversion of $Ce^{3+} \leftrightarrow Ce^{4+}$ in an aqueous solution of Br^-, $BrO_4^=$ and H_2SO_4. The reactor system was also simulated using the Noyes-Field-Thompson model.

(*Continued*)

TABLE 11-1 Miscellaneous Reactions (*cont'd*)

Author(s)	Objective	Reaction & Reaction Conditions	Modulated Variable	Observations	Comments
Schinor et al. (1999)	Investigation of the modulation of a bistable reaction and a thermokinetic system	Minimal bromate reaction in a CSTR but with a region of bistability rather than Hopf bifurcations and either types of thresholds for the thermokinetic system. Chemicals as in the Hohmann et al. (1997) study, but in 3 M H_2SO_4.	Sinusoidal variation of volumetric feed rate, also step and pulse experiments	Modulation forced the reaction system to remain in the high rate branch resulting in a huge enhancement of production over that possible under steady-state operation. Production rates in thermokinetic reactions were not significantly enhanced through modulation.	At some reactant concentrations, the minimal bromate reaction exhibited bistability as flow rate changed without passing through a region of oscillations. Thermokinetic reactions exhibit a reactant activation step that is T independent.
Golay et al. (1997, 1998, 1999a, b)	Investigation of the stop-effect	Ethanol dehydration to ethene and diethyl ether over γ -Al_2O_3 catalyst. $T = 180°C$; $P = 140$ kPa, total flow-rate of 200 mL /min (NTP). $(\overline{C}_{EtOH})_0 = 0.013$ mol/m^3, the feed concentration that maximized the steady-state rate.	Bang-bang cycling between Ar and ethanol for various periods.	Maximum enhancement was achieved for short non-symmetric periods, with most of the cycle under Ar (relaxed steady state). A 6-step, 2-site dehydrogenation model for the stop-effect qualitatively predicted the influence of all cycle parameters.	Stop-effect is due to reactant inhibition, leading to the existence of an optimal inlet concentration of ethanol at steady state.
Buechler et al. (1999)	Investigation of a photocatalytic decomposition	Decomposition of formate over a TiO_2 surface using black light illumination. Formate solution flowed through a rotating illuminated disc at room temperature	Mechanical shutter used to control duration of light and dark periods	4-fold increase in photonic efficiency was observed under periodic illumination. Efficiency decreased with increasing light intensity.	
Rouge and Renken (2001a, b)	Operation of a stop-effect reaction in a microchannel reactor	Isopropanol dehydration to $C_3^{=}$ and dipropyl ether over γ-Al_2O_3. $T = 200°C$, $P = 1.3$ bar, $V_{reactor} = 0.5$ cm^3, catalyst mass = 0.075 g	Isothermal bang-bang cycling between i-C_3OH in Ar and Ar alone, $\tau_{cycle} = 30$ s.	Significant enhancement of dehydration rate was observed under modulation. Authors' model based on a modified 2-step mechanism closely describes concentrations exiting the microreactor.	Rate enhancement of many stop-effect reactions requires rapid cycling of concentration or temperature. Fast modulation of temperature is possible only in low-mass microreactors.

Reshetnikov et al. (2003)	Modeling of the stop-effect and study of the stop-effect using a circulating fluidized bed	An arbitrary $A \rightarrow B + C$ reaction considered over a catalyst possessing two different active sites. Authors used Koubek I model in (**TABLE 11-3**)	Bang-bang switching between a carrier gas and the gas containing reactant A.	Considering isothermal, bang-bang switching between an inert and inert + reactant, highest enhancement was observed for relaxed steady-state operation.	A two-site catalyst with each site producing a different product and temperature-dependent interconversion of sites also considered. Authors showed that combining on-off reactant and temperature modulation improved selectivity.
Wang and Ku (2006)	Exploration of the effect of chamber design and illumination on dye decomposition	Degradation of the reactive dye red 22 over TiO_2 surfaces under UV radiation from an LED source. Dye solution at $pH = 2$ flowed through planar reactor with the TiO_2 coating on one surface.	Variables studied were light pulse frequency and intensity, dye strength and location of the TiO_2 surface	Photonic efficiency was much higher under periodic illumination and when the TiO_2 surface was placed below the flowing fluid for low light intensity. At high intensity, position of the catalyst did not affect efficiency.	Red 22 is an azo dye (substituted phenyl azonaphthol) that is degraded by an oxidative attack on the azo group (-N=N-).
Sakurai et al. (2007)	Circumvention of a reaction equilibrium limitation using a spray-pulse method	$CH_3CH(OH)CH_3$ (2-propanol) $\rightarrow H_2 + (CH_3)_2CO$ (acetone). A Pt/Al_2O_3 catalyst spread on an Al plate was used. Liquid 2-propanol was pulse sprayed onto a heated catalyst coated plate where vaporization and reaction occurred. Dehydrogenation at 80°C was used.	A series of non-steady operations (spraying, vaporizing and reacting) was repeated cyclically at various time intervals.	Evaporation and reaction rates were evaluated under conditions of 1) controlled power and 2) constant temperature. For both, evaporation and reaction rates decreased as the pulse interval increased. The ratio of evaporation/reaction rates under condition 2 was larger than that under condition 1.	The spray-pulse method was used to increase the low gas-phase equilibrium conversion of this endothermic reaction. A comparison with a more conventional reactor such as a catalytic fixed bed reactor was not made.
Gogova and Hanika (2009b)	Modulation of the oxygen feed rate to a 3-phase slurry reactor to avoid catalyst deactivation	Glucose selective oxidation to gluconic acid over Pd/C catalyst, semicontinuous stirred tank reactor was used. The system model assumed five elementary surface reactions and a 2-step reversible deactivation of the catalyst.	Bang-bang modulation of O_2 and N_2 at 101 kPa and $T = 303$ K.	Glucose oxidation is inhibited by O_2. Raising O_2 concentration in the liquid phase causes greater formation of inactive oxygen species responsible for catalyst deactivation. Under composition modulation, the catalyst activity is maintained for a significantly longer time.	Rate equations were based on best fit of experiments. Physisorption, chemisorption and inactive catalytic sites were posited. The proposed model described experimental data very well. Conditions for the highest CSTR productivity were identified.

side-reactions (polymerization, hydrogenation and oxidation to form CO_2) as well as acid reduction to SO_2 do not proceed to any significant extent so selectivity is not an issue. Even so, products of side-reactions, although small, are strongly adsorbed and act as poisons.

Modulation experiments were carried out around a time-average HAc mol fraction of 0.0825, corresponding to the steady-state mol fraction for which EtAc production exhibits a maximum. The reaction was run in excess ethene. *Leupold and Renken* defined the split as the fraction of the cycle during which HAc passed through the reactor. They observed that cycle period, τ_{cycle}, seemed less important than split, s, provided τ_{cycle} exceeded about 5 min. **Figure 11-1** shows the effect of both on the enhancement of the EtAc concentration for the

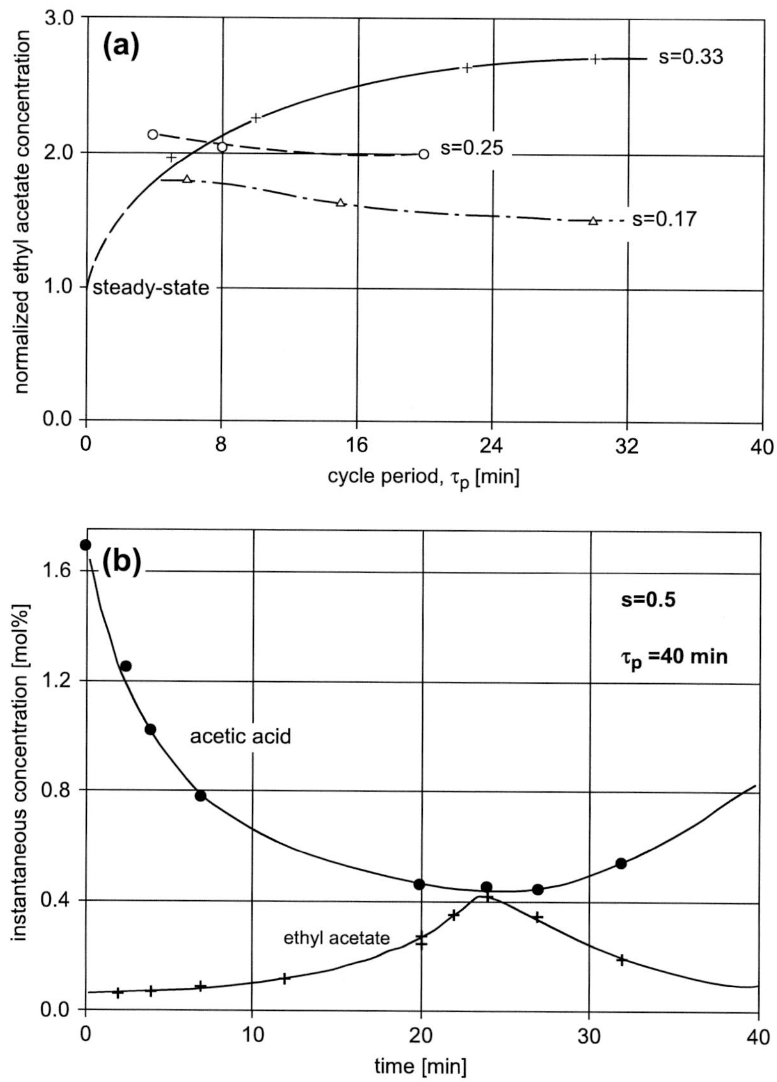

FIGURE 11-1 Ethyl acetate production under on-off cycling of the HAc feed to a fixed bed reactor packed with H_2SO_4 loaded SiO_2 catalyst operating at 6 bar and 140°C with a continuous flow of $C_2H_4(g)$: (a) enhancement or normalized EtAc concentration versus cycle period for various cycle splits, (b) variation of reactant and product mole fractions during a cycle for $\tau_p = 40$ minutes and $s = 0.5$. *(Figure reproduced from* Leupold and Renken (1978) *with permission. © 1978 Verlag Chemie.)*

range of cycle splits. As the time-average the HAc mol fraction used corresponded to the optimal steady-state mol fraction, it is the global enhancement, Ψ^*, that is given in the figure.

The concentration measurements made under symmetrical cycling ($s = 0.5$) and shown in **Figure 11-1(b)** explain the enhancement seen in **(a)**. A stationary cycling state had been reached in the experiment. HAc flow was discontinued at $t = 0$ in the figure and resumed at $t = 20$ min. The slow decrease over the 20 min half-period of HAc concentration and the equally slow build-up over the same duration suggest a slow displacement of absorbate in the catalytic layer. EtAc concentration behaves oppositely and reaches a maximum when HAc attains a minimum. The ester formation at the HAc minimum is about seven times the maximum achieved under steady-state operation. These observations suggest HAc inhibition of the esterification reaction. Thus, composition modulation improved reactor performance by providing ethene access into the catalyst layer.

Leupold and Renken (1978) also observed that on-off modulation of the HAc feed to the reactor prevented deactivation of the catalyst as illustrated in **Figure 11-2**. Ethyl acetate concentrations in the product stream leaving the reactor have been normalized relative to the product concentration under steady state after deactivation ended at about 500 h after start-up. Activity increased for both steady-state and periodic operation initially and the difference in activity between the two operations was small. Once production under steady state reached a maximum, EtAc concentration in the product stream dropped because of HAc inhibition. A further discussion of the *Leupold* papers is to be found in *Composition Modulation of Catalytic Reactors* (*Silveston, 1998*).

Renken and co-workers eventually developed a dynamic model to predict differential reactor performance under composition modulation (*Dettmer and Renken, 1983a, b; Walker and Renken, 1983, 1984; Renken et al. 1984*). From this effort, *Renken* concluded that ethene absorption in the supported acid phase is rate-limiting and inhibited by HAc. The mechanism is complicated and could involve ethyl sulfate, protonated ethene, or the ethyl ester of a sulfonic acid. One or the other of these solute species reacts with HAc dissolved in the acid phase to form the ester which then desorbs. Measurement of the sorption rate of ethene as a function of acetic acid loaded per mass of catalyst indicated inhibition was a strong function of the HAc concentration (*Walker and Renken, 1984*). Thus *Renken et al. (1984)* modeled inhibition as an exponential function of the HAc concentration. In this model the ethene intermediates in the acid phase were assumed to be in equilibrium with dissolved ethene. *Silveston (1998)*

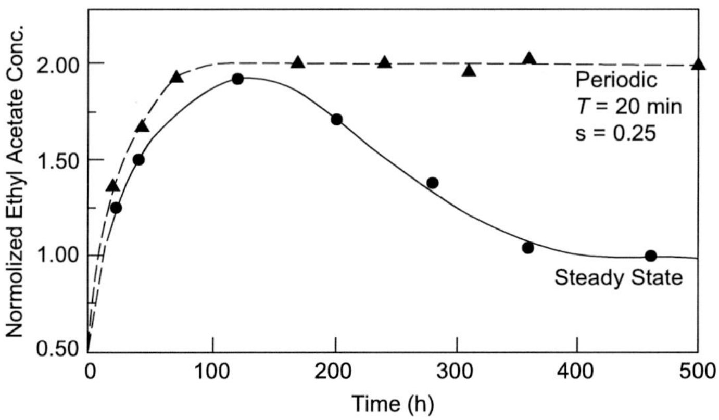

FIGURE 11-2 Comparison of normalized ethyl acetate production with time after start up for steady state and on-off modulation of the HAC feed at 6 bar and 140°C over a H_2SO_4 loaded SiO_2 catalyst with SV = 3050 h^{-1}. *(Figure reproduced from* Leupold and Renken (1978) *with permission. © 1978 Verlag Chemie.)*

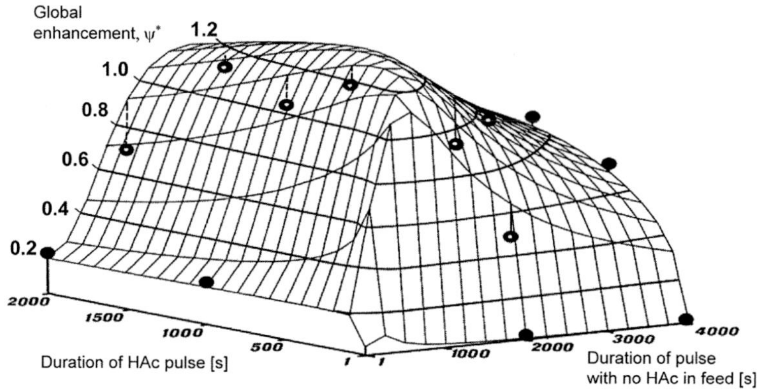

FIGURE 11-3 Comparison of experimental data (•) with surface of model enhancement predictions for modulation of HAc flow to a differential packed bed reactor for the formation of ethyl acetate. *(Figure reproduced from* Truffer and Renken (1986) *with permission. © 1986 by the AIChE.)*

discusses this model and compares it with experimental measurements.

Truffer and Renken (1986) undertook both transient and cycling experiments on the acid catalyzed HAc-C_2H_4 system using a differential fixed bed reactor and introducing HAc in an N_2 diluent. By diverting the N_2 diluent stream to the reactor without first passing through a saturator, space velocity was held constant in the step-change and modulation experiments. *Truffer and Renken* improved the model as discussed by *Silveston (1998)* and with these changes they were able to simulate the startup of their experimental reactor under on-off modulation reasonably well. Simulating composition modulation is a severe challenge for a reactor model.

The *Truffer and Renken* model produced a dome-shaped global enhancement surface as a function of acetic acid and inert gas pulses for the operating conditions of their reactor. That surface fitted experimental measurements reasonably well (**Figure 11-3**). Short exposures to HAc, of the order of 500 s, offered the largest enhancements, about 30%, for the conditions studied. Enhancement was less sensitive to the exposure duration of the low HAc portion of the cycle, e.g., for a feed rich in ethene. In the region of highest enhancement, the *Truffer-Renken* model over-predicted enhancement by just 10%. The largest overprediction was about 20%.

Truffer and Renken (1986) found that performance under HAc modulation depended strongly on the sorption rate of HAc. At high rates, Ψ^* could reach 1.75. Enhancement also depended upon the time-average HAc partial pressure. Increasing the mean partial pressure from 10 mbar to 740 mbar raised Ψ^* from 30 to about 50%.

11.2 CLAUS REACTION

Composition modulation experiments on this reaction over a commercial bauxite catalyst were undertaken by *El Masry (1985)*. The H_2S:SO_2 ratio was modulated at cycle periods between $2 < \tau_{cycle} < 12$ min; while the flow of an N_2 diluent was counter-modulated to keep the space velocity constant. Other variables in the *El Masry* study were the cycle split, defined as the fraction of the cycle period which the feed is H_2S-rich, and the time-average mol fraction of H_2S in the feed. *Hudgins et al. (1988)* re-analyzed *El Masry's* experimental data but corrections made did not change the *El Masry* conclusion that composition modulation of the Claus reaction led to a 35% enhancement of the reaction rate. This may be seen in **Figure 11-4**. The

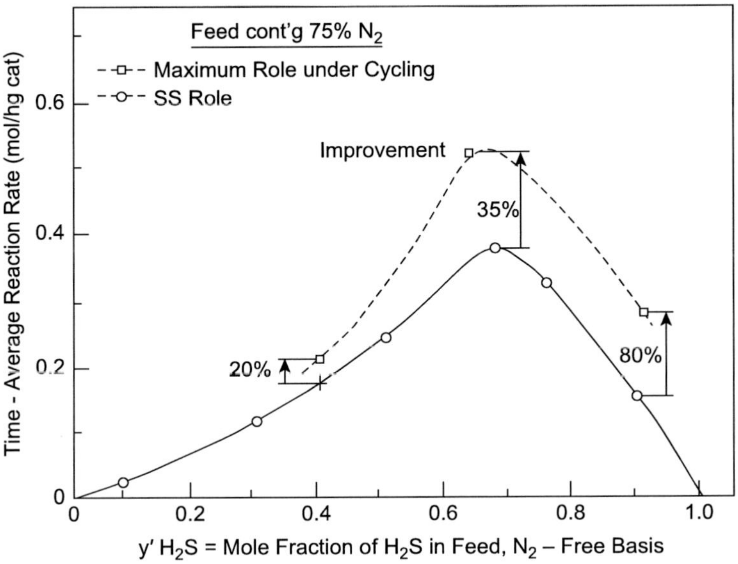

FIGURE 11-4 Comparison of time-average and steady-state H_2S conversion rates versus time-average H_2S mol fraction in the reactor feed for composition modulation of the H_2S:SO_2 ratio: Claus reaction over a bauxite catalyst in an isothermal, differential reactor at 107.54 kPa and 483 K with time-averaged 75% N_2 diluent and a H_2S mol fraction of 0.625 in the feed. *(Figure reproduced from* Hudgins et al. (1988) *with permission. © 1988 Elsevier Science Publishers.)*

maximum under modulation appears to coincide with the mol fraction of H_2S that maximizes the steady-state rate. Enhancement, however, reached about 80% in the H_2S-rich region. Time-average rates were found to depend on τ_{cycle}. Quasi-steady state was approached at cycle periods greater than 8 min, while the periodical optimal required cycle periods under 2 min. The Claus reaction is believed to be poisoned by sulfur dioxide. *Hudgins et al.* show data for prior exposure of the catalyst to pure SO_2 which support this view. Periodic flow direction switching seems to be a more efficacious way of promoting the Claus reaction (see **Chapter 22**).

11.3 DEHYDROGENATION OF METHANOL

This reaction, producing formaldehyde and hydrogen, is catalyzed by sodium carbonate supported on activated carbon or on sodium aluminate. *Renken* and co-workers (*Su et al., 1992; 1994; Zaza et al., 1994*) considered the reaction as an alternative to the commercial partial oxidation process. Step-change methods were used by *Zaza et al. (1994)* to explore reaction mechanism and to identify rate limiting steps. Formaldehyde appears to form from an adsorbed methoxy species, CH_3O-s, created by the dissociative adsorption of CH_3OH. Atomic hydrogen, formed in the dissociation, is strongly adsorbed so that its desorption limits the rate of formaldehyde formation.

The dehydrogenation reaction would seem to be a good candidate for composition modulation using air or another oxidant to periodically scavenge adsorbed H_2. In their experiments, *Zaza et al. (1994)* employed a circulating fluidized bed reactor and observed that passing the catalyst between regions of high and low methanol concentrations raised both selectivity to formaldehyde and methanol conversion above those measured under steady-state in a fixed bed reactor. Temperature measurements were made and demonstrated that the improvements could not be the result of a higher operating temperature in the fluidized bed or temperature non-homogeneity.

11.4 DEAMINATION AND ALCOHOL DEHYDRATION REACTIONS

At temperatures above 250°C over various simple or mixed oxides (e.g., alumina, thoria or chromia) or decationized zeolites, primary amines evolve NH_3 while yielding the corresponding olefin. A disproportionation reaction also occurs in which a diamine is formed. These reactions occur at about the same rate at 325°C under a steady state operation. Over the same catalysts, but above 150°C, analogous reactions occur with alcohols. Alcohols split out water and form either an olefin or an ether. Because both the disproportionation and dehydration or deamination reactions proceed over the oxide catalysts, a selectivity problem arises. These two reactions are discussed here rather than in **Chapter 12** because published modulation studies consider just the single reactions. Selectivity effects, though certainly present, do not seem to have been treated in the literature.

11.4.1 Experimental

Discussing an abrupt increase of two to three times in the rate of propylene formation from the deamination of diisopropylamine over alumina after the flow of the amine-N_2 mixture to the catalyst was replaced by N_2 alone, *Koubek et al. (1980a, b)* referred to this behavior as the "stop-effect." They observed similar phenomena using C_3 to C_4 amines and alcohols over decationized zeolites as well as over alumina, chromia and thoria oxides, and binary mixtures of those oxides. There are other mentions of the stop-effect in the literature (*Koubek et al., 1980a*), but only *Koubek et al.* seem to have given the "stop-effect" thorough study.

Figure 11-5 shows deamination rates with successive start-stop steps for propylamine added to an N_2 carrier using a differential reactor packed with a de-cationized 13X zeolite catalyst. The figure illustrates the remarkable magnitude of the stop-flow effect. After removing the amine from the feed, the rate of propene formation increases almost 20-fold during a "stop" period of 10 min. The figure suggests that the time-average rate enhancement must have been about 10-fold.

Table 11-2 shows the magnitude of the stop-effect for different amines and alcohols. For the

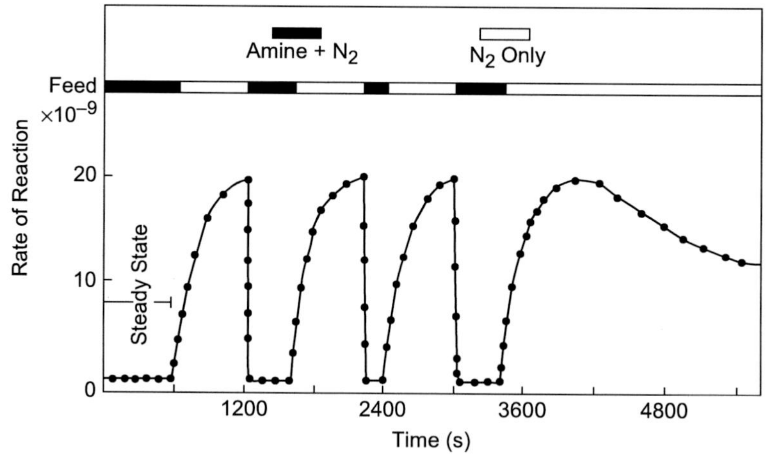

FIGURE 11-5 Variation of the rate of deamination for successive pulses of propylamine (P_{amine} = 11 kPa) in a N_2 carrier gas feeding a differential reactor packed with a decationized 13X-zeolite catalyst at 573 K. *(Figures reproduced from* Koubek et al. (1980b) *with permission. © 1980 Elsevier Science Publishers.)*

amines, rates of olefin production at steady-state and an estimate of site blockage responsible for those low rates are also given. The magnitude is the maximum transient rate of olefin formation divided by the rate under a steady state. The stop-effect is observed only for primary and secondary amines. Steric effects also occur. The magnitude of the stop-effect and the time needed to reach the maximum rate decreased as temperature increased. On the other hand, both were independent of amine partial pressure.

Disproportionation reactions were also observed, but rates decreased after the amine or alcohol was eliminated from the reactor feed. The stop-effect was observed over a chromia catalyst for ethyl alcohol and propylamine, for silica and zeolites only for propylamine, but the effect was not observed for the thoria catalyst. On an alumina-silica catalyst, the maximum rate of propylene formation was 50 times the steady-state rate after interruption of the propylamine feed—the largest stop-effect seen. The decay of the propylene concentration to a negligible level in the gas flowing through the reactor lasted up to 6 h. When ethanol was fed in place of the primary amine, the maximum rate, after halting the alcohol flow, was about 5 times the steady-state rate of ethene formation.

Koubek et al. (1980b) suggest commercially operating dehydration or deamination reactions by introducing a pulse of alcohol or amine just after olefin formation reaches a maximum. That pulse duration would depend on reactant partial pressure and the amount of reactant consumed during diluent flow through the reactor.

The stop-effect was employed by *Koubek* and co-workers to investigate the deamination and dehydration mechanisms. Through the use of selective poisoning of the Lewis acid and basic sites on alumina, *Koubek et al. (1980b)* concluded that the deamination reaction proceeded via the adsorption of the NH_2 or NH end of the molecule on an acid site (a surface Al) and the subsequent or simultaneous adsorption of the opposite end of the molecule on an adjacent basic site (O^{2-} or OH^-). Rearrangement, probably involving H migration from the β-carbon to give the olefin precursor, was followed by cleavages of the C-N bond at the acid site and the C-H bond at the basic site and then olefin

TABLE 11-2 Magnitude of the Stop-Effect for Various Amines and Alcohols

Amine/Alcohol	Magnitude of the Stop-Effect	Steady-State Rate of Olefin Formation in $mol/cm^3 \cdot s$	Estimated Site Blockage %
Propylamine*	11.5	1.14×10^{-9}	95
Butylamine	10.5	0.99	96
Isobutylamine	18.0	0.32	96
Isopropylamine	9.5	10.9	92
N-methylisopropyl amine	1.7	25.5	52
N,N-dimethyl isopropyl amine	1.0	50.5	12
Ethyl alcohol**	6.1	-	-
Propyl alcohol	6.5	-	-
Isopropyl alcohol	1.5	-	-

* *Data reproduced at 300°C.*
** *at 187°C.*

desorption. The stop-effect arises because the amine moiety also adsorbs on basic sites via hydrogen bonding. Adsorption is rapid but weak and reversible. Nevertheless, the basic site is blocked in this way and can no longer participate with the Lewis acid site in concerted deamination. Abruptly cutting off the amine feed results in desorption of the amine end of the molecule from the basic sites, freeing those sites for the concerted deamination of molecules adsorbed on adjacent Lewis acid sites. However, removing amine from the feed decreased the amine adsorbed on the acid sites only slightly as these are strongly bound. Desorbing olefin frees both the acid and basic sites. Indeed, *Koubek et al. (1980b)* hypothesized that Lewis acid sites store amine and the deamination rate is controlled by the availability of basic sites. They suggested a similar mechanism for alcohol dehydration to explain the stop-effect observed with that reaction. Variations of the stop-effect for different oxides as well as for different amines on the same oxide indicate a reaction sensitive to site geometry and/or distribution.

A Swiss team followed up the Koubek work using composition modulation (*Golay et al., 1997; 1998; 1999a, b*). Ethanol was dehydrated over a γ-Al_2O_3 catalyst to ethene and diethyl ether in a differential fixed bed reactor at 180°C. Modulation was found to greatly enhance ethene production. The dehydration reaction could be modeled by either of the two-site mechanisms in **Table 11-3** as expected. Indeed, both qualitatively predicted results for bang-bang modulation of ethanol pulses in Ar gas, as shown in **Figure 11-6**. Periods of less than 80 min and splits below 0.5 significantly enhanced ethene production under the given experimental conditions. Earlier *Thullie and Renken (1991)* showed that the optimal flow interruption duration for the stop-effect depended on model kinetic parameters. *Golay et al. (1999a)* observed that the optimal s and τ_{cycle} for dehydration change with the chemical and catalyst system used.

Application of temperature modulation to raise reactor performance is constrained by reactor mass that increasingly dampens imposed temperature variations as the mass gets larger. Recognition of this has led to development of miniature reactors with flow through microchannels. *Rouge et al. (2001)* chose to explore the performance of one of the microchannel reactor designs utilizing isopropanol dehydration over γ-alumina at 200°C but with concentration rather than temperature forcing. *Morávek (1992a)* had shown experimentally some years earlier that the side-reaction forming an ether interfered minimally with dehydration for the conditions chosen.

Figure 11-7 shows the structure of this miniature reactor. The narrow 300 × 240 μm

TABLE 11-3 Mechanisms for the Stop-Effect

Koubek I	Koubek II	Modified Two-Site Model
Nowobilski and Takoudis (1986)	Thullie and Renken (1991)	Rouge et al. (2001)
$A + s_1 \rightarrow A\text{-}s_1$	$A + s \rightarrow A\text{-}s$	$A + C\text{-}s_1 \rightarrow A\text{-}s_1 + C$
$A + s_2 \rightarrow A\text{-}s_2$	$A\text{-}s + A \rightarrow A\text{-}s\text{-}A$	$A + s_2 \leftrightarrow A\text{-}s_2$
$A\text{-}s_1 + s_2 \rightarrow B + C + s_1 + s_2$	$A\text{-}s \rightarrow B + C + s$	
		$C + s_2 \leftrightarrow C\text{-}s_2$
		$A\text{-}s_1 + s_2 \rightarrow B + C\text{-}s_1 + s_2$
		$A\text{-}s_1 + A\text{-}s_2 \rightarrow D + C\text{-} s_1 + s_2$

where A = amine or alcohol, s = acid-base site, B = olefin, C = NH_3 or H_2O, D = dipropyl ether

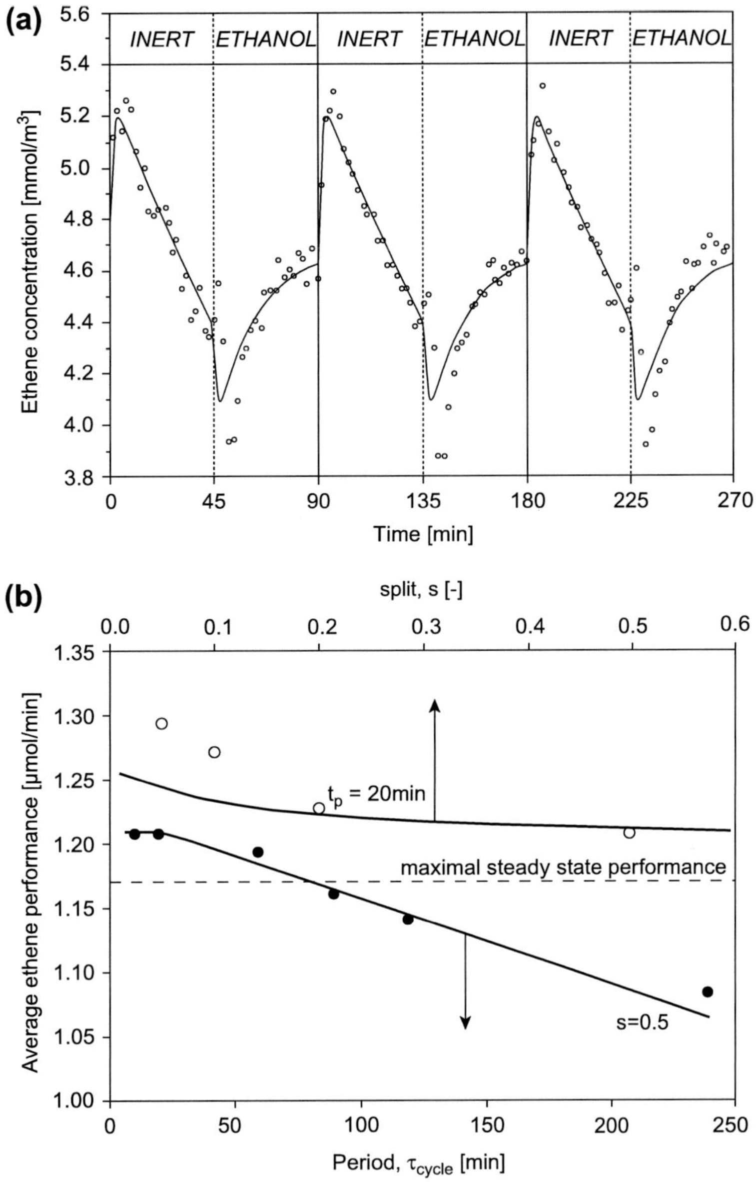

FIGURE 11-6 Modeling of ethanol dehydration under on-off ethanol pulsing versus experimental measurements. Solid lines represent the two-site catalyst model (Table 11-3), while black dots show experimental measurements: (a) $C_2^{=}$ concentration vs. time in successive cycles (180°C, 200 NTP mL/min, 510 mg catalyst, ethanol pulse concentration = 0.013 mol/m^3, τ_{cycle} = 90 min., s = 0.5), (b) influence of cycle split, s, on average ethene production under pulsing for conditions similar to (a) except for τ_{cycle} = 20 min., and influence of τ_{cycle} on ethene production for conditions of (a) with s = 0.5. *(Figures reproduced from* Golay et al., 1999b *with permission. © 1999 Pergamon Press PLC.)*

channels result in plug flow behavior because of the high rate of radial diffusion despite the laminar flow profile of the fluid forced through a channel. Channels were machined from a thin stainless steel plate and anodized to roughen the surface for catalyst adherence. The catalyst, γ-Al_2O_3, was deposited by wash coating with an aqueous boehmite γ-Al_2O_3 slurry. The reactor used in the *Rouge et al.* experiments consisted of five microchannel plates cemented together and held in a heated holder as can be seen in the figure.

Limited experimental results are given in **Figure 11-8**. These show that the dehydration rate under on-off modulation of isopropanol flow to the reactor increased the rate of dehydration over that obtained for a comparable steady-state operation.

11.4.2 Modeling

Koubek and co-workers proposed a model for the sequence described above which adequately represented the rapid increase in rate after

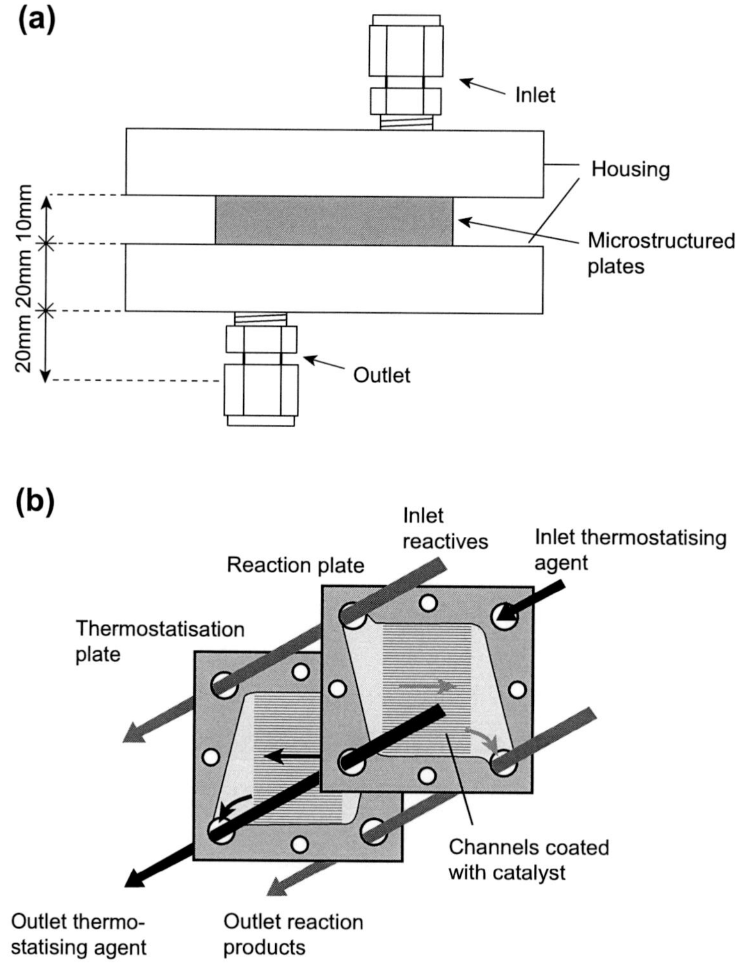

FIGURE 11-7 **Miniature microchannel reactor used by *Rouge et al.* *(2001)*; (a) shows the holder for the machined plates with reactant and product outlets (inlet and outlet of the thermostatting fluids are not shown); (b) shows a pair of microchannel plates: the upper plate conducts the reacting gas and its channels are wash coated with catalyst, the bottom plate conducts the heating fluid for the endothermic reaction.** *(Figure reproduced from* Rouge et al. (2001) *with permission. © 2001 by Elsevier Science Ltd.)*

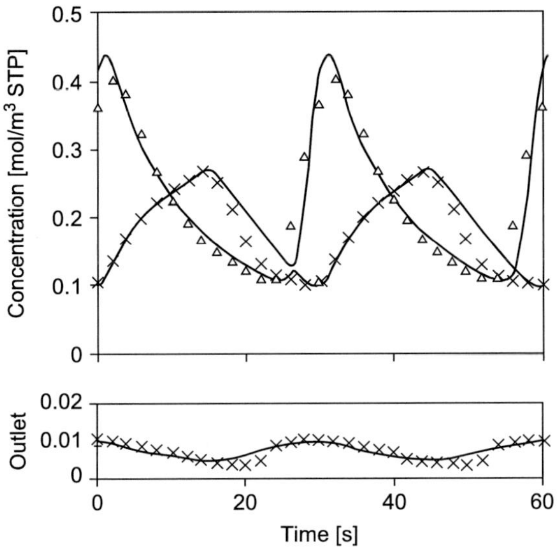

FIGURE 11-8 **Concentration variations over successive cycles under bang-bang modulation of isopropanol concentration in the feed to a 5 plate, microchannel reactor shown in Figure 11-7. Solid lines are model based simulations. Symbols: Δ = isopropanol, x = propene, * = diisopropyl ether. Reaction conditions: $T = 200°C$, $P = 1.3$ bar, $0 < C_{isopropanol} < 0.86$ mol/m³ at $\tau_{cycle} = 30$ s.** *(Figure reproduced from* Rouge et al. (2001) *with permission. © 2001 by Elsevier Science Ltd.)*

removing reactant from the feed and later the gradual drop in rate as only an inert flowed through the reactor (*Koubek et al., 1980a*). A simpler form results if the acid-base site pair is treated as a site itself. *Koubek* and his co-workers found that the simpler model gave an equally good representation of the stop-effect.

The two-site mechanism of *Koubek* was formulated as a model by *Nowobilski and Takoudis (1986)* who used the model to argue that a steady-state operating condition can be found for which rates of reaction will equal or exceed any time-average rate under composition modulation. Their assumed mechanism is given in **Table 11-3** as Koubek I along with an alternate mechanism with a single site, Koubek II, proposed a few years later. Koubek II was actually used in a slightly different form by *Koubek et al. (1980a)*. *Morávek (1992)* provided supporting evidence for the Koubek II model using step-change and pulse experiments. *Koubek et al. (2000)* discussed mechanism and the models arising therefrom.

Based on the Koubek II and the Koubek I mechanism, *Thullie and Renken (1990, 1991, 1993)* developed models to describe the steady-state, transient and the periodic behavior of differential reactors. Results under steady state or from a step-change of the reactants are described quantitatively by both models. But different behavior is predicted under periodic operation. *Golay et al. (1997, 1998)* experimentally discriminated between the two models for the catalytic dehydration of ethanol over alumina and showed that a model based on the Koubek I mechanism allowed a quantitative description of their experimental results. Model and predictions are discussed in more detail by *Silveston (1998)*. Generally, fast cycling, i.e., operating at the relaxed steady state, results in a large enhancement of olefin production.

Reshetnikov and Ivanov (1999) also explored the Koubek I model and concluded that the relaxed steady state provides the largest rate enhancement under bang-bang modulation of reactants fed to a CSTR. These authors also suggested periodic operation can be replaced by a continuous one using two fluidized beds operating at the same temperature with one reactor fed with just carrier gas and the other fed with a mixture containing the reactant. Catalyst, of course, circulates continuously between the reactors. Performance of this two bed system depends on hold-up in the beds and the circulation rate because adsorption and desorption are rate controlled processes.

The Koubek I mechanism was the basis of the modified two-site model (*Rouge et al., 2001*) derived from *Morávek's (1992a)* observation that water was produced in step 1 (**Table 11-3**) and that it was adsorbed on the s_2 site. *Rouge et al.* allowed as well for ether formation which arose through interaction of isopropanol on the s_1 and s_2 sites. Using rate and equilibrium measurements made independently, a model based on the modified two-site mechanism reproduced the experiments closely as can be seen in **Figure 11-8**.

11.5 PHOTOCATALYTIC DEGRADATION OF AZO DYES

Shortcomings of conventional methods for treating widely used azo dyes prior to disposal have led to examination of photolytic processes for their degradation. The azo bond (>N = N<) in the dye molecule can be destroyed by UV light in the presence of a TiO_2 catalyst forming more easily treated organic and inorganic compounds (*Hoffman et al., 1995*). A key issue in photocatalytic degradation has been the photonic efficiency, the ratio of molecules decomposed to the number of photons incident on the catalyst surface during a unit of time.

Sczechowski et al. (1993) reported that the photonic efficiency increased when the photo illumination changed from continuous to periodic. Several years later this finding was confirmed by *Buechler et al. (1999)* who, like *Sczechowski et al.*, studied formate decomposition, but used a cell bounded by a TiO_2 surface rather than a catalyst suspension. The Buechler team observed a four-fold increase in efficiency under periodic illumination.

More recently, the photonic efficiency for the degradation of an azo dye, RR 22, was investigated by *Wang and Ku (2006)* using a rectangular steel cell with a clear quartz surface on one side and a TiO_2 coated surface on the other side. Illumination employed an array of UV-LEDs emitting at 405 nm. Light intensity was varied, but the cycle under periodic operation was 0.1 s on and 1.0 s off (dark). **Figure 11-9** illustrates the improved photonic efficiency through an on-off modulation. The efficiency shown in the figure is quite small and decreases with increasing light intensity. UV absorption is small through the clear quartz cover but becomes large as the radiation passes through the TiO_2 coating and the 2 mm thick flow channel containing the flowing dye solution. *Wang and Ku* concluded from their work that photocatalytic destruction of the RR 22 dye is technically feasible.

11.6 THE MINIMAL BROMATE REACTION

This reaction, oxidation of Ce^{3+} to Ce^{4+} in an acidic solution of bromine, bromate and sulfate ions, is part of the widely studied spontaneously oscillating Belousov-Zhabotinsky reaction. *Hohmann et al. (1997)* investigated this reaction experimentally under flow rate modulation. They employed a stirred spectrophotometric cell that functioned as a CSTR. **Figure 11-10** gives their steady-state measurements of the Ce^{4+} concentration in the cell outlet and shows that the reaction system displays a threshold effect, namely a bifurcation at which the Ce^{4+} concentration may abruptly increase as the flow rate decreases. Decreasing flow rate further leads to oscillatory behavior so the bifurcation is a subcritical Hopf bifurcation (HB). An even greater decrease in flow rate results in restoration of a single operating state, thus the reaction displays a supercritical Hopf bifurcation. Modulating the flow rate with sufficient amplitude can entrain the spontaneous oscillation and lead to a higher Ce^{4+} formation rate. Part **(b)** of the figure plots the steady-state formation rate of Ce^{4+} ions as mol/s vs. the flow rate. The maximum formation rate occurs at the Hopf subcritical point.

Hohmann et al. forced their system sinusoidally and with non-symmetrical waves around a space velocity of 0.004 s^{-1}, which is just above the velocity at HB_2 in **Figure 11-10**. Amplitude of the wave was chosen so that the flow rate crossed from the single state to the oscillatory state region and stayed in the oscillatory region for a portion of a cycle. **Figure 11-11(a)** shows the enhancement achieved as a function of τ_{cycle} and the amplitude

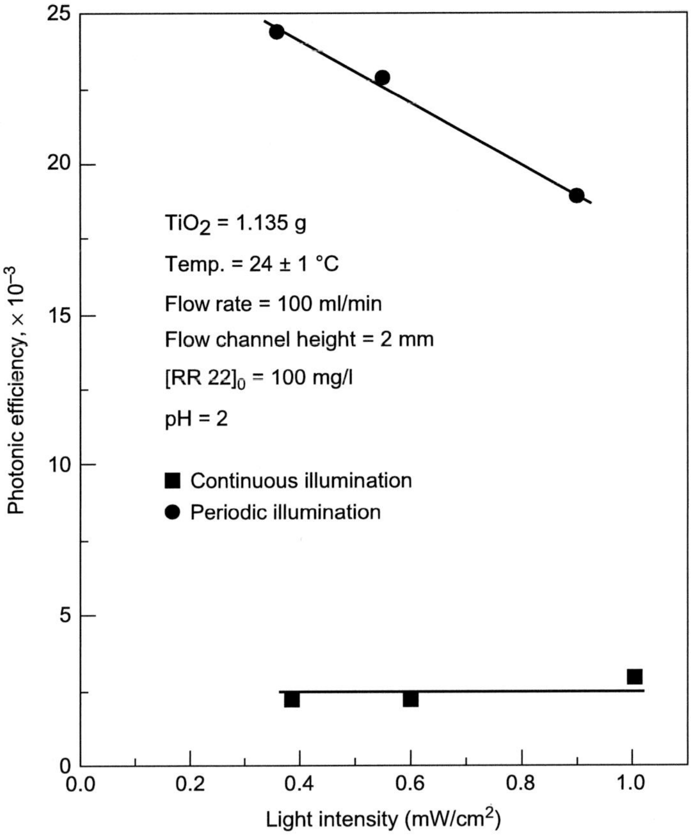

FIGURE 11-9 Comparison of the photonic efficiency for the photocatalytic decomposition of an azo dye, RR 22, as a function of UV light intensity under continuous and intermittent illumination (τ_{cycle} = 1.1 s) in a flat channel reactor with a TiO_2 coated surface. *(Figure reproduced from* Wang and Ku (2006) *with permission. © 2006 by Elsevier Ltd.)*

expressed in relation to the flow rate at the Hopf bifurcation. The enhancement is large. In **(b)**, a model assuming isothermal operation of a CSTR and the Noyes-Field-Thompson mechanism for the minimal bromate reaction has been used. This mechanism is:

$$BrO_3^- + Br^- + 2\,H^+ \rightarrow HBrO_2 + HOBr$$
$$HBrO_2 + Br^- + H^+ \rightarrow 2\,HOBr$$
$$HOBr + Br^- + H^+ \leftrightarrow Br_2 + H_2O$$
$$BrO_3^- + HBrO_2 + H^+ \leftrightarrow 2\,BrO_2^* + H_2O$$
$$Ce^{3+} + BrO_2^* + H^+ \leftrightarrow Ce^{4+} + HBrO_2.$$

At a lower pH, the minimal bromate reaction displays a bistability in which, below a specific flow rate through the reactor, the Ce^{4+} concentration shifts abruptly to a significantly higher level representing a different branch of the concentration vs. flow rate relationship. This shift takes place without transiting a region of spontaneous oscillation. This second type of bistability, also a threshold behavior, is given in **Figure 11-12** based on experimental measurements. The end of each concentration vs. flow rate curve in **(a)** of the figure is termed a turning point (tp). Note that the behavior of the formation rate of Ce^{4+} is different from the one seen in **Figure 11-12(b)**. In the lower (SS_2) branch, the formation rate increases with flow rate.

Schinor et al. (1999) showed that the formation rate of the Ce^{4+} ion can be increased by up to

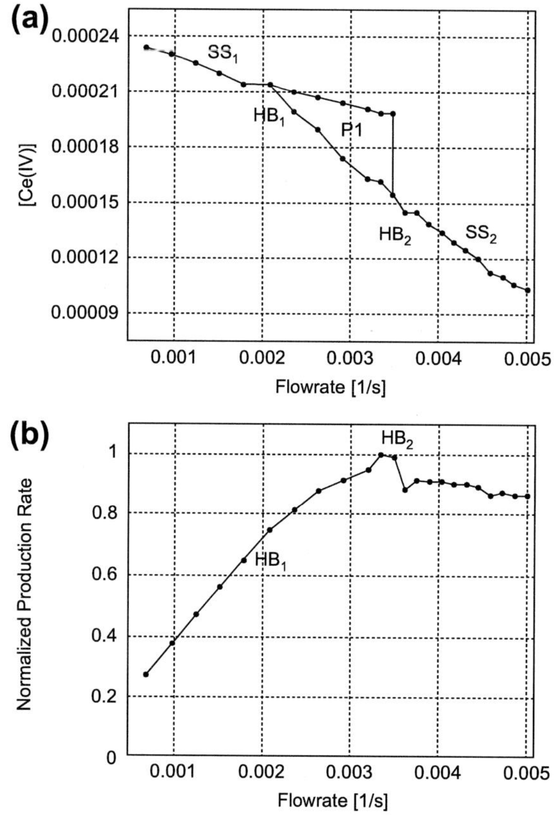

FIGURE 11-10 (a) is the bifurcation diagram for the Ce^{4+} concentration in the minimal bromate reaction plotted against the flow rate through the reactor. SS_1 and SS_2 are single state regions, while P_1 is oscillatory operation between the boundaries shown. HB_1 and HB_2 are the supercritical and subcritical Hopf bifurcations; (b) shows the Ce^{4+} formation rate normalized by the maximum mean formation rate at the subcritical Hopf bifurcation. *(Figure reproduced from* Hohmann et al. (1997) *with permission. © 1997 by the American Chemical Society.)*

about 30% by square-wave flow modulation around the tp_1 turning point in **Figure 11-12**. Sinusoidal modulation was also tested and observed to enhance the Ce^{4+} concentration in about the same way as step-modulation. For the former, the optimal duration of the zero or low-flow step was 30 s. Unlike the result with the Hopf bifurcation, the largest enhancement occurred when the low flow was small, but not zero. *Schinor et al.* also undertook simulation of the modulation behavior using the Noyes-Field-Thompson mechanism given above. They found that a model incorporating the mechanism accurately reproduced the experimental measurements.

The authors examined what they referred to as a thermokinetic reaction under modulation assuming a stirred tank reactor. A thermokinetic system may be represented as: $X_0 \rightarrow X \rightarrow P$ in which the rate of activation of X_0 to form X is not a function of temperature and no heat is evolved or consumed. The second step, the formation of the product P, is exothermic and its rate is governed by the Arrhenius expression. Such a system exhibits a threshold. It has a low temperature branch in which reaction temperature increases slowly with flow rate. At a turning point, this branch stops and the system jumps abruptly to a high temperature branch, where evidently activation becomes fast and temperature rises sharply with increasing flow rate. Under other operating conditions, a Hopf, rather than a turning point bifurcation, arises.

The *Schinor et al.* simulation of a thermokinetic system showed that sinusoidal flow rate modulation around either the bistable point or the Hopf bifurcation caused just a small enhancement of product formation. Other investigators (*Lazar and Ross, 1990; Schell et al., 1987*) reported that flow rate modulation caused at best a small production rate enhancement for reactions that do not exhibit a threshold, such as those first- and second-order reactions obeying the Arrhenius relation.

11.7 PROPANOL DEHYDROGENATION

The reversible dehydrogenation of 2-propanol to acetone and hydrogen was examined by *Sakurai et al. (2007)*. The reaction has been considered for a chemical heat pump or an H_2 storage system. However, the equilibrium conversion of the dehydrogenation reaction is low at

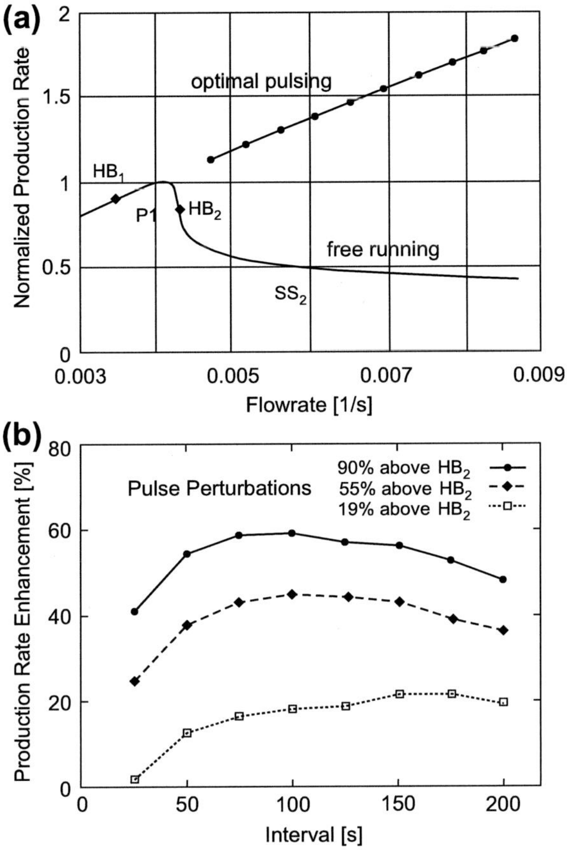

FIGURE 11-11 **(a) Enhancement of the Ce^{4+} formation rate through square-wave pulsing of flow rate as a function of τ_{cycle} and the flow rate in the high-flow step expressed in terms of flow at HB_2 for non-symmetrical cycles with near-zero flow for 10 s. (b) comparison of the formation rates of Ce^{4+} (normalized as in Figure 11-10) under optimal square-wave modulation as in (a) as a function of space velocity.** *(Figure reproduced from* Hohmann et al. (1997) *with permission. © 1997 by the American Chemical Society.)*

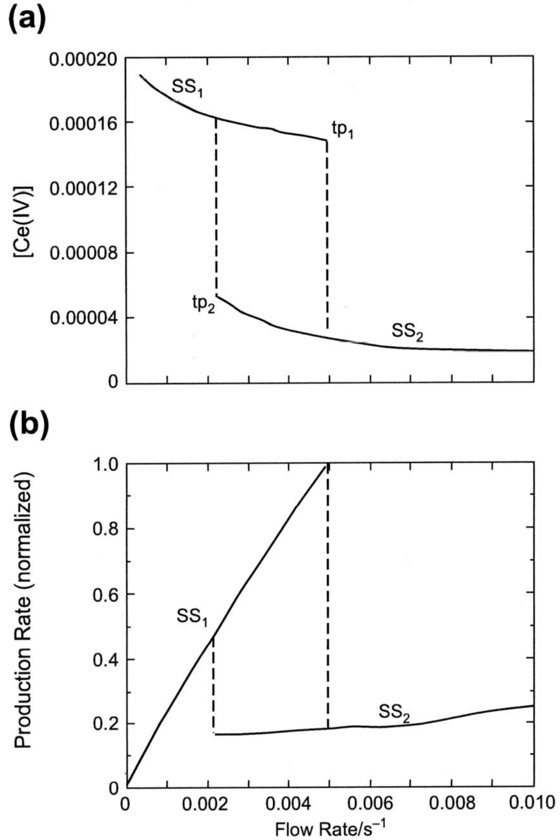

FIGURE 11-12 **(a) is the experimental bifurcation diagram for the Ce^{4+} concentration in the minimal bromate reaction plotted against the flow rate through the reactor. SS_1 and SS_2 are single state branches. At the turning points, tp_1 and tp_2, there is a discontinuity and the formation rate switches to the other branch. (b) shows the Ce^{4+} formation rate normalized by the maximum mean formation rate at the tp_1 turning point.** *(Figure reproduced from* Schinor et al. (1999) *with permission. © 1997 by the American Chemical Society.)*

practical temperatures. To sidestep this limitation, the reaction was carried out by *Sakurai et al.* so as to avoid having reactants and products equilibrate in the gas phase. Liquid reactant, 2-propanol, was periodically sprayed through a nozzle onto a heated $Pt/Al_2O_3/Al$ plate, where it reacted on vaporization with acetone adsorbed on the Pt catalyst and gaseous H_2 was withdrawn. In a subsequent step, acetone was released by heating. This series of unsteady operations was repeated cyclically (**Figure 11-13**).

The authors used Langmuir–Hinshelwood kinetics to describe dehydrogenation. Evaporation and reaction rates were evaluated under operating conditions of 1) constant heating rate or 2) constant temperature. Both evaporation and reaction rates decreased as the time between pulses increased. Under either operating condition, the ratio of rates of evaporation to reaction remained constant at various feed rates.

However, the ratio for constant temperature was larger than that for controlled heating rate at similar cycle periods. For this study, the lowest ratio of evaporation to reaction was 3.2 at constant heating rate. Unfortunately the only comparison used by the authors was an internal one of the rates of reaction at constant heating rate versus those at constant temperature. The conversions achieved were not related to an equilibrium limit. Until such a comparison is addressed, the pulse spray operation remains an interesting technique but one that lacks a suitable assessment.

11.8 GLUCOSE OXIDATION

Gogová and Hanika (2009b) chose glucose (Glc) oxidation with oxygen to gluconic acid (Glcac) as a model reaction typical of many commercial gas–liquid reactions catalyzed by a solid catalyst, in their case, palladium on a commercial activated carbon. Typically, such reactions are undertaken in a 3-phase slurry reactor; reactions in slurry reactors under composition modulation, however, have not been previously studied. Motivation for the research was that glucose oxidation is limited by fast, reversible deactivation of the catalyst. The authors acted on a reported beneficial effect of composition modulation on maintaining constant catalytic activity in an electrochemical redox system (*Markusse et al., 2001*).

Gogova and Hanika performed experiments in a 1 L well-stirred, semi-batch reactor filled initially with an aqueous glucose solution and a suspension of 45 μm Pd on activated carbon catalyst particles. A gas stream containing N_2 or a mixture of O_2–N_2 was fed continuously. The O_2 content of that mixture was modulated by on/off control of the O_2 stream. **Figure 11-14**

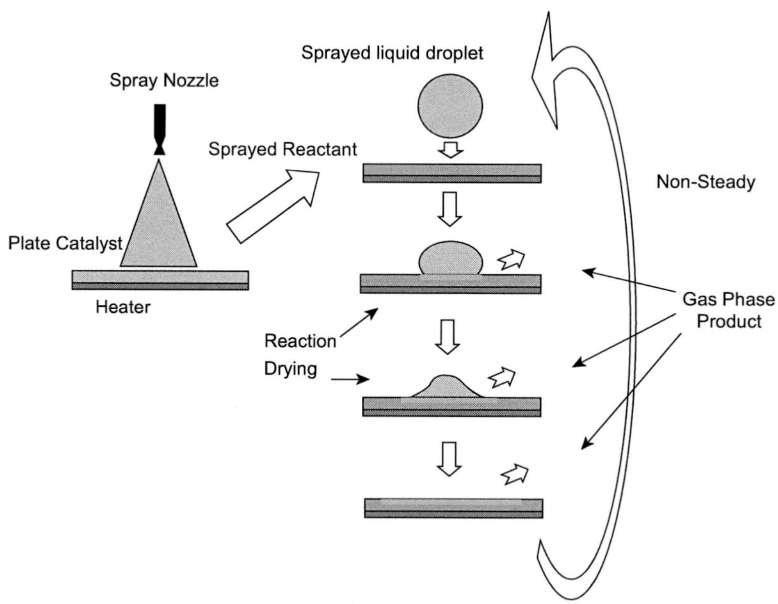

FIGURE 11-13 Diagram showing the concept of the spray-pulse operation. *(Figure reproduced from* Sakurai et al. (2007) *with permission. © 2006 International Association for Hydrogen Energy. Published by Elsevier Ltd.)*

shows a portion of one of their experiments. The solid points are experimental measurements of the rate of glucose (Glc) consumption. The decay is caused by catalyst deactivation because of the build-up of an oxide on the Pd surface. This build-up is indicated by the gray curve in the figure. That oxide is unstable and is removed once O_2 is cut off from the gas bubbled into the reactor. This procedure initiates a brief rebound of the rate, which quickly decays to zero because of the absence of O_2. The open points in the figure have been corrected for deactivation and thus illustrate the effect of this process.

To interpret their results, *Gogova and Hanika* constructed a kinetic model based on the complex mechanism given in **Table 11-4**. Mass balances on three types of catalytic sites were made, based on their earlier studies modified so as to satisfactorily fit the data in **Figure 11-14** and adjusted to account for the extent of catalyst deactivation during a cycle.

In the model, it was assumed that oxygen was chemisorbed on active sites in the form of O^- and that the surface reaction between adsorbed O_2 and physisorbed alcohol was rate determining. With this model, the authors went on to calculate an optimal periodic operations surface for the particular conditions of the oxidation experiment. This surface can be seen in **Figure 11-15** which plots the time average rate of glucose oxidation against cycle period and cycle split, defined as the relative duration of O_2 in gas flowing to the reactor. A CSTR has been assumed. The effect of choosing the optimal modulation parameters is dramatic. There is a two-fold increase of rate between the optimal split and period vs. 1200 s period and a small split. For a steady-state experiment, **Figure 11-15** suggests the oxidation rate becomes vanishingly small. *Gogová and Hanicka* noted that under composition modulation, even though some catalyst activity was lost during a cycle, the long-term activity remained constant.

11.9 OVERVIEW

The first reaction considered in this chapter, acid catalyzed addition of C_2H_4 to HAc to make the ester, could be commercially interesting. Modulation appears to be a possible

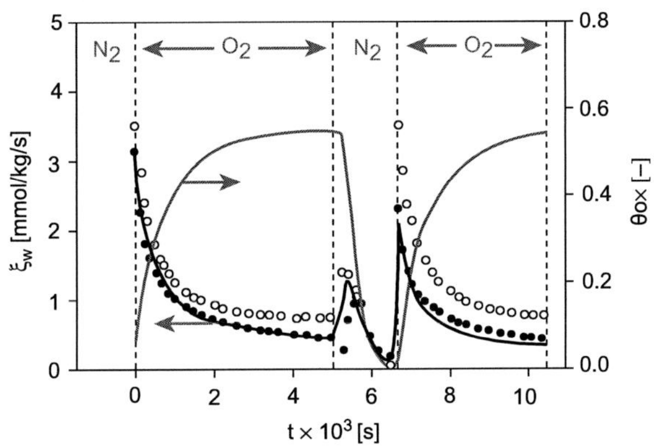

FIGURE 11-14 Change in the measured rate of glucose consumption and the calculated extent of oxide coverage of the Pd surface as functions of time in successive cycles after start up when O_2 is introduced into the gas feed to the reactor. Solid points are measured data; open points have been corrected for catalyst deactivation. *(Figure reproduced from* Gogová and Hanika (2009b) *with permission. © 2009 Elsevier B.V.)*

TABLE 11-4 Proposed Mechanism for the Oxidation of Glucose to Gluconic Acid in an Aqueous Solution

1)	R-COH + s_p ↔ R-COH·s_p
2)	O_2 + s ↔ 2 O·s
3)	R-COH·s_p + H_2O ↔ R-CH$(OH)_2$·s_p
4)	R-CH$(OH)_2$·s_p + O·s + s → R-C(OOH) ·s_p + H_2O + 2 s
5)	R-C(OOH) ·s_p + OH^- ↔ $RCOO^-$ + H_2O + s_p
6)	O·s + s_{Pd} → PdO + s
7)	PdO + s → O·s + s_{Pd}

where s = chemisorption site, s_p = physisorption site and s_{Pd} = active metal atom

alternative to existing technology. Since the early research was industrially sponsored, one must wonder why a modulated process was not pursued further over the last four decades.

Olefins are less costly for the most part than amines or alcohols so there seems to be no incentive to develop the stop-effect for olefin production. It appears that the modulated reaction may be useful in mechanistic studies. The technique, however, is an extension of step-change methods and hardly offers any advantages over that technique, although behavior seems to be quite sensitive to slow secondary processes such as deactivation. Perhaps its most promising use is as a laboratory demonstration of surface fundamentals in catalysis.

The outlook for the application of composition modulation to the industrially important Claus reaction is poor. Modern Claus plants now operate at high conversions. Emissions are the primary concern. Periodic flow reversal, examined in **Chapter 18**, addresses the emission problem.

Alcohol dehydrogenation was included in this chapter even though it has not been tested under composition modulation. The spray-pulse system devised by *Sakurai et al.* is a separative reaction system combining adsorption for separation

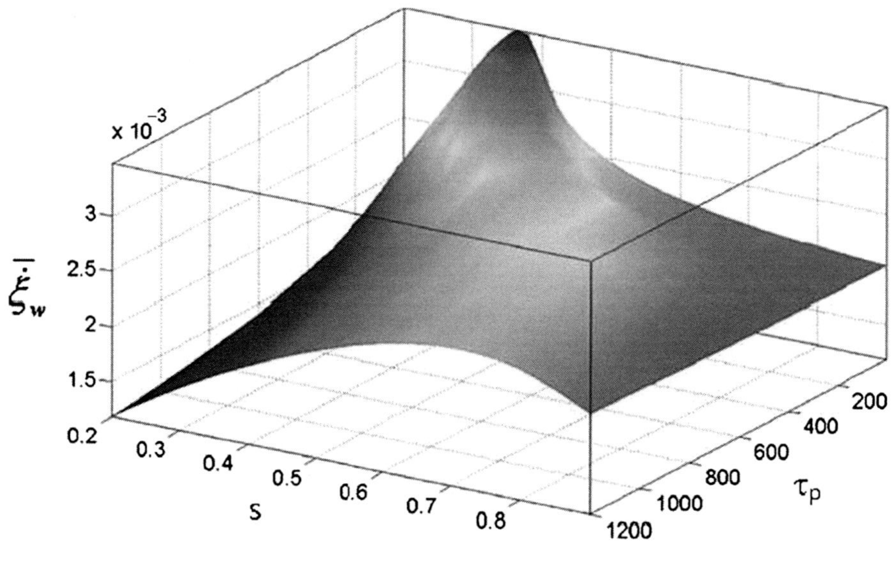

FIGURE 11-15 **Simulation of the effect of cycle period, τ_{cycle}, and cycle split, *s*, on the time-average rate of glucose oxidation to gluconic acid in an O_2 on-off modulated gas flow to a CSTR with a slurry Pd/act. carbon catalyst. Catalyst density = 1000 g/m^3. y_{O2} = 0.21, $C_{Glucose}$ = 100 mol/m^3, $X_{Glucose}$ = 2%.** *(Figure reproduced from* Gogová and Hanika (2009b) *with permission. © 2009 Elsevier B.V.)*

with a chemical reaction. Such systems are industrially attractive but not in the rotating plate configuration discussed. **Chapters 20** to **22** examine chromatographic and pressure/temperature swing systems that are now under rather intensive development for promoting reaction advancement through adsorption.

Threshold reactions discussed in this chapter are probably of academic interest only. It is difficult to conceive of reactions with regions of instability operating commercially. However, the final system addressed in this chapter, a 3-phase bubbling slurry reactor, showed dramatic rate improvement through modulation. 3-phase reactors are used in various organic chemical and biochemical processes. They were discussed in the previous chapter and will be revisited in **Chapter 17** which deals with trickle bed processes. Modulation has been effective for enhancing rates in such reactors.

Nomenclature

A,B,C	=	reactant or product species
$\underline{C_i}$	=	concentration of species i (mol/m^3)
$\overline{C_i}$	=	time average concentration of species i
CSTR	=	continuous stirred tank reactor
d_p	=	particle diameter (mm)
EtAc	=	ethyl acetate
Glc	=	glucose
HAc	=	acetic acid
HB_i	=	Hopf bifurcation point
i.d.	=	inner diameter (cm)
k_n	=	rate constant for the nth reaction (various units)
LED	=	light emitting diode
NTP	=	normal temperature and pressure
n	=	reaction index or number of intervals
P	=	pressure (Pa, bar, atm)
	=	product species
P_i	=	partial pressure of ith species, oscillatory region
SS	=	steady state
SLPC	=	supported liquid phase catalyst
SV	=	space velocity
STP	=	standard temperature and pressure (25°C, 101.5 kPa)
s	=	cycle split (-)
s_i	=	active site, surface reaction site in mechanistic equations
T	=	temperature (°C, K)
t	=	time (s, min)
tp	=	turning point
$V_{reactor}$	=	reactor volume (L, cm^3)
X	=	thermokinetic reactant
y_i	=	mole fraction of species i

Greek

θ	=	dimensionless time (-)
θ_i	=	fraction of surface sites occupied by species i
τ, τ_{cycle}	=	cycle or forcing period (s, min)
ξ	=	rate of reaction
ψ^*	=	global enhancement

Subscript

i	=	species
p	=	physisorption
0	=	entering, initial

CHAPTER

12

Modulation of Multiple Reactions

Robert Ross Hudgins, Peter Lewis Silveston*, Albert Renken*[†]

*Waterloo, Ontario, Canada and [†]Lausanne, Switzerland

OUTLINE

12.1 Introduction 341
12.1.1 Selectivity and Yield 342
12.1.2 Classes of Multiple Reactions 349

12.2 Homogeneous Reactions 349
12.2.1 Ethyl Adipate Saponification 351
12.2.2 Formation of Diethanolamine 352

12.3 Solids Catalyzed Reactions 353
12.3.1 Catalytic Reactions of Ammonia and Ethylene Oxide 353
12.3.2 Butadiene Hydrogenation 353
12.3.2 Acetylene Hydrogenation 355
12.3.3 Total Oxidation 357

12.4 Competitive Reactions 359
12.4.1 Methanol Synthesis 359
12.4.2 Ethanol Dehydration 359
12.4.3 Parasitic Reaction Systems 360

12.5 Methane Homologation 361
12.5.1 Methane Homologation at a High Temperature 361
12.5.2 Methane Homologation at a Low Temperature 363

12.6 Oligomerization of Ethene 365

12.7 Modulation of Multiple Inputs 366

12.1 INTRODUCTION

The presence of secondary reactions in a process system often causes operating problems. Parasitic reactions consume feedstock and may complicate recovery of products. Secondary reactions may cause catalyst fouling or have large heat effects that compromise reactor stability or require extra heat-transfer surface. With multiple reactions, an objective of reactor design becomes minimizing the extent of the secondary reactions without throttling the desirable ones.

Early studies of multiple reactions used arbitrary models and employed numerical analysis, variational or relaxed steady-state techniques. *Horn and Bailey (1968)*, for example, considered two competitive reactions of different orders to demonstrate improved selectivity through composition modulation. Several years later, *Renken (1972)* examined a system of three consecutive-competitive reactions in a continuous stirred tank reactor and found the selectivity to an intermediate product could be improved by composition modulation. An important theoretical and experimental literature dealing with selectivity improvement through feed composition modulation has emerged since these publications. Discussion of the application of composition modulation to multiple reactions has been split between **Chapters 4** (partial oxidation and oxidative dehydrogenation of hydrocarbons), **Chapter 7** (polymerization processes) and this chapter. Many of the published studies on the application of modulation to multiple reactions are summarized in **Table 12-1**.

12.1.1 Selectivity and Yield

Although mentioned in **Chapter 1**, these measures were not defined. Overall selectivity ($S_{i/j}$) is calculated from molar flows across a reactor. It is an absolute quantity and corresponds to the ratio of the moles of the desired product "i" formed and the moles of the key reactant "J" transformed:

$$S_{i/j} = \frac{(n_i)_0 - n_i}{(n_j)_0 - n_j} \frac{\nu_j}{\nu_i}. \tag{12-1}$$

Normally, (n_i) is a species molar flow rate (mol/s); it could be also moles in the reactor if a batch reactor is considered. The ratio of the stoichiometric factors ν_j/ν_i is introduced to get selectivity values between zero and unity. It should be mentioned that the stoichiometric factor is always negative for the disappearing species (left hand side of a stoichiometric equation).

Yield is the moles of product "i" formed per mole of key reactant "j" fed to the reactor. It is also an absolute quantity and a product species and the key reactant must be indicated. Using the symbols given above,

$$Y_{i/j} = \frac{(n_i)_0 - n_i}{(n_j)_0} \frac{\nu_j}{\nu_i}. \tag{12-2}$$

Thus, the yield, $Y_{i/j}$, is the product of the selectivity, $S_{i/j}$, and the conversion, X_j, of the key reactant.

An instantaneous selectivity, $s_{i/j}$, may be also used. It is defined as the ratio of the rate of formation of product "i" to the rate of consumption of key reactant "J." Occasionally a selectivity ratio, $\varsigma_{i/k}$, is used in place of the instantaneous selectivity. It is the ratio of rates of formation, usually the ratio of the rate for the desired product "i" to the rate for an important unwanted product "K." Selectivity ratio comparisons often lead to large differences and thus dramatize the effect of changes in operating conditions. Instantaneous selectivity and the selectivity ratio at steady state reflect local concentrations and temperatures and, of course, the reaction network and kinetics. Selectivity, as an overall quantity, reflects in addition the reactor type and the conversion achieved.

Enhancement can be defined in terms of either conversion, selectivity or yield. The subscript designates the enhancement considered:

$$\psi_X = \frac{X_j}{(X_j)_{SS,}} \tag{12-3}$$

$$\psi_S = \frac{S_{i/j}}{(S_{i/j})_{SS,}} \tag{12-4}$$

$$\psi_Y = \frac{Y_{1/j}}{(Y_{i/j})_{SS,}} \tag{12-5}$$

where X_j, $S_{i/j}$ and $Y_{i/j}$ are the time-average quantities under composition modulation while the quantities with a SS subscript are those measurements taken under steady state at the time-average composition of the feed used in the modulated operation.

Global enhancements are defined as well by the relation just given under the condition that

TABLE 12-1 Modulation of Multiple Reactions

References	Objective	Reaction and Reaction Conditions	Modulated Variable	Observations	Comments
Renken (1972)	Effect of composition modulation on the intermediate yield in a consecutive-parallel reaction system	Arbitrary reaction: $A + B \rightarrow C + D$, $C + B \rightarrow E + D$, $E + B \rightarrow F + D$. Stoichiometry is typical of the reaction of Et_2O with MEA. Simulation assumed a CSTR	Bang-bang cycling of reactants A and B	Modulation improved the yield of the "C" intermediate.	Systems of competitive parallel reactions were also studied by *Farhadpour and Gibilaro (1975)* and by *Lee and Bailey (1980)*.
Wandrey and Renken (1973)	Effect of composition modulation on yield of an intermediate product	Oxidation of butane and cyclohexene; Pt gauze catalyst. $T_0 = 415°C$; $P = 103$ kPa; $SV = 0.86\ s^{-1}$	1) In-phase cycling of both reactants; 2) cycling of O_2 flow with constant C_4 or C_6 flow; 3) constant O_2 flow with cycling of C_4 or C_6 flow. Diluent flow modulated to maintain a constant *SV*	Strategy (2) increased the CO yield and selectivity more than strategy (1). Both yield and selectivity were strong functions of period.	Two CO peaks were observed; the second peak resulted from combustion of C deposits on the catalyst surface. Authors explained increased selectivity and yield by temperature changes arising during modulation. The presence of a reservoir of C on the Pt surface was suggested.
Renken et al. (1974)	Proper comparison of composition modulation and with steady-state results	Formation of DEA and TEA from Et_2O and MEA. $T = 50°C$; $SV = 0.033$ and $0.071\ min^{-1}$	Bang-bang cycling of Et_2O and MEA concentrations	Yield increase of 9% and was strongly dependent on period for symmetrical cycling. For asymmetrical cycling, there was a small enhancement of yield.	Experimental system was modeled and closely reproduced the experimental measurements.
Wandrey and Renken (1974)	Effect of composition modulation on the yield of an intermediate	Oxidation of propane; Pt gauze catalyst. $T_0 = 530°C$; $P = 103$ kPa	1) In-phase cycling of both reactants; 2) cycling of O_2 with constant C_3 flow; 3) constant O_2 with cycling of C_3 flow. Diluent flow modulated to maintain a constant *SV*	With strategy (1), an order of magnitude increase in the yield of CO from propane combustion was observed.	

(*Continued*)

TABLE 12-1 Modulation of Multiple Reactions (*cont'd*)

References	Objective	Reaction and Reaction Conditions	Modulated Variable	Observations	Comments
Renken et al. (1974)	Effect of composition modulation on competitive reactions	Dehydration of ethanol to ethene or diethyl ether; γ-alumina catalyst. $T = 240°C$; $P = 1$ atm; packed bed reactor	Cycling of ethanol flow to the system	Interruption of feed increased the yield of ethylene and decreased the yield of diethyl ether. Both cycle split and period affect yield.	Observations were assumed to result from reactant or water blockage of sites for adsorption. Model based on steady-state data failed to predict experimental observations.
Leupold and Renken (1977, 1978); Leupold et al. (1979)	Effect of modulation on ethyl acetate from the acid catalyzed reaction of ethylene and acetic acid	$C_2H_4 + CH_3COOH = CH_3COOC_2H_5$; H_2SO_4/SiO_2 as catalyst; $SV = 3050\ h^{-1}$; $T = 140°C$	C_2 was fed continuously; HAc was vaporized and cycled in an on-off mode. Experiments were carried out around a time-average $y_{HAc} = 0.0825$; $4 < \tau_{cycle} < 30$ min; $0.167 < s < 0.333$ based on HAc	At $\tau_{cycle} \geq 5$ min, ψ increased with s, and exceeded 2.5 with $\tau_{cycle} = 32$ min and $s = 0.33$. Under symmetrical cycling, with $\tau_{cycle} = 40$ min, ψ was 7 times greater than at steady state. On-off cycling of HAc prevented the catalyst deactivation observed at steady state.	$C_2^{=}$ participates in several minor side-reactions such as polymerization, hydrogenation, oxidation to form CO_2, and acid reduction to form SO_2. Products of these side-reactions are strongly adsorbed and poison the catalyst.
Al-Taie and Kershenbaum (1978)	Exploration of the effect of composition modulation on consecutive reactions	Butadiene hydrogenation. Catalyst was 23 wt% Ni on kieselguhr	System modulated by adding $C_4^{2=}$ or H_2 to a steady flow of the mixture of the two in H_2	Enhancement of $C_4^{2=}$ conversion required that $\tau_{cycle} < 2$ s. Amplitude of the concentration change was important. Selectivity to and yield of $C_4^{=}$ were enhanced by 20% at the largest amplitude for $\tau_{cycle} = 2$ s.	Enhancement was attributed to adsorption of $C_4^{2=}$ on the catalyst surface. A model qualitatively predicted experimental observations.
Bilimoria and Bailey (1978); Lee and Bailey (1980)	Determine level of modeling needed to characterize performance under composition modulation	Acetylene hydrogenation; Harshaw Ni-0707T; Girdler G-52 catalysts; CSTR (Berty-type), $T = 166°C$; $P = 108$ kPa	Modulation of H_2 flow with constant acetylene flow at constant SV. $\tau_{cycle} = 66$ s	Yields of $C_2^{=}$ and C_2 increased under modulation. Yields depended primarily on s rather than τ_{cycle}. Modulating the H_2 concentration did not increase selectivity to C_2 but suppressed conversion of acetylene.	Simulation of the system using steady-state data to evaluate model parameters failed to predict experimental performance.

Lee et al. (1980)	Effect of composition modulation on a homogeneous reaction	Reaction of diethyl adipate with NaOH to form disodium adipate. Diester concentration held constant at 0.023 mol/m^3; solvent was 50/50 wt% mixture of isoproanol and water. $T = 25°C$	Bang-bang cycling between diethyl adipate and NaOH; $\tau_{cycle} = 22.5$ min	Composition modulation increased yield of monoester by 10%. Performance depended on the ratio of diethyl adipate/NaOH.	A model was developed and successfully simulated experimental results.
Dettmer and Renken (1983b); Walker and Renken (1983, 1984); Renken et al. (1984)	Model for ethyl acetate from the acid catalyzed reaction of ethylene and acetic acid	$C_2H_4 + CH_3COOH = CH_3COOC_2H_5$; catalyst was H_2SO_4/SiO_2. Authors evaluated model parameters from experiments on equilibrium adsorption and step-change responses	Transient mass balances on reactants in acid phase and reactants and product in the gas phase	Comparisons of predicted and measured responses to step-changes in HAc partial pressure were made for start-up and shut-off.	The model was deemed good considering that no fitting of the experimental data was used to evaluate model parameters.
Truffer and Renken (1986)	Transient and cycling experiments on the acid catalyzed reaction of $C_2^=$ and HAc	$C_2H_4 + CH_3COOH = CH_3COOC_2H_5$; Experimental measurements in a differential reactor; catalyst was H_2SO_4/SiO_2. Conditions: $T = 110°C$; $P = 1.05$ bar	Data of Dettmer and Renken were fitted to an Elovich isotherm to re-evaluate adsorption and desorption constants for a better fit to sorption data.	Measured partial pressures were about 20% below predicted values just before the step-up in HAc partial pressure. The model predicted a global enhancement of 75%.	Authors' model reproduced the measured data. When applied to their pulsing experiments, the model over-predicted enhancement by about 10%.
Vamling (1987); Liden and Vamling (1989)	Exploration of composition modulation using stop-flow	NH_3 and Et_2O to produce MEA, DEA and TEA; tubular reactor with an Amberlite 200 ion exchange catalyst. $T = 87°C$ inlet; $P = 313$ MPa; near-adiabatic operation	Continuous NH_3 flow with stop-flow of Et_2O	No enhancement of MEA production was observed.	Authors observed that the amplitude of the modulation was probably insufficient to affect performance in the presence of mass transfer resistance. Simulation was also undertaken.

(Continued)

TABLE 12-1 Modulation of Multiple Reactions (*cont'd*)

References	Objective	Reaction and Reaction Conditions	Modulated Variable	Observations	Comments
Koerts and van Santen (1991,1992, 1993); Koerts et al. (1992)	Effect of temperature and composition modulation on consecutive reactions	CH_4 homologation over transition metals supported on silica. CH_4 exposed to the catalyst at 450°C, cooled to 150°C and hydrogenated	Complex temperature and composition cycle	Ru provided the highest yield of C_2^+ products. 21% of C deposited was converted to paraffinic products. Conversion to higher hydrocarbons increased with the temperature of the hydrogenation step.	Only parts of the cycles were investigated. Observations were used to develop a mechanism for hydrocarbon formation.
Belgued et al. (1991, 1992a, b); Amariglio et al. (1994,1995a, b)	Effect of composition modulation on consecutive reactions	CH_4 homologation; catalysts were Euro-Pt 1 (a standardized material with 6.3 wt% Pt on SiO_2), Ru (4.7 wt% on SiO_2), and Co (2 wt% on kieselguhr); $150 < T < 300°C$	Cycling between methane and diluent and cycling of diluent alone	Main products with Pt and Ru catalysts were C_2 and C_5. With Co catalyst, main products were C_2, C_4 and C_5. The duration of CH_4 exposure affected the product yield and selectivity.	Yields of hydrocarbons were low. C deposition was mostly refractive and was hydrogenated back to methane.
Rao et al. (1992)	Investigation of composition modulation	Hydrogenation of benzene; $T = 172°C$, $P = 2$ MPa	1) Modulation of benzene flow at constant H_2 flow; 2) modulation of H_2 flow at constant benzene flow. $20 < \tau_{cycle} < 60$ min	20 to 25% enhancement in benzene conversion for $\tau_{cycle} = 20$ min. For a 60-min period, conversions were below comparable steady-state values.	
Wang et al. (1993)	Effect of temperature and composition modulation on consecutive reactions	CH_4 homologation; ZSM-5 zeolite catalyst promoted with Mo or Zn	Complex temperature and composition cycle involving CH_4 exposure to the catalyst at 700°C, with hydrogenation on cooling	Selectivity to benzene approached 100% at a 7.2% CH_4 conversion.	

Lefort et al. (1994)	Effect of composition modulation on consecutive reactions	Oligomerization of ethene; 6.3 wt% Pt on silica catalyst: $100 < T < 150°C$	Cycling between ethene and diluent and diluent alone; asymmetrical cycles	At 100°C, 42% of the ethene adsorbed appeared as C_3+ products. At 150°C, 64% appeared as C_3+ products	Ethyne formation is believed to occur on ethene adsorption.
Visser et al. (1994)	Investigation of composition modulation	Hydrogenation of 1% phenylacetylene in toluene; catalyst was Pd/alumina; $20 < T < 100°C$, $P = 1$ atm	Bang-bang modulation between H_2 and N_2	No significant differences between modulated and steady-state selectivities and conversions to styrene were found.	Failure to find an improvement was attributed to attenuation of square-wave disturbance by mixing in the catalyst bed.
Solymosi and Cserényi (1994, 1995); Solymosi et al. (1992, 1994, 1995)	Effect of temperature and composition modulation on consecutive reactions	Methane homologation using 2 wt% MoO_3 on silica as a catalyst	Complex temperature and composition cycles involved CH_4 exposure to the catalyst at 700°C, with hydrogenation on cooling below 700°C	Main products were benzene and C_2 components. CH_4 conversion reached 5.4%; selectivities to benzene up to 56% were observed.	Graphitic carbon formed on exposure. This carbon was hydrogenated at temperatures below 700°C.
Creaser et al. (1999a)	Composition modulation of the oxidative dehydrogenation of propane	$C_3H_8 + O_2 \rightarrow C_3H_6 + H_2O$, $C_3H_6 + O_2 \rightarrow CO_2 + H_2O$ (not balanced) over a V-Mg-O catalyst in a micro-packed bed at 510°C	Bang-bang switching between 6 vol% C_3 and 6 vol% O_2 for $10 < \tau_{cycle} < 400$ s; also used switching between C_3 and a C_3-O_2 mixture	Bang-bang switching significantly increased dehydrogenation selectivity and $C_3^=$ yield. Further selectivity improvement came from adding C_3 to the O_2 half-cycle.	*Creaser* study is discussed in **Chapter 4**.

(*Continued*)

TABLE 12-1 Modulation of Multiple Reactions (*cont'd*)

References	Objective	Reaction and Reaction Conditions	Modulated Variable	Observations	Comments
Kiwi-Minsker et al. (2002)	Investigation of dehydrogenation in a membrane reactor with a filamentous catalyst	$C_3H_8 \rightarrow C_3H_6 + H_2 + C$; $C + O_2 \rightarrow CO_2$; catalyst was Pt/Sn impregnated on alumina wash coated silica fibers. Fibers packed in a tubular reactor with two concentric compartments separated by an H_2 permeable Pd- Ag membrane; $T = 823$ K, $P = 0.14$ MPa	Concentric compartments were fed either C_3 or air, and feed to each compartment was switched after 85 min	Membrane-separated concentric reactor system with flow cycling resulted in a 97% $C_3^=$ selectivity at a C_3 conversion of 22%.	Slow deterioration of performance was observed.
Ye et al. (2005); Munder et al. (2007)	Investigation of butane partial oxidation in an electrochemical membrane reactor	$C_4 + O_2 \rightarrow MA + CO_2$ using a vanadia phosphate (VPO) catalysts under anaerobic operation with electrical potential O_2 pumping with a yttria-stabilized zirconia membrane	Duration of air and C_4 flows were 40 min followed by electrical O_2 pumping for 120 min	Anaerobic operation with electrical pumping of O_2 improved the selectivity to MA to 92 from 63 at steady state for 75% C_4 conversion. Models fitted to periodic and steady-state data show MA yield enhancement through periodic operation of 1.25 to 1.40 depending on N_{Da} and the reactor model used.	Rate of C_4 conversion was quite small and throughput very limited.

the molar reactant feeds are identical at steady state and periodic operation. For a meaningful comparison, the denominator of each term corresponds to the maximum steady-state conversion, selectivity or yield in the range of feed composition covered by the modulations.

12.1.2 Classes of Multiple Reactions

Competitive or parallel reactions are those that share one or more reactants:

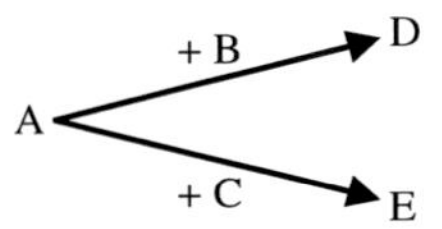

(12-6)

Or, for two reactants:

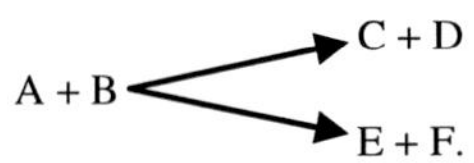

(12-7)

Consecutive, sequential or serial reactions are rather rare:

$$A \longrightarrow B \longrightarrow C \longrightarrow D \longrightarrow E. \quad (12\text{-}8)$$

A great many systems fall into the class of consecutive-competitive multiple reactions:

$$\begin{aligned} A + B &\longrightarrow C + D \\ C + B &\longrightarrow E + D \\ E + B &\longrightarrow F + D. \end{aligned} \quad (12\text{-}9)$$

Chlorination of hydrocarbons, forming successively the monochloro, dichloro and trichloro products, represent this class. The common reactant "B" is Cl_2 and the common product "D" is HCl. When the reactant "B" in **Eq. (12-9)** is in large excess with respect to "A," the consecutive class is encountered, either as **Eq. (12-8)** or one in which each step has a common product that does not react further. Hydrogenation of acetylene through ethene to ethane is an example of a consecutive reaction class.

12.2 HOMOGENEOUS REACTIONS

Renken (1972) found through a numerical simulation of a CSTR that selectivity to an intermediate product "C" in a homogeneous, consecutive-competitive reaction network represented by **Eq. (12-9)** can be increased by in-phase concentration cycling of reactants "A" and "B." Of course, kinetics of the reaction must be globally nonlinear. In general, a first-order reaction relating to A and B is found. Others who have looked at consecutive-competitive reaction networks have reached the same conclusion (*Farhadpour and Gibilaro, 1975; Lee and Bailey, 1980*). The concept behind in-phase modulation is simple: as the first reaction proceeds, building up the concentration of "C," the rate of the second reaction forming "E" accelerates. By cutting off the feed of "A" and "B," the second reaction can be throttled before the concentration of "C" and the rate of formation of "E" reach maxima.

Renken (1972) was concerned with a proper comparison of modulation and steady state with respect to global enhancement. He concluded it is not possible to carry out comparable experiments under the same nominal time-average operating conditions. Thus, the concentration range used in a modulation study must be searched for a maximum selectivity or yield under steady state to see if modulation has achieved global enhancement. He proposed using the specific reactor performance, i.e., the amount of product obtained per reactor volume and time, as a criterion for comparison.

For the consecutive competing reactions considered the maximum attainable yield of the intermediate product is only a function of the ratio of rate constants, k_1/k_2, and the reactor type. Therefore, under steady state, maximum product formation will be obtained for constant reactant feed at the highest inlet concentration. Imposing concentration modulations of the reactant leads to a decrease of the time average concentration and, therefore, a decrease of the molar feed. To get the same molar feed as under optimal steady-state conditions the volumetric flow must be increased to compensate for the lower average concentration. The following conditions were applied in the study:

$$n_j = Q_{ss}C_{j,o,ss} = Q_{per}\,\overline{C}_{j,0} = Q_{per}\,C_{j,o,ss}\cdot s \tag{12-10}$$

with n_j the molar feed (mol/s) of reactant "A" or "B," Q_{ss}, Q_{per} are the volumetric flow under steady state and periodic operation, respectively, and $C_{j,o,ss}$ is the steady-state concentration with $\overline{C}_{j,0}$ the mean concentration under periodic operation, which depends on the period split, s.

Of course, this strategy leads to an important reduction of the mean residence time and, consequently, to a decrease of the conversion of the key reactants "A" and "B." *Renken* showed, however, that the yield of "C," $Y_{C/A}$, actually increased even under those conditions, if a suitable choice of cycle period and split are used, although the increase is small. Much more important is a selectivity increase in the order of 20 to 30% and, therefore, a much better use of the key reactants for the same specific production of the intermediate (kmol/m^3·s). Renken's results are summarized in **Figure 12-1**.

Two experimental studies were conceived to test the above predictions, recognizing that kinetics were imperfectly known, mixing can give rise to inhomogeneity, and mechanically stirred tanks were not perfectly isothermal or fully backmixed. Each of these studies used a different reaction system and imposed different restrictions on comparing modulated and steady-state operation.

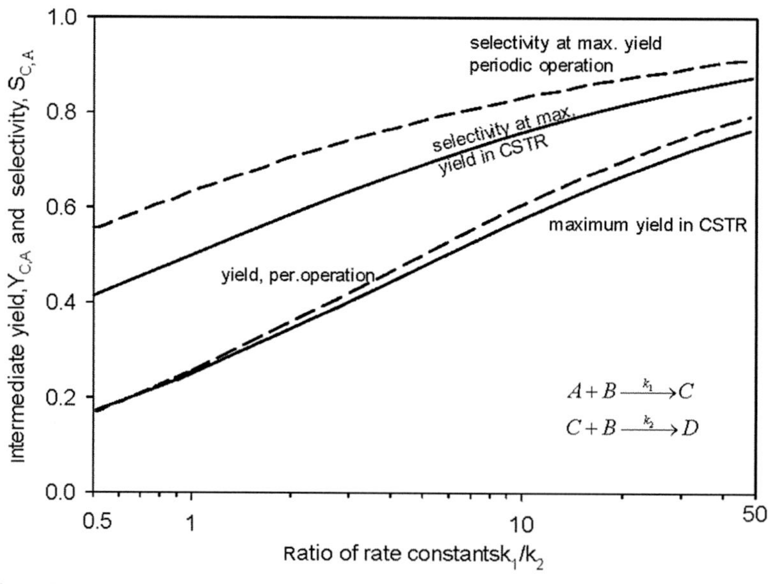

FIGURE 12-1 Achievable yield and selectivity under periodic operation in a CSTR for optimal cycle time and split compared to the maximum yield and the corresponding selectivity to "C." under steady state operation. Constant molar reactant feed: $Q_{ss}C_{A,o,ss} = Q_{per}\overline{C}_{A,0}$; $C_{A,0}/C_{B,0} = 1$. *(Data reproduced from* Renken (1972) *with the author's permission. © 1972 by Elsevier Science Publishers.)*

12.2.1 Ethyl Adipate Saponification

Ethyl adipate is the ethyl diester of adipic acid. It saponifies in the presence of a strong base to the adipate salt:

$$\begin{aligned} Et_2A + NaOH &\rightarrow NaEtA + EtOH \\ NaEtA + NaOH &\rightarrow Na_2A + EtOH. \end{aligned} \quad (12\text{-}11)$$

This reaction was studied by *Lee et al. (1980)* whose objective was to measure the global enhancement of the selectivity to the intermediate, sodium monoethyl adipate (NaEtA) under composition modulation.

The experimental system employed by *Lee et al. (1980)* was chosen to model a standard industrial stirred autoclave and dealt only with the effect of reactant ratio on yield enhancement. Amplitude was at a maximum as the flow was switched from the reactant-solvent mixture to solvent only. For comparison purposes, steady-state measurements were made over the same concentrations used for the modulation study.

Figure 12-2 presents experimental time-average yields of the monoester (NaEtA) under in-phase modulation versus the diester/saponification agent ratio. These data points lie above their steady-state counterparts. Maxima are apparent for both periodic and steady-state data so that global enhancement of yield was achieved. *Lee et al. (1980)* modeled their experiments assuming an isothermal CSTR, irreversible reactions and rates that are first order in each reactant. Separate, steady-state experiments were performed to evaluate the rate constants and activation energies for each of the saponification reactions.

The curves shown in **Figure 12-2** are simulated yields based on the steady-state experiments. Except for the feed reactant ratio of 1.6, the agreement between model and experiment is good. From the simulated yields, the global yield enhancement observed by *Lee et al.* was about 10%. Because the modulation employed decreased conversion substantially at a split of 0.1, a yield improvement means that selectivity increased enormously through cycling.

Unfortunately, the authors did not compare the achieved yield under periodic operation with the maximums obtainable in continuous stirred tank reactors, which can easily be calculated from the presumed kinetics. The maximum yield is independent of the inlet concentration as long as $\overline{C}_{B,0}/\overline{C}_{A,0} \geq 1$. In addition, experiments

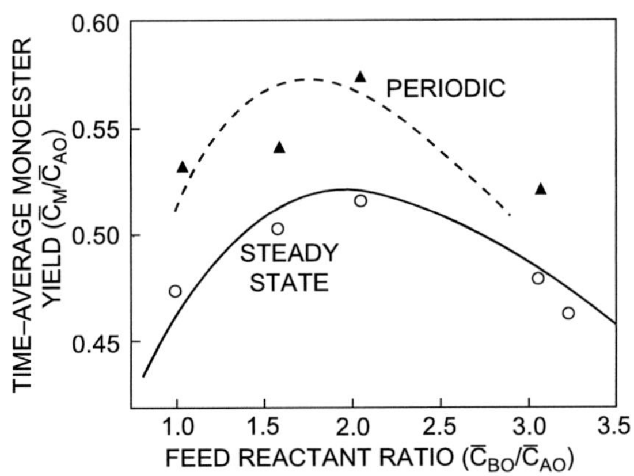

FIGURE 12-2 **Comparison of time-average sodium monoethyl adipate yields with steady-state yields at various time-average diethyl adipate/sodium hydroxide concentration ratios at 298 K and $\tau_{cycle} = 22.5$ min for diester saponification using a 50:50 wt% solvent mixture of isopropanol and water.** *(Figure reproduced from* Lee et al. (1980) *with permission of the authors.)*

were carried out at constant mean residence time leading to a drastic decrease of specific productivity for the intermediate of around 50%.

12.2.2 Formation of Diethanolamine

Renken (1974b) studied the formation of diethanolamine (DEA) and triethanolamine (TEA) from the reaction of ethylene oxide (Et_2O) and monoethanolamine (MEA) in an aqueous solution. The reaction network is given by **Eq. (12-12)** along with kinetic constants. Both reactions proceed at moderate rates around 50°C. Experimental studies under steady-state conditions confirmed the classical kinetic model for consecutive competing reactions.

$$MEA + Et_2O \xrightarrow{k_1} DEA$$

$$k_1 = 4.63 \cdot 10^{-3} m^3 kmol^{-1} s^{-1}$$

$$DEA + Et_2O \xrightarrow{k_2} TEA$$

$$k_2 = 2.97 \cdot 10^{-3}\, m^3 kmol^{-1} s^{-1}. \quad (12\text{-}12)$$

Parasitic reactions of Et_2O are not significant at this temperature. *Renken* employed in-phase modulation of reactant concentrations to investigate the effect of cycle period and split on selectivity to DEA and its yield.

Symmetrical and asymmetrical in-phase cycling of MEA and Et_2O concentrations were undertaken for conditions given in **Figure 12-3**. For all experimental runs, the maximum reactant concentration and the molar feeds were kept constant. As concentration modulation leads to lower average inlet concentrations, the volumetric flow rate was adapted for each run (see **Eq. 12-10**). As mentioned earlier, this modulation strategy reduces the mean residence time and depresses conversion but

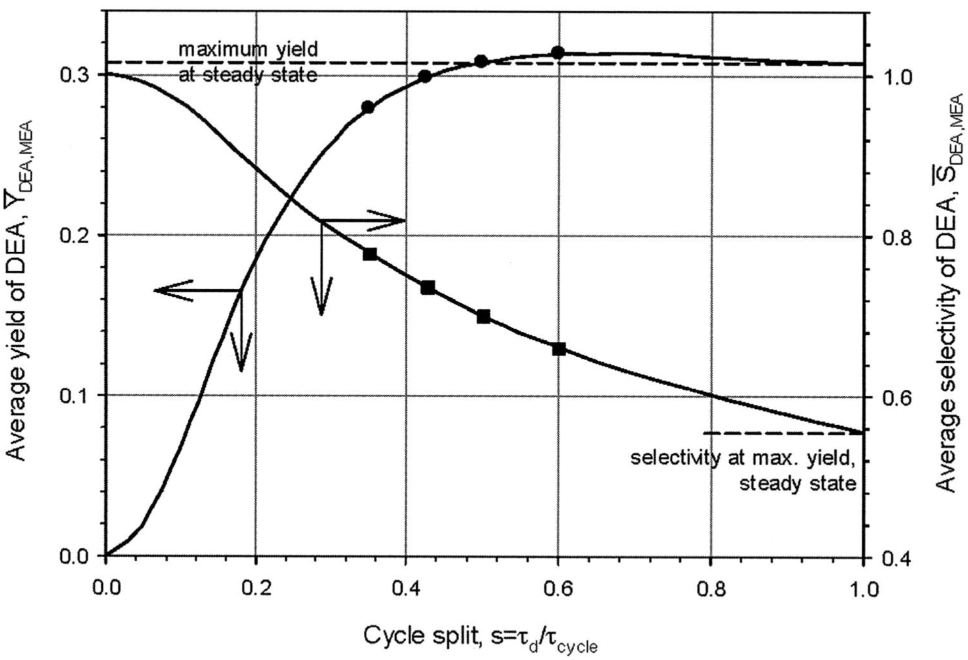

FIGURE 12-3 Experimental and predicted time-average yield of DEA and selectivity to DEA versus the cycle split, $\tau_{cycle} = 3\tau$ (τ = residence time), concentration modulation of the reactant feed concentrations of MEA (0 to 0.47 mol/L) and Et_2O (0.5 to 0 mol/L) in a CSTR at 323 K. *(Figure reproduced from* Renken (1974b) *with permission of the author.)*

augments selectivity, so the effect on yield is uncertain.

For symmetric cycling, yield decreased in the range of variables studied by 9.1% at the smallest period used to 2.6% at the highest. The selectivity, $S_{DEA/MEA}$, changed in the opposite direction, but it was roughly 27% higher than the steady-state selectivity at maximum yield. The time-average yield of DEA can be increased slightly above the maximum yield at steady state by letting the $s > 0.5$ as shown in **Figure 12-3**. For asymmetrical forcing where space velocity is not constant, the experimental time-average yield exceeded the steady-state yield by 1.5% for a normalized cycle period of 3.0 at $s = 0.6$. This is a global yield enhancement in the range of variables investigated. Note that $s = 1.0$ represents steady state, while at $s \rightarrow 0$, the mean residence time $\tau \rightarrow 0$; in consequence of this, the yield goes to zero. The selectivity, $S_{DEA/MEA}$, between 1 and 0.56, varies considerably with s. The experimental results, shown in **Figure 12-3**, agree with model predictions.

The two experimental studies discussed in this section illustrate that yield and selectivity improvement can be achieved for consecutive-competitive reaction systems through a suitable composition modulation strategy. Other strategies, such as modulation with a phase shift, could be usefully explored using homogeneous reaction systems. This would extend our still limited knowledge of how modulation and modulation strategy affects performance. Good agreement between model and experiment suggests that simulation would be satisfactory for such studies. See *Silveston (1998)* for further discussion.

12.3 SOLIDS CATALYZED REACTIONS

Many heterogeneous catalytic reaction systems, including some of industrial significance, are also consecutive-competitive reactions. Examples are most hydrogenations, many partial oxidations, and polymerization reactions. However, mechanisms occurring on the catalyst surface added to transport steps needed to bring reactants to the surface introduce complications such that observations for homogeneous reactions cannot be confidently extended to heterogeneous catalytic reactions.

12.3.1 Catalytic Reactions of Ammonia and Ethylene Oxide

Ethanolamines are produced commercially by an acid-catalyzed, liquid phase reaction between ethylene oxide (Et_2O) and ammonia. *Vamling (1987)* explored ways to increase Et_2O conversion and selectivity to MEA using on-off modulation of the Et_2O flow to a tubular, near-adiabatic packed bed reactor. Further details of *Vamling's* work are given by *Silveston (1998)*. Modulation employed switching between NH_3 and an NH_3-Et_2O mixture. *Vamling* observed no difference in MEA production per mol of Et_2O consumed between composition modulated and steady-state operation. A smaller number of experiments with a zeolite 13-X catalyst also failed to show a production difference.

Liden and Vamling (1989) believed that the lack of enhancement through modulation occurred because the reaction system was not sufficiently disturbed by an on-off operation, since 99.9% Et_2O conversion caused the amplitude to vanish by the reactor outlet. Furthermore, use of a liquid-phase reaction reduced transport rates into the porous catalysts and lowered the amplitude significantly within particles. Their observations suggested a cycle frequency limit with solid catalyzed, liquid phase reactions for any rate or selectivity enhancement.

12.3.2 Butadiene Hydrogenation

Al-Taie and Kershenbaum (1978) and *Bilimoria and Bailey (1978)* used hydrogenation as a - model reaction to explore selectivity improvement under modulation and to test predictions of a theoretical analysis (*Bilimoria and Bailey, 1978*). Objectives of the former study were to measure selectivity and yield improvement through composition modulation and to find optimal conditions. Results were compared at the same C_4H_6 feed rate and a constant space velocity. The maximum selectivity for the first product in a consecutive-competitive reaction system occurs at zero conversion; consequently, selectivities to butene were compared for forcing conditions that gave yields identical to the maximum steady-state yield.

Cycle period and amplitude exert a large influence on butadiene conversion as shown in **Figure 12-4(a)**. Fast modulation at $\tau_{cycle} = 2$ s or less affords conversions that equal or slightly exceed the maximum attained under steady state. Increasing the period, however, suppresses conversion. Amplitude, λ, defined by the variation of the C_4H_6 concentration between the half-periods, is important. $\lambda = 1$ represents switching between H_2/N_2 and C_4H_6/N_2. In **(b)** a substantial increase in selectivity (about 20%) can be seen. Apart from modulation at the amplitude extreme, neither amplitude nor cycle period substantially influenced selectivity. A curious result is that operating at $\tau_{cycle} = 2$ s and $\lambda = 1$ increased conversion by about 8% over steady state but provided the smallest improvement in selectivity. **Figure 12-4(c)** shows that the selectivity enhancement leads to a yield enhancement at shorter cycle periods, despite the negative effect of cycling on conversion. The yield reaches 20% at $\tau_{cycle} = 2$ s, the shortest period used. Amplitude was an important variable: A small amplitude, corresponding to switching between C_4H_6-H_2-N_2 mixtures, resulted in improved butene yields at τ_{cycle} up to 30 s.

Al-Taie and Kershenbaum (1978) attributed improvement under periodic forcing to the dynamics of butadiene adsorption. They formulated a plug flow model for their reactor assuming competitive adsorption of C_4H_6, C_4H_8 and C_4H_{10}, but negligible H_2 adsorption.

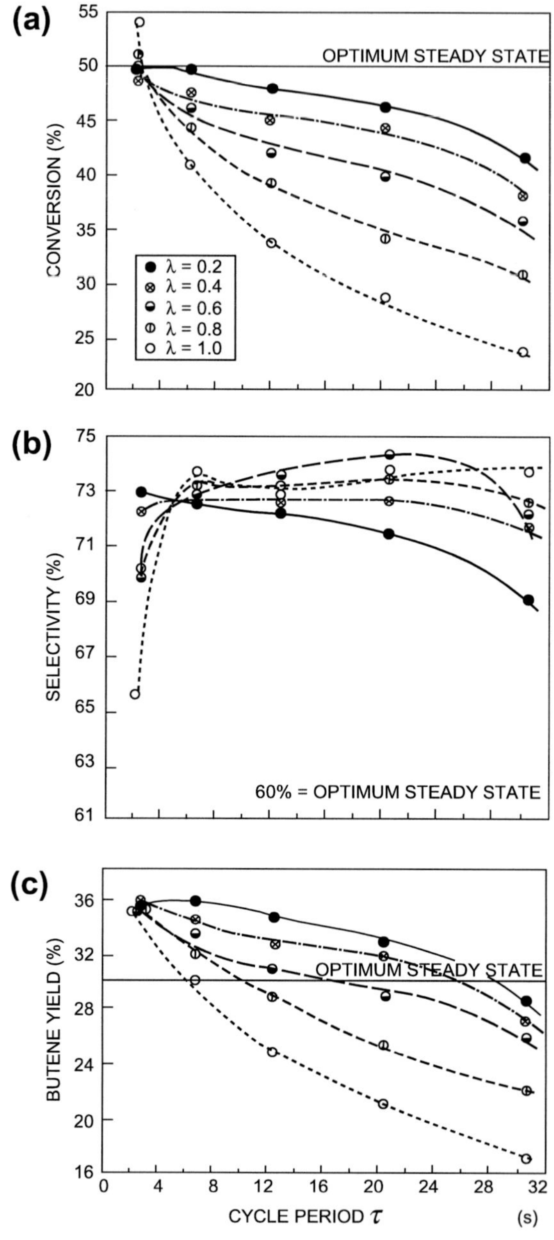

FIGURE 12-4 Variation of reactor performance under symmetrical modulation as a function of cycle period and amplitude for the consecutive hydrogenation of butadiene to butene and butane at 343 K and 1.07 bar over a Ni/kieselguhr catalyst in a near isothermal packed bed reactor: (a) conversion, (b) selectivity and (c) yield of butene. *(Figure adapted from* Al-Taie and Kershenbaum (1978) *with permission. © 1978 American Chemical Society.)*

They further assumed that hydrogenation occurred via a Rideal mechanism. *Silveston (1998)* provides further discussion.

12.3.2 Acetylene Hydrogenation

The experimental study of this reaction by *Bilimoria and Bailey (1978)* dealt with enhancement by composition modulation and the level of modeling needed to predict performance. A later simulation of this reaction (*Lee and Bailey, 1980*) employed steady-state data but assumed the same reactor configuration and operating conditions used earlier. The question addressed was whether steady-state data used in a simulation properly predict reactor performance under composition modulation.

Experiments of *Bilimoria and Bailey* and those of *Lee and Bailey* used two commercial catalysts: Harshaw Ni-0707T and Girdler G52. The former was used by *Komiyama and Inoue (1968)* and furnished the kinetics adopted by *Lee and Bailey (1980)*. These experiments showed that the product yields increased under composition modulation for both catalysts. **Figure 12-5(a)** plots the mole fractions of all the hydrocarbons leaving the reactor, whereas **Figure 12-5(b)** shows only the ethane mole fractions leaving. The ethene mole fractions exhibit the same shape, but are higher. A comparison of the ethane results indicates the Girdler catalyst is about 10 times more active than the Harshaw catalyst despite fouling and deactivation. *Bilimoria and Bailey* handled deactivation by running modulation experiments between steady-state ones. Because of fouling, only a qualitative interpretation of the cycle period effect can be made. Short periods, ca. 60 s, had just a small influence on yield. Yield depended primarily on cycle split, *s*. The largest H_2/C_2H_2 ratio in the figures corresponds to the highest *s* used. For the Harshaw catalyst, increasing τ_{cycle} raised yields of both C_2H_4 and C_2H_6; however, for the Girdler catalyst the optimum τ_{cycle} for yield enhancement appeared to be about 180 s. Longer periods depressed the ethane yield. An optimum τ_{cycle} for the Harshaw catalyst probably exists but it was not identified by the experiments. Increasing *s* increased the ethene yield for both catalysts.

In contrast to butadiene hydrogenation, where a different cycling strategy was used, modulating the H_2 mole fraction did not increase selectivity to the ethene intermediate. For the Harshaw catalyst, calculations revealed that the steady-state selectivity to C_2H_4, $S_{C2H4/C2H6}$ is 91% at $H_2/C_2H_2 = 1$, but under modulation $S_{C2H4/C2H6}$ dropped to 87% at $\tau_{cycle} = 60$ s and to 82% at 300 s. For $H_2/C_2H_2 = 3$, the C_2H_4 selectivity is 72% at 60 s, but fell to 71% at $\tau_{cycle} = 180$ s and to 64% at 300 s. Modulating the H_2 mole fraction raised the acetylene conversion and selectivity to the final product, ethane. However, out-of-phase forcing of both butadiene and hydrogen mole fractions increased $S_{C2H4/C2H6}$, but raised $Y_{ethene/ethane}$ only at the highest frequencies. Conversion was suppressed by modulation except at high frequency where the improvement was small.

Simulation of acetylene hydrogenation in a continuous stirred tank reactor by *Lee and Bailey (1980)* used Langmuir-Hinshelwood rate expressions for acetylene and ethane hydrogenation, derived by *Komiyama and Inoue (1968)*, in which the hydrocarbon concentration appears in the denominator. These are convex with respect to H_2 indicating that H_2 forcing should improve $S_{C2H4/C2H6}$. Indeed, when these terms were used in a model neglecting adsorbate storage on the nickel surface, the behavior given in **Figure 12-5** was obtained for $\tau_{cycle} = 60$ s and a space time of 50 s. With the assumptions made, H_2 modulation yielded a 46% improvement in $Y_{ethene/ethyne}$ at $H_2/C_2H_2 = 0.5$. A 30% decrease in $Y_{ethane/ethyne}$ was predicted as well so that the simulation showed enhanced selectivity to ethene. However, a comparison of ethane yields from **Figures 12-5a** with those predicted from steady-state data by *Komiyama and Inoue (1968)* show that they are badly in error

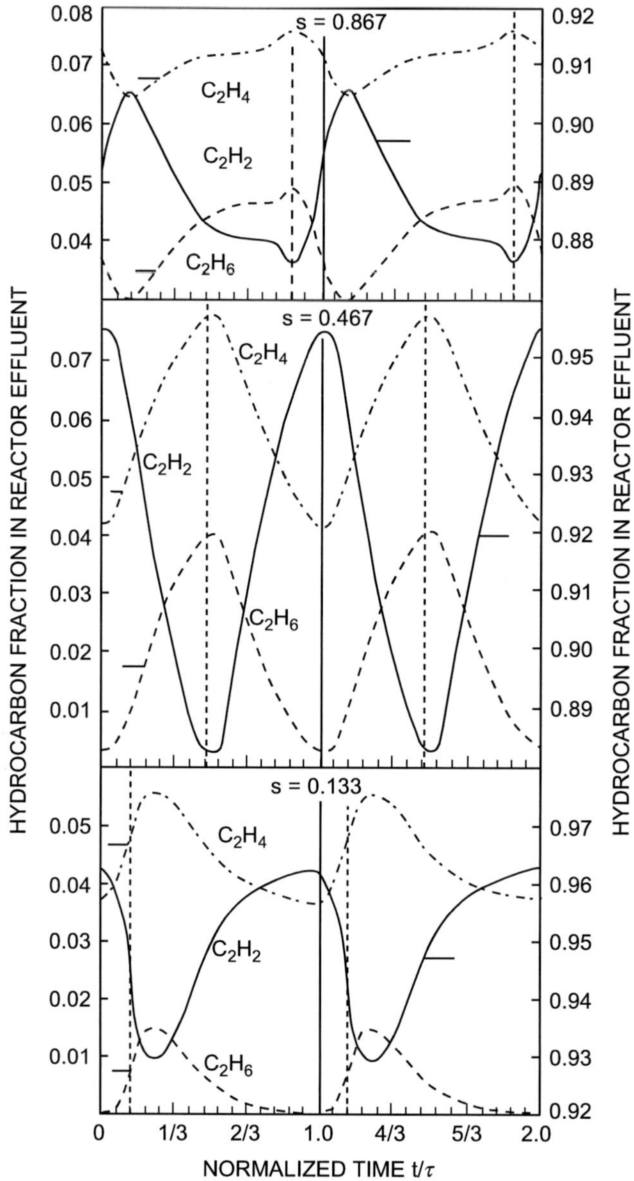

FIGURE 12-5 Variation of hydrocarbon mole fractions in the reactor off-gas with time within a cycle under modulation of the H_2 concentration in the feed to a backmixed reactor employing a Harshaw Ni-0707T catalyst and operating at 439 K, 108 kPa and a space velocity of 0.015 s^{-1}. Modulation used $\tau_{cycle} = 300$ s and $s = 0.867$ in (a), 0.467 in (b) and 0.133 in (c). *(Figure adapted from* Bilimoria and Bailey (1978) *with permission. © 1978 American Chemical Society.)*

in magnitude as well as in the influences of τ_{cycle} and s on yield and selectivity.

Failure of the model resulted from neglecting C_2H_2 and C_2H_4 adsorption dynamics on the catalyst surface. Adsorption dynamics also explains *Bilimoria and Bailey's* experimental data. **Figure 12-5** for the Harshaw catalyst shows concentration measurements during a cycle. For $s = 0.867$, wrong-way behavior appears: Ethene and ethane mole fractions drop rather than increase after the H_2 partial pressure is stepped up at $t/\tau_{cycle} = 0$ or 1. For $s = 0.133$, wrong-way behavior again appears. Wrong-way behavior is explained by strong C_2H_2 adsorption relative to other species and by slow adsorption compared with other reaction steps. The importance of C_2H_2 adsorption is consistent with the effect of the H_2/C_2H_2 ratio on product concentrations for steady-state operation shown in **Figure 12-5(a)**. Clearly, adsorbate storage must be included if a model is to accurately represent composition modulation.

Phenylacetylene hydrogenation was investigated by *Visser et al. (1994)* in a cocurrent, upflow packed bed reactor with a constant flow of the liquid phase and switching between H_2 and an N_2 diluent in the gas phase over a range of cycle periods and cycle splits. A surface-impregnated alumina supported Pd catalyst was used. The *Visser* study encompassed both experiments and simulation using a dynamic model based on steady-state data. The authors observed no significant improvement through this inlet composition modulation experimentally. Their simulation assuming a non-porous catalyst particle confirmed this result. The absence of any improvement was attributed to mass transfer resistance between gas-liquid and solid phases in the three-phase reactor that attenuated the step-change in H_2 concentration. Results of the *Visser* experiments are surprising in view of the significant effect of flow modulation on the performance of trickle bed reactors as discussed in **Chapter 17**.

12.3.3 Total Oxidation

In an early modulation study of catalyzed, multiple reactions, *Wandrey and Renken (1973, 1974)* measured conversion and product distribution for hydrocarbon oxidation over a Pt gauze (**Table 12-1**). They assumed that the reaction network is represented by **Eq. (12-8)**. The authors investigated the effect of modulation on selectivity to CO, an intermediate in the consecutive reaction. Their experiments were less comprehensive than the hydrogenation ones just considered, but they represent a different reaction system and are interesting as well because they were carried out under near-adiabatic conditions.

Wandrey and Renken used a reactor consisting of Pt gauze fitted horizontally in an open tube. The feed system allowed for in-phase cycling of O_2 and hydrocarbons or for on-off cycling of one of the reactants with the other held steady. An N_2 or He diluent was used, the flow rate of which was adjusted to maintain constant space velocity in a cycle. Butane and cyclohexene were used in the 1973 study, while propane was used in 1974. Symmetrical cycles were used and the effect of cycle period was examined.

For in-phase cycling used with propane, the largest increase in selectivity to CO was found at $\tau_{cycle} = 5.4$ s. Modulation increased the yield to 6.45% compared to 0.25% under steady-state operation. The primary CO peak appeared well before the maximum temperature on the gauze and also led the CO_2 peak, while the H_2O peak after the switch to He preceded the CO_2 peak. This was explained by the initial formation of CO followed by its oxidation to CO_2. The secondary CO maximum coincided with the temperature maximum after the O_2 flow had been halted and suggests either burn-off of residual carbon on the Pt surface with adsorbed oxygen or CO desorption from the surface. The decline in CO_2 production during the time when no reactants pass over the catalyst suggested there must be residual matter and adsorbed O_2 on the Pt. The temperature lags in the figure are likely due to the endothermic cracking of C_3H_8 on the gauze surface after the reactant mixture is reintroduced and to continuing oxidation after the mixture flow has terminated. Evidently, the reaction network is more complicated than the sequence given by **Eq. (12-8)**.

A much larger increase in CO yield and selectivity was achieved by cycling just the oxygen flow (*Wandrey and Renken, 1973*). For the shortest τ_{cycle} used, 16 s, mixing in the fluidized bed damped the variation of the O_2 mol% in the feed so that the system behaved almost as though it was at steady state. The effect of raising the cycle period is shown in **Figure 12-6**. Cyclohexene conversion increased from 9.5 to 45.3% for change in τ_{cycle} from 16 to 120 s, while the selectivity to CO went from about 8.5 to 55%. However, if the selectivity change is viewed from an enhancement standpoint, in-phase switching between diluent and reactants appeared to be about as effective as cycling O_2. Either modulation strategy enhanced the CO yields by almost 25-fold.

Two explanations were offered for these enhancements. The first was that O_2 modulation caused oxidation to proceed at a time-average temperature that could not be attained in steady-state operation. Composition modulation induced a time-average gauze temperature in between a high temperature associated with

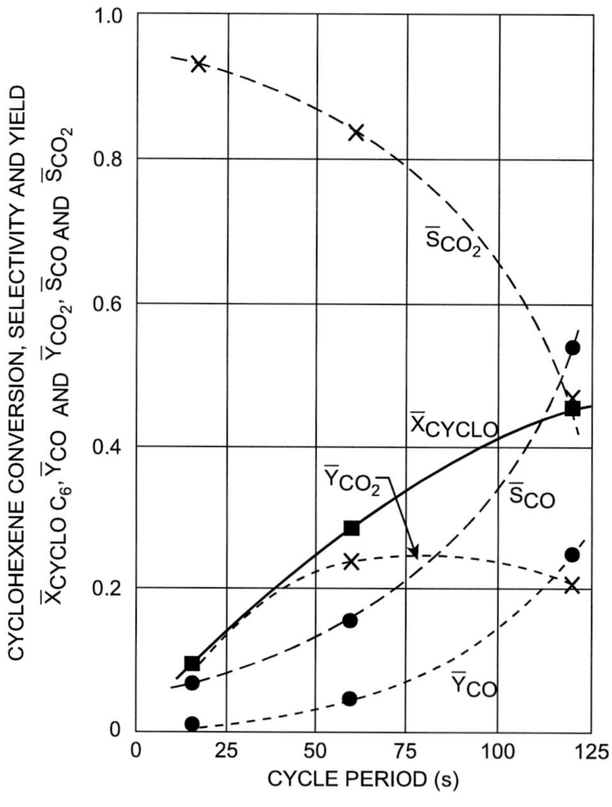

FIGURE 12-6 Influence of cycle period on the time-average cyclohexene conversion, carbon oxide yields and selectivities for the oxidation of cyclohexene over a Pt gauze with a preheater temperature = 415°C, pressure = 103 kPa, and time-average C_6H_{10} and O_2 mole fractions = 0.025 and 0.088 respectively, $SV = 0.86\ s^{-1}$. O_2 mole fraction in the feed modulated symmetrically between 0 and 0.175. *(Figure adapted from* Wandrey and Renken (1973) *with permission. © 1973 Verlag Chemie GmbH.)*

the ignited state and a low temperature observed in the extinguished state. *Wandrey and Renken (1973)* demonstrated stabilization of a time-average temperature experimentally using butane oxidation where the C_4 flow rate was cycled while the O_2 flow rate remained constant.

Their second explanation was that the strong adsorption of hydrocarbons and CO on Pt provided a reservoir of carbon and CO on the surface. With high carbon or CO coverage, O_2 adsorption is limited so the surface is oxygen-deficient, encouraging CO formation. CO is retained on the surface and desorbs as the surface temperature rises, but this takes place after the O_2 flow has stopped so that further oxidation to CO_2 does not proceed. Measurements of changing product concentrations within a cycle for $\tau_{cycle} = 120$ s showed the mole fraction of the carbon oxides in the gas leaving the reactor reached a maximum some 30 s after O_2 was reintroduced. This long delay can be explained only through strong adsorption.

Experiments discussed in this section demonstrate that the selectivity to an intermediate product in solid catalyzed, consecutive-competitive reaction systems can be profoundly altered through composition modulation provided the forcing amplitude is adequate and the frequency is low enough so that amplitude is not significantly reduced by mixing. Selectivity enhancement arises from nonlinear reaction kinetics on surfaces, differences in adsorption/desorption equilibria and sorption rates for reactants and products. The dual source of improvements makes it risky to adopt a cycling strategy before measurements are made or a reliable model exists.

12.4 COMPETITIVE REACTIONS

12.4.1 Methanol Synthesis

This synthesis is only rarely recognized as a competitive reaction. Methane was a side-product with the early catalysts employed. Modern copper-zinc mixed oxide catalysts yield higher alcohols along with methanol, but little methane.

Modulation of this synthesis, studied by *Nappi et al. (1985)* and later by *Chanchlani et al. (1992, 1994)* over copper-zinc mixed oxide catalysts, was discussed in **Section 2.4**. For the conditions they employed, methane was formed. Selectivity to methane was heavily influenced by composition modulation but selectivity to methanol was not affected. However, the system is more complicated than a simple competitive reaction. If water is present, a shift reaction takes place forming CO_2 which also may be hydrogenated to methanol. *Chanchlani et al. (1992)* explored hydrogenation of the mixed carbon oxides under composition modulation over a Cu/ZnO catalyst. They observed a large effect of modulation on the methanol synthesis rate. Their system involved competitive as well as consecutive reactions.

12.4.2 Ethanol Dehydration

This reaction proceeds over alumina at moderate temperatures. Two products are formed: ethene ($C_2^=$) and diethyl ether (DEE):

$$CH_3CH_2OH \rightarrow CH_2{=}CH_2 + H_2O$$
$$CH_3CH_2OH \rightarrow CH_3CH_2OCH_2CH_3 + H_2O. \qquad (12\text{-}13)$$

Renken et al. (1974b) studied this system in a tubular reactor packed with a γ-alumina. The system was modulated by replacing ethanol (EtOH) in the feed by a diluent. Split was defined as the cycle fraction with EtOH in the feed stream. *Renken et al.* mentioned that diethyl ether may decompose to $C_2^=$ and water or disproportionate to the olefin and the alcohol. Furthermore, the reactions given by **Eq. (12-13)** are reversible. However, Renken et al. assumed the two side reactions and the reverse reactions did not proceed under the conditions used and they were neglected in the system model.

Periodic interruption of the EtOH flow to the catalyst bed enhanced $Y_{C2=/EtOH}$. The maximum increase was about 25% for symmetrical cycling and was obtained at $\tau_{cycle}/t' = 20$ where t' is the space time. For $\tau_{cycle}/t' < 10$, the yield enhancement decreased towards zero, probably because mixing in the bed diminished the amplitude of the EtOH concentration variation. Results suggested that as the dimensionless cycle period approached the quasi-steady-state limit, the yield enhancement vanishes.

Increasing the relative duration of the flow interruption portion of the cycle raised selectivity to $C_2^=$ and $C_2^=$ yield even further. At $s = 0.25$, the lowest used, and $\tau_{cycle}/t' = 73$, $Y_{C2=/EtOH}$ reached 70%, about a 1.8-fold enhancement over a steady-state operation. Water and to a lesser extent ethanol are relatively strongly adsorbed on the alumina surface and these species inhibit the formation of $C_2^=$ more than they do the formation of DEE. Evidence that $C_2^=$ selectivity enhancement arises through interference with adsorption/desorption processes came from measuring the product mole fractions leaving the fixed bed as well as from modeling. The $C_2^=$ mole fraction fell immediately when EtOH is re-introduced into the reactor feed, while there was a delay in the DEE mole fraction before it rose rapidly. When the EtOH flow was shut off, the DEE mole fraction fell and the $C_2^=$ mole fractions rose at rates that appear to be in a ratio of 2:1.

In modeling, *Renken et al.* used measured adsorption equilibrium constants. Adsorption rate constants were assumed to be in ratios of $(k_{ad})_{DEE}:(k_{ad})_{EtOH}:(k_{ad})_W = 6:3:1$, while kinetic rate constants were drawn from steady-state measurements assuming their reactor behaved as a cascade of 10 CSTRs-in-series. Although the general behavior could be qualitatively predicted, the increase in $Y_{C2=/EtOH}$ was over-predicted and the optimal τ_{cycle} was under-predicted by an order of magnitude. Nevertheless, modeling supported the suggestion that the selectivity improvement arose from interference in the adsorption/desorption processes on the alumina surface. *Silveston (1998)* provides a fuller discussion of this early work.

12.4.3 Parasitic Reaction Systems

Leupold and Renken (1978) observed slow deactivation of the acid catalyst during composition modulation of the formation of ethyl acetate (EtAc) from ethene and acetic acid (HAc). They attributed this to parasitic reactions: $C_2^=$ polymerization and oxidation. Composition modulation of this system was discussed in **Chapter 11** (see **Table 11-1**). Periodically interrupting the HAc feed prevented catalyst deactivation, as illustrated by **Figure 12-7**. This figure shows a nearly 50% decline in catalyst activity over 400 h under steady-state operation. Unfortunately, in this and subsequent publications (*Dettmer and Renken, 1983b*; *Renken et al., 1984*; *Truffer and Renken, 1986*), the influence of the cycling variables on the parasitic reactions was not explored.

Rabl et al. (1986), *Rabl and Renken (1986)* and *Renken et al. (1989)* investigated the acetoxidation of ethene over a 2 wt% Pd/SiO_2 catalyst promoted by K, Cd and Mn compounds using composition modulation as well as other dynamic methods. Parasitic polymerization and oxidation occur in this reaction, causing catalyst deactivation. However, the effect of modulation on catalyst deactivation was not discussed.

The large reduction in catalyst deactivation through composition modulation shown in **Figure 12-7** is not an isolated observation. *Chanchlani et al. (1994)* reported a two-fold reduction in the rate of deactivation in methanol synthesis. The effect of composition modulation on deactivation by parasitic reactions deserves further study.

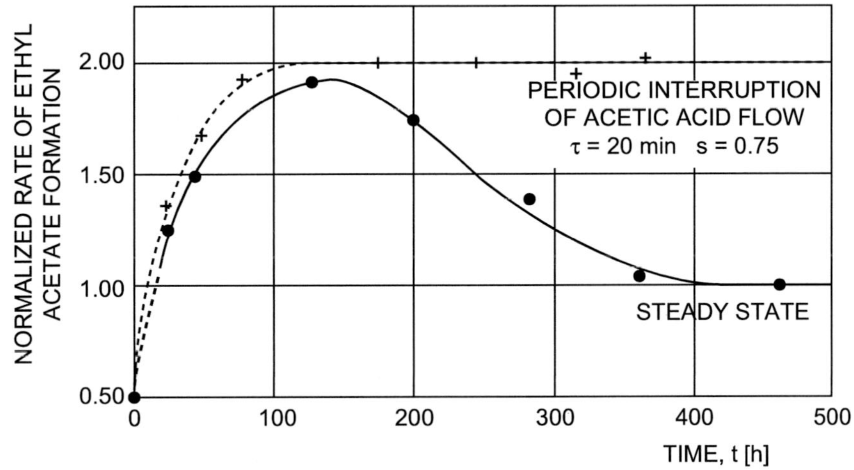

FIGURE 12-7 Rates of ethyl acetate formation under steady state and modulation of the HAc flow rate. Rates have been normalized using the steady-state rate after 400 h. Modulation results are for τ_{cycle} = 20 min and s = 0.75. Both modes of operation employed a 25 wt% H_2SO_4 on SiO_2 catalyst at 413 K and 6 bar, and SV = 4880 h^{-1}. *(Figure adapted from* Leupold and Renken (1978) *with permission. © 1978 Verlag Chemie GmbH.)*

12.5 METHANE HOMOLOGATION

Several authors have considered methane homologation under cyclic operation because the reaction appears to proceed in two steps: 1) dissociation of CH_4 to yield a carbide or carbon residue on the catalyst surface and 2) hydrogenation of that residue from hydrogen released in the decomposition step. *Belgued et al. (1991, 1992a, b), Amariglio et al. (1994, 1995b)* and *Mielczarski et al. (1993)* studied homologation on Pt, Ru and Cu catalysts at temperatures between 300 and 600 K (**Table 12-1**). *Koerts and van Santen (1991, 1992)* and *Koerts et al. (1992)* worked at 723 K. *Solymosi et al. (1992)* also studied the homologation reaction. A review paper by *Amarilgio et al. (1995a)* summarizes the studies undertaken to that date.

Koerts et al. (1992) commented on the thermodynamic barrier for methane decomposition and the subsequent hydrogenation of carbon to ethane. Temperatures below 350 K are needed for the reaction to proceed. Consequently, temperature as well as composition modulation are required.

12.5.1 Methane Homologation at a High Temperature

Van Santen and co-workers employed Group VIII transition metals supported on SiO_2. The catalyst was exposed to a 3 min pulse of 0.5 vol% CH_4 in He at 450°C, then quickly cooled to 150°C or less and exposed to pure H_2 at a pressure of 1 bar. The cycle ended by reheating the catalyst back to 450°C. Rapid cooling was found necessary to prevent transformation of the carbide or carbon residue into an unreactive state, possibly the C_γ state that is widely discussed in the literature (*McCarty and Wise, 1979; Winslow and Bell, 1984, 1985*). Of the metals considered by *Koerts and van Santen (1991, 1992)* and by *Koerts et al. (1993)*, ruthenium and cobalt seemed to be the most promising. Only the hydrogenation part of the cycle was investigated by *van Santen* and co-workers.

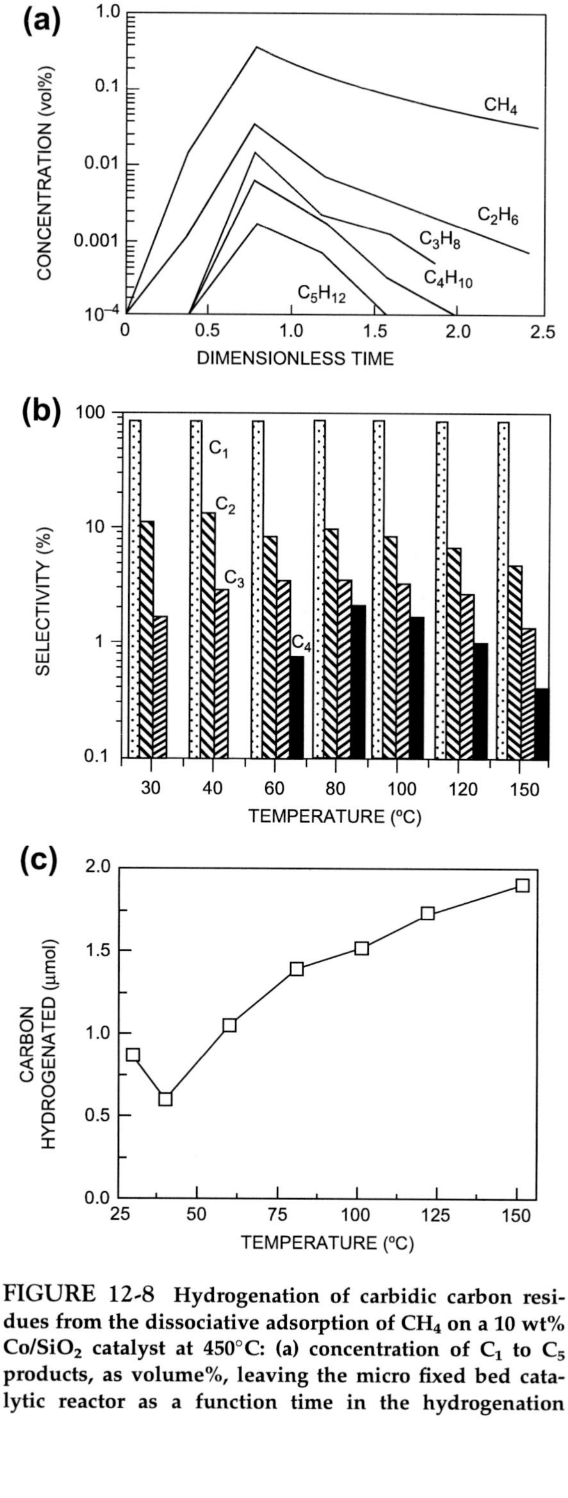

FIGURE 12-8 **Hydrogenation of carbidic carbon residues from the dissociative adsorption of CH_4 on a 10 wt% Co/SiO_2 catalyst at 450°C: (a) concentration of C_1 to C_5 products, as volume%, leaving the micro fixed bed catalytic reactor as a function time in the hydrogenation portion of a multipart cycle for methane homologation, (b) selectivity to the C_1 to C_4 hydrocarbon and (c) the amount of surface carbon residues converted at various temperatures in a multi-part cycle.** *(Figure adopted from* Koerts et al. (1992) *with permission. © 1992 Academic Press.)*

Highest yield of the C_2+ products was 13% and arose with a 5 wt% Ru/SiO_2 catalyst at a fractional CO surface coverage of 0.18 and a hydrogenation temperature of 368 K. About 8% of the carbon deposited was converted back to methane. *Van Santen* and co-workers noted that hydrogenation was the slow step in the cycle. Thus, they investigated the effect of process variables on this step. **Figure 12-8** shows in part **(a)** the variation of the hydrocarbon concentrations with H_2 exposure duration. The rapid appearance of products in the first seconds of H_2 exposure was attributed by *Koerts et al.* to hydrogenation of a reactive, probably carbidic carbon known in the literature as C_o (e.g., *McCarty and Wise, 1979; Winslow and Bell, 1984*).

Figure 12-8(a) shows that most of the deposited carbon is hydrogenated back to CH_4. Part **(b)** of the figure gives the selectivity to the C_1 to C_4 products as a function of temperature in the hydrogenation step. Temperature has evidently a large influence on the product distribution and on the amount hydrogenated (**Figure 12-8(c)**). Temperature strongly influenced product distribution and also the amount hydrogenated as seen in **Figure 12-8(b)**. Increasing temperature shifts the products towards methane. Temperatures above 150°C result in just CH_4. Indeed, above 200°C, formation of C_2+ hydrocarbons is thermodynamically impossible according to *Koerts et al. (1992)*. Not all of the deposited carbon can be recovered through hydrogenation. For example, at 95°C, only about half of the carbon deposited is hydrogenated. This means that a carbon burn-off step would have to be inserted in a homologation cycle.

Wang et al. (1993) also pursued composition modulation for high-temperature homologation. They demonstrated that methane can be forced to benzene at temperatures above 700°C using commercial ZSM-5 zeolites that incorporated either molybdenum or Zn. MoZSM-5 was the most active zeolite, providing 7% CH_4 conversion and a selectivity to benzene of almost 100%, despite some catalyst deactivation. *Solymosi et al. (1995)* achieved benzene selectivities up to 56% using a 2 wt% MoO_3/SiO_2 catalyst.

12.5.2 Methane Homologation at a Low Temperature

Formation of small amounts of ethane or ethene on adsorption of methane on Group VIII transition metals and the ease of the CH_4-D_2 exchange led *Amariglio* and co-workers to consider a relatively low temperature two-step cyclic process (*Belgued et al., 1991*). In this cycle, methane is dissociatively adsorbed on a transition metal catalyst, splitting off H_2. Carbon-carbon bond formation takes place between adsorbed CH_X fragments so when the catalyst is flushed with hydrogen, higher hydrocarbons are recovered.

With a Pt catalyst, *Belgued et al. (1992a, b)* observed an equilibrium-controlled adsorption step with C_2 formation. C_2 was not observed in the adsorption step for the Ru and Co catalysts. In the second part of the cycle at 1 bar, hydrogenation was so rapid that *Belgued et al. (1992a)* were able to measure only the cumulative amounts of hydrocarbons formed. With a constant temperature over a cycle, an optimal operating window was found for each catalyst: 225–275°C for Pt, 140–180°C for Ru and 260 to 320°C for Co. Temperature had a large influence on selectivity. Higher temperatures decreased $S_{C2/C1}$, but raised selectivity to the higher carbon numbers for Pt. With the Ru and Co catalysts, the effect of temperature was reversed for $S_{C2/C1}$ and $S_{C5/C1}$. Further details of the *Belgued* study are given by *Amariglio et al. (1995b)* and are discussed in *Silveston (1998)*.

Belgued et al. (1992a) observed that the CH_4 space velocity and duration of exposure affected the amount of methane converted and product selectivity. Increasing both generally increased CH_4 homologation, although that increase became quite small for the Pt catalyst after a 1.5 min exposure. Higher CH_4 flow rates increased C_{2+} production and shifted the product distribution toward higher carbon numbers. Longer CH_4 exposures increased $S_{C5/C1}$ as well as the CH_4 conversion. Selectivity, $S_{C4+/C1}$, jumped almost 25-fold when the exposure duration increased from 5 to 20 min (*Amariglio et al., 1995a*). Higher H_2 flow rates also increased selectivity to the higher hydrocarbons.

Using a lower temperature in the hydrogenation portion of the cycle produced a dramatic change. Dropping the temperature to 100 from 160°C in the adsorption part of the cycle raised the CH_4 homologation by almost 50% and $S_{C5/C1}$ to about 55% (*Belgued et al., 1992b*). With the Pt catalyst, CH_4 converted decreased with the decreasing temperature of the hydrogenation portion of the cycle. Selectivity, however, shifted to C_5 just as with the Ru catalyst. Deactivation was investigated over 20 successive cycles by *Belgued et al. (1992a)*. No apparent deactivation was observed for the Pt and Ru samples under the conditions used. However, the Co catalyst exhibited about a 25% activity loss.

Amariglio et al. (1995b) explored inserting a helium flush between the CH_4 and H_2 exposure steps for the Eurocat Pt catalyst they employed. Some CH_4 was swept from the Pt surface and this reduced the amount of C_{2+} material formed in the subsequent H_2 exposure step. This suggests either adsorbed CH_4 remains on the surface after the methane flow is discontinued, or that there is a dynamic equilibrium among the hydrogen-deficient hydrocarbons on the Pt surface. Replacing H_2 by CO at 1 bar produced about 55% of the homologation achieved in two-part CH_4 (30 s)–H_2 (15 min) cycle for the same CH_4 exposure duration. The product distribution, however, changed dramatically. Hydrocarbons above C_{3+} were primarily olefins and carbon numbers as high as C_8 were detected. Both alkanes and alkenes were branched and small amounts of cyclopentene were found. No aromatics, alkynes or oxygenates were seen. The amount of CO or the duration of the exposure did not significantly affect the homologation or the product distribution. Diluting CO with He reduced homologation. Following the CO dosage with an extended exposure to H_2 increased CH_4 homologation to 76% of that obtained in the two-part cycle. Only alkanes were recovered during the hydrogenation step.

Belgued et al. (1992a) speculated that the relatively low temperature adsorption generated adsorbed CH_3, CH_2, CH and H species on the surface. The H-species combined and desorbed as H_2. Neighboring unsaturated hydrocarbons dimerized primarily to adsorbed C_2H_6, CH_2-CH_2, and CH=CH species; these reacted in turn to form the higher carbon number precursors that desorbed on hydrogenation. Mild temperatures limited dehydrogenation to carbon in the adsorption step. Surface populations were probably similar to those obtained from the adsorption of C_2 and C_3 olefins on the same metal surfaces. Product distribution did not follow the Anderson-Schulz-Flory distribution probably because 1) dimerization was the primary process for C-C bond formation, and 2) hydrogenolysis occurred during the hydrogenation step. The higher selectivity to C_5 and the different selectivity trend with temperature found for the Ru catalyst was explained by its lower hydrogenolysis activity. This lower activity arose because of the lower adsorption temperature with the Ru catalyst. Adsorption rates were higher for the Ru catalyst than for the Pt one. *Amariglio et al. (1995b)* explained their CO dosing results by displacement of the olefinic precursors on the surface by CO. A more complete summary of these homologation studies is given by *Silveston (1998)*.

How attractive is the *Amariglio* cyclic process for methane upgrading? There has been considerable industrial interest in upgrading for many years. Alternative routes would be oxidative coupling or the Fischer-Tropsch synthesis. Key considerations are methane conversion per pass and the yield of C_2+ hydrocarbons. The former was very low ($< 1\%$) in the *Belgued et al. (1991, 1992a, b)* experiments. *Amariglio et al. (1994, 1995a)* argued that these can be substantially improved. They demonstrated that yields of about 40% can be achieved by using a batch reactor with external circulation of CH_4 and by adding a Pd/Al_2O_3 trap to recover H_2. In a continuous process, two fluidized beds could be used with catalyst transport between the beds. The first bed would operate at a higher temperature and dissociatively adsorb methane. The off-gas would flow to a membrane separator to strip out H_2 so that the unconverted methane could be recycled. The relatively pure H_2 would be compressed and fed to the second fluidized bed, operated at a lower temperature and a higher pressure, to produce hydrocarbons. Further process details are given by *Amariglio et al. (1994, 1995a)* and discussed by *Silveston (1998)*.

Choice of catalyst will also be important if C_{2+} yields are to be improved. *Solymosi et al. (1992)* investigated methane decomposition on supported noble metal catalysts. At 250°C, rhenium proved to be more active than platinum for CH_4 dehydrogenation and the rates of C_{2+} formation were only somewhat smaller than those observed for Pt. The Pd catalyst had lower activity than Rh. *Solymosi et al.* suggested that carbon mobility on the metal surface and possibly spillover onto the support were important. They also proposed that methane is stepwise-dehydrogenated and only the CH_3 fragments are loosely enough bonded to dimerize and desorb. Step-experiments indicated the presence of an active carbon, referred to as C_α, which readily hydrogenated below 250°C. Removal of catalytic carbon deposits, except on Rh, required temperatures above 300°C. This observation suggested the existence of refractory carbons known as C_β and C_γ (e.g., *McCarty and Wise, 1979*; *Winslow and Bell, 1984, 1985*).

Solymosi et al. (1994) demonstrated that the catalyst support is important. They suggested the performance improvement using Pd was largely due to better dispersion of Pd on the support. *Solymosi and Cserényi (1994)* working with an Ir catalyst also remarked that the support was important. Ir dispersion was thought to explain their observations. Indications in the literature that copper promotes coupling of CH_3 species and suppresses dehydrogenation caused *Solymosi and Cserényi (1995)* to explore CH_4 homologation on Cu-Rh catalysts supported on a silica aerogel. The Cu/Rh ratio was important. Cu/Rh between 0.5 and 1 dramatically increased $S_{C5/C1}$ in the hydrogenation step. Since Cu/SiO_2 cannot decompose CH_4 below 350°, *Solymosi and Cserényi* suggested that CH_4 decomposes on Rh and the CH_3 and CH_2 species formed migrate to Cu sites where dimerization also occurs. Significant amounts of ethene are formed in both the decomposition and hydrogenation steps on this catalyst. In general, the experiments of *Solymosi* and co-workers supported the results and mechanistic implications reported by *Belgued et al. (1991, 1992b)*. Decomposition of adsorbed CH_4 proceeds stepwise with concurrent dimerization of the partially dehydrogenated CH_4. Hydrogenation of the C_α species may produce C_{2+}, but much of the carbon produced even at 250° was refractory and hydrogenated only slowly to CH_4.

Erdöhelyi et al. (1992) explored CH_4 homologation using CO_2. They found that the presence of CO_2 in the CH_4 feed suppressed formation of the higher hydrocarbons. Reforming of CO_2 to CO and H_2 resulted.

From an engineering standpoint, the disadvantage of the *van Santen* homologation process (**Section 12.5.1**) is the more than a 300° temperature difference between the adsorption and hydrogenation parts of the cycle. Further difficulties are the accumulation of graphitic carbon and the low $S_{C2+/C1}$. On the other hand, the yield of the C_{2+} products can reach 13% in a flow reactor, an order of magnitude greater than the yields measured for the moderate temperature *Amariglio* process. That process, however, has the advantage of lower operating temperatures. A research target for the *Amariglio* process must be increasing CH_4 adsorption in the CH_4 exposure portion of the cycle and achieving a higher selectivity to the C_{2+} products during methane decomposition. This is a matter of finding a better catalyst. *Solymosi* and co-workers have demonstrated that the choice of catalyst and support has a large effect on product distribution.

12.6 OLIGOMERIZATION OF ETHENE

Following their work on methane homologation, *Amariglio* and co-workers extended their research to the oligomerization of ethene (*Lefort et al., 1994*). For this, they employed the reactor system and procedures used by *Belgued et al. (1992a)*, but the Pt catalyst was EUROPT-2 (6.3 wt% Pt on silica). With a cycle consisting of a 1 min exposure to 1 vol% C_2H_4 in Ar followed by a 30 s exposure to H_2, they found about 42% of the C_2H_4 adsorbed at 100°C was oligomerized to C_3+ products, while at 150°C the percentage increased to 64%. Co-feeding C_2H_4 and H_2 at these temperatures resulted in only C_2H_6. **Figure 12-9** shows the products formed in terms of product selectivity.

Hydrogenation of C_2H_4 occurred as well and the presence of CH_4, propane and pentane means hydrogenolysis also took place. About 18% of the C_2H_4 adsorbed is not removed in the H_2 portion of the cycle. *Lefort et al.* also concluded from their observations that strongly adsorbed ethyne (C_2H_2) and H_{ad} form on adsorption of C_2H_4. The adsorbed H_2 is available to hydrogenate some of the adsorbed C_2H_4 to C_2H_6 species detected in the off-gas.

Precursors of the higher hydrocarbons resulted from the oligomerization of adsorbed ethyne species.

The experiments of *Lefort et al. (1994)* were exploratory. Selectivities to higher hydrocarbons were substantial so that further examination of this modulation process would seem worthwhile.

12.7 MODULATION OF MULTIPLE INPUTS

Only single inputs, such as composition or a flow rate, have been modulated in the experiments considered in this chapter and, indeed, in the other chapters of this monograph. It is possible to modulate more than just one input. That opportunity has been explored analytically by *Parulekar (2000, 2003)* (see **Chapters 1** and **10**) for a competitive bioreaction and for a series-parallel reaction such as encountered with butadiene and acetylene hydrogenation or hydrocarbon oxidation, discussed in **Sections 12.3** or with methane homologation examined in **Section 12.5**. *Parulekar* extended the generalized π-criterion of *Sterman and Ydstie (1990a,b; 1991)* to multiple inputs, deriving conditions for the appropriateness of periodic operation, relations for optimal phase and amplitude differences between manipulated inputs and predictions of the maximum enhancement possible over steady-state operation. Although the π-criterion is obtained from a perturbation analysis and thus is valid for only small amplitudes, *Parulekar* suggested that larger amplitudes extend the region of variable space in which

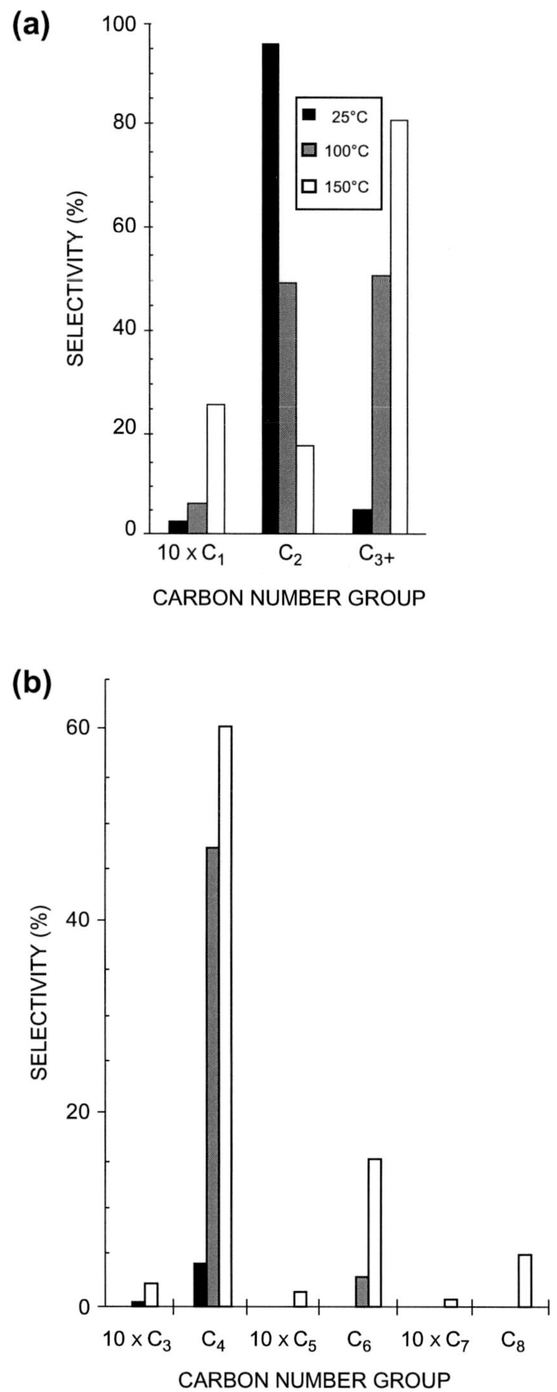

FIGURE 12-9 **Selectivity to different hydrocarbon products for a two-step, cyclic process over a 6.3 wt% Pt/SiO_2 catalyst with 12.5 vol% C_2H_4 in Ar for 60 s followed by 100% H_2 for 30 s: (a) Breakdown of the products among C_1 to C_{3+}, (b) breakdown of the C_{3+} products.** *(Figure adopted from* Lefort et al. (1994) *with permission. © 1994 by J.C. Baltzer AG.)*

enhancement may be expected, if not the magnitude of the enhancement.

A numerical example for a series-parallel reaction in a CSTR with separate feed streams for the reactants (*Parulekar, 2003*) demonstrated that simultaneously modulating the flow rates of both feed streams extends the modulation span in which conversion enhancement through periodic operation occurs. The magnitude of the enhancement depends, of course, on reaction kinetics and operating conditions.

Nomenclature

A,B	=	reactant species
C,D	=	usually intermediate species
E,F	=	product species
C	=	concentration
	=	carbon
C_i	=	concentration of species i
$C_{j,o,ss}$	=	initial concentration of species j under steady state operation
C_M, C_m	=	monomer or monoester concentration
$\overline{C_i}$	=	time average concentration of species i
$\overline{C_{j,o}}$	=	mean concentration under periodic operation
CSTR	=	continuous stirred tank reactor
DEA	=	diethanolamine
DEE	=	diethylether
EtAc	=	ethyl acetate
Et_2A	=	diethyl adipate
Et_2O	=	diethyl ether
EtOH	=	ethyl alcohol
HAc	=	acetic acid
i	=	product species
j	=	reactant species
k	=	unwanted product species
	=	reaction rate constant
k_{ad}	=	adsorption rate constant
MA	=	maleic anhydride
MEA	=	monoethanolamine
N_{Da}	=	Damkoehler number
Na_2A	=	disodium adipate
NaEtA	=	sodium monoethyl adipate
n_i, n_j	=	molar flux of species i, J
$(n_i)_0$	=	entering molar flux of species i
P	=	pressure (KPa, MPa)
Q_{per}	=	volumetric flow rate under periodic operation
Q_{ss}	=	volumetric flow rate under steady state operation
$S_{i/j}$, S_i	=	overall selectivity to product "i" based on a key reactant "j"(based on reactor outlet) (-)
$\overline{S}_{i/j}$	=	time average overall selectivity to product "i" based on key reactant "j"
$s_{i/j}$	=	instantaneous selectivity to product "i" for key reactant "j" (based on reaction rates (-)
s	=	cycle split (-)
SV	=	space velocity (s^{-1})
T	=	temperature (K)
T_0	=	entering or initial temperature
t	=	time (s, min.)
TEA	=	triethanolamine
$\bar{t}$	=	residence time
t'	=	space time
X_j	=	conversion of reactant "j" (-)
$\overline{X}_I$	=	time average conversion of "i"
$Y_{i/j}$, Y_i	=	yield of "i" referred to reactant "j"
$\overline{Y}_{i/j}$	=	time average yield of "i" referred to reactant "j"
y_i	=	mol fraction of species "i"
ZSM 5	=	zeolite

Greek

λ	=	amplitude
υ_I	=	stoichiometric coefficient for species i
τ, τ_{cycle}	=	period, cycle period (s, min.)
	=	residence time
τ_d	=	duration of partial cycle
ψ	=	enhancement (-)
ψ_X	=	conversion enhancement
ψ_S	=	selectivity enhancement
ψ_Y	=	yield enhancement
$\zeta_{i/k}$	=	selectivity ratio

Subscripts

DEE	=	diethyl ether
$C_{2=}$	=	ethene
EtOH	=	ethyl alcohol
HAc	=	acetic acid
i	=	product
j	=	reactant
k	=	unwanted product
S	=	selectivity
SS	=	steady state
v	=	state of carbon deposit
W	=	water
X	=	conversion
Y	=	yield
α	=	state of carbon deposit
0	=	initial or inlet

CHAPTER

13

Use of Modulation in Mechanistic Studies

Albert Renken, Robert Ross Hudgins*†*, Peter Lewis Silveston*†*

*Lausanne, Switzerland and †Waterloo, Ontario, Canada

OUTLINE

13.1 Introduction 369

13.2 Qualitative Applications 370

13.3 Quantitative Applications 371

13.4 Modulation of Light Intensity 374

13.5 Application of Modulation to the Testing of Rival Models 375

13.6 Overview 385

13.1 INTRODUCTION

Modulation is one of several dynamic methods for studying reaction mechanism. Other widely used methods are response to a single pulse, or to a step-change of a reactant or promoter (*Kobayashi and Kobayashi, 1974; Kobayashi, 1982; Müller and Hofmann, 1987; Hinrichsen, 2008*). Temperature programmed desorption (TPD) or temperature programmed reaction (TPR) are variants of the step-change technique (*Falconer and Schwarz, 1983*). The technique perhaps originated from successive pulsing of a reactant as practiced by *Hattori and Murakami (1968)*.

Kobayashi and Kobayashi (1974) and *Bennett (1976)* have published comprehensive reviews of the application of dynamic methods in mechanistic studies. A review by *Renken (1990a, b)* discussed the use of modulation in this context. *Silveston (1998)* devoted a chapter to this topic.

The advantage of all dynamic methods is that they generally are able to detect in a reaction sequence relatively rapid steps that are obscured in steady-state measurements. Dynamic methods also offer a sensitive means for distinguishing among rival kinetic models. Models based on steady-state measurements do not predict the performance of periodically forced catalytic reactors.

13.2 QUALITATIVE APPLICATIONS

Resonance frequencies in modulation of a reaction variable are those frequencies that maximize a response or observed variable. They often suggest the nature of the rate-controlling step in a catalytic reaction and identify in this way a portion of the reaction mechanism. In an early experimental study of composition modulation applied to SO_2 oxidation over a potassium promoted vanadia catalyst, *Unni et al. (1973)* observed that a diffusion step, probably of O_2 or SO_2 in a melt or glassy phase occurring in the working catalyst, likely controlled the rate of SO_2 oxidation based on resonance observed at frequencies between 35 and 280 μHz for this reaction over the catalysts used. Consequently, oxidation probably occurs in a melt phase. At the other end of the frequency spectrum, *Silveston (1995)* reasoned from the relatively high frequencies at which resonance is observed as well as temperature dependency of the magnitude of the resonance effect that CO adsorption and/or surface reaction are rate-controlling for CO oxidation over noble metal catalysts.

Chiao et al. (1987) found that the time-average rate of NH_3 synthesis increased slowly the with time after switching on N_2-H_2 modulation, but when modulation was switched off hours later, the steady-state rate was much higher than previously. They used this observation to account for the failure of models to predict synthesis performance under periodic forcing, concluding that composition modulation activates the catalyst probably by surface reorganization. *Labastida-Bardales et al. (1989)* investigated composition modulation of propene partial oxidation over a copper-molybdate catalyst. They concluded that a surface transformation occurred with the formation of new copper molybdate crystal phases.

The role of surface residues and the participation of oxygen in ethene epoxidation were explored by *Park et al. (1983)* using four- and eight-part cycles. The four-part cycles consisted of successive exposure of the catalyst to C_2H_4 or a C_2H_4-O_2 mixture, followed by an N_2 flush, O_2 exposure and a second N_2 flush. The eight-part cycle added H_2 exposure (to hydrogenate residues), another O_2 exposure and two more N_2 flushes. *Park et al.* varied the ethene and O_2 exposure duration. They concluded from these modulation experiments that three different oxygen adsorbates participate in the epoxidation reaction. Of these species, adsorbed O is responsible for total oxidation of ethane and other adsorbed hydrocarbons. Surface residues appeared to suppress O_2 dissocation on adsorption.

The mechanism of propene partial oxidation was investigated using two- and three-part cycles in one of the earliest studies of composition modulation by *Niwa and Murakami (1972a)*. They formed their catalyst into disks and measured electrical conductivity as well as products formed during successive pulses of different composition. From the resistance measurements they identified weak and strong adsorption of propene on different partial oxidation catalysts and related the adsorption strength to catalyst performance under steady-state and cyclic operation. Bismuth in oxide catalysts was found to contribute to catalyst reoxidation and to provide high lattice-oxygen mobility.

The question of intermediates in the reaction pathway under transient conditions was examined by *Zhou and Gulari (1986c)* using a near-transparent wafer of silica impregnated with Pt mounted in a flow-through IR cell. The flow of CO and O_2 was modulated in a bang-bang mode and the absorption spectrum measured. **Figure 13-1** shows the variation in the percentage absorbed with wavelength under steady state and when the flow is modulated

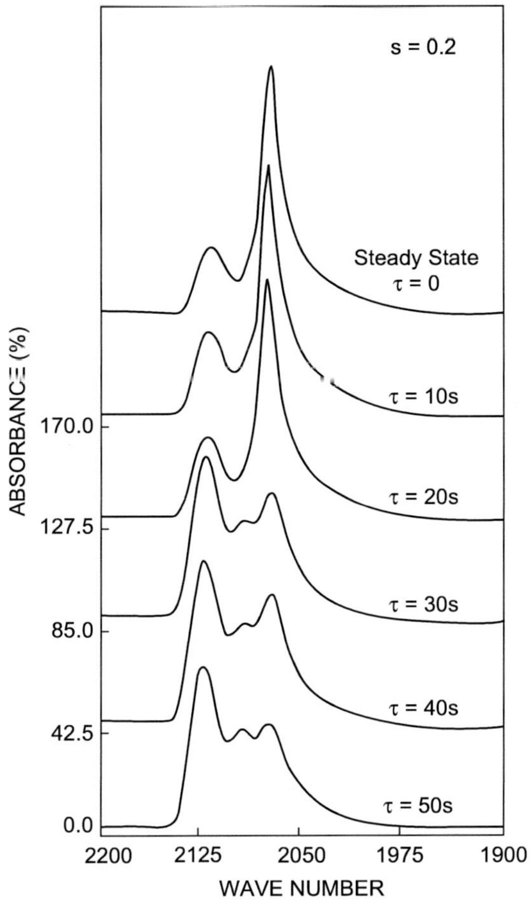

FIGURE 13-1 Variation of the absorption spectra of chemisorbed CO on Pt in the presence of O_2 with cycle period for the bang-bang modulation of composition. A cycle period of zero is equivalent to steady state. *(Figure adapted from* Zhou and Gulari (1986c) *with permission. © 1986 Pergamon Press.)*

with different cycle periods. The broad peak centered at 2125 cm^{-1} under steady-state operation is linearly adsorbed CO on the Pt surface, while the narrow peak at 2080 cm^{-1} that breaks down when the flow is modulated and is initially bridge-bonded CO. Under modulation, the peak splits and overlaps the linear-bonded CO. *Zhou and Gulari* attributed this transformation to CO bridge-bonded to a chemisorbed oxygen and Pt. The changing amplitude of the peaks suggests a shift in the reaction intermediate as the cycling frequency decreases from 0.1 to 0.02 Hz from linearly bonded CO to CO bridge-bonded to O and a Pt site.

13.3 QUANTITATIVE APPLICATIONS

An elegant use of modulation was carried out by *Renken* and co-workers (*Marwood et al., 1994, 1996*). Simultaneous measurements of the time variation of gas phase and adsorbate concentrations were made for the methanation of CO_2 over a 2 wt% Ru on TiO_2 catalyst at 383°C and near-atmospheric pressure. A fixed bed of the supported Ru catalyst was placed in a recycle loop and a product stream was withdrawn through a diffuse reflectance, infrared Fourier transform spectrometer (DRIFTS) cell containing a small amount of the Ru catalyst maintained at the same temperature as the fixed bed. **Figure 13-2** shows a schematic of the *Marwood* experimental system. The recycle loop assures a nearly uniform composition in the fixed bed, while just a small amount of catalyst in the DRIFTS cell keeps the gas phase composition in the cell the same as the composition in the reactor so that adsorbate concentrations represent those in the reactor. Placing the cell in the exit line rather than in the recycle loop greatly reduced the pressure drop in the DRIFTS cell.

Marwood et al. (1994, 1996) used this system to identify intermediate species on the catalyst surface. In earlier investigations, they had observed adsorbed CO (CO_{ad}), carbon (C_α) and a formate on the surface. The reasoning for their approach was that during feed composition modulation the response of a surface intermediate in the hydrogenation reaction should have the same time profile as that of the product, but with an amplitude smaller than that of the reactant and greater than that of the product. Also, the intermediate should

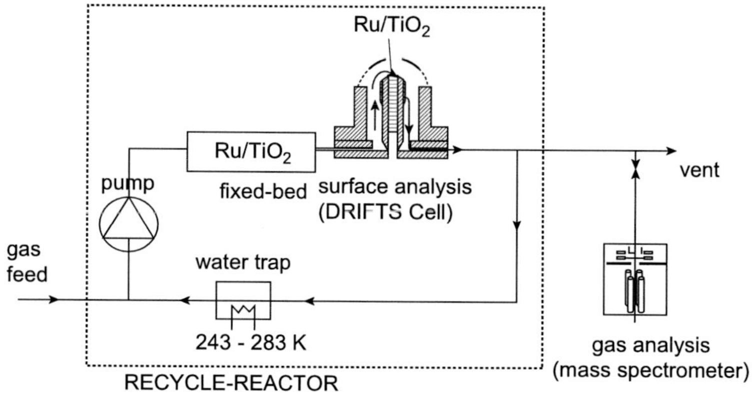

FIGURE 13-2 Schematic of an apparatus for studying composition modulation of a fixed bed catalytic reactor and simultaneously observing the time variation of adsorbates on the catalyst surface. *(Figure reproduced from* Marwood et al. (1996) *with permission. © 1994 Elsevier Science Publishers.)*

exhibit a phase lag less than that of the product. Furthermore, a surface species not in the reaction sequence should have the same amplitude as the modulated reactant and just a small phase lag. With multiple surface intermediates, the phase lag should indicate the reaction sequence.

Experiments were conducted using constant flows of H_2 and He by periodically switching the CO feed on and off so as to maintain a time-average composition of $CO:H_2:He = 1:4:5$ (10 vol% CO). Gas phase and adsorbate responses for a 20 min, symmetrical cycle are shown in **Figure 13-3**. The curves have been normalized with steady-state measurements at the time-average feed composition.

The DRIFTS-recycle-reactor was simulated as a continuous stirred tank reactor with a residence time of 1.58 min and a simplified kinetic model with an intermediate adsorbed on the surface, $(CO)_{ad}$, and a reservoir in the form of adsorbed formates as summarized in the following equation:

$$CO_2 \overset{r1}{\rightarrow} (CO)_{ad} \overset{r2}{\rightarrow} CH_4(CO)_{ad} \underset{r4}{\overset{r3}{\rightleftarrows}} formates. \tag{13-1}$$

As the surface concentrations are not quantitatively known, as a result of the DRIFTS technique, the curves were normalized by dividing the experimental transient data with the values obtained at steady state. The dynamic developed model describes the steady state as well as the dynamic behavior under periodic operation as can be seen in **Figure 13-3**. In addition to the simplified lumped model used for the reactor simulations, a detailed mechanistic model was developed and is discussed by *Marwood et al. (1996)*.

There are several examples in the engineering literature in which modulation has been employed to measure adsorption/desorption rates. *Li et al. (1989)* used sinusoidal modulation of CO concentration in a carrier gas to measure sorption rates on Pt/SiO_2. The catalyst, 1 wt% Pt/SiO_2, was incorporated in a SiO_2 wafer that was placed in a temperature-controlled IR flow cell so that the gas-phase and adsorbate signals at the appropriate wavelengths could be monitored simultaneously. **Figure 13-4** shows the time-dependent IR signals for CO in the gas and adsorbed phases after a stationary cycling state was reached. These signals have the same frequency but are phase-shifted and have different amplitudes. A dynamic model was used to extract adsorption rate and equilibrium constants. Experiments were conducted at different cycling frequencies.

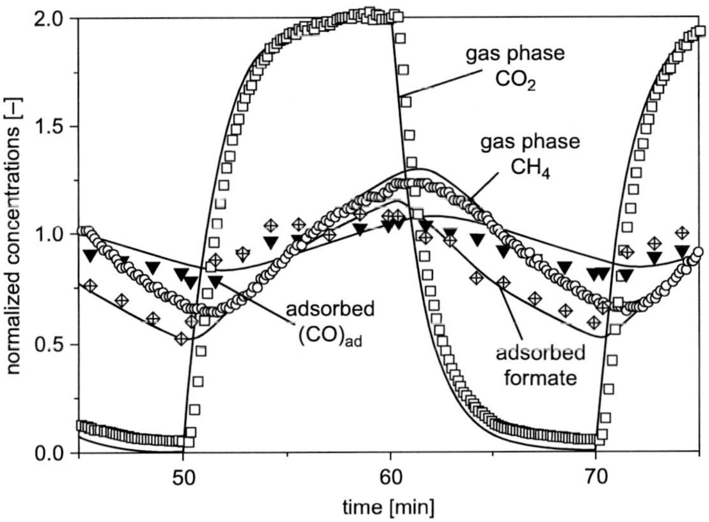

FIGURE 13-3 Measured response of pro-duct and surface species to the symmetrical bang-bang modulation of CO_2 feed rate to a recycle reactor containing a packed bed of 2 wt% Ru on TiO_2 catalyst operating at 383 K, about 105 kPa, time-average feed composition as CO:H_2:He = 1:4:5 with τ_{cycle} = 20 min. *(Figure reproduced from* Marwood et al. (1996) *with permission of the authors.)*

The kinetics of CO oxidation over a Cu/Al_2O_3 catalyst were studied by *van Neer et al. (1997)* who employed successive oxidation and reduction cycles. However, their investigation at 493 K did not rely solely on composition modulation to suggest a mechanistic model of this reaction. They also carried out experiments with isotopically labeled oxygen ($^{18}O_2$) and monitored surface species during transients in an FTIR flow cell. One important inference from their isotopic work was to note that CO and O_2 did not adsorb competitively on the catalyst. The result of their comprehensive study was a three-step mechanistic model for the reoxidation of the catalyst. This model described their experimental results qualitatively and in part

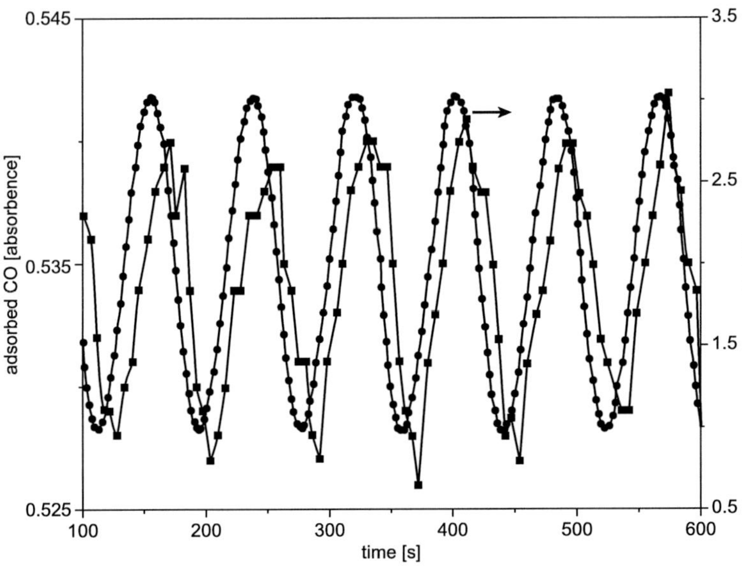

FIGURE 13-4 Experimental response to the sinusoidal forcing of CO flow to an IR cell containing a wafer of 1 wt% Pt/SiO_2 at 343 K and 101.3 kPa. Gas-phase CO is given by the right axis, while adsorbed CO at λ = 2080 cm^{-1} is given by the left axis. *(Figure reproduced from* Li et al. (1989) *with permission. © 1989 AIChE.)*

quantitatively. The authors concluded that composition modulation offered a powerful tool for inferring mechanisms in heterogeneous catalysis.

Adsorption rates on α-Al_2O_3 for CH_4, CO and CO_2 were measured by *Lynch and Walters (1990)* using sinusoidal forcing with a packed bed of adsorbent, gas recycle and an IR flow cell in the recycle stream. Both phase shift and amplitude ratio were used to obtain adsorption rate constants. The technique employed by these researchers used a model of the adsorption process which was fitted to experimental measurements in the Fourier domain. Sinusoidal modulation was found to be more reliable than pulse methods, which suffered from parameter correlation. *Silveston (1998)* also discusses this research.

In chemical physics, composition modulation has been widely used in conjunction with molecular beams to measure the rate of fundamental surface processes such as surface diffusion, desorption or surface reactions. The technique is referred to as fast chopping rather than composition modulation. One or more beams impinging on a catalyst surface, often a single crystal face, are interrupted, usually at high frequencies. At the catalyst surface, however, the interruptions are perceived as square-wave variations of the rate of molecular collisions of a species with the surface. Collision frequency is directly proportional to the partial pressure or concentration of a species. Chopping, therefore, has the same effect as bang-bang composition modulation, except for the higher energy in the molecular beam. Beams, too, are used under high vacuums. Because of these differences, fast chopping, though similar to periodic forcing, is beyond the scope of this monograph. It is discussed only briefly below. *Schwarz and Madix (1974)* describe the method.

Fast chopping has been applied to measuring the rate of carbon deposition on a Pt filament under pressures of about 0.1 kPa using a system that did not employ molecular beams (*Halpern and al Mutaz, 1986*). This application illustrates the methodology used with molecular beams. Chopping was accomplished using a rapidly swinging syringe that passed a razor edge separating a bypass to the vacuum pump inlet from a flow channel containing the hot Pt filament. The system is shown schematically in **Figure 13-5**. Two swinging syringes were used, one feeding hydrocarbon and the other oxygen (not shown in the figure). These were mounted so that they swung past the razor edge moving in opposite directions. This arrangement produced alternating square-wave exposure of the Pt filament to the two reactants. Thermionic electron emission from the heated filament was very sensitive to carbon on the Pt surface so changes of the emission with time measured the deposition rate. The O_2 pulse cleaned the Pt surface prior to the next exposure. *Halpern and al Mutaz* operated their flow modulator below 20 Hz and so were able to observe the rise time of the emission signal. From their data, they obtained the probability of carbon deposition per molecular impact and demonstrated that this relationship was a logarithmic function of the carbon number of the paraffin used.

13.4 MODULATION OF LIGHT INTENSITY

Modulation of light intensity, primarily in a bang-bang mode, to the gas phase free radical reactions is a classic application of modulation to the investigations of mechanism and kinetics. The technique employs a brief, intense illumination of the contents of a reactor. The decay of reactants or formation of products is measured in the dark period by a variety of methods. The systems studied are noncatalytic and thus fall outside the scope of this monograph. There is a large literature on modulation techniques for measurements of elementary reactions. The technique has also been applied to free radical

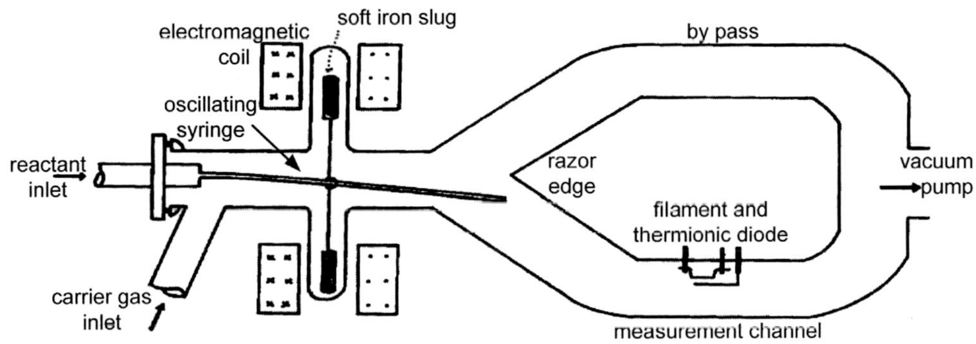

FIGURE 13-5 Fast chopping apparatus for the study of carbon deposition on a hot Pt filament using a swinging gas syringe.

polymerization reactions. Unlike the classical application, consecutive pulses of light are used and the polymer properties, such as number chain length, μ_n and polydispersity, D, are related to frequency and duration of the dark or light phase. Originally, an open sector in a rotating disk was used for pulsing, but this has been replaced by pulsed laser illumination of the reaction cell (*Olaj et al., 1985, 1987*). Light photons generate free radicals that react with monomer to initiate chains, but the vastly increased number of radicals also cause chain termination by elementary reactions discussed in **Chapter 7**. Application of the technique is discussed by *Silveston (1998)*.

Davis et al. (1989a) used light-pulsing to measure the propagation constant for the homopolymerization of styrene and methyl methacrylate in various solvents with benzoin or azobisisobutyronitrile initiators. The constants were found to be independent of chain length, initiator and solvent. Such was not true for the co-polymerization of the two monomers; k_p depended on composition in this reaction (*Davis et al., 1989b*). Light-pulsing to measure k_p for other co-polymerization systems and use of results to explore the composition effect are described in further papers (*Davis et al., 1990, 1991*). Measurements were also used to explain the variations observed (*Sanayei et al., 1994*).

13.5 APPLICATION OF MODULATION TO THE TESTING OF RIVAL MODELS

Several investigators (*Li et al., 1985a, b; Renken, 1990a, b*) report that modulation data offer a discriminating test for rival mechanisms if those mechanisms can be formulated as models. In the early 1980s, the literature on ammonia synthesis over a promoted iron catalyst claimed NH_3 was formed by the dissociative adsorption of H_2 and N_2 and the stepwise hydrogenation of the adsorbed N moiety. It was uncertain whether adsorption was competitive or H_2 and N_2 adsorbed on separate sites. *Li et al. (1985a)* constructed dynamic models of the synthesis for each of the proposed adsorption routes. These were tested by step-change experiments and later by composition modulation (*Li et al., 1985b*). A two-site model survived the step-change testing but failed to predict the resonance seen as the cycling frequency varied (**Figure 13-6**). By modifying the two-step model

FIGURE 13-6 Predicted time-average rate of ammonia synthesis for two-site models as a function cycle period, τ_{cycle}, and comparison with CSTR data collected at 400°C and 2.39 MPa. *(Figure reproduced from* Li et al. (1985b) *with permission of the authors.)*

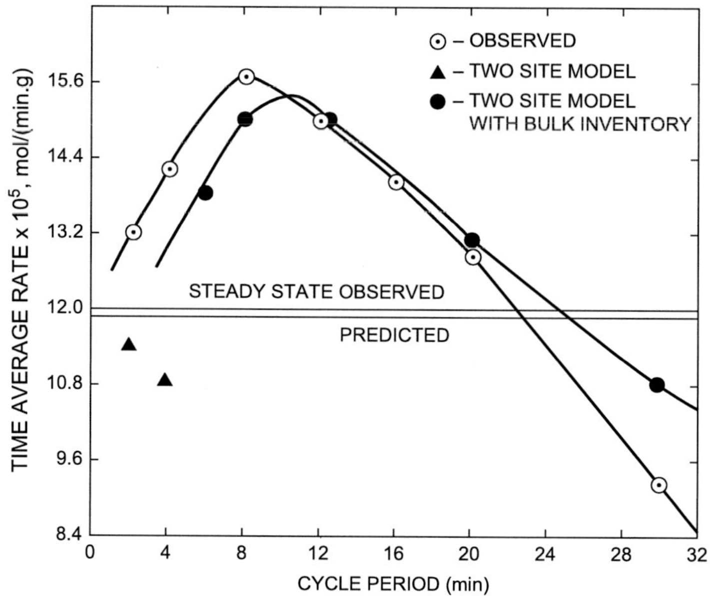

to allow for N_2 storage in the bulk of the catalyst, *Li* was able to fit the resonance results as **Figure 13-6** demonstrates. Nevertheless, the two-sites with bulk storage model failed to adequately reproduce the variation of the ammonia synthesis rate during a forcing cycle.

Nam et al. (1990) extended *Li's* work by examining two classes of storage models: 1) A class assuming the surface contained spillover regions where reaction could not occur but where adsorbate storage was possible and 2) models that stored a reactant below the catalyst surface through dissolution when the reactant was present in the gas phase at a high concentration. These reservoirs, either on or below the surface, supplied the active surface with one of the reactants when that reactant was absent or in low concentration in the gas phase. **Figure 13-7** shows the two classes as well as variants dealing with how transfer occurred between storage and active sites on the surface. Models representing the two classes of mechanisms may be seen in the Nam paper or in *Silveston (1998)*. Following *Li et al.*, the models assume non-competitive adsorption and that the multi-step hydrogenation of atomic nitrogen is lumped into a single rate expression. Although the experiments of *Jain et al. (1982b)* observed storage of both N_2 and H_2, the storage of H_2 was much less than that of N_2, so that only N_2 storage was considered in the models. In both models, the Elovich equation was used for adsorption.

All of the models tested by *Nam et al.* described, the steady-state ammonia synthesis data of *Jain (1981a)* most closely reproduced the step-change measurements adequately. However, only two of the models predicted resonance under variation of the modulation frequency. These models were the surface storage model (a) and the bulk dissolution model (b) in **Figure 13-7**. To make the surface storage model satisfactory, it was necessary to assume that the inactive surface had to have greater than 10 times more capacity for N_2 than the active surface where reaction takes place. **Figure 13-8** compares predicted normalized rates of ammonia production for four

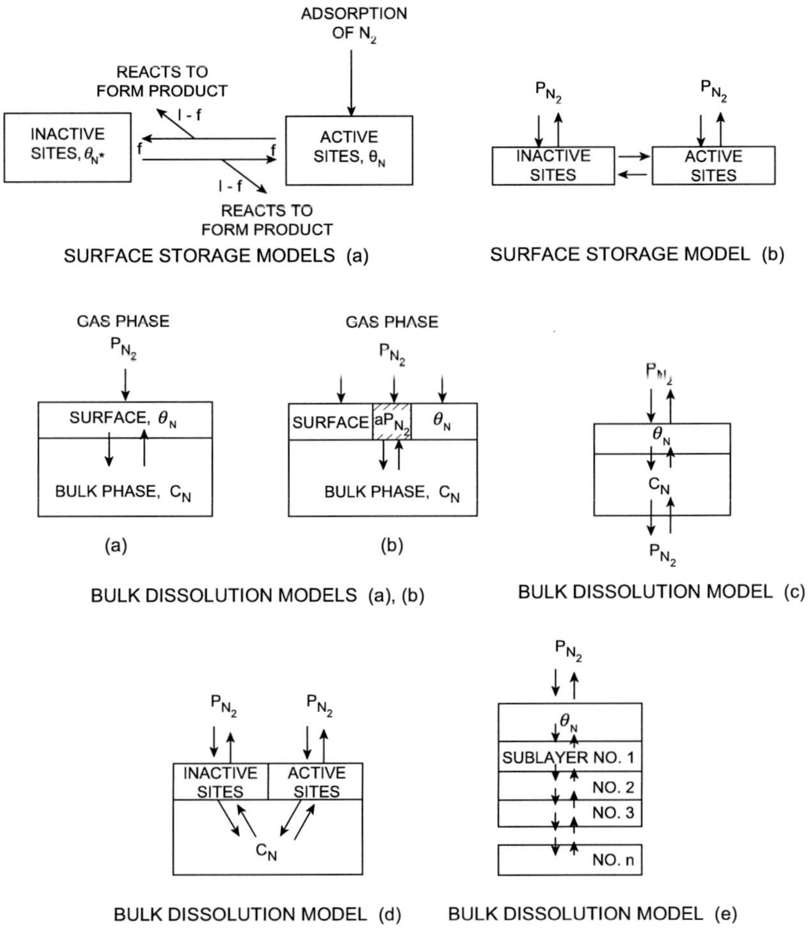

FIGURE 13-7 Storage models for ammonia synthesis tested using modulation. *(Figure adapted from* Nam et al. (1990) *with permission. © 1990 Pergamon Press.)*

models with the measured normalized rates. The bulk dissolution model, assuming separate sites for N_2 adsorption and reaction and for N_2 dissolution (bulk dissolution model (b) in **Figure 13-7**), predicted the observed results closely. The single dissociation site model (bulk dissolution, model (a) in **Figure 13-7**) did not predict resonance for a feed containing 25 vol% N_2. The surface storage model (a) (**Figure 13-7**) predicted resonance only when the N_2 capacity on the inactive surface was about 10 times the capacity on the active surface. Although the bulk dissolution model provides a superior fit to the data, uncertainty associated with fitting the nine parameters used in the model makes it difficult to choose that model over the surface storage model.

A weakness of the model discrimination studies of *Li et al. (1985b)* and *Nam et al. (1990)* is the use of the same data for parameter estimation and model discrimination. *Renken* and coworkers take pains to use separate data for these two tasks. In one of their earlier studies, the Pd/SiO_2 catalyzed acetoxidation of ethene to form ethyl acetate, separate measurement of the adsorption kinetics of acetic acid and ethyl acetate were undertaken (*Rabl et al., 1986*). These showed adsorption was slow and suggested

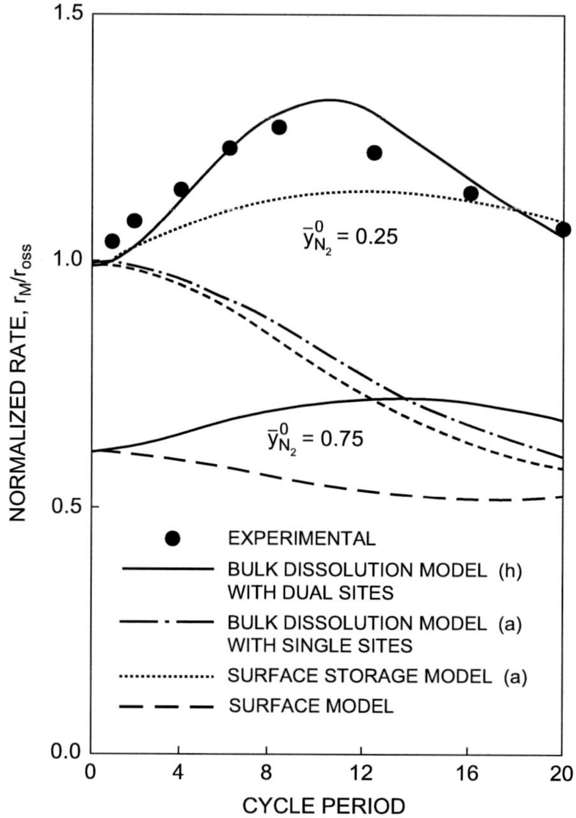

FIGURE 13-8 **Comparison of experimental observations of the normalized time-average rate of ammonia synthesis over a commercial iron catalyst at $T = 400°C$, $P = 2.38$ MPa with the predictions of models for different cycle periods and $s = 0.5$. Physical models are shown in Figure 13-7. The surface storage model assumes the ratio of the N_2 adsorption capacities on the inactive and catalytically active surfaces = 10.** *(Figure adapted from* Nam et al. (1990) *with permission. © 1990 Pergamon Press.)*

that the step could be modeled by assuming the activation energy for adsorption was a linear function of the adsorbate coverage. Several models for the acetoxidation were considered. All assumed ethyl acetate forms from the surface reaction of the acetic acid and ethene adsorbate. Both a differential reactor and a CSTR were considered (*Truffer and Renken, 1986*). Four models were examined: 1) Fast adsorption that implied equilibrium was achieved, 2) slow adsorption with rate constants independent of temperature, 3) slow adsorption/desorption with temperature-dependent rate constants and 4) slow and inhibited adsorption/desorption of ethane. The last model represents the step-change response data well and **Figure 13-9** shows that the model satisfactorily depicted the start-up of a modulation experiment.

The "stop effect" in which interrupting the flow of reactants leads to a rapid but brief acceleration of the reaction has been used skillfully by *Thullie and Renken (1993)* to demonstrate the use of modulation for model discrimination. The literature provides two competing

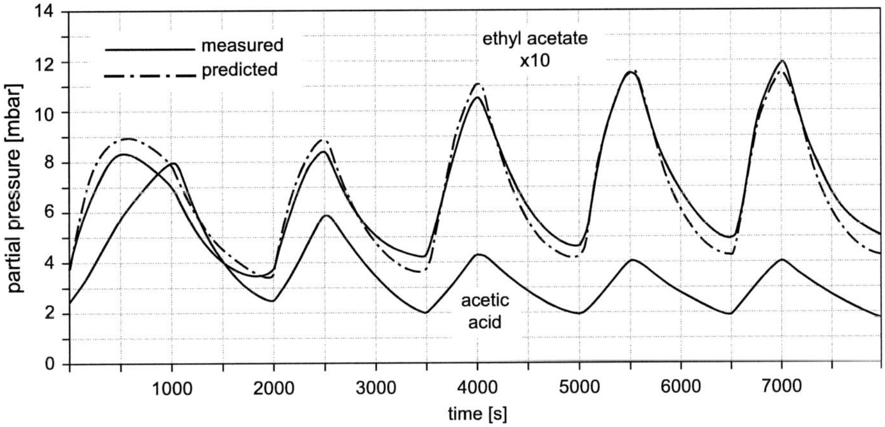

FIGURE 13-9 Comparison of measured (——) and predicted (- - - -) partial pressure of ethyl acetate leaving a differential reactor under modulation of the acetic acid flow to the reactor for $s = 0.5$ and $\tau_{cycle} = 1500$ s. *(Figure adapted from* Truffer and Renken (1986) *with permission. © 1986 AIChE.)*

explanations or models for the effect (*Koubek et al., 1980a, b*). The first, Model 1, assumes that different active surface sites, s_1, and, s_2, on the alumina catalyst participate. The s_1 site is acidic and strongly adsorbs an alcohol or an amine. Both reactants can adsorb as well on the basic site, s_2, on the alumina so as to block this site for H extraction to give the olefin. Reducing the reactant concentration frees the basic site, allowing it to react with adsorbate on the s_1 site to extract an H atom and form the olefin. *Nowobilski and Takoudis (1986)* developed and analyzed a model based on this proposal. Model 2 treats the acid and basic sites on alumina as an ensemble. Adsorption of a second alcohol or amine on a site blocks its ability to extract H. Such a blockage is a form of educt inhibition and, assuming adsorptive equilibrium, leads to the substrate inhibition model that is widely used to describe enzyme kinetics. A third model has been proposed by *Morávek (1992b)* based on the Ipatiev mechanism for bimolecular substitution. Some simplification gives this model the same mathematical formulation as Model 2.

Thullie and Renken (1993) demonstrate that in steady-state operation when adsorptive equilibrium can be assumed, the Langmuir-Hinshelwood models that result are indistinguishable. By examining the predicted rise in reaction rate after the flow of reactant is interrupted, these authors show that the rise is proportional to the steady-state rate predicted by the model. Consequently, if the models cannot be distinguished using steady-state data, it will be difficult to choose between them using step-change response data.

Composition modulation, however, can discriminate between the models at the asymptotic limit of the relaxed steady state. *Thullie and Renken* show that two models give different rates of dehydrogenation, as is illustrated in **Figure 13-10**. Discussion of this discrimination is summarized by *Silveston (1998)* and given in detail by *Thullie and Renken (1993).*

Away from the relaxed steady-state limit, as the cycle period increases, the difference between the calculated time-average rates decrease, as may be seen in **Figure 13-10** and it becomes more difficult to discriminate between the models. As the discussion of *Li et al. (1985b)* and *Nam et al. (1990)* suggests, variation of rate or a product concentration with time within

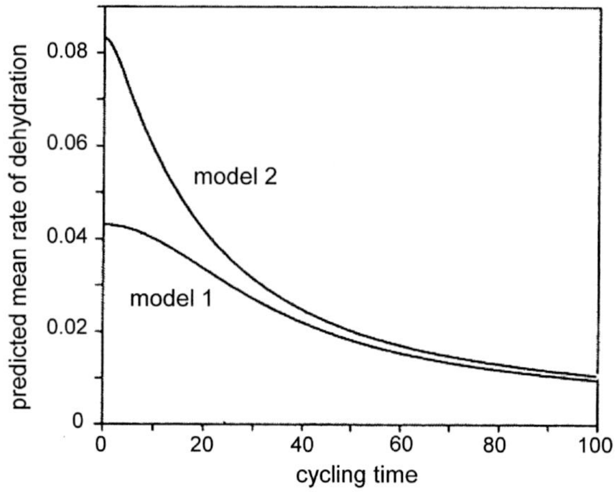

FIGURE 13-10 Calculated time-average rate of dehydration (mol/g_{cat}·s) for the two models versus cycle period (in seconds) for a time-average reactant concentration, the same set of rate constants and $s = 0.9$. *(Figures adapted from* Thullie and Renken (1993) *with permission. © 1993 Pergamon Press.)*

a cycle is a more sensitive test of a model than the prediction of resonance or the time-average behavior. Thus, it should be possible to choose between the models using realistic cycle periods, but instantaneous measurements of the dehydration rate or of the olefin concentration must be made.

This was demonstrated by *Golay et al. (1997)* by simultaneous measurements of gas phase and surface concentration applied to the dehydration of ethanol under periodic concentration modulation. Strongly adsorbed ethoxide species are thought to be reaction intermediates. They are characterized by C-H stretching vibrations of the ethyl group in the range of 2970, 2930 and 2870 cm^{-1}. The surface concentration of ethoxides were quantitatively determined at 2970 cm^{-1}.

The experimentally obtained concentration-time curves under periodic stop age of the ethanol feed are shown in **Figure 13-11** together with the predicted curves. The simulations are based on the same two models proposed by *Thullie and Renken (1990)*. The response of ethylene becomes reproducible after the first cycle, whereas the infrared signal increases slowly during several periods and becomes invariant after ca. three periods. **Figure 13-11** clearly shows that the experiments cannot be described by model 2, whereas model 1 is able to predict the experimental observations quantitatively.

Randall et al. (1996) employed modulation as proof of mechanism acceptability rather than for discrimination. Discrimination among three alternative mechanisms for NO reduction by CO over a silica-supported iron catalyst was handled satisfactorily using step response. Mechanisms considered by *Randall et al.* were 1) a homogeneous reaction and 2) a heterogeneous reaction involving the adsorption of NO and a further reaction with gas phase NO to form an O adatom on the surface and N_2O. The N_2O adsorbed to form another O adatom and gaseous N_2, while CO reacted with an O adatom to form CO_2. A third mechanism involved a redox reaction in which O vacancies on the catalyst successively extract O from NO to yield gaseous N_2. A final step was the reaction of the catalyst with CO to yield CO_2 and an O vacancy. When tested against steady-state rate data, the rate models for all three mechanisms were found to be adequate. Step-change experiments disqualified Model 1 and suggested that Model 2 was more likely than Model 3. This result was confirmed by a modulation experiment utilizing a 4-part cycle. Details are given by *Randall et al. (1996)*. *Silveston (1998)* also discusses the study.

Competing mechanisms for N_2O reduction by CO were investigated by *Sadhankar and Lynch (1994)* using composition modulation. Their experiments examined a Pt/Al_2O_3 catalyst and were performed at 499°C. Their model assumed oxygen extraction from N_2O to yield N_2 and an O adatom, the reaction of CO with this adatom, and the adsorption of CO to form a CO adatom. The last step was found necessary by

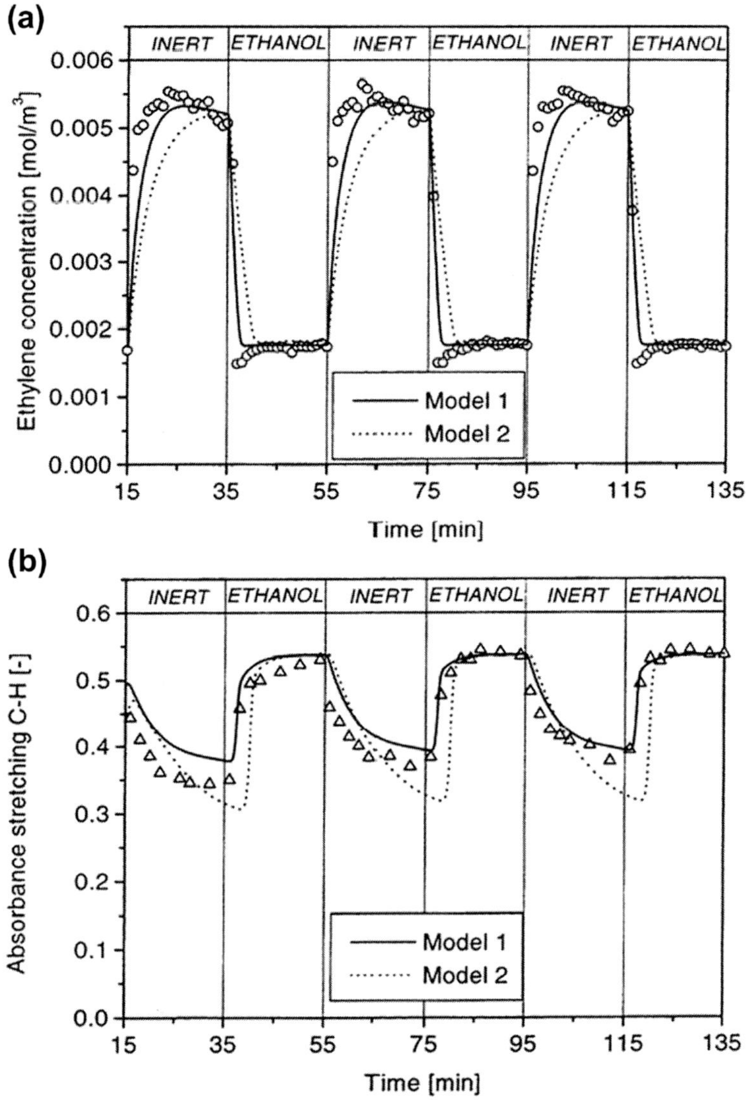

FIGURE 13-11 Comparison of the measured and predicted gas phase and surface concentration under periodic operation *(Golay et al., 1997, used with the authors permission.)*

Randall et al. (1996) to make their heterogeneous NO reduction model acceptable. Reduction of the Pt surface involved adsorbed CO rather than gas phase CO. Two models considered for steady-state experiments were: 1) Adsorption equilibrium assuming a Langmuir isotherm that gives the familiar Langmuir-Hinshelwood model and 2) CO adsorption with self-exclusion. In the latter, it was assumed that an adsorbed CO molecule excludes adsorption of CO on surrounding Pt sites. N_2O, however, could adsorb and react on those sites.

Steady-state data were adequately fitted by both models. To resolve this problem, *Sadhankar and Lynch (1994)* chose bang-bang modulation through varying the concentration of each reactant between 0 and 2.4 vol% and introduced phase lead (or lag) as an additional modulation parameter. In terms of phase lead, most applications of periodic forcing have utilized a 180° lead. When *Sadhankar and Lynch* used a 180° phase lead they found that the Langmuir-Hinshelwood and CO exclusion models fitted the low frequency data equally well, but that only the latter model matched their higher frequency data. In between the high and low frequencies, both models failed. **Figure 13-12** shows their observations. To resolve the poor fit, *Sadhankar and Lynch* followed *Chiao and Rinker (1989)* and introduced a factor for transformation of the surface under bang-bang cycling. They referred to this surface state as an adsorbate-induced phase transformation. **Figure 13-12** shows that including the factor in the CO exclusion model now fits the experimental CO_2 data satisfactorily. CO_2 rather than N_2O or N_2 data were used because more accurate measurements can be made of CO_2. *Sadhankar and Lynch (1994)* undertook forcing with different phase leads to further test their model. Predicting rate or concentration behavior as frequency changes is a sensitive test of a model.

Systematizing model discrimination using instantaneous rates or concentrations measured at points within a packed bed of catalyst during a cycle was undertaken by *Yuan* and co-workers using the partial oxidation of benzene to maleic anhydride over a vanadia catalyst as a test system. *Tan et al. (1994)* applied the Karhunen-Loeve expansion. $C(z,t)$ is the predicted concentration of one of the competing models at a point z in the catalyst bed at time t, subtracting the time-average value gives a deviation $V(z,t)$. Assuming the predicted deviations are part of a stationary time series, deviations at n different points for m different times form matrices $\boldsymbol{V(z,t)}\cdot$ $\boldsymbol{C} = \boldsymbol{VV}^T$ and $\boldsymbol{D} = \boldsymbol{V}^T\boldsymbol{V}$. These can be shown to be non-negative, real symmetrical matrices sharing a set of common coefficient λ_i. The Karhunen-Loeve expansion sets the vector $\boldsymbol{V}$ to be:

$$V(z,t) = \sum_{i=1}^{r} \lambda_i \phi_i(z) \xi_i^T(t), \qquad (13\text{-}2)$$

where $\varphi_i(z)$ and $\xi_i^T(t)$ are terms in the eigenfunctions of $\boldsymbol{C}$ and $\boldsymbol{D}$ respectively and are orthogonal. The sum in **Eq. (13-2)** can be truncated to h terms where h is related to some target error. The eigenfunction terms $\varphi_i(z)$ can be evaluated from the model once its parameters are determined and measured eigenfunction terms $\varphi_i(z)$ can be calculated from the experimental

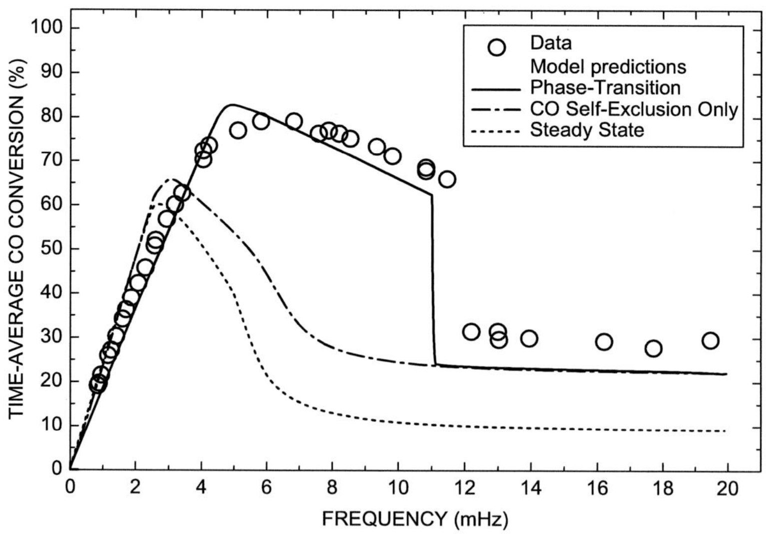

FIGURE 13-12 Comparison of calculated and experimental CO conversion as a function of frequency for the symmetrical bang-bang modulation of N_2O and CO from 0 to 2.4 vol% over a Pt/Al_2O_3 catalyst at 499°C. The three models are: Equilibrium adsorption of CO, adsorption of CO with CO exclusion and CO exclusion modified for phase transformation of the Pt surface. *(Figure adapted from* Sadhankar and Lynch (1994) *with permission. © 1994 Academic Press.)*

observations through **Eq. (13-2)**. The procedure involves recording instantaneous rates or concentrations at n points in the bed at m different times. These values are used to establish time-average values at the n points and are then used to find $V(z,t)$ for the catalyst bed.

To demonstrate the discrimination procedure, *Tan et al.* collected data at atmospheric pressure and three temperatures (633, 653 and 673 K) in a 12 mm (i.d.) × 200 mm bed of 40–60 mesh commercial V_2O_5 for the well-studied partial oxidation of benzene to maleic anhydride. The catalyst bed was kept isothermal by a thick copper sheath outside of the reactor shell. This sheath and the reactor shell were pierced by four sampling tubes located 42, 100, 158 and 196 mm from the top of the bed. Feed composition modulation was achieved by solenoid valves that switched flow from an air-N_2 stream containing 1.4 vol% Bz to an air-N_2 stream without Bz. Cycles were symmetrical with $\tau_{cycle} = 60$ s. A quadrupole mass spectrometer monitored the gas composition of the samples that were collected and this data was used to construct the deviation matrix $V(z,t)$, step-change measurements were also made for parameter estimation.

Tan et al. (1994) considered detailed mechanisms for the partial oxidation and from these derived rate expressions. They used an isothermal, one-dimensional plug flow model with conventional boundary conditions. Initial conditions assumed either an empty reactor or steady state and depended on whether the experiment modeled was periodic or involved a step change. Parameters for each model were estimated by fitting the step-change data. Four of the six models considered were discarded at this stage because parameters were either negative or the residual errors could not be reduced to a target level.

The two remaining models (2 and 3) were tested against modulation data using the Karhunen-Loeve expansion. **Figure 13-13** shows the test results. In **(a)** using the maleic anhydride data, model discrimination is poor, although it is evident that Model 3 is preferred. However, if the benzene data are employed, the model predictions are dramatically different and there is no question that Model 3 is superior. *Silveston (1998)* summarizes the work of *Tan et al. (1994)* to which the reader is referred for details.

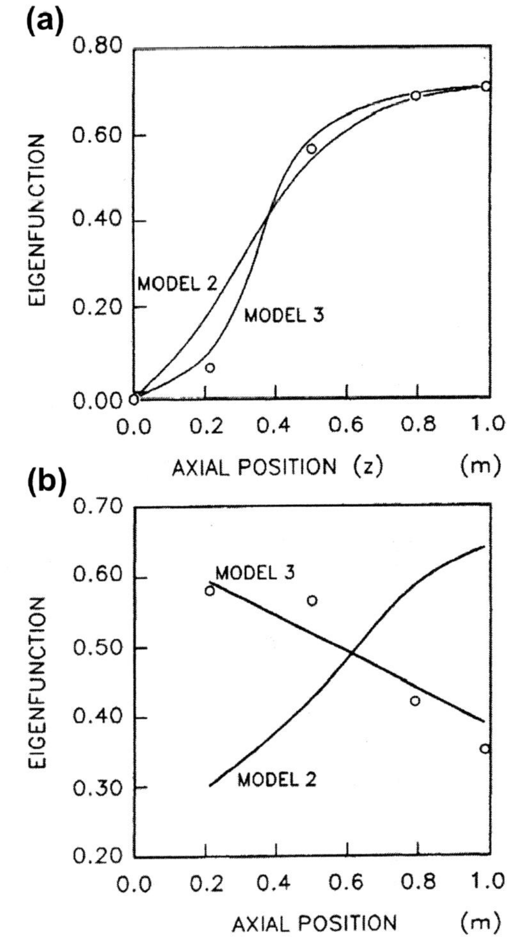

FIGURE 13-13 Testing of competing mechanistic models for the partial oxidation of benzene over a V_2O_5 catalyst at 673 K and 1 atm using eigenfunctions calculated from composition modulation data with $\tau_{cycle} = 60$ s and $s = 0.5$: a) maleic anhydride measurements, b) benzene measurements. *(Figure adapted from* Tan et al. (1994) *with permission. © 1995 Elsevier Science Ltd.)*

A variation of this approach was reproduced by *Tomilov et al. (1999)*, who used a model to predict how composition modulation could improve the production of butadiene from butene-1. This team of researchers collected data on oxidative dehydrogenation of butene-1 to butadiene over a Sb-Sn oxide catalyst in order to predict the outcome of operation under square-wave cycling of the reactants: Butene-1 and O_2 at average feed concentrations of 0.05% and 0.5% respectively and a temperature of 400°C. From their own transient (relaxation) trials and results from the literature, they inferred that: 1) A molecule of butene-1 uses two surface oxygen sites for adsorption and one additional site for oxidative elimination of hydrogen atoms from butene; 2) oxygen reactivity is not a linear function of the concentration but depends upon the degree of catalyst surface reduction/oxidation; and 3) water surface concentration is practically equal to zero. These assumptions led to a five-step mechanism that modeled their data well. From this model, they concluded that, if the reaction were carried out in an ideal, isothermal CSTR environment under composition modulation, process performance would be enhanced relative to steady state, resulting in increased selectivity and yield of butadiene. This result was restricted to residence times from 0.1 to 2.5 s and cycle periods from 1 s to 30 min, but these were considered to be feasible from the perspective of industrial practice. Unfortunately, no experimental follow-up of this research appears to have been undertaken.

Creaser et al. (1999a, b, c, d) presented a significant study of the oxidative dehydrogenation of propane over a V–Mg–O catalyst using various transient and modulation techniques. Although they did not develop a mathematical model for their system, their results supported a redox mechanism in which propane reduced the catalyst, which was reoxidized by gas-phase O_2. Only lattice oxygen participated in propene formation. Desorbable oxygen was a major source of poor selectivity, although lattice oxygen also caused total oxidation. Consequently, propene selectivity in the absence of gas-phase O_2 was found to be superior to co-feed, steady-state selectivity at the same C_3 conversion. Propene selectivity was further improved by increasing the degree of reduction of the catalyst.

Na-Ranong et al. (2002) used data from the composition modulation of the reaction between NO and CO over a Pt/Al_2O_3 catalyst as a means of developing a model that could serve not only for dynamic but also for steady-state operation. Employing an isothermal reactor at 150°C, the authors undertook a series of bang-bang experiments switching between pure NO and pure CO to produce a square-wave input. Having observed that the dissociation of adsorbed NO could be neglected at 150°C, they assumed a five-step, single-site, Langmuir-Hinshelwood mechanism in which equilibrium adsorption of both NO and CO were assumed. At the reactor outlet, deformed square waves of NO and CO concentrations were used to extract kinetic parameters. The resulting model was sensitive to all kinetic parameters except adsorption rate constants. The model predicted experimental results under periodic operation reasonably well even for conditions not employed in estimating the parameters.

Na-Ranong et al. (2009) revisited the reaction between NO and CO using composition modulation but this time with an Rh/Al_2O_3 catalyst. As previously, the authors used an isothermal reactor at 150°C with bang-bang experiments between pure NO and pure CO to produce a square-wave input. The shape of the deformed C_{NO} wave indicated that both reactants were strongly adsorbed on the catalyst's active sites. Out of three prospective models, the only one that could describe the experimental results with a single set of parameters in both steady and composition modulation modes was one based on an Eley-Rideal mechanism.

The kinetics of the oxidative dehydrogenation of alkanes has been studied by several research teams since the *Creaser et al.* work discussed above. *Del Rosso et al. (1998)* explored the effect of deep reduction of the catalyst on catalyst activity and reaction kinetics for the dehydrogenation of propane with a β-$NiMoO_4$ catalyst. A four-part cycle with successive exposure of the catalyst to a C_3 stream, an He flush, an O_2 stream, followed by another He flush. The variables were τ_{cycle} and T_o; He flushes were of constant duration. Increasing the cycle period at $T_o > 420°C$ beyond 180 s led to a sharp activity loss, a change in the C_3 reaction pathway and carbon whisker growth.

13.6 OVERVIEW

This chapter demonstrates that composition modulation, or indeed, other types of modulation have a place in research apart from investigating whether dynamic operation or systems with circulating catalyst beds can improve selectivity, yield or throughput. Although the use of modulation for mechanism discrimination is a relatively new technique, it is clear that this approach has promise. Furthermore, modulation equipment is relatively simple, at least for square-wave cycling, and measurements require no more effort than following concentrations after a step change. Consequently there is no experimental barrier for this application of modulation.

The examples considered suggest that modulation may be attractive when step-change experiments are unable to discriminate among rival models. Modulation over a range of frequencies should be especially useful for model discrimination when relaxation times of the catalytic processes are relatively fast so step-change experiments are limited by finite rates of data collection of analytical instruments employed. Indeed, modulation might be able to stabilize a surface species that might otherwise be unobservable so that the species' response to changing mean concentrations, temperature or other operating conditions can be determined and related to product behavior.

Nomenclature

C	= matrix formed from the deviation matrix V
C_i	= concentration of the ith species in the gas phase ($gmol/m^3$)
C_α	= carbon deposit, α form
C(z,t)	= concentration at position z and at time t
C_N	= concentration of nitrogen atoms in the catalyst bulk ($gmol/cm^3$)
CSTR	= continuous stirred tank reactor
D	= polydispersity
D	= matrix formed from the deviation matrix V
DRIFTS	= diffuse reflectance infra-red Fourier transform spectrometer
FTIR	= Fourier transform infra-red spectrometer
f	= fraction of spillover species not consumed by reaction
h	= number of terms in an expansion
i.d.	= inner diameter (mm)
k	= rate constant for reaction or adsorption (various units)
k_p	= propagation rate constant
MA	= maleic anhydride
m	= time interval indicator
N_i	= concentration of ith species on spillover or support surface
n	= measurement point, position interval indicator
P	= total pressure (kPa)
p_i	= partial pressure of the ith species
r	= rate (usually $gmol/cm^2s$)
s	= cycle split (-)
s_i	= ith surface site
s_s	= surface site for adsorption on catalyst support
T	= temperature (K)
t	= time (s)
V(z,t)	= deviation from a time-average concentration at position *z* and time *t*
V, V(z,t)	= deviation matrix
y_i	= mole fraction in the gas phase of the ith component
z	= axial position (m)

Greek

θ_i	= fraction of sites occupied by species i
λ_i	= coefficient in an eigenfunction expansion
μ_p	= number chain length of a polymer
$\xi_i T$	= term of an eigenfunction of the matrix D

τ, τ_{cycle} = cycle period (s, min.)
φ_i = term of an eigenfunction of the matrix C

Subscripts

ad = adsorbed
i = species or model index, index for a term in an expansion
M = mean
N = nitrogen
o, 0 = initial or entrance
RSS = relaxed steady state
SS = steady state

Superscripts

o = intercept or base value
T = vector or matrix transpose

CHAPTER

14

Evaluation of Periodic Processes

*Menka Petkovska**, *Andreas Seidel-Morgenstern*†

*Belgrade, Serbia and †Magdeburg, Germany

OUTLINE

14.1 Introduction 387

14.1.1 Periodic Processes 388

14.1.2 Early Approaches to Identifying Enhancement 388

14.2 Nonlinear Frequency Response and Higher Order Frequency Response Functions 389

14.2.1 Frequency Response Functions 390

14.2.1.1 Relation between the Output and the Frequency Response Functions 390

14.3 Estimation of the Time Average Performance of Periodic Processes Using Nonlinear Frequency Response Analysis 393

14.3.1 Simple Homogeneous Reactions 393

14.3.2 Simple Heterogeneous Reactions 401

14.3.3 Comparison of Homogeneous and Heterogeneous Reactions 405

14.4 Application of Nonlinear Frequency Response Analysis for the Estimation of the Periodic Steady States of Cyclic Processes 405

14.4.1 Mathematical Foundation of Approximate Calculations 405

14.4.2 Procedure for Approximate Calculation of the Periodic Quasi-Steady State 407

14.4.2.1 Example 407

14.5 Summary 412

Acknowledgment 412

14.1 INTRODUCTION

Throughout this review, examples have been given that demonstrate that periodic operation of chemical reactors may offer a significant improvement in performance. Naturally, it would be useful to have generally applicable methods of identifying catalytic reaction systems that are enhanced by periodic operation and to be able to estimate the enhancement possible. In this chapter, we review past work on such methods and discuss a recent proposal that appears to be admirably suited for this purpose.

The three questions of identification of candidate systems for enhancement through periodic operation, estimation of the magnitude of such enhancements, and what system properties result in improvements, have occupied researchers since the investigation by *Horn (1965)* and the simulation study of *Douglas and Rippin (1966)*. These questions were addressed tangentially by *Horn and Lin (1967)* and *Horn and Bailey (1968)*. They are succinctly discussed in the fine review by *Bailey (1973)*. The issues of identification and estimating have been dealt with by *Watanabe et al. (1981)*, *Farhadpour and Gibilaro (1981)* and *Sterman and Ydstie (1989)*. *Nowobilski and Takoudis (1986)* and *Thullie et al. (1986)* have discussed the magnitude question.

When a stable system is perturbed by a periodic input change, after some time the transient response vanishes and the system reaches a quasi-stationary or periodic steady state, defined by an output time variation which identically repeats itself. In process dynamics and process control, this quasi-stationary response is known as frequency response (*Stephanopoulos, 1984*), and is one of the most common and useful tools for process analysis, identification and synthesis.

14.1.1 Periodic Processes

Among processes operated periodically, one can distinguish two different classes. The first comprises processes which can operate either under steady-state conditions or under periodic forcing. In this chapter, those processes will be addressed as deliberate periodic processes. In analyzing such processes, it is of interest to determine whether the deliberate periodic operation imposed by forced modulation of certain process inputs leads to better performance.

The second class, in contrast, comprises processes which cannot be operated under steady-state conditions so that periodic operation is essential. These are processes where a finite storage capacity is required in order to operate. Examples are adsorption processes, which after a specific process time require regeneration of the sorbent (*Ruthven et al., 1994*; *Yang, 1987*). A number of different cyclic adsorption processes have been developed in the last 30 years, such as temperature swing adsorption, pressure swing adsorption (first introduced by *Skarstrom, 1960*), parametric pumping (first introduced by *Wilhelm et al., 1966*), cyclic zone adsorption (first introduced by *Pigford et al., 1969*), as well as the simulated moving bed (first introduced by *Broughton and Gerhold, 1961*). All these processes are cyclic. In this chapter such processes are referred to as inherent periodic processes.

The efficient calculation of the cyclic steady state, which is established after a certain, often long, transient duration, is an important task in evaluating and optimizing inherent periodic processes.

14.1.2 Early Approaches to Identifying Enhancement

Experimentally identifying catalytic reaction systems that are candidates for periodic operation and estimating enhancement possible has proved to be time-consuming and expensive. To reduce this effort, mathematical modeling and simulation are often used. However, models are frequently complex, usually in the form of coupled nonlinear partial differential equations that can be solved, generally, only numerically. Simulating the periodic (cyclic) steady-state operation of the process requires running calculations until the transient response completely relaxes. For slow processes, this can be very long. Direct calculation of the periodic steady state, without going through the transient solution of the model equations would be very desirable. Some attempts in this direction have been reported by *Unger (1999)* and *LeVan* and coworkers (*Croft and LeVan, 1994a, b; Ding and LeVan, 2001; Ding et al., 2002*). This approach will be developed further in this chapter.

The theoretical problems related to identification and estimation of enhancement have been analyzed based on control criteria, using for instance Pontryagin's maximum principle and relaxed steady-state analysis (*Horn and Lin, 1967; Bailey and Horn, 1971*). The optimal steady-state condition was used as a reference for evaluating the optimal periodic operation. Guardabassi's π-criterion (Guardabassi et al., 1974; Noldus, 1977; Schädlich et al., 1983) was used for frequencies close to the critical frequencies which delimit the intervals where performance improvement may be achievable.

However, the literature of the past several decades shows that these analytical techniques have not been applied. The reason for this is probably the complexity of their application and some lingering uncertainty about their reliability. There is still a need for a simple and reliable method to quantitatively evaluate the possibility of process improvements through periodic operation. In **Section 14.3** of this Chapter, we will present such a method which gives a fast answer as to whether working under periodic conditions is advantageous with respect to average reactor performance. It is based on the Volterra series, nonlinear frequency response and the concept of higher order frequency response functions (*Weiner and Spina, 1980*). Results obtained with this simple method indicate possible process improvements, a promising starting point for further analysis and optimization.

In **Section 14.4**, a new method for fast, approximate calculation of quasi-steady states of cyclic processes is given. The method is based on higher order frequency response functions. The system input is represented in the form of a Fourier series, while the output is presented as a Volterra series. For practical applications, both the input and the output series are approximated by finite length sums. In this way, the approximate periodic quasi-steady state of the system output can be calculated directly, without tedious numerical integration.

14.2 NONLINEAR FREQUENCY RESPONSE AND HIGHER ORDER FREQUENCY RESPONSE FUNCTIONS

Kinetics and temperature effects for most catalytic reaction systems result in system nonlinearity. The quasi-stationary response of a nonlinear system to a periodic (sinusoidal or co-sinusoidal) input around a steady state is referred to as a nonlinear frequency response (NLFR). One of the most convenient tools for mathematical treatment of an NLFR is the concept of higher order frequency response functions (FRFs) (*Weiner and Spina, 1980*) which is based on a Volterra series and the generalized Fourier transform.

Consider a stable system with a single input x and a single output y. Dynamic response of a linear system to an arbitrary input $x(t)$ can be expressed by the convolution integral:

$$y(t) = \int_{-\infty}^{\infty} g(\tau)\; x(t-\tau)\; d\tau \tag{14-1}$$

where $g(\tau)$ is the so-called impulse-response function of the system or its kernel. On the other hand, the response of a weakly nonlinear system, for which the system nonlinearity has a polynomial form (or can be expanded into a Taylor series), can be represented in the form of a Volterra series (*Volterra, 1959*):

$$y(t) = \sum_{n=1}^{\infty} y_n(t) \tag{14-2}$$

in which the nth element of the series is defined as:

$$y_n(t) = \int_{-\infty}^{\infty} \cdots \int_{-\infty}^{\infty} g_n(\tau_1, \cdots, \tau_n)\ x(t - \tau_1)\ \cdots\ x(t - \tau_n) d\tau_1 \cdots d\tau_n \tag{14-3}$$

with $g_n(\tau_1,\ldots,\tau_n)$ being the nth order Volterra kernel, or the generalized impulse response function of order n. It is assumed here that the catalytic reaction systems are generally weakly nonlinear.

The first element of the Volterra series y_1 corresponds to the linearized model, while y_2, y_3, ... are the "correction" functions of the 1st, 2nd, 3rd... order. Similar to a Taylor series expansion, a Volterra series of infinite length is needed to represent exactly a nonlinear system. For practical applications, however, a series of finite length can be used.

14.2.1 Frequency Response Functions

By applying the Fourier transform to the function $g(\tau)$ in **Eq. (14-1)** where τ represents time, the frequency-response function, or frequency transfer function is obtained:

$$G(\omega) = \int_{-\infty}^{\infty} g(\tau)\ e^{-j\omega\tau}\ d\tau. \tag{14-4}$$

This function is directly related to the amplitude and phase of the quasi-stationary response to a single harmonic input:

$$\begin{aligned} x &= A\cos(\omega t)\ \Rightarrow\ \ t \to \infty\ : \\ y(t) &= A|G(\omega)|\cos(\omega t + \arg(G(\omega)). \end{aligned} \tag{14-5}$$

On the other hand, applying the multidimensional Fourier transform to the function $g_n(\tau_1,\ldots,\tau_n)$, the nth order FRF, or the nth order generalized transfer function is obtained (*Weiner and Spina, 1980*):

$$G_n(\omega_1, \cdots, \omega_n) = \int_{-\infty}^{\infty} \cdots \int_{-\infty}^{\infty} g_n(\tau_1, \cdots, \tau_n)\ e^{j(\omega_1\tau_1 + \cdots \omega_n\tau_n)} d\tau_1 \cdots d\tau_n. \tag{14-6}$$

In this way, a nonlinear model of the system can be replaced by an infinite sequence of functions of the 1st, 2nd, 3rd, ... order. The relation between x and y is illustrated schematically in **Figure 14-1**.

14.2.1.1 Relation between the Output and the Frequency Response Functions

If the input is a periodic function of the general form:

$$x(t) = \sum_{k=1}^{N} A_k e^{j\omega_k t} \tag{14-7}$$

the nth element of the Volterra series defined in **Eq. (14-2)** would be:

$$y_n(t) = \sum_{k_1=1}^{N} \sum_{k_2=1}^{N} \cdots \sum_{k_n=1}^{N} A_{k_1} A_{k_2} \cdots A_{k_n}\ G_n(\omega_{k_1}, \omega_{k_2}, \cdots, \omega_{k_n})\ e^{j(\omega_{k_1} + \omega_{k_2} + \cdots + \omega_{k_n})t}. \tag{14-8}$$

For a single harmonic input:

$$x(t) = A\ \cos(\omega t) = \frac{A}{2} e^{j\omega t} + \frac{A}{2} e^{-j\omega t} \tag{14-9}$$

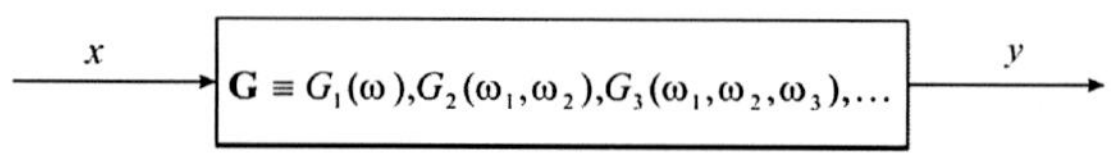

FIGURE 14-1 Block diagram of a weakly nonlinear system.

the first three elements of the Volterra series become:

$$y_1(t) = G_1(\omega)\frac{A}{2}e^{j\omega t} + G_1(-\omega)\frac{A}{2}e^{-j\omega t} \quad (14\text{-}10)$$

$$y_2(t) = G_2(\omega,\omega)\left(\frac{A}{2}\right)^2 e^{2j\omega t} + 2G_2(\omega,-\omega) \times \left(\frac{A}{2}\right)^2 e^0 + G_2(-\omega,-\omega)\left(\frac{A}{2}\right)^2 e^{-2j\omega t}, \quad (14\text{-}11)$$

$$y_3(t) = G_3(\omega,\omega,\omega)\left(\frac{A}{2}\right)^3 e^{3j\omega t} + 3G_3(\omega,\omega,-\omega) \times \left(\frac{A}{2}\right)^3 e^{j\omega t} + 3G_3(-\omega,-\omega,\omega) \times \left(\frac{A}{2}\right)^3 e^{-j\omega t} + G_3(-\omega,-\omega,-\omega) \times \left(\frac{A}{2}\right)^3 e^{-3j\omega t}. \quad (14\text{-}12)$$

By collecting the terms of the same frequency, it can be easily shown that the response of a weakly nonlinear system to a single harmonic input is obtained as a sum of the basic harmonic (of the same frequency as the input), a DC (nonperiodic) term and an infinite number of higher harmonics. The DC component of the output is obtained by collecting the nonperiodic terms (the terms with e^0):

$$y_{DC} = 2(A/2)^2 G_2(\omega,-\omega) + 6(A/2)^4 G_4(\omega,\omega,-\omega,-\omega) + \cdots \quad (14\text{-}13)$$

the first harmonic by collecting the terms of frequency ω (the terms with $e^{j\omega t}$ and $e^{-j\omega t}$):

$$y_I = B_I\cos(\omega t + \varphi_I) = \{(A/2)G_1(\omega) + 3(A/2)^3 G_3(\omega,\omega,-\omega) + \cdots\}\, e^{j\omega t} + \{(A/2)G_1(-\omega) + 3(A/2)^3 G_3(\omega,-\omega,-\omega) + \cdots\}\, e^{-j\omega t} \quad (14\text{-}14)$$

the second harmonic by collecting the terms of frequency 2ω (the terms with $e^{2j\omega t}$ and $e^{-2j\omega t}$):

$$y_{II} = B_{II}\cos(2\omega t + \varphi_{II}) = \{(A/2)^2 G_2(\omega,\omega) + 4(A/2)^4 G_4(\omega,\omega,\omega,-\omega) + \cdots\}\, e^{2j\omega t} \quad (14\text{-}15)$$

the third by collecting the terms of frequency 3ω (the terms with $e^{3j\omega t}$ and $e^{-3j\omega t}$):

$$y_{III} = B_{III}\cos(3\omega t + \varphi_{III}) = \{(A/2)^3 G_3(\omega,\omega,\omega) + 5(A/2)^5 G_5(\omega,\omega,\omega,\omega,-\omega) + \cdots\}\, e^{3j\omega t} + \{(A/2)^3 G_3(-\omega,-\omega,-\omega) + 5(A/2)^5 G_5(\omega,\omega,-\omega,-\omega,-\omega) + \cdots\}\, e^{-3j\omega t} \quad (14\text{-}16)$$

and so on.

For weakly nonlinear systems, contributions of the higher harmonics and higher FRFs decrease with their increasing order. Different harmonics of the output can be estimated directly by harmonic analysis of the output signal. On the other hand, as can be seen from **Eqs (14-13)** to **(14-16)**, the first-order function $G_1(\omega)$ corresponds to the dominant term of the first harmonic, the second-order functions $G_2(\omega,\omega)$ and $G_2(\omega,-\omega)$ to the dominant terms of the second harmonic and the DC component, respectively, and the third-order function $G_3(\omega,\omega,\omega)$ to the dominant term of the third harmonic, and so on.

If the nonlinear system has multiple inputs that can be modulated, its dynamic model consists of several sets of FRFs. A block diagram, representing a nonlinear system with two inputs, is shown in **Figure 14-2**. It should be noted that, for this case, to state the complete model, it is necessary to identify three sets of FRFs: Two of them relating the output to each of the inputs, and one cross-function, relating the output to both inputs. This third set contains only functions of the second and higher orders.

In **Figure 14-2**, the FRFs are described in the following way: G_{n,x^n} and G_{n,z^n} are n^{th} order FRFs corresponding to the individual inputs x and z,

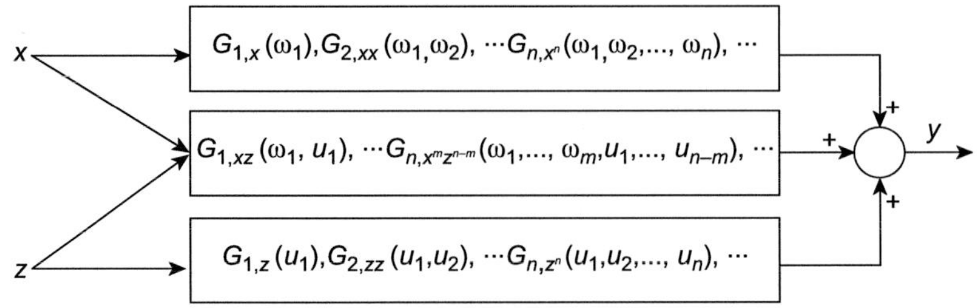

FIGURE 14-2 Block diagram of a nonlinear system with two inputs and one output.

respectively, while $G_{n,x^m z^{n-m}}$ is the nth order cross-function, with order m in respect to the input x and order n-m, with respect to the input z.

The output of this system is obtained as a sum of the contributions of the inputs x and z separately (via the G_x- and G_z-functions), as well as the additional contribution corresponding to the cross-effect of both inputs (via the G_{xz}-functions). Each of these contributions can be presented in Volterra series form as:

$$y(t) = y_x(t) + y_z(t) + y_{xz}(y) = \sum_{n=1}^{\infty} y_{x,n}(t) + \sum_{n=1}^{\infty} y_{z,n}(t) + \sum_{n=1}^{\infty} y_{xz,n}(t). \tag{14-17}$$

If the inputs are defined as general periodic functions:

$$x(t) = \sum_{k=1}^{N} A_k e^{j\omega_k t}, \quad z(t) = \sum_{k=1}^{M} B_k e^{ju_k t}. \tag{14-18}$$

The nth elements of the Volterra series corresponding to single inputs $y_{x,n}$ and $y_{z,n}$ can be represented in an analogous way as in **Eq. (14-8)**. The nth element corresponding to the cross effect, on the other hand, would be:

$$y_{xz,n}(t) = \sum_{m=1}^{n} \sum_{k_1=1}^{N} \cdots \sum_{k_n=1}^{M} A_{k_1} \cdots A_{k_m} B_{k_{m+1}} \cdots B_{k_n} \times G_{n,x^m z^{n-m}}(\omega_{k_1}, \cdots, \omega_{k_m}, u_{k_{m+1}}, \cdots, u_{k_n}) \times e^{j(\omega_{k_1} + \cdots + \omega_{k_m} + u_{k_{m+1}} + \cdots + u_{k_n})t}. \tag{14-19}$$

This analysis could be extrapolated to nonlinear systems with 3 or more input variables. Thus, for a nonlinear system with three inputs, it would be necessary to define 7 sets of HFRFs: 3 of them relating the output to each of the inputs, three sets of cross-functions relating the output to each combination of two inputs and one set of cross-functions relating the output to all three inputs. This last set would contain only the FRFs of the third and higher orders. On the other hand, for a single input, **Figure 14-2** reduces to **Figure 14-1**, while $G_{n,x^n}(\omega_1, \omega_2, \cdots, \omega_n) \equiv G_n(\omega_1, \omega_2, \cdots, \omega_n)$.

Although a powerful tool, frequency response has not found broad application in Chemical Engineering, so far. Up to now, it has been used for developing new experimental techniques for investigating equilibria and kinetics of adsorption systems (*Petkovska, 2006; Ilić et al., 2008*), as well as for identification of kinetic mechanisms in adsorption and electrochemical reaction systems (Petkovska and Do, 2000; Petkovska and Petkovska, 2003; Petkovska, 2006; Bensmann et al., 2010).

In **Section 14.3**, higher order FRFs are used for direct, although approximate evaluation of periodic processes, using the property that the frequency response functions are directly related to the quasi-stationary behavior or frequency response of a system under the influence of forced periodic input changes.

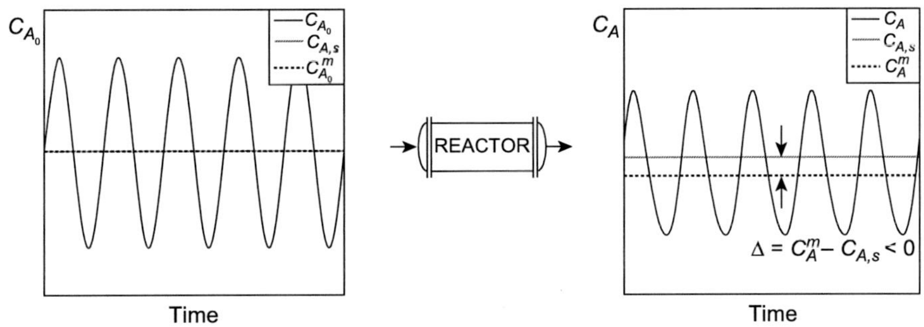

FIGURE 14-3 **Representation of a favorable periodic reactor operation in which A is a reactant.**

14.3 ESTIMATION OF THE TIME AVERAGE PERFORMANCE OF PERIODIC PROCESSES USING NONLINEAR FREQUENCY RESPONSE ANALYSIS

In some cases the time-average process performance under periodic operation is superior to performance under steady-state operation, i.e., conversion or/and selectivity increases. **Figure 14-3** illustrates the increase of conversion for a simple reaction $A \rightarrow$ products.

Assuming that, $C_{A_0,s}$ and $C_{A,s}$ are the steady-state input and output concentrations of reactant A, respectively. If the input concentration is modulated around its steady-state value, the outlet concentration will also oscillate. The mean value of the outlet concentration C_A^m for a nonlinear system will be different from $C_{A,s}$. The difference $\Delta = C_A^m - C_{A,s}$ can be negative, zero, or positive, depending on the type of nonlinearity. If $\Delta < 0$, periodic operation will be favorable, as it corresponds to increased conversion compared with steady state.

Testing whether a periodic operation leads to better reactor performance generally demands a tedious experimental and/or numerical effort. Application of the relations presented in **Section 14.2.2** avoids numerical integration of the model equations and reduces substantially the analytical effort required.

In what follows, the use of higher order FRFs for investigating the average performance of periodic processes is developed. Only the DC component, corresponding to Δ in **Figure 14-3**, and the asymmetrical second-order FRF $G_2(\omega, -\omega)$, corresponding to its dominant term (see **Eq. (14-13)**), are of interest. The sign of the function $G_2(\omega, -\omega)$ will define the sign of the DC component. In order to decide on the advantage of a particular periodic operation relative to the corresponding steady state, only derivation and evaluation of the function $G_2(\omega,\omega)$ is needed.

14.3.1 Simple Homogeneous Reactions

Consider a simple homogeneous isothermal n^{th} order reaction of the type $A \rightarrow$ products and determine the average periodic performance of three reactor types: a continuous stirred tank reactor (CSTR), a plug flow reactor (PFR) and a dispersed flow tubular reactor (DFTR). The nonstationary material balance equations are listed in **Table 14-1**. Notation is conventional and given at the end of the Chapter.

It is common practice to render the model equations dimensionless by writing the concentration variables as relative deviations from their steady-state values. Definitions of these dimensionless variables are given in **Table 14-2**.

TABLE 14-1 Model Equations for a Homogeneous nth Order Reaction in Standard Reactor Types

Reactor	Material Balance		Boundary Conditions	
CSTR	$V\frac{dC_A}{dt} = F(C_{A_0} - C_A) - kVC_A^n$	(14-20)		
PFR	$\frac{\partial C_A}{\partial t} + v\frac{\partial C_A}{\partial z} + kC_A{}^n = 0$	(14-21)	$z = 0:\ C_A(0,t) = C_{A_0}(t)$	(14-21a)
DFTR	$\frac{\partial C_A}{\partial t} + v\frac{\partial C_A}{\partial z} + kC_A^n = D_{ax}\frac{\partial^2 C_A}{\partial z^2}$	(14-22)	$z = 0: C_A(0,t) = C_{A_0}(t) + D_{ax}\left.\frac{\partial C_A}{\partial z}\right\rvert_{z=0}$	
			$z = L: \left.\frac{\partial C_A}{\partial z}\right\rvert_{z=L} = 0$	(14-22a)

Subscript s denotes the steady-state value of the corresponding concentration. The contact time τ_c used for defining the dimensionless time is $\tau_c = \frac{V}{F}$.

The steady-state concentrations that define the dimensionless concentrations are obtained by solving the corresponding steady-state models. The resulting dimensionless equations for the three reactor types under consideration are listed in **Table 14-3**. To evaluate higher order FRFs, all model nonlinearities need to be stated in polynomial form. For that reason, the nonlinear terms in **Table 14-3** have been replaced by their Taylor series expansions (details are given by *Marković et al. (2008)*. N_{TP} used in **Eq. (14-25)** is the number of theoretical plates of the DFTR and defined as $N_{TP} = \frac{vL}{2D_{ax}}$.

TABLE 14-2 Definitions of the Dimensionless Variables

Reactant concentration	$C = \frac{C_A - C_{A,s}}{C_{A,s}}$
Input reactant concentration	$C_i = \frac{C_{A_0} - C_{A_0,s}}{C_{A_0,s}}$
Time	$\tau = \frac{t}{\tau_c}$
Axial coordinate	$x = \frac{z}{L}$

Derivation of higher order FRFs is standard and can be found in (*Petkovska and Do, 1998, 2000; Petkovska, 2001, 2006; Petkovska and Marković, 2006*). The basic steps of this procedure, applied to the system under consideration, are:

1) Define the input concentration $C_i(\tau)$ in the form of a co-sinusoidal function (**Eq. (14-9)**).
2) Express the output concentration $C(\tau)$ as a Volterra series (**Eq. (14-7)** with **Eqs (14-10)** to **(14-12)**).
3) Substitute the expressions for $C_i(\tau)$ and $C(\tau)$ into the reactor model equations.
4) Apply harmonic probing to the equations obtained in Step 3 (collect the terms with the same power of the input amplitude and same frequency and equate them to zero).
5) Solve the equations obtained in Step 4. For the CSTR model these equations are algebraic, while for the PFR and DFTR they are linear ODEs. The derivation procedure is recurrent, i.e., the first-order FRF has to be derived first, then the second, and so on.

TABLE 14-3 Dimensionless Form of the Reactor Models (see **Table 14-1**)

Reactor	Dimensionless model	
CSTR	$\frac{dC}{d\tau} = (1 + k\tau_c C_{A,s}{}^{n-1})C_i - C - k\tau_c C_{A,s}{}^{n-1} nC - \frac{1}{2}k\tau_c^{n-1} C_{A,s} n(n-1)C^2 - \ldots$	(14-23)
PFR	$\frac{\partial C}{\partial \tau} + \frac{\partial C}{\partial x} + \tau_c k C_{A,s}{}^{n-1}(n-1)C + \frac{1}{2}\tau_c n(n-1)kC_{A,s}{}^{n-1}C^2 + \ldots = 0$	(14-24)
	$x = 0: \; C = C_i(0, \tau)$	
DFTR	$\frac{\partial C}{\partial \tau} + \frac{\partial C}{\partial x} + \tau_c k C_{A,s}{}^{n-1}(n-1)C + \frac{1}{2}\tau_c n(n-1)kC_{A,s}{}^{n-1}C^2 + \cdots = \frac{1}{2N_{TP}}\frac{\partial^2 C}{\partial x^2}$	(14-25)
	$x = 0: \quad C(0, \theta) = C_i(\tau) + \frac{1}{2N_{TP}}\frac{\partial C}{\partial x}\Big\vert_{x=0}, \; x = 1: \; \frac{\partial C}{\partial x}\Big\vert_{x=1} = 0$	

For evaluating the average reactor performance, it is sufficient to estimate the DC component. The sign of the DC component is determined by the sign of the asymmetrical second-order FRF $G_2(\omega,-\omega)$ (e.g., **Eq. (14-13)**). Consequently, derivations and analysis are limited to the first ($G_1(\omega)$) and asymmetrical second-order FRFs ($G_2(\omega,-\omega)$). Details of the procedure can be found in *Marković et al., (2008)*. Only the final expressions for the asymmetrical second-order FRFs of the three reactor models considered are given below.

CSTR model

$$G_2(\omega,-\omega) = -\frac{1}{2}n(n-1)\frac{k\tau_c C_{A,s}{}^{n-1}(1 + k\tau_c C_{A,s}{}^{n-1})^2}{(1 + k\tau_c C_{A,s}{}^{n-1}n)((1 + k\tau_c C_{A,s}{}^{n-1}n)^2 + \omega^2)}. \quad (14\text{-}26)$$

PFR model

$$G_2(\omega,-\omega) = -\frac{1}{2}n(n-1)\frac{k\tau_c C_{Ai,s}^{n-1}}{2(1 + k\tau_c(n-1)C_{Ai,s}^{n-1})^2}. \quad (14\text{-}27)$$

DFTR model

The derived expression for the second-order FRF corresponding to the DC component for the DFTR model is cumbersome. The result can be put into the following form:

$$\begin{aligned} G_2(\omega,-\omega) = {} & D_1 + D_2e^{2N} + d_1e^{\alpha_1(\omega)+2N} \\ & + d_2e^{\alpha_2(\omega)+2N} + d_3e^{\alpha_1(-\omega)+2N} \\ & + d_4e^{\alpha_2(-\omega)+2N} + f_1(1)e^{\alpha_1(\omega)+\alpha_1(-\omega)} \\ & + f_2(1)e^{\alpha_1(\omega)+\alpha_2(-\omega)} \\ & + f_3(1)e^{\alpha_1(-\omega)+\alpha_2(\omega)} \\ & + f_4(1)e^{\alpha_2(-\omega)+\alpha_2(\omega)}, \end{aligned} \quad (14\text{-}28)$$

where $\alpha_{1,2}$ are characteristic values, D_1, D_2, and d_1 to d_4 are integration constants and $f_1(x)$ to $f_4(x)$ are auxiliary functions. Their definitions can be found in **Table 14-4**.

Since a dimensionless time is used in **Eqs (14-23)** to **(28)**, the frequency, ω, in **Eqs (13-26)** to **(28)** is also dimensionless.

Analysis of the expressions for the asymmetrical second-order FRF $G_2(\omega,-\omega)$ for the CSTR and PFR leads to the following conclusions:

1. $G_2(\omega,-\omega) < 0$, for $n < 0$ and $n > 1$. This corresponds to improved reactor performance through periodic operation.

TABLE 14-4 Integration Constants for **Eq. (14-28)**

$\alpha_{1,2}$ are characteristic values of the corresponding second-order ODE, and can be obtained using:

$$\alpha_{1,2}(\omega) = N_{TP} \pm \sqrt{N_{TP}^2 + 4N_{TP}j\omega} \tag{14-28a}$$

Integration constants $d_1 - d_4$ in **Eq. (14-28)** are defined by:

$$d_1 = \frac{2N_{TP}k\tau(n-1)nC_{A,s}^{n-1}C_1(\omega)}{(\alpha(\omega)_1 + 2N_{TP})\alpha_1(\omega)}, \quad d_2 = \frac{2N_{TP}k\tau(n-1)nC_{A,s}^{n-1}C_2(\omega)}{(\alpha(\omega)_2 + 2N_{TP})\alpha_2(\omega)}, \quad d_3 = \frac{2N_{TP}k\tau(n-1)nC_{A,s}^{n-1}C_1(-\omega)}{(\alpha(-\omega)_1 + 2N_{TP})\alpha_1(-\omega)},$$

$$d_4 = \frac{2N_{TP}k\tau(n-1)nC_{A,s}^{n-1}C_2(-\omega)}{(\alpha(-\omega)_2 + 2N_{TP})\alpha_2(-\omega)} \tag{14-28b}$$

while the integration constants D_1 and D_2 are obtained from the following set of linear algebraic relations:

$$\begin{aligned}
&D_1 + D_2 + d_1 + d_2 + d_3 + d_4 + N_{TP}n\left(\frac{C_1(\omega)C_1(-\omega)}{(\alpha_1(\omega)+\alpha_1(-\omega))(\alpha_1(\omega)+\alpha_1(-\omega)-2N_{TP})} + \frac{C_1(\omega)C_2(-\omega)}{(\alpha_1(\omega)+\alpha_2(-\omega))(\alpha_1(\omega)+\alpha_2(-\omega)-2N_{TP})}\right.\\
&\left.+\frac{C_1(-\omega)C_2(\omega)}{(\alpha_1(-\omega)+\alpha_2(\omega))(\alpha_1(-\omega)+\alpha_2(\omega)-2N_{TP})} + \frac{C_2(\omega)C_2(-\omega)}{(\alpha_2(\omega)+\alpha_2(-\omega))(\alpha_2(\omega)+\alpha_2(-\omega)-2N_{TP})}\right)\\
&= \frac{1}{2N_{TP}}\left\{k\tau C_{A_0s}^{n-1}(n-1)D_1 + 2N_{TP}D_2 + d_1(\alpha_1(\omega)+2N_{TP}) + d_2(\alpha_2(\omega)+2N_{TP}) + d_3(\alpha_1(-\omega)+2N_{TP})\right.\\
&+ d_4(\alpha_2(-\omega)+2N_{TP}) + 2N_{TP}n(n-1)k\tau C_{A_0s}^{n-1} \times (C_1(\omega)C_1(-\omega) \times \left[C_1(\omega)C_1(-\omega)\left(\frac{\alpha_1(\omega)+\alpha_1(-\omega)-(n-1)k\tau C_{A_0s}^{n-1}}{f_1(0)}\right.\right.\\
&\left.+(n-1)k\tau C_{A_0s}^{n-1}(\alpha_1(\omega)+\alpha_1(-\omega))(\alpha_1(\omega)+\alpha_1(-\omega)-2N_{TP})\right) + C_1(\omega)C_2(-\omega)\left(\frac{\alpha_1(\omega)+\alpha_2(-\omega)-(n-1)k\tau C_{A_0s}^{n-1}}{f_2(0)}\right.\\
&\left.+(n-1)k\tau C_{A_0s}^{n-1}(\alpha_1(\omega)+\alpha_2(-\omega))(\alpha_1(\omega)+\alpha_2(-\omega)-2N_{TP})\right) + C_1(-\omega)C_2(\omega)\left(\frac{\alpha_1(-\omega)+\alpha_2(\omega)-(n-1)k\tau C_{A_0s}^{n-1}}{f_3(0)}\right.\\
&\left.+(n-1)k\tau C_{A_0s}^{n-1}(\alpha_1(-\omega)+\alpha_2(\omega))(\alpha_1(-\omega)+\alpha_2(\omega)-2N_{TP})\right) + C_2(\omega)C_2(-\omega)\left(\frac{\alpha_2(\omega)+\alpha_2(-\omega)-(n-1)k\tau C_{A_0s}^{n-1}}{f_4(0)}\right.\\
&\left.\left.+(n-1)k\tau C_{A_0s}^{n-1}(\alpha_2(\omega)+\alpha_2(-\omega))(\alpha_2(\omega)+\alpha_2(-\omega)-2N_{TP})\right)\right\}
\end{aligned} \tag{14-28c}$$

$$
\begin{aligned}
& k\tau(n-1)C_{A_0s}^{n-1}D_1 + 2N_{TP}\,e^{2N_{TP}} + d_1(\alpha_1(\omega)+2N_{TP})\;e^{(\alpha_1(\omega)+2N_{TP})} + d_2(\alpha_2(\omega)+2n)\;e^{(\alpha_2(\omega)+2N_{TP})} + d_3(\alpha_1(-\omega) \\
& \quad + 2N_{TP})\;e^{(\alpha_1(-\omega)+2N_{TP})} + d_4(\alpha_2(\omega)+2N_{TP})\;e^{(\alpha_2(\omega)+2N_{TP})} + \frac{2N_{TP}k\tau(n-1)C_{A_0s}^{n-1}}{1+k\tau(n-1)C_{A_0s}^{n-1}} \\
& \quad \times \Bigg\{ C_1(\omega)C_1(-\omega)e^{(\alpha_1(\omega)+\alpha_1(-\omega))}\left(\frac{1}{f_1(1)}\left(\alpha_1(\omega)+\alpha_1(-\omega)-\frac{k\tau(n-1)C_{A_0s}^{n-1}}{1+k\tau(n-1)C_{A_0s}^{n-1}}\right)+k\tau(n-1)C_{A_0s}^{n-1}(\alpha_1(\omega)+\alpha_1(-\omega))\right. \\
& \quad \left.\times\,((\alpha_1(\omega)+\alpha_1(-\omega))-2N_{TP}\vphantom{\frac{1}{1}}\right) + C_1(\omega)C_2(-\omega)e^{(\alpha_1(\omega)+\alpha_2(-\omega))}\left(\frac{1}{f_2(1)}\left(\alpha_1(\omega)+\alpha_2(-\omega)-\frac{k\tau(n-1)C_{A_0s}^{n-1}}{1+k\tau(n-1)C_{A_0s}^{n-1}}\right)\right. \\
& \quad \left. + k\tau(n-1)C_{A_0s}^{n-1}(\alpha_1(\omega)+\alpha_2(-\omega))((\alpha_1(\omega)+\alpha_2(-\omega))-2N_{TP}\vphantom{\frac{1}{1}}\right) + C_1(-\omega)C_2(\omega)e^{(\alpha_1(-\omega)+\alpha_2(\omega))} \\
& \quad \times \left(\frac{1}{f_3(1)}\left(\alpha_1(-\omega)+\alpha_2(\omega)-\frac{k\tau(n-1)C_{A_0s}^{n-1}}{1+k\tau(n-1)C_{A_0s}^{n-1}}\right)+k\tau(n-1)C_{A_0s}^{n-1}(\alpha_1(-\omega)+\alpha_2(\omega))((\alpha_1(-\omega)+\alpha_2(\omega))-2N_{TP}\right) \\
& \quad + C_2(\omega)C_2(-\omega)e^{(\alpha_2(\omega)+\alpha_2(-\omega))}\left(\frac{1}{f_4(1)}\left(\alpha_2(\omega)+\alpha_2(-\omega)-\frac{k\tau(n-1)C_{A_0s}^{n-1}}{1+k\tau(n-1)C_{A_0s}^{n-1}}\right)+k\tau(n-1)C_{A_0s}^{n-1}(\alpha_2(\omega)+\alpha_2(-\omega))\right. \\
& \quad \left.\times\,((\alpha_2(\omega)+\alpha_2(-\omega))-2N_{TP})\vphantom{\frac{1}{1}}\right)\Bigg\} = 0
\end{aligned}
\tag{14-28d}
$$

$C_1(\omega)$ and $C_2(\omega)$ used in these equations are obtained from:

(Continued)

$$C_1(\omega) + C_2(\omega) = 1 + \frac{1}{2N_{TP}}(\alpha_1(\omega)C_1(\omega) + \alpha_2(\omega)C_2(\omega) + f + 2N_{TP}) \tag{14-28e}$$

$$C_1(\omega)\alpha_1(\omega)e^{\alpha_1(\omega)} + C_2(\omega)\alpha_2(\omega)e^{\alpha_2(\omega)} + (f + 2N_{TP}(1+f))e^{2N_{TP}} = 0 \tag{14-28f}$$

The functions $f_1(x)$ to $f_4(x)$ were introduced to simplify **Eq. (14-28)** and **Eqs. (14-28c)** and **(14-28d)**. They are defined as:

$$\begin{aligned} f_1(x) &= (\alpha_1(\omega) + \alpha_1(-\omega))(2k\tau(n-1)C_{A_0s}^{n-1} + (1 + k\tau(n-1)C_{A_0s}^{n-1}x)(\alpha_1(\omega) + \alpha_1(-\omega) - 2N_{TP}))f_2(x) \\ &= (\alpha_1(\omega) + \alpha_2(-\omega))(2k\tau(n-1)C_{A_0s}^{n-1} + (1 + k\tau(n-1)C_{A_0s}^{n-1}x)(\alpha_1(\omega) + \alpha_2(-\omega) - 2N_{TP}))f_3(x) \\ &= (\alpha_1(-\omega) + \alpha_2(\omega))(2k\tau(n-1)C_{A_0s}^{n-1} + (1 + k\tau(n-1)C_{A_0s}^{n-1}x)(\alpha_1(-\omega) + \alpha_2(\omega) - 2N_{TP}))f_4(x) \\ &= (\alpha_2(\omega) + \alpha_2(-\omega))(2k\tau(n-1)C_{A_0s}^{n-1} + (1 + k\tau(n-1)C_{A_0s}^{n-1}x)(\alpha_2(\omega) + \alpha_2(-\omega) - 2N_{TP})) \end{aligned} \tag{14-28g}$$

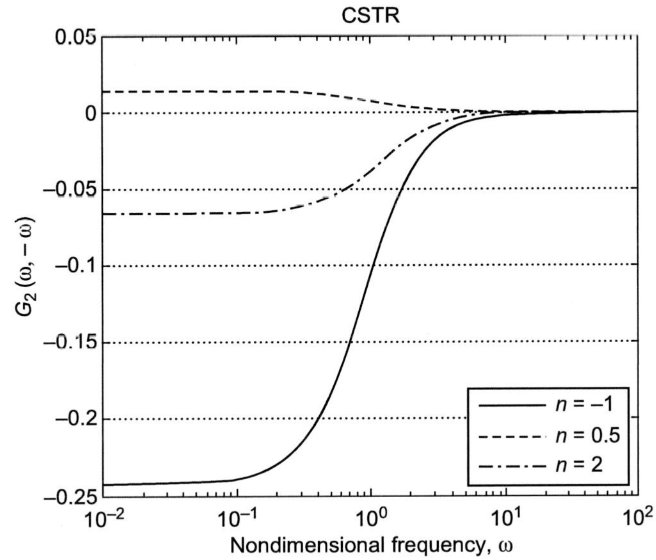

FIGURE 14-4 **The second order functions $G_2(\omega,-\omega)$ for a reaction in a homogeneous CSTR with different orders of reaction.**

2. $G_2(\omega,-\omega)=0$, for $n=0$ and $n=1$. This means that there is no influence of periodic operation on reactor performance.
3. $G_2(\omega,-\omega)>0$, for $0<n<1$. This corresponds to poorer reactor performance through periodic operation.

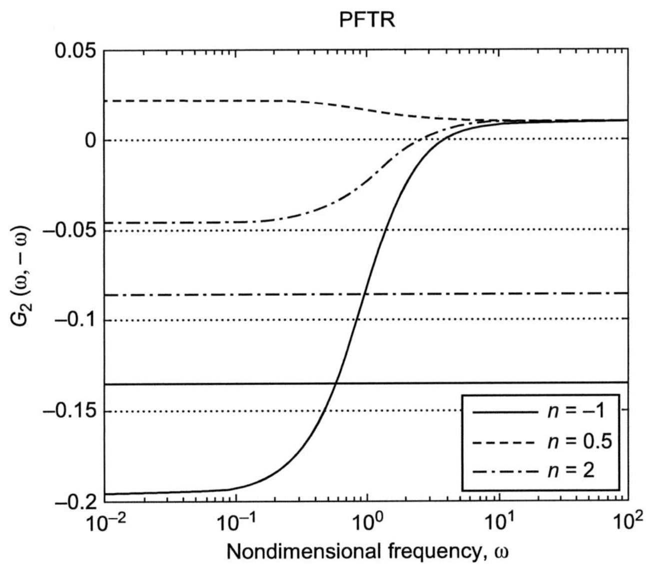

FIGURE 14-5 **The second order functions, $G_2(\omega,-\omega)$, for a homogeneous PFR with different reaction orders.**

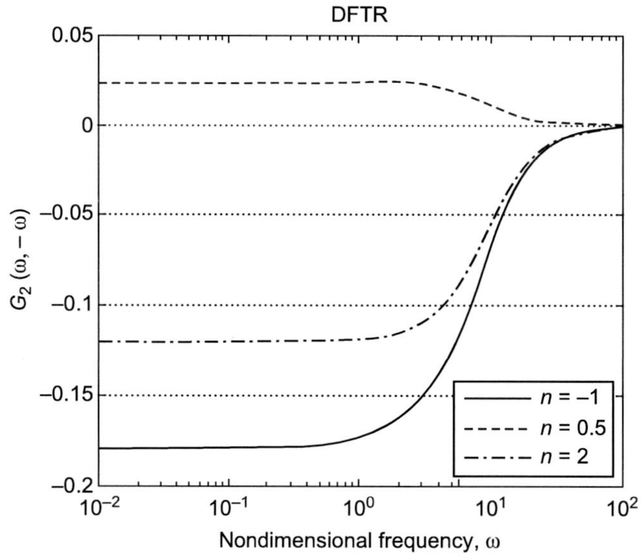

FIGURE 14-6 **The second order functions, $G_2(\omega,-\omega)$, for a homogeneous DFTR with different reaction orders.**

These results agree with those of *Schädlich et al. (1983)*, obtained using the Hoelder inequality, the π criterion and the Legendre-Clebsch inequality.

Identical results were obtained for both the CSTR and PFR. Analysis of **Eq. (14-28)**, defining the function $G_2(\omega,-\omega)$ for the DFTR, is not so simple. Nevertheless, it can be shown by numerical analysis that the same conclusions are valid for this reactor type as well. This is important as the analysis for the CSTR is much simpler than for the other two reactor types because the model equation in the frequency domain is a linear algebraic equation. Thus, the conclusion concerning the advantage of periodic operation obtained for the CSTR model can be applied to other types of reactor.

To illustrate this observation, the corresponding functions were calculated for the three reactors using three reaction orders ($n = -1$, 0.5 and 2) employing **Eqs (14-26)** to **(14-28)**. These results are shown in **Figures 14-4** to **14-6** for the three types of reactors. Parameter values used for calculations are given in **Table 14-5**. Calculations were performed with the same values of contact time, steady-state inlet concentration and rate constant for each reactor type.

Results presented in **Figures 14-4** to **14-6** show that over the whole frequency range, values of $G_2(\omega,-\omega)$) for $n = -1$ ($n < 0$) and $n = 2$ ($n > 1$) are negative, while for $n = 0.5$ ($0 < n < 1$), the $G_2(\omega,-\omega)$ value is positive.

Figure 14-4 demonstrates that the function $G_2(\omega,-\omega)$ for a CSTR is frequency dependent, with different low-frequency plateaus for different orders of the reaction rate. On the other hand, for any reaction order, $G_2(\omega,-\omega)$ approaches zero at high frequencies. This implies that low-frequency modulation of inlet concentration (for the parameters used) can

TABLE 14-5 Model Parameters Used for Calculations

Rate constant, k	**$0.001\ s^{-1}mol^{1-n}$**
Mean inlet concentration of reactant A, C_{A_0s}	1 mol m^{-3}
Contact time, τ_c	100 s
Number of theoretical plates, N_{TP} (DFTR model)	100

considerably improve or reduce the reactor performance compared with steady-state operation, while high-frequency modulations ($\omega > 10$) make practically no difference as $G_2(\omega,-\omega) \approx 0$. This means that for a periodically operated CSTR an optimal range of modulation frequency exists.

On the other hand, as can be seen from **Figure 14-5**, the $G_2(\omega,-\omega)$ functions for a PFR are constant and independent of frequency. This is a result of plug flow, which, from a dynamic point of view, represents a pure time-delay. The PFR does not change the shape and intensity of the input signal, but only translates it in time. For such a system, the amplitude of the output is determined solely by its static gain, i.e., it is frequency independent. Consequently there is no optimal frequency range for a periodically operated PFR.

Figure 14-6 shows that $G_2(\omega,-\omega)$ for the DFTR model is again frequency dependent. This was expected as a result of back mixing introduced into the reactor model through the dispersive term. The resulting function, $G_2(\omega,-\omega)$, for the DFTR model has shape similar to that for the CSTR model. The differences are that the $G_2(\omega,-\omega)$ functions approach zero for higher frequencies ($\omega > 100$) while the optimal frequencies for concentration modulation are also higher than those for the CSTR model ($\omega < 1$).

To illustrate behavior under periodic forcing, a quasi-steady-state segment of the output concentration from a CSTR, obtained by numerical solution of the nonlinear model (**Eq. (14-20)**), is shown in **Figure 14-7** for a sinusoidal input. The following parameters were used for the calculation: $n = -1$, $\tau_c = 100$ s, $k = 0.001$ mol^2/s, $C_{A0,s} = 1$ mol/m^3, $\omega^* = 0.01$ rad/s and $A = 75\%$. The corresponding steady-state concentration ($C_{A,s} = 0.8873$ mol/m) and the mean value of the outlet concentration ($C_A^m = 0.8176$ mol/m) are also shown in **Figure 14-7**.

The reactor performance improvement owing to the periodic operation is:

$$\Delta = C_A^m - C_{A,s} = -0.0697 \text{ mol/m}^3.$$

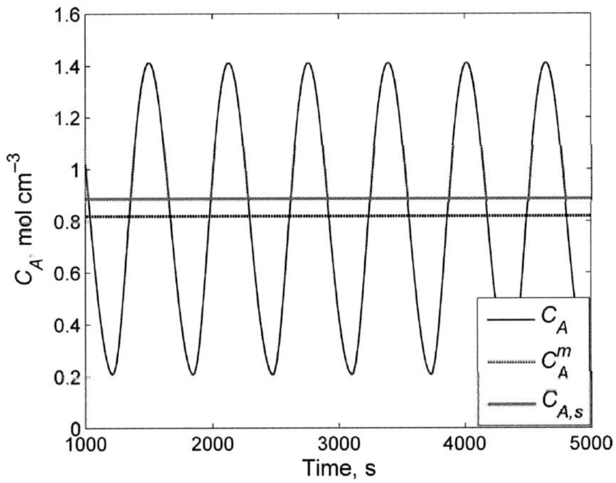

FIGURE 14-7 **Numerical simulation of the CSTR outlet concentration for a sinusoidal input concentration (A = 75%, ω^* = 0.01 rad/s) showing the difference between the mean and steady-state values for reaction order $n = -1$.**

The approximate value of the DC component, calculated based on the second-order FRF, $G_2(\omega,-\omega)$, i.e., using only the first term in **Eq. (14-13)**, is:

$$y_{DC} \approx 2\left(\frac{A}{2}\right)^2 G_2(\omega,-\omega)C_{A,s}$$
$$= -0.0605 \text{ mol/m}^3$$

which is close to the value of Δ obtained from the numerical solution. For a more precise estimation of the DC component, contributions of the asymmetrical fourth-order FRF, $G_4(\omega,\omega,-\omega,-\omega)$, and possibly higher order FRFs, would have to be taken into account.

14.3.2 Simple Heterogeneous Reactions

The second example is a simple, heterogeneous, catalytic n^{th} order reaction of the type $A \leftrightarrow A_{ads} \rightarrow Products$ carried out in a CSTR. The model could describe a catalytic fluidized bed reactor.

The material balance for the reactant A is:

$$V_f\frac{dC_A}{dt} + V_s\frac{dq}{dt} = F(C_A - C_A) - kV_sq^n. \quad (14\text{-}29)$$

In this equation, q is the adsorbate concentration in the solid phase, and V_f, and V_s are the volumes of the fluid and of the solid phases, respectively.

If mass transfer resistances are negligible, concentrations in the two phases are related by an adsorption isotherm which is generally nonlinear:

$$q = \phi(C_A). \tag{14-30}$$

Using the dimensionless variables defined in **Table 14-2**, and $Q = (q - q_s)/q_s$, the model (**Eq. (14-29)**) may be transformed into dimensionless form, which, after using Taylor series expansions of all nonlinear terms, becomes:

$$\begin{aligned}\frac{dC}{d\tau} + f\frac{dQ}{d\tau} &= \left(1 + k\tau_{c,s}q_s^{n-1}\frac{q_s}{C_{A,s}}\right)C_i - C \\ &- k\tau_{c,s}\frac{q_s}{C_{A,s}}q_s nQ - \frac{1}{2}\frac{q_s}{C_{A,s}} \\ &\times n(n-1)k\tau_{c,s}q_sQ^2 - \ldots\end{aligned} \tag{14-31}$$

In this equation, the subscript s once again indicates the concentrations at steady state; the contact time used to normalize time is defined as: $\tau_c = \dfrac{V_f + V_s}{F}$ and f and $\tau_{c,s}$ are auxiliary parameters: $f = \dfrac{V_s}{V_f}\dfrac{q_s}{C_{A,s}}$ and $\tau_{c,s} = \dfrac{V_s}{F}$.

The nonlinear adsorption isotherm, **Eq. (14-30)**, may be also converted into its dimensionless form and replaced by a Taylor series expansion around the steady state:

$$Q = \Phi(C) = \tilde{a}C + \tilde{b}C^2 + \ldots \tag{14-32}$$

where $\tilde{a}$ and $\tilde{b}$ are proportional to the local isotherm derivatives:

$$\tilde{a} = \left.\frac{d\phi}{dC_A}\right|_{C_A=C_{A,s}}\frac{C_{A,s}}{q_s}, \quad \tilde{b} = \frac{1}{2}\left.\frac{d^2\phi}{dC_A^2}\right|_{C_A=C_{A,s}}\frac{C_{A,s}^2}{q_s}. \tag{14-33}$$

Following the derivation given in **Section 14.3.1**, the resulting expression for the asymmetrical second-order FRF will be obtained for the heterogeneous reaction carried out in a CSTR:

$$\begin{aligned}G_2(\omega, -\omega) &= \frac{n\left(\frac{1}{2}(1-n)\tilde{a}^2 - \tilde{b}\right)}{(1+n\beta\tilde{a})} \\ &\times \frac{\beta(1+\beta)^2}{(1+n\beta\tilde{a})^2 + \omega^2(1+f\tilde{a})^2}\end{aligned} \tag{14-34}$$

where

$$\beta = k\tau_{c,s}\frac{q_s^n}{C_{A,s}} \tag{14-35}$$

is an auxiliary parameter (derivation details given by *Petkovska et al., 2010*).

The advantage of periodic operation depends, of course, on the sign of the asymmetrical second-order FRF, $G_2(\omega,-\omega)$. However, the sign of the $G_2(\omega,-\omega)$ function is determined by the sign of the constant term $\dfrac{n\left(\frac{1}{2}(1-n)\tilde{a}^2 - \tilde{b}\right)}{(1+n\beta\tilde{a})}$ which, in turn, depends on the reaction order n, as well as on the values of the adsorption isotherm derivatives $\tilde{a}$ and $\tilde{b}$. Analysis of this term shows that when the reaction order n changes there are three possible changes of sign of the $G_2(\omega,-\omega)$ function. The values of n for which these changes occur are:

$$\begin{aligned}n_1 &= 0 \\ n_2 &= 1 - \frac{2\tilde{b}}{\tilde{a}^2} \\ n_3 &= -\frac{1}{\beta\tilde{a}}.\end{aligned} \tag{14-36}$$

With respect to the isotherm derivatives $\tilde{a}$ and $\tilde{b}$, two cases arise $\tilde{b} < \tilde{a}^2/2$ and $\tilde{b} > \tilde{a}^2/2$.

Case 1: $\tilde{b} < \tilde{a}^2/2$. This case corresponds to all favorable and many of the unfavorable isotherms. For these circumstances, n_2 is positive and the sign of $G_2(\omega,-\omega)$ changes with n in the following way:

$$\begin{aligned}
&n < n_3 \Rightarrow G_2(\omega,-\omega) > 0\\
&n_3 < n < 0 \Rightarrow G_2(\omega,-\omega) < 0\\
&n = 0 \Rightarrow G_2(\omega,-\omega) = 0\\
&0 < n < n_2 \Rightarrow G_2(\omega,-\omega) > 0\\
&n = n_2 \Rightarrow G_2(\omega,-\omega) = 0\\
&n > n_2 \Rightarrow G_2(\omega,-\omega) < 0.
\end{aligned} \qquad (14\text{-}37)$$

The value of n_2 depends on the adsorption isotherm derivatives $\tilde{a}$ and $\tilde{b}$, while the value of n_3 depends on $\tilde{a}$ and the auxiliary variable β, which depends on the reaction order (see **Eq. (14-35)**). Accordingly, the value of n_3 will be determined by solving the following set of nonlinear equations:

$$\begin{aligned}
&n_3 k\tau_{c,s}\frac{q_s^{n_3}}{C_{A,s}}\tilde{a} + 1 = 0\\
&C_s\left(1 + k\tau_{c,s}\frac{q_s^{n}}{C_{A,s}}\right) = C_{A_0 s}\\
&q_s = \phi(C_{A,s})
\end{aligned} \qquad (14\text{-}38)$$

This set of equations has no solution. Consequently, three ranges of the reaction order n may be defined in which the sign of the $G_2(\omega,-\omega)$ function needs to be considered: $n < 0$, $0 < n < n_2$ and $n > n_2$. Thus,

$$\begin{aligned}
&G_2(\omega,-\omega) < 0 \quad \Leftrightarrow \quad n < 0 \quad \vee \quad n > 1 - \frac{2\tilde{b}}{\tilde{a}^2}\\
&G_2(\omega,-\omega) > 0 \quad \Leftrightarrow \quad 0 < n < 1 - \frac{2\tilde{b}}{\tilde{a}^2}\\
&G_2(\omega,-\omega) = 0 \quad \Leftrightarrow \quad n = 0 \quad \vee \quad n = 1 - \frac{2\tilde{b}}{\tilde{a}^2}.
\end{aligned} \qquad (14\text{-}39)$$

Case 2: $\tilde{b} > \tilde{a}^2/2$ (possible only for unfavorable isotherms). For this case the value n_2 is negative, and the following situations are possible:

$$\begin{aligned}
&G_2(\omega,-\omega) < 0 \quad \Leftrightarrow \quad 1 - \frac{2\tilde{b}}{\tilde{a}^2} < n < 0\\
&G_2(\omega,\ \ \omega) > 0 \quad \Leftrightarrow \quad n < 1 \quad \frac{2\tilde{b}}{\tilde{a}^2} \quad \vee \quad n > 0\\
&G_2(\omega,-\omega) = 0 \quad \Leftrightarrow \quad n = 0 \quad \vee \quad n = 1 - \frac{2\tilde{b}}{\tilde{a}^2}.
\end{aligned} \qquad (14\text{-}40)$$

To illustrate the results that are obtained, the asymmetrical second-order FRFs are calculated for the set of model parameters given in **Table 14-2**, a ratio of the solid to fluid phase volumes $(V_s/V_f) = 0.3$ and for the Langmuir isotherm: $q = 20\frac{0.2C_A}{1+0.2C_A}$, then n_2, defined in **Eq. (14-36)**, is 1.235.

The simulated $G_2(\omega,-\omega)$ functions corresponding to six different values of n are shown in **Figure 14-8**. The values of n were chosen from all three regions defined by **Eq. (14-39)**. As expected, the value of $G_2(\omega,-\omega)$ is negative for negative values of n (−3, −1 and −0.5) and for $n > 1.235$ (3), while it is positive for $0 < n < 1.235$ (0.5 and 1). It can be seen also that for all values of n, the values of the $G_2(\omega,-\omega)$ function become negligibly small at frequencies higher than 1 and tend to constant values at frequencies lower than 0.01.

In **Figure 14-9**, the calculated values in the low-frequency, asymptotic range of $G_2(\omega,-\omega)$ are shown as a function of the reaction order n, over a wide range between −10 and 10. As can be observed, the limiting value of $G_2(\omega,-\omega)$ is negative for $n < 0$, but it becomes very close to zero for $n < -5$. There is a local minimum in this range of n, that corresponds to $n \approx -0.9$ and the $\lim_{\omega\to 0} G_2(\omega,-\omega) \approx -0.1$ (representing a 10% improvement, when compared with a steady-state operation). Within the range $0 < n < 1.235$, $G_2(\omega,-\omega)$ is positive with a maximum in this range

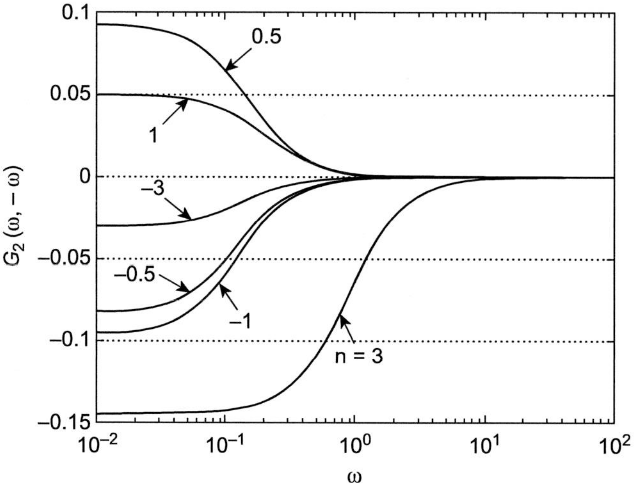

FIGURE 14-8 **Calculated $G_2(\omega,-\omega)$ functions for a heterogeneous CSTR for different values of the reaction order n and for $\tilde{b} < \tilde{a}^2/2$.**

corresponding to $n \approx 0.6$ and $\lim_{\omega\to 0} G_2(\omega,-\omega) \approx +0.1$ (performance reduced by 10%). Finally, for $n > 1.235$, $G_2(\omega,-\omega)$ is again negative, and there is a minimum in this interval, corresponding to $n \approx 2.8$ and $\lim_{\omega\to 0} G_2(\omega,-\omega) \approx -0.15$ (representing a 15% improvement). These results were obtained for a favorable isotherm and with the model parameters defined in **Table 14-2**. For another set of parameters, the values corresponding to the

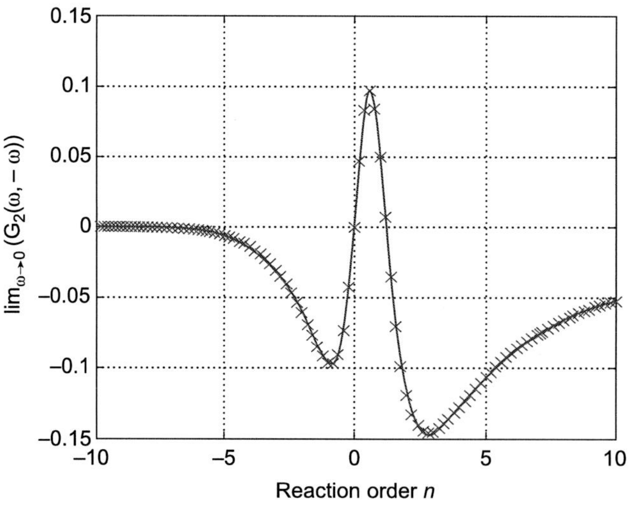

FIGURE 14-9 **The low-frequency asymptotic value of $G_2(\omega,-\omega)$ for a heterogeneous reaction carried out in a CSTR as a function of the reaction order n for $\tilde{b} < \tilde{a}^2/2$.**

local minimums and maximums, of course, will change.

14.3.3 Comparison of Homogeneous and Heterogeneous Reactions

Comparing results for the homogeneous and heterogeneous nth order reactions (**Sections 14.3.1** and **14.3.2**), it is evident that:

1. There are two values of the reaction order for which there is no influence of inlet concentration modulation on conversion. One of those values is $n_1 = 0$. For the homogeneous reaction case, the second value is $n_2 = 1$; while for the heterogeneous case, this value depends on the local adsorption isotherm: $n_2 = 1 - \frac{2\tilde{b}}{\tilde{a}^2}$.
2. The two values of the reaction order divide the complete range into three intervals, in which the sign of the $G_2(\omega,-\omega)$ function is either negative or positive, corresponding to a favorable or an unfavorable periodic operation. For the homogeneous case, a periodic operation is favorable for all negative reaction orders and for reaction orders higher than 1, while it is unfavorable for reaction orders between 0 and 1. The situation is similar for the heterogeneous reaction if the isotherm derivatives fulfill the condition $\tilde{b} < \tilde{a}^2/2$. In this case, the value n_2 is positive, and the only difference is that the periodic operation is favorable for $n > 1 - \frac{2\tilde{b}}{\tilde{a}^2}$ and unfavorable for $0 < n < 1 - \frac{2\tilde{b}}{\tilde{a}^2}$. On the other hand, if $\tilde{b} > \tilde{a}^2/2$, the value n_2 is negative, and the situation is completely reversed, compared with the homogeneous reaction. Periodic operation is favorable for reaction orders between n_2 and 0 and unfavorable for $n < n_2$ and $n > 0$.

14.4 APPLICATION OF NONLINEAR FREQUENCY RESPONSE ANALYSIS FOR THE ESTIMATION OF THE PERIODIC STEADY STATES OF CYCLIC PROCESSES

14.4.1 Mathematical Foundation of Approximate Calculations

In this section, a method for fast approximate calculations of the complete behavior of periodic processes is given. This method is also based on the use of higher order frequency response functions, described in **Section 14.2.2**. It replaces the time-consuming numerical solutions of ordinary or partial differential equations needed to calculate the periodic steady state of inherently periodic processes. For increased generality, the method is developed for a nonlinear system with two modulated inputs, which can be represented by the block diagram shown in **Figure 14-2**. The method applies to arbitrary shapes of the input modulations and assumes the same frequency for both.

Even though any periodic function can be represented as an indefinite sum of harmonic functions, i.e. as a Fourier series, in practice, the indefinite Fourier series can be approximated by finite sums, containing only the most significant harmonics. Accordingly, the input functions x and z can be written as:

$$x(t) = \sum_{k=-\infty}^{\infty} \frac{1}{2} A_k e^{jk\omega t} \approx \sum_{k=-K}^{K} \frac{1}{2} A_k e^{jk\omega t} \quad (14\text{-}41)$$

$$z(t) = \sum_{k=-\infty}^{\infty} \frac{1}{2} B_k e^{jk\omega t} \approx \sum_{k=-K}^{K} \frac{1}{2} B_k e^{jk\omega t} \quad (14\text{-}42)$$

where $A_{-k} = conj(A_k)$, $B_{-k} = conj(B_k)$, ω is the basic frequency of the inputs x and z and K is the number of harmonics used for the approximation. The modules of A_k and B_k are the amplitudes and their arguments the phases of the kth harmonics of x and z for the frequency $k\omega$.

For two periodic inputs, the periodic quasi-steady state of the output can be represented as the sum of three terms (see **Eq. (14-17)**), each of them corresponding to the contribution of a set of FRFs, as given in **Figure 14-2**.

The Volterra series, given by **Eq. (14-17)**, with indefinite sums, represents the periodic quasi-stationary state of the output exactly. In practice only finite sums are calculated, leading to an approximation of the periodic output. When the inputs are also approximated by finite sums of only the first K harmonics, the approximate output is:

$$\begin{aligned} y_{app}(t) &= y_{x,app}(t) + y_{z,app}(t) + y_{xz,app}(t) \\ &= \sum_{n=1}^{N} y_{x,n,app}(t) + \sum_{n=1}^{N} y_{z,n,app}(t) \\ &\quad + \sum_{n=1}^{N} y_{xz,n,app}(t) \end{aligned} \tag{14-43}$$

where:

$$\begin{aligned} y_{x,n,app}(t) &= \frac{1}{2^n} \sum_{k_1=-K}^{K} \sum_{k_2=-K}^{K} \cdots \sum_{k_n=-K}^{K} A_{k_1} \\ &\quad \times A_{k_2} \cdots A_{k_n} G_{n,x^n}(k_1\omega, k_2\omega, \cdots, k_n\omega) \\ &\quad \times e^{j(k_1\omega + k_2\omega + \cdots + k_n\omega)t} \end{aligned} \tag{14-44}$$

$$\begin{aligned} y_{z,n,app}(t) &= \frac{1}{2^n} \sum_{k_1=-K}^{K} \sum_{k_2=-K}^{K} \cdots \sum_{k_n=-K}^{K} B_{k_1} B_{k_2} \cdots B_{k_n} \\ &\quad \times G_{n,z^n}(k_1\omega, k_2\omega, \cdots, k_n\omega) \\ &\quad \times e^{j(k_1\omega, k_2\omega, \cdots, k_n\omega)t} \end{aligned} \tag{14-45}$$

$$\begin{aligned} y_{xz,n,app}(t) &= \frac{1}{2^n} \sum_{m=1}^{n} \sum_{k_1=-K}^{K} \cdots \sum_{k_n=-K}^{K} A_{k_1} \cdots A_{k_m} \\ &\quad \times B_{k_{m+1}} \cdots B_{k_n} G_{n,x^m z^{n-m}}(k_1\omega, k_2\omega, \cdots, k_n\omega). \\ &\quad \times e^{j(k_1\omega, k_2\omega, \cdots, k_n\omega)t} \end{aligned} \tag{14-46}$$

The final output, obtained with only the FRFs up to N^{th} order, and only the first K harmonics of the time varying input will contain N x K harmonics:

$$y(t) = \sum_{n=0}^{K \times N} |Y_n| \cos(n\omega t + \arg(Y_n)) \tag{14-47}$$

where Y_n is the Fourier transform of the nth harmonic (of frequency $n\omega$) in the calculated approximate output, which can be calculated using complex algebra.

$$\begin{aligned} Y_n(n\omega) &= \sum_{\substack{k=-K \\ |k|=n}}^{K} (G_{1,x}(k\omega)A_k + G_{1,z}(k\omega)B_k) \\ &\quad + \frac{1}{2} \sum_{k_1=-K}^{K} \sum_{\substack{k_2=-K \\ |k_1+k_2|=n}}^{K} (G_{2,xx}(k_1\omega, k_2\omega)A_{k_2}A_{k_2} \\ &\quad + G_{2,xz}(k_1\omega, k_2\omega)A_{k_1}B_{k_2} \\ &\quad + G_{2,zz}(k_1\omega, k_2\omega)B_{k_1}B_{k_2}) + \cdots \\ &\quad + \frac{1}{2^{N-1}} \sum_{m=0}^{N} \sum_{k_1=-K}^{K} \sum_{k_2=-K}^{K} \cdots \\ &\quad \sum_{k_N=-K}^{K} G_{N,x^m z^{N-m}|k_1+k_2+\ldots k_N|=n}(k_1\omega, k_2\omega, \ldots, k_N\omega) \\ &\quad A_{k_2} \cdots A_{k_m} B_{k_{m+1}} \cdots B_{k_N}. \end{aligned} \tag{14-48}$$

This equation defines the nth harmonic of the output for the simultaneous modulation of both inputs. If, however, a periodic operation with only one modulated input is considered, derivation of the appropriate expression is trivial and based on setting all the unnecessary FRFs to zero, e.g., if x is the only modulated input, all G_z- and G_{xz}-functions in **Eq. (14-48)** should be set to zero. The quality of the approximate solution obtained in this way, i.e., its closeness to the exact solution, increases with both K and N.

14.4.2 Procedure for Approximate Calculation of the Periodic Quasi-Steady State

This procedure consists of the following steps:

1. Postulating a mathematical model for the process considered.
2. Deriving the FRFs up to the Nth order, based on the model.
3. Defining the periodic input or inputs and their approximation by finite sums, taking into account only the first K harmonics.
4. Calculating the approximate output, using **Eqs (14-47)** and **(14-48)**.

In this procedure, the most complex and time consuming is the second step, i.e., derivation of the FRFs. Nevertheless, this step has to be performed only once for each system investigated. Once derived, these functions can be used for any shapes of the input modulation and for any set of model parameters.

Steps 3 and 4, of course, have to be repeated for each particular case or study. Step 3 is easily performed by applying the Fourier transform to the periodic input functions, while the last step, the calculation of the periodic output in its quasi-steady state, reduces to simple algebra in the complex domain.

14.4.2.1 Example

To illustrate the procedure, the cyclic operation of an adsorption column with modulation of the inlet concentration and/or the temperature of the adsorbent bed will be predicted. Since adsorption in fixed beds has to be performed in a periodic manner, it is an inherently periodic process. The example also is of practical importance. The inputs that are periodically modulated—inlet concentration and temperature of the adsorbent bed—correspond to the so-called standing-wave mode of operation (*Pigford et al., 1969*). The periodic behavior of the adsorption column is treated for the simplest case of a single adsorbing component and for a constant flow rate of the mobile phase.

The starting point for deriving the FRFs is the equilibrium-dispersive model of an adsorption column (*Guiochon et al., 1994*). In this model, the only nonlinearity is the adsorption isotherm, a nonlinear function of the concentration and the temperature. A dimensionless form of the model equations was used, for convenience. Details will be found in *Petkovska and Seidel-Morgenstern (2005) and Petkovska and Marković (2006)*.

Based on this model, three series of FRFs, were derived: Functions that correlate the outlet and the inlet concentrations (the G_c - functions), the FRFs that correlate the outlet concentration and the temperature of the adsorbent bed (the $G_\theta -$ functions), and the mixed functions (the $G_{c\theta} -$ functions). The procedure for derivation of the higher order FRFs is standard and similar to that given in Section 14.3.1. Although, in principle, the FRFs can be derived analytically up to any desired order, the complexity of both the solution and the derivation procedure increases considerably with an increase of the order of the FRFs. FRFs up to the third order can be found in *Petkovska and Marković, (2006)*, together with some details of the derivation procedure.

The mathematical expressions obtained, which are cumbersome and space consuming, have been omitted. Instead, just the graphical representation of the FRFs up to the third order and the resulting approximate outlet concentration changes are given.

The FRFs were calculated for a laboratory scale packed column of the following dimensions and characteristics:

- column length $L = 10$ cm
- fluid phase velocity $v = 0.0737$ cm/s
- bed porosity $\varepsilon = 0.602$
- number of theoretical plates $N_{TP} = 1000$

and for equilibrium described by a Langmuir adsorption isotherm:

$$q = \frac{q_0 b C_A}{1 + bC_A}, \quad b = b_0 \exp\left(\frac{k}{T}\right)$$

with the following parameters: $q_0 = 0.0885$ mol/dm^3, $b_0 = 0.00765$ dm^3/mol and $k = 2647$ K, corresponding to a microcrystalline cellulose triacetate adsorbent, a Tröger's base (-) enantiomer as adsorbate and ethanol as the carrier fluid (*Seidel-Morgenstern and Guiochon, 1993*). The FRFs were calculated for the steady-state concentration, $C_{A,s} = 0.005$ mol/dm^3 and temperatures: $T_s = 315$ K.

Three sets of the calculated FRFs up to the third order are presented in **Figures 14-10** (the G_c-functions), **14-11** (the G_θ-functions) and **14-12** (the $G_{c\theta}$-functions). All FRFs are presented in standard Bodé plot form (amplitude versus frequency in log-log, and phase versus frequency in semi-log diagrams). A dimensionless frequency is used in **Figures 14-10** to **14-12**.

All second and third-order FRFs have very small amplitudes both for low and high frequencies, while in the middle frequency range their values become substantial. As a consequence, the importance of system nonlinearity will be different for different frequencies of the inlet modulations.

Oscillatory behavior of the G_θ- and $G_{c\theta}$-function amplitudes, with a number of resonant frequencies and a number of frequencies for which the amplitudes tend to zero, is evident. As a consequence, a small change of the input frequency in the $2 < \omega < 30$ range can lead to large differences in system nonlinearity. This behavior is not unexpected, as these FRFs correspond to periodic change of temperature imposed on the entire adsorbent bed. Such behavior has already been reported for shell and tube heat exchangers with sinusoidal change of the heating fluid temperature (*Douglas, 1972*).

Based on the functions shown in **Figures 14-10** to **14-12** and using **Eqs (14-47)** and **(14-48)**, approximate periodic outlet concentration can be calculated for any form of the input. Some of these results are presented in **Figures 14-13** for sinusoidal inputs and **14-14** for rectangular ones with a 3:2 ratio of the cold and hot portions of a cycle.

In both figures, the top row of responses corresponds to inlet concentration modulation with an amplitude of 25%, the middle row to temperature modulation with an amplitude of 3.2% (~20 K) and the bottom row to simultaneous, out of phase changes of inlet

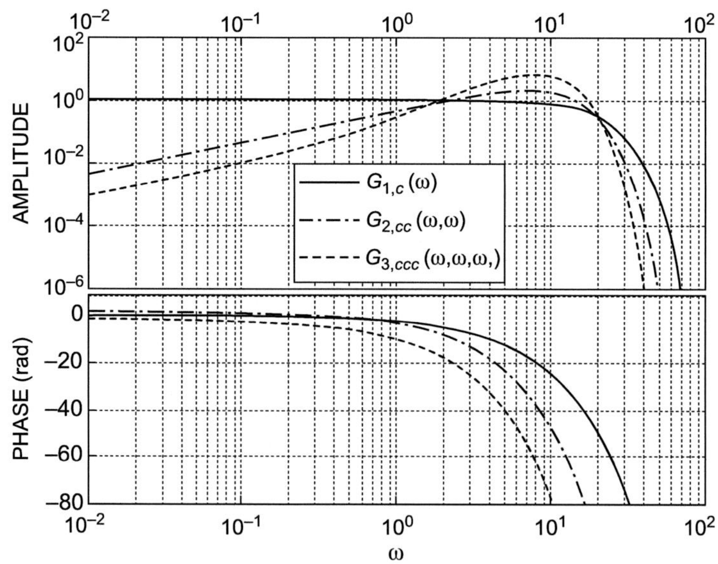

FIGURE 14-10 The simulated first, second and third order G_c-functions of an adsorption column.

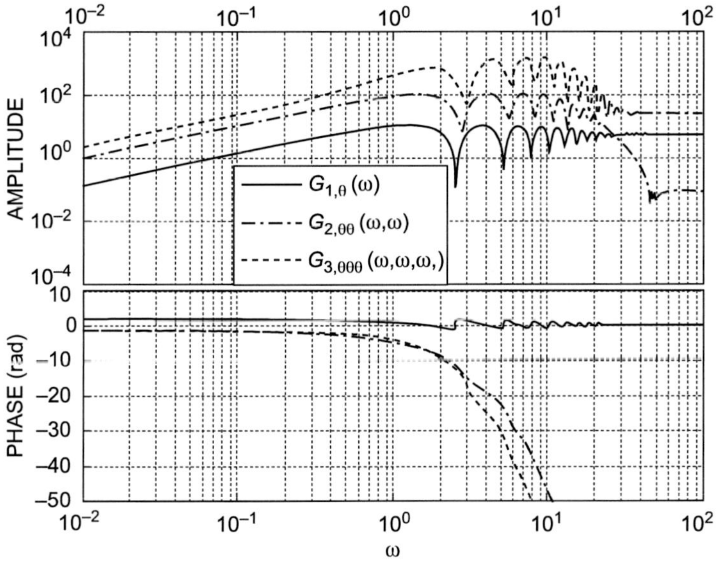

FIGURE 14-11 The calculated first, second and third order FRFs G_q-functions of an adsorption column.

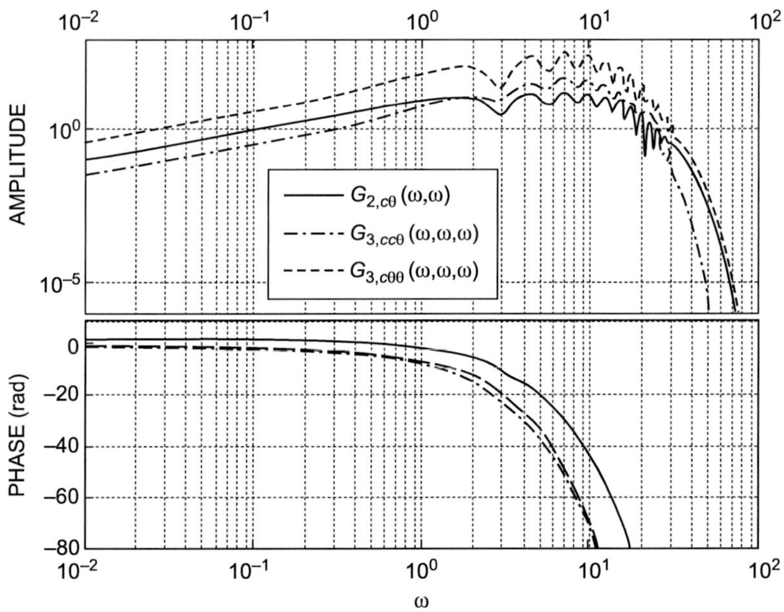

FIGURE 14-12 The calculated second and third order G_{cq}-functions of an adsorption column.

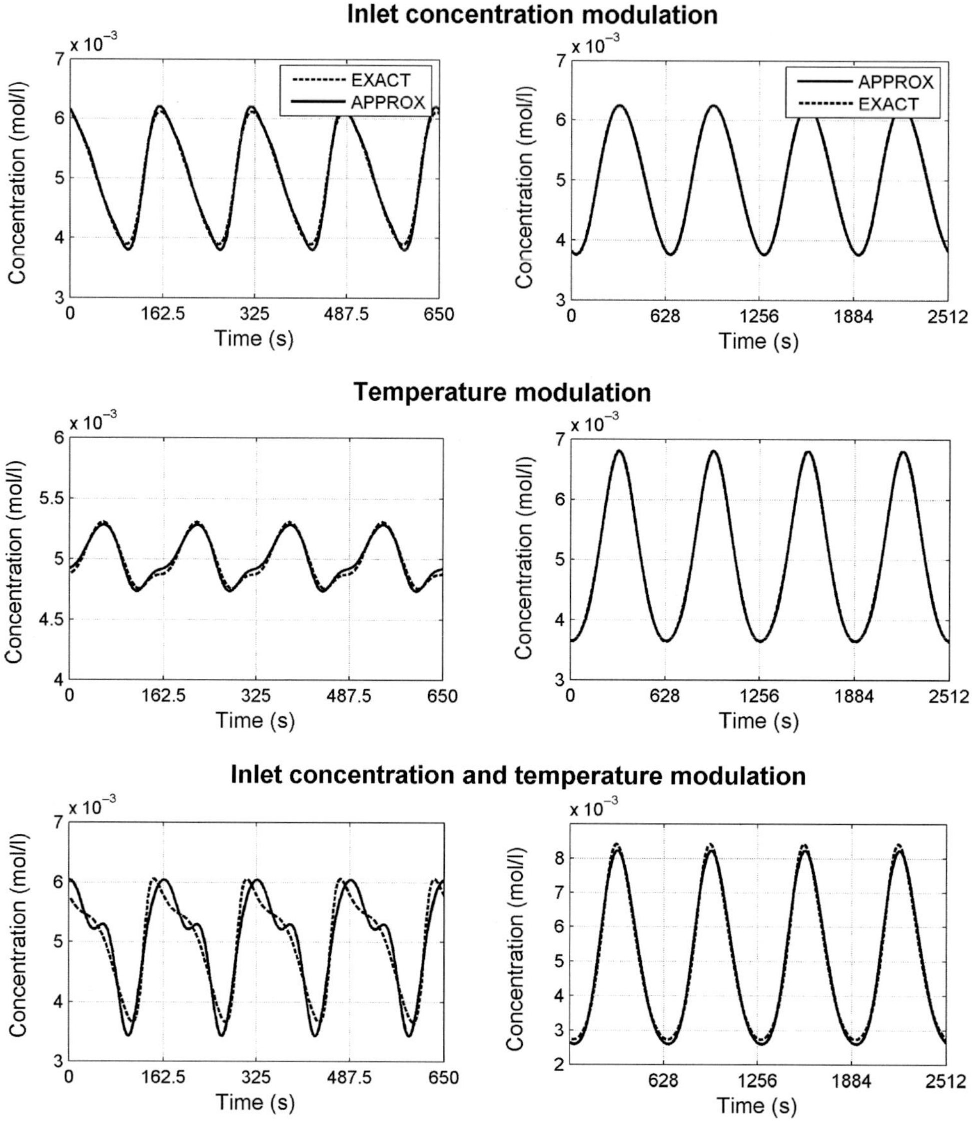

FIGURE 14-13 **Approximate (solid) and exact (dashed) outlet concentrations obtained for sinusoidal input changes: Top: inlet concentration only, Middle: temperature only, Bottom: inlet concentration and temperature; Left: period = 162.5 s, Right: period = 628 s.**

concentration (25%) and column temperature (3.2%). These out-of-phase changes of concentration and temperature are representative of the real operation of adsorption columns, which are cooled during adsorption, and heated during desorption. The graphs on the left correspond to an input modulation period of 162.5 s (dimensionless frequency = 5.25) while those on the right to 628 s (dimensionless frequency = 1.36). For comparison, the exact solution of the model equations is also shown in **Figures 14-13** and **14-14**. The "exact" solution was obtained by numerical integration, using the fast converging algorithm of *Seidel-Morgenstern et al. (1998)*. All approximate solutions presented in these figures were obtained based on the FRFs up to the third order. For the case of rectangular inputs (**Figure 14-14**), only the first four input harmonics were used.

Figures 14-13 and **14-14** show that very good agreement was obtained between the approximate and the "exact" outlet concentrations. In

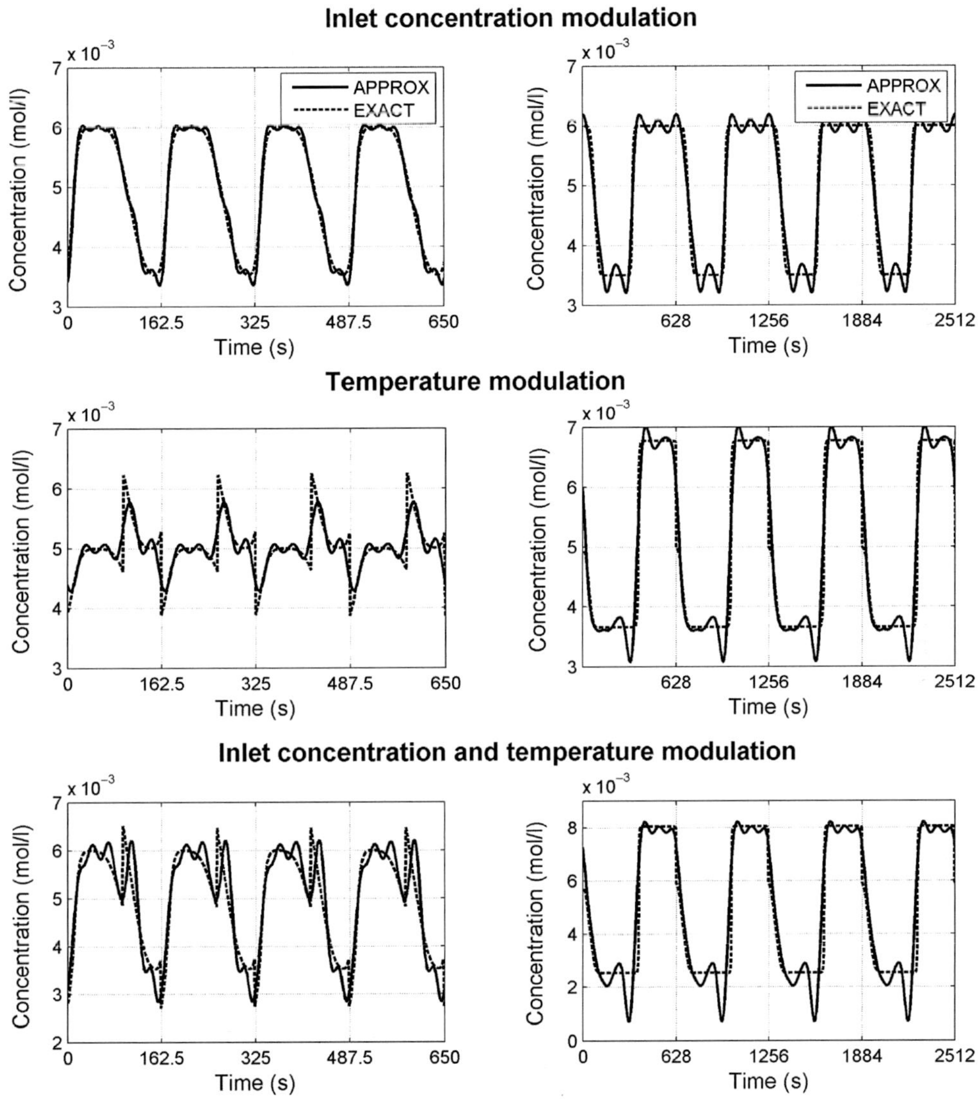

FIGURE 14-14 **The approximate (solid) and exact (dashed) outlet concentrations obtained for rectangular cyclic input changes: Top: inlet concentration only; Middle: temperature only; Bottom: inlet concentration and temperature; Left: period = 162.5 s, Right: period = 628 s.**

some cases, the dashed lines, corresponding to the exact solution, are not even visible. As expected, the agreement is better for the sinusoidal input modulations at longer periods, which correspond to weaker nonlinearity (see **Figures 14-10** to **14-12**), and for modulation of only one input.

A detailed analysis, investigating the influence of the number of FRFs and the number of input harmonics used in the calculations on the quality of the approximate solution, can be found in *Petkovska and Marković (2006)*.

14.5 SUMMARY

Analytical methods of assessing advantages of the periodic operation of chemical processes have been discussed in this chapter and it is suggested that a recently introduced method of frequency response analysis is an efficient way of evaluating the periodic steady-state behavior of nonlinear processes. The method was illustrated for two important generic problems: Rapid direct estimation of the time-averaged performance of periodic processes and the approximate estimation of the cyclic stationary state.

The first situation arises when periodic operation is proposed as a means to enhance the performance of stationary processes. In this case, the analysis can give a fast answer as to whether or not the periodic operation is advantageous compared with the corresponding steady-state one. Not only does the analysis indicate whether periodic operation is attractive, it can also indicate the optimal input modulation frequencies that should be used.

The second situation considered concerns questions arising in the appplication of periodic processes. In this case, nonlinear frequency response analysis provides an approximation of the complete periodic behavior. Results can be used to optimize the periodic process with respect to the shapes, amplitudes and frequencies of the input modulations. In either situation, the analysis replaces tedious numerical calculations that are generally needed to simulate periodic behavior.

Of course, the examples used in this chapter were rather simple to facilitate understanding of the technique. The methodology can be used, in principle, for the analysis of any periodic process.

Acknowledgment

The authors are pleased to acknowledge the assistance of Dr. A. Marković and Ph.D. student D. Nikolić in the preparation of this chapter.

Notation

A	=	amplitude, in general, and for the dimensionless inlet concentration change; reacting species, reactant
$\tilde{a}$	=	first order coefficient of the Taylor series expansion of the dimensionless adsorption isotherm
B	=	amplitude
b	=	temperature dependent Langmuir isotherm coefficient (dm^3/mol)
b_0	=	Langmuir isotherm coefficient (dm^3/mol)
$\tilde{b}$	=	second order coefficient of the Taylor series expansion of the dimensionless adsorption isotherm
C	=	dimensionless concentration in the fluid phase
C_Am	=	mean concentration of species A
$C_{A,s}$	=	steady state concentration of species A
$C_{A_0,s}$, $C_{A_i,s}$	=	inlet concentration of species A at steady state; normalizing concentration, normalizing concentration
C_i	=	dimensionless inlet concentration
D_j	=	constants of integration in **Eq. (14-28)**
D_{ax}	=	axial dispersion coefficient (cm^2/s)
d	=	constants of integration in **Eq. (14-28)**
F	=	flow-rate (cm^3/s)
f	=	auxiliary parameter in **Eq. (14-31)**
f_j	=	j^{th} auxiliary function in **Eq. (14-28)**
G_n	=	n^{th} order frequency response function
$G(\omega)$	=	frequency transfer function
$G_1(\omega)$	=	1st order, asymmetric FRF
$G_2(\omega, -\omega)$	=	asymmetric 2nd order FRF
G_c	=	1st order FRF correlating outlet and inlet concentrations
G_θ	=	1st order FRF correlating bed temperature and outlet concentration

$G_{c\theta}$ = mixed FRF correlating inlet concentration and bed temperature with outlet concentration
g_n = n^{th} order impulse response function (Volterra kernel)
g = impulse response function
K = number of input harmonics
k = reaction rate constant ($s^{-1}mol^{1-n}$); temperature coefficient of the Langmuir isotherm (K); index in **Eq. (14-41)**, harmonic index in **Eq. (14-42)**
L = reactor or column length (cm)
N = number of frequency response functions or harmonics
N_{TP} = number of theoretical plates
n = reaction order, equation order, order in Volterra series
n_j = values of reaction order for a sign change to occur (**Eq. (14-36)**)
Q = dimensionless concentration in the solid phase
q = concentration in the solid phase (mol/dm^3)
q_0 = concentration in the solid phase at maximal coverage (Langmuir isotherm) (mol/dm^3)
T = temperature (K)
T_s = steady state temperature
t = time (s)
u = frequency, general
V = volume (cm^3)
v = fluid velocity (cm/s)
x = input (general), dimensionless axial coordinate
$x(t)$ = input function (**Eq. (14-41)**)
Y_n = Fourier transform of the nth harmonic (**Eq. (14-48)**)
y = output (general), response
y_{DC} = DC component of response or output
y_n = n^{th} element of a Volterra series
y_I, y_{II}, y_{III} = response or output harmonics
z = input (general); axial coordinate (cm)
$z(t)$ = input function

Greek

$\alpha, \alpha_1, \alpha_{1,2}$ = characteristic values (in Table 14-4, **Eqs (14-28), (14-28a)**)
β = auxiliary parameter, **Eq. (14-35)**
Δ = difference between the mean and the steady-state concentration
ε = porosity
Φ = dimensionless isotherm relation
φ = phase shift
ϕ = adsorption isotherm relation
θ = dimensionless temperature in **Eq. (14-25)**
τ = time, dimensionless time, variable in convolution integral
τ_c = contact time (s)
$\tau_{c,s}$ = solids contact time (s)
ω = frequency, dimensionless frequency
ω^* = frequency in rad/s

Subscripts

A = component A or adsorbate
A_i, A_0 = component at reactor inlet
DC = DC component
ads = adsorbed
app = approximate
ax = axial
c = with respect to concentration, contact as in contact time
$c,$ = with respect steady-state concentration
f = fluid
i = inlet
k = harmonic index
n = element number
s = steady-state; solid
x = with regard to input x
z = with regard to input z
0 = inlet
θ = with regard to temperature
I, II, III = harmonic index for output

Superscripts

m = mean

Abbreviations

CSTR = continuous stirred tank reactor
DC = non oscillating portion of input or output
DFTR = dispersive flow tubular reactor
FRF = frequency response function
HFRF = higher order frequency response function
NLFR = nonlinear frequency response
ODE = ordinary differential equation
PFR = plug flow tubular reactor

CHAPTER

15

Pressure Modulation

Peter Lewis Silveston, Robert Ross Hudgins
Waterloo, Ontario, Canada

OUTLINE

15.1 Introduction 415

15.2 Acceleration of Mass Transfer 416
15.2.1 Gaseous Reactants 416
15.2.2 Gas-Liquid Systems 424

15.3 Sonocatalysis 427

15.4 Periodic Pressure Reduction 430

15.5 Combined Compression and Reaction 431

15.6 Application to Rate and Equilibrium Measurements 432

15.7 Assessment and Research Opportunities 433

15.1 INTRODUCTION

The use of total pressure modulation to improve reactor performance has attracted attention for about three decades in several different fields of Engineering. Ultrasonic excitation of both mass transfer and reaction systems has been discussed in the literature extensively. Because ultrasonic pressure waves are low-amplitude, high-frequency, total-pressure oscillations, this review considers such excitation just peripherally. Developments in ultrasonic excitation are covered in two annual publications: *Ultrasonics* and *Ultrasonics Sonochemistry*.

Attention in this chapter is focused on modulation at higher amplitudes and much lower frequencies. Much of the research to date has been directed at enhancing mass transfer in porous catalyst particles. High-amplitude forcing has been discussed in the preceding chapters. Temperature modulation is the subject of the following chapter.

Rates of heterogeneously catalyzed reactions can be affected by pressure modulation in at least two ways. The first way is that with porous catalysts and gas or gas-liquid systems, pressure modulation can significantly increase the rate of transport within the structure or at the particle surface. This is perhaps the most widespread application. The second way is that reactions involving adjacent active sites on the catalyst surface or inhibited by a strongly adsorbing reactant or product may be sensitive to pressure changes.

Pressure modulation of heterogeneous catalytic reactions has been discussed in the scientific and engineering literature since the 1970s, but this application of periodic operation has only begun to attract substantial attention in the last decade (see **Table 15-1**). *Silveston and Hudgins (2004a)* have reviewed the published research on pressure modulation; this chapter is an expansion of that review.

15.2 ACCELERATION OF MASS TRANSFER

Porous catalyst particles are often designed with a bidispersed pore structure consisting of large pores that provide transport into the catalyst interior and a network of fine pores that contain most of the solid surface of the catalyst. With high intrinsic rates of reaction on the surface, transport of reactants through the porous solid may sometimes control reaction rate. Diffusion of products may influence selectivity when multiple reactions occur. With small, highly porous catalyst particles and/or low external flow velocities, diffusion through an irregular, liquid hydrodynamic boundary layer surrounding the particle may also retard the overall rate of reaction. If the reaction forms a gas, bubbles of gas can be retained within the structure. This condition is frequently encountered in anaerobic fermentation in which methane and carbon dioxide form. In addition to diffusional transport, non-equimolar reactions induce bulk or convective flow. In rare cases, the pressure drop across a particle can cause a small convective flow contribution. Under steady-state operation, convective flow within a porous catalyst is usually not important.

The effect of a pressure change on transport depends on the fluid phases present in the porous catalyst. With compressible fluids, such as gases, compression or rarefaction will cause a density change and convective flow in or out of the particle. The amount of flow or the region of the particle affected depends on the magnitude of the pressure change. With incompressible fluids, like most liquids, convective flow on a pressure change will be negligible. An exception to this occurs when bubbles are present inside the solid or are formed or destroyed through the pressure change. These bubbles expand or contract during the pressure change, inducing convective flow of the liquid phase. Evaporation and condensation can also occur. *Pismen (1976)* points out that gases collect in the larger pores as a result of capillary forces and can form bubbles. Bubble movement caused by reaction or external pressure change can induce liquid flow in the fine liquid-filled pores in the opposite direction. Such flows may dominate mass transfer.

15.2.1 Gaseous Reactants

Investigations of total pressure modulation to enhance mass transfer in porous solids began in the 1970s. *Hamer and Cormack (1978)* appear to be the first who applied modulation to catalytic reactions. They considered a reaction A → n B in which n = 1 would indicate isomerization and n ≥ 2 cracking. The porous solid was represented by a dusty gas model with a slab geometry of depth 2L. Pore dimensions exceeding the mean free path length of the diffusing reactant were assumed so that only bulk diffusion occurs. For n = 1, assuming a negligible pressure gradient within the porous solid, the dusty gas model becomes:

$$\varepsilon_p P \frac{\partial y_A}{\partial t} = P D_{AB} \frac{\partial^2 y_A}{\partial z^2} - P v \frac{\partial y_A}{\partial z} + \nu_A r. \quad (15\text{-}1)$$

TABLE 15-1 Pressure Modulation Literature

Authors	Objective	Reaction	Modulated Variable	Reaction Conditions	Observations	Comments
Hamer and Cormack (1978)	Investigation of modulation on rate and selectivity in a 1st order consecutive reaction in a porous catalyst pellet	$A \rightarrow B \rightarrow C$ to represent cracking of light gas oil in a cracking catalyst	Total pressure	1st order reaction controlled by diffusion of reactant within the particle; 4/10 mesh molecular sieve catalyst; $P = 100$ kPa; $A = 50$ kPa; $f = 1$ Hz.	For the authors' choice of "typical" conditions under diffusion control, the effectiveness factor for reaction was almost doubled by pressure modulation. Selectivity to intermediate B was enhanced.	The model has no diffusional contribution to mass transport under modulation, nor does it allow for the disruption of the hydrodynamic boundary layer on the particle surface from rapid changes in velocity.
Sohn and Chaubal (1984)	Simulation of modulation on rate for a 1st order reaction in a porous catalyst pellet	Reduction of a porous iron oxide pellet with hydrogen	Total pressure	Authors assumed a low-permeability solid and a dusty gas model for intra particle diffusion.	Simulation showed rate enhancement through pressure modulation.	Knudsen diffusion would likely be present to some extent in such a system, but if predominant as the diffusion limiting step, it would likely curtail enhancement.
Sohn and Aboukheshem (1992)	Experimentally investigated and modeled the effect of modulation on reaction rate using a dusty gas model	Hydrogen reduction of fine nickel oxide powder	Total pressure	$T = 653$ K; time-average H_2 pressure $= 86.1$ kPa; $d_p = 1$ cm. Permeability was sufficiently high that bulk diffusion was present in catalyst pores.	As conversion increased from exterior to interior, modulation enhanced the conversion rate.	Experimental results were consistent with model predictions. Enhancement increased with the Thiele modulus at any cycle frequency.

(Continued)

TABLE 15-1 Pressure Modulation Literature (*cont'd*)

Authors	Objective	Reaction	Modulated Variable	Reaction Conditions	Observations	Comments
Baiker and Richarz (1976)	Measure the effect of total pressure modulation on the rate of a catalytic gas phase reaction.	Hydrogenation of ethene over a commercial supported Ni catalyst	Compressor and a throttling valve used to provide pressure modulation.	Cycle periods from 0.3 to 8 s	Modulation considerably reduced ethene conversion compared with steady-state conversion at the time-average pressure. Increasing amplitude decreased conversion.	Results were explained by the interference of mass transfer at the external surface of the particle and through secondary effects, all of which were neglected in their model.
Leucker et al. (1978)	Investigation of mass transfer under modulation from a porous sphere	Mass transfer from a porous sphere filled with naphthalene into a flowing air stream	Modulation was produced by alternately evacuating and pressurizing a vessel containing a porous sphere. This also produced variations in the gas velocity.	Variables were the period and the cycle split, the time-average flow rate of air and the depth of the straight, regular cylindrical pores in the solid. Amplitudes were from 15 to 45 kPa.	Mass transfer rates increased from 17 to 35% for periods from 1.5 to 4 s. Largest increases arose for a cycle split $s = 0.5$. Tripling modulation amplitude increased the mass transfer rate. Main effect of modulation appeared to be on mass transfer through the particle boundary layer.	Higher transport rates likely result from boundary layer thinning by flow-direction switching in the pores of the solid caused by external pressure fluctuations.
Van den Heuvel et al. (1996, 1997a)	Investigation of high methane production rates observed in fermentation within a loop reactor	Fermenting biomass containing granules of a methanogenic organism.	Pressure was varied in a triangular waveform.	Amplitude of the variation was about 100 kPa and the frequency was 60 Hz.	About a 22 to 26% increase occurred in methane generation.	Results were interpreted in terms of granule "breathing." Activity of the organism remained high after modulation was halted, suggesting bubble movement alters the granule structure.

Van den Heuvel et al. (1996)	Investigations to explore the particle "breathing" concept	Catalytic decomposition of H_2O_2 over Pt.	Reactor vessel was subjected to time-varying O_2 pressure.	Pt deposited on 0.5 mm i.d. × 4 mm sintered glass cylinders, mounted on a rotating shaft in a reactor vessel subjected to time-varying O_2 pressure. Average pore size for the sintered glass was 20 μm; particle porosity was 0.3. Thiele modulus = 7; $T = 35°C$.	Large enhancement of the decomposition rate was measured for the system.	Predicted enhancement was observed experimentally.
Wang et al. (2003)	Measurement of adsorption equilibria and kinetics using frequency response methods	Adsorption of O_2 and N_2 on a carbon molecular sieve.	Gas pressure was varied sinusoidally.	Adsorber contained 6.6 g of carbon, $T = 298$ K, $P_{mean} = 0.067, 27$ and 1 bar; $5 \times 10^{-5} < \omega < 0.2$ Hz, $A = 0.5\%$ of P_{mean}, $Q = 1\ cm^3/min$	Adsorption was described by a Langmuir isotherm. Mass transfer coefficient for O_2 was about 30 × the coefficient for N_2. Difference was attributed to different rate controlling processes.	The authors offer a simple method for rate and equilibrium measurements in porous solids.

In this relation, z is the dimension in the flow direction, y is a mole fraction, v the superficial velocity and P is the total pressure. Pressure is a time-dependent function:

$$P = P_0(1 + A\sin\omega t). \qquad (15\text{-}2)$$

The continuity equation for compressible flow is:

$$v = -\frac{\varepsilon_p z}{P}\frac{dP}{dt}. \qquad (15\text{-}3)$$

When these equations are rendered dimensionless, the following dimensionless groups appear and become parameters for the effect of pressure modulation on reaction rate: φ (Thiele Modulus), $K = \frac{k}{\varepsilon_p \omega}$ (dimensionless reaction rate constant) and λ (a dimensionless frequency). Only two of these three parameters are independent. From the solution to the material balance and continuity equation, *Hamer and Cormack* calculated an effectiveness factor for a first-order isomerization reaction. The behavior of the effectiveness factor as a function of the modulation parameters may be seen in **Figure 15-1**.

The three parts of **Figure 15-1** show the effect of normalized amplitude, R, on the effectiveness factor. The upper figure represents a small amplitude, while the bottom figure shows the effect of a large amplitude. The maximum amplitude for an atmospheric operation would be 1 atm. The abscissa in the figures is the dimensionless reaction rate constant, K; however, from the definition above, K varies with 1/ω, namely the effect of modulation frequency. The curves in the figure show that increasing the Thiele modulus retards the reaction because of greater mass transfer resistance. These curves approach a limit of η_{osc}. This limit comes from a solution to the model that allows just for convective flow caused by pressure pulsations. The RHS of the figure shows the steady-state effectiveness factor for the choice of model parameters. It is evident that, at an amplitude of 0.1 atm, R=0.1, pressure modulation produces only a modest increase in the rate of reaction, ca. 30 to 40% at high frequencies when φ = 20. As the modulation amplitude increases, rate increases become very large. It can also be seen that the curves of constant φ

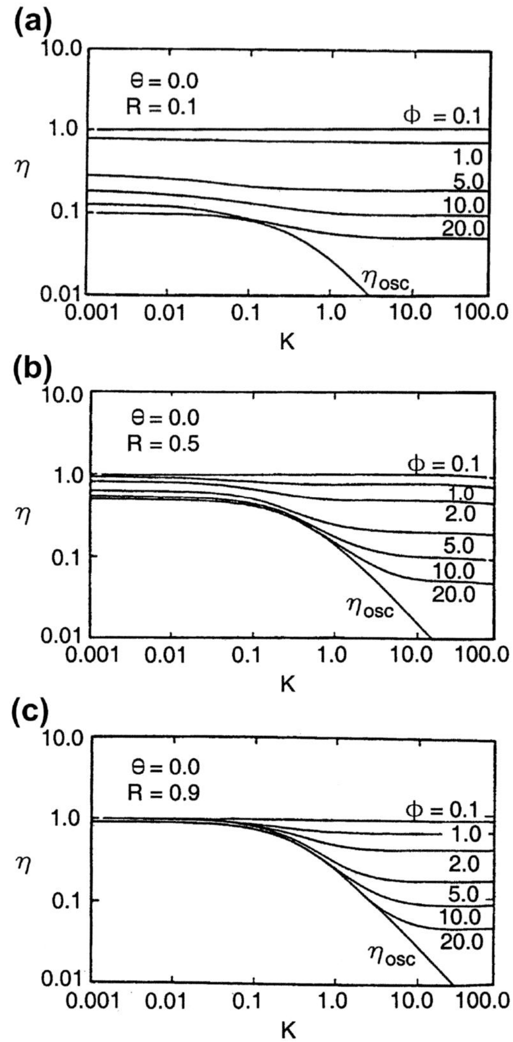

FIGURE 15-1 The effect of sinusoidal pressure modulation amplitude and reciprocal frequency on the effectiveness factor for a first-order gas phase reaction occurring in a porous slab under bulk diffusion conditions. R is the dimensionless amplitude and θ is the fractional volume change on reaction. *(Figure adapted from* Hamer and Cormack (1978) *with permission of the publisher. © 1978 by Elsevier Science Ltd.)*

are flat at high and low modulation frequencies. Thus, at these extremes, frequency has little effect on mass transfer. Of course, in the case of large amplitudes ($R = 0.9$), the flatness arises from effectiveness factors that approach unity. Nevertheless, the simulation results suggest that there is a critical K or, for any given system, a critical frequency for pressure modulation to affect mass transport. **Figure 15-1** suggests this critical frequency is set by the condition of $\eta_{osc} = \eta_{steady\ state}$. When rate is strongly limited by mass transfer, this condition becomes: $\eta_{osc} = 1/\varphi$. *Hamer and Cormack (1978)* obtained an analytical solution for η_{osc}:

$$\eta_{osc} = \frac{A}{2\pi K}\left[2 - \left(\frac{1 + e^{-2\pi K}}{4K^2 + 1}\right)\right]. \qquad (15\text{-}4)$$

A change in volume on reaction reduces the improvement possible through pressure modulation if there is a volume increase, but there is nevertheless a significant increase of rate for a dimensionless amplitude > 0.5 within a suitable range of frequency. There is also a critical K or $1/\omega$ for improvement through modulation. *Hamer and Cormack* show that

$$K_{crit} = \frac{\varphi A}{\sqrt{2}(1+\theta)\pi}\left[\frac{1}{\theta} - \frac{1}{\theta^2}\ln(1+\theta)\right]^{-1/2} \qquad (15\text{-}5)$$

where $\theta = (n - 1)(y_A)_0$. They also show that selectivity to the intermediate product B in the consecutive reaction $A \rightarrow B \rightarrow C$ is improved also by pressure modulation through increasing the rate of mass transfer of B out of the catalyst particle.

According to *Hamer and Cormack*, modulation amplitude affects the fraction of the particle volume where reaction occurs, while frequency mainly influences the steepness of the concentration gradient in the particle and thus the mean reactant concentration in the particle. The gradient increases with more frequent pumping of gas into and out of the particle.

The frequencies that are required for large changes in η can be estimated from a cracking catalyst example given by *Hamer and Cormack*. Assuming a 4/10 mesh molecular sieve catalyst operating on a light gas oil, the authors estimate a Thiele modulus of 10.6 and a steady-state effectiveness factor $= 0.094$. With $\varepsilon_p = 0.28$ and $\theta = 2.4$, modulating the total pressure with an amplitude of about 50 kPa and a frequency of 1 Hz raises the oscillating effectiveness factor, η_{osc}, to 0.17, indicating an 80% increase in the rate of cracking. Other factors could raise the effectiveness factor even further. The rapid oscillatory gas movement will cause mixing and raise the effective diffusivity. The *Hamer-Cormack* analysis does not allow for a diffusional contribution to mass transport under pressure modulation, nor does the analysis allow for the destruction of the hydrodynamic boundary layer on the outer surface of the particle by the rapidly changing velocities at this surface. Loss of the boundary layer could also raise the mass transport rate if that resistance is significant in the system.

Modulating pressure by opening and closing a conventional throttling valve to achieve a 100 kPa change at 1 Hz is probably not practical for an industrial-scale reactor. Yet, it is difficult to conceive of modulation on a large scale without using throttling valves. Application may have to wait on development of new valve designs.

Sohn and Chaubal (1984) revisited mass transport in a porous solid particle bathed in a flowing gas system. They considered the reduction of a porous iron oxide pellet by flowing hydrogen gas. The permeability of the solid they considered was rather low so Knudsen diffusion was probably encountered. Their simulation using the dusty gas model applied to a spherical particle showed rate enhancement through pressure modulation. Such an increase would not occur if Knudsen diffusion was predominant.

Reduction of a granular particle composed of a fine nickel oxide powder was examined by *Sohn and Aboukheshem (1992)* experimentally and through modeling. Permeability of the pellets used was sufficiently high that bulk diffusion was present in the transport pores. A spherical particle was assumed and the dusty gas model was used. Reduction converts the oxide to a metal so some changes in pore size and porosity would be expected, but such changes were ignored. **Eqs (15-1)** and **(15-3)** state the model except that the spherical geometry changes the forms of the terms. Normalization of the model variables generates the dimensionless groups mentioned above, modified by introducing the particle radius in place of the slab half-depth and the solid density. Because of the particular kinetics, another group arises,

$$\varsigma = \frac{\varepsilon_p y_{A_0} P_0}{RT(1 - \varepsilon_p)\rho_p}. \quad (15\text{-}6)$$

The effect of pressure modulation on the extent of oxide reduction was similar to that *Hamer and Cormack*. Conversion enhancement through pressure modulation at any modulation frequency, ω, increases with the Thiele modulus for the particle. Enhancement at any ω and φ increases also with the modulation amplitude. *Sohn and Aboukheshem*, however, did not observe critical values of K and frequency. They did note that because the particle conversion increases from the exterior to the interior, enhancement of the conversion rate through modulation increased. This result is consistent with the effect of φ on enhancement. As conversion of the oxide increases, the reduction front moves towards the center of the particle and the diffusional resistance to transport grows.

An attempt to measure the effect of total pressure modulation on the rate of a catalytic gas phase reaction was undertaken by *Baiker and Richarz (1976)* using a compressor and a throttling valve to provide pressure modulation. These authors considered the hydrogenation of ethene over a commercial supported nickel catalyst. The experimental results are given in an abstract and an unpublished manuscript *(Baiker and Richarz, 1976)*. Modeling of the system was mentioned, but the model was not given. Because of the operation mode, the throughput or contact time of ethene in the packed bed of catalyst varied along with the total pressure. For this situation, *Baiker and Richarz* found that modulation considerably reduced ethene conversion compared with conversion under steady state at the time-average pressure for cycle periods of from 0.3 to 8 s (or cycle frequencies of 0.12 to 3 Hz). Surprisingly, increasing amplitude decreased conversion. However, conversion under modulation closely approached steady-state conversion as the time-average gas flow rate increased. These results contradicted the prediction of *Hamer and Cormack* as well as those of *Sohn and Aboukheshem*. Those models predicted a conversion enhancement through modulation. *Baiker and Richarz* explain the lack of agreement by the interference of mass transfer at the external surface of the particle and through secondary effects, all of which were neglected in their model. Their explanation is supported by the effect of gas flow rate on the approach of conversion under modulation to conversion at steady state. In addition, pore size distribution data in the authors' unpublished manuscript indicate that for about 35% of the pore volume in the catalyst, Knudsen diffusion would be expected, whereas for another 35% diffusion would be in the transition region between Knudsen and bulk diffusion. Pressure modulation should not affect internal mass transport when Knudsen diffusion predominates.

Leucker et al. (1978) investigated mass transfer from a porous sphere filled with naphthalene into a flowing air stream under pressure modulation. Alternately evacuating and recharging the system with compressed air caused a trapezoidal variation of pressure in a cycle and therefore variations in the gas velocity past the sphere. This action increased the rate of mass transfer from about 17 to 35% with cycle periods from 1.5 to 4 s. Increasing the modulation amplitude from about 15 to 45 kPa increased the mass transfer rate. The primary effect of pressure modulation appears to be on mass transfer through the boundary layer encircling the particle and can be explained through distortion or thinning of the external boundary layer by flow reversal in the pores of the solid as a result of external pressure fluctuations.

The experimental portion of the *Sohn and Aboukheshem* study of nickel oxide reduction was carried out in the reactor shown in **Figure 15-2**. The catalyst pellet was formed from nickel oxide powder and a binder by agglomeration and shaping.

After this the pellet was dried and calcined. Reduction of the oxide was followed by weighing at the end of an experiment. Variables in the study were modulation frequency, amplitude and reactor temperature.

Both frequency and amplitude had a significant influence on the rate of oxide reduction as **Figure 15-3** illustrates. Increasing the frequency increased the fractional conversion of the oxide to the metal at any time point in the reduction process (**Figure 15-3(a)**). For the frequency experiment, the modulation amplitude was 67.9 kPa. **Figure 15-3(b)** shows that increasing the amplitude of the modulation also increased the fractional conversion. These experiments

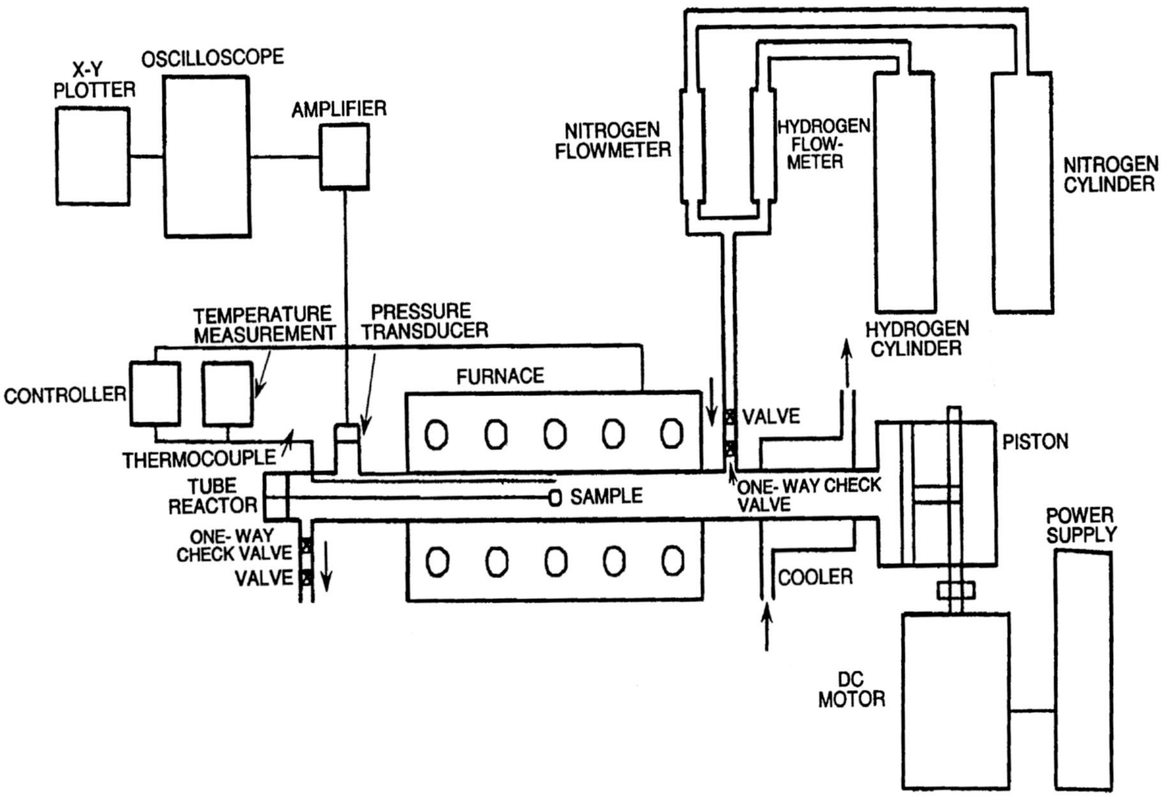

FIGURE 15-2 Experimental system used in the study of the influence of pressure modulation on the rate of reduction of a porous sphere of nickel oxide in flowing hydrogen. *(Figure reproduced from* Sohn and Aboukheshem (1992) *with permission of the copyright holder. © 1992 by Metallurgical Transactions.)*

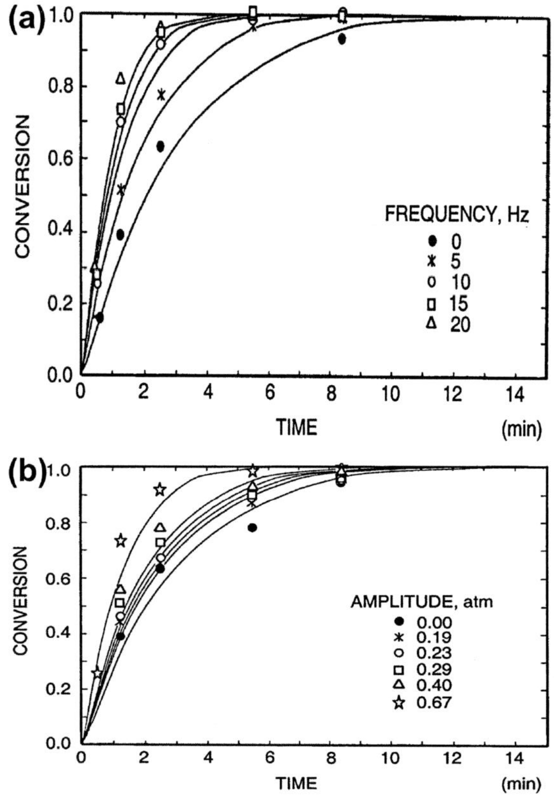

FIGURE 15-3 **Fractional conversion of a 1 cm porous nickel oxide sphere as a function of time on stream under pressure modulation at 653 K, 86.1 kPa: (a) The effect of modulation frequency at an amplitude of 67.9 kPa, (b) the effect of modulation amplitude at a frequency of 10 Hz.**

were done at a frequency of 10 Hz. The solid lines in both figures are model predictions at the respective frequencies and amplitudes used experimentally. Parameters for the authors' model (discussed above) were obtained from steady-state experiments. With further experiments done at constant pressure, the agreement between model and data in **Figure 15-3** is just satisfactory.

The *Sohn and Aboukheshem* study demonstrated that the dusty gas model of a porous solid used by them and by *Hamer and Cormack* satisfactorily represents the influence of pressure modulation on solid catalyzed gas phase reactions and gas-solid reactions. Experimentally, *Sohn and Aboukheshem* used an acoustic generator in their study of nickel oxide reduction. Acoustic generators produce pressure waves of low amplitude but high frequency. Modeling studies mentioned above predicted that the largest increases in rates of internal mass transport are found at the highest cycling frequencies. Thus, *Sohn and Aboukheshem* examined pressure modulation at a frequency of 100 Hz and an amplitude of 6.9 kPa for the 1 cm diameter, porous nickel oxide sphere bathed in flowing hydrogen. Their simulation showed that with an acoustic generator the time needed to achieve a 90% reduction of the oxide was just one-third of that needed without pressure modulation. Acoustic generators are readily available and can be attached to most types of flow reactors, so this means of overcoming a transport limitation in solid catalyzed, gas-phase reactions is promising.

15.2.2 Gas-Liquid Systems

Remarkably high methane production rates observed in fermentation experiments undertaken in a loop reactor have been attributed by *Van den Heuvel et al. (1996)* to "breathing" by granules of a methanogenic organism as they circulate from top to bottom in a loop reactor. These fibrous granules are relatively rigid and porous. Free space within the granules will contain small bubbles of methane and CO_2 produced by the organisms feeding on the substrate that diffuses into the granules. As suggested in **Figure 15-4**, the gas bubbles are compressed under the higher hydrostatic pressure at the bottom of the loop reactor. Thus, the aqueous phase containing substrate is drawn into the granule. When the granule reaches the top of the loop where the hydrostatic pressure is lower, the bubbles

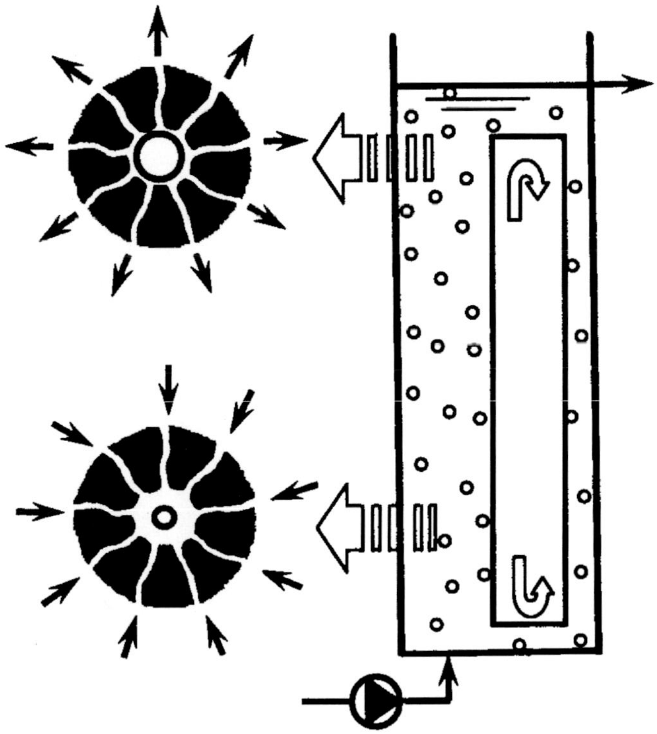

FIGURE 15-4 **Induction of particle breathing through particle circulation in a loop reactor. Liquid inflow and outflow at lower and upper points within the reactor.** *(Figure reproduced from* Van den Heuvel et al. (1997a) *with permission of the copyright holder. © 1997 by Elsevier Science Ltd.)*

expand, driving the aqueous phase out of the granules.

To verify the breathing mechanism with bioflocs, *Van den Heuvel et al. (1997a)* undertook a series of batch experiments in which methanogenic organisms suspended as a biofloc in a stirred reactor were subjected to time-varying pressure. These experiments demonstrated about a 22 to 26% increase in methane generation once the biomass was acclimatized to the substrate. Vertical movement of the suspended biofloc was evident as the external pressure changed. Floc particles rose as the pressure was reduced and water expelled. They settled as the pressure increased and water flowed inward. Activity of the organism remained high after pressure modulation was discontinued, suggesting bubble movement alters the floc structure. Some disintegregation of the floc was observed. *Van den Heuvel et al. (1997b)* explored flow variations within the bioflocs caused by the pressure fluctuations.

Van den Heuvel et al. (1996) undertook further experiments designed to demonstrate the particle "breathing" concept using the catalytic decomposition of H_2O_2 over Pt deposited on sintered glass. The experimental system appears in **Figure 15-5**. There was no interference from mass transfer external to the particle. Reactor temperature was maintained at 35°C.

The researchers treated the pumping action due to pressure variation as unsteady flow between "mixers" of constant volume representing radial points within the particle.

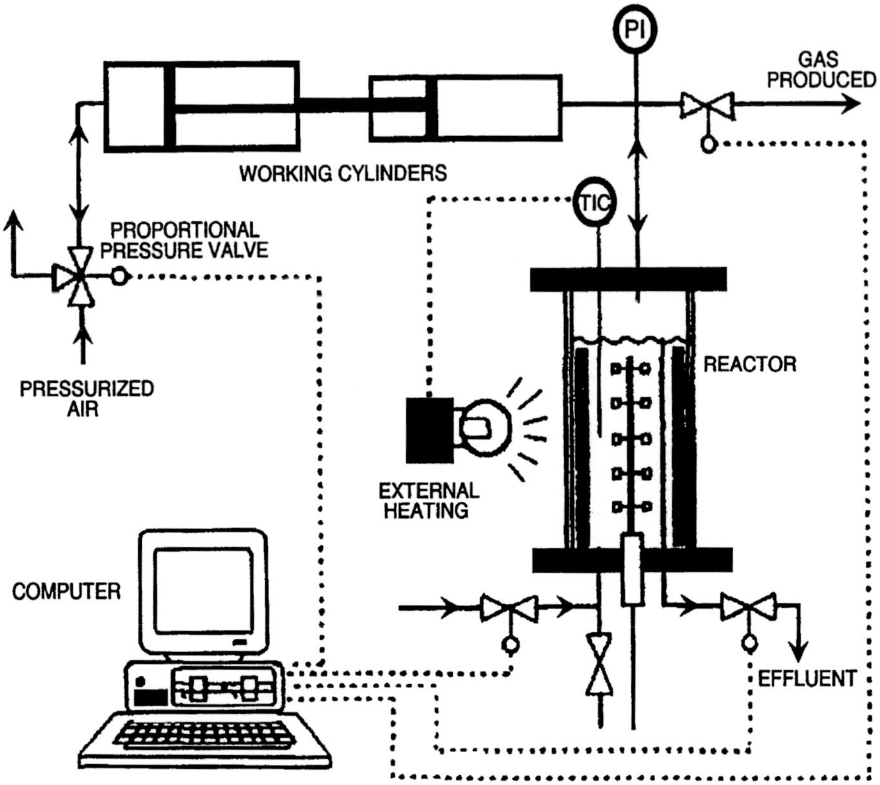

FIGURE 15-5 Experimental system for testing the effect of external pressure modulation on the rate of H_2O_2 decomposition over a supported Pt catalyst with a large mean pore size and narrow pore size distribution. *(Figure reproduced from* Van den Heuvel et al. (1996) *with permission of the copyright holder. © 1996 by Elsevier Science Ltd.)*

To simplify the model, the gas phase was assumed to reside at the particle center. Intraparticle velocity of the exchange flow was assumed to be constant so that the volumetric flow decreased in the inward, radial direction. The transient "mixer" model is described by:

$$V_n \frac{dC_n}{dt} = q_n(C_{n-1,t} - C_{n,t}) - q_{n+1,t} \times (C_{n,t} - C_{n+1,t}) - V_n k C_{n,t}. \tag{15-7}$$

In this model, V_n is the volume of mixer corresponding to the nth shell in the particle and q_n is the volumetric flow from the nth to the $(n\text{-}1)^{th}$ mixer or shell. The decomposition was assumed to be first order in H_2O_2. *Van den Heuvel et al.* assumed the cylindrical catalyst particle could be modeled with eight mixers or shells. Presumably the pumping flow, q_n, was a function of modulation frequency and the gas holdup in the particle. Published data were used to obtain model parameters.

Figure 15-6 shows the predicted enhancement of the decomposition rate by pressure modulation for $\varphi = 4$. At a frequency of 1 Hz, and a gas hold up, $\varepsilon_g = 0.05$, the decomposition rate increases about 3.5-fold. The predicted effect, a large enhancement of the decomposition rate, was observed experimentally. Measurements of rates of H_2O_2 decomposition in the

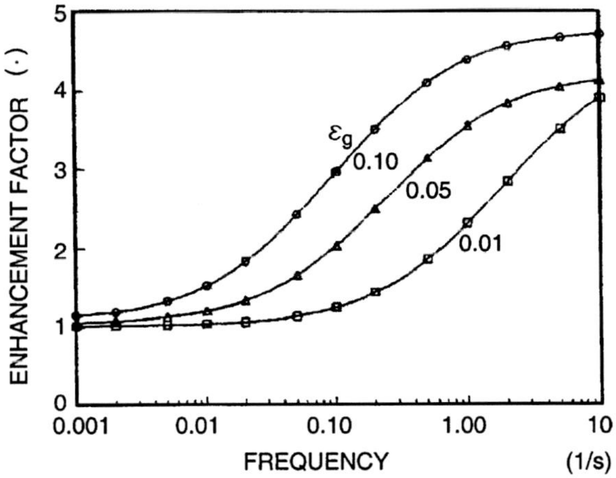

FIGURE 15-6 **Enhancement of a first-order decomposition reaction generating a gas product through external pressure modulation at φ = 4.** *(Figure reproduced from* Van den Heuvel et al. (1996) *with permission of the copyright holder. © 1996 by Elsevier Science Ltd.)*

system described by **Figure 15-5** are shown in **Figure 15-7**. Their variations with modulation frequency are compared with the prediction by the model discussed above in the same figure

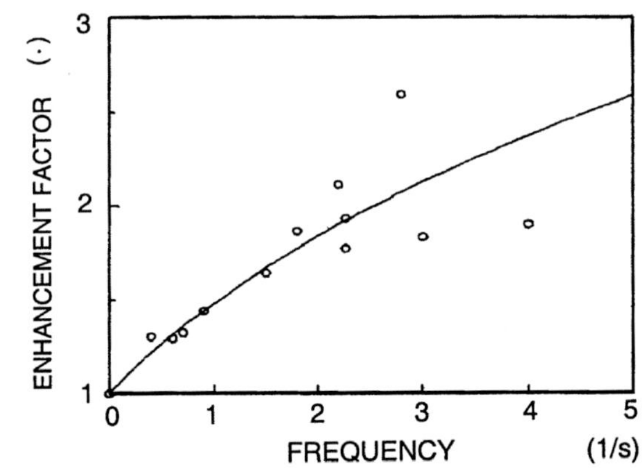

FIGURE 15-7 **Enhancement of the mass transport limited rate of H_2O_2 decomposition over a Pt catalyst supported on sintered glass by total pressure modulation as a function of frequency (see Figure 15-5). The solid curve is the prediction of the model given by Eq. (15-6) with $\varepsilon_g = 0.042$.** *(Figure reproduced from* Van den Heuvel et al. (1996) *with permission of the copyright holder. © 1996 by Elsevier Science Ltd.)*

for an assumed gas holdup of 0.042. The *Van den Heuvel* model reproduced the experimental observations reasonably well.

An alternate model describing the effect of pressure modulation on mass transfer in the presence of gas pockets trapped in the porous solids has been reported *(Zirkwa et al., 2001)*.

15.3 SONOCATALYSIS

Pressure modulation at frequencies several orders of magnitude higher than those just considered can have profound effects on the rates and selectivity of certain heterogeneously catalyzed reactions. This subject is referred to as ultrasonic excitation or sonochemistry. Several monographs have appeared on sonochemistry with extended discussions of applications to heterogeneous catalysis (*Suslick, 1988; Mason and Lorimer, 1988; Mason, 1990, 1991, 1993; Price, 1993* and *van Eldik and Hubbard, 1997*). Two annual book series, *Ultrasonics* and *Ultrasonics Sonochemistry,* are devoted to sonochemistry.

The frequencies of pressure modulation in sonochemistry range from 20 kHz to 2 MHz. As comparison, acoustic frequencies are 16 Hz to 16 kHz. Even the high frequencies in ultrasonic excitation are several orders of magnitude below the vibrational frequencies associated with molecular bonds; thus, sonic radiation is not absorbed at the molecular level. Excitation operates through several mechanisms. Ultrasonic irradiation causes cavitation in liquids if the power input is sufficient and the modulation frequency is suitable. Tiny short-lived bubbles are generated by the rarefaction portion of the modulation. These bubbles eventually collapse in the compression portion, but growth and deformation can occur between these events as illustrated in **Figure 15-8**. It is during the collapse of the cavities shown in the figure that phenomena influencing reaction rates arise. Cavity collapse in the bulk liquid causes abrupt adiabatic compression of the vapor within the

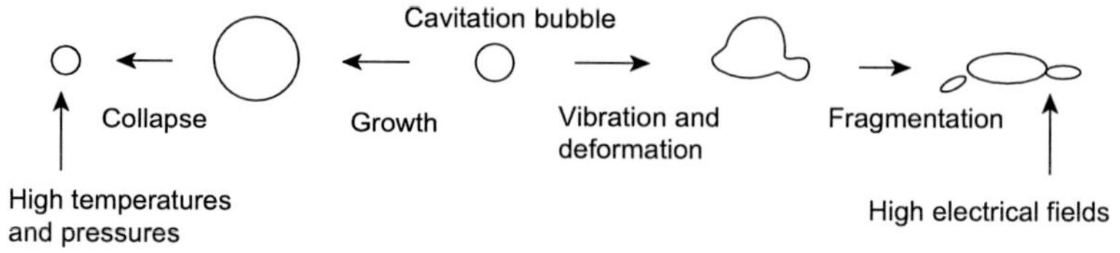

FIGURE 15-8 Physical events in cavitation through ultrasonic irradiation of liquids. *(Figure reproduced from* van Eldik and Hubbard (1997) *with permission of the copyright holder. © 1997 John Wiley & Sons.)*

bubble that generates temperatures sufficient to create free radicals. The in-rushing liquid during collapse spawns intense shear forces and turbulence locally. **Figure 15-9** summarizes these effects.

Free radicals formed in the collapsing cavity react or may migrate into the surrounding liquid. This effect is the phenomenon observed in homogeneous reaction systems. In water, for example, H^* and OH^* radicals form in the cavities. The OH^* radicals react to form H_2O_2. In the presence of organic solutes, such as esters or halogenated hydrocarbons, hydrolysis or solvolysis is observed when sonification begins.

With heterogeneous systems, the intense shear forces stemming from bubble collapse disrupt liquid–liquid phase boundaries. Emulsions can be formed depending on the surface tensions in the two liquid phases. Cavity distortion near a phase boundary apparently leads to

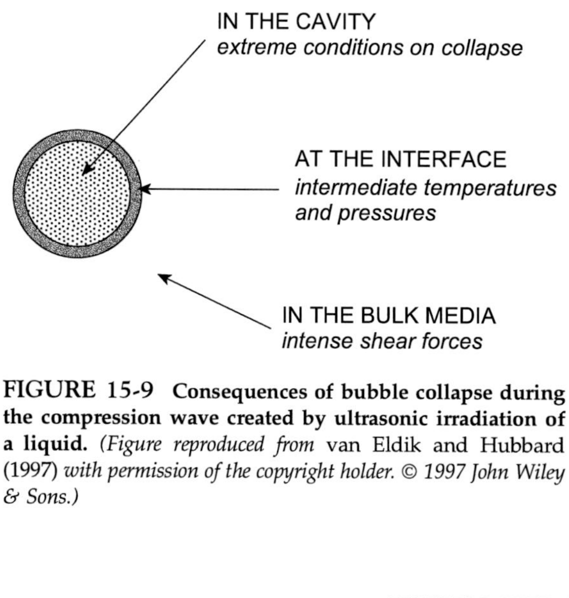

FIGURE 15-9 Consequences of bubble collapse during the compression wave created by ultrasonic irradiation of a liquid. *(Figure reproduced from* van Eldik and Hubbard (1997) *with permission of the copyright holder. © 1997 John Wiley & Sons.)*

a liquid jet in the direction of the interface. Jet impingement is the mechanism involved in sonic cleaning baths that are now widely used commercially. It is hypothesized that intense jet impingement can penetrate passivating oxide layers on metal surfaces such as nickel and iron. Creation of active metal sites by this mechanism is the explanation of the activation of metal catalysts through ultrasonic irradiation. With coarse particles, cavitation bubbles can form in surface imperfections. Forces generated on collapse of such bubbles may fragment the particle. On the other hand, with very fine particles, high temperatures arising from bubble collapse may induce particle fusion and thus agglomeration (*Mason and Luche, 1996*). Different phenomena occurring in the compression wave during ultrasonic irradiation are illustrated in **Figure 15-10**.

Many examples of the application of ultrasonic irradiation exist in the literature. The Cannizzaro reaction, disproportionation of an aryl aldehyde to the acid and alcohol, is one of many that are enormously accelerated by irradiation.

Disproportionation of benzaldehyde to benzoic acid and benzyl alcohol over a barium hydroxide base catalyst does not proceed significantly at room temperature, but with irradiation at 20 kHz the reaction goes to completion (*Fuentes and Sinisterra, 1986*). *Ando et al. (1984)* studied the formation of aromatic cyanides from the corresponding bromide over an activated alumina using toluene as the solvent and a small quantity of water. Benzyl cyanide is formed in high yield from benzyl bromide in this system when ultrasonic irradiation is

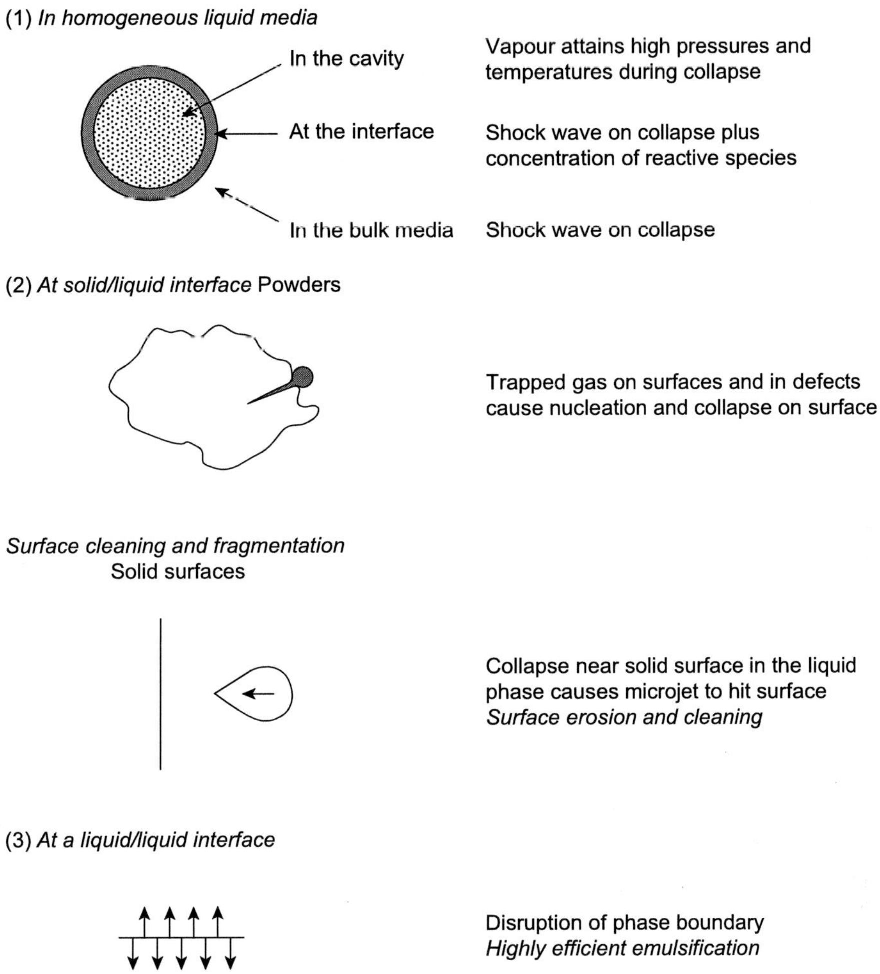

FIGURE 15-10 Phenomena associated with collapsing bubble cavities during ultrasonic irradiation in heterogeneous systems. *(Figure reproduced from Practical Sonochemistry: Power Ultrasound Uses and Applications (Horwood Chemical Science Series), 9781898563839, Mason and Peters, Woodhead Publishing Limited).*

used. The rate is very slow in the absence of irradiation. When 4-methylbenzyl bromide is used in place of benzyl bromide, ultrasonic irradiation changes the product formed. Irradiation results in 4-methylbenzyl cyanide in high yield, whereas, if ultrasound is not used, the Friedel-Crafts adduct of toluene and 4-methyl toluene forms. The explanation in both cases appears to be that the adsorption of KCN by alumina is promoted by irradiation. With 4-methyl benzyl bromide in toluene, KCN adsorption suppresses the Friedel-Crafts activity of alumina.

Ultrasonic irradiation activates nickel powder in alkene hydrogenation and provides a hydrogenation activity comparable to that displayed by Raney nickel. The explanation for this activation is the enrichment of the passivated nickel surface with Ni through the jet effect mentioned above, rather than a surface area enhancement (*Suslick and Casadonte, 1987*).

Hydrogenation of soybean oil employing ultrasonic irradiation at 20 kHz is performed on a commercial scale at 180°C and 115 psig hydrogen pressure using a 1 wt% copper chromite catalyst. Hydrogen is dispersed in the oil and irradiated in a tank attached to a tubular reactor packed with catalyst (*Moulton et al., 1983*).

15.4 PERIODIC PRESSURE REDUCTION

Low-frequency pressure modulation, the other extreme to ultrasonic irradiation, apparently has not been investigated. Experiments undertaken on the "stop effect" (see **Chapter 11**) suggest that pressure modulation at low frequency and high amplitude should significantly increase production rates over steady state for reaction systems exhibiting reactant inhibition. This effect, its mechanism and the use of bang-bang flow control to increase rates are succinctly discussed by *Silveston (1998)*. There is an explosive increase in the rate of reaction when reactant flow is replaced by a flow of an inert carrier gas. The behavior is illustrated in **Figure 15-11** for the deamination of propylamine over a 13X zeolite. Similar behavior has been reported for butylamine, isopropylamine and isobutylamine. It is also observed for the dehydration of ethyl, propyl

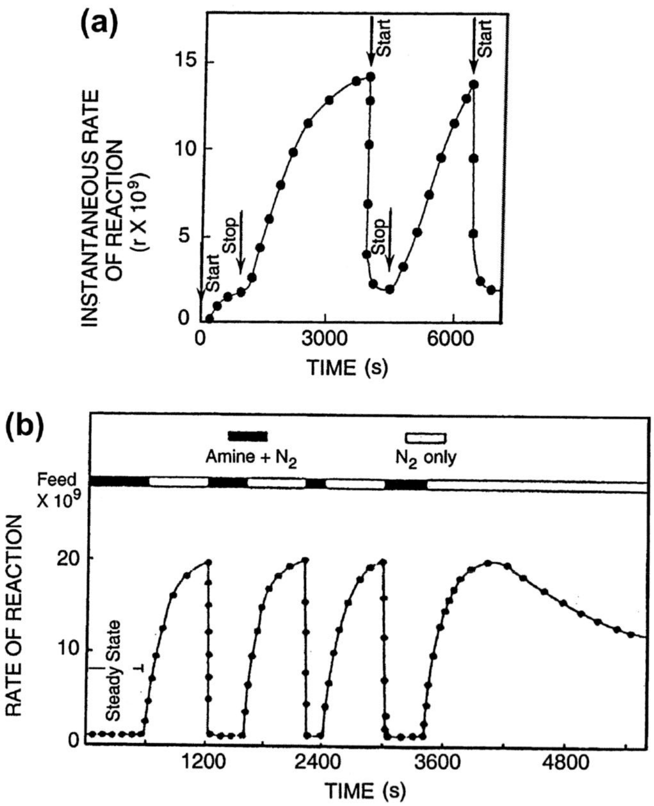

FIGURE 15-11 **Variation of the rate of deamination for successive addition and removal of propylamine from a N_2 carrier gas fed to a differential reactor packed with decationized 13X zeolite. T = 573 K, p_{amine} in the feed = 11 kPa.** *(Figure adapted from* Koubek et al. (1980b) *with permission of the copyright holder. © 1980 by Elsevier Science Publishers.)*

and isopropyl alcohols over alumina and 13X zeolite.

As **Figure 15-11** suggests, periodically halting the mixing of propylamine into a N_2 carrier gas for 600 s out of the 1000 s cycle raises the production rate of propylene 10-fold. Decreasing the duration of propylamine addition would probably increase production even more. Indeed, rate increases of 50-fold have been reported through periodically halting the addition of a reactant to the carrier gas flow.

Explanation of the stop effect is that deamination or dehydration requires vacant adjacent surface sites to proceed, but the amine or alcohol reactant is so strongly adsorbed that unoccupied adjacent sites are rare on the surface during steady-state operation except when the reactants are extremely dilute. Eliminating reactant from the feed periodically allows a few unoccupied sites to arise that are adjacent to adsorbed amine or alcohol. Reaction at these locations frees additional sites and this event leads to an avalanche of product. **Chapter 11** discusses the stop effect.

Pressure modulation should achieve similar increases in production rates. Feed to a catalytic deamination or dehydration reactor would be held constant by a flow-control valve. A positive displacement vacuum pump would be placed immediately downstream of the reactor. To emulate the first cycle in **Figure 15-11**, the vacuum pump would operate for 600 s and then rest for 400 s. This procedure would result in a modulation frequency of 0.001 Hz and an amplitude of about 50 kPa, namely a high-amplitude, low-frequency application of pressure modulation.

15.5 COMBINED COMPRESSION AND REACTION

Undertaking a gaseous chemical reaction while compressing a reactant mixture is quite an old concept, going back almost 100 years in the patent literature. Whereas in a combustion engine, the piston-cylinder system extracts energy from a spontaneous exothermic reaction yielding waste products, the same system can be used to provide energy to an endothermic reaction or to carry out a reaction yielding desired products and some of the available energy with exothermic reactions. Combining compression and reaction is, perhaps, an extreme example of pressure modulation. However, pressure in such systems is changing repetitively with time.

Roestenberg et al. (2010) discussed the application of a free piston system with opposed compression chambers to an apparent homogeneous, gas-phase reaction. **Figure 15-12** illustrates the system considered. A number of different designs exist (*Mikalsen and Roskilly, 2007*). As the figure shows, upward movement of the piston closes the intake conduit, trapping and then compressing the charge until it ignites. Piston movement allows flow to the bottom chamber flushing out product gas and recharging that chamber. Start-up and attainment of

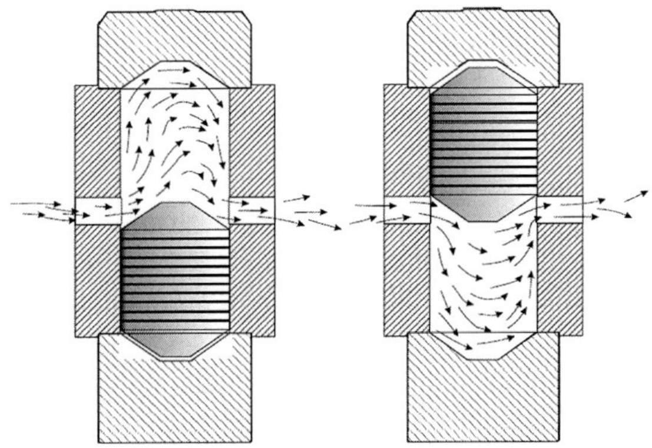

FIGURE 15-12 **Operating principle of a free piston operating reactor with two reaction chambers. Upward movement of the piston adiabatically compresses and heats the reactants in the upper chamber to their ignition temperature. Exhaust and feed cut-off allows reaction to take place and forces piston downwards to compress gas in the lower chamber.** *(Figure reproduced from* Roestenberg et al. (2010) *with permission. © 2010 Elsevier Ltd.)*

a stable operation were examined experimentally using a 10 vol% CH_4 − 5 vol% O_2 mixture in N_2-Ar by *Roestenberg et al.* Simulation of the free piston system with a CSTR-based model extended the measurements. The researchers used just a single compression-expansion stroke in their experiments but concluded, nevertheless, that stable operation was possible. Runaway that would destroy the system could occur if the energy release through reaction exceeds energy loss through gas expansion and heat loss through the chamber walls.

Although the application was apparently created for a homogeneous reaction, the Roestenberg free piston design could be used for an exothermic heterogeneously catalyzed reaction by coating a portion of the compression chamber head and/or the piston head with a catalytically active washcoat interspersed with solid metal piers to protect the wash-coat at the top of the piston stroke.

15.6 APPLICATION TO RATE AND EQUILIBRIUM MEASUREMENTS

Pressure modulation has been and continues to be widely used in frequency response methods for equilibrium and rate measurement in reacting as well as non-reacting systems. The subject is somewhat distant from our topic in this monograph. However, frequency response studies will be examined briefly as an illustration of a common use of pressure modulation. A large literature exists. Several good references are available on the application (*Jordi and Do, 1993*; *Reyes et al., 1997*; *Wang et al., 2003*).

Wang et al. (2003) considered adsorption of a pure gas onto a microporous carbon with sieve properties. The experimental system employed is typical of those used for rate measurement in reacting as well as non-reacting systems. It is shown schematically in **Figure 15-13**. The initial mass flow controller insures a constant flow rate of the adsorbing gas. Low-amplitude pressure fluctuations (sinusoidal or square-wave) are created by the pressure transducer and the responses to these disturbances are measured by the mass flow meter. At the start of an experiment, gas is introduced and pressure perturbation begun. From the Fourier series transformation of the flow measurements and the pressure forcing, the amplitude ratio and the phase lag are determined. Mass transfer coefficients and the amount adsorbed are obtained by fitting the amplitude ratio and phase lag drawn from a Fourier transform of the model to the transformed experimental values. Several models were examined: surface diffusion control

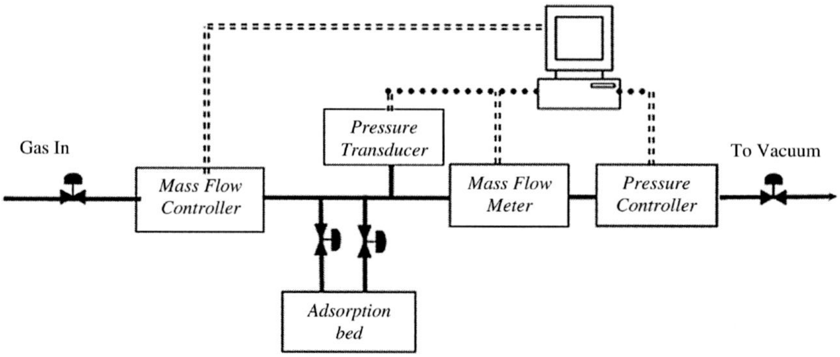

FIGURE 15-13 Experimental system for measuring adsorption equilibria and mass transfer rates in a porous adsorbent. *(Figure adapted from* Wang et al. (2003) *with permission of the publisher. © 2003 by the American Chemical Society.)*

of mass transfer, control by transfer through a gas phase boundary layer at the pore mouth, and combined mass transfer control. All of these models are described in the research literature.

Wang et al. show that the adsorption equilibrium data are well fitted for O_2 and N_2 by a Langmuir isotherm. These isotherms are quite similar for the two gases, the only ones considered by the experimenters in their study. However, O_2 diffuses through the porous carbon much more rapidly than N_2. This observation is explained by the ratio of the mean pore mouth diameter to the gas molecular diameter which is larger for O_2 than for N_2. Consequently, the pore mouth resistance is more important for N_2 than for O_2. Indeed, *Wang et al.* observed that mass transfer of N_2 into the carbon adsorbent is controlled by pore mouth resistance, whereas, for O_2, this resistance as well as that due to surface diffusion is important.

15.7 ASSESSMENT AND RESEARCH OPPORTUNITIES

Many biochemical reactions, particularly fermentation, take place in flocs large enough that the enzymatic reaction occurring is rate-limited by mass transfer through the floc. Consequently, increasing mass transport rates without otherwise affecting the floc or operating conditions would be of commercial interest. Experiments by the *Van den Heuvel* team suggest that pressure modulation at sub-acoustic frequencies substantially raises the bioreaction rate. Improvement depends on the gas entrapment in the floc. However, there is evidence that modulation with large amplitudes destroys a floc. Currently, little is known about the limits on amplitude and the interplay between amplitude and modulation frequency with respect to enhanced mass transport. More cell culture systems need study because of differences in floc structure and strength as well as differences in respiration rates of the biomass. Systematic experiments varying amplitude and frequency should be undertaken up to the limit of floc disintegration. The design of equipment for such studies appears to have been solved by the University of Amsterdam team as may be seen in **Figure 15-5**. It is evident that experimental studies do not require expensive, sophisticated equipment.

Rate enhancement through pressure modulation at sub-acoustic frequencies seems limited to special situations with respect to heterogeneous catalysis. Although the H_2O_2 decomposition experiments demonstrate that significant rate increases can be achieved, few industrially important solid-catalyzed reactions generate a gas phase from liquid-phase reactants. There are, of course, a great many reactions with reactants in separate phases, e.g., hydrogenation, but wetting of the catalyst by the liquid phase and reactor operation precludes gas bubbles from the liquid-filled catalyst interior. The oxide reduction experiments discussed show that modulation at sub-acoustic frequencies does increase reaction rates in porous solids only when a gas phase is present, but the rate enhancement is modest. The same benefit is more easily achieved by reducing the size of the solid particle. Consequently, pressure modulation at sub-acoustic frequencies appears of interest for solid-catalyzed gaseous reactions only when reducing the particle size is unworkable, perhaps because of pressure-drop considerations.

Pressure modulation at ultrasonic frequencies of heterogeneously catalyzed liquid phase reactions may be quite attractive. The rates of these reactions are almost always mass-transfer limited. It is evident from ultrasonic irradiation experiments briefly mentioned that irradiation generates considerable small-scale mixing that could take place within catalyst transport pores. Such mixing would increase effective diffusivities and thereby raise transport rates. Surface damage is a concern with ultrasonic irradiation, but this can be controlled by mediating bubble collapse through operating temperature, total pressure, and irradiation amplitude. This possibility has not been considered in the literature. Ion exchange systems would seem to be good candidates and experiments could be undertaken in several of the tubular reactor systems described in the sonochemistry literature (*Mason and Lorimer, 1988*).

Low-frequency pressure modulation may offer a means of substantially raising reaction rates where adjacent sites are needed for a reaction to proceed in situations in which the reactants or possibly products are strongly adsorbed. It is an application of modulation that has not been treated in the literature.

Nomenclature

A, B,	=	chemical species
A	=	amplitude
C	=	concentration (kmol/m^3)
$C_{n,t}$	=	concentration in the nth mixer or shell of a catalyst particle
D_{AB}	=	effective or binary diffusivity (m^2/s)
K	=	dimensionless rate constant
k	=	reaction rate constant
n	=	products to reactants molar ratio
P	=	pressure (kPa)
P_0	=	median pressure (kPa)
q_n	=	volumetric exchange flow between shells in a catalyst particle (mL/s)
R	=	dimensionless amplitude of pressure modulation
r	=	rate of reaction (mol/m^{3}s)
t	=	time (s)
Vn	=	volume of the nth mixer or shell in a catalytic particle (mL)
v	=	superficial velocity (m/s)
y_i	=	mole fraction of species i
z	=	dimension along flow path (m)

Greek

ε_p	=	particle void fraction
η	=	effectiveness factor
η_{osc}	=	effectiveness factor for convective oscillations
φ	=	Thiele modulus
λ	=	dimensionless frequency
ν_i	=	stoichiometric coefficient of species i
θ	=	increase in volume on reaction factor
ρ	=	density (kg/m^3)
ω	=	rotational frequency (s^{-1})

CHAPTER

16

Temperature Modulation

Jürgen J. Brandner, Peter Lewis Silveston†, Robert Ross Hudgins†*

*Karlsruhe, Germany and †Waterloo, Ontario, Canada

OUTLINE

16.1 Introduction 435

16.2 Theoretical Studies 436

16.3 Simulation Studies 437

16.4 Experimental Studies with Conventional Laboratory Equipment 439

16.5 Temperature Modulation of Trickle Beds 444

16.6 Experimental Studies with Microreactors 444

16.6.1 Micro Devices for Temperature Modulation 445

16.6.2 Experimental Procedure with Microreactors 452

16.6.3 Experimental Observations 453

16.6.4 How Temperature Modulation Increases Reaction Rate 459

16.7 Overview and Comments 462

16.1 INTRODUCTION

The exponential relationship of the rate constant to temperature has served as a magnet to attract the attention of researchers to periodic temperature modulation. Early theoretical studies suggested that significant improvements in rate and selectivity could be expected. Failure of the early experiments to confirm these expectations caused a long lapse in research interest on temperature modulation. Progress in the development of microreactors with their greatly reduced thermal mass has revived research activity during the last decade.

Modulating the reactor temperature was discussed in the earliest studies on periodic operation (*Horn and Lin, 1967; Denis and Kabel, 1970; Bailey et al., 1971; Dorawala and Douglas, 1971*). A simple calculation explains this early interest. Assuming a point in a reactor operating at a temperature of 773 K were cycled symmetrically with an amplitude of 10 K, the time-average rate would be increased by 7.4% over the steady-state rate at 773 K for a reaction with a low activation energy of about 83.6 kJ/mol. For a large activation energy of 418 kJ/mol the increase would be 266%. If the amplitude were increased to 20 K, rate improvements for the range of activation energies would become 30 and 2300% respectively (*Silveston and Hudgins, 2004b*).

From a practical point of view, such an enhancement cannot be achieved in reality because of the large thermal inertia of conventional process equipment. Early attempts to carry out temperature modulation failed in the laboratory because of the thermal inertia of catalyst beds and reactor vessels used so that there has been little research activity on this topic for about three decades. The amplitudes necessary to see a modulation effect would be in the range of some 10 K, and the required frequency would be in the range of a few Hz or more (*van Neer et al., 1996*). Such levels of performance could not be attained with any equipment available at that time. At frequencies below a few Hz, the reaction system operates in a quasi-steady state, an appropriate average of operation at each of the two temperatures used for modulation. Frequencies higher than a few Hz will force the system into the relaxed steady state discussed in **Chapter 1**.

The appearance of microreactors with relatively low mass in the last few years has reawakened interest in temperature forcing. Furthermore, some remarkable increases in rates of reaction in three-phase trickle beds observed for periodic interruption of liquid phase flow now appear to be caused by temperature variations (*Haure et al., 1989*).

16.2 THEORETICAL STUDIES

Early studies on periodic temperature forcing, mentioned in the previous section, were theoretical and were summarized in a review (*Bailey, 1974*). Two types of operations were considered: 1) Quasi-steady state and 2) relaxed steady state. In the former, the time needed to achieve steady behavior within a half-cycle is negligible relative to the duration of the half-cycle. For the relaxed steady state, however, switching is so fast that there is a negligible change in state of the manipulated system with time. Improved performance for quasi-steady state requires convexity in the equations describing the reactor system. For an isothermal CSTR in which a first-order reaction is occurring, the system is simply described as:

$$x_1 = \frac{1}{(1 + N_{Da}(-y/x_2))} \qquad (16\text{-}1)$$

where x_1, x_2 are the dimensionless concentration and temperature in the reactor. If temperature is an independent variable, the system exhibits convexity. Indeed, using the numbers given in the introduction, a bang-bang periodic change in reactor temperature, x_2, can significantly increase conversion. Convexity is exhibited also by a plug flow, packed bed catalytic reactor provided the characteristic time for a change in surface concentration is small relative to the cycle period.

The relaxed steady state arises when the variation of the manipulated variable forces a reactor output into a limit cycle that shrinks as the cycle frequency increases into effectively a single state. It will be different from a steady state of the system if the dynamics of the reactor are nonlinear or the relation between an objective function to be optimized, such as conversion, and the state and manipulated variables are nonlinear. In temperature modulation, temperature is the manipulated variable, so that the condition for a relaxed steady state that differs from a steady state is met. Whether that relaxed steady state results in improvement can be established by application of an optimization routine to the system of descriptive equations. The first application to a reactor system was in the classical paper by *Horn and Lin (1967)* considering a system of parallel reactions of different order in a CSTR under bang-bang temperature cycling. Applying Pontryagin's Maximum Principle, *Horn* and Lin found a set of temperature states for which the objective function, the concentration of a product from one of the competing reactions, was contained in a small limit cycle located outside of the range of steady states, as shown in **Figure 16-1**. The plot assumes a second-order reaction to the desired product and a first-order reaction to a waste product. The activation energy for the second reaction is 3/4 of the activation energy for the reaction to the desired product. The cycle switches between two temperatures. In the plot, the concentration of the desired product, C_B, is plotted against the exit concentration of the reactant C_A. It can be seen that the variation of C_B with C_A is concave under steady-state operation so a quasi-steady state cannot improve upon steady state (maximum selectivity to C_B at I). Fast cycling of the CSTR temperature leads to at least two relaxed steady-state operations. The operation designated by II shows a modest improvement over the optimal steady state, while the second region just above I is another relaxed steady state but one giving negligible improvement over the best steady-state operation.

Dorawala and Douglas (1971) considered a consecutive reaction in a CSTR with possible temperature cycling. In this system, the intermediate product was desired. They demonstrated that a relaxed steady state could be found under temperature cycling that offered a selectivity to the intermediate superior to that of the optimal steady state. *Sterman and Ydstie (1990a, 1991)* considered both parallel and consecutive reactions in a CSTR with temperature modulation. In their 1991 paper, they developed an optimization scheme to determine relaxed steady states that improved upon the optimal steady state.

Sinčić and Bailey (1977) discovered that temperature modulation in a CSTR can lead to some surprising dynamic behavior. For first-order parallel reactions with different activation energies, a single limit cycle for the reactor was observed. There was also a complex cycle that was 2τ-periodic, with two stable limit cycles. Several of the cycles exhibited irregular temperature trajectories. One such cycle is illustrated in **Figure 16-2**.

A comprehensive investigation of the dynamic phenomena that may arise was undertaken by *Kevrekidis et al. (1986)*.

16.3 SIMULATION STUDIES

The earliest published study of temperature modulation was a simulation undertaken by

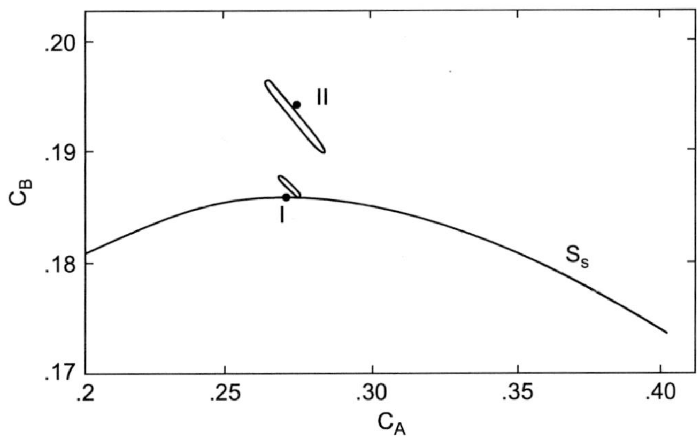

FIGURE 16-1 **Exit concentrations under steady-state and periodic operation with temperature forcing of a parallel reaction (A→B, A→C) in a CSTR. S_S = curve of steady-state operation, I, II = relaxed steady operations.** *(Figure reproduced from* Bailey, 1974.*)*

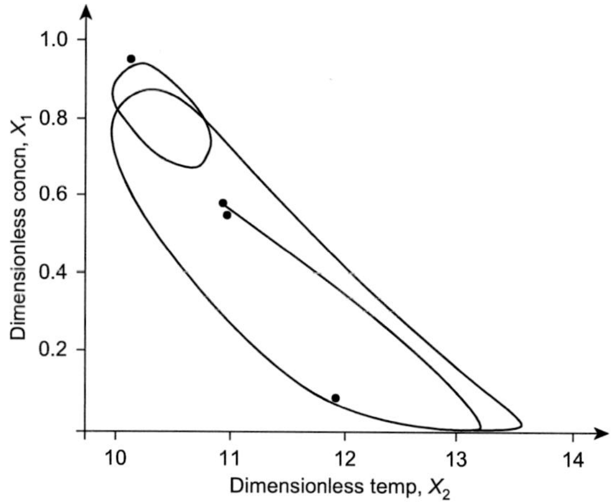

FIGURE 16-2 Concentration-temperature trajectory for an A → B, A → C reaction system in a CSTR forced by modulation of the feed temperature. Points in the figure represent steady operating states. The two loops indicate 2τ-periodicity. *(Figure reproduced from* Sinčić and Bailey (1977) *with permission. © 1977 by Elsevier Science Publishers.)*

Chang and Bankoff (1968), who examined a jacketed packed bed reactor. These investigators found that conversion was optimal in a quasi-steady-state operation. Later, *Denis and Kabel (1970)* simulated a temperature modulation of a packed bed catalytic reactor running the vapor-phase dehydrogenation of ethanol over a cation exchange resin. Improved conversion over steady-state operation was found. An interesting aspect of their work was the use of the response to step-change variations in the reactor jacket temperature to obtain parameters for their dynamic packed bed model. That model allowed for adsorption/desorption from the catalyst surface and considered variations of velocity in the reactor.

Cyclic temperature variations in a jacketed packed bed reactor were considered in a simulation study by *Kim and Hulburt (1972)*. The authors also examined simultaneous modulation of concentration and feed rate along with temperature. They observed improvements that they claimed were due primarily to temperature cycling.

Temperature modulation of a CSTR was simulated by *Dorawala and Douglas (1971)* who assumed square-wave changes of coolant flow rate in the CSTR jacket. They also examined feed temperature modulation. Generally, these simulation studies showed that temperature cycling with the proper frequency and cycle split improved reaction selectivity. Many of these studies are summarized in a review by *Bailey (1977)*. In 1980, *Lee and Bailey (1980)* considered consecutive-parallel reactions in a CSTR under modulation and demonstrated a significant increase in selectivity to the desired intermediate product in the reaction chain. Unfortunately, these encouraging early simulations as well as theoretical studies discussed above have not stood up to the rigor of experiment.

Luther et al. (2008a, b) combined simulation with experimental studies using a microreactor of the oxidation of CO. This oxidation, one of the most studied catalytic reactions, is known to show self-sustained oscillations under certain conditions (*van Neer and Bliek, 1999; Yuranov et al., 2000*). Although the mechanism explaining these self-oscillations is still not certain, the basic sequence described by *Engel and Ertl (1978), Rinnemo et al. (1997)*, and *Nibbelke et al. (1997)* is widely accepted. The reaction model is of Langmuir-Hinshelwood type assuming reversible adsorption of CO and O_2 on the catalyst surface followed by a surface reaction between the adsorbed CO molecule and oxygen atoms. Adsorption for CO and O_2 is competitive with only one type of active site. The CO_2 formed interacts only weakly with the surface (*Engel and Ertl, 1978; Dulaurent and Bianchi, 2000*) and thus desorbs immediately from the catalyst surface. Thus, its surface coverage may be set to zero. To explain self-oscillations, it is necessary to include a relatively slow and nonlinear feedback step in the mechanism (*van Neer and Bliek, 1999*) as the Langmuir-Hinshelwood mechanism itself is not sufficiently nonlinear to show complex dynamic behavior.

However, in the case of temperature, the strongly nonlinear temperature dependence of the rate constants for the different reaction and transport steps allow unforced oscillations to occur.

To perform simulations, numerical values are needed for each model parameter. Of course, a range of values exists in the literature, but since kinetic parameters depend on the experimental conditions, these can be seldom used. Moreover, parameters determined under stationary conditions may not be applicable for transient systems (*Renken et al., 1989*). Consequently, the *Luther et al.* dynamic model (*Luther, 2008; Luther et al., 2008a, b*) was tested under constant temperature and found to reproduce well the experimentally measured product concentrations. Additional parameters were evaluated for the variable temperature terms in the model so that *Luther et al. (2008a, b)* assumed it was appropriate for investigating temperature modulation. When the model was applied to experiments, it turned out to be inaccurate (*Luther, 2008*). Surface transients were fast compared to the period of the temperature modulation applied. To overcome this problem, *Luther et al.* introduced a feedback

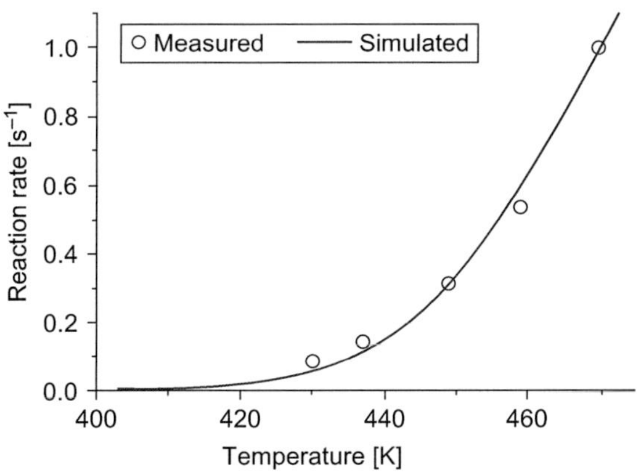

FIGURE 16-3 Comparison of simulated reaction rates calculated from an enhanced Langmuir-Hinshelwood model and experimentally measured values. *(Figure reproduced from* (Luther, 2008) *with author's permission.)*

step into their model similar to ones used by *Yuranov et al. (2000)* and *van Neer et al. (1996)* to explain the self-oscillations they observed. This feedback step assumed a kind of rearrangement of the adsorbed species on the active site. Details are given in *Luther (2008)*. This modification was used to generate the curve in **Figure 16-3**. From the figure, it can be seen that the modified model fits the experimental data closely.

Using their model, *Luther (2008)* simulated the CO oxidation rates under temperature modulation that are shown in **Figure 16-4**. This figure demonstrates that the time-average reaction rate under temperature modulation is higher than the rate at steady state.

16.4 EXPERIMENTAL STUDIES WITH CONVENTIONAL LABORATORY EQUIPMENT

Experimental investigations of temperature modulation began within a decade of the *Horn and Lin (1967)* paper. One of the first dealt with small-amplitude, low-frequency perturbations intended to test dynamic packed bed reactor models (*Hansen and Jorgensen, 1974*). Oxidation of H_2 on a Pt/Al_2O_3 catalyst in lab-scale equipment was used. Oxidation went to completion in the reactor so the authors were unable to gain information about the effect of temperature modulation on rate. They observed wrong-way behavior for a step-change in feed temperature. Also, they reported a large amplification of internal temperature waves when a small-amplitude, periodic, square-wave temperature disturbance was imposed upon the reactor feed.

The temperature modulation question was addressed by *Abdul-Kareem et al. (1980b)* using CO oxidation over a vanadia catalyst in a laboratory packed bed reactor immersed in a fluidized sand bath to maintain the reactor at sand-bath temperature. Modulation was produced by

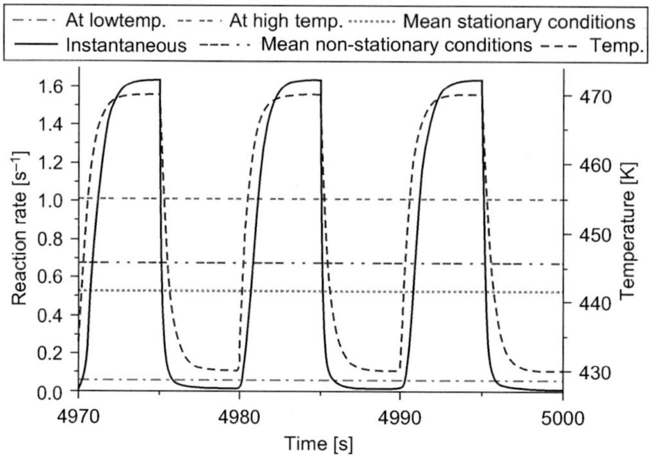

FIGURE 16-4 Comparison of simulated reaction rates under temperature modulation and steady-state operation. Curved lines show instantaneous rate under modulation (- - - - -) and instantaneous temperature during modulation (- - -) Horizontal lines (from the top): Show high-temperature extremum, time-average rate under modulation, rate under steady-state operation, and low temperature extremum. *(Figure reproduced from* Luther (2008) *with author's permission.)*

switching the temperature in the sand-bath using high-output electrical heaters as well as by adjusting the air flow rate. Because of thermal inertia, temperature variations were trapezoidal, but approached a square-wave shape for long cycle periods (*Abdul-Kareem et al., 1980b; Silveston and Hudgins, 2004b*). Steady-state rate measurements over the temperature range of the modulation furnished an activation energy of 103 ± 5 kJ/mol, in good agreement with published values. Cycle periods of the order of hours were needed to switch temperatures between the two levels chosen. Thus, with a 60 min cycle period, the bed resided at each temperature level for just 20 min in each partial cycle.

Abdul-Kareem et al. observed that the time-average rate of CO oxidation under temperature modulation increased with increasing cycle period for both of the amplitudes they used (10°C and 20°C). For cycle periods greater than six hours, the quasi-steady-state rate was obtained. Low-frequency modulation, thus, increased the reaction rate above that possible at steady state, but the quasi-steady-state limit was not exceeded. Because the oxidation rate versus temperature curve is convex, the quasi-steady-state rate must exceed the steady-state rate at any time-average temperature. What is surprising about the measurements is that the time-average rate lies below the quasi-steady-state rate at cycle periods up to 6 h. *Abdul-Kareem et al.* pointed out that the sand bath reached a constant temperature in 10 min after a change, while their calculations showed that the catalyst particles' temperature should follow the bed temperature with a lag of just 60 s (*Abdul-Kareem et al., 1980b; Silveston and Hudgins, 2004b*). Thus, particle temperature in the catalyst bed follows the sand-bath temperature closely on a scale of hours. The remarkably long time of 6 h needed to reach the reaction rate at quasi-state indicates that the changing temperature on the catalyst surface retards rather than enhances the catalytic reaction rate. Experimental results with cycle periods less than 2 h and a 10°C amplitude reached only the steady-state rate at the time-average temperature. This rate is well below the observed quasi-steady state and so supports this interpretation. Such an idiosyncratic result merits further investigation.

Bang-bang temperature modulation was imposed on the fluid flowing through the jacket of a CSTR reactor by *Lee and Bailey (1980)*. In order to verify the simulation discussed in **Section 16.3**, *Lee et al (1980)* used a small jacketed CSTR to study the saponification of diethyl adipate with sodium hydroxide, a consecutive-parallel reaction. The experimental yields under modulation were below simulated values and close to those measured under steady-state. Dynamic measurements on the jacketed CSTR indicate a relaxation time of about 2.5 min. Thus, the cycle period used was so short that the temperature change in the reactor was greatly attenuated. Damping through heat exchange and mixing in the heating jacket seems to explain the small improvement compared with steady state.

Temperature modulation experiments have been extended in order to investigate temperature amplification, a phenomena first reported by *Padberg and Wicke (1967)*. The concern in these experiments was not so much rate increase as possible damage to the catalyst through local overheating and consequent deactivation. *Jaree et al. (2001b, 2003a, b)* employed CO oxidation over a Pt catalyst while *Na-Ranong et al. (2003)* worked with the same reaction over a CuO catalyst. Experiments were carried out in a laboratory tubular reactor packed with a catalyst designed to minimize heat loss as shown in

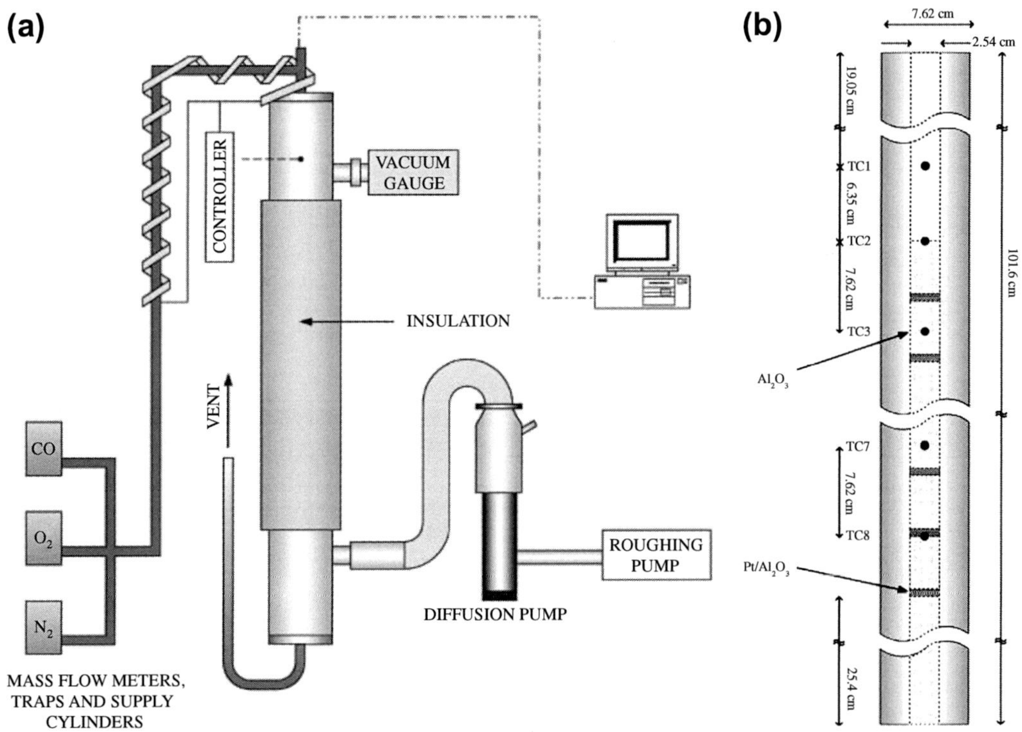

FIGURE 16-5 Experimental system and reactor used to observe input wave amplification in well-insulated packed bed reactors: (a) Overall experimental system showing reactant flow and method of temperature variation, (b) thin-walled reactor showing thermocouple placement. *(Figure adapted from* Jaree et al. (2001b*) with the author's permission.)*

Figure 16-5. Eight uniformly-spaced thermocouples were installed to follow triangular temperature variations carried into the catalyst bed by the reactant mixture. These variations were created through a computer-controlled heating-tape.

Both *Jaree et al. (2001b, 2003a, b)* and *Na-Ranong et al. (2003)* observed significant amplification of these periodic temperature disturbances. Amplification set in shortly after the entrance to the catalyst bed and jumped to a maximum that depended on the catalyst chosen. **Figure 16-6(a)** shows that the 1 K inlet disturbance amplitude grows to about 9 K inside the catalyst bed. Such a temperature rise could be serious if the reactor operated near a parametric sensitivity boundary. The amplification exhibited resonance with respect to the temperature disturbance frequency as well as saturation as the amplitude of the inlet disturbance increased. Axial variations in catalyst activity strongly affected the gain in amplitude

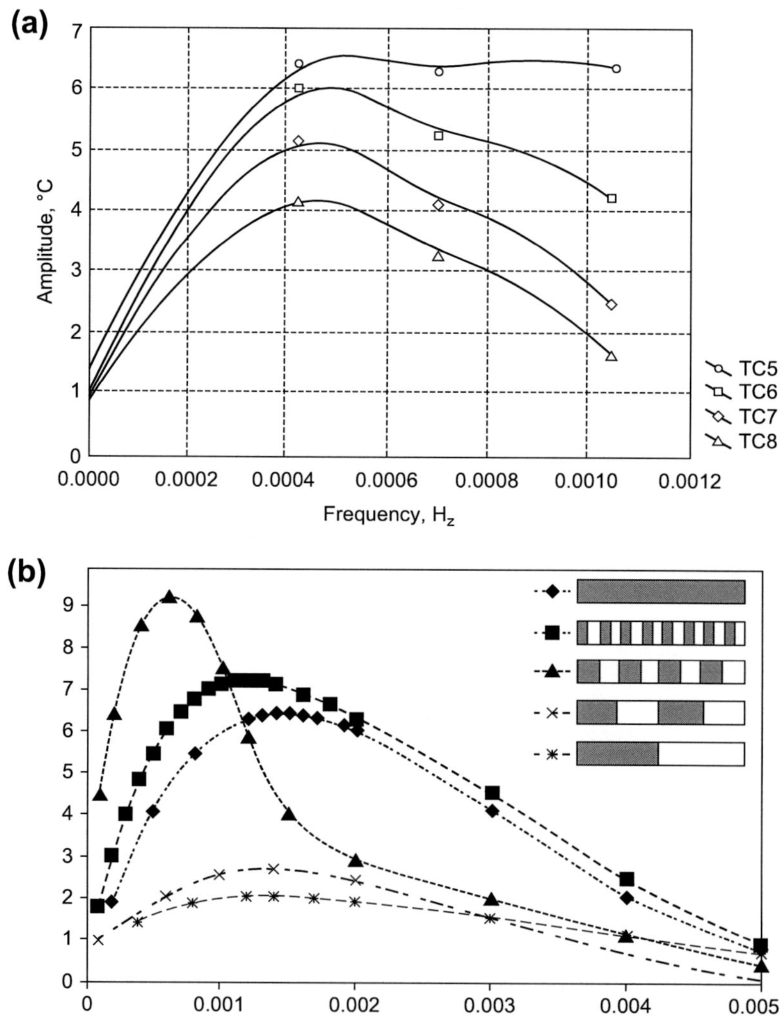

FIGURE 16-6 Gain in amplitude of temperature variations within a packed bed reactor vs. modulation frequency with triangular temperature waves in the inlet reacting gas entering a packed bed of catalyst. (a) Effect of position in the bed (Experimental for CO oxidation): (b) effect of bed structure in terms of catalytically active and inert packing (see inset). Simulation for CO oxidation. *(Figures adapted from* Jaree et al. (2003a) *and* Jaree et al. (2004) *respectively, with the authors' permission.)*

and the frequency at which resonance was observed. *Jaree et al. (2001b, 2004)* explored this behavior by simulating CO oxidation in beds of alternating catalyst and inactive carrier. **Figure 16-6(b)** shows the catalyst-carrier configurations examined. The amplitude could be almost doubled by alternating sections of catalyst with sections of inert carrier with the same density, thermal conductivity and heat capacity. **Figure 16-6(b)** illustrates these observations. Curiously, the maximum gain observed is well below the gain measured by *Hansen and Jorgensen (1974)*.

Yakhnin et al. (1995a, b) simulated amplification in packed bed reactors assuming oxidation of CO in air and a widely used packed bed reactor model. Their work showed that large amplification of small, periodic disturbances was possible. With the proper choice of model parameters, their simulations closely modeled the response to periodic temperature variation measured experimentally by *Jaree et al.* for a Pt catalyst in **Figure 16-6(a)** and by *Na-Ranong et al.* for a CuO catalyst. With their model, *Yakhnin and Menzinger (2008)* explored the effect of CO concentration on disturbance amplification. The CO reactant strongly inhibits oxidation on the Pt catalyst but has just a small effect with the CuO catalyst. **Figure 16-7** summarizes the Yakhnin results. Their simulations indicated that amplification increases with rising reactant concentration, while a comparison of their simulations for the Pt and CuO catalysts showed that reactant inhibition suppresses amplification.

Explanation of the amplification phenomenon is the interaction of concentration and temperature waves moving through the packed bed reactor at different velocities. Faster moving concentration waves overtake the slower-moving temperature waves, causing alternating bursts and sharp drops in energy generation with time. This is the same phenomenon that is associated with wrong-way behavior. As the waves move upstream, heat exchange between gas and the solid catalyst as well as heat loss from the reactor attenuate the temperature variations. A more detailed discussion of this phenomenon is available (*Posarac et al., 2003*).

Reaction went to completion in the experimental studies of *Jaree et al. (2003a, b)* and *Na-Ranong et al. (2003)*, just as was the case for

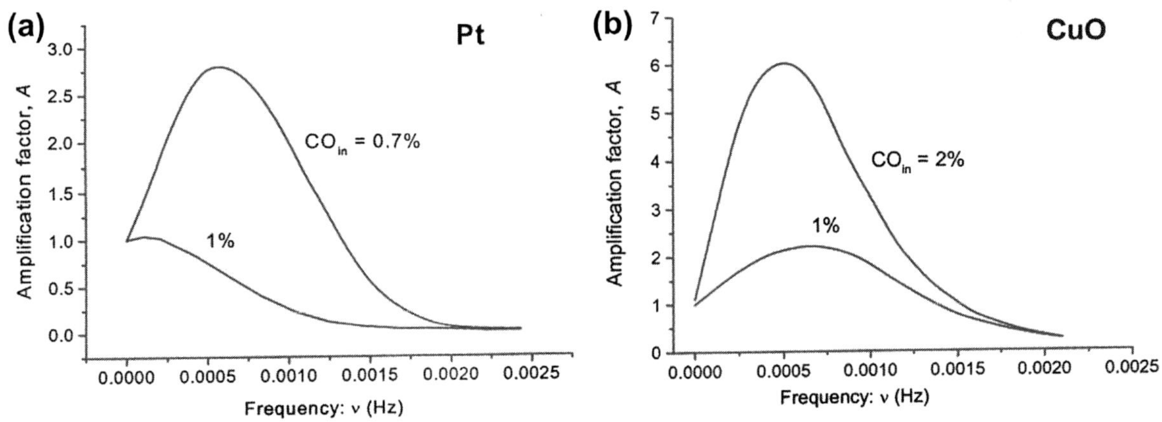

FIGURE 16-7 **Effect of reactant inhibition and reactant concentration on amplification of periodic temperature disturbances in packed beds of catalyst assuming CO oxidation in 10 vol% O_2 at about 1 bar and 400 K: (a) Pt/Al_2O_3 catalyst for which the reaction is strongly CO inhibited, and (b) CuO/Al_2O_3 catalyst exhibiting just weak inhibition by CO and CO_2.** *(Figure adapted from* Yakhnin and Menzinger (2008) *with the authors' permission.)*

the Hansen and *Jorgensen* (1974) experiments. Consequently, the question of whether inlet temperature modulation increases the oxidation rate cannot be determined. Increasing the flow rate, while raising catalyst activity so that the reaction does not go to completion in the reactor, should provide an answer to the rate improvement question. Further experiments are needed.

16.5 TEMPERATURE MODULATION OF TRICKLE BEDS

Haure et al. (1989) observed that periodic interruption of the liquid flow through a trickle bed dramatically increases the reaction rate in the bed. The authors demonstrated that part of the rate increase arises from a temperature increase in the trickle bed after the flow has stopped. It is for this reason that flow interruption is mentioned here. The subject is discussed in **Chapter 17**.

Liquid flow is the major carrier of heat from a trickle bed in which an exothermic reaction takes place. Flow interruption was used by *Haure* and co-workers to generate temperature excursions in their studies of the catalytic hydrogenation of α-methylstyrene in a co-current downflow trickle bed. Exothermicity of this reaction coupled with the cessation of heat removal through the flowing liquid causes the excursion. This periodic temperature jump is one source of the increase in the time-average rate of reaction *(Haure et al., 1989)*. At the shortest τ_{cycle}, there was about a four-fold increase in the rate of hydrogenation. This remarkable rate increase arises from the partial wetting of the external surface of a catalyst particle. When particle wetting decreases, the more rapid diffusion through the gas phase than through the liquid phase means higher transport of H_2 to the catalyst interior. Indeed, the hydrogenation rate for a gas-phase pathway can be as much as an order of magnitude greater than rates through partially liquid-filled pathways. Periodically interrupting the liquid flow reduces wetting of the catalyst surface and augments the contribution of the gas-phase transport pathway, increasing the rate of hydrogenation. This rate is further enhanced by the temperature jump produced by the exothermicity of reaction. The reaction is not conducted in the gas phase because of the high heat generation that can lead to runaway. Consequently, periodic temperature pulsing caused by flow interruption provides a means of avoiding runaway while taking at least partial advantage of the more rapid gas-phase reaction.

Temperature pulses can also be created by periodically altering the feed composition to the reactor, as is described by *Hanika et al. (1990)*. Of course with a three-phase reactor, feed temperature modulation can be undertaken very simply without resorting to flow interruption or feed composition switching.

Any beneficial effect of feed temperature modulation on trickle bed performance, indicated by the experiments of *Gabarain et al. (1997)*, is clouded by interference of flow interruption that profoundly changes reactor operation. Further experimental study seems warranted. Also, the effect of temperature modulation on both rate of reaction and selectivity deserves attention. A suitable candidate system for this study would be the catalytic hydrogenation of crotonaldehyde. In this process, there are two parallel reactions depending on the catalyst used: Hydrogenation of the double bond to produce butyraldehyde (butanal) and hydrogenation of the aldehyde group to give crotyl alcohol. *Stradiotto et al. (1999)* used the crotonaldehyde reaction in their study of flow modulation in trickle beds.

16.6 EXPERIMENTAL STUDIES WITH MICROREACTORS

Although the experimental results discussed earlier were not encouraging, they did not rule out an effect of temperature modulation on the reaction rate. After a lapse of some 20 years, interest in periodic temperature forcing resurfaced, driven by the availability of microreactors with fast heating and relatively low thermal inertia that make high cycle frequencies accessible. In 2000, a suggestion for a microchannel device suitable for fast heating and cooling was published by *Alépée et al. (2000)*. Brandner's PhD thesis (*Brandner, 2003*) introduced the first micro-channel device capable of fast periodic temperature changes for heterogeneously catalyzed gas phase reactions, a so-called fast temperature cycling (FTC) system.

Microstructure devices improve heat transfer, overpressure resistance, and flexibility over conventional laboratory equipment. However, the applicability of these tiny devices is restricted by corrosion, possible flow blockage, and mass flow limitations that cannot easily be overcome. A detailed description of the opportunities for, as well as the advantages and disadvantages of, microstructure devices are discussed by *Rouge et al. (2001)* and *Brand et al. (2006)* and exhaustively by *Quiram et al. (2007a)*.

16.6.1 Micro Devices for Temperature Modulation

Even though the thermal inertia of micro devices is very low compared with conventional laboratory equipment, the design of these devices has to be undertaken with care to achieve clean sine or square-wave modulation at reasonable frequencies. *Van Neer et al. (1996)* suggested there is a frequency window for rate enhancement through temperature modulation. At low frequencies, the reaction will simply operate in the quasi-steady state, while at sufficiently high frequencies it will operate in the relaxed steady state or at steady state. Thus, micro reactors must be designed to fit the frequency window for the reaction in question. Recommendations on design, manufacturing and testing of microstructure devices are offered by *Brand et al. (2006)*.

The original microstructure device for temperature modulation of a reaction was improved by *Rouge et al. (2001)* who exchanged heating and cooling fluids to produce temperature modulation. A description of their micro device is given in **Table 16-1**. **Figure 16-8(a)** shows the stack structure and flow routing. Parallel channels machined into the wafers served as flow channels; they are depicted in **Figure 16-8(b)**.

An improved stainless steel body, shown in **Figure 16-9**, was designed for heterogeneously catalyzed gas-phase reactions. It was continuously heated by high power resistor cartridges and periodically cooled with de-ionized water. It consisted of a central block with insertions for the heater cartridges and microchannel foils for passage of reactant gases and cooling fluid, arranged in alternating order. A top plate and a base plate completed the device. Details on the design and performance of the device can be found in (*Brandner, 2003, 2008*; *Brandner et al., 2001, 2002, 2003*).

Microchannel foils were designed to be short for the cooling fluid and perpendicular to the reactant channels to reduce pressure drop and provide adequate cooling. Reactant channels were sized to provide a long residence time for the gas reactants. Thus, they had a path length of about 812 mm. The reactor itself was only 44 mm long, 35 mm wide and about 8 mm high.

To carry out temperature modulation, two magnetic valves, under computer control, switched cooling water flows on and off. Temperatures were measured by fine thermocouples inserted into small holes in the stainless steel body. To overcome water hold-up in the cooling passages that limited the heating rate in that part of the cycle, a third valve was added to the switching setup to initiate the flow of a short high-pressure pulse (100 to 200 ms) of air through the cooling channels at the

TABLE 16-1 Temperature Modulation Experiments

Authors	Research Objective	Reaction	Reaction Conditions	Modulated Variable	Observations	Comments
Abdul-Kareem et al. (1980b)	Investigation of temperature modulation using catalytic CO oxidation.	CO oxidation over a commercial vanadia catalyst.	Packed bed, differential reactor, 6.35 mm i.d. immersed in a fluidized sand bath that was electrically heated.	Temperature of the fluidized sand bath was modulated using a timer on current to heating coils in the sand bath. $1 < \tau_{cycle} < 8$ h	$\tau_{cycle} > 5$ h brought the time average oxidation rate = to the quasi-steady-state reaction rate. This rate was greater than the time average of the steady-state rates at the upper and lower temperatures.	Because of thermal inertia, temperature waves in the sand bath were trapezoidal rather than square. At $\tau_{cycle} < 2$ h, time average oxidation rate is less than the mean of the steady-state rates at the upper and lower temperatures.
Lee et al. (1980)	Investigation of temperature modulation for a series reaction.	Saponification of diethyl adipate with NaOH in an isopropanol/water mixture.	Reaction carried out in a jacketed CSTR. Modulation undertaken by sending alternately a fluid at 15°C and one at 90°C through the jacket.	Bang-bang temperature modulation by switching between hot and cooler fluids in the reactor jacket. $\tau_{cycle} = 5$ min; $0.1 \leq s \leq 0.9$ based on the flow of the 90° fluid.	Yield of the intermediate increased only 1.5 to 22%, i.e., much less than predicted by simulation.	Cycle period was apparently too short to achieve the desired temperature amplitude.
Jaree et al. (2001b, 2003a, b); Na-Ranong et al. (2003)	Investigate amplification of a temperature disturbance using CO oxidation in a packed bed of catalyst.	CO oxidation over Pt and Cu catalysts.	2.54 cm i.d. packed bed reactor was surrounded by a high vacuum jacket to reduce heat loss. Seven thermocouples, evenly spaced axially, provided transient temperatures.	Inlet temperature triangular waves were generated by electrical heaters.	Amplification of the temperature variation was observed, while periodicity of the inlet modulation was preserved. Maximum gain was 4- to 5-fold.	Resonance, saturation and extinction phenomena were also observed.

Rouge et al. (2001)	Development of a microchannel reactor with a narrow residence time distribution for investigating modulation.	Dehydration of isopropanol to propene over a γ-alumina catalyst.	20 mm × 20 mm stack consisting of 9 ceramic wafers and 9 steel plates for flow control, each wafer containing 34 quadrangular microchannels of 300 μm width, 240 μm depth and 20 mm length.	Symmetric square-wave inlet concentrations of iPrOH were fed between 0 and 0.86 mol/m^3.	Experiments showed reactor performance under modulation exceeded the maximum possible performance under steady state.	This was an early version of a microreactor intended eventually for temperature modulation.
Ohmori et al. (2001, 2003, 2011)	Investigation of the modulation of a tubular packed bed reactor mainly by simulations.	Conversion of methanol to light olefins.	In some studies, certain products were recycled back to the inlet or middle of the reactor.	Reactor wall temperature was varied sinusoidally along its axis.	Small improvements in yield and selectivity were predicted along with global reaction rate enhancement.	A synergetic effect was observed between temperature modulation and recycling of some products.
Brandner (2003); Brandner et al. (2004)	Investigation of the modulation of a microreactor using periods in the second to subsecond range.	Oxidation of CO over Pt deposited on alumina by sol-gel technique.	Online FTIR spectrometer analyzed the concentration of the CO_2 generated.	Fast temperature-modulation through continuously heating by six 1.05 kW resistor cartridges and periodic cooling by a fluid stream.	100 K temperature excursions were observed in a 5 s cycle period, as well as a significant increase in CO conversion to CO_2.	One of the early microreactors with integrated product analysis and temperature measurement.
Carlsson et al. (2004)	Development of a microreactor for use in low-temperature oxidation.	Total oxidation of propane.	Monolith samples (L=15 mm, diameter =12 mm, 69 channels) cut from a honeycomb of cordierite were impregnated with Pt solution and calcined. FTIR spectroscopy and mass spectrometry were combined to analyze the catalyst surface and product composition.	Step- and pulse-changes of feed gas composition.	Rate enhancement was observed.	One of the early microreactors with integrated product analysis and temperature measurement.

(*Continued*)

TABLE 16-1 Temperature Modulation Experiments (*cont'd*)

Authors	Research Objective	Reaction	Reaction Conditions	Modulated Variable	Observations	Comments
Hansen et al. (2006)	Investigation of the fast temperature modulation of a catalytic reaction.	Oxidation of CO over a supported Pt/Al_2O_3 catalyst.	Feed gas composition was 5% CO, 2.5% O_2, 2.5% Ar and 90% He. Flow of 5 mL/s ≡ *SV* of 1 s^{-1}.	Temperature was modulation via computer control of heaters at frequencies between $0.03 < \omega < 0.1$ Hz.	Reaction rates increased monotonically as function of frequency but displayed a maximum as a function of the mean temperature and the amplitude.	Phase shifts of CO and O_2 relative to CO_2 varied monotonically with the mean reactor temperature, amplitude and frequency.
Sotowa et al. (2006)	Investigation of the fast temperature modulation of a catalytic reaction.	Oxidation of propylene over Pt/γ-alumina catalyst.	$C_3^=$ and O_2 concentrations in the feed were 10 % and 5 %, respectively. Carrier gas was He. $1 < \tau_{cycle} < 10$ s and $0.1 < s < 0.9$.	Temperature of the catalyst surface through the periodic heating of the reactor cavity.	Time-average conversion of $C_3^=$ under modulation was greater than under steady-state. The reaction approached a relaxed steady state. Quasi-steady state was not reached at $\tau_{cycle} = 10$ s.	It was not possible to measure the reaction temperature, which was estimated only indirectly from the heater current.
Jensen et al. (2007, 2008b)	Development of a reactor with integrated heating and measurement systems to operate at high frequencies.	Oxidation of CO over Pt/Al_2O_3 catalyst.	Silicon microreactor chip (0.9 mm × 16 mm × 20 mm) built for rapid heat conduction and compatibility with semiconductor sensors and actuators. Reactor chamber was separated from the heater by 200 nm SiO_2 and 30 μm of silicon to speed heat transfer.	Reactor temperature was varied by applying a sinusoidally varying voltage across the resistance heater element.	Rate enhancements up to 70% were observed for ω_{cycle} up to 1 Hz; these decreased up to 2.5 Hz, the maximum used. The decrease was attributed to amplitude loss at higher frequencies.	
Luther et al. (2008a, b)	Development of a microreactor for catalytic gas-phase reactions under temperature modulation.	Oxidation of CO over Pt/Al_2O_3 catalyst.	Temperatures up to 500 K were used.	Temperature modulation achieved by electrically heating the microreactor continuously and periodic cooling with water passing through cooling coils.	Temperature oscillations up to 60 K at $\omega = 0.06$ Hz were demonstrated with gradients < 15 K within the reactor.	

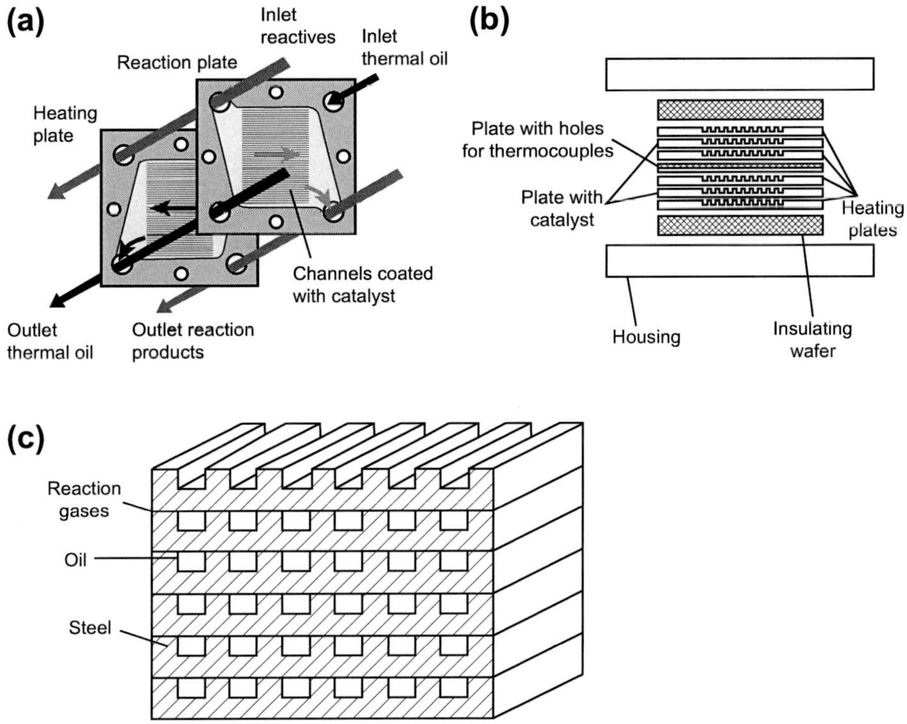

FIGURE 16-8 Micro device for temperature modulation of a reaction system: (a) schematic of directional masks for fluids containing reactants (reacting fluids) and heat transfer fluids, (b) micro reactor-heat transfer stack, (c) stacking of the machined out wafers showing the flow channels. *(Figures adapted from* Rouge and Renken (2001a) *with kind permission. © 2001 by Springer Science+Business Media.)*

beginning of each heating cycle. A maximum frequency of about 4 Hz with an amplitude of 100 K was possible with this setup (*Brandner, 2003*).

Brandner et al. found that the cycles were reproducible and could be precisely controlled. The temperature distribution inside the FTC reactor was examined using infrared thermography which disclosed a fairly uniform temperature in the central region of the microreactor during the cooling portion of the cycle, but much higher temperatures persisted at the entrance and exit regions. *Brandner et al.* observed that the temperature variations imposed could be passed on to the reactant flow.

Several microcreactors of different design for heterogeneously catalyzed gas phase reactions have been built since 2001. Details of these devices are given in **Table 16-2**. Microreactors have been a popular topic since 1990 and a modest literature exists discussing their design and application. A group of three publications by *Quiram et al. (2007a, b, c)* provide a good introduction to this literature.

Because of the non-homogeneous temperature distribution in *Brandner's* first microreactor, *Brandner* and co-workers re-designed the reactor. Several versions were developed. One of these proved suitable for the CO oxidation experiments by achieving a more

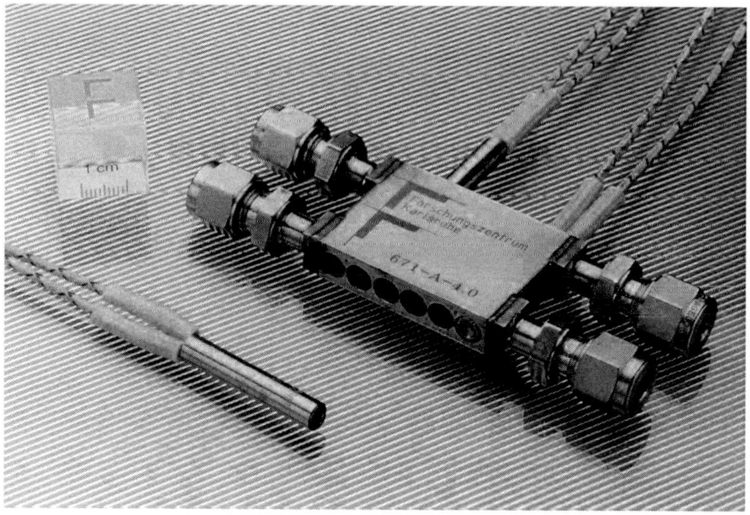

FIGURE 16-9 A stainless steel micro-channel reactor for fast periodic temperature changes, the so-called fast temperature cycling (FTC) device. *(Photograph reproduced from* Brandner (2008) *with the author's permission.)*

uniform temperature distribution in the microreactor. That design is shown in **Figure 16-10**. The device is held together with screws and sealed with a polymer O-ring. Thus the structure inside the reactor can be interchanged.

Two thermocouples were placed inside the reactor to measure the temperature of the metal plate used as the catalyst carrier. The temperature change of the carrier depended on the cycle period. However, for short periods of about 16 s, a temperature difference

TABLE 16-2 Microreactor Designs and their Performance Characteristics

Reference	Reactor Material, Heating / Cooling	Size (mm)	Maximum Temperature Amplitude (K)*	Maximum Cycle Frequency (Hz)	Maximum Heating / Cooling rate (K/s)
Hansen and Jorgensen (1974)	Electrical heating; fluid cooling	10 × 20	60 (20)	0.03 (0.1)	150
Rouge et al. (2001)	Fluid heating and cooling	10 ϕ × 200	10	0.1	2
Brandner et al. (2004)	Steel body; electrical heating; fluid cooling		50+	4	140
Jensen et al. (2007, 2008b)	Electrical heating; convective cooling	16 × 20	60 +	> 1 Hz	> 150
Sotowa et al. (2008)	Pulse electrical heating; cooling by heat loss	2 ϕ × 50	250 (200)	0.1 (1)	n.a.
Luther et al. (2008a)	Electrical heating; fluid cooling	53 × 35	100 (60)	0.1 (0.06)	n.a.

* *Amplitude at the higher frequency in ( ).*

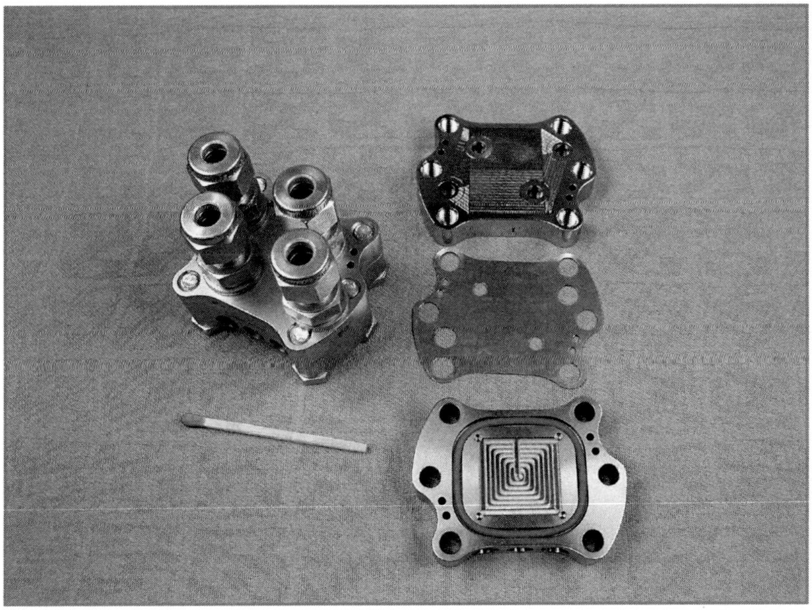

FIGURE 16-10 **Photo of the modified microreactor before assembly. The foil shown in the middle was welded to the upper frame (top) to seal the cooling passage from the reaction chamber.** *(Photograph reproduced from* Luther (2008) *with the author's permission.)*

of 14 K between inlet and outlet streams was observed at an amplitude of 50 K in the heating cycle.

Catalysts can be introduced into microreactors in several ways (*Hessel et al., 2009; Brand et al., 2006*). One way is to make the reactor out of the catalyst itself, e.g., platinum (*Sotowa et al., 2008*). However, most researchers used ceramic materials such as silica and employed wet impregnation (*Hansen et al., 2006; Quiram et al., 2007a; Brooks et al., 2002*). *Rouge et al. (2001)* used ceramic wafers and applied a washcoat to the channel walls of the wafers and then impregnated the washcoat with the active catalyst. *Brandner et al. (2004)* and *Luther et al. (2008b)* used a sol-gel technique to create an alumina support with sufficient porosity to be wet-impregnated with a Pt-salt solution (*Haas-Santo et al., 2002*). *Luther et al.* substituted a sintered metal fibre plate for a micro-channel system and covered the plate with an alumina layer deposited by a sol-gel-method (*Luther, 2008; Haas-Santo et al., 2002*).

The chip-based microreactor used by *Jensen et al. (2007, 2008b)* and *Hansen et al. (2006)* was modeled on silicon chips used in the semiconductor industry. It was fabricated from a silicon wafer, and flow micro-channels as well as channels for resistance heating and thermometry were prepared by etching methods. A μm hole near the end of the flow channel leaks into a line to a mass spectrometer. A side view of the microreactor is given in **Figure 16-11**. An aluminum housing contains the chip and provides flow connections. The gas inlet is shown in the figure. At the bottom of the silicon chip, a very thin NiSi strip has been laid down to serve as an electric resistance heater. Another strip of the same material but with a much lower electrical resistance serves as a surface thermometer. As may be seen in the figure, spring-loaded connectors bring in power and measure

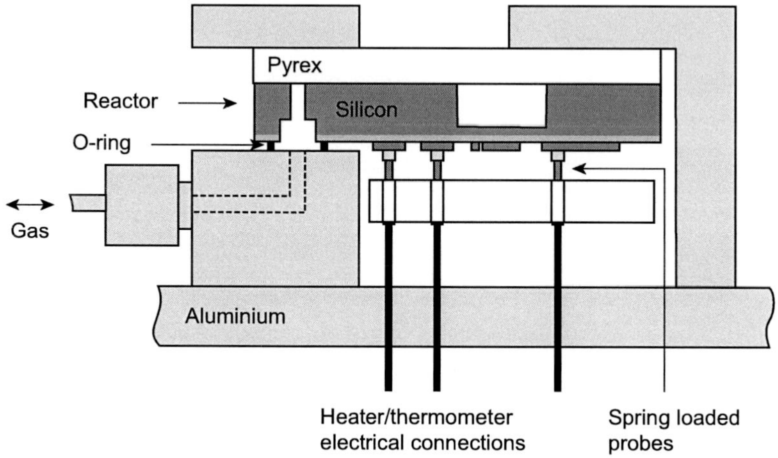

FIGURE 16-11 Side view of the Jensen Reactor showing its position in an aluminium housing and the flow and instrument connections. *(Figure reproduced from* Jensen et al. (2008b) *with permission. © 2008 by Elsevier Inc.)*

ohmic resistance. A Pyrex glass plate seals the top of the silicon chip. Further details of this interesting design are given in the *Jensen* papers mentioned above. A washcoat plus an impregnation procedure was used to introduce a Pt catalyst. The BET surface area was found to be 130 m^2/g. Temperature modulation for the *Jensen* microreactor was provided by on-off operation of the heater supplemented by blowing air over the device. Temperature was not uniform along the reactor microchannel etched in the silicon chip.

Sotowa et al. (2008) developed a cylindrical microreactor with an α-alumina shell mounted in a stainless steel T-junction (**Figure 16-12**). Temperature cycling was generated by on-off heating using a platinum resistance heater located axially in the alumina shell as shown in the figure. Flow occurred parallel to the heating tube so that the temperature was not uniform across the reaction surface. Indeed, the authors report variations reaching as much as 100 K. Heating was provided by micro-pulses of power as illustrated in the lower half of the figure. The system provided an amplitude of about 45 K at a frequency of 1.2 Hz. The inside of the alumina tube was washcoated with a boehmite suspension. Calcining converted the washcoat to γ-alumina that was then impregnated with a chloroplatinate solution. The *Sotowa* paper contains details of the microreactor design, temperature modulation, and catalyst preparation.

16.6.2 Experimental Procedure with Microreactors

Under fast temperature modulation, conventional analytic systems such as gas chromatography and some mass spectrometry systems cannot follow the rapidly changing output from a microreactor. *Brandner et al.* used a fast FTIR spectrometer, providing four full scans per second, to analyze the gas leaving the reactor (*Brandner, 2003, 2008; Brandner et al., 2004*). The main disadvantage of this system is that reactant species must have IR absorption lines. *Rouge et al. (2001)* used a fast GC-MS system for their analysis. Although this is probably the favored way to analyze all components in the product stream, the dead volume between the microreactor and the analytical system smooths the time changes, averaging the effects under study. A solution to the smoothing problem was developed by *Quaade et al. (2004, 2005)*. Their microreactor was fitted with a very small nozzle in the outlet, having a diameter of just 3 μm, that led to a mass spectrometer. This device

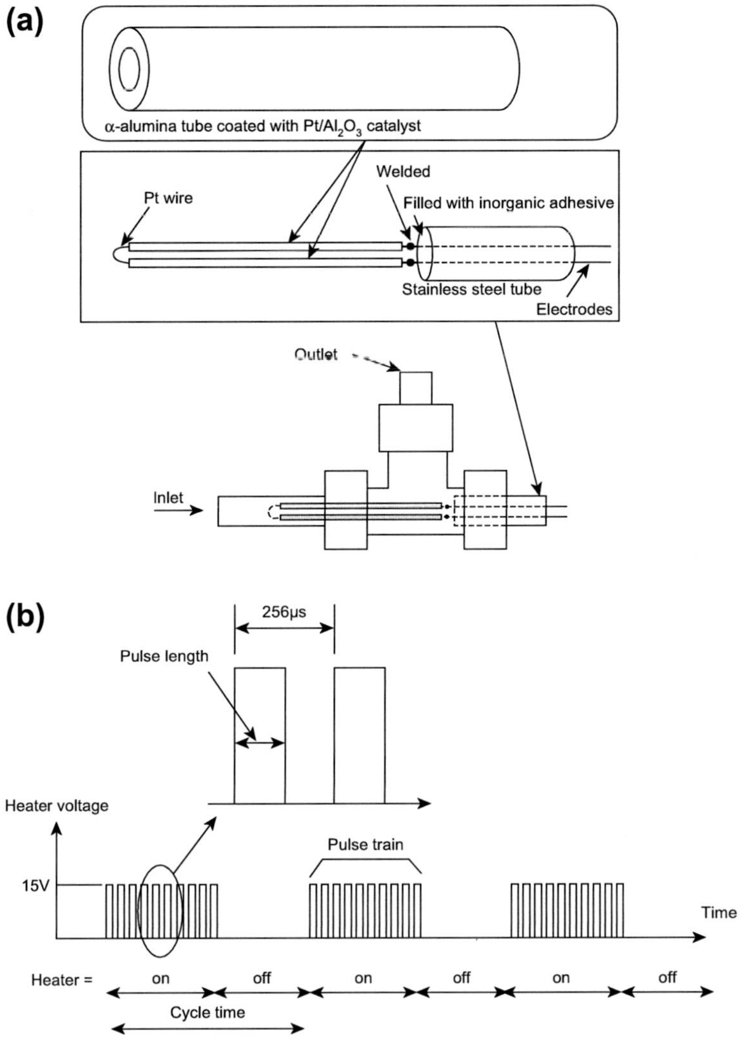

FIGURE 16-12 Sotowa Reactor: Reactor and temperature modulation employed by Sotowa et al. (2008): (a) T-junction reactor containing two 50 mL sections of an α-alumina tube whose outer surface had a Pt-impregnated, γ-alumina coating. Pt wires mounted along the center-line of the tube provided heating; (b) temperature modulation generated by electrical heating in an on-off mode (overheating of the Pt wire prevented by using micropulses of current). *(Figure adapted from* Sotowa et al. (2008) *with permission. © 2008 by Elsevier Inc.)*

facilitated online analysis without major dead volumes. *Luther et al. (2008a, b)* adapted this technology for their metal microreactor.

The systems briefly described above follow reaction under temperature modulation reasonably well. However, the best currently available analytical devices have time constants of the same order as those of the rapid dynamic systems under study. This results in some irreducible uncertainty.

16.6.3 Experimental Observations

Rouge and Renken (2001a) investigated the dehydration of isopropanol over alumina using modulation as well as step-changes in temperature. They employed a 20 s period with a 10 K amplitude. Dehydration is both product- and reactant-inhibited. Adjacent sites are needed for the surface reaction. Even at low isopropanol concentrations, isopropanol and the water product on the surface severely reduce the reaction rate. However, both species are reversibly adsorbed, so the reaction rate can be increased several orders of magnitude by periodically flushing the catalyst with a non-adsorbing carrier gas (*Morávek, 1992a; Silveston, 1998*). *Rouge and Renken* found from step-change observations that an abrupt increase in temperature had a similar effect. Lags in reaction rate ranging from 5 to 10 s were also seen. These results led *Rouge and Renken (2001a)* to undertake modulation measurements at $\tau_{cycle} = 20$ s for the dehydration reaction. Their data are reproduced in **Figure 16-13**. A 5 s phase shift may be seen between the peaks of the isopropanol and propene concentrations. Further calculations show at best a small increase of not more than 1 to 2% in the time-average propene concentration due to modulation. Ether is a reaction by-product.

Brandner and co-workers used the oxidation of CO to CO_2 on an alumina-supported Pt catalyst for their first experiments on temperature modulation (*Brandner, 2003; Brandner et al., 2001, 2002, 2003*). **Figure 16-14** was obtained using the microreactor shown in **Figure 16-10** with an FTIR to measure CO_2 in the off-gas. Measurements made under modulation are compared with those made at both steady state and quasi-steady state in the figure. This comparison shows a significant increase in yield due to modulation at $\tau_{cycle} = 10$ s and $A \approx 100$ K. There is about a three-fold yield enhancement. A convenient way of visualizing the enhancement is to note that the relative CO_2 concentration in the product gas at 411 K under steady-state operation can be reached at 355 K through temperature modulation.

Brandner (2003) attributed the enhancement under temperature modulation either to increased time-average isopropanol coverage of the catalyst surface or to an improved distribution of isopropanol and vacancies on the surface. Either of these explanations is

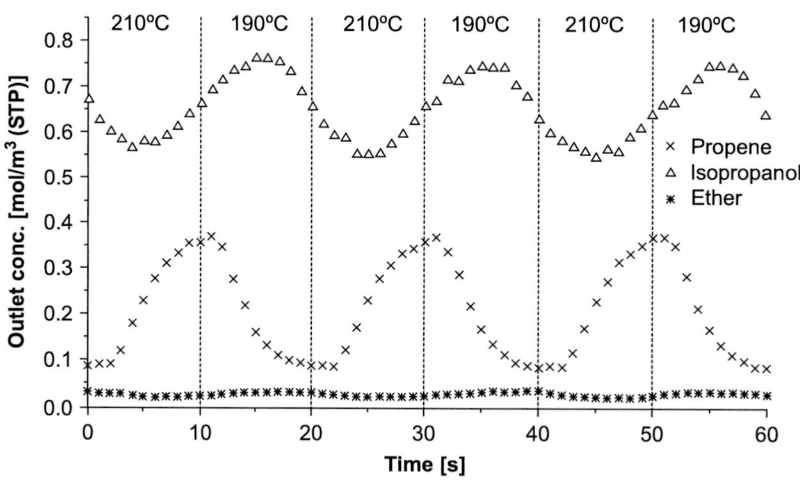

FIGURE 16-13 Response of isopropanol, propene and ether concentrations to coolant temperature modulation in a microreactor. *(Figure reproduced from* Rouge et al. (2001a) *with permission. © 2001 by Elsevier Inc.)*

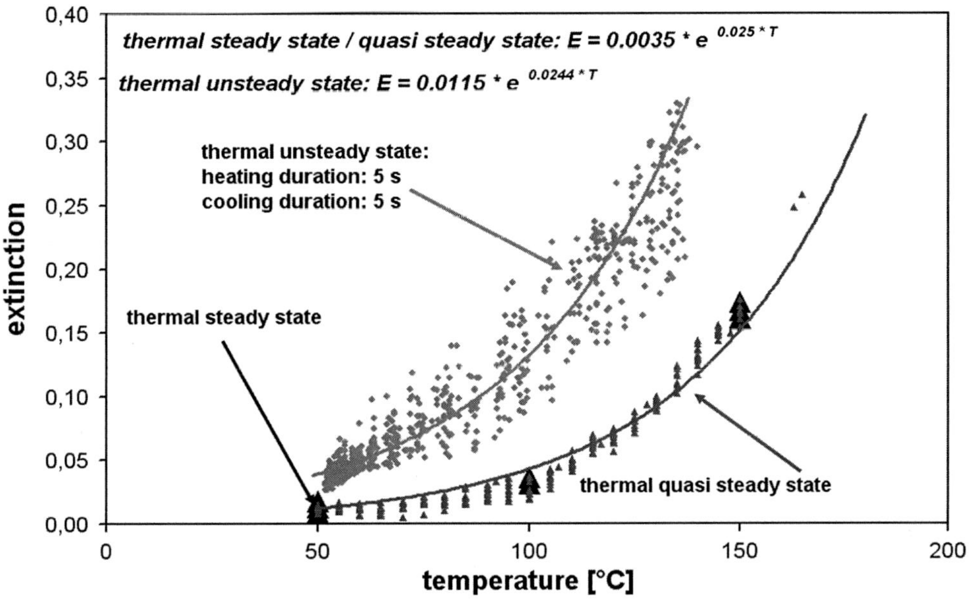

FIGURE 16-14 Comparison of CO_2 concentrations leaving the miocroreactor, expressed as FTIR extinction coefficient, under different modes of operation (Δ steady state; ▲ quasi-steady state; ◆ unsteady state) for the catalyzed oxidation of CO to CO_2. Unsteady-state and quasi-steady-state measurements were made under temperature modulation. Frequencies for the unsteady-state measurements were more than an order of magnitude greater than those used for quasi-steady state. *(Figure reproduced from* Brandner (2003, 2008) *with the author's permission.)*

consistent with adsorption/desorption changes that are slower than the temperature changes imposed on the catalyst.

Luther et al. (2008a) investigated the oxidation of CO under temperature modulation using a FTIR spectrometer as well as a mass spectrometer to measure product gas composition. They too observed a large increase in the oxidation rate as shown in **Figure 16-15**. This figure plots the instantaneous CO_2 concentration measured vs. the temperature reading in the reactor. Different symbols identify readings made with increasing and decreasing temperature during a modulation cycle. An isola is present and must represent some type of hysteresis behavior on the catalyst surface. Steady-state measurements are also shown in this figure for comparison purposes. The concentration of CO_2 is significantly higher under modulation than at steady state.

Figure 16-16(a) compares measurements at two additional frequencies and amplitudes. It is evident that performance is affected by these variables. If we assume that only frequency is important, the enhancement of the time-average rate of CO oxidation increases with increasing modulation frequency, as plotted in **Figure 16-16(b)**. This plot suggests that the claim by *van Neer et al. (1996)* is correct, namely, that conversion increases with increasing frequency. Of course, eventually a relaxed steady state is attained whereupon there can be no further effect of frequency on conversion. As frequency decreases, the quasi-steady state is reached where again there is no effect of frequency on conversion.

Comparison of **Figure 16-14** with **Figures 16-15 and 16-16(a)** discloses a contradiction. The latter figures show that CO conversion under temperature modulation approaches

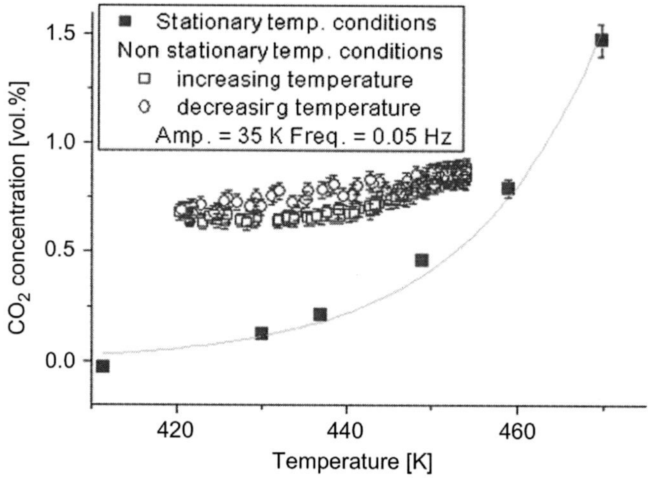

FIGURE 16-15 **Comparison of instantaneous off-gas CO_2 concentrations during modulation with steady-state values. ■ Steady state; temperature modulation with $A = 35$ K and $\omega = 0.05$ Hz: □ Portion of the cycle in which temperature increases, ○ portion of the cycle in which temperature decreases.** *(Figure reproduced from* Luther (2008) *with the author's permission.)*

conversion under steady-state operation as the mean reaction temperature approaches 480 K. **Figure 16-14**, however, shows a divergence of the curves. The amplitude of forcing is greater for the results given in this figure and measurements were made over a somewhat lower temperature range.

The results obtained by *Hansen et al. (2006)* are shown in **Figure 16-17**. Part **(a)** of the figure shows the instantaneous variation of CO_2 in the microreactor off-gas with time, while in **(b)** the measured concentrations are plotted versus the observed temperature. Also shown in **(b)** is the steady-state CO_2 concentration exiting the microreactor plotted against temperature. This behavior was measured by very slowly changing the reactor temperature. Although obtained for the same reaction and a similar Pt/Al_2O_3 catalyst to that used by *Brandner et al.* and *Luther et al.* and covering about the same range of mean temperatures, **Figure 16-17** indicates a much lower enhancement of the rate of CO oxidation.

The instantaneous points in (b) trace an isola suggesting a hysteresis phenomenon. Further experiments disclosed that the oxidation rate enhancement relative to the rate under quasi-steady state goes through a maximum at about 425 K. The maximum arises from a saturation phenomenon, i.e., nearly complete conversion at the maximum temperature excursion in a cycle. A maximum was also observed with modulation amplitude for the same reason. However, above $\omega = 0.05$ Hz, the enhancement increases almost linearly with modulation frequency. Use of a mass spectrometer allowed determination of the phase lags between products and reactants. Reactant changes were in-phase, but their maxima lagged the CO_2 minima by less than a second. This lag decreased with increasing mean temperature and amplitude and reached zero at the highest frequencies used. The lag switched to a phase lead at low frequency.

Jensen and co-workers used a modified version of the Hansen microreactor mentioned above to explore the effect of higher modulation

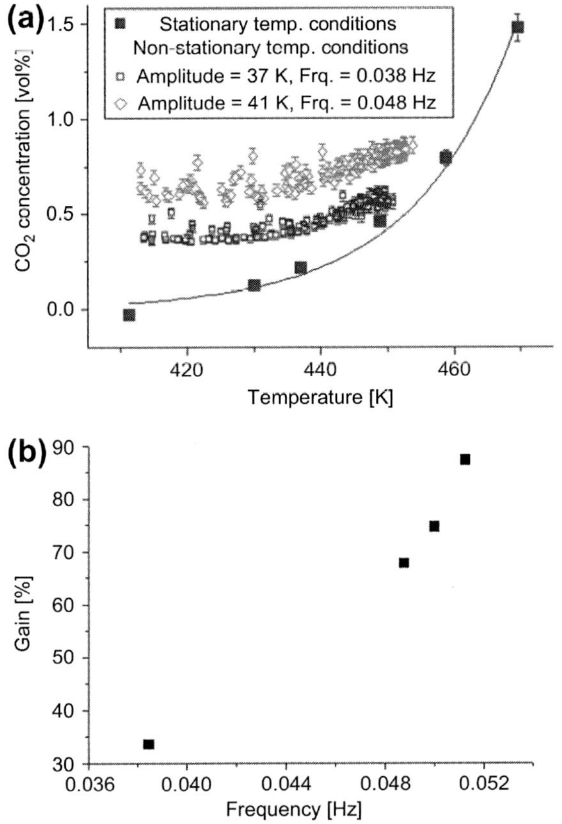

FIGURE 16-16 **Effect of amplitude and cycling frequency on CO_2 production and rate enhancement during temperature modulation: (a) □ experiments with $A = 37$ K and $\omega = 0.038$ Hz; experiments ◇ with $A = 41$ K and $\omega = 0.048$ Hz; corresponding steady-state concentrations (■ and full line); (b) ■ percentage increase in the exiting CO_2 concentration above steady state (gain) as a function of modulation frequency.** *(Figure reproduced from* Brandner (2008); Luther, (2008) *with the authors' permission.)*

frequencies. **Figure 16-18** shows the results of those experiments. In the figure, fractional enhancement is compared to that at a quasi-steady state reached by cycling at a low frequency, about 0.02 Hz. It can be seen that enhancement increases with increasing amplitude. The dip at frequencies above 1 Hz was caused by a decrease in amplitude of the temperature change at the catalyst surface. Modeling by the investigators disclosed that the conventional surface model for CO oxidation on a Pt surface (see **Section 3.2.5** in **Chapter 3**) cannot account for enhancement. However, if the model is modified to allow for the change in surface coverage by CO and O atoms with temperature, a good fit of the data can be obtained except at the highest amplitude used experimentally. The result is shown by the solid lines in **Figure 16-18**. Details of the model are discussed by *Jensen et al. (2007)*.

Based on further measurements using the same experimental system, *Jensen et al. (2008b)* observed that rate enhancement from temperature modulation depended on the mean temperature of the cycle and the reactant space time. *Jensen et al.* related enhancement to conversion and in this way were able to isolate the effect of the modulation amplitude on reaction rate (**Figure 16-19**). It is evident that, at low conversion, enhancement depends strongly on amplitude; whereas at the largest amplitude used, conversion dominates the enhancement observed. The shape shown in the figure results from the limit on enhancement as conversion goes to completion and the exponential effect of temperature on rate that arises when the amplitude is doubled. The joint effect of amplitude and frequency on rate enhancement is best seen at low conversions and at intermediate temperature. **Figure 16-20** shows these effects. Both amplitude and frequency strongly affect rate enhancement over steady-state.

Sotowa et al. (2008) investigated the oxidation of propene over a Pt/γ-Al_2O_3 catalyst coated onto the outer surface of an alumina tube as described in **Section 16.6.1**. The micropulse system of heating allowed a maximum temperature of about 1070 K and a modulation amplitude that could be varied from 500 to 600 K. **Figure 16-21** illustrates the strong effect of τ_{cycle} on conversion. Cycle split, the fraction of the cycle at high temperature, also strongly influenced conversion. Measurements in the

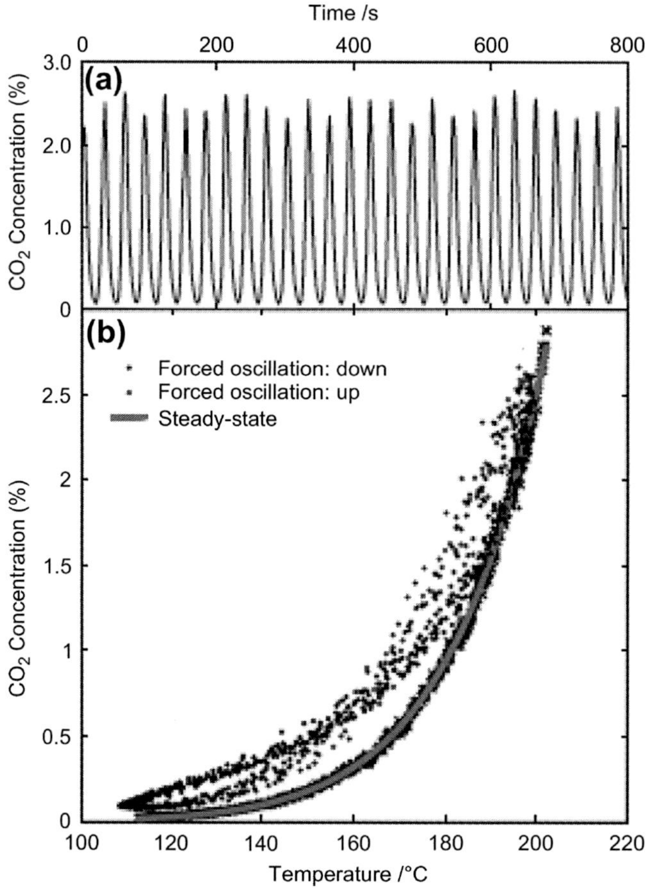

FIGURE 16-17 Temperature modulation of CO oxidation by O_2 over a Pt/Al_2O_3 catalyst at a mean temperature of 426 K, a space velocity of 1 s^{-1} and a forcing frequency of 0.03 Hz: (a) Variation of CO_2 concentration with time, (b) instantaneous CO_2 concentration vs. instantaneous temperature. Steady-state measurements are shown around the solid grey line which is an Arrhenius fit of the data points. *(Figure reproduced from* Hansen et al. (2006) *with permission. © 2006 by Elsevier Inc.)*

absence of reaction indicated that the maximum temperature in a cycle was 1070 K and was independent of period: The minimum temperature in a cycle decreased slowly with period and was about 550 K at a period of 10 s. Figure 16-21 shows that about 40% conversion could be reached through temperature modulation at a cycle split of 0.5 (i.e., equal on-off durations) using the space velocity and the micropulse heating described. Steady-state measurements with the same unit gave about 42% conversion at 1070 K, the highest temperature in the cycle. From these measurements, conversion under quasi-steady state would be 27%. Thus temperature modulation raises the oxidation rate significantly. The authors estimated that a relaxed steady state is attained at a cycle period of 1 s. **Figure 16-21** indicates that $C_3^=$ conversion for this state is about 30%.

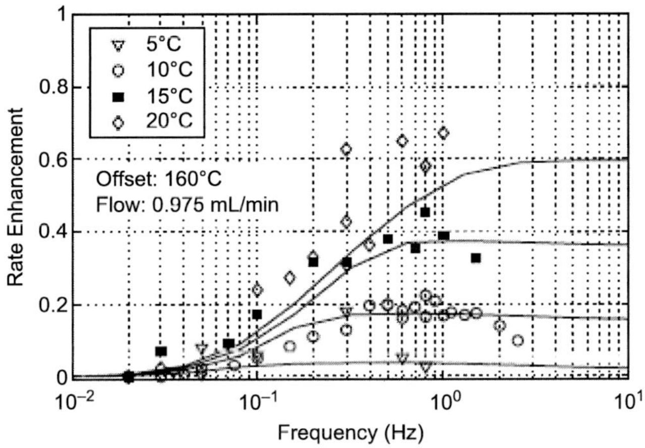

FIGURE 16-18 The effect of modulation frequency on the fractional enhancement of the CO oxidation rate above the quasi-steady-state rate at 433 K and different modulation amplitudes. Curves are from a simulatiojn. *(Figure adapted from* Jensen et al. (2008a *with permission. © 2008 by Elsevier Inc.)*

16.6.4 How Temperature Modulation Increases Reaction Rate

Only in the case of trickle beds with periodic flow interruption is there some consensus on the mechanism for rate improvement. In such systems, temperature modulation is initiated by the flow rate change. When flow is interrupted, more of the catalyst surface becomes accessible to gas phase reactants so it is the increase in

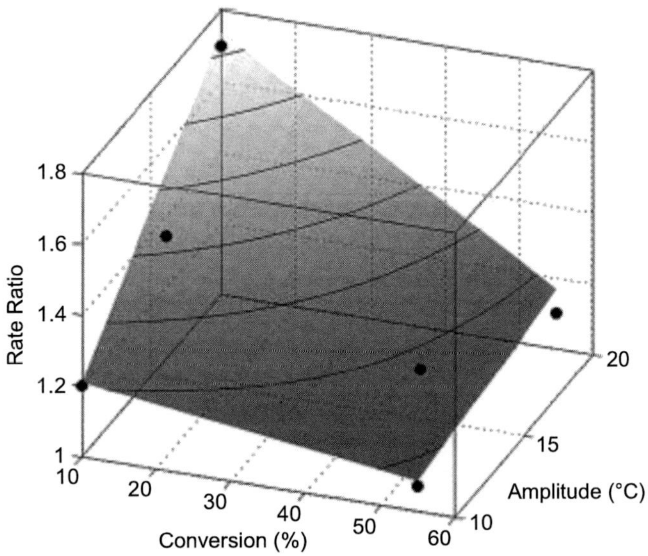

FIGURE 16-19 Surface of CO oxidation rate enhancement through temperature modulation as a function of CO conversion and modulation amplitude. *(Figure adapted from* Jensen et al. (2008a) *with permission. © 2008 by Elsevier Inc.)*

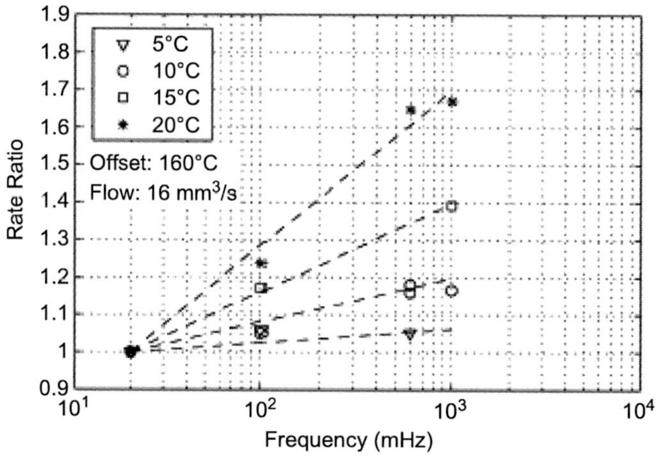

FIGURE 16-20 Dependence of the enhancement of the rate of CO oxidation through temperature modulation on frequency and amplitude for a time-average reactor temperature of 433 K and a gas flow rate of 16 mL/s corresponding to a CO conversion of 4%. *(Figure adapted from* Jensen et al. (2008) *with permission. © 2008 by Elsevier Inc.)*

reactant transport rate along with higher surface temperature that spawns higher reaction rates.

This does not appear to be the explanation for the strong temperature modulation effect on reactor performance described above. Most experimental studies of temperature modulation have utilized Pt-catalyzed oxidation of CO. Composition forcing or modulation of this reaction has been studied for decades (see **Chapter 3**). It is reasonable therefore to

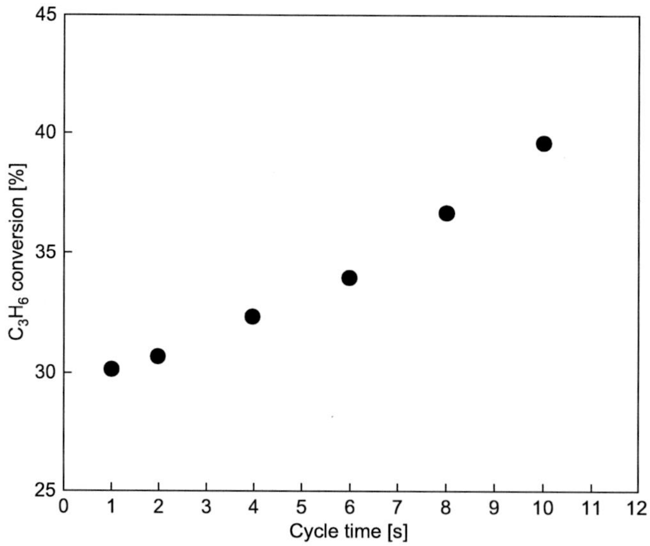

FIGURE 16-21 Dependence of $C_3^{=}$ conversion on modulation period for oxidation over a Pt/γ-Al_2O_3 catalyst at $s = 0.5$. *(Figure adapted from* Sotowa et al. (2008) *with permission.)*

look to such past studies for an explanation of the temperature modulation effect. Physical processes that would seem to allow for a modulation effect include: 1) Changes in adsorbate coverage through competitive adsorption and desorption, 2) surface restructuring induced by adsorbate concentration change, or 3) reactant adsorption as "islands" with reaction taking place only at the perimeters of islands of each adsorbed reactant.

Van Neer et al. (1996) invoked a Langmuir-Hinshelwood model of a surface reaction involving adsorbates and observed through simulation that, for some values of the model constants, temperature modulation rate enhancement occurs if the proper modulation frequencies are chosen. *Jensen et al. (2007)* adopted this analysis for CO oxidation on a Pt surface. They chose a widely used model:

$$CO + s \rightleftarrows CO - s \qquad (16\text{-}2)$$

$$O_2 + 2\,s \rightleftarrows 2\,O - s \qquad (16\text{-}3)$$

$$CO - s + O - s \rightleftarrows CO_2 + 2\,s. \qquad (16\text{-}4)$$

Letting θ represent an adsorbate species and assuming that the stoichiometric relations **(16-2 to 16-4)** designate elementary reactions, the dynamic behavior of the CO and O adsorbates are given by:

$$\frac{d\theta_{CO}}{dt} = k_1 P_{CO}\theta_s - k_{-1}\theta_{CO} - k_3\theta_{CO}\theta_O \qquad (16\text{-}5)$$

$$\frac{d\theta_O}{dt} = k_2 P_{O_2}\theta_s^2 - k_{-2}\theta_O^2 - k_3\theta_{CO}\theta_O. \qquad (16\text{-}6)$$

A material balance on a microreactor assuming complete backmixing yields:

$$\frac{VdP_{CO}}{k_B dt} + N_s\frac{d\theta_{CO}}{dt} = \frac{Q}{k_B\bar{t}}((P_{CO})_0 - P_{CO}) - k_3 N_s\theta_{CO}\theta_O \qquad (16\text{-}7)$$

$$\frac{VdP_{O_2}}{k_B dt} + N_s\frac{d\theta_O}{dt} = \frac{Q}{k_B\bar{t}}((P_{O_2})_0 - P_{O_2}) - k_3 N\theta_{CO}\theta_O. \qquad (16\text{-}8)$$

In the above balance, k_B is the Boltzmann constant, V is the volume of the microreactor, Q is the volumetric flow rate, $\bar{t}$ is the fluid residence time and N_s is the total number of active sites for adsorption, desorption or surface reaction in the reactor, represented by "s" in **Eqs (16-2)** to **(16-4)**. *Jensen et al.* took model parameters from the literature and integrated **Eqs (16-5)** to **(16-8)**. Instantaneous temperature in the backmixed reactor was given by a sinusoidal function of time $A\,\sin(2\pi\omega t)$ where ω is the frequency and t represents time.

The solid curves in **Figure 16-18** give the model predictions for the amplitudes used experimentally. It can be seen that the experimental measurements are quite well represented by the model to the extent that rate enhancement requires frequencies greater than 0.02 Hz and ceases to increase the CO oxidation rate above 1 Hz. Unfortunately, the model does not establish the enhancement mechanism. **Eqs (16-5)** and **(16-6)** are consistent with any of the three physical phenomena mentioned earlier.

In unpublished modeling, *Luther (2008)* assumed that there is a slow reaction step in the CO adsorption-oxidation sequence. *Yuranov et al. (2000)*, based on earlier work (*Engel and Ertl, 1978*; *Sales et al., 1982*; *Bassett and Imbihl, 1990*), proposed adding oxidation and reduction of the metal catalyst surface to the reaction sequence given by **Eqs (16-2)** to **(16-4)**. Oxidation of the metal deactivates the catalyst. Incorporating the extra steps in the kinetic equation for CO oxidation and using the kinetic data given by *Sales et al. (1982)*, *Luther (2008)* simulated the CO oxidation under temperature modulation and found good agreement with experiment. He concluded that the reason for improvement under modulation is a decrease in the number of oxidized sites on the catalyst surface blocked by O_2 or CO adsorption. Decrease in the density of these oxidized sites is important primarily in the low-temperature portion of the modulation cycle.

16.7 OVERVIEW AND COMMENTS

Experiments discussed in this chapter demonstrate that temperature modulation can enhance the reaction rates of at least some catalytic reactions provided the modulation amplitude and frequency are properly chosen. However, further study is necessary. Microreactors appear to be well-suited to any future studies. There are many candidate systems; among these are the dehydration reactions of amines and alcohols. These reactions are reactant- and product-inhibited and often exhibit a "stop" effect. Periodic desorption has been shown to increase reaction rates. Thus periodic temperature pulses providing desorption should enhance reaction rates. Dehydration reactions also have consecutive type side-reactions so that effects on selectivity could be explored. The saponification reaction, discussed by *Lee et al. (2003)*, could also be a candidate for selectivity studies under temperature modulation. Partial oxidation of hydrocarbons over metal oxide catalysts also have those characteristics.

However, for commercial application, the microreactor results may have just limited currency, since only at the microscale can the thermal inertia of the vessel and support structures be overcome. A task for future research is to investigate how the rate enhancement possible through modulation can be exploited at a production scale. It should be noted that for biotechnology, there are several reactions running under thermal unsteady-state conditions like the rapid polymerasis chain reaction for multiplying DNA fragments *(Silveston, 1998)*.

Notation

Ci	=	concentration of species I (mol/m^3)
CSTR	=	continuous stirred tank reactor (fully back mixed reactor
Da	=	Damköhler number
STP	=	standard temperature and pressure
s	=	cycle split
x_1	=	dimensionless educt concentration
x_2	=	dimensionless temperature
γ	=	dimensionless activation energy
τ	=	cycle period (s, min)
θ	=	dimensionless time

Subscripts

A	=	reactant
B	=	product
P	=	product
o	=	initial

CHAPTER

17

Flow Interruption in Trickle Beds

Patricia Haure, Jiri Hanika†, Peter Lewis Silveston***

*Mar del Plata, Argentina, †Prague, Czech Republic and **Waterloo, Ontario, Canada

OUTLINE

17.1 Introduction 463

17.2 Steady-State Operation of A Trickle Bed Reactor 464

17.2.1 Spatial Segregation 474

17.3 Periodic Operation of Trickle Bed Reactors 474

17.3.1 Rate or Conversion Enhancement 475

17.3.2 Base-Peak Liquid Flow Rate Modulation 478

17.3.3 Liquid Flow Modulation Under Liquid-Reactant Limiting Conditions 479

17.3.4 Effect of Liquid Flow Modulation on Catalyst Activity or Stability 479

17.3.5 CYCLOP Project 480

17.4 Liquid Flow Modulation with Multiple Reactions 481

17.5 Hydrodynamics Under Liquid Flow Modulation 486

17.6 Modeling of the Periodic Operation of Trickle Bed Reactors 487

17.7 Summary 493

17.1 INTRODUCTION

Trickle beds can be problematic because of the presence of two fluid phases. The liquid phase trickles down through a bed of catalyst particles helped along by a co-current gas phase. Typically, a trickle bed is made up of porous catalyst particles in an unordered, random packing. Open channels through which the fluid can flow are irregular in shape and cross section, tortuous and of wide variability in length. Channels can separate and also intersect. Wetting of the particles and entrapment of the liquid phase between particles affects channel shape. Environments within these fluid filled spaces can be quite dissimilar when one or more reaction occurs because of diffusion into surface wetted particles, liquid filled particles or particles devoid of liquid. Both gas and liquid compositions as well as temperature can differ significantly among nearby channels. As a result of this situation, rates of a catalytic reaction (typically exothermic ones: Hydrogenation, hydrotreatment, hydrocracking, oxidation, etc.) may be enhanced in some regions of the packing (often with an increase in the local temperature) and reduced in others. Some regions of the bed may have temperatures well above the average. These may be designated as "hot spots." Other regions may suffer from so little liquid that they may be viewed as "dry zones." In addition, flow distribution becomes poorer over the bed cross section with time as preferred flow channels develop because of temperature changes at different points within the bed. These changes can cause bed hot spots or dry regions to fade and reappear, and even wander with time.

Because of this inherent complexity, the steady-state operations of trickle-bed reactors (TBRs) will be briefly considered before examining their operation under periodic operation.

17.2 STEADY-STATE OPERATION OF A TRICKLE BED REACTOR

Co-current, downflow trickle beds are used extensively in the petroleum refining industry for hydrotreating and hydro-desulfurization and for hydrogenation, oxidation and hydro-denitrogenation in the chemical, biochemical and waste treatment industries. Catalysts employed are often characterized by high reactivity; hence, internal and external mass transport rates are frequently rate-limiting so high operating pressures and temperatures are generally employed to obtain desirable transport and reaction behavior.

A common feature of reactions carried out in TBRs is that gaseous reactants must first dissolve in and then pass through the liquid phase to reach the catalyst surface. Understanding the reaction-transport-wetting phenomena and their interactions is complicated by the presence of different catalyst wetting conditions caused by two-phase flow. These affect reactor performance differently depending on whether the reaction is gas- or liquid-reactant-limited (*Mills and Dudukovic, 1980a, b*). For a steady-state system with negligible external mass transport limitations, the limiting reactant can be determined by a criterion proposed by *Khadilkar et al. (1996)*:

$$\gamma = D_S C_S^0 / b D_A C_A^* \tag{17-1}$$

where D_A is the effective diffusivity of the reactant in the catalyst particle, C_A^* is the gas reactant concentration in the liquid phase at saturation, C_S^0 is the initial liquid reactant concentration and b is the stoichiometric coefficient. The parameter γ indicates the relative availability of species A at the reaction site. When $\gamma >> 1$, the reaction is gas-reactant limited and downflow TBR operation is preferred for high conversion. Conversely, the reaction is liquid-limited if $\gamma < 1$ and upflow operation would be chosen.

The overall reaction rate in a partially wetted pellet can be higher or lower than that achieved for a fully wetted particle, depending on the limiting reactant. If it is the liquid phase, higher mass transport to the catalyst is obtained at high wetting. For a gas-limited reaction, higher reaction rates are obtained at lower wetting, whereby gas can access the particle interior without dissolution (*Beaudry et al., 1987*; *Iliuta et al., 1999*). In this case, TBR performance enhancement may take advantage of competition between the fluid phases in reactant supply to the particle interior. In a few cases, a decrease in wetting may improve the supply of the gas reactant to the particle, increasing reaction rate, conversion or selectivity. Spatial or temporal segregation of the phases may be used.

Spatial segregation can be achieved through reactor design or by carrying out the reaction over hydrophobic or waterproofed catalysts for oxygenation or hydrogenation in aqueous media. Temporal segregation can be attained by forced gas or liquid flow modulation. Gas flow modulation (GFM) has been only rarely considered (e.g., *Xiao et al., 2001*); liquid flow modulation (LFM) has generally been chosen (see **Table 17-1**).

TABLE 17-1 Studies on the Periodic Operation of Trickle Beds

Reference	Topic or Objective	System/Reaction Description	Observations
Haure et al. (1989, 1990a, b)	Influence of LFM on SO_2 oxidation in a trickle bed of activated carbon catalyst	Gaseous O_2 and SO_2 were fed continuously into a TBR packed with an activated carbon catalyst and operated with on-off LFM. Mean liquid superficial velocities of 0.86 and 1.65 mm/s were used.	For $u_L = 1.65$ mm/s, SO_2 oxidation rate increased by 30 to 45% for $2 < \tau_{cycle} < 80$ min and a bed temperature increase of up to 7°C occurred. On-off liquid flow causes a rising then falling temperature front in the flow direction. These moving waves were predicted by a dynamic TBR model. Even without fitted constants, the model satisfactorily represented the experimental data.
Castellari and Haure (1995)	Experimental study of LFM on α- methylstyrene hydrogenation in a trickle bed	Shallow trickle bed (2.54 cm i.d.× 1.5 cm) packed with 0.5 wt% Pd/Al_2O_3 spheres (2–3 mm) under constant flow of H_2 and AMS flow modulated around a mean of 2.3 mL/s.	About a 4-fold increase of hydrogenation rate observed for $10 < \tau_{cycle} < 15$ min at $0.3 < s < 0.5$. Enhancement attributed to partial wetting of the catalyst and temperature raise through on-off LFM as well as contribution of the gas phase hydrogenation reaction.
Metzinger et al. (1992, 1994); Gangwal et al. (1993)	Application of on-off gas flow modulation for SO_2 removal from stack gas	5 cm i.d. × 19 cm deep TBR used to test activated carbon and Pt impregnated carbon for SO_2 absorption from an air stream (2,500 ppm SO_2) into dilute acid and oxidation to SO_3.	First set of experiments used water and found that $\tau_{cycle} < 20$ min and a low s gave best results. Dilute sulfuric or sulfurous acid reduces solubility of SO_2 so investigators used on-off modulation of gas flow substituting water for acid solution in the liquid phase. Pt impregnated activated carbon gave best SO_2 removal but activated carbons were also very good.
Stegasov et al. (1994)	Model development for periodic operation of a TBR for catalytic oxidation of SO_2	Trickle bed was represented by liquid film flowing down along a catalytically active slab with cocurrent gas flow. Transient, adiabatic model was used with parameters and kinetics evaluated from literature correlations.	Model predicted the effect of liquid feed rate on SO_2 conversion; also the effect of cycle split was closely predicted. However, the prediction of cycle period and temperature rise was poor. The latter was attributed to heat loss from the TBR.
Lange et al. (1994a, b)	Experimental study of base-peak periodic operation for hydrogenation of cyclohexene and for α-AMS	Glass tube reactor (3 cm i.d. × 35 cm) packed with Pd/charcoal catalyst for cyclohexene and a 2 cm i.d. × 70 cm TBR packed with Pd/α Al_2O_3 for α-AMS.	Cycling mitigated temperature rise and avoided a 2-phase reaction for cyclohexene hydrogenation. Within $1 < \tau_{cycle} < 25$ min, both base-peak and on-off LFM substantially increased α-AMS conversion to styrene. Cycle split was also important.

(*Continued*)

TABLE 17-1 Studies on the Periodic Operation of Trickle Beds (*cont'd*)

Reference	Topic or Objective	System/Reaction Description	Observations
Lee et al. (1995, 1996a, b)	Investigation of the removal of low concentrations of SO_2 from stack gas using a TBR packed with activated carbon and LFM	Experiments conducted in a glass reactor (5 cm i.d. × 60 cm) at 80°C with counter-heating to maintain near isothermal operation. Different activated carbons used, some with Pt impregnation. Effect of cycle parameters was studied.	Activated carbons differed significantly in activity, however, for the most active carbon impregnation with 0.1 wt% Pt did not further increase activity. Both τ_{cycle} and s affected SO_2 capture and conversion to sulfuric acid. These effects are explained by wetting of the activated carbon and reduced SO_2 solubility as the acid concentration in the liquid increases.
Gabarain et al. (1997)	Analytical and experimental study of LFM on AMS hydrogenation in a trickle bed	System identical to that used by *Castellari and Haure* above. Additional experiments were analyzed by a model assuming rate and temperature rise controlled by heat and mass transfer between the catalyst particle and the flowing gas and liquid phases.	Experimental enhancement of AMS hydrogenation rates were > 4 at $\tau_{cycle} = 10$ min and $s = 0.3$, but dropped rapidly for $\tau_{cycle} < 10$. Cycle split, s, had an important effect. Researcher's simple model predicted the effect of τ_{cycle} on enhancement and on the temperature rise in the catalyst bed with time.
Yamada and Goto (1997)	Application of periodic operation to a 4-phase TBR system with product precipitation	Carbobenzoxyphenylalanine soluble in organic media when hydrogenated to phenylalanine in the presence of water precipitates on the catalyst. Authors studied periodic washing with water to re-dissolve the phenylalanine.	Deactivation of the catalyst through phenylalanine precipitation was prevented by periodic operation of the TBR through switching between the water – organic and just water using $\tau_{cycle} = 150$ min and $s = 0.2$ where s is the relative duration of the water wash.
Kouris et al. (1998)	Catalyst particles with constant and periodically changing degree of external wetting.	Model developed for reaction between one gaseous and two non-volatile reactants on partially wetted catalysts with time-dependent wetting to simulate a TBR in pulsing flow.	Model showed that wetting, location of wetted zones, and modulation frequency must be specified to calculate time-average effectiveness factors.
Boelhouwer et al. (1999)	Enlargement of the pulsing flow regime by periodic operation of a trickle-bed reactor.	Experiments were carried out in a 1.0 m deep column packed with 6 mm glass beads by alternately cycling water and air to a trickle bed.	Experiments demonstrated that it is possible to shift the transition from trickling to pulsing flow toward lower average gas and liquid flows through forced cycling.
Khadilkar et al. (1999)	Parametric study of flow modulation in trickle-bed reactors.	LFM was used for hydrogenation of α-methylstyrene in a hexane solvent over 0.5% Pd on alumina spheres, under both gas- and liquid-reactant-limited conditions.	LFM may result in reactor performance better than that obtained under steady flow. Improved performance depended on the limiting reactant, cycle period and split, liquid mass velocity, and extent of liquid maldistribution.

Stradiotto et al. (1999)	Hydrogenation of crotonaldehyde under steady and periodic liquid flow.	Hydrogenation of crotonaldehyde over a Pt catalyst. On-off liquid flows at 253°C and 1.1 MPa pressure with periods of 5 and 20 min; various cycle splits.	For higher liquid flow rates, conversion rates under periodic operation were less than those under comparable steady operation. However, for lower liquid flow rates (0.04-0.4 kg/(m^2·s)), periodic operation showed up to a 50% increase in the crotonaldehyde conversion rate.
Lange et al. (1999)	Experiments on AMS hydrogenation under periodic modulation of AMS flow	Experiments employed a 2 cm i.d. × 70 cm steel reactor packed with a crushed Pd/α-Al_2O_3 catalyst as a TBR with constant gas flow and modulated liquid flow. Authors' model assumed rate and particle temperature controlled by mass and heat transfer to the external fluid.	Large rate enhancement observed experimentally for $5 < \tau_{cycle} < 10$ min for $s = 0.5$; enhancement was much smaller for base-peak modulation. The authors' model predicted much higher rate enhancement. Catalyst wetting was believed to be the dominant effect on time-average hydrogenation rate.
Boelhouwer et al. (2001a, b)	Induction of pulses in a TBR by LFM	Air/water flow in a trickle bed (11 cm i.d. × 1.0 m) packed with 3 or 6 mm glass beads.	Modulation of liquid or gas flow rates in a TBR can result in pulsing flow at average throughputs of liquid usually found in steady-state trickle flow operation. This liquid-induced pulsing flow can improve reactor performance and eliminate hot spots. Liquid-induced pulsing flow allows tuning of the pulse frequency. During periodic operation, continuity shock waves are initiated and divide the TBR into regions of high and low liquid holdup. A critical dimension is the available bed length for disturbances to grow into pulses.
Xiao et al. (2001)	Hydrodynamic behavior of a TBR under LFM	Tests were done in a Plexiglas cylinder 1.0 m × 0.10 m by varying the air flow in an air/water flow system over glass beads at a frequency of 0.135 Hz.	Axial and radial liquid distributions were found to be more uniform than those in spontaneously pulsing flow. The induced liquid flow sharply reduced the thickness of the liquid film over the particles. For liquid holdup, LFM frequency was the most influential parameter.
Turco et al. (2001)	Experimental study of periodic flow modulation of either gas or liquid phases using AMS hydrogenation	Experiments were undertaken in an isothermal TBR (4.0 cm i.d. × 11 cm). Liquid and H_2 feeds were separately modulated at 40°C and near-ambient pressure. $\tau_{cycle} = 10$ min, but two values of s were used.	Liquid modulation produced a large increase in AMS conversion probably caused by a shift into the pulsing regime from the trickling regime. Gas phase flow modulation suppressed conversion.

(*Continued*)

TABLE 17-1 Studies on the Periodic Operation of Trickle Beds (*cont'd*)

Reference	Topic or Objective	System/Reaction Description	Observations
Boelhouwer et al. (2002a, b)	Nature and characteristics of pulsing flow in trickle-bed reactors	Experiments were carried out in an 11 cm i.d. × 3.2 m column packed with 6 mm glass beads using an air/water system.	Properties of pulsing flow (liquid holdup, velocity and duration) do not depend on superficial liquid velocity at constant gas flow. However, the pulse frequency increases with increasing superficial liquid velocity. Liquid holdup in the parts of the bed between pulses equals the liquid holdup at the transition to pulsing flow at all gas flow rates.
Tukač et al. (2003)	Catalytic wet air oxidation of phenol under LFM	A 1.8 cm i.d. × 50 cm steel trickle bed was used packed with 4 × 5 mm activated carbon catalyst and used a 0.5 vol% aqueous phenol solution and O_2 at 1-7 MPa. $125 < T < 170°C$.	Liquid holdup increased with τ_{cycle}. Optimal conversion of phenol occurred at $\tau_{cycle} = 60$ s and $s = 0.5$ for conditions used and represented a 10% increase over steady-state operation. Optimal cycle time closely matched drainage time from the TBR.
Wilhite et al. (2003)	Experimental and modeling study of a parallel hydrogenation reaction under LFM	Experiments used a 2.54 cm i.d. × 40 cm pyrex tube packed with 0.5 wt% Pt/γ Al_2O_3 pellets to an 8 cm depth. TBR fed with dilute phenylacetylene in n-tetradecane. Operation was isothermal. Results were compared with detailed model.	Base-peak LFM switched steady-state regime between trickling and pulsing flow; however, LFM suppressed phenylacetylene conversion and selectivity to styrene under gas reactant limiting. Authors' model which closely predicted experimental measurements confirmed suppression for gas phase limiting, but predicted a small conversion and a large selectivity enhancement when the liquid reactant is limiting.
Borremans et al. (2003, 2004)	Influence of periodic operation on liquid flow distribution and particle-fluid heat transfer in TBRs	Experiments used saturated air and water in a 10 cm i.d. × 3.1 m trickle bed packed with either glass spheres or cylindrical extrudates.	Modulation did not significantly enhance liquid distribution at the two spatial scales examined. Radial thermal conductivity and liquid flow distribution were not notably modified compared with steady-state operation at comparable average feed rates.
Banchero et al. (2004)	Experimental investigation of fast-mode LFM	Hydrogenation of α-methyl styrene on a Pd/C catalyst in an isothermal TBR.	Fast-mode modulation increased conversion by up to 60%. Isothermal operation meant that conversion improvement was due only to enhanced mass transfer.

Fraguío et al. (2004); Muzen et al. (2005)	LFM effect on TBR performance	Oxidation of aqueous ethanol and butanol solutions with O_2. TBR packed with 1 wt% Pt/γ-Al_2O_3 and operated at 1 atm and 70°C using on-off LFM at $10 \leq \tau_{cycle} \leq 10^3$ s.	Higher conversions attained for short τ_{cycle}, especially for low s and lower alcohol concentrations. Conversion and product distribution were affected by past operational history. Longer dry periods improved selectivity to acetic acid.
Lange et al. (2004)	Catalytic hydrogenation in a laboratory-scale TBR under steady and periodic operation.	Hydrogenation of α-methylstyrene over a supported Pd catalyst.	LFM increased the time-average α-methylstyrene conversion relative to steady-state operation. Higher conversions were explained by variation in liquid holdup and catalyst wetting. Available reactor models were insufficiently detailed to simulate transient operation.
Urseanu et al. (2004)	Hydrodynamics and reactor performance for induced pulse operation of a high-pressure TBR	Hydrogenation of α-methyl styrene over 2 wt% Pd/C in a TBR at a feed temperature of 40°C and a pressure of 0.2 MPa.	Comparison of the steady-state and modulated modes of operation showed a significant improvement in reaction rate under modulation.
Ayude et al. (2005)	Phenomenological study of the effect of trickle bed, liquid flow modulation at the scale of a single particle	Liquid flow modulation was modeled for a 1st order reaction in a spherical porous particle in a trickle bed. Reaction between a gas component and a nonvolatile liquid species was considered.	Dynamic reactant profiles inside a catalytic particle were obtained for different cycling and system conditions. Enhancement factors for periodic operation were calculated for a range of conditions.
Liu and Mi (2005); Liu et al. (2005)	Experimental study of 2-ethylanthraquinone under LFM	Hydrogenation of 2-EAQ to 2-ethylanthrahydroquinone undertaken in an isothermal TBR (2.1 cm i.d. × 0.65 m) over a Pd/Al_2O_3 catalyst.	LFM modulation at $\tau_{cycle} < 200$ s and $0.2 < s < 0.6$ significantly increased 2-EAQ conversion and selectivity to the 2-ethylanthrahydroquinone. Increasing liquid feed rate had little effect on conversion enhancement through periodic operation but eliminated the selectivity enhancement.
Massa et al. (2005)	Experimental study of phenol oxidation under LFM	Phenol oxidation was undertaken in a 2.1 cm i.d.x 40 cm isothermal TBR at 140°C and 7 atm over a CuO/γ-Al_2O_3 catalyst.	Periodic operation offers a small enhancement of phenol conversion but a larger increase in selectivity to total oxidation.
Dietrich et al. (2005)	Modeling of periodically wetted catalyst particles	Modeling approaches examined were: 1) A general model, 2) a simplified model with two different wetting zones and 3) as in 2) but considering particle interior as a sink.	Investigators point out that the problem of simulating diffusion into a wetted particle with periodically changing wet and dry surface conditions is probably intractable without large simplifying assumptions.

(*Continued*)

TABLE 17-1 Studies on the Periodic Operation of Trickle Beds (*cont'd*)

Reference	Topic or Objective	System/Reaction Description	Observations
Iliuta and Larachi (2005)	Deposition of suspended solids in TBRs under steady flow and LFM	Modeling entrapment and release of fine suspended solids in trickle beds for steady and pulsating liquid flow.	Induced pulsing reduces the entrapment of fine, non-Brownian particles and increases detachment of deposited particles.
Giakoumakis et al. (2005)	Investigation of induced pulsing in trickle beds	Tests were conducted with an air-water system in a 1.24 m $\times$ 0.14 m i.d. bed packed with glass spheres, irrigated intermittently with water at frequencies between 0.05 and 10 Hz to identify conditions for improved fluid distribution and transport rates.	Liquid holdup measurements were used to determine propagation and attenuation of induced pulses in fast-mode LFM. Calculated attenuation factors depended on dynamic holdup, pressure drop, pulse magnitude and modulation frequency. Time-average holdup and pulse shape were almost constant along the bed and independent of the modulation frequency. Conditions were identified to maximize benefits from induced pulsing.
Sicardi et al. (2005)	Application of a quasi-steady-state model to slow mode LFM in TBRs	Model tested on hydrogenation of α-methyl styrene over a Pd/C catalyst.	The mean conversion rate in the pulsed mode was assumed equal to the weighted mean of the two steady-state conversion rates at the minimum and maximum liquid flow rates. This assumption satisfactorily reproduced the experimental data.
Gunjal et al. (2005)	Application of CFD to pulsed liquid flow	Model developed from CFD study of pulsing regime in steady operation of TBRs was applied to LFM.	Variation of pressure drop and liquid holdup with time in a modulation cycle adequately predicted by CFD model. These hydrodynamic properties were found to be insensitive to the forcing frequency.
Trivizadakis et al. (2006b); Trivizadakis and Karabelas (2006)	Particle shape and size effects on pulse characteristics for induced pulsing in trickle beds, effect of LFM on mass transfer	Porous spherical and cylindrical extrudates were employed with fast mode, bang-bang liquid feeding to measure dynamic holdup, pressure drop, pulse shape and amplitude as functions of gas and liquid feed rates and cycle frequency.	Measurements were made in the trickle flow regime. For extrudates of comparable size, the cylindrical particles provided larger dynamic holdup and pressure drop along with greater pulse attenuation. Nonetheless, spherical extrudates offered advantages over cylindrical ones.

Aydin et al. (2006)	Slow-mode LFM in TBRs at elevated temperatures	Air-water used for Newtonian flow study of shockwave breakthrough; decay time, plateau height and amplitude at breakthrough were measured. Experiments were repeated for a non-Newtonian liquid: Water +0.25 wt% CMC. TBR was 4.8 cm i.d. × 1.07 m and was packed with 3 mm glass beds.	Temperature and pressure affected shockwave decay rate and breakthrough time. Collapse of the liquid pulses was observed with increasing *T* or *P* caused by reduction of liquid holdup in pulse due to seepage into base flow and dispersive effects in flow through the packed bed.
Borremans et al. (2007)	Cyclic variation of the liquid residence time in periodically operated trickle-bed reactors	Experiments with salt tracer in water were compared with a residence time model based on the theory of continuity shock waves.	Periodic operation results in repetitive variations in reactor residence time and can affect conversion of the liquid reactant. Adjustment of cycling variables can yield a constant and predictable residence time.
Ayude et al. (2007a)	Effect of LFM on liquid holdup	7 cm i.d. × 22 cm tube packed with 3 mm alumina spheres was used for experiments at atmospheric pressure and 200°C.	Three types of holdup behavior were identified depending on on-off pulse frequency: Fast ($\tau_{cycle} < 10$ s), intermediate and slow. Holdup in fast pulses changed little in a cycle and in passage through the bed. With slow pulsing, holdup resembles a deformed square wave, draining occurs and holdup changes with depth. In off portion, holdup decays exponentially with time.
Ayude et al. (2007b); Larruy et al. (2007)	Investigation of wet air oxidation in a periodically operated TBR	Long duration experiments were conducted in a laboratory trickle bed at 140 to 160°C packed with activated carbon on oxidation of dilute aqueous phenol under modulation of gas flow and switching between O_2 and N_2.	Either periodic switching between O_2 and N_2 or on-off gas flow reduces oxidation of activated carbon in the TBR and thus extends carbon life. Extended durations of up to 144 h on stream were used.
Brkljac et al. (2007)	Modeling of unsteady-state hydrodynamics in periodically operated TBRs	Influence of the liquid-phase physical properties during base-peak and on-off LFM was evaluated using a model of two-phase pressure drop and dynamic liquid holdup.	By selecting suitable permeability parameters in their model, using steady-state measurements, unsteady-state hydrodynamics in different periodic operating modes could be predicted.
Gladden et al. (2007)	Flow patterns in periodically operated trickle-bed reactors	Magnetic resonance imaging was applied to a 4.5 cm i.d. bed packed with 1 to 3 mm particles.	MRI images show non-uniform liquid flow over bed cross section with many liquid-filled channels.

(*Continued*)

TABLE 17-1 Studies on the Periodic Operation of Trickle Beds (*cont'd*)

Reference	Topic or Objective	System/Reaction Description	Observations
Tukač et al. (2007a, b); Hanika et al. (2007)	Investigated the behavior of a pilot trickle-bed reactor under periodic operation	Hydrogenation of styrene and/or dicyclopentadiene on a Pd egg-shell type catalyst served as model reactions to test the use of LFM in olefin and diene hydrogenation.	Productivity of styrene hydrogenation was enhanced by 30% under LFM modulation compared with steady operation. Catalyst wetting was affected by modulation for pilot-scale reactors.
Hamidipour et al. (2007; 2010)	Effect of LFM on suspended solids deposition in TBRs; application of tomography to explore hydrodynamics in TBRs	Experimental study of entrapment of suspended solids in the liquid feed in trickle beds under steady and induced pulsing flow. The 2010 paper demonstrated the use of capacitance tomography to characterize hydrodynamic properties under LFM.	Induced pulsing through feed LFM cannot prevent the deposition/entrapment of suspended solids in a TBR but can reduce the rate of solids build-up. Capacitance tomography found to be an effective tool for measuring liquid holdup and pulse properties such as breakthrough and decay times. Alternating gas-liquid feed gave the best pulsing performance.
Ayude et al. (2008)	Explored the effect of liquid flow modulation on catalytic activity	Slow bang-bang liquid flow modulation on the oxidation of ethanol in aqueous solutions using a 0.5% Pd/Al_2O_3 commercial egg-shell catalyst in a laboratory TBR.	Significant improvements over continuous operation were obtained when the catalyst was exposed to a short excess of oxygen after a longer duration in a mass-transfer limited portion of the cycle with high liquid flow rates and moderate gas flow rates to ensure complete catalyst wetting.
Liu et al. (2008)	Cyclic TBR operation for a multi-step exothermic reaction	A comprehensive unsteady-state TBR model for hydrogenation of (DCPD) over a Pd/Al_2O_3 catalyst that assumed gas and liquid plug-flow, stagnant liquid behavior, three-zone partial wetting, vapor-liquid phase equilibrium and allowed for heat effects was tested against experiments.	The authors' detailed model adequately predicted experimental enhancement for five different modulation strategies.
Skála and Hanika (2008)	Experimental study of dicyclopentadiene (DCPD) hydrogenation in a liquid flow modulated TBR	DCPD hydrogenation was chosen as a model reaction to explore LFM under isothermal conditions in a laboratory reactor using an egg-shell impregnated Pd catalyst.	Hydrogenation rate at atmospheric pressure was raised by 12% over steady-state operation by symmetrically modulating the DCPD feed rate at a period of 60 s. Selectivity decreased because of the higher levels of adsorbed H_2 on the catalyst.

Stanek et al. (2009)	Modeling and experimental investigation of styrene hydrogenation in a pilot scale TBR	Model consisted of mass and heat balances assuming all phases at a point in the reactor were at the same temperature. Experiments were conducted in a 10 cm i.d. × 2.0 m steel reactor.	Experimental enhancement of styrene hydrogenation was found in the low-liquid portion of a cycle. The authors' model predicted experimental results only if catalyst wetting changed with liquid flow, thus demonstrating that LFM enhancement depends on catalyst wetting.
Lysova et al. (2009)	MRI imagining of 1-octene hydrogenation under steady flow and on-off periodic flow modulation	Laboratory scale TBR (1 cm i.d.) used for hydrogenation of 1-octene over 1 wt% Pd/Al_2O_3 with multilayer MRI imaging. Experiments were undertaken for steady flow and symmetrical on-off LFM at constant gas flow.	Fast on-off LFM, $\tau_{cycle} = 6$ s, 1 mm ϕ catalyst particles remain liquid filled and T_{bed} varies by about 10° around 162°C while at $\tau_{cycle} = 60$ s, particles dry out during off-periods and T_{bed} varies by about 35° around 140°C. Catalyst particle size affects optimal τ_{cycle}.
Ayude et al. (2009)	Model development for TBRs operating under LFM	Authors' earlier surface wetted particle model combined two different trickle bed models: 1) Draining with transient reduction of holdup and liquid velocity in the "off" portion of a cycle, 2) square-wave model in which only two holdup and liquid velocity states occur in a cycle.	The liquid draining model in which holdup, surface wetting and liquid velocity vary with time and axial position in flow-off portion of a cycle predicts lower enhancement under periodic operation than the simpler square wave which assumes constant holdup, wetting and liquid velocity in each part of a cycle.

17.2.1 Spatial Segregation

Three-phase catalytic reactor performance may be improved by using ordered (rather than random) packing and segregating the liquid and gas. In monolithic reactors, liquid flows on one side of the catalytic packing and gas on the other. Because of this, segregated flow reactors are an attractive alternative to both slurry and trickle-bed reactors. Advantages of monoliths are the avoidance of filtration to remove precipitants or other solids in the liquid, low pressure drop (two orders of magnitude less than for TBRs), high external surface area, safer operation, and simpler scale-up. The main disadvantage is the difficulty of achieving uniform contact times for reactants flowing within the reactor. Discussion of the application of monolithic catalysts to three-phase processes are presented by *Roy et al. (2004)* and *Cybulski and Moulijn (2006)*.

Another approach to controlling wetting is to modify the catalyst surface. Although hydrophobic catalysts were initially designed for deuterium enrichment and tritium separation by H_2 $-H_2O$(liq.) isotopic exchange, their use in the oxidation of organic compounds and hydrogenation reactions in aqueous media has been reported (*Hatziantoniou et al., 1986; Chuang et al., 1994; Lavelle and McMonagle, 2001; Horowitz et al., 1999; Omota et al., 2005*). With these catalysts, a microlayer of gas surrounds each particle immersed in water, thereby avoiding the resistance added by gas-liquid mass transfer. Hydrophobic catalysts can be prepared either by treating the catalyst surface (*Berruti et al., 1984; Horowitz et al., 1999; Massa et al. 2009; Cheng and Chuang, 1992*) or by supporting the active phase on a hydrophobic support (*Lavelle and McMonagle, 2001; Omota et al., 2005*).

Horowitz et al. (1999) compared ethanol oxidation in a TBR packed either with a hydrophilic catalyst, a hydrophobic one or mixtures of a hydrophobic catalyst and a hydrophilic inert support. The improvement in reactor performance was remarkable for the hydrophobic catalyst. Improved performance persisted for almost all the liquid flow rates used, except for the lowest one in which the ethanol conversion was high and the ethanol/oxygen ratio decreased markedly. Similar results were reported by *Massa et al. (2009)* for the oxidation of phenol in a TBR at low liquid flow rates. The influence of O_2 mass transfer disappears without a large excess of liquid reactant and the performance of hydrophilic and hydrophobic catalysts become similar. Temporal separation is possible only in periodic operations.

17.3 PERIODIC OPERATION OF TRICKLE BED REACTORS

Literature on periodic operation of trickle beds continues to expand. **Table 17-1** summarizes much of this literature.

1. *Schädlich et al. (1983)* demonstrated that enhanced TBR performance through cycling is possible only if the system is nonlinear. The chief source of nonlinearity in TBRs lies in the reaction kinetics, heat effects and their coupling through the Arrhenius relation. Coupling of reaction rate with mass transfer and catalyst wetting contribute.
2. Performance enhancement of a TBR has been obtained with forced unsteady-state operation, as **Table 17-1** demonstrates, by temporal, input segregation of the phases which lessens gas-liquid mass transfer resistance and also improves liquid flow distribution through the reactor packing.

With liquid flow modulation (LFM), inlet liquid volumetric flow rate periodically changes between two set levels, while the volumetric gas flow rate remains constant. Cycle period gives the time between repetitions of the same input flow rate. On-off operation is the extreme situation, where the cycle split, s, is the fraction of a period during which liquid flushes the bed. Thus, the TBR is forced to run continuously in a transient mode in which the liquid surface coverage of a catalyst particle varies periodically. Periodic thinning of the liquid film encasing a catalyst particle enhances the rate of gaseous reactant supply. **Figure 17-1** from *Boelhouwer et al. (2002a)* shows wetting as a function of the liquid flow rate for steady-state operation (concave curve SS). By cycling the liquid feed rate between u_{l_b} and u_{l_p}, the average wetting during periodic operation is located somewhere on the line USS1 (unsteady state), depending on the relative durations of low and high liquid flow rates. If on-off operation is used, average wetting is located on the line USS2. Average wetting during an on-off cycle is at all times less than during a high-low cycled feed. Thus, the average wetting is reduced and the mass transfer rate of the gaseous reactant is increased.

17.3.1 Rate or Conversion Enhancement

The application of periodic operation to a TBR was first suggested by *Gupta (1985)* in a US patent. Independent experimental work by *Bellut* at about the same time with follow-up work by *Haure* resulted in US and Canadian patents *(Haure et al., 1988, 1992)*. In both the patents and other studies carried out at the University of Waterloo, *Haure et al. (1989, 1990a, b)* studied the oxidation of sulfur dioxide over activated carbon in an integral TBR. Reactants, dilute SO_2 and air, were fed continuously to the reactor in the gas phase, while the liquid phase input, water, was switched on and off. This sequence was continued over many cycles so that the production rate of sulfuric acid could be calculated from pH measurements in the liquid draining from the bed and the total liquid collected. Steady-state measurements were performed under identical conditions. A comparison of the two sets of measurements showed that slow bang-bang cycling resulted in about a 40% increase in acid production as illustrated in **Figure 17-2**.

Scatter of data is rather high but all points lie well above 1.0. Indeed, in some experiments the normalized rate is greater than 1.65 meaning the rate increased by 65% with periodic flow interruption. Additionally, during the off-portion of the cycle period, heat removal from the bed was reduced, so that the bed temperature increased and the flow interruptions also led to traveling temperature waves. Measurements

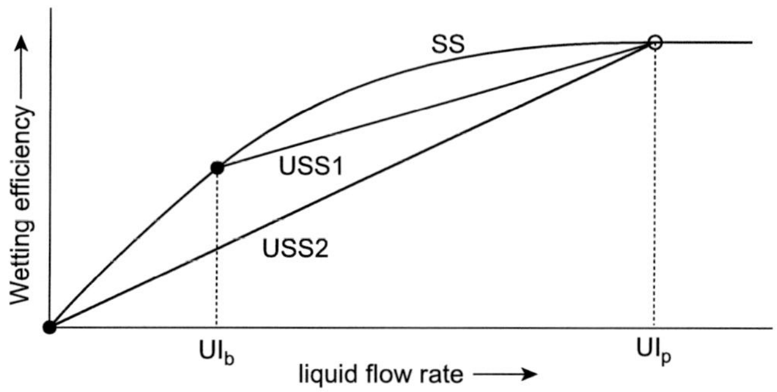

FIGURE 17-1 The effect of liquid flow rate on catalyst wetting efficiency illustrating the potential advantage for periodic liquid flow rate variation on gas-liquid mass transfer rates in a TBR (USS1 = high-low flow variation; USS2 = on-off variation) and steady-state operation (SS). *(Figure adapted from* Boelhouwer et al. (2002a) *with permission. © 2002 by Elsevier Science Ltd.)*

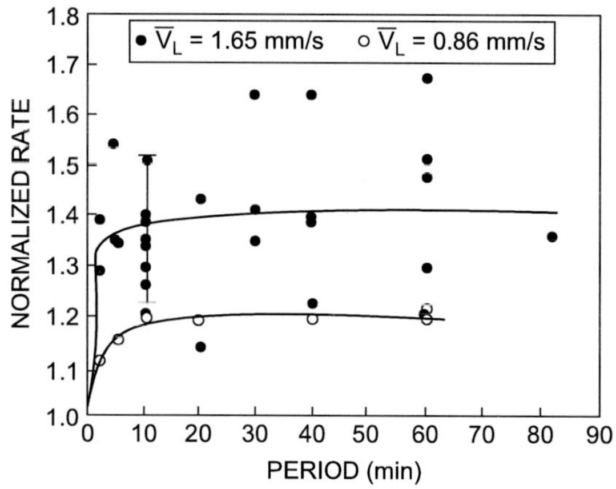

FIGURE 17-2 **Normalized rate of H_2SO_4 production vs. liquid flow cycle period for SO_2 oxidation over activated carbon in a trickle bed with gas (air and SO_2) and water in cocurrent flow.** *(Figure adapted from* Haure et al. (1989) *with permission of authors.)*

of these waves are shown in **Figure 17-3** (*Haure et al., 1990b*).

Assuming that feeding and draining were instantaneous, *Haure et al. (1990b)* used a series of first-order partial differential equations (PDEs) to predict the observed traveling thermal fronts. Enhanced mass transfer and higher average bed temperature were shown to be responsible for the improvements observed. The SO_2 oxidation system was further studied by several researchers (*Metzinger et al., 1992, 1994; Lee et al., 1995*) using a Pt-impregnated activated carbon catalyst. These investigators also observed enhancement of acid production. They attributed higher rates to more rapid mass transport of O_2 through a periodic disruption of an external water film encasing catalyst particles of the trickle bed.

Based on the work of *Haure et al. (1990b), Stegasov et al. (1994)* developed a dynamic model for the SO_2 oxidation under periodic operation, taking into account partial wetting but neglecting axial dispersion and other hydrodynamic effects, such as changes in liquid distribution that are found after liquid flow begins in trickle beds. The *Stegasov* model crudely predicted the observed transient temperature profiles in the bed and changing acid concentration leaving the reactor.

These early experimental results suggested that temperature effects for exothermic reactions contribute importantly to the remarkable enhancements of the overall reaction rate observed. *Castellari and Haure (1995)* and *Gabarain et al. (1997)* explored this suggestion by studying the hydrogenation of α-methyl styrene (AMS) in a laboratory TBR packed with 2 to 3.2 mm spheres of a shell-impregnated 0.5% Pd /Al_2O_3 catalyst. An increase in the AMS hydrogenation rate, reaching four-fold for short periods, was observed through periodic flow interruption, as may be seen in **Figure 17-4**.

Additionally, when the liquid flow rate to the reactor was reduced to zero, a sharp increase in temperature was observed, as shown in **Figure 17-5**. Although the average trickle bed temperature is about 12°C higher under on-off cycling than at steady state, the observed rate enhancement can be explained only by events occurring during the off-portion of the cycle. Once liquid flow is interrupted, the bed partially drains and reaction proceeds between AMS held up in the packing and the flowing H_2. Heat generated by the exothermic reaction cannot be completely removed by the flowing gas, so the temperature and reaction rate increase. Indeed, under these conditions, vaporization of some AMS occurs. Then, a gas phase reaction contributes to AMS consumption. The rapid temperature rise in "B" of **Figure 17-5** is due to this sequence. Eventually, depletion of AMS in both liquid and gas phases slows the reaction and the reaction zone is cooled by the incoming gas and liquid phase evaporation. This is shown by region "C" in the figure. During the on-portion of the cycle, liquid flow returns and replenishes the bed with AMS and removes the hydrogenated product. Consequently, the observed enhancement arises from an increased rate of mass transfer, the formation of controlled hot spots that result in higher

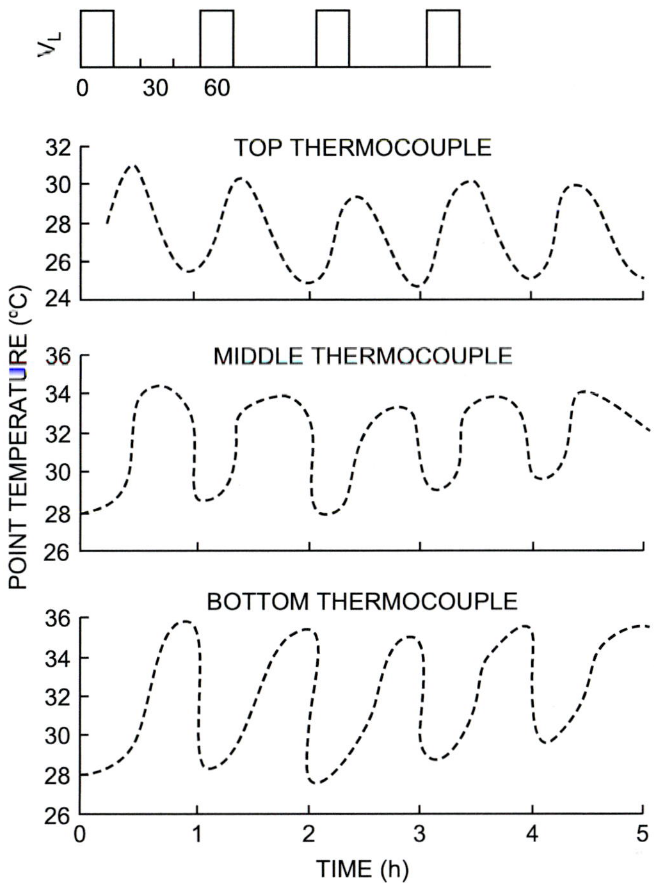

FIGURE 17-3 Temperature profiles at different reactor depths for the air oxidation of SO_2 in a TBR packed with activated carbon. *(Figure adapted from* Haure et al. (1990b) *with permission of authors.)*

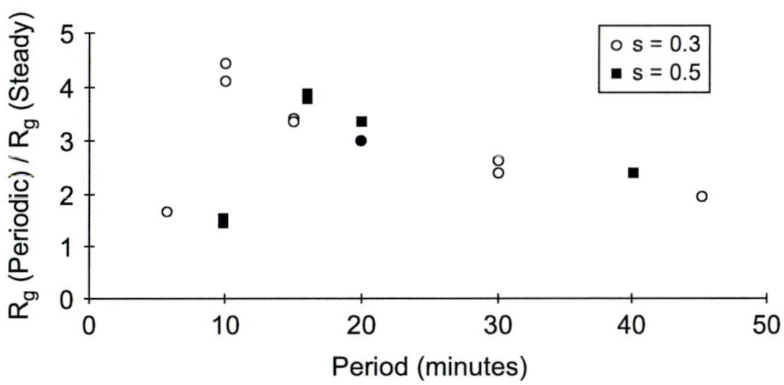

FIGURE 17-4 Normalized rate of AMS hydrogenation in a TBR packed with a supported Pd catalyst vs cycle period for bang-bang on-off liquid flow. *(Figure adapted from* Castellari and Haure (1995) *with permission of authors.)*

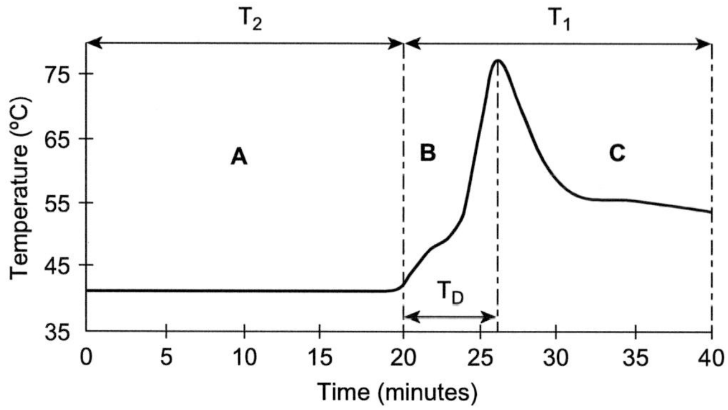

FIGURE 17-5 Temperature profile within an on-off flow cycle for the operation shown in Figure 17-4. *(Figure adapted from* Castellari and Haure (1995) *with permission of authors.)*

catalyst temperatures, and the appearance of a gas-phase reaction over a dry catalyst. The enhancement depends on the duration of wet and dry cycles, i.e., cycle period and split. *Gabarain et al. (1997)* developed a one-dimensional pseudo-homogeneous model for AMS hydrogenation under LFM that satisfactorily represented their experimental results after determining several model parameters from separate experimental data.

LFM increased AMS hydrogenation rates even when there was little or no temperature variation, according to experiments undertaken by *Lange et al. (1992)*. These authors also used a Pd catalyst but introduced AMS into the TBR in a paraffin solvent so as to avoid large temperature fluctuations. In related work, *Tukač et al. (2007a, b)* studied the hydrogenation of toluene solutions of styrene and dicyclopentadiene over a palladium egg-shell catalyst under LFM in bench- and pilot-scale trickle-bed reactors operated with constant H_2 flow. Productivity enhancements of 30% were reported. *Urseanu et al. (2004)* compared the performance of periodically and continuously operated TBRs for hydrogenation of AMS over 2-wt% Pd/C catalyst in a pilot scale reactor at 40°C and 0.2 MPa. They measured a reaction rate increase of 50% through periodic operation. *Stradiotto et al. (1999)* investigated in some detail the improvement of wetting through flow interruption, choosing hydrogenation of an aqueous solution of crotonaldehyde for their study. Heat effects were negligible and there was no gas phase reaction as was the case for AMS hydrogenation. Rate enhancement was found, but only at very low time-average liquid loading, just as in the *Khadilkar et al. (1999)* study. A tracer experiment showed that incomplete wetting was encountered at liquid loadings well above those employed in the flow interruption experiments. *Fraguío et al. (2004)* and *Muzen et al. (2005)* studied the oxidation of ethanol and benzyl alcohol in a TBR with on-off LFM. They concluded that alcohol conversion can be improved through cycling for many combinations of cycle split and period.

17.3.2 Base-Peak Liquid Flow Rate Modulation

There have been several LFM studies employing square-wave switching of liquid feed between high and low rates (*Haure et al., 1989; Lange et al., 1992; Lange et al., 1999; Khadilkar et al., 1999*). These are now described as Base-Peak modulation. All of the above studies observed that this type of flow modulation increased the reaction rate; however, for the same amplitude, increases in the reaction rate were much smaller than those observed for flow interruption. Typically, rate or conversion enhancements ranged from 10 to 20%. On the other hand, flow interruption, in cases of strongly exothermic reactions, could lead to the development of dangerous hot spots in trickle bed during the no feed portion of an LFM cycle. This, potentially, limits the use of on-off cycling.

As liquid holdup drains from the bed or while operating at very low liquid feed rates, the base portion of a cycle, catalyst in the trickle bed may be insufficiently wetted so that liquid maldistribution and/or liquid channelling (rivulets) arise in the bed. In such cases, important temperature excursions within a trickle bed can occur if the reaction carried out is highly exothermic. For example, at low feed rate during the start up of nickel catalyzed thymol hydrogenation in a trickle bed reactor, *Dudas and Hanika (2009)* reported hot spot formation.

Explanations of the improvement through base-peak flow modulation differ according to the system studied. In one case, enhancement appeared to result from higher mass transfer rates when the liquid film flowing over particles was thinned after the liquid loading was reduced. In another, enhancement was attributed to improved wetting after the feed rate to the trickle bed was raised from the low to the high level. In the third study, enhancement was thought to arise from better distribution of the liquid across the trickle bed at the high feed rate which persisted after the rate was dropped to the lower level. Nonlinearity of the liquid film thickness, particle wetting or flow distribution dependence on the liquid feed rate seems to be involved. The effect of LFM parameters as well as the Thiele modulus value (comparing gaseous reactant diffusion through flowing liquid film and reaction rate) on enhancement has been investigated by *Skála et al. (2001)*.

17.3.3 Liquid Flow Modulation Under Liquid-Reactant Limiting Conditions

Most investigations reported in the trickle bed periodic operation literature have examined just gas-reactant-limiting systems. LFM under gas-reactant-limiting conditions provides periodic partial dry-out of the catalyst particles and thus direct access of gaseous reactants to the catalytic sites. This is followed in the next half-cycle by replenishment of the liquid reactant, periodic removal of reaction products and quenching of any local temperature rise. *Khadilkar et al. (1999)* studied the effect of the limiting reactant on periodic forcing of AMS hydrogenation in a TBR. Under gaseous reactant limiting conditions, low H_2 partial pressure and high AMS concentrations in the liquid phase, periodic flow interruption produced conversion enhancements of 20–40%. This enhancement was strongly influenced by the mean liquid flow rate. As flow rate decreased, equivalent to greater residence time, enhancement became quite large. *Lange et al. (1992, 1994)* as well as *Khadilkar et al. (1999)* observed resonance with respect to cycle period so that under gas-limiting conditions there is a region of significant conversion improvement. However, when the liquid is rate limiting, so that the hydrogenation rate is limited by the AMS concentration, periodic flow interruption results in lower AMS conversions than under steady-state operation.

17.3.4 Effect of Liquid Flow Modulation on Catalyst Activity or Stability

The effect of LFM on catalyst activity/stability has been ignored until recently. Most studies have assumed there would be no changes in either activity or stability. However, such an assumption is not always correct. In oxidation reactions, if the catalyst is a semiconductor, it may participate in an intermediate stage of the reaction. For instance, surface properties of a metal oxide catalyst may be altered by exposure to reactants, so its activity may be modified as the catalyst is exposed to different oxygen concentrations. Liquid-phase oxidation reactions with semiconductor catalysts often proceed through a stepwise mechanism in which O_2 acts as an oxidizing agent and adsorbed oxygen may also augment the concentration of active sites. Thus, a boost of O_2 concentration could be doubly favorable (*Massa et al., 2005*).

The effect of on-off modulation of the flow of aqueous solutions of ethanol on its oxidation with air in a TBR was investigated by *Ayude et al. (2008)* using a 0.5% Pd/Al_2O_3 commercial egg-shell catalyst. Catalyst activity was monitored through continuous experiments. Some overoxidation of the catalyst was observed depending on the cycling conditions used, but activity was recovered by exposing the catalyst to flowing N_2 in the presence of the ethanol solution. A significant improvement over steady-state operation was obtained when the catalyst was exposed to a short pulse of excess oxygen after an interval of operating in the mass-transfer-limiting regime. Indeed, high liquid flow rates and moderate gas flow rates must be employed to ensure complete wetting of the catalyst during the on-portion of a cycle and, to minimize overoxidation, during the off-portion. Even though, at the conditions studied, periodic operation did not improve catalyst stability, it was evident that proper selection of cycle duration and an appropriate start-up protocol are crucial to maintain long-term catalytic activity. The *Ayude* study indicated that to understand the performance under LFM the catalyst's dynamic behavior must be taken into account.

17.3.5 CYCLOP Project

Figure 17-1 is qualitative and shows how flow modulation enhances TBR productivity, focusing on just a wetting efficiency and using two static measurements to explain an inherently complex, dynamic process. A comprehensive analysis of flow modulation was undertaken from 2000 to 2003 by a European Consortium for Cycling Operation (CYCLOP)— Cordis Project. This project involved 10 teams from both academia and industry, coordinated by DSM Research BV, Geelen, the Netherlands and consisting of Centre National de la Recherche Scientifique, Nancy, France; Foundation for Research & Technology Hellas, CPERI, Thessaloniki, Greece; Martin Luther University of Halle-Wittenberg, Halle, Germany; Institute of Chemical Technology, Prague, Czech Republic; Politecnico di Torino, Turin, Italy; Solvay SA, Bruxelles, Belgium; Gesellschaft für Heuristisch-Numerische Beratungssyteme, GmbH, Dortmund, Germany; Sulzer AG, Winterthur, Switzerland, and Akzo Nobel, Amsterdam, The Netherlands. The ambitious targets for the project were exploring optimal designs for TBRs, devising control systems for safe operation, demonstrating that periodic can compete or surpass steady operation, development of heuristics for process development, elucidating mechanisms for improvement under cyclic operation, preparing design correlations and software, and, of course, examining applications. For this effort, pilot plants were constructed in Geleen, The Netherlands, and in Litvinov, Czech Republic.

Not all of the research results are published. Those that were have been presented in a special symposium as a part of the 15th International Congress of Chemical and Process Engineering (CHISA) held in Prague during August, 2002. Because the CYCLOP effort covered a wide range of research activities, results of the project are discussed in different sections of this chapter rather than in one place.

Investigations undertaken during the European project, however, identified two distinct modes of periodic operation. The first is a slow mode in which the period is long enough that the entire catalyst bed will have the same mean liquid velocity. The cycle period, τ, is then comparable to the liquid residence time, $\bar{t}$, in the reactor, which is often several minutes. The second mode, a fast mode, occurs under pulsing of the liquid fed to the bed over just a few seconds. In this case liquid pulses, or shock waves, are generated that travel through the reactor at velocities usually ranging from 0.1 to 0.3 m/s. The cycle period is so short that the length of a single liquid pulse is well below the length of the reactor. The existence of at least one pulse in the packed bed identifies the fast mode.

Fast mode cycling may be considered an extension of the high-interaction regime observed under steady-state TBR operation at high gas and liquid flow rates. This regime exhibits much more rapid mass and heat transfer. *Boelhouwer et al. (1999, 2001a, b, 2002)* have demonstrated similar heightened transport during fast-mode cycling operation. More recent work on fast-mode cycling shows that particle shape and size effects induced pulsing (*Trivizadakis et al., 2006a, b*).

17.4 LIQUID FLOW MODULATION WITH MULTIPLE REACTIONS

The effect of LFM on selectivity when more than a single reaction occurs has not been fully investigated (*Boelhouwer, 2001*). Oxidation of hydrocarbons typifies systems where multiple reactions arise. *Tukač et al. (2003)* discussed the oxidation of phenol over extrudates of active carbon in a TBR operated with fast LFM. Both phenol and rates of total oxidation increased compared with the steady-state operation, the former by more than 10%. Selectivity towards total oxidation was enhanced, especially for a 60 s cycle period at a split, *s*, of 0.5.

The often used phenylacetylene hydrogenation reaction to styrene and then ethylbenzene has been employed by *Wilhite et al. (2003)* under gas reactant and liquid reactant limiting conditions to explore LFM induced pulsing flow. Their isothermal experiments were undertaken through adjusting liquid feed rates so that, at base flow, the TBR was operating in the trickling hydrodynamic regime while at peak flow, the bed was in the pulsing regime. These investigators found that LFM suppressed conversion and selectivity to styrene under gas reactant limiting conditions. A detailed model developed by the authors predicted this observation. When applied to a liquid limiting condition, the model predicted a small enhancement of phenylacetylene conversion and a large enhancement in selectivity to styrene. Enhancements were confined to short peak flow times < 5 s.

Fraguío et al. (2004) examined the oxidation of ethanol over a Pt/Al_2O_3 catalyst in an aqueous solution in which reaction products were acetaldehyde, acetic acid and CO_2. They found that LFM increased the selectivity to acetic acid with perhaps a small increase of this selectivity as τ_{cycle} was extended. The effect of period was opposite to that on EtOH conversion.

Massa et al. (2005) also studied the oxidation of phenol in solution but over a CuO/Al_2O_3 catalyst, using slow-mode cycling. Over the range of operating conditions investigated for a partially wetted TBR, they observed that on-off LFM had just a small effect on phenol conversion but a much larger effect on selectivity to total oxidation, especially at longer τ_{cycle}. When the liquid feed to the bed is halted, the nonvolatile intermediate products remain inside the catalyst particles where they can be further oxidized. Average gas phase and dissolved oxygen concentrations inside the catalyst particles during LFM are higher than those attained under steady-state operation. The effect is more pronounced for extended τ_{cycle}, where phenol oxidation is greater.

Skála and Hanika (2008) explored the hydrogenation of dicyclopentadiene (DCPD) on an egg-shell type palladium catalyst in laboratory-scale, pseudo isothermal and adiabatic TBRs. Experiments were performed at atmospheric pressure using $s = 0.5$ and $\tau_{cycle} = 60$ s. These researchers found a 12% rate increase over steady state at comparable operating conditions. However, selectivity of the hydrogenation to tetrahydrodicyclopentadiene (THDCPD) decreased. This effect is similar to selectivity suppression of consecutive hydrogenation of dienes or acetylenic substrates at higher partial pressure of hydrogen. This selectivity observation did not agree with *Liu and Mi (2005)* and *Liu et al. (2005)*, who reported a small increase in the selectivity to an intermediate over a broad range of conditions for the consecutive hydrogenation of 2-ethylanthraquinones in a periodically operated, isothermal TBR. *Liu et al. (2008)* returned to the exothermic hydrogenation of DCPD to THDCPD also using a Pd/Al_2O_3 catalyst. The influence of operating strategies on the reactor performance were studied and compared with their steady-operation counterparts, including on-off and peak–base LFM or DCPD concentration modulation. A novel hybrid mode involving modulation of both liquid flow rate and DCPD concentration was also tested.

Figure 17-6 summarizes the effect of τ_{cycle} and cycle split, s, on DCPD conversion, X_{DCPD}, THDCPD yield, Y_{THDCPD} and global hydrogenation rate, R_H. Maxima were observed of up to 20% over steady state at $\tau_{cycle} = 240$ s ($s = 0.75$), but these measures drop below steady-state values for cycle periods greater than 400 s. LFM improves hydrogenation rates from 10 to 20% and causes temperature spikes of ca. 15 K. *Liu et al. (2008)* ascribed this performance enhancement to an increase in the H_2 mass transfer rate. Modulation of the liquid phase DCPD concentration in the feed to the bed produced less than a 5% conversion and yield increase and just a 10 K temperature rise. Enhancement under the hybrid modulation of about 15% was higher than the peak-base modulation of either flow or DCPD concentration but the temperature rise was less than 3K. Even though only a few of the possible operating conditions for the different modulation strategies were considered, the *Liu et al.* results suggest modulation is promising for exothermic reactions.

Hydrogenation of pyrolysis gasoline is widely applied in industry to prevent gum-formation and catalyst deactivation caused by diene oligomerization. Advantages of trickle

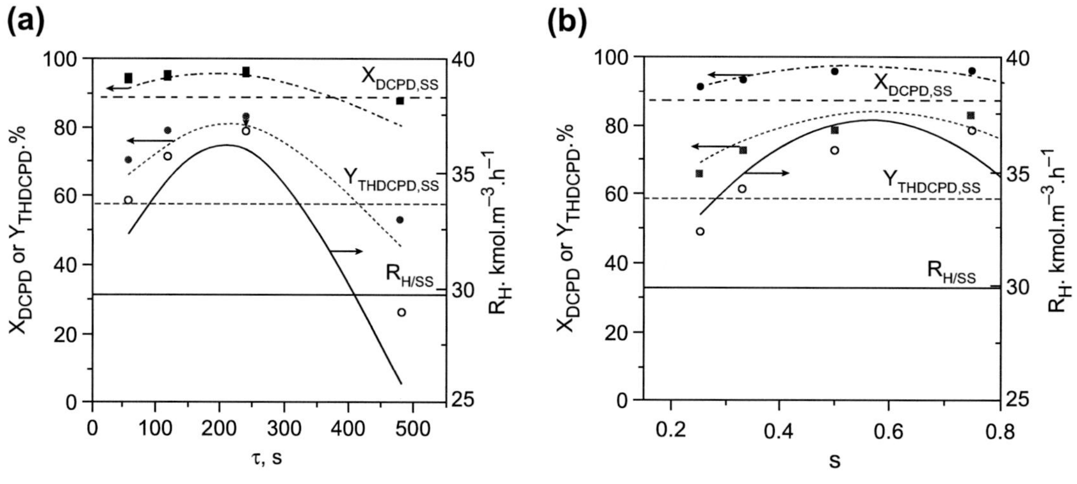

FIGURE 17-6 The effect of (a) cycle period, τ, and (b) cycle split, s, on TBR performance for the hydrogenation of dicyclopentadiene (DCPD) using a palladium catalyst in pseudo isothermal and adiabatic TBRs operating at 1 atm, (■) X_{DCPD}; (●) Y_{THDCPD}; (○) R_H; lines show calculated results; (a) s = 0.75; (b) τ = 240 s; V_{LA} = 15 mL/min. *(Figure adapted from* Liu et al. (2008) *with permission of authors.)*

bed reactors usually adopted for this exothermic process are accompanied by drawbacks such as poor external heat exchange, liquid maldistribution and rate-limiting mass transfer of hydrogen to the catalyst surface. LFM offers a solution to these disadvantages by creating a pulsing flow pattern throughout the catalyst bed that provides periodic renewal of liquid wetting of the catalyst surface, more intensive interaction between the gas and liquid phases, higher mean bed temperature and more rapid heat transport during the liquid rich part of the cycle.

Hanika et al. (2007) and *Tukač et al. (2007a, b)* undertook a comprehensive study of the apparent advantages of LFM by comparing the hydrogenation rate of pyrolysis gasoline with hydrogenation rates of typical components, i.e., styrene and dicyclopentadiene (DCPD) under steady-state and periodic modulation of the liquid feed rate. Experiments were carried out in a pilot scale TBR at the R & D Center of the Chemopetrol Co. (Litvínov, Czech Republic). The reactor, with a 10 cm i.d. × 2 m height, had automatic control and a data acquisition system. Through thick insulation, reactor behavior was close to adiabatic. Catalyst, pelleted 0.1 wt% Pd/ Al_2O_3 (Degussa Noblyst 1505, $d_p = 3$ mm), was randomly packed to a depth of 1 m. During the test program, catalyst deactivation and hydrodynamic hysteresis was observed. Tests determined parametric sensitivity of the hydrogenation rate to the inlet temperature and pressure, and to liquid feed modulation parameters, like τ_{cycle}, s as well as the ratio of peak to base liquid feed rates. Of course, the hydrogenation reactions employed exhibited a significant adiabatic temperature rise which resulted in a measurable axial temperature profile.

Results for styrene hydrogenation (*Hanika et al., 2007*) are shown in **Figures 17-7 to 17-9**. In the first of these, variation of the mol fraction of ethylbenzene (EtBz) in the liquid product with time in a cycle appears. "A" in the figure corresponds to a steady-state feed rate, while "B" gives the mean EtBz mol fraction for the time modulated feed rate of styrene in the reactor inlet. LFM imparts a ca. 10% increase of EtBz in the liquid leaving the TBR compared with a continuous constant feed rate mode. The higher conversion is due to a higher mean temperature in the TBR through periodic operation (see **Figure 17-8**).

Any change of feed rate, of course, may modify the hydrodynamic regime in the TBR

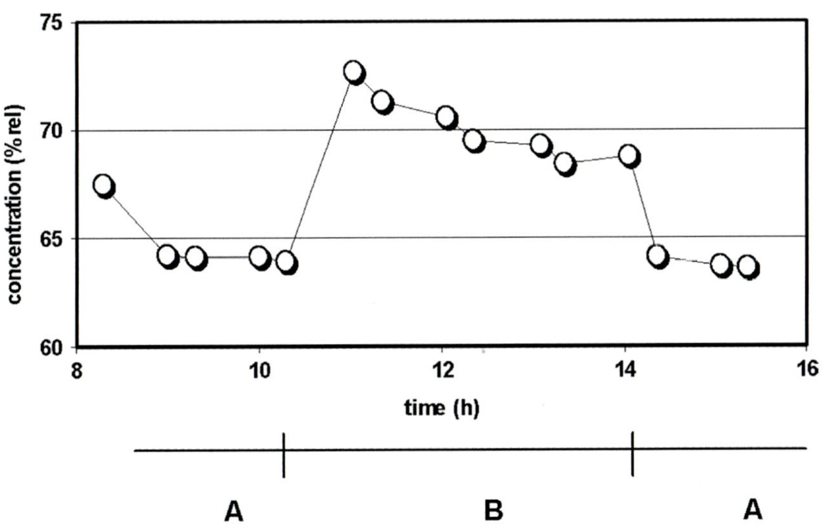

FIGURE 17-7 Relative concentration of ethylbenzene in reaction product leaving a pilot scale trickle bed during styrene hydrogenation over 0.1% Pd/Al_2O_3 during (A) steady state and (B) LFM operations: $T_{feed} = 30°C$; $P_{feed} = 0.2$ MPa; $\tau_{cycle} = 60$ s; $s = 0.15$; q_{feed} (liq.) = 0.125 m^3/h; liquid feed = 10% styrene in toluene. *(Figure reproduced from* Hanika et al. (2007) *or* Tukač et al. (2007a, (b) *with permission of the authors.)*

FIGURE 17-8 Transient axial temperature profiles in adiabatic trickle-bed reactor during styrene hydrogenation (See Figure 17-7 for experimental conditions). *(Figure reproduced from* Hanika et al. (2007) *or* Tukač et al. (2007) *with the authors' permission.)*

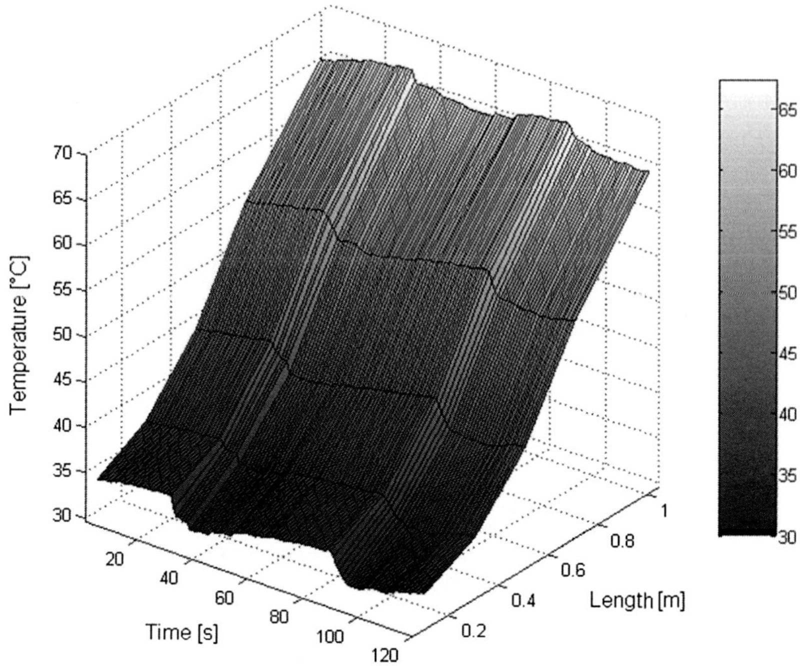

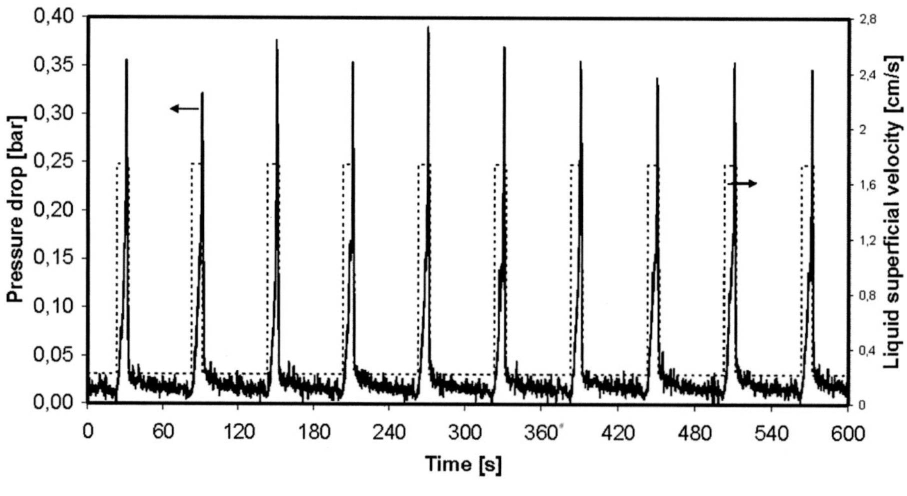

FIGURE 17-9 Transient pressure drop and liquid superficial velocity during liquid feed rate modulation (See Figure 17-7 for experimental conditions). ← belongs to the solid lines and refers to the scale on the left ordinate; → belongs to the dotted lines and refers to the scale on the right ordinate. *(Figure reproduced from* Hanika et al. (2007) *or* Tukač et al. (2007) *with the authors' permission.)*

and will be reflected by altered liquid holdup and consequently pressure drop over the catalyst bed (see **Figure 17-7**). With a reduced feed rate, liquid holdup and thus the thickness of liquid films enveloping catalyst pellets decreases so that the hydrogen transfer rate to catalyst external surface is increased. It is this effect that accelerates the progress of the reaction.

In the case of pyrolysis gasoline hydrogenation, the situation is more complicated. Various reactions proceed so that many reactants compete at the active centres on the catalyst surface. However, LFM leads to a higher mean temperature in the TBR in comparison with a constant feed rate at steady state (see **Figure 17-10**). Mean reaction rate over the entire TBR, namely reactor productivity, is significantly higher, by ca. 30%, compared with steady-state operation with identical feed conditions.

Mean hydrogenation rates over the catalyst bed of pyrolysis gasoline components are also enhanced through LFM, but as might be expected the enhancement is not uniform: Thus they vary from 7% for cyclopentadiene and isoprene to 20% for indene. **Table 17-2** lists the LFM enhancements over steady-state operation for the major unsaturated components of pyrolysis gasoline measured by either *Hanika et al. (2007)* or *Tukač et al. (2007)*.

Cycle split, *s*, exerted an important effect on the production rate enhancement of the gasoline components. Lower values of *s*, i.e. shorter splits raised the hydrogenation enhancement of individual components by about 10 to 30% over $0.15 \lesssim s \lesssim 0.5$ except for dicyclopentadiene the enhancement of which differed little from the factor given in **Table 17-2**.

From their studies *Hanika et al. (2007)* and *Tukac et al. (2007)* concluded that hydrogenation was confined to the diene compounds for the Pd catalyst and the operating conditions used. Aromatic rings were not attacked. The throughput enhancement for LFM operation, seen in the experiments as higher reaction rates, arose from short-term excursions into the pulsing regime from the film (trickle) regime resulting in more intense interaction of the gaseous and liquid phases. Higher rates are accompanied by a ca. 2°C mean temperature rise for the TBR. These investigators found

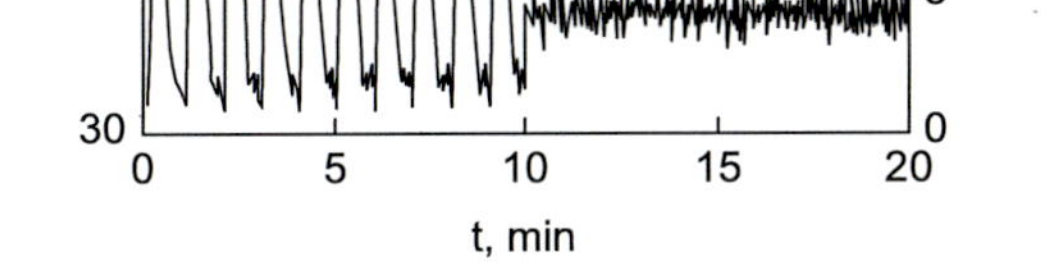

FIGURE 17-10 Liquid outlet temperature and TBR pressure drop during transition from periodic (LFM) to steady-state operation for T_{feed} = 60°C, mean feed rate = 124.7 L/h, τ_{cycle} = 60 s, s = 0.25. *(Figure reproduced from* Tukač et al. (2007b) *with the authors' permission.)*

TABLE 17-2 Enhancement Factors for Pyrolysis Gasoline Components ($\bar{q}$ = 125 L/h, τ_{cycle} = 60 s, s = 0.25, T_{feed} = 60°C)

Compound	Enhancement Factor
styrene	1.07
allylbenzene	1.13
3-methylstyrene	1.14
2- methylstyrene	1.16
4- methylstyrene	1.17
cyclopentadiene + isoprene	1.07
endo-dicyclopentadiene	1.09
indene	1.20
endo-dihydrodicyclopentadiene	1.08

that the optimum cycle period corresponds to the time it takes for liquid to drain from the trickle bed after the liquid feed is switched off. Period split, $s = 0.25$ or less, offered the best enhancement.

17.5 HYDRODYNAMICS UNDER LIQUID FLOW MODULATION

Important reactor properties, such as pressure drop, liquid holdup, and wetting efficiency, are affected by the external surface coverage of the catalyst pellets. If LFM is imposed on a TBR, these properties change as the liquid feed rate to the bed increases or decreases. Even so, only a handful of studies examining the influence of LFM on bed hydrodynamics have been reported. Indeed, the effect of periodic operation on liquid holdup, mass and heat transfer coefficients, and wetting efficiency needs more investigation. **Table 17-1** summarizes publications on this subject up to 2010. Most experimental studies of TBR hydrodynamics were undertaken without an accompanying chemical reaction and were part of the CYCLOP project (*Boelhouwer et al., 2001a, 2002a, b; Borremans et al., 2004, 2007; Giakoumakis et al., 2005*). This effort identified the importance of gas-liquid mass transfer, but found that the average rate under LFM did not differ from rates interpolated from steady-state rates at the base-peak flows in a cycle. A similar observation was made for heat transfer. Pulse decay occurred and could be modeled using a slight modification of published correlations. Also liquid holdup could be calculated similarly. Flow maps derived for the model systems were applicable to these pulse induced systems.

Research has continued after the end of the CYCLOP effort. *Brkljac et al. (2007)* applied the two-phase flow concept of relative permeability to model the hydrodynamics of a periodically operated TBR. Relative permeabilities based on steady-state measurements accurately predicted liquid holdup and pressure drop. Liquid relative permeabilities depended mainly on liquid viscosity, whereas those for the gas depended on liquid surface tension, viscosity and the flow operation, i.e. whether flow was increasing or decreasing. For the latter, hysteresis is involved arising from changes in flow pattern in the packing. With on-off LFM, the filament structure of liquid flow through a TBR is disrupted by flow stoppage and builds back slowly when liquid flow is restored. MRI imaging (*Gladden et al., 2007)* supports the relative permeability concept as images of a TBR cross section show numerous channels filled with liquid and thus impermeable to gas phase flow, while other channels contain just gas and so impede liquid flow. From a different standpoint, CFD based models were used by *Gunjal et al. (2005)* to predict pressure drop and liquid holdup under externally imposed modulation.

A 5 cm (i.d.) × 1 m bed filled with a non-porous packing was used to study the behavior and stability of a LFM induced pulse moving downward in a TBR using Newtonian and non-Newtonian liquids at $25°C < T < 75°C$ and $P < 0.7$ MPa (*Aydin et al., 2006*). An idealized pulse is shown in **Figure 17-11**. The effect of a narrow range of temperature and pressure on the parameters describing the induced pulse was explored for an air-water system and for an air-0.25 wt% CMC solution representing a non-Newtonian liquid. The investigators found that the pulse parameters B_t and D_t became shorter while P_t lengthened with increasing temperature and pressure. The breakthrough amplitude, B_a, proceeded to collapse as it moved down the bed and this was abetted by higher fluid temperature and gas pressure. Draining of liquid holdup in the pulse to the base flow holdup occurred through dispersive phenomena in the TBR. Behavior was similar in Newtonian and non-Newtonian liquids, but was much slower in the latter. These results indicate bed depth limits for periodic flow forcing.

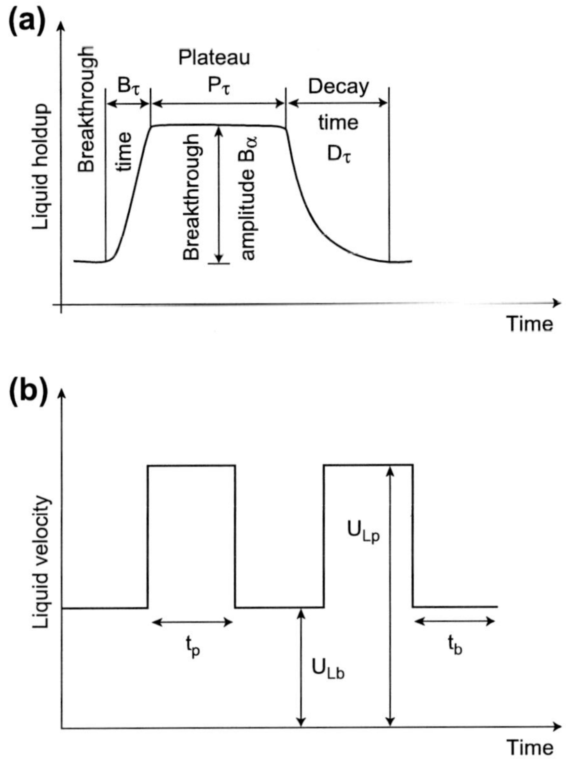

FIGURE 17-11 **Shape and descriptive parameter of the slow-mode liquid pulses (a) induced by modulation of the liquid feed to the TBR shown in (b).** *(Figure reproduced from* Aydin et al. (2006) *with permission. © 2006 by the AIChE.)*

Electrical capacitance tomography was applied by *Hamidipour et al. (2010)* to measure the variation with time of liquid distribution over a TBR cross-section. Such measurements enabled calculation of liquid holdup and pulse characteristics such as pulse duration, break through and decay times. An unexpected result of these measurements was that a new mode of modulation, on-off modulation of liquid than gas flow, yielded a closer approximation of the square-wave pulse than just LFM and also the most uniform distribution of liquid flow across the trickle bed.

NMR imaging of LFM in trickle beds was applied by *Lysova et al. (2009)* to spherical Pd/Al_2O_3 catalyst particles in a shallow 1 cm i.d. trickle bed. Particle size was a variable and sizes from 1 mm to 4.2 mm were used. Hydrogenation of 1-octene was the model reaction. **Figure 17-12** shows bed temperatures under on-off modulation in **(a)** and base-peak modulation in **(b)**. H_2 flow is continuous and enters the bed at 77°C.

Temperature measurements in **(a)** of the figure indicate a large increase in the hydrogenation rate once LFM begins. Some of the increase results from an increase in bed temperature. MRI measurements show catalyst particles retain some liquid even though temperatures exceed the normal boiling point of both octene and octane. The cycle period is 60 s in **(b)** of the figure. Large temperature fluctuations in the base-peak modulation are consistent with dry-out of the catalyst particles during the liquid-off portion of a cycle that was seen from the MRI results.

Despite the discussion above, hydrodynamic studies under periodic operation are still limited and have been carried out mostly for non-porous packing. *Dietrich et al. (2005)* have tackled the problem of modeling diffusion and reaction in catalyst particles, pointing out that practical models will require simplifying assumptions about geometry and internal conditions. Of course, the influence of the cycling parameters has been examined only within a narrow range. Further experiments employing a broader range of operating conditions and using porous packing are required.

17.6 MODELING OF THE PERIODIC OPERATION OF TRICKLE BED REACTORS

Several investigators have attempted to model the modulation of TBRs. *Stegasov et al. (1994)* and later *Lange et al. (1999, 2004)* proposed dynamic models for a periodically operated TBR, although they neglected the dynamics inside catalyst pellets making up the

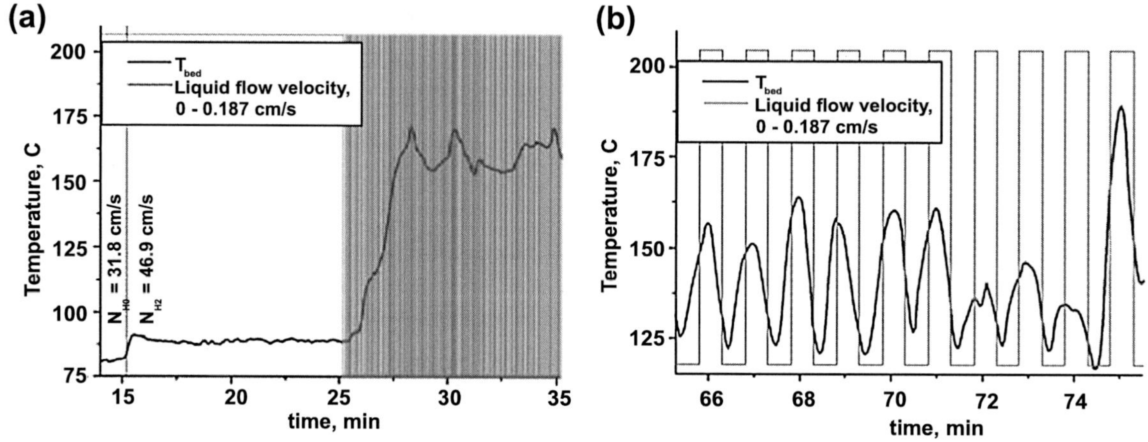

FIGURE 17-12 **Temperature variations with time during on-off LFM for 1-octene hydrogenation in a shallow trickle bed with 1mm Pd/Al_2O_3 catalyst particles: (a) Shows T_{bed} with He. At 15 min, H_2 replaces He and the TBR operates under steady liquid flow until 25 min when modulation begins at $\tau_{cycle} = 6$ s, (b) T_{bed} under modulation at $\tau_{cycle} = 60$ s.** *(Figure reproduced from* Lysova et al. (2009) *with permission. © 2009 by Springer Science + Business Media.)*

bed. *Boelhouwer (2001)* solved dynamic mass balances allowing for internal diffusion assuming a wash-coated catalyst pellet with an impermeable core, but mass transfer between the dry and wet zones was not considered. Reactant and product profiles were obtained during fast cycling, base-peak modulation; however, reaction enhancement was not evaluated. *Boelhouwer* commented that the rate of internal diffusion largely determines the optimal cycle period.

A dynamic model has been presented by *Kouris et al. (1998)* to reproduce the behavior of a catalytic pellet in a TBR pulsing-flow regime, where regions enriched in gas and liquid rapidly pass through the reactor. As the period tends to zero, the pellet is unable to follow the rapid changes in wetting and reaches a pseudo-stationary state, namely the relaxed steady state described by *Bailey (1977)*. *Kouris et al.* noted that catalyst performance is improved in the pulsing-flow regime. Experiments discussed earlier have shown that pulsing flow is created by fast cyclic operation of a TBR. However, intermediate and/or slow flow modulation has different characteristics and was not examined by *Kouris et al. (1998)*.

Modeling slow flow modulation was treated by *Sicardi et al. (2005)* who showed, using data from the literature, that if temperatures are held constant, performance under periodic operation can be closely predicted from steady-state results at the base and peak flow rates using the cycle split, *s*, to weigh the contribution from each steady state.

Ayude et al. (2005) modeled the transient behavior of a catalyst pellet in a TBR under on-off cycling. Effectiveness factors for stationary and transient operation were calculated from the model. An enhancement factor for LFM was defined from these effectiveness factors and evaluated for different operating conditions. In this way the influence of cycling and system parameters on performance was analyzed over a wide range of conditions. The authors concluded that LFM will not significantly improve reactor performance if mass transfer is not rate limiting or if the external wetting is low in the reference steady-state. In addition, they found that the ratio of reactant concentrations significantly affects the possible enhancement. Experimental comparisons among different cycling strategies must consider this parameter. If reaction stoichiometry has the reactant ratio changing with conversion, extrapolating results obtained in a differential reactor could be misleading.

Recognizing that base-peak or on-off LFM raises conversion above that at steady-state if reaction rate is controlled by gas-liquid mass transfer, *Hanika et al. (2007)* devised a simple, pseudo-steady state, isothermal model describing the transient concentration of a gaseous reactant dissolved within a moving liquid film with a time-varying thickness flowing down along an inclined catalytically active wall. Mass and momentum balances were made on a space element within the laminar film assuming a parabolic transversal velocity profile. A pure gaseous reactant was assumed making a gas phase mass balance unnecessary. Simulations were carried out for an irreversible, monomolecular, first-order reaction with respect to the gas phase reactant. In the simulation, periodic on-off forcing was applied by specifying a 10-fold thinner film during the liquid off portion of the cycle to represent static liquid holdup in the trickle bed.

Pellet-to-pellet variation of the depth of liquid films enveloping the catalyst external surface and thus also the variation of liquid dynamic holdup is typical in trickle beds in most operating regimes. With LFM, large temporal variations arise. Under such conditions, lowering the mean base flow, and thus liquid film thickness markedly raises the gas reactant's mass transfer rate to the external surface of a catalyst pellet. The situation is shown schematically in **Figure 17-13**. The left-hand side of this figure represents square wave, on off liquid feed modulation while the right-hand side illustrates the continuous feed mode, first formulated, solved, and experimentally verified by *Hanika et al. (1970, 1971)*.

Applying this model to a selected τ_{cycle} allowing cycle split and reaction-diffusion parameters to vary, expressed as the Thiele Modulus, provided a comparison of reaction rates under periodic and continuous operation. This was expressed by *Hanika et al. (2007)* as a ratio of external efficiency. Their calculated ratio, showing dependence on cycle split value and a modified Thiele modulus, appears as **Figure 17-14**. Ratios > 1 indicate rate improvement through on-off feed rate modulation. That improvement is sensitive to s and becomes quite large when there is a significant diffusional reaction barrier. An optimum occurs at $s = 0.3$ for that condition.

Liu et al. (2008) developed a dynamic non-isothermal TBR model taking into account

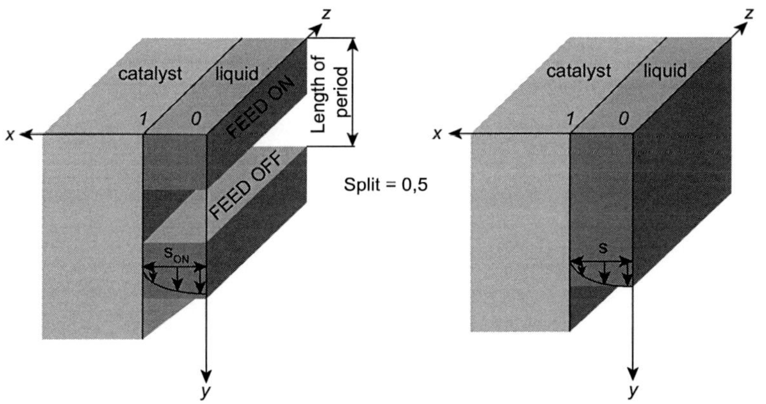

FIGURE 17-13 Simplified particle scale model of a catalyst pellet external surface in a trickle bed with periodically interrupted and continuous liquid feed. *(Figure reproduced from* Hanika et al. (2007) *with the authors' permission.)*

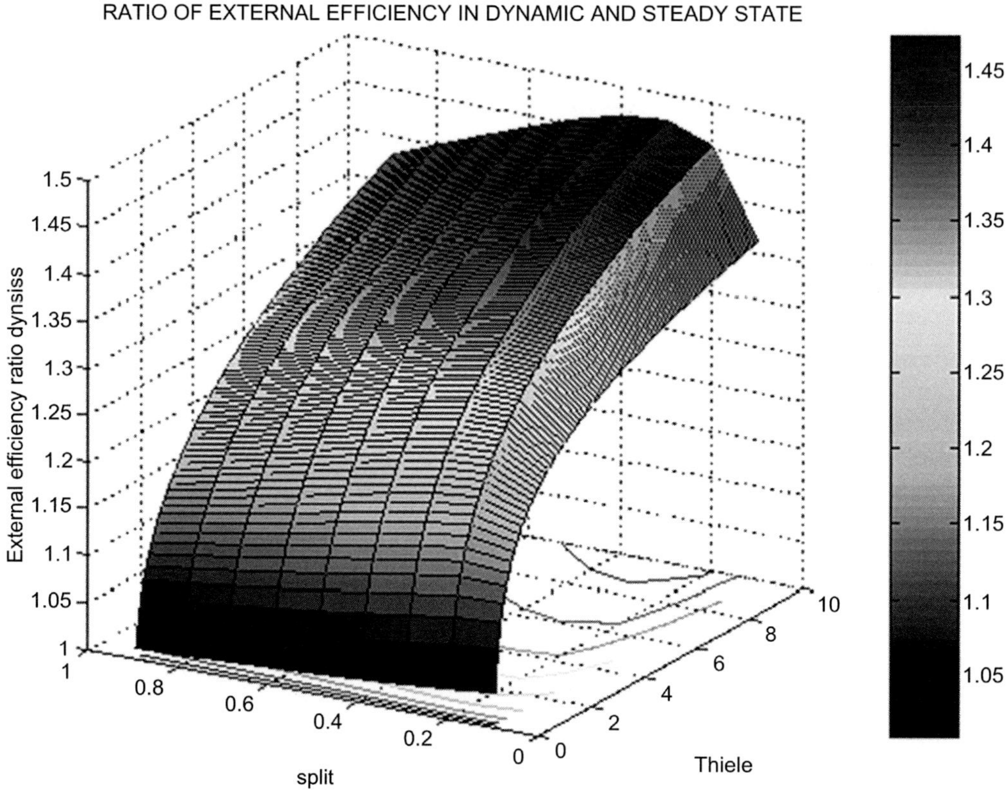

FIGURE 17-14 Enhancement of external efficiency, a comparison of production rates independent of kinetic model, for periodic liquid feed interruption and continuous feed for an idealized trickle bed as a function of cycle split and the balance of reaction and diffusion through a liquid film (Thiele Modulus). *(Figure reproduced from* Hanika et al. (2007) *with permission of the authors.)*

time-dependent variations of reactant concentrations in the gas phase and the contributions of dynamic and static liquid zones in a trickle bed. Liquid holdup variations encountered with LFM were approximated following the approach suggested by *Lange et al. (2004)*. In a recent contribution, *Ayude et al. (2009)* examined the response of an isothermal TBR to on-off liquid flow modulation employing a reactor scale-model. The reaction was gas reactant limited and first order with respect to reactants in both phases. Liquid phase hydrodynamic behavior was introduced into the model through two different approaches: 1) "Liquid draining," based on experimental results and 2) ideal "square waves." The model allowed evaluation of, among other variables, liquid holdup, liquid velocity and the time variation of liquid reactant conversion at different axial positions within the reactor. An enhancement factor for periodic operation was defined by computing a temporal average, liquid reactant conversion during an invariant cycling state and referring this to its corresponding steady-state conversion. For all conditions investigated, enhancements were lower when the liquid draining approach was assumed.

A mathematical model of catalytic hydrogenation in a TBR under LFM of the feed rate has been formulated by *Stanek et al. (2009)* and the performance predicted by that model has been compared with pilot plant experiments for the catalytic hydrogenation of styrene. The computed mean conversion of the reactor under forced modulation of the liquid feed is plotted in **Figure 17-15** as a function of cycle split. Constant liquid feed is represented by $s = 1$. Thus, the figure shows a considerably higher conversion at low values of the split. Numerical results indicate that the principal source of improvement under modulation is an increase in the wetted surface of the catalyst.

The improved conversion predicted above was observed experimentally at low splits with LFM. The base u_L temporarily brought the TBR regime close to the transition from the trickling to the natural pulsing regime. The experiments of *Stanek et al.* were carried out on a pilot plant trickle bed reactor described earlier in **Section 17.4**. A 2 m long reactor was designed to sustain temperatures up to 200°C and pressure to 2.5 MPa. The unit was equipped with two pumps (maximum feed rate 120 and 600 L/h), two storage tanks of 1.1 m^3 volume, a 40-L phase-separator and a ControlWeb 2000 SP9 data logger and control system. The function of the pumps was controlled by a computer program which created modulation of the liquid feed to the reactor. Thermocouples and pressure ports located axially through the catalyst bed along with pump speed provided operating data that were logged at the frequency of 10 Hz.

Stanek et al. ran the hydrogenation of styrene in a toluene solution (10 wt%) in their pilot unit at 0.2 MPa and a 30°C inlet temperature using a Noblyst R 1505 catalyst (egg shell impregnated, 0.1 wt% Pd/Al_2O_3 on 2.5 mm spheres). **Figure 17-16** compares predicted (line) and experimental (points) conversion in the TBR expressed as the ratio of conversion under LFM to that under continuous operation. The reactor was started with continuous liquid feed, but between 12:45 and 16:28 h, operation switched to the forced periodic modulation with: $s = 0.15$, $\tau_{cycle} = 60$ s, $u_G = 0.098$ m/s, base $u_L = 0.0021$ m/s and $\overline{u_L} = 0.0044$ m/s.

The figure shows that the relative periodic performance of the reactor is predicted reasonably well by the model of *Stanek et al.*, but the experimental performance deteriorates with time. This was attributed to the observed rapid

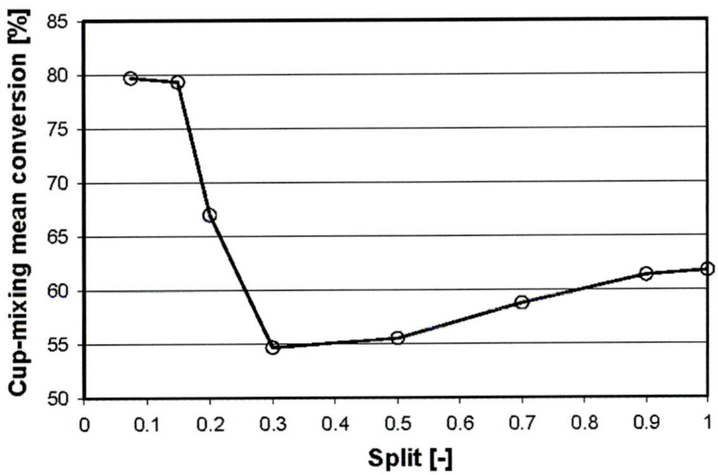

FIGURE 17-15 **Predicted mean conversion as a function of the split under periodic modulation of the liquid feed: $\tau_{cycle} = 60$ s; $u_G = 0.098$ m/s; base $u_L = 0.0021$ m/s; $\overline{u_L} = 0.0044$ m/s.** *(Figure reproduced from* Stanek et al. (2009) *with the authors' permission.)*

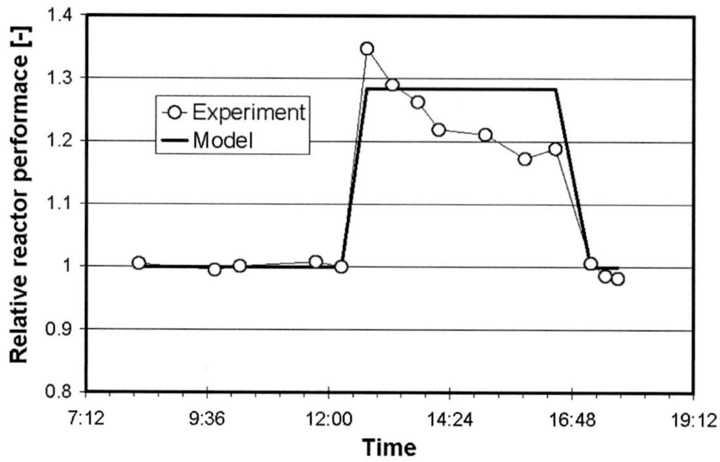

FIGURE 17-16 Predicted (line) and experimental (o) performance of the reactor relative to steady-state operation when operated under LFM feed between 12:45 and 16:28 h. Operating conditions are given in Figure 17-15. *(Figure reproduced from* Stanek et al. (2009) *with the authors' permission.)*

deactivation of the catalyst primarily as a result of styrene polymerization.

Figure 17-17 compares the predicted transient temperatures at different depths within the trickle bed with the experimental observations for LFM at $\tau_{cycle} = 60$ s and $s = 0.15$. The observations correspond to the later stages of an experiment when the effects of TBR start up have died down. Thus, zero time does not coincide with the beginning of the experiment. The figure shows fair agreement of the predictions of the temperature waves and the measurements at the various depths from the top of the bed. A systematic deviation can be seen in that the experimental data exhibit the peak temperatures on each bed level more or less at the same time instant, while the model predicts the peaks at different depths mutually shifted in time. A probable source is that the model assumed the same temperature for all three phases while the bed thermocouples reacted fastest to the temperature of the gas for which the time shift between levels indeed may be small.

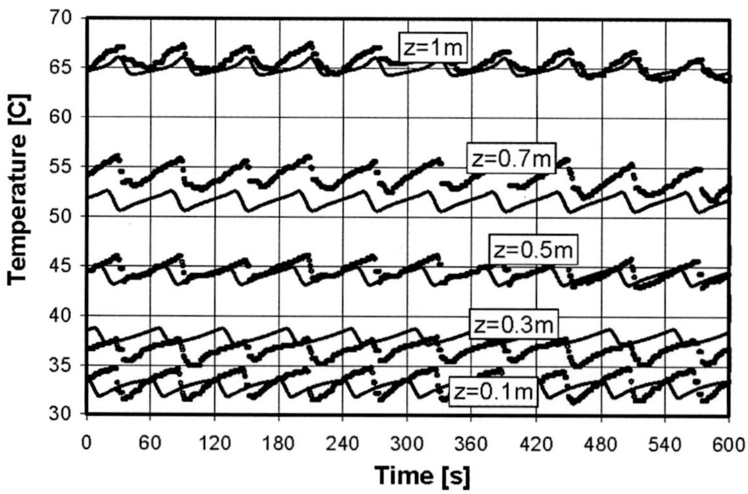

FIGURE 17-17 Predicted (lines) and experimental (circles) temperatures in a pilot scale TBR under periodic modulation of the liquid feed. For operating conditions, see Figure 17-16. *(Figure reproduced from* Stanek et al. (2009) *with permission of the authors.)*

Both computed and experimental results indicated that LFM has a much greater impact on the transient concentration profiles than on the temperature profiles. Increased conversion appeared at lower end cycle splits where the forced modulation of the liquid feed brings the reactor temporarily into the transition zone between the trickling and pulsing flow regimes (*Stanek et al., 2009*).

17.7 SUMMARY

This chapter has demonstrated that LFM affects reaction rates and thus also product distribution. The latter effect has not been studied. Indeed changes in product distribution for consecutive reactions under LFM will require a more detailed analysis to establish optimal cycling conditions. Furthermore, the impact of LFM on hydrodynamic properties in TBRs needs to be addressed in more detail, such as the effect of cycling on liquid holdup, mass transfer and catalyst wetting. Results presented in the literature to date cover only a narrow range of experimental conditions. Certainly, further experimental measurements are required.

Nomenclature

A = reactant species
AMS = α-methylstyrene
B_a = pulse breakthrough amplitude
B_t = pulse breakthrough time (s)
C = concentration (mol/m^3)
C^0 = initial or entrance concentration
$C*_i$ = concentration of gas species i in the liquid phase at saturation
C_S = liquid reactant concentration
CFD = computational fluid dynamics
CMC = carboxymethylcellulose
D_i = effective diffusivity of species i in a catalyst particle
D_t = pulse decay time (s)
DCPD = dicyclopentadiene
EAQ = ethylanthraquinone
EtBz = ethylbenzene
i.d. = inner diameter
MRI = magnetic resonance imaging
LFM = liquid or liquid feed flow modulation
P = pressure (kPa, MPa)
P_t = pulse plateau duration (s)
ΔP = pressure drop (kPa)
q = volumetric flow rate (L/s)
$\bar{q}$ = mean or time average volumetric flow rate
s = cycle split (-)
T = temperature (°C, K)
THDCPD = tetrahydrodicyclopentadiene
t = time (s)
$\bar{t}$ = mean residence time (min)
u = liquid superficial velocity (cm/s)
v = liquid velocity (cm/s)
$\bar{v}$ = mean or time average velocity
X_i = conversion of species i
x = dimension, usually depth (m)
Y_i = yield of species i
y = dimension (m)
z = dimension (m)

Greek

γ = criterion for the limiting reactant
τ = period, usually cycle period (min, h)
τ_{cycle} = cycle period

Subscripts

b = base
G, g = gas
L, l = liquid
p = peak
ss = steady sate
uss = unsteady state

CHAPTER

18

Periodic Flow Reversal

Grigorii Bunimovich, Hristo Sapoundjiev†*

*Matros Technologies, Inc., St. Louis, USA and †Montreal, Canada

OUTLINE

18.1 The Heat-Trapping Concept 496

18.2 Theoretical Aspects 500

18.2.1 Modeling 500

18.2.1.1 Simplifications 504

18.2.1.2 Complications 504

18.2.2 High Switching Frequency Reactor Model 505

18.2.3 Stationary Heat Front Movement 507

18.2.4 Dynamic Behavior 507

18.2.5 Simulation and Preliminary Reactor Design 509

18.3 Oxidation of Volatile Organic Compounds 511

18.3.1 Methane Combustion with Heat Recovery 515

18.3.2 Reverse Flow Reactor Control 519

18.4 Other Applications of Reverse Flow Reactors 521

18.5 Thermal Reactors 523

18.6 Endothermic Processes 528

18.7 Mass Trapping Reactors 534

18.8 Biofilters 537

18.9 Miscellaneous Applications 537

18.10 Commercial Applications 538

18.10.1 VOC Mitigation 538

18.10.2 Sulfuric Acid Plants 539

18.10.3 Other Commercial Applications 540

Periodic flow reversal is fundamentally a heat or mass trapping technique that makes it possible to increase temperatures or accumulate adsorbate in a bed of packing. The operation efficiently integrates chemical reaction and heat or mass exchange in a fixed bed reactor.

The concept goes back about 80 years, when the first commercial unit for removal of noxious organic emissions through thermal oxidation was apparently built by *Cottrell* (*Cameron, 1952*). *Matros* rediscovered the design in the 1970s and together with co-workers from the Boreskov Institute of Catalysis in Novosibirsk (Russia) undertook extensive mathematical modeling and experimental studies as well as commercialization of the concept for a number of catalytic processes. English language reviews began appearing in the 1980s (*Boreskov and Matros,*

1983a, b; Matros, 1989, 1996; Matros and Bunimovich, 1996; Matros et al., 1988b, 1993a, b), and the operation has been discussed in considerable detail in a book (*Matros, 1989*). More recent reviews on heat-integrated reverse flow reactors have been published by *Kolios* and co-workers (*Kolios et al., 2000, 2005a, b*). The literature on the operation is quite large and is summarized in **Tables 18.1 to 18.14**.

Periodic flow reversal is the only periodic operation that has been widely adopted commercially. Thousands of thermal and catalytic reactors, known also as regenerative oxidizers, control emissions of volatile organic compounds in the wood-working, automobile, and chemical industries. Dozens of reactors have been applied to selective catalytic reduction of NO_x, catalytic decomposition of N_2O, oxidation of SO_2 to SO_3, and combustion of coal bed ventilation exhaust.

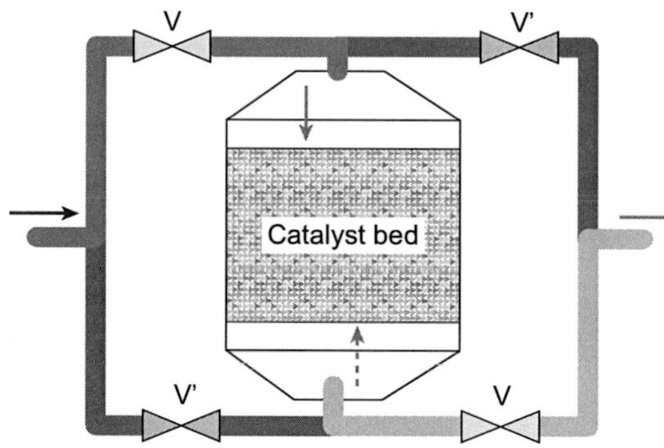

FIGURE 18-1 **Schematic of the basic flow reversal reactor: the pair of one-way valves V and V′ are opened and closed simultaneously to change the flow direction through a catalyst bed.** *(Figure adapted from* Matros (1989) *with permission. © 1989 by the Elsevier Science Publ. Co.)*

18.1 THE HEAT-TRAPPING CONCEPT

Figure 18-1 shows the basic reverse flow system diagram developed by *Matros (1977)* consisting of a single fixed bed reactor and switching valves that provide for periodic reversal of flow direction through the bed.

The reactor shown in the figure is well-insulated and would be used for an exothermic gas-phase reaction process. If the catalyst bed is brought to reaction temperature initially by some external means and fed a cold gas, that feed will be heated by the hot packed bed. If the bed is long enough, the gas will reach ignition temperature and undergo reaction. Indeed, the gas and catalyst bed can be brought to a higher temperature through heat released by an exothermic reaction. A temperature front created by the cold entering feed moves in the flow direction cooling the catalyst bed. Temperature and reactant conversion profiles developed from a mathematical model for SO_2 oxidation are illustrated in **Figure 18-2**. Reactant conversion in the moving front is high, achieving almost equilibrium conversion.

Time of front movement through the bed is determined primarily by the ratio of the heat capacities of solid and gas phases. It can be several minutes or even hours at a gas residence time in the order of several seconds. Eventually, the moving hot front reaches the end of the bed where the hot exiting gas would carry heat out of the reactor unless the flow direction is reversed. **Figure 18-2(b)** shows temperature and conversion profiles after flow reversal has taken place, some 40 minutes after startup. Now temperature at the opposite end of the bed drops through cooling by the cold entering gas, while temperatures at the other end increase. This pattern repeats each time the flow direction changes. The result of multiple flow reversals is the high temperature plateau clearly seen in **Figure 18-2(b)**.

Periodic flow reversal at a sufficiently high frequency confines the high temperature plateau to a region around the midpoint of the bed. Due to high heat capacity of the catalyst bed, both ends of the bed remain at a low temperature set by the temperature of the gaseous feed. After several flow reversals, a steady pattern is attained. Profiles at any time after a flow reversal

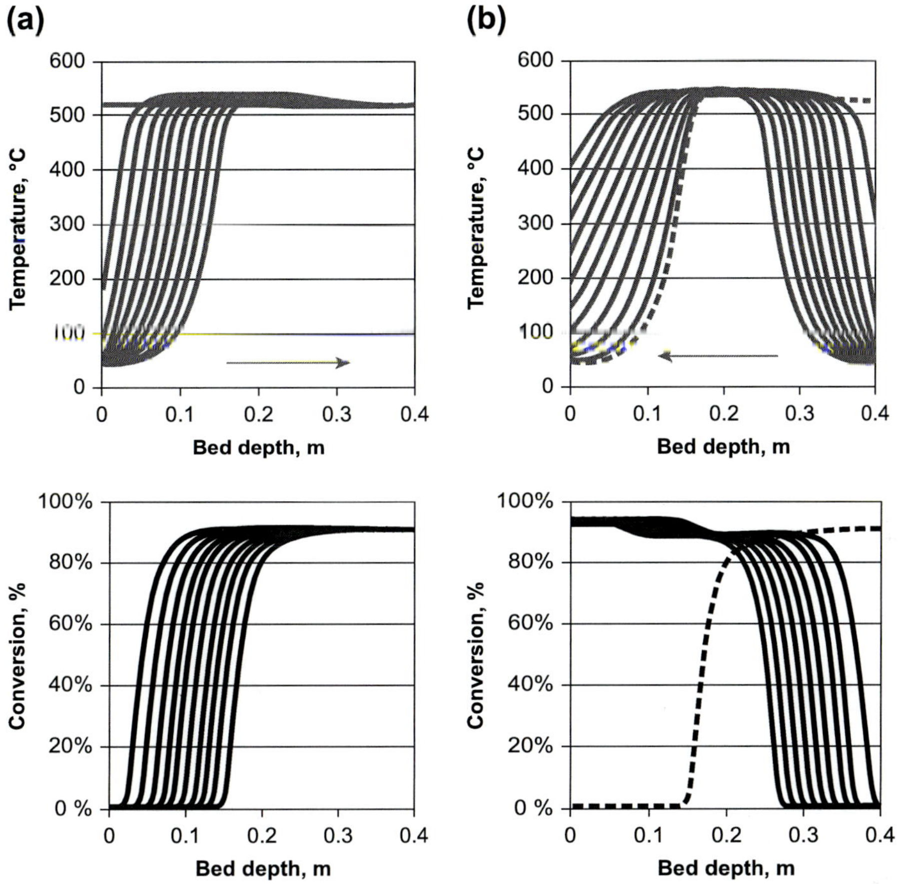

FIGURE 18-2 **Temperature and conversion profiles along the length of a catalyst bed at different times: (a) Passing the combustible feed through an initially heated catalyst bed; (b) evolution after flow reversal. Profiles are given for 4 min intervals. Arrows indicate direction of gas stream through the catalyst bed and direction of the profile movement. Dashed lines show profiles just before flow reversal. System: Oxidation of SO_2 over a vanadia catalyst for experimental conditions given by *Boreskov et al. (1982b)*.** *(Figure adapted from* Matros and Bunimovich (1996) *with the authors' permission.)*

are symmetrical in half-cycles with opposite flow directions. Temperature in the plateau depends on the adiabatic temperature rise of the feed, linear velocity through the bed, size and shape of the catalyst particles, and kinetic parameters. It depends little on the inlet temperature of the gas mixture and the cycle duration.

Often the difference between the maximum and inlet temperatures is much higher than the adiabatic temperature rise. Thus, the catalyst not only accelerates combustion, but also accumulates heat. Interphase heat transfer rates in packed beds can be quite high, especially for high surface area packing, so that steep temperature fronts can be achieved for reactions with a low ΔT_{ad}. This offers an energy efficient operation for dilute and cold gases common in air pollution mitigation.

Periodic flow reversal with its unique temperature profile along the catalyst bed is important for reversible exothermic reactions such as SO_2 oxidation allowing high conversion in a single catalyst bed. Initially conversion increases occur along the reaction front developed as the gas pushes into the catalyst bed. It increases further in the falling temperature region at the bed outlet (see **Figure 18-2(b)** (conversion)).

Another feature of the temperature profiles in **Figure 18-2(b)** is that the ends of a bed do not contribute to the reaction during most of the cycle so that the catalyst at the ends can be replaced by a less costly heat-storage material. This is routinely done in regenerative catalytic oxidizers, the flow reversal reactors used for mitigating emissions of highly dilute volatile organic compounds. Use of catalytically inert material at bed ends prevents loss of conversion for some reversible processes such as the reforming of methane.

Hundreds, if not thousands, of thermal reactors are now used by industry for the destruction of dilute gaseous organic wastes. These regenerative thermal oxidizers first described in the *Cottrell (1938)* patent combine regenerative heat exchange using ceramic media beds with homogeneous oxidation. The basic system, **Figure 18-3**, includes two ceramic heat-exchange beds and a combustion chamber with a burner installed in its middle. This system operates with a cold start-up. The burner is fired to heat and combust the cold feed gas and flow reversal is initiated. The flow reversal cycle is rather short, from 1 to 3 minutes, for maximum thermal efficiency.

Figure 18-4 illustrates the evolution of temperature profiles during the oxidizer start-up. Temperatures in the upper portion of the ceramic beds and in the combustion chamber increase every switching cycle but remain low in the bottom fraction of the ceramic media. There is a small temperature rise in the middle of the chamber caused by burner firing. The process reaches steady state at about 800°C, a temperature quite sufficient for complete oxidation of organic compounds at residence times in the combustion chamber of less than 1 s. Profiles in this figure as well as the next are from a mathematical simulation using a model assuming a small temperature rise due to VOC oxidation. The height of each bed = 1.2 m; linear velocity = 1 m/s. The period between flow reversals was 1 min, inlet temperature = 40°C, residence time in the combustion chamber was 2 s at 273 K. Burner operation involves steady firing at combustion chamber temperatures below 780°C and a modulated firing between 850°C and 780°C.

Figure 18-5 illustrates temperature evolution in the combustion chamber and at bed boundaries. At rather brief burner firings, a stationary oscillatory operation is reached in about 3 h.

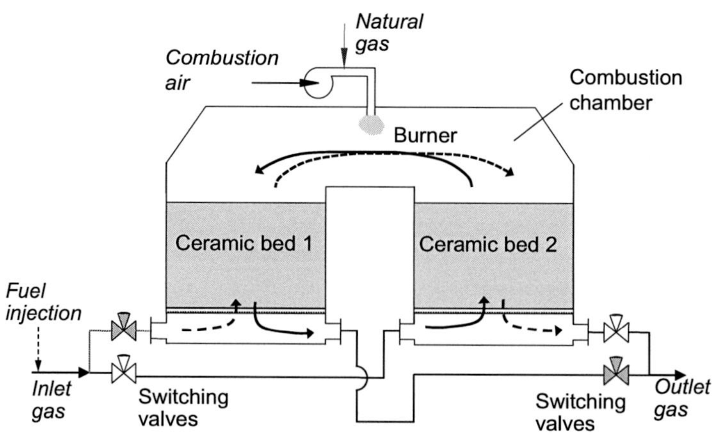

FIGURE 18-3 Schematic diagram of a thermal flow-reversal reactor. *(Figure adapted from* Bunimovich and Matros (2010a) *with permission. © 2010 by McGraw-Hill Corp.)*

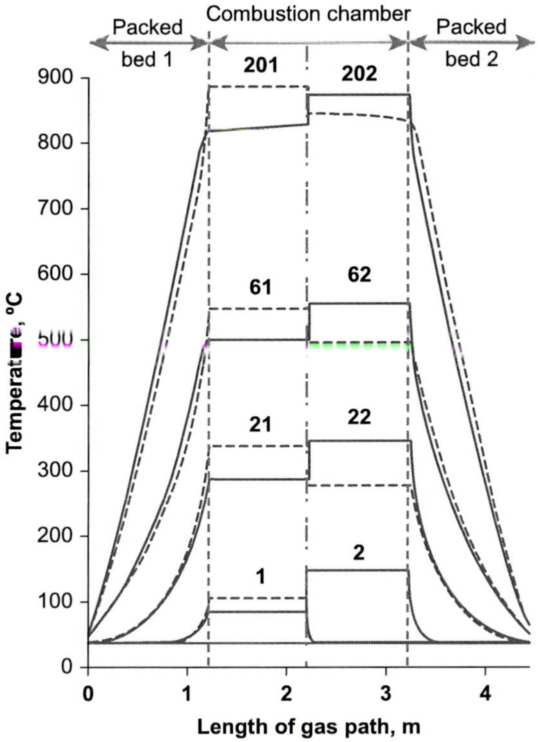

FIGURE 18-4 Evolution of temperature profiles during start-up of a regenerative thermal oxidizer. Solid lines: Profiles obtained by the end of a half-cycle for a flow direction from left to right, dashed lines: similar but from right to left. Number of flow reversals after the reactor start-up is shown above the profiles. Dotted vertical lines indicate upper boundaries of packed beds in the oxidizer canisters. Vertical dotted line indicates location of the fuel burner in the middle of the combustion chamber. *(Figure adapted from* Bunimovich and Matros (2010b) *with permission © Matros Technologies, Inc.)*

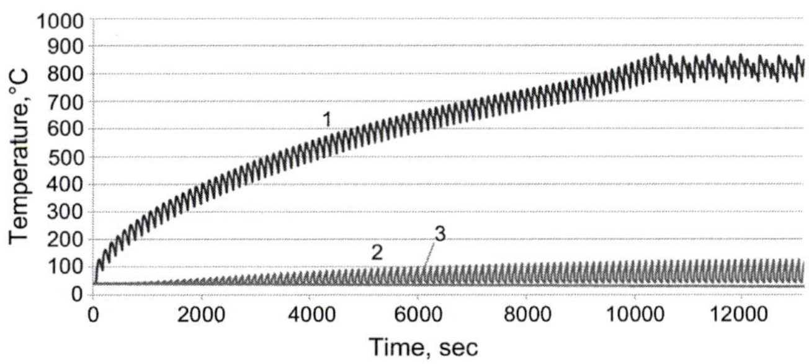

FIGURE 18-5 Temperature changes during a start-up of a regenerative thermal oxidizer. 1: Temperature in the middle of the combustion chamber, 2: outlet temperature, 3: inlet temperature. *(Figure adapted from* Bunimovich and Matros (2010b) *with permission © Matros Technologies, Inc.)*

The rapid oscillations indicate flow reversal, while the longer, 2-peak oscillations observed during stationary operation arise in the figure from periodic burner firings, designed to maintain high temperature in the combustion chamber at very low organics concentration in the combustible gas.

18.2 THEORETICAL ASPECTS

Table 18.1 lists most of the publications on the principles of heat-integrated reversed-flow reactors. Most works concern catalytic reactors though several discuss noncatalytic thermal processes.

18.2.1 Modeling

Commonly used one-dimensional models have been discussed by many investigators (see **Table 18.1**). A general heterogeneous, transient model for a single catalytic reaction, A → B, can be represented by the following system of differential equations:

Heat balance in the solid phase:

$$(1-\varepsilon_b)\left(\rho_s c_s(1-\varepsilon_p)+\rho_g c_g \varepsilon_p\right)\frac{\partial T_s}{\partial t} = \lambda_{s,eff}(1-\varepsilon_b)\frac{\partial^2 T_s}{\partial z^2} + ha_v(T_g - T_s) + (-\Delta H_R)r(C_s, T) \tag{18-1}$$

Heat balance in the gas phase:

$$\varepsilon_b \rho_g c_g \frac{\partial T_g}{\partial t} = \lambda_{g,eff}\frac{\partial^2 T_g}{\partial z^2} - \rho_g c_g u(t)\frac{\partial T_g}{\partial z} + ha_v(T_s - T_g) + h_w \frac{4}{d_R}(T_w - T_g). \tag{18-2}$$

TABLE 18-1 Contributions to Basic Theory and Mathematical Simulation of Heat-integrated Catalytic Reverse Flow Reactors

Reference	Contribution
Boreskov et al. (1977b), Matros (1977)	The connection between the steady-state movement of reaction fronts in catalytic beds and the reverse flow reactor. Concept of RFR for exothermic reversible reactions and discussion of temperature and conversion profiles based on a mathematical model. Mathematical simulation of temperature and conversion profiles for reversible exothermic reaction using a heterogeneous model.
Boreskov et al. (1983)	Original development and analysis of a high frequency switching model based on a pseudo-homogeneous model; also consideration of state multiplicity and stability for an adiabatic fixed bed reactor.
Eigenberger and Nieken (1988)	Mathematical simulation and experimental study of an RFR for the oxidation of volatile organic compounds. Experiments were performed in 50 mm reactor loaded with ceramic honeycomb monoliths with both inert and catalytic sections. Temperature control using hot gas bypass (hot gas withdrawal) was introduced.
Gawdzik and Rakovski (1988)	Mathematical simulation of an adiabatic RFR based on a heterogeneous model.
Sapundzhiev et al. (1988)	Mathematical simulation based on a two-dimensional cylindrical model; high temperature and reactant conversion in the reactor core can coexist with low temperature and zero conversion near the reactor wall.
Matros (1989)	Fundamentals of reactor modeling and simulation; analytical estimation of process parameters. Simulation study of a reversible exothermic reaction in an adiabatic bed. Analysis of selected catalytic processes in an isothermal reactor based on model reaction mechanisms.

(*Continued*)

TABLE 18-1 Contributions to Basic Theory and Mathematical Simulation of Heat-integrated Catalytic Reverse Flow Reactors (*cont'd*)

Reference	Contribution
Chumakova and Zolotarskii (1990); Chumakova and Matros (1991)	Analysis of a high frequency switching model; proof of profile symmetry for an adiabatic process; existence of isolated high temperature solutions at various linear velocities and bed lengths.
Bhatia (1991)	Simplified solution obtained using regular perturbations around a high frequency steady state.
Gupta and Bhatia (1991)	Rapid numerical method for RFR mathematical simulations. The original dynamic system for two-phase adiabatic fixed bed reactor model is transformed to a boundary value problem representing symmetrical profiles for stationary cyclic operation. The boundary problem is further solved using spatial finite collocation and Newton's iterative algorithm. The method was demonstrated for a 2nd order exothermic reaction.
Zagoruiko et al. (1992)	Mathematical simulation of a model for a selective catalytic process with three reactions. At low inlet temperatures reverse-flow operations showed increased product yield and selectivity compared with steady-state operation.
Řeháček et al. (1992, 1998)	Analysis of a wall-cooled reactor model indicated complex reactor behavior. Asymmetric and aperiodic regimes were discovered along with a normal one for adiabatic symmetric steady states.
Young et al. (1992)	Mathematical simulation for an exothermic reversible reaction. For the bed parameters used in the model, the reverse-flow reactor required a slightly lower amount of catalyst than a steady-state single-bed reactor operated at an optimal inlet temperature. Two RFRs placed in series required less catalyst than an optimally operated double bed reactor with intercooler.
Van de Beld and Westerterp (1994)	Mathematical simulation using 1-D heterogeneous and quasi-homogeneous models; also a high frequency switching model was developed.
Nieken et al. (1995)	Applications of stationary front and countercurrent reactor models for estimating RFR performance based on heterogeneous and quasi-homogenous systems. Analytical estimation of maximum temperature. Discussion of multiplicity and stability of the cyclic steady state for an adiabatic reactor.
Wallace and Viljoen (1995)	Mathematical simulation using a 1-D model; estimation of maximum bed temperatures.
Haynes et al. (1995)	Short-cut technique for reactor design based on a high frequency switching model.
Thullie and Burghardt (1995)	A procedure for estimating catalyst bed depth and a maximum flow reversal cycle combining calculations of reactor steady states and estimation of a reaction zone shift during a cycle based on the model of a stationary moving front.
Sun et al. (1996)	Analyzed a simple reversible reaction using a high frequency switching model.
Salinger and Eigenberger (1996a, b)	Computational procedure proposed by *Gupta and Bhatia (1991)* was combined with a numerical continuation technique and Floquet stability analysis for studying complex reactor behavior. The parameter regions with different numbers of solutions (up to 5) were mapped for the example of two independent exothermic reactions. Calculations for a wall-cooled reactor model confirm the existence of asymmetric and aperiodic states found by *Řeháček et al. (1992, 1998)*. The asymmetric and symmetric states can be obtained for the same reactor parameters depending on initial conditions.

(*Continued*)

TABLE 18-1 Contributions to Basic Theory and Mathematical Simulation of Heat-integrated Catalytic Reverse Flow Reactors (*cont'd*)

Reference	Contribution
Khinast and Luss (1997)	Study of RFR complex dynamic behavior based on a systematic methodology for mapping parameter regions with qualitatively different bifurcation diagrams. The method involves a solution for a boundary value problem for the cyclic reactor state together with adjoining sets of conditions defining singularities. The maps constructed for a pseudo-homogeneous adiabatic reactor model for a single and two independent exothermic first order reactions disclose single and double hysteresis loops, and isolated solution patterns discovered in earlier works of other authors. Generally an adiabatic RFR attains only symmetric states and q+1 of the 2q+1 states are stable (q is the number of independent exothermic reactions).
Khinast et al. (1998)	Systematic methodology for mapping parameter regions with qualitatively different dynamic behavior is extended to a 1-D pseudo-homogeneous model of a wall-cooled RFR. The map, constructed in Δ - τ_f plane, where Δ is reactor cooling capacity, and τ_f is flow reversal period, shows presence of asymmetric and quasi-periodic states at short flow reversal periods. Quasi-periodic states usually exist at lower cooling capacity than asymmetric ones. There is a narrow area of parameters where stable symmetric and asymmetric states coexist.
Khinast et al. (1999)	Continued study of the dynamic behavior of a wall-cooled RFR considering heterogeneous and pseudo-homogeneous reactor models. In most situations, both models yield almost identical dynamic behavior and parametric maps. Increasing complexity for dynamic behavior, from complex aperiodic to chaotic motions, was demonstrated for both models within a narrow range of parameters. The two models can lead to rather different predictions of the dynamic features when there are large differences between feed and coolant temperature.
Ramdani et al. (2001)	Derivation of a simple linear model for an RFR comprising a catalyst bed flanked by two beds of heat-exchange ceramic material. The model assumes instantaneous exothermic reaction at the boundaries of the catalyst bed and is intended for use in an automatic control system. The transients obtained with the simplified model generally agreed with the detailed model and experiments.
Cittadini et al. (2001)	Simplified procedure for design of a single bed catalytic RFR based on a 1-D pseudo-homogeneous model.
Keith et al. (1999); Keith (2003); Chan and Keith (2006)	Theory, mathematical modeling and simulation of the effect of insertion of metal rods or pins in a packed bed. It was speculated that the bed thermal dispersion would increase very much due to the Taylor-Aris mechanism and so prevent temperature runaway after a reactant concentration increase. The optimal bed design with the inserts was evaluated.
Snyder and Subramaniam (1999)	Mathematical simulation and experimental study of the discrete heating of gas in an RFR operated at an insignificant input of energy generated by the reaction. The external energy input can be efficiently exploited for manipulating temperature profiles while achieving efficient energy integration.
Smith and Bobrova (2002)	Mathematical simulation and simplified theory for a simple reaction model of ammonia oxidation over a platinum catalyst.
Budhi et al. (2004)	Study of the effect of catalyst dynamics on conversion and selectivity using mathematical simulation of NH_3 oxidation over a Pt catalyst.
Balaji and Lakshminarayanan (2006); Viecco and Caram (2006)	Comparison between flow reversal and a simulated moving bed reactor.

(*Continued*)

TABLE 18-1 Contributions to Basic Theory and Mathematical Simulation of Heat-integrated Catalytic Reverse Flow Reactors (*cont'd*)

Reference	Contribution
van de Rotten et al. (2006)	Development of an efficient limited memory iterative method for simulation of periodically forced processes. A wall-cooled RFR was simulated for a 2-D model accounting for radial temperature gradients. Reactor dynamics may be different for 1-D and 2-D reactor representations.
Balaji et al. (2008)	Equations for estimating RFR temperatures and other parameters were derived based on dimensional scaling groups. Mathematical simulation and parametric analysis for a first order exothermic reaction was undertaken.

mass balance in the solid phase:

$$(1-\varepsilon_b)\varepsilon_p\frac{\partial C_s}{\partial t} = k_m a_v(C_g - C_s) + r(C_s, T) \tag{18-3}$$

mass balance in the gas phase:

$$\varepsilon_b\frac{\partial C_g}{\partial t} = D_{eff}\frac{\partial^2 C_g}{\partial z^2} - u(t)\frac{\partial C_g}{\partial z} + k_m\, a_v(C_s - C_g). \tag{18-4}$$

Boundary conditions:

$$\begin{aligned} &\text{At } z = 0:\ \lambda_{s,eff}\frac{\partial T_s}{\partial z} = 0,\\ &\lambda_{g,eff}\frac{\partial T_g}{\partial z} = \rho_g c_g u(t)\kappa_+(t)(T_g - T_{in}),\\ &D_{eff}\frac{\partial C_g}{\partial z} = u(t)\kappa_+(t)(C_g - C_{in}) \end{aligned} \tag{18-5}$$

$$\begin{aligned} &\text{at } z = L:\ \lambda_{s,eff}\frac{\partial T_s}{\partial z} = 0,\\ &\lambda_{g,eff}\frac{\partial T_g}{\partial z} = \rho_g c_g u(t)\kappa_-(t)(T_g - T_{in}),\\ &D_{eff}\frac{\partial C_g}{\partial z} = u(t)\kappa_-(t)(C_g - C_{in}) \end{aligned} \tag{18-6}$$

and initial conditions:

$$\begin{aligned} &\text{at } t = 0:\ T_s = T_s^0(z),\ \ T_g = T_g^0(z),\\ &C_s = C_s^0(z),\ \ C_g = C_g^0(z). \end{aligned} \tag{18-7}$$

Unique in periodic flow reversal relative to other periodic operations considered in this monograph is the change in flow direction. With reference to the modeling, this feature is reflected through a change of sign of the convective transport derivatives in the gas phase heat and mass balances, **Eqs (18-2)** and **(18-4)**:

$$u(t) = \text{sgn}(t)U, \tag{18-8}$$

where $\text{sgn}(t) =$

$$\begin{cases} +1 \text{ at } t\in\left[(\text{n-1})t_c,\ \left(\text{n-}\frac{1}{2}\right)t_c\right] \\ -1 \text{ at } t\in\left[\left(\text{n-}\frac{1}{2}\right)t_c,\ nt_c\right] \end{cases}$$

and U is the absolute velocity; $n = 1,2,3$, etc. represents a count of flow reversal cycles.

The boundary conditions, **Eqs (18-5) and (18-6)**, depend on functions of the flow direction:

$$\kappa_+(t) = \frac{1+\text{sgn}\,(t)}{2},\qquad \kappa_-(t) = \frac{1-\text{sgn}\,(t)}{2} \tag{18-9}$$

If the reactor attains a periodic steady state, it follows from a reactor heat balance that:

$$T_{in} + \Delta T_{ad}\,\overline{x_{out}} = \overline{T_{out}} \tag{18-10}$$

where

$$\Delta T_{ad} = \frac{(-\Delta H_R)C_{in}}{\rho_g c_g}\overline{x_{out}} = \frac{1}{2t_c}\int_0^{t_c}\left[2 - \frac{\kappa_+(t)C_g(Z,t) + \kappa_-(t)C_g(0,t)}{C_{in}}\right]dt,$$

and

$$\overline{T_{out}} = \frac{1}{2t_c}\int_0^{t_c}\left[\kappa_+(t)T_g(Z,t) + \kappa_-(t)T_g(0,t)\right]dt,$$

ΔT_{ad} is the adiabatic temperature rise, $\overline{x_{out}}$ and $\overline{T_{out}}$ are the cycle average outlet conversion and temperature respectively.

Most theoretical papers consider just first-order kinetics:

$$r\left(C_s, T_s\right) = -k_\infty e^{-\frac{E}{RT_s}}C_s. \quad (18\text{-}11)$$

Heat and mass transfer parameters $h, k_m, \lambda_{eff,s}, \lambda_{eff,g}$ and D_{eff} can be found in the literature, for example, *Salomons et al. (2004)* and *Fissore et al. (2005)* for randomly packed beds or *Liu et al. (2001b)* for structured beds.

18.2.1.1 Simplifications

Typical simplifications of the heterogeneous model, **Eqs (18-1)** to **(18-7)**, omit wall cooling for large diameter insulated reactors ($h_w\frac{4}{d_R}\rightarrow 0$). Another simplification is neglecting the temporal derivatives in **Eqs (18-2)** to **(18-4)**, i.e., assuming quasi-steady state in the gas phase. Temperature transients in the solid phase actually determine the overall process transients because of the much higher density of the solid phase. A quasi-steady state is assumed for cycles that substantially exceed the gas residence time in the catalyst bed. Heat conductivity through the solid phase is low for a randomly packed bed, although it must be accounted for sometimes with structured monolith packing. At high linear velocities, heat conductivity and diffusivity terms in **Eqs (18-1)**, **(18-2)** and **(18-4)** can be omitted. The model then turns into a two-phase plug flow model, which works particularly well for catalyst and inert media structured as straight channel monoliths.

Use of a pseudo-homogeneous model is common even for simulating a reactor running at rather high linear velocities. Widely used relations assume a high rate of heat transfer between phases and fast mass transfer within a catalyst pellet:

$$(1-\varepsilon_b)\rho_s c_s\frac{\partial T}{\partial t} = \lambda_{eff}\frac{\partial^2 T}{\partial z^2} - u(t)\rho_g c_g\frac{\partial T}{\partial z} + (-\Delta H_R)f(C,T) - h_w\frac{4}{d_R}(T-T_w) \quad (18\text{-}12)$$

$$\varepsilon_b\frac{\partial C}{\partial t} = D_{eff,A}\frac{\partial^2 C}{\partial z^2} - u(t)\frac{\partial C}{\partial z} - f(C,T) \quad (18\text{-}13)$$

where C is the reactant concentration in the gas phase. For a first-order reaction, the combined kinetic/mass transfer rate is:

$$f(C,T) = \frac{k_m a_v k_\infty e^{-\frac{E}{RT}}}{k_m a_v + k_\infty e^{-\frac{E}{RT}}}C, \quad (18\text{-}14)$$

which reduces to **Eq. (18-11)** for a high rate of interphase mass-transfer.

In many situations, a pseudo-homogeneous model can replace the heterogeneous model by employing the *Vortmeyer and Shaeffer (1974)* equivalence:

$$\lambda_{eff} = (1-\varepsilon_b)\lambda_{s,eff} + \lambda_{g,eff} + \frac{\left(U\rho_g c_g\right)^2}{h a_v}. \quad (18\text{-}15)$$

Using this relation was satisfactory for a complex wall-cooled RFR studied by *Khinast et al. (1999)*.

18.2.1.2 Complications

Of course, the basic RFR model, **Eqs (18-1)** to **(18-7)** can be modified to encompass complex process kinetics. For example, *Matros (1989)*, *Matros and Bunimovich (1996)* and *Young et al. (1992)* introduced reversible exothermic reactions. **Table 18-10** provides a list of contributions considering exo- and endothermic processes. Contributions simulating homogeneous thermal processes (**Table 18-9**) adds a chemical reaction to gas phase heat and mass balance **Eqs (18-2)** and **(18-4)**. Except for several studies, simplified global reaction models are used instead of employing detailed chain-branch reaction kinetics.

Modeling of many catalytic processes (see *Matros, 1989*) introduces an effectiveness factor to account for diffusion within a catalyst pellet. *Matros (1989)* and *Quinta Ferreira et al. (1999)* also accounted for the dynamics of intrapellet heat transfer. Several studies (*Bunimovich et al., 1995; Smith and Bobrova, 2002; Budhi et al., 2004*) allowed for catalyst surface dynamics. The dynamics of adsorbed NH_3 are particularly important when representing the selective catalytic reduction of NO_x over vanadia-titania catalysts (*Agar and Ruppel, 1988a, b; Noskov et al., 1993*).

A two-dimensional radial-symmetric model was used by *Sapoundzhiev et al. (1988), Bunimovich et al. (1990b)* and *Matros and Bunimovich (1996)* to explore temperature nonuniformity in large diameter RFRs running weakly exothermic processes. They found that even a large diameter reactor can exhibit substantial radial temperature gradients induced by increased void fraction in a thin peripheral zone near the reactor wall.

Heat losses and possibly radial temperature nonuniformity should be taken into account in simulating small diameter laboratory reactors. Canadian research groups (*Aubé and Sapoundjiev, 2000; Salomons et al.*, 2003) validated the two-dimensional radial-symmetric model by comparing calculated and experimental temperature profiles measured in pilot-scale RFRs for the catalytic oxidation of methane. In other studies, *van de Beld and Westerterp (1996), Fissore et al. (2005), Sheintuch (2005)* and *Marín et al. (2008)* demonstrated that a one-dimensional model that included an effective wall heat transfer coefficient, h_w, quantitatively represents experimental results for the oxidation of VOCs in laboratory reactors with diameters between 50 and 150 mm. These studies added to their model a heat balance on the reactor wall to account for wall axial thermal conductivity and heat capacity. Temperature differences along the reactor radius found in the experiments by *Kushwaha et al. (2004)* were moderate, at least much lower than the overall temperature difference along the reactor axis. Perhaps, in this situation, using a one-dimensional model is justified.

Van de Rotten et al. (2006) developed a numerical solution for the two-dimensional radial-symmetric model and demonstrated differences between the dynamic behavior of a wall-cooled reactor using one- and two-dimensional models. Only *Liu et al. (2007)* considered three-dimensional CFD modeling for an RFR. They investigated an automotive converter for combusting CH_4 and CO described by *Zheng et al. (1999)*. Even though the CFD modeling clearly demonstrates the three-dimensional nature of the temperature distribution within the converter, it is uncertain how these nonuniformities affected converter behavior. Earlier, *Liu et al. (2001b)* showed that a one-dimensional heterogeneous, plug-flow model, even without a heat loss term, provided a reasonably accurate representation of the same reverse-flow converter.

18.2.2 High Switching Frequency Reactor Model

The high switching frequency model (HSFM) first suggested by *Boreskov et al. (1983)* is actually a model for intermediate switching cycles:

$$\varepsilon_b \tau_c << \frac{t_c}{2} << \frac{(1-\varepsilon_b)\rho_s c_s}{\rho_g c_g}\tau_c, \qquad (18\text{-}16)$$

where $\tau_c = \dfrac{Z}{U}$ is contact time of gas flowing through the catalyst bed calculated at NTP.

At high reversal frequencies, temperature profiles oscillate just narrowly as shown in **Figure 18-6**. Cycle period is substantially shorter than the time needed for the temperature front to travel through the catalyst bed. On the other hand, t_c is substantially longer than the gas flow residence time, τ_c. This provides continuous flow though the bed and avoids conversion loss due to periodic back flushing of unreacted feed at the reactor inlet. The intermediate t_c satisfying

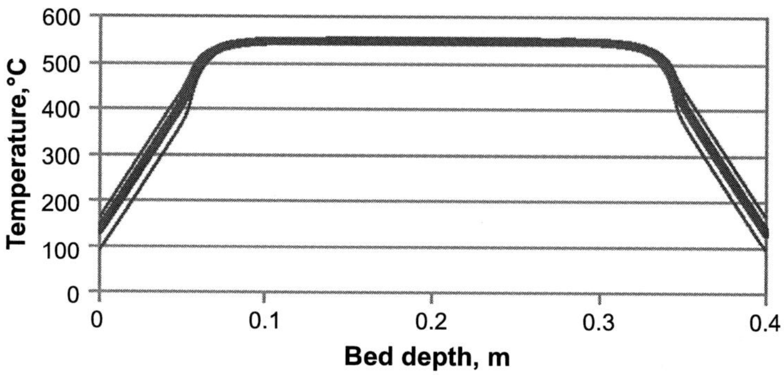

FIGURE 18-6 Temperature profile at high reversal frequencies. *(Figure redrawn from* Matros and Bunimovich (1996) *with permission. © 1996 by Marcel Decker Inc.)*

both requirements can be of the order of several minutes because of substantial differences between solid and gas densities.

For an adiabatic reactor with negligible axial dispersion ($D_{eff} = 0$), *Boreskov et al. (1983)* proposed a pseudo-homogeneous model consisting of:

$$(1-\varepsilon_b)\rho_s c_s \frac{\partial T}{\partial t} = \lambda_{eff}\frac{\partial^2 T}{\partial z^2} + \frac{1}{2}(-\Delta H_R) \times [f(C_{\rightarrow}, T) + f(C_{\leftarrow}, T)] \quad (18\text{-}17)$$

$$-U\frac{\partial C_{\rightarrow}}{\partial z} - f(C_{\rightarrow}, T) = 0, \quad (18\text{-}18)$$

$$U\frac{\partial C_{\leftarrow}}{\partial z} - f(C_{\leftarrow}, T) = 0. \quad (18\text{-}19)$$

With fast, symmetric cycling, the convective terms in the heat balance equation cancel and the reaction term is averaged over a cycle as **Eq. (18-17)** shows. The material balance equation splits into **Eqs (18-18)** and **(18-19)** for the different flow directions assuming negligible temperature change during a cycle. The boundary conditions for the heat balance are averaged between two flow directions:

$$\lambda_{eff}\frac{\partial T}{\partial z} = \frac{1}{2}U\rho_g c_g (T - T_{in}) \text{ at } z = 0$$

and

$$\lambda_{eff}\frac{\partial T}{\partial z} = -\frac{1}{2}U\rho_g c_g (T - T_{in}) \text{ at } z = Z. \quad (18\text{-}20)$$

The boundary conditions for the mass balances are:

$$C_{\rightarrow} = C_{in} \text{ at } z = 0 \text{ and } C_{\leftarrow} = C_{in} \text{ at } z = Z \quad (18\text{-}21)$$

Bhatia (1991) and *Haynes et al. (1995)* derived **Eqs (18-17)** to **(18-20)** as a zero-order expansion of the temperature profile at small increments of t_c around zero.

Nieken et al. (1992, 1994a, b) noted that the HSFM is equivalent to a model of a countercurrent reactor that comprises two separate streams going in different directions through two compartments with a common catalytic wall that permits heat exchange. The flow rate of the exchanging streams should be equal to half of the overall flow rate and every stream should interact with only half of the total catalyst volume (*Nieken et al. 1995*). Indeed, the material balances, **Eqs (18-18)** and **(18-19)** consider separate streams in two flow directions, omitting any effect of dynamic changes of composition and temperature on the reactor performance. Still the reduced model reflects the essential dynamic properties of the reactor that are determined by the slow heat transfer process in the solid phase. The solution at $\frac{\partial T}{\partial t} = 0$ corresponds to the cyclic steady state. *Matros (1989), Nieken et al. (1995), Haynes et al. (1995)* and *van de Beld and Westerterp (1994)* showed that the HSFM adequately represents the shape of the RFR temperature profiles. It quantitatively agrees with experimental data obtained in laboratory VOC oxidation reactors (*Züfle and Turek, 1997a, b; Ben-Tullilah et al. 2003b; Sheintuch, 2005*) and in a commercial reactor for SO_2 oxidation (*Kunitski et al., 1984; Bunimovich et al., 1990a; Matros, 1989*).

18.2.3 Stationary Heat Front Movement

As may be seen in **Figure 18-2**, the successive temperature and conversion profiles in the forward portion of a catalyst bed have almost unchanged shapes and move at almost constant speed. In a deep bed the profiles can be approximated by heat and mass balances derived for a coordinate system moving with a constant velocity, U_f. These are referred to as creeping waves. Creeping waves in fixed bed reactors is a well-studied subject (*Wicke and Vortmeyer, 1959; Padberg and Wicke, 1967; Wicke and Padberg, 1968; Feiguth and Wicke, 1971; Vortmeyer and Yahnel, 1972; Gilles, 1974; Rhee et al., 1974*). A comprehensive one-dimensional analytical theory was developed at the end of the 1970s by *Boreskov et al. (1979), Kiselev and Matros (1980, 1987), Pinjala et al. (1988), Matros (1989), Il'in and Luss (1992, 1993), Kiselev (1993)* and *Burghardt et al. (1999)*.

A fundamental heat balance (*Wicke and Vortmeyer, 1959*) connects the front velocity and the maximum temperature:

$$U_F = \frac{U}{\gamma}\left(1 - \frac{\Delta T_{ad}}{T_F - T_{in}}\right)\frac{1}{1 + \dfrac{\varepsilon_g}{\gamma}\dfrac{\Delta T_{ad}}{T_F - T_{in}}}, \tag{18-22}$$

where $\gamma = \dfrac{(1-\varepsilon_b)c_s\rho_s}{c_g\rho_g}$ is the ratio between the heat capacities of solid and gas. In many practical cases: $\dfrac{\varepsilon_g}{\gamma}$ <<1 and the final fraction in the RHS of **Eq. (18-22)** can be eliminated. For a small ΔT_{ad}, the front moves in the same direction as the gas. An opposed direction or a standing wave are possible for highly exothermic processes. Of course, the maximum period between flow reversals cannot be longer than the time for front movement through catalyst bed, e.g., $\frac{t_c}{2} < \frac{Z}{U_F}$. Then, **Eq. (18-22)** can be used to determine the reactor length. Actually, the bed length available for front movement is shorter than the total length. However, if the bed is too short, a high temperature plateau may not exist.

Matros (1989), Matros and Bunimovich (1996), Nieken et al. (1995) and *Somani et al. (1997)* found similarities between maximum temperatures achieved in a steady-state front and those predicted by the HSFM. It was also observed that the bed temperature in an RFR is rather insensitive to cycle duration.

18.2.4 Dynamic Behavior

Both the HSFM and the dynamic one-dimensional models discussed earlier have been used to represent reactor dynamics. If the temperature at the RFR inlet is too low, the process cannot be ignited without an external heater. This case represents a low temperature steady-state solution as opposed to a high temperature solution illustrated in **Figure 18-2** arising after preheating the catalyst bed. An intermediate state between these two solutions is unstable and cannot be realized without external control. The two stable steady states are pertinent for systems combining an exothermic reaction and heat exchange such as in conventional stationary reactors with external heat exchanger or in countercurrent reactors. Steady state multiplicity is important for autothermal systems where the energy for maintaining the high temperature state is obtained from the heat generated during the reaction. Various autothermal processes are described by *Eigenberger* and co-workers (*Kolios et al., 2000, 2005a, b*).

Presence of three steady states for an adiabatic RFR was demonstrated by *Boreskov et al. (1983)* using their high switching frequency model. This model provides an efficient algorithm for calculating multiple solutions. *Matros (1989)* and *Matros and Bunimovich (1996)*, using the HSFM for an exothermic reaction in an adiabatic reactor, observed hysteresis. Thus the reactor maximum temperature decreased abruptly for a decrease in inlet reactant concentration below a critical value. On the other hand, the system self-ignited at an increase of feed concentration above another critical value. Hysteresis behavior was also observed for the dependency of maximum temperature and conversion on residence time, inlet temperature, length of catalyst bed and axial thermal conductivity (*Boreskov et al., 1983; Matros, 1989; Young et al., 1992; Snyder and Subramaniam, 1993; Nieken et al., 1995; Matros and Bunimovich, 1996*).

The dependence on linear velocity is more complex as this parameter affects both the thermal conductivity and residence time in the catalyst bed. Relations between maximum temperature and velocity obtained using the HSFM are shown in **Figure 18-7**.

According to the figure, high temperature operations are possible at intermediate velocities (see **Figure 18-7(b)** and **(c)**). At high velocity, the residence time in the catalyst bed is too short to achieve sufficient conversion for igini-tion. At very low velocity (less than 0.01 m/s in **Figure 18-7(e)**), thermal dispersion via axial heat conductivity is so high that the RFR approaches extinction. The lower line 5 in **Figure 18-7(e)** shows hysteresis at a very low linear velocity.

Ivanov et al. (1992) found up to five steady-state solutions using the HSFM with two independent exothermic reactions. Three states were stable, two of them ignited, while the other states were

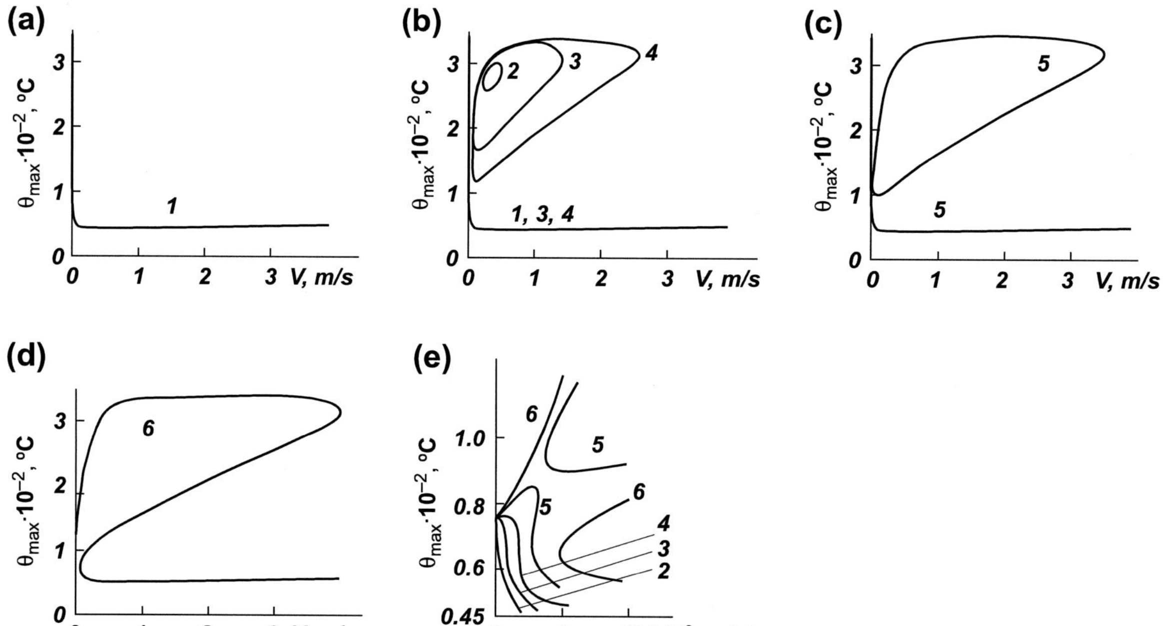

FIGURE 18-7 **Dependence of maximum temperature on linear gas velocity at different depths of a catalyst bed: (a) Low temperature solution at a shallow bed depth, (b) isolated solutions at intermediate depths, (c) and (d) disappearance of isolated solutions as bed depth increases, (e) dependences at small linear velocities (0–0.02 m/s). Line 1: z = 1 m, 2: z = 1.1 m, 3: z = 2.2 m, 4: z = 3.9 m, 5: z = 5.15 m, 6: z = 6 m.** *(Figure adapted from* Chumakova and Matros (1991) *with the authors' permission.)*

unstable. Evidently, reactions occur in different temperature regions to yield this complex behavior. *Nieken et al. (1994a, b)* observed three stable steady states experimentally during the oxidation of an air-diluted mixture of propane and propene over a platinum catalyst. The reactions were truly separated as propene oxidation reached high conversion at about 280°C while propane oxidation required about 500°C for completion. From a numerical study, *Salinger and Eigenberger (1996a, b)* constructed maps of possible entering propane and propene concentrations where ignited states can coexist.

Dynamic behavior becomes much more complex for a wall-cooled reactor whose model includes a heat sink term (see **Eq. (18-2)**) which increases linearly with increasing reactor temperature. The wall-cooled RFR was first studied by *Řeháček et al. (1992, 1998)* who showed the presence of asymmetric and aperiodic states at intermediate flow reversal frequencies. *Salinger and Eigenberger (1996a, b)* were able to calculate the unstable asymmetric and symmetric solutions, confirming the results of *Řeháček et al.* A comprehensive study by *Khinast and Luss (1997)* and *Khinast et al. (1998, 1999)* showed that simple symmetric period-1 operations occupy a major area of the high temperature states (**Figure 18-8**). Quasi-periodic and asymmetric states exist mainly at rather short flow reversal periods.

Kolios et al. (2000) have discussed the states identified for the wall-cooled RFR. **Figure 18-9** from their work illustrates aperiodic operation. A low heat removal rate and a high residence time over part of a bed where the temperature gradient is low, for example, in the left half of the bed during the first period, causes a gradual increase in temperature. A higher heat removal rate and lower residence time narrows the temperature peak in the right half of the bed during the first cycle. Apparently between 26 and 31 periods are needed to produce similar but mirrored changes in the profiles.

Van de Rotten et al. (2006) used a two-dimensional model to demonstrate that dynamic

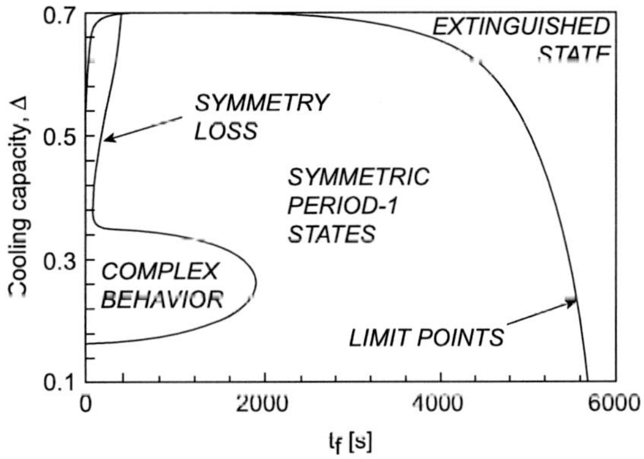

FIGURE 18-8 Map of regions with qualitatively different dynamic behavior of a wall-cooled RFR. *(Figure adapted from* Khinast et al. (1999) *with permission. © 1999 by the American Institute of Chemical Engineers.)*

features of a wall-cooled RFR can substantially change with a change of reactor diameter. For example, quasi-periodic behavior turns into stable asymmetric states for a large reactor diameter. *Gosiewski and Warmuzinski (2007)* showed that asymmetric operations also arise for an adiabatic reactor with cooling at a single point in the middle of the catalyst bed. This asymmetry impairs the heat utilization efficiency of the intercooler. Evidently, complex dynamic behavior including non-symmetric, aperiodic and chaotic states can arise for cooled RFRs, including wall-cooled reactors, small units with relatively high heat losses and large diameter adiabatic reactors with intermediate heat removal. However, complex dynamics have been studied using simulation only. No experimental verification has been presented so far.

18.2.5 Simulation and Preliminary Reactor Design

In most simulation routines, flow reversal is actually replaced by a reversal of the reactor. In **Eqs (18-1)** to **(18-6)**, the velocity $u(t) = U$ is kept constant and positive. Profiles for the process variables $v = \{T_s, T_g, C_sC_g\}$ are calculated through several solutions of the original

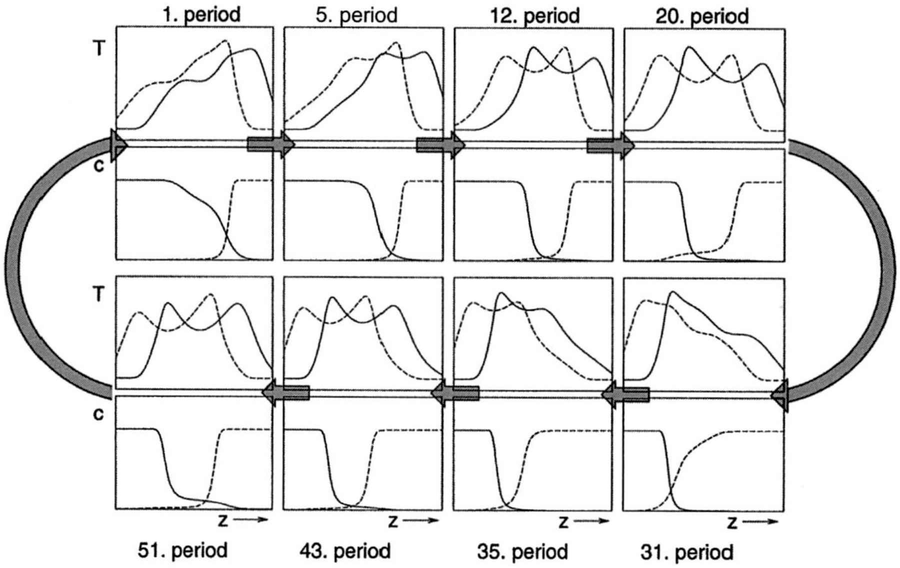

FIGURE 18-9 Evolution of temperature (T) and concentration (c) profiles in a wall-cooled RFR. Solid lines mark the end of the period with flow from the left, dotted lines show the end of the period with flow from the right. *(Figure reproduced from* Kolios et al. (2000) *with permission. © 2000 by Elsevier, Inc.)*

system within intervals equal to the time between flow reversals: $0 \le t \le t_c/2$. The initial data for every subsequent solution $v^{(j+1)}(0,z)$ depends on the preceding solution, i.e., $v^{(j+1)}(0,z) = v^{(j)}\left(t_c/2, Z - z\right)$. Numerical techniques used for these calculations included implicit finite difference schemes (*Eigenberger and Niekin, 1988; Řeháček et al., 1992; van de Beld and Westerterp, 1994; Haynes et al., 1995; van Sint Annaland et al., 2002; Ben-Tullilah et al., 2003b*) and orthogonal collocation (*De Groote et al., 1996; Neophytides and Tripakis, 1996*). *Gupta and Bhatia (1991)* proposed a numerical procedure for identifying both stable and unstable reactor steady states. The original system of equations was converted in their procedure into a boundary value problem, with boundary conditions for every subsequent solution represented by the cyclic condition: $v^{(j+1)}(0,z) = v^{(j)}(t_c, z)$. The solution to the boundary value problem was found using a shooting method. This procedure was further developed by *Salinger and Eigenberger (1996a, b), Khinast and Luss (1997)* and *Khinast et al. (1999)* and used for simulating the complex dynamic behavior of two independent exothermic reactions in an adiabatic RFR or of a single exothermic reaction in a wall-cooled RFR.

Table 18-1 references several papers published during the 1990s developing "short-cut" techniques for estimating the reverse flow reactor design parameters, employing analytical approximations and "ready to use" design charts. These were limited to a single first-order exothermic reaction in an adiabatic bed and adequately described only the oxidation of dilute VOCs. One of the important design tasks involved finding process parameters to locate the boundary between extinguished and ignited states. In early work on the subject, *Boreskov et al. (1983)* suggested a simple estimate for the residence time along the ignited state boundary for a single reactor bed operated with frequent flow reversals:

$$\tau_{c,\min} = 4\frac{\lambda_{eff}}{U^2 c_p \rho_g}\left(\frac{T_{max} - T_{in}}{\Delta T_{ad}} - 1\right). \quad (18\text{-}23)$$

Eq. (18-23) follows from a linear approximation of the temperature profile in an RFR and assumes that the temperature plateau disappears at the point of extinction while the maximum temperature is still equal to T_{max}. The equation was also derived by *Ramdani et al. (2001)* and by *Ben-Tullilah et al. (2003b)*. It is analogous to the heat transfer relation for an autothermal reactor combining a catalyst bed and a countercurrent recuperative heat exchanger.

18.3 OXIDATION OF VOLATILE ORGANIC COMPOUNDS

Complete oxidation of highly diluted volatile organic compounds (VOCs), including methane, is the simplest application of reverse flow reactors and the most widely used in industry. An overview of publications on VOC destruction is given in **Table 18-2** starting with the theoretical and experimental studies at the beginning of the 1980s by *Matros* and co-workers in Russia.

The basic RFR design for VOC oxidation consists of a bed of catalyst flanked by beds of ceramic packing. The high temperature zone is trapped within the catalyst bed while the inert packed beds are held at low temperature and serve solely as regenerative heat exchangers. Laboratory units tested so far (**Table 18-2**) have reactor diameters from 5 to 200 cm and employed several different noble-metal and base metal catalysts supported on either pellets or monoliths. Most experimental units (*Boreskov et al., 1984; Nieken et al., 1992, 1994a; Chaouki et al., 1994; van de Beld and Westerterp, 1996; Ben-Tullilah et al., 2003b; Marin et al., 2008*) utilized a single reactor (see **Figure 18-10(a)**) heated from one end only during startup. In those experiments, autothermality was maintained by feeding sufficiently high inlet concentrations of VOCs.

Commercial designs (*Schoubye and Nielsen, 1993; Matros and Bunimovich, 1995; Bunimovich and Matros, 2010a*) separate the bed into two parts (**Figure 18-10(b)**) and insert a central heater designed to maintain temperature at very low VOC concentration. Some of the laboratory units described in the literature were equipped with compensatory heating or cooling to make the laboratory reactor function adiabatically (*Fissore et al., 2005*).

Many of the papers listed in **Table 18-2** studied the influence of operating parameters on reactor performance. These disclosed autonomous control features such as an increase in bed temperature upon a decrease in catalyst activity. Autothermal operation was possible at very low VOC concentrations corresponding to a ΔT_{ad} as low as 15–20°C. *Nieken et al. (1994b)* noted that while process extinction caused by a dilute feed was easily prevented, it was much more difficult to limit the maximum temperature when the feed was too rich. *Sapundzhiev et al. (1994)* investigated a simultaneous oxidation of two or more VOC mixtures in an RFR. An operation in which all substances were combusted depended on the properties of the substances and their ratio in the feed as well as on the ratio between inert and catalyst and their thermo-physical properties.

Industry uses burners installed in the center of the reactor or direct fuel injection to the inlet of the system (*Bunimovich and Matros, 2010a*) for maintaining temperature at low VOC concentrations. Intuitively, overheating when running rich VOCs may be prevented by installing an internal heat exchanger in the mid-section of the reactor (see **Figure 18-10b**), inlet mixture dilution or hot gas bypass from the bed center (see **Figure 18-10c**). According to *Eigenberger and Nieken (1988, 1994)*, hot bypass results in a contraction of the temperature plateau in the

TABLE 18-2 Oxidation of Volatile Organic Compounds in Reverse Flow Reactors

Reference	Contribution
Matros et al. (1981); Lugovskoy et al. (1982); Boreskov et al. (1984)	Concepts; experimental study in a 175 mm o.d. reactor on the oxidation of n-C_3 and n-C_4 over a Cr catalyst using various combinations of inert heat-exchange ceramics and catalyst.
Wojciechowki and Haber (1982)	Described the Swingtherm VOC oxidation reactor comprising a catalyst bed flanked between two beds of ceramic heat-exchange material with an electric heater installed in the catalyst bed center. Operating data provided for the unit treating 5000 m^3/h of gas discharge from a phenol laminate plant.
Eigenberger and Nieken (1988, 1991, 1994)	Process parameters estimations; discussion of an experimental study employing a 5 cm o.d. reactor, including the effect of gas withdrawal (hot bypass) from the reactor center.
Matros (1989)	Concepts, mathematical modeling and experiments; estimations of design parameters for commercial catalytic RFRs.
Matros et al. (1991, 1993a, b)	Reviewed the RFR mathematical simulation and presented operating results for a pilot unit (30–70 m^3/hr) and commercial units (300 m^3/h and 6000 m^3/hr). Comparison between calculated and measured temperature profiles and VOC destruction efficiency. They also showed the applicability of a 1-D model.
Sapundzhiev et al. (1991)	The effect of the type of VOCs compound, its initial feed concentration, linear gas velocity, inert to catalyst ratio and inert thermo-physical characteristics on the temperature and concentration profiles in an RFR. Autothermal behavior was found at $\Delta T_{ad} > 20°C$.
Nieken et al. (1992, 1994a)	Detailed experimental study of VOC oxidation in an 8.5 cm o.d. reactor packed with a platinum honeycomb monolith catalyst. Mathematical simulation of the effects on reactor performance of cycle parameters, feed concentration, catalyst activity for the oxidation of different VOCs, catalyst deactivation, inert boundary sections and hot gas withdrawal from the reactor center. Presence of two stable ignited steady states was demonstrated experimentally for the oxidation of a mixture of $C_3^=$ and C_3.
Schoubye and Nielsen (1993)	Discussed the Haldor Topsoe REGENOX technology with a comparison between pilot tests and a mathematical model; description of a purge system for mitigating the bypass of VOCs to the reactor exit immediately after a flow reversal.
Vanin et al. (1993)	Design and operation of a commercial RFR treating a waste gas containing nitrogen bearing VOCs, acrylonitrile and cyanic acid. A 5000 Nm^3/h oxidizer showed high oxidizing efficiency with a selectivity toward formation of nitrogen oxides of less than 5%.
Nieken et al. (1994b)	Mathematical simulation of, and experiments on, reactor temperature control at low and high VOC concentrations. Heat removal from the reactor center using either a heat-exchanger, cold gas injection or hot gas withdrawal were found to have just a limited control of T_{max} with high concentrations of VOCs in the gas feed. Control via hot bypass can be improved for a structured bed containing catalyst in the bed center and inert heat exchange material at the ends of the bed.
Sapundzhiev et al. (1994)	Investigated simultaneous oxidation of two or more VOCs. Achieving complete combustion depends on characteristics and inlet concentrations of the components.
Purwono et al. (1994)	Experiments on reactor dynamic behavior using a wall-cooled experimental reactor with a 1 cm o.d. Focus was on extinction and runaway.

(Continued)

TABLE 18-2 Oxidation of Volatile Organic Compounds in Reverse Flow Reactors (*cont'd*)

Reference	Contribution
van de Beld et al. (1994)	Experimental study of $C_2^=$, C_3 and CH_3OH oxidation over a Pd/Al_2O_3 catalyst in a 145 mm o.d. reactor. An evacuated jacket around the reactor provided relatively low heat loss. Effects of adiabatic temperature rise, different fuels, pressure and flow rate were studied.
van de Beld and Westerterp (1994)	Performed mathematical simulations of an RFR. 1-D quasi-homogeneous and heterogeneous models show almost identical behavior. This behavior can be described by a simplified countercurrent reactor model.
van de Beld and Westerterp (1996)	1-D heterogeneous model corrected for heat loss and supplemented with heat conductivity and heat retention in the reactor wall agreed well with the experimental measurements.
Matros and Bunimovich (1995)	Presented an overview of VOC oxidation in RFRs with comparison to regenerative flow diagrams, mathematical simulation, estimations of maximum temperature and bed length, economic optimization and a discussion of commercial applications. Also a discussion of the advantages of catalytic compared with thermal systems.
Budman et al. (1996)	Parametric study and process control study of CO oxidation in a wall-cooled RFR.
Matros et al. (1996)	Economic evaluation of a catalytic RFR based on mathematical simulation. The overall cost for a catalytic RFR has a distinct minimum as a function of fuel consumption and linear gas velocity. The catalytic RFR has an advantage over a thermal RFR for a wide range of low VOC concentrations, but its advantages disappear at very high destruction efficiency requirements.
Grozev and Sapundzhiev (1997)	Catalytic decontamination of waste VOC gases in RFRs. Utilization of a significant part of the heat release in the reactor while exceeding 99.5% conversion was demonstrated, even with a severe reduction of catalyst activity.
Cunill et al. (1997)	Experimental study of a pilot catalytic RFR equipped with an electric heater for oxidation of $C_2^=$ and C_3. Continuous heating provided for VOC removal from very dilute mixtures. 1-D heterogeneous model agreed well with test results except under extreme conditions.
Züfle and Turek (1997a, b)	Study of CO and C_3 oxidation over a platinum catalyst in a laboratory RFR with just a single catalyst bed. A simplified countercurrent reactor (high switching frequency) model was applied that appeared to agree well with the measured variables such as maximum temperature and conversion.
Fissore and Barresi (2002)	Comparison between an RFR and a SMBR using mathematical simulation. The SMBR had a narrower stability range than the reverse-flow reactor.
Cittadini et al. (2002)	Dynamic simulation of the start-up of a catalytic RFR. High initial temperatures were required to prevent a temporary conversion decrease during the start-up.
Ben-Tullilah et al. (2003b)	Experiments with oxidation of $C_2^=$ over a Pt/Al_2O_3 catalyst in a 47 mm o.d. reactor. 1-D homogeneous and heterogeneous models of a wall-cooled RFR agreed well with the experimental data. Reactor model was based on an instantaneous reaction in the middle of the catalyst bed.
Strots et al. (2003)	Developed an RFR for treatment of nitrogen-containing VOCs. Use of a base metal catalyst reduced formation of NO_x. Results of kinetic studies and mathematical modeling were used in the design of a commercial VOC oxidizer.
Fissore et al. (2005)	Studied the effect of reactor walls on RFR performance. The outer wall temperatures in a small (5.2 cm) o.d. reactor were controlled using several electric band heaters and cooling air to simulate adiabatic operation.

(*Continued*)

TABLE 18-2 Oxidation of Volatile Organic Compounds in Reverse Flow Reactors (*cont'd*)

Reference	Contribution
Barresi et al. (2007)	Comparison between an RFR and a simulated moving bed reactor. Also a review of 2001–2007 research at Politecnico di Torino.
Bos et al. (2007)	Employed a catalyst bed flanked by beds of adsorbent for removal of water, other reaction inhibitors or catalyst poisons. Concept was proven by experiments on CH_4 oxidation over a palladium catalyst.
Zagoruiko (2008b)	Modification of an RFR model for VOCs to allow for VOC desorption equilibria.
Marín et al. (2008)	Oxidation of highly diluted mixtures of toluene and hexane in air in a 5 cm o.d. RFR with a bed filled with spherical platinum catalysts and adjacent beds of ceramic material. A 1-D heterogeneous model accounting for heat transfer in the reactor wall showed good agreement with temporal temperature and concentration measurements.
Lou and Huang (2008)	A synthesized copper-cobalt catalyst supported on honeycomb ceramic monolith was tested in a 15 cm o.d. x 15 cm regenerative oxidizer and demonstrated a high oxidation rate for isopropyl alcohol without formation of side products at 400–450°C.
Marín et al. (2009a, d)	Comparison between three different simplified techniques for RFR design indicated that the method of *Cittadini et al. (2001)* is suitable for preliminary design. Economic optimization confirmed earlier results indicating optimal linear velocities between 0.6 and 1 m/s for VOC destruction.
Bunimovich and Matros (2010a)	Discussion of commercial regenerative thermal and catalytic VOC RFRs and conversion of thermal units to catalytic units. Several case studies and examples of large commercial units utilizing nonprecious metal catalysts are given.

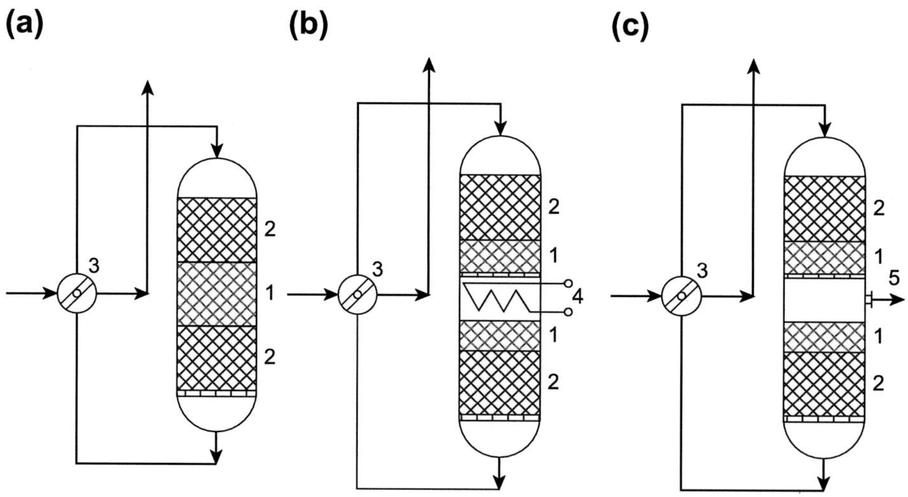

FIGURE 18-10 Common configurations for a VOC oxidation reactor - 1: Catalyst bed, 2: ceramic packing, 3: switching valve, 4: heater or cooler, 5: hot bypass.

middle of the catalyst bed without much change in T_{max}. Heat can be recovered from the hot bypass, but the issue of catalyst overheating remains. *Nieken et al. (1994b)* suggested inserting metal wires into the catalyst bed to increase the effective axial thermal conductivity. This technique has never been used in practice albeit *Keith et al. (1999)*, *Keith (2003)* and *Chan and Keith (2006)* predicted a reduced T_{max} through the Taylor-Aris relation.

Matros et al. (1988b, 1993b) demonstrated the applicability of the one-dimensional RFR model for predicting the general performance of commercial units. *Van de Beld and Westerterp (1994)*, *Ben-Tullilah et al. (2003b)* and *Marín et al. (2008)* compared calculated and experimental temperature profiles as well as maximum temperatures over a wide range of operating conditions and confirmed the adequacy of the simple one-dimensional model.

18.3.1 Methane Combustion with Heat Recovery

A number of studies (**Table 18-3**) have discussed the catalytic oxidation of dilute CH_4 streams. Methane is released in huge quantities with coal mine ventilation air. It is a much more potent greenhouse gas than CO_2 so that its complete oxidation contributes significantly to greenhouse gas mitigation. If the methane concentration is high enough, it is possible to use the waste air to produce high pressure steam and electricity. The reverse flow combustor with

TABLE 18-3 Catalytic Combustion of Dilute Methane Streams

Reference	Contribution
Boreskov et al. (1985); Matros et al. (1988a); Matros (1989)	Original numerical simulation and experimental studies of RFRs with a mid-bed heat exchanger (**Figure 18-10(b)**). Simulation observed an optimal heat recovery with respect to heat exchanger temperature, switching cycle and other parameters. Experiments with dilute methane, 0.45–1.2 vol%, in a 17.5 cm i.d. unit loaded with Raschig rings and a copper chromite catalyst demonstrated a heat recovery of about 50%, less substantial heat losses to the surrounding environment.
Chaouki et al. (1994); Sapundzhiev et al. (1993)	Application of an RFR to the combustion of natural gas was investigated by numerical simulation. The appropriate choice of catalyst and inert size and shape, the volumetric ratio between catalyst and inert material in the bed, allowed methane combustion at much lower reactor temperature than in a steady-state reactor. Under lower temperatures, no nitrogen oxides are produced in combustion. Experimental study was undertaken in a 5 cm. o.d. reactor packed with a perovskite catalyst. Process was simulated by a heterogeneous 1-D model.
Sapoundjiev et al. (1999)	Economic assessment of using the heat produced by catalytic combustion of coal mine ventilation air in a flow reversal reactor to produce electricity.
Aubé and Sapoundjiev (2000)	2-D model adequately represented temperature profiles in 200 mm and 500 mm o.d., pilot scale catalytic RFRs with an embedded heat exchanger or a hot bypass. Accounting for the properties of the reactor, insulation was necessary for quantitative agreement between model and experiment.
Gosiewski (2004)	Comparison between different techniques for numerical simulation of the stationary performance of a double bed, catalytic RFR with an intercooler. Direct iterative methods did not provide a computational advantage compared with dynamic simulation.

(Continued)

TABLE 18-3 Catalytic Combustion of Dilute Methane Streams (*cont'd*)

Reference	Contribution
Hayes (2004)	Catalytic RFR investigated as one of the possible techniques for mitigating fugitive CH_4 emissions.
Salomons et al. (2003)	Experimental investigation in a 200 mm o.d. RFR for the catalytic destruction of lean mixtures of methane in air. Space velocity, cycle period and CH_4 concentration were parameters that determined reactor performance. Control of these parameters is important to the amount of energy that can be usefully extracted from the reactor.
Salomons et al. (2004)	Development of a 2-D heterogeneous model for simulating the catalytic combustion of dilute CH_4 streams. The external heat transfer resistance was shown to be an important parameter in the operation as evidenced by the relative failure of a pseudo-homogeneous model to adequately capture the experimental results. The 2-D model was in good agreement with the 200 mm o.d. reactor experiments.
Kushwaha et al. (2004)	Experimental study in a 200 mm o.d. reactor with different inert material/catalyst configurations. The effect of the type of inert on the axial and radial temperature profiles, the rate of heat accumulation and the reactor stability was evaluated for the combustion of lean CH_4-air mixtures. Inert packing type had a significant influence on the temperature profiles, especially the radial ones. A metal monolith reduced the magnitude of the radial gradients as well as T_{max}. On the other hand, stable operation of the reactor with metal monolith packing required the use of shorter cycles.
Gosiewski (2005)	Numerical simulation of a hot bypass, double bed RFR with a comparison of palladium and metal oxide catalysts.
Kushwaha et al. (2005)	Experiments on the oxidation of a dilute CH_4 stream (0.6% CH_4 in air) in an RFR comprising two interconnected vessels 0.1 $\times$ 0.1 m cross-section area, loaded with inert heat exchange material and a catalyst. A fraction of the process gas at 600–800°C was removed between the beds for potential heat utilization. The tests showed viability of heat recovery using hot bypass. Testing various heat-exchange media indicated advantages of ceramic monolith versus metallic monolith and ceramic balls.
Marin et al. (2005)	Simulation of spherical particle and honeycomb packed bed RFRs with an inert-palladium catalyst-inert configuration for the combustion of a lean CH_4 stream. At the same bed depths, a particle bed reactor required longer cycle times, while the monolith packing offered much lower pressure drops.
Litto et al. (2006)	Numerical simulation of a catalytic CH_4 oxidation reactor employing a heterogeneous 2-D model that included heat losses and heat conduction through the insulation. The wall losses in the insulated reactor influenced the bed temperature and other properties at reactor diameters up to 1 m.
Balaji and Lakshminarayanan (2006)	Comparison between an RFR and a simulated moving bed reactor, possibly combining an RFR with a SMBR.
Hevia et al. (2006)	Test study on bench-scale (5 cm o.d.) pyrex and stainless steel reactors coupled with a mathematical simulation. Thermal conductivity and heat capacity of the reactor wall material affect the bench-scale RFR behavior especially at low linear velocities.
Hevia et al. (2007a, b)	Comparison between an RFR and two and three bed SMBRs for catalytic combustion of 0.1 to 1 vol% CH_4 in air. The RFR is preferred due to a wider stability range and a narrower temperature variation in the catalyst bed. Testing and modeling of an autothermal reactor with palladium and manganese oxide catalysts was undertaken.

(*Continued*)

TABLE 18-3 Catalytic Combustion of Dilute Methane Streams (*cont'd*)

Reference	Contribution
Gosiewski and Warmuzinski (2007)	Mathematical simulation comparing two different modes of intermediate heat removal in a double bed RFR.
Gosiewski et al. (2008)	Experiments and simulation of the thermal combustion of CH_4 in an RFR packed with pelletized and monolith heat-exchange ceramics.
Marín et al. (2009a, b)	Comparison between four different flow arrangements for heat recovery after combustion of dilute CH_4 streams in an inert-catalyst-inert RFR. Extraction of hot gas without its return to the reactor is preferred to a process with cooled gas return as it provided greater reactor stability as well as more efficient heat utilization. Extraction of hot gas from the outlet of a catalyst bed offers stability benefits compared with the more widely used extraction in the middle of the bed.

either an internal heat exchanger (**Figure 18-10(b)**) or heat removal from a side stream (**Figure 18-10(c)**) is suitable for such a task.

An RFR for CH_4 combustion would not differ from those discussed above except for a device to remove energy. Of course, catalytic combustion of methane requires higher temperatures than those for VOC, so the thermal process might be preferable. The first studies on CH_4 combustion were performed during the 1980s in the Boreskov Institute of Catalysis. A simulation study of the catalytic process with intermediate heat removal (**Figure 18-10(b)**) using CO oxidation kinetics (*Boreskov et al. 1985; Matros et al. 1988a; Matros, 1989*) showed that the optimum heat recovery depended on cycle duration and the gas temperature after the heat exchanger. Proof-of-concept experiments using oxidation of methane in a 175 mm o.d. catalytic reactor demonstrated a potential heat recovery greater than 50% for CH_4 concentrations of 0.4 to 1.2% (*Matros et al., 1988a*). Maximum temperature in the catalyst bed reached 800 to 1000°C depending on the inlet concentration of methane. Later, *Chaouki et al. (1994)* found a maximum temperature as low as 650°C for an experimental RFR operating with a palladium catalyst and methane concentrations as high as 1.4 vol%.

Figure 18-11 compares temperature profiles in reactors with a mid-bed heat exchanger with those with a hot bypass. In the latter reactor the outlet temperature is usually almost equal to the inlet temperature to the reactor (*Eigenberger and Nieken, 1988, 1991, 1994; Nieken et al., 1992, 1994a, b*). As a result, the fraction of the heat of reaction recovered can reach almost 100%, even at a relatively low overall quantity of packed bed heat-exchange material. The energy-generation part of a unit can be quite small as only a fraction of the total gas stream is utilized. Indeed, the maximum fraction, $\delta_{\max} = \frac{\Delta T_{ad} x_{tot}}{T_{\max} - T_{in}}$, where x_{tot} is overall conversion of reactant. Thus, δ_{max} makes up 10–20% of the total stream which results in a rather compact heat recovery unit built separately from the combustor. An RFR with an internal heat exchanger (**Figure 18-10(b)**) generally offers lower utilization of reaction heat, still it can offer a higher energy yield compared with an RFR with hot bypass because the entire stream is passing through the heat recovery unit.

Aubé and Sapoundjiev (2000) compared the two flow arrangements, **Figure 18-10(b) and 18-10(c)**. The hot bypass system created a significant temperature decrease in the mid-section of the reactor but also reduced the range of stable

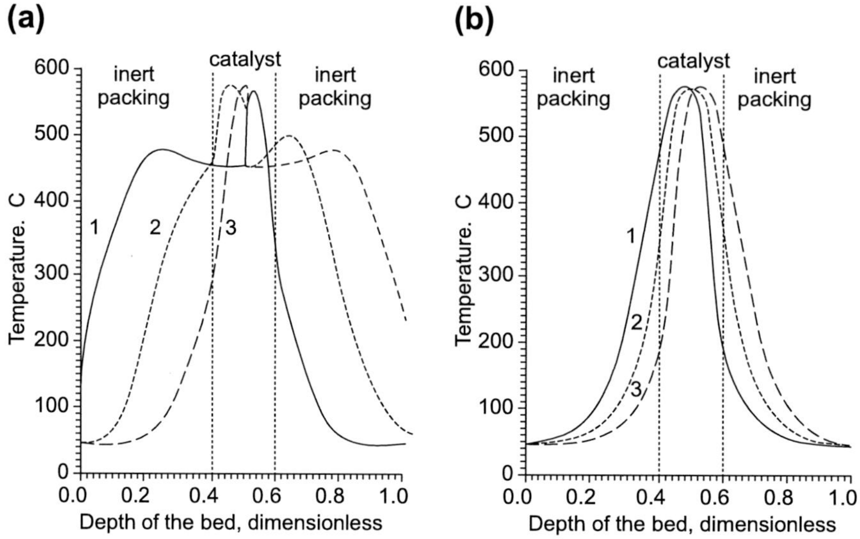

FIGURE 18-11 Comparison of temperature profiles for RFRs using center-of-the-bed (a) heat exchange and (b) hot gas bypass. *(Figure reproduced from* Matros and Bunimovich (1996) *with permission. © 1996 by Elsevier Inc.)*

operation. However, the system is preferable to mid-bed heat exchange because it still provides a more stable operation and better heat utilization without asymmetric temperature profiles. *Gosiewski and Warmuzinski (2007)* reached the same conclusion using simulation. Their more detailed study confirmed that internal heat removal created large half-cycle asymmetry that could result in extinction. *Hevia et al. (2006)* showed, through simulation, that cooling of the gas stream at two points along the bed, such as between the inert and catalytic materials, increased stability relative to cooling at the center. Some commercial ventilation air-methane reactors and pilot combustion reactors utilize two imbedded heat exchangers instead of just one.

Canadian publications (*Aubé and Sapoundjiev, 2000; Salomons et al., 2003, 2004; Kushwaha et al., 2004, 2005; Litto et al., 2006*) describe the application of catalytic RFRs. Pilot scale units with diameters of 200 mm and 500 mm were used. Two different configurations were tested: A vertical configuration (**Figure 18-10(b) or 18-10(c)**) and a configuration similar to **Figure 18-3**, where on top of the two ceramic beds, a catalyst bed was added. The latter performed better and had a simpler design. The experimental program (*Sapoundjiev, 2000*) was designed to test different catalysts under field conditions and so was undertaken at a mine site with actual mine ventilation air over a number of months. Catalyst performance has been discussed by *Gosiewski (2005)* and *Hevia et al. (2007b)*. Less active metal oxides increase the operating window with respect to inlet CH_4 concentrations.

Salomons et al. (2004) and *Litto et al. (2006)* developed a heterogeneous two-dimensional model allowing for the dynamics of reactor wall insulation. This model agreed well with the experimental data reported by *Salomons et al. (2003)* and *Sapoundjiev (2000)* as may be seen in **Figure 18-12**. These researchers found that their RFR system remained autothermal at about 0.1 vol% CH_4 in air. At 0.6 vol% and high Q_0, T_{max} was around 900°C, just within the thermal stability limit for the metal oxide catalyst used in the experiments. To avoid overheating during CH_4 concentration excursions, *Kushwaha et al. (2004)* tested various heat storage materials as well as the RFR cycle period. Their experiments showed that by proper choice of packing and RFR operating parameters almost total CH_4 conversion and high heat recovery was possible for coal mine ventilation air. Even with fluctuations in flow rate and CH_4 concentration, a typical situation in coal mining, stable operation could be maintained. *Sapoundjiev (2000)* reported that for $0.3 < CH_4 < 1.0$ vol% the heat recovery efficiency was between 50 and 95%. Typical coal mine ventilation air is 0.4 to 0.5 vol% CH_4. A typical experimental axial temperature profile for the 500 mm o.d. pilot unit with mid-point, hot gas withdrawal is shown in **Figure 18-12**.

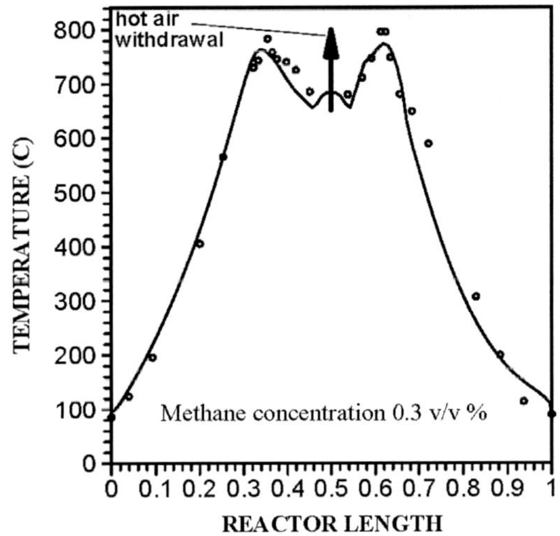

FIGURE 18-12 **Axial temperature profile at the end of the half-cycle in a pilot-scale RFR with hot gas withdrawal, flow direction is from left to right. Experimental results (circles) compared with model predictions (line) for a feed CH_4 concentration = 0.3 vol% and T_{in} = 20°C.** *(Figure reproduced from* Sapoundjiev (2000) *with the author's permission.)*

18.3.2 Reverse Flow Reactor Control

Table 18-4 summarizes the publications on automatic control systems for reverse flow reactors. These have been limited so far to oxidation of VOC and dilute methane streams. The function of control was to maintain a suitable switching period: short enough to avoid displacement of the reaction zone but long enough to prevent inlet gas bypassing. However, if inlet conditions change only slightly the cycle duration (t_c) can be fixed. Normally set are short cycles which result in highest thermal efficiency at low feed concentrations. A thermocouple-based control option was applied in SO_2 and VOC oxidation reactors operated in Russia from the early 1980s. This system increases t_c for increased feed concentration and decreased flow rate. Recently *Marín et al. (2010)* undertook a study of the thermocouple system. These authors varied locations of the control thermocouples and the control logic using a 5 cm o.d. laboratory reactor packed with a honeycomb precious metal catalyst. They found that flow reversal based on setting a minimum temperature in just a single thermocouple in the reactor center provided adequate control comparable to the use of two thermocouples. Apparently, this single point controller cannot be used with short periods when temperature oscillations in the high temperature plateau are very small.

A further function of a control system is to achieve optimal operation under variable inlet conditions. *Budman and Silveston (2008)* reviewed the control of periodically operated reactors, including RFRs. *Balaji et al. (2007)* also considered RFR control. *Matros (1989)* pointed out that an RFR can withstand substantial fluctuations in operating parameters. Its behavior is not sensitive to fluctuations occurring at times shorter than $\frac{t_c}{2}$ provided the period does not exceed the time for movement of the temperature front through the catalyst bed. Besides,

TABLE 18-4 Control of Reverse Flow Reactors

Reference	Contribution
Budman et al. (1996)	Worked on feedback and feed-forward control strategies for CO-oxidation reactor based on the CO outlet concentration.
Ramdani et al. (2001)	Studied VOC oxidation in an RFR consisting of two central catalyst beds and two boundary heat exchange beds controlled using a simplified dynamic model devised for high frequency switching and assuming fast reaction limited by interphase mass transfer.
Dufour et al. (2003); Dufour and Touré (2004)	Developed a model predictive control for an RFR studied initially by *Budman et al. (1996)*.
Edouard et al. (2004, 2005a)	Studied reactor modeling, simulation and experiments leading to the development of a model-based controller using a "high gain observer" and linear quadratic regulator control.
Edouard et al. (2005b)	Ran a comparison between LQR and MPC control strategies.
Fissore et al. (2006a)	A high-gain observer algorithm suggested by *Edouard et al. (2004)* was further developed for a kinetically controlled reaction. Besides temperature profile and inlet concentration of the fuel, the observer reconciles fuel conversion even for uncertain kinetic parameters.
Balaji et al. (2007)	Model predictive control strategy was tested for a CH_4 oxidation RFR. Application of a repetitive model control strategy is discussed.
Fuxman et al. (2008)	Presented a discussion of a model-based linear quadratic regulator applied to CH_4 oxidation in an RFR with temperature control using hot gas bypass.
Fissore and Barresi (2008)	Developed a robust control system for the VOC oxidation reactor considered by *Ramdani et al. (2001)*. Their feedback control system accounted for possible uncertainties in the reactor outputs and controller design.
Marín et al. (2010)	Flow reversal frequency in experimental unit was controlled based on specified minimum temperatures at one or two reference thermocouples in the catalyst bed.

the RFR is inherently stable to rather slow fluctuations potentially leading to extinction. The need for a control system arises with slow and large variations in feed concentration, or with high combustible concentrations that may cause catalyst bed overheating, or very low concentrations that lead to extinction. *Budman et al. (1996)* studied a laboratory wall-cooled RFR operated at high concentrations of CO and controlled via the coolant or t_c. Maps of stationary states were constructed showing process extinction and temperature runaway boundaries. Outlet CO concentration, measured easily, was selected as the controlled variable. The authors concluded that an optimal control strategy combines a conventional feedback PID controller with feed-forward model predictive control. The feedback controller performs well during start-up and operation at relatively low concentrations of CO, while the feed-forward controller maintains a safe operating window at high concentrations of CO.

Edouard and *Dufour* and their co-workers considered a 5 cm o.d. reactor described by *Ramdani et al. (2001)* that used a high cell density catalyst and inert monoliths and 20 s long switching cycle so that the HSFM applies. An electric heater mounted between two catalyst beds served to avoid low catalyst temperatures while adding air dilution before the catalyst bed prevented catalyst overheating. Flow rate was stabilized by replacing feed with air at high VOC concentrations, a substitution unacceptable in practice. Nevertheless, their experimental unit was more realistic than that considered by *Budman et al. (1996)*. *Edouard et al. (2004, 2005a, b)* suggested control systems involving a "high gain observer," an algorithm that reconstructs the reactor temperature profiles and inlet concentration of VOC based on a limited number of actual temperature measurements. *Edouard et al. (2005a)* considered two control strategies: a linear quadratic regulator (LQR) and model predictive control (MPC). The LQR used the observer's estimated temperature profile to compare with reference profiles obtained off-line representing the extinguished and run-away states. With these, the controller used a linearized and discretized representation of the reactor model to minimize a cost criterion employing state and manipulated variables. The required incremental changes in the electric power or air dilution rate are determined quickly on-line when the system approaches one of the reference states. The method was validated experimentally by *Edouard et al. (2005b)*.

Dufour et al. (2003) developed an open-loop MPC strategy for cyclically operated systems that has been extended to an RFR (*Dufour and Touré, 2004*; *Edouard et al., 2005a*). The controller receives as an input the inlet feed concentration; maximum temperature and outlet concentration are calculated and controlled based on a model incorporated into the controller. The MPC system allows for deviations in the controlled variables outside of the reference limits while values and changes for manipulated variables are kept within the limits. To reduce controller action time during relatively short sampling periods, an approximate, linearized model that continuously calculates variations in the output and controlled parameters was used. The original nonlinear model was used off-line to calculate a reference trajectory, around which the approximate solution varies. The authors observed that the MPC system required accurate tuning to keep operations within the reference window and avoid undesirable outcomes when both heating and cooling controls work simultaneously. *Edouard et al. (2005a)* showed that the LQR strategy works better than MPC in terms of control cost.

Baluji et al. (2007) and *Fuxman et al. (2008)* described a repetitive MPC that periodically resets the model variables along with the predictive control action. This system was simulated for CH_4 oxidation in a mid-bed heat exchanger RFR and inlet air dilution or fuel injection to avoid process extinction at low CH_4 concentrations. *Fissore and Barresi (2008)* considered the same VOC reactor and control objectives. They applied an H-infinity based robust control technique described by *Fissore (2008)*. Although quite sophisticated control systems have been proposed for VOC/methane oxidation RFRs, industry currently relies on empirically tuned PID regulators.

18.4 OTHER APPLICATIONS OF REVERSE FLOW REACTORS

Research has been undertaken on the application of RFRs to mitigating automotive exhaust pollution. The effort is summarized in **Table 18-5**. **Chapter 6** is devoted to the use of periodic processes for automotive exhaust treatment; however, the use of reverse flow reactors was left to this section. A decade of research on this application has demonstrated that periodic reverse flow can be used for emission control after natural gas fueled diesel engines. Using heat-integrated reactors makes sense in this case because of high temperatures required for oxidation of CH_4 over common noble-metal catalysts and the low temperature of diesel engine exhaust.

TABLE 18-5 Application to Automotive Emissions

Reference	Contribution
Zheng et al. (1998); Matros et al. (1999a); Strots et al. (1998)	Concept and potential configurations of a reverse-flow catalytic converter (RFC) for automotive emission control. A simple single bed converter was examined for the destruction of HCs and CO in the exhausts of diesel engine fueled with natural gas as well as NO_x reduction using lean-NO_x HC-SCR technology. Reduction of HC emissions during engine start-up was examined also. Numerical simulation and proof-of-concept experiments demonstrated RFC advantages such as lower tail pipe emissions at low temperature.
Zheng et al. (1999)	Developed a compact RFC to reduce HC and CO emissions for natural gas fueled diesel engines. The tested converter design combined a catalyst monolith and rotating valve in one casing. In both dynamometer and on-road tests, the RFC demonstrated substantially better performance during various engine transients then the traditional unidirectional converter.
Konstandopoulos and Kostoglou (2000)	Numerical simulation showed that reversed flow operation increased the rate and improved the uniformity of diesel soot filter regeneration.
Liu et al. (2001b)	Experimental study of 152 x152 mm unit installed after a natural gas engine dynamometer demonstrated that an RFC can maintain high temperature for a light engine load. A 1-D model accurately represented the operation.
Zheng and Reader (2004)	Mathematical modeling of an RFC demonstrated satisfactory energy management.
Liu et al. (2007)	Developed a CFD FLUENT code for modeling an RFC for removal of CO and CH_4 from a dual fuel diesel engine. The simulation established the 3-D temperature distribution in a 250 mm diam. converter and agreed well with experimental results. According to experiments and simulations, an RFC can be ignited at lower inlet temperature than the usual unidirectional unit, however the latter converter ignites much faster at high inlet temperatures.
Zheng and Banerjee (2009)	Carried out modeling of a combined diesel soot filter and oxidation catalyst bed in different flow configurations.

A compact converter-prototype design combining a catalytic monolith and a rotating valve in one casing developed by *Zheng et al. (1999)* has been tested both in an engine dynamometer and on-road. Commercialization has been held up, probably due to low market demand and the absence of regulations on CH_4 emission.

Table 18-6 provides a list of publications on selective catalytic reduction of NOx. The heat-integrating reversed-flow reactor for selective catalytic reduction of nitrogen oxides was developed by *Matros* and co-workers in the 1980s. *Noskov et al. (1993)* discussed the design for the first commercial unit including the development of the flow diagram, effect of operating parameters on performance, potential hazard from buildup of ammonia salts, experiments with monolithic catalyst, and the parameters chosen for the commercial unit. These authors indicated the beneficial effect of ammonia adsorption, which was explored by *Agar and Ruppel (1988a, b)* for their isothermal mass-trapping reactor.

The development of RFRs for SO_2 oxidation, summarized in **Table 18-7**, has led to

TABLE 18-6 Application to the Selective Catalytic Reduction of NO_x with Ammonia

Reference	Contribution
Bobrova et al. (1988)	Process concept, reactor design and an estimate of the operating parameters for a commercial unit employing injection of NH_3 in the middle of the reactor.
Agar and Ruppel (1988a, b)	Concept of an isothermal reverse-flow adsorptive reactor with interruption of the NH_3 feed at the end of a cycle.
Kolter and Lintz (1991); Kolter et al. (1992)	Tests on laboratory and small-scale commercial units showed that a Ljungström rotating heat-exchanger fitted with an SCR catalyst supported on the exchanger surface provided substantial NO reduction without requiring major power plant modifications.
Noskov et al. (1993)	Overview of the developments of reverse-flow reactors for the selective catalytic reduction of NO_x. Discussion of the flow arrangement, effect of operating parameters, potential hazard from buildup of ammonia salts, experiments with monolith catalysts and adsorbents, identification of key commercial unit parameters and an indication of the beneficial effect of NH_3 adsorption.
Snyder and Subramaniam (1998)	Numerical simulation of an isothermal reactor with mid-reactor NH_3 injection demonstrated the effects of various parameters on NH_3 consumption.
Galle et al. (2001); Nalpantidis et al. (2006)	Concept of a thermal RFR for decomposition of N_2O. Thermal and catalytic destruction mechanisms can be distinguished based on the analysis of reactor dynamic behavior.
Fissore et al. (2006b)	Offered a comparison of an RFR and a 3-bed SMBR for the simultaneous introduction of reactants to the adiabatic units. NH_3 outlet concentration was reduced in the SMBR even though that reactor had a smaller stable operating window compared with an RFR.
Botar-Jid et al. (2009)	A steady-state countercurrent reactor model adequately approximated a dynamic SCR-RFR model.

commercialization with a number of installations in Russia, Bulgaria and China as will be reviewed in **Section 18.10.2**. **Table 18-8** lists investigations of RFR applications to methanol and ammonia synthesis. The incentive for this work was potential process diagram simplification as well as improved H_2 utilization and air pollution mitigation.

18.5 THERMAL REACTORS

The heat trapping feature of periodic flow reversal has been applied, not surprisingly, to noncatalytic reactions. Indeed, this development preceded its use with catalytic systems that has been the primary topic of this chapter. Because the principles of reverse flow application to noncatalytic, thermal reactors are essentially the same as what has been discussed until now, the examination of these reactors will be brief.

Commercial thermal reverse-flow reactors or regenerative thermal oxidizers (*Lewandowski (2000)*, the operation of which is illustrated by **Figures 18.4** and **18.5**, achieve 95–96% thermal efficiency such that the difference between maximum and inlet temperature is about 20 times greater than the total adiabatic temperature rise from combustion of both VOC and the added fuel. Further substantial savings in the added fuel can be achieved via the

TABLE 18-7 Application to the Oxidation of Sulfur Dioxide

Reference	Contribution
Boreskov et al. (1982a)	Earliest experimental study using a 0.18 m o.d. pilot-scale reactor.
Boreskov et al. (1982b); Balashev et al. (1987)	Pilot scale tests employing 1.6 and 2.8 m o.d. reactors.
Sapundzhiev et al. (1984).	The effect of pellet size on the SO_2 oxidation kinetics in a flow reversal operation was studied. Despite a decrease in the effectiveness factor with increasing size, the temperature profile created in the bed increased conversion and lowered pressure drop in the reactor. Nevertheless, small catalyst pellets improve reactor stability. Simulation results were tested in an industrial reactor.
Matros et al. (1984)	Summary of research on, and testing of, RFRs for sulfuric acid production.
Isozaki et al. (1990)	Pilot scale tests of a double contact system with 1.9 m diam. RFRs in both contact stages
Sapundzhiev et al., (1990)	Numerical simulation of an RFR showed that the amount of catalyst can be substantially reduced by substitution of a suitable inert material at each end of the reactor.
Matros (1989); Bunimovich et al. (1990a); Gosiewski (1993); Snyder and Subramaniam (1993); Huixiong et al. (1996)	Numerical simulations using steady-state kinetics and a heterogeneous model of the catalyst bed were used to study the optimal catalyst size and operation of an RFR.
Matros (1989); Bunimovich et al. (1990a)	Process design for various SO_2 gas concentrations. Summary of pilot and commercial reactors. Treatment of moderately concentrated gases required large catalyst oellet size and intermediate heat removal.
Bunimovich et al. (1995)	Numerical simulation employing a dynamic kinetic model. Non-stationary behavior of oxide intermediates explained deviations between experiments and calculations that employed steady-state kinetic models, specifically an increase in SO_2 conversion at high inlet concentration of SO_2 and decreases at low concentrations.
Kunitski et al. (1984); Matros (1989); Bunimovich et al. (1990a)	Summary of operating data for a single-bed commercial, 7.2 m o.d. reactor installed in a copper smelter and treating a dilute SO_2 stream; results for long-term operation and a comparison with predictions from a mathematical model of the unit.
Matros and Bunimovich (1996)	Summary of RFR performance for commercial units operated in Russia.
Grozev et al. (1994)	Operating data for a 6 m o.d. triple bed RFR treating dilute concentrations of SO_2 emitted by a lead smelter.
Sorin et al. (1994)	Calculation of exergy flows for an RFR reactor. Exergy efficiency closely follows the fractional conversion of SO_2.
Xiao and Yuan (1996)	Test results for a 33 mm o.d. laboratory reactor and a pilot scale 0.8 m o.d. triple bed RFR.
Schoubye (1999)	Description of the Haldor-Topsoe REGESOX process combining regenerative heat transfer with catalytic oxidation of SO_2, and condensation of H_2SO_4 in a two vessel RFR. REGESOX was tested in a 60 Nm^3/h reactor. Primary application of this design is to revamping of a single contact system to a double contact one.
Matros et al. (1999b)	Numerical simulation of the application of an RFR to producing sulfuric acid from dilute H_2S and COS emissions in viscose production.

(*Continued*)

TABLE 18-7 Application to the Oxidation of Sulfur Dioxide (*cont'd*)

Reference	Contribution
Vernikovskaya et al. (1999)	SO_2 reactor modeling allowing for dynamic properties of the complex vanadia catalyst.
Xiao et al. (1999b)	Numerical simulation of double and triple bed reactors with mid-bed heat exchange; testing of converter designs for various feed gas concentrations through simulation.
Xiao et al. (1999c)	Numerical simulation of a triple bed RFR with a mid-bed cold feed gas quench.
Xiao et al. (1999a, d)	Operating description of a commercial 6 m o.d. triple bed converter with two inter-bed heat exchangers; description of the control system for treatment of a highly fluctuating SO_2 concentration in the reactor feed.
Dimitrov et al. (2003)	Description of a periodic DC/DA acid system operated at a non-ferrous metallurgical plant. Operating data show stable autothermal operation in the presence of inlet SO_2 concentration fluctuations. SO_2 conversion in the system was over 99%.

substitution of regenerative catalytic oxidizers (*Bunimovich and Matros, 2010a*).

The thermal reverse-flow reactor has been studied over the years by quite a number of research teams for a variety of applications (**Table 18-9**). There has been recent interest in RFRs as compact combustion devices for small-scale boilers. **Figure 18-13** shows a laboratory version of such a system developed at the University of Ilinois in Chicago. Generation of steam with reverse flow combustion in porous media has advantages in exceptionally low emissions of NO_x and CO, and production of steam from low-grade fuels.

TABLE 18-8 Application to the Syntheses of Methanol and Ammonia

Reference	Contribution
Matros (1989); Gerasev and Matros (1991)	Numerical simulation of an RFR for the synthesis of NH_3. At low pressure, the efficiency of an RFR is equivalent to 2 to 3 beds of a steady-state converter. A larger catalyst size is required to avoid catalyst overheating. Purge gases can be utilized in an RFR.
Matros (1989)	Numerical simulation of MeOH synthesis. Possible applications of RFRs include MeOH production from dilute purge gases.
Neophytides and Froment (1992)	Studied MeOH synthesis in a bench scale RFR; a substantial effect of catalyst dynamics was identified.
Vanden Bussche et al. (1993); Vanden Bussche and Froment (1996)	Numerical simulation of MeOH synthesis; analysis indicated that gas phase heat capacity and density change must be taken into account in any model. A triple bed reactor is suggested for alleviating conversion losses just after flow reversal.
Egyhàzy et al. (1998)	Experiments on NH_3 synthesis over iron catalyst in a 50 mm o.d. reactor were conducted at 240 to 300 bar. The experimental NH_3 production rate in an RFR is higher than the rate calculated based on steady-state reaction kinetics and temperature profiles obtained in the RFR.

TABLE 18-9 Application to Various Noncatalytic Combustion Processes

Reference	Contribution
Cottrell (1938)	Concept of an RFR for removal of noxious and odorous components by combining thermal destruction and regenerative heat exchange. Design used a burner and hot bypass for temperature control. Use of catalysts was discussed but not implemented.
Gilbert and Daniels (1948); Hendrickson and Daniels (1953)	Experimental study of the oxidation of N_2 to NO through heating of air and cooling of the product gas in a regenerative oxidizer with two pebble beds. Fuel was injected in a combustion chamber between the beds. Extremely high temperatures, up to 2200°C, and about 1 vol% of NO in the reactor outlet were achieved in a test unit with 0.45 m o.d. beds.
Ermenc (1956)	Operation of 40 t/day nitric acid pilot plant using a regenerative pebble furnace. Design and operating problems were encountered with respect to combustion, pebble attrition and support grates. The plant functioned for over a year until it was abandoned due to poor economics and equipment difficulties.
Bixler and Coberly (1952)	Concept and operation of a small-scale industrial plant for acetylene production through thermal cracking of hydrocarbons.
Hanamura et al. (1993)	Numerical simulation of single bed, reversed flow autothermal combustor employing a 1-D heterogeneous model accounting for thermal radiation in a packed bed.
Hoffmann et al. (1997)	Bench-scale study of periodically reversing super-adiabatic combustion in a porous media. Experimental conditions: 100 mm × 100 mm × 200 mm deep bed filled with ceramic foam having a 0.17 to 0.8 mm pore size. Used combustion of city gas at an A/F ratio = 0.03 to 0.4, linear velocities = 0.08 to 0.33 m/s, flow reversal periods = 5 to 100 s. Emission of NO_x was very low, about 1 ppm, while emission of CO depended on operating conditions with a minimum for the larger pore size ceramic foam.
Nijdam and van der Geld (1997)	1.6 m o.d. reactor filled with a structured bed of Intallox ceramic saddles was tested for combustion of low CH_4 concentrations. A 1-D model generally reproduced test results after accounting for thermocouple transients.
Nijdam and van der Geld (1999)	Numerical and semi-analytical techniques for the simulation of thermal reverse flow reactors.
Choi and Yi (2000)	CFD modeling of processes in the combustion chamber of a regenerative thermal oxidizer.
Cheng et al. (2002)	Testing of three pilot, gravel-packed RTOs, operating at flow rates of about 1000 Nm^3/h, demonstrated destruction of various VOCs without use of a catalyst at 300 to 440°C.
Contarin et al. (2003)	Numerical study of a reverse-flow filtration combustor with two heat exchangers embedded in the ends of packed beds. Heat utilization efficiency was between 50% and 80%, increasing with higher A/F ratio and linear velocity in the packed bed.
Jugjai and Sawanon (2004)	Experimental study of a 26 × 30 cm^2 RF packed bed combustor with a water-fed tube bank at the bed center and additional tube banks at the bed ends. Tuning τ_{cycle} (10 to 30 s), the A/F ratio (0.38 to 0.49) and the fuel flow rate for a ceramic $d_p = 16$ mm stabilized a high temperature zone near the central tube bank and achieved high energy utilization (80%) at low exit concentrations of NO_x and CO.
Dobrego et al. (2008c)	Study of the lean combustibility limit (LCL) (minimum ΔT_{ad} for autothermal operation) for an RFR with heat losses using estimations based on an HSFM and numerical results for a heterogeneous 1-D model. A minimum in the LCL versus linear velocity was observed.

(*Continued*)

TABLE 18-9 Application to Various Noncatalytic Combustion Processes (*cont'd*)

Reference	Contribution
Dobrego et al. (2008a, 2008b)	Numerical simulation of the noncatalytic partial oxidation of CH_4 in an RFR. A 1-D heterogeneous model allowed for heat losses from the reactor walls. CH_4 conversion of about 80% with 25% H_2 in the reactor outlet was predicted for mixtures of CH_4 and air. Addition of water to the feed increased the H_2 content in the product gas.
Barcellos et al. (2011)	Experimental study of a prototype RF porous burner with an embedded steam boiler for CH_4 combustion at A/F ratios = 0.2 to 1 resulted in ultra low NO_X and CO emissions at a heat utilization efficiency of 50 to 80%.
Koznacheev et al. (2011)	Study of water purification from organic inclusions through combustion with air in porous beds: minimum concentration of acetone in water for achieving autothermal operation was about 14%.

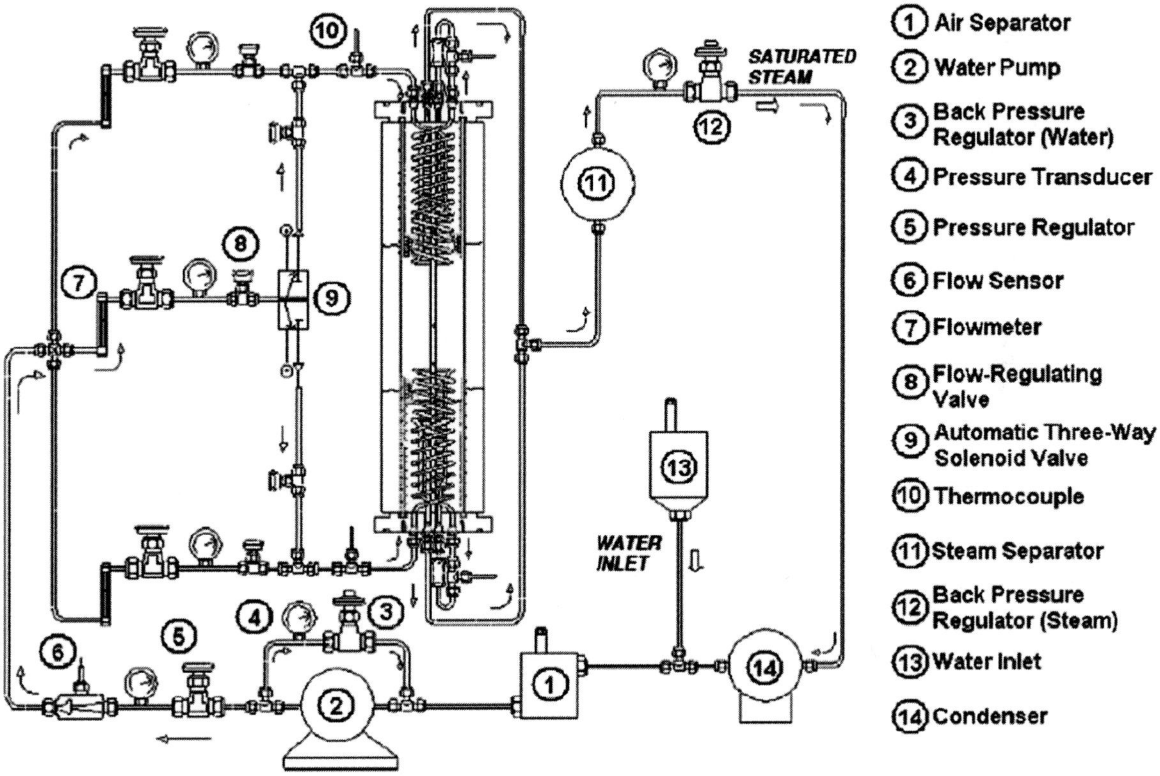

FIGURE 18-13 **Flow diagram of the water side of a reverse-flow combustor with an embedded steam boiler. The RFR is contained in the inner cylinder with its three compartments: a central combustion chamber and two packed bed recuperators containing the boiler heating coils. Air and combustion gas switch flow direction together with the boiler feed or use an adjustable off-set.** *(Figure adapted from* Barcellos et al., (2011), *with permission. © 2011by Elsevier, Inc.)*

18.6 ENDOTHERMIC PROCESSES

Several highly endothermic processes, such as reforming of natural gas or cracking of ethane to ethylene, employ externally-fired tubular reactors that consume much fuel and emit high volumes of CO_2 per tonne of product. Direct coupling between endothermic and exothermic processes still requires rather high fuel consumption when high temperature must be maintained at the end of the reaction zone. RFRs offer high energy recuperation and can reduce fuel consumption. They have long been considered as an attractive alternative to conventional operation (see **Table 18-10**).

Three ways for introducing heat into endothermic processes may be used in RFRs: 1) heating of the catalyst bed by an inert fluid, 2) combustion of coke accumulated on the catalyst during the process and 3) combining an endothermic process with an exothermic one. In 1) and 3), the heating fluid or fuel for the exothermic process can be introduced together with the reactants at the reactor inlet or injected into a high temperature zone in the middle of the reactor.

The inert fluid heating system proposed by *Heggs and Abdullah (1986)* for the catalytic dehydrogenation of ethylbenzene combined passing ethylbenzene through the bed in one flow direction and a heating fluid, steam, in the other. Continuous operation required doubling the catalyst beds and four-way valves. Steam, flowing countercurrent to the reactants, creates a desirable rising temperature in the direction of reactant flow that is advantageous for equilibrium-limited endothermic reaction. Heat recovery from the product stream can be undertaken by preheating the feed with the product in a shell and tube heat exchanger or by using a pair of cyclically operated packed bed exchangers. An ethylbenzene dehydrogenation system considered by *Snyder and Subramaniam (1994)* used one or more steam injection points in the center of the catalyst bed.

Gartside et al. (1994) and *Ercan and Gartside (1996)* described a modified Catofin Houdry process for propane or isobutane dehydrogenation. The original Houdry process periodically switched the reactor feed between hydrocarbons and air. Coke accumulated on the catalyst during dehydrogenation was combusted during the air portion of the cycle regenerating the catalyst and heating the bed to the reaction temperature (ca. 600°C). Unlike the original process that fed hydrocarbons and air at the same point, reverse flow, counter-current sequential feeding of the paraffins and air resulted in higher conversions and lower energy demand. This improvement was attributed to a more favorable axial temperature profile (*Ercan and Gartside, 1996*).

Most of the other studies listed in **Table 18-10** relate to combinations of gas phase endothermic and exothermic reactions. *Kolios et al. (2000)* distinguished between simultaneous, asymmetric and symmetric operations. **Table 18-11** illustrates these operating modes with flow diagrams and process examples. Simultaneous endothermic and exothermic operation involves feeding of the reactants for both processes at the same time to separate locations or at different times to the same feed point. This can be done, for example, with a combined feed of natural gas, air and steam in the partial oxidation or autothermal reforming of natural gas, a process studied in the pioneering work of *Blanks et al. (1990)* and more recently by *Veser* and coworkers from the University of Pittsburg (*Veser, 2010; Neumann and Veser, 2005; Liu et al., 2009*). The process would be the simplest and most straightforward option if not for the problem caused by differences between the rates of the endothermic and exothermic processes. **Figure 18-14** illustrates the types of temperature profiles that can arise through rate mismatching using a simple example of a pair of independent, irreversible reactions. The adiabatic temperature rise achieved due to the exothermic reaction ($(\Delta T_{ad})_{exothermic}$ = 1200°C) is only slightly higher than the

TABLE 18-10 Publications on Reverse-Flow Heat Supply to Endothermic Processes

Reference	Contribution
Heggs (1986); Heggs and Abdullah (1986)	Conception and numerical simulation of an RFR with an endothermic reaction in one flow direction and bed reheat with an inert gas in the opposite direction. Modeling and experiments undertaken for the dehydrogenation of ethylbenzene using steam as the reheating media. RFR achieved higher conversion and yield compared with steady-state operation with co-feed of the heating media and reactants.
Blanks et al. (1990)	Conception, pilot-scale testing and numerical simulation of CH_4 partial oxidation. The 0.6 m o.d. reactor was packed with Ni catalyst sandwiched between beds of inert material. The pilot unit operated for over a year on mixtures of CH_4 and air with $\tau_{cycle} \approx 1$ h. Operating procedure included steam flushing at the beginning of every flow direction switch. About 10% of the CH_4 feed was combusted before entering the reactor to prevent high catalyst temperatures and coke formation. A 1-D heterogeneous model represented the test results quite well.
Haynes et al. (1992)	Numerical simulation of the dehydrogenation of ethylbenzene using an RF operation alternating between dehydrogenation and bed reheating with a hot inert gas. Researchers concluded RF operation was superior to steady-state at a high inert gas flow rate.
Snyder and Subramaniam (1994)	Numerical simulation of ethylbenzene dehydrogenation using reverse flow of superheated steam for bed reheat. The RF operation produced higher temperature and product yield at lower steam flow rates compared with the conventional process. Uniform temperature distribution was achieved by using multiple steam injection points along the reactor length.
De Groote et al. (1996)	Conducted numerical simulation of CH_4 partial oxidation with O_2 and air based on detailed kinetics including formation of carbon. Dynamic profiles were characterized by steep temperature gradients. At every flow reversal, high temperature peaks occurred in the inlet of the catalyst bed but dissipated by the outlet. Coke formed continuously in the high temperature zone in one flow direction but was gasified during reverse flow with air and O_2.
Gartside et al. (1994); Ercan and Gartside (1996)	The KATOFIN process alternates C_3-C_4 paraffin dehydrogenation and air oxidation of by-product coke. Countercurrent operation with feeding of hydrocarbon and air in different directions in a catalyst bed creates a favorable rising temperature profile that increases selectivity compared with the conventional co-current operation.
Kulkarni and Duduković (1996, 1997, 1998)	Numerical simulation of the RE-GAS process (*Levenspiel, 1988*) using a heterogeneous reactor model with simplified kinetics combining CH_4 steam reforming and oxidation. The asymmetric RFR operations involved reforming in one flow direction and combustion in the other. Introduced the fuel gas at low concentrations of fuel as well as reactant. Temperature control can be improved by heating the combustion mixture above ignition temperature before entering the catalyst bed.
Poirier and Sapundzhiev (1997)	Catalytic decomposition of natural gas to H_2 for fuel cell application was undertaken in an asymmetric flow reversal mode (**Table 18-11**). Cracking of CH_4 over the catalyst produces H_2 and deposits coke on the catalyst. In the other part of a cycle, the coke is burnt off by air. H_2 purity is 95% (compared with 75% for conventional steam reforming) and contains ppm levels of CO and CO_2.

(Continued)

TABLE 18-10 Publications on Reverse-Flow Heat Supply to Endothermic Processes (*cont'd*)

Reference	Contribution
Kolios and Eigenberger (1999)	Simulation and experiments on the dehydrogenation of ethylbenzene. A symmetrical RFR mode was used with injection of hot gas in the middle of the reactor. This provided higher conversion and heat recuperation compared with an asymmetrical operation with opposed flows for feed and reheat gases. High volume of injected reheating gas reduced heat recuperation. Injection of air only increases the thermal efficiency, but decreases the selectivity to styrene.
Borio et al. (1999)	Numerical simulation of the dehydrogenation of 1-butene over a chromia catalyst in an RFR using air for catalyst regeneration and burn out of coke deposits. Results showed greater coke burnout, higher butadiene yield, reduced regeneration temperatures and longer catalyst life compared with the conventional technology.
Agar (1999)	Developed the concept of a reverse-flow Andrussow HCN process.
van Sint Annaland et al. (2001)	Simulation and experiments for an RFR consisting of catalyst with an inert monolith packing found that there was no gas-phase back flash during switching between propane and air indicating that switching between oxidating and reducing conditions does not require intermediate flushing with inert gas in an RFR.
van Sint Annaland et al. (2002)	Numerical simulations of various reactor symmetric sequential operations alternating endothermic and exothermic reactions in the same catalyst bed or in two contiguous beds exchanging heat. Study explored a design strategy involving the matching of the operating conditions for the two reactions with respect to energy required. For the reversible dehydrogenation of C_3, provision must be made to suppress hydrogenation at low temperature in the catalyst bed outlet.
Van Sint Annaland and Nijssen (2002)	Experimental dehydrogenation of C_3 in a 5.5 cm $\times$ 5.5 cm reactor containing inert and catalytically active honeycomb ceramic packing using periodic switching between a combustible CH_4–O_2 mixture and C_3 to give a symmetric endothermic-exothermic cycle. Design was developed from a numerical simulation of a high space velocity CH_4-air system with 0.7 vol% CH_4 in the combustion mixture. Experiments used ceramic packing in the ends of the bed to avoid hydrogenation and obtained about a 20% yield of $C_3^=$ and $C_2^=$. O_2 vol% in the combustion mixture must be reduced to prevent temperature spikes from the combustion of coke deposited during dehydrogenation. Also, there must be a periodical increase in τ_{cycle} to bake out slowly accumulated coke deposits.
Gosiewski (2001)	Simulation of CH_4 partial oxidation in an RFR with a simultaneous feed of CH_4/O_2 mixture at a 2/1 ratio with and without addition of steam. Without steam, a substantial overheating of the catalyst bed along with possible coke formation occurs.
Kikas et al. (2003); Kaisare et al. (2005a, 2005b)	Experiments on CH_4 partial oxidation in a microreactor with 0.5 mm x 11 cm channels containing Pt/Pd wires as a catalyst. Experimental results and a numerical simulation found that an RFR achieved marginally higher H_2 yield at lower temperature compared with uni-directional operation. The yield improvement resulted from more favorable conditions for the water-gas shift reaction at the lower outlet temperature attained in an RFR.

(*Continued*)

TABLE 18-10 Publications on Reverse-Flow Heat Supply to Endothermic Processes (*cont'd*)

Reference	Contribution
Glöckler et al. (2003)	Employed an asymmetric RF reactor combining endothermic steam reforming of methane in one flow direction and bed reheat through combustion in the other. During the combustion portion, the gaseous fuel is distributed to several points along the reactor length.
Glöckler et al. (2004)	Overview of various countercurrent recuperative and regenerative reactors for combining endothermic and exothermic processes. Asymmetric operation requires spatial distribution of the feed for the exothermic reaction to prevent excess temperatures and ensure stable operation.
Glöckler et al. (2004, 2007, 2009); Tellaeche et al. (2009)	Experiments using a 50 mm o.d. RFR operated asymmetrically. Reheating during the regeneration part cycle employed three point fuel injection along the catalyst bed. Operational problems encountered were overheating and catalyst sintering. Later reactor designs utilized flameless homogeneous combustion with simultaneous injection of air and fuel in combustion zones and a continuous stream of inert gas passing through the reactor. This distributed combustion reheating permitted operating at higher concentrations of CH_4 and longer τ_{cycle}.
Hershkowitz et al. (2004, 2005, 2009)	An asymmetric RFR, a so-called pressure swing reformer, was studied by simulation and experiment. The reactor switches between high-pressure steam reforming and atmospheric combustion using a bed filled with a reforming catalyst at one end and an inert packing at the other end. 93% heat recovery and high reactor volumetric productivity was obtained with a honeycomb packing and operating at 10–30 s reversal durations. Tests with CH_4 and some liquid fuels (up to 1000 h duration) demonstrated high conversion and heat recovery with no loss in catalyst activity. Applications to generating H_2 for automotive fuel cells and H_2 fired turbines are possible.
Neumann and Veser (2005); Neumann et al. (2004a, b); Mitri et al. (2004); Liu et al. (2009); Veser (2010)	Experimental studies of partial oxidation or autothermal reforming of CH_4 in a 1.7 cm o.d. reactor filled with a bed of noble metal catalyst supported on a ceramic foam monolith flanked between two inert honeycomb blocks. RFR operation provided substantially greater yield of synthesis gas compared with a conventional reactor. Benefits included extending space velocities up to 240,000 h^{-1} and compensating for catalyst deactivation through self-sustaining temperature increases. Explanation for the improvements is the high catalyst bed entrance temperature even with a small adiabatic temperature rise.
Smit et al. (2004, 2005a, b, c; 2007)	Concept of a reverse-flow membrane reactor for partial oxidation of CH_4. Concentric tubular reactor has the inner tube filled with an oxidation catalyst that has a porous wall in its central part. The outer shell contained a combustion catalyst. 100% methane is fed to the inner tube and air enters the reaction zone from the shell uniformly through the porous tube wall. Air and feedstock are separately heated in beds of inert packing flanking the catalyst beds in the reactor tube and packed in the shell through periodic flow reversal. A uniform high temperature is maintained in the inner tube reaction zone by adding steam. Numerical simulation as well as experiments with 8 mm o.d. stainless steel tube demonstrated high selectivity toward synthesis gas having an almost equilibrium composition in the reactor outlet at $H_2/CO \approx 2$.

TABLE 18-11 Flow Reversal Systems for Endothermic Processes

Type of operation	Simultaneous	Asymmetric	Symmetric with heating the reactor center	Symmetric with separate heat and production cycles
Flow Diagrams	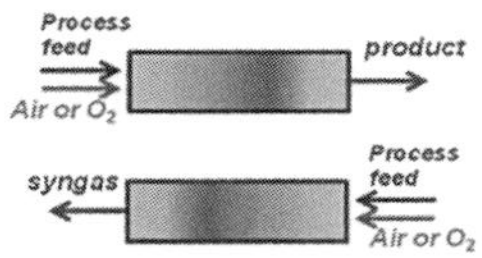	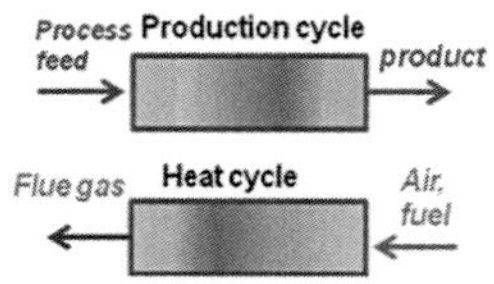	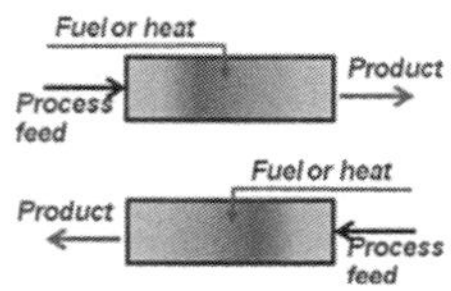	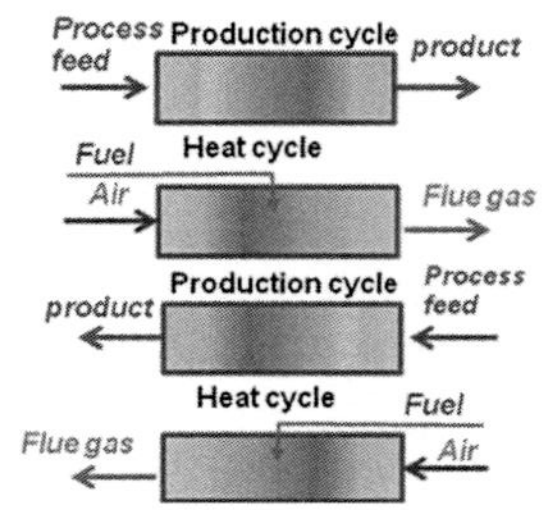
Number of Reactors	1	2	1	2
Number of 4-way Valves	1	2	2	4
Examples with References	Partial oxidation of methane (*Blanks et al., 1990*) Autothermal reforming and partial oxidation of CH_4 (*Veser, 2010*)	Cracking of CH_4 to H_2 for fuel cells (*Poirier and Sapundzhiev, 1997*) Coal gasification (*Levenspiel, 2005*) CH_4 reforming (*Tellaeche et al., 2009; Hershkowitz et al., 2009*) Dehydrogenation of paraffin (*Ercan and Gartside, 1996; Borio et al., 1999*) Dehydrogenation of ethylbenzene (*Heggs and Abdullah, 1986; Kolios and Eigenberger, 1999*)	High temperature oxidation of N_2 (*Gilbert and Daniels, 1948*) Dehydrogenation of ethylbenzene (*Snyder and Subramaniam, 1994; Kolios and Eigenberger, 1999*)	Thermal dehydrogenation of alkanes to dienes (*Bixler and Coberly, 1952*) Catalytic dehydrogenation of C_3 (*van Sint Annaland and Nijssen, 2002*)

temperature decrease $(\Delta T_{ad})_{endothermic} = -1120°C$) from the other reaction. When the reaction rates are the same, the profile (Line 1, **Figure 18-14**) is identical to a single exothermic process with a moderately high adiabatic temperature rise equal to the sum of the two ΔT_{ad}'s, e.g. 80°C. The shape of the profile remains the same when the rate of the endothermic reaction exceeds that of exothermic one (Line 2, **Figure 18-14**). In this situation the plateau temperature increases because the higher local heat liberation from the exothermic reaction is not fully compensated for by the consumption of the endothermic process that takes place at prior points in the bed. Another profile is observed if the rate of the exothermic process is higher. The rapid exothermic reaction results in a distinct peak followed by a temperature decrease due to the trailing endothermic process. High temperature in the peak area can affect the catalyst stability while the low temperature in the middle of the bed can hinder completion of the endothermic reaction, and, for the particular example of methane reforming, cause the formation of coke. Enclosing a short catalyst bed between two beds of inert materials can help to increase the temperature and flatten the profile. Nevertheless, the process may be difficult to control. *Blanks et al. (1990)* reported successfully operating a 0.6 m diameter pilot reactor combining catalyst and ceramic balls for over a year. Still Amoco, the company

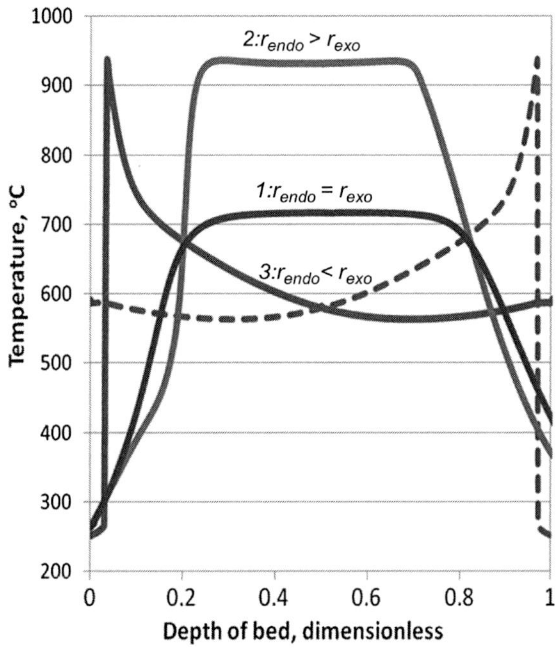

FIGURE 18-14 **Temperature profiles for simultaneous reverse-flow operations combining endothermic (r_{endo}) and exothermic (r_{exo}) reactions. Solid profiles are for the middle of a flow reversal period with gas direction from left to right. Dotted profile is for the middle of a half-cycle with gas direction from right to left. The profiles assume two independent first-order reactions. Activation energies for both reactions are 105 kJ/mole and rates are limited by intra-phase mass transfer. Reaction rate constants for the endothermic—exothermic curves 2: and 3: were set to achieve approximately the same maximum temperatures.** *(Figure adapted from* Bunimovich and Matros (2010b) *with permission © Matros Technologies, Inc.)*

undertaking the research, abandoned further development, supposedly because of problems with temperature control (*Kolios et al., 2000*).

The problem of temperature control for combined endothermic-exothermic systems has been addressed by *Eigenberger* and co workers (*Glöckler et al., 2003, 2004, 2007, 2009*) who proposed distributing combustion air for reheating at points along the RFR axis. The asymmetric system investigated and the intended performance is indicated in **Figure 18-15**. Methane reforming proceeds generating the temperature profiles on the left hand side of the figure while reheating of the catalyst appears on the right hand side. A stationary system has been assumed in the figure so a cooling front moves from left to right in the LHS of the figure accompanied by the endothermic reaction front seen at the bottom of the figure. On the RHS, cold fuel enters from the right and is heated to ignition (1000°C) by the hot inert packing. Air entering through holes in the tube wall combusts the fuel resulting in a leftward moving hot front that reheats the bed back to 1000°. Experiments reported by *Glöckler et al. (2007)* found homogeneous combustion at the air injection points caused a sharp local temperature rise that increased with time on stream. The research team handled this by nonuniform air distribution, adding CO_2 and steam to the fuel to suppress the homogeneous combustion and decreasing the combustion air flow with time. These modifications resulted in the time-varying temperature profiles given in **Figure 18-16**. Fuel for reheating was a pressure swing adsorber discharge containing steam and CO_2. Temperature spikes were not eliminated but were limited in magnitude and do not grow with time on stream. Stable operation over an extended time was demonstrated.

18.7 MASS TRAPPING REACTORS

Periodic reverse flow can be employed in adsorptive separations in situations where the

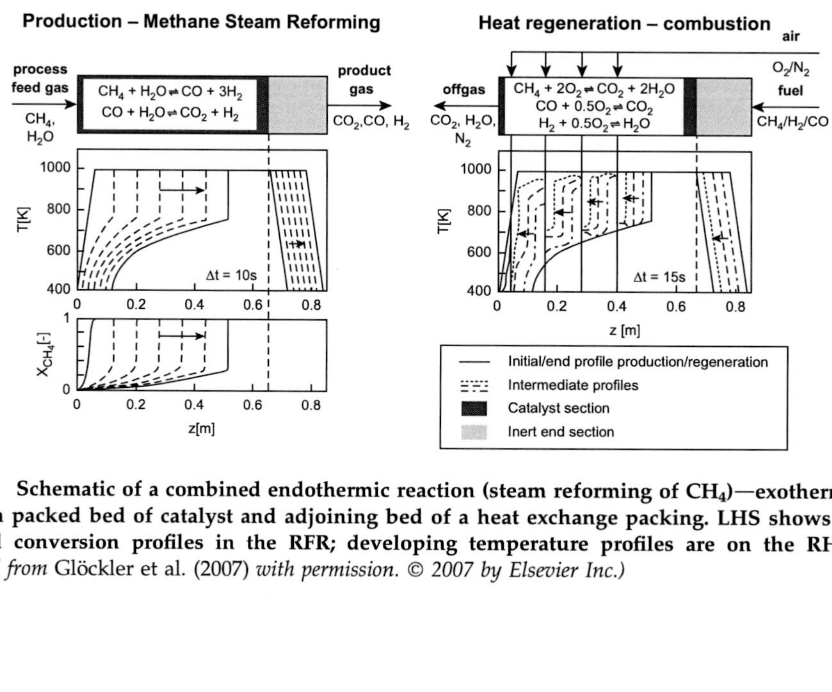

FIGURE 18-15 Schematic of a combined endothermic reaction (steam reforming of CH_4)—exothermic reaction (CH_4 combustion) in a packed bed of catalyst and adjoining bed of a heat exchange packing. LHS shows the time varying temperature and conversion profiles in the RFR; developing temperature profiles are on the RHS. $\tau_{cycle} = 120$ s. *(Figure reproduced from* Glöckler et al. (2007) *with permission. © 2007 by Elsevier Inc.)*

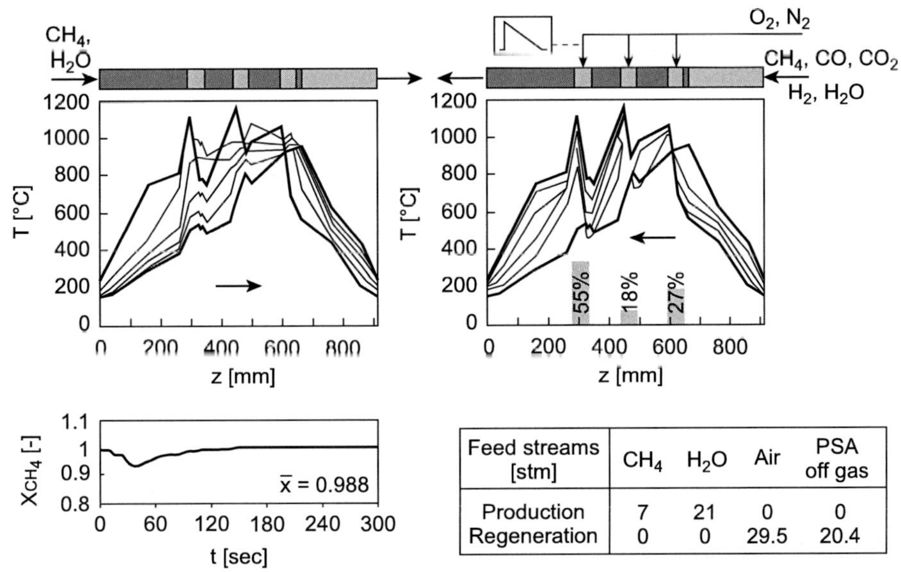

Feed streams [stm]	CH_4	H_2O	Air	PSA off gas
Production	7	21	0	0
Regeneration	0	0	29.5	20.4

FIGURE 18-16 Operation of the single bed periodic flow reversal reactor using an industrial discharge for fuel with distributed combustion air (see Figure 18-15). Figure shows change in the measured axial temperature profiles with time in the reforming and reheating half-cycles (60 s between each profile) *(Figure reproduced from* Glöckler et al. (2007) *with permission. © 2006 by Elsevier Ltd.)*

adsorbate is to be recovered as well as concentrated. Periodic reverse flow was cleverly applied to adsorptive decontamination of a waste gas containing NO_x by *Agar and Ruppel (1988a, b)*. The system they devised consisted of a bed packed with a NO_x reduction catalyst operating with periodic flow reversal (**Figure 18-17**). NH_3, the reductant, was mixed with the waste gas upstream from the catalyst bed. The reductant is strongly adsorbed by noble metals and reacts on these metals with weakly adsorbed NO_x. Just prior to when the NO_x front

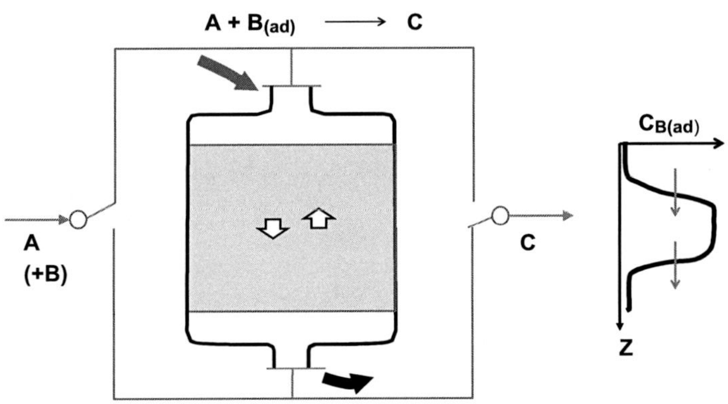

FIGURE 18-17 (a) Schematic of the operation of a periodic flow reversal reactive adsorber for the catalytic decontamination of a NO_x containing waste gas. NH_3 is stopped for a fraction of cycle time at the end of every reversal cycle. (b) upward and downward moving wave of adsorbed NH_3. *(Figure adapted from* Agar and Ruppel (1988a) *with permission. © 1988 by VCH Verlagsgesellschaft MbH.)*

reaches the end of the bed and would break through, the flow direction in the figure changes from downward to upward. A small excess of the stoichiometric NH_3 for reaction leaves NH_3 adsorbed on the catalyst throughout the reactor. NH_3 is not added to the upflow stream. Nevertheless, NO_x in the now upflowing waste gas reacts with adsorbed NH_3 on contact. A stationary cycling state is reached after several cycles with a decontaminated waste stream containing a low level of NH_3. It can be seen from the sketch that a near-square wave of adsorbed NH_3 moves alternately upwards and downwards in the catalyst bed. The reactor in the figure contains a multi-channel block composed of metal oxides. A variation of the process used NH_3 injection in the middle of the reactor. Other explorations of reverse flow adsorptive trapping are summarized in **Table 18-12**.

Agar and Ruppel (1988a, b) simulated their reverse flow system using a one-dimensional, plug flow pseudo-homogeneous model that assumed NH_3 adsorption and the reaction with NO_x was kinetically controlled and that

TABLE 18-12 Publications on Reverse-Flow Adsorptive Reactors

Reference	Contribution
Agar and Ruppel (1988a, b)	Concept and modeling of a NO_x mass-trapping adsorptive reactor using periodic flow reversal and periodic interruption of the NH_3 reactant. Strongly adsorbed ammonia on a metal catalyst coupled with flow reversal reduces NH_3 loss.
Falle et al. (1994)	General theory of a chromatographic RFR for a process including adsorption of reactant followed by reaction of the adsorbed species with another reactant. The adsorbed reactant is injected in the middle of the catalyst bed. Analytical estimates of the cycle duration and the outlet concentration of adsorbed reactant were obtained. Efficiency can be increased in an RFR by a high adsorption capacity and short cycle duration.
Jeong and Luss (2003)	See *Falle et al. (1994)* above. Development of a methodology for finding the operating conditions that would lead to the minimal level of the weighted emission of both reactants. A continuous feed of strongly adsorbed reactant (e.g. NH_3) is superior to an operation in which the same amount is intermittently fed during each half-cycle.
Elsner et al. (2002a, 2003)	Operation of the Claus process using a mixed bed of alumina catalyst and a water adsorbent. Selection of adsorbent, mathematical modeling and proof-of-concept experiment. A double bed RFR was proposed for a continuous process combining reaction and adsorbent regeneration.
Grünewald and Agar (2004)	Use of thermal desorption of an inert gas during an exothermic reaction for temperature control. A mixed bed of catalyst and adsorbent used with reverse flow for adsorption of the inert in a partial cycle and for reaction and desorption in the remainder of the cycle. The concept was proved using water vapor adsorption and desorption and the exothermic oxidation of CO.
Lorkovic et al. (2004); Breed et al. (2005)	Concept of a reactor for the Br_2-mediated coupling of methane using injection of CH_4 in the center of the reactor and periodic reversal of air flow through two packed beds filled with catalyst and metal oxide. Hydrogen bromide formed in the central zone is trapped as metal bromide in the outlet packed bed. Molecular bromine is released from the inlet bed when regenerated back to the metal oxide by air.
Bos et al. (2007)	Adsorption and desorption of water in RF reactor boundary beds for improving the rate of VOC oxidation.

the reactor was isothermal. A similar model was used for the same system several years later by *Snyder and Subramaniam (1998)*. A heterogeneous model was applied to the system by *Salden and Eigenberger (2001)*, while *Jeong and Luss (2003)*, following *Falle et al. (1994)*, chose a two-dimensional model. Their interest was optimizating the NO_x capture. They observed that the reverse flow operation substantially reduced the use of NH_3 over that needed for a uni-directional operation. Since the Agar-Ruppel work was carried out by industry, it seems likely that, given the predicted performance, the mitigation process has been commercialized. However, there is no record of an application in the Engineering literature. *Aida and Silveston (2005)* discuss the simulation studies mentioned above in more detail.

18.8 BIOFILTERS

Biofilters employ living organisms to capture and destroy pollutants. They are three-phase reactors with a porous solid phase to support the organisms, usually as a slime coating, a trickling liquid phase that supplies nutrients to maintain the biomass slime, and a gas phase that is to be decontaminated. Different structures are used, ranging from randomly packed ceramic shapes to ordered monolithic surfaces with uniform flow pathways. Most slimes must be acclimatized to the pollutant to be removed. That pollutant is actually utilized by the biomass as food.

Biofilters have been used for off-gas treatment on a commercial scale for decades. Just recently, however, have they become possible applications for periodic flow reversal. The operation seems capable of improving the slime or biomass distribution in the bed and/or controlling the decontamination activity of the slime by affecting the growth phase or the slime condition. Improvement of biofilter performance under abrupt changes in pollutant concentration, so-called shock loading, has been reported also. **Table 18-13** summarizes several recent publications.

18.9 MISCELLANEOUS APPLICATIONS

Table 18-14 lists reverse flow publications that fit poorly in the various sections of this chapter. All involve one or more chemical reactions.

TABLE 18-13 Publications on the Use of Reverse-Flow Biofilters

Reference	Contribution
Song and Kinney (2000a, b); Park and Kinney (2001); Woertz et al. (2001)	Experimental study of a reverse-flow biofilter operating on air containing toluene. These authors found that $\tau_{cycle} = 6$ days achieved a more even biomass accumulation, higher biofilm activity and a lower pressure drop through the slime bed than a similar biofilter operating with unidirectional flow.
Wright et al. (2005); Wright (2005)	RF biofilters handled variations in VOC concentration and air flow rates much better than unidirectional units and provided as much as twice the removal rate. Filters used $\tau_{cycle} = 24$ h. Periodic flow reversal mimics the feast/famine of microbial processes described by *Irvine and Moe (2001)* that enhances the ability of a microbial community to adsorb and store VOC during a transient shock loading event.
Znad et al. (2007)	Experimental study showing that periodically alternating the flow direction with a period of several days improved the performance of a laboratory biofilter enabling its operation under VOC (toluene) inlet concentrations up to 13 g/m^3.

TABLE 18-14 Publications on the Use of Reverse Flow for Miscellaneous Reactions

Reference	Contribution
Seiler and Emig (1997, 1998, 1999)	Experimental study and numerical simulation of a steam-iron process using periodic reduction of iron oxide by CO or HCs and reoxidation by steam to generate high purity H_2.
Quinta Ferreira et al. (1999)	Modeling of a walled cooled tubular reactor for the partial oxidation of o-xylene. Reverse-flow improved reactor performance by lowering the operating temperature and marginally increasing selectivity to phthalic anhydride.
Neophytides and Tripakis (1996); Neophytides (1999); Paloukis and Neophytides (2007)	Numerical simulation of a solid oxide fuel cell operating with periodic flow reversals. Cell designs with co-current and cross-flow streams of fuel and oxidant configurations were considered. A co-current CH_4 fuel cell combining generation of electricity and synthesis gas was also studied. RF provides high cell efficiency at low fuel and oxidant temperatures in a compact system without a heat exchanger.
Zagoruiko and Matros (2002)	Numerical simulation of the catalytic Claus H_2S process for producing sulfur allowing for periodic sulfur condensation. An RFR increased conversion compared with a steady-state operation in a single bed reactor. A double bed RFR including sulfur removal between the beds provided higher conversion compared with its steady-state equivalent. A brief description of a pilot unit operated at a gas flow rate of 5000 m^3(NTP)/h is contained in the publication.
Hung et al. (2003)	Conceptual study of an RFR application to liquid phase processes with a homogeneous catalyst adsorbed onto a solid support. Reverse flow operation enabled catalyst immobilization without requiring strong bonding between catalyst and support that lowers catalyst solubility and thus activity.
Logist et al. (2007)	Demonstrated that wall-cooled catalytic reactors can approach the optimal trapezoidal temperature profile for endothermic reactions using periodic flow reversal.
Marín et al. (2009c)	Modeling and numerical simulation of the water gas shift reaction over an iron chromium catalyst on pellets or monolith support. Use of an RFR consisting of a single catalyst bed has a lower conversion than a uni-direction bed reactor. Adding inert materials at the ends of the bed improved RFR conversion.

However, these reactions and the reverse flow application are listed for purposes of thoroughness and will not be discussed further.

18.10 COMMERCIAL APPLICATIONS

Periodic flow reversal is the only periodic operation to date that has seen wide commercial adoption with perhaps the exception of several bioprocesses discussed in **Chapter 10**. Concepts and technology have been discussed earlier in this chapter. Commercial design and operation are the focus of this section.

18.10.1 VOC Mitigation

There are thousands of reverse flow VOC incinerators in operation globally. Both types of systems described in **Section 18.3**, regenerative thermal oxidizers (RTO) and regenerative catalytic oxidizers (RCO), are in use. The basic design is given by **Figure 18-3**. The central burner is necessary in catalytic units when the ΔT_{ad} of the waste stream is less than ~30°C and for start-up regardless of the heating value of the stream. *Lewandowski (2000)* provides a good overview of RTO technology evolution, mechanical design and operation. Catalytic designs place beds of catalyst on top of each of the heat exchange beds seen in the figure. Bed depth depends primarily on the catalyst employed, but apart from the catalyst, the RFR design is quite close to that of the thermal units. Catalysts are generally metal oxides, such as manganese, chromium or copper, and platinum/palladium noble-metal catalysts supported on alumina. Of course, the thermal units operate at a considerably higher temperature, up to 1000°C, whereas the catalytic units run between 280 and 500°C. Cycle periods are generally 60 to 180 s.

RFR control is relatively simple because of the high heat capacitance of the recuperative beds. A sensor installed in the catalyst bed or combustion chamber above the bed signals if dropping temperatures require use of the burner and auxiliary fuel, while high temperature initiates bypassing of the recuperator beds. Design and operation of these units are discussed by *Bunimovich and Matros (2010a)*. Examples of installations are given.

18.10.2 Sulfuric Acid Plants

The sulfuric acid reversed flow reactor, developed in the Boreskov Institute of Catalysis at the end of the 1970s, was enthusiastically promoted so that large pilot and commercial scale units were built uncharacteristically fast for the Soviet chemical industry. During 1979–1985 the process was tested by BIC and the Moscow Fertilizer Institute at a pilot plant operated at Voskresensk, Russia. Initially a single 2.8 m diameter reactor was tested; then two reactors were combined to build a miniature double-contact—double absorption plant. The first commercial reactor was built in 1982 in the Krasnouralsk copper-melting plant in the Urals. A single bed reactor, diameter 7 m, treated about 30,000 Nm^3/hr of sulfurous gases leaving the copper smelters with SO_2 concentrations between 1 and 2.5%. *Kunitski et al. (1984)* and *Matros (1989)* provided detailed information about the unit start-up and operation. Matros and Bunimovich (1996) provided summary of operations of other commercial units.

By the end of the 1980s, the technology was licensed to Bulgarian, Chinese and Japanese companies. The operation of a triple bed converter designed by the Boreskov Catalysis Institute for a Bulgarian Plovdiv lead-zinc smelter was described by *Grozev et al. (1994)*. The reactor operated at a gas flow rate of 20,000–45,000 m^3(NTP)/h with a 1.4–4.8 vol% inlet SO_2. A similar flow design was used for a Chinese reactor operated at the Shenyang copper smelter for several years. UNILAB, in Shanghai, designed another converter built at the Jiyuan lead smelter (Henan Province, China) in 1987. As per *Xiao et al.* (1999a), this 6.5 m diameter converter treated 33,500 m^3/h and included two additional beds of ceramic pellets besides the three beds used in the previous design. The ceramic beds served solely to reduce the amplitude of the outlet gas temperature oscillation. Buffering the outlet temperature avoided cooling the switching valves and so reduced risks of corrosion and acid mist formation. The same result was achieved in some Russian reversed flow converters operated with a small external heat exchanger. Another distinctive feature of the UNILAB design was on-off control for the intermediate heat-exchangers instead of gradual modulation used earlier. The converter performance at an average SO_2 concentration of 3% reached 97% but was reduced later to 93.5% due to the effect of catalyst deactivation arising from CO and hydrocarbon contamination according to *Xiao et al. (1999d)*. The authors also reported a gas channeling phenomenon near the reactor walls that affected the overall performance during start up but was later corrected by a special distributor.

After 10 years of successful operation, the RFR at the Plovdiv lead-zinc smelter was incorporated into the first stage (R1) of a double conversion (DC)—double absorption (DA) acid plant operated at the same facility. A schematic flow diagram is shown in **Figure 18-18**. The added second stage (R2) also used flow reversal. Partially converted SO_2, after the first stage (R1) and heat exchanger (HE 4), enters the intermediate absorber (A1) where the SO_3 is recovered, and proceeds to the second RFR (R2) for further conversion and recovery (A2). The DC/DA system offered stable autothermal operation even in the presence of inlet SO_2 fluctuations. The final SO_2 conversion was over 99% and SO_2 tail gas emissions were below 600 ppm (*Dimitrov et al., 2003*).

18.10.3 Other Commercial Applications

Pilot reversed flow reactors for reduction of NO_X with ammonia were built during the 1980s in the Boreskov Institute of Catalysis. A commercial unit for 10,000 m^3/h of tail gas leaving NO_X absorbers was put into operation as early as 1985 (*Bobrova et al. (1988); Noskov et al. (1993)*). Recently Babcock Power, Inc. developed commercial reverse flow reactor technology for NO_X removal for boilers fired by wood chip and tree biomass. Four or five commercial units have been successfully built and operated at a maximum unit power of 54 MW that corresponds to about 160,000 m^3/h (NTP) of treated gas.

Based on the available literature, the thermal reversed flow reactor is the most competitive technology for mitigation of methane in coal mine ventilation air. US EPA (*Somers and Schultz, 2008*) lists several companies, RTO and RCO equipment manufacturers, which have been involved in the development of thermal reversed flow reactors for methane utilization. In 2007, the first large-scale commercial unit using the technology designed by MEGTEC Systems, De Pere, Wisconsin was put into operation at West Cliff Colliery, Australia following

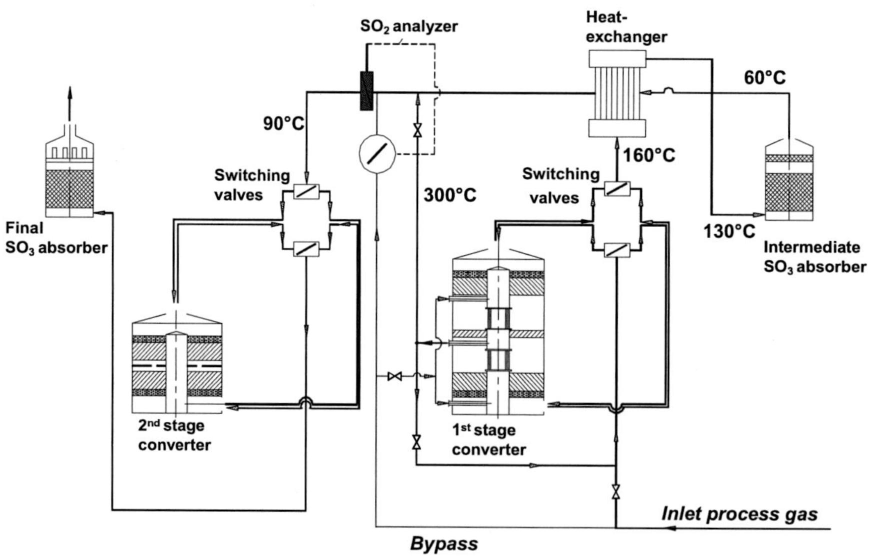

FIGURE 18-18 **Flow diagram of a DC/DA sulphuric acid plant using periodic flow reversal built in Plovdiv, Bulgaria in 2000. The plant handled 45,000–50,000 m^3(NTP)/h with a 1.2–6.5 vol% SO_2 content.** *(Figure adapted with permission of* Dimitrov et al. (2003).*)*

testing of a pilot unit at the same mine site. There is not very much detail available about the system operation except that it continuously treats about 20% of the mine VAM at a total flow rate of 250,000 m^3/h, simultaneously generating 6 MW of electricity or about 25% from the total heating value of the inflow methane. The concentration of methane is controlled at 0.9 vol% via adding a highly concentrated stream of mine drainage methane available at the site. There are several other large demonstration units operating in the US and China. The transition from demonstration phase to widespread deployment is bottle-necked by economic and regulatory factors noted by *Somers and Schultz (2008)*. Currently available oxidizers offer appreciable return on investments at inflow VAM concentration 0.9–1%. Usually ventilation air has lower concentrations and the system needs addition of more concentrated methane (for example from drainage gas), which can affect the economics. At current development of the European carbon market, carbon credit prices at less than $10/ton of CO_2 limit the economically feasible VAM treatment to concentrations higher than 0.6%. In addition to market constraints the technology needs to secure approval of mine safety regulators in leading coal mining countries.

Catalytic decomposition of N_2O is another commercial application of the RFR directed to the abatement of greenhouse gas emissions. The technology designed by Japanese company Sumiko Eco-Engineering makes use of the high thermal efficiency of the RFR for the treatment of low concentrated and low temperature N_2O gases emitted from caprolactam and other chemical plants. From available sources the system comprises a double-canister reactor resembling a RCO where the palladium catalyst supported over alumina-magnesia is loaded above the bed of heat-exchange ceramics. The catalyst operates at 550°C whereas the inlet temperature of the process gas is close to ambient. Perhaps three or four reversed flow N_2O decomposition reactors built by Sumiko operate in caprolactam and fertilizers production plants in South East Asia.

Nomenclature

A, B, C	=	reacting species, usually A, B are reactants
A/F	=	air fuel ratio
a_v	=	volumetric heat transfer surface area (cm^{-1})
C_g	=	concentration in the gas phase (mol/cm^3)
C_{in}	=	entrance concentration
C_s	=	concentration on the catalyst surface
$C_{\rightarrow}$	=	concentration in the direction of increasing z
$C_{\leftarrow}$	=	concentration in the direction of decreasing z
C_g°, C_s°	=	initial concentrations in gas and on the catalyst surface
CFD	=	computational fluid dynamics
c_g	=	heat capacity of the gas phase (cal/g·°C)
c_s	=	heat capacity of the solid phase
$D_{eff,I}$	=	effective diffusivity for reactant "i"
DC/DA	=	double conversion-double adsorption (sulfuric acid plant)
d_p	=	particle or pellet diameter (m)
d_R	=	reactor diameter or characteristic dimension of the external heat transfer surface (cm)
E	=	activation energy (cal/mol)
$f(C_{\rightarrow},T)$	=	rate of reaction in the direction of increasing z
HC	=	hydrocarbon
HC-SCR	=	hydrocarbon selective catalytic reduction
HFSM	=	high frequency switching model
h	=	fluid-particle heat transfer coefficient (cal/cm^2 °C s)
h_w	=	fluid or bed heat transfer coefficient at the external heat exchange surface
i.d.	=	inner diameter
k, k_{∞}	=	kinetic rate constant
k_m	=	fluid-solid mass transfer coefficient (cm/s)
L	=	bed or reactor length (m)
LCL	=	lean combustibility limit
LHS	=	left hand side (of figure)
LQR	=	linearized quadratic regulation
MeOH	=	methanol
MPC	=	model predictive control
NTP	=	normal or standard temperature and pressure (0°C, 1 bar)
n	=	cycle count (1, 2, 3, etc.)
o.d.	=	outer diameter
PID	=	proportional-integral-derivative (control mode)
PSA	=	pressure swing adsorber
Q_0	=	volumetric flow rate at reactor entrance
q	=	number of independent reactions
R	=	gas constant (various units)

RF = reverse flow
RFC = reverse flow converter or catalytic converter
RFR = reverse flow reactor
RHS = right hand side (of figure)
RTO = regenerative thermal oxidizer
r = rate of reaction (mol/s cm^3)
$r(C_s,T)$ = rate of reaction for the surface concentration of the key reactant at temperature T
r_{endo} = rate of endothermic reaction
r_{exo} = rate of exothermic reaction
SCR = selective catalytic reduction
SMBR = simulated moving bed reactor
$sgn(t)$ = flow direction (takes on + or − values depending on time t)
T = temperature (°C, K)
T_F = maximum temperature in stationary front
T_g = gas phase temperature (°C)
T_g° = initial gas temperature
T_{in} = entrance temperature
T_{max} = maximum temperature in reactor or catalyst bed
T_s = temperature of the solid (°C)
T_w = wall temperature
$\overline{T}_{out}$ = time average reactor exit temperature
t = time (s, min, h)
t_c = switching duration or time
U = fluid superficial velocity (cm/s)
U_F = thermal front velocity
$u(t)$ = superficial axial velocity (cm/s)
VOC = volatile organic compound
X, x = conversion (-)
$\overline{X}$ = time average conversion
$\overline{x}_{out}$ = time average conversion at the reactor exit
x_{tot} = total conversion
Z = reactor exit, reactor length
z = axial coordinate, distance (cm)
1-D = one-dimensional

Greek

Δ = reactor cooling capacity
ΔH_R = heat of reaction (cal/mol)
ΔT_{ad} = adiabatic temperature rise (K, °C)
δ_{max} = maximum fractional heat recovery
ε_b = bed void fraction (-)
ε_p = pellet or particle porosity
γ = ratio of solid and gas heat capacities
κ = flow direction variable, takes on either 1 or −1
λ_{eff} = reactor effective thermal conductivity (see **Eq. (18-15)**)
$\lambda_{g,\ eff}$ = effective thermal conductivity of the gas phase (cal/s cm^3)
$\lambda_{s,\ eff}$ = effective thermal conductivity of the solid phase (cal/s cm^3)
ρ_g = gas density (g/m^3)
ρ_{g0} = gas density of the feed stream
ρ_s = solid phase density (kg/m^3)
τ_c = residence time in catalyst bed
τ_{cycle} = cycle period (s, min, h)
τ_f = time between flow reversal
υ = temperature or concentration profile solution

Subscripts

ad = adiabatic
b = bed
c = cycle, contact
eff = effective
endo = endothermic
exo = exothermic
g = gas
i = species indicator
in = reactor inlet
out = reactor outlet
p = particle
R = reaction, reactor
s = solid, surface

Superscripts

j = iteration counter
0 = initial or entrance

CHAPTER

19

Control of Periodically Operated Reactors

Hector Budman, Peter Lewis Silveston
Waterloo, Ontario, Canada

OUTLINE

19.1 Formulation of an Optimal Control Problem for a Periodically Operated Reactor 545
19.2 Chattering Controls 546
19.3 Controls for Stirred Slurry and Fluidized Bed Reactors 548
19.4 Controls for Packed Bed Reactors 553
19.5 Control of Packed Bed Reactors with Flow-Direction Switching 553
19.6 Control of Simulated Moving Bed Chromatographic Reactors 562
19.7 Other Control Schemes for Periodically Operated Reactors 564
19.8 Comments and Research Needs 565

Just as for reactors operating at steady state, the purpose of control for periodically operated reactors is to minimize the effect of random disturbances on product composition and yield or reactant conversion. Nevertheless, the time-varying nature of variables under periodic operation poses additional challenges. As a result, simple feedback control employing feedback PID regulators is rarely adequate because of problems with tracking a time-varying output, sensing and controlling system lags and various nonlinear effects.

The literature on the control of periodically operated reactors is not very large. In recent years, model predictive control studies have been conducted especially for reactors operated under periodic flow-reversal.

As illustrated in earlier chapters, input variables in periodic reactor operation, such as reactant concentration, vessel temperature or feed flow rate are manipulated to follow a periodic function of specific frequency and amplitude. For the most part, just a single variable, such as a reactant flow rate is periodically manipulated and a single variable, typically conversion, has been controlled. These are single input-single output (SISO) systems. The variable that is switched periodically is referred to as the manipulated variable (MV) which is also the term used for variables manipulated in closed loop feedback control. The concern in this review is with the control of the class of periodically operated reactors mentioned above. The class contains several different types of reactors as suggested in **Table 19-1**. Batch reactors, of course, are periodically operated but their periodicity is trivial and will not be discussed. Only the first five types in the table will be considered, since the remaining three types have not been tackled as yet by process control researchers.

The discovery that the performance of a packed bed reactor may be improved by intentional periodic manipulation of an input variable came out of the process control literature (e.g., *Horn and Lin, 1967*). Hence, it is surprising that the control of periodically operated reactors has only attracted significant research in just the last decade. Nevertheless, there are some early studies by pioneers of periodic operation that did explore the control aspects of this operating mode, recognizing that a periodically changing input results in a time-varying output and thus the controller has to be able to cope with such time-varying behavior (*Liaw and Bailey, 1974*).

Several early studies on periodically operated reactors approached the problem as an open loop control strategy (*Horn and Lin, 1967; Douglas, 1967*) and it can indeed be viewed as such. In what follows, mainly control strategies that are designed to counteract or minimize the effect of incoming disturbances using closed loop configurations will be considered. Nevertheless, the use of open-loop periodic forcing inputs to suppress the effect of zero mean disturbances or to stabilize an operation around an unstable steady state or an unstable limit cycle will be discussed briefly. This mode of periodic operation is often referred to as chattering or vibrational control.

In examining control strategies, discussion will be limited to disturbances in uncontrolled and/or unmeasured variables. The objective is to describe the types of control proposed and the test results when those controls have been implemented. Details will be found in the references given. These will contain the mathematical formulation of the control strategy and set out advantages and disadvantages. An elementary understanding of feedback control

TABLE 19-1 Periodically Operated Reactors

Type of Reactor	Variable(s) Manipulated
Catalytic Packed Beds	Feed Composition Feed Temperature
Stirred Tank Reactors	Coolant or Heating Fluid Temperature
Stirred Polymerization Reactor	Initiator Flow Rate or Radiation Intensity, Feed or Coolant Flow Rate
Catalytic Trickle Beds	Liquid Flow Rate (On-off)
Catalytic Packed Beds (Heat Recapture)	Flow Direction
Simulated Moving Bed Chromatographic Reactors	Location of Feed Point and Product Removal Points
Pressure Swing Adsorptive Reactors	Pressure, Flow Direction, and Feed Composition
Temperature Swing Adsorptive Reactors	Temperature, Flow Direction, and Feed Composition

is assumed in what follows. At the end of the chapter, topics that warrant further study will be highlighted.

19.1 FORMULATION OF AN OPTIMAL CONTROL PROBLEM FOR A PERIODICALLY OPERATED REACTOR

The control objective for a periodically operated reactor is to manipulate a reactor input variable u (t) based on feedback information about a reactor output variable $y(t)$ so that the temporal mean of the output, $\overline{y}$ remains close to some desired value or so any instantaneous value of y is maintained under some specified limit. In some circumstances, however, the objective may be to maximize conversion or selectivity. The formulation of an optimal control strategy usually involves the minimization of a cost function, $J()$. Maximizing a quantity or maintaining an output below some limiting value can often be transformed into a minimization problem. Thus, optimal control strategies are developed according to the following general minimization problem:

$$\min J\,(x(0),\ u(\bullet),\ \tau) \tag{19-1}$$

where $J(x(0),\ u(\bullet),\ \tau)$ depends on the initial conditions. $x(0)$, the control strategy $u(\bullet)$ and the cycle period, τ, as:

$$J(x(0),\ u(\bullet),\ \tau) = \frac{1}{\tau}\int_0^{\tau} g\,(y(t))dt, \tag{19-2}$$

$y(t)$ is the reactor output and is obtained from the solution of an ODE if the reactor can be described by a lumped parameter model or from the solution of a PDE if it is a distributed parameter system where the general nonlinear dynamic model can be stated as:

$$\begin{aligned}\frac{\partial x}{\partial t} &= f(x(t),\ d(t),\ u(t), z)\\ y &= h(x(t), d(t), u(t), z).\end{aligned} \tag{19-3}$$

In **Eq. (19-3)**, x is the time and position dependent vector of state variables, z represents the axial position in the reactor, if radial gradients are neglected, $d(t)$ are disturbances and $u(t)$ is the time-varying manipulated variable. For a lumped parameter system the equations in **(19-3)** will be independent of z. In a dynamic distributed model of the reactor, $d(t)$ and $u(t)$ may appear in the boundary conditions and not explicitly in the differential equation model. Because the system is periodic, a required condition for the undisturbed system is that:

$$x(\tau) = x(0). \tag{19-4}$$

The problem in **Eq. (19-1)** can be restated by using **Eq. (19-3)** in terms of states as an optimal state feedback control where the deviations in all the states with respect to a reference are used to calculate the following cost function:

$$J(x(0),\ u(\bullet),\ \tau) = \frac{1}{\tau}\int_0^{\tau} L(x(t),\ u(t))dt. \tag{19-5}$$

In general, y(t), x(t), etc. are vector quantities that may become scalar for specific simple cases. $J(x(0),\ u(\bullet),\ \tau)$ is usually scalar.

Minimization of the objective function leads to the determination of the control strategy u(•). Because disturbances are random in magnitude, duration and incidence, in some optimal control strategies the minimization stated in **Eq. (19-5)** is recalculated over time. This is the case for predictive controllers and it has been applied to the control of periodically operated reactors for which the optimization is redone at each time interval.

Several assumptions must be made for solving the optimal control problem stated in **Eq. (19-5)**. First, the functions g, f and L exist as do their partial derivatives up to second order in x and u. The existence of the derivatives is needed because often the solution of the optimal control problem relies on linearization of the original nonlinear model. Also, the functions are autonomous and the period τ is such that $x(t) \neq x(0)\ \forall\ t \in (0, \tau)$. The control strategy, u(•), must be piece-wise continuous in the (0, τ) interval. For an optimal solution, a time-varying Lagrange multiplier must exist and there must be a real, symmetric periodic solution to a Riccati differential equation.

Constraints on manipulated variables and/or controlled variables are frequently encountered. These constraints can be tackled either as soft or hard constraints. In the soft-constraints approach, a deviation of the variable from its constraint value is minimized by including this deviation, multiplied by a large weight, in the cost function $J(x(0), u(\bullet), \tau)$ to be minimized. When a hard-constraint approach is chosen, a constrained optimization, based on the use of Lagrange multipliers for linear systems, must be used.

19.2 CHATTERING CONTROLS

This type of control, also referred to as vibrational control, has a history as long as that of reactor periodic operation (*Fjeld, 1972, 1974; Meerkov, 1973, 1980*). It is an open-loop technique, since measurement of a controlled variable for the purpose of feedback is not needed. The technique was motivated by the observation that by periodically modulating an input variable a steady-state reactor system will be forced into a limit cycle with the input's periodicity. Even an autonomously oscillating system will be forced into the forcing input's periodicity by chattering control, although peculiar phenomena such as period doubling and resonant horns may occur. Under this type of control, the resulting time-average state of the reactor may differ significantly from the steady state of the unforced (and nonlinear) reactor. Because of process inertia, as the forcing frequency increases, the space in the state plane occupied by the limit cycle progressively decreases until becoming a point which is referred to as the relaxed steady state. Thus, with increasing forcing frequency, the average state approaches the relaxed steady state.

The rationale behind the use of chattering control for some operating situations can be best explained by **Figure 19-1**. The figure shows the familiar "S" shaped locus of steady states for a CSTR defined in terms of feed temperature values for different residence times where the latter are determined by the feed flow rates. Three possible residence times are shown as vertical lines. Different steady states are defined by the intersections of these vertical lines with the curve. At short residence times, the intersection occurs on the lower branch and low conversion occurs corresponding to a low temperature

FIGURE 19-1 Temperature-residence time plane for an equilibrium-limited, first-order, exothermic reaction occurring in an adiabatic CSTR.

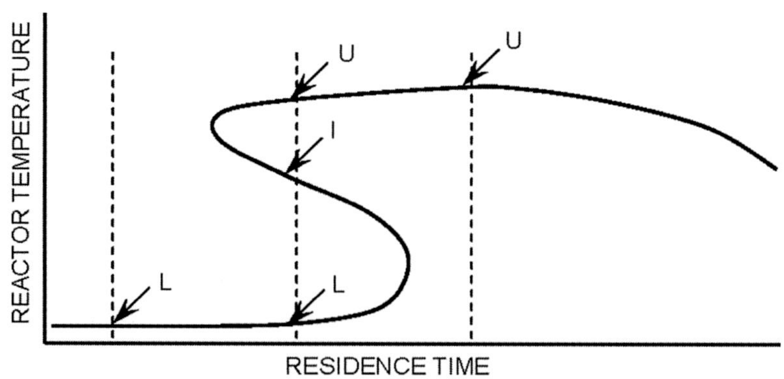

value, assuming an exothermic reaction. On the other hand, at long residence times, the conversion is very high as indicated by the corresponding high temperature at steady state. Between the two cusps of the curve, three intersections are possible. Although stability around a particular steady state should be formally assessed by analyzing the eigenvalues of the linearized model of the system around that state (*Shacham et al., 1994*), as a first approximation it is generally possible to predict stability from the sign of the slope of the curve. Where the curve has a positive slope at the intersection, the state, in this case the reactor temperature, is generally conditionally stable while, when the slope is negative, the state is generally unstable. Then, when the system starts from the conditionally stable point, even an infinitesimal disturbance will draw the reactor towards either one of the stable states, depending on the disturbance direction. The states corresponding to the different intersections between the steady-state locus and the vertical lines are labeled in the diagram as L, I, U, for lower, intermediate and upper, respectively. A stability analysis employing a dynamic model of the reactor linearized around an operating point will generally verify the previous description.

Modulating an input variable, such as feed flow rate around its steady-state value forces the CSTR output, namely the reactor's temperature and its corresponding conversion, into a limit cycle. The time-averages of the limit cycle outputs differ from those of the unforced system at steady state. At a chosen forcing frequency, the plot of the average temperature, as a function of residence time, will still appear as an "S" shaped curve; however, the upper branch is shifted downward to lower temperatures while the lower branch is shifted to higher temperatures resulting in a shrinking of the multiplicity region. The latter can be eliminated altogether through a proper choice of frequency and amplitude (*Çinar et al., 1987a, b, c*). Furthermore, stable average steady states can be shown to progressively converge, for increasing frequencies of the forcing, toward the unstable intermediate state of the unforced steady-state CSTR. Accordingly, chattering control can cause the CSTR to operate on average at the unstable steady-state value of the original unforced system. Thus, this may be a possible mode of operation in situations where the desired conversion value corresponds to the intermediate steady state (I).

As a consequence of the behavior of a system under chattering control, two applications of this controller have been reported: 1) Stabilization of a desirable, but unstable operating state for a reactor and 2) extending the stability of a reactor operating at steady state near a region of steady-state multiplicity. The first problem was explored by *Bellman et al. (1983)*, *Çinar et al. (1987a, b)* and *Shu et al. (1989)* using simulation. Experiments have been also reported by *Çinar et al. (1987a)* and by *Shu et al. (1989)*. In the latter, the effect of simultaneous forcing of feed flow rate, inlet concentration and cooling temperature was investigated for a CSTR. The phase shift between the different forcing inputs was found to be of paramount importance to determine whether the periodic operation enhanced the performance. An interesting extension of the stabilization property of vibrational controllers involves the avoidance of chaotic behavior by forcing a periodic input. The use of small periodic perturbations to force systems with chaotic dynamic behavior to a desired periodic state was initially examined by *Shinbrot et al. (1993)* for general nonlinear dynamic systems. This particular stabilization method is explained by a principle referred to as the "butterfly effect." The idea is that, in a nonlinear system, there may be an infinite number of unstable limit cycles coexisting with chaotic motion. Small periodic perturbation in a particular process parameter may help to stabilize the system around one particular limit cycle which would be unstable in the absence of those perturbations. This stabilization principle has been recently applied to chemical reactors in two numerical studies (*Ajbar, 2001; Cordoba et al., 2006*). For example, using a Cellular-Automata based model, *Cordoba* found that by forcing the system at one of its natural frequencies, it is possible to drive the process from an unwanted chaotic behavior to a preferred periodic operation. Neither industrial nor experimental applications of this stabilization method have been reported. Of course, an unstable state in a CSTR can also be stabilized by a suitably designed feedback controller (*Bruns and Bailey, 1977; Sterman and Ydstie, 1990a*) but this type of stabilization is not of interest for discussing open loop vibrational control.

The second application that involves extending the stability of a steady-state operating point is motivated by the observation that the S curve shown in **Figure 19-1** changes when chattering control is applied. Typically, a disturbance that temporarily brings the CSTR deep enough into the multiplicity region destabilizes the reactor which will not return to the original state once the disturbance is removed. Chattering control, as discussed above, distorts the "S" shaped curve representing the time-average behavior by shrinking the multiplicity region. Thus, with proper choice of forcing frequency and amplitude, chattering control can be used to reject random disturbances since the system will be maintained within the bounds of the reduced multiplicity region. Evidence for this disturbance attenuation property can be found in vibrational control studies that consider simultaneous forcing of two variables mentioned above (*Shu et al. 1989*) where one variable could be viewed as the one that is purposefully forced whereas the second one could be viewed an unmeasured zero mean disturbance entering the process. On the other hand, this control is ineffective for step-disturbances or long-pulse disturbances that will alter the final steady state. It is, thus, a controller that provides robustness in the face of zero mean disturbances only. Because of this limitation, the application of this open loop control technique to industrial reactors seems not to have been pursued. To ensure the success of open-loop chattering control based methodologies in the presence of unmeasured disturbances, they must be implemented in conjunction with conventional feedback controllers. Some of the benefits of implementing such combined control strategies have been previously identified (*Liaw and Bailey, 1974; Çinar et al., 1987a, b, c*). For example, *Çinar et al.* used the combination to keep feedback control actions away from saturation limits. Later, the same research group proposed using nonlinear predictive control together with forced periodic fluctuations of feed flow rate and reactant inlet concentration to maximize product yield for ethylene oxidation (*Özgülsen et al., 1993*). They demonstrated through simulation that this strategy maintained improvements in reactor performance in the face of parametric uncertainties and unmeasured disturbances.

19.3 CONTROLS FOR STIRRED SLURRY AND FLUIDIZED BED REACTORS

Well-mixed reactors are frequently chosen for control studies because as lumped parameter systems their dynamics may be described using ordinary differential equations (ODEs). Accordingly, their open- and closed-loop behavior can be described by the responses of a small set of manipulated and controlled variables.

Multivariable proportional and proportional integral feedback controllers with time-varying tuning parameters were initially proposed by *Liaw and Bailey (1974)* for a cooled CSTR, running exothermic parallel reactions. As mentioned earlier, *Liaw and Bailey,* however, combined those conventional control strategies with vibrational controllers. They observed that in the absence of disturbances to the input variables, open loop, square wave modulation of the heat removal rate, with period τ, increased reaction selectivity. Since *Liaw and Bailey* assume that all external disturbances are undesirable, conventional P or PI controllers were chosen to reject these disturbances and maintain the reference process behavior corresponding to periodic heat removal. Researchers examining steady-state processes often assume that any disturbances are undesirable and should be rejected by the control system. However, recognizing the benefits of forcing periodic deviations in certain variables, Liaw and Bailey suggested that modulation should be imposed only on those variables that can improve reactor performance, namely the feed flow rate or the heat flux to the coolant in their system.

For the design of optimal linear control strategies, it can be assumed for relatively small disturbances that the state equations describing the reactor can be linearized around the desired operating state. This allows standard procedures for linear systems to be used for controller design. Optimal tuning parameters of the multivariable controllers were obtained by *Liaw and Bailey* by minimizing a quadratic performance index containing a term penalizing state deviations for the P controllers or both state deviations and the derivatives of control actions for the PI controller. Following this, a periodic, symmetric K matrix arising from the solution to a Riccati equation was calculated. This provides the control strategy $u(\bullet)$ or $u(t)$ schematically represented by the block diagram in **Figure 19-2**. A and B are the state space matrices of the linearized model multiplying the state and control action vectors respectively where the prime on B represents its transpose. As shown in the figure, the regulator for the periodic CSTR is a multivariable proportional controller with time dependent gains resulting from the periodic nature of the solution.

To test the efficacy of their design, *Liaw and Bailey (1974)* examined the resulting simulated closed loop responses for sinusoidal, random pulse and step-disturbances. They observed that the latter provided the severest test of controller performance. To further explore the closed loop performance in extreme situations, they used a simultaneous 10% step-down in feed concentration and feed temperature, assuming the manipulated variable was a change in either the heat removal rate, with respect to a pre-specified periodic trajectory, or the feed rate to the reactor. Heat removal rate, as the modulated variable, maintained selectivity much better than the manipulation of flow rate to the reactor. In that case, the flow rate was maintained as a constant. Without control, the step changes drove the selectivity towards zero.

For comparison purposes, *Liaw and Bailey* also examined conventional time-invariant proportional control, i.e., the K matrix was

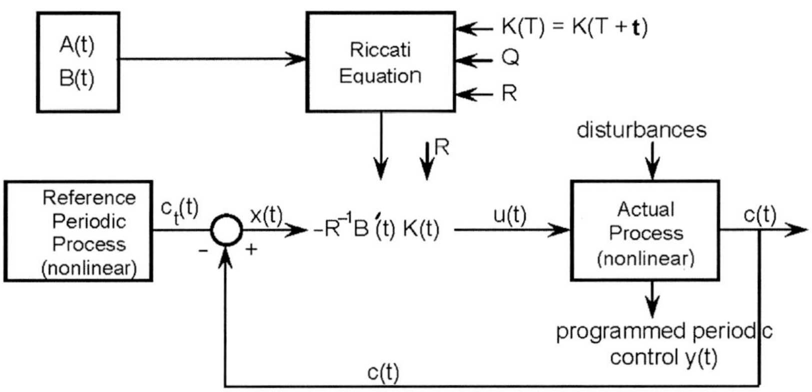

FIGURE 19-2 Optimal PI controller for a periodically operated CSTR Note that *c(t)* is the state variable (vector) *x(t)* in **Eqs (19-3)** and **(19-5)**. *x(t)* in the figure is a deviation variable (vector). *(Figure reproduced from Liaw and Bailey (1974) with permission. Copyright 1974 by the AIChE.)*

held constant in **Figure 19-2**. Curiously, for the A matrix of coefficients used in their study, there was just a small loss in control performance. This is an important, recurring observation in this and subsequent studies undertaken in control of periodic process, i.e., simple control strategies provide in many cases similar closed loop performance as the performance obtained with more sophisticated control strategies. **Figure 19-3** shows the variation in the heat removal rate during the first cycle after the step change when this rate is used as the manipulated variable. The thin line shows the square-wave pulse in heat flux used as the reference process, while the triangular symbols trace the heat flux evolution for a multivariable PI controller and the heavy solid line shows the evolution for a multivariable P controller. The step change causes a distortion of the control variable in the first cycle. The original square-wave modulation will be gradually restored in subsequent cycles, but the mean value of the wave will change to counteract the effect of the step disturbance. The P and PI controllers resulted in similar controller performance in terms of controlled variable errors but the P controller resulted in larger fluctuations in the manipulated variable than the PI device as shown in **Figure 19-3**.

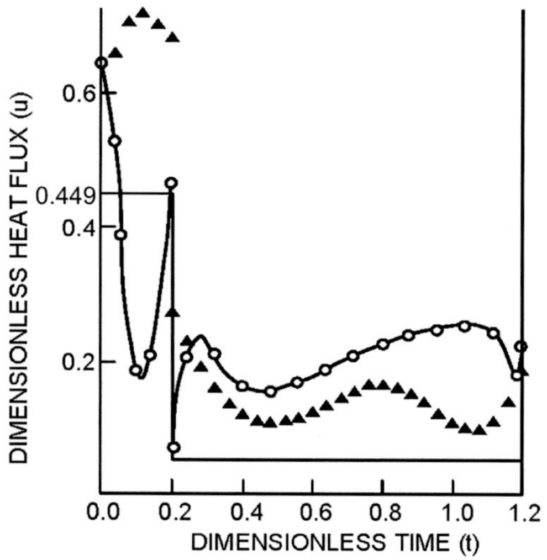

FIGURE 19-3 **Variation of the heat removal rate with time in the first cycle after a 10% step-down in feed concentration and temperature in a simulation of the controlled periodically operated reactor. ——— = prior to step-change (reference process), o—o—o = P control, ▲▲▲▲ = PI control after a step-change.** *(Figure reproduced from* Liaw and Bailey (1974) *with permission. © 1974 by the AIChE.)*

In the decades following the work of *Liaw and Bailey,* the use of model-based control for chemical systems expanded rapidly (viz. *Bequette (1991)*). For example, Model-Predictive Control or MPC is a type of model-based control algorithm that is widely used in industry for multivariable systems with constraints (*Camacho and Bordons, 1999; Maciejowski, 2001*). MPC calculates optimal control moves based on predictions of the controlled variable along a time horizon obtained using a model of the process. An advantage of the MPC algorithm is that it accounts for constraints in the solution of the optimization problem. Both linear and nonlinear models have been used for prediction where the corresponding algorithms are referred to as linear or nonlinear MPC (NMPC). Adaptive versions of MPC exist where model parameters are tuned online based on differences between model predictions and measured values of the manipulated and controlled variables. The nonlinear version is appropriate for processes displaying significant nonlinearity whereas the adaptive version is suitable for either time-varying processes or for nonlinear processes that can be efficiently represented by time-varying models. *Özgülsen et al. (1993)* proposed nonlinear model-predictive control (NMPC) to regulate yield for a single reaction carried out in a jacket-cooled CSTR. *Özgülsen et al.* considered the set point y_{sp} as a mean value of the periodic reference and used coolant flow rate as the manipulated variable. Their strategy is shown schematically in **Figure 19-4**.

Özgülsen et al. use a multiple of the period, assumed to be constant, as the prediction

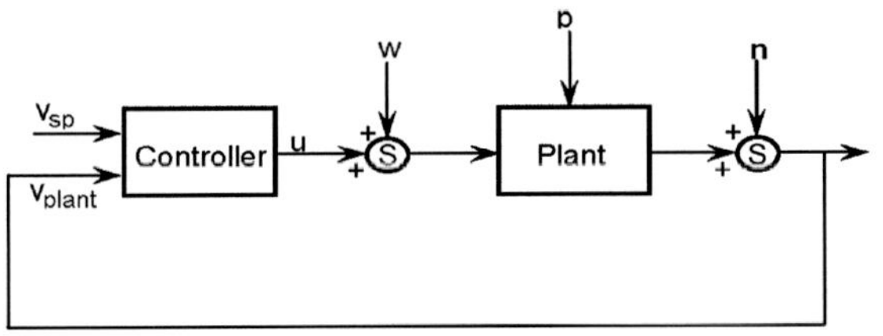

FIGURE 19-4 Closed loop control system considered where p is a vector of the modulated variable which forces a periodic trajectory, w is an error introduced in the control action while v is an error in the measured variable y_{plant}, u is the manipulated variable and n is measurement error and/or output disturbance. *(Figure reproduced from Ozgulsen et al. (1993) with permission. © 1993 by the AIChE.)*

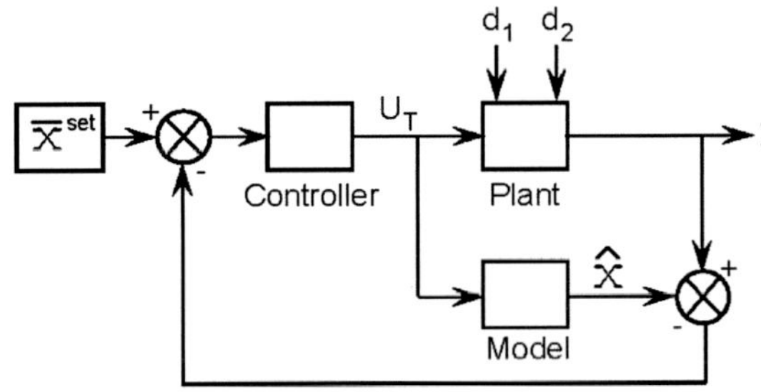

FIGURE 19-5 Operation of internal model control in the NMPC procedure. $\overline{X}^{set}$, X, $\widehat{X}$ are controlled variables set-points, measurements and predictions respectively. *(Figure reproduced from* Ravi Kumar et al. (1991) *with permission. © 1991 by The Royal Society of Chemistry.)*

horizon to predict the output y, using values from the end of the previous step and the disturbance. w and n are unmeasured input and output disturbances that may enter the process but their values are not needed for the control action calculation. This calculation through the predictive control algorithm involves the solution of a nonlinear programming problem using a suitable optimization routine. As in standard model-predictive control, the control actions are kept constant between the end of the control horizon and the end of the prediction horizon. Since the disturbances entering the system cannot be measured, the disturbance vector is estimated at each sampling interval as the difference between the plant output, y_{plant}, and the model output, y. It is introduced into the model as feedback as shown in **Figure 19-5**. Both **Figures 19-4** and **19-5** are schematic representations of the same controller in the sense that the combination of the internal model and the controller in **Figure 19-5** is equivalent to the controller shown in **Figure 19-4**. The difference between the diagrams is that **Figure 19-4** represents a conventional feedback block diagram description of the algorithm while **Figure 19-5** illustrates the explicit use of the internal model used for prediction. Reduction of the offset is accelerated by using just the first part of the predicted control action. At any given time when a new measurement of the output is made available to the control system, the estimated disturbance vector is re-calculated, and the corresponding control actions are ordered by the nonlinear optimization program.

For a 5% step-change disturbance, deviations from the set point were largely eliminated in three cycles. The control horizon was selected as a third of a cycle period, while the prediction horizon was chosen as two cycle periods. Constraints were considered on the magnitude of the control variable change as well as on the upper and lower limits of that variable.

Özgülsen et al. also assessed the robustness of their algorithm to model errors and concluded that an accurate mathematical model of the reactor and the reaction kinetics was not necessary. For prediction, they used an empirical nonlinear autoregressive model with exogenous inputs (a NARX model). In this technique, the predicted output is calculated by a polynomial function involving input and output of the previous time steps, the prediction error in those steps and time lags for different variables. The coefficients of the NARX polynomial model are obtained from plant data. The researchers concluded that using a shorter prediction horizon resulted in better performance and accordingly they recommended using 1/3 τ as the horizon.

In a later study, *Özgülsen et al.* showed that non-adaptive versions of MPC fail for steady-state, non-adiabatic CSTRs in the region of steady-state multiplicity for the presence of parametric variation (*Ravi Kumar et al., 1991*). Of interest in that paper, is the failure of a controller applied to a CSTR operating around an unstable steady state. This situation was considered earlier in the context of vibrational controllers. *Ravi Kumar et al.* demonstrated that if a model-based controller with parameter adaptation is used, a reactor can operate in a stable limit cycle around a steady state that would be unstable in an open loop system even in the presence of a wide range of disturbances. **Figure 19-6** shows the controller with the addition of an adaptive control element that generates a control correction which depends on the deviation of the measured variable from its set point.

Another novel control technique (*Pan and Lee, 2003*) particularly suited for periodic operation formulates a linear, stochastic, periodically-varying model for the forced reactor of the form:

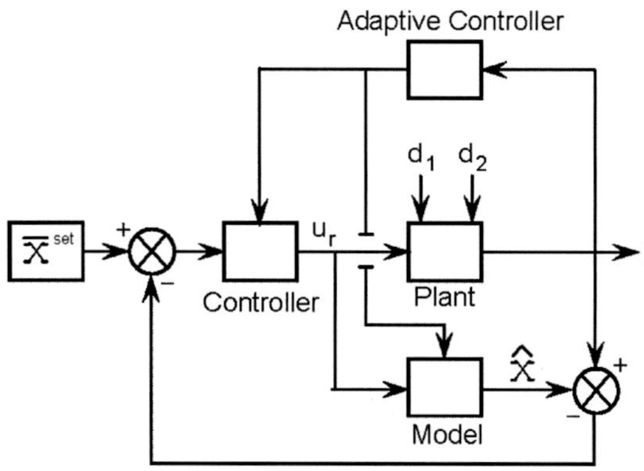

FIGURE 19-6 Schematic of the closed loop controller combining internal model and adaptive elements. *(Figure reproduced from* Ravi Kumar et al. (1991) *with permission. © 1991 by The Royal Society of Chemistry.)*

$$\begin{aligned} x_k(t+1) &= A(t)x_k(t) + B(t)u_k(t) + K(t)e_k(t) \\ y_k(t) &= C(t)x_k(t) + D(t)u_k(t) + e_k(t) \\ for \quad t &= 0, 1, \ldots\ldots N-1 \\ x_{k+1}(0) &= x_k(N) \end{aligned} \tag{19-6}$$

where x and y represent the state and output vectors, while u is a vector of manipulated variables, and e is the prediction error vector, assumed to be a white noise sequence. k is the period index and t is a time index within the sampling horizon. Each period is discretized into N sampling times. The state space matrices A to D and K are time-varying. **Eq. (19-6)** can be identified from the open loop response to step or pulse disturbances or, alternatively, can be obtained from linearization of a first principles model (*Pan and Lee, 2003*).

With a linear time-varying model, such as **Eq. (19-6)**, the controller can be designed by well-established procedures. To eliminate the offset for step changes in set point or disturbances, controllers with integrators are needed. Thus, to eliminate offsets in processes with periodic set points or disturbances a controller that performs integration over each period will be needed (*Lee et al., 2001*). Period-wise integrators are embedded in the model to eliminate cycle to cycle offsets by redefining the states in the model to be equal to differences of states values at time t within a period with respect to states values at the same time t during the previous cycle. Then the concept of "lifting" is used where the vector of the redefined states at a specific interval during a cycle is given as a function of the vector of redefined states at the same sampling time interval of the previous cycle. The newly defined states can be estimated with the help of a time-varying Kalman filter. The output disturbance is assumed to be periodically integrated white noise. Then, based on the redefined states and the assumed periodic disturbance, a periodic linear quadratic regulator using a Riccati finite difference equation can be designed for unconstrained systems or a model-predictive controller may be applied when constraints must be satisfied on manipulated and/or controlled variables. *Pan and Lee (2003)* give details of these techniques and their use with **Eq. (19-6)**. They also provide results for a simulation of a polyamine CSTR under square-wave, feed flow rate modulation during start-up. In addition, they illustrated the ability of their controller to reject a step-change disturbance in the temperature of the feed stream.

19.4 CONTROLS FOR PACKED BED REACTORS

Both non-adiabatic and adiabatic (or rather near-adiabatic) packed bed reactors are widely used by the chemical industry. These are distributed parameter systems, described mathematically by partial differential equations (PDEs). Nevertheless, the variables regulated are usually only those at the reactor outlet.

Control of distributed parameter systems is more challenging than control of lumped parameters ones because of the higher dimensionality of the state vectors and interactions existing between states. The problem has been addressed in the literature in two ways: 1) Early lumping, where the distributed parameter equation system is discretized initially into an approximate model consisting solely of a set of ordinary differential equations in time, or 2) late-lumping that allows for the distributed parameter nature but uses lumping of states for computing the controlled variable of interest as a function of the states, e.g., the average temperature as a function of the temperature profile in the reactor (*Ray, 1981*).

In early lumping applications spatial discretization of the PDEs was accomplished by either finite-difference or finite-element methods resulting in a lumped parameter system of ODEs that can be treated by methods such as Laplace transforms. However, spatial discretization is only an approximation of the original distributed parameters system and this may lead to inaccuracies in the conditions of controllability of the closed-loop. It may even lead to an incorrect prediction of stability of the open-loop and/or closed-loop system (*Ray, 1981*; *Christofides, 2000*). This inaccuracy in the prediction of stability motivated a significant research effort on the late-lumping approach whereby systems modeled by PDEs can be formulated in a state space form similar to that for lumped parameter systems. Theoretical results for the control of linear time-invariant systems of infinite dimensions, arising from the infinite dimension of the vector of states in problems modeled by PDEs, have been reported (*Curtain and Zwart, 1996*). However, applications of these results to the control of chemical reactors have not been found.

Moving temperature and concentration fronts are generated for periodically operated reactors that may produce unexpected behavior. Only isothermal systems have been studied using feed composition forcing. Although the literature on open loop operation is extensive (e.g., *Silveston (1998)*), this type of periodic operation has not been pursued in industry. Consequently, closed loop control studies of reactors under feed composition modulation, although challenging, has not been studied. On the other hand, several industrial applications have been reported for the control of packed bed reactors with flow-direction switching.

19.5 CONTROL OF PACKED BED REACTORS WITH FLOW-DIRECTION SWITCHING

This class of periodically operated reactors has seen widespread commercial application (see **Chapter 18**). The reactor consists of a catalyst bed situated between beds of an inert, heat storage material. Once started up, the ambient temperature feed containing a low concentration of a combustible waste in air is heated to at least ignition temperature by the one of the beds of inert packing already hot from heat transferred by exiting gas in the previous cycle. The waste-air mixture combusts on reaching the catalytic packing. This hot, combusted stream flows across the second inert bed where it is cooled back to almost ambient temperature by heat transfer to the inert packing. After a chosen time, the direction of the flow is reversed just before the moving hot front leaves the catalyst bed and the cold feed now enters the second, now hot, bed of inert packing, flowing in the opposite direction. Periodic reversal of the flow direction, thus, recovers most of the heat of reaction in the gas and uses it to bring the incoming cold feed to the catalytic ignition temperature.

Heat trapping and re-use is an example of autothermal operation, a type of operation known to increase parametric sensitivity and reduces the region of stable operation. *Matros* and colleagues at the Boreskov Institute of Catalysis (*Matros, 1989*) recognized this, but concluded that the large heat capacity of the catalyst and the inert packing relative to the heat capacity of the flowing gas would stabilize the periodic operation with respect to pulse disturbances in feed flow rate, temperature and combustible concentration as long as those disturbances had a zero mean and the duration of the pulses, t_{pulse}, was short relative to the cycle period. These two conditions may be expressed as,

$$\int_{t}^{t+\tau} (m(t) - m_m)(x(t) - x_m)dt = 0 \textit{ and } t_{pulse} < \frac{1}{2}\tau \tag{19-7}$$

where m represents the molar flow rate of the feed, while x is the mole fraction of combustible matter. *Matros (1990)* discusses stability bounds for an adiabatically operated, periodic flow-reversal reactor. *Nieken et al. (1994a)* found that multiple cyclic steady states were possible when fuel mixtures were fed to the reactor. *Salinger and Eigenberger (1996b)* applied stability analysis to an adiabatically operated reactor and verified the possibility of multiplicity.

The stability range for flow reversal with respect to in flow and concentration disturbances can be extended significantly by allowing for heat withdrawal or heat addition. This can be accomplished by withdrawing hot gas from the middle of the catalyst bed or by adding a combustible mixture at that point. A heat exchanger also can be used. The exchanger can be located either within the bed or placed by dividing the bed in half and inserting the exchanger between the halves.

In one of the few experimental investigations of stability of a flow-reversal reactor, *Purwono et al. (1994)* tested a bench-scale reactor combined with a double-pipe heat-exchanger using cool or hot air flowing in an outer tube around a catalyst bed. The reaction studied was the air oxidation of CO over a Pt/Al_2O_3 catalyst. These researchers found that the reactor was stable in just a narrow range of operating conditions. Outside of this range, temperature runaway or flame-out occurred. Heating or cooling extended the range of stable operation.

Following up on Purwono's work, *Budman et al. (1996)* studied the open loop stability and closed loop control of the same flow-reversal system. *Budman et al.* used a one-dimensional, pseudo-homogeneous dynamic model for a packed bed to establish the region of stability for a flow-reversal reactor with a fixed switching frequency during open loop operation. They used an experimental rate model for CO combustion in air and literature values for axial dispersion and heat transfer to a cooling or heating medium. The rate of reaction parameters was slightly adjusted to fit Purwono's experimental data. The three parts of **Figure 19-7** show the stability region for different combinations of cycle period and the overall heat transfer coefficient. The purpose of the study was to assess the system's open loop properties as the first step towards the design of a closed loop control strategy. The two variables used to define the stability range were chosen assuming a closed loop strategy where the heat exchanged

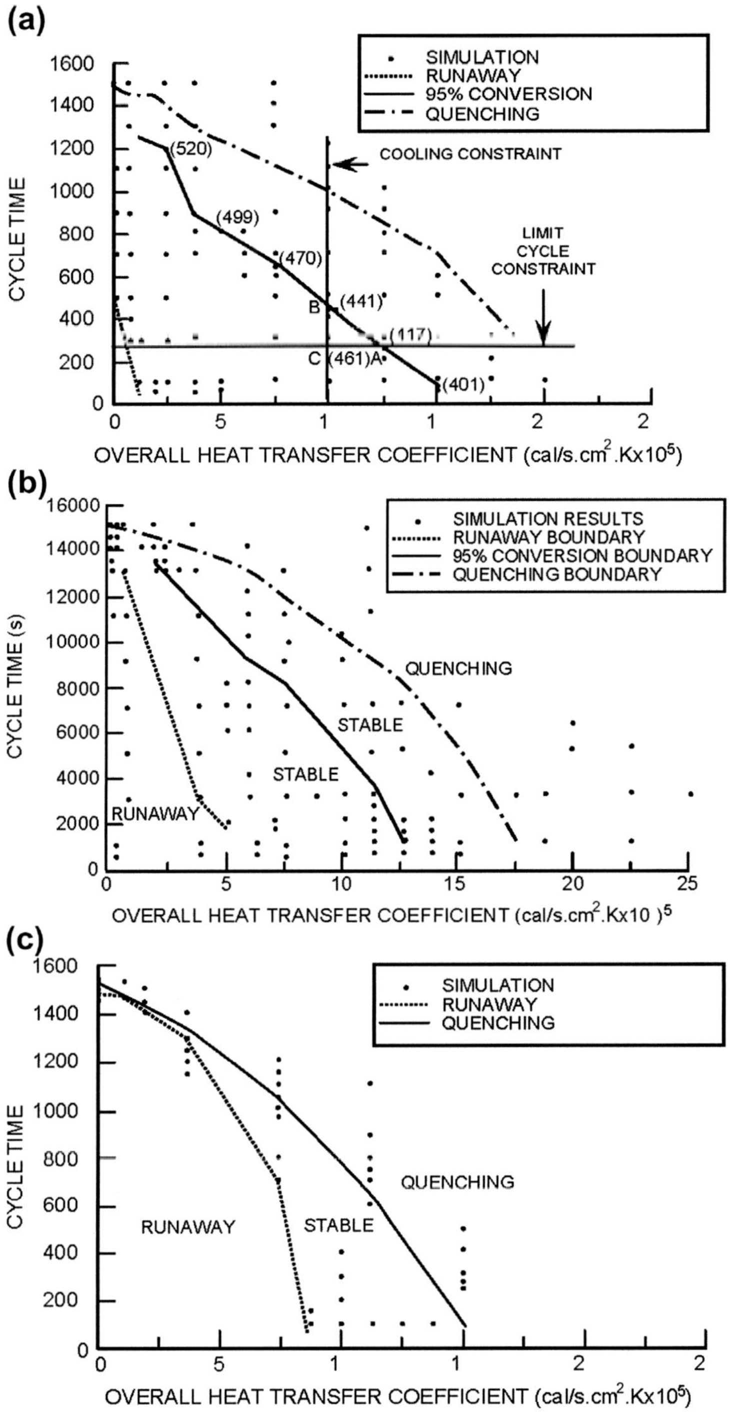

FIGURE 19-7 **Stability boundaries for the catalytic combustion of CO fed to a periodic flow reversal reactor at 25°C as a function of cycle period and overall heat transfer coefficient: (a) CO_{feed} = 1 vol%, (b) CO_{feed} = 2 vol%, (c) CO_{feed} = 3 vol%.** *(Figure reproduced from* Budman et al. (1996).*)*

was the manipulated variable and conversion functioned as the controlled variable. Cycle time was fixed at an optimal value.

In the various parts of **Figure 19-7**, the separate regions are designated as normal or stable operation, quenching, and runaway. Runaway is defined as the condition where the maximum bed temperature exceeds 600°C in 100 cycles or less, whereas quenching is the condition where the minimum bed temperature reaches 80°C (coolant temperature). In this latter case the reaction is eventually extinguished. A typical target for VOC elimination is 95% conversion of combustible matter. This target boundary is shown as a solid line in **Figures 19-7(a)** and **19-7(b)** indicating that 95% conversion is achieved to the left of and below the line. Also, in real applications, constraints on the heat transfer coefficient and on the cycle period exist. These, of course, depend on the design. These constraints are shown only in **Figure 19-7(a)** where a minimum τ_{cycle} and limitations on the heat transfer coefficient were considered. For instance, in **Figure 19-7(a)**, $\tau_{cycle} > 2700$ s. Pump choice or media selection limits the heat transfer coefficient so only regions to the left of the vertical line in **Figure 19-7(a)** are feasible. There is a large region of stable operation for a feed containing 1 vol% CO in air but it may not always be possible to reach 95% conversion. The feasible region shrinks in **Figure 19-7(b)** at 2 vol% CO in air indicating that runaway is now possible. Finally, if the CO concentration in the feed reaches 3 vol%, the region of open loop stable operation becomes quite narrow and consequently some sort of closed loop control will be needed.

Additional stability limitations can arise when using a heat exchanger. For example, simulations carried out by *Řeháček et al. (1992)* for a feed rich in combustible matter so that large amounts of heat are removed through a wall heat exchanger, nonsymmetrical temperature profiles arise in the forward and reverse directions. A subsequent stability analysis showed that the system becomes unstable under some conditions (*Salinger and Eigenberger, 1996b*).

To address the problem of a very narrow range of stable operation under certain conditions in the presence of disturbances, *Budman et al. (1996)* investigated the use of feedback control to stabilize the reverse flow reactor. They explored the use of PID feedback control with anti-windup maintaining the cycle period constant while manipulating the heat transfer coefficient through the flow rate of the heat transfer fluid. *Budman et al.* chose conversion as the controlled variable because it is indicative of the stability of the process and easier to measure compared with other variables such as the hot-spot temperature. Temperature is especially difficult to measure since its location shifts with time while sensors are located in fixed positions. Tuning of the PID controller was initially done using a linearized version of the open loop nonlinear model given in **Table 19-2** and the IMC tuning rules proposed by *Morari and Zafiriou (1989)*.

The controller was tested with a start-up scenario and for a step-change disturbance in feed concentration. Although the PID controller performed adequately for small step-changes and for start-up with about 1 vol% CO in the feed, it failed badly at 3 vol% CO because at that concentration a small change in the heat transfer coefficient switched conversion from complete to zero or vice versa. Even at the lower CO concentration, the closed loop response showed brief, sharp spikes in the exit CO concentration when there was a switch in flow direction. Although these spikes initiated controller action, they had little effect on the average conversion.

To improve performance, *Budman et al.* also investigated using a model-based feed-forward controller that could be operated as a stand-alone controller or in combination with the PID controller mentioned above. This combined feedback-feed-forward control strategy is shown schematically in **Figure 19-8**. This system is given a feed-concentration (x_0), which subsequently goes to the reactor model. Then, the

TABLE 19-2 Model for a Packed Bed, Catalytic Reactor with Flow Reversal (*Budman et al.*, 1996)

Material balances:

$$\frac{\partial C}{\partial t} = 0 = -v_z \frac{\partial C}{\partial z} + \varepsilon_B D_z \frac{\partial^2 C}{\partial z^2} - \rho_B R. \tag{19-8}$$

Where R′ = 0 for the recuperator.
Heat balances:
Recuperator

$$\frac{\partial T}{\partial t} = \frac{v_z \rho_g (C_p)_g}{[\rho_B (C_p)]_{rec}} = \frac{\partial T}{\partial z} + \frac{(k_z)_{rec}}{[\rho_B (C_\rho)]_{rec}} \frac{\partial^2 T}{\partial z^2}. \tag{19-9}$$

Catalyst bed

$$\frac{\partial T}{\partial t} = -\frac{v_z \rho_g (C_\rho)_g}{[\rho_B (C_p)]_{cat}} \frac{\partial T}{\partial z} + \frac{(k_z)_{rec}}{[\rho_B (C_p)]_{cat}} \frac{\partial^2 T}{\partial z^2} - \frac{2U}{[\rho_B (C_p)]_{cat} r}(T - T_c) - \frac{\Delta H\, R}{(C_p)_{cat}}. \tag{19-10}$$

Initial conditions:

$$t = 0, all\ z: \quad | \quad C = 0, \quad T = 200^\circ \mathrm{C}. \tag{19-11}$$

Boundary conditions:

$$\mathrm{z} = 0,\ \mathrm{t} > 0: \quad \left(\frac{\partial C}{\partial z}\right)_{z=O} = 0 \tag{19-12}$$

$$-(k_z)_{rec}\left(\frac{\partial T}{\partial z}\right)_{z=0} = v_z \rho_g (C_\rho)_g (T_o - T_{z=0}) \tag{19-13}$$

$$\mathrm{z} = \mathrm{L},\ \mathrm{t} > 0: \quad \left(\frac{\partial C}{\partial z}\right)_{z=L} = 0 \tag{19-14}$$

$$-(k_z)_{rec}\left(\frac{\partial T}{\partial z}\right)_{z=o} = v_z \rho_g (C_p)_g (T_O - T_{z=0}) \tag{19-15}$$

$$\mathrm{z} = \mathrm{z_{cat}},\ \mathrm{t} > 0: \quad -\varepsilon_B D_z \left(\frac{\partial C}{\partial z}\right)_{z_{cat}} = v_z (C_O - C_{z=0}) \tag{19-16}$$

$$-(k_z)_{rec}\left(\frac{\partial T}{\partial z}\right)_{z=z_{cat}} = -(k_z)_{cat}\left(\frac{\partial T}{\partial z}\right)_{z=z_{cat}.} \tag{19-17}$$

(Continued)

TABLE 19-2 Model for a Packed Bed, Catalytic Reactor with Flow Reversal (*Budman et al., 1996*) (*cont'd*)

Entrance to catalyst bed:

$$\begin{aligned}
&(T)_{z=z_{cat}}\ (recuperator) = T_{z=z_{cat}}(catalyst)\\
&z = z_{rec},\ t > 0:\\
&-\varepsilon_B D_z\left(\frac{\partial C}{\partial z}\right)_{z_{rec}} = 0\\
&-(k_z)_{cat}\left(\frac{\partial T}{\partial z}\right)_{z=z_{rec}} = -(k_z)_{rec}\left(\frac{\partial T}{\partial z}\right)_{z=z_{rec}}.
\end{aligned} \tag{19-18}$$

Exit from catalyst bed:

$$(T)_{z=z_{rec}}(catalyst) = T_{z=z_{rec}}(recuperator). \tag{19-19}$$

feed-forward controller, based on the reactor model, determines the combination of τ_{cycle} and overall heat transfer coefficient that achieves the target VOC conversion in the reactor. *Budman et al.* conjectured that the combination could reject unmeasured disturbances of the process by manipulating deviations in the heat transfer coefficient with respect to the nominal value generated by the feed-forward controller. This combined feed-forward-feedback strategy functioned well for a step-change disturbance, but it controlled poorly on start-up when the system experienced large changes in operating conditions. A drawback of the strategy was that the feed-forward control

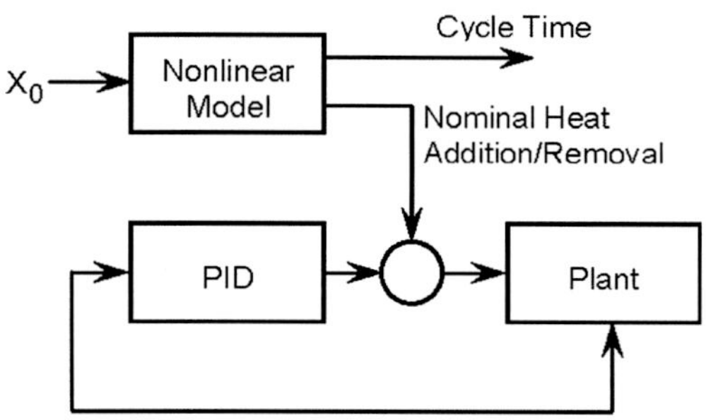

FIGURE 19-8 Schematic of a PID controller supplemented by a feed-forward controller. *(Figure reproduced from Budman et al. (1996).)*

calculations were time-consuming since they involved integrating the nonlinear PDE model given in **Table 19-2**. *Budman et al.* proposed the use of look-up tables calculated off-line to facilitate real-time implementation.

Recently, several applications of MPC, discussed earlier, to flow-reversal reactors have been reported (*Dufour et al., 2003; Dufour and Toure, 2004; Balaji et al., 2007; Fuxman et al., 2006*). From these contributions, the main challenges for the reverse flow operation are the use of distributed parameters in, and the inherent nonlinearity of, the mathematical models required.

Dufour et al. (2003) considered a monolithic design in their control application with a gas residence time in seconds and a fixed constant cycle period of 20 s. They proposed minimizing a performance index and treating any statutory limit on an emission as a constraint as per the following equation:

$$\frac{1}{n\tau}\int_{t}^{t+n\tau} y(t)dt \le y_{statute}. \tag{19-20}$$

Their performance index was the cost of heat addition or removal, an important component of the overall operating costs of the combustor. To simplify the solution, the hard constraint on

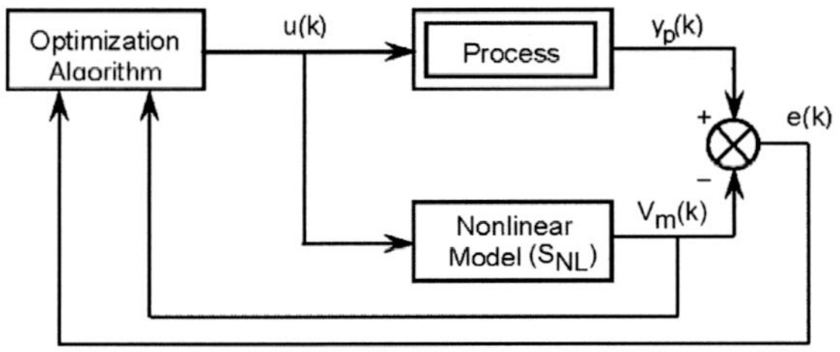

FIGURE 19-9 NMPC for a periodically operated catalytic combustor. *(Figure reproduced from* Dufour et al. (2003) *with permission. © 2003 by the IEEE.)*

VOC emission given by **Eq. (19-20)** was converted into a soft constraint by adding a severe penalty function to the performance index $J(x(0), u(\bullet), \tau)$ for emissions that exceed the limit. Additional constraints were maximum and minimum limits on the manipulated variable, heat transferred or removed, as well as a constraint on the rate of change of that variable.

The controller investigated by *Dufour et al.* is a version of the nonlinear MPC design discussed earlier for a CSTR. **Figure 19-9** is a schematic of that NMPC scheme showing the process model used to predict future outputs. Disturbances and observational errors are not shown in the schematic for clarity, but they do appear in **Figure 19-4**, another schematic of an NMPC regulator. Control action is initiated by an error signal e(k) at the kth sampling time obtained by subtracting the model output from the plant output. The controller uses this signal as feedback to estimate the unmeasured disturbances as explained previously. Using this estimate and the internal model the optimizer block generates a vector of the future manipulated variable u(•), the heat transfer rate, over the control horizon. It is conventional in predictive control practice to select a control horizon shorter than the prediction horizon. The error signal e(k) accounts for model error as well as disturbances or sampling errors.

A high computational burden results from repeated integration of a nonlinear reactor model. In real-time implementation this burden is somewhat alleviated by conducting the control actions partially offline. This is done by replacing the nonlinear PDE by a nonlinear model formulated for the desired trajectory of the system plus a time-varying linearized model around this reference trajectory with parameters that are continuously updated. Then, to save computation time, calculations using the nonlinear reference model are conducted off line; only predictions from the linearized model are done on-line. These are added to the offline calculated values from the nonlinear reference model. The nonlinear model consisted of energy balances for the solid and the gas and a material balance for the gas. Axial diffusion was neglected, but axial conduction was accounted for. A block diagram of this controller showing the blocks executing the offline and online calculations mentioned above is shown in **Figure 19-10**.

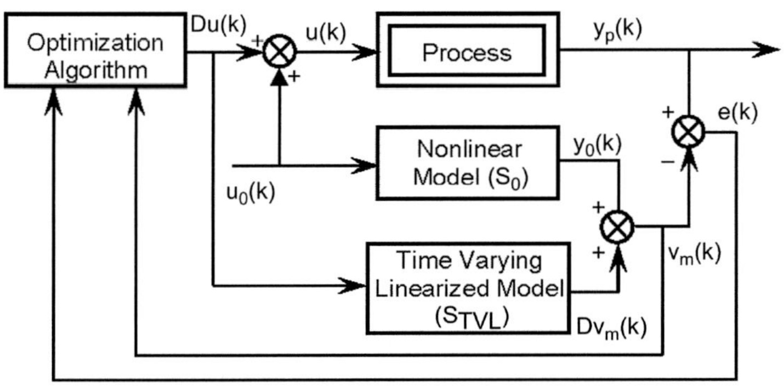

FIGURE 19-10 Schematic of an NMPC controller with linearized and non-linear models embedded in the controller. *(Figure reproduced from* Dufour et al. (2003) *with permission. © 2003 by the IEEE.)*

Simulations to test the control procedure were carried out by discretizing each monolith section into 200 elements. Accordingly, this approach represents an early-lumping following the classification discussed earlier. Sampling time was 10 s or twice in a flow-reversal cycle. Prediction and control horizons were 20 s. Stochastic step-disturbances were used to test the efficacy of the controller. Mean levels of combustible matter were so low that heat addition was necessary. Controller performance was judged on the ability of the controller to keep the VOC emissions below limits and to provide smooth heat input changes while remaining well under the maximum constraint. The *Dufour et al.* strategy has not been compared with others; consequently, it is difficult to assess the advantages of their algorithm compared with other potentially simpler control techniques.

Fuxman et al. (2006) also applied MPC to a flow-reversal reactor. In contrast to *Dufour et al.*, these investigators used a late lumping procedure whereby the method of characteristics for the solution of hyperbolic PDEs was embedded within an MPC scheme. The selection of this technique was motivated by *Shang et al. (2004)* who found that a high-performance MPC, which does not require substantial online computations to provide high accuracy prediction of output variables, can be designed based on the method of characteristics. This method requires the assumption that diffusion is negligible. The mass and energy balances model used by *Fuxman et al.* assumed plug flow. The controller was designed to drive towards a stability region and to keep the reactor within that region.

Formulation of the control scheme was conceived for reactors with slow flow switching frequencies. An average of the temperature profile along the reactor was used as the control variable to avoid reaction extinction. In addition to average temperature, a maximum temperature constraint was added to avoid catalyst overheating and deactivation. Through simulations of the closed loop system, *Fuxman et al.* showed that either the average or the maximum temperature can be used to keep the reactor stable in the presence of disturbances in the inlet reactant mole fraction or gas velocity. They also showed that an MPC scheme using heat extraction by means of removing mass can provide effective control. Although the control appears promising, further studies and comparisons are needed to assess the advantage of this late-lumping strategy over other algorithms such as early-lumping proposed by *Dufour et al.* and discussed earlier.

Balaji et al. (2007) applied the repetitive model-predictive controller of *Lee et al. (2001)* to a flow-reversal reactor with intermediate heat removal that was similar to the process studied by *Fuxman et al.* The *Lee et al.* controller was developed for stirred reactors as explained in the previous section. *Balaji et al.* found that diluting the feed along with heat extraction from the middle of the catalyst bed under rich-feed or the addition of reactants under poor-feed conditions were effective for maintaining the maximum catalyst bed temperature within a specified range. Early-lumping was used in this study involving the discretization of a homogeneous reactor model. Numerical simulation of the closed loop system showed that an RMPC strategy performed well compared with a conventional MPC strategy which did not use the repetitive control concept even though the same model was utilized in both the RMPC and MPC controllers.

Linear-Quadratic (LQ) optimal control is another technique that has been applied to flow reversal reactors. LQ is a special case of the optimal control problem described by **Eqs 19-1** to **19-5** given at the beginning of this chapter and later used by *Liaw and Bailey (1974)* where the argument of the integral in **Eq. (19-5)** is made up of a superposition of the quadratic cost of the states and the quadratic cost related to changes in the manipulated variable (*Aksikas et al., 2007a, b*). A limitation of the LQ algorithm is that it is based on a linear model of the process while the actual process is highly nonlinear. The LQ-feedback control problem for a reverse-flow reactor was treated in *Edouard et al. (2005a)* by using the traditional LQ solution for a finite dimension system obtained by early-lumping. The LQ controller was used to maintain the temperature at a single spatial point along the axis of the reactor between limits specified by manipulating either the dilution rate or the heating input. Although the LQ approach is based on a linear model, *Edouard et al.* claim that any nonlinear behavior of the reactor around a specific operating condition is not very significant which justifies use of a linear approximation. Implementation of the LQ controller necessitates the use of an observer to estimate the temperature profile along the reactor as well as feed disturbances from a few available temperature measurements. The researchers compared the LQ system with an MPC one for their control problem. A simulation study found the LQ to be superior to the MPC system as the result of a better estimate of the disturbance.

Fuxman et al. (2007) followed up the MPC research mentioned above by investigating the use of an LQG algorithm. They assumed a plug-flow reactor model with unidirectional flow and developed an LQ controller based on an infinite dimension representation of the model equations. The LQ-controller was formulated for a linearized model and was designed to keep the temperature along the reactor axis at stationary state values. *Fuxman et al.* assumed that the spatially distributed input variable, namely gas flow velocity, could be manipulated to control the temperature distribution along the axis of the reactor Although manipulation of flow velocity along the axis at a large (or infinite) number of points is not realistic, it gives the best achievable control performance and thus it can be used as a best case scenario for comparison purposes. For implementation, the authors proposed approximating the optimal distribution of the manipulated variable by using an average of this distribution. *Fuxman et al.* concluded from simulations of the closed loop system that a proportional controller could be used to drive the reverse flow reactor to a stationary state and to maintain stability in face of inlet disturbances. However, control objectives such as offset elimination or avoidance of temperature excursions above a threshold could not be achieved.

Eigenberger and co-workers (*Eigenberger and Nieken, 1988*; *Nieken et al., 1994a, b*) conducted a comprehensive empirical investigation of the control of a periodic flow-reversal reactor combusting a waste gas. Their research differed from the aforementioned control studies in that it considered different reactor designs rather than control strategies. They compared a catalytic bed, a catalytic bed faced on either end with beds of inert material, and a three-part bed in which the axial conductivity of the inert and catalytic sections were different. Heat exchange throughout the catalytic bed, splitting the catalytic bed and placing a heat exchanger between the section, as well as heat withdrawal through sucking off hot gas at different points in the catalyst bed or heat addition through supplying combustible material were considered. Heating or cooling of the feed was also examined. No formal optimization of design was pursued; instead, they examined the different alternatives through simulation and experiment. Constraints were assumed for a maximum bed temperature and the mean level of combustible matter in the bed exhaust. Step-change disturbances were implicitly assumed for both combustibles concentration and gas composition. Their primary focus was the stability of the flow-reversal reactor.

Nieken et al. (1994a) observed that a three-section reactor with low axial conductivity in the heat exchange material before and after the catalyst bed or a split catalyst bed employing a catalyst support with high axial conductivity

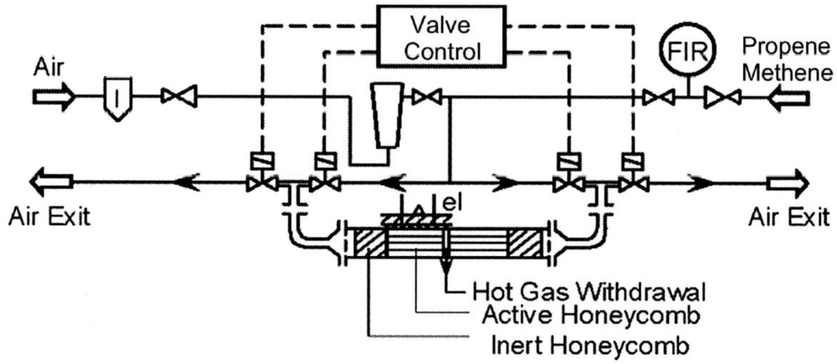

FIGURE 19-11 Schematic of the experimental periodic flow-reversal reactor used experimentally by *Eigenberger and Nieken (1988). (Figure reproduced from Eigenberger and Nieken (1988) with permission. © 1988 by Elsevier Inc.)*

were the best of the designs studied. Either choice resulted in a smooth temperature profile in the front of the catalyst bed receiving fresh feed. It also reduced the region of multiple steady states when a mixture of combustibles was used. Withdrawal of hot gas between the sections of the catalyst bed when a rich feed (high $\Delta T_{adiabatic}$) was used or heat addition through burning a fuel and adding the hot gas between the sections of the catalyst bed resulted in the widest range of feed conditions for stable operation within the pre-specified constraints. **Figure 19-11** shows their experimental unit. In their reactor, *Eigenberger* and co-workers used beds of monolith, wash-coated with a catalytically active noble metal salt. The flow-reversal reactor was fed with various mixtures of propylene and propane in air.

19.6 CONTROL OF SIMULATED MOVING BED CHROMATOGRAPHIC REACTORS

Chromatographic reactors and pressure or temperature swing reactors are examples of separating reactors incorporating adsorption. These reactors must be operated periodically or use moving beds in order to regenerate the adsorbent used. However, only periodically operated systems are used industrially because of the operating problems associated with moving beds. Simulated moving bed chromatographic reactors (SMBCR) have been heavily studied during the last two decades (*Aida and Silveston, 2005*). SMBCRs are referred to as hybrid dynamic systems because they involve both continuous and discrete changes of the dependent variables. Separation rather than conversion is usually the limiting process in SMBCRs; thus the objective function for such systems is formulated to maximize separation. Constraints are many and severe, including limits on product purity, switching time and on fluid velocity in the packing in each section of the reactor.

An SMBCR usually has two inlet ports, one for the reactant and the second for a solvent or carrier gas, and two outlet ports where the extract and raffinate are removed. Flow to each port is switched periodically so a port serves consecutively in a cycle as a feed inlet, a raffinate outlet, a carrier inlet and an extract outlet. The cycle is in the direction of fluid flow, but fluid velocity in a bed section between ports may change after a switch. This velocity is controlled by the feed and carrier flow rates.

Kloppenburg and Gilles (1999) applied the method of exact input/output linearization for the automatic control of a simulated moving bed process used for the separation of C_8 aromatics. The linearization method consisted in finding a mathematical transformation of the manipulated variables (inputs) so that a linear relation resulted between the transformed inputs and the controlled variables (outputs). The transformation was based on a nonlinear model of the process. The authors employed a mathematical model of a true moving bed (TMB) because it is simpler than the SMB model while providing similar results. The nonlinear transformation requires knowledge about the states vector which has infinite dimensions because of the distributed parameter model. Accordingly, the nonlinear model was discretized using orthogonal collocations. Since it would be prohibitive time-wise to measure states at all collocation points, a nonlinear Kalman filter-based estimator was used to estimate those states from an available smaller set of measured states. The strategy was multivariate involving two controlled variables, purity of the extract and yield of p-xylene, and two manipulated variables, extract and raffinate flow. Since the controlled variable's values were extracted from the discretized model of the process, the approach can be considered to be late lumping type as per the classification given earlier.

To further simplify the control problem, the process outputs were sampled only once at the middle of the switching period, thus avoiding the complication of controlling variables which are changing in a periodic fashion. Then, using the linear relation between inputs to outputs resulting from application of the model-based transformation, reference trajectories were selected so as to force the outputs to converge to their set-points in an exponential fashion. Simulation showed that the controller could be used to control the SMBCR in the presence of disturbances such as changes in feed composition or recycle flow rate.

Natarajan and Lee (2000) applied a combination of repetitive and model-based predictive control to an SMBCR, similar to the strategy that they previously applied to stirred reactors as discussed earlier, claiming that the cyclic nature of an SMBCR makes it well suited to repetitive control. Since the original repetitive control algorithm cannot handle multivariable systems, process delays and constraints, *Natarajan and Lee* suggested combining repetitive with predictive control. As mentioned in a previous section, this concept was later applied by *Balaji et al. (2007)* to control flow-reversal reactors. The structure and operational details of combined repetitive and predictive model-based control were developed in a separate paper (*Lee et al., 2001*).

The controller proposed by *Natarajan and Lee* is similar to that illustrated by **Figure 19-9** with the modification that the nonlinear model is replaced by a time-varying linear one. The latter was obtained by assuming isothermal operation, linear adsorption isotherms and then by employing spatial finite difference to eliminate the second-order terms in the mass balances for the fluid streams. Several remaining nonlinear terms were linearized around steady-state values. The manipulated variable in each section is the fluid velocity which was expressed as an algebraic function of the feed and carrier flow rates. The linear equations for each section are in state space form. Manipulated variables were the four zone velocities; key disturbances considered were the feed concentrations. With each switch of ports, the matrices of the linearized model change. Thus they change four times per cycle and resume their original values with the start of a new cycle. *Natarajan and Lee* used a balanced model reduction methodology to significantly reduce the dimensions of the model system in order to simplify the calculations. As in their previous application of the algorithm, the period-wise integration is embedded into the control algorithm by assuming that the model error is evolving according to a periodically integrated white noise relation. The controller employs prediction errors with the model through linear programming chooses the optimal feed and carrier flow rates to use over the prediction horizon. It also uses the differences between model predictions and measurements with a periodic Kalman filter to update the model.

This is similar to the procedure illustrated in **Figure 19-10**.

In a test of their controller, *Natarajan and Lee* used the duration of a cycle as the prediction horizon and considered four sampling times within the cycle for calculating the control actions. Parameters for their model were obtained from a study of the separation of phenylalanine and tryptophan. The time between port switching was 50 min, so the cycle duration was 200 min. After comparing the closed loop operation with open loop operation during start-up by simulation, they observed that combined repetitive and predictive control reduced the number of cycles needed to reach a cyclic steady state from four to two. Thus, they concluded that the combination of repetitive and predictive control should be applicable to many types of periodically operated reactors.

19.7 OTHER CONTROL SCHEMES FOR PERIODICALLY OPERATED REACTORS

Earlier in this chapter the necessity of adjusting control model parameters to improve closed loop performance for step or time-varying (oscillating) disturbances with a time trend in their means was discussed. The reason for this was that a change in the mean value of a disturbance forces the system to shift to a new steady state. Then, if the process is nonlinear and a linear model is used for control, this linear model may be inaccurate when the system is operated around a different steady state. Consequently, following the introduction of a step-disturbance, the linear model on which the controller is based should be updated. Other situations that may require the updating of model parameters are those where the model is not known a priori or where it is applied to time-varying systems. Control strategies that involve online updating of model parameters used in model-based control are referred to as adaptive strategies. This type of control has been discussed in the context of periodic operations by *Sinha and Ramaswamy (1990)*.

Adapting or updating model parameters as a function of the feedback errors involves using differential equations of the following form:

$$\frac{da}{dt} = \varepsilon(x - x_{setpoint}), \qquad (19\text{-}21)$$

where a is either a single parameter or a parameter vector that is to be adapted over time and ε is the stiffness coefficient or adaptation gain. The control action is subsequently calculated from the updated model. A few examples of parameter adaptation algorithms have been reported for periodically operated reactors, especially in combination with predictive control. For example, the controller proposed by *Ravi Kumar et al. (1991)*, discussed earlier and illustrated in **Figure 19-6**, involved combining an adaptive element and a predictive controller. A few applications of pure adaptive control have been reported for periodic processes. For example, *Sinha and Ramaswamy (1990)* have controlled a limit cycle by using a characteristic radius of the cycle to force changes in parameters. **Figure 19-12** shows the

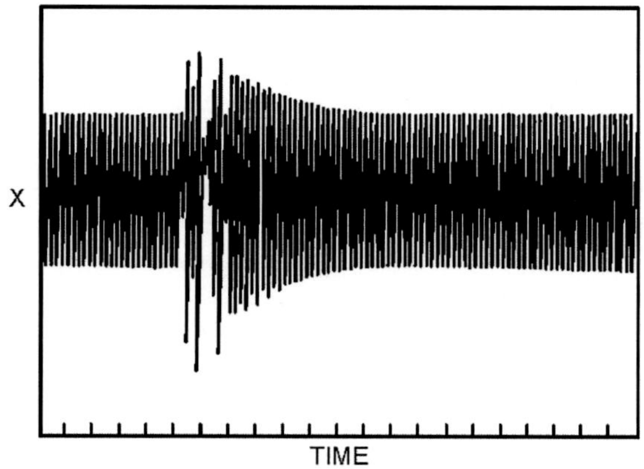

FIGURE 19-12 **Response of a limit cycle process under adaptive control to a step-disturbance.** *(Figure reproduced from* Sinha and Ramaswamy (1990) *with permission. © 1990 by Elsevier Science.)*

response of a limit cycle process to a step-disturbance that increases the cycle radius.

Jirát et al. (2001) applied adaptive control to a periodically operated catalytic converter for diesel emissions. Gas cooling is needed for the catalyst to function properly. The *Jirát et al.* cooler combined countercurrent heat exchange between cold air and hot combustion gas with flow direction switching. The cooler outlet temperature was a proxy for converter conversion since it was easier to measure. A feed-forward adaptive controller was used to control the temperature at the inlet to the converter. This controller employed a neural network model whose coefficients were learnt by the controller algorithm.

Although not applied as yet to reactors, *Bauer and Dourdoumas (2004)* proposed a controller specifically for cyclically operated systems that may be potentially useful for reactors. In their scheme parameter adaptation is driven by the time-varying open loop output of the undisturbed system. Designating the reference signal of the closed loop as r(t) and assuming that the cycle has a period τ, the Fourier transform of the reference signal is:

$$r(t) = \sum_{m=-\infty}^{\infty} R_m e^{2\pi jm\frac{t}{\tau}}. \qquad (19\text{-}22)$$

This signal can be fed into a controller as shown in **Figure 19-13** that manipulates a reactor input. If the system is linear with a transfer function P(s), then, for the output to track the reference signal exactly, the controller output must be:

$$u(t) = \sum_{m=-\infty}^{\infty} \frac{R_m}{P\left(\frac{2\pi jm}{\tau}\right)} e^{\frac{2\pi jmt}{\tau}}. \qquad (19\text{-}23)$$

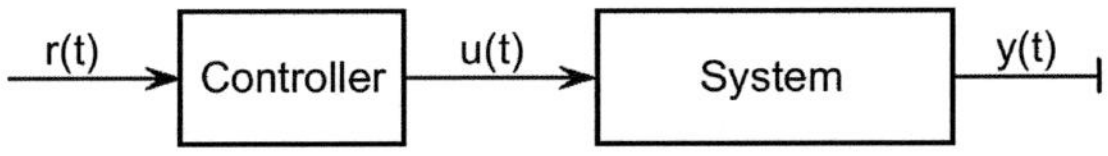

FIGURE 19-13 Schematic of a controller for a cyclic linear system with a known transfer function.

Because of disturbances that require feedback to eliminate offset, the use of a sampled rather than a continuous output, the existence of limits on the number of harmonics for real-time use, and possibly imperfect knowledge of the system or error introduced by linearization, the implementation of the apparently simple scheme in **Figure 19-13** requires several more steps. Sampled signals r(t) and y(t) must be transformed into Fourier series and the number of harmonics reduced. The predicted values of the Fourier coefficients of the transfer function must be calculated and updated based on a feedback measurement. Since noise is present in the feedback, filtering is needed in that calculation step. *Bauer and Dourdoumas* suggested using separate low pass filters on phase and amplitude respectively. Finally a time-varying controller output, u(t), must be generated from the Fourier transform. **Figure 19-14** shows the components of the control strategy proposed by *Bauer and Dourdoumas*. The controller was tested only on a simple repetitive mechanical operation but not yet on periodically operated reactors.

19.8 COMMENTS AND RESEARCH NEEDS

It has been shown in this chapter that model-based predictive controllers can be readily adapted for use with periodically operated CSTRs and packed beds employing periodic flow direction changes. Strategies combining a predictive control algorithm and repetitive control have been successfully applied to CSTRs, reactors with flow reversal and simulated moving bed chromatographic reactors. But research is lacking on the application of model-based predictive controllers to packed beds under flow or composition modulation or to separating reactors such as pressure- and temperature-swing adsorptive reactors.

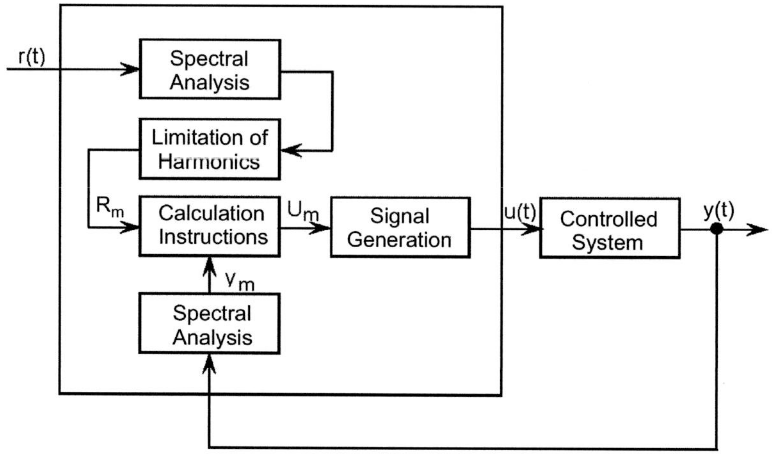

FIGURE 19-14 Schematic of the *Bauer and Dourdoumas (2004)* controller using a reference trajectory from the undisturbed system. *(Figure reproduced from Bauer and Dourdoumas (2004) with permission. © 2004 by Elsevier Inc.)*

So far, controllers have been tested only by simulation and that with just limited disturbance frequencies and magnitudes. Therefore, it will be important to test the model-based schemes experimentally and assess their relative advantages and disadvantages. Such tests have not been found in the literature. Furthermore, studies by *Liaw and Bailey (1974)* and *Budman et al. (1996)* suggest that for certain operating conditions and reactor designs, simple controllers may provide acceptable closed loop performance. Hence, it is important to compare sophisticated versus more conventional control strategies to assess whether or not added algorithmic complexity is justified. The work of *Eigenberger* and co-workers suggests that reactor design may render systems that are easier to control or which possess wide regions of stability in open loop operation. Integration of control and design to achieve an operation that is simpler to control is of great importance. Further research on applying adaptive controllers to periodically operated reactors seems warranted.

Nomenclature

A = matrix of coefficients
= time-varying matrix of state space
a = parameter or vector of parameters
B = matrix of coefficients
C = concentration (mol/m^3)
C_p = specific heat
CSTR = continuous stirred tank (back mixed) reactor
c(t) = state vector in **Figure 19-2**
D = time-varying matrix of state space
Du(k) = Deviation of control action with respect to nominal trajectory calculated by optimization as per **Figure 19-10**
Dv_m(k) = Deviation of in output prediction with respect to predicted output of nonlinear model calculated by a linearized model as per **Figure 19-10**
D_z = axial diffusion or dispersion
d(t) = time-varying disturbance
e = error or vector of errors
e(k) = error signal or error vector at the kth sampling time
f( ) = function
g( ) = function
h( ) = function
IMC = Internal Model Control
J(x, u, τ) = cost or objective function, performance function
j = complex number $\sqrt{-1}$
K = time-varying matrix of state space
k = sampling time index
k_z = axial heat conductivity (J/cm^2 s)
L = reactor or bed length (m)
L( ) = function
LQ = Linear Quadratic control

MPC = Model Predicative Control
MV = manipulated variable
m = running summation index in **Eqs. 19-22** and **19-23**
m(t) = time-varying molar flow rate
N = integer
NARX = non-linear auto regressive with exogenous inputs
NMPC = Non-linear Model Predicative Control
n = unmeasured input disturbance or vector of unmeasured input disturbances
= integer
ODE = ordinary differential equations
P(s) = transform function
P = proportional mode
PDE = partial differential equations
PI = proportional-integral mode
PID = proportional-integral-derivative mode
p = vector of modulated variables
R = rate of reaction
R_m = coefficients of Fourier transform of reference trajectory in **Eq. 19-22**
RMPC = Repetitive Model Predictive Control
r(t) = closed loop reference signal
S_{NL} = non-linear system model
SMB = simulated moving bed
SMBCR = simulated moving bed chromatographic reactor
r = reactor or bed radius (cm)
T = temperature (°C, K)
T_c = coolant temperature
t = time (s, min)
t_{pulse} = pulse duration (s)
U = overall heat transfer coefficient between the coolant and the reactor (J/cm^2·s·K)
U_m = Fourier transform of control actions in **Figure 19-14**
u = manipulated or controlled variable or vector of variables
u(•) = control strategy (vector of manipulated variables)
u(t) = time-varying controller output or output vector
VOC = volatile organic compound
v = axial flow velocity (cm/s)
= error in measured variable or vector of errors in measured variables
= set point or set point vector
v_z = axial flow velocity
w = error in control action or vector of errors in control action, unmeasured disturbance or vector of unmeasured disturbances
X = measured controlled variables in **Figure 19-5**
$\hat{X}$ = predicted controlled variables in **Figure 19-5**
$\overline{X}^{set}$ = set-points in **Figure 19-5**
x = state variable or vector of state variables
= mol fraction of combustible matter (**Eq. 19-7**)
x_0 = feed state variable
x(0) = initial condition or vector of initial conditions
x(t) = deviation vector in **Figure 19-2**
y = output variable or vector of variables or measured output variable or vector
= model output or model output vector
y_{plant} = plant output or plant output vector
$y_{statute}$ = legal or specified limit (constraint)
$y_m(k)$ = model output or output vector at the kth sampling time
$y_p(k)$ = process or plant output or output vector at the kth sampling time
y(t) = time-varying output variable or vector of variables
$\overline{y}$ = mean output variable or vector of variables
z = axial position (cm, m)

Greek

ΔH = heat of reaction (J/mol)
$\Delta T_{adiabatic}$ = adiabatic temperature rise (°C)
ε = void fraction (-)
= stiffness (adaption) coefficient
ε_B = catalyst bed void fraction
ρ = density (kg/m^3)
ρ_B = bed density
ρ_g = gas density
τ = period, cycle period (s, min.)
τ_{cycle} = cycle period

Subscripts

B = bed (catalyst bed)
Cat = catalyst
g = gas
k = period index
m = molar
o = initial
rec = recuperator
set point = specified variable or vector of variables
t = time index for sampling horizon
0 = entering

CHAPTER

20

Chromatographic Reactors

Peter Lewis Silveston, Kenji Hashimoto†, Motoaki Kawase†*

*Waterloo, Ontario, Canada and †Kyoto, Japan

OUTLINE

20.1 Introduction 569

20.2 Concept and Types 570

20.3 General Models 573

20.3.1 Distributed Systems 573

20.3.2 Lumped Models 578

20.4 Cyclic Steady State 579

20.5 Pulse Chromatographic Reactor 580

20.6 Countercurrent Moving Bed Chromatographic Reactor 587

20.7 Continuous Rotating Annular Chromatographic Reactor 590

20.8 Stepwise, Countercurrent Multi-Stage Fluidized Bed Chromatographic Reactor 591

20.9 Fixed Bed Chromatographic Reactor With Flow Direction Switching 592

20.10 Extractive Reactor Systems 593

20.11 Centrifugal Partition Chromatographic Reactor 593

20.1 INTRODUCTION

With this chapter and the two that follow, a discussion of separating reactors begins. Such reactors draw together a solid catalyst and an adsorbent, a solid that traps one of the reaction products. The adsorbent serves to prevent a reverse reaction or a further reaction of that product, and to initiate its separation from the reaction mixture. Chromatographic reactors employ such a mixture of catalyst and adsorbent. Swing reactors, considered in **Chapter 22**, do so as well. Using an adsorbent requires regeneration to release the product and to re-use the solid, thus mandating periodic operation. Separating reactors differ from one another only in the way regeneration is accomplished.

Mathematical models for these systems are virtually the same and their initial conditions and boundary conditions are often similar. Consequently, modeling and calculational considerations for both chromatographic and swing reactors will be discussed in this chapter. *Aida and Silveston (2005)* discuss chromatographic and swing reactors in much more detail. These authors also examine other types of separation that have been tested in combination with chemical reaction such as extraction, vaporization (catalytic distillation), and sedimentation. Separation of the second phase in such systems is simple and continuous operation is routine.

The chromatographic reactor is the oldest of the reaction and separation combinations. Not long after the introduction of chromatography as a separation technique, *Roginskii et al. (1962)* observed that an equilibrium-limited reaction could be forced toward completion by carrying out the reaction in a chromatographic column. Shortly thereafter, *Magee (1963)* proposed a simplified model for a reaction in such a column. Promoting reaction through separating products from reactants is an even older concept. *Glueckauf and Kitt (1957)* and *Thomas and Smith (1959)* described a pulse reactor in which product adsorption shifted the reaction towards completion. *Hattori and Murakami (1968)* theoretically derived the conversion and yield in a chromatographic reactor and compared them with those in a tubular reactor at steady state, while in a succeeding paper (*Murakami et al., 1968*) they studied the cracking of cumene over a silica-alumina catalyst using a rectangular pulse. The effect of pulse width on the conversion predicted in their first paper was verified. Shortly before *Murakami, Matsen et al. (1965)*, using the dehydrogenation of cyclohexane, experimentally verified that conversion in a chromatographic reactor was greater than that in a continuous tubular reactor under the same temperature and pressure. *Langer et al. (1969)* discussed the overall potential of the gas chromatographic column as a chemical reactor and derived a simple equation for estimating the irreversible first-order rate constant from chromatographic measurements. Disadvantages and limitations of chromatographic reactors were also discussed.

Since these exploratory efforts, a large literature on versions of chromatographic reactors, models for such reactors, and their application to chemical systems have appeared. Reviews of work prior to 1990 have been published by *Barker et al. (1987a,b)*, *Barker and Ganetsos (1988)*, *Barker et al. (1992a, 1990b)* and by *Carta (1991)*. Unfortunately, the emphasis of these reviews is on biotechnology applications. A comprehensive review of the published literature up to 2005 is contained in a book by *Aida and Silveston (2005)*.

20.2 CONCEPT AND TYPES

Figure 20-1 illustrates the principle of a chromatographic reactor (CR) using as an example the catalytic dehydrogenation of cyclohexane. A pulse of C_6H_{12} (A) in a nitrogen carrier gas is introduced into a fixed bed of mixed catalyst and adsorbent in **(a)**. Above 200°C, C_6H_{12} chemisorbs on the catalyst and breaks down to benzene (B) and atomic hydrogen. The latter rapidly combines to form H_2 and desorbs. C_6H_6 is strongly adsorbed. The adsorption difference inaugurates separation between products as well as reactant. Weakly adsorbed, H_2 moves through the reactor at almost the velocity of the carrier gas. C_6H_{12}, more strongly adsorbed, moves more slowly through the reactor and continues to break down. This is the situation shown in **(b)**. The sequence continues as the C_6H_{12} pulse moves through the reactor causing spreading of the C_6H_6 and H_2 peaks evident in **(c)**. Once C_6H_6 reaches the end of the bed, a new pulse of C_6H_{12} is introduced at the entrance (**Figure 20-1d**). Nearly complete conversion may be obtained in pulse operation provided the bed is long

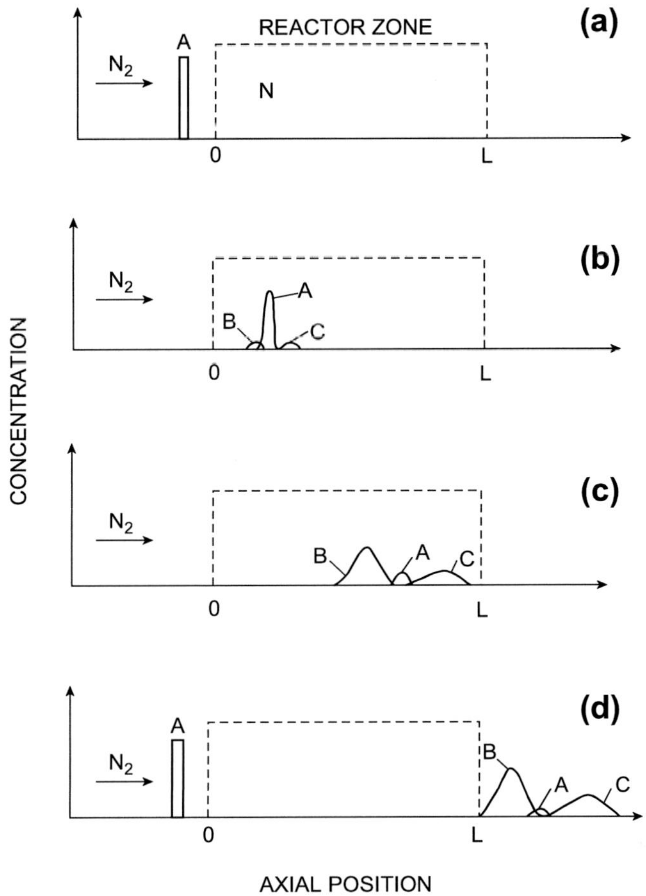

FIGURE 20-1 **Concentrations of the C_6H_{12} reactant (A), C_6H_6 product (B) and H_2 product (C) in successive times as a pulse of C_6H_{12} in an N_2 carrier gas moves through a mixed bed of catalyst and adsorbent and undergoes catalytic dehydrogenation.** *(Figure reproduced from* Aida and Silveston (2005) *with permission. © 2005 by Blackwell Publ. Ltd.)*

enough. However, there is a throughput penalty because reactant is not fed continuously to the reactor.

Increasing throughput is made possible by maintaining a constant feed rate and physically moving the catalyst and adsorbent bed past the feed point. This is the countercurrent moving bed chromatographic reactor (CMCR) operation shown in **Figure 20-2**. Gaseous reactant in a carrier gas moves upward, while the solid bed of catalyst and adsorbent moves downward. Dehydrogenation of C_6H_{12} again serves as the example. C_6H_{12} is fed at the middle of the reactor, while a carrier gas enters from the bottom. Reaction products, H_2 and C_6H_6 are separated by adsorption differences. Weakly adsorbing H_2 is driven upward through the falling bed of catalyst and adsorbent by the carrier gas and emerges in the effluent at the top of the reactor. Catalyst and adsorbent particles are fed at the top and are withdrawn from the bottom. C_6H_{12} and C_6H_6 adsorb on the catalyst and are carried downward. If the bed is deep enough, cyclohexane is fully converted

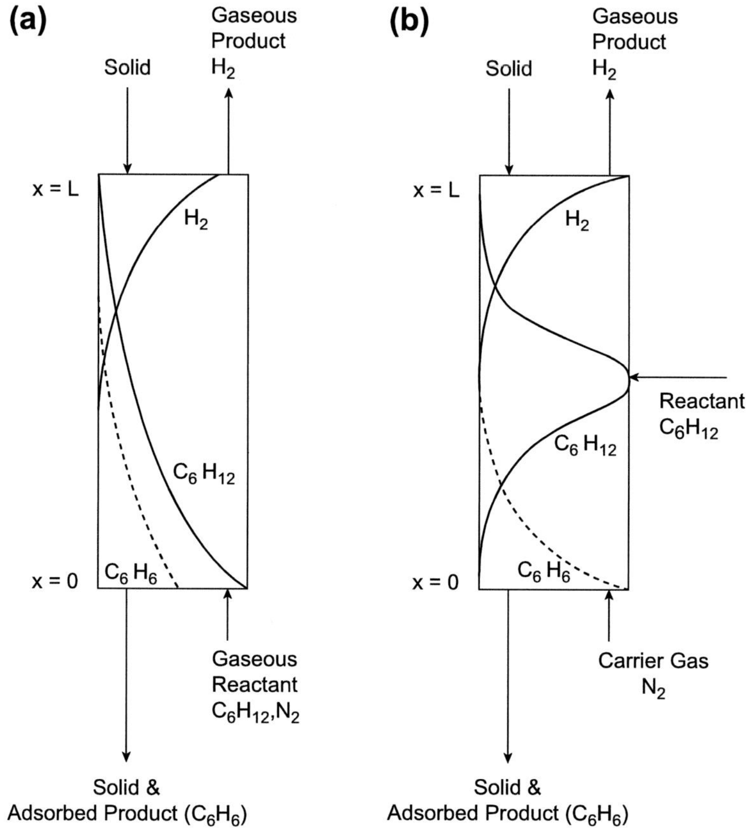

FIGURE 20-2 Steady-state concentration profiles in a countercurrent moving bed chromatographic reactor (CMCR) packed with a homogeneously mixed bed of catalyst and adsorbent. *(Figure reproduced from* Aida and Silveston (2005) *with permission. © 2005 by Blackwell Publ. Ltd.)*

and only C_6H_6 adsorbed on the catalyst emerges from the bottom of the bed.

A variation of this contacting technique is the continuous rotating annular chromatographic reactor (CRAC). The principle is the same, but in the rotating bed the catalyst/adsorbent particles are motionless relative to the bed. The bed itself is carried past the feed inlet and product outlet.

In these systems, all flows are continuous. A countercurrent moving bed of catalyst and adsorbent converts the reactor system from an unsteady state to steady state. Of course, catalyst and adsorbent circulating through the reactor encounter periodically varying fluid environments.

A third version is the simulated countercurrent moving bed chromatographic reactor (SCMCR) in which the bed is stationary but the feed entrance and product exits shift with time. Fluid flow through the bed is continuous as well. **Chapter 21** is devoted to simulated moving bed systems, so we reserve further discussion until that chapter.

A great many reaction systems, such as hydrogenation, chlorination or oxidation, involve more than a single reactant. For these systems, periodic flow reversal can be introduced into the chromatographic reactor. Called reverse-flow chromatographic reactors, such systems can either improve performance if the size of a mixed bed of catalyst and adsorbent is restricted, or reduce bed size for a specified conversion. Reverse flow chromatographic reactors will be considered at the end of this chapter.

In chromatographic reactors, a fluid desorbent, called an eluent, strips adsorbed product and residual reactant from the adsorbent. It is also possible to use pressure reduction or a higher temperature to force desorption. Reactors using pressure change are referred to as pressure swing systems, while those employing a temperature increase are called temperature swing ones. **Chapter 22** deals with both reactor types.

There are important constraints on the choice of adsorbent. Weak adsorption of the key product increases bed length and thus capital expense, while strong adsorption raises the carrier gas or solvent requirement per unit of reactant processed and results, also, in a more dilute product. This increases the cost of product concentration and recovery. A further constraint arises for single reactions with multiple reactants. Significantly different affinities of the adsorbent for the reactants can cause their separation and a retardation of the reaction rate. This is not a problem when one reactant serves as the carrier or solvent. Screening of candidate adsorbents is an important step in development of chromatographic reactor systems. The topic is discussed by *Aida and Silveston (2005)*.

20.3 GENERAL MODELS

20.3.1 Distributed Systems

The purpose of **Figure 20-3** is to define the variables used in the model presented in this section. The separating reactor shown schematically in the figure could be any one of a CR, a swing reactor, or a section between ports of a CMCR or a SCMCR. In the first two cases the solid phase is stationary and the bottom line representing flow of solids should be removed. The variables q_i and T_s now refer to conditions at a fixed plane in the bed. For a SCMCR, the solid phase is stationary and U_s represents virtual motion. The schematic is broken into parts to indicate internal variables in the bed. Symbols at the ends indicate input variables, some of which will be functions of time. Reactor parameters, treated as constant in the model, are the symbols shown in the boxes. A single solid phase serving as catalyst and adsorbent is assumed. Dispersive fluxes resulting from concentration or temperature gradients are not shown in the schematic.

Because isotropic beds of catalyst and adsorbent are generally used, separating reactors, chromatographic as well as swing, are described by the same set of heat and material balances. A momentum balance should be part of the set, but in practice it is replaced by a phenomenological equation. **Table 20-1** states the pseudo-homogeneous, one-dimensional model for a separating reactor containing a packed bed made up of a single solid functioning as both adsorbent and catalyst or a homogeneous, fully mixed solids bed of separate adsorbent and catalyst. The model does vary with the application as indicated in

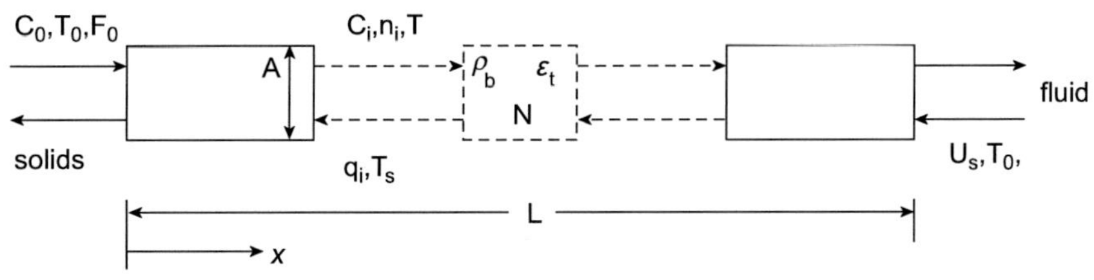

FIGURE 20-3 Schematic of the catalyst-adsorbent bed of cross-section *A* and length *L* for a separating reactor showing bed parameters, inlet and bed variables. *(Figure reproduced from* Aida and Silveston (2005) *with permission. © 2005 by Blackwell Publ. Ltd.)*

TABLE 20-1 Partial Differential Equations of Pseudo-Homogeneous Models for Chromatographic and Swing Reactors

Mass balance for component i

$$\varepsilon_t \frac{\partial C_i}{\partial t} + \rho_b \frac{\partial q_i}{\partial t} = D_x \frac{\partial^2 C_i}{\partial x^2} - \frac{1}{A}\frac{\partial n_i}{\partial x} + U_s \rho_b \frac{\partial q_i}{\partial x} + \rho_b \nu_i r. \quad (20\text{-}1)$$

Overall mass balance for components

$$\varepsilon_t \frac{\partial C}{\partial t} + \rho_b \sum_{i=1}^{N_C} \frac{\partial q_i}{\partial t} = D_x \frac{\partial^2 C}{\partial x^2} - \frac{1}{A}\frac{\partial n}{\partial x} + U_s \rho_b \sum_{i=1}^{N_C} \frac{\partial q_i}{\partial x} + \rho_b \sum_{i=1}^{N_C} \nu_i r. \quad (20\text{-}2)$$

In both equations: $U_s = 0$ for a CR, SCMCR, PSR, or TSR; while

$$\varepsilon_t \frac{\partial C}{\partial t} + \rho_b \sum_{i=1}^{N_C} \frac{\partial q_i}{\partial t} = 0 \text{ for a CMCR at steady state,}$$

energy balance

$$\left(\varepsilon_t C_{p_g} C + \rho_b C_{p_s}\right) \frac{\partial T}{\partial t} = k_x \frac{\partial^2 T}{\partial x^2} - C_{p_g} \frac{n}{A}\frac{\partial T}{\partial x} + U_s \rho_b C_{p_s} \frac{\partial T}{\partial x} - \frac{P}{A}\frac{\partial}{\partial x}\left(\frac{n}{C}\right) - \rho_b \sum_{i=1}^{N_C} H_{a_i} \frac{\partial q_i}{\partial t}$$
$$+ \rho_b \Delta H_R r - \frac{4h_0}{d_c}\left(T - T_a\right). \quad (20\text{-}3)$$

In this equation: $U_s = 0$ for a CR, SCMCR, PSR or TSR;

$$\frac{\partial}{\partial x}\left(\frac{n}{C}\right) = 0 \text{ for a CR, CMCR or SCMCR;}$$

$$\left(\varepsilon_t C_{p_g} C + \rho_b C_{p_s}\right) \frac{\partial T}{\partial t} = 0 \text{ for a CMCR at steady state.}$$

Flow model: Ergun correlation (instead of momentum balance)

$$\frac{dP}{dx} = -J_v\, u - J_k u^2 \quad (20\text{-}4)$$

where

$$J_v = \alpha \frac{\mu_g [\lambda_s (1 - \varepsilon_b)]^2}{d_p^2\, \varepsilon_b^3} \quad (20\text{-}5)$$

$$J_k = \frac{\lambda_s (1 - \varepsilon_b) \rho_g}{d_p\, \varepsilon_b^3}. \quad (20\text{-}6)$$

Table 20-1. These changes arise in the case of a CR or SCMCR or the swing reactors because the solids are stationary. For a CMCR, the changes occur because this reactor operates at steady state. Assumptions of the model are that: 1) The bed is isotropic, 2) intraparticle diffusion proceeds so rapidly that it can be neglected, 3) only a single reaction occurs,

4) particles are isothermal and uniformly bathed by the fluid phase, 5) the bed has no radial gradients, 6) all motion in the bed is plug flow, 7) fluids, simple or mixtures, exhibit ideal behavior, 8) all transport and thermodynamic properties do not vary as pressure or temperature change, 9) mass and heat transfer between fluid and solid is fast enough so that concentration and temperature differences between the phases are negligible and, finally, 10) both adsorption and desorption are rapid so that the adsorbate is at equilibrium with the fluid phase.

The pseudo-homogeneous model is the simplest model for chromatographic and swing reactors, but it presents a dilemma: the rate constant, k, in the rate term, r, can be extracted only with difficulty from independent rate measurements on the catalyst itself because the term in the model is formulated in terms of the bulk density of a combined catalyst and adsorbent. It is preferable to evaluate the rate constant from measurements with the catalyst-adsorbent mixture employing the same pseudo-homogeneous model.

In **Table 20-1**, the total void fraction $\varepsilon_t = \varepsilon_b + (1 - \varepsilon_b)\varepsilon_p$, where ε_b and ε_p are the bed or bulk and particle porosity respectively. Usually, an intimate mixture of catalyst and adsorbent, two solid phases, will be used. When this is the case, ρ_b modifying the adsorbate density q_i terms must be replaced by $(\rho_b)_{ads}$ and the ρ_b modifying the reaction rate term must be replaced by $(\rho_b)_{cat}$. These "bulk" densities are the product of a volume fraction and the particle density. Thus, $(\rho_b)_{ads} = \varepsilon_{ads}\ (\rho_p)_{ads}$. The total void fraction term changes too: $\varepsilon_t = \varepsilon_b + \varepsilon_{ads}\varepsilon_{p_{ads}} + \varepsilon_{cat}\varepsilon_{p_{cat}}$. The dispersion term in **Table 20-1** is based on the fluid volume. It may also be based on the reactor volume. When this definition is used D_x should be replaced by $\varepsilon_t D_x$. In **Eq. (20-3)**, specific heat and density may not be the same for catalyst and adsorbent. Thus, $\rho_b(C_p)_s$ should be replaced by $\rho_{ads}(C_p)_{ads} + \rho_{cat}(C_p)_{cat}$. The two other ρ_b terms also change because they become specific to either catalyst or adsorbent.

If **Eqs (20-1)** and **(20-2)** are broken into separate balances for the fluid and solid phases, a heterogeneous model results. When this is done, the balances must be coupled by rate of transport and/or rate of adsorption terms. Of course, separate balances for each phase are necessary when transport between phases and/or adsorption is rate controlling and when measurements of the transport or adsorption steps are to be made. When the solid phase is stationary, the adsorbate flux terms disappears from all models, including **Eqs (20-1)** and **(20-2)** in the pseudo-homogeneous model.

The heterogeneous model is presented in **Table 20-2**. This model assumes that fluid-particle transport is slow compared with adsorption rates. The Ergun relation, **Eqs (20-4)** to **(20-6)**, is used to determine the pressure drop for the heterogeneous model, just as it is with the homogeneous model.

Heat and material balances in **Table 20-2** again assume a single solid phase serving both as catalyst and adsorbent. As discussed for **Table 20-1**, different materials are normally used. The versions presented were chosen because they offer a reduced set of dimensionless groups. Generally, heat transfer between phases is rapid so that the pseudo-homogeneous assumption is appropriate for the energy balance. Lumping catalyst and adsorbent causes problems in defining a rate of reaction term, as mentioned above, because the surface concentration, q_i, represents an adsorbate concentration rather than the concentration on the surface of the catalyst. A similar problem arises with the flux terms. Although the mass and heat transfer coefficients for catalyst and adsorbent will be similar, the driving forces will not be. An advantage of distinguishing between catalyst and adsorbent is that the rate expression and constants for r can be taken from independent rate measurements on the catalyst. Lumping both materials into a single solid phase means

TABLE 20-2 Partial Differential Equations of Heterogeneous Models for Chromatographic and Swing Reactors

Mass balance for component i in the fluid phase

$$\varepsilon_t \frac{\partial C_i}{\partial t} = D_{x_i} \frac{\partial^2 C_i}{\partial x^2} - \frac{1}{A} \frac{\partial n_i}{\partial x} - k_{m_i} a_m \rho_b \left(C_i - \frac{q_i}{K_i} \right). \tag{20-7}$$

Overall mass balance for components in the fluid phase

$$\varepsilon_t \frac{\partial C}{\partial t} = D_x \frac{\partial^2 C}{\partial x^2} - \frac{1}{A} \frac{\partial n}{\partial x} - a_m \rho_b \sum_{1=1}^{N_c} k_{m_i} \left(C_i - \frac{q_i}{K_i} \right). \tag{20-8}$$

In both equations, $\frac{\partial C_i}{\partial t} = 0$ for a CMCR at steady state.

Mass balance for component i in the solid phase

$$\rho_b \frac{\partial q_i}{\partial t} = U_s \rho_b \frac{\partial q_i}{\partial x} + k_{m_i} a_m \rho_b \left(C_i - \frac{q_i}{K_i} \right) + \rho_b \nu_i r. \tag{20-9}$$

Overall mass balance equation for components in the solid phase

$$\rho_b \sum_{i=1}^{N_C} \frac{\partial q_i}{\partial t} = U_s \rho_b \sum_{i=1}^{N_C} \frac{\partial q_i}{\partial x} + a_m \rho_b \sum_{i=1}^{N_c} k_{m_i} \left(C_i - \frac{q_i}{K_i} \right) - \rho_b \sum_{i=1}^{N_C} \nu_i r. \tag{20-10}$$

In both equations, $U_s = 0$ for a CR, SCMCR, PSR or TSR;

$$\frac{\partial q_i}{\partial t} = 0 \text{ for a CMCR at steady state,}$$

Energy balance for the fluid phase

$$\left(\varepsilon_t C_{p_g} C \right) \frac{\partial T}{\partial t} = k_x \frac{\partial^2 T}{\partial x^2} - C_{p_g} \frac{n}{A} \frac{\partial T}{\partial x} - \frac{P}{A} \frac{\partial}{\partial x} \left(\frac{n}{C} \right) + a_m h \rho_b \left(T_s - T \right). \tag{20-11}$$

In this equation, $\frac{\partial}{\partial x} \left(\frac{n}{C} \right) = 0$ for a CR, CMCR and SCMCR;

$$\frac{\partial T}{\partial t} = 0 \text{ for a CMCR at steady state.}$$

Energy balance for the solid phase

$$\rho_b C_{p_s} \frac{\partial T_s}{\partial t} = k_s \frac{\partial^2 T_s}{\partial x^2} + U_s \rho_b C_{p_s} \frac{\partial T_s}{\partial x} - a_m h \rho_b \left(T_s - T \right) - \rho_b \sum_{i=1}^{N_C} H_{a_i} \frac{\partial q_i}{\partial t} - \rho_b \Delta H_R r. \tag{20-12}$$

In this equation, $U_s = 0$ for a CR, SCMCR, PSR, or TSR;

$$\frac{\partial T_s}{\partial t} = 0 \text{ for a CMCR at steady state.}$$

that the rate constants must be extracted from measurements on the adsorbent-catalyst mixture using an appropriate model.

Instead of being listed in another table, the equations are amended below assuming catalyst and adsorbent are different materials. Volume fractions and particle densities are used in place of "bulk" densities.

Fluid phase material on component i:

$$\varepsilon_t \frac{\partial C_i}{\partial t} = D_{x_i} \frac{\partial^2 C_i}{\partial x^2} - \frac{1}{A}\frac{\partial n_i}{\partial x} - k_{m_i} a_m (\varepsilon_{ads} \rho_{p_{ads}}) \left(C_i - \frac{q_i}{K_i} \right) + \varepsilon_{cat} \rho_{p_{cat}} \nu_i r. \tag{20-13}$$

Solid phase material balance on component i:

$$\varepsilon_{ads} \rho_{p_{ads}} \frac{\partial q_i}{\partial t} = U_s \varepsilon_{ads} \rho_{p_{ads}} \frac{\partial q_i}{\partial x} + k_{m_i} a_m \varepsilon_{ads} \rho_{p_{ads}} \left(C_i - \frac{q_i}{K_i} \right) \tag{20-14}$$

considers just the adsorbent because adsorption on the catalyst has been assumed to be negligible. Changes to the total mass balances, **Eqs (20-9)** and **(20-10)** must also be made. These changes may be readily seen from the amended component material balances above. The pseudo-homogeneous assumption for the energy balance is made by simple changes in **Eq. (20-3)**. Thus,

$$\left(\varepsilon_t C_{p_g} C + \varepsilon_{cat} \rho_{p_{cat}} C_{p_{cat}} + \varepsilon_{ads} \rho_{p_{ads}} C_{p_{ads}} \right) \frac{\partial T}{\partial t} = k_x \frac{\partial^2 T}{\partial x^2} - C_{p_g} \frac{n}{A} \frac{\partial T}{\partial x} + U_s \left(\varepsilon_{ads} \rho_{p_{ads}} C_{p_{ads}} + \varepsilon_{cat} \rho_{p_{cat}} C_{p_{cat}} \right) \frac{\partial T}{\partial x} - \frac{P}{A} \frac{\partial}{\partial x} \left(\frac{n}{C} \right) - \varepsilon_{ads} \rho_{p_{ads}} \sum_{i=1}^{N_C} H_{a_i} \frac{\partial q_i}{\partial t} + \varepsilon_{cat} \rho_{p_{cat}} \Delta H_R r - \frac{4 h_0}{d_c} (T - T_a). \tag{20-15}$$

Eqs (20-1 to 20-3), **(20-7)**, **(20-9)** and **(20-11)** represent the fluid velocity by the total molar flow rate for the fluid, n. Velocity is affected by temperature and pressure. If isothermal and isobaric conditions are assumed and the fluid phase is dilute, fluid velocity, u, can replace the molar flow rate.

For many years, the balances given in **Tables 20-1** and **20-2** were used in normalized or dimensionless form for calculations. This practice has lost favor in recent years. Nonetheless, dimensionless groups that arise from normalization are important for correlated experimental data. These groups are summarized in **Table 20-3**. Dimensionless forms of the pseudo-homogeneous and heterogeneous separating reactor models are given by *Aida and Silveston (2005)* and their application discussed. Model variables in those sets of equations have been rendered dimensionless by characteristic length, space time, mean velocity and adsorbent capacity. In the energy balance the temperature departure (T-T_0) is used and made dimensionless by the inlet temperature (T_0). The normalized models continue to assume a single solid acting as both catalyst and adsorbent.

Table 20-3 defines the dimensionless variables that arise from normalization. Following the suggestion of *Aida and Silveston (2005)*, the solid-fluid velocity ratio, so important in characterizing moving bed chromatographic reactors, is called the Aris Number (N_{Ar}) to honor his important contributions to the theory of chromatographic reactors. Use of the Damköhler number is well-established. However the first form shown in **Table 20-3** is strongly temperature-dependent. Frequently the rate terms in dimensionless models are represented by a dimensionless Damköhler number, N'_{Da}, containing β, the Prater number.

For simulated moving bed chromatographic reactors, the Danckwerts boundary conditions are used and apply to a segment of the bed between j and j+1 or j-1 and j in **Figure 21-1**.

TABLE 20-3 Definition of Dimensionless Groups Used with Chromatographic and Swing Reactors

Bodenstein number for mass : $N_d = \frac{D_x A}{LF_0}.$

Bodenstein number for heat (gas) : $(N_h)_g = \frac{k_x}{C_{p_g}\rho_b N}\frac{A}{LF_0}.$

Bodenstein number for heat (solid) : $(N_h)_s = \frac{k_s}{C_{p_s}\rho_b N}\frac{A}{LF_0}$ *or*

$$= \frac{k_s}{C_{p_{cat}}\varepsilon_{cat}\rho_{p_{cat}} + C_{p_{ads}}\varepsilon_{ads}\rho_{p_{ads}}}\frac{A}{NLF_0}.$$

Damkoehler number for component i assuming an irreversible, 1storder rate model : $N_{Da\ i} = \frac{\varepsilon_t ALk_{r_i}}{F_0}$ or $k_{\gamma_i}\bar{t}$.

Temperature independent Damkoehler number : $N_{Da\ i} = \frac{\varepsilon_t ALk_i}{F_0}.$

Prater number : $\beta = \frac{E}{RT_0}.$

Aris Number(N_{Ar}) or the velocity ratio of solid and fluid : $\sigma = \frac{\varepsilon_t AU_s}{F_0}.$

Boundary conditions characterize the different types of chromatographic reactors and will be set forth, if necessary, when each type of reactor is discussed in this chapter.

Fluid flows downward in the figure so at j where fluid enters the j-1 to j segment,

$$D_{x_i}\frac{\mathrm{d}}{\mathrm{d}x}(C_i)\bigg|_{j=0^-} = \frac{1}{A}\left((n_i)|_{j=0^-} - (n_i)|_{j=0^+}\right) \quad (20\text{-}16)$$

$$k_x\frac{\mathrm{d}T}{\mathrm{d}x}\bigg|_{j=0^-} = \frac{C_{p_g}}{A}\left((nT)|_{j=0^-} - (nT)|_{j=0^+}\right). \quad (20\text{-}17)$$

At j-1 where fluid leaves the segment,

$$\frac{dT}{dx}\bigg| = \frac{dC_i}{dx}\bigg| = 0. \quad (20\text{-}18)$$

If the boundary, j, between two segments is a junction, the flow, n_i, in **Eqs (20-16)** and **(21-17)** must be augmented by $(n_i)_f$ if it is a feed point or diminished by $(n_i)_p$ if it is a product withdrawal point. The position at which the boundary conditions apply varies with time.

20.3.2 Lumped Models

A cell model can be a useful, alternative representation of a packed bed of mixed adsorbent and catalyst. In such models, a cascade of cells, each with a uniform temperature and composition, is assumed. Thus, each cell can be described by a lumped model. Cascade models are best suited to fixed bed CRs and SCMCRs. Isothermal operation is usually assumed as a representation of heat dispersion would require the use of pseudo-heat transfer coefficients to represent conductive flows between cells. Variables in a cascade model are shown in given by **Figure 20-4**. In this figure, the boxes are cells, *j*-1, *j* and *j*+1. There are junctions around each cell denoted by *j*-1 and *j* that allow the model to be applied to a SCMCR.

Cell models are used mainly with dilute systems under isothermal and isobaric conditions. Assuming a pseudo-homogeneous system and using the assumption listed for **Table 20-1**, the material balance for the ith component is:

$$V\left(\varepsilon_t\frac{dC_i^j}{dt}+\rho_{b_{ads}}\frac{dq_i^j}{dt}\right)=F(C_i^{j-1}-C_i^j)+\rho_{b_{cat}}\nu_i Vr. \quad (20\text{-}19)$$

Superscripts identify the cell or junction. The overall material balance is:

$$V\left(\varepsilon_t\frac{dC^j}{dt}+\rho_{b_{ads}}\sum_{i=1}^{N_c}\frac{dq_i}{dt}\right)=F(C^{j-1}-C^j)-\rho_{b_{cat}}Vr\sum_{i=1}^{N_c}\nu_i. \quad (20\text{-}20)$$

Mass balances at junctions between cells, *j-1, j, j+1,* in **Figure 20-4** are algebraic relations.

It is necessary to estimate the number of cells to apply the cell model. This number, N_{cell}, can be estimated from the chromatographic peak characteristic using the following relationship (*Falk and Seidel-Morgenstern, 1999*):

$$N_{cell}=5.54(t_R/w_{0.5})^2 \quad (20\text{-}21)$$

where t_R is the retention time, and $w_{0.5}$ is the corresponding peak width at half-height.

20.4 CYCLIC STEADY STATE

With the exception of the simple, pulse version, chromatographic reactors run continuously and eventually reach what can be called a cyclic steady state. In this state, concentration of any component, C_i, temperature, T, velocity, u, quantity of an adsorbate on a surface, q_i, are constant at identical times measured from the start of the cycle at a specific position in the reactor. That condition can be written for a cycle of period, τ, for $n\tau \le t \le (n+1)\tau$ as:

$$C_i(x,t)=C_i(x,t+\tau),\quad q_i(x,\tau)=q_i(x,t+\tau),\quad u(x,\tau)=u(x,t+\tau),\quad T(x,t)=T(x,t+\tau), \quad (20\text{-}22)$$

where n is a large integer. Of course, the condition applies as well to the reaction, adsorption and transport rates and thus it implies the absence of fouling and deactivation processes. **Figure 20-5** illustrates the cyclic steady state with respect to a reactant concentration.

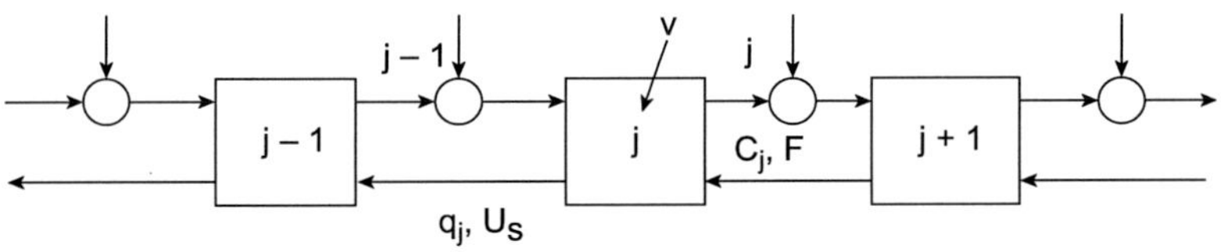

FIGURE 20-4 **Cascade representation of a separating reactor. Each cell, (*j* -1), *j*, (*j* + 1), is assumed to have a uniform concentration and temperature. Junctions on either side of all but the first and final cells permit fluid addition or withdrawal.** *(Figure reproduced from* Aida and Silveston (2005). *© 2005 by Blackwell Publ. Ltd.)*

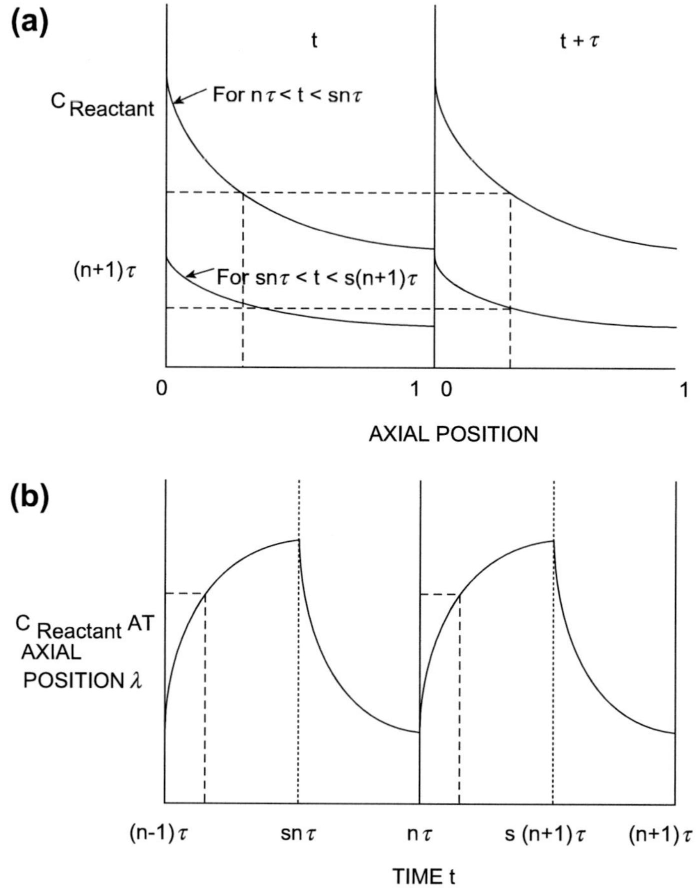

FIGURE 20-5 Schematic representation of the cyclic steady state for a reactant at an axial point λ within the reactor. Two cycles are shown. *(Figure reproduced from* Aida and Silveston (2005) *with permission. © Blackwell Publ. Ltd.)*

20.5 PULSE CHROMATOGRAPHIC REACTOR

Of the various types of chromatographic reactors, the simplest introduces a pulse of reactant into a bed of intimately mixed catalyst and adsorbent in a continuously flowing carrier fluid. The principle of the pulse chromatographic reactor (CR) was illustrated in **Figure 20-1**. The operation was the first to be proposed (*Roginskii et al., 1962*) as well as the first to be modeled (*Magee, 1963*) and studied experimentally by several research teams (*Bassett and Habgood, 1960; Gaziev et al., 1963; Roginskii et al., 1962; Matsen et al., 1965*). Pulse chromatographic reactors are now used for analyzing the reaction kinetics and testing adsorbents and catalysts. *Aida and Silveston (2005)* offer a thorough discussion of modeling as well as application of pulse chromatographic reactors. **Table 20-4** summarizes many of the theoretical and experimental contributions for this type of reactor.

The pulse chromatographic reactor was first applied to kinetic studies of heterogeneous catalytic reactions because this device uses

TABLE 20-4 The Pulse Chromatographic Reactor (CR) Literature

Investigators	Type of Study	Reaction	Observations
Klinkenberg (1961)	Calculational	General	Reaction effect on chromatographic band width
Gaziev et al. (1963)	Experimental		Dehydrogenation of C_6H_{12} Demonstrated C_6H_6 yield in excess of equilibrium
Roginskii et al. (1961)	Calculational	General	Interpretation of peak shape
Roginskii et al. (1962)	Experimental	As above	As above
Roginskii and Rozental (1962)	Calculational	General	Model development
Magee (1963)	Simulation	General reversible reaction	Proposed simplified model
Semenenko et al. (1964)	Simulation		Dehydrogenation of n-C_4H_{10} Yields in excess of equilibrium
Roginskii and Rozental (1964)	Calculational	General	Derivation of kinetic models
Matsen et al. (1965)	Experimental	C_6H_{12}	Dehydrogenation Conversion in excess of equilibrium
Saito et al. (1965)	Calculational	General	Model development
Gore (1967)	Calculational	General	Represented operation of a CR in terms of sine and error function waves
Kocirik (1967)	Simulation	General	Analysis of CR using moments of chromatographic curves
Hattori and Murakami (1968)	Calculational	General	Analysis of pulse reactors and CRs
Chu and Tsang (1971)	Simulation	General	Effect of operating parameters on CR performance
Wetherold et al. (1974)	Experimental & Simulation	Hydrolysis of methyl formate	Conversion in excess of equilibrium; model validation
Unger and Rinker (1976)	Experimental	Ammonia synthesis	Demonstrated increased conversion
Schweich and Villermaux (1978)	Experimental & Simulation	C_6H_{12}	Dehydrogenation Model development and validation
Sardin and Villermaux (1979)	Experimental	Esterification	Increased conversion with no separation of reactants
Schweich et al. (1980)	Calculational	General	Calculation method

(Continued)

TABLE 20-4 The Pulse Chromatographic Reactor (CR) Literature (*cont'd*)

Investigators	Type of Study	Reaction	Observations
Schweich and Villermaux (1982a, b, c)	Simulation	C_6H_{12}	Dehydrogenation Assumption of local reaction equilibrium questionable
Antonucci et al. (1978)	Experimental	C_2H_6	Dehydrogenation Conversion in excess of equilibrium
Cho and West (1986)	Experimental	CO oxidation	Chromatographic effects within a catalyst pellet
Zafar and Barker (1988)	Experimental & Simulation	Enzymatic sucrose polymerization	Validation of a plug flow model
Liden and Vamling (1989)	Simulation	Consecutive reactions	Improved selectivity over PFR
Sad et al. (1996)	Experimental		Dehydroisomerization of n-C_4 Use of composite catalysts and different adsorbents
Mazzotti et al. (1997a)	Experimental & Simulation	Esterification of alcohol with HAc	Back-mixing due to density differences in displacement front
Wu (1998)	Experimental	Enantioselective esterification	Analytical method of measuring enantioselectivity
Falk and Seidel-Morgenstern (1999)	Simulation	Hydrolysis of methyl formate	Conversion improvement due to a dilution effect
Wu and Liu (1999)	Experimental	Esterification of racemic naproxen	Analytical method of measuring enantioselectivity
Migliorini et al. (2000)	Experimental & Simulation	Enzymatic diol esterification	Importance of water content; model validation
Silva and Rodrigues (2002)	Experimental & Simulation	Diethyl acetal synthesis	Model validation and parameter measurement
Falk and Seidel-Morgenstern (2002)	Experimental & Simulation	Hydrolysis of methyl formate	Product separation and recovery more important than conversion
Gelosa et al. (2003)	Experimental & Simulation	Triacetine synthesis	Conversion greatly in excess of equilibrium
Yu et al. (2004)	Experimental & Simulation	MeOH - HAc esterification	Measurement of adsorption and kinetic parameters
Vu et al. (2005)	Simulation	Hydrolysis of methyl formate and acetate	Model for equilibrium and reaction transformed into an equivalent equilibrium system

Note: Calculational = theoretical study, usually a simulation.

only a small amount of reactants and requires less experimental time than classical steady-state procedures. Since the reactor operates under unsteady state, to obtain the kinetic data for reactor design, it was necessary to relate conversion and yield in the two systems. This was done by *Saito et al. (1965)* and *Hattori and Murakami (1968)*. They found that in the case of linear reactions, the results from a pulse chromatographic reactor are in agreement with those from a steady-state reactor, while for nonlinear reactions there is a remarkable disagreement.

In pulse chromatographic reactors, the reactants are fed periodically to the reactor inlet in a carrier fluid. The shape of the pulse, the amount of injection, the cycle time and the interval between injected pulses are important operating variables. *Matsen et al. (1965)* experimentally examined the performance of the pulse chromatographic reactor for the dehydrogenation of cyclohexane to benzene and hydrogen with a Pt/Al_2O_3 catalyst, where alumina acted as the adsorbent. Rectangular pulses of C_6H_{12} were fed into the reactor in a helium carrier gas. The thermal conductivity of the effluent gas was recorded potentiometrically. Using single pulses in a relatively long reactor, *Matsen et al.* observed well-separated, fairly sharp peaks of H_2 and C_6H_6 with a small reactant sample. As the sample size increased, these peaks tailed toward each other. With a sufficiently large sample, a sharp peak of unreacted cyclohexane appeared between H_2 and C_6H_6. Since the effect of carrier gas flow rate on conversion was not significant, adsorption, reaction and desorption were rapid under the conditions studied so dehydrogenation was equilibrium-limited. As the pulse frequency increases, the H_2 peak from a fresh pulse catches up with the C_6H_6 peak from the previous pulse and a reconverted C_6H_{12} peak appears in the product chromatogram. Thus, the optimum pulsing operation is to use a sample pulse as large as possible and to pulse as frequently as possible without allowing a peak of unconverted C_6H_{12} to arise in the exit stream. **Figure 20-6** demonstrates such an optimum mode of operation. The experimental conditions are given in the figure caption. Conversion in the pulse chromatographic reactor was 96.4%. This was about 30% greater than that under equilibrium for the reactor operating conditions.

Schweich and Villermaux (1982b) also dealt with the dehydrogenation of cyclohexane in

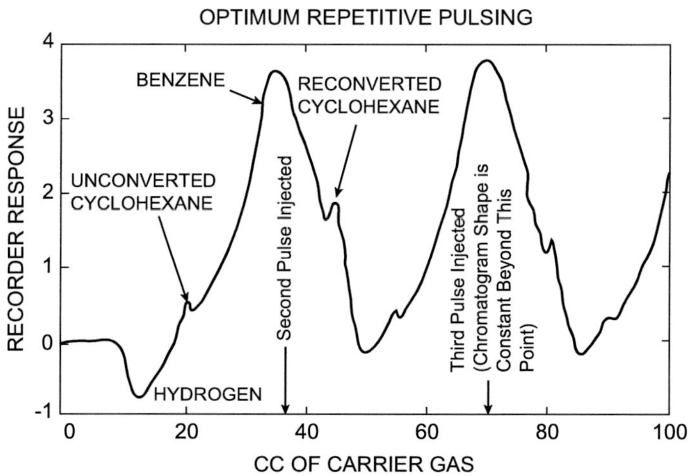

FIGURE 20-6 Product chromatogram for an optimal repetitive pulsing. Measurement conditions: $T_{Reactor} = 225°C$, $L_{Reactor} = 50$ cm. Catalyst Wt. = 8 g, $U_{Carrier} = 10$ cm³/min, pulse volume = 12 μL, pulse separation = 3.5 min. *(Figure adapted from* Matsen et al. (1965) *with permission. © 1965 by the American Chemical Society.)*

the pulse chromatographic reactor. Conversion X was measured as a function of the period of the injections. The authors observed an optimal sequence of injections that depended on the reaction temperature. At $T_{Reactor} = 220°C$, the optimal time between injections was about five minutes. Experimental results showed that conversion in the pulsed chromatographic reactor exceeded the equilibrium conversion.

A mixed cell model was used by *Falk and Seidel-Morgenstern (1999)* to simulate the performance of the pulse chromatographic reactor. Porosity of the bed, the number of cells and equilibrium constants were estimated from the elution profiles for the hydrolysis of methyl formate over an acidic ion exchange resin. Reaction kinetics were assumed to be reversible and second-order; kinetic parameters were determined from batch experiments. Simulated elution profiles agreed well with separate experimental measurements. The authors addressed the question of whether a chromatographic reactor provides a higher conversion than a continuously operated fixed bed reactor under the same operating conditions. Thus, they compared the performance of both reactors for the same amount of reactant per unit time and the same volumetric flow rate. The authors found that, in the range of the conditions considered, the conversion for the fixed bed reactor with the diluted reactant exceeded that of the pulsed chromatographic reactor. However, as the injection is extended, the two reactors perform similarly. Of course, the chromatographic reactor has the advantage of directly collecting separated products.

Falk and Seidel-Morgenstern (2002) extended this earlier work introducing a dilution ratio, φ, which specifies the rectangular pulse illustrated in **Figure 20-7**:

$$\varphi = 1 - t_{\text{inj}}/\tau_{\text{cyc}} = t_{\text{reg}}/\tau_{\text{cyc}}, \qquad (20\text{-}23)$$

where inj τ_{cyc} and t_{reg} represent, respectively, the injection duration, the cycle period and the regeneration time between successive injections.

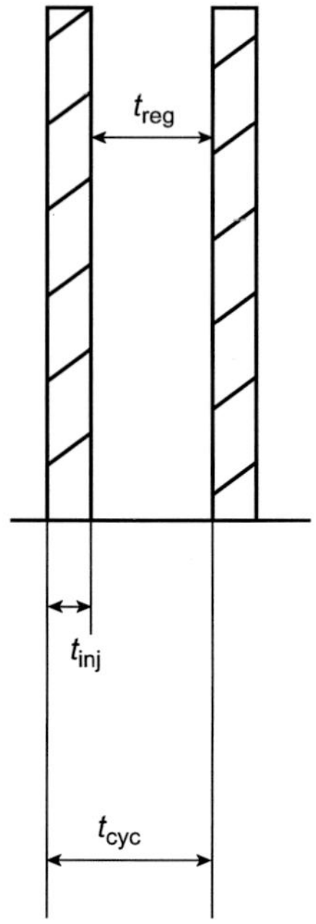

FIGURE 20-7 Illustration of injection, regeneration and cycle time for a pulse chromatographic reactor. *(Figure reproduced from* Falk and Seidel-Morgenstern (2002). *© 2002 by Elsevier Publishing Co.)*

When the cycle period is only slightly less than t_{inj} the dilution ratio is small and the operation of the chromatographic reactor functions like a conventional fixed bed reactor. Longer regeneration times lead to larger values of φ. *Falk and Seidel-Morgenstern (2002)* discussed the performance of chromatographic reactors in terms of three performance parameters: Productivity of the desired product, conversion of reactant and recovery of injected reactant in the pure product. The production rate showed maximum values of $\varphi = 0.8$. An increase of φ increased conversion and product separation. Recovery decreased as the dilution ratio drops and feed flow rate increases.

In the pulse chromatographic reactor, the catalyst is continuously regenerated by the carrier fluid between successive pulses. However, when one of the products is strongly adsorbed, regeneration is difficult for short

t_{reg} between pulses. An example arises in esterification using a cationic exchange resin as both catalyst and adsorbent because water is strongly adsorbed on the resin. Reactive chromatography is a solution of such a problem. In this operation the solute reacts with the adsorbate to regenerate the adsorbent. A reactive chromatographic reactor is also referred to as a fixed bed adsorptive reactor.

Mazzotti et al. (1997a) demonstrated experimentally the operation of reactive chromatography for the synthesis of ethyl acetate (EtAc) from ethanol (EtOH) and acetic acid (HAc) over an Amberlyst 15 resin. At start-up a mixture of reactants is continuously fed to the resin bed initially saturated with the EtOH solvent. As reaction occurs, water is trapped by the resin, whereas EtAc is carried by the fluid stream along the column. Since a product (EtAc) is removed from the resin surface, esterification proceeds until the limiting reactant (HAc) is consumed. At each location in the bed, this process continues until water saturates the resin. Thereafter, the composition remains constant and at equilibrium. **Figure 20-8** illustrates the time evolution of the outlet composition. First, EtAc breaks through together with ethanol. After water saturates the bed of resin, water and acetic acid break through together and the steady-state equilibrium composition is attained. In between these breakthroughs, a solution of EtAc in alcohol can be collected. Before a new feed pulse can be introduced, the resin must be regenerated. This can be accomplished using EtOH as the desorbent, as illustrated in **Figure 20-9**. The weakly adsorbed components, EtAc and HAc, are quickly desorbed resulting in a rather sharp transition. On the other hand, water elutes much more slowly. In both figures the curves were calculated by using a pseudo-homogeneous dispersion model with a rate term employing component activities rather than concentrations. Agreement between model and experimental and theory is satisfactory.

Silva and Rodrigues (2002) discussed in more detail the dynamic behavior of a fixed bed adsorptive reactor for the synthesis of diethylacetal from EtOH and acetaldehyde over a cationic resin catalyst (Amberlyst 18) with water as a byproduct. For a simulation they employed a mathematical model allowing for axial dispersion as well as external and internal mass transfer resistances. Adsorption was

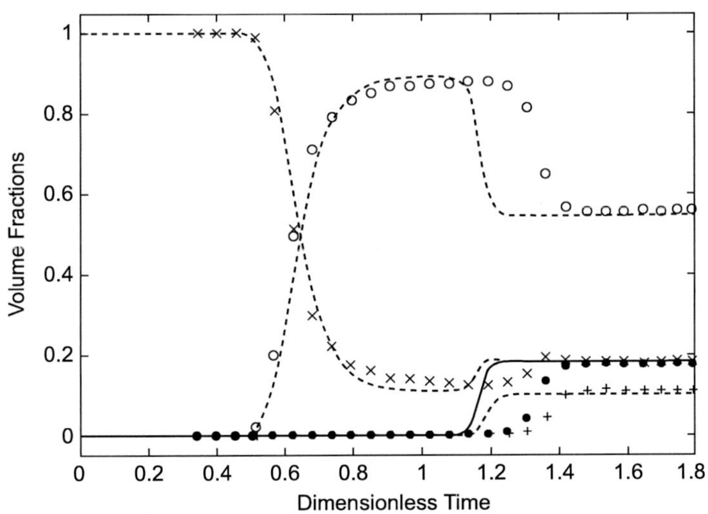

FIGURE 20-8 Reaction experiment with upflow through the bed; the plot shows the measured composition as volume fraction of species exiting the reactor: (●) acetic acid; (×) ethanol; (+) water; (○) ethyl acetate. All curves are calculated by the simulation model. *(Figure reproduced from* Mazzotti et al. (1997a). © *1997 American Chemical Society.)*

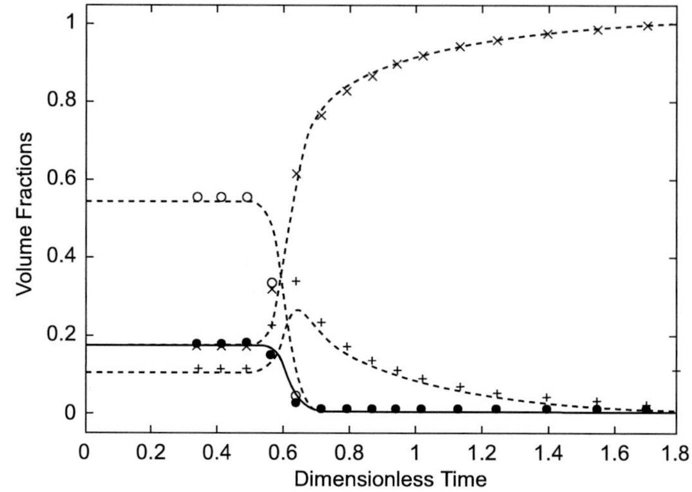

FIGURE 20-9 Regeneration experiment with downflow through the bed; the plot shows the measured composition as volume fraction of species exiting the reactor: (●) acetic acid; (×) ethanol; (+) water; (○) ethyl acetate; all curves are calculated by the simulation model. *(Figure reproduced from* Mazzotti et al. (1997a). © *1997 American Chemical Society.)*

described by a multi-component Langmuir isotherm and an experimental reaction rate expression was used. The experimental concentration vs. time profiles in their reactor for the reaction and the regeneration steps were similar to those shown in **Figures 20-8** and **20-9**. In order to better understand the behavior of the concentration-time profiles, the internal concentration profiles of each species in the fluid phase inside the reactor at different times were determined through simulation.

Gelosa et al. (2003) investigated the synthesis of a commercial product glycerol triacetate (triacetine) by esterification of glycerol with acetic acid using reactive chromatography and the approach of *Mazzotti et al. (1997a)*. A chromatographic reactor can produce high-purity triacetine at high conversions, but the presence of water in the resin at the end of the regeneration step has a detrimental effect on the final triacetine purity. The reaction proceeds consecutively through a series of three esterification steps, each producing water, with glycerol monoacetate and glycerol as intermediate products. The experimental apparatus consisted of a single column with an internal diameter of about 1.5 cm and a length of 44 cm that was packed with the resin Amberlyst 15. Transient changes in the concentrations of components at the reactor outlet in both reaction and regeneration experiments were compared with predictions of a mathematical model, the kinetics and equilibrium parameters of which were estimated from batch experiments. Agreement was considered to be good.

In addition to being used as a screening tool, pulse chromatographic reactors have been used to examine the behavior of catalysts that swell in the presence of reactants or products so that the trapping capacity of the solid changes with fluid composition. This behavior is observed with polymer-based ion exchange resins such as Amberlyst 15. Researchers in Zurich developed a model for the esterification of ethanol by acetic acid over a cationic exchange resin that allowed for a variation of the resin void fraction and in this way its adsorption capacity. For this reaction, the rate of diffusion in the resin is affected by swelling. However, the effect on reactor performance is small (*Mazzotti et al., 1997a*).

Extraction of rate of reaction and adsorption equilibrium constants was demonstrated by *Yu et al. (2004)*. Good agreement was found with constants obtained through other techniques. In a new development, *Seidel-Morgenstern* and colleagues use matrix-based adsorption modeling assuming adsorptive equilibria whose behavior in a fixed bed is well-known and show how a reactive system can be transformed into an adsorptive one provided the reactions are linear. This approach was applied to hydrolysis reactions of different stoichiometries to determine whether complete conversion and product separation is possible (*Vu et al., 2005*). In addition, these investigators present a predictive model that introduces dispersion and is not restricted to linear kinetics.

20.6 COUNTERCURRENT MOVING BED CHROMATOGRAPHIC REACTOR

Production limitations of pulse operation can be overcome by allowing the catalyst and adsorbent solids to flow through the reactor countercurrent to the flow of reactants and products. When this is done, the feed to the reactor and withdrawal of products are continuous rather than periodic. The reactor operates at steady state and only ordinary differential equations describe the reactor performance. Catalyst and adsorbent, however, are not at steady state as they move through the reactor space and come into contact with different fluid environments. These solids experience cyclically changing environments.

The throughput of a chromatographic reactor may be substantially increased by using a countercurrent moving bed. Product yield per weight of catalyst or adsorbent, however, may not change greatly because a much larger weight of solids will be needed when they circulate through the reactor. The moving bed alternative exchanges the complexities and problems of periodic operation for those of moving solids. These problems are nonuniform flow and attrition of the solids. Attrition means solids replacement and the continuous removal of fines.

The countercurrent moving bed chromatographic reactor (CMCR) is often associated with the University of Minnesota research group directed initially by *Aris* and later by *Carr*. However, the concept was proposed at about the same time by a Japanese team (*Takeuchi and Uraguchi, 1976a*) who studied separations possible with moving beds and demonstrated throughput advantages. **Table 20-5**, redrawn from *Aida and Silveston (2005)*, summarizes the countercurrent moving bed chromatographic literature.

There appears to be little current interest in the type of countercurrent moving bed chromatographic reactors shown in **Figure 20-2** with downward cascading solid catalyst and adsorbent because of the well-known problems encountered with flowing solids. However, past work, summarized in **Table 20-5**, has been useful for understanding concentration distributions in practical reactive chromatographic systems.

Tables 20-1 and **20-2** contain pseudo-homogeneous and heterogeneous models for CMCRs. Danckwert's boundary conditions are used for this type of reactor.

Using a pseudo-homogeneous model and assuming plug flow, *Aris* and co-workers searched for concentration discontinuities (shock fronts) that arise in CMCRs. They did so using various mathematical procedures: 1) Method of characteristic where results are plotted as phase planes or as hodographs (*Viswanathan and Aris, 1974; Cho et al., 1982; Fish and Carr, 1989*) and 2) stability analysis (*Petroulas et al., 1985a,b*). Top discontinuities where the solid phase enters the CMCR, as well as feed point and bottom discontinuities where fluids enter, are always present and internal discontinuities can occur if the fixed bed of catalyst and adsorbent is long enough. The critical parameters for discontinuity location are κ, the relative adsorptivities of product and reactant,

TABLE 20-5 Countercurrent Moving Bed Chromatographic Reactor Literature

Investigators	Type of Study	Reaction	Observations
Viswanathan and Aris (1974)	Calculation and simulation	General irreversible reaction	No improvement in conversion was found, but existence of shock fronts was observed.
Takeuchi and Uraguchi (1976a, b)	Calculation	General	Calculation of separation based on moments.
Takeuchi and Uraguchi (1977a, b)	Experimental and simulation	General, CO oxidation	No improvement in conversion was observed. Model validation undertaken
Takeuchi et al. (1978)	Simulation	General consecutive reactions	Selectivity improvement over a PFR was found.
Cho et al. (1982)	Simulation	General reversible reaction	Discontinuity identified and discussed.
Altshuller (1983)	Simulation	General	Undertook treatment of a complex isotherm.
Petroulas et al. (1985a, b)	Experimental and simulation	Hydrogenation of mesitylene	Improved product purity and conversion over equivalent PFR.
Fish et al. (1986)	Experimental and simulation	As above	Description of reactor design and troubleshooting of operational problems.
Fish and Carr (1989)	Experimental and simulation	As above	Model validation.
Lode et al. (2003a)	Simulation	General	Comparison of CMCR and SCMCR design.

and the Aris Number, σ, the ratio of solid velocity to fluid velocity in the CMCR. This term arises in the dimensionless mass balances and is defined in **Table 20-3**. To describe the movement of a reaction species i, another form of the Aris No. can be used,

$$\sigma_i = \frac{U_s}{u_f}\frac{1-\varepsilon_b}{\varepsilon_b}\frac{NK_A}{1+K_AC_A+K_BC_B} \quad (20\text{-}24)$$

where a binary reaction A $\rightleftarrows$ B and adsorptive equilibrium given by a Langmuir isotherm is assumed. The presence of an internal discontinuity is essential to performance of the CMCR. In the region of the bed below the discontinuity, CMCR performance is given by the reaction system, namely reaction equilibrium, whereas, above the discontinuity, adsorptive equilibria determine the overhead composition. **Figure 20-10** illustrates discontinuities in a CMCR with a stripping section for an A $\rightleftarrows$ B isomerization in which A is the more strongly adsorbed component. The recycled adsorbent contains neither A nor B. The feed contains just A, but adsorbed B is carried past the feed point by the adsorbent. A is trapped by the adsorbent and converted to B so it does not leave in the product stream. The reaction system is isothermal and the equilibrium relation is given in the hodograph, **(c)** in the figure.

Mixing cannot be avoided in flow through a porous media. Thus, more accurate mass and heat balance models must contain dispersion terms, such as the $D_{x_i}\frac{\partial^2 C_i}{\partial x^2}$ term in **Eq. (20-1)** and the $k_x\frac{\partial^2 T}{\partial x^2}$ term in **Eq. (20-3)** for the pseudo-homogeneous model of **Table 20-1** or in **Eqs (20-7)**, **(20-8)**, **(20-11)** and **(20-12)** for the heterogeneous model in **Table 20-2**. Allowing

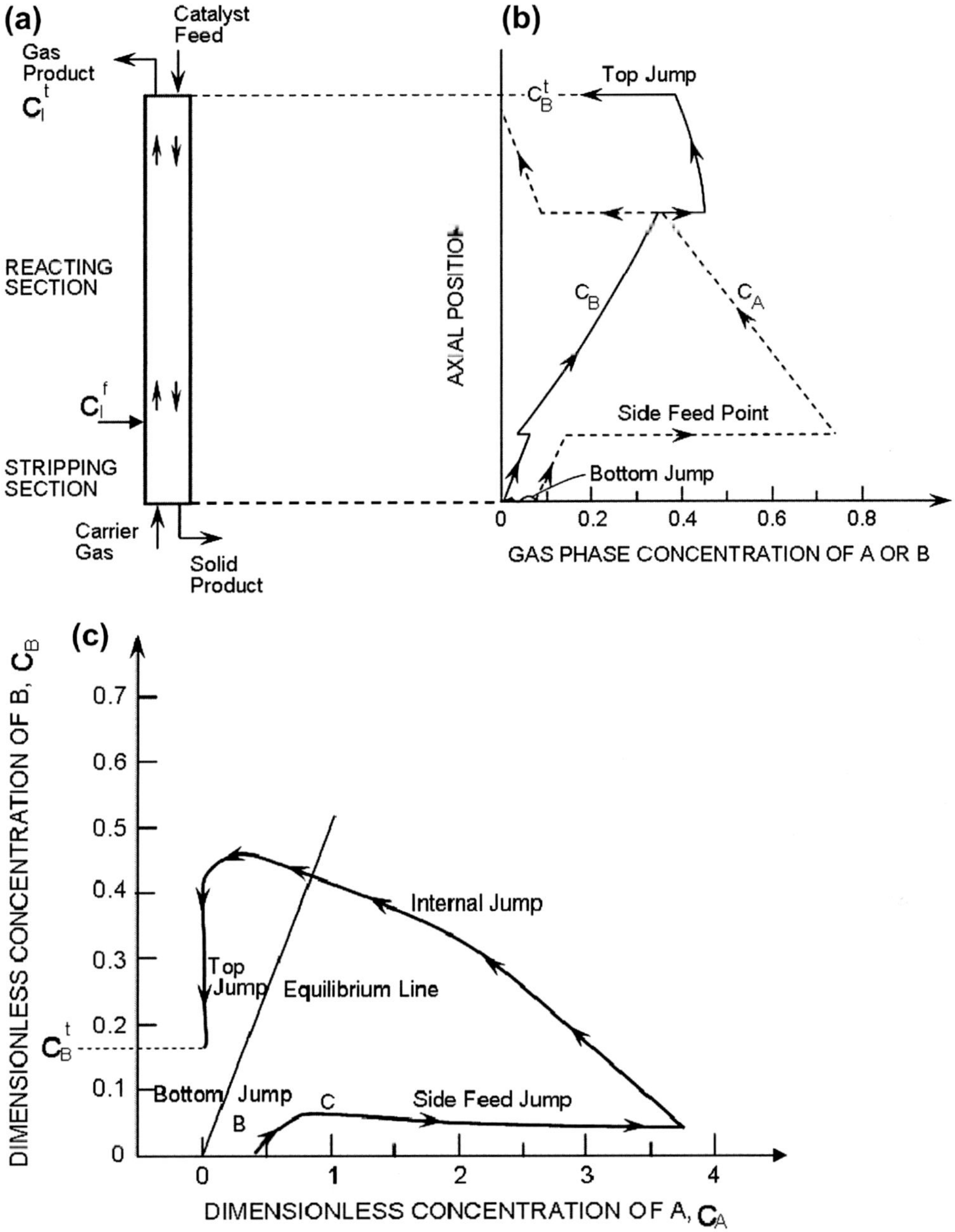

FIGURE 20-10 Discontinuities in a simulation of an A ↔ B equilibrium controlled reaction in a countercurrent moving bed chromatographic reactor: (a) Schematic of the CMCR with a side feed point and stripping gas fed at the bottom, (b) reactant "A" and product (B) trajectories within the reactor bed and (c) hodographic representation showing discontinuities. *(Figure copied from* Petroulas et al. (1985a) *with permission. © 1985 Pergamon Press Ltd.)*

for dispersion replaces the discontinuities by steep concentration and temperature gradients. Their steepness depends on the magnitude of the dispersion coefficient. Indeed, steep gradients are predicted in models for simulated moving bed chromatographic reactors. Examples of measured or predicted steep gradients in SCMCRs may be seen, for example, in **Figures 21-4**, **21-7**, or **21-19**.

20.7 CONTINUOUS ROTATING ANNULAR CHROMATOGRAPHIC REACTOR

Moving the bed instead of individual catalyst/adsorbent particles solves the attrition and nonuniform flow problems of moving beds mentioned earlier. Such a moving bed can be realized by rotating an annular bed of catalyst and adsorbent past feed and carrier fluid injection points in a top plane and product and carrier fluid removal points along a bottom plane as shown in **Figure 20-11**. Although the figure shows a biochemical application, the continuous rotating annular chromatographic reactor (CRAC) can be used for any reaction with a small heat. Passing fluids through rotating packed beds of solids is a well-established technology in energy conservation. Rotating beds are commercially available in various sizes for recovering heat continuously from hot exhausts and transferring the captured heat to combustion air. They are described and their performance analyzed in the regenerative heating literature.

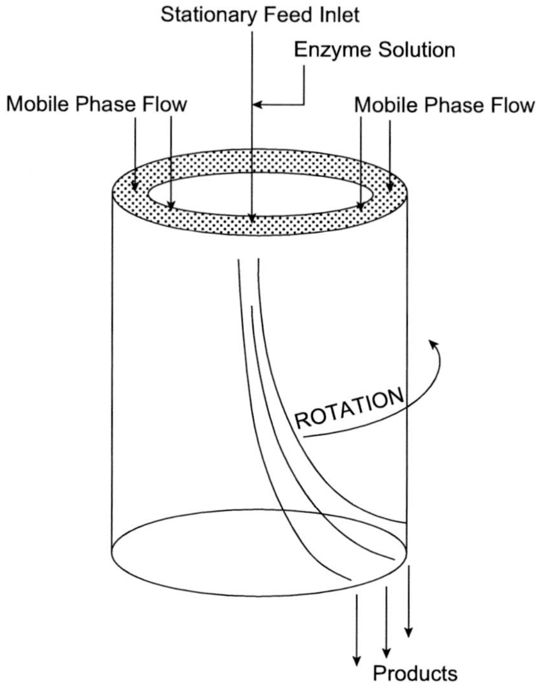

FIGURE 20-11 **Operating principle of a CRAC reactor for an enzyme-catalyzed reaction. The rotating annular volume is packed with a mixture of catalyst and adsorbent while the mobile phase is the eluent. The reaction species carried furthest from the feed point in the angular direction is the species most strongly adsorbed.** *(Figure reproduced from* Sarmidi and Barker (1993a) *with permission. © 1993 by the Society for the Chemical Industry.)*

CRAC systems have been used in preparative chromatography and are being explored for reactive chromatography as indicated in **Table 20-6**. Interest seems to have lapsed since 1995 despite their simple operation and demonstrated performance. CRAC is perhaps the only moving bed chromatographic reactor that is ready for commercial application. The only limitation at the present state of development is throughput, arising from mechanical limitations on rotating a heavy body of solids in the form of an annulus. Laboratory scale devices are discussed by *Aida and Silveston (2005)*.

With the exception of an early paper (*Wardwell et al., 1982*), CRAC reactors have been used for liquid systems. These systems, particularly biochemical ones, are isothermal so they can be described by just material balances. They are also continuous so the time derivative disappears. Their geometry, however, requires modification of their model equations; the position variable, x, from the material

TABLE 20-6 Continuous Rotating Annular Chromatographic Reactor Literature

Investigators	Type of Study	Reaction	Observations
Cho et al. (1980a, b)	Simulation and experiment	Catalytic hydrolysis of methyl formate	Complete conversion was obtained; model predicted experiments closely.
Wardwell et al. (1982)	Simulation and experiment	Catalytic dehydrogenation of C_6H_{12}	Conversion greatly exceeded equilibrium limit. Simulation predicted separation poorly.
Sarmidi & Barker (1993a)	Experiment	Enzymatic saccharification of starch	Conversion to maltose very much higher than in a batch reactor.
Sarmidi & Barker (1993b)	Simulation and experiment	Enzymatic inversion of sucrose	Complete conversion of sucrose; simulation closely predicted experimental performance.
Herbsthofer & Bart (2003)	Simulation and experiment	Fe^{3+} reduction by Ir^{4+}	CRAC reactor gave 5 x higher conversion than a PFR. Model verified experimentally.

balances in **Tables 20-1** and **20-2** is retained, but a second variable, ζ, reflecting angular position from the point of feed injection, must be introduced. Thus,

$$\varepsilon\omega\frac{\partial C_i}{\partial \varsigma}+\rho_{ads}\frac{\partial q_i}{\partial \varsigma}+\frac{1}{A}\frac{\partial n_i}{\partial x}-D_{x_i}\frac{\partial^2 C_i}{\partial x^2}=\rho_{cat}\nu_i r, \quad (20\text{-}25)$$

where ω is the angular velocity and A is the x-section of the mixed bed of catalyst and adsorbent. Molecular diffusion in the direction of rotation is neglected.

Sarmidi and Barker's experiments are compared with a simulation using **Eq. (20-25)** in **Figure 20-12**. Data were obtained from a 1.2 × 135 cm X-section annular bed packed with a Dowex 50W-X4 adsorbent and an α-amylase maltogenase. The enzyme was bled into the eluent stream. It can be seen that the simulation slightly overestimated the product concentrations in the take-off streams.

Herbsthofer and Bart (2003) using the reduction of ferric ion by an iridium ion as their test system also observed good agreement of simulation and experiments at 50°C. They also found an optimal angular velocity for Fe^{3+} reduction. *Aida and Silveston (2005)* provide a further discussion of this contribution as well as some earlier work.

20.8 STEPWISE, COUNTERCURRENT MULTI-STAGE FLUIDIZED BED CHROMATOGRAPHIC REACTOR

Dutch investigators (*van der Wielen et al., 1990,1998; Vos et al., 1990a,b,c*) have investigated a true moving bed chromatographic reactor employing periodic fluidization of a mixed

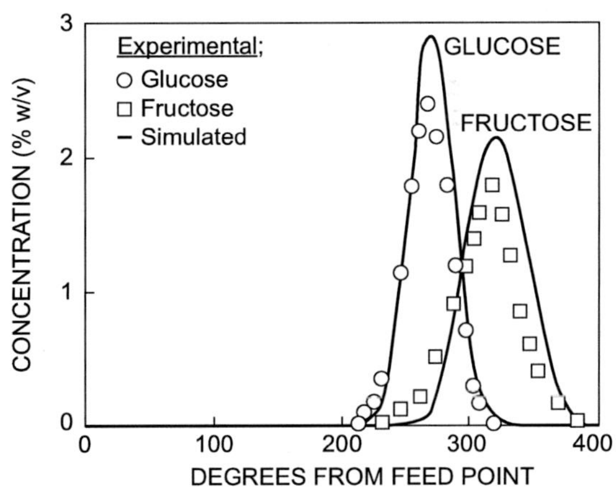

FIGURE 20-12 Comparison of a CRAC reactor simulation with experimental measurements for the enzymatic inversion of sucrose at 100°C. Eluent flow rate = 8000 cm^3/h, liquid feed rate = 230 cm^3/h containing 25 wt% sucrose; ω = 240°/h. *(Figure reproduced from* Sarmidi and Barker (1993b) *with permission. © 1993 by Pergamon Press Ltd.)*

catalyst and adsorbent bed. By using particles of a different size and density for adsorbent and catalyst, these researchers caused the catalyst particles to be held stationary in the reactor vessel, while the adsorbent particles separated from the mixed bed and flowed under gravity downward to a lower stage. Thus the adsorbent moved countercurrent to the reactant. The means for accomplishing segregation and downflow was adjustment of the fluid upflow velocity and periodic flow direction reversal. Operation of the *Vos-van der Wielen* system is depicted in **Figure 20-13**. *Van der Wielen et al. (1998)* discuss the application of this system to the preparation of 6-aminopenicillanic acid from penicillin G. Research on this novel moving bed system appears to have ended. *Aida and Silveston (2005)* provide a lengthier discussion of the *van der Wielen* work.

20.9 FIXED BED CHROMATOGRAPHIC REACTOR WITH FLOW DIRECTION SWITCHING

A strongly adsorbed reactant or product can be carried out of the bottom of a countercurrent moving bed chromatographic reactor if the eluent flow is too small, diminishing the yield or product separation. Raising the eluent flow corrects this difficulty but with the penalty of a diluted overhead product and the extra eluent cost. *Caram and Viecco (2004)* demonstrate that periodically switching the flow direction through the catalyst-adsorbent bed can trap the strongly adsorbed component to ensure its complete conversion. This rearrangement avoids higher eluent consumption and product dilution. **Figure 20-14** illustrates the operation of the flow reversal alternative. The authors investigated the performance of their system for first-order, reversible and irreversible isomerization reactions, for an irreversible consecutive reaction using cell (lumped) models (see **Section 20.2.2**). They also undertook a simulation of the hydrogenation of mesitylene in order to compare performance to that of a CMCR (*Petroulas et al. 1985a,b; Fish & Carr, 1989*) and observed that conversion and product purity achieved by the reverse flow chromatographic reactors are lower than those achieved in a CMCR for the same assumptions.

Reverse flow operation is the subject of **Chapter 18**.

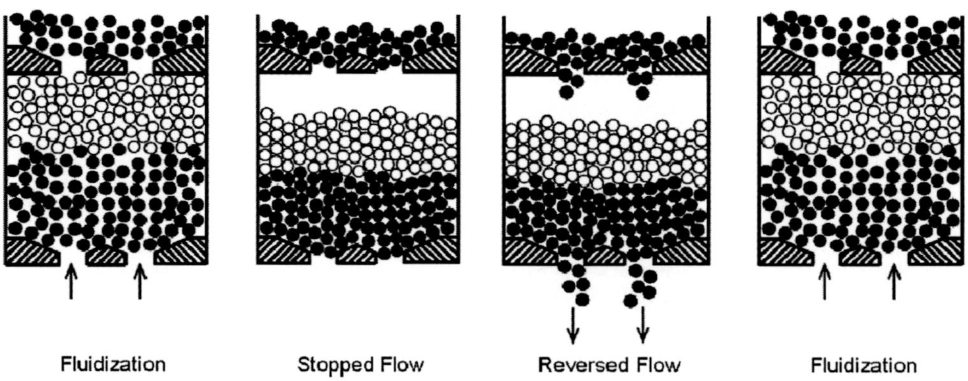

FIGURE 20-13 **Sequence of fluid flow operations in a liquid flow pulsed, fluidized bed chromatographic reactor with countercurrent flow of liquid phase reactant and adsorbent. Filled circles = adsorbent particles; open circles = catalyst particles.** *(Figure reproduced from* van der Wielen et al. (1998) *with permission. © 1998 by the AIChE.)*

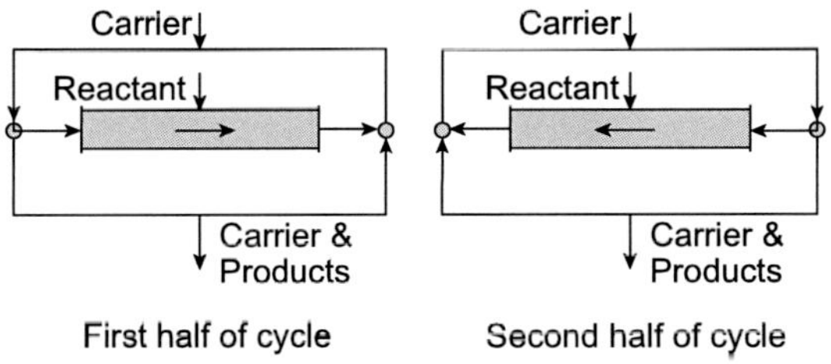

FIGURE 20-14 Schematic of the operation of a reversing flow chromatographic reactor showing the two parts of a cycle. *(Figure reproduced from* Caram and Viecco (2004) *with permission. © 2004 AIChE.)*

20.10 EXTRACTIVE REACTOR SYSTEMS

Extractive reactors are liquid-liquid systems that can function as separating reactors for reactants available as liquids. Likewise absorbers can also perform both reaction and separation for gas phase reactants. We do not consider these operations in this book because they are generally operated continuously. Furthermore, contacting immiscible streams of liquids or gases and liquids are well-developed operations in industry.

There are exceptions for several processes employing reactive extraction and absorption. For these, periodic operation is either advantageous or simply necessary. One of these processes is the air oxidation of SO_2 over an activated carbon catalyst. Several activated carbons function at room temperature, but they strongly adsorb the SO_3 product. In the past, investigators have shown that the adsorbed SO_3 can be quantitatively stripped from the carbon surface by water to form sulfuric acid. Oxidation can be performed in a trickle bed packed with carbon granules with a continuous, co-current feed of air, SO_2 and water (*Hartman and Coughlin, 1972; Mata and Smith, 1981*); however, considerably higher oxidation rates are attained by periodically switching between liquid and gas flow (*Haure et al., 1989; Metzinger et al., 1994; Lee et al. 1996b*). **Chapter 17** deals with periodic operation of three phase reactors and SO_2 oxidation is discussed there in some detail.

A similar technique was used by *Yamada and Goto (1997)* for the hydrogenolysis of carbobenzophenylalanine to phenylalanine over a Pd/carbon catalyst employing a dichloroethane solvent. Phenylalanine is insoluble in this solvent and deactivates the catalyst by crystallizing on its surface. Water, however, dissolves phenylalanine so these researchers carried out the hydrogenolysis in a countercurrent trickle bed by periodically switching the liquid feed at the top of the bed between the reactant dissolved in dichloroethane and water. The latter stream contained the product. *Aida and Silveston (2005)* discuss the *Yamada and Goto* experimental work.

20.11 CENTRIFUGAL PARTITION CHROMATOGRAPHIC REACTOR

We end this chapter with a brief discussion of the application of a centrifugal partition chromatographic reactor (CPCR) for a liquid phase reaction. The CPCR employs reactant pulses and operates with two liquid phases, one of which is held stationary at the centrifuge outer wall by the vessel design and a density difference between it and the second, circulating liquid. The stationary phase contains a catalyst, generally an enzyme, as applications have been to biochemical systems. Products of the reaction must have different partition coefficients for the two liquid phases.

When a pulse is carried into contact with the stationary liquid, a reactant enters that phase. Products are formed and diffuse back into the circulating phase, partially separated by differing partition coefficients. The mobile phase is moved through the centrifuge for repeated interchange with a different portion of the stationary phase, thus affecting reactant conversion and separation of the reaction products as may be seen by separate peaks in the circulating liquid. *DenHollander et al. (1998)* describe the application of CPCR to the chirally selective enzymatic hydrolysis of N-acetyl-L-methionine into L-methionine and by-product acetic acid. A second paper dealt with hydrolysis of a racemic mixture of N-acetyl-methionine.

Notation

A = bed or reactor cross sectional area (cm^2)
A = component
a = specific surface area
a_m = specific surface area for mass transfer (cm^2/g_{solid})
B = component
CMCR = continuous moving bed chromatographic reactor
CPCR = centrifugal partitioning chromatographic reactor
CR = chromatographic reactor
CRAC = continuous rotating annular-bed chromatographic reactor
CSTR = continuous stirred tank reactor
C = concentration (mol/cm^3)component
C_i = concentration of i
C_p = specific heat (cal/mol·K)
D_{x_i} = axial dispersion or diffusivity of species i
d = diameter (cm)
d_c = vessel diameter (cm)
d_p = particle diameter (mm)
E = activation energy (various units)
EtAc = ethyl acetate
EtOH = ethanol
F = volumetric flow rate (m^3/h)
F_0 = entering volumetric flow rate
H_a = heat of adsorption (kJ/mol)
HAc = acetic acid
ΔH_R = heat of reaction (kJ/mol)
h = heat transfer coefficient (various units)
h_0 = heat transfer coefficient
i = chemical species indicator
J_k = kinetic coefficient in the Ergun Eq.
J_v = viscous coefficient in the Ergun Eq.
j = position, stage or step indicator, cell number
K = equilibrium constant
K_i = adsorption equilibrium constant for species i (L/g)
k = position, stage or step indicator species or component indicator
k = rate constant (various units)
k′ = temperature independent rate constant (coefficient in the Arrhenius term)
k_m = mass transfer coefficient ($mol/cm^2 \cdot s$)
k_r = rate constant
k_s = conductivity of the solid phase ($J/cm^2 \cdot s$)
k_x = axial conductivity ($J/cm^2 \cdot s$)
k_γ = defined in **Table 20-3**
L = bed or column length (m)
MeOH = methanol
m = position, stage, step indicator
N = adsorption capacity, number of adsorption sites (1/g adsorbent)
N_{Ar} = Aris number defined in **Table 20-3**
N_c = Number of components
N_{cell} = number of cell
N_{Da} = Damköhler No.
N'_{Da} = temperature independent Damköhler No.
N_d = Bodenstein No. for mass
N_h = Bodenstein No. for heat
N_m = ratio of volumetric mass transfer coefficient to space velocity
n = integer
n = molar flow rate (mol/h)
n_i = molar flux of component I, molar flow rate for species I (mol/h)
n_s = number of segments or stages in a column or bed
PDE = partial differential equation
PFR = plug flow reactor
PSR = pressure swing reactor
P = pressure (kPa)
q_i = adsorbate loading ($mol/g_{adsorbent}$)
r = rate of reaction (various units)
SCMCR = simulated continuous moving bed chromatographic reactor
s = counting integer
TSR = temperature swing reactor
T = temperature (K)
T_a = ambient temperature
T_s = solid or surface temperature
t = time (s, h)
t_{inj} = injection time (s)
t_R = retention time (s, min)
t_{reg} = regeneration duration (s, min)
$\bar{t}$ = residence time (h)
U_s = solids velocity
u = superficial velocity (cm/s)
u_f = fluid superficial velocity (cm/s)
VOC = volatile organic compound

V = volume, usually of bed or reactor (m^3)
Vj = volume of jth bed or column
$w_{0.5}$ = width at half height
X_i = conversion of reactant i
x = axial position, also stage no. counting from top or bottom of a column

Greek Symbols

α = coefficient in the Ergun Eq.
β = Prater No.
ε = void fraction ( )
ε_{ads} = adsorbent volume fraction
ε_b = void fraction in bulk of solid, also bed void fraction
ε_{bed} = void fraction in bed
ε_{cat} = catalyst volume fraction
ε_p = particle (adsorbent) void fraction
ε_t = total void fraction, usually referred to the bed volume, but also to particle volume
φ = dilution ratio
κ = ratio of adsorptivities
λ = parameter in Ergun relation
μ_g = gas viscosity (various units)
ν = dimensionless total molar flow rate
ν_I = stoichiometric coefficient for component i dimensionless molar flow rate for component i
θ = dimensionless time
ρ = density
ρ_{ads} = adsorbent density (g/cm^3)
ρ_b = bulk density of adsorbent or catalyst-adsorbent (g/cm^3)
ρ_{cat} = bulk density of catalyst
ρ_g = gas density
ρ_s = solid density
σ = Aris No. or flow ratio of solid and fluid
τ = period (s, h)
τ_{cycle} = cycle time or period (s, h)
τ_s = switching time, duration between movement of inlet/outlet location (s, h)
ω = angular velocity (rad/s)
= ratio of solid to fluid heat capacities
ψ = ratio of pressure and thermal energies
ζ_v = normalized viscous pressure drop contribution in the Ergun Eq.
ζ_k = normalized kinetic pressure drop contribution in the Ergun Eq.

Superscripts

f = feed
j = cell or position or step indicator
k = position or step indicator
m = time-step indicator
t = top
* = flow ratio, equilibrium quantity

Subscripts

A, B = chemical species
a = ambient, adsorbent adsorption
ads = adsorbent
b = bulk
cat = catalyst
cyc = cycle
Da = Damköhler
f = feed
g = gas
i = reaction component or species
inj = injection
j = j^{th} section, zone or bed, also reaction species or reaction number
k = kth section, zone or bed, also reaction species indicator
L = outlet or exit
m = mass
n = stage number
ns = number of top stage
p = particle, bead, pellet or withdrawal point
R = reaction
reg = regeneration
s = solid, or switch (switching)
t = total
x = axial
0 = feed or inlet
0^- = just before inlet
0^+ = just after inlet
0.5 = half height
1,2 = position or time indicators

CHAPTER

21

Simulated Moving Bed Chromatographic Reactors

Kenji Hashimoto, Motoaki Kawase, Peter Lewis Silveston

Kyoto, Japan and Waterloo, Ontario, Canada

OUTLINE

21.1 Operation and Application 598

21.2 Modeling and Simulation 599

21.2.1 Isothermal Systems 599

21.2.2 Modeling of SCMCRs 599

21.2.3 Numerical Simulation of SCMCRs 609

21.2.4 Optimization 614

21.2.5 Non-isothermal Systems 616

21.2.6 Separate Catalyst and Adsorbent Beds 616

21.3 Experimental Studies 620

21.3.1 Gas–Solid Systems 620

21.3.2 Liquid–Solid Systems 624

21.3.3 Biochemical Systems 627

21.4 Other Reactor Applications of Simulated Moving Beds 632

The simulated countercurrent moving bed chromatographic reactor (SCMCR), introduced in the previous chapter, employs a stationary bed, as with a pulse chromatographic reactor, but with feed and product removal points that shift locations with time. This operation can be carried out with a single fixed bed, multiple beds in a cascade or with an annular bed. The reactor is an adaption of the simulated countercurrent moving bed chromatographic separator described by *Liapis and Rippin (1979)* and *Barker* and co-workers (*Barker and Deeble, 1975; Barker et al., 1983; Barker and Ching, 1980*). Operation of the separator and several applications have been discussed by *Ruthven and Ching (1989)*. The extension to reacting systems seems to have been disclosed in a U.S. patent by *Zabransky and Anderson (1977). Aida and Silveston (2005)* discuss research on SCMCRs in considerable detail.

21.1 OPERATION AND APPLICATION

The operation of a single bed reactor is illustrated in **Figure 21-1**. The bed is divided into n_s segments. For each segment, a feed or a stripping fluid (eluent) can be added or a product stream withdrawn. At the mth time interval, reactant is fed into the jth segment and product is taken out from the kth segment. The entrances or exits of other segments are closed. After a time period, the switching time τ_s, the feed and product segments are shifted from j^{th} to $(j\text{-}1)^{th}$ and from k^{th} to $(k\text{-}1)^{th}$ ports respectively. The shifting of these ports is periodic and moves from position to position along the axis of the bed. An SCMCR consequently retains all the characteristics of a countercurrent moving bed chromatographic reactor (CMCR) without actual movement of the solid phase. Switching solves the difficult mechanical problems of continuously introducing and removing solids from a vessel, uniform plug flow movement, as well as catalyst attrition and loss of inventory. The design shown in the figure is feasible for small diameter columns, up to 4 cm, but in larger installations, consecutive beds would be employed instead of segments. Each bed is then connected to the succeeding bed through a conduit which contains a valved tee connection that can be used to introduce feed or an eluent or to remove product. With multiple beds, catalyst or adsorbent could be changed, or volume or temperature of a bed altered. The SCMCR mode provides probably the widest operational flexibility for a chromatographic reactor.

The SCMCR essentially functions as a countercurrent moving bed chromatographic reactor

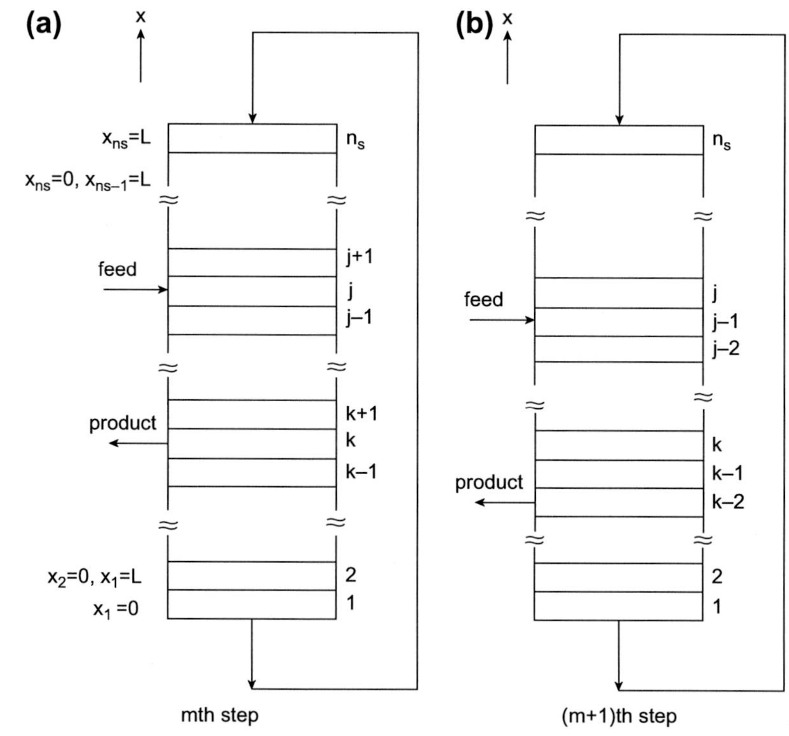

FIGURE 21-1 Principle of the simulated countercurrent moving bed chromatographic reactor in which the feed and product removal ports in a single-bed reactor change periodically: (a) Location of the ports in the mth time step, (b) location of the ports in the $(m+1)^{th}$ time step. (*Figure adapted from* Aida and Silveston (2008) *with permission. © 1990 Pergamon Press Plc.*)

(CMCR) and should provide a similar performance. Indeed, equivalence relations have been by devised by *Lode et al. (2001)*. The more beds a section contains, the closer is the approximation to a CMCR. For a section in an SCMCR:

$$V_j = n_j V_{bed} \quad (21\text{-}1)$$

$$\tau_s = (1 - \varepsilon_{bed}) \frac{V_{bed}}{(Q_s)_{CMCR}} \quad (21\text{-}2)$$

$$(Q_{fluid})_{SCMCR} = (Q_{fluid})_{CMCR}(1 - \varepsilon_{bed}) \frac{V_{bed}}{\tau_s}. \quad (21\text{-}3)$$

In the first relation, V_j is the volume of a section in either an SCMCR or a CMCR, while V_{bed} is the volume of a column or a separate bed in that section of the SCMCR. τ_s is the switching time or the time between changes of port location in the SCMCR and Q_s is the volumetric rate of solids flow in the equivalent CMCR. Q_{fluid} is the flow rate of the fluid in the same section.

Since the original SCMCR concept in the late 1970s, there have been many studies of SCMCRs through simulation and experiment. **Table 21-1** and **Table 21-2** summarize the experimental studies; many of these also consider simulation.

Simulated countercurrent moving bed chromatographic reactors usually consist of several sections with different functions, so that running numbers, such as 1,2,3,4, are attached to each section. Unfortunately, this numbering differs among researchers. In what follows, the numbering system introduced by *Hashimoto et al. (1983)* is used.

21.2 MODELING AND SIMULATION

21.2.1 Isothermal Systems

A pseudo-homogeneous, isothermal mass balance for adsorption and reaction in an SCMCR was introduced by *Ray et al. (1990, 1994)*. It is a PDE with independent variables t (time) and x (position) given as **Eq. (20-1)** in **Table 20-1**. *Ray et al. (1990)* left out the dispersion term in **Eq. (20-1)** because local mixing is usually small and barely affects reactor performance. Other investigators, however, retain the term because it allows for error introduced by neglecting mass transfer and property changes along the flow path.

An SCMCR contains J beds or columns; also **Eq. (21-4)** must be written for all reactants. With I independent species, usually reactants, an SCMCR must be described by at least $I \times J$ PDEs. Finding a solution numerically is often a formidable undertaking. The PDEs must be integrated after each switching duration, τ_s, for the initial conditions ($t = 0$):

$$C_{i_j}(x) = C_{i_{j-1}}(x, \tau_s), \quad q_{i_j}(x) = q_{i_{j-1}}(x, \tau_s),$$
$$T(x) = T(x, \tau_s), \quad (21\text{-}4)$$

where T, $C_{i_{j-1}}$, $q_{i_{j-1}}$ are the temperature, concentration, and the adsorbate loading at the same position, x, in the upstream bed just prior to switching the feed and withdrawal points. **Eq. (21-4)** applies as well to non-isothermal operation.

Although the pseudo-homogeneous model has been widely employed, some applications must allow for fluid-solid mass transfer. This requires use of a heterogeneous model such as given by the PDEs in **Table 20-2**. Initial and boundary conditions for the heterogeneous model follow those given above for the pseudo-homogeneous application.

21.2.2 Modeling of SCMCRs

Ray et al. (1990, 1994) considered a reversible reaction: Mesitylene (MES) hydrogenation. Parameters examined were switching time, τ_s, and a switching velocity, ζ. The latter represented the hypothetical velocity of the solid phase, defined as $\Delta x/\tau_s$, where Δx is the spacing

TABLE 21-1 Application of Simulated Countercurrent Moving Bed Chromatographic Reactors to Gas–Solid Systems

Authors	Study Objectives	Reactor Configuration	Reaction	Operating Conditions/ Variables	Observations	Comments
Tonkovich et al. (1993)	Increasing C_2 yield	Separate reactor followed by a heat exchanger and a separate adsorber	Oxidative coupling: $2\,CH_4 + O_2 \rightarrow C_2H_6 + H_2O$	SmO_2 catalyst held at $823\text{ K} \leq T \leq 1048$ K, activated charcoal served as adsorbent and operated at 373 K, $20\text{ s} \leq \tau_s \leq 33$ s, $2{:}1 \leq CH_4/\,O_2 \leq 3{:}1$	CH_4 conversion per reactor was between 2 and 3%. Selectivity to C_2 + products close 100% at $\tau_s = 20$ s.	Experiments were designed to find optimal operating conditions.
Tonkovich and Carr (1994a)	Increasing C_2 yield	As above; SmO_2 catalyst packed in 13 mm o.d. × 500 mm quartz tubes with 70 mg of catalyst; activated. charcoal adsorbent in 7 mm o.d. × 76–110 mm steel tubes	Oxidative coupling: $2\,CH_4 + O_2 \rightarrow C_2H_6 + H_2O$	As above	Selectivity reaches 80% at 1023 K and $\tau_s = 28$ s, CH_4 conversion is 65%.	C_2 yields were ca. 50%, well in excess of the 30% target for a commercial operation.
Ray and Carr (1995a)	Obtaining conversion in excess of equilibrium	5 packed bed reactors with valving for mixed C_9 and H_2 feed and a N_2 flush	Hydrogenation of mesitylene (MES) to trimethylcyclohexane (TMC)	13 mm o.d. x 300 mm beds packed with 10 wt% Pt/ Al_2O_3 and 90 wt% Chromosorb 106, $T = 463$, 473 K, Eluent = N_2; variables were τ_s and T	Excellent product/reactant separation with 83% MES conversion to TMC; 96% TMC in product stream.	Fair agreement with system simulation, offset due to inadequacy of the Langmuir adsorption isotherm for MES.
Bjorklund and Carr (2002)	Increasing MeOH yield	Single tubular reactor feeding 3 separation columns packed with a Supelcorport adsorbent	Non-catalytic partial oxidation of CH_4 with O_2	Non catalytic, with unconverted CH_4 and O_2 recycled, P = 100 atm, feed $CH_4{:}O_2 = 2{:}1$; variables were T and τ_s	At 750 K, $\tau_s = 440$ s, CH_4 conversion was 50%, selectivity to $CH_4 = 50\%$ for an MeOH yield of 25%.	Recycle of unconverted gas mixed with fresh feed gave a $CH_4{:}O_2 = 16{:}1$ at reactor inlet. Cyclic steady state reached in 60 min, major problem was loss of CH_4.
Kundu et al. (2009)	Increasing C_2 yield	4 section reactor with 8 columns, τ_S = 30s	Oxidative coupling: $2\,CH_4 + O_2 \rightarrow C_2H_6 + H_2O$	Columns were 0.64 cm id and 51cm in length, reaction temp. = 750°C, separator temp. = 100°C, τ_s = 20 to 33 s	At 750 K, $\tau_s = 33$ s, methane conversion was 54.5%, selectivity to $C_2 = 65.6\%$ for a C_2 yield of 36%.	Conversion and yield go through maximum values with switching time. Selectivity decreases. Simulation showed good agreement with experimental results.

TABLE 21-2 Application of Simulated Countercurrent Moving Bed Chromatographic Reactors to Liquid–Solid Systems

Authors	Study Objectives	Reactor Configuration	Reaction	Operating Conditions/ Variables	Observations	Comments
Hashimoto et al. (1983)	Demonstration of increased fructose yield, model verification	3-section rotating system with fixed inlet and take off valves consisting of 7 reactors and 16 adsorbers	Glucose isomerization over a solid immobilized glucose isomerase; Ca^{2+} exchanged Y zeolite served as adsorbent.	Atmospheric pressure, $T = 323$ K, $120 < \tau_s < 180$ s, water served as eluent; experimental variables were glucose and eluent feed rates and glucose concentration.	Target conversion of 55% was achieved.	SCMCR achieved target conversion at a much lower energy demand than existing processes. SCMCR and a CMCR model represent experimental measurements adequately.
Barker et al., (1987a); Barker and Ganetsos (1988)	Investigation of SCMCR performance.	Enzyme catalyzed reaction with 12 adsorption columns, 5.4 cm i.d. × 75 cm, with a Ca^{2+} ion exchange resin	Formation of dextran from sucrose over a Ca^{2+} exchanged polystyrene resin.	$T = 298$ K, continuous feed of enzyme at 0.8 DSU/cm^3, water served as eluent.	Fructose product of sucrose polymerization is trapped by the resin. This allows polymerization to proceed.	Increased pulse volume or sucrose concentration suppressed dextran formation.
Barker et al., (1987b)	As above.	As above.	As above.	Effect of sucrose concentration, single vs. multiple pulsing explored.	High molecular weight dextran was possible.	Pulse frequency is important with multiple pulsing.
Akintoye et al. (1990, 1991)	Enzyme catalyzed reaction with 12 adsorption columns, 5.4 cm i.d. × 75 cm, with a Ca^{2+} ion exchange resin.	Formation of dextran from sucrose over a Ca^{2+} exchanged polystyrene resin.	Sucrose inversion using an immobilized invertase enzyme	$T = 25°C$, throughput at 16 kg sucrose/m^3 resin·h	SCMCR forced inversion to near completion.	Much lower loss of enzyme reported by authors.
Barker et al., (1992a, 1990b)	As above.	As above.	Saccharification of a starch over an immobilized maltogenase	Starch feed rate = 116 g/h	Starch conversion of 60% with maltose product that was 96%	
Barker et al., (1992b)	As above.	As above.	Biosynthesis of dextran from sucrose.	$T = 25°C$, pH = 5.3, $\tau_s = 30$ min	Complete conversion of sucrose, but dextran product contaminated by levan, a polyglucose.	SCMCR performance deteriorated after 50 h on stream.

(Continued)

TABLE 21-2 Application of Simulated Countercurrent Moving Bed Chromatographic Reactors to Liquid–Solid Systems (*cont'd*)

Authors	Study Objectives	Reactor Configuration	Reaction	Operating Conditions/ Variables	Observations	Comments
Shieh and Barker (1995)	Effect of operating variables on performance.	As above, but with α-amylase maltonase added to the eluent.	Starch conversion; hydrolysis of the maltotroise to maltose and glucose.	$T = 25°C$; variables were Q_{eluent}, τ_s, starch concentration, enzyme activity.	Eluent flow rate has only a small affect on starch conversion, but affects maltose purity.	Limit in starch concentration set by viscosity effects that appear to limit mixing of the enzyme with the starch feed.
Shieh and Barker (1996)	Comparison of an SCMCR with a CRAC reactor	As above, but with a lactase enzyme co fed with lactose; additional column added to desorb galactose.	Hydrolysis of lactose by the enzyme lactase to form glucose and galactose.	$T = 25°C$; variables were Q_{eluent}, lactose concentration, τ_s, enzyme activity.	Hydrolysis driven to completion by galactose adsorption, SCMCR performance controlled by product separation.	Use of an SCMCR substantially reduced enzyme requirement, but simple model failed to predict product purity.
Kawase et al. (1996)	Determination of SCMCR performance and model testing	Four-section SCMCR consisting of 6 columns, 1 cm i.d. x 30 cm, packed with an Amberlyst 15 ion exchange resin that acted as catalyst and adsorbent.	Esterification of β-phenetyl alcohol with acetic acid.	Variables were the ratio of withdrawal rates of extract and raffinate, column temperature.	Conversion of acetic acid reached 99%, ester purity = 99% at optimal operating conditions. Stationary, cyclic operation attained after 300 min on stream.	Authors' model closely predicted experimental results. SCMCR performance was sensitive to the operating temperature.
Mazzotti et al. (1996)	Demonstration of the use of an SCMCR to force a reaction to completion	Three-section SCMCR consisting of 8 thermostated columns, 7 cm in length, each packed with 3.1 g of an Amberlyst 15 ion exchange resin.	Esterification of ethyl alcohol with acetic acid with the alcohol serving as eluent.	$\tau_s = 6$ min, HAc fed at 15 g/h with alcohol to acid feed rate = 20:1	Complete conversion of acetic acid achieved, raffinate EtAc mole fraction = 0.10 with100% recovery of the ester; performance sensitive to reactant feed rate and eluent/ reactant feed ratios.	Both CMCR and SCMCR models predicted experimental performance, but the moving bed model was easier to use. Density difference driven flow seriously compromised performance.

Meurer et al. (1996)	Comparison of SCMCR with a separate reactor and SMB separator	Four-section SCMCR consisting of 8 columns (2-2-2-2) with 2.5 cm i.d. × 12.8 m.	Irreversible enzymatic inversion of sucrose with excess water over a Dowex 99 resin.	Switching time = 630 s	SCMCR performed better than a single reactor followed by SMB separators, with higher product purity at the same feed flow rate. Maximum purity was 99% vs. 97% for use of an SMB separator.	SCMCR gives higher yield and better separation for reversible as well as irreversible reactions.
Mensah and Carta (1999)	Investigation of a two-section SCMCR and different operating modes	Two-section SCMCR consisting of 4 columns packed with a mixture of an immobilized lipozyme and an ion exchange resin.	Esterification of isoamyl alcohol by propionic acid using an enzyme catalyst	Reactants fed as hexane solutions, column being regenerated is removed and flushed separately by isoamyl alcohol in hexane at a high flow rate. Variables studied were number of columns, enzyme to adsorbent ratio and τ_s.	High ester concentrations were measured; model predicted concentration variations with time, but this was not observed. Increasing columns in a section from 1 to 2 with no change in immobilized enzyme greatly increased productivity.	A problem occurred with irreversible enzyme deactivation when the stream flowing through a column was almost only water. SCMCR system showed much higher productivity than use of the immobilized enzyme adsorbent.
Kawase et al. (2001)	Investigation of the application of an SCMCR to a multi-reaction system	4-section unit consisting of 12 columns, 1.2 cm i.d. × 19 cm, packed with Amberlite CR-1310, enzyme catalyst was introduced with reactant.	Production of lactosucrose from lactose and sucrose employing the β-fructofuranosidase enzyme.	$\tau_s = 10$ min, enzyme activity = 100 activity units/m^3	Lactosucrose yield ≅ 60% at sucrose conversions > 50% with complete separation of reaction products; cyclic stationary state achieved in 900 min.	Performance greatly exceeded that obtained in a batch reactor without using an adsorbent.
Lode et al. (2001)	Evaluation of SCMCR performance and model verification	Model C-920 moving bed chromatographic separator consisting of 10 2.4 cm i.d. × 30 cm columns packed with Amberlyst 15 ion exchange resin.	Esterification of MeOH with HAc with the alcohol serving as reactant and eluent.	$\tau_s = 24$ min, $T = 25°C$; Variables were reactant ratios and MeOH feed rate.	Complete conversion of acetic acid along with separation of the ester where achieved, but at low ester productivity/m^3 of resin and high use of MeOH solvent.	Downflow versus upflow tested to determine the effect of mixing caused by fluid density differences. Only a small effect was observed.

(Continued)

TABLE 21-2 Application of Simulated Countercurrent Moving Bed Chromatographic Reactors to Liquid–Solid Systems (*cont'd*)

Authors	Study Objectives	Reactor Configuration	Reaction	Operating Conditions/ Variables	Observations	Comments
Silva and Rodrigues (2005)	Development of an SCMCR process with simulation, lab. and pilot-scale experiments	4-section pilot scale SCMCR with12 columns (3-3-3-3), 2.6 cm i.d × 23 cm length. Resin radius = 400 μm.	Synthesis of diethylacetal (fragrance) from acetaldehyde and ethanol over an Amberlyst 15 resin.	Flow rates (for reference case) were $Q_F = 10$, $Q_R = 25$, $Q_D = 50$, $Q_X = 35$, $Q_{Rec} = 20$ mL/min, $\tau_s = 3.7$ min, $T = 10°C$	Simulated purities were raffinate = 97.8%, extract = 99.9%, Conversion = 99.7%.	Performance parameters and concentration profiles were simulated using true moving-bed and SCMCR models. Both models gave good agreement.
Da Silva et al. (2005)	Application of SCMCR to fructose production	Simulation of a 4 reactor, 8 adsorber system (see above) using experimentally determined model parameters.	Isomerization of glucose to fructose using glucose isomerase and Mg^{2+} exchanged Y zeolite.	Ideal operation at 55°C assumed with 1 molar glucose feed.	Simulations indicated 70 to 90% fructose in the extract depending on operating conditions with an Mg^{2+} exchanged resin.	Experimental verification needed.
Pereira et al. (2009)	Test of models. Examination of the effects of operating parameters	4-section pilot scale SCMCR with12 columns (3-3-3-3) or (3-3-4-2), 2.6 cm i.d. × 23 cm length. Resin radius = 343 μm	Synthesis of ethyl lactate (biodegradable solvent) from lactic acid and ethanol over a wet Amberlyst 15 resin	Flow rates were $Q_F = 1.8$, $Q_R = 8.8$, $Q_D = 27.4$, $Q_{rec} = 24.7$ mL/min in different experiments with $\tau_s = 2.9$ and 3 min, $T = 50°C$	Experimental purities (%): Raffinate = 72.8 to 75.2, extract = 95.5 to 97.8, conversion = 99.1 to 99.7%. These results showed fairly good agreement with simulation.	Performance parameters and experimental profiles were predicted with good accuracy using an SCMCR model.

between inlet and outlet of a bed. With this definition, the switching velocity can be related to the critical CMCR parameter, σ_I, by:

$$\sigma_i = \frac{1-\varepsilon}{\varepsilon} N K_{eq} \frac{\zeta}{u_g} \qquad (21\text{-}5)$$

where N and K_{eq} represent the number of adsorption sites and the equilibrium constant respectively. The parameter σ_I is proportional to the ratio of the switching and gas velocities. By adjusting u_g, and ζ, the condition $\sigma_A < 1$ and $\sigma_B > 1$ necessary to achieve separation and high conversion for a reversible reaction can be obtained.

Ray et al. (1994) used their model to examine the effect of inlet and outlet port location on reactor performance and on the time-varying concentration profiles in the catalyst and adsorbent beds. Their results are summarized by *Aida and Silveston (2005)*. For the SCMCR they considered, MES conversion was 97.7% after a stationary cycling state had been reached, whereas equilibrium conversion in a comparable PFR is 62%. **Figure 21-2** demonstrates the superiority of an SCMCR to a PFR.

The effect of the process parameters was further examined by *Fricke et al. (1999)* for a reversible decomposition, A ⇄ B + C, in the four-section SCMCR system (**Figure 21-3**). The net fluid flow and the simulated solid flow are shown as solid and dashed lines, respectively. Reactant A is fed into the system between Sections 2 and 3. Product B was assumed to be more strongly adsorbed than C. Therefore, C propagates towards the raffinate node and can be obtained at a high concentration between Sections 3 and 4 whereas B moves toward the extract node. In Section 1, a high concentration of desorbent forces B to be desorbed where it is then recovered as an extract.

In their model, *Fricke et al.* considered axial dispersion and interphase mass transfer but assumed that reaction takes place only in the liquid phase. The mass balances employed are given by **Eqs (20-7)** and **(20-9)** in **Table 20-2**. The volumetric mass transfer coefficient, however, was based on just the surface area of the adsorbent. The reaction, A ⇄ B + C, was assumed to be elementary. Initial conditions after a switch in port location are given by **Eq. (21-4)** and boundary conditions are discussed thereafter.

Performance of their SCMCR was judged by the maximum feed flow rate to achieve a product purity of 99.75% and the specific solvent consumption for that rate. *Fricke et al.* examined the effect on performance of the reactant adsorption constant (K_A), the separation factor for the products, namely K_B/K_C, the reaction

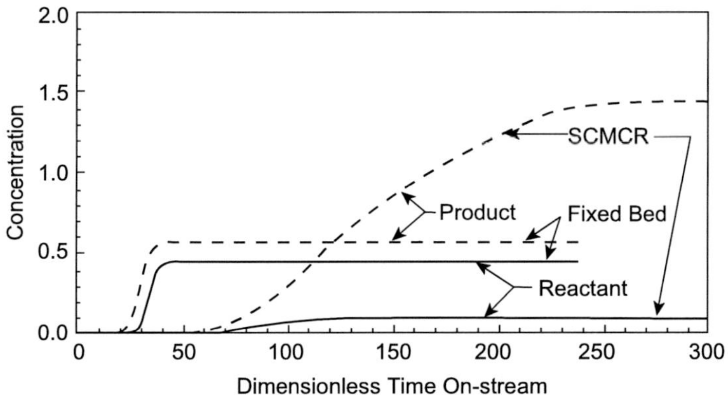

FIGURE 21-2 Transient concentrations of reactant and product for a fixed-bed reactor (i.e., a PFR) and for an SCMCR at a switching time of 10 s and a carrier flow rate of 10 cm $\cdot s^{-1}$. *(Figure adapted from* Ray et al. (1994) *with permission. © 1994 by Elsevier Science Ltd.)*

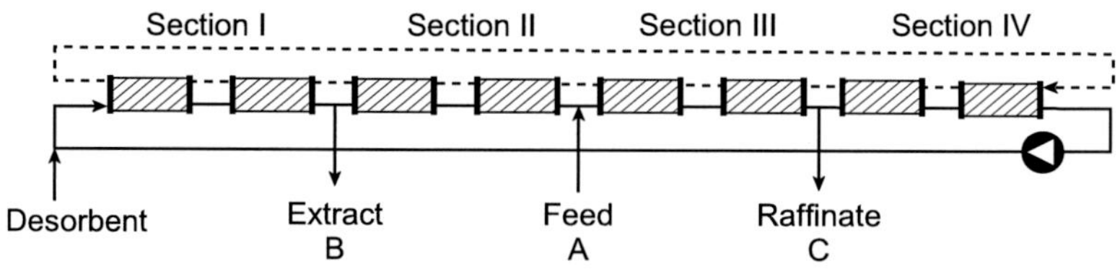

FIGURE 21-3 Schematic diagram of the multi-bed SCMCR studied by *Fricke et al. (1999)*. (*Figure adapted from* Fricke et al. (1999) *with permission of the copyright holder. © 1999 by Elsevier Science Ltd.*)

rate and equilibrium constants. **Figure 21-4** presents an example of their calculated results. Clearly, reactant A is present only in Sections 2 and 3, so C can be obtained with high purity in the raffinate and nearly pure B can be collected in the extract. The conditions which provide the performance shown in the figure are discussed by *Aida and Silveston (2005)*.

A comprehensive exploration of factors influencing SCMCR performance for an A + B ⇄ C + D reaction has been published by *Migliorini et al. (1999a, b)* and *Lode et al. (2001)* using HAc esterification of EtOH over an ion exchange resin as a model reaction. EtOH in this system acts both as reactant and solvent, while the resin serves as catalyst and absorbent for the water. Absorption occurs through swelling of the resin which is component- and concentration-dependent. Thus the void fraction shifts within a bed and differs between beds making up the SCMCR. EtOH, HAc, and reaction products have quite different densities, so convective mixing driven by density gradients was possible. *Migliorini et al. (1999a)* based their analysis on a heterogeneous PDE model, while *Lode et al. (2001)* used a CMCR model and applied the equivalence relations, **Eqs (21-1)** to **(21-3)**.

Both teams considered the SCMCR system treated by *Fricke et al. (1999)* and shown in **Figure 21-3**. Employing their studies of simulated moving bed chromatographic separators

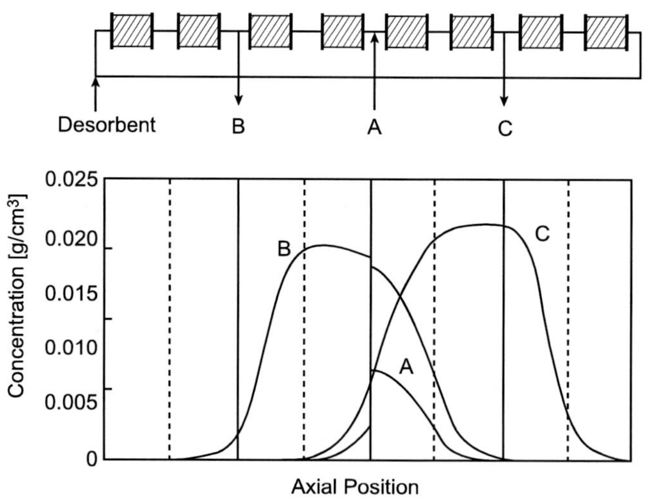

FIGURE 21-4 Concentration profiles of reactant and products in an SCMCR shown in Figure 21-3 at the end of a cycle. (*Figure adapted from* Fricke et al. (1999) *with permission. © 1999 by Elsevier Science Ltd.*)

(*Storti et al., 1988; Storti et al. 1993; Mazzotti et al., 1997a, b*), they identified Sections 2 and 3 in **Figure 21-3** as critical for performance, provided Sections 1 and 4 were operated to completely recover the raffinate and regenerate the adsorbent. In the previous chapter, we observed that performance of a moving bed chromatographic separator was found to depend on the absorptivity ratio, κ, and flow rate ratio, m_j. This remains true for an SCMCR. The flow rate ratios in Sections 2 and 3 control the separation. For the SCMCR, these ratios are

$$m_i = \frac{\textit{net fluid flow rate}}{\textit{adsorbed phase flow rate}} = \frac{Q_j\tau_s - V_j(1-\varepsilon_b)\varepsilon_p}{V_j(1-\varepsilon_b)(1-\varepsilon_p)}. \tag{21-6}$$

The index j represents a column or section in **Figure 21-3**, V_j is the volume of the column and τ_s is the switching time. Parameter m_j is inversely proportional to σ_j defined by **Eq. (21-5)** and is thus equivalent to $\beta_{kj} K_k$ (K_k is a distribution coefficient) defined by *Hashimoto's* group (*Hashimoto et al. 1983, 1993a, 1993b*) and to K/γ defined by *Ruthven's* group (*Ruthven and Ching, 1989*).

With complete regeneration of the adsorbent, reaction occurs only in Sections 2 and 3. Consequently, the range of acetic acid conversion and separation between ethyl acetate and water must be given by the m_2–m_3 plane. Separation performance depends on the switching time, τ_s, the volumetric flow rate, Q_j, and bed volume, V_j. Feed composition in terms of the HAc to EtOH ratio is important. EtOH is a reactant as well as the carrier so pure HAc can be fed to the reactor. Diluting the acid with EtOH, however, improves conversion and product purity, that is, it enlarges the region in the m_2–m_3 plane in which uncontaminated products can be found (*Migliorini et al., 1999a, b*).

Normalizing **Eq. (20-1)** introduces the Damköhler number. This important parameter is affected by residence time so that each of the four sections in the SCMCR may have a different number. Fluid flow rates are usually different and the number of beds in each section can be varied even if the total volume of adsorbent and catalyst is kept constant. Since esterification is usually confined to Sections 2 and 3, it is the residence times or the Damköhler numbers in these sections that matter. *Lode et al. (2001)* showed that if most of the esterification occurs in Section 3, increasing the Damköhler number in this section increased conversion, if not already complete, and raised productivity.

With respect to design, an SCMCR is like a simulated moving bed chromatographic separator. Usually both have four sections where the first and last provide regeneration of the adsorbent. Best performance of either system is achieved when reaction or adsorption does not occur in these sections. As a consequence, separation of products and the extent of an equilibrium-limited reaction will be governed by the middle sections. In these sections, the flow ratios given by **Eq. (21-6)**, m_2 and m_3, or by **Eq. (21-5)**, σ_2 and σ_3, and the appropriate Damköhler numbers, $(N_{Da})_2$ and $(N_{Da})_3$ determine the SCMCR performance. Considering just separation, the choice of eluent flow rate and τ_s for a bed of a specified length should place the reactor within the triangle defined by the diagonal in the m_2–m_3 plane shown in **Figure 21-5(a)** if adsorption can be represented by a linear isotherm. The boundaries of the triangular region are given by the Henry's Law constants for the adsorbed species. The abscissas of the points "a" and "b" are H_A and H_B, respectively. If competitive adsorption occurs, mandating the use of a multi-component Langmuir isotherm, the non-diagonal boundaries of the region for complete conversion in the SCMCR are curved as shown in **Figure 21-5 (b)**. In either case, the more strongly adsorbed product will be recovered at the extract port, while the weakly adsorbed product exits through the raffinate port.

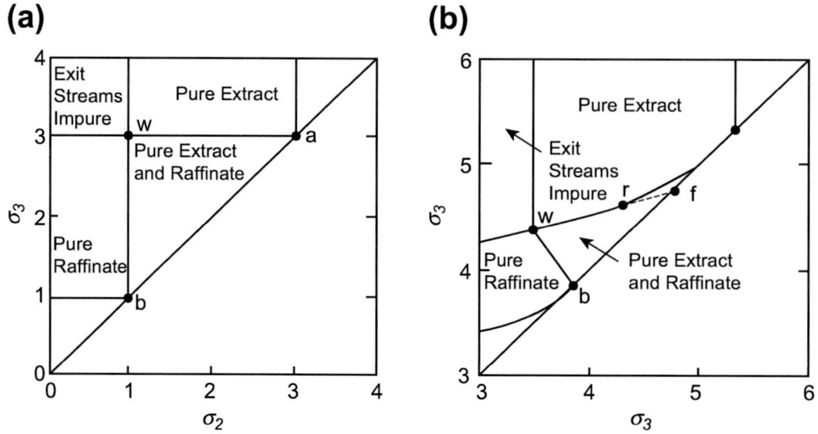

FIGURE 21-5 Triangular diagram representation of an SCMCR using flow rate ratios in Sections 2 and 3 surrounding the feed point: (a) for linear adsorption isotherms, (b) for multi-component Langmuir isotherms. (*Figure adapted from* Mazzotti et al. (1997c) *with permission. © 1997 Elsevier Science B.V.*)

Figure 21-5 applies to just equilibrium-limited reactions of stoichiometry A + B ⇄ C + D where one of the reactants functions as eluent. If equilibrium is not attained in either Section 2 or 3, the size of the triangular region in **Figure 21-5 (a)** and **(b)** shrinks as the Damköhler numbers in the sections decrease. Examination of the m_2–m_3 plane should indicate the expected performance even if neither reactant serves as eluent. However, in that case, the extract or raffinate could be contaminated by a reactant.

Although the simulated moving bed does not include movement of the adsorbent particles, its operation is more easily understood in terms of a moving bed. Thus, simplified models, called true countercurrent moving bed models, based on this consideration, are often employed. Computational effort is greatly reduced compared with that for a dynamic model which follows the actual switching within an SCMCR. However, the validity of the substitution must be examined. *Lode et al. (2003a)* compare true countercurrent moving beds and SCMCRs and discuss the interplay among the process design parameters. Analytical solution of the mass balance PDEs for the continuous countercurrent process was developed, using as a model system the reaction A ⇄ B + C with each species exhibiting linear adsorption behavior. Based on the solution, criteria were derived for the optimal process design with respect to productivity and solvent consumption. Comparison of these results with the numerical simulations of an SCMCR shows that a true moving bed model does not apply to an actual SCMCR with a finite number of columns per section, because the two units exhibit different residence time distributions and hence different degrees of conversion.

Strohlein et al. (2005) discussed effects of the equilibrium constant K_{Eq} for the reaction A ⇄ B + C on the triangular shaped region in the m_2–m_3 plane in **Figure 21-5 (b)** using an equilibrium model. They demonstrated that the operating regions for specific equilibrium constants approached the one for complete separation. They also examined experimentally the validity of the proposed method using the MeAc synthesis from HAc and MeOH over an ion exchange resin catalyst. Their measurements were consistent with the relative position in the m_2–m_3 plane with respect to complete separation and full conversion regions as predicted by their model.

21.2.3 Numerical Simulation of SCMCRs

A pseudo-homogeneous model was chosen by *Mensah et al. (1998a, b)* and *Mensah and Carta (1999)* for the adsorber-reactor in the two-section, four-column SCMCR they simulated. A two-section SCMCR can be used when just one reaction product adsorbs. The reaction considered was the esterification of propionic acid with isoamyl alcohol over a lipase immobilized on a macroporous anion exchange resin. Alcohol served as eluent, while the carrier fluid was hexane. A cationic exchange resin acted as the adsorbent for the water product. For their simulation, the various authors assumed the adsorption of water was mass transfer controlled, while esterification was kinetically controlled. However, adsorbed water deactivates the enzyme so an inhibition rate model was employed. Model parameters were determined experimentally (*Mensah et al., 1998b*). Simulation results are discussed together with the authors' experiments in **Section 21.3.2**.

Kawase et al. (1999) simulated the production of bisphenol A (BPA) from acetone and phenol in their SCMCR study. An ion-exchange resin catalyzes this liquid phase reaction and functions as adsorbent. The authors' simulation utilized experimental data. The SCMCR system examined by *Kawase et al.* had three sections as illustrated in **Figure 21-6**. Unlike the system used by *Fricke et al. (1999)*, there was no Section 1 so there was no recycle flow from Section 2 to Section 4. Recycle could be dispensed with because purification of desorbent was unnecessary due to the weak adsorption of BPA by the resin. Phenol functioned both as reactant and solvent so the reaction was conducted in excess phenol. Only acetone was fed between Sections 2 and 3. Phenol entered after Section 4.

In modeling, *Kawase et al.* neglected axial dispersion and assumed isothermal operation. As in the work of *Fricke et al. (1999)*, mass transfer between the fluid and solid phases was considered. Because there is no homogeneous reaction and the solid phase functions as both adsorbent and catalyst, **Eq. (20-9)** was used. The product, BPA, was not adsorbed in their model. A complication was that the adsorption of acetone depended on adsorption of water. A Langmuir isotherm was used for water while for acetone a linear one was assumed with an adsorption constant that depended inversely on water concentration. To obtain a rate expression for the reaction, a carbonium ion mechanism with decomposition of an intermediate species was assumed and taken to be rate

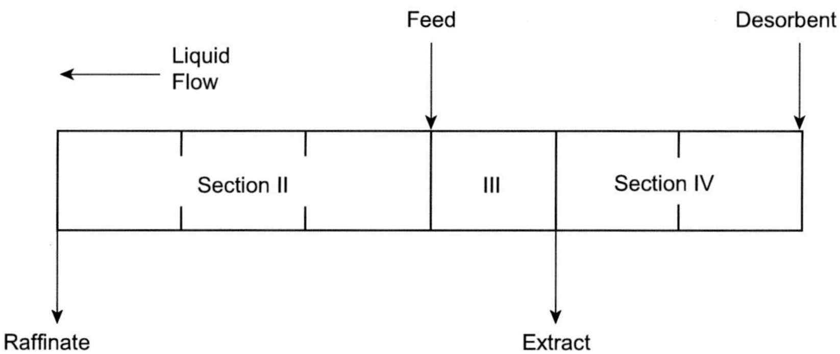

FIGURE 21-6 Schematic diagram of a simulated moving-bed reactor for production of bisphenol A. (*Figure adapted from* Kawase et al. (1999) *with permission. © 1999 by Elsevier Science Ltd.*)

controlling. Under the conditions used, the rate term simplified to the concentration of acetone divided by that of water. Initial and boundary conditions were typical of those used for SCMCRs.

Figure 21-7 shows an example of their simulation results. From **(a)** in this figure, it can be seen that water is found in all the sections whereas BPA and acetone are restricted to Sections 2 and 3. BPA can be obtained at high purity in the raffinate as shown in **Figure 21-7 (b)**. At the same time, phenol and water are withdrawn as extract. That stream is much larger than the raffinate stream because it is primarily phenol. *Kawase et al.* discuss the effect of parameters such as switching time, the number of beds in Section 2 and liquid velocity in Section 4 on acetone conversion.

Some years earlier, *Kawase et al. (1996)* simulated the esterification of HAc and β-phenetyl alcohol using a combined catalyst and adsorbent in an SCMCR system shown as **Figure 21-17**. Their simulation employed the model discussed above. Langmuir isotherms described the equilibrium adsorption of reactants and products, while a conventional Langmuir-Hinshelwood model was employed for reaction kinetics. *Kawase et al.* conducted a series of experiments in fixed beds to determine isotherm parameters and measure liquid–solid mass transfer coefficients. Batch experiments determined parameters for the kinetic model. Simulation results are compared with experimental data in **Section 21.3.2**.

The simulation work discussed so far in this section deals with model validation or assessing the influence of operating variables on SCMCR performance. *Zhang et al. (2001)* nicely illustrate the effect of these operating variables. Their system was the commercially important production of methyl tertiary butyl ether (MTBE) from tertiary butyl alcohol (TBA) and MeOH over an ion exchange resin. MeOH also served as the solvent. A four-section SCMCR was considered with two columns in each section (see **Figure 21-3**). Flow rate ratios (**Eq. (21-5)**) for each section were set so that

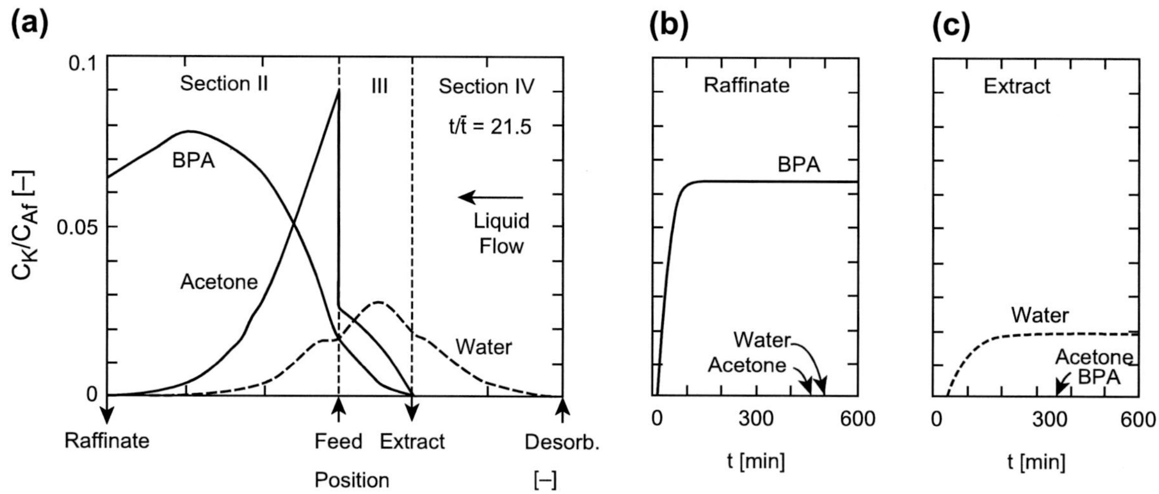

FIGURE 21-7 **Stationary concentration profiles in an SCMCR system (a), transient concentrations in the raffinate (b) and in the extract (c) for typical operating conditions.** (*Figure adapted from* Kawase et al. (1999) *with permission. © 1999 by Elsevier Science Ltd.*)

$\sigma_{MBTE} < 1$, while $\sigma_W > 1$. Thus, MTBE was the raffinate and water the extract.

Zhang et al. systematically studied the effects of switching time, solvent and raffinate flow rates, and number of columns on the conversion of TBA (the limiting reactant) as well as on yield, purity of and selectivity to MTBE. These authors simulated an operation using a feed of 20% TBA in MeOH, with raffinate withdrawal at 20% of the total flow at the feed port, and solvent introduction after Section 3 at 50% of the feed flow rate. Their base case assumed a switching time, $\tau_s = 840$ s, a column length of 25 cm and feed flow rate of 0.0167 cm^3/s. Experimentally determined adsorption equilibrium constants were used. The authors found that their SCMCR model was cyclically stationary after 75 switching periods (about 9 cycles). See *Aida and Silveston (2005)* for a summary of results. The predicted concentration profiles for all components except MeOH are shown in **Figure 21-8 (a)**. The adsorbed water removed in the extract stream controlled the time needed to reach the stationary state. This may be seen in **Figure 21-8 (c)**.

Our discussion of the *Migliorini et al. (1999a, b)* and *Lode et al. (2003a, b)* contributions pointed out that performance depends critically on the switching time. This is illustrated by comparing **Figure 21-8(b)** or **(c)** with **Figure 21-8(a)**. Only τ_s is changed in these figures. Reducing τ_s by about 30% lowers the conversion a small amount but results in very large decreases in yield and purity of MTBE in the raffinate. Much lower purity occurs because σ_{MTBE} increases with a smaller τ_s and MTBE is carried into Section 4 as **Figure 21-8(b)** shows. Increasing τ_s by about 30% (from 840 to 1080 s) should have raised conversion, but it decreased slightly. The explanation for this is that product separation is poorer, allowing the reverse reaction to proceed in Section 2. Increasing the solvent flow rate only strips water from the ion exchange resin and it leaves with the extra solvent at the extract port. Because water in Sections 2 and 3 is much lower, TBA conversion and MTBE purity at the raffinate port increase. Although higher solvent usage improves the SCMCR performance with respect to conversion and raffinate purity, Y_{MTBE} is not increased. Recovery of solvent and recycle are important costs. Increasing the withdrawal rate at the raffinate port changes the flow in Section 3 and 4, improving the separation in these sections and causes more of the reaction to occur in Section 2 as the TBA curves confirm. Both Y_{MTBE} and P_{MTBE} increase significantly. Raising the withdrawal rate at the raffinate port to 75% of the feed flow causes a slight reduction in performance because the solvent flow is insufficient to sweep all the water from the resin so water breaks through into Sections 2 and 3.

Could performance be further improved by adding or removing a column in Section 2, where almost all the reaction occurs? Removing a bed from Section 2 and adding one to Section 1, so the total number in the SCMCR is unchanged, decreased performance because the residence time in Section 2, being reduced by half, meant that TBA broke through into Section 3. This lowered conversion and MTBE purity in the raffinate stream. Increasing columns in Section 2 by one at the expense of Section 1 still reduced performance because the residence time in Section 1 was too short and water broke through into Section 2.

The above discussion of the *Zhang et al.* and *Lode et al.* contributions indicates that the best SCMCR performance is obtained for a large number of columns and high solvent-to-feed ratios. Cost considerations, however, limit bed numbers and solvent-to-feed ratios. In some reaction systems, it is difficult to express all design considerations in monetary terms. Safety, for example, is one such consideration. In other systems, insufficient cost data prevent the formulation of a single objective function. Often this is the situation in the early stages of process design. In such a case, optimization must be undertaken with multiple variables whose

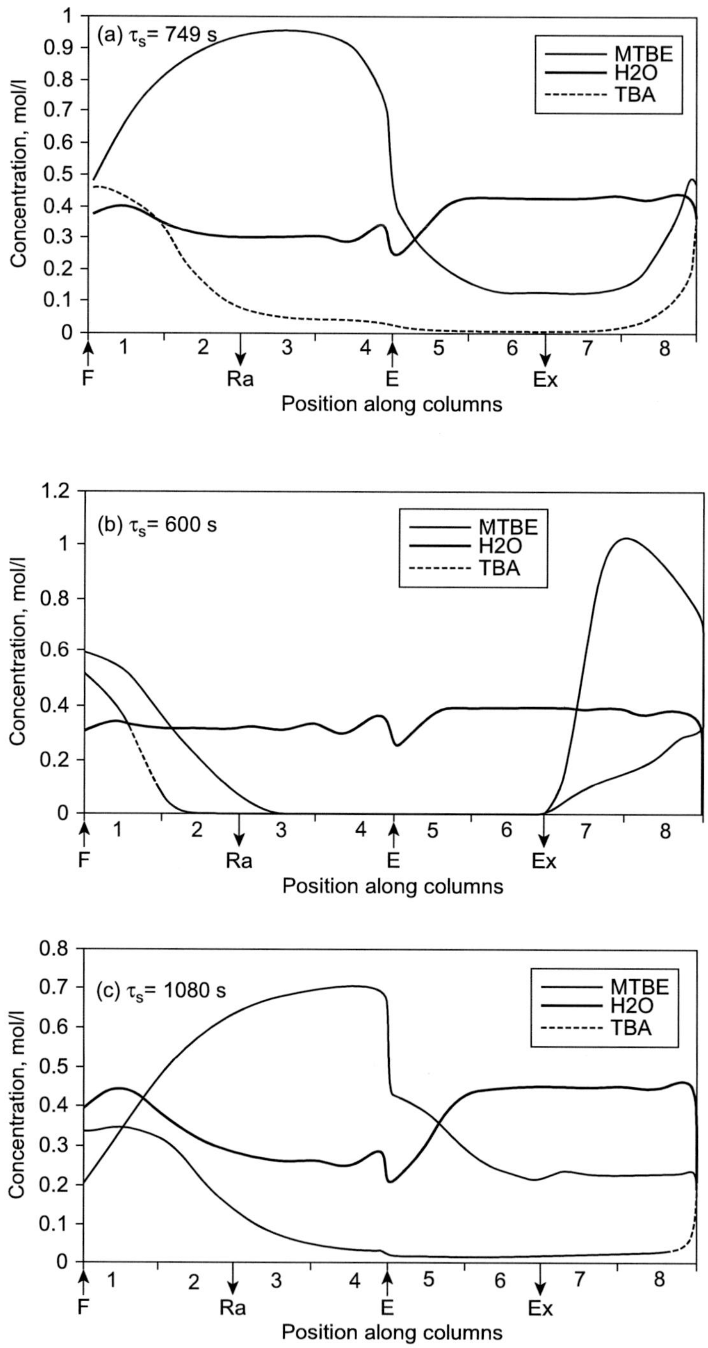

FIGURE 21-8 Effect of switching time on concentration profiles in a four-section, eight-column SCMCR with an ion exchange catalyzed reaction between TBA and MeOH to form MTBE: (a) base case, τ_s = 749 s, (b) τ_s = 600 s, (c) τ_s = 1080 s. (*Figure adapted from* Zhang et al. (2001) *with permission. © 2001 by the American Chemical Society.*)

maxima or minima are sought. There are several ways these variables may be expressed. *Zhang et al. (2002)* suggested for a reaction A $\rightleftarrows$ B + C that these variables are X_A, conversion of A in the SCMCR, Y_B, yield of the desired product B at the raffinate port, Γ_B, purity of B at that port, and Ξ_B, the selectivity to the desired product at that port. Normally, yield is the product of selectivity and conversion. However, for an SCMCR, yield and selectivity are defined in terms of products collected at the raffinate port, while conversion is based on products collected at both the extract and raffinate ports. In most of the examples considered so far, water, a product of no value, has been the extract. Consequently, only what is recovered at the raffinate port is important. Of course, Y_B and Ξ_B behave much alike when an independent variable is changed, provided X_A changes little, so it will not always be necessary to consider four functions in optimization. In this formulation, the solvent or extractant-to-feed ratio is not a dependent variable. It is independent and in all cases would be constrained, just as are the number of beds, the bed length, and the allocation of beds to the three or four sections of the SCMCR.

A multi-reaction simulation of a sugar system was undertaken by *Kawase et al. (2001)* as part of an experimental study. They considered the formation of lactosucrose from lactose and sucrose. Side reactions were the hydrolysis of sucrose to yield glucose and fructose, and the hydrolysis of lactosucrose, forming fructose and lactose. Transfer of the fructosyl moiety of sucrose to lactose is an enzymatic reaction occurring in the liquid phase. The hydrolysis of sucrose and the lactosucrose product are also enzyme-catalyzed. Lactosucrose is weakly adsorbed on the ion exchange resin compared with fructose and glucose. Equilibrium adsorption was assumed, so mass transfer from the liquid phase to the solid adsorbent was rate controlling. For their isothermal system, *Kawase et al.* employed **Eqs (20-7)** and **(20-9)** but assumed plug flow. Concentration C_i^* for the mass transfer driving force term was the mean value in the resin and assumed to be in equilibrium with the adsorbate. *Kawase et al. (2001)* found from their simulation that application of an SCMCR to lactosucrose formation would substantially increase conversion over that obtainable in a single phase, batch reactor.

The application of the SCMCR to the production of a high-purity biodiesel fuel has been reported by *Kapil et al. (2010)*. Biodiesel (fatty acid methyl ester, FAME) is produced by the transesterification of waste vegetable oils (triglycerides, FA) with MeOH. Vegetable oils, however, contain free fatty acid (FFA), which act as a catalyst poison for the homogeneous alkali-catalyzed reaction in transesterification. The concentration of FFA can be reduced by esterification with MeOH over solid catalysts. This reaction is equilibrium limited producing water as a byproduct. The catalyst also acts as an adsorbent for water, which is more strongly adsorbed than FAME or FFA. When the reaction is carried out in an SCMCR, high-purity FAME is recovered in the raffinate, while water is eluted with MeOH in the extract. *Kapil et al.* selected the esterification of acid dissolved in commercial sunflower oil over a silica-supported Nafion resin as their model reaction. Reaction kinetics were represented by a nonlinear rate expression; adsorption followed Henry's law. Parameter values were estimated from published experimental results. The typical *Kapil* SCMCR had four sections. Dynamic simulations showed a more than 90% conversion of FA and a FAME purity of 80% were achieved in the SCMCR.

Meurer et al. (1996) have shown that an SCMCR offers higher yields and better separation for reversible as well as irreversible decomposition reactions, such as the enzymatic inversion of sucrose to glucose and fructose. Their simulation model considered axial dispersion, and mass transfer resistances. Adsorption isotherms of all three components over a Dowex

Monosphere 99 resin were approximately linear, while the reaction kinetics were represented by the Michaelis–Menten equation and included substrate inhibition. The authors compared the performance of an SCMCR with two conventional processes for sucrose inversion. The first comprised an inversion reactor and chromatographic separation of glucose and fructose. The second process consisted of the same reactor and a simulated moving bed separator. Complete conversion could not be achieved in the reactor and residual sucrose as well as other impurities entered the chromatographic separator and thus decreased product purity in all sections of the unit. However, the SCMCR process was able to convert sucrose completely.

21.2.4 Optimization

An alternate set of dependent variables for the A ⇄ B + C reaction is productivity, ϑ_A, and the eluent requirement:

$$\vartheta_A = \frac{n_A}{V_{reactor}} \tag{21-7}$$

$$\Delta = \frac{n_{.solvent} + C_{solvent}Q_f}{n_A} \tag{21-8}$$

where Δ is the eluent requirement in terms of the feed rate, recognizing that eluent may enter after Section 3 and with the feed. *Lode et al. (2001)* used these dependent variables for the case of complete conversion of A and separation of B and C that they considered in their study. For optimization without complete conversion or product separation, productivity and solvent requirement could be added to X_A, $Y_B{}^*$, Γ_B and Ξ_B^*, thereby reducing optimizing constraints.

For their system, *Lode et al. (2001)* demonstrated that the residence time or Damköhler number in Section 3 and the solvent-to-reactant ratio in the feed stream control the size and shape of the complete conversion and separation region in the m_2–m_3 plane. They also showed that the solvent ratio is inversely proportional to the switching time and the productivity defined by **Eq. (21-7)**. Productivity, in turn, is inversely proportional to the switching time and the number of columns in the SCMCR. Higher productivity through smaller switching times implies a higher solvent ratio. There is thus a trade-off between productivity and the solvent requirement.

Optimization is usually an iterative operation for SCMCRs. Column length and total number of columns must be specified; then an arrangement of columns into sections can be chosen. If sufficient cost data are available, a single objective function can be fashioned from feed cost, ϑ_A, and Δ. A suitable two-dimensional optimization routine can be applied to find the optimal τ_s and solvent to feed ratio. Another column arrangement may then be tested to see if the objective function can be reduced further. Once the optimal arrangement of bed into sections has been identified, the influence of the number of columns and column length can be investigated.

If a single objective function cannot be formulated, the trade-off between productivity and solvent ratio, or, more generally, between conversion, selectivity and purity leads to Pareto-optimal solutions in which pairs of variables, such as τ_s and feed composition give ostensibly acceptable performance. Choice of the pair or set depends on other non-quantifiable information. *Zhang et al. (2002)* discussed optimization in this case using MTBE synthesis as their example. Two objective functions were considered: yield, Y_{MTBE}, and purity, Γ_{MTBE}. Conversion, X_{TBA}, also important, was treated as a constrained variable rather than as an objective function.

Figure 21-9 illustrates that as long as $Y_{MTBE} < 0.88$, high values of Γ_{MTBE} and Ξ_{MTBE}^* can be achieved. X_{TBA} is greater than 0.98 for all values of the independent variables chosen within the constraint limits. All of these solutions are satisfactory so that further information is needed to select from among them.

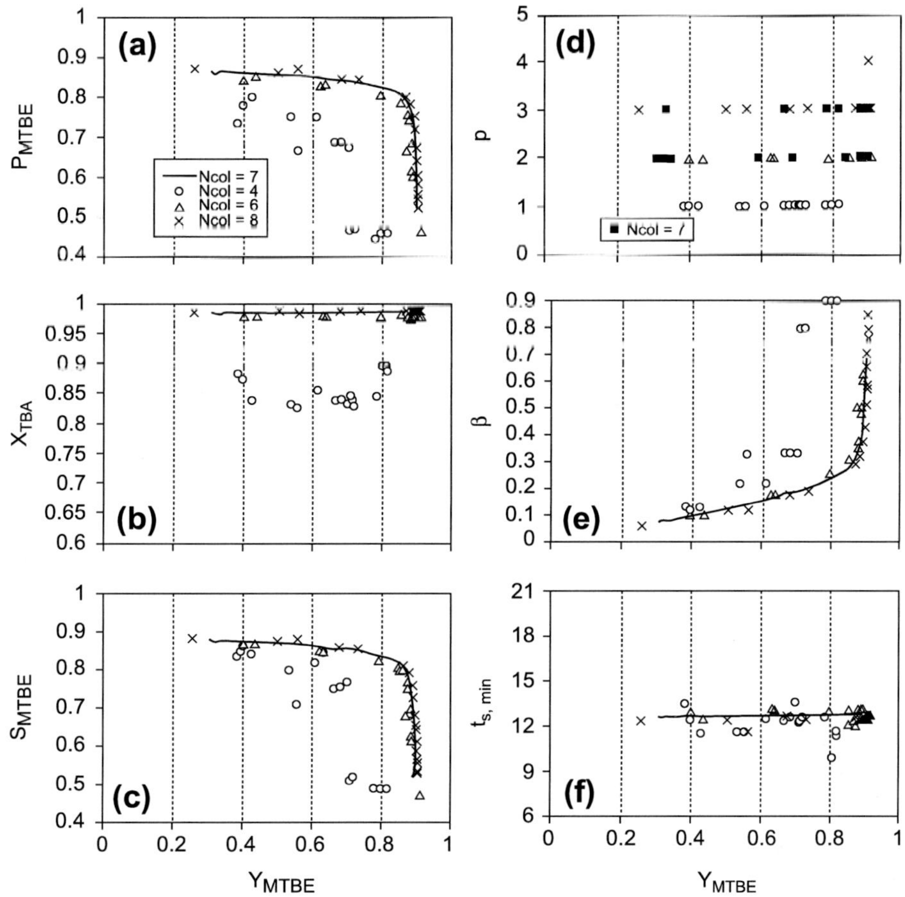

FIGURE 21-9 Pareto optimal sets and values of the search variables for optimization of MTBE production in a four-section SCMCR. (*Figure adapted from* Zhang et al. (2002) *with permission. © 2002 by the American Chemical Society.)*

Raffinate withdrawal of greater than 20% of the feed is necessary to achieve $Y_{MTBE} > 0.88$, but at this withdrawal rate, water appears in the raffinate port causing purity and selectivity to drop drastically. Symbols in the figure show that above six, the number of columns in the SCMCR becomes unimportant, that is, the Pareto-optimal solutions are unaffected. However, as the total number increases, the best performance requires adding columns to Section 2. When there are just four columns in the four-section SCMCR, Section 2 can have just one column so the achievable performance drops regardless of the independent variables and the conversion constraint cannot be met.

Dünnebier et al. (2000) present a practical optimization and design strategy for an SCMCR. They used a general design model considering axial dispersion, mass transfer between fluid and solid phases, particle diffusion, nonlinear adsorption and reaction kinetics. A single objective function for optimization was chosen avoiding competing and contradictory targets. The actual optimization task was minimization of the specific separation costs, which were the sum of absolute separation costs divided by the amount of each product. Absolute separation cost consisted of the costs of solvent, adsorbent, and plant operation. The capability of the proposed approach was illustrated using two sets of experimental results reported previously. The first was the inversion of sucrose to produce fructose and glucose (*Meurer et al., 1996*), which is a fast, homogeneous, enzyme catalyzed reaction with linear adsorption isotherms. The second was the production of phenethyl acetate (*Kawase et al., 1996*), a reversible heterogeneous reaction with nonlinear adsorption isotherms. In both applications, the *Dünnebier* strategy proved to be efficient.

21.2.5 Non-isothermal Systems

Kruglov (1994) investigated MeOH synthesis from syngas in an SCMCR. He was interested in the use of an adsorbent to trap MeOH so that the reaction would be shifted toward higher conversion. The concept had been applied earlier by *Westerterp* and co-workers (*Kuczynski et al., 1987b*; *Westerterp and Kuczynski, 1987a*; *Westerterp et al., 1989*). *Kruglov* presented two different configurations of an SCMCR for the synthesis. In the first, catalyst and adsorbent are jointly placed in fixed sections, whereas in the second, the catalyst and adsorbent are packed in beds alternately as shown in **Figure 21-10**. The first arrangement corresponds to mixed catalyst and adsorbent. The operation is carried out adiabatically, so a heat balance, **Eq. (20-3)** in **Table 20-1,** and appropriate boundary conditions must be introduced. In the second, reaction and separation are carried out isothermally. *Kruglov* analyzed and compared the performance of these two configurations. For his model, *Kruglov* allowed for mass transport into the porous adsorbent but assumed a linear adsorption isotherm. He also assumed uniform initial conditions in the bed, but pointed out that initial conditions are not

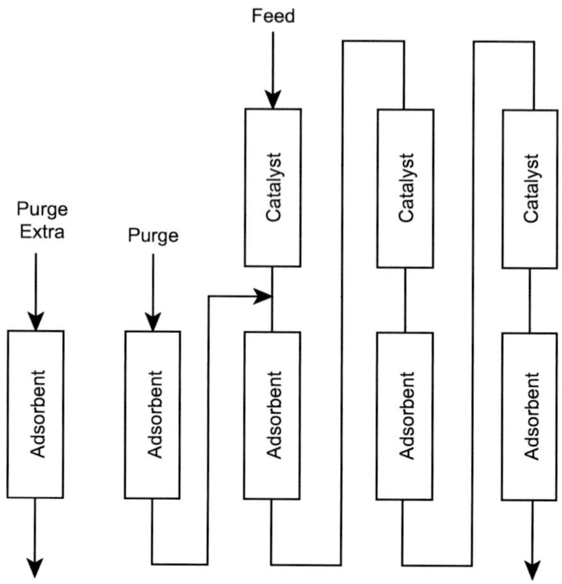

FIGURE 21-10 SCMCR cascade with separation of adsorbent and catalyst into different beds. *(Figure adapted from* Kruglov (1994) *with permission. © 1994 by Elsevier Science Ltd.)*

important because the cyclic stationary state is independent of these conditions. His model was converted to finite differences for the spatial derivatives and the Thomas algorithm employing a Newton-Raphson procedure handled the sparse matrix that resulted. Because of the steep gradients that arise, an adaptive grid was necessary. Model parameters were developed from known properties of the solids, while kinetic and adsorption parameters were obtained from the literature. Further details of the operation and simulation results are given by *Aida and Silveston (2005)*.

21.2.6 Separate Catalyst and Adsorbent Beds

For the SCMCR configuration shown in **Figure 21-10**, mass and heat balances for the beds in which adsorption or reaction takes place are easily derived from those given in **Table 20-1**. Boundary conditions are not changed as far as the feed into one bed is the effluent from the upstream one. *Kruglov (1994)* demonstrated that in this separation arrangement conversion greater than 98% could be attained if the operation was isothermal. Switching frequency had a strong effect on conversion and an optimum existed. Separation of adsorbent and catalyst was proposed much earlier by *Hashimoto et al. (1982, 1983)*. One of the interesting features of an SCMCR is its configurational flexibility compared with a CMCR. Several variations of the arrangement in **Figure 21-10** have been reported.

Hashimoto et al. (1983) studied the isomerization of glucose to fructose by immobilized glucose isomerase in an SCMCR consisting of seven reactors (R1—R7) and sixteen adsorption columns (A1—A16). The arrangement of the beds is shown in **Figure 21-11**. Adsorption columns, A2—A7, were placed in between the reactors, while the other columns were connected consecutively. The reactors are fixed but the adsorption columns "move" in the manner of an SCMCR operation from left to right while the fluid flows from right to left. It is the beds that move; the adsorbent within each bed is fixed. Appropriate valve changes replace the bed movement. Section 1 corresponds to Section 2 in **Figure 21-3**, Section 2 to Section 3 and Section 3 to Section 4. The model of the system, as mentioned above, differs from the material balances in **Table 20-1** because the reactors have only a reaction term with no storage on the solid phase whereas the adsorbers have no reaction term.

Hashimoto et al. proposed an alternate CMCR model that assumed the adsorbent particles were moving countercurrent to the liquid flow. In this model, steady state can be assumed so there are no time derivatives. Other researchers (*Ching and Lu, 1997; Storti et al., 1988; Lode et al., 2001*) also used such a model for a combined catalyst-adsorbent. Both models were tested against experimental data and both represented the experimental results well. Simulations showed that fructose content in excess of 60% could be obtained by an SCMCR using much lower amounts of desorbent than a chromatographic separator following the enzymatic reactor.

With a limited number of reactors, glucose is not completely converted to fructose in **Section 1**, and the remaining glucose is transported to **Section 3** with the recycled liquid flow, followed by mixing with the product stream. This resulted in a reduction in fructose content in the product. To increase the content of fructose, *Hashimoto et al. (1982)* proposed a modified system in which another section, consisting of adsorption columns only, was added to the left of Section 1 in **Figure 21-11** and a portion of the liquid flow was withdrawn as raffinate. As **Figure 21-21(b)** shows, adding extra adsorption columns to create a four-section SCMCR increased the concentration of fructose in the extract stream considerably.

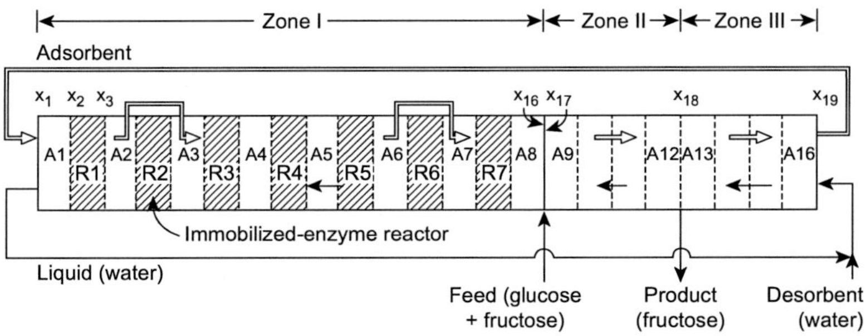

FIGURE 21-11 **Schematic of the operation of a three-section SCMCR with separate beds of catalyst and adsorbent.** (*Figure adapted from* Hashimoto et al. (1983) *with permission. © 1983 by John Wiley & Sons.*)

Zhang et al. (2007b) proposed two different modifications to the SCMCR system. In the first modification, an additional reactor is installed alongside the SCMCR, and the raffinate stream is fed to this reactor and recycled back with the fresh feed. In the second modification, an additional reactor is added in the section between the raffinate and eluent streams in the first modification. *Zhang et al.* claimed that the proposed modifications led to higher productivity and fructose purity in the product.

Minceva et al. (2008) have applied a system combining adsorption columns and a reactor similar to **Figure 21-11** to produce liquid phase *p*-xylene (PX) by isomerization of a xylene mixture. They combined five isomerization reactors and six absorption columns in a section between the feed and raffinate streams. The simulation was carried out using an equivalent moving bed reactor model, in which axial dispersion, external and internal mass transfer, and multi component Langmuir isotherms were considered. The influences of switching time, reactor length, and operating temperature on the SCMCR performance were evaluated. This performance was defined by the extract and raffinate purities, PX productivity, desorbent consumption, and PX deviation from equilibrium. Calculated performances were sensitive to small changes of switching time, especially near the optimal operating points.

Bergeot et al. (2010) also simulated PX production in the same type of SCMCR used by *Mineceva et al.* Commercially, PX is separated from o-xylene (OX) and m-xylene (MX) and remaining isomers, which are recycled to an external isomerization reactor where more PX is produced. Fresh isomer feed is added to this recycle flow and both streams are fed to the separator. In the Bergeot process, the separator is replaced by an SCMCR. This modification reduced both the amount of recycle and equipment size, resulting in lower operating costs. An example of the simulation results for this modification was a reduction of 41.6% in the feed flow rate, 32% in the raffinate flow rate, 54% in the isomerization reactor feed, and 58% of the feed to the distillation column separating xylenes.

Another variation on separate reactors and absorbers, shown in **Figure 21-10**, was employed by *Tonkovich and Carr (1994b)* for their simulation of the oxidative coupling of methane (OCM). Separation of adsorbers and reactors is essential because adsorbents do not exist for the high reaction temperatures employed in an OCM reactor. Methane coupling is a consecutive reaction with C_2's as intermediate products. Rates of CH_4 conversion and C_2 consumption are proportional to O_2 partial pressure so the selectivity to C_2 drops as conversion of CH_4 increases. Thus, at low conversion, high selectivity to C_2 products can be realized. Continuous separation of these intermediates to prevent their further oxidation seems the most promising route to increasing C_2 yield.

The configuration used by *Tonkovich and Carr (1994b)* is shown in **Figure 21-12**. Low-temperature adsorption columns separating C_2 products and CH_4 follow the short high-temperature reactors, operating near 1000 K. One-pass conversion in these reactors is low so selectivity to C_2s it is high. Initially feed is introduced into the reactor located second to the right. This reactor discharges into a serially connected separation column. Carrier gas enters the reactor on the far right, passes through both reactor and adsorption column and then mixes with feed and passes through the second reactor. Some adsorbed C_2s are stripped from the adsorption column by the carrier gas. Unreacted CH_4 is weakly adsorbed so it passes through the far right adsorption column more rapidly than C_2 and is swept into the second-to-the-right reactor by the carrier gas. Effluent from this adsorption column contains little C_2 initially and CH_4 is heavily depleted. It is

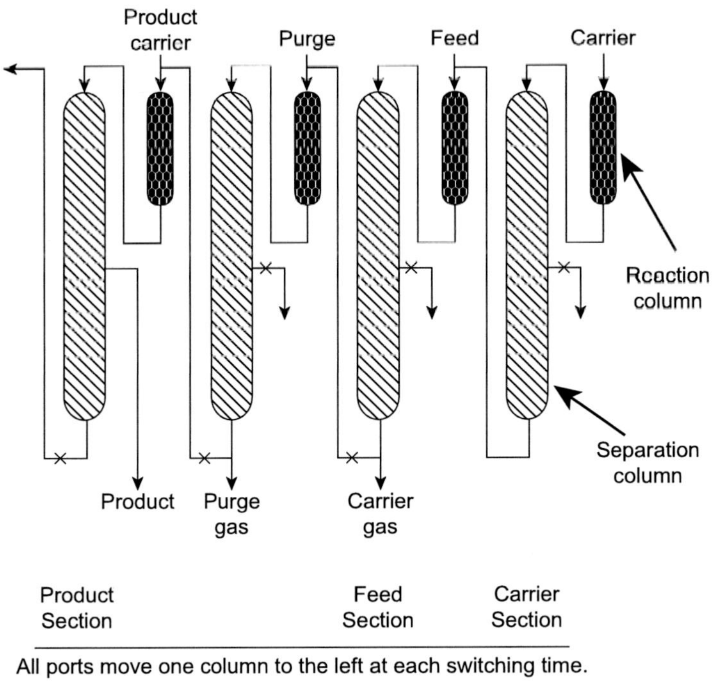

FIGURE 21-12 Separate adsorbent and catalyst bed SCMCR used to achieve high product yields in the oxidative coupling of methane. Adsorbent beds and reactors operate at widely different temperatures. (*Figure adapted from* Tonkovich and Carr (1994b) *with permission. © 1994 by Elsevier Science Ltd.*)

mainly carrier gas, so the stream is recycled. Well before C_2 breakthrough occurs in the adsorption column second on the right, feed is switched to the second-to-the-left reactor that had been purged in the previous step. Purge continues in this reactor, second on the left, just up to the C_2 breakthrough. Before breakthrough becomes substantial, carrier gas flushes the adsorption column. The effluent is the product stream and contains negligible CH_4. Product recovery occurs in the far left reactor-adsorption column sequence in **Figure 21-12**. Further operational details are given by *Aida and Silveston (2005).*

In modeling their system, *Tonkovich and Carr* used first-order, reversible kinetics for the OCM reaction. Solution of the model equations followed *Kruglov* except that integration employed a fourth-order Runge-Kutta algorithm. Model parameters were based on experimental data. The authors calculated concentration profiles as a function of position at different times within a cycle, but agreement with experiments for realistic switching periods of less than 30 s was poor.

Kundu et al. (2009) examined in more detail the effects of different operating variables on the performance of an SCMCR for the oxidative coupling of methane. The reaction rate and adsorption isotherm parameters required for the SCMCR model were estimated from data published previously. Transient model equations were solved numerically to obtain concentration profiles in the SCMCR, conversion of CH_4, selectivity, and yield of the desired products, ethylene and ethane. Their predictions were found to be in agreement with the experimental results of *Tonkovich and Carr (1994a)*. In an SCMCR, separation of the products is accomplished by appropriate selection of operating variables such as τ_s and flow rates in each section. *Kundu et al.* examined these effects using simulation and found that conversion could reach 93.3%, while a selectivity of 99.9%

and a yield of 58.5% were possible, although not simultaneously.

21.3 EXPERIMENTAL STUDIES

21.3.1 Gas–Solid Systems

An experimental investigation by *Ray and Carr (1995a)*, see **Table 21-1**, followed up the simulation work of *Ray et al. (1994)* discussed above. Catalyst and reaction temperature in the study were the same as those used by *Fish et al. (1986)*; however, Chromosorb 106 was chosen as the adsorbent instead of alumina. A mixture of 10 wt% catalyst and 90 wt% adsorbent was employed. Their SCMCR system is shown in **Figure 21-13**. Solenoid valves performing the feed, flush, and product removal functions are shown in the figure. Each column contained an equal amount of the catalyst/adsorbent mixture. The columns were heated and temperature controlled to insure the same temperature in each column. With the adsorbent employed, the product, TMC, broke through the bed before MES. Responding to this situation, Ray and Carr used switching times between 240 and 300 s. Experiments employed 25% excess H_2 on a volume basis and a feed flow rate of 4.95×10^{-6} mol/s. Nitrogen carrier gas flow was 1.32×10^{-4} mol/s.

Figure 21-14 shows the mole fraction of MES and TMC leaving ports B and A of the previous figure. Measurements were taken after one hour of continuous cyclic operation and appear to represent a cyclic stationary state. Port B is located after the column receiving feed while port A is located after the column being purged by the N_2 carrier gas. That port is located two columns behind the feed column. It is evident from these figures that product and reactant separation is excellent. TMC is mainly in the stream leaving through port B while MES exits port A with negligible TMC. Temperature and switching time are important. **Figure 21-14(a)** used a switching time of 300 s and a column temperature of 463 K while the bottom figure employed a switching time of 240 s and a column temperature of 473 K. The conversion at 463 K reached 79%. MES recovery in port A is dictated by its adsorption isotherm. Because of adsorption, not all MES is removed by purging. Results are discussed further by *Aida and Silveston (2005)*.

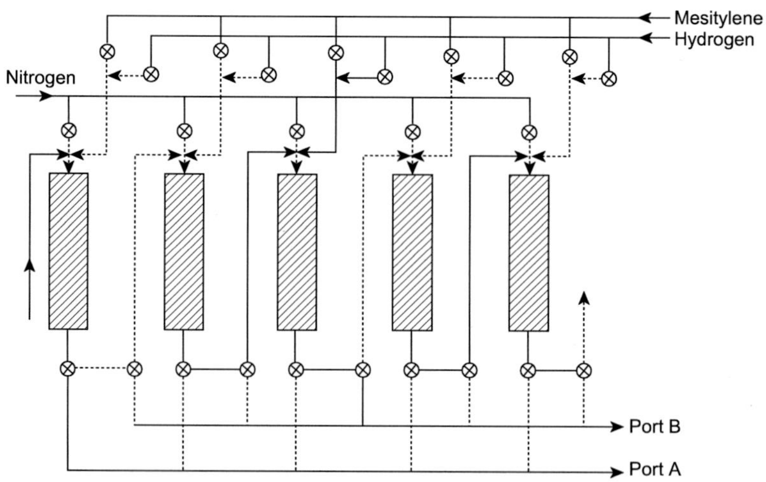

FIGURE 21-13 Schematic of a multi-bed, experimental SCMCR built for hydrogenation of mesitylene. (*Figure adapted from* Ray and Carr (1995a) *with permission. © 1995 by Elsevier Science Ltd.*)

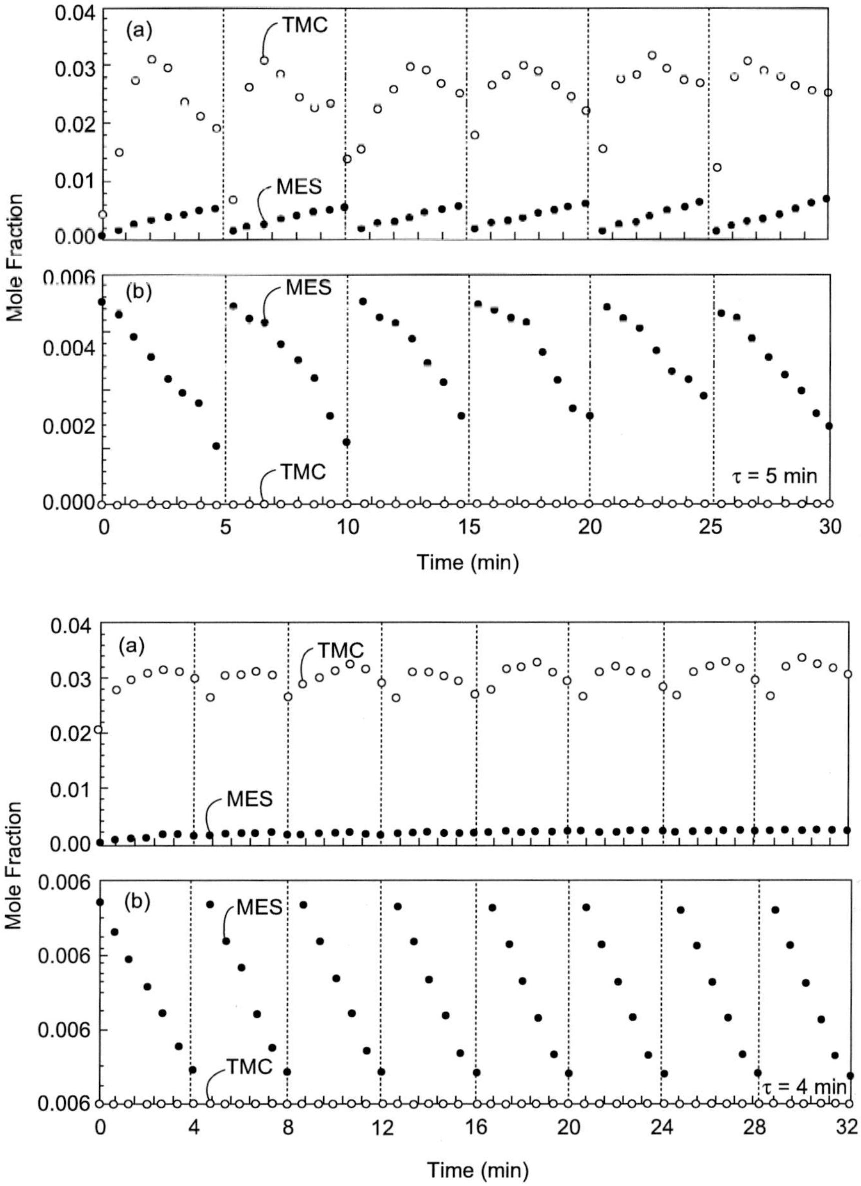

FIGURE 21-14 Mole fractions of reactant (MES) and product (TMC) leaving port B (upper) and port A (lower) at (a) 463 K and τ_s = 300 s and (b) 473 K and τ_s = 240 s. (*Figure adapted from* Ray and Carr (1995a) *with permission. © 1995 by Elsevier Science Ltd.*)

A comparison of experimental and simulation results given by *Ray and Carr (1995b)* showed that their mean MES conversion was 83% and the mean TMC purity in the effluent leaving port B was 96%, while the SCMCR simulation gave 97 and 98% for this stream and the CMCR simulation gave 97 and 100% respectively. The agreement of the simulated values is not surprising as an SCMCR is represented closely by a CMCR. The lack of agreement, however, between prediction and experiment is significant and suggests a model inadequacy. See *Aida and Silveston (2005)* for further discussion.

Because of the high temperatures needed, the SCMCR system had to be modified for methane coupling. That system, shown in **Figure 21-12,** was discussed earlier. *Tonkovich and Carr* used 70 mg of pure Sm_2O_3 in each reactor. The adsorbent beds of activated charcoal were packed into two lengths of steel tubing located one after the other. This permitted take-off within an adsorbent bed during flushing. N_2 carrier gas flowing at 100 mL/min was used in all experiments. *Aida and Silveston (2005)* describe the experiments further, summarize results and discuss commercial prospects for CH_4 coupling. Methane conversion per reactor varied between 2 and 3%. Switching time, τ_s, and reactor temperature were important as can be seen in **Figure 21-15**. The times, τ_s, are close to the optimum. The molar reactant/oxidant ratio also affects performance significantly above a ratio of 2.6.

Bjorklund and Carr (1995) carried out further oxidative coupling experiments using an SCMCR similar to the unit shown in **Figure 21-12,** also with an Sm_2O_3 catalyst and an activated charcoal adsorbent. Their SCMCR contained four sections, each of which consisted of a high-temperature reactor followed by a low-temperature separation column. Operation was essentially similar to that used by *Tonkovich and Carr (1994a)*. Prior to CH_4 breakthrough from the separators, the feed was switched and CH_4 desorbed by the carrier gas (N_2) during the previous cycle was added to the new feed stream. Products were removed from the first separation column several switches later. Conversion of over 65% was obtained with a yield higher than 50%.

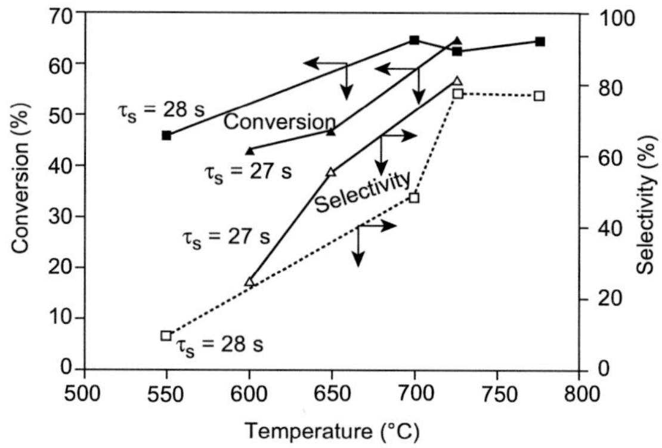

FIGURE 21-15 Temperature effects on a modified SCMCR performance for oxidative coupling of methane over Sm_2O_3. (*Figure adapted from* Tonkovich and Carr (1994a) *with permission. © 1994 by Elsevier Science Ltd.*)

Kruglov et al. (1996) found that the methane loss and the decreased conversion observed were caused by incomplete desorption of CH_4 in the separation column of the SCMCR. A hydrophobic carbon molecular sieve, examined along with activated carbon and zeolite, was found to be better for CH_4 storage and efficient separation. Three catalysts, Sm_2O_3, $Y_1Ba_2Zr_3O_{9.5}$ and $Y_1Ba_2Ge_1O_{3.5}$ were tested. A combination of the hydrophobic sieve and $Y_1Ba_2Zr_3O_{9.5}$ catalyst gave the best performance with a C_2+ yield of 55% at a conversion of 75%. The authors suggested that adaptive flow switching and use of a non-uniform make-up feed were promising ways to improve performance.

An alternative for converting natural gas to higher value products is partial oxidation. Selectivity to MeOH is a key consideration because MeOH readily oxidizes to water and the carbon oxides as temperature increases. *Bjorklund and Carr (2002)* investigated the non-catalytic partial

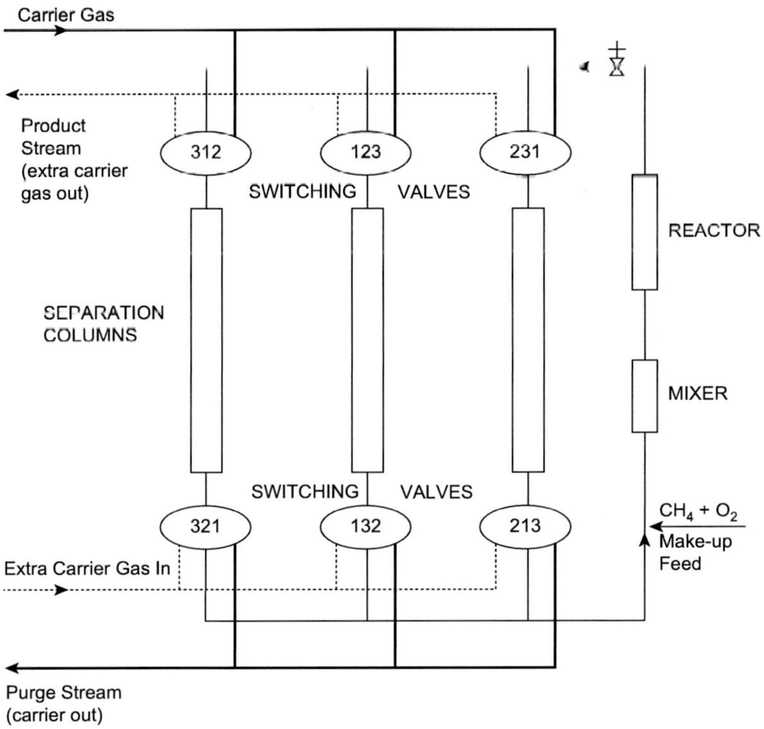

FIGURE 21-16 Schematic of a simulated moving bed chromatographic reactor with a separate high temperature reactor and low temperature adsorbers designed for the partial oxidation of methane to methanol. (*Figure adapted from* Bjorklund and Carr (2002) *with permission. © 2002 by the American Chemical Society.*)

oxidation of CH_4 using a three-section SCMCR consisting of a single, open-tube reactor and three adsorption columns (See **Table 21-1**). **Figure 21-16** provides a schematic of their experimental unit. Ovals at either end of the adsorption columns represent switching valves. The numbers within the ovals show which valve is open in each step of the three-step cycle. The reactor used a fused silica-lined steel tube wrapped with heating tape. At a pressure of 100 atm and CH_4:O_2 = 16:1, a temperature of 735 K initiated the reaction and gave the highest MeOH yield. Conversion was 6% and selectivity to MeOH was about 50%. Thus, the 16:1 ratio was fed to the SCMCR initially. However, their make-up ratio used the stoichiometric 2:1 ratio. Experiments were run at 750 K and at this temperature, a 2:1 feed ratio and $\tau_s = 440$ s were optimal for partial oxidation. Steel tubes packed with a 80/100 mesh Supelcoport™ particles served as adsorbents. This material provided a retention sequence at 375 K: H_2O > MeOH >> CO ≈ CO_2 > CH_4. Ratio of reactant:-product breakthrough times in the adsorption columns was MeOH:CH_4 ≈ 2.5:1, quite close to the 2:1 target for good separation.

In the first of three steps, effluent from the reactor went to the first adsorption column on the LHS of **Figure 21-16**. Helium eluent entered the third adsorber (counting from the left), while additional He, designated as "extra" in the figure, entered the second column. The effluent from the third column mixed with fresh feed before passing into the reactor. Effluent from the second column was the product stream. Just before CH_4 breakthrough in the first adsorber column, valves labeled "1" in the ovals were closed and those labeled "2" were opened.

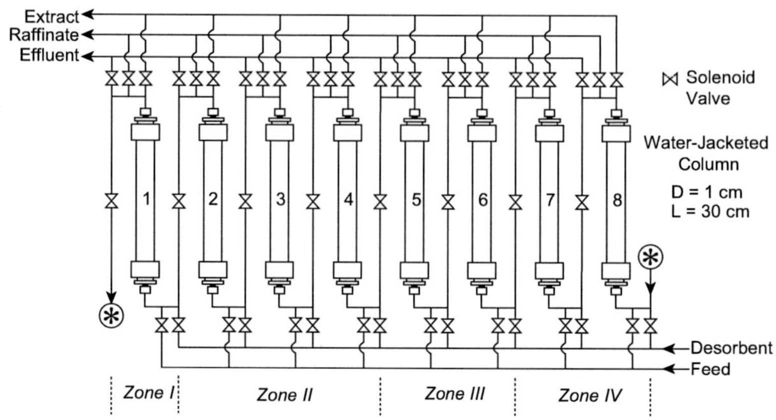

FIGURE 21-17 Schematic of an experimental SCMCR used for the esterification of acetic acid and β-phenetyl alcohol. (*Figure adapted from* Kawase et al. (1996) *with permission. © 1996 by Elsevier Science Ltd.*)

Reactor effluent now entered Column (2), while He flowed into Column (1) and extra He entered the third column. Effluent from Column (1) mixed with fresh feed and flowed into the reactor. Effluent from Column (2) passed into the purge stream shown in the figure, while the product stream was the effluent from the third column. Prior to CH_4 breakthrough from Column (2), the valves were changed again. Valves labeled "2" in the ovals were closed and those labeled "3" were opened. Flows can be worked out from **Figure 21-16**. The cycle repeats just before CH_4 breakthrough from Column (3) on the right hand side of the figure. Trapping of CH_4 by the adsorption column, recycle and mixing with fresh feed creates the optimal 16:1 mixture entering the reactor.

For the operation just described, CH_4 conversion and selectivity to MeOH reached about 50% resulting in a 25% yield of the alcohol. Selectivity declines with increasing temperature so higher yields through higher reactor temperature is not possible.

21.3.2 Liquid–Solid Systems

Kawase et al. (1996) undertook an experimental study of the liquid phase esterification of β-phenetyl alcohol with acetic acid using an Amberlyst 15 ion-exchange resin that functioned as both catalyst and adsorbent for the water product. 1,4-Dioxane served as the eluent. **Figure 21-17** shows a schematic of the four-section apparatus used.

Figure 21-18 compares the predicted stationary cyclic concentration profile with the experimental results after 7.7 h of operation. There is good agreement. The *Kawase* model was discussed in **Section 21.2.3**. The figure shows that the ester and water exist in Sections

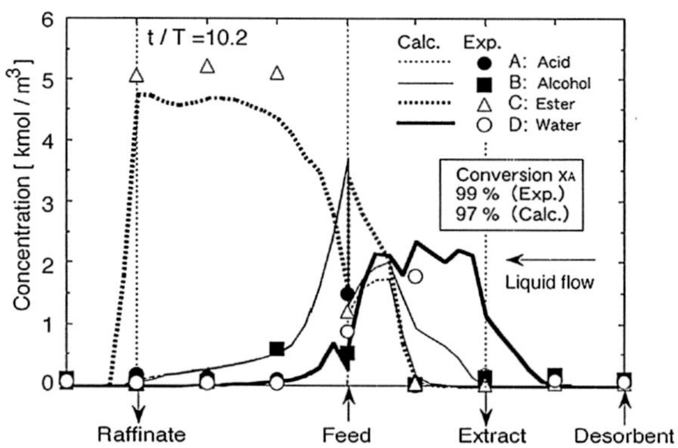

FIGURE 21-18 Predicted and experimental stationary concentration profiles for the esterification of acetic acid and β-phenetyl alcohol in an SCMCR under optimal operating conditions. (*Figure adapted from* Kawase et al. (1996) *with permission. © 1996 by Elsevier Science Ltd.*)

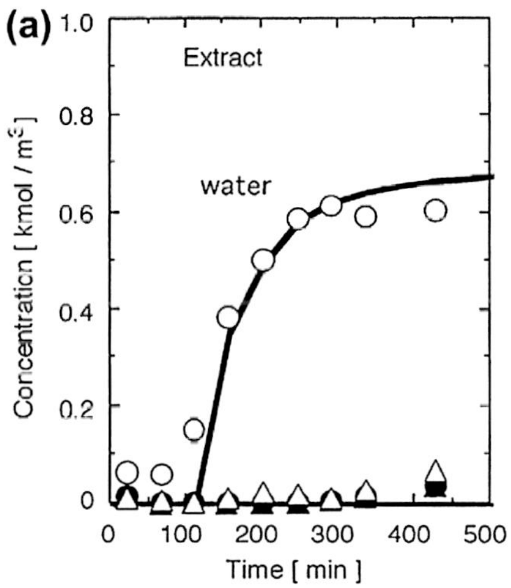

FIGURE 21-19 Predicted and experimental transient concentrations in the extract stream (a) and in the raffinate stream (b) under optimal operating conditions. (*Figure adapted from* Kawase et al. (1996) *with permission. © 1996 by Elsevier Science Ltd.*)

2 and 3 but the amount of water is low in Section 2. Reactants are present near the feed point. At the raffinate withdrawal, both water and alcohol are virtually absent, while at the extract point, water is the only contaminant.

Figure 21-19 shows the changes of product concentrations in the raffinate and extract streams after start-up. A cyclic stationary state was attained after about five hours. Agreement between simulation and experiment was good. Excluding solvent, purity of the ester in the raffinate was 99%. Overall conversion was 99% experimentally, considerably exceeding the equilibrium conversion of 63%. *Kawase et al.* demonstrated that SCMCR performance critically depends on the relative flow rates of the extract and raffinate streams. Performance was also sensitive to temperature.

Esterification of EtOH was used by *Mazzotti et al. (1996)* to demonstrate that a properly operated SCMCR can achieve complete conversion of a key reactant and separation of reaction products. Amberlyst 15 served as the esterification catalyst and product adsorbent. For modeling, equilibrium distribution of water between resin and the fluid phase was assumed. Activities for the solution were predicted by the UNIFAC group contribution method, while an extended Flory-Huggins model was used for the polymer phase. The *Mazzotti* model required swelling data, which was acquired experimentally. Interaction parameters in the model were set equal to zero. Reaction rate was found to be first order for each reactant. Details of the experiments are summarized by *Aida and Silveston (2005)*. Close prediction of the time varying concentrations using the equations discussed in **Section 21.2.2** verified the investigator's model and the parameters used.

For their SCMCR experiments, *Mazzotti et al.* used eight resin packed columns with five columns in Section 3, one in Section 2 and two in Section 1. Port movement employed a rotary valve consisting of a stationary upper disk connected to the inlet and outlet lines and a rotating lower disk that gave connections to the columns making up the SCMCR. Operating with $\tau_s = 6$ min and a EtOH:HAc volumetric ratio of 20:1, complete conversion of acid was obtained with only ester and EtOH in the raffinate. There was no ester in the extract stream. *Aida and Silveston* give further results.

Lode et al. (2001) employed a Model C-920 chromatographic separation unit (Advanced Separation Technologies Inc.) in their study of the esterification of HAc and MeOH over an Amberlyst 15 ion exchange resin. For details of this unit and its use by *Lode et al.*, see *Aida and Silveston*. A limited number of parametric experiments were undertaken: for a specific eluent rate, a low acid feed rate and fast switching. 100% conversion of the acid was obtained along with complete separation of the ester and water.

Productivity, however, was low. Increasing the eluent rate improved product separation at full conversion, but diluted the products. Increasing the acid feed rate more than doubled productivity and reduced the required methanol/ester ratio by 1/3; conversion, however, dropped to below 90% and separation was poorer. One of the experiments by *Lode et al.* was undertaken with fluid downflow in the columns in place of upflow. Their purpose was to test the effect of mixing on performance caused by density differences. Only a small effect was observed.

As mentioned in **Section 21.2.3**, a two-section SCMCR can be used if just one reaction product adsorbs. This is the situation for immobilized enzyme catalyzed esterification of propionic acid with isoamyl alcohol in a hexane carrier. The enzyme support and adsorbent were ion exchange resins that take up water. *Mensah and Carta (1999)* studied this system using a two-section, four-column SCMCR. The reactant mixture passes consecutively through three columns packed with the immobilized enzyme and ion-exchange adsorbent, while the fourth column strips the water product from the solid phase with the isoamyl alcohol eluent. Measurements showed that the SCMCR produced a hexane solution of the ester contaminated by a small amount of eluent. Pure water, which deactivated the enzyme irreversibly, was never present according to the simulation. The authors observed that the simulation reproduced the effluent from the SCMCR closely. A much higher productivity was found using a 2:1 mixture of lipozyme and adsorbent than with using the lipozyme alone. Increasing the number of mixed immobilized enzyme-adsorbent columns, while holding the amount of enzyme constant, increased productivity.

Yu et al. (2003) evaluated numerically and experimentally the performance of an SCMCR for the synthesis of methyl acetate catalyzed by an Amberlyst 15 ion-exchange resin. Concentration profiles in their SCMCR and conversion, purity and selectivity were computed using a rigorous mathematical model. Indeed, this model closely predicted their experimental results. The effects of various process parameters such as τ_S, feed and eluent flow rates on the behavior of the SCMCR were thoroughly examined. The researchers found that switching time, τ_S, determined performance, provided suitable operating conditions were selected. Although the *Yu et al.* paper should be consulted for the effect of different design parameters on SCMCR performance, the authors' observations supported what has been heretofore discussed.

Lode et al. (2003b) also examined the influence of the operating parameters, especially the influence of the feed stream composition, on SCMCR performance for applications with nonlinear equilibrium relationships. Their study combined both simulation and experiment. Synthesis of MeAc from HAc and MeOH over an Amberlyst 15 resin served as their model reaction. Unlike other experiments considered in this chapter, these were carried out on a mini-plant scale. Reliability of their simulation model was established from a comparison of their simulation results with experimental data. A notable finding was that optimal performance in terms of productivity and eluent consumption is achieved with feed compositions composed of equimolar mixtures of the two reactants.

Silva and Rodrigues (2005) evaluated a novel SCMCR process for diethylacetal (acetal) experimentally and by simulation. Acetal is an important raw material for fragrances and pharmaceuticals. Production occurs through a liquid phase, reversible reaction of acetaldehyde with EtOH. Water is a byproduct. An Amberlyst 15 resin serves as catalyst as well as a selective adsorbent. The authors calculated the SCMCR performance using a CMCR model that considered axial dispersion, assumed a linear driving force for fluid-particle mass transfer, and multi-component adsorption equilibria. Reaction rate was given by a Langmuir-Hinshelwood model. Following the usual practice, model parameters were evaluated experimentally. The best experimental performance obtained was a raffinate purity of 87% and an acetaldehyde conversion of about 98%. Simulation showed that high acetal purity and complete conversion of acetaldehyde could be achieved by increasing the eluent flow rate. Influences of feed composition, τ_S, and mass transfer resistance on SCMCR performance was obtained by simulation.

Performance of an SCMCR process for ethyl lactate was studied by *Pereira et al. (2009)* using the approach and equipment employed by *Silva and Rodrigues (2005)*. Ethyl lactate is a biodegradable green solvent with many applications as a food additive. It is produced through the liquid phase esterification of lactic acid with EtOH over an Amberlyst 15 resin used as both catalyst and adsorbent. In the *Pereira et al.* model, mass transfer between the solid and fluid phases was represented by an overall mass transfer coefficient k_L; while the reaction rate and the adsorption isotherms were expressed by nonlinear functions. Concentration profiles in the SCMCR were calculated using both SCMCR and equivalent CMCR models. Only small differences in the calculated profiles were found. The authors introduced new parameters, γ_j in the jth section, defined as the ratio of the fluid interstitial velocity, u_j, to the simulated solid velocity, U_S. This is the column length, L, divided by τ_S. Separation performance was discussed in terms of γ_3 versus γ_2. This representation is similar to **Figure 21-5(b)**. Performance was evaluated based on the purities of raffinate and extract, conversion of lactic acid, raffinate productivity, and eluent consumption. Agreement between calculated values of these process parameters and experimental ones was good. Since the rate of the ethyl lactate formation is low, its production in an SCMCR is determined by the reaction rate. Thus complete conversion of lactic acid is critical to optimal SCMCR performance.

21.3.3 Biochemical Systems

In perhaps the earliest experimental investigation of an SCMCR, *Hashimoto et al. (1983)* demonstrated that high fructose syrups can be made from a glucose feed. The rotating system of adsorption columns and stationary enzymatic reactor beds used in their experiments is illustrated by **Figure 21-20**. The system contained 2.5 times as many adsorption columns as enzymatic reactors because of the functions performed by the columns: 1) Trapping of reaction products, 2) separation of glucose and fructose by their different adsorptivities, 3) desorption of the mixed fructose-glucose product. These functions are indicated in the figure by Sections 1–3. Connections between adsorption columns, reactor beds and the feed or eluent sources take place through a combination of rotating and fixed discs shown in the figure. The reactor beds are stationary while the adsorbent columns, mounted on the rotating disc, move counterclockwise. After each switching interval, a column moves one step counterclockwise. Thus, the final adsorbent column in Section 1 moves into Section 2 and carries adsorbents heavily loaded with the fructose-glucose mixture into this section where the adsorbent is washed countercurrently with water containing fructose initially at the product

FIGURE 21-20 Experimental SCMCR with rotating adsorbent columns and stationary reactor columns used for glucose isomerization. (*Figure adapted from* Hashimoto et al. (1983) *with permission.* © *1983 by John Wiley & Sons.*)

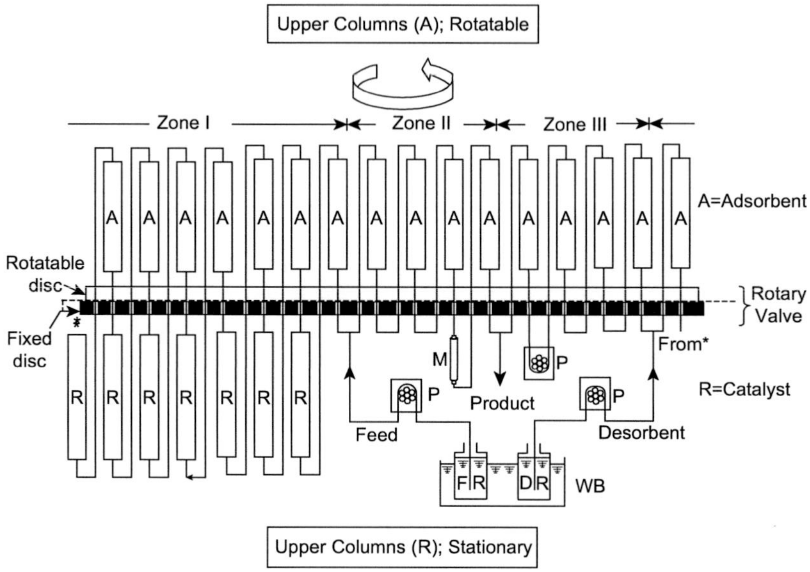

concentration. By the end of Section 2, the adsorbent contains a high ratio of fructose to glucose. On the switch, the final adsorbent column in Section 2 moves into Section 3 where glucose and eventually fructose are stripped from the adsorbent by the counter-current flowing eluent. Finally, the last column in Section 3 moves into Section 1 returning stripped adsorbent to that section.

Hashimoto et al. employed an immobilized glucose isomerase as a catalytic substrate for the conversion of glucose. The adsorbent was the Ca^{2+} form of Y-zeolite. The columns had the same diameter, but differing lengths (see **Table 21-2**). Isothermality was maintained by circulating thermostated water at 323 K through the jackets of the columns. Rotation of the upper disc was in 22.5 degree steps and τ_S was either 120 or 180 s.

An objective of *Hashimoto et al.* was to verify their model. For this purpose, they varied feed, eluent and withdrawal flow rates, and feed concentrations of fructose and glucose. **Figure 21-21(a)** shows the measured change of glucose and fructose in the adsorption columns. Conversion of glucose takes place in Section 1 where the concentration goes from about 0.3 in the feed to 0.05 kmol/m^3 in the stream leaving the last reactor. Separation of glucose and fructose is also shown. The sugars are about equal at 0.3 in the feed, whereas in the stream leaving the last adsorbent bed fructose is 0.12 and glucose is 0.07 kmol/m^3. Adsorbates are further depleted in Section 3 and drop to 0.04 kmol/m^3 in the stream leaving the last adsorbent column. Good agreement between results and predictions demonstrated that the model successfully represents sugar concentrations in the SCMCR despite experimental scatter in Section 1. Discontinuities at the feed and eluent introduction points are seen quite well; product compositions are close to those measured. *Hashimoto et al.* also discovered that the approximate CMCR model represented the discontinuous SCMCR satisfactorily. A steady-state model makes calculations easier; however, such a model does not indicate discontinuities.

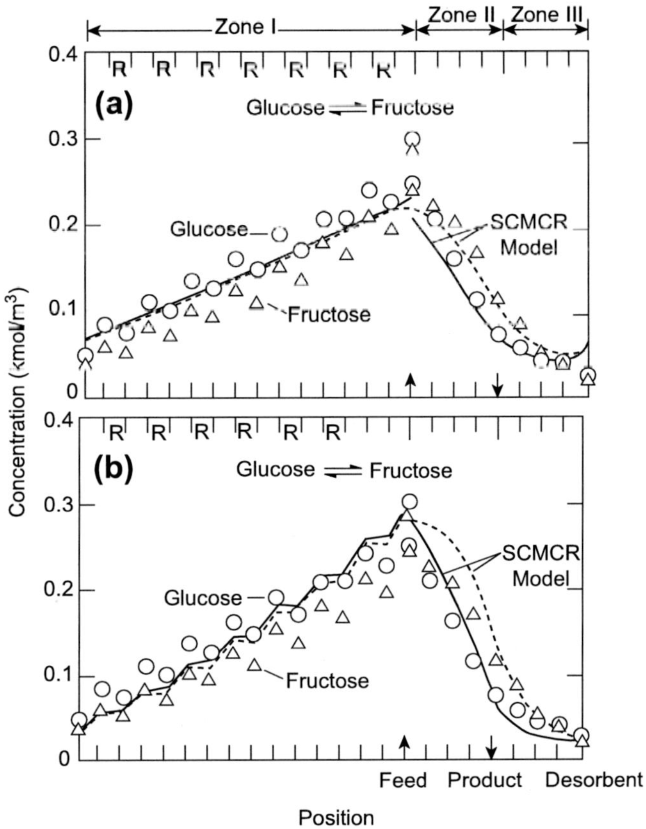

FIGURE 21-21 **(a) Experimental and predicted composition profiles in a three-section modified SCMCR for glucose isomerization employing a periodic switching model; (b) experimental and predicted composition profiles in a four-section modified SCMCR employing a continuous moving-bed model.** (*Figures adapted from* Hashimoto et al. (1982, 1983) *with permission. © 1982 by The Asahi Glass Foundation, © 1983 by John Wiley & Sons.*)

Hashimoto et al. (1983) demonstrated that their modified SCMCR achieved a fructose content of about 55% with a significantly lower eluent to feed ratio than a simple SCMCR process or a fixed bed enzymatic reactor followed by a fixed bed adsorber with recycle of unconverted glucose. **Figure 21-21(b)** demonstrates that the concentration of fructose in the extract stream in a four-section modified SCMCR increased significantly compared with that in the three-section modified SCMCR shown in **Figure 21-21(a)**. The concentration profiles in the four-section system were computed using the CMCR model. They compare well with the experimental concentration profiles.

Hashimoto et al. (1982, 1983) employed the Ca^{2+} form of Y-zeolite as the adsorbent and added Mg^{2+} to the eluent to stabilize the enzyme. *Da Silva et al. (2005)* reported simulations which showed that a fructose purity over 80% could be attained employing zeolite in the Mg^{2+} form, which is inferior to Ca^{2+} form for separation, but more durable in an Mg^{2+}-containing eluent.

Barker and co-workers discussed the enzymatic conversion of sugars using the Ca^{2+}

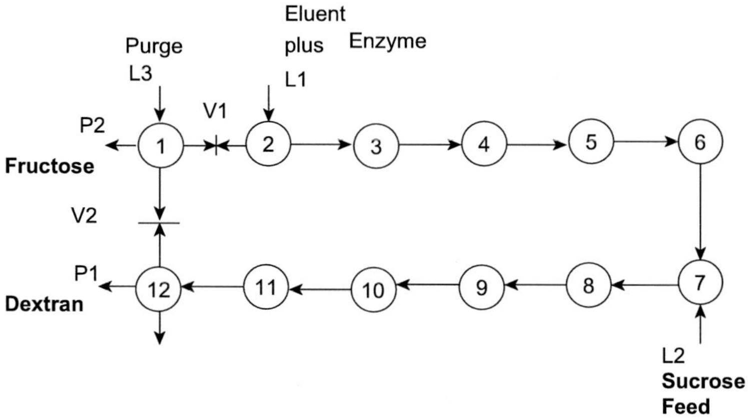

FIGURE 21-22 Operating scheme of a 12-column SCMCR employed in the biosynthesis of dextran from sucrose. V1 and V2 are on-off valves used to isolate Column (1) in order to strip fructose from the resin using a separate purge stream. Numbers represent columns in the three-section SCMCR. (*Figure adapted from* Barker et al. (1992a, 1990b) *with permission. © 1993 by International Adsorption Society.*)

form of an ion exchange resin as adsorbent (*Barker and Ching, 1980; Zafar and Barker, 1988; Barker et al.,1988; Barker et al., 1990a; Akintoye et al., 1990, 1991; Barker et al., 1992a, 1990b*). A 12-column, preparative scale unit, built originally for chromatographic separations, was used for their studies (see **Table 21-2**). Six on-off valves were attached to each column, three at each end. These valves were connected to one of the inlet or outlet streams (raffinate, extract and the succeeding column) and were timer-operated. A water purge was introduced just upstream of the eluent and enzyme feed point to remove fructose from the resin. System operation is shown schematically in **Figure 21-22**.

Test reactions used were sucrose inversion using an invertase enzyme (*Akintoye et al., 1990, 1991*), the biosynthesis of a polyglucose, dextran, employing the enzyme dextransucrase (*Barker et al., 1992b; Shieh and Barker, 1995, 1996*) and the saccharification of a modified starch using maltogenase (*Barker et al., 1992 a, 1990b*). In all cases the adsorbent was the calcium form of either a Purolite ion exchange resin (PCR 563) or a Korela resin (V07C). The inversion rate is apparently inhibited by the educt or substrate for the sucrose system in the presence of products. Dextran production is equilibrium limited. Fructose, a product of both sugar reactions, is preferentially adsorbed by the resin. The enzymatic reaction occurred in the aqueous phase so enzymes, at low concentrations, are added to the SCMCR with the water eluent.

Akintoye et al. (1990, 1991) found that sucrose inversion went readily to completion at 25°C in their SCMCR system even for concentrated sucrose solutions. Separation of the glucose and fructose products was virtually complete at throughputs of 16 kg sucrose/m^3 resin·h with glucose in the raffinate. Enzyme usage was just 20% of that consumed in a stirred, batch reactor in the absence of adsorbent.

Biosynthesis of dextran was less successful (*Barker et al., 1993*). Although initially, complete conversion of sucrose was obtained at a pH of 5.2, $T = 25°C$ with $\tau_s = 30$ min and a dextran product free of fructose was recovered, the separation of dextran and fructose deteriorated after 50 h on stream even at low sucrose throughputs. Levan, a polyglucose, was found in the dextran and glucose was present in the fructose extract. The problem was traced to the displacement of Ca^{2+} from the resin by small amounts of K^+ and Na^+ in the enzyme. Periodically regenerating the resin with calcium hydroxide permitted much longer run times, but incomplete separation persisted. *Aida and Silveston (2005)* discuss these experiments in more detail.

A *Barker* study of the production of maltose from modified starch found a conversion of 60% at a feed rate of 116 g starch/h with 96% maltose purity. Enzyme consumption was substantially reduced compared with using just a CSTR for this reaction. There was no deterioration of performance with time (*Barker et al., 1990b*). Further experiments reported by *Shieh and Barker (1995)* used an exo-acting α-amylase that hydrolysed the maltotriose of the starch. A Ca^{2+} exchanged resin preferentially adsorbed the maltose formed. There were strong interactions among the variables examined. Increasing the eluent flow rate decreased conversion because of its effect on residence time, but this was small because of the extra enzyme the eluent brought into the system. Maltose purity improved. Higher starch concentrations in the feed lowered conversion; however, a larger effect was higher viscosity that reduced mixing of starch and enzyme. Pressure drop was also affected by viscosity so this set a limit on the starch concentration fed to the SCMCR. The effect of τ_s on performance was similar to that of the eluent flow rate. Results of their experiments were compared with those obtained in a rotating annular chromatographic (CRAC) reactor (*Sarmidi and Barker, 1993a, b*). For essentially the same enzyme, adsorbent and operating conditions, the SCMCR used a significantly smaller amount of enzyme for a larger starch conversion. The maltose product, however, contained a small amount of glucose.

Shieh and Barker (1996) investigated the hydrolysis of lactose to galactose and glucose by the enzyme lactase using the SCMCR shown schematically in **Figure 21-22**, but modified by isolating a column in the adsorbent regeneration step and further stripping that column with the water eluent. This was necessary because of the strong adsorption of galactose by the resin and inhibition of hydrolysis by that product. Experimentally, galactose adsorption drove the hydrolysis reaction to completion even with a less active enzyme so the SCMCR performance was governed by separation of the products. A strong interaction among the variables investigated was also found for this hydrolysis. Switching time was the most important variable, followed by the eluent flow rate. Enzyme activity had little influence on performance. Use of a modified SCMCR reduced the enzyme needed for hydrolysis by a factor of 3 compared with a conventional batch reactor. The SCMCR produced 99.9% pure glucose, but galactose purity was lower and that sugar was quite dilute.

Kawase et al. (2001) took up the application of an SCMCR to the production of lactosucrose from lactose and sucrose. Simulation using experimentally determined model parameters indicated that a substantial improvement in throughput and yield over that of a batch reactor could be obtained. Lactosucrose formation is catalyzed by the enzyme β-fructofuranosidase. However this enzyme also catalyzes the hydrolysis of sucrose to yield glucose and fructose as well as the hydrolysis of the lactosucrose product to produce lactose and fructose. Glucose is also produced in the primary reaction. All these reactions are reversible. Use of an adsorbent specific to glucose increased lactose conversion by forcing the primary reaction and limiting hydrolysis of the lactosucrose. Further details of these experiments are given by *Aida and Silveston (2005)*. The experimental rotating SCMCR of *Kawase et al.* consisted of 12 columns arranged in four sections with two columns in Section 1, four in each of Sections 2 and 3 and two in Section 4. Several experiments used a 2-6-2-2 arrangement. **Figure 21-23** gives a schematic of their unit. Note that the enzyme is introduced with the lactose–sucrose feed stream and undergoes some hydrolysis that might be avoided by immobilizing the enzyme in some way. *Kawase et al.* carefully adjusted feed and withdrawal rates so that lactosucrose and glucose were separated in Sections 2 and 3.

With $\tau_s = 10$ min, a cyclic steady state was attained after 7.5 cycles and resulted in high lactosucrose yields at sucrose conversions in excess of 50%. Separation of lactosucrose and glucose was complete, as **Figure 21-24** demonstrates.

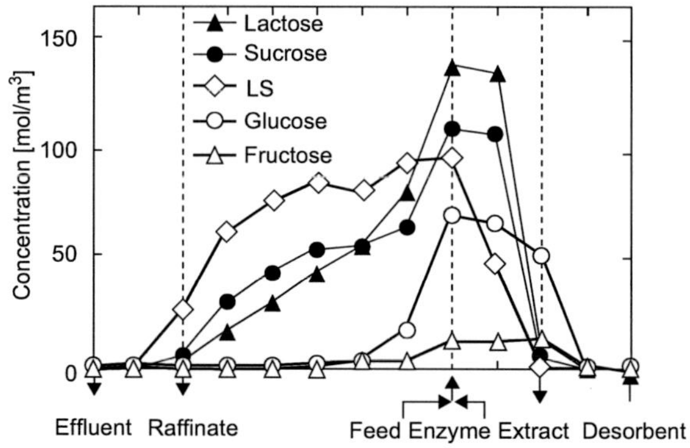

FIGURE 21-24 Concentration profiles in a 12-column SCMCR for the formation of lactosucrose from sucrose at $(C_{sucrose})_0$ = 500 mol/ m^3 and lactose at $(C_{lactose})_0$ = 530 mol/m^3 at τ_s = 10 min. Feed concentration of the β-fructofuranosidase was 100 enzyme activity units/m^3. (*Figure adapted from* Kawase et al. (2001) *with permission of the authors.*)

21.4 OTHER REACTOR APPLICATIONS OF SIMULATED MOVING BEDS

Exothermic solid catalyzed reactions undertaken in a packed bed reactor usually discharge hot gases. Of course, it is advantageous to capture this "waste" heat to preheat the entering fluids. Heat recovery can be performed through separate heat exchangers or by periodic flow reversal.**Chapter 18** examines the latter option. Another method replaces a single packed bed reactor by two, three or more separate, interconnected, catalyst beds with their feed inlets and product outlets shifting periodically to simulate a moving bed. *Haynes and Caram (1994)* proposed a two-unit, simulated moving bed catalytic reactor as an alternative to periodic flow reversal using just a single packed bed catalytic reactor and showed that higher productivity results. Interconnection forms a loop of reactors so such schemes are often referred to as loop reactors. *Haynes and Caram*

FIGURE 21-23 Schematic of the 12-column SCMCR showing the operation of the rotary valving. (*Figure adapted from* Kawase et al. (2001) *with permission of the authors.*)

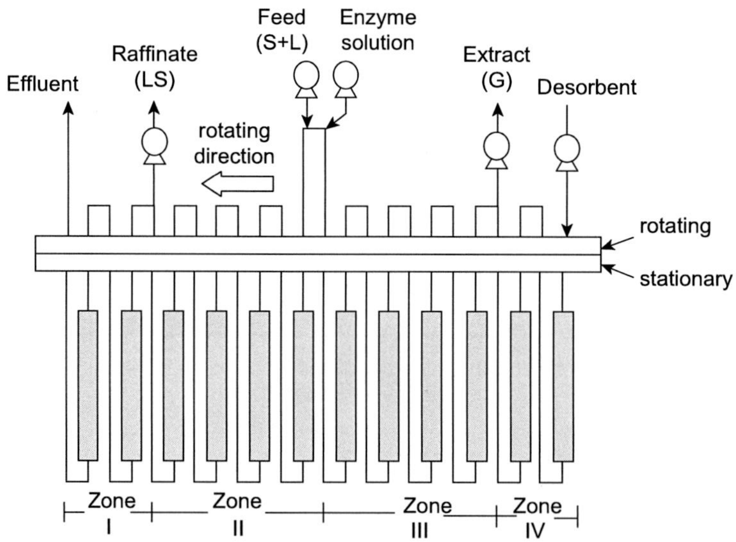

point out that the loop arrangement can significantly reduce washout, a problem that arises for periodic flow direction switching. Prior to *Haynes and Caram*, a ring arrangement using a system of interconnected reactors with periodically shifting inlets and outlets had been shown to also avoid washout (*Matros, 1985; Vanden Bussche and Froment, 1996*). Simpler three-bed systems have been proposed. *Brinkmann et al. (1999)* considered a three-bed simulated continuous moving bed reactor and numerically examined VOC destruction with such a system. *Velardi and Barresi (2002)* applied that system to MeOH synthesis **(Figure 21-25)**. They found through simulation that the loop reactor provided higher syngas conversion to MeOH than a single reactor of equal volume operating with periodic flow reversal. However, stable operation was possible in just two narrow regions of switching times: $\tau_S < 25$ s and $160 < \tau_S < 210$ s.

In a multi-bed loop reactor, a temperature front arises that moves through the catalyst bed at a velocity more slowly than the fluid velocity by some three orders of magnitude. Stable operation requires that the advancing thermal front remain in the cooler portion of a catalyst bed. Switching times that are too long or too short extinguish the reaction or result in instability. *Velardi and Barresi (2002)* and *Velardi et al. (2004)* have explored stable and unstable loops of three reactors. *Sheintuch and Nekhamkina (2004, 2005)* examined the limits of fast switching for a large number of beds in a loop and, in a later paper (*Nekhamkina and Sheintuch, 2008*), discussed rotating pulse solutions to the equations describing loop reactors. Computational requirements for exploring the behavior of loop reactors with respect to system variables such as switching time, inlet temperature, heat loss, heat capacity of the packing or maximum temperature rise led *Zahn et al. (2009, 2010)* to propose using a CMCR model in place of a detailed description of the loop reactor. *Altimari et al. (2006)* have applied bifurcation analysis to the three-bed loop system to identify the various domains that arise and their boundaries. A two-bed system was examined by members of the same group (*Continillo et al., 2006*), while *Mancusi et al. (2007)*

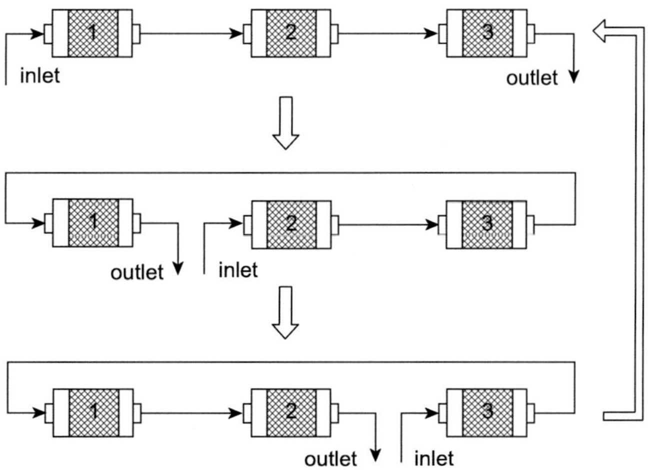

FIGURE 21-25 Cyclic operation of a three-bed simulated moving bed reactor proposed for methanol synthesis using a cold feed stream. (*Figure adapted from* Velardi and Barresi (2002) *with permission. © 2002 by Elsevier Science Ltd.*)

considered the dynamic behavior of the three-bed loop for two different switching strategies. A detailed discussion of these contributions is beyond the scope of this monograph.

Nomenclature

A,B,C, D = reacting species
BPA = Bisphenol A
CMCR = continuous moving bed chromatographic reactor
CRAC = continuous rotating annular bed chromatographic reactor
CSTR = continuous stirred tank reactor
C_i = concentration of species i (mol/L)
$(C_i)_0$ = feed concentration of species i
E = entrance
Ex = extract port
EtAc = ethyl acetate
EtOH = ethanol
FA = fatty acids (triglycerides)
FAME = fatty acid methyl ester
FFA = free fatty acid
H_i = Henry's Law constant for solute i
HAc = acetic acid
I = number of chemical species in reaction system
i.d. = inside diameter
J = number of beds or columns in a section or reactor
j = position, stage or step indicator
K_{eq} = adsorption equilibrium constant
K_i = adsorption constant for species i (L/g)
K_K = distribution coefficient
k = position, stage or step indicator species or component indicator
L = bed or column length (m)
MeAc = methyl acetate
MeOH = methanol
MES = mesitylene
MTBE = methyl tertiary butyl ether
MX = m-xylene
m = time step (integer)
m_j = flow-rate ratio in section j
N = number of adsorption sites
N_{Da} = Damköhler No.
n_i = molar flow rate for species i (mol/h)
n_j = number of beds or columns in the jth section or section
n_s = number of segments or stages in a column or bed
OX = o-xylene
o.d. = outer diameter (mm, cm)
P = pressure (kPa, bar)
PDE = partial differential equation
PFR = plug flow reactor
PX = p-xylene
P_i = purity of product i
P_{MTBE} = purity of methyl tertiary butyl ester
p = number of columns in Section P
Q = volumetric flow rate (L/min)
Q_{eluent} = eluent or desorbent flow rate
Q_{fluid} = fluid flow rate
Q_f = fluid flow rate
Q_s = solid phase volumetric flow rate (cm^3/min)
q_i = adsorbate loading (mol/$g_{adsorbent}$)
Ra = raffinate port
SCMCR = simulated countercurrent moving bed chromatographic reactor
SMB = simulated moving bed
T = temperature (K) period in **Figure 21-18**
TBA = tertiary butyl alcohol
TMC = 1,3,5-trimethyl cyclohexane
t = time (s, h)
$\bar{t}$ = residence time (h)
U_s = solids velocity
u = superficial velocity (cm/s)
u_f = fluid superficial velocity
u_g = gas velocity
u_i = interstitial velocity
VOC = volatile organic compound
V = volume, usually of bed or reactor (m^3)
V_{bed} = bed volume (m^3)
V_j = volume of jth bed or column
X_i = conversion of reactant i
x = axial position, also stage no. counting from top or bottom of a column (cm, m)
Y_i = yield of product i
Y_{MTBE} = yield of methyl tertiary butyl ether

Greek

β = fraction of feed flow rate withdrawn at raffinate port
β_{kj} = flow rate ratio defined by Hashimoto et al. (1983, 1993a,b)
Δ = eluent requirement in terms of reactant feed rate
Δx = spacing (distance) between take off points
ε = void fraction
ε_b = void fraction in bulk of solid or bed
ε_{bed} = void fraction in bed
ε_p = particle (adsorbent) void fraction
γ = eluent/feed ratio
γ_j = interstitial velocity/simulated solids velocity
κ = adsorptivity ratio

Γ_I	=	purity of product i
Γ_B^*	=	purity of the desired product at the raffinate port B
ϑ_I	=	productivity of product i in SCMCR
ζ	=	switching velocity (1/h)
σ_I	=	flow ratio for reactant or product I, critical CMCR parameter
τ_s	=	switching time, duration between movement of inlet/outlet location (s, h)
Ξ_I'	=	selectivity to the desired product at the raffinate port
Ξ_B	=	selectivity to the desired product at port B

Subscripts

A,B,C	=	species
Da	=	Damköhler
G	=	gas
F, f	=	feed
i	=	reaction component or species
j	=	jth section, section or bed, also reaction species or reaction number
rec	=	recycle
s	=	solid, or switch (switching), no. of columns in S section of SCMCR
0	=	feed or inlet
2,3	=	sections in SCMCR

Superscripts

j	=	position or step indicator
k	=	position or step indicator
m	=	time step indicator
*	=	equilibrium quantity

CHAPTER

22

Pressure and Temperature Swing Reactors

Alirio Rodrigues[*], *Peter Lewis Silveston*[†]

[*]Oporto, Portugal and [†]Waterloo, Ontario, Canada

OUTLINE

22.1 Concepts and Types of Pressure Swing Reactors 638

22.2 Models for Swing Reactors 640

22.3 Computational Considerations 642

22.4 Simulations of Pressure Swing Systems 644

22.4.1 Isothermal Simulation 644

22.4.2 Non-isothermal Simulations 649

22.5 Experimental Studies 652

22.6 Temperature Swing Reactors 657

22.6.1 Types of Reactors 657

22.6.2 Modeling 658

22.7 Simulation of Temperature Swing Systems 660

22.8 Temperature Swing Reactor Networks 663

22.9 Experimental 663

22.10 Combined Pressure and Temperature Swing Reactors 671

22.10.1 Simulation Studies 671

22.11 Overview and Assessment 674

Acknowledgment 676

Swing reactors employ a mixed bed of catalyst and adsorbent and utilize either a pressure or temperature change for regenerating that adsorbent. Because of the adsorbent, they must operate in a cyclic mode. Separation is often limited so that recovery of a reaction product in high purity is unusual. Often the primary justification for choosing a swing system is that product separation, even if incomplete, extends conversion when a reaction is equilibrium limited.

Development of swing reactors appears to be less advanced than chromatographic reactors, even though their separation counter parts are well established technology. Pressure swing reactors (PSRs) were the first to be considered (*Vaporciyan and Kadlec, 1987, 1989*). Temperature swing reactors (TSRs) garnered attention about seven years later. Consideration of combined pressure and temperature swing operation has been taken up much more recently.

A compelling reason for considering swing reactors is that they can operate at lower temperatures for equilibrium-limited endothermic reactions than a single pressure reactor. Dehydrogenation would be an example of such an application. Lower dehydrogenation temperatures reduce the interference of secondary reactions and lower capital and operating costs. On the other hand, for equilibrium-limited exothermic reactions with large activation energies, a swing reactor might increase reaction yield without requiring a reduction in reaction temperature.

22.1 CONCEPTS AND TYPES OF PRESSURE SWING REACTORS

There are two design alternatives for pressure swing systems: A single catalyst-adsorbent bed cycled through feed, delivery purge and exhaust flow or a continuous two vessel moving or fluidized bed with one vessel for reaction and adsorptive separation operating at high pressure and a second vessel for adsorbent regeneration operating at lower pressure. A continuous operation could also use a rotating annular bed so that the catalyst and adsorbent sequentially pass the feed, exhaust and delivery connections. In principle, the rotating bed is just like a discontinuously operated fixed bed, that is, the set of equations describing the two systems are the same except for the interpretation of the time variable. In this chapter, we consider only the fixed bed alternative.

PSR designs follow those used for adsorption systems. **Figure 22-1** illustrates two PSA

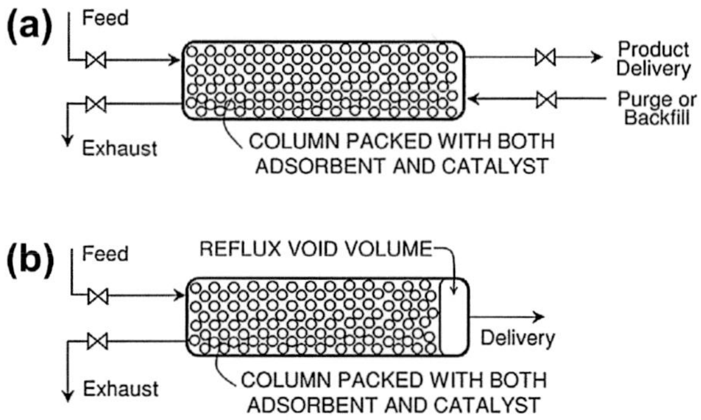

FIGURE 22-1 Connections and bed arrangements for pressure swing adsorbers and reactors: (a) standard design, (b) design with reflux volume. *(Reproduced from* Vaporciyan and Kadlec (1987) *with permission. © 1987 AIChE.)*

designs that have been adopted for reactors. In **(a)**, the reactor is completely filled with catalyst and adsorbent, while in **(b)** a void volume follows the bed. This latter volume provides purge. It could also be an attached empty vessel. The design with attached void volume **(b)** was intended for rapid cycling (*Turnock and Kadlec, 1971; Kowler and Kadlec, 1972*). With deeper beds, built without a reflux volume, cycle times of several minutes are used. Purge is provided from an external source. In recent years, further variations have been introduced: The reactor bed can be divided into zones and maintained at different temperatures by heating or cooling the wall for that zone. Also, stream withdrawal or purge may be carried out between zones.

Pressure swing reactors can be operated in several different modes. **Figure 22-2** shows the basic pressure swing cycles employed: **(a)** Two steps, **(b)** three steps with a delivery step, **(c)** four steps with an added purge or **(d)** with added backfill and **(e)** six steps with two purges and repressurization. For industrial operations, at least two PSR units would be used in parallel to generate quasi-continuous flow of the feed, product and exhaust streams. There are many

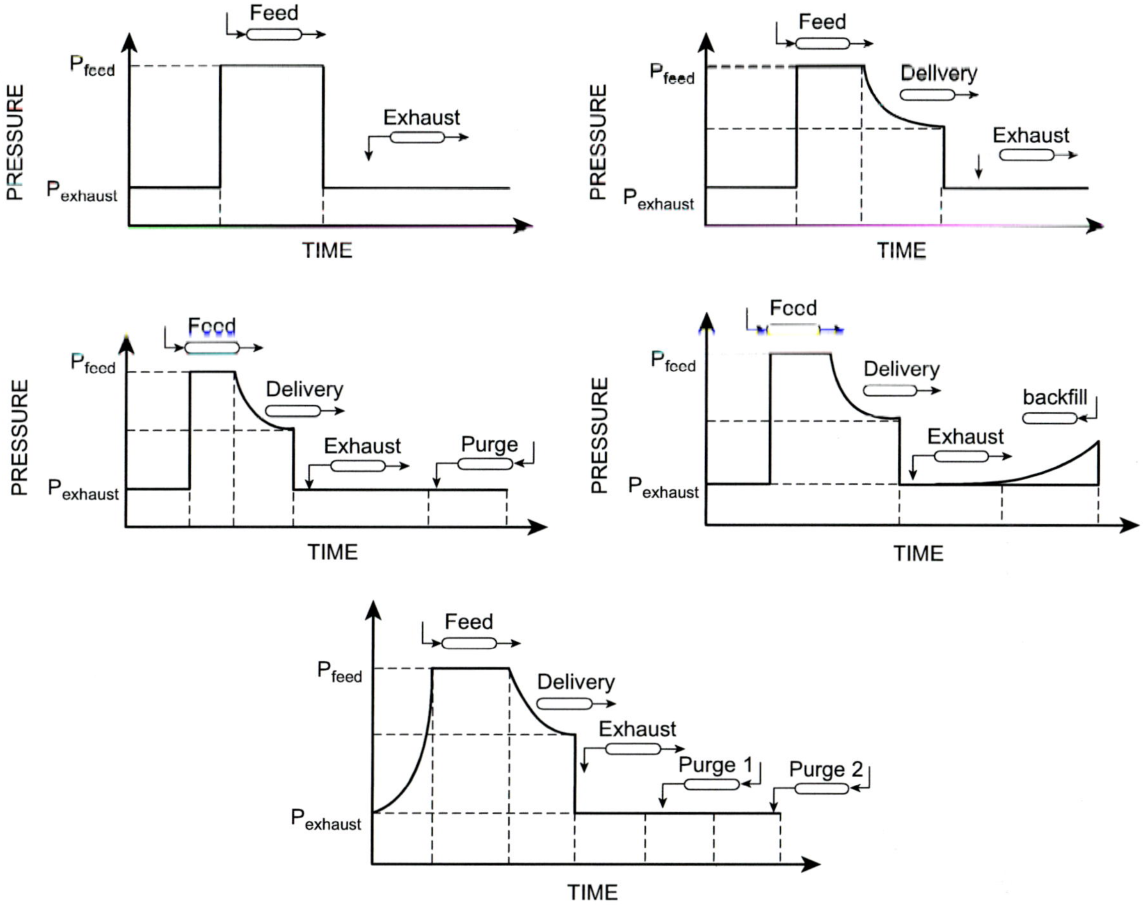

FIGURE 22-2 Basic operational cycles for pressure swing reactors. *(Adapted from* Vaporciyan and Kadlec (1987) *with permission. © 1987 AIChE.)*

variations of the basic cycles shown in **Figure 22-2**. The number of PSR units used and their arrangement in a network also has variations. As a result, the patent literature is large.

The pressure swing vessel in **Figure 22-1** has one inlet for the feed. This inlet is held at the highest pressure in the system. Normally the adsorbent bed is vertically oriented with an inlet at the bottom of the bed. The product or discharge stream is removed through an outlet at the top of the bed. It is withdrawn at an intermediate pressure. The second outlet, the exhaust stream, is located at the feed end of the bed and is held at the lowest pressure used in the cycle.

The basic PSR cycle consists of a pressurization or feed step (**Figure 22-2(b)**) in which feed is introduced to the reactor-adsorbent bed at the upper pressure limit. This brief step is followed by a discharge, sometimes called a delivery step, in which the discharge pressure is reduced to an intermediate value. During this step, the discharge port is open. The third and final step is exhaust or depressurization, when the feed side outlet is opened and the feed and discharge ports are closed. Pressure in the exhaust outlet is held constant at the lowest pressure in the cycle. Bed pressure falls to this level by the end of the step. Step duration varies with particle size of adsorbent, bed length and adsorption properties of the mixture to be separated. Usually the discharge step is the longest.

A purge from an external source adds a fourth step following depressurization. As can be seen in **(c)**, the discharge port is opened and the product is allowed to flow back into the bed. The product flow in **Figure 22-1** moves from right to left providing a flush of the mixed bed of catalyst and adsorbent. The exhaust port remains open. Purge cycles improve separation. They are used in rapid pressure swing systems but at the cost of either a smaller throughput or a longer cycle period if the same feed amount is maintained.

Figure 22-2(d) illustrates a backfill cycle. In this four-step operation, the exhaust port is closed after depressurization and the product stream from the discharge port partially repressurizes the bed. The cycle is also employed to improve separation. With both four-step cycles, it is assumed that product is discharged into a well-mixed vessel which provides either purge, backfill or both. A variant of the backfill cycle employs product or a non-adsorbing feed component to repressurize the bed. In this case, shown in **Figure 22-2(e)**, the feed in step 1 forces the repressurizing gas out of the reactor. The figure also shows use of a second purge with a composition different from purge 1.

Discussion of the various modes of operation for pressure swing systems are given by *Ruthven (1984)*, *Wankat (1986)*, *Yang (1987)* and *Suzuki (1990)*. Rapid cycles seem to be most suited to reaction applications. These generally employ short, low-aspect-ratio beds of fine particles and cycle periods on the order of seconds.

22.2 MODELS FOR SWING REACTORS

The material, momentum and energy balances for pressure or temperature swing reactors are shown in **Table 22-1** for the case of a one-dimensional, pseudo-homogeneous model widely used for swing reactor design. This model adequately describes reactor behavior at the limits of adiabatic or isothermal operation. In packed beds, temperature differences between phases are of the order of a degree or less so the pseudo-homogeneous energy balance can be used even when adsorption equilibrium cannot be assumed. The Ergun equation is given in place of a momentum balance in these tables. This equation reduces to Darcy's law if J_k in the equation is set to zero. Equations in **Table 22-1** are one dimensional. Two-dimensional models must be used if the reactor is not adiabatic. Packed beds are isotropic implying a uniform mixture of catalyst and adsorbent. The equations in the table assume that a single solid acts both as catalyst and adsorbent.

Partial pressures are sometimes used in place of concentration in material balances because adsorption equilibria are stated in terms of that variable. Usually gas properties are assumed to be independent of composition and temperature, and mixtures are taken to behave ideally. Catalyst and adsorbent are treated as well mixed and uniformly distributed in the reactor. Inlet/outlet effects, such as expansion, are neglected.

Extension of the models to situations in which the adsorbent and catalyst cannot be lumped are discussed in **Chapter 20**.

Models are now often used in dimensionless form. These have been discussed in **Chapter 20**. In that chapter, **Table 20-9** defines the

TABLE 22-1 Pseudo-Homogeneous, One-Dimensional Model Equations for Pressure, Temperature and Combined Swing Reactors

Component Mass Balance

$$\varepsilon_t \frac{\partial C_i}{\partial t} + \rho_{b_{ads}} \frac{\partial q_i}{\partial t} = \varepsilon_b \frac{\partial}{\partial z}\left(D_{ax_i} \frac{\partial C_i}{\partial z}\right) - \frac{\partial u C_i}{\partial z} - \rho_{b_{cat}} \sum_{j=1}^{N} \nu_{ij} \eta_j R_j. \tag{22-1}$$

Total Mass Balance

$$\varepsilon_t \frac{\partial C}{\partial t} + \rho_{b_{ads}} \frac{\partial q}{\partial t} = -\frac{\partial u C}{\partial z} - \rho_{b_{cat}} \sum_{i=1}^{n} \sum_{j=1}^{N} \nu_{ij} \eta_j R_j. \tag{22-2}$$

Momentum Balance as an Empirical Relation (Ergun Eq.)

$$\frac{\partial \rho_g u}{\partial z} = \frac{d P}{d z} = -J_v u - J_k u^2. \tag{22-3}$$

Energy (Thermal) Balance

$$\varepsilon_t C_{p_g} \frac{\partial (\rho_g T)}{\partial t} + (\rho_{b_{ads}} + \rho_{b_{cat}}) C_{p_s} \frac{\partial T}{\partial t} = k_z \frac{\partial^2 T}{\partial z^2} - C_{p_g} \frac{\partial (u \rho_g T)}{\partial z} - \rho_{b_{ads}} \sum_{i=1}^{N_C} (-\Delta H_{a_i}) \frac{\partial q_i}{\partial t}$$
$$- \rho_{b_{cat}} \sum_{i=1}^{N_C} \Delta H_j \eta_i R_j + \frac{4 h_W}{d_r} (T_W - T). \tag{22-4}$$

Adsorption Rate Model (Linear Driving Force Assumption)

$$\frac{\partial q_i}{\partial t} = k_i (q_{s_i} - q_i). \tag{22-5}$$

dimensionless groups employed. For swing reactors, other dimensionless groups arise and the use of reflux leads to more. These are given in **Table 22-2**.

In this table, cycle fractions define the duration of the different parts of a swing cycle. **Table 20-9** defines the Damköhler number for a first-order reaction. The number depends, of course, on the kinetic model. A power law model is used for the definition in **Table 22-2**. There are separate definitions for forward and reverse reactions.

Aida and Silveston (2005) suggested naming two of these dimensionless groups, N_{RU} and N_{Ka}, to honor Professors Douglas Ruthven and Robert Kadlec, pioneers in the study and modeling of pressure swing adsorbers. Separation achieved in either a PSA or PSR is given by a separating factor, also defined in **Table 22-2**. Separation can also be described by the yield of a product removed in either the discharge or exhaust port.

Cell or lumped models of a packed bed were once used because they simplified programming. With the advent of reactor computing packages, cell models are seldom advantageous. The concept of a cell model is illustrated by **Figure 20-4**. They are discussed in **Section 20.2.2.**

Boundary and initial conditions complete the PDE models of **Table 22-1**. When a port is closed, of course, boundary conditions are not needed. However, if there is no open inlet or outlet at an end of the reactor bed, an end of bed boundary condition must be specified. The latter are the standard zero flux conditions, namely,

$$\frac{d T}{d z} = \frac{d P}{d z} = \frac{d y_i}{d z} = 0. \tag{22-6}$$

TABLE 22-2 Definition of Specific Dimensionless Variables, Parameters and Groups used in Swing Reactor Models

Cycle fractions:	
Feed	$f_f = \frac{t_f}{\tau}$
Discharge	$f_d = \frac{t_d}{\tau}$
Exhaust	$f_e = \frac{t_e}{\tau}$
Damköhler number for the forward reaction of order β:	$N_{Da_f} = k_j P_f^{\beta_i - 1} \rho_c RT\tau$
Damköhler number for the reverse reaction of order α:	$N_{Da_r} = \frac{k_j}{K_{eq_j}} P_f^{\alpha j - 1} \rho_c RT\tau$
Pressure Ratio	$\frac{P_e}{p_f}$
Kadlec number	$N_{Ka} = \frac{n_d RT\tau}{P_f V_R}$
Ruthven number	$N_{Ru} = \frac{\tau P_f k}{\mu L^2}$
Volume Ratio	$\frac{1}{f_R} = \frac{V_R}{A_x L}$
Capacity	$\gamma_i = \varepsilon + \rho_A RT\alpha_i$
Separation Factor (i/j)	$\frac{(y_i/y_j)_{discharge}}{(y_i/y_j)_{exhaust}}$

Table 22-3 sets out boundary and initial conditions for pressure swing reactors. Note that these equations are independent of the models employed for adsorption and reaction, and also of reaction stoichiometry.

Phenomenological models for reaction and adsorption vary among investigators so these will be introduced as necessary when we discuss specific contributions.

22.3 COMPUTATIONAL CONSIDERATIONS

Periodic operation of reactors results in moving reaction and temperature fronts. These fronts manifest abrupt changes in concentration and temperature that lead to unstable behavior in many commonly chosen integration routines. Such instability can be managed only by employing inordinately small space and time steps. In the presence of both adsorption and reaction, the instability problem appears to be exacerbated. *Vaporciyan and Kadlec (1987)* found that integration routines which functioned satisfactorily with PSA systems became unstable when an equilibrium reaction was introduced. Several investigators have explored efficient computation methods employing spatial discretization to convert the reactor model into a set of ordinary differential equations (ODEs) using: 1) Orthogonal collocation (OC), 2) orthogonal collocation on finite elements (OCFE), 3) double collocation on finite elements (DOCFE) to convert the PDEs to a set of algebraic equations, and 4) by assuming the reactor to consist of well-mixed cells-in-series (CIS). (*Alpay et al., 1993; Chatsiriwech et al., 1992, 1994; Cheng et al., 1998b; van Noorden et al., 2002*). *Cheng et al. (1998b)* were concerned with the computational effort for optimizing PSR variables. They compared discretizing the spatial variables only and solving non-linear dynamic equations with discretization of both temporal and spatial variables and solving a large system of non-linear algebraic equations. *Aida and Silveston* summarize these important studies and provide their conclusions in their 2005 book, "Cyclic Separating Reactors."

Engineering of pressure swing reactors requires knowledge of the cyclic stationary state. The problem of rapidly estimating such states has attracted much attention. Indeed, it is a concern for all types of periodic processes so it is not surprising that a large research literature has arisen on the "shooting" methods that are used (*Croft and LeVan (1994a), van Noorden et al. (2002, 2003a, 2003b)*. *Aida and Silveston (2005)* offer a highly condensed review of various "shooting" methods.

TABLE 22-3 Boundary and Initial Conditions for Pressure Swing Reactors

Bed Initial Condition for All Cycle Parts

$$\zeta(z, \tau_p) = \zeta(z, \tau_p)^{+} \quad (22\text{-}7)$$

Once a cyclic stationary state is achieved, a further initial condition may be used:

$$\zeta(z, t) = \zeta(z, t + \tau_p) \quad (22\text{-}8)$$

where ζ is any dependent variable (y_i, P, T, for example)

Inlet Port Boundary Condition

Pressurization: $p = p_f$ **(22-9)**

$$D_z \frac{d}{dz}(Cy_i) = \frac{n}{A_x}(y_i - y_{i_f}) \quad (22\text{-}10)$$

$$\lambda_z \frac{dT}{dz} = \frac{nC_{p_g}}{A_x}(T - T_f) \quad (22\text{-}11)$$

Exhaust/Purge/Backfill (This port is usually closed):

$$dP/dz = dy_i/dz = dT/dz = 0 \quad (22\text{-}12)$$

Discharge Port/Purge or Backfill Inlet Port

Pressurization/Delay/Discharge:

$$dP/dz = dy_i/dz = dT/dz = 0 \quad (22\text{-}13)$$

$$n = C_v\sqrt{C(P_{z=L} - P_d)} \quad (22\text{-}14)$$

or,

$$n = n_d \text{ (constant delivery)} \quad (22\text{-}15)$$

$$\frac{d}{dz}(Cy_i) = \frac{dT}{dz} = 0 \quad (22\text{-}16)$$

Exhaust (Two operations are possible):

1) the discharge port is open and conditions given by and/or **Eqs (22-14)** to **(22-16)** apply, or 2) the discharge port is closed so the condition given by **Eq. (22-13)** applies)

Purge:

$$P = P_p \quad (22\text{-}17)$$

and $n = C_v\sqrt{C(P_p - P_{z=L})}$ **(22-18)**

or

$n = n_p$ (constant delivery) **(22-19)**

Also,

$$D_z \frac{d}{dz}(Cy_i) = \frac{n}{A_x}(y_i - y_{ip}), \text{ where } y_{ip} = y_{i_{mix}} \quad (22\text{-}20)$$

$$\lambda_z \frac{dT}{dz} = \frac{nc_{p_g}}{A_x}(T - T_p), \quad \text{where } T_p = (T)_{mix} \quad (22\text{-}21)$$

Backfill:

These are the same as for the purge step, except P_b replaces P_p and n_b replaces n_p

(Continued)

TABLE 22-3 Boundary and Initial Conditions for Pressure Swing Reactors (*cont'd*)

Exhaust (Depressurization) Port

Feed/Delay/Discharge:

Discharge port is shut at the end of bed so a boundary condition must be used. (**Eq. (22-13)**)

Exhaust:

$$P = P_V \tag{22-22}$$

$$n = C_v\sqrt{C(P_{z=0} - P_v)} \tag{22-23}$$

and

$$\frac{d}{dz}(Cy_i) = \frac{dT}{dz} = 0 \tag{22-24}$$

Backfill:

Discharge port is closed so end of bed conditions apply. (**Eq. (22-13)**)

End Of Packed Bed (With Reflux Void Volume)

Pressurization/Discharge:

$$\frac{dP}{dz} = \frac{d}{dz}(Cy_i) = \frac{dT}{dz} = 0 \tag{22-25}$$

Exhaust:

$$D_z\frac{d}{dz}(Cy_i) = \frac{n}{A_x}(y_i - y_{i_R}), \text{ where } y_{i_R} = y_{i_{mix}} \tag{22-26}$$

$$\lambda_z\frac{dT}{dz} = \frac{nc_{p_g}}{A_x}(T - T_R), \text{ where } T_R = (T)_{mix} \tag{22-27}$$

22.4 SIMULATIONS OF PRESSURE SWING SYSTEMS

22.4.1 Isothermal Simulation

Early simulation of PSR operations assumed isothermality. These important investigations are summarized in **Table 22-4**.

In an early study of a PSR, *Vaporciyan and Kadlec (1987)* examined fast reactions under conditions of adsorption equilibria. Because equilibrium is attained as the feed enters the reactor and is maintained at all points inside this vessel, pressure swing operation cannot force the reaction to exceed its equilibrium limit. Thus, the authors were interested in how the presence of reaction affected the separation performance of a PSA. Changes in the number of moles on a reaction mean that the equilibrium partial pressures are affected by reactor pressure. *Vaporciyan and Kadlec (1987)* observed a restricted range for cycle periods so as to maintain a specified discharge flow rate. Within this range, the cycle period had just a small effect on reactant conversion. With very rapid cycling, the separation factor approached unity so no separation occurred. Once the cycle period exceeded a critical value, there was good separation which became independent of period as period increased. *Vaporciyan and Kadlec* noted that their results are applicable to other pressure swing cycles. The presence of reaction in the adsorbent bed changes the separation performance of the pressure swing system in terms of the exit steams in which reactants and products appear, but the extent of separation is not greatly altered.

TABLE 22-4 Simulations of Pressure Swing Reactors

Authors	Reactor System	Reaction Considered	Reactor Operation	Integration Method	Observations	Comments
Vaporciyan and Kadlec (1987)	Mixed bed isothermal PSR with attached reflux compartment	General, equilibrium reactions: A ↔ C; A ↔ 2 C; 2 A ↔ C: A ↔ ½ B + C	3-step, rapid cycle; constant discharge rate; $P_{discharrge} = 1/3\ P_{feed}$, $t_{exhaust} = t_{feed}$, $t_{discharge} \approx 0$	Cells in series models with solving a set of ODEs	Equilibrium conversion cannot be exceeded. There is no separation unless the number of moles changes during reaction. In such cases separation depends on τ_{cycle} and f_{feed}.	Equilibrium constraint is often valid for FSA operation, but is not realistic for a PSR.
Vaporciyan and Kadlec (1989)	As above with $V_{reactor} = 10.9$ V_{reflux}	General, irreversible reactions: A → ½ B + ½ C, A → B, C ← A → B	3-step, rapid cycle; $\tau_{cycle} = 5$ s, $t_{exhaust} = t_{feed}$, $t_{discharge} \approx 0$, $P_{feed} = 1.68$ bar, $P_{exhaust}/P_{feed} = 0.595$	As above	If reactant is not adsorbed, $X_{PSR} > X_{PFR}$ provided that a reaction product adsorbs, separation performance is similar to PSA for the same products.	Key performance parameters are $P_{exhaust}/P_{feed}$, τ_{cycle}, f_{feed}.
Lee and Kadlec (1989)	4-zone isothermal PSR with different catalyst-adsorbent mixtures in each zone	General, equilibrium reaction B ↔ A ↔ C	3-step rapid cycle; zone composition and length adjusted to maximize product purities at discharge and exhaust		Catalyst must be placed in the front of the reactor and adsorbent for the reactant A should be uniformly distributed.	Distribution of catalyst and adsorbent just weakly affected the performance.
Alpay et al. (1993)	3- and 4-step isothermal PSR using a portion of the discharge as purge	General: A → B	$L_{bed} = 1$ m, $(y_A)_{feed} = 0.05$, $P_{feed} = 10$ bar, $P_{exhaust} = 1$ bar, $\tau_{cycle} = 24$ s, $t_{discharge} = 12$ s, t = 6 s for other steps	Method of characteristics resulted in set of ODEs integrated numerically (R-K-M routine)	$Y_{PSR} < Y_{PFR}$ for slow reaction well removed from equilibrium; $Y_{PSR} > Y_{PFR}$ for fast reactions approaching equilibrium. Use of purge increases product recovery.	PSR behaves like a PFR for fast reactions approaching equilibrium.
Alpay et al. (1994)	As above with an equal mixture of catalyst and adsorbent	General: A → B, Methylcyclohexane (MCH) dehydrogenation	As above for isomerization; $\tau_{cycle} = 36$ s, $t_{discharge} = 18$ s for dehydrogenation	As above	Without purge, toluene yield in MCH dehydrogenation less for a PSR than for a PFR. Purge improves PSR yield.	Model parameters used values that were determined experimentally.
Chatsiriwech et al. (1994)	3-step isothermal PSR with portion of well mixed discharge as purge	General: 2 A ↔ B + C with B adsorbed		Method of characteristics resulted in set of ODEs integrated numerically (R-K-M routine)	Optimal purge amount maximized Y_B. Both depend on magnitude of K_B. $(Y_B)_{PSR} > (Y_B)_{PFR}$ with adsorption of B, Amount of discharge affects performance.	Separation factors for products also depend on K_B.
Kirkby and Morgan (1994)	Isothermal PSR with 3-step cycle and 4-step cycle with purge, also with backfill of constant composition	General, irreversible: A → B	$\tau_{cycle} = 90$ s with $t_{feed} = t_{discharge} = t_{purge} = t_{backfill}$, $P_{feed} = 1$ MPa, $P_{exhaust} = 0.1$ Mpa, $L_{Bed} = 1$ m	Method of characteristics resulted in set of ODEs integrated numerically (R-K-M routine)	Adsorbed product B recovered at exhaust, but low product discharge inversion takes B to discharge.	Several factors were found to affect inversion: type of cycle, product draw-off, amount of purge. Backfill and purge cycles give similar performance.

(*Continued*)

TABLE 22-4 Simulations of Pressure Swing Reactors (*cont'd*)

Authors	Reactor System	Reaction Considered	Reactor Operation	Integration Method	Observations	Comments
Kodde and Bliek (1997)	3 step, isothermal PSR	General, $A \rightarrow B \rightarrow C$ Only B adsorbed			PSR operation improves selectivity to B with high conversion of A.	
Cheng et al. (1998b)		General, reversible: $A \rightarrow B + C$				Objective was optimization of control variables.
Kodde et al. (2000)	4 step, isothermal PSR with a portion of the delivery stream used as purge	General: $A + B \rightarrow C$, $C + B \rightarrow D$	$P_{feed} : P_{exhaust} = 3$	2nd order upwind method used to discretize reactor axially, resulting ODEs solved by gPROMS	Selectivity to C depended strongly on adsorption. Maximum conversion is achieved when reaction and adsorption are balanced.	Performance of PSR was better than a stand-alone PFR or a PFR followed by a PSA.
Ding and Alpay (2000a, b)	Non-isothermal PSR with a 5-step cycle: 1) Feed and production, 2) depressurization, 3) H_2 purge, 4) steam purge and 5) pressurization	Steam reforming of CH_4 using a hydrotalcite for CO_2 capture	See Figure 22-2	2nd order collocation procedure dividing the reactor into 100 elements. Resulting array of algebraic equations solved using gPROMS	Authors observed a low temperature wave travelling through the reactor. They found a literature kinetic model for steam reforming satisfactory.	Study was limited to the production step and compared model simulations with experiment.
Sheikh et al. (2001)	2-step isothermal PSR, with depressurization through only the discharge port or through both discharge and exhaust ports	General, equilibrium: $A \leftrightarrow B + C$ (1 butene dehydrogenation to butadiene)	Rapid operation with $t_{pressurization} = t_{depressurization} = 1$ s		Depressurization using both ports increases B concentration in discharge and $(Y_B)_{PSR} > (Y_B)_{PFR}$. Depressurization through discharge only increases conversion.	Model parameters were determined experimentally.
Gomes and Yee (2002)	4 step, isothermal PSR	Disproportionation of $C_3^=$ with $C_4^=$ strongly adsorbed	$100 < \tau_{cycle} < 450$ s	Collocation procedure applied to a PDE model	Cycle time, τ_{cycle}, affected $C_2^=$ purity and recovery in the discharge strongly and in opposite ways. Feed rate effects purity and recovery similarly.	Simulation was verified experimentally.
Van Noorden et al. (2003a)		General, $A \leftrightarrow B + C$			PSR cycle efficiently optimized by a numerical gradient method.	
Al-Juhani and Loughlin (2003)	4 step, isothermal PSR with catalyst placed in front half of reactor and adsorbent in the back half	Isomerization of n-C_5, n-C_6 in the presence of H_2. Operation near reaction equilibrium. Model parameters chosen to represent 0.5 wt% Pd on Y and 5A zeolite adsorbents	Steps in cycle: 1) pressurization with feed, 2) feed at constant P with discharge open, 3) draw down through exhaust, 4) purge with H_2. $P_{feed} = 15$ bar, $P_{discharge} = 2$ bar	Orthogonal collocation of finite element model deduced from PDE	Increasing the catalyst activity or space time raised conversion through their effect on the Damköhler No. Increasing the bed temperature decreased conversion.	High paraffin concentration discharge suggests this stream should be used for purging the catalyst bed.

Xiu et al. (2002a, b)	5 step, non-isothermal PSR with two purge steps and repressurization with H_2	Steam reforming of CH_4 over an equivolume mixture of catalyst and adsorbent	$L_{reactor} = 2$ m, $t_1 = 1000$ s, $t_3 = 900$ s, $t_2 = t_4 = t_5 = 100$ s, Steam:$CH_4 = 6{:}1$, $T_f = T_p = T_W = 723$ K		Increasing the packing density, adsorbent/catalyst and $\bar{f}$ increased the CH_4 conversion and H_2 purity.	Model parameters were drawn from *Hufton et al. (1999).*
Xiu et al. (2003a, b)	As above	As above, however with a 2:1 adsorbent: catalyst mixture	As above with 3 mm spherical catalyst and adorbent particles		Diffusion within the catalyst particles was important for reforming and the water gas shift reaction: $0.4 < \eta_{reforming} < 0.95$; $0.7 < \eta_{shift} < 0.95$.	Only the production step was examined.
Reijers et al. (2006)	4-step PSR-TSR system	Steam reforming of methane	$P_{feed} = 17$ bar, $T_{reactor} = 600°C$, $(Steam/CH_4)_{feed} = 3{:}1$; $P_{regenerator} = 17$ bar, $T_{regenerator} = 1000°C$, using steam as regenerant		Simulation gave 93% methane conversion, overall energy efficiency was 56% with 85% CO_2 capture.	
Ochoa-Fernandez et al. (2007)	5-step, combined PSR-TSR with a pressurization to permit comparison with conventional multi-step reforming	Steam reforming of natural gas in fixed bed	$P_{feed} = 10$ bar, $T_{reactor} = 575°C$, $(Steam/CH_4)_{feed} = 3{:}1$; $P_{regenerator} = 1$ bar, $T_{regenerator} = 870°C$		Hydrogen products were 99.9% H_2. H_2 recovery was 90% just as in conventional process, but carbon capture was 100%.	The swing process had a net energy efficiency of 79% compared with 71% for conventional reforming.
Hershkowitz et al. (2009)	2-step, non-isothermal PSR combined with a reverse flow heat recuperator and integrated into a gas turbine power cycle	Steam reforming of natural gas and light distillates	$P_{feed} = 10$ atm, $P_{regenerator} = 1$ atm, $T_{feed} = 250°C$, $\tau_{cycle} = 12$ s, steam:$CH_4 = 3{:}2$	Subdivided reactor into 30 well-mixed cells and integrated ODE model (using Xu and Froment kinetics)	Conversion approached 100% with 85–90% recovery of the combustion energy.	Some experiments discussed, but no results or details were given.

Vaporciyan and Kadlec (1989) continued investigating an isothermal, rapid cycling PSR but focused on systems containing a reflux compartment. Considering an equilibrium limited parallel reaction, B ↔ A ↔ C, they observed that the pressure ratio, P_e/P_f, caused selectivity to B and conversion of A to increase as the ratio decreased. Cycle period and the feed fraction of a cycle influenced selectivity and exhibited optima.

A design variant was considered by *Lee and Kadlec (1989)* for parallel equilibrium reactions, B ↔ A ↔ C, using a three-step pressure swing cycle. Three adsorbents, each specific for a reactant or a product were each mixed with catalyst and distributed in one to three sections of a four-zone reactor with zones of variable length. The concept was to control the concentration profile of the components so as to maximize the purity of one of the products, in this case product B, in the discharge stream. *Lee and Kadlec* found that catalyst should be placed in the first 10% of the bed. The adsorbent for A should be uniformly distributed in the remaining 90%, whereas the adsorbents for B and C should be distributed uniformly throughout the reactor. Nevertheless, their results indicated that distribution of catalyst and adsorbents had a small effect on conversion and selectivity to B in the product stream.

European researchers (*Alpay et al., 1993, 1994; Chatsiriwech et al., 1994; Cheng et al., 1998b; Yongsunthon and Alpay, 1998a, b, 1999, 2000; Xiu et al., 2002a, b, 2003a, b*) have dominated research on PSRs for the last twenty years. *Alpay et al. (1994)* employed parameters taken from experimental kinetic and adsorption data. A simple pressure swing system with a four-step cycle (**Figure 22-2(c)**) was considered in which a well mixed portion of the product stream was used as purge. Catalyst and adsorbent were separate phases. A linear driving force model **(Eq. 22-5)** was used to represent the combined resistance of adsorption and mass transfer: relations for the coefficient k_i in the model have been proposed by *Alpay et al. (1993)* and *Ding and Alpay (2000b)*.

Simulated yields for simple, reversible isomerization, A ↔ B, were compared as a function of the forward rate constant for pressure swing and plug flow reactors. Yields for the PFR were at 10 bar, the feed pressure for the PSR. Well away from equilibrium, a PFR outperformed the PSR because the reverse reaction is not important. However, as equilibrium is approached, the situation is reversed. Improvement depended strongly on the magnitude of the adsorption constant. At high reaction rates, the PSR behaved as a pressure swing adsorber. Adding a purge cycle further improved the separation of reactant and product. Indeed, *Alpay et al.* showed that purge increased yields near equilibrium and provided greater product and reactant separation. Simulation of a disproportionation reaction, 2A ↔ B + C, in a PSR and a PFR for rate and adsorption constants similar to those used for isomerization gave the behavior just discussed.

Alpay et al. (1994) suggest that a PSR is not the best choice for equilibrium reactions with an increase in moles through reaction. Their simulation of methylcyclohexane dehydrogenation showed that the yield of a PSR operated at a feed pressure of 2 bar and an exhaust pressure of 1 bar is always below the PFR yield regardless of the adsorption constant and the purge amount. Nevertheless, a PSR can exceed PFR yields for dehydrogenation reactions in which one mole of H_2 is formed for each mole of product, e.g., ethane dehydrogenation to ethene.

In closely related work, *Chatsiriwech et al. (1994)* examined the 2A ↔ B+C reaction. Yield depended on the adsorption of B, varying the purge duration alters that yield, but fractional durations, $f_{delivery}$ and f_{purge}, had just a small effect. Cocurrent and countercurrent operations of a two-step PSR were considered by *Sheikh et al. (2001)* for 1-butene dehydrogenation. In the former mode, depressurization occurred only through the discharge port while in the latter both discharge and exhaust port were opened. Cocurrent depressurization increased conversion of $C_4^=$ dramatically but at the expense of reactant and product separation.

Kirkby and Morgan (1994) observed an inversion in their study of an irreversible isomerization with a non-adsorbing reactant. The product was concentrated in the exhaust rather than in the discharge as the amount of product drawn off per cycle decreased. Their study suggests that PSR performance with respect to operating parameters is similar regardless of operating mode.

The equilibrium-limited disproportionation of $C_3^=$ was investigated by *Gomes and Yee (2002)* through simulation as well as experimentation. A four-step isothermal pressure swing cycle was considered. Weakly adsorbed $C_2^=$ was recovered primarily in the discharge. Their simulation explored the effect of cycle period and reactant feed rate on performance. The effect of the period is shown in **Figure 22-3** and that of feed rate is similar.

Al-Juhani and Loughlin (2003) explored through simulation separate beds of catalyst and adsorbent as a way of reducing the capital outlay for equilibrium-limited reactions, such as isomerization, in which recycle of unreacted feed is necessary. In this process, H_2 is added to prevent catalyst deactivation. A four-step cycle was considered with: 1) Bed pressurization with feed, 2) feed step with recovery of the isomerate product in H_2, 3) depressurization producing a waste stream of mixed reactant, product and H_2, and 4) a low pressure purge using H_2. The liquid isomerate product is flashed to recover H_2. The investigators explored the influence of catalyst activity and temperature on performance. Of particular interest was the optimal fraction of the bed occupied by catalyst. Published models were used for isomerization kinetics and adsorption equilibria.

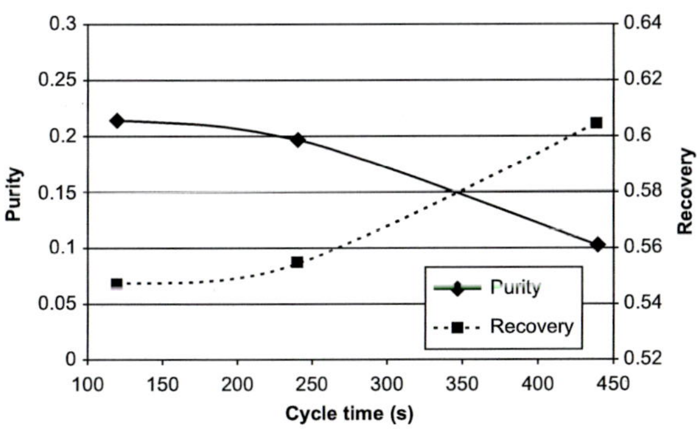

FIGURE 22-3 Ethene product purity and total recovery as a function of cycle period for a four-step pressure swing cycle imposed on the disproportionation of propene. *(Figure adapted from* Gomes and Yee (2002) *with permission. © 2002 by Elsevier Science Publishers.)*

22.4.2 Non-isothermal Simulations

Adsorption and desorption exhibit significant heat effects. Only on a laboratory scale can these heats be balanced by external heating or cooling to realize isothermal operation. In vessels whose diameter exceeds 4–5 cm, radial as well as axial temperature variations must be addressed. Nevertheless, non-isothermal operation of PSRs has been addressed only recently (*Yongsunthon and Alpay, 1998a, b, 1999; Xiu et al., 2002a, b; 2003a, b; Reijers et al., 2006; Ochoa-Fernandez et al., 2007*).

The operating principle of all combined reaction and adsorption systems is that product adsorption reduces the rate of the reverse reaction thereby raising conversion as equilibrium is approached. With endothermic reactions, *Yongsunthon and Alpay (1998b)* note that adding an adsorbent changes the temperature conversion trajectory favorably with respect to conversion. Defining conversion to include reactant trapped by the adsorbent gives the following expression:

$$X_A = 1 - \frac{C_A(1 + (\varphi_A))}{C_{A_0}(1 + (\varphi_{A_0}))} \tag{22-28}$$

where φ_A is the volumetric capacity of the adsorbent for reactant, A, per unit volume of fluid in the reactor. The subscript 0 indicates a feed or initial condition. For a well-mixed continuous flow reactor at steady state, the material and heat balances of **Table 22-1** combine to give:

$$X = -\frac{1}{(\Delta H_r)_f}\left(\frac{(\gamma)_f(T-T_0)}{Q_0 C_{A_0}} + \frac{(\varphi_A)_f}{1+(\varphi_A)_f}\Delta H_{a_A}\right) \tag{22-29}$$

the slope of which is:

$$\frac{dX}{dT} = -\frac{1}{(\Delta H_r)_f}\left(\frac{(\gamma)_f}{Q_0 C_{A_0}} + \frac{(\varphi_A)_f}{1+(\varphi_A)_f}\frac{d\Theta_i}{dT}\right) \tag{22-30}$$

where $(\Delta H_r)_f$, Θ_i and $(\gamma)_f$ are, respectively, the effective heat of reaction, a dimensionless adsorbate capacity and an effective thermal capacity defined as:

$$(\Delta H_r)_f = \Delta H_r + \sum_i \frac{\nu_i}{|\nu_A|}\frac{(\varphi_i)_f}{1+(\varphi_i)_f}\Delta H_{a_i} \tag{22-31}$$

$$\Theta_i = \frac{(\varphi_i)_f}{1+(\varphi_i)_f}\frac{\Delta H_{a_i}}{T}\left(1+\frac{\Delta H_{a_i}}{RT}\right) \tag{22-32}$$

$$(\gamma)_f = \rho_s(C_p)_s + \varepsilon\sum_i (C_p)_g M_i C_i. \tag{22-33}$$

The f subscript indicates a flowing adsorbent added to the reactor in the *Yongsunthon and Alpay* analysis. For an endothermic reaction, the slope, $dX/_{dT}$ increases as T increases. Thus, the reactor temperature is greater for the same degree of conversion so the rate accelerates. Conversely, if the reaction is exothermic, non-isothermal operation decreases conversions.

Ding and Alpay (2000a) simulated the steam reforming of methane in the presence of a hydrotalcite adsorbent for CO_2 assuming a non-isothermal PSR and using a pseudo-homogeneous one-dimensional model for their mass balance. Their energy balance allowed for heat loss through the reactor walls. The authors adopted the steam reforming kinetics proposed by *Xu and Froment (1989a, b)*. These kinetics have been widely used, but are without validation for dynamic processes. Adsorption data for hydrotalcite was obtained in a separate study and fitted to a Langmuir isotherm (*Ding and Alpay, 2000b*). Their simulation is compared with experiments in the next section.

A comprehensive simulation of a non-isothermal pressure swing reactor for the steam reforming of methane has been published by *Xiu et al. (2002a, b; 2003a, b)*. Their first paper (see **Table 22-4**) examined the five-step cycle used experimentally by *Carvill et al. (1996)* and *Hufton et al. (1999)*. Feed, purge and repressurization temperatures were the same as the wall temperature of the reactor. For their model and, indeed, for other systems considered in later papers, *Xiu et al.* followed *Ding and Alpay (2000a)* except that they assumed that adsorption was governed by the rate of mass transfer, so that their model was a heterogeneous one. They employed the Xu and Froment kinetic model, but different adsorption equilibria relations were used in the presence or absence of steam. *Xiu et al. (2002a)* simulated the entire PSR cycle. Conditions and properties used represented the *Hufton et al. (1999)* experiments. **Figure 22-4** shows the simulated pressure, temperature and superficial velocities profiles within the 2 m reactor at the end of the cycle steps for the first to fifteenth cycles. Variations disappear after the third cycles except for step one where they persist for more than five cycles. By then, $\bar{y}_{H_2} = 0.81$ and remains constant. CO_2 concentrations in the gas phase are highest at the end of the second step and drop continuously in the third to fifth steps as expected. At the end of the feed step, the front 2/3 of the reactor is saturated with CO_2. Some CO_2 remains in the front half of the reactor even after the two purge steps.

The *Xiu* simulation indicated that increasing the adsorbent to catalyst ratio and the residence

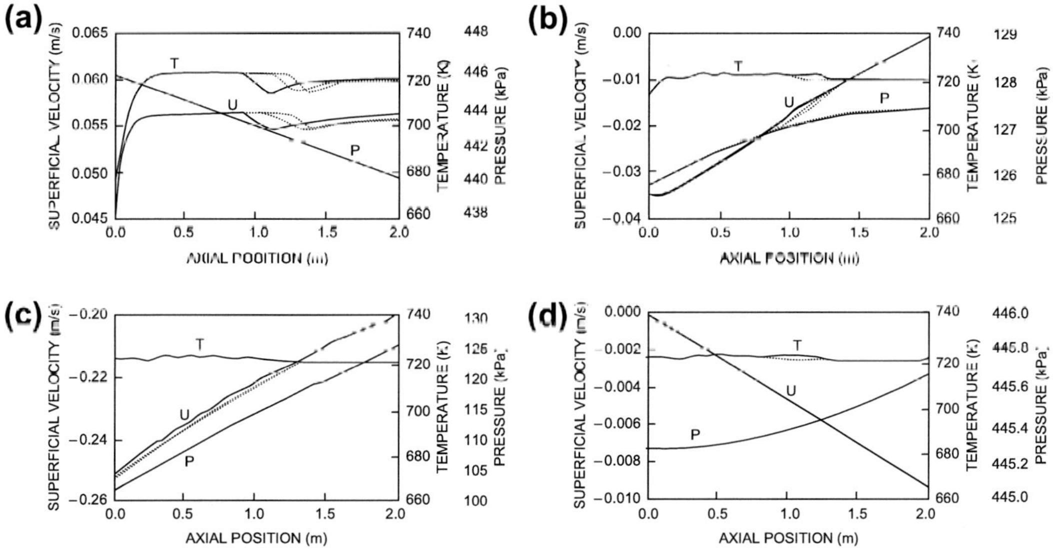

FIGURE 22-4 **Pressure, temperature and superficial velocity profiles in a five-step PSR system running the steam reforming of methane.** Operating conditions are stated in the text. The solid lines are profiles at the end of the first cycle, dashed lines are profiles after the second, fifth and fifteenth cycles. *(Figure adapted from* Xiu et al. (2002b) *with permission. © 2002 by the AIChE.)*

time enhanced performance. Using a longer reactor for this lengthens the regeneration time and increases pressure drop. In an extension, *Xiu et al. (2003a)* investigated the effect of catalyst and adsorbent particle size on PSR performance. They considered just the production step and not the entire cycle. However they had to add to their model material balances on a shell within the catalyst and adsorbent particle assuming that catalytic and adsorption sites were isotropically distributed. *Xiu et al.* assumed rapid gas-solid heat transfer and modified their material balances to include storage for component i within a particle.

From their simulations, *Xiu et al. (2003a)* observed that mass transfer is important. Reaction rates are high at the reactor entrance but fall deeper into the catalyst bed as methane is consumed. Adsorption rates are low initially because little CO_2 is available. They increase downstream so that by $z = 0.1$ they are comparable to rates of reaction at $t = 100$ s. Adsorption becomes important then in maintaining the rates of reaction. Adsorbent at the entrance is saturated with CO_2, but this does not affect reaction rate. After 700 s on stream, reaction rates at the entrance remain high, but drop by $z = 0.1$. Adsorption rates are low at this point as the adsorbent approaches saturation. The adsorption front has moved downstream to $z = 1$, increasing the adsorption rate which again becomes comparable to reaction rates. Adsorption now contributes to CH_4 conversion.

Time and position variation of rates causes erratic variations of effectiveness factors (**Figure 22-5**). These are well below unity so intraparticle diffusion and external mass transfer are important. Low η values occur at the front end of the reactor. Consequently the effect of concentration gradients on rate of reaction is important only in the first 10% of bed length. Because rate is high in this region, intraparticle diffusion does not have a large effect on the length of the reactor needed to meet the H_2 purity or the CO

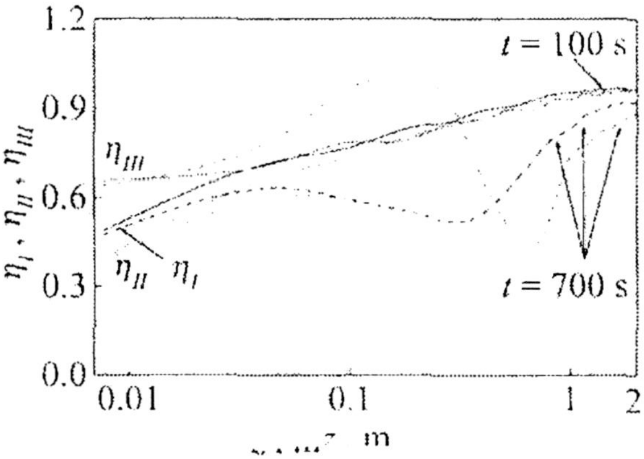

FIGURE 22-5 **Variation with axial position of catalyst effectiveness factors for reactions I, II and III at two different times after the start of the feed step for steam reforming of methane in a PSR. Operating conditions are the same as in Figure 22-4. Subscripts refer to reactions in the Xu and Froment kinetic model for steam reforming.** *(Figure reproduced from* Xiu et al. (2003a) *with permission of the copyright holder, © 2003 by Elsevier Science B.V.)*

constraint on the product gas. Recognizing the computing burden imposed by considering intraparticle diffusion, *Xiu et al. (2003a)* recommended assuming constant effectiveness factors for the rate expressions. They suggested calculating these factors at the beginning and end of the feed step and using an average for simulation.

Studies of PSR systems have now extended to integration into other processes. For example, *Hershkowitz et al. (2009)* have investigated introducing a non-isothermal PSR into power generation from natural gas and light distillates using a gas turbine. The role of the PSR is to convert the fuel into H_2 and CO_2. The former is combusted for power production while the latter stream is sequestered. The authors discuss a simulation and indicate some pilot scale experiments have been undertaken.

22.5 EXPERIMENTAL STUDIES

Table 22-5 summarizes experimental studies on pressure swing reactors. *Vaporciyan and Kadlec (1989)* examined the oxidation of dilute CO by O_2 over an eggshell Pt catalyst in the presence of a molecular sieve adsorbent. A short time cycle configuration was used with a reflux of variable volume following the catalyst-adsorbent bed. Objectives were to understand PSR behavior and establish system advantages. The effect on PSR performance of several operating variables was examined, but not exhaustively.

Vaporciyan and Kadlec observed lower rates of CO_2 production when the concentration of CO in the reactor feed increased caused by strong adsorption of CO on Pt and consequent inhibition of the oxidation rate. PSR operation increased the residence time of CO in the bed. Inversion was observed as O_2 and CO concentrations in the reactor feed changed. These investigators found good agreement between measurements and predictions for feed CO concentrations up to 1 vol%. Between 2 and 4 vol%, prediction of the critical CO_2 concentration displayed considerable error. This was attributed to the failure to reach a cyclic stationary state. They reported, however, temperature variations of up to 5°C along their reactor as well as moving temperature fronts so their experiments were not performed at constant temperature. *Vaporciyan and Kadlec's* measurements suggest that, for CO oxidation, a PSR operates primarily as a reactant-product separator.

Carvill et al. (1996) discuss a six-step PSR cycle illustrated in **Figure 22-2(e)** for CO production from CO_2 by the reverse water gas shift reaction. Feed is introduced into a pressurized reactor. The discharge port is open and CO, the weakly adsorbed product, is removed through that port. The remaining steps are depressurization, countercurrent flushing with carrier gas, followed by a second purge using the collected CO. In the final step, the reactor is repressurized by the CO product. This system was demonstrated at a bench scale by the authors using a commercial low temperature shift catalyst and a NaX zeolite as the adsorbent for water. A 36% conversion of CO_2 to CO was

TABLE 22-5 Experimental Studies of Pressure Swing Reactors

Authors	Reactor System	Reaction	Operating Conditions	Variables Examined	Observations	Comments
Vaporciyan and Kadlec (1989)	Near isothermal rapid PSR with an empty end-of-bed reflux volume. Catalyst: egg-shell Pt impregnated Al_2O_3. Adsorbent was zeolite 5A (32/48 mesh)	Oxidation of dilute CO by O_2	$T = 430$ K, $(CO)_0 = 1.6$ vol%, $0.1 < \varpi < 0.5$ Hz	O_2 and CO concentrations, delivery flow rate, P_{feed}, τ_{cycle}, t_{feed}, $t_{exhaust}$	PSR overcame CO inhibition of oxidation at higher CO concentrations but otherwise functioned primarily as product-reactant separator.	CO_2 concentration at the delivery port was poorly represented by the simulation. Better agreement occurred at the exhaust port.
Goto et al. (1993)	Non-isothermal packed bed with a 2-step cycle using a mixture of 5 wt% Pt/Al_2O_3 as catalyst and a $CaNi_5$ alloy as an H_2 adsorbent	Dehydrogenation of C_6H_{12}	Wt. of adsorbent : wt. of catalyst = 5, He served as carrier gas, $423\text{ K} < T < 463$ K, $Q_{purge}:Q_{feed} = 20:1$, $\tau_{cycle} = 40$ min, $s = 0.5$	$\bar{t}$ (space time), T, t_{purge}, P_{purge}	Complete conversion achieved.	Experiments were better described as periodic pulsing than PSR, because the flow was always cocurrent.
Goto et al. (1995)	As above with Zn exchanged ZSM-5 as catalyst and $Mg_{51}Zn_{20}$ as adsorbent	Dehydroaromatization of n-C_6	Wt. of adsorbent:wt. of catalyst = 13:1, $T = 573$ K	Adsorbent alloy, catalyst:adsorbent mixture, segregated catalyst and adsorbent	Ca. 50% increase in conversion occurred in the first 10 min, slowly decreasing as adsorbent became saturated. Activity was restored after He purge.	As above
Goto et al. (1996)	As above with a Ti adsorbent	Dehydroaromatization of C_3	Wt. adsorbent:wt. catalyst = 6:5, $T = 793$ K, $\tau_{cycle} = 130$ min, $t_{feed} = 90$ min	Adsorbent alloy, catalyst:adsorbent mixture, segregated mixture; particle size	As above	As above
Carvill et al. (1996)	6-step isothermal PSR with pressurization to P_{feed} in the final step with a commercial, low temperature shift catalyst and an NaX zeolite as adsorbent	Reverse water gas shift: $CO_2 + H_2 \leftrightarrow H_2O + CO$	Wt. adsorbent:wt. catalyst = 1:1, $T = 523$ K, $P_{feed} = 480$ kPa, $P_{purge} = 129$ kPa, $P_{exhaust} = 115$ kPa, feed was an equimolar mixture of CO_2 and H_2	Bench scale demonstration	Demonstration unit achieved 99% CO purity delivered at $P_{delivery} = P_{feed} = 480$ kPa.	Object of the demonstration was the production of high-pressure and high-purity CO.

(*Continued*)

TABLE 22-5 Experimental Studies of Pressure Swing Reactors (*cont'd*)

Authors	Reactor System	Reaction	Operating Conditions	Variables Examined	Observations	Comments
Hufton et al. (1999)	5-step, near isothermal, demonstration scale PSR with repressurization to P_{feed} using H_2 in the 5th step. An Ni reforming catalyst was used with a hydrotalcite adsorbent	Steam reforming of CH_4: $2\,H_2O + CH_4 \rightarrow CO_2 + 4\,H_2$	1:1 wt. adsorbent:wt. catalyst, $T = 723$ K, $P_{feed} = 480$ kPa, $P_{purge} = P_{exhaust} =$ 115 kPa, $(y_{steam}/y_{methane})_{feed} = 6$	Bench scale demonstration	Product was 96 vol% H_2 with just 50 ppmv of carbon oxides at 82% CH_4 conversion. Adsorbent capacity for CO_2 decreased with time-on-stream until 15 cycles, then remained constant.	Authors concluded that hydrotalcite was a suitable CO_2 adsorbent.
Sheikh et al. (2001)	2-step Rapid PSR with depressurization through the exhaust or the discharge port. Catalyst was CrO_2/Al_2O_3 with KY zeolite as adsorbent	Dehydrogenation of 1-butene to butadiene	$V_{catalyst}:V_{adsorbent} =$ 1:1, $P_{feed} = 2$ bar, $\tau_{cycle} = 2$ s with $t_{pressurization} = t_{depressurization}$	Q_{feed}	Depressurization through the discharge port increased conversion. Also, catalyst deactivation caused by coke deposition was decreased.	Experimental measurements did not agree with simulations, perhaps because of incorrect adsorption data.
Ding and Alpay (2000a, b)	5-step laboratory scale, non-isothermal PSR using an MgO-Na2Si2O4-Ni-NiO catalyst and a hydrotalcite adsorbent	Steam reforming of CH_4 to make H_2	$Wt_{catalyst}:wt_{adsorbent} = 0.5$, $y_{steam}:y_{methane} =$ 3:1, $T_{feed} = 523$ K, $P_{feed} = 455$ kPa, $T_{repressurization} =$ 753 K, repressurization with H_2	$y_{steam}:y_{methane}$, T_{feed}, P_{feed}	A low-temperature front caused by endothermic reaction moved through reactor during the feed step. Kinetic model of *Xu and Froment (1989)* was adequate for the simulation.	Only the feed step studied experimentally.
Waldron et al. (2001)	4-step, demonstration scale PSR using a single superheated steam purge and steam repressurization. Catalyst was Pt/Al2O3 and adsorbent was K promoted hydrotalcite	Steam reforming of CH_4 to make H_2	Wt. adsorbent:wt. catalyst = 1/2, $630 < \tau_{cycle} < 1130$ s, $T = 763$ K, $179 < P_{feed} < 455$ kPa, $P_{depressurization} =$ 103 kPa, $P_{purge} = 69$ kPa (763 K steam), $P_{exhaust} = 34$ kPa, $(y_{steam}/y_{methane})_{feed} = 6$	τ_{cycle}, t_{feed}, t_{purge}, $t_{exhaust}$, $T_{reactor}$, P_{feed}, Q_{steam}	Through control of the step duration, feed rates, and purge/feed ratios, H_2 purity ranged from 88.6 to 94.4% with CO present only in sub-ppm levels.	Superheated steam provided heat input for hydrotalcite regeneration making the process a hybrid pressure-temperature swing system.
Gomes and Yee (2002)	4-step PSR with a RhO/γ-Al_2O_3 catalyst and a 13X zeolite adsorbent	Metathesis of $C_3^=$ to $C_2^=$ and 2-$C_4^=$	$Wt._{catalyst}$: $Wt._{adsorbent} = 1:1$, $P_{feed} = 3$ bar, $P_{purge} = 1$ bar, $Q_{purge}:Q_{feed} = 0.8$, $y_{feed} = 0.333$ for $C_3^=$, $T = 298$ K	Demonstration and measurement of reactor performance	Authors observed a substantial improvement over PFR performance. $C_2^=$ concentration was 5-fold greater than PFR concentration; yield was 4-fold higher.	Rapid catalyst deactivation compromised study results. Experiments verified simulation results.

obtained in the PSR compared with an equilibrium conversion of 9.8% at 523 K. The PSR product was essentially pure CO at 480 kPa. *Carvill et al.* pointed out that conversion measured in their reactor depended on bed length relative to the length of the reactive front moving through the reactor. Higher conversion could have been achieved by using a longer reactor. **Figure 22-6** shows the time-variation of CO in the discharge during the feed step of the cycle. The CO produced up to 30 min (hatched area in the figure) was CO stored in the bed during the pressurization step. The second rectangular area, from 30 to 67 min, is the CO produced in the cycle. Afterwards, reactants break through to the discharge. An experiment with either CO_2 or H_2 backfill showed that high purity CO could not be produced. The *Carvill* process offers a low temperature, single vessel system for a relatively pure product delivered at a high pressure. Disadvantages are that a mixture of reactants and products are discharged in the exhaust step and a non-adsorbing carrier gas is required.

Sheikh et al. (2001) investigated the dehydrogenation of 1-C_4 to $C_4^{2=}$ using a mixed bed of CrO_2/Al_2O_3 catalyst and KY zeolite adsorbent. A simple two-step rapid PSR system was used to study the flow direction of the depressurization step. Improved performance under cocurrent

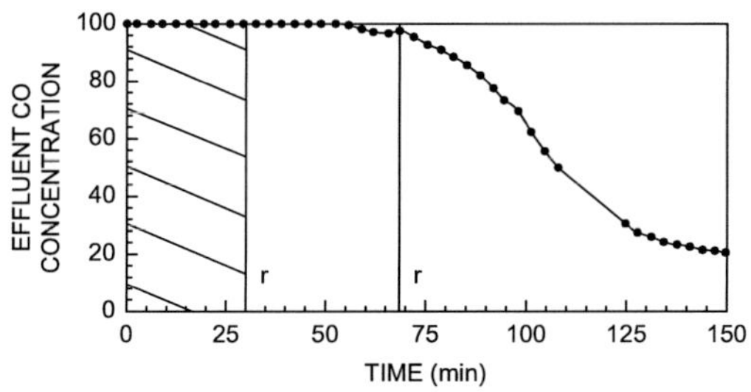

FIGURE 22-6 **Time-variation of CO volume% in the PSR discharge for the production of CO from CO_2 using the reverse water gas shift reaction with selective adsorption of water.** *(Figure reproduced from* Carvill et al. (1996) *with permission. © 1996 by the AIChE.)*

depressurization resulted from the formation of butadiene primarily in the latter part of their reactor. Countercurrent depressurization remixed this product with unreacted 1-C_4. Experiments failed to confirm the simulations discussed in **Section 22.4.1.** This was attributed to an error in the adsorption equilibrium measurement.

Gomes and Yee (2002) used a continuously operated, two-column PSR unit to test their simulation of propene metathesis. Both simulation and experiments indicated about six cycles are needed to reach a cyclic stationary state for $\tau_{cycle} = 240$ s. Because of its weak adsorption, $C_2^=$ elutes first and its concentration goes through a maximum during start-up. The simulation did not represent experimental behavior closely. Catalyst deactivation appeared to be the cause. PSR operation nevertheless substantially improved reactor performance. Concentration of $C_2^=$ in the product increased by about five-fold, while the yield rose by about four-fold compared with a packed bed reactor without adsorbent.

Section 22.4.2 briefly discussed the *Ding and Alpay (2000a, b)* simulation of methane steam refoming over a Ni catalyst and a hydrotalcite adsorbent. These researchers also undertook an experimental investigation, but just the feed step of the cycle was studied. Temperature measurements showed a low temperature wave associated with the endothermic reforming reaction developing and moving downstream in the reactor. Endothermicity and heat loss dropped temperature in the front of the reactor by more than 60°C.

The *Ding and Alpay* experiments demonstrated that the steady-state kinetic model of *Xu and Froment (1989a, b)* for steam reforming of methane is adequate for simulation. The conversion enhancement predicted by the PSR model using those kinetics agreed with the measured enhancement, allowing for the inevitable mixing and sampling problems. *Ding* and *Alpay* observed that a PSR enhancement of conversion improves as space time and pressure increase, but drops off as the steam to CH_4 ratio rises. Temperature affects equilibrium conversion and kinetics as well as adsorption suggesting that there is an optimal temperature profile for a mixed catalyst-adsorbent bed.

In the 1990s Air Products and Chemicals Inc. began developing a PSR system for hydrogen production. Results are described by *Hufton et al. (1999)* and *Waldron et al. (2001)*. The former contribution deals primarily with the properties of the K_2CO_3 promoted hydrotalcite developed as a sorbent for CO_2. This and similar oxide/hydroxide materials are not strictly adsorbents. The base component reacts to form a carbonate, which readily decomposes during regeneration. The gas-solid reactions are so rapid that they can be deemed as being at equilibrium. *Hufton et al.* found this equilibrium could be represented by a Langmuir isotherm. The promoted hydrotalcite exhibited a capacity of about 0.45 mol CO_2/kg of hydrotalcite at 673 K at a partial pressure of 30 kPa CO_2 after undergoing about 15 adsorption/desorption cycles.

Several experiments were carried out using the equipment shown in **Figure 22-7**. The reactor consisted of a 38 mm $\phi \times 1.07$ m tube packed with uniformly mixed 3 mm ϕ catalyst and adsorbent. A five-step cycle was used with two consecutive, countercurrent purges of CH_4 or steam at 115 kPa and a portion of the H_2 product. These were followed by repressurization with the product. *Hufton et al.* reported a product containing 96 vol% H_2 at a CH_4 conversion of 82%. Productivity was 0.8 mole of H_2 /kg catalyst. A PFR operating under the same conditions produced a gas

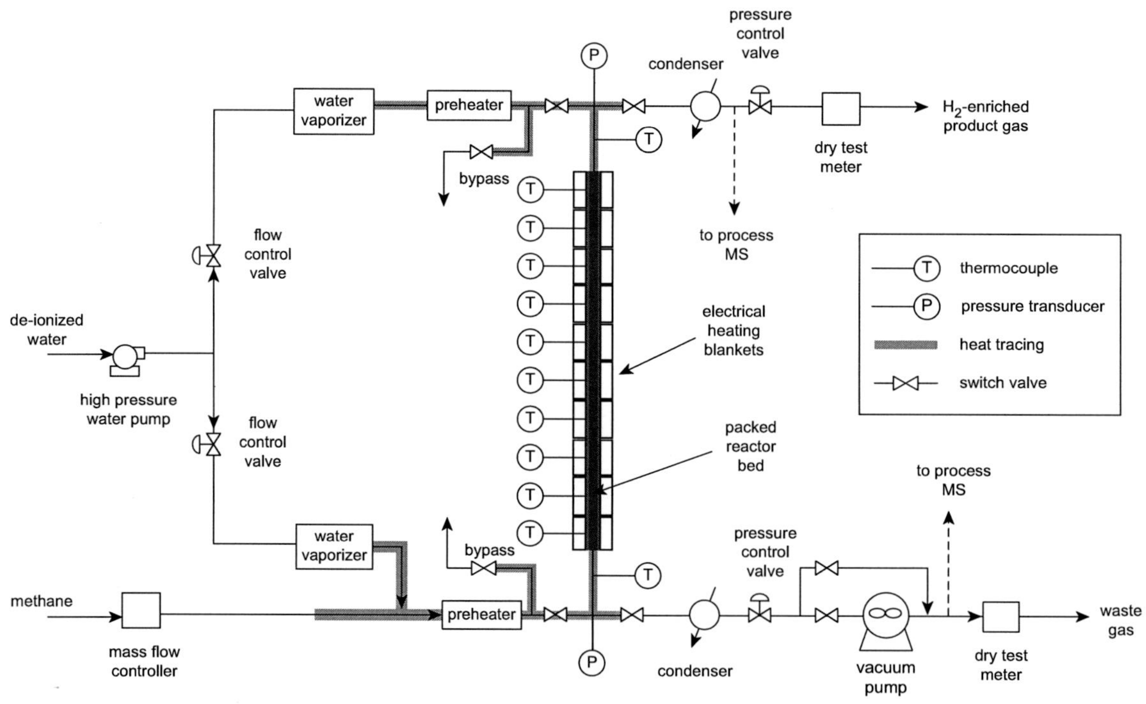

FIGURE 22-7 Schematic diagram of the hydrogen pilot plant for the steam reforming of methane employing a four-step isothermal PSR system. *(Figure redrawn from* Waldron et al. (2001) *with permission. © 2001 Air Products and Chemicals, Inc.)*

with 53 vol% H_2 at a conversion of just 28%. Loss of sorbent capacity is an important consideration. Capacity loss was observed; however, a plateau was attained after about 15 cycles at about 50% of the initial level.

Waldron et al. (2001) operated a four-step PSR having just a single low pressure countercurrent purge with superheated steam at 69 kPa. Repressurization also used steam. Experiments were performed in the small continuously operated pilot unit shown in **Figure 22-7**, but using a 25 mm i.d. × 61 mm steel tube. Step durations were variable; however, the cycle period was kept between 630 and 1130 s. Regeneration of the hydrotalcite was carried out with superheated steam. Thus, the Air Products process was a hybrid, non-isothermal pressure swing system. Pilot plant results were impressive with H_2 purities between 89 and 94%, CO_2 content usually under 100 ppm and no detectable CO in the product gas. Raising the amount of purge increased the purity and improved conversion significantly. *Aida and Silveston (2005)* present further details.

22.6 TEMPERATURE SWING REACTORS

Industrial adsorbers generally use thermal regeneration. Adsorption occurs at low temperature and the adsorbent is then flushed with a hot carrier gas to strip off the adsorbate, essentially a temperature swing operation. Temperature swing reactors (TSR) have been discussed in the engineering literature since the late 1980s. A barrier to application is that some of the heat supplied for regeneration is retained by the adsorbent thereby increasing the energy demand. Because temperatures are low in many adsorption systems, the heat expended is difficult to recover and much is wasted. Circulating adsorbent or periodic flow reversal offers a way to reduce energy loss. The latter technique was discussed in **Chapter 18**.

22.6.1 Types of Reactors

Cyclically operated fixed bed systems such as those discussed earlier in this chapter can be used for temperature swing systems. It is also possible to use moving beds. In these, a mixture of catalyst and adsorbent flows through a vessel fed by reactants where products are formed and some are adsorbed. The solids then flow to a regenerator in which adsorbate is recovered. The operation is then continuous rather than periodic. These systems are discussed at length in **Chapter 8** as circulating fluidized beds or chemical looping. They are mentioned briefly in **Chapter 20**.

Figure 22-8 shows a schematic of a circulating fluidized bed for producing H_2 from CH_4. It must be borne in mind that fluidized beds are well mixed and not advantageous for equilibrium limited reactions. The contacting pattern, however, is not disadvantageous for regeneration. An alternative to reactors with moving particles is to move the bed as a whole. The simplest way is to rotate the bed around an axis through the centerline of the vessel. **Figure 22-9** shows how such a rotating bed would operate for SO_2 reduction by H_2S. Rotating chromatographic reactors were considered in **Chapter 20**. *Agar (2003)* and *Menge et al. (2003)* discuss rotating temperature swing systems, but descriptions of the laboratory units are not given.

Segregation of catalyst and adsorbent is an alternative to a well-mixed bed. Segregation has been studied for SMBCRs (see **Chapter 21**), but only *Agar (2003)* and *Menge et al. (2003)* mention this possibility for swing systems. A possible arrangement is given in **Figure 22-10**. Here reactants flow consecutively into a bed of catalyst and then into a bed of adsorbent. At least two catalyst beds are necessary to take advantage of product adsorption.

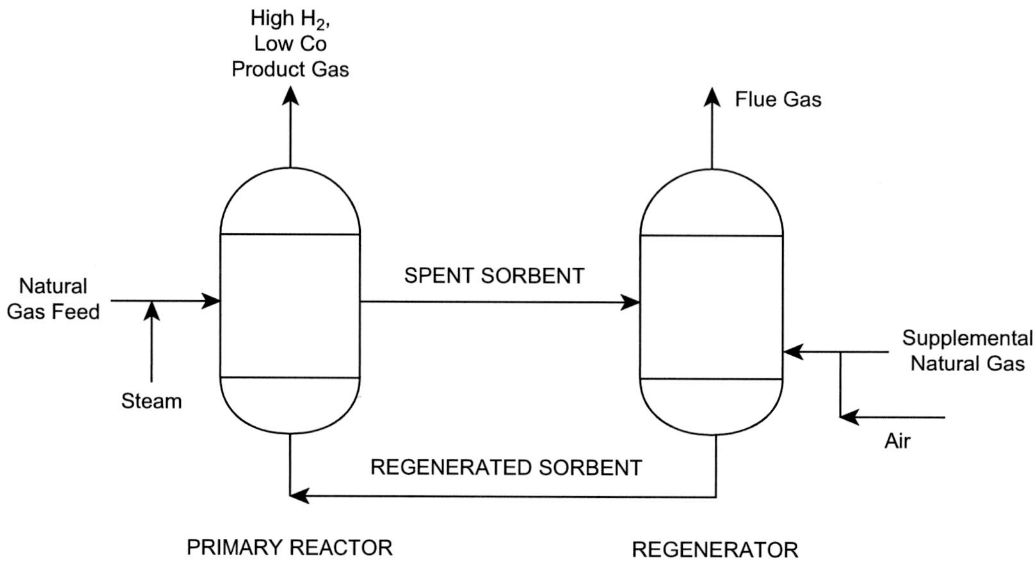

FIGURE 22-8 Schematic of a circulating fluidized bed TSR system for the steam reforming of methane. *(Figure reproduced from* Harrison and Peng (2003) *with permission. © 2003 by Berkeley Electronic Press.)*

Three catalyst beds are shown in the figure. Segregation provides operating flexibility at the expense of capital outlay. For example, with separate beds, the adsorbent could be regenerated at temperatures that would deactivate the catalyst. Or, different temperatures or even catalysts might be used in successive catalyst beds.

Menge et al. (2003) proposed a rotating reactor with separate catalyst and adsorbent compartments. A schematic of their design is shown in **Figure 22-11**. There are entrance and exit catalyst beds but just a single adsorbent bed. Hot carrier gas for regeneration passes only through the adsorbent bed. Although **Figures 22-10** and **22-11** suggest two-step processes, any number of steps is possible by adding feed and take-off connections.

Rotating bed reactors and segregated catalyst-adsorbent systems have been introduced in this section rather than earlier because they are impractical, though not impossible, with large pressure changes encountered in pressure swing systems.

22.6.2 Modeling

Most simulations of temperature swing reactors assume a one-dimensional, pseudo-homogeneous model for an isotropic mixture of adsorbent and catalyst. Reactors are not isothermal so an energy balance must be invoked. Appropriate equations are given in **Table 22-1**. If mass or heat transfer to or from the solid phase must be taken into account, a heterogeneous model must be used (see **Table 20-2**). When energy is supplied or removed through vessel walls or internal heat transfer surfaces, a two-dimensional model must be considered. However, energy transfer through surfaces is unusual in swing reactors and heat storage in vessel walls is generally neglected. With TSRs, energy is removed or brought into the reactor through a flowing fluid.

For reactor design, a pseudo-homogeneous model is usually adequate. The heterogeneous model is probably preferable for simulation.

Initial condition for all models, continuity, is stated in **Table 22-3** as **Eqs (22-7** or **22-8)** and

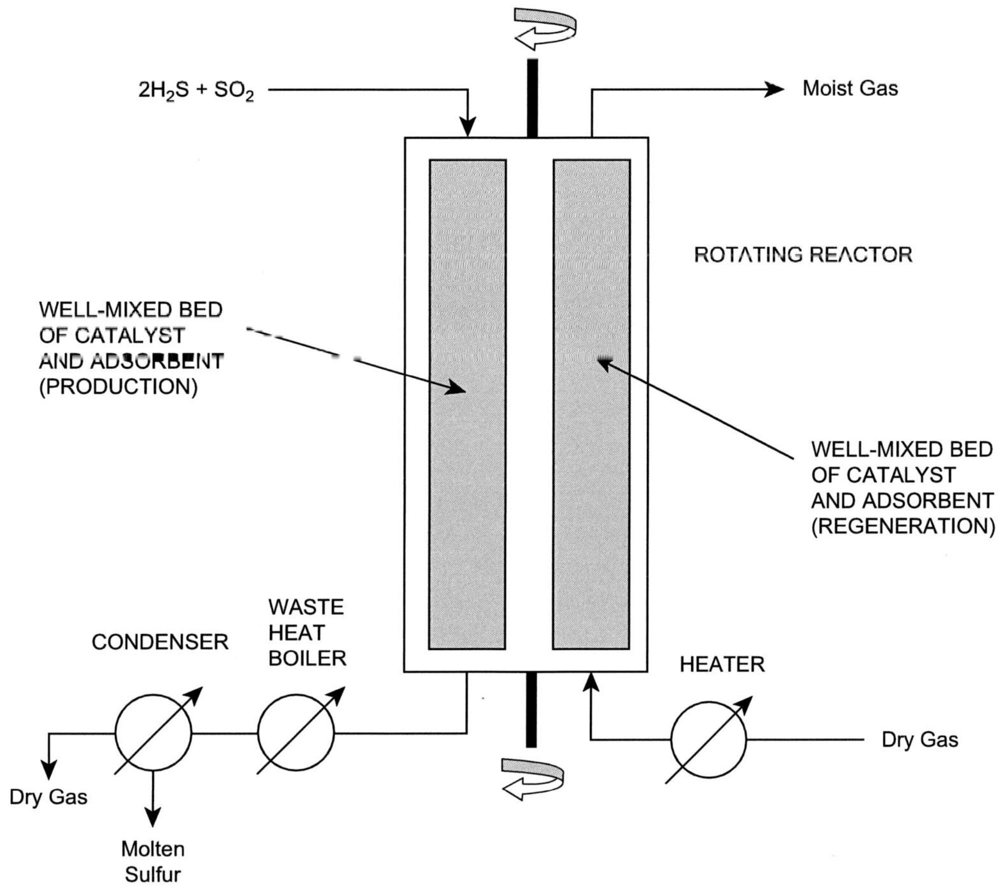

FIGURE 22-9 Schematic operation of a two-section, rotating bed TSR for an A ↔ B + C reaction in which product B is adsorbed. *(Figure reproduced from Agar et al. (2003) with permission.)*

must be applied at the beginning of each step in the cycle. It is the initial and boundary conditions at each step and port that distinguish each type of separating reactor. Some of these are given in **Table 22-3**. Various boundary conditions are possible for pressure. It is often convenient to set an inlet pressure, such as P_f, or specify a superficial or nozzle velocity.

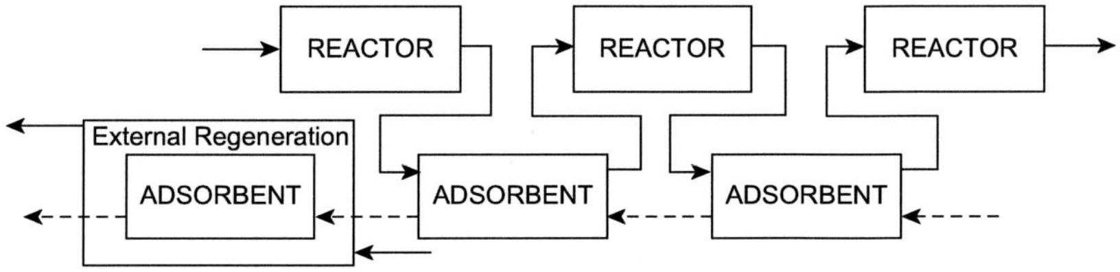

FIGURE 22-10 Schematic of a TSR with catalyst and adsorbent isolated in different beds. *(Figure reproduced from* Menge et al. (2003) *with permission. © 2004 by Elsevier Science Ltd.)*

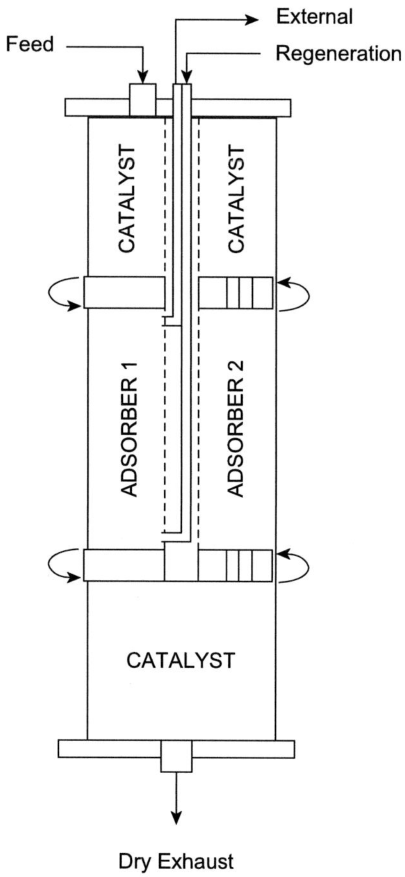

FIGURE 22-11 **Schematic of a rotating bed TSR with catalyst and adsorbent isolated in different beds.** *(Figure reproduced from* Menge et al. (2003) *with permission. © 2004 by Elsevier Science Ltd.)*

22.7 SIMULATION OF TEMPERATURE SWING SYSTEMS

Simulations have been under taken to explore design of reactor beds, the use of multiple reactors and their arrangement in a network and to investigate the effect on performance of cycle parameters. Many of the published studies are summarized in **Table 22-6**.

The important Claus reaction oxidizes $S^{=}$ to S^0 with S^{4+} with stoichiometry:

$$2H_2S + SO_2 \leftrightarrow 3/n\ S_n + 2H_2O.$$

The reaction proceeds rapidly above 230°C. However, equilibrium limitations require that several stages must be used in the conventional process to achieve the high levels of conversion necessary in an industrial operation. Use of flow reversal together with temperature swing appears to make a single stage feasible (*Elsner et al., 2002a*). Such a reactor is shown in **Figure 22-12**. The reactor bed consists of a mixture of alumina catalyst and zeolite 3A molecular sieve. The upper part of the figure shows the forward operating direction where reactants are fed to the front of the hot bed and exit through a condenser that removes molten sulfur and cools the gas. The effluent is then reheated to a temperature exceeding the reaction temperature of about 250°C and passes through a second bed where the hot gas strips water vapor or the adsorbent. When either adsorbent in the first bed is saturated or water is fully stripped from the adsorbent of the second bed, the flow direction switches and then cold reactor feed enters from the right end of the diagram shown in the figure. Consequently, most of the energy supplied to strip water from the adsorbent is recovered and heats the cold feed to reaction temperature. The Claus reaction is exothermic so some low-pressure steam can be recovered in a heat exchanger before the gas flows to the condenser.

The temperature swing reactor with flow reversal illustrated previously was proposed several years earlier by *Yongsunthon and Alpay (1999)*. Indeed, flow reversal is used widely in temperature swing adsorbers (TSA).

Operation of the Claus reaction system shown in **Figure 22-13** has been simulated by *Elsner et al. (2002a, b)*. Kinetics were evaluated from experiments, but formation of carbonyl sulfide from the parasitic reaction of H_2S with

TABLE 22-6 Simulations of Temperature Swing Reactor Systems

Authors	Reactor System	Reaction Considered	Reactor Operation	Integration Method	Observations	Comments
Yongsunthon and Alpay (1999)	2 step, network of flow splitters, mixing junctions and adiabatic TSRs with a Pt/Al_2O_3 catalyst and a zeolite 5A adsorbent	Dehydrogenation of methylcyclohexane to toluene: $C_7H_{14} \rightarrow C_7H_8 + 3\ H_2$, MCH weakly adsorbed	Up to 5 reactors, $1 < L_{reactor} < 5$ m, $298 < T < 623$ K, $0.2 < U_{feed} < 10$ m/s	Optimization of splits and operating parameters through quadratic programming	Within the constraints imposed, a 5-reactor network outperformed a single reactor; a mix of serial and parallel connections was better than serial only.	Authors assumed energy input came from heating all feed streams to a temperature sufficient for adiabatic reactor operation.
Yongsunthon and Alpay (2000)	As above, but with reverse (recycle) flow permitted in the network	As above	As above	As above, extended to include flow direction in regeneration and recycle	Complexity that was introduced by allowing recycle and reverse flow in regeneration offered only a small improvement in performance.	Multiple reactor networks improved performance.
Elsner et al. (2002a, b)	2 step, adiabatic and isobaric TSR using an Al_2O_3 catalyst and a zeolite 3A adsorbent for water	Claus reaction: $2\ H_2S + SO_2 \rightarrow ½\ S_6 + 2\ H_2O$	$(y_{H2S})_{feed} = 0.10$, $(y_{SO2})_{feed} = 0.05$, $(SV)_{feed} = 100$ $m^3_{(STP)}/m^3_{catalyst} \cdot h$		43 vol% of adsorbent in the reactor maximized the breakthrough time in the reactor and production/volume of the reactor	Authors employed experimental reaction kinetics and adsorption isotherm.
Menge et al. (2003)	As in *Elsner et al. (2002a, b)* above, but allowed for partial segregation of catalyst and adsorbent	Claus reaction: $2\ H_2S + SO_2 \rightarrow ½\ S_6 + 2\ H_2O$	$(y_{H2S})_{feed} = 0.10$, $(y_{SO2})_{feed} = 0.05$, $(SV)_{feed} = 100$ $m^3_{(STP)}/m^3_{catalyst} \cdot h$		Continuous or stepwise variation of adsorbent/catalyst ratio in reactor reduces τ_{cycle} for the reactor performance.	Authors investigated performance improvement through variation of catalyst/adsorbent ratio along reactor length.

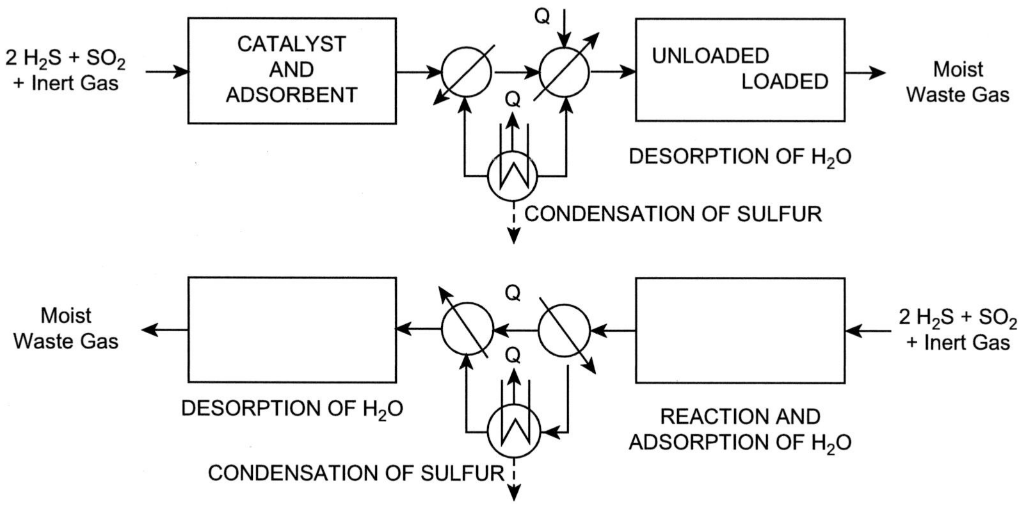

FIGURE 22-12 Schematic of the operation of a two-bed temperature swing reactor (TSR) for the Claus reaction. *(Figure reproduced from* Elsner et al., 2002b *with permission. © 2002 Elsevier Science Ltd.)*

CO_2 was not included. Experiments with zeolite 3A produced Freundlich isotherms for water. The rate of adsorption was represented by the linear driving force model, **Eq. (22-5)**. Based on

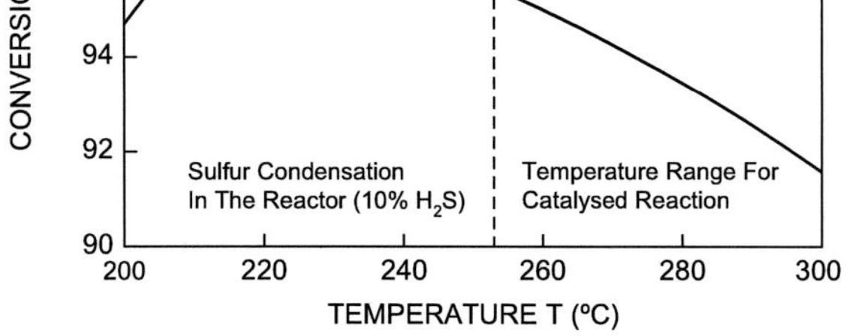

FIGURE 22-13 Equilibrium conversion of H_2S for isothermal operation with and without adsorbent as a function of temperature assuming a feed space velocity of 100 Nm^3/m^3catalyst·h and 10% H_2S and 5% SO_2 in the gas feed. Dashed horizontal line indicates required conversion. *(Figure reproduced from* Elsner et al. (2002a) *with permission, © 2002 Elsevier Science Ltd.)*

equilibrium and assuming isothermal operation, **Figure 22-13** shows the removal. Thus, an adsorptive reactor, such as a TSR, must be used. *Elsner et al. (2002b)* explored the effect of the adsorbent to catalyst ratio on the H_2S breakthrough time in the first step of the cycle. They observed that a 3A zeolite at 43 vol% maximized the breakthrough time. With lower ratios, the reverse reaction becomes more important leading to earlier breakthrough, while at higher ratios, lower catalyst density reduces the reaction rates.

Menge et al. (2003) extended the *Elsner* work examining the effect of varying the volume fraction of adsorbent continuously or stepwise in the bed. Continuous variation of the fraction from zero to one with a mean of 0.45 or a two-section bed of equal lengths with the first section containing a 0.2 adsorbent fraction and the next section 0.8 reduced the cycle period by about 20%. This bed structure allowed a higher throughput. The *Menge* study extended to particles formed from mixtures of adsorbent and catalyst. Layered particles with adsorbent centers or catalyst centers were also considered.

The HCN synthesis via the slightly endothermic gas-phase reaction, 2 CO + NH_3 ↔ HCN + CO_2, using a two-step cycle was also simulated by *Elsner et al. (2002a)* and *Menge et al. (2003)*. An isothermal, one-dimensional, pseudo-homogeneous model that incorporated axial dispersion was used. Kinetics for the catalyst, Fe_3O_4/Cr_2O_3, were measured and modeled, while the adsorption isotherm and the linear driving force rate expression developed by *Ding and Alpay (2000b)* was used for CO_2 adsorption on potassium-modified hydrotalcite. Further details on the research reviewed above are given by *Aida and Silveston (2005)*.

22.8 TEMPERATURE SWING REACTOR NETWORKS

A comprehensive analysis of multi-bed networks was undertaken by *Yongsunthon and Alpay (1999)* to examine what improvements in performance might be possible. They considered a TSR cycle consisting of pressurization with the removal of products at the discharge port and a regenerant step at a higher temperature to desorb product and sweep it out through the exhaust port. For a network, a serial arrangement is shown in the upper portion of **Figure 22-14** with purge added to the first bed or split between beds. In the lower portion, some of the flow can be removed between stages in both the pressurization and regeneration steps. Feed or regenerant to beds other than the first are discharges from upstream beds when serial connections are used. With serial-parallel connections, a mixer is located at the entrance to the bed in the flow direction so the feed or purge composition will be a mixture of discharge from the previous bed and fresh feed or purge. Of course, splits of fresh feed or regenerant to the various beds and splits for the discharge of each bed to the product line must be specified.

Aida and Silveston (2005) examine over several pages the *Yongsunthon and Alpay* analysis: optimization routine, models employed, assumptions, adsorption isotherms and the kinetics of their model system. Only results for five stages were calculated by *Yongsunthon and Alpay*. They observed that a single TSR or a network substantially decreased the reactant feed rate needed to produce the target amount of product. Multiple reactors increased conversion significantly but at a cost of slightly lower toluene recovery in the purge stream eluted from the beds as exhaust. The purity of the toluene in all the TSR networks was greater than 80% showing that both good separation and an increase in conversion can be achieved. The TSR networks also offered a small energy saving. A 10% energy saving over a PFR with a single TSR can be realized by regenerating the bed with countercurrent flow.

A follow-up paper (*Yongsunthon and Alpay, 2000*) explored all possible flow arrangements by adding flow mixers and splitters to the network. Optimization began with the performance of the optimal unidirectional series-parallel network, that is, the minimum feed rate subject to the constraints considered, and attempted to find a connection scheme that would decrease that feed rate. The search required a huge expenditure of computing resources and only a small improvement in system performance was found.

22.9 EXPERIMENTAL

Surprisingly, experimental investigations preceded simulation studies for TSRs. These contributions are summarized in **Table 22-7**. *Han and Harrison (1994)* studied CO_2 uptake by calcined dolomite as a means of intensifying the water gas shift (WGS) reaction for H_2 production. At temperatures above 823 K, the shift reaction is forced towards H_2 by CO_2 adsorption. These researchers proposed a two-fluidized bed system for the WGS reaction. Others (*Han and Harrison, 1997;*

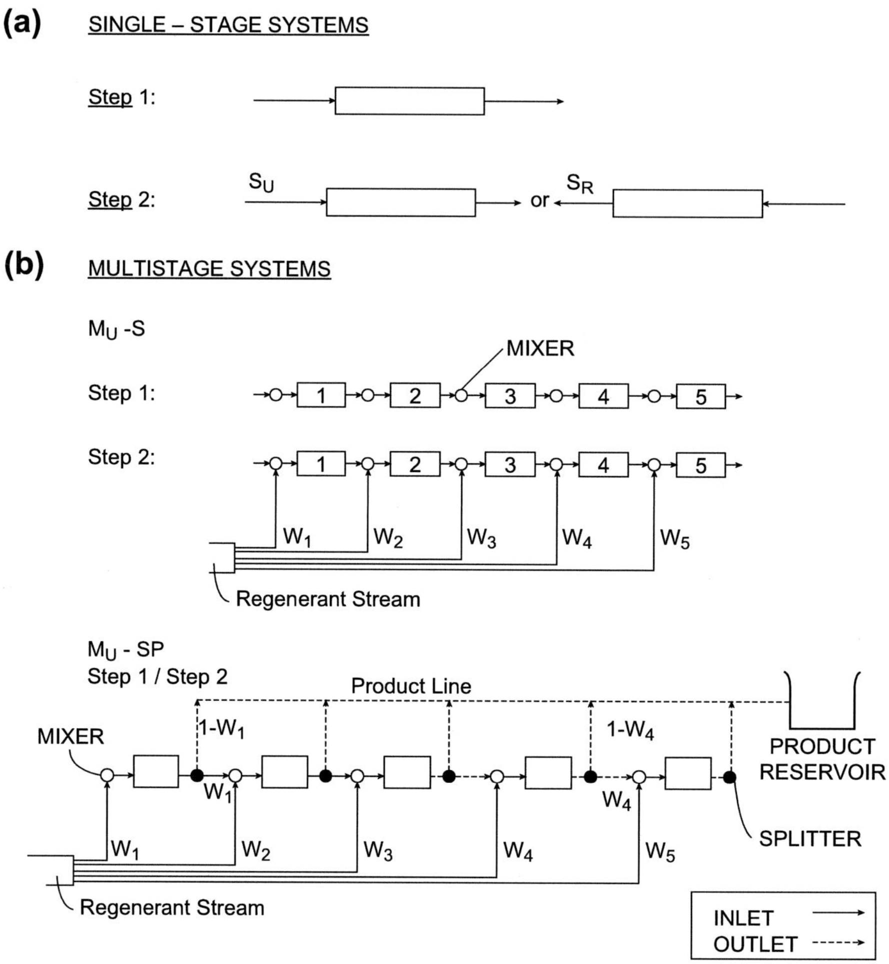

FIGURE 22-14 Mixed catalyst-adsorbent bed and flow arrangements considered in the optimization study of a multi reactor TSR. *(Figure reproduced from* Yongsunthon and Alpay (1999) *with permission. © 1999 by Elsevier Science Ltd.)*

Balasubramanian et al., 1999; Lopez Ortiz and Harrison, 2001; Harrison and Peng, 2003) proposed such a two-bed system for the steam reforming of methane.

A packed bed reactor placed in a constant temperature furnace was employed by *Han and Harrison (1994)* for their experiments. Controlled amounts of pure gases were fed into the reactor; these were heated in up-flow around the bed. Reaction products were cooled to condense out water; pressure was reduced upstream from the analysis section. At temperatures between 773 and 873 K, CO conversion at 15 bar reached 99.5%. Carbon oxide concentration in the reactor off-gas was below 400 ppm on a dry basis.

Economic viability of a CO_2 acceptor process for H_2 production through the WGS reaction critically depends on maintaining the adsorption capacity of the calcined dolomite over many cycles. This constraint was explored by *Han and Harrison (1997)* using a commercial

TABLE 22-7 Experimental Studies of Temperature Swing Reactors

Authors	Reactor System	Reaction	Operating Conditions	Variables Examined	Observations	Comments
Han and Harrison (1994)	2-step TSR with a mixed bed of catalyst and adsorbent; reaction at 823 K and regeneration at 1023 K	Water gas shift: $CO + H_2O \rightarrow CO_2 + H_2$ using a shift catalyst and a dolomite adsorbent	$P_{feed} = 15$ bar, SV = 1425 h^{-1}, $(y_{CO})_{feed}$ = 5.5 vol%, $(y_{H2O})_{feed}$ = 26.4 vol%, $(y_{CO2})_{feed}$ = 3.1 vol%, $(y_{H2})_{feed}$ = 4.5 vol% (in N_2); regenerant is N_2	$T_{reactor}$, P_{feed}, $P_{regeneration}$, feed gas composition, SV_{feed}, dolomite preparation (properties)	CO conversions were up to 99.6%. CO + CO_2 were below 400 ppmv on dry basis. Minimum CO in product gas was 30 ppmv.	Dolomite adsorption capacity declined with time. Rate of decline depended on $T_{regeneration}$ and the presence of steam.
Machocki (1996)	Used separate catalyst and adsorbent beds operating at 1073 and 300 K respectively; effluent from adsorber recycled to reactor	Oxidative coupling of CH_4: $4\ CH_4 + 2\ O_2 \rightarrow 2\ C_2H_4 + 4\ H_2O$ over a calcined mixture of Na_2CO_3/$CaCO_3$ with a zeolite 5A adsorbent	$(Q_{CH4}/Q_{O2})_{feed} = 20$, $\bar{t} = 0.4$ s/g·cm^3		Only short runs were undertaken with an average CH_4 conversion of 94%. Yield of C_2+ was 66%.	Adsorber was regenerated off-line at 573 K so the system is actually a reactor + PSA.
Han and Harrison (1997)	2-step TSR with a mixed bed of catalyst and adsorbent	Water gas shift: $CO + H_2O \rightarrow CO_2 + H_2$ using a shift catalyst and a dolomite adsorbent	$P_{feed} = 15$ bar, $T_{reaction}$ = 823 K, $T_{regeneration}$ = 1023 K, SV = 1425 h^{-1}, $(y_{CO})_{feed}$ = 5.5 vol%, $(y_{H2O})_{feed} = 26.4$ vol%, $(y_{CO2})_{feed} = 3.1$ vol%, $(y_{H2})_{feed} = 4.5$ vol% (in N_2); regenerant is N_2		Dolomite deactivation observed over successive cycles, but was independent of syngas composition.	Authors investigated 1) breakthrough of carbon oxides and 2) adsorbent deactivation.
Balasubramanian et al. (1999)	As in *Han and Harrison (1997)* above, but with reaction at 723 K	Steam reforming of CH_4 with a commercial NiO catalyst (18 wt% NiO/Al_2O_3) and calcined lime as adsorbent	$750 < T_{reactor} < 970$ K, $P_{feed} = P_{regeneration}$ = 15 bar, 3 < Steam/CH_4 < 5, 23 wt% and 50 wt% catalyst in reactor	$T_{reactor}$, $(y_{CH4})_{feed}$, Q_{steam}/Q_{CH4}	TSR cycle provides higher H_2 purity and lower CH_4 and steam consumption than conventional steam reforming for the same H_2 production. Increasing Q_{steam}/Q_{CH4} and $T_{reactor}$ increased CH_4 conversion.	Only the feed step was investigated.
Lopez Ortiz and Harrison (2001)	As above	As above but with calcined dolomite as adsorbent	$T_{reactor}$ = 923 K, 1073 < $T_{regeneration}$ < 1223 K, $P_{feed} = P_{regeneration}$ = 15 bar, steam/CH_4 = 4	$T_{regeneration}$, composition of regenerant gas	The presence of O_2 in the regenerant oxidized the catalyst but was reduced in feed step. The use of CO_2 as a regenerant dropped adsorbent capacity. NiO crystallite growth indicated catalyst deactivation with time.	Authors investigated the regeneration cycle primarily. Results indicated continual or periodic replacement of dolomite was necessary.

(*Continued*)

TABLE 22-7 Experimental Studies of Temperature Swing Reactors (*cont'd*)

Authors	Reactor System	Reaction	Operating Conditions	Variables Examined	Observations	Comments
Elsner et al. (2002a, b)	2-step isothermal TSR cycle using a WGS catalyst	HCN synthesis: $2\ CO + NH_3 \leftrightarrow HCN + CO_2 + H_2$ with a WGS catalyst (Fe_3O_4/Cr_2O_3) with a K-hydrotalcite adsorbent		$T_{reactor}/T_{regeneration}$, $P_{reactor}$, wt. adsorbent/wt. catalyst, feed composition, reactant/carrier gas ratio	Experiments failed: no enhancement of the reaction was observed. The failure was attributed to the consumption of CO by the Boudouard reaction or reactant adsorption.	Catalyst deactivation was also postulated by the authors.
Harrison and Peng (2003)	2 step TSR with a mixed bed of catalyst and adsorbent; regeneration at 1023 K	Steam reforming of CH_4 with a commercial NiO catalyst and calcined lime as adsorbent	$750 < T_{reactor} = 775$ K, $P_{feed} = P_{regeneration} =$ 5 bar, $3 <$ Steam/CH_4 $<$ 5, 23 wt% and 50 wt% catalyst in reactor	$T_{reactor}$, Q_{CH4}, $(y_{CH4})_{feed}$	Changing the operating variable reduced CO by less than 10 ppmv. H_2 purity reached 98%.	Authors explored the effect of operation on CO content of H_2 product. Only the feed step was investigated.
Kato et al. (2003)	2-step isothermal TSR with a mixed bed of catalyst + adsorbent; with reaction at 923 K, regeneration at 1023 K	As in *Harrison and Peng (2003)* above, but with lime adsorbent made from high purity $CaCO_3$	$823 < T < 1023$ K, $P_{reactor} = P_{regeneration} =$ 1 bar, $(Q_{steam}/Q_{CH4})_{feed} =$ 4, wt. catalyst/wt. adsorbent ≈ 0.5	Mixed or segregated catalyst and adsorbent	A mixed bed yields higher H_2 content and a lower CO ppm in the product gas than a PFR containing just catalyst.	Authors attempted to make an H_2 feed for a fuel cell vehicle. Only the feed step was investigated.
Menge et al. (2003)	Pilot scale, 2-step near-isothermal TSR	Claus reaction: $2\ H_2S + SO_2 \leftrightarrow 3/n\ S_n + 2\ H_2O$ with a Al_2O_3 catalyst and a zeolite 3A adsorbent	$T_{reactor} = 623$ K, $P_{feed} = 1$ bar	Demonstration	Authors observed different conversions of H_2S and SO_2 in the reaction step. This was attributed to adsorption of reactants.	Experimental measurements validated the model used for simulation.
Satrio et al. (2005)	2 step TSR with a mixed bed of catalyst at 923 K and adsorbent; regeneration at 1023 K	Steam reforming of C_3H_8 with a commercial NiO catalyst and calcined dolomite as adsorbent	$T_{reactor} = 923$ K, $T_{regeneration} = 1023$ K, $(Q_{steam}/Q_{propane})_{feed} = 3$	Catalyst and adsorbent preparation	Authors prepared particles incorporating catalyst and adsorbent. Loss of adsorptive capacity with successive cycles was observed.	Authors claim to have formed composite particles contained in an abrasion-resistant shell.
Wiltowski et al. (2008)	2-step TSR with separate but consecutive beds of catalyst and adsorbent	Steam reforming of a $CO\text{-}CO_2\text{-}H_2\text{-}CH_4$ synthesis gas using Fe_2O_3 as catalyst and solid reactant	$750° < T_{raector} < 850°C$, $0 < P_{feed} < 150$ psig, catalyst (Fe_2O_3) loading, Wt. catalyst:wt. adsorbent = 1:2, 1.3	$T_{reactor}$, Fe_2O_3 loading, P_{syngas}, CaO loading, $(y_{steam})_{feed}$	Using a syngas feed with $T_{reactor} > 600°C$, the Boudouard reaction proceeds. A TSR gasifies carbon producing a CO_2 rich stream in the regeneration.	Objective of this study was H_2 production from coal. Several experiments reported combined coal gasification and a TSR cycle for making H_2.

dolomite calcined at 1023 K in N_2. Their experimental dolomite bed exhibited an abrupt breakthrough of both CO and CO_2. Han and Harrison observed a steady deactivation of dolomite during successive cycles of exposure to a synthesis gas at 823 K and regeneration in N_2 at 1023 K. Up to breakthrough, the removal of carbon oxides was about 99.6% while the CO level in the product gas was about 30 ppm. Deactivation was independent of gas composition and space velocity but depended strongly on temperature. The addition of steam in the regeneration step reduced the rate of deactivation.

The Harrison team completed several studies on hydrogen production via the catalytic steam reforming of methane using a commercial nickel catalyst, NiO/Al_2O_3 mixed with calcined lime (*Balasubramanian et al., 1999; Harrison and Peng, 2003*) or calcined dolomite (*Lopez Ortiz and Harrison, 2001*) as CO_2 adsorbers. Commercial steam reforming incorporates three reaction, two separation as well as several heat exchange and compression steps to produce high purity hydrogen. The endothermic reforming reaction proceeds at 1100 to 1150 K so, like temperature swing, heat must be supplied at the highest temperature in the system. Consequently, the high temperature penalty for regeneration of the CO_2 acceptor in the two-step TSR may not be an obstacle in this application. The Harrison work is summarized by *Aida and Silveston (2005)*.

Harrison and Peng (2003) used a different nickel catalyst and changed the procedure for making the lime adsorbent. Their effort focused on CO content in the product gas. Only the production step was considered. Pressure and temperature were lowered to 5 bar and 750 K respectively. These investigators found that CO levels under 10 ppmv could be achieved, while H_2 purity just under 98% was possible. In the range of variables studied, the CO content was independent of the volumetric feed rate and volume% CH_4 in the feed. It was, however, affected by temperature. Somewhat lower temperatures and pressures were used by *Yi and Harrison (2005)* in a packed bed to achieve 11 ppmv CO in the product H_2 stream. Experiments by *Johnsen et al. (2006)* with a mixture of a Ni reforming catalyst and dolomite at 600°C and 1 bar gave a 98% H_2 stream on CO breakthrough.

Carbonaceous materials, such as bituminous coal, are another source of H_2, in this case via a water gas shift. *Wiltowski et al. (2008)* experimented with applying a TSR to reforming syngas. In place of mixing catalyst with their lime adsorbent, these materials were split into two contiguous beds with adsorbent following the catalyst. *Wiltowski et al.* observed disproportionation of CO depositing carbon in the catalyst bed. The CO_2 produced by both the Boudouard and WGS reactions was trapped by the lime adsorbent in the second bed producing a 98% H_2 stream at 800°C. Once carbon oxides break through the adsorbent bed, feed to the catalyst bed switches to O_2-or oxygen enriched air to regenerate the reduced Fe_2O_3 catalyst, an exothermic reaction, and decompose the $CaCO_3$ in the adsorption bed through an endothermic reaction. The CO_2-rich stream emerging is suitable for sequestration. The investigators explored process variables and concluded that the optimal operating temperature for both of the swing steps was about 650°C at a catalyst to adsorbent weight ratio between 1:2 and 1:3. *Wiltowski et al. (2008)* also reported some preliminary experiments using a mixed bed of coal and catalyst (at a 22:1 wt. ratio) in their consecutive catalyst-adsorbent bed system with O_2-steam gasification at 100 psig and 650°C. They observed a product containing 98% H_2.

Key steps in a TSR process for carbonation of lime or calcined dolomite and their regeneration in a hot gas were studied by *Silaban and Harrison (1995)* and *Silaban et al. (1996)* using an electro balance. CO_2 uptake was examined at 1 to 15 bar, 823 to 1023 K with a feed containing up to 15 vol% CO_2, while regeneration was studied in the same pressure and CO_2 concentration range, but at temperatures of 1023 to 1173 K. *Silaban and Harrison* observed that at 1023 K calcination drops the weight of the solid by almost 50% in the initial cycle, but this loss is not recovered in the subsequent carbonation. Indeed, weight gain is initially rapid but soon slows. Pore volume expands greatly on calcination, but the gain is largely lost in the next step. Only about half reappears at the end of the first regeneration. Despite volume loss, the mean pore diameter increases suggesting irreversible closure of the fine micropores. A hot zone moving inward from the particle surface appears to be retarded by the endothermic decarbonation. A layer of carbonate forms on the surface and causes a drop in the rate as diffusion through that layer becomes controlling. This behavior continues to occur in subsequent cycles. Eventually, a cyclic stationary state is attained, but dolomite capacity becomes just a small fraction of the initial capacity. Thus, continual replacement of the lime adsorbent will be required.

Dolomite is a physical mixture of calcium and magnesium carbonate with less than 1 wt% of silicon, iron and aluminum oxides. Magnesium oxide, MgO, in calcined dolomite is not carbonated under the conditions expected in hydrogen production using a TSR process. This is evident from **Figure 22-15** that follows weight loss and gain in two calcination-carbonation cycles for a commercial dolomite. Recarbonation proceeds rapidly initially, then abruptly changes and becomes very slow suggesting that a different material forms.

The microscopic processes of pore opening and closure from carbonate formation are the same for lime and dolomite. Presence of MgO, however, prevents closure of some of the pore structure in dolomite. Even though calcined dolomite has only 42% of the CO_2 capacity of lime, because of MgO, the stability of this lower capacity over repeated cycles makes it a better adsorbent choice. *Silaban et al. (1996)* show that

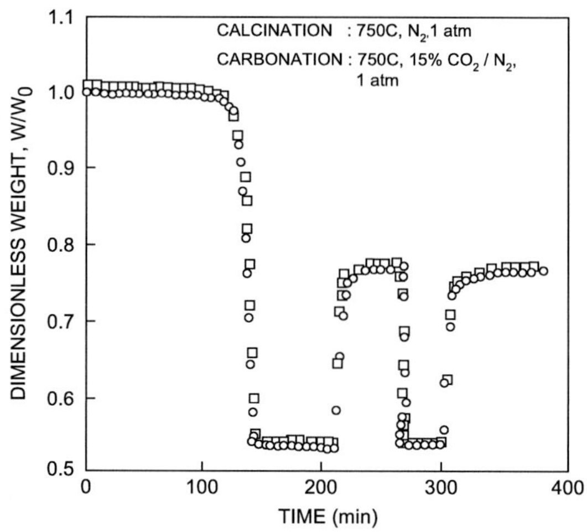

FIGURE 22-15 Change in adsorbent weight over two calcination-carbonation cycles starting with raw dolomite. Measurements made at 1 bar, 1023 K with only N_2 in the calcination and 15 vol% CO_2 in N_2 for the carbonation steps. *(Figure reproduced from* Silaban et al. (1996) *with permission. © 1996 by Gordon and Breach Science Publishers SA.)*

carbonation as a fraction of the ultimate CO_2 capacity of the solid reached 97% after five minutes and did not change over 10 carbonation-calcination cycles. These authors report that 1023 K is about the lowest temperature that can be used for regeneration of calcined dolomite. Adding steam during regeneration increases the CO_2 capacity. Recognizing the gradual decline in the adsorptive properties, a TSR must incorporate adsorbent replacement and probably some type of fluidized bed must be used. Dolomite, nevertheless, emerges as the CO_2 acceptor of choice for H_2 production.

Multi-cycle performance of a TSR process for H_2 based on dolomite has been explored by *Lopez Ortiz and Harrison (2001)*. Minimal loss of sorbent capacity on calcination was observed with an N_2 carrier gas at 1073 and 1123 K. After 25 cycles, there was a ca. 30% loss in carbonation capacity. Increasing the regeneration temperature to 1223 K accelerated the capacity loss. When CO_2 was used in place of N_2, there was an abrupt drop in capacity after the first cycle. Thereafter the loss was about the same as with N_2. *Lopez Ortiz and Harrison (2001)* further investigated sorbent durability using an electro balance with cycling between 1073 and 1223 K in CO_2 at 1 bar. Chance in fractional carbonation for experiments with up to 144 cycles agreed well with data for a 25 cycle fixed bed experiment on H_2 production at 923 K followed by regeneration in pure CO_2 at 1223 K. Thus, their experiments adequately represented the decline in sorption capacity.

Evidently, continuous replacement of sorbent will be necessary even with relatively stable dolomite. Roughly a 5 wt% exchange of sorbent per cycle would maintain the dolomite sorption capacity at about 50% of its initial value. Replacement at this rate would mandate use of a fluidized bed or a transport reactor. Loss of carbonation capacity seems to be associated with sintering. Addition of steam to the carrier gas may suppress sintering, but this requires evaluation. In addition to sorbent capacity loss, Ni crystallite growth was observed at 1,073 K. This was relatively rapid in the first five cycles on stream, but slowed appreciably in the succeeding 25 cycles. Regeneration at 1223 K caused continued rapid crystallite growth that would be manifested in a reactor as a loss in catalyst activity.

Development of sorbents for temperature swing reactors is discussed by *Satrio et al. (2005)*. Compressive strength and abrasion resistance are essential. *Satrio et al.* proposed achieving this by forming an alumina shell around a dolomite core. They then impregnated the shell with an Ni salt solution so that on calcination a single particle contains both catalyst and sorbent. Strongest particles were formed when 5 to 10 wt% limestone was added to the alumina. The authors hypothesize that a calcium aluminate is formed that bonds both alumina and the alumina shell to dolomite. The impregnated particle was tested for the steam reforming of propane using an electrobalance. Reversible CO_2 adsorption by dolomite was observed, but there was a continual loss of capacity with successive cycles.

In the last several years, interest in adsorbent development has continued. Activity is summarized by *Harrison (2008, 2009)*. Restoration of adsorbent capacity by removing material from the mixed catalyst-adsorbent bed and treating adsorbent particles to reactivate them is emerging as a research theme. *Manovic and Anthony (2007)* steam-treated a lime at 475 K that had deactivated over 20 production-regeneration cycles to retention of just 35% of its fresh capacity. Treatment increased capacity to 65–75% of that of fresh lime. Apparently, steam exposure forms $Ca(OH)_2$ which decomposes under reactor conditions forming small pores of a high surface area. *Harrison (2009)* reports other investigators mention the use of hydration for restoring adsorption capacity.

Kato et al. (2003) discuss vehicle-based H_2 production for a fuel cell employing steam reforming of methane in a TSR. The reforming catalyst was a supported Ni material and the sorbent was lime, generated in situ from $CaCO_3$. These solids were intimately mixed and used only to collect data for the feed step of a TSR cycle. Results were compared with those for a conventional PFR packed only with catalyst. Prior to use, the packed bed was calcined at 1,103 K in Ar for four hours. Isothermal conditions were maintained for reforming. The feed was a steam/CH_4 mixture at 1 bar and temperatures used ranged from 823 to 1,023 K. The concentration of H_2 leaving the reactor was significantly higher for the TSR, while the concentration of CO in the effluent was substantially reduced. From their results, *Kato et al.* estimated that about 7.5 kg of CaO would be required in a vehicle to generate sufficient H_2 to travel 100 km. For this amount of H_2, 2.1 kg of CH_4 and 4.8 kg of water would be necessary. The sorbent would be removed from the vehicle in a service station and regenerated there while collecting the CO_2 formed. Thus, there would be no carbon oxide emissions during travel.

Oxidative coupling was considered as an application of SCMCRs in **Chapter 21**. The temperature swing has been explored as an alternative for removing ethylene produced in oxidative coupling before it is further oxidized (*Machocki, 1996*). Separate catalyst and adsorbent beds operating at widely different temperatures were used. A zeolite 5A adsorbent bed followed a reactor bed of calcined sodium and calcium carbonates. *Aida and Silveston (2005)* discuss this partial temperature swing system in their book.

A pilot scale reactor using a mixture of catalyst and 3A zeolite was employed by *Menge et al. (2003)* to validate their model for the Claus reaction simulations discussed in Section 22.7. The adequacy of the model for the first step of the two-step cycle is clearly indicated by **Figure 22-16**. *Menge et al.* observed that the conversions of the two reactants, H_2S and SO_2, were not the same despite Claus reaction stoichiometry. Lower conversion of SO_2 was explained by differences in the adsorption of the reactants on the alumina surface, rather than by the parasitic formation of carbonyl sulfide, COS, from traces of CO_2 in the reactor feed.

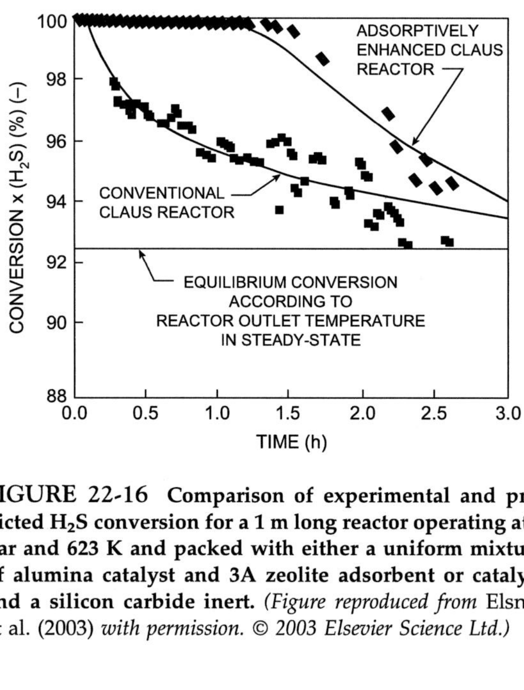

FIGURE 22-16 **Comparison of experimental and predicted H_2S conversion for a 1 m long reactor operating at 1 bar and 623 K and packed with either a uniform mixture of alumina catalyst and 3A zeolite adsorbent or catalyst and a silicon carbide inert.** *(Figure reproduced from* Elsner et al. (2003) *with permission. © 2003 Elsevier Science Ltd.)*

HCN synthesis from ammonia and carbon monoxide proceeds via a gas phase reaction with stoichiometry: $2\ CO + NH_3 \leftrightarrow HCN + CO_2$. The equilibrium conversion is only 17% for this reaction at 673 K with a stoichiometric ratio of reactants. *Elsner et al. (2002a)* found that the conventional high temperature water gas shift catalyst, Fe_3O_4/Cr_2O_3, catalyzes the synthesis and that an adsorbent for CO_2 could drive the reaction to completion. However, with a reaction temperature greater than 523 K, the CO_2 capacity for virtually all adsorbents is limited so the investigators turned to a gas-solid reaction for CO_2 capture. Their early choice, the exothermic formation of lithium carbonate from the zirconate, proceeded too slowly in the range of operating conditions studied so, instead, a hydrotalcite was chosen.

Using an isothermal, reactor, *Elsner et al.* explored just the production step of a two-step temperature swing cycle over a wide range of operating conditions and found that it was not possible to enhance the equilibrium conversion. Their analysis suggested the unexpected failure was caused by the Boudouard reaction, $2\ CO \leftrightarrow C + CO_2$, which is also forced in the forward direction by CO_2 capture. Carbon deposition and co-adsorption of HCN and CO_2 by hydrotalcite also contributed to the lack of enhancement.

This negative outcome illustrates pitfalls that can occur in developing PSR or TSR applications. For the HCN synthesis, these were: 1) An important side reaction, 2) "capture" of both reaction products by the adsorbent, 3) fouling of the catalyst by a side reaction and 4) low rates of adsorption compared with the rate of reaction, at least for the zirconate adsorbent.

22.10 COMBINED PRESSURE AND TEMPERATURE SWING REACTORS

Regeneration of adsorbent is often the challenge in the development of swing reactor applications. Instead of just depressurization or a temperature jump, it is possible to combine these methods. An example of such an application is the production of H_2 for use in fuel cells. For this purpose H_2 must have a CO content less than 30 ppm. Research undertaken in the late 1990s indicated that deep beds would be needed in either a PSR system or a more conventional steam reforming process. However, regeneration of deep beds is difficult. Simple depressurization is no longer adequate.

Production of high purity H_2 by steam reforming is an industrially important multistep process. The primary reaction is endothermic. Under typical industrial operating conditions, 6:1 H_2O/CH_4 and 446 kPa, at least 1,070 K is necessary for complete conversion of methane. If CO_2 can be removed from the equilibrium mixture, the temperature can be reduced substantially. Even at the above conditions, 99% removal of CO_2 reduces CO in the mixture to less than 10 ppm and CO_2 to about 30 ppm, and H_2 purity in the dry gas approaches 100%. Achieving nearly pure H_2 is probably not possible through a PSR alone. *Xiu et al. (2002b)* suggested that adsorbed CO_2 could be removed in a PSR through conversion to CH_4 which requires reducing the bed temperature. Implementing this implies a combined pressure-temperature swing system.

22.10.1 Simulation Studies

Figure 22-17 shows a five-step reactive regeneration process, involving combining pressure and temperature swing (*Xiu et al., 2002b*). It differs from a four-step PSA cycle by reducing the temperature in the depressurization step and replacing step three by an H_2 purge. The second purge with H_2O becomes step four. Justification is that reforming is strongly endothermic and lowers temperature, whereas methanation taking place during the second purge is exothermic and raises temperature. For the simulation, the model given in **Table 22-1** was used. Initial and boundary conditions are set forth in **Table 22-3**. A bed filled with a non-adsorbing gas at the temperature and pressure of the first cycle step was assumed at start-up. Walls were assumed to be at constant temperature. Duration of each step in the cycle was 500, 150, 400, 50 and 100 s respectively. *Xiu et al.* assumed that cooling of the bed is instantaneous when the wall temperature is lowered and at the end of the purge period when the wall temperature rises.

Xiu et al. used *Xu and Froment (1989a, b)* for the kinetics of the two reforming reactions and the water-gas shift reaction. A constant effectiveness factor of 0.8 was assumed for all reactions. Adsorption of CO_2 on the hydrotalcite employed the isotherms proposed by *Ding and Alpay (2000b)*. *Xiu et al. (2003a)* simulated the experimental results of *Waldron et al. (2001)* reasonably well for their assumptions (see the discussion in *Aida and Silveston, 2005*).

Reactive regeneration (combined swing) is compared with just a steam purge at the feed temperature for a four-step PSR in **Figure 22-18**. The figure shows the adsorbate concentration on the hydrotalcite at the end of the first three or four steps for the 2 m long bed assumed. The first 0.5 m of the bed is saturated at the end of the first step. Depressurizing, step two, desorbs only a small amount of CO_2. Adsorbate concentration drops sharply in the purge step as the four bottom curves in the figure show. Clearly, the reactive purge is more effective than the steam purge at 723 K. Comparing the hydrogen purge curves at three different temperatures, there appears to be an optimal temperature for this purge of about 673 K. At the back end of the reactor, adsorbate concentrations by the end of step two are low so

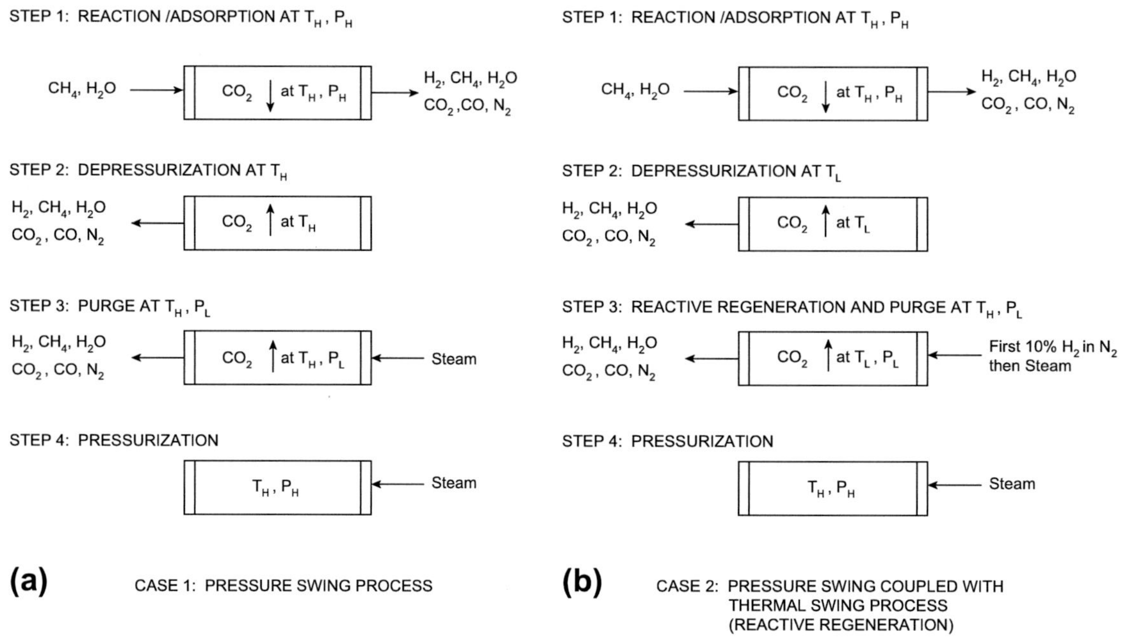

FIGURE 22-17 Multi-step, non-isothermal, combined swing process for producing high purity H_2 by steam reforming of CH_4. *(Figure reproduced from* Xiu et al., (2003b) *with permission. © 2003 by Elsevier Ltd.)*

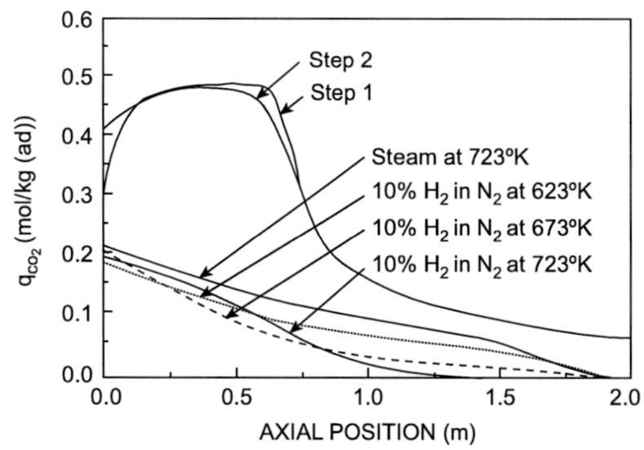

FIGURE 22-18 Adsorbate concentration on hydrotalcite at the end of cycle steps as a function of axial position for a 2 m long reactor operating with $u_{1_0} = 0.08m/s$, $u_{3_0} = 0.3m/s$ and cycle durations given in the text. Steam curves are for a four-step PSR while !0% H_2 curves are for reactive regeneration *(Figure reproduced from* Xiu et al. (2002b) *with permission. © 2002 by Elsevier Science B.V.)*

CO_2 desorption is slow making a steam purge, or preferably an H_2 purge, necessary to fully regenerate the adsorbent.

The *Xiu* simulation showed that a 2 m bed was too short to reduce the CO concentration to the target level of 30 ppm. Compared with a four-step PSR cycle, CH_4 conversion rose from 47.5 to 51%, H_2 purity from 78 to 80.5% and CO in the product dropped from 150 ppm to 110 ppm for the combined swing operation. Because some of the H_2 is used as the purge, production for the 2 m bed decreased. Increasing bed length, however, should solve those shortcomings so *Xiu et al.* tested the effect of bed length. A 6 m bed brought the CH_4 conversion to 62% and the H_2 purity to 86.5% for the combined swing cycle. CO contamination of the product fell to about 50 ppm, but the H_2 production rate dropped drastically to 0.17 $mol/kg_{solids} \cdot cycle$.

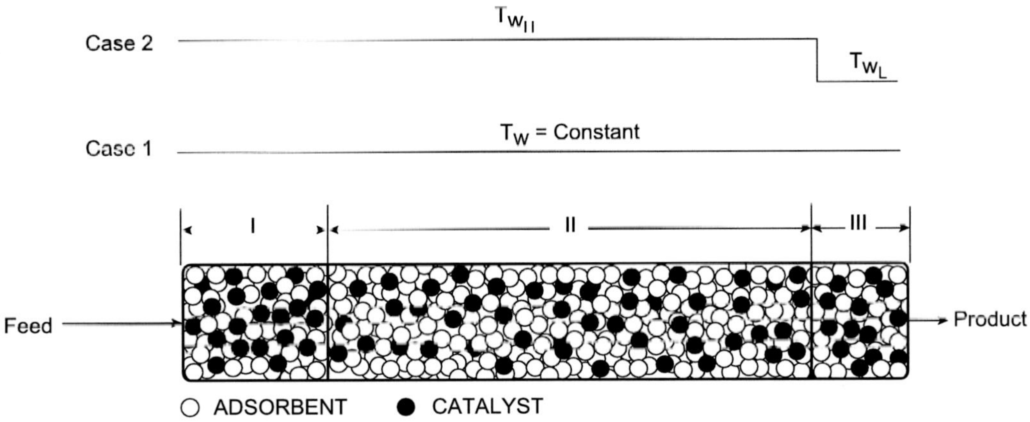

FIGURE 22-19 Schematic of a three-section, mixed catalyst and adsorbent bed investigated with the proposed wall temperature profile for the combined swing case studied by *Xiu et al. (2003b)*. *(Figure modified from* Xiu et al. (2003b) *with permission. © 2002 by Elsevier Science B.V.)*

Lengthening the mixed catalyst and adsorbent bed, however, creates a much larger region that has low adsorbate levels at the discharge end of the bed. Indeed this region expands from about one m for the two m long bed (see **Figure 22-19**) to almost four m for the six m deep bed. For either bed length most of the CO_2 is held in the first meter of length at the end of step 1 and saturates the adsorbent in this region. An alternative is to change the cycle period or the duration of the individual steps. Reducing the duration of the feed step to 350 s from 500 s increased conversion, H_2 purity and produced a gas that met the CO constraint for fuel cell use. Production rate, however, fell to 0.11 $mol/kg_{solids} \cdot cycle$. Reaching a cyclic steady state required about eight cycles or about 2.5 h for both cycles considered.

A problem of the combined cycle is the poor utilization of the catalyst and adsorbent. Methane conversion occurs primarily in the front portion of the bed, close to the entrance. *Xiu et al. (2003b)* suggest that the problem can be solved by dividing the reactor into sections and operating with different temperatures and catalyst to adsorbent ratios in each section. The concept underlying subdividing the bed is that different processes predominate at different points. Catalytic reactions occur in the front section, while in the middle section adsorption becomes important and hinders reverse reactions that consume product. At the end of the bed, adsorption, though slow, operates to reduce concentrations of undesired components in the product gas. *Xiu et al. (2003b)* proposed enhancing these different roles by increasing the catalyst to adsorbent ratio in the front section of the bed, Section I, and operating at the maximum temperature to force the endothermic reforming reactions to completion. In the middle of the bed, Section II, the catalyst to adsorbent ratio would be sharply decreased but temperature kept high to remove carbon oxides thereby stifling methanation and the reverse water gas shift. At the rear of the bed, Section III, the temperature would be lowered by withdrawing heat through the wall and the catalyst to adsorbent ratio increased. Lowering temperature shifts the equilibrium from reforming to methanation and forces CO towards CO_2 by the reverse water gas shift. This change reduces the CO level in the product gas without adding to the adsorption burden and serves to make the adsorbent regeneration step less critical. **Figure 22-19** shows the division of the bed into sections proposed by *Xiu et al.* and the wall temperature profile for the combined swing case studied numerically.

The objective of *Xiu et al. (2003b)* was to reduce the carbon oxide content of the product gas and increase H_2 productivity. Methane conversion, however, was less important because CH_4 is tolerated by PEM fuel cells and waste gas from the cells is consumed to supply the heat demand. Their improved cycle resembles that shown in **Figure 22-17** except that the steam purge and repressurization is at a lower temperature.

For their simulation, *Xiu et al.* retained from their previous study the 2:1 adsorbent to catalyst weight ratio in bed sections I and III but doubled the ratio in section II. Total bed depth was taken to be 6 m. Depths of section I, II and III were 0.1 m, 4.3 m and 0.6 m respectively, and T_H was mainly 763 K while T_L was 693 K. Other variables were unchanged. The authors explored T_H, T_L, P_H, and t_1, the duration of the production step. *Xiu et al.* commented that acceptable results can be obtained at least for the critical CO level over a fairly wide range of operating conditions for the combined swing cycle.

Lowering the wall temperature in section III and purging at the lower temperature were critical for keeping CO levels under 30 ppm. This is illustrated by **Figure 22-20**, a comparison of simulations for the combined cycle and a PSR cycle with the same set of operating conditions, cycle period, bed and bed section dimensions. The difference between this simulation and the previous one for the operation just discussed is the wall and purge temperatures in step three, and the repressurization wall temperatures in step four. It can be seen from **Figure 22-20** that the discharge composition is virtually the same in the two cases except in the last 2000 s of the feed step where reducing the wall temperature by 40°C in Section III substantially reduces the level of the carbon oxides. The wall temperature effect can also be seen at the end of reactor, the last meter, in **Figures 22-20(b)**, **(c)** and **(d)**. The lack of difference in the profiles from zero to five m or in the profile at t = 472 s suggest that the temperature changes in the purge and repressurization steps do not have an important effect on behavior. Purge temperature strongly affects superficial velocity. Wall temperature propagates rapidly through the 12.5 mm i.d. bed assumed in their study. *Xiu et al.* found that about 10 cycles or about three hours are needed to reach a cyclic stationary state.

Simulations of combined PSR-TSR have continued (see **Table 22-4**) and have been summarized by *Harrison (2008, 2009)*. Unlike the *Xiu* work discussed above, recent work has looked at swing cycles embedded in processes. Thus, *Reijers et al. (2006)* examined power generation via turbines, while *Ochoa-Fernandez et al. (2007)* considered an H_2 production plant supplying a fuel cell installation. Both teams compared the swing cycles with the conventional multi-step reforming alternative and concluded that the swing cycles were distinctly better with respect to energy use and carbon capture.

22.11 OVERVIEW AND ASSESSMENT

Swing reactors when applied to equilibrium limited reactions can increase conversions often significantly. Nevertheless the main advantage these reactors offer is to provide reactant conversion along with separation of reactants and products in the same vessel. Experimental studies, although limited, show some but hardly remarkable conversion increases; they do support simulations demonstrating that substantial product separation can be accomplished.

The pressure swing reactors have commercial promise, perhaps because pressure swing adsorbers have found wide application. Design of a PSR and scale-up of experimental units rely on the well-understood fluid dynamics of packed

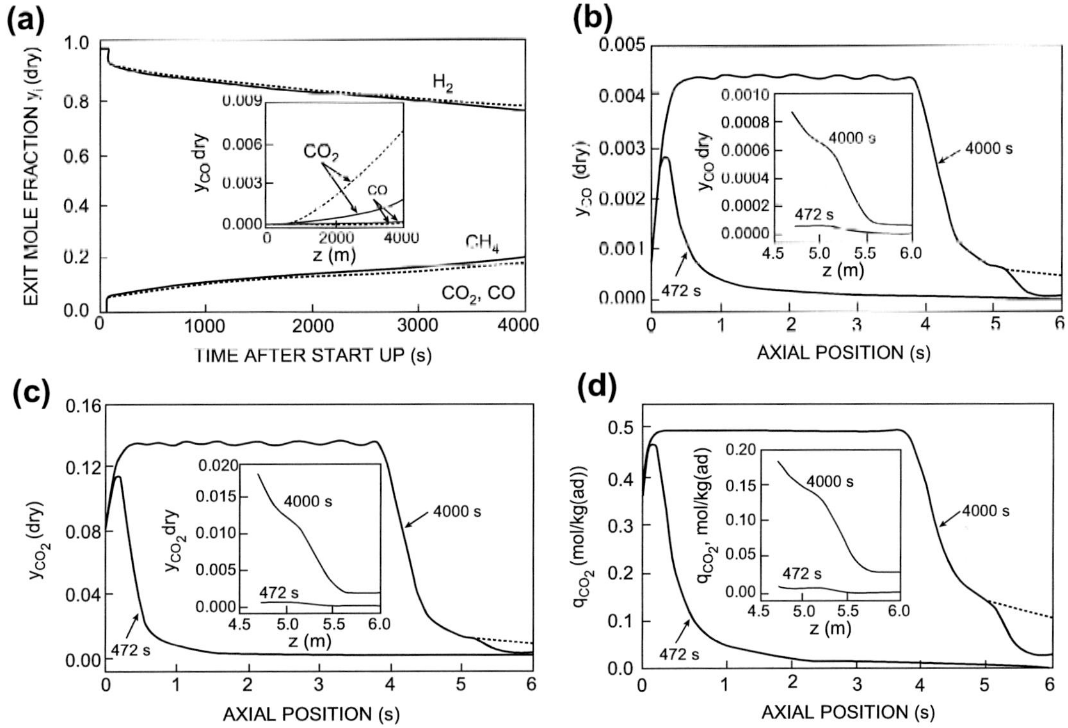

FIGURE 22-20 **Product gas composition leaving the discharge port at the end of the reactor: (a) And the axial distribution of CO (b) and CO_2 (c) in the gas phase and CO_2 adsorbate loading (d) at different times, 472 s and 4000 s, in the production step for H_2.** Solid line represents the combined temperature and pressure swing reactor for section III wall temperature case shown in Figure 22-19. Dashed lines = PSR with constant wall temperature. *(Figure reproduced from* Xiu et al. (2003b) *with permission. © 2003 by Elsevier Ltd.)*

beds so the engineering of isothermal units, at least, is straightforward. Decades of PSA applications provide a firm foundation of practical design knowledge. Perhaps the only uncertainty in PSR design is provision for the potentially large heat effect that will arise for systems with reactant concentrations of commercial interest.

Experimental work on pressure swing reactors has lagged behind theoretical studies. There are only a few experimental investigations of PSR systems. Just one (*Vaporciyan and Kadlec, 1989*) compared simulation and experiment and observed that the model simplifications employed are not satisfactory, particularly the isothermality assumption. The widely used pseudo-steady-state assumption also needs investigation. *Yongsunthon and Alpay (1999)* examined non-isothermal systems through simulation, but made no comparison between isothermal and non-isothermal models, nor was there a comparison with experiment, so it is still uncertain under what conditions the simpler isothermal representation is satisfactory. Indeed, significant heat effects for adsorption/desorption as well as for reaction suggest that most commercial applications will employ a combined pressure and temperature swing cycle. Experiments are needed to verify the simulation of *Xiu et al. (2002 a, b; 2003a, b)*. Experimental ingenuity will be required because of problems of heat loss in laboratory-scale equipment. Energy flows for steps in a cycle, indeed, energy integration, needs further investigation. *Yongsunthon and Alpay (1999)* made only a start on this potentially important problem.

Integration of a PSR or TSR with other separation operations seems worth examining. *Esteves and Mota (2007)* discuss combining a PSA with membranes, although the arrangement *Esteves and Mota* employed is difficult to visualize for a reacting system.

Only single reactions in PSRs have been studied experimentally or through simulation. Expansion to systems with multiple reactions should be undertaken. Work on defining the characteristics of reaction/separation systems suitable for PSR or TSR applications is needed. The use of a well-mixed reactor to explore candidate systems, choose adsorbents and suggest operating conditions has been suggested by *Sheikh et al. (1998)*. This application, however, has yet to be undertaken.

Because of its commercial promise, identification and development of PSR/TSR application needs to be pursued. It must be largely an experimental activity.

Acknowledgment

The editors wish to acknowledge the advice of professor David Agar (Universitaet Dortmund) for this chapter.

Notation

A	=	reacting species, usually a reactant
A_x	=	cross sectional area (cm^2)
B	=	reacting species
CIS	=	cells in series
C	=	total concentration (mol/m^3) proportionality constant, reacting species, often a product
C_i	=	concentration of the ith species (mol/m^3)
C_v	=	valve or throttling coefficient
C_p	=	specific heat (kJ/g·°C)
c_p	=	molar specific heat (cal/mol·K)
DOCFE	=	double orthogonal collocation on finite elements
D	=	reacting species, usually a product
D_{ax}	=	axial dispersion coefficient
D_z	=	axial dispersion coefficient
d_R, d_r	=	reactor diameter
f_i	=	cycle fraction for the ith step in a cycle ( )
f_R	=	reflux volume ratio for PSR with reflux
h	=	heat transfer coefficient
h_w	=	heat transfer coefficient at the tube or reactor wall
J_k	=	kinematic coefficient in the Ergun equation
J_v	=	viscous coefficient in the Ergun equation
K, K_{eq}	=	equilibrium constant
K_i	=	adsorption equilibrium constant or Langmuir constant for the ith component
k	=	rate constant, conductivity
k_f, k_j	=	rate constant of the forward reaction
k_i	=	adsorption rate constant for species i
k_z	=	axial conductivity
L	=	length (cm, m)
L_{bed}	=	bed length
M_i	=	molecular weight of the ith species
N	=	number of reacting species
N_c	=	number of reactions in system
N_{Da}	=	Damköhler number ( )
N_{Ka}	=	Kadlec number ( )
N_{Ru}	=	Ruthven number ( )
n	=	molar flow (mol/s), species number
n_b	=	backfill molar flow
n_d	=	discharge (delivery) molar flow
n_p	=	purge molar flow
OC	=	orthogonal collocation
ODE	=	ordinary differential equation
OCFE	=	orthogonal collocation on finite elements
PDE	=	partial differential equation
PFR	=	plug flow reactor
PSA	=	pressure swing adsorber
PSR	=	pressure swing reactor
P	=	pressure (kPa, bar)
P_b	=	backfill pressure
P_d	=	discharge pressure
P_e	=	exhaust pressure
P_f	=	feed pressure
P_H	=	high pressure in pressure swing cycle
P_i	=	partial pressure of ith component
P_L	=	low pressure in pressure swing cycle
P_p	=	purge pressure
P_v	=	exhaust pressure
R	=	gas constant
R_i	=	rate of the ith reaction, also rate of reaction kinetics for ith reaction
Q	=	volumetric flow (m^3/min), or heat quantity (cal.)
Q_0	=	entering volumetric flow
q	=	adsorbate concentration (mole/g adsorbent)
q_i	=	concentration of adsorbed species i on the adsorbent
q_s	=	concentration of adsorbed species at saturation

SV = space velocity
TCAR = temperature swing reactor
TSA = temperature swing adsorber
TSR = temperature swing reactor
T = temperature (K)
T_f = feed temperature
T_p = purge temperature
T_H = High temperature in a temperature swing cycle
T_L = Low temperature in a temperature swing cycle
$(T)_{mix}$ = mixed or mixture temperature
T_W = wall temperature
t = time (s, min)
t_i = duration of the ith step in a cycle (s, min)
$\bar{t}$ = residence time (s, min)
U = velocity
u = superficial velocity (cm/s)
V = volume (cm^3, m^3)
V_R = reflux volume
v = interstitial velocity
WGS = water gas shift
w = flow splitting variable
X_i = conversion of the ith reactant
y = mole fraction
y_i = mole fraction of the ith component
$\bar{y}$ = mean mole fraction
z = axial position or distance (cm, m)

Greek Letters

α, β = exponents
α_i = adsorption constant for species i
ΔH_{ad} = heat of adsorption
ΔH_a = heat of adsorption
ΔH_i = heat of adsorption for ith component heat of reaction for ith reactant
ΔH_r, ΔH_R = heat of reaction
ε = void fraction ( )
ε_b = void fraction in packed bed
ε_t = total void fraction
γ_f = total heat capacity of flowing fluid
γ_i = dimensionless capacity for component i
ϕ_i = volumetric capacity of adsorbent for species i
η = effectiveness factor ( )
λ = conductivity
λ_z = axial conductivity
μ = viscosity
Θ = dimensionless adsorption capacity
ρ = density
ρ_A = adsorbent density
ρ_a = catalyst density
ρ_B, ρ_b = bulk or bed density (g/cm^3(bed))
ρ_g = gas density
σ = exponent
ν = stoichiometric coefficient ( )
ν_{ij} = stoichiometric coefficient of component i in the jth reaction
τ, τ_{cycle} = cycle period
τ_p = period
ζ = system variable

Subscripts

A = adsorbent
a = adsorption
ads = adsorbent
ax = axial
b = backfill
c, cat = catalyst
d = discharge, delivery
e = exhaust
eq = equilibrium
f = feed, flow, forward
g = gas
H = high
I = reaction index
i = index, step index, species index
j = reaction index, species index
k = step index (in cycle)
L = low
max = maximum
mix = mixture
o = entering (feed), initial
p = period
r = reverse, reaction, reactor
s = solid
t = total
v = exhaust
W, w = wall
z = axial
0 = initial, sometimes feed
1,2,3 = step numbers in a cycle
+ = downstream of boundary
- = upstream of boundary

Superscripts

α, β = exponents
* = equilibrium
\- = downstream
\+ = upstream

CHAPTER

23

New Directions—Research and Development Challenges

Kenji Hashimoto, Motoaki Kawase*, Menka Petskovska*[†]*, Jiri Hanika*[‡]*, Cheng-Yue Li*[¶]*, Adesoji Adesina*[§]*, Hristo Sapoundjiev*[$]*, Jeno Scharer**, Peter Lewis Silveston***

***Kyoto, Japan, [†]Belgrade, Serbia, [‡]Prague, Czech Republic, [¶]Beijing, China, [§]Sydney, Australia, [$]Montreal, Canada and **Waterloo, Canada**

OUTLINE

23.1 Challenges 680
- *23.1.1 Identifying Attractive Reactions* 680
- *23.1.2 Identifying Alternate Separating Techniques for Reaction-Separation Applications* 681
- *23.1.3 Research on New Cycles* 681
- *23.1.4 Biochemical Applications* 681
  - 23.1.4.1 Product-inhibited Biochemical Reactions 681
  - 23.1.4.2 Temperature Modulation / Temperature Modulation Combined with Inducer Modulation to Enhance Recombinant Production 682
  - 23.1.4.3 Pressure Modulation for Biochemical Reaction within Bioflocs 683
- *23.1.5 Adoption of Electrochemical Techniques to Non-electro Chemical Reactions* 683
- *23.1.6 Use of Supercritical Fluids in Simulated Moving Bed Chromatographic Reactors* 683
- *23.1.7 Separation of Multicomponent Mixtures Using Modified Simulated Moving Bed Chromatographic Reactors* 684
- *23.1.8 Development of Control Systems for Periodic Processes* 684
- *23.1.9 Use of Modulation to Control Catalyst Activity* 684
- *23.1.10 Circulating Solids vs. Fixed Beds with Composition and Flow Direction Modulation* 685
- *23.1.11 Modulation as a Tool for Mechanistic Studies* 685
- *23.1.12 Swing Reactors* 686

23.2 New Directions 686
- *23.2.1 Applications to Systems with Multiple Objectives* 686

23.2.2 *Control of Carbon Deposition* 686
23.2.3 *Modulation of Multiple Reactor Inputs* 687
23.2.4 *Improvement of Photocatalysis* 688
23.2.5 *Optimization of Periodic Processes* 688
23.2.6 *Engineering of Periodically Operated Processes* 688

Forty years of research and development on periodic flow reversal have resulted in widespread adoption of the technology for mitigation of low levels of hydrocarbon and other volatile organics compounds in industrial exhausts. As discussed in **Chapter 18**, reverse flow catalytic mitigation systems have begun to even replace reverse flow thermal oxidizers that over recent decades largely replaced once-through combustion oxidizers.

Many of the early chapters have demonstrated the use of composition modulation in elucidating reaction mechanism. This application, indeed, was the subject of **Chapter 13**. Composition modulation has now emerged as a useful tool for mechanism research and monitoring the role of a catalyst in accelerating reactions. From an opposite perspective, **Chapter 14** shows how an assessment of the benefits of periodic operation on reactor performance can be made through the application of nonlinear frequency response provided a kinetic model of the reaction of interest is available.

Recent vigorous research on applying chemical looping, a type of periodic operation, to sequestering CO_2 generated from the combustion of carbon containing fuels has made periodic reactor operation an important topic of Chemical Engineering research. Besides chemical looping, discussions in previous chapters point to other applications of periodic operation that warrant attention.

An objective of this chapter is to specify what many current researchers believe to be the R & D challenges for periodic operation of chemical reactors. These are separated in what follows as: 1) challenges from research already underway and 2) identifying new directions that research might take. Indeed, this discussion was begun earlier. **Section 10.10** at the end of **Chapter 10** addresses such challenges for biochemical systems. The chapters following this also consider research needs in their final sections. Those needs or challenges will be stated just summarily in what follows. Of course, an important purpose of this chapter is to encourage further expansion of research, development and the application of periodic operations.

23.1 CHALLENGES

Ordering of topics in the following subsections should not imply importance or ease in achievement.

23.1.1 Identifying Attractive Reactions

Research on periodic operations can serve several purposes: 1) Reducing the capital or operating cost, or safety of a commercial operation, 2) identifying operating strategies or optimal operating conditions and 3) identifying chemical systems that might benefit if operated periodically. In general terms for 3), identifying the characteristics of reactions that are suitable for periodic operation.

The framework for identification has been laid through the application of nonlinear frequency response, NLFR, introduced in **Chapter 14**. Evaluation of the dominant component — the second-order frequency response function, $G_2(\omega, -\omega)$, or just its sign is needed. For this, a kinetic model as well as a reactor model is required. How reliable does that model have to be? How should the value of the response be interpreted? What needs to be undertaken is an application of NLFR to a chemical system that has been studied experimentally. The existing NLFR publications consider just an adsorption system or a rather elementary, isothermal reaction system. An important application of periodic operations is enhancing selectivity. Certainly NLFR should be studied for such an application.

23.1.2 Identifying Alternate Separating Techniques for Reaction-Separation Applications

Reaction-separation systems examined in this monograph have dealt with just adsorptive separation simply because it is the only system discussed in the literature. Of course, the other separation system readily compatible with chemical reactions is absorption/extraction in which one or more liquid phases are involved. Both of these operations offer flexibility because a liquid trapping phase can be readily moved; on the other hand, temperature restrictions arise because the fluid must remain liquid throughout the operation. Probable applications of extractive separation are with biochemical reactions that operate mainly at temperatures between 20 and 80°C. Combined extractive separation and reaction systems are discussed in **Section 20.9**.

Alternate separating techniques, such as chemical binding of products, or phase change (vaporization/solidification) appear to have very limited commercial application.

23.1.3 Research on New Cycles

Coupling exothermic and endothermic reactions has been considered in **Chapter 18**. In this operation, the exothermic reaction produces just heat and a waste gas. An intriguing possibility is to choose an exothermic reaction system that produces a product as well. One possibility is to combine two exothermic reactions such as the total and partial oxidation of CH_4 using a low quality oxygen in cycles with steam and dry (CO_2) reforming of CH_4. Both systems produce a synthesis gas of different quality or fuels for anaerobic combustion and power generation.

Early work discussed in **Section 12.6** suggested that homologation of $C_2^=$, $C_3^=$ was a potentially attractive route to higher olefins if conducted periodically in a flow reversal system with burning off of the coke deposition in homologation by air. There are insufficient data in the original homologation studies of *Lefort et al. (1994)* to undertake modeling of a possible system, so further study will be required.

23.1.4 Biochemical Applications

It is evident from **Chapter 10** that processes employing prokaryotic or eukaryotic cells offer a rich opportunity for exploiting periodic operations. Only a few of these can be highlighted in this section. Further opportunities are discussed in **Section 10.10**.

23.1.4.1 Product-inhibited Biochemical Reactions

The acid hydrolysis of starch and its derivatives, such as maltose, to glucose is a process that is used extensively in the food industry. For improved selectivity, a biocatalyst, glucoamilase is used. However, glucose, a hydrolysis product, inhibits the reaction and must be continuously separated. An approach considered by *Yankov et al. (1995)* was to immobilize the glucoamilase enzyme on a polyacrylamide gel and carry out the hydrolysis using periodic flow reversal. Modeling undertaken by *Yankov et al.* suggested that gel particle size and the concentration of maltose in the reactor feed were important variables. Periodical reversal of the feed stream created more favorable concentration profiles in the reactor, increasing the selectivity to glucose considerably over that possible by conventional acid hydrolysis.

23.1.4.2 Temperature Modulation / Temperature Modulation Combined with Inducer Modulation to Enhance Recombinant Production

Biological processes employing prokaryotic (bacteria) and eukaryotic (mammalian cells, yeasts, fungi) platforms are known to be strong functions of temperature. For example, gene expression (transcription and translation) has been shown to be up-regulated by a temperature increase (*Hasan and Shimizu, 2009*) or, in some cases, by a temperature decline (*Underhill and Smales, 2007*). Stability of the synthesized protein may also be temperature dependent. Recognizing that specific gene expression, growth and product stability are optimal at different temperatures, temperature shift operations have been employed over three decades (*Lemaux et al., 1978; Chohji et al., 1983*). Systematic studies of temperature shift effects on the molecular genomics and proteomics level are more recent. *Sun et al. (2007)* reported a programmed three-step temperature increase as being optimal for human preproinsulin production by *E. coli*. Temperature shifts, however, are usually single step operations where the temperature change is introduced once only during the course of cell culture. Periodic temperature changes in either fed-batch or fully continuous flow operations are rarely performed and would represent a new opportunity. Model-based optimization may be carried out once the temperature dependence of the key metabolic networks has been established (*Nolan and Lee, 2011*) as has been done recently for a Chinese hamster ovary (CHO) cell line. *Clarke et al. (2011)* constructed a model for recombinant gene expression with temperature-shifted CHO cells.

A particularly fruitful approach would be the combination of temperature shift with the periodic addition of precursors or inducers for recombinant protein pharmaceutical production. The combination of precursor addition (l-cysteine) with a temperature downshift from a growth temperature of 30° to 26°C enhanced the production of glutathione in a *Candida utilis* culture (*Wei et al., 2008*). *Toru et al. (2010)* reported enhanced porcine adenylate kinase expression by coupling temperature up-shift (35° to 42°C) with amino acid (valine and leucine) supplement in a recombinant *E. coli* harboring the pUC plasmid under control of the *trp* promoter.

A promising, as yet unexplored, endeavor would be the combination of periodic temperature shifts with periodic induction of the desired recombinant gene expression for pharmaceutical protein production. Inducers such as the non-metabolizable inducer IPTG (isopropyl-1-thio-β-G-galactopyranoside) and TMG (thyomethyl-β-galactoside) of the recombinant genes residing within the *Lac* operon (*Marbach and Bettenbrock, 2011*) function in the methanol inducing recombinant protein synthesis in the methylotrophic yeast, *Pichia pastoris*. The inducer concentrations need to be carefully controlled, since high levels can cause unwanted cell death and lysis. On the other hand, inducers at low concentrations tend to lose their efficacy after a few hours since daughter cells are exposed to ineffective concentrations of the inducer. The experience has been so far that, using either temperature shift or induction alone, the concentration of product and cell specific productivity cannot be maximized simultaneously under batch, fed-batch or continuous growth conditions. High cell-specific productivity reduces production costs by shortening the bioreactor operating time. On the other hand, high product concentration is desirable for downstream purification of the recombinant biopharmaceuticals. Combining periodic temperature shift with periodic induction in fully continuous or fed-batch culture provides a challenging task for simultaneously optimizing both cell specific productivity and the concentration of the recombinant product. Model-based optimization may be carried out by establishing the temperature dependence of the key metabolic networks (*Nolan and Lee, 2011*) and recombinant gene expression (*Clarke et al., 2011*) as has been done recently for specific Chinese hamster ovary (CHO) cell lines as mentioned above.

23.1.4.3 *Pressure Modulation for Biochemical Reaction within Bioflocs*

Section 15.7 discusses studies that show that pressure modulation enhances gas phase mass transfer through slimes and flocs, but that it also destroys the structure of these masses. These dual effects need experimental study to identify the influence of amplitude and frequency on both transfer and floc strength. Ultrasonic irradiation has been found to enhance mass transfer in gas phase, heterogeneous systems. Irradiation causes bubble formation and collapse that appear to be the mechanism for more rapid mass transfer. Pressure modulation, perhaps bordering the ultrasonic frequencies, should also increase mass transport in biochemical flocs. Conversion rates in such systems are often mass transfer controlled.

23.1.5 Adoption of Electrochemical Techniques to Non-electro Chemical Reactions

Modulation of electrochemical processes, according to the discussion in **Chapter 9**, function by interfering with the transport, adsorption/desorption or chemical processes on or near electrode surfaces. Modulation available in catalytic systems discussed elsewhere in this monograph cannot operate directly on near surface processes. Only sonic or radiation processes may be able to influence surface/near surface operations. Thus, apart from such irradiation, there is little that can be carried over from the many important periodic operations applications in electrochemical processes into conventional catalytic process operations. Indeed, another characteristic of periodic operations used in electrochemical processes, complex periodic cycles, does not seem applicable to processes of interest in other chapters because of inertial and/or mixing effects. An exception to this is perhaps the application of irradiation techniques, such as electromagnetic or ultrasonic, that can function on surfaces. Such applications could be a potentially attractive research direction.

23.1.6 Use of Supercritical Fluids in Simulated Moving Bed Chromatographic Reactors

Solvents, such as water or alcohol, generally serve as carrier fluids in liquid-phase chromatographic reactors. However, it is often costly to remove the solvent from the product stream leaving a unit. An attractive solution might be to employ a supercritical fluid as a solvent. Carbon dioxide in its supercritical state appears to be an ideal choice for such a mobile phase: It is an excellent solvent, relatively inexpensive, nontoxic and nonflammable.

Supercritical chromatography was first used for analytical purposes and only later extended to preparative chromatographic separations. The latest extension, supercritical simulated moving bed chromatography (SCF-SMB), has proven to be feasible. It offers the advantage that the separation of a solvent from products and control of adsorption and desorption in each zone of a unit are easily carried out by changing pressures. The SCF-SMB has already been applied to the separation of enantiomers in pharmaceutical processes. A review (*Taylor, 2009*) deals with the characteristics of SCF as well as applications to preparative scale chromatography and simulated moving beds.

Several investigators (*Di Giovanni et al., 2001*; *Peper et al., 2002, 2007*; *Zhang et al., 2007c*) have shown that the models and design procedure developed for the conventional SMB processes are applicable to SCF-SMB. These designs and the experimental examinations of the models were carried out for the separation of enantiomers using supercritical carbon dioxide (*Schulte and Strube, 2001*; *Freund et al., 2009*). Extension of this technology is a challenge offering important benefits if successful.

23.1.7 Separation of Multicomponent Mixtures Using Modified Simulated Moving Bed Chromatographic Reactors

SMBCR applications to complicated reactions have been limited to reactions in which the same product, namely water, is produced in each reaction and must be almost entirely removed to reach high yields of the target product. *Silva et al. (2011)* investigated a simulated moving bed membrane reactor (PermSMBR) that combined an SMBR with hydrophilic membranes. This novel reactor was applied to the ethyl lactate and 1,1-diethoxyethane syntheses. The unit studied consisted of a set of columns connected in series, each column containing a membrane with a catalyst/adsorbent mixture packed either inside or outside of the membrane depending on the position of the membrane active layer. Water produced by the two parallel reactions investigated was effectively removed through the membranes, resulting in a high yields of products. The success of this early work needs confirmation from studies of other multiple reaction systems.

Consecutive reactions are another important class of multiple reaction systems in which the intermediate product is desired. SMBRs have not been applied, apparently, to such systems for yield enhancement. Finding a chemical system suitable for SMBR application and demonstrating benefits from the application is a worthy challenge.

Seidel-Morgenstern et al. (2008) have reviewed modifications and/or extensions of SMB technology proposed during the last decade. Gradient operations (*Wei et al., 2011*) and ternary and pseudo-ternary separations (*Borges and Rodrigues, 2008*) offer promise. Although these modifications were originally applicable to separations in SMBs, the concepts should be applicable to simulated moving bed chromatographic reactors (SMBRs).

23.1.8 Development of Control Systems for Periodic Processes

This is an R & D topic still active as is clear from **Chapter 19**. Conventional feedback PID schemes have been dismissed as unsuitable for controlling periodic processes because of the time-varying observer signal. Model-based predictive control, already well-established for steady-state control in the chemical and petrochemical industries, has also been studied for control of periodic processes (see **Sections 19.6** and **19.7**). Although simulations suggest these model based systems are robust, they have not been tested in an application. Many involve several levels of control and/or undertake corrective action over several cycles. The question arises as to whether these more sophisticated control schemes offer enough improvement over the simpler PID controllers that they can be made to function with a time-varying input. Although simulation continues to be useful, there is a pressing need for experimental investigations.

23.1.9 Use of Modulation to Control Catalyst Activity

Some catalysts deactivate so rapidly that they are uneconomical despite their initial high activity. An example of this is palladium, a possible catalyst for glucose partial oxidation. Czech investigators (*Gogová and Hanika, 2009a, b, c*; *Gogová et al., 2010*) have found that oxygen depletion in the liquid phase initiates deactivation. They have proposed and demonstrated catalyst activity control by periodic modulation of the gas phase oxygen concentration during liquid phase oxidation of glucose using both an aerated CSTR and a gas-lift reactor. These investigators attempted to find, for any reactor input, conditions for the highest reactor productivity, i.e., conditions where the reaction and catalyst reactivation times are shared optimally. Glucose partial oxidation is discussed in **Section 11.7**.

23.1.10 Circulating Solids vs. Fixed Beds with Composition and Flow Direction Modulation

Research and development on CO_2 capture and sequestration has been confined in recent years primarily to circulating solids systems, those discussed in **Chapters 5** and **8**. Although circulating solids seem better suited to large operations such as coal burning power stations, attrition of the oxygen carrier and fines separation and handling are serious problems.

Replacing a dense or dilute phase fluidized bed by a fixed bed would substantially reduce, but not eliminate, the attrition problem and the associated difficulties of fines separation and continual catalyst replacement. Gas mixing is reduced so there would be a closer approach to plug flow that would offset some of the increased transport resistance arising from the much larger carrier particle size. Less purge gas would be required for stripping between oxidative and reducing steps. Perhaps of most importance, the power cost of circulating beds of fine solids is avoided. There are probably sufficient data on attrition losses for circulating solids to conduct an economic comparison of the two alternatives. Such a comparison would be an important first step before undertaking an experimental program on periodic operation.

Section 4.7 discussed the DuPont maleic anhydride process that attempted the use of a circulating solids catalyst but could not overcome the catalyst attrition problem. Perhaps the process should be re-evaluated with consideration of periodic flow reversal with fixed catalyst beds replacing circulating catalyst.

23.1.11 Modulation as a Tool for Mechanistic Studies

Examples considered in **Chapter 13** suggest that modulation may be attractive when step-change experiments are unable to discriminate among rival models. Modulation over a range of frequencies should be useful for model discrimination when relaxation times of the catalytic processes are relatively fast so that step-change experiments are limited by the finite rates of data collection of available analytical instruments. Indeed, modulation might be able to stabilize an otherwise unobservable surface species so that its response to changing operating conditions can be determined and related to product behavior.

Frequency response methods have been widely used to characterize electrical networks and some mechanical systems. The characterization is often stated as a transfer function. **Chapter 14** discusses interpreting the suitability of periodic operation for a chemical reaction based on the availability of a model incorporating kinetic and reactor flow characteristics. Could an experimentally derived transfer function be used in the same way? Derivation of a transfer function experimentally with a packed bed of catalyst involves determining the response at varying frequencies. These measurements, of course, answer the suitability question if a response maximum is observed for the frequencies tested. However, reactor performance also depends on the reactor residence time distribution. Measurements are made, generally, in just a plug flow situation. Experiments needed are to show how plug flow FR measurements can be used to assess the suitability of PO for different reactor flow patterns.

23.1.12 Swing Reactors

Pressure and temperature swing reactors still remain research topics at this time. The state of development as per the published literature is summarized in the preceding chapter. Experimental studies are limited so that most advances come through simulation. More experimental work is needed to support existing simulations. **Section 22.11** discusses specific research needs.

23.2 NEW DIRECTIONS

23.2.1 Applications to Systems with Multiple Objectives

Virtually all of the systems considered in the literature on periodic operations have dealt with improving a single performance measure such as selectivity to a desired product or reactant conversion. In an industrial environment, optimization problems may focus on situations involving multiple objectives. This subject has drawn a good deal of attention for steady-state systems and sometimes leads to a Pareto set, i.e., a problem in which not all of the objectives can be optimized under the same choice of operating conditions. There is a growing literature on this topic. Recently, researchers at Petrochina Company Ltd working together with Tsinghua University (*Gao et al., 2008a*) investigated the production of lower olefins from naphtha pyrolysis. The problem specifically was to optimize the production of either ethylene or propylene. Pyrolysis of naphtha becomes a periodic process on a timescale of days because the cracking furnace must periodically be shut down for coke removal. In an earlier paper, *Gao et al. (2006)* considered the periodic burn-off of coke in attempting to optimize the yield of just ethylene. Consideration of yields of both ethylene and propylene generates a Pareto set because yields of both products cannot be maximized at the same reactor state.

The Tsinghua researchers developed a model for the naphtha process and attempted to use conventional multi-objective algorithms to find an optimal solution. The algorithms used turned out to be extremely slow, so the authors modified one of them to produce what they termed a "parallel hybrid algorithm." Details of this algorithm and multi-objective optimization are too lengthy to discuss here. *Gao et al. (2008a)* offer a well-written treatment.

The system considered by *Gao et al. (2005, 2008a)*, and earlier by *Xu et al. (2001)*, introduces a type of periodic operation rarely examined in the literature. Each part of the cycle considered had a duration of days so that investigators, including *Gao et al.*, replaced dynamics models with pseudo-steady-state ones. Nevertheless, future problems with multiple objectives will surely arise in systems described by dynamic models, such as those considered in this monograph. Devising optimization techniques for problems in which dynamic models cannot be avoided and for which there are multiple objectives as well as multiple constraints poses a difficult challenge.

23.2.2 Control of Carbon Deposition

Despite claims in the literature that types of periodic operation extend catalyst life, there has been little study of this application. Catalyst deactivation through carbon deposition has been tackled by *Jaeger et al. (1992)* in an examination of CO methanation on a Ni-faujasite catalyst. The authors used a vibrationally fluidized bed in which the reactants CO and H_2 could be added separately or combined in a single feed. In the separate-feed operation, H_2- and CO-rich gas regions exist in the fluidized bed. Catalyst particles circulate through these regions. In an H_2-rich region, hydrogen reacts with the active carbon deposited by the Boudouard reaction to form methane, thereby removing carbon and regenerating the Ni catalyst. *Jaeger et al.* demonstrated that the catalyst activity can be maintained indefinitely with a separate feed design. This results in a greater methanation rate for long on-stream times. This interesting experimental study should be pursued further.

Recently, *Alenazey et al. (2009)* found that carbon deposition during the steam reforming of propane in a fluidized bed reactor operating under at 823 K and an atmosphere could be reduced by over 50% through forced cycling with CO_2. Deactivation was greatly retarded. The main gaseous products were H_2, CO and CH_4 in both parts of the cycle. H_2 and CO were formed from the steam gasification of coke deposited during the dehydrogenation part of the cycle and CO_2 gasification of the surviving surface deposit in the remainder of the cycle. The Co-Ni catalyst used by *Alenazey et al.* exhibited stable activity over a three day run under periodic operation. Artificial neural network modeling of the data obtained showed that cycle split had the greatest impact (*Arcotumapathy et al., 2011*). The coupled reactions are not heat balanced in general. Could O_2 be co-fed with CO_2 in the partial cycle to address the heat balance or could a hot CO_2 stream be used? Is a similar operation feasible with a natural gas feed?

Control of carbon deposition in methane cracking for fuel cell H_2 production through periodic flow reversal was reported by *Poirier and Sapundzhiev (1997)*. Their process employed the catalytic decomposition of natural gas. Methane decomposes releasing H_2 and depositing coke on the catalyst. As the catalyst bed fills and constrains gas flow, the partial cycle ends, the flow direction switches and the catalyst bed is regenerated by combusting the coke with air in the other partial cycle. The operation creates a moving heat front in the reactor, causing a more even coke deposition that retards plugging. The combination of exothermic coke combustion and endothermic CH_4 decomposition approaches autothermal operation and produces a greater than 95% H_2 stream. Further experimental study of this reverse flow application is necessary. Temperature control will probably be important. Coke reactivity is sensitive to the thermal history of a coke deposit. Process optimization will be a challenge.

23.2.3 Modulation of Multiple Reactor Inputs

The current literature on periodic operation has considered just modulation of a single input such as feed composition or reactant concentration, reactor temperature or reactant flow rate. *Parulekar (2003)* appears to be the first to examine modulating more than one reactor input simultaneously. He provides, in this important paper, the conditions that must be fulfilled for an improvement to occur in reactor performance using the π-criterion, which was discussed briefly in **Chapters 10** and **14**. *Parulekar's* analysis found that the interaction between the simultaneously modulated inputs dominates the magnitude of the reaction improvement achieved.

Two examples are given in the 2003 paper to illustrate the improvement possible with the simultaneous modulation of two inputs. The first of these deals with the competition between recombinant microorganisms and nonrecombinant ones. This competition was dealt with by *Lee and Parulekar (1996)* and others. It is discussed further in **Chapter 11**. *Lee and Parulekar* examined the use of antibiotics in a continuously operated stirred tank reactor to control the nonrecombinant cells and demonstrated through modeling that by modulating different inputs to the reactor they could maintain a higher concentration of the desirable recombinant organisms. These inputs were substrate concentration and dilution rate. Improvements also occurred when the antibiotic concentration was modulated. *Parulekar* considered two different types of antibiotics in his analysis.

A second example utilized a series-parallel reaction in which the intermediate product was desired. This system had been analyzed under periodic operation by *Watanabe et al. (1981)*. Modulation of flow rates of reactants was examined. Those authors observed that the frequency region in which periodic operation improved performance was extended significantly when both of the feed rates were modulated. *Parulekar's* analysis was applied to small variations in the forced variables, a situation which allows linearization of the descriptive equations. Multiple reactions under periodic operation are discussed in **Section 12.7**.

The immediate need is to follow up this extension of periodic operations with experimental confirmation.

23.2.4 Improvement of Photocatalysis

Photocatalytic processes using a TiO_2 catalyst have attracted environmental attention for decomposing organic contaminants into simple inorganic species. *Wang et al. (2006)* examined the application of periodic illumination using the ultraviolet light from a light-emitting diode to the photocatalytic decomposition of Reactive Red 22 dye. The dye was assumed to represent an undesirable contaminant of a water stream. The UV-LED was used with TiO_2 in a rectangular planar fixed-film reactor operated with recirculation of a water stream of low pH (= 2). Results indicated that the photonic efficiency with periodic illumination was much higher than with continuous illumination. However, the photonic efficiencies for these two arrangements became nearly identical at higher light intensities. Experiments are needed to see if this is an isolated phenomenon, or is periodic illumination generally more efficient. If it is more efficient, what is the explanation?

23.2.5 Optimization of Periodic Processes

Some additional important issues for the industrial application of periodic operations are: 1) Finding the optimal time profiles that maximize productivity and/or selectivity and 2) designing a control system to ensure the optimal performance. More sophisticated mathematical methods are needed to optimize a periodic process compared with standard methods used for finding optimal steady states.

Optimization of periodic processes has relied on simulation. Despite algorithm improvements, simulation of periodic processes is a notorious consumer of computing resources. This situation is unlikely to change. Some relief will be achieved for relatively simple systems that can be reduced to mathematical models by the use of nonlinear frequency response analysis that enables direct estimation of the periodic steady states. **Chapter 14** presents a synopsis of theory and an application example. Wider use of the technique will require more application examples. A challenge will be to find performance improvement from an experimentally determined transfer function in systems for which models are unavailable.

23.2.6 Engineering of Periodically Operated Processes

Besides identifying attractive reaction systems for which the periodic operation would be more efficient than conventional stationary processes, a series of practical problems must be considered to move this operation mode into industrial use, for example, how to efficiently integrate steady-state operating up-stream and down-stream units with a periodically operated apparatus? Large hold-up tanks are a simple solution when a plant is in a stationary operation. But, how can plant start-up and shut down be carried out? Valves are the primary means for flow switching. These must be large, fast-acting and leak-free. Such valves are expensive, require maintenance and will certainly affect process economics.

Notation

CHO	= Chinese hamster ovary
FR	= frequency response
NLFR	= non-linear frequency response
PID	= proportional-integral-derivative (controller)
SCF	= supercritical fluid
SMB	= simulated moving bed
SMBCR	= simulated moving bed chromatographic reactor
UV-LED	= ultra-violet light emitting diode

References

Abdolahi, F., Mortazavi, Y., Khodadadi, A., Hudgins, R.R., Silveston, P.L., 2005. Modeling of methane oxidative coupling under periodic operation by neural network. Chem. Eng. Technol. 28 (5), 581–586 [4].

Abdul-Kareem, H.K., Hudgins, R.R., Silveston, P.L., 1980b. Forced cycling of the catalytic oxidation of CO over a V_2O_5 catalyst. II Temperature cycling. Chem. Eng. Sci. 35, 2085–2088 [16].

Abdul-Kareem, H.K., Hudgins, R.R., Silveston, P.L., 1981. Differences in behavior in forced periodic catalytic oxidation of CO. Proc. 2nd World Congr. Chem. Eng. Montreal, Canada [3].

Abdul-Kareem, H.K., Jain, A.K., Hudgins, R.R., Silveston, P.L., 1980c. Harmonic behaviour of the rate of catalytic oxidation of CO under cycling conditions, Chem. Eng. Sci. 35, 273–282.

Abdul-Kareem, H.K., Silveston, P.L., Hudgins, R.R., 1980a. Forced cycling of the catalytic oxidation of CO over a V_2O_5 catalyst: I. Concentration cycling. Chem. Eng. Sci. 35, 2077–2084 [1] [3].

Abon, M., Bere, K.E., Delichere, P., 1997. Nature of active oxygen in the n-butane selective oxidation over well-defined V-P-O catalysts: an oxygen isotopic labelling study. Catal. Today 33, 15–23 [4].

Abulesz, E.M., Lyberatos, G., 1987. Periodic optimization of continuous microbial growth processes. Biotechnol. Bioeng. 29, 1059–1065 [10].

Abulesz, E.M., Lyberatos, G., 1989. Periodic operation of a continuous culture of Baker's yeast. Biotechnol. Bioeng. 34, 741–749 [10].

Acharya, B., Dutta, A., Basu, P., 2009. Chemical looping gasification of biomass for hydrogen-enriched gas production with in-process carbon dioxide capture. Energy Fuels 23, 5077–5083 [8].

Adànez, J., de Diego, L.F., Garcia-Labiano, F., Gayan, P., Abad, A., Palacios, J.M., 2004. Selection of oxygen carriers for chemical-looping combustion. Energy Fuels 18 (2), 371–377 [8].

Adams, W.A., Gardner, C.L., 2005. Enhancement of the performance of direct methanol fuel cells by periodic pulsed oxidation. 207th Mtg. Electrochem. Soc. (Abstract) [9].

Agar, D.W., 1999. Multifunctional reactors: old misconceptions and new dimensions. Chem. Eng. Sci. 54, 1299–1305 [18].

Agar, D.W., 2003. Dos and don'ts of adsorptive reactors, 4th Intern. Confer. Unsteady-State Processes in Catalysis, Oct. 26–29, 2003. Montreal, Canada [22].

Agar, D.W., Ruppel, W., 1988a. Extended reactor concept for dynamic deNOx design. Chem. Eng. Sci. 43, 2073–2078 [18].

Agar, D.W., Ruppel, W., 1988b. Multifunktionale reaktoren für die heterogene katalyse. Chem. Ing. Tech. 43, 731–741 [18].

Aida, T., Kawakami, R., Niiyama, H., 1997b. Periodic operation effects in selective catalytic reduction of nitrogen monoxide with propane over alumina. J. Chem. Eng. Japan 30 (2), 321–327 [6].

Aida, T., Kobayashi, R., Niiyama, H., 1997a. Periodic operation of NO-CO reaction over Pt/Al_2O_3—effect of intraparticle diffusion. Kagaku Kogaku Ron. 23, 962–968 [3].

Aida, T., Na-Ranong, D., Kobayashi, R., Niiyama, H., 1999. Effect of diffusion and adsorption-desorption on periodic operation performance of NO-CO reaction over supported noble metal catalysts. Chem. Eng. Sci. 54, 4449–4457 [3].

Aida, T., Silveston, P.L., 2005. Cyclic Separating Reactors. Blackwell Science Publishers, Oxford, U.K. [10] [18] [19] [20] [21] [22].

Aidoun, Z., Ternan, M., 2004. The unsteady state overall heat transfer coefficient in chemical heat pump reactor: the NH_3—$CoCl_2$ system. Chem. Eng. Sci. 59, 4023–4031 [8].

Aiello, R., Fiscus, J.E., zur Loye, H.C., Amaridis, M.D., 2000. Hydrogen production via the direct cracking of methane over Ni/SiO_2: catalyst deactivation and regeneration. Appl. Catal. A-Gen. 192, 227–234 [4].

Ajbar, A., 2001. Stabilization of chaotic behavior in a two-phase autocatalytic reactor. Chaos Soliton Fract. 12, 903–918 [19].

Akintoye, A., Ganetsos, G., Barker, P.E., 1990. Preparative scale chromatographic systems as combined biochemical reactor-separators in Advances in Separation Processes. Inst. Chem. Eng. Sympos. Ser. 118, 17–24 [21].

Akintoye, A., Ganetsos, G., Barker, P.E., 1991. The inversion of sucrose on a semi-continuous countercurrent chromatographic bioreactor-separator. Trans. Inst. Chem. Eng. 69C, 35–44 [21].

Aksikas, I., Winkin, J.J., Dochain, D., 2007a. Asymptotic stability of infinite-dimensional semilinear systems: application to a nonisothermal reactor. Syst. Control Lett. 56 (2), 122–132 [19].

Aksikas, I., Winkin, J.J., Dochain, D., 2007b. Optimal LQ-feedback regulation of a nonisothermal plug flow reactor model by spectral factorization. IEEE Trans. Autom. Control 52 (7), 1179–1193 [19].

Alépée, Ch., Mauer, R., Paratte, L., Vulpesu, L., Renaud, Ph., Renken, A., 2000. Fast heating and cooling for high temperature microreactors, Proc. 3rd Int. Conf. Microreaction Technology IMRET3. Springer-Verlag, Berlin, Germany, pp. 514–525 [16].

Alenazey, F., Dave, C., El-Nashaie, S.S.E.H., Susu, A.A., Adesina, A.A., 2009. Simultaneous coke reduction with improved syngas production during propane steam reforming using forced CO_2 cycling. Chem. Prod. Process Model. 4 (5). Article 6. (http://en.scientificcommons.org/48952634) [23].

Al-Harbi, M., Epling, W.S., 2009. The effects of regeneration-phase CO and/or H_2 amount on the performance of a NOX storage and reduction catalyst. Appl. Catal. B-Environ. 89, 315–325 [6].

Ali, M.A.-H., Ajbar, E.H.A., Alhumaizi, K., 2010. Control of molecular weight distribution of polyethylene in gas-phase fluidized bed reactors. Korean J. Chem. Eng. 27 (1), 364–371 [7].

Ali, M.A.-H., Stroomer, J., Betlem, B., Weickert, G., Roffel, B., 2008. Molecular weight distribution broadening of polypropylene by periodic switching of hydrogen and catalyst additions. J. Appl. Polym. Sci. 108, 2446–2457 [7].

Al-Juhani, A.A., Loughlin, K.F., 2003. Simulation of a combined isomerization reactor & pressure swing adsorption unit. Adsorption 9, 251–264 [22].

Alpay, E., Chatsiriwech, D., Kershenbaum, L.S., Hull, C.P., Kirkby, N.F., 1994. Combined reaction and separation in pressure swing processes. Chem. Eng. Sci. 49 (24B), 5845–5864 [22].

Alpay, E., Kenney, C.N., Scott, D.M., 1993. Simulation of rapid pressure swing adsorption and reaction processes. Chem. Eng. Sci. 48 (18), 3173–3186 [22].

Al-Taie, A.S., Kershenbaum, L.S., 1978. Effect of periodic operation on the selectivity of catalytic reactions. In: Weekman Jr., V.W., Luss, D. (Eds.), Chemical Reaction Engineering-Houston, ACS Symp.Series, 65, pp. 512–525 [12].

Altimari, P., Maffettone, P.L., Crescetelli, S., Russo, L., Mancusi, E., 2006. Nonlinear dynamics of a VOC combustion loop reactor. AIChE J. 52 (8), 2812–2822 [5] [21].

Altshuller, D., 1983. Design equations and transient behaviour of the counter-current moving bed chromatographic reactor. Chem. Eng. Commun. 19, 363–375 [20].

Amariglio, A., Pareja, P., Belgued, M., Amariglio, H., 1994. The possibility of obtaining appreciable yields in methane homologation through a two-step reaction at 250°C on a platinum catalyst. J. Chem. Soc. Chem. Comm., 561–562 [12].

Amariglio, H., Belgued, M., Pareja, P., Amariglio, A., 1995b. Carbon monoxide induced desorption of alkanes and alkenes up to C8 after chemisorption of methane on platinum. Catal. Lett. 31, 19–26 [12].

Amariglio, H., Pareja, P., Amariglio, A., 1995a. Periodic operation of a catalyst as a means of overcoming a thermodynamic constraint. The case of methane homologation on metals. Catal. Today 25 (2), 113–125 [12].

Amariglio, H., Rambeau, G., 1977. Kinetic manifestations of reversible changes induced by ammonia synthesis upon iron catalysts. In: Bond, G.C., Wells, P.B., Tomplins, F.C. (Eds.), Proc. 6th Int. Congr. Catal. 2, pp. 1113–1124 [2].

Amberntsson, A., Fridell, E., Skoglundh, M., 2003. Influence of platinum and rhodium composition on NOx storage and sulphur tolerance of barium based NOx storage catalysts. Appl. Catal. B-Environ. 43, 429–439 [6].

Amenomiya, Y., Birss, V.I., Goledzinowski, M., Galuszka, J., Sanger, A., 1990. Conversion of methane by oxidative coupling. Catal. Rev. 32, 163–227 [4].

Ando, T., Sumi, S., Kawate, T., Ichiharra, J., Hanafusa, T., 1984. Sonochemical Switching of Reaction Pathways in Solid Liquid 2-Phase Reactions. J. Chem. Soc. Chem. Commun. 7, 439–440 [15].

Anthony, E.J., 2008. Solid looping cycles: a new technology for coal conversion. Ind. Eng. Chem. Res. 47, 1747–1754 [1] [8].

Antonucci, F., Giordano, N., Bart, J.C.J., 1978. Chemical reactions in chromatographic columns. Dehydrogenation of ethane over cadmium-exchanged zeolite 4A. J. Chromatogr. 150, 309–317 [20].

Arcotumapathy, V., Alenazey, F., Adesina, A.A., 2011. Artificial neural network modelling of forced cycling operation between propane steam reforming and CO_2 carbon gasifier. Catal. Today 164, 275–281 [23].

Arcuri, E.J., Nicols, T.S., Brix, J.R., 1983. Thienamycin production by immobilized cells of Streptomyces cattleya in a bubble column. Biotechn. Bioeng. 28 (10), 2399–2411 [10].

Arnold lll, E.W., Sundaresan, S., 1989. Dynamics of packed-bed reactors loaded with oxide catalysts. AIChE J. 35 (5), 746–754 [4].

Aroyo, M., Tzonev, N., 2002. Pulse periodic reverse plating – new possibilities for electrodepositionof metal coatings with improved properties: Part 1. Plat. Surf. Finish. July, 48–53 [9].

Aroyo, M., Tzonev, N., 2003. Pulse periodic reverse plating – new possibilities for electrodepositionof metal coatings with improved properties: Part 2. Plat. Surf. Finish. February, 50-54 [9].

Assabumrungrat, S., Charoenseri, S., Laosiripojana, N., Kiatkittipong, W., Praserthdam, P., 2009. Effect of oxygen addition on catalytic performance of Ni/SiO_2-MgO toward carbon dioxide reforming of methane under periodic operation. Int. J. Hydrogen Energy 34, 6211–6220 [4].

Asua, J.M. (Ed.), 2007. Polymer Reaction Engineering, Blackwell Publishing Co. Oxford, U.K. [7].

Atasoy, I., Yuceer, M., Oguz-Utker, E., Berber, R., 2007. Neural network based control of the acrylonitrile polymerization process. Chem. Eng. Technol. 30 (11), 1525–1531 [7].

Aubé, F., Sapoundjiev, H., 2000. Mathematical model and numerical simulations of catalytic flow reversal reactors for industrial applications. Comput. Chem. Eng. 24, 2623–2632 [18].

Avila, A.J., 1986. Pulsed Electrodeposition of Alloys. In: Puippe, J.-C., Leaman, F. (Eds.), Theory and Practice of Pulse Plating. American Electroplaters & Surface Finishers Society, Washington, D.C. 1986 [9].

Avila, A.J., Brown, M.J., 1970. Design factors in pulse plating. Plating 57 (11), 1105–1108 [9].

Aydin, B., Fries, D., Lange, R., Larachi, F., 2006. Slow-mode induced pulsing in trickle bed reactors at elevated temperature. AIChE J. 52 (11), 3891–3901 [17].

Ayude, M.A., Cassanello, M.C., Martínez, O.M., Haure, P.M., 2005. Phenomenological approach to interpret the effect of liquid flow modulation in trickle bed reactors at the particle scale. Chem Eng. Sci. 60, 6262–6269 [17].

Ayude, M.A., Cechini, J., Cassanello, M., Martínez, O.M., Haure, P., 2008. Trickle bed reactors: effect of liquid flow modulation on catalytic activity. Chem. Eng. Sci. 63 (20), 4969–4973 [17].

Ayude, M.A., Haure, P.M., Martínez, O.M., Cassanello, M.C., 2009. Towards a comprehensive model for liquid flow modulation in trickle bed reactors. Chem. Eng. J. 154 (1–3), 162–167 [17].

Ayude, M.A., Martínez, O.M., Cassanello, M.C., 2007a. Modulation of liquid holdup along a trickle bed reactor with periodic operation 62, 6002–6014 [17].

Ayude, M.A., Rodriguez, T., Font, J., Fortuny, A., Bengoa, C., Fabregat, A., Stuber, F., 2007b. Effect of gas feed flow and gas composition modulation on activated carbon performance in phenol wet air oxidation. Chem. Eng. Sci. 62 (24), 7351–7358 [17].

Baiker, A., Bergougnan, M., 1985a. Investigation of a fixed-bed pilot plant reactor by dynamic experimentation. Part 1. Apparatus and experimental results. Can. J. Chem. Eng. 63, 138–145 [2].

Baiker, A., Bergougnan, M., 1985b. Investigation of a fixed-bed pilot plant reactor by dynamic experimentation. Part 2. Simulation of reactor behaviour. Can. J. Chem. Eng. 63, 146–154 [2].

Baiker, A., Casanova, R., Richarz, W., 1980. Transient behaviour of an adiabatic fixed-bed catalytic reactor in the range of multiple steady states. Ger. Chem. Eng. 3 (2), 112–117 [3].

Baiker, A., Richarz, W., 1976. Behavior of solid bed reactors on periodic variations of gas throughput and total pressure variations. Chem. Ing. Tech. 48 (12) 1203–1203 [2] [15].

Bailey, J.E., 1974. Periodic operation of chemical reactors: a review. Chem. Eng. Commun. 1, 111–124 [1] [10] [14] [16].

Bailey, J.E., 1977. Periodic phenomena. In: Lapidus, L., Amundson, N.R. (Eds.), Chemical Reactor Theory: A Review. Prentice-Hall, Englewood Cliffs, NJ, pp. 758–813 [1] [16] [17].

Bailey, J.E., Horn, F.J.M., 1971. Comparison between two sufficient conditions for improvement of an optimal steady state process by periodic operation. J. Optim. Theory Appl. 7, 378–384 [14].

Bailey, J.E., Horn, F.J.M., Lin, R.C., 1971. Cyclic operation of reaction systems: effects of heat and mass transfer resistance. AIChE J. 17, 818–825 [16].

Balaji, S., Fuxman, A., Lakshiminarayan, S., Forbes, J., Hayes, R., 2007. Repetitive model predicative control of a reverse flow reactor. Chem. Eng. Sci. 62, 2154–2167 [18] [19].

Balaji, S.A., Lakshminarayanan, S., 2006. Performance comparison of autothermal reactor configurations for methane combustion. Ind. Eng. Chem. Res. 45, 3880–3890 [18].

Balaji, S.A., Lakshiminarayan, S., Krantz, W.B., 2008. Scaling and sensitivity analysis of a reverse flow reactor. Chem. Eng. Sci. 63, 342–355 [18].

Balashev, A.A., Filatov, Yu.V., Kozlov, V.P., Epifanov, V.S., Popov, A.E., Smirnova, O.N., Glukhova, T.Yu., 1987. A non-stationary method of sulfur dioxide oxidation in sulfuric acid production. Khim. Prom. Chem. Ind. 780–733 (in Russian) [18].

Balasubramanian, B., Lopez Ortiz, A., Kaytakoglu, S., Harrison, D.P., 1999. Hydrogen from methane in a single-step process. Chem. Eng. Sci. 54, 3543–3552 [22].

Baltzis, B.C., Tsangaris, D.M., Wang, K.W., 1996. Utilization of substitutable substrates in a bioreactor under cycling: a study on process dynamics and optimization. Chem. Eng. Sci. 51 (15), 3801–3811 [10].

Balzhinimaev, B.S., Ivanov, A.A., Lapinaa, O.B., Mastikhin, M., Zamaraev, K.I., 1989. The mechanism of SO_2 oxidation over supported vanadium catalysts. Faraday Disc. Chem. Soc. 87 (8), 1–15 [3].

Balzhinimaev, B.S., Park, D.W., Gau, G., 1984. Non steady state oxidation of propylene on supported silver catalysts. React. Kinet. Catal. Lett. 24, 59–64 [4] [8].

Banchero, M., Manna, L., Sicardi, S., Ferri, A., 2004. Experimental investigation of fast-mode liquid modulation in a trickle-bed reactor. Chem. Eng. Sci. 59, 4149–4154 [17].

Bandermann, F., 1971. Optimierende reaktionsführung von polyreaktionen. Angew. Makromol. Chem. 18, 137–147 [7].

Bär, M., Falcke, M., Zülicke, C., Engel, H., Eiswirth, M., Ertl, G., 1992. Reaction fronts and pulses in the CO oxidation on Pt—theoretical analysis. Surf. Sci. A 270, 471–475 [3].

Barcellos, W.M., Souza, L.C.E.O., Saveliev, A.V., Kennedy, L.A., 2011. Ultra-low-emission steam boiler constituted of reciprocal flow porous burner. Exp. Therm. Fluid Sci. 35, 570–580 [18].

Bard, A.J., Faulkner, L.R., 1980. Electrochemical Methods: Fundamentals and Applications, 1st Edition. John Wiley and Sons, Somerset, New Jersey [9].

Bard, A.J., Faulkner, L.R., 2001. Electrochemical Methods: Fundamentals and Applications, second ed. John Wiley and Sons, Somerset, New Jersey [9].

Barker, P.E., Ajongwen, N.J., Shieh, M.T., Ganetsos, G., 1992c. Simulated counter-current chromatographic bioreactor-separators, In: Fundamentals of Adsorption. 17–22 May. Proc. 4th Intern. Confer. Fund. Adsorption, Kyoto, Japan, pp. 35–44 [21].

Barker, P.E., Ching, C.B., 1980. Continuous liquid chromatographic process for separation of fructose-glucose mixtures. Eur. Fed. Chem. Eng. Kemtek 5, Copenhagen [21].

Barker, P.E., Deeble, R.E., 1975. Sequential chromatographic equipment for separation of a wide range of organic mixtures. Chromatographia 8 (2), 67–79 [21].

Barker, P.E., England, K., Ganetsos, G., 1988. Fractionation of dextran using repetitive batch chromatography. J. Chem. Technol. Biot. 41, 61–68 [21].

Barker, P.E., England, K., Vlachogiannis, G., 1983. Mathematical model for the fractionation of dextran on a semi-continuous counter-current simulated moving bed chromatograph. Chem. Eng. Res. Des. 61, 241–247 [21].

Barker, P.E., Ganetsos, G., 1988. Chemical and biochemical separations using preparative and large scale batch and continuous chromatography. Sep. Purif. Methods 17 (1), 1–65 [1] [20] [21].

Barker, P.E., Ganetsos, G., Ajongwen, N.J., 1993. A novel approach to the production of clinical-grade dextran. J. Chem. Technol. Biot. 57 (1), 21–26 [21].

Barker, P.E., Ganetsos, G., Ajongwen, A., Akintoye, A., 1992a. Bioreaction-separation on continuous chromatographic systems. Chem. Eng. J. 50, B23–B28 [20] [21].

Barker, P.E., Ganetsos, G., Ajongwen, A., Akintoye, A., 1992b. Continuous chromatographic bioreaction-separation. In: Pyle, D. (Ed.), Separations for Biotechnology. Elsevier Science Publishers, London, pp. 549–557 [20] [21].

Barker, P.E., Knoechelamnn, A., Ganetsos, G., 1990. Simulated countercurrent moving chromatography used in the continuous separation of carbohydrate mixtures. Chromatographia 29, 161–166 [21].

Barker, P.E., Zafar, I., Alsop, R.M., 1987a. A novel method for the production of dextrans and fructose. In: Moody, G.W., Baker, P.B. (Eds.), Bioreactors and Biotransformations. Elsevier Science Publishers, Amsterdam, pp. 141–157 [20] [21].

Barker, P.E., Zafar, I., Alsop, R.M., 1987b. Production of dextran and fructose in a chromatographic reactor-separator. In: Verral, M.S., Hudson, M.J. (Eds.), Separations for Biotechnology, Ellis-Horwood. London (Chapter 9), pp. 127–151 [20] [21].

Barresi, A.A., Baldi, G., Fissore, D., 2007. Forced unsteady-state reactors as efficient devices for integrated processes: case histories and new perspectives. Ind. Eng. Chem. Res. 46 (25), 8693–8700 [18].

Barshad, Y., Gulari, E., 1985a. A dynamic study of CO oxidation on supported platinum. AIChE J. 31, 649–658 [3].

Barshad, Y., Gulari, E., 1985b. A novel catalytic reactor system for transient response and its use in CO oxidation on Pd/Al_2O_3. J. Catal. 94, 468–477 [3].

Barshad, Y., Zhou, X., Gulari, E., 1985. CO oxidation under transient conditions: a Fourier transform infra red transmission spectroscopy study. J. Catal. 94, 128–141 [3].

Bartholomew, C.H., 1982. Carbon deposition in steam reforming and methanation. Catal. Rev.-Sci. Engng. 24 (1), 67–112 [8].

Bartsch, S., Falkowski, J., Hofmann, H., 1989. Catalyst development for oxidative methane coupling. Catal. Today 4, 421–431 [4].

Bassett, D.W., Habgood, H.W., 1960. A gas chromatographic study of the catalytic isomerization of cyclopropane. J. Phys. Chem. 64, 769–773 [20].

Bassett, M.R., Imbihl, R., 1990. Mathematical modeling of the kinetic oscillations in the catalytic CO oxidation on Pd(110): the subsurface oxygen model. J. Chem. Phys. 93 (1), 811–821 [16].

Bauer, R., Dourdoumas, N., 2004. An approach to the control of periodically cyclic processes. Int. J. Electron. Commun. 58, 37–40 [19].

Beaudry, E.G., Dudukovic, M.P., Mills, P.L., 1987. Trickle-bed reactors: Liquid diffusional effects in a gas-limited reaction. AIChE J. 33, 1435–1447 [17].

Becker, E., Carlsson, P.-A., Skogslundh, M., 2009. Methane oxidation over alumina and ceria supported platinum. Top. Catal. 52, 1957–1961 [5].

Beden, B., Lèger, J.M., Lamy, C., 1992. Modern Aspects of Electrochemistry, 22. In: Bockris, J.O'M., Conway, B.E. (Eds.), Plenum Press, New York, p. 97 [9].

Belgued, M., Amariglio, H., Pareja, P., Amariglio, A., Saint-Just, J., 1992a. Low temperature catalytic homologation of methane on platinum, ruthenium and cobalt. Catal. Today 13, 437–445 [12].

Belgued, M., Monteverdi, S., Pareja, P., Amariglio, H., Amariglio, A., Saint-Just, J., 1992b. Homologation of methane on metallic surfaces: consideration of reaction pathways, Preprint, Div. Petrol. Chem. Am. Chem. Soc. April, 1992 Meeting, San Francisco, 37, 324–335 [12].

Belgued, M., Pareja, P., Amariglio, A., Amariglio, H., 1991. Conversion of methane into higher hydrocarbons on platinum. Nature 352, 789–790 [12].

Belkhiria, S., Meyer, T., Renken, A., 1994. Styrene maleic anhydride copolymerization in a recycle tubular reactor: reactor stability and product quality. Chem. Eng. Sci. 49 (24B), 4981–4990 [7].

Bellman, R., Bentsman, J., Meerkov, S.M., 1983. Vibrational control of systems with Arrhenius dynamics. J. Math. Anal. Appl. 91 (1), 152–191 [19].

Belo, I., Pinheiro, R., Mota, M., 2003. Fed-batch cultivation of Saccharomyces cerevisiae in a hyperbaric bioreactor. Biotechnol. Prog. 19, 665–671 [10].

Bennett, C.O., 1976. The transient method and elementary steps in heterogeneous catalysis. Catal. Rev. 13, 121–148 [13].

Bensmann, B., Petkovska, M., Vidaković, T., Hanke-Rauschenbach, R., Sundmacher, K., 2010. Nonlinear frequency response of electrochemical methanol oxidation: a theoretical analysis. J. Electrochem. Soc. 157, B1279–B1289 [14].

Ben-Tullilah, M., Alajem, E., Gal, R., Sheintuch, M., 2003a. Comparison of flow reversal and internal recirculation reactors: experiments, simulations and approximations. AIChE J. 49, 1849–1858 [5].

Ben-Tullilah, M., Alajem, E., Gal, R., Sheintuch, M., 2003b. Flow-rate effects in flow-reversal reactors: experiments, simulations and approximations. Chem. Eng. Sci. 58, 1135–1146 [18].

Bequette, B.W., 1991. Nonlinear control of chemical processes: a review. Ind. Eng. Chem. Res. 30, 1391–1413 [19].

Bergeot, G., Leinekugel-Le Cocq, D., Wolff, L., Muhr, L., Bailly, M., 2010. Intensification of paraxylene producing using a simulated moving bed reactor. Oil & Gas Sci. and Technol. - Rev. JEP Energies nouvelles 65 (5), 721–733 [21].

Berguerand, N., Lyngfelt, A., 2008. Design and operation of a 10 kWth chemical-looping combustor for solid fuels — testing with a South African coal. Fuel 87 (12), 2713–2726 [8].

Berruti, F., Hudgins, R.R., Rhodes, E., 1984. Oxidation of sulfur dioxide in a trickle-bed reactor: a study of reactor modelling. Can. J. Chem. Eng. 62, 644–650 [17].

Bhatia, S.K., 1991. Analysis of catalytic reactor operation with periodic flow reversals. Chem. Eng. Sci. 46, 361–367 [18].

Bi, Y.-L., Zhen, K.-J., Jiang, Y.-T., Teng, C.-W., Yang, X.-G., 1988. Catalytic oxidative coupling of methane over alkali, alkaline earth and rare earth metal oxides. Appl. Catal. 39 (1-2), 185–190 [4].

Bielinski, J., Ibl, N., Angerer, H., 1979. Porosity of hard gold coatings. Oberfläche Surf. 20 (1), 2–4 [9].

Bilimoria, M.R., Bailey, J.E., 1978. Dynamic studies of acetylene hydrogenation on nickel catalysts. In: Weekman Jr., V.W., Luss, D. (Eds.), Chemical Reaction Engineering—Houston. ACS Symp. Series, 65, pp. 526–536 [12].

Bixler, G.H., Coberly, C.W., 1952. Wulff process acetylene. Ind. Eng. Chem. 45, 2596–2606 [18].

Bjorklund, M.C., Carr, R.W., 1995. The simulated countercurrent moving bed chromatographic reactor: a catalytic and separative reactor. Catal. Today 25, 159–168 [21].

Bjorklund, M.C., Carr, R.W., 2002. Enhanced methanol yields from the direct partial oxidation of methane in a simulated countercurrent moving bed chromatographic reactor. Ind. Eng. Chem. Res. 41, 6528–6536 [21].

Blanks, R.F., Wittrig, T.S., Petterson, D.A., 1990. Bidirectional adiabatic synthesis gas generator. Chem. Eng. Sci. 45, 2407–2413 [18].

Bobrova, L.N., Slavinskaya, E.M., Noskov, A.S., Matros, Yu.Sh., 1988. Unsteady-state performance of NOx catalytic reduction by NH_3. React. Kinet. Catal. Lett. 37, 267–272 [18].

Bockris, J.O'M., Reddy, A.K.N., 1998a. Modern Electrochemistry: vol. 1, second ed. Plenum Publishers, New York, New York [9].

Bockris, J.O'M., Reddy, A.K.N., 1998b. Modern Electrochemistry: vol. 2B, second ed. Kluwer Academic/Plenum Publishers, Dordrecht, The Netherlands [9].

Bodega, P.S., Alonso, S., Rotermund, H.H., 2009. Effects of external global noise on the catalytic CO oxidation on Pt(110). J. Chem. Phys. 130, 084704 [3].

Bodega, P.S., Kaira, P., Beta, C., et al., 2007. High frequency periodic forcing of the oscillatory catalytic CO oxidation on Pt (110). New J. Phys. 9, 61–79 [3].

Boelhouwer, J.G., 2001. Nonsteady operation of trickle-bed reactors: hydrodynamics, mass and heat transfer. Ph.D. Thesis. Technische Universiteit Eindhoven, Eindhoven, The Netherlands [17].

Boelhouwer, J.G., Piepers, H.W., Drinkenburg, A.A.H., 1999. Enlargement of the pulsing flow regime by periodic operation of a trickle-bed reactor. Chem. Eng. Sci. 54, 4661–4667 [17].

Boelhouwer, J.G., Piepers, H.W., Drinkenburg, A.A.H., 2001a. The induction of pulses in trickle-bed reactors by cycling the liquid feed. Chem. Eng. Sci. 56, 2605–2614 [17].

Boelhouwer, J.G., Piepers, H.W., Drinkenburg, A.A.H., 2001b. Non-steady state operation of trickle-bed reactors. Stud. Surf. Sci. Catal. 133, 231–238 [17].

Boelhouwer, J.G., Piepers, H.W., Drinkenburg, A.A.H., 2002a. Liquid-induced pulsing flow in trickle-bed reactors. Chem. Eng. Sci. 57 (16), 3387–3399 [1] [17].

Boelhouwer, J.G., Piepers, H.W., Drinkenburg, A.A.H., 2002b. Non-steady state operation of trickle-bed reactors: nature and characteristics of pulsing flow in trickle-bed reactors. Chem. Eng. Sci. 57 (22–23), 4865–4876 [17].

Boreskov, G.K., Bunimovich, G.A., Matros, Yu.Sh., Ivanov, A.A., 1982a. Catalytic processes under non-steady state conditions. II. Switching the direction for the feed of the reaction mixture to the catalyst bed, experimental results. Kinet. Katal. 23, 402–406 [18].

Boreskov, G.K., Bunimovich, G.A., Matros, Yu.Sh., Zolotarski, Y.S., Kiselev, O.V., 1983. Cyclic regimes in fixed bed reactor operated at flow reversal of gas mixture. Dokl. Acad. Nauk SSR (Reports of the USSR Acad. Sci. Chem. Technol.) 268, 646–650 [18].

Boreskov, G.K., Matros, Yu.Sh., 1983a. Unsteady-state performance of heterogeneous catalytic reactions. Catal. Rev. 25 (4), 551–590 [1] [18].

Boreskov, G.K., Matros, Yu.Sh., 1983b. Flow reversal of reaction mixture in a fixed bed—a way to increase the efficiency of chemical processes. Appl. Catal. 5, 337–343 [18].

Boreskov, G.K., Matros, Yu.Sh., Bunimovich, G.A., Balashov, A.A., Filatov, Yu.V., Kozlov, V.P., Khripunov, N.F., Epifanov, V.S., Popov, A.E., Soloviev, Yu.M., Safonov, A.V., Smirnova, O.N., 1982b. Pilot study of unsteady-state method of sulfuric acid oxidation. In: Unsteady-State Processes in Chemical Reactors, Institute of Catalysis, Siberian Branch of the USSR Academy of Sciences. Novosibirsk, Russia [18].

Boreskov, G.K., Matros, Yu.Sh., Ivanov, A.G., 1985. Utilization of heat energy of catalytic combustion of low-calorific value gases and fuels in the reactor with periodic switching of the direction of gas. Dokl. Chem. Technol. 288, 429–433 [18].

Boreskov, G.K., Matros, Yu.Sh., Ivanov, A.G., 1986. Utilization of heat energy of catalytic combustion of low calorie gaseous fuel mixtures by reversing the direction of their input. Dokl. Chem. Technol. 286–288, 55–59 [18].

Boreskov, G.K., Matros, Yu.Sh., Kiselev, O.V., Bunimovich, G.A., 1977. Realization of heterogeneous catalytic processes under unsteady conditions (Transl. from Russian). Dokl. Acad. Nauk USSR 237, 160–163 [1] [18].

Boreskov, G.K., Matros, Yu.Sh., Kiselev, O.V., 1979. Catalytic processes under nonsteady state conditions. I. Heat front in fixed catalyst bed. Kinet. Katal. 20, 773–780 [18].

Boreskov, G.K., Matros, Yu.Sh., Lugovskoy, V.I., Bunimovich, G.A., Puzhilova, V.I., 1984. Unsteady-state process of complete oxidation in reactor with fixed catalyst bed. Teor. Osn. Khim. Technol. 18, 328–334 [18].

Borges da Silva, E.A., Rodrigues, A.E., 2008. Design methodology and performance analysis of a pseudo-simulated moving bed for ternary separation. Sep. Sci. Technol. 43, 533–566 [23].

Borio, D.O., Schbib, N.S., Gatica, J.E., 1999. Reversal flow in fixed bed reactors operated under reaction-regeneration cycles. Chem. Eng. Sci. 54, 1313–1318 [18].

Borremans, D., Rode, S., Carré, P., Wild, G., 2003. The influence of periodic operation on the effective radial thermal conductivity in trickle bed reactors. Can. J. Chem. Eng. 81, 795–801 [17].

Borremans, D., Rode, S., Wild, G., 2004. Liquid flow distribution and particle-fluid heat transfer in trickle-bed reactors: the influence of periodic operation. Chem. Eng. Proc. 43, 1403–1410 [17].

Borremans, D., Rode, S., Wild, G., 2007. Cyclic variation of the liquid flow residence time in periodically operated trickle bed reactors. Chem. Eng. Sci. 62, 1230–1238 [1] [17].

Borzani, W., Gregori, R.E., Vairo, M.L.R., 1976. Response of a continuous anaerobic culture to periodic variations of the feeding mash concentration. Biotechnol. Bioeng. 18, 623–632 [10].

Bos, A.N.R., Lange, J.-P., Kabra, G., 2007. A novel reverse flow reactor with integrated separation. Chem. Eng. Sci. 62, 5661–5662 [18].

Botar-Jid, C.C., Agachi, P.S., Fissore, D., 2009. Comparison of reverse flow and counter current reactors in the case of selective reduction of NOx. Comput. Chem. Eng. 33, 782–787 [18].

Bouzek, K., Roušar, I., 1993. The study of electrochemical preparation of ferrate (VI) using alternating current superimposed on the direct current: Frequency dependence of current yields. Electrochim. Acta 38 (13), 1717–1720 [9].

Bouzek, K., Roušar, I., 1996. Influence of anode material on current yields during ferrate (VI) production by anodic iron dissolution. Part I: current efficiency during anodic dissolution of grey cast iron to ferrate (VI) in concentrated alkali hydroxide solutions. J. Appl. Electrochem. 26, 919–923 [9].

Bradley, J.-C., Babu, S., Mittal, A., Ndungu, P., Carroll, B., Samuel, B., 2001. Pulsed bipolar electrodeposition of palladium onto graphite powder. J. Electrochem. Soc. 148 (9), C647–C651 [9].

Brand, O., Fedder, G.K., Hierold, C., Korvink, J.G., Tabata, O., Kockmann, N. (Eds.), 2006, Micro Process Engineering, Advanced Micro and Nano Processes, vol. 5. Wiley-VCH, Weinheim, Germany [16].

Brandner, J.J., 2003. Entwicklung von Mikrostrukturreaktoren zum thermisch instationären Betrieb chemischer Reaktionen. Forschungszentrum Karlsruhe, Wissenschaftliche Berichte FZKA 6891, Dissert. Universität Karlsruhe (TH) [16].

Brandner, J.J., 2008. Micro Process Engineering: Effects of Miniaturization for a new Technology. Habilitation Thesis. Technical University Dresden, Department of Mechanical Engineering, Institute for Process Engineering and Environmental Engineering, Dec. 2008 [16].

Brandner, J.J., Emig, G., Liauw, M.A., Schubert, K., 2002. Fast temperature cycling with microstructure devices. Proc. 6th Int. Conf. on Microreaction Technology IMRET 6, Atlanta, GA, USA, pp. 275–280 [16].

Brandner, J.J., Emig, G., Liauw, M.A., Schubert, K., 2003. Fast temperature cycling of CO oxidation in microstructure devices. Proc. 7th Int. Conf. on Microreaction Technology IMRET 7, Lausanne, Switzerland, pp. 30–32 [16].

Brandner, J.J., Emig, G., Liauw, M.A., Schubert, K., 2004. Fast temperature cycling in microstructure devices. Chem. Eng. J. 101 (1–3), 217–224 [1] [16].

Brandner, J.J., Fichtner, M., Liauw, M.A., Emig, G., Schubert, K., 2001. A new microstructure device for fast temperature cycling. Proc. 5th Int. Conf. on Microreaction Technology IMRET 5, Springer-Verlag, Berlin, Germany, pp. 164–174 [16].

Brazdil, J.F., Suresh, D.D., Grasselli, R.K., 1980. Redox kinetics of bismuth molybdate ammoxidation catalysts. J. Catal. 66, 347–367 [4].

Brétagnol, F., Lejeune, M., Papadopoulou-Bouraoui, A., Hasiwa, M., Rauscher, H., Ceccone, G., Colpo, P., Rossi, F., 2006. Fouling and non-fouling surfaces produced by plasma polymerization of ethylene oxide monomer. Acta Biomater. 2, 165–172 [7].

Breed, A., Doherty, M.F., Gadewar, S., Grosso, P., Lorkovic, I.M., McFarland, E.W., Weiss, M.J., 2005. Natural gas conversion to liquid fuels in a zone reactor. Catal. Today 106, 301–304 [18].

Briggs, J.P., Hudgins, R.R., Silveston, P.L., 1977. Composition cycling of an SO_2 oxidation reactor. Chem. Eng. Sci. 32, 1087–1092 [3].

Briggs, J.P., Kane, D.M., Hudgins, R.R., Silveston, P.L., 1978. Reduction of SO_2 emissions from an SO_2 converter by means of feed composition cycling to the final stage. In: Moo Young, M.W., Farquhar, G. (Eds.), Proc. 1st Int. Waste Treatment and Utilization Conf. Univ. Waterloo, Waterloo, ON, Canada, pp. 521–533 [3].

Brinkmann, M., Barresi, A.A., Vanni, M., Baldi, G., 1999. Unsteady-state treatment of very lean waste gases in a network of catalytic burners. Catal. Today 47 (1–4), 263–277 [5] [21].

Brkljac, B., Bludowsky, T., Dietrich, W., Grünewald, M., Agar, D., 2007. Modelling of unsteady-state hydrodynamics in periodically operated trickle-bed reactors: influence of the liquid-phase physical properties. Chem. Eng. Sci. 62, 7011–7019 [1] [17].

Brooks, K.P., Rassat, S.D., Wegeng, R.S., Stenkamp, V.S., Tegrotenhuis, W.E., Caldwell, D.D., 2002. Component development for a microchannel in situ propellant production system. Proc. 6th Int. Conf. on Microreaction Technology IMRET 6, Atlanta, GA, USA, pp. 219–229 [16].

Broughton, D., Gerhold, C., 1961. Continuous sorption process employing fixed bed of sorbent and moving inlets and outlets, US patent 2,985,589 [14].

Brown, T.A., Dennis, J.S., Scott, S.A., Davidson, J.F., Hayhurst, A.N., 2010. Gasification and chemical looping combustion of a lignite char in a fluidized bed of iron oxide. Energy Fuels 24, 3034–3048 [8].

Bruns, D.D., Bailey, J.E., 1977. Nonlinear feedback control for operating a nonisothermal CSTR near an unstable steady state. Chem. Eng. Sci. 32, 257–264 [19].

Budhi, Y.W., Jaree, A., Hoebink, J.H.B.J., Schouten, J.C., 2004. Simulation of reverse flow operation for manipulation of catalyst surface coverage in the selective oxidation of ammonia. Chem. Eng. Sci. 59, 4125–4135 [18].

Budman, H., Kzyonsek, M., Silveston, P.L., 1996. Control of a nonadiabatic packed bed reactor under periodic flow reversal. Can. J. Chem. Eng. 74, 751–759 [18] [19].

Budman, H., Silveston, P.L., 2008. Control of periodically operated reactors. Chem. Eng. Sci. 63, 4942–4954 [18].

Buechler, K.J., Nam, C.H., Zawistowski, T.M., Noble, R.D., Koval, C.A., 1999. Design and evaluation of a novel-controlled periodic illumination reactor to study photocatalysis. Ind. Chem. Eng. Res. 38, 1258–1263 [11].

Bulushev, D.A., Ivanov, E.A., Reshetnikov, S.I., Renken, A., 2005. Transient kinetics of toluene interaction with V/Ti-oxides in anaerobic conditions. Chem. Eng. J. 107 (1–3), 147–155. SI [4] [8].

Bulushev, D.A., Kiwi-Minsker, L., Renken, A., 2000a. Vanadia/titania catalysts for gas phase partial toluene oxidation: spectroscopic characterization and transient kinetic study. Catal. Today 57 (3–4), 231–239 [4].

Bulushev, D.A., Kiwi-Minsker, L., Renken, A., 2000b. Transient kinetics of toluene partial oxidation over V/Ti oxide catalysts. Catal. Today 61 (1–4), 271–277 [4].

Bulushev, D.A., Kiwi-Minsker, L., Zaikovskii, V.I., Lapina, O.P., Ivanov, E.A., Reshetnikov, S.I., Renken, A., 2000c. Effect of potassium doping on the structural and catalytic properties of V/Ti-oxide in selective toluene oxidation. Appl. Catal. A-Gen. 202, 243–250 [8].

Bulushev, D.A., Rainone, F., Kiwi-Minsker, L., 2004. Partial oxidation of toluene to benzaldehyde and benzoic acid over model vanadia/titania catalysts: role of vanadia species. Catal. Today 96 (4), 195–203 [4] [8].

Bunimovich, G.A., Matros, Yu.Sh., 2010a. Saving energy in regenerative oxidizers. Chem. Eng. 117 (3), 26–32 [18].

Bunimovich, G.A., Matros, Yu.Sh., 2010b. Reversed-flow reactors: potential and realized, key-note presentation at XIX International Conference on Chemical Reactors "CHEMREACTOR-19." Austria, Vienna. 6 Sept. 2010 [18].

Bunimovich, G.A., Sapundjiev, H., Matros, Yu.Sh., Drobyshevich, V.I., Yaushva, L.Ya, Elenkov, D., 1990b. Mathematical simulation of temperature inhomogeneities in a flow reversal reactor. In: Proc. Intern. Conf. Unsteady State Processes in Catalysis, Proc. Novosibirsk, USSR, VSP Publishers, Utrecht, The Netherlands, 1990 [18].

Bunimovich, G.A., Strots, V.O., Goldman, O.V., 1990a. Theory and industrial applications of SO_2 oxidation reverse process for sulfuric acid production, Proc. Int. Conf. Unsteady-State Processes in Catalysis, Novosibirsk, USSR. VSP Publishers. Utrecht, The Netherlands, pp. 7–24 [1] [18].

Bunimovich, G.A., Vernikovskaya, N.V., Strots, V.O., Matros, Yu.Sh., Balzhinimaev, B.S., 1995. SO_2 oxidation in a reverse-flow reactor: influence of vanadium catalyst dynamic properties. Chem. Eng. Sci. 50, 565–580 [18].

Burghardt, A., Berezowski, M., Jacobsen, E.W., 1999. Approximate characteristics of a moving temperature front in a fixed bed catalytic reactor. Chem. Eng. Sci. 38, 19–34 [18].

Bussard, A., Dooley, K.M., 2008. Polymer hydrogenation in pulsed flow systems. AIChE J. 54 (4), 1064–1072 [7].

Butler, M., 2007. Cell Culture and Upstream Processing. Taylor & Francis Science Publishers, New York [10].

Çinar, A., Deng, J., Meerkov, S.M., Shu, X., 1987a. Vibrational control of an exothermic reaction in a CSTR: theory and experiment. AIChE J. 33, 353–365 [19].

Çinar, A., Deng, J., Meerkov, S.M., Shu, X., 1987c. Vibrational stabilization of a chemical reactor: an experimental study. IEEE Trans. Autom. Control, AC-32, 348–352 [19].

Çinar, A., Rigopoulos, K., Shu, X., Meerkov, S.M., 1987b. Vibrational control of chemical reactors: stabilization and improvement in an exothermic CSTR. Chem. Eng. Commun. 59, 299–308 [19].

Cail, R., Rickard, P.A.D., Rogers, P.L., 1979. Maximization of Candida-Utilis cytochromes by pulse additions of ethanol to continuous cultures. Biotechn. Bioeng. 21 (5), 803–820 [10].

Callahan, J.L., Grasselli, R.K., Milberger, E.C., Strecker, H.A., 1970. Oxidation and ammoxidation of propylene over bismuth molybdate catalyst. Ind. Eng. Chem. Prod. Res. Dev. 9 (2), 134–142 [1] [4].

Camacho, E.F., Bordons, C., 1999. Model Predictive Control, second ed. Springer Verlag, Berlin, Germany [19].

Cameron, F., 1993. Cottrell: Samaritan of Science. The biography of scientist and inventor F.G. Cottrell. Reprint Edition Published by Research Corporation Arizona, Tucson. 414 [18].

Cao, Y., Casenas, B., Pan, W.P., 2006. Investigation of chemical looping combustion by solid fuels: 2. redox reaction kinetics and product characterization with coal, biomass and solid waste as solid fuels and CuO as an oxygen carrier. Energy Fuels 20 (5), 1845–1854 [8].

Cao, Y., Pan, W.P., 2006. Investigation of chemical looping combustion of solid fuels. 1. Process analysis. Energy Fuels 20 (5), 1836–1844 [8].

Capsaskis, S.C., Kenney, C.N., 1986. Subharmonic response of a heterogeneous catalytic oscillator, the Cantabrator, to a periodic input. J. Phys. Chem. 90 (19), 4631–4637 [3].

Caram, H.S., Viecco, G.A., 2004. Analysis of the reverse flow chromatographic reactor. AIChE J. 50 (9), 2266–2275 [20].

Cardinal, L.C., Stenstrom, M.K., 1991. Enhanced biodegradation of polyaromatic hydrocarbons in the activated sludge process. Res. J. Water Environ. Fed. 63, 950–957 [10].

Carlsson, P.-A., Fridell, E., Skogslundh, M., 2007. Methane oxidation over Pt/Al_2O_3 and Pd/Al_2O_3 catalysts under transient conditions. Catal. Lett. 115 (1-2), 1–7 [5].

Carlsson, P.-A., Mollner, S., Arnby, K., Skoglundh, M., 2004. Effect of periodic operation on low-temperature activity for propane oxidation over Pt/Al_2O_3 catalysts. Chem. Eng. Sci. 59, 4313–4323 [6] [16].

Carta, G., 1991. Simultaneous reaction and chromatography. In: Costa, C.A., Cabral, J.S. (Eds.), Chromatographic and Membrane Processes in Biotechnology. Kluwer Academic Publishers, The Netherlands, pp. 429–442 [20].

Cartwright, T., 1994. Animal Cells as Bioreactors. Cambridge University Press, Cambridge U.K. [10].

Carvill, B.T., Hufton, J.R., Anand, M., Sircar, S., 1996. Sorption-enhanced reaction process. AIChE J. 42 (10), 2765–2772 [22].

Casanova, R., Baiker, A., Richarz, W., 1979. Dynamisches verhalten eines festbettreaktors im bereich instabiler betriebszustände. Chem. Ing. Tech. 51 (10), 978–979 [3].

Castellari, A.T., Haure, P.M., 1995. Experimental study of the periodic operation of a trickle bed reactor. AIChE J. 41 (6), 1593–1597 [17].

Cavallotti, P.L., Vallauri, R., Vicenzo, A., 2000. Electrodeposition of Cu/CoNi(Cu) GMR superlattices, Proc. 5th Intern. Pulse Plating Symp., Amer. Electroplater & Surface Finishers Soc. Florida, Orlando. 109–117 [9].

Chakrabarty, T., Hudgins, R.R., Silveston, P.L., 1982. Influence of diluent gases on CO oxidation over platinum-alumina catalyst. J. Catal. 77, 527–538 [3].

Chalakov, L., Rihko-Struckmann, L.K., Munder, B., Sundmacher, K., 2007. Reaction induced current generation by butane oxidation in high temperature electrochemical membrane reactor. Chem. Eng. J. 131, 15–22 [4].

Chan, F.L., Keith, J.M., 2006. Designing reverse-flow packed bed reactors for stable treatment of volatile organic compounds. J. Environ. Manag. 78, 223–231 [18].

Chanchlani, K.G., Hudgins, R.R., Silveston, P.L., 1992. Methanol synthesis from H_2, CO and CO_2 over Cu/ZnO catalysts. J. Catal. 136, 59–75 [1] [2] [12].

Chanchlani, K.G., Hudgins, R.R., Silveston, P.L., 1994. Methanol synthesis under periodic operation: An experimental investigation. Can. J. Chem. Eng. 72, 657–671 [1] [2] [12].

Chang, K.S., Bankoff, S.G., 1968. Oscillatory operation of jacketed tubular reactors. Ind. Eng. Chem. Fund. 7, 633–639 [16].

Chaouki, J., Gui, G., Sapundzhiev, C., Kusohorsky, D., Klvana, D., 1994. Combustion of methane in a cyclic catalytic reactor. Ind. Eng. Chem. Res. 33, 2957–2963 [18].

Charon, O., Jouvard, D., Toledo, E., Genies, B., 1991. New technique for NOx emission reduction, Preprint 1st Intern. Conf. Combustion Technologies for a Clean Environment. Villamoura, Portugal Sept. 1991 [5].

Chatsiriwech, D., Alpay, E., Kershenbaum, L.S., Hull, C.P., Kirkby, N.F., 1994. Enhancement of catalytic reactions by pressure swing adsorption. Catal. Today 20, 351–366 [22].

Chatsiriwech, D., Kershenbaum, L.S., Kirkby, N.F., 1992. The enhancement of catalytic reactor performance by pressure swing operation, Internal Report. Imperial College, London, England [22].

Chen, B.-H., Dai, Q.-L., Lu, D.-W., 1996. Development and modeling of a loop fluidized bed reactor with baffle for propylene ammoxidation. Chem. Eng. Sci. 51, 2983–2988 [4].

Chen, B.-H., Huang, X.-F., Li, C.-Y., Liang, R.-Z., Zhao, B.-R., 2002. Mass spectrometric studies of selective oxidation of n-butane over a VPO catalyst. Chin. J. Chem. Eng. 10, 177–182 [4].

Chen, H.T., Kuan, C.N., Setthachayanon, S., Chartier, P.A., 1980. Photopolymerization in a continuous stirred-tank reactor-experiment. AIChE J. 26, 672–675 [7].

Chen, L.Z., Nguang, S.K., 2002. On-line identification and optimization of feed rate profiles for high productivity fed-batch culture of hybridoma cells using genetic algorithms. ISA Trans. 41, 409–441 [10].

Chen, L.Z., Nguang, S.K., Chen, X.D., Li, X.M., 2004. Modelling and optimization of fed-batch fermentation processes using dynamic neural networks and genetic algorithms. Biochem. Eng. J. 22, 51–61 [10].

Cheng, Y.S., Alpay, E., Kershenbaum, L.S., 1998b. Simulation and optimization of a rapid pressure swing reactor. Comput. Chem. Eng. 22, S45–S52 [22].

Cheng, C.W., Cheng, F.T., Man, H.C., 1998a. Improvement of protective coating on Nd-Fe-B magnet by pulse nickel plating. J. Appl. Phys. 83 (11), 6417–6419 [9].

Cheng, W.-H., Chou, M.-Sh., Lee, W.-Sh., Huang, B.-J., 2002. Applications of low-temperature regenerative thermal oxidizers to treat volatile organic compounds. J. Environ. Eng. April, 313-319 [18].

Cheng, S., Chuang, K.T., 1992. Simultaneous methanol removal and destruction from wastewater in a trickle-bed reactor. Can. J. Chem. Eng. 70, 727–733 [17].

Chevalot, I., Visvikis, A., Nabet, P., Engasser, J.M., Marc, A., 1994. Production of a membrane-bound protein, the human gamma-glutamyl transferase, by CHO cells cultivated on microcarriers, in aggregates and in suspension. Cytotechn. 16, 121–129 [10].

Chiao, L., Rinker, R.G., 1989. A kinetic study of ammonia synthesis: modeling high pressure, steady state forced-cycling behavior. Chem. Eng. Sci. 44, 9–19 [2] [13].

Chiao, L., Zack, F.K., Thullie, J., Rinker, R.G., 1987. Concentration forcing in ammonia synthesis: plug flow experiments at high temperature and pressure. Chem. Eng. Comm. 49, 273–289 [2] [13].

Ching, C.B., Lu, Z.P., 1997. Simulated moving bed reactor: application in bioreaction and separation. Ind. Eng. Chem. Res. 36, 152–159 [21].

Cho, B.K., 1983. Dynamic behavior of a single catalyst pellet. I. Symmetric concentration cycling during CO oxidation over Pt/Al_2O_3. Ind. Eng. Chem. Fund. 22, 410–420 [3].

Cho, B.K., 1994. Mechanistic importance of intermediate N_2O + CO reaction in overall NO + CO reaction system. J. Catal. 148, 697–708 [3].

Cho, B.K., Aris, R., Carr, R.W., 1982. The mathematical theory of a countercurrent catalytic reactor. Proc. Roy. Soc. Lond. A383, 147–189 [1] [20].

Cho, B.K., Carr, R.W., Aris, R., 1980a. A new continuous flow reactor for simultaneous reaction and separation. Sep. Sci. Technol. 15, 679–696 [20].

Cho, B.K., Carr, R.W., Aris, R., 1980b. A continuous chromatographic reactor. Chem. Eng. Sci. 35, 74–81 [20].

Cho, B.K., Shanks, B.H., Bailey, J.E., 1989. Kinetics of NO reduction by CO over supported rhodium catalysts: isotopic cycling experiments. J. Catal. 115, 486–499 [3].

Cho, B.K., West, L.A., 1986. Cyclic operation of Pt/Al_2O_3 catalysts for CO oxidation. Ind. Eng. Chem. Fund. 25 (1), 158–164 [3] [20].

Chohji, T., Sawada, T., Kuno, S., 1983. Effect of temperature shift on growth rate of *Escherichia coli*. Biotechnol. Bioeng. 25, 2991–3003 [23].

Choi, K.Y., Ray, W.H., 1985. Dynamic behavior of fluidized bed reactors for solid catalysed gas phase olefin polymerization. Chem. Eng. Sci. 40 (12), 2261–2279 [7].

Choi, B.-S., Yi, J., 2000. Simulation and optimization on the regenerative thermal oxidation of volatile organic compounds. Chem. Eng. J. 76, 103–114 [18].

Christofides, P.D., 2000. Nonlinear and Robust Control of PDE Systems. Birkhauser Publishing, Boston, Massachusetts [19].

Chu, C., Tsang, L.C., 1971. Behavior of a chromatographic reactor. Ind. Eng. Chem. Proc. Des. Dev. 10 (1), 47–53 [20].

Chuang, K.T., Zhou, B., Tong, S.M., 1994. , kinetics and mechanism of catalytic-oxidation of formaldehyde over hydrophobic catalysts. Ind. Eng. Chem. Res. 33 (7), 1680–1686 [17].

Chudoba, J., Grau, P., Ottova, V., 1973. Control of activated sludge filamentous bulking: selection of microorganisms by means of a selector. Water Res. 7, 1384–1406 [10].

Chumakova, N.A., Matros, Yu.Sh., 1991. Isolating laminar-flow regimes with periodic sweep reversal in a fixed catalyst layer. Theor. Found. Chem. Eng. 25, 304–308 [18].

Chumakova, N.A., Zolotarskii, I.A., 1990. Reverse-flow operation of a fixed-bed catalytic reactor. Analysis of relaxed steady-state regimes. In: Matros, Yu.Sh. (Ed.), Unsteady State Processes in Catalysis, VPS, Utrecht, The Netherlands [18].

Ciambelli, P., Cimino, S., Lasorella, G., Lisi, L., De Rossi, S., Faticanti, M., Minelli, G., Porta, P., 2002. CO oxidation and methane combustion on $LaAl_{1-x} Fe_xO_3$ perovskites solid solutions. Appl. Catal. B-Environ. 37, 231–241 [4].

Cittadini, M., Vanni, M., Barresi, A.A., 2002. Transient behavior and start up of periodic flow reversal reactors for catalytic decontamination of waste gases. Chem. Eng. Process. 41, 437–443 [18].

Cittadini, M., Vanni, M., Barresi, A.A., Baldi, G., 2001. Simplified procedure for design of catalytic combustors with periodic flow reversal. Chem. Eng. Process. 40, 255–262 [18].

Clarke, C., Doolan, P., Barron, N., Meleady, P., O'Sullivan, F., Gammel, I.P., Melville, M., Leonard, M., Clynes, M., 2011. Predicting cell specific productivity from CHO gene expression. J. Biotechnol. 151, 159–161 [23].

Claybaugh, B.E., Griffon, J.R., Watson, A.T., 1969. Process for broadening the MWD in polymers, U.S. Patent 3,472,829 [7].

Collins, N.R., Twigg, M.V., 2007. Three-way catalyst emission control technologies for spark-ignition engines – recent trends and future developments. Top. Catal. 42-43, 323–332 [6].

Constantinides, A., Mehta, N., 1991. Periodic operation of immobilized live cell bioreactor for the production of candicidin. Biotechnol. Bioeng. 37, 1010–1020 [10].

Contarin, F., Saveliev, A.V., Fridman, A.A., Kennedy, L.A., 2003. A reciprocal flow filtration combustor with embedded heat exchangers: numerical study. Int. J. Heat Mass Tran. 46, 949–961 [18].

Continillo, G., Grabski, A., Mancusi, E., Russo, L., 2006. Bifurcation analysis of a periodically forced pair of tubular catalytic combustors. Combust. Theor. Model. 10 (6), 1023–1035 [5] [21].

Contractor, R.M., 1987. Improved vapor phase catalytic oxidation of butane to maleic anhydride, U.S. Patent 4,668,802 (May 26) [4].

Contractor, R.M., 1999. DuPont's CFB technology for maleic anhydride. Chem. Eng. Sci. 54, 5627–5632 [4].

Contractor, R.M., Bergna, H.E., Horowitz, H.S., Blackstone, C.M., Chowdhry, U., Sleight, A.W., 1988. Butane oxidation to maleic anhydride in a recirculating solids reactor. In: Ward, J.W. (Ed.), Proc. 10th North American Meeting of the Catalysis Society. Elsevier Science Publishers B.V., Amsterdam, pp. 645–654 [4] [8].

Contractor, R.M., Bergna, H.E., Horowitz, H.S., Blackstone, C.M., Malone, B., Torardi, C.C., Griffiths, B., Chowdhry, U., Sleight, A.W., 1987. Butane oxidation to maleic anhydride over vanadium phosphate catalysts. Catal. Today 1, 49–58 [1] [4].

Contractor, R.M., Ebner, J., Mummey, M.J., 1990. Butane oxidation in a transport bed reactor—redox characteristics of the vanadium phosphorus oxide catalyst. In: Centi, G., Trifiro F., (Eds.), New Developments in Selective Oxidation. Elsevier Science Publishers B.V., Amsterdam, pp. 553–562 [4] [8].

Contractor, R.M., Garnett, D.I., Horowitz, H.S., Bergna, H.E., Patience, G.S., Schwartz, J.T., Sisler, G.M., 1994. A new commercial scale process for n-butane oxidation to maleic anhydride using a circulating fluidized-bed reactor. In: Corberan, V.C., Bellon, S.V. (Eds.), New Developments in Selective Oxidation II. Stud. Surf. Sci., 82. Elsevier Science Publishers B.V., Amsterdam, pp. 233–242 [4].

Contractor, R.M., Sleight, A.W., 1988. Selective oxidation in a riser-reactor. Catal. Today 3, 175–184 [4].

Cooper, B.J., Evans, W.D.J., Harrison, B., 1987. Aspects of automotive catalyst preparation, performance and durability. In: Crucq, A., Frennet, A. (Eds.), Catalysis and Automotive Pollution Control. Stud. Surf. Sci. Elsevier Science Publ, Amsterdam, pp. 117–141 [6].

Cordoba, A., Lemos, M.C., Jimenez-Morales, F., 2006. Periodical forcing for the control of chaos in a chemical reaction. J. Chem. Phys. 124 article no. 014707 [19].

Cordova, V.R., Gau, G., 1983. Redox kinetics of benzene oxidation to maleic anhydride. Can. J. Chem. Eng. 61, 200–207 [4] [8].

Corliss, J.M., Putnam, A.A., Murphy, M.J., Locklin, D.W., 1984. NOx Emissions from Several Pulse Combustors. ASME Paper 85-JPGC-APC-2, 105th ASME Winter Annual Meeting, New Orleans [5].

Cottrell, F.G., 1938. Purifying gases and apparatus therefore. US Patent 121, 733. US Patent Office, Washington, DC. [18].

Couso, D.A., Alassia, L.M., Meira, G.R., 1985. Molecular weight distribution control in a semi batch "living" anionic polymerization. 1. Theoretical study. J. Appl. Poly. Sci. 30 (8), 3249–3265 (7).

Couso, D.A., Meira, G.R., 1984. Living anionic polymerizations in a continuous stirred-tank reactor under sustained feed oscillations. Polymer Eng. Sci. 24 (6), 391–397 [7].

Creaser, D., Andersson, B., Hudgins, R.R., Silveston, P.L., 1999a. Cyclic operation of the oxidative dehydrogenation of propane. Chem. Eng. Sci. 54, 4437–4448 [4] [8] [12] [13].

Creaser, D., Andersson, B., Hudgins, R.R., Silveston, P.L., 1999b. Transient study of oxidative dehydrogenation of propane. Appl. Catal. A-Gen. 187, 147–160 [4] [8] [13].

Creaser, D., Andersson, B., Hudgins, R.R., Silveston, P.L., 1999c. Transient kinetic analysis of the oxidative dehydrogenation of propane. J. Catal. 182, 264–269 [4] [8] [13].

Creaser, D., Andersson, B., Hudgins, R.R., Silveston, P.L., 1999d. Oxygen partial pressure effects on the oxidative dehydrogenation of propane. Chem. Eng. Sci. 54, 4365–4370 [4] [13].

Creaser, D., Andersson, B., Hudgins, R.R., Silveston, P.L., 2000. Kinetic modelling of oxygen dependence in oxidative dehydrogenation of propane. Can. J. Chem. Eng. 78, 182–193 [4].

Crittendon, B., Thomas, W.J., 1998. Adsorption Technology and Design. Butterworth-Heinemann Science Publ., Burlington, Massachusetts [1].

Croft, D.T., LeVan, M.D., 1994a. Periodic states of adsorption cycles—I. Direct determination and stability. Chem. Eng. Sci. 49 (11), 1821–1829 [14] [22].

Croft, D.T., LeVan, M.D., 1994b. Periodic states of adsorption cycles—II. Solution spaces and multiplicity. Chem. Eng. Sci. 11, 1831–1841 [14].

Crone, G., Renken, A., 1979a. Der Einfluss periodischer Proczessführung auf radikalische Polymerisationsreaktionen. Chem. -Ing. -Tech. 51, 42–43 [7].

Crone, G., Renken, A., 1979b. Influence of cyclic operation on free radical polymerization. Ger. Chem. Eng. 2, 337–342 [7].

Croughan, M.S., Wang, D.I.C., 1990. Reversible removal and hydrodynamic phenomena in CHO microcarrier cultures. Biotechnol. Bioeng. 36, 316–319 [10].

Crucq, A. (Ed.), 1991. Catalysis and Automotive Pollution Control II, Proc. 2nd Int. Sympos. (CAPOC 2). Brussels, Belgium, *Brussels*, pp. 10–13 [6].

Crucq, A., Frennet, A. (Eds.), 1987. Catalysis and Automotive Pollution Control, Studies in Surface Science, 30. Elsevier Science Publ., Amsterdam, The Netherlands [6].

Cunill, F., van de Beld, L., Westerterp, K.R., 1997. Catalytic combustion of very lean mixtures in a reverse flow reactor using an internal electric heater. Ind. Eng. Chem. Res. 36, 4198–4206 [18].

Curtain, R.F., Zwart, H., 1996. Riccati equations and normalized coprime factorizations for strongly stabilizable infinite-dimensional systems. Syst. Control Lett. 28 (1), 11–22 [19].

Cuthrell, J.E., Biegler, L.T., 1989. Simultaneous optimization and solution methods for batch reactor control profiles. Comput. Chem. Eng. 13, 49–46 [10].

Cutlip, M.B., 1979. Concentration forcing of catalytic surface rate processes. AIChE J. 25, 503–508 [3].

Cutlip, M.B., Hawkins, C.J., Mukesh, D., Morton, W., Kenney, C.N., 1983. Modelling of forced periodic oscillations of carbon monoxide over platinum catalyst. Chem. Eng. Commun. 22, 329–344 [3].

Cybulski, A.A., Moulijn, J.A., 1994. Monoliths in heterogeneous catalysis. Catal. Rev. 36 (2), 179–270 [6].

Cybulski, A., Moulijn, J., 2006. Structured Catalysts and Reactors, second ed. CRC Press, New York, New York [17].

Da Silva, E.A.B., Ulson de Souza, A.A., de Souza, S.G., Rodrigues, A.E., 2005. Simulated moving bed technology in the reactive process of glucose isomerization. Adsorption 11, 847–851 [21].

Dai, X.-P., Li, R.-J., Yu, C.-C., Hao, Z.-P., 2006a. Unsteady-state direct partial oxidation of methane to synthesis gas in a fixed-bed reactor using $AFeO_3$ (A=La, Nd, Eu.), perovskite-type oxides as oxygen storage. J. Phys. Chem. B 110 (45), 22525–22531 [4].

Dai, X.-P., Wu, Q., Li, R.-J., Yu, C.-C, Hao, Z.-P., 2006b. Hydrogen production from a combination of the water-gas shift and redox cycle process of methane partial oxidation via lattice oxygen over $LaFeO_3$ perovskite catalyst. J. Phys. Chem. B 110 (51), 25856–25862 [4].

Dai, X.-P., Yu, C.-C., Li, R.-J., Wu, Q., Hao, Z.-P., 2008. Synthesis gas production using oxygen storage materials as oxygen carrier over circulating fluidized bed. J. Rare Earths 26, 76–80 [4].

Dambrowsky, N., Bade, K., Schulz, J., Köster, F., 2005. Electroforming of thick layers from sulfite gold alloys in the LIGA process, Proc. 2005 SUR/FIN Conf., Amer. Electroplating & Surface Finishing Soc. 343–349 [9].

Dath, J. P., Fink, Th., Imbihl, R., Ertl, G., 1992. Periodic and random perturbation of a system exhibiting damped kinetic oscillations—Pt(100)/NO+CO. J. Chem. Phys. 92, 1582–1589 [3].

Datta, M., Landolt, D., 1981. Electrochemical machining under pulsed current conditions. Electrochim. Acta 26 (7), 899–907 [9].

Datta, M., Landolt, D., 1982. High rate transpassive dissolution of nickel with pulsating current. Electrochim. Acta 27 (3), 385–390 [9].

Davis, T.P., O'Driscoll, K.F., Piton, M.C., Winnik, M.A., 1989a. Determination of propagation rate constants using a pulsed laser technique. Macromolecules 22, 2785–2788 [13].

Davis, T.P., O'Driscoll, K.F., Piton, M.C., Winnik, M.A., 1989b. Determination of propagation rate constants for the copolymerization of methylmethacrylate and styrene using a pulsed laser technique. J. Polym. Sci. Part C Polym. Lett. 27, 181–185 [13].

Davis, T.P., O'Driscoll, K.F., Piton, M.C., Winnik, M.A., 1990. Copolymerization propagation kinetics of styrene with alkyl methacrylates. Macromolecules 23, 2113–2119 [13].

Davis, T.P., O'Driscoll, K.F., Piton, M.C., Winnik, M.A., 1991. Copolymerization propagation kinetics of styrene with alkyl methacrylates. Polym. Int. 24, 65–70 [13].

Davison, B.H., Stephanopoulos, G., 1986. Effect of pH oscillations on a competing mixed culture. Biotechnol. Bioeng. 28, 1127–1137 [10].

De Alwis, D.M., Dutton, R.L., Scharer, J., Moo-Young, M., 2005. Statistical method of media optimization for batch and fed-batch animal cell culture. Bioproc. Biosyst. Eng. 30, 107–113 [10].

De Groote, A.M., Froment, G.F., Kobylinski, Th., 1996. Synthesis gas production from natural gas in a fixed bed reactor with reversed flow. Can. J. Chem. Eng. 74, 735–742 [18].

Del Rosso, R., Kaddouri, A., Fumagalli, D., Mazzocchia, C., Gronchi, P., Centola, P., 1998. Deactivation of alkane oxidative dehydrogenation catalyst by deep reduction in periodic flow reactor. Catal. Lett. 55, 93–95 [4].

Dementhon, J.B., Collious, T., Martin, B., Bouchez, M., Guyon, M., Messaoudi, I., Noirot, R., Michon, S., Gérentet, C., Pierron, L., 2003. Application of NOx adsorber to diesel depollution: performances and durability. Oil Gas Sci. Technol.- Rev. Inst. Fran. Petro. 58 (1), 129–149 [6].

den Hollander, J.L., Stribos, B.I., van Buel, M.J., Luyben, K.Ch., van der Wielen, L.A.M., 1998. Centrifugal partition chromatographic reaction for the production of chiral amino acids. J. Chromat. B 711, 223–235 [20].

Denis, G.H., Kabel, R.L., 1970. The effect of temperature change on tubular heterogeneous catalytic reactors. Chem. Eng. Sci. 25, 1057–1071 [16].

Dennis, J.S., Mueller, C.R., Scott, A.S., 2010. In situ gasification and CO_2 separation using chemical looping with a Cu-based oxygen carrier: performance with bituminous coals. Fuel 89, 2353–2364 [8].

Dennis, J.S., Scott, A.S., 2010. In situ gasification of a lignite coal and CO_2 separation using chemical looping with a Cu-based oxygen carrier. Fuel 89, 1623–1640 [8].

Dettmer, M., Renken, A., 1983a. Zur kinetik der ethylacetatbildung unter stationaeren und instationaeren bedingungen. Chem. Ing. Tech. 55 (2), 146–147 [11].

Dettmer, M., Renken, A., 1983b. Kinetic studies on the formation of ethyl acetate under steady-state and non-steady conditions. Ger. Chem. Eng. 6, 356–365 [11] [12].

Dietrich, W., Grünewald, M., Agar, D.W., 2005. Dynamic modelling of periodically wetted catalyst particles. Chem. Eng. Sci. 60, 6254–6261 [17].

Di Giovanni, O., Mazzotti, M., Morbidelli, M., Denet, F., Hauck, W., Nicoud, R.M., 2001. Supercritical fluid simulated moving bed chromatography: Langmuir isotherm. J. Chromatogr. A. 919, 1–12 [23].

Dimitrov, D., Konarev, N., Sapoundjiev, H., 2003. Unsteady-state double conversion- double absorption method for sulphuric acid production. In: Proc. 4th Intern. Conf. Unsteady-State Processes in Catalysis, Montreal, Canada. October 2003 [18].

Ding, Y., Alpay, E., 2000a. Adsorption-enhanced steam-methane reforming. Chem. Eng. Sci. 55, 3929–3940. [22].

Ding, Y., Alpay, E., 2000b. Equilibria and kinetics of CO_2 adsorption on hydrotalcite adsorbent. Chem. Eng. Sci. 55, 3461–3474 [22].

Ding, Y., Croft, D.T., LeVan, M.D., 2002. Periodic states of adsorption cycles IV. Direct optimization. Chem. Eng. Sci. 57, 4521–4531 [14].

Ding, Y., LeVan, M.D., 2001. Periodic states of adsorption cycles III. Convergence acceleration for direct determination. Chem. Eng. Sci. 56, 5217–5230 [14].

Dionisi, D., Majone, M., Tandoi, V., Beccari, M., 2001. Sequencing batch reactor: influence of periodic operation on performance of activated sludges in biological waste treatment. Ind. Eng. Chem. Res. 40 (23), 5110–5119 [10].

Dobrego, K.V., Gnezdilov, N.N., Lee, S.H., Choi, H.K., 2008a. Methane partial oxidation reverse flow reactor scale up and optimization. Int. J. Hydr. Energy 33, 5501–5509 [18].

Dobrego, K.V., Gnezdilov, N.N., Lee, S.H., Choi, H.K., 2008b. Partial oxidation of methane in a reverse flow porous media reactor. Water admixing optimization. Int. J. Hydr. Energy 33, 5535–5544 [18].

Dobrego, K.V., Gnezdilov, N.N., Lee, S.H., Choi, H.K., 2008c. Lean combustibility limit of methane in reciprocal flow filtration combustion reactor. Int. J. Heat Mass. Tran. 51, 2190–2198 [18].

Dobrev, D., Vetter, J., Angert, N., Neumann, R., 1999. Electrochemical growth of copper single crystals in pores of polymer ion-track membranes. Appl. Phys. A: Material Sci. Process 69, 233–237 [9].

Dobrev, D., Vetter, J., Angert, N., Neumann, R., 2000. Periodic reverse current electrodeposition of gold in an ultrasonic field using ion-track membranes as templates: growth of gold single-crystals. Electrochim. Acta 45 (19), 3117–3125 [9].

Doedel, E., Champneys, A.R., Fairgrieve, T.F., Kuznetsov, Y.A., Sandstede, B., Wang, X., 1997. AUTO 97: Continuation and bifurcation software for ordinary differential equations—User's Guide. Concordia University, Montreal, Canada [7].

Dorawala, T.G., Douglas, J.M., 1971. Complex reactions in oscillating reactors. AIChE J. 17 (4), 974–981 [16].

Dorka, P., Fischer, H., Budman, H., Scharer, J.M., 2009. Metabolic flux based modeling of MAb in batch and fed-batch culture. Bioprocess Biosyst. Eng. 32, 182–196 [10].

Doroshenko, V.A., Shapovalova, L.P., Dolya, L.P., 1986. Preparation and properties of magnesium-molybdenum catalysts for oxidative dehydrogenation of n-butane. Zhurn. Priklad. Khimi. 59 (5), 1176–1179 (Plenum Publ. Corp. Translation) [4].

Dossenbach, O., 1986. Current distribution in pulse plating. In: Puippe, J.-C., Leaman, F. (Eds.), Theory and Practice of Pulse Plating, American Electroplaters & Surface Finishers Society, Washington, D.C. (1986) [9].

Douglas, J.M., 1967. Periodic reactor operation. Ind. Eng. Chem. Prod. Res. Dev. 6 (5), 42–49 [1] [19].

Douglas, J.M., 1972. Process Dynamics and Control. Prentice-Hall, Englewood Cliffs, New Jersey [14].

Douglas, J.M., Rippin, D.W.T., 1966. Unsteady state process operation. Chem. Eng. Sci. 21, 305–315 [1] [14].

Dünnebier, G., Fricke, J., Klatt, K.-U., 2000. Optimal design and operation of simulated moving bed chromatographic reactors. Ind. Eng. Chem. Res. 39, 2290–2304 [21].

Dünnebier, G., van Hessam, D., Kadam, J.V., Klett, K.-U., Schlegel, M., 2005. Optimization and control of polymerization processes. Chem. Eng. Technol. 28 (5), 575–580 [7].

Dudas, J., Hanika, J., 2009. Design, scale up and safe piloting of thymol hydrogenation and menthol racemization. Chem. Eng. Res. Des. 87, 83–90 [17].

Dufour, P., Couenne, F., Touré, Y., 2003. Model predictive control of a catalytic reverse flow reactor. IEEE Trans. Control Syst. Technol. 11 (5), 705–713 [18] [19].

Dufour, P., Touré, Y., 2004. Multivariable model predicative control of a catalytic reverse flow reactor. Comput. Chem. Eng. 28 (11), 2259–2270 [18] [19].

Dulaurent, O., Bianchi, D., 2000. Adsorption isobars for CO on a Pt/Al_2O_3 catalyst at high temperatures using FTIR spectroscopy: isosteric heat of adsorption and adsorption model. Appl. Catal. A-Gen. 196, 271–280 [16].

Dutton, R., Scharer, J.M., 2007. Advanced Technologies in Biopharmaceutical Processing. Wiley - Blackwell Publishing Inc., New York, New York [10].

Eberhart, R.C., Kennedy, J., 1995. A new optimizer using particle swarm theory. Proc. 6th Intern. Symp. on Micro Machine and Human Sci., Nagoya, Japan, 39–43 [10].

Eckler, T.A., Manty, B.A., McDaniel, P.L., 1980. Pulse plating of chromium-molybdenum coatings. Plat. Surf. Finish. 57 (9), 60–64 [9].

Edouard, D., Dufour, P., Hammouri, H., 2005a. Observer based multivariable control of a catalytic reverse flow reactor: comparison between LQR and MPC approaches. Comput. Chem. Eng. 29, 851–862 [18] [19].

Edouard, D., Hammouri, H., Zhou, X., 2005b. Control of a reverse flow reactor for VOC combustion. Chem. Eng. Sci. 60, 1661–1672 [18].

Edouard, D., Schweich, D., Hammouri, H., 2004. Observer design for reverse flow reactor. AIChE J. 50, 2155–2166 [18].

Egyházy, T., Kovács, J., Scholtz, J., 1998. Experimental study of flow reversal ammonia synthesis. Chem. Eng. Technol. 21 (12), 967–974 [18].

Eigenberger, G., Kolios, G., Nieken, U., 2007. Thermal pattern formation and process intensification in chemical reaction engineering. Chem. Eng. Sci. 62, 4825–4841 [6].

Eigenberger, G., Nieken, U., 1988. Catalytic combustion with periodic flow reversal. Chem. Eng. Sci. 43 (8), 2109–2115 [18] [19].

Eigenberger, G., Nieken, U., 1991. Katalytische Abluftreinigung: Verfahrens-technische Aufgaben und neue Lösungen. Chem. Ing. Tech. 63, 781–791 [18].

Eigenburger, G., Nieken, U., 1994. Catalytic cleaning of polluted air: reaction engineering problems and new solutions. Int. Chem. Eng. 34, 4–16 [18].

Eiswirth, M., Ertl, G., 1988. Dynamic order in a surface process. Appl. Phys. A 47, 91–94 [3].

Eiswirth, M., Möller, P., Wetzl, K., Imbihl, R., Ertl, G., 1989. Mechanism of spatial self-organization in isothermal kinetic oscillations during carbon monoxide oxidation on platinum single crystal surfaces. J. Chem. Phys. 90, 510–521 [3].

Eliçabe, G.E., Frontini, G.L., Meira, G.R., 1986. Adaptive model-following control of a certain class of nonlinear processes. Application to a continuous periodic reactor. Chem. Eng. Sci. 41 (7), 1833–1841 [7].

Eliçabe, G.E., Meira, G.R., 1988. Estimation and control in polymerization reactors. A review. Polymer Eng. Sci. 28 (3), 121–135 [7].

Eliçabe, G.E., Meira, G.R., 1989. Model reference adaptive control of a continuous polymerization reactor under periodic operation. Polym. Eng. Sci. 29 (6), 374–382 [7].

El Masry, H.A., 1985. The Claus reaction; effect of forced feed composition cycling. Appl. Catal. 16, 301–313 [11].

El-Sherik, A.M., Erb, U., Page, J., 1997. Microstructural evolution in pulse plated nickel electrodeposits. Surf. Coat. Tech. 88 (1–3), 70–78 [9].

Elsner, M.P., Dittrich, C., Agar, D.W., 2002a. Adsorptive reactors for enhancing equilibrium gas-phase reactions - two case studies. Chem. Eng. Sci. 57, 1607–1619 [18] [22].

Elsner, M.P., Menge, M., Agar, D.W., 2002b. Assessment of adsorption enhanced reaction exemplified by the Claus process, preprint CAMURE Conference. Lausanne, Switzerland. Sept. 2002 [22].

Elsner, M.P., Menge, M., Müller, C., Agar, D.W., 2003. The Claus process: teaching an old dog new tricks. Catal. Today 79-89, 487–494 [18].

Emig, G., Uihlein, K., Haecker, C.J., 1994. Separation of catalyst oxidation and reduction - an alternative to the conventional oxidation of n-butane to maleic anhydride. Stud. Surf. Sci. Catal. 82, 243–251 [4].

Engel, T., Ertl, G., 1978. A molecular beam investigation of the catalytic oxidation of CO on Pd(111), J. Chem. Phys. 69, 1267–1281 [16].

Epling, W.S., Campbell, L.E., Yezerets, A., Currier, N.W., Parks II, J.E., 2004. Overview of the fundamental reactions and degradation mechanisms of NOx storage/reduction catalysts. Cataly. Rev. 46 (2), 163–245 [6].

Ercan, C., Gartside, R.J., 1996. Reactor performance and stability in an alternating reaction-reheat paraffin dehydrogenation system. Can. J. Chem. Eng. 74, 626–637 [18].

Erdöhelyi, A., Cserényi, J., Solymosi, F., 1992. Activation of CH_4 and its reaction with CO_2 over supported Rh catalysts. J. Catal. 141, 287–299 [12].

Ermenc, E.D., 1956. Wisconsin process. Pebble furnace fixes atmospheric nitrogen. Chem. Eng. Progr. 52, 149–153 [18].

Ertl, G., 1990. Oscillatory catalytic reactions at single crystal surfaces. Adv. Catal. 37, 213–277 [3].

Esteves, I.A.A.C., Mota, J.P.B., 2007. Gas separation by a novel hybrid membrane/pressure swing adsorption process. Ind. Eng. Chem. Res. 46, 5723–5733 [22].

Ett, G., Pessine, E.J., 1999. Pulse current plating of TiB_2 in molten fluoride. Electrochim. Acta 44 (17), 2859–2870 [9].

Fahidy, T.Z., 1985. Principles of Electrochemical Reactor Analysis. Elsevier, Amsterdam, The Netherlands [9].

Falconer, J.L., Schwarz, J.A., 1983. Temperature-Programmed Desorption and Reaction: Application to Supported Catalysts. Catal. Rev. -Sci. Eng. 25 (2), 141–227 (13).

Falk, T., Seidel-Morgenstern, A., 1999. Comparison between a fixed-bed reactor and a chromatographic reactor. Chem. Eng. Sci. 54, 1479–1485 [20].

Falk, T., Seidel-Morgenstern, A., 2002. Analysis of a discontinuously operated chromatographic reactor. Chem. Eng. Sci. 57, 1599–1606 [20].

Falle, S.A.E.G., Kallrath, J., Brockmüller, B., Schreieck, A., Giddings, J.R., Agar, D.W., Watzenberger, O., 1994. The dynamics of reverse flow chromatographic reactors with side stream feed. Chem. Eng. Commun. 135, 185–211 [18].

Fan, L.-S., 2010. Chemical Looping Systems for Fossil Energy Conversion. AIChE-John Wiley & Sons, New York [1] [8].

Fan, L.-S., Li, F., 2010. Chemical looping technology and its fossil energy conversion applications. Ind. Eng. Chem. Res. 49, 10200–10211 [1] [8].

Fan, L.-S., Li, F., Ramkumar, S., 2008. Utilization of chemical looping strategy in coal gasification processes. Partic. 6, 131–142 [8].

Fann, C.H., Guirgis, F., Chen, G., Lao, M.S., Piret, J.M., 2000. Limitations to the amplification and stability of human tissue-type plasminogen activator expression by Chinese hamster ovary cells. Biotechnol. Bioeng. 69, 204–212 [10].

Farhadpour, F.A., Gibilaro, L.G., 1975. Continuous unsteady state operation of a stirred tank reactor. Chem. Eng. Sci. 30, 997–999 [12].

Farhadpour, F.A., Gibilaro, L.G., 1981. On the optimal unsteady state operation of a continuous stirred tank reactor. Chem. Eng. Sci. 36, 143–147 [1] [14].

Fathi, M., Bjorgum, E., Viig, T., Rokstad, O.A., 2000a. Partial oxidation of methane to synthesis gas: Elimination of gas phase oxygen. Catal. Today 63, 489–497 [4].

Fathi, M., Monnet, F., Schuurman, Y., Holmen, A., Mirodatos, C., 2000b. Reactive oxygen species on platinum gauzes during partial oxidation of methane into synthesis gas. J. Catal. 190, 439–445 [4].

Fauser, H., Zielonka, A., 2004. Microelectroforming for medical application, Proc. 2004 SUR/FIN Conf., American Electroplaters & Surface Finishers Society 394–410 [9].

Feiguth, P., Wicke, E., 1971. Der Übergang vom Zünd/Lösch-Verhalten zu stabilen Reaktionszuständen bei einem adiabatischen Rohrreaktor. Chem. Ing. Tech. 43, 604–608 [18].

Feimer, J.L., Hudgins, R.R., Silveston, P.L., 1985. Influence of forced cycling on the Fischer-Tropsch synthesis. Part II Response to periodic operation. Can. J. Chem. Eng. 63, 86–92 [1].

Feldhues, M., Schäfer, H.J., 1985. Selective mixed coupling of carboxylic acids (I.) Electrolysis, thermolysis and photolysis of unsymmetrical diacyl peroxides with acyclic and cyclic alkyl groups. Tetrahedron 41 (19), 4195–4212 [9].

Ferrer-Miralles, N., Domingo-Espin, J., Corchero, J.L., Vázquez, J., Villaverde, A., 2009. Microbial factories for recombinant pharmaceuticals. Microbial Cell Fact. 8 Article # 17, Mar. 24, 2009 [10].

Field, R.J., Noyes, R.M., 1974. Oscillations in chemical systems. IV. Limit cycle behaviour in a model of a real chemical reaction. J. Chem. Phys. 60 (5), 1877–1884 [1].

Fink, Th., Dath, J.-P., Bassett, M.R., Imbihl, R., Ertl, G., 1991. The mechanism of the explosive NO+CO reaction on Pt(100): experiments and mathematical modeling. Surf. Sci. 245, 96–110 [3].

Fiolitakis, E., Hofmann, H., 1983. Wave front analysis, a specific analysis of transient responses for the investigation of continuously operated distributed heterogeneous reaction systems. Catal. Rev. 24, 113–157 [4] [8].

Fiolitakis, E., Schmid, M., Hofmann, H., Silveston, P.L., 1983. Investigation of the catalytic oxidation of benzene to maleic anhydride by wave front analysis with respect to the application of periodic operation. Can. J. Chem. Eng. 61, 703–709 [4] [8].

Fish, B.B., Carr, R.W., 1989. An experimental study of the countercurrent moving-bed chromatographic reactor. Chem. Eng. Sci. 44 (9), 1773–1783 [20].

Fish, B., Carr, R.W., Aris, R., 1986. The continuous countercurrent moving bed chromatographic reactor. Chem. Eng. Sci. 41 (4), 661–668 [1] [20] [21].

Fissore, D., 2008. Robust control in presence of parametric uncertainties: observer-based feedback controller design. Chem. Eng. Sci. 63, 1890–1900 [18].

Fissore, D., Barresi, A.A., 2002. Comparison between the reverse-flow reactor and a network of reactors for the oxidation of lean VOC mixtures. Chem. Eng. Technol. 25, 421–426 [5].

Fissore, D., Barresi, A.A., 2003. On the influence of the catalyst physical properties on the stability of forced unsteady-state after-burners. Chem. Eng. Res. Des. Trans Inst. Chem. Eng. A. 81, 611–617 [18].

Fissore, D., Barresi, A.A., 2008. Robust control of a reverse-flow reactor. Chem. Eng. Sci. 63 (7), 1901–1913 [18].

Fissore, D., Barresi, A.A., Baldi, G., Hevia, M.A.G., Ordóñez, S., Diez, F.V., 2005. Design and testing of small-scale, unsteady-state afterburners and reactors. AIChE J. 51 (6), 1654–1664 [18].

Fissore, D., Barresi, A.A., Botar-Jid, C.C., 2006b. NOx removal in forced unsteady-state chromatographic reactors. Chem. Eng. Sci. 61, 3409–3414 [18].

Fissore, D., Eduard, D., Hammouri, H., Baressi, A.A., 2006a. Nonlinear soft-sensors design for unsteady-state VOC afterburners. AIChE J 52 (1), 282–291 [18].

Fjeld, M., 1972. Comments on "A controllability theory for nonlinear-systems." IEEE Trans. Autom. Contr., AC17 (4), 579–580 [19].

Fjeld, M., 1974. Relaxed controls in asynchronous quenching and dynamical optimization. Chem. Eng. Sci. 29 (4), 921–933 [19].

Fleischmann, M., Goodridge, F., 1968. Anodic oxidation under pulse conditions. Discuss. Faraday Soc. 45, 254–260 [9].

Fleischmann, M., Pletcher, D., 1969. The electrosynthesis of organic compounds. Platinum Met. Rev. 13, 46–52 [9].

Fleischmann, M., Mansfield, J.R., Wynne-Jones, Lord, 1965a. The anodic oxidation of aqueous solutions of acetate ions at smooth platinum electrodes. Part I. the non-steady state oxidation of platinum. J. Electroanal. Chem. 10, 511–521 [9].

Fleischmann, M., Mansfield, J.R., Wynne-Jones, Lord, 1965b. The anodic oxidation of aqueous solutions of acetate ions at smooth platinum electrodes. Part II: the non-steady state of the Kolbe synthesis of ethane. J. Electroanal. Chem. 10, 522–537 [9].

Fleury, T.A., Meyer, T., Renken, A., 1992. Solution polymerization of methyl- methacrylate at high conversion in a recycle tubular reactor. Chem. Eng. Sci. 47 (9–11), 2597–2602 [7].

Forzatti, P., Nova, I., Tronconi, E., 2009. Enhanced NH_3 selective catalytic reduction for NOx abatement. Angew. Chem. Intern. Edition 48, 8366–8369 [6].

Fraguío, M.S., Muzen, A., Martínez, O., Bonelli, P., Cassanello, M., 2004. Catalytic wet oxidation of ethanol in an integral trickle-bed reactor operated with liquid flow modulation. Lat. Am. Appl. Res. 34 (1), 11–16 [17].

Freedman, R.B., Greenall, C., Jenkins, N., Tuite, M.F., 1995. Protein folding in the secretory pathway of animal cells. Cytotechnology 18, 77–82 [10].

Fregeau, M., Hermanson, J.C., 2009. CO/NOx emissions of strongly-pulsed jet diffusion flames. Combust. Sci. Technol. 181 (3), 536–542 [5].

Freund, E., Abel, S., Huthmann, E., Lill, J., 2009. Chiral chromatography in the early phases of pharmaceutical development. Chim. Oggi 27 (5), 62–64 [23].

Fricke, J., Meurer, M., Dreisörner, J., Schmidt-Traub, H., 1999. Effect of process parameters on the performance of a simulated moving bed chromatographic reactor. Chem. Eng. Sci. 54, 1487–1492 [21].

Frontini, G.L., Eliçabe, G.E., Couso, D.A., Meira, G.R., 1986. Optimal periodic control of a continuous living anionic polymerization. J. Appl. Polym. Sci. 31 (4), 1019–1039 [7].

Frontini, G.L., Eliçabe, G.E., Meira, G.R., 1987. Optimal periodic control of a continuous living anionic polymerization II. New Theoretical Results. J. Appl. Polym. Sci. 33, 2165–2177 [7].

Fuentes, A., Sinisterra, J.V., 1986. Single electron-transfer mechanism of cannizzaro reaction in heterogeneous phase, under ultrasonic conditions. Tetrahedron Lett. 27, 2967–2970 [15].

Fukase, S., Suzuka, T., 1993. Residual oil cracking with generation of hydrogen—deactivation of iron-oxide catalyst in the steam iron reaction. Appl. Catal. A-Gen. 100, 1–17 [4].

Furrow, S.D., 1995. Comparison of several substrates in the Briggs-Rauscher oscillating system. J. Phys. Chem. 99 (28), 11131–11140 [1].

Fuxman, A.M., Aksikas, I., Forbes, J.F., Hayes, R.E., 2007. Control of a catalytic flow reversal reactor model by linear quadratic regulator. 8th International IFAC Symposium on Dynamics and Control of Process Systems (DYCOPS), Cancun, Mexico [19].

Fuxman, A., Aksikas, I., Forbes, J., Hayes, R., 2008. LQ-feedback control of a reverse flow reactor. J. Proc. Control 18, 654–662 [18].

Fuxman, A.M., Forbes, J.F., Hayes, R.E., 2006. Model predictive control of a catalytic flow reversal reactor with heat extraction, Proceedings of ADCHEM 2006 Conference. Gramado, Brazil, 1009–1014 [19].

Gabarain, L., Castellari, A.T., Cechini, J., Tobolski, A., Haure, P., 1997. Analysis of rate enhancement in a periodically operated trickle bed reactor. AIChE J. 43 (1), 166–172 [16] [17].

Galle, M., Agar, D.W., Watzenberger, O., 2001. Thermal N_2O decomposition in regenerative heat exchanger reactors. Chem. Eng. Sci. 56, 1587–1595 [18].

Galvita, V., Sundmacher, K., 2005. Hydrogen production from methane by steam reforming in a periodically operated two-layer catalytic reactor. Appl. Catal. A-Gen. 289, 121–127 [4].

Gangwal, S.K., Howe, G.B., Spivey, J.J., Silveston, P.L., Hudgins, R.R., Metzinger, J.G., 1993. Low-temperature carbon-based process for flue-gas cleanup. Environ. Prog. 12 (2), 128–132 [17].

Gao, X.D., Chen, B.Z., He, X.R., 2005. Study on the optimization of periodic operation for the naphtha pyrolysis furnace, in PSE ASIA 2005 Conference Proceedings vol. P027, 240–244 [1] [23].

Gao, X., Chen, B., He, X., 2006. An agent oriented architecture for modeling and optimization of naphtha pyrolysis process. In: Marquardt, W., Pantelides, C. (Eds.), 16th European Symposium on Computer Aided Process Engineering 21C 475-481 [23].

Gao, X., Chen, B., He, X., Qiu, T., Lib, J., Wang, C., Zhang, L., 2008a. Multi-objective optimization for the periodic operation of the naphtha pyrolysis process using a new parallel hybrid algorithm combining NSGA-II with SQP. Comput. Chem. Eng. 32, 2801–2811 [1] [23].

Gao, Z.-P., Shen, L.-H., Xiao, J., Qing, C.-J., Song, Q.-L., 2008b. Use of coal as a fuel for chemical-looping combustion with Ni-based oxygen carrier. Ind. Eng. Chem. Res. 47, 9279–9287 [5].

Garcia-Labiano, F., Adanez, J., de Diego, L.F., Gayan, P., Abad, A., 2006. Effect of pressure on the behaviour of copper-, iron-, nickel-based oxygen carriers for chemical looping combustion. Energy Fuels 20 (1), 26–33 [8].

Garcia-Labiano, F., de Diego, L.F., Adanez, J., Abad, A., Gayan, P., 2005. Temperature variations in the oxygen carrier particles during their reduction and oxidation in a chemical looping combustion system. Chem. Eng. Sci. 60 (3), 851–861 [8].

Gartside, R.J., Feldman, R.J., Ercan, C., Dautzenberg, F., 1994. New CATAFIN™ process provides higher conversion and selectivity, Proc. World Conf. Ref. Proc. Reform. Gasoline. San Antonio, Texas, March 22–24 [18].

Gawdzik, A., Rakovski, L., 1988. Dynamic properties of the adiabatic tubular reactor with switch flow. Chem. Eng. Sci. 43, 3023–3030 [18].

Gaziev, G.A., Roginskii, S.Z., Yanovskii, M.I., 1963. USSR Patent, 149, 398 [20].

Gelosa, D., Ramaioli, M., Valente, G., Morbidelli, M., 2003. Chromatographic reactors: esterification of glycerol with acetic acid using acidic polymeric resins. Ind. Eng. Chem. Res. 42, 6536–6544 [20].

Genser, F., Pietrzyk, S., 1999. Oxidative dehydrogenation of propane on V_2O_5/TiO_2 catalysts under transient conditions. Chem. Eng. Sci. 54, 4315–4325 [4] [8].

Gerasev, A.P., Matros, Yu.Sh., 1991. Unsteady-state process of synthesis of ammonia. Theor. Found. Chem. Eng. 25, 821–827 [18].

Ghasem, N.F., 1999. Effect of polymer particle size and inlet gas temperature on industrial fluidized bed polyethylene reactors. Chem. Eng. Technol. 22, 777–783 [7].

Ghazali, S., Park, D.W., Gau, G., 1983. Kinetics of ethylene epoxidation on a silver catalyst. Appl. Catal. 6, 195–208 [4].

Ghose, T.K., Roychoudhury, P.K., Ghosh, P., 1984. Simultaneous saccharification and fermentation (SSF) of lignocellulose to ethanol under vacuum cycling and step feeding. Biotechnol. Bioeng. 26, 377–381 [10].

Ghosh, S.K., Grover, A.K., Dey, G.K., Totlani, M.K., 2000. Nanocrystalline Ni-Cu alloy plating by pulse electrolysis. Surf. Coat. Tech. 126 (1), 48–63 [9].

Giakoumakis, D., Kostoglou, M., Karabelas, A.J., 2005. Induced pulsing in trickle beds—characteristics and attenuation of pulses. Chem. Eng. Sci. 60, 5813–5197 [1] [17].

Gilbert, N., Daniels, F., 1948. Fixation of atmospheric nitrogen in a gas heated furnace. Ind. Eng. Chem. 40 (9), 1719–1723 [18].

Gilks, W.R., Richardson, S., Spiegelhalter, D.J., 1998b. Markov Chain Monte Carlo Practice. Chapman & Hall, London, U.K. [10].

Gilks, W.R., Roberts, G.O., Suhu, S.K., 1998a. Adaptive Markov chain Monte Carlo through regeneration. J. Amer. Stat. Assoc. 93 (443), 1045–1054 [10].

Gilles, E.D., 1974. Quasi-stationäres Verhalten von wandernden Brenzonen. Chem. Eng. Sci. 29, 1211–1216 [18].

Gilley, J.W., Bungay, H.R., 1968. Frequency response analysis of dilution rate effects on yeast growth. Biotechn. Bioeng. 10 (1), 99–101 [10].

Gladden, L.F., Anadon, L.D., Dunckley, C.P., Mantle, M.D., Sederman, A.J., 2007. Insights into gas-liquid-solid reactors obtained by magnetic resonance imaging. Chem. Eng. Sci. 62, 6969–6977 [17].

Glöckler, B., Gritsch, A., Morillo, A., Kolios, G., Eigenberger, G., 2004. Autothermal reactor concepts for endothermic fixed-bed reactions. Chem. Eng. Res. Des. 82 (A2), 148–159 [18].

Glöckler, B., Dieter, H., Eigenberger, G., Nieken, U., 2007. Efficient reheating of a reverse flow reformer - an experimental study. Chem. Eng. Sci. 62, 5638–5643 [18].

Glöckler, B., Kolios, G., Eigenberger, G., 2003. Analysis of a novel reverse flow concept for autothermal methane steam reforming. Chem. Eng. Sci. 58, 593–601 [18].

Glöckler, B., Kolios, G., Tellaeche, C., Nieken, U., 2009. A heat-integrated reverse-flow reactor concept for endothermic high-temperature synthesis. Part I: Fundamentals—short-cut theory and experimental verification of a traveling endothermic reaction zone. Chem. Eng. Technol. 32, 1339–1347 [18].

Glueckauf, E., Kitt, G.P., 1957. Gas chromatographic separation of hydrogen isotopes. In: Desty, D.H. (Ed.), Vapor Phase Chromatography. Butterworths, London, pp. 422–427 [20].

Goertz, V., Elie, F., Spinner, B., 1993. The structure and performance of single effect solid-gas chemical heat pumps. Heat Recov. Sys. & Chem. Heat Pumps 13 (1), 79–96 [8].

Gogová, Z., Hanika, J., 2009a. Reactivation of a palladium catalyst during glucose oxidation by molecular oxygen. Chem. Pap. 63 (5), 520–526 [23].

Gogová, Z., Hanika, J., 2009b. Dynamic modelling of glucose oxidation with palladium catalyst deactivation in multifunctional CSTR; benefits of periodic operation. Chem. Eng. J. 150 (1), 223–230 [11] [23].

Gogová, Z., Hanika, J., 2009c. Model aided design of three-phase gas-lift reactor for oxidation accompanied with catalyst reversible deactivation. Chem. Eng. Technol. 32 (12), 1929–1940 [23].

Gogová, Z., Hanika, J., Markos, J., 2010. Optimal design of multifunctional reactor for catalytic oxidation of glucose with fast catalyst deactivation (Chapter 12). In: Brito, A. (Ed.), Dynamic Modelling, Intech, Vukovar, Czech Republic, pp. 209–232 [23].

Golay, S., Doepper, R., Renken, A., 1998. In-situ characterisation of the surface intermediates for the ethanol dehydration reaction over γ-alumina under dynamic conditions. Appl. Catal. A-Gen. 172, 97–106 [11].

Golay, S., Doepper, R., Renken, A., 1999b. Reactor performance enhancement under periodic operation for the ethanol dehydration over γ-alumina, a reaction with a stop-effect. Chem. Eng. Sci. 54, 4469–4474 [11].

Golay, S., Kiwi-Minsker, L., Doepper, R., Renken, A., 1999a. Influence of the catalyst acid/base properties on the catalytic ethanol dehydration under steady-state and dynamic conditions—in-situ surface and gas phase analysis. Chem. Eng. Sci. 54 (15–16), 3593–3598 [11].

Golay, S., Wolfrath, O., Doepper, R., Renken, A., 1997. Model discrimination for reactions with stop-effect. Stud. Surf. Sci. Catal. 109, 295–304 [11] [13].

Gomes, V.G., Yee, K.W.K., 2002. A periodic separating reactor for propene metathesis. Chem. Eng. Sci. 57, 3839–3850 [22].

Goodman, M.G., Cutlip, M.B., Kenney, C.N., Morton, W., Mukesh, D., 1982. Transient studies of CO oxidation over platinum catalyst. Surf. Sci. 120, L453–L460 [3].

Goodridge, F., Scott, K., 1995. Electrochemical Process Engineering: A Guide to the Design of Electrolytic Plants. Plenum Publishers, New York [9].

Gordon, D.L., Weidner, K.R., 1981. Control of particle size distribution through emulsifier metering based on rate of conversion. In: Bassett, D.R., Hamielec, A.E. (Eds.), Emulsion Polymers and Emulsion Polymerization, ACS Symp. Series, No. 165, pp. 515–532 [7].

Gore, F.E., 1967. Performance of chromatographic reactors in cyclic operation. Ind. Eng. Chem. Proc. Des. Dev. 6 (1), 10–22 [20].

Gosden, R.G., Auguste, S., Edwards, H.G.M., Johnson, A.F., Meszena, Z.G., Mohsin, M.A., 1995. Living polymerization reactors. Part I. Modelling and simulation of flow reactors operated under cyclical-steady-state feed conditions for the control of molecular weight distribution. Polym. React. Eng. 3 (4), 331–359 [7].

Gosden, R.G., Meszena, Z.G., Mohsin, M.A., Auguste, S., Johnson, A.F., 1997a. Living polymerization reactors. Part II. Theoretical and experimental tests on an algorithm which predicts MWDs from CSTRs with perturbed feeds. Polym. React. Eng. 5 (1–2), 45–81 [7].

Gosden, R.G., Meszena, Z.G., Mohsin, M.A., Johnson, A.F., 1997b. Living polymerization reactors. Part III. Towards molecular weight distribution control in a CSTR through periodic feed perturbation. Polym. React. Eng. 5 (4), 205–232 [7].

Gosiewski, K., 1993. Dynamic modeling of industrial SO_2 oxidation reactors. Part II. Model of a reverse-flow reactor. Chem. Eng. Process. 32, 233–244 [18].

Gosiewski, K., 2001. Simulations of non-stationary reactors for the catalytic conversion of methane to synthesis gas. Chem. Eng. Sci. 56, 1501–1510 [18].

Gosiewski, K., 2004. Effective approach to cyclic steady state in the catalytic reverse-flow combustion of methane. Chem. Eng. Sci. 59, 4095–5101 [18].

Gosiewski, K., 2005. Efficiency of heat recovery versus maximum catalyst temperature in a reverse-flow combustion of methane. Chem. Eng. J. 107, 19–25 [18].

Gosiewski, K., Matros, Yu.Sh., Warmuzinski, K., Jaschik, M., Tanczyk, M., 2008. Homogeneous vs. catalytic combustion of lean methane—air mixtures in reverse-flow reactors. Chem. Eng. Sci. 63, 5010–5019 [18].

Gosiewski, K., Warmuzinski, K., 2007. Effect of the mode of heat withdrawal on the asymmetry of temperature profiles in reverse-flow reactors. Catalytic combustion of methane as a test case. Chem. Eng. Sci. 62, 2679–2689 [18].

Goto, S., Tagawa, T., Ohkawara, H., 1995. Promotion effect of hydrogen storage alloys on the catalytic dehydroaromatization of n-hexane. Sekiyu Gakkaishi 38 (3), 167–172 (in Japanese) [22].

Goto, S., Tagawa, T., Ohkawara, H., 1996. Promotive effect of hydrogen storage metals on catalytic dehydroaromatization of propane in a PSA type reactor. Sekiyu Gakkaishi 39 (6), 389–394 (in Japanese) [22].

Goto, S., Tagawa, T., Oomiya, T., 1993. Dehydrogenation of cyclohexane in a PSA reactor using hydrogen storage alloy. Kagaku Kogaku Ronbunshu 19 (6), 978–983 (in Japanese) [22].

Grabmueller, H., Hoffmann, U., Schädlich, K., 1985. Prediction of conversion improvements by periodic operation for isothermal plug-flow reactors. Chem. Eng. Sci. 40, 951–960 [1].

Grabowski, R., Pietrzyk, S., Soczynski, J., Genser, F., Wcisco, K., Grzybowska-Swierkosz, B., 2002. Kinetics of the propane oxidative dehydrogenation on vanadia/titania catalysts from steady-state and transient experiments. Appl. Catal. A-Gen. 232, 277–288 [4] [8].

Graham, W.R.C., Lynch, D.T., 1984. Model validation through an experimental investigation of resonant behaviour for the catalytic oxidation of carbon monoxide on platinum. In: Kaliaguine, S., Mahay, A. (Eds.), Catalysis on the Energy Scene. Stud. Surf. Sci. Catal., 19, pp. 197–204 [3].

Graham, W.R.C., Lynch, D.T., 1987. CO oxidation on Pt: model discrimination using experimental bifurcation behaviour. AIChE J. 33 (5), 792–800 [3].

Graham, W.R.C., Lynch, D.T., 1988. Forced cycling of catalytic reactors: systems with two inputs. Stud. Surf. Sci. Catal. 38, 693–703 [3].

Graham, W.R.C., Lynch, D.T., 1990. CO oxidation on Pt: variable phasing of inputs during forced composition cycling. AIChE J. 36 (12), 1796–1806 [3].

Grasselli, R.K., Stern, D.L., Tsikoyiannis, G.G., 1999. Catalytic dehydrogenation (DH) of light hydrocarbons combined with selective hydrogen combustion (SHC) II. DH+SHC catalysts mixed (redox process mode). Appl. Catal. A Gen. 189, 9–14. [4] [8].

Grasselli, R.K., Stern, D.L., Tsikoyiannis, J.G., 2000. Strategies for combining light paraffin dehydrogenation (DH) with selective hydrogen combustion. Stud. Surf. Sci. Catal. 130A, 773–778 [8].

Greger, M., Ihme, B., Kotter, M., 1984. Nonequilibrium phase-transitions with hysteresis in solid catalysts—The system copper-oxygen-propene. Berich. Bunsen Ges. Phys. Chem. Chem. Phys. 88 (4), 427–433 [4].

Grozev, G., Sapundzhiev, H., 1997. Modeling of the reverse flow fixed bed reactor for catalytic decontamination of waste gases. Chem. Eng. Technol. 20, 378–383 [18].

Grozev, G.G., Sapundzhiev, H., Elenkov, D.G., 1994. Unsteady-state SO_2 oxidation: Practical results. Ind. Eng. Chem. Res. 33, 2248–2250 [18].

Grünewald, M., Agar, D., 2004. Intensification of regenerative heat exchange in chemical reactors using desorptive cooling. Ind. Eng. Chem. Res. 43, 4733–4779 [18].

Grube, G., Gmelin, H., 1920. The influence of superimposed alternating current on the anodic ferrate formation. Z. Elektrochem. Angew. P. 26, 153–161 [9].

Güthenke, A., Chatterjee, D., Weibel, M., Waldbusser, M., Kočí, P., Marek, M., Kubíček, M., 2007. Development and application of a model for a NOx storage and reduction catalyst. Chem. Eng. Sci. 62, 5357–5363 [6].

Gu, M.B., Todd, P., Kompala, D.S., 1993. Foreign gene expression (β-galactosidase) during the cell cycle phases in recombinant CHO cells. Biotechnol. Bioeng. 42, 1113–1123 [10].

Guardabassi, G., Locatelli, A., Rinaldi, S., 1974. Status of periodic optimization of dynamical systems. J. Optim. Theory Appl. 14, 1–20 [14].

Gugliotta, L.M., Meira, G.R., 1986. Control of particle size distribution in lattices through periodic operation of continuous emulsion polymerization reactors, Macromol. Chem. Macromol. Symp. 2, 209–218 [7].

Gugliotta, L.M., Meira, G.R., 1988. Control de la distribution de pesos moleculaires y de la distribution de tamanos de particulas en reactures continuous de polimerizacion en emulsion. Preprint 1er Simposio Latinoamericano de Polimeros, Porlamar, Venezuela (July, 1988) [7].

Guiochon, G., Shirazi, S.G., Katti, A.M., 1994. Fundamentals of Preparative and Nonlinear Chromatography. Academic Press, London [14].

Gunjal, P.R., Kashid, M.N., Ranade, V.V., Chaudhari, R.V., 2005. Hydrodynamics of trickle bed reactors: experiments and CFD modeling. Ind. Eng. Chem. Res. 44, 6278–6294 [17].

Gupta, R., 1985. Pulsed flow, vapor-liquid reactor. US Patent 4,526,757. US Government Patent Office, Washington, D.C [17].

Gupta, V.K., Bhatia, S.K., 1991. Solution of cyclic profiles in catalytic reactor operation with periodic flow reversal. Comput. Chem. Eng. 15 (4), 229–237 [18].

Gupta, S.K., Nath, S., Kumar, A., 1985. Forced oscillations in continuous flow stirred tank reactors with nonlinear step growth polymerization. J. Appl. Polym. Sci. 30, 557–569 [7].

Gupta, P., Velazquesz-Vargas, L.G., Fan, L.-S., 2007. Syngas redox (SGR) process to produce hydrogen from coal derived syngas. Energy Fuels 21 (5), 2900–2908 [8].

Haas-Santo, K., Görke, O., Pfeifer, P., Schubert, K., 2002. Catalyst coatings for microstructure reactors. Chimia 56, 605–610 [16].

Haber, J., 1980. Dynamics of surface reconstruction and its influence on catalytic properties. Izves. Khim. 13, 65–75 [4].

Haber, J., Stoch, J., Wiltowski, T., 1980a. Surface transformations of copper molybdate catalysts. In: Seiyama, T., Tanabe, K. (Eds.), New Horiz. Catal. B, pp. 1402–1403 [4].

Haber, J., Stoch, J., Wiltowski, T., 1980b. Surface transformations of copper oxidation catalysts. React. Kinet. Catal. Lett. 13, 161–165 [4].

Hacker, V., Fankhauser, R., Faleschini, G., Fuchs, H., Friedrich, K., Muhr, M., Kordesch, K., 2000. Hydrogen production by steam-iron process. J. Power Sources 86 (1–2), 531–535 [4].

Haggin, J., 1992. Direct conversion of methane to fuels, chemicals still intensely sought. Chem. Eng. News 70 (17), 33–35 [4].

Halpern, B., al Mutaz, I., 1986. A chopped fast flow technique for transient surface reaction rate measurements: carbon deposition and oxidation on hot Pt. Chem. Eng. Sci. 41 (6), 899–904 [13].

Hamer, J.W., Akramov, T.A., Ray, W.H., 1981. The dynamic behavior of continuous polymerization reactors. II Nonisothermal solution homopolymerization and copolymerization in a CSTR. Chem. Eng. Sci. 36 (12), 1897–1914 [7].

Hamer, J.W., Cormack, D.E., 1978. Influence of oscillating external pressure on gas-phase reactions in porous catalysts. Chem. Eng. Sci. 33, 935–944 [15].

Hamidipour, M., Larachi, F., Ring, Z., 2007. Cyclic operation strategy for extending cycle life of trickle beds under gas-liquid filtration. Chem. Eng. Sci. 62, 7426–7435 [17].

Hamidipour, M., Larachi, F., Ring, Z., 2010. Cyclic operation strategies in trickle beds and electrical capacitance tomography imaging of filtration dynamics. Ind. Eng. Chem. Res. 49, 934–952 [17].

Hamielec, A.E., Tobita, H., 1992. Polymerization Processes. In: Ullmann's Encyclopedia of Industrial Chemistry. vol. A21. VCH Publishers Inc., Weinheim, Germany [7].

Hamielec, A.E., Tobita, H., 1998. Polymerization Processes. In: Wilks, E.S. (Ed.), Industrial Polymers Handbook. John Wiley Publishers, New York, New York [7].

Han, C., Harrison, D.P., 1994. Simultaneous shift reaction and carbon dioxide separation for the direct production of hydrogen. Chem. Eng. Sci. 49, 5875–5882 [1] [22].

Han, C., Harrison, D.P., 1997. Multicycle performance of a single step process for H_2 production. Separ. Sci. Technol. 32 (1–4), 681–697 [22].

Hanamura, K., Echigo, R., Zhdanok, S.A., 1993. Super-adiabatic combustion in porous medium. Int. J. Heat Mass Tran. 36, 3201–3209 [18].

Hanaoka, T., Yoshida, T., Fujimoto, S., Kamei, K., Harada, M., Suzuki, Y., Hatano, H., Yokoyama, S.-Y., Minowa, T., 2005. Hydrogen production from woody biomass by steam gasification using a CO_2 sorbent. Biomass Bioenergy 28, 63–68 [8].

Hanika, J., Jiřičný, V., Karnetová, P., Kolena, J., Lederer, J., Skála, D., Staněk, V., Tukač, V., 2007. Trickle bed reactor operation under forced liquid feed rate modulation. Chem. Ind. Chem. Eng. Quart. 13 (4), 192–198 [17].

Hanika, J., Lange, R., Turek, F., 1990. Computer-aided control of a laboratory trickle bed reactor. Chem. Eng. Process. 28, 23–27 [16].

Hanika, J., Sporka, K., Růžička, V., 1970. Investigation of Hydrogenation in Liquid Phase XVI. Theoretical Models of Hydrogenation in Liquid Film; Flat Plate. Collect. Czech. Chem. Commun. 35, 2111–2123 [17].

Hanika, J., Sporka, K., Růžička, V., 1971. Investigation of Hydrogenation in Liquid Phase XVIII. Experimental Verification of Theoretical Model of Trickle Bed Reactor. Collect. Czech. Chem. Commun. 36, 2903–2913 [17].

Hansen, K.W., Jorgensen, S.B., 1974. Experimental investigations of the dynamics of a catalytic fixed bed reactor. Adv. Chem. Ser. 133, 505–518 [16].

Hansen, H.A., Olsen, J.L., Jensen, S., Hansen, O., Quaade, U.J., 2006. Rate enhancement in microfabricated chemical reactors under forced temperature oscillations. Catal. Commun. 7 (5), 272–275 [16].

Harmsen, J.M.A., Hoebink, J.H.B.J., Schouten, J.C., 2000. Transient kinetic modeling of ethylene and carbon monoxide oxidation over a commercial automotive exhaust gas catalysis. Ind. Eng. Chem. Res. 39, 599–609 [6].

Harmsen, J.M.A., Hoebink, J.H.B.J., Schouten, J.C., 2001a. Acetylene and carbon monoxide oxidation over a Pt/Rh/CeO_2/γ-Al_2O_3 automotive exhaust gas catalyst: kinetic modeling of transient experiments. Chem. Eng. Sci. 56, 2019–2035 [6].

Harmsen, J.M.A., Hoebink, J.H.B.J., Schouten, J.C., 2001b. NO reduction by CO over automotive exhaust gas catalysts in the presence of O_2. Catal. Lett. 71, 81–90 [6].

Harmsen, J.M.A., Hoebink, J.H.B.J., Schouten, J.C., 2001c. Kinetic modeling of transient NO reduction by CO in the presence of O_2 over an automotive exhaust gas catalyst. Stud. Surf. Sci. Catal. 133, 349–356 [3].

Harrison, D.P., 2008. Sorption-enhanced hydrogen production: a review. Ind. Eng. Chem. Res. 47, 6486–6501 [22].

Harrison, D.P., 2009. Calcium enhanced hydrogen production with CO_2 capture. Energy Procedia 1, 675–681 [22].

Harrison, D.P., Peng, Z.-Y., 2003. Low-carbon monoxide hydrogen by sorption-enhanced reaction. Intern. J. Chem. React. Eng. 1, A37 [22].

Hartman, M., Coughlin, R., 1972. SO_2 oxidation in a trickle bed reactor packed with carbon. Chem. Eng. Sci. 27, 867–880 [20].

Hasan, C.M.M., Shimizu, K., 2008. Effect of temperature upshift on fermentation and metabolic characteristics in view of gene expressions in *E. coli*, Microb. Cell Factor. On-line publication 7, 35 [23].

Hashimoto, K., Adachi, S., Nojima, H., Ueda, Y., 1983. A new process combining adsorption and enzyme reaction for producing higher-fructose syrup. Biotechnol. Bioengng. 25, 2371–2393 [1] [21].

Hashimoto, K., Adachi, S., Nojima, H., Shimono, M., Miyahara, M., 1982. Development of a new process for producing a higher fructose syrup by using a system combining adsorption columns and immobilized reactors. Rep. Asahi Glass Found. Ind. Technol. 41, 265–277 [21].

Hashimoto, K., Adachi, S., Shirai, Y., 1993b. Development of new bioreactors of a simulated moving-bed type. In: Ganetsos, G., Barker, P.E. (Eds.), Preparative and Production Scale Chromatography. Marcel Dekker, New York, pp. 395–419 [21].

Hashimoto, K., Adachi, S., Shirai, Y., Morishita, M., 1993a. Operation and design of simulated moving-bed adsorbers. In: Ganetsos, G., Barker, P.E. (Eds.), Preparative and Production Scale Chromatography. Marcel Dekker, New York, pp. 273–300 [21].

Hashimoto, S., Kawakami, W., Akehata, T., 1976a. A study on control of radiation polymerization by dose rate control. Ind. Eng. Chem. Proc. Des. Dev. 15, 244–249 [7].

Hashimoto, S., Kawakami, W., Akehata, T., 1976b. Experimental study on control of radiation polymerization by dose rate control. Ind. Eng. Chem. Proc. Des. Dev. 15, 549–552 [7].

Hattori, T., Murakami, Y., 1968. Study on the pulse reaction technique I. Theoretical study. J. Catal. 10, 114–122 [1] [13] [20].

Hatziantoniou, V., Andersson, B., Larsson, T., Schöön, N.-H., Carlsson, L., Schwartz, S., Wideen, K.-B., 1986. Preparation, characterization and testing of a new type of porous catalytic plates usable for liquid phase hydrogenation at enhanced mass transfer conditions. Ind. Eng. Chem. Proc. Des. Dev. 25, 143–150 [17].

Haure, P.M., Bogdashev, S.M., Bunimovich, M., Stegasov, A.N., Hudgins, R.R., Silveston, P.L., 1990b. Thermal waves in the periodic operation of a trickle-bed reactor. Chem. Eng. Sci. 45, 2255–2261 [17].

Haure, P.M., Hudgins, R.R., Silveston, P.L., 1989. Periodic operation of a trickle-bed reactor. AIChE J. 35, 1437–1444 [1] [16] [17] [20].

Haure, P.M., Hudgins, R.R., Silveston, P.L., 1990a. Steady-state models for SO_2 oxidation in a trickle-bed reactor. Chem. Eng. J. 43, 121–125 [17].

Haure, P.M., Silveston, P.L., Hudgins, R.R., Bellut, M., 1992. Conversion Efficiency in Trickle Bed Reactors 1988 US Patent 5011675, issued 15 Mar. 1988; Can. Patent 1299843, issued 5 May 1992 [17].

Hayes, R.E., 2004. Catalytic solutions for fugitive methane emissions in the oil and gas sector. Chem. Eng. Sci. 59, 4073–4080 [18].

Haynes, T.N., Caram, H.S., 1994. The simulated moving bed chemical reactor. Chem. Eng. Sci. 49, 5465–5472 [5] [21].

Haynes, T.N., Georgakis, Ch., Caram, H.S., 1992. The application of reverse flow reactors to endothermic reactions. Chem. Eng. Sci. 47, 2927–2932 [18].

Haynes, T.N., Georgakis, Ch., Caram, H.S., 1995. The design of reverse flow reactors for catalytic combustion systems. Chem. Eng. Sci. 50 (3), 401–416 [18].

Hayter, P.M., Curling, E.M.A., Gould, M.L., Baines, A.J., Jenkins, N., Salmon, I., Strange, P.G., Bull, A.T., 1993. The effect of the dilution rate on CHO cell physiology and recombinant interferon-γ production of glucose-limited chemostat culture. Biotechnol. Bioeng. 42, 1077–1085 [10].

Hegedus, L.L., Chang, C.C., McEwen, D.J., Sloan, E.M., 1980. Response of catalyst surface concentrations to forced concentration oscillations in the gas phase. The NO, CO, O_2 system over Pt alumina. Ind. Eng. Chem. Fund. 19, 367–373 [6].

Hegedus, L.L., Oh, S.H., Baron, K., 1977. Multiple steady states in an isothermal, integral reactor: the catalytic oxidation of carbon monoxide over platinum alumina. AIChE J. 23, 632–642 [3].

Hegedus, L.L., Summers, J.C., Schlatter, J.C., Baron, K., 1979. Poison resistant catalysts for the simultaneous control of hydrocarbon, carbon monoxide and nitrogen oxide emission. J. Catal. 56, 321–335 [6].

Heggs, P.J., 1986. Modelling of the dynamics of regenerative catalytic reactors. Trans. Inst. Math. Comput. 8 (3), 115–122 [18].

Heggs, P.J., Abdullah, N., 1986. Experimental investigation of a transient catalytic reactor for the dehydrogenation of ethylbenzene to styrene. Chem. Eng. Res. Des. 64, 258–265 [18].

Helmrich, H., Renken, A., Schügerl, K., 1974. Control of effective rate of heterogeneous catalytic reactions by forced variation of concentration. Chem. Ing. Tech. 46 (15), 647 (Synopsis) [1] [2].

Hendrickson, W.G., Daniels, F., 1953. Fixation of atmospheric nitrogen in a gas heated furnace. Ind. Eng. Chem. 45 (12), 2613–2615 [18].

Herbsthofer, R., Bart, H.-J., 2003. Influence of reaction kinetics on the performance of a chromatographic reactor. Chem. Eng. Technol. 26, 874–879 [20].

Herguido, J., Menéndez, M., Santamaría, J., 2005. On the use of fluidized bed catalytic reactors where reduction and oxidation zones are present simultaneously. Catal. Today 100, 181–189 [4] [8].

Herman, R.G., Klier, K., Simmons, G.W., Finn, B.P., Bulko, J.B., Kobylinski, T.P., 1979. Catalytic synthesis of methanol from carbon monoxide/hydrogen, I. Phase composition, electronic properties and activities of the copper/zinc oxide/M_2O_3. J. Catal. 56, 407–429 [2].

Hermanson, J.C., Gaehm-Maghami, E., Johari, H., 2004a. CO/unburned hydrocarbon emissions of strongly-pulsed turbulent diffusion flames. Combust. Sci. Technol. 176 (11), 1855–1866 [5].

Hermanson, J.C., Johari, H., Stocker, D.P., Hegde, U.G., 2004b. Buoyancy effects in strongly pulsed turbulent diffusion flames. Combust. Flame 139, 61–76 [5].

Hershkowitz, F., Berlowitz, P.J., Deckman, H.W., Marucchi-Soos, E., Gurciullo, C.S., Frederick, J.W., Rados, N., Agnihotri, R., 2004. A Breakthrough process for the production of hydrogen, AIChE Annual meeting. Fuel Cell Topical Conference November, Paper 23D [18].

Hershkowitz, F., Berlowitz, P.J., Socha, R.F., Marucchi-Soos, E., Frederick, J.W., 2005. Hydrogen production via steam reforming in a reverse-flow reactor. Proc. AIChE Annual Meeting., New York, New York [18].

Hershkowitz, F., Deckman, H.W., Frederick, J.W., Fulton, J.W., Socha, R.F., 2009. Pressure swing reforming: a novel process to improve cost and efficiency of CO_2 capture in power generation. Energy Procedia 1, 683–688 [18] [22].

Herz, R.K., 1981. Dynamic behavior of automotive catalysts: 1. Catalyst oxidation and reduction. Ind. Eng. Chem. Prod. Res. Dev. 20, 451–457 [6].

Herz, R.K., 1982. The dynamic behavior of three way automotive catalysts. In: Bell, A.T., Hegedus, L.L. (Eds.), Catalysis Under Transient Conditions. American Chemical Society, Washington, D.C., ACS Symp. Series pp. 178 59-78 [6].

Herz, R.K., 1987. Dynamic behavior of automotive three-way emission control systems. In: Crucq, A., Frennet, A. (Eds.), Catalysis and Automotive Pollution Control. Elsevier Science Publ. Amsterdam, pp. 427–445 [6].

Herz, R.K., Kiela, J.B., Sell, J.A., 1983. Dynamic behavior of automotive catalysts: 2. Carbon monoxide conversion under transient air/fuel ratio conditions. Ind. Eng. Chem. Prod. Res. Dev. 22, 387–396 [6].

Herz, R.K., Sell, J.A., 1985. The dynamic behavior of automotive catalysts. 3. Transient enhancement of water gas shift over rhodium. J. Catal. 94, 166–174 [6].

Herz, R.K., Shinouskis, E.J., 1985. Dynamic behavior of automotive catalysts. 4. Impact of air/fuel ratio excursions during driving. Ind. Eng. Chem. Prod. Res. Dev. 24, 385–390 [6].

Hess, St., Freund, H., Liauw, M.A., Emig, G., 2001a. Butane oxidation to maleic anhydride over a VPO catalyst following the riser regenerator approach. In: Froment, G.F., Waugh, K.C. (Eds.), Proc. 3rd Intern Sympos. Reaction Kinetics and the Development and Operation of Catalytic Processes (2001), Oostende, Belgium [4].

Hess, St., Freund, H., Liauw, M.A., Emig, G., 2001b. Butane oxidation to maleic anhydride over a VPO catalyst following the riser regenerator approach. Stud. Surf. Sci. Catal. 133, 205–210 [4].

Hessel, V., Renken, A., Schouten, J.C., Yoshida, J. (Eds.), 2009. Micro Process Engineering: A Comprehensive Handbook Vols. 1-3, Wiley-VCH, Weinheim, Germany [16].

Hevia, M.A.G., Fissore, D., Ordóñez, S., Díez, F.V., Barresi, A.A., 2007a. Combustion of medium concentration CH_4–air mixtures in non-stationary reactors. Chem. Eng. J. 131, 343–349 [18].

Hevia, M.A.G., Ordóñez, S., Díez, F.V., 2006. Effect of wall properties on the behavior of bench-scale reverse flow reactors. AIChE J. 52 (9), 3203–3209 [18].

Hevia, M.A.G., Ordóñez, S., Díez, F.V., 2007b. Effect of the catalyst properties on the performance of a reverse flow reactor for methane combustion in lean mixtures. Chem. Eng. J. 129, 1–10 [18].

Hinrichsen, K.-O., 2008. Transient Catalytic Studies. In: Ertl, G., Knözinger, H., Schüth, F., Weitkamp, J. (Eds.), Handbook of Heterogeneous Catalysis, Wiley-VCH, pp. 1543–1560 [13].

Hirata, H., Kishita, K., Nagai, Y., Dohmae, K., Shinjoh, H., Matsumoto, S., 2011. Characterization and dynamic behavior of precious metals in automotive exhaust gas purification catalysts. Catal. Today 164, 467–473 [6].

Hoebink, J.H.B.J., Harmsen, J.M.A., Balenovic, M., Bacx, A.C.P.M., Schouten, J.C., 2001. Automotive exhaust gas conversion: from elementary step kinetics to prediction of emission dynamics. Top. Catal. 16–17 (1–4), 319–327 [6].

Hoebink, J.H.B.J., Nievergeld, A.J.L., Marin, G.B., 1999. CO oxidation in a fixed bed reactor with high frequency cycling of the feed. Chem. Eng. Sci. 54 (20), 4459–4468 [6].

Hoedemakers, G.F., Thoenes, D., 1990. Continuous emulsion polymerization in a pulsed packed column. In: Kleintjens, L.A. (Ed.), Integration of Fundamental Polymer Science and Technology. Elsevier Science Publishers, London, U.K. pp. 183–191 [7].

Hoffman, M.R., Martin, S.T., Choi, W., 1995. Environmental applications of semiconductor photocatalysis. Chem. Rev. 95, 69–96 [11].

Hoffmann, J.G., Echigo, R., Youshida, H., Tada, S., 1997. Experimental study on combustion in porous media with a reciprocating flow system. Combust. Flame 111, 32–46 [18].

Hohmann, W., Lebender, D., Müller, J., Schinor, N., Schneider, F.W., 1997. Enhancement of the Production Rate in Chemical Reactions with Thresholds. J. Phys. Chem. A 101, 9132–9136 [11].

Horn, F.J.M., 1965. Attainable and unattainable regions in chemical reaction technique. Proc. 3rd Europ. Symp. Chem. React. Eng. Pergamon Press, Oxford, U.K. pp. 292–303 [14].

Horn, F.J.M., Bailey, J.E., 1968. An application of the theorem of relaxed control to the problem of improving catalyst selectivity. J. Opt. Theory Appl. 2 (6), 441–449 [12] [14].

Horn, F.J.M., Lin, R.C., 1967. Periodic processes: a variational approach. Ind. Eng. Chem. Proc. Des. Dev. 6 (1), 21–30 [1] [14] [16] [19].

Horowitz, G., Martinez, O.M., Cukierman, A.L., Cassanello, M., 1999. Effect of the catalyst wettability on the performance of a trickle-bed reactor for ethanol oxidation as a case study. Chem. Eng. Sci. 54, 4811–4816 [17].

Houck, C., Joines, J., Kay, M., 1995. A genetic algorithm for function optimization: a MATLAB implementation. North Carolina State University. NCSU-IE TR 95–09 [10].

Hu, Y., Zhao, F., Wei, F., Jin, Y., 2007. Ammoxidation of propylene to acrylonitrile in a bench-scale circulating fluidized bed reactor. Chem. Eng. Process. 46, 918–923 [4].

Huang, X.-F., 1999. Dynamic kinetic model of butane selective oxidation to maleic anhydride and its application. Ph.D. thesis. Beijing Univ. Chem. Technol. Beijing (China) [4].

Huang, X.-F., Chen, B.-H., Liu, B.-J., Silveston, P.L., Li, C.-Y., 2002b. Reoxidation kinetics of a VPO catalyst. Catal. Today 74 (2), 121–130 [4].

Huang, X.-F., Li, C.-Y., Chen, B.-H., Qiao, C.-Z., Yang, D.-H., 2001. Investigation of the unsteady state oxidation of n-butane to maleic anhydride in fixed-bed reactors. Ind. Eng. Chem. Res. 40, 768–773 [4].

Huang, X.-F., Li, C.-Y., Chen, B.-H., Silveston, P.L., 2002a. Transient kinetics of n-butane oxidation to maleic anhydride over a VPO catalyst. AIChE J. 48 (4), 846–855 [4].

Huang, H.-J., Wu, G.-B., Yang, J., Dai, Y.-C., Yuan, W.-K., Lu, H.-B., 2004. Modeling of gas-solid chemisorption in chemical heat pumps. Sep. Purif. Tech. 34, 191–200 [8].

Hudgins, R.R., Silveston, P.L., Edvinsson, R., 1988. Claus reaction: effect of forced feed composition cycling. Appl. Catal. 50, 303–306 [11].

Hufton, J.R., Mayorga, S., Sircar, S., 1999. Sorption-enhanced reaction process for hydrogen production. AIChE J. 45 (2), 248–256 [22].

Hugo, A.J., Jakelski, D.M., Stanitsas, G., Sullivan, G.R., Hudgins, R.R., Silveston, P.L., 1986. Long-term transients in a reactor under forced concentration cycling. Can. J. Chem. Eng. 64, 349–351 [3].

Huixiong., W., Shuzeng, Zh., Li C.-Y., 1996. Study of unstready-state catalytic oxidation of sulfur dioxide by periodic flow reversal. Can. J. Chem. Eng. 74, 766–771 [18].

Hung, K.G.W., Papadias, D., Björnbom, P., Anderlung, M., Åkermark, B., 2003. Reverse-flow operation for application of imperfectly immobilized catalysts. AIChE J. 49 (1), 151–167 [18].

Hungenberg, K.D., 1992. Strategies for PP-homo- and co polymers with broadened MWD in stirred-bed gas phase reactors. In: DECHEMA - Monographien, VCH Verlaggesel, Frankfurt, Germany, 127 pp. 501–509 [7].

Hyanek, I., Zacca, J., Teymour, F., Ray, W.H., 1995. Dynamics and stability of polymerization process flow sheets. Ind. Eng. Chem. Res. 34, 3872–3877 [7].

Iggland, M., Leion, H., Mattisson, T., Lyngfelt, A., 2010. Effect of fuel particle size on reaction rate in chemical looping combustion. Chem. Eng. Sci. 65, 5841–5851 [8].

Ihssen, J., Kowarik, M., Dilettoso, S., Tanner, C., Wacker, M., Thony-Meyer, L., 2010. Production of glycoprotein vaccines in *Escherichia coli,* Microb. Cell Fact. 9 Article # 61 [10].

Ilić, M., Petkovska, M., Seidel-Morgenstern, A., 2008. Estimation of single solute adsoprtion isotherms applying the nonlinear frequency response method using non-optimal frequencies. J. Chromat. A. 1200, 183–192 [14].

Il'in, A., Luss, D., 1992. Wrong-way behavior of packed-bed reactors: influence of reactant adsorption on support. AIChE J. 38, 1609–1617 [18].

Il'in, A., Luss, D., 1993. Wrong-way behavior of packed-bed reactors: Influence of an undesired consecutive reaction. Ind. Eng. Chem. Res. 32, 247–252 [18].

Iliuta, I., Larachi, F., 2005. Stretching operational life of trickle bed filters by liquid-induced pulse flow. AIChE J. 51 (7), 2034–2047 [17].

Iliuta, I., Leclerc, A., Larachi, F., 2010. Allothermal steam gasification of biomass in cyclic multi-compartment bubbling fluidized-bed gasifier/combustor—new reactor concept. Biores. Technol. 101, 3194–3208 [8].

Iliuta, I., Ortiz-Arroyo, A., Larachi, F., Grandjean, B., Wild, G., 1999. Hydrodynamics and mass transfer in trickle-bed reactors: an overview. Chem. Eng. Sci. 54, 5329–5337 [17].

Irvine, R.L., Davis, W.B., 1971. Use of sequencing batch reactors for waste treatment, 26th Annual Industrial Waste Treatment Conference. Purdue University, West Lafayette, Indiana [10].

Irvine, R.L., Moe, W.M., 2001. Periodic biofilter operation for enhanced performance during unsteady-state loading conditions. Water Sci. Technol. 43 (3), 231–239 [18].

Irvine, R.L., Wilderer, P.A., Flemming, H.-C., 1997. Controlled unsteady state processes and technologies—an overview. Water Sci. Technol. 35 (1), 1–10 [10].

Irvine, R.L., Yocum, P.S., Early, J.P., Chozick, R., 1993. Periodic processes for in-situ and on site remediation of leachates and soils. Water Sci. Technol. 27, 97–104 [10].

Ishihara, T., Yamada, T., Akbay, T., Takita, Y., 1999. Partial oxidation of methane over fuel cell type reactor for simultaneous generation of synthesis gas and electric power. Chem. Eng. Sci. 54 (10), 1535–1540 [4].

Isozaki, C., Katagiri, T., Nakamura, Y., Hirabayashi, S., Yabe, S., Yamaki, T., 1990. Demonstration of DC/DA sulphuric acid plant based on unsteady-state process. In: Matros, Yu.Sh. (Ed.), Unsteady-State Processes in Catalysis. VPS Publishers BV, Utrecht, The Netherlands [18].

Ito, T., Wang, J.-X., Lin, C.-H., Lunsford, J.H., 1985. Oxidative dimerization of methane over a lithium promoted magnesium oxide catalyst. JACS 107, 5062–5068 [4].

Ivanov, A.A., Balzhinimaev, B.S., 1987. New data on kinetics and reaction-mechanism for SO_2 oxidation over vanadium catalysts. React. Kinet. Catal. Lett. 35, 413–424 [3].

Ivanov, Yu.V., Noskov, A.S., Matros, Yu.Sh., Bunimovich, G.A., 1992. Multiplicity of states in catalytic reactors with periodic flow reversal of gas mixture flow. Teor. Osn. Khimich. Tekhnol. 26, 228–235 [18].

Iyer, M.V., Gupta, H., Sakadjian, B.B., Fan, L.-S., 2004. Multicyclic study on the simultaneous carbonation and sulfation of high-reactivity CaO. Ind. Eng. Chem. Res. 43, 3939–3947 [8].

Jadot, F., Bastin, G., Van Impe, J.F., 1998. Optimal adaptive control of a bioprocess with yield-productivity conflict. J. Biotechnol. 65, 61–68 [10].

Jaeger, N.I., Plath, P.J., Svensson, P., 1990. Concentration forcing of the oscillating heterogeneous catalytic oxidation of carbon monoxide. In: Gray, P., Nicolis, G., Baras, F., Borckmans, P., Scott, S.K. (Eds.), Spatial Inhomogeneities and Transient Behaviour in Chemical Kinetics. Manchester University Press, Manchester, U.K, pp. 593–603 [2] [3].

Jaeger, N.I., Schulz-Eckloff, G., Kapicka, J., 1992. In situ perpetual regeneration of a Ni-faujasite methanation catalyst. Catal. Lett. 14, 331–337 [1] [2] [23].

Jagtap, S.B., Wheelock, T.D., 1996. Regeneration of sulfided calcium-based sorbents by a cyclic process. Energy Fuels 10, 821–827 [8].

Jahn, R., Šnita, D., Kubíček, M., Marek, M., 2001. Evolution of spatiotemporal temperature patterns in monolithic catalytic reactor. Catal. Today 70, 393–409 [6].

Jain, A.K., 1981a. A study of forced feed composition cycling in the catalytic synthesis of ammonia. PhD Thesis. Dept. Chem. Eng. Univ. of Waterloo, Waterloo, Ontario, Canada [2] [13].

Jain, A.K., Hudgins, R.R., Silveston, P.L., 1981b. A model to Unni's experiments on cyclic catalytic oxidation of SO_2. Chem. Eng. Sci. 36, 233–234 [3].

Jain, A.K., Hudgins, R.R., Silveston, P.L., 1982a. Forced composition cycling experiments in a fixed bed ammonia synthesis reactor, Proc. 7th ISCRE. ACS Symp. Ser. 196, 97–107 [2].

Jain, A.K., Hudgins, R.R., Silveston, P.L., 1982b. Evidence for bulk-phase nitrogen dissolution in iron in catalytic ammonia synthesis. Can. J. Chem. Eng. 60, 809–811 [13].

Jain, A.K., Hudgins, R.R., Silveston, P.L., 1983. Influence of forced feed composition cycling on the rate of ammonia synthesis over an industrial iron catalyst: 1—Effect of cycling parameters and mean composition. Can. J. Chem. Eng. 61, 824–832 [2].

Jain, A.K., Hudgins, R.R., Silveston, P.L., 1985. Effectiveness factor under cyclic operation of a reactor. Can. J. Chem. Eng. 63, 165–169 [2].

Jaree, A., Budman, H., Hudgins, R.R., Silveston, P.L., Yakhnin, V., Menzinger, M., 2001b. Temperature excursions in reactors packed with segregated layers of catalysts. Catal. Today 69, 137–146 [16].

Jaree, A., Hudgins, R.R., Budman, H., Silveston, P.L., Menzinger, M., 2004. Numerical investigation of resonance behavior of a tubular packed bed reactor with non-uniform activity. Can. J. Chem. Eng. 82, 387–391 [3] [16].

Jaree, A., Hudgins, R.R., Budman, H., Silveston, P.L., Yakhnin, V., Menzinger, M., 2003a. Amplification of inlet temperature disturbances in a packed-bed reactor for CO oxidation over Pt/Al_2O_3. Chem. Eng. Sci. 58, 833–839 [3] [16].

Jaree, A., Hudgins, R.R., Budman, H., Silveston, P.L., Yakhnin, V., Menzinger, M., 2003b. Hysteresis and extinction waves in catalytic CO oxidation caused by reactant concentration perturbations in a packed bed reactor. Ind. Eng. Chem. Res. 42, 1662–1673 [3] [16].

Jaree, A., Silveston, P.L., Hudgins, R.R., Budman, H., Menzinger, M., 2001a. Temperature response to reactant concentration perturbation in a packed-bed reactor. Can. J. Chem. Eng. 79, 823–827 [3].

Jayaraman, V.K., Kulkarni, B.D., Gupta, K., Rajesh, J., Kusumaker, H.S., 2001. Dynamic optimization of fed-batch bioreactors using the ant algorithm. Biotech. Prog. 17, 81–88 [10].

Jensen, J.D., Gabe, D.R., Wilcox, G.D., 1998. The practical realization of zinc-iron CMA coatings. Surf. Coat. Tech. 105 (3), 240–250 [9].

Jensen, S., Olsen, J.L., Thorsteinsson, S., Hansen, O., Quaade, U.J., 2007. Forced thermal cycling of catalytic reactions: Experiments and modeling. Catal. Commun. 8, 1985–1990 [16].

Jensen, S., Olsen, J.L., Thorsteinsson, S., Hansen, O., Quaade, U.J., 2008b. Forced thermal cycling of catalytic reactions: Experiments and modeling. Chem. Eng. J. 135S, S237–S241 [16].

Jensen, S., Thorsteinsson, S., Hansen, O., Quaade, U.J., 2008a. Parametric investigation of rate enhancement during fast temperature cycling of CO oxidation in microreactors. Chem. Eng. J. 135S, S237–S241 [1].

Jeong, Y.O., Luss, D., 2003. Pollutant destruction in a reverse-flow chromatographic reactor. Chem. Eng. Sci. 58, 1095–1102 [18].

Jin, H.G., Ishida, M., 2002. Reactivity study on natural-gas-fueled chemical-looping combustion by fixed bed reactor. Ind. Eng. Chem. Res. 41 (16), 4004–4007 [8].

Jirát, J., Štěpánek, F.M., Marek, M., Kubíček, K., 2001. Comparison of design and operation strategies for temperature control during selective catalytic reduction of NO_X. Chem. Eng. Technol. 24, 35–40 [19].

Johnsen, K., Ryu, H.J., Grace, J.R., Lim, C.J., 2006. Sorption-enhanced steam reforming of methane in a fluidized-bed reactor with dolomite as CO_2-acceptor. Chem. Eng. Sci. 61, 1195–1202 [22].

Jones, C.A., Leonard, J.J., Sofranko, J.A., 1987a. The oxidative conversion of methane to higher hydrocarbons over alkali-promoted Mn/SiO_2. J. Catal. 103, 38–319 [4].

Jones, C.A., Leonard, J.J., Sofranko, J.A., 1987b. Fuels for the future: remote gas conversion. Energy Fuels 1, 12–16 [4].

Jordi, R.G., Do, D.D., 1993. Analysis of the frequency-response method for sorption kinetics in bidispersed structured sorbents. Chem. Eng. Sci. 48, 1103–1130 [15].

Joy, G.J., Lester, G.R., Molinaro, F.S., 1979. The influence of sulfur species on the laboratory performance of automotive three component control catalysts. SAE Tech. Paper Series, Paper # 790943, Society of Automotive Engineers, Warrendale, PA [6].

Jugjai, S., Sawanon, A., 2004. The surface combustor-heater with cyclic flow reversal combustion embedded with water tube bank. Fuel 83, 2369–2379 [18].

Kaisare, N.S., Lee, J.H., Fedorov, A.G., 2005a. Hydrogen generation in a reverse-flow microreactor: 1. Model formulation and scaling. AIChE J. 51 (8), 2254–2264 [18].

Kaisare, N.S., Lee, J.H., Fedorov, A.G., 2005b. Hydrogen generation in a reverse-flow microreactor: 2. Simulation and analysis. AIChE J. 51 (8), 2265–2272 [18].

Kalantary, M.R., Wilcox, G.D., Gabe, D.R., 1995. The production of compositionally modulated alloys by simulated high speed deposition from a single source. Electrochim. Acta 40 (11), 1609–1616 [9].

Kaneko, Y., Kobayashi, H., Komagoma, R., Hirako, O., Nakayama, O., 1978. Effect of the air-fuel ratio modulation on conversion efficiency of three-way catalysts. SAE Tech. Series Paper #780607, Soc. Automot. Eng. Warrendale, PA [6].

Kapička, J., Marek, M., 1989a. Oscillations on individual catalytic pellets in a packed bed: CO oxidation on Pt/Al_2O_3. J. Catal. 119, 508–511 [3].

Kapička, J., Marek, M., 1989b. Transition to chaos in the oscillating CO oxidation. Surf. Sci. 222, L885–L889 [3].

Kapil, A., Bhatand, S.A., Sadhukhan, J., 2010. Dynamic simulation of sorption enhanced reaction processes for biodiesel production. Ind. Eng. Chem. Res. 49, 2326–2335 [21].

Kato, Y., Ando, K., Yoshizawa, Y., 2003. Study on a regenerative fuel reformer for a zero-emission vehicle system. J. Chem. Eng. Jpn. 36 (7), 860–866 [22].

Kawase, M., Pilgrim, A., Araki, T., Hashimoto, K., 2001. Lactosucrose production using a simulated moving bed reactor. Chem. Eng. Sci. 56 (2), 453–458 [21].

Kawase, M., Suzuki, T.B., Inoue, K., Yoshimoto, K., Hashimoto, K., 1996. Increased esterification conversion by application of the simulated moving-bed reactor. Chem. Eng. Sci. 51, 2971–2976 [21].

Kawase, M., Inoue, Y., Araki, T., Hashimoto, K., 1999. The simulated moving-bed reactor for production of bisphenol A. Catal. Today 48, 199–209 [21].

Keil, W., Wicke, E., 1980. Über die kinetischen instabilitäten bei der CO-oxidation an platin-katalysatoren. Ber. Bunsen Ges. Phys. Chem. 84, 377–383 [3].

Keith, J.M., 2003. Controlling reverse-flow reactors via multiscale transient thermal dispersion. Adv. Environ. Res. 7, 521–535 [18].

Keith, J.M., Leighton Jr., D.T., Chang, H.-C., 1999. A new design of reverse-flow reactors with enhanced thermal dispersion. Ind. Eng. Chem. Res. 38 (3), 667–682 [18].

Keller, G.E., Bhasin, M.M., 1982. Synthesis of ethylene via oxidative coupling of methane. J. Catal. 73, 9–19 [4].

Kevrekidis, I.G., Schmidt, L.D., Aris, R., 1984. On the dynamics of periodically forced chemical reactors. Chem. Eng. Comm. 30, 323–330 [3].

Kevrekidis, I.G., Schmidt, L.D., Aris, R., 1986. Some common features of periodically forced reacting systems. Chem. Eng. Sci. 41, 1263–1276 [16].

Khadilkar, M.R., Al-Dahhan, M.H., Dudukovic, M.P., 1999. Parametric study of unsteady-state flow modulation in trickle-bed reactors. Chem. Eng. Sci. 54 (13–14), 2585–2595 [17].

Khadilkar, M.R., Al-Dahhan, M.H., Dudukovic, M.P., Colakyan, M., 1996. Comparison of trickle-bed and upflow reactor performance at high pressure: model predictions and experimental observations. Chem. Sci. Eng. 51 (10), 2139–2148 [17].

Khazai, B., Vrieland, G.E., Murchison, C.B., 1995. Process of oxidizing aliphatic hydrocarbons in the presence of a solid heterogeneous catalyst. European Patent, B1 0 409 355 [4].

Khinast, J., Gurumoorthy, A., Luss, D., 1998. Complex dynamic features of cooled reverse flow reactor. AIChE J. 44, 1128–1135 [18].

Khinast, J., Jeong, Y.O., Luss, D., 1999. Dependence of cooled reverse-flow reactor dynamics on reactor model. AIChE J. 45 (2), 299–309 [18].

Khinast, J., Luss, D., 1997. Mapping regions with different bifurcation diagrams of a reverse-flow reactor. AIChE J. 43 (8) 2937–2047 [18].

Kiatkittipong, W., Tagawa, T., Goto, S., Assabumrungrat, S., Praserthdam, P., 2004. Oxidative coupling of methane in LSM/YSZ/LaAlO SOFC reactor. J. Chem. Eng. Japan 37, 1461–1470 [4].

Kikas, T., Bardenshteyn, I., Williamson, C., Ejimofor, C., Puri, P., Fedorov, A.G., 2003. Hydrogen production in a reverse-flow autothermal catalytic microreactor: from evidence of performance enhancement to innovative reactor design. Ind. Eng. Chem. Res. 42, 6273–6279 [18].

Kim, D.K., Hulburt, H.M., 1972. Preprint, Proc. 64th Annual AIChE Meeting, San Francisco, CA, USA [16].

Kim, K.J., Choi, K.Y., 1991. Continuous olefin copolymerization with soluble Ziegler-Natta catalysts. AIChE J. 37 (8), 1255–1260 [7].

Kiparissides, C., MacGregor, J.F., Hamielec, A.E., 1979. Continuous emulsion polymerization. Modeling oscillations in vinyl acetate polymerization. J. Appl. Polym. Sci. 23, 401–411 [7].

Kirkby, N.F., Morgan, J.E.P., 1994. A theoretical investigation of pressure swing reaction. Trans. I. Chem. E. 72 (A), 541–550 [22].

Kiselev, O.V., 1993. Theoretical study of the phenomena of heat waves movement in catalyst bed, Monograph, Russian Academy of Sciences. Institute of Catalysis, Novosibirsk, Russia (in Russian) [18].

Kiselev., O.V., Matros, Yu.Sh., 1980. Propagation of the combustion front of a gas mixture in granular bed of catalyst. Combust. Explo. Shock 16, 152–157 [18].

Kiselev, O.V., Matros, Yu.Sh., 1987. Heat output from the hot region in a thermal front propagating in a catalyst layer. Combust. Explo. Shock 23 (5), 655–662 [18].

Kiss, I.Z., Zhai, Y.M., Hudson, J.L., 2008. Resonance clustering in globally coupled electrochemical oscillators with external forcing. Phys. Rev. E 77 (4) 046204 Part 2 [3].

Kiwi-Minsker, L., Wolfrath, O., Renken, A., 2002. Membrane reactor microstructured by filamentous catalyst. Chem. Eng. Sci. 57, 4947–4953 [4] [12].

Klahn, J., Bandermann, F., 1979. Optimising procedure of reaction of living polymer. Theoretical considerations about molecular weight distribution of living polymers in a periodically stirred tank reactor. Angew. Makromol. Chem. 83, 1–19 [7].

Klinkenberg, A., 1961. Chromatography of substances undergoing slow reversible chemical reaction. Chem. Eng. Sci. 15, 255–259 [20].

Kloppenburg, E., Gilles, E.D., 1999. Automatic control of the simulated moving bed process for C_8 aromatics separation using asymptotically exact input/output linearization. J. Process Contr. 9, 41–50 [19].

Knödler, A., 1986. Pulsed electrodeposition of precious metals. In: Puippe, J.-C., Leaman, F. (Eds.), Theory and Practice of Pulse Plating. American Electroplaters & Surface Finishers Society, Washington, D.C. [9].

Knödler, A., Ch.Raub, J., Raub, E., 1985. Die galvanische Abscheidung der Nickel-Zink, Nickel-Zinn und Blei-Zinn Legierungen mit pulse-plating Strom. Metalloberfläche 39 (1), 21–25 [9].

Kobayashi, M., 1982. Characterization of transient response curves in heterogeneous catalysis. Chem. Eng. Sci. 37, 393–401 [13].

Kobayashi, M., Kobayashi, H., 1974. Transient response method in heterogeneous catalysis. Catal. Rev. 23, 1–15 [13].

Kočí, P., Kubíček, M., Marek, M., 2004a. Periodic forcing of three-way catalyst with diffusion in the washcoat. Catal. Today 98, 345–355 [6].

Kočí, P., Marek, M., Kubíček, M., Maunula, T., Härkönen, M., 2004b. Modelling of catalytic monolith converters with low- and high-temperature NOx storage compounds and differentiated washcoat. Chem. Eng. J. 97, 131–139 [6].

Kocirik, M., 1967. A Contribution to the theory of chromatography; Linear non-equilibrium chromatography with a chemical reaction of the first order. J. Chromatography 30, 459–468 [20].

Kodde, A.J., Bliek, A., 1997. Selectivity enhancement in consecutive reactions using the pressure swing reactor. Stud. Surf. Sci. Catal. 109, 419–428 [22].

Kodde, A.J., Fokma, I.S., Bliek, A., 2000. Selectivity effects on series reactions by reactant storage and PSA operation. AIChE J. 46 (11), 2295–2304 [22].

Koerts, T., Deelen, M.J.A.G., van Santen, R.A., 1992. Hydrocarbon formation from methane by a low-temperature, two-step reaction sequence. J. Catal. 138, 101–114 [12].

Koerts, T., van Santen, R.A., 1991. A low temperature reaction sequence for methane conversion. J. Chem. Soc. Chem. Comm., 1281–1283 [12].

Koerts, T., van Santen, R.A., 1992. Transient-response study of CO insertion into CHx surface intermediates on a vanadium-promoted rhodium catalyst. J. Catal. 134 (1), 13–23 [12].

Koerts, T., van Santen, R.A., 1993. Reaction routes for methane conversion on transition-metals at low-temperature. Stud. Surf. Sci. Catal. 75, 1065–1078 [12].

Koerts, T., van Santen, R.A., Ross, J.R.H., Solymosi, F., Moyes, R.B., Bhasin, M.M., Woo, S.I., Vanhardeveld, R., Wang, D., Machocki, A., Foger, K., Goodwin, J.G., Sarkany, A., 1993. Reaction routes for methane conversion on transition-metals at low-temperature. Stud. Surf. Sci. Catal. 75, 1065–1078 [12].

Kolbitsch, P., Bolhàr-Nordenkramf, J., Pröll, T., Hofbauer, H., 2010. Operating experience with chemical looping combustion in a 120kW dual circulating fluidized bed (DCFB) unit. Intern. J. Greenh. Gas Control 4, 180–185 [8].

Kolbitsch, P., Pröll, T., Bolhàr-Nordenkramf, J., Hofbauer, H., 2009. Operating experience with chemical looping combustion in a 120kW dual circulating fluidized bed (DCFB) unit. Ener. Procedia 1, 1465–1472 [8].

Kolios, G., Eigenberger, G., 1999. Styrene synthesis in a reverse-flow reactor. Chem. Eng. Sci. 54, 2637–2646 [18].

Kolios, G., Frauhammer, J., Eigenberger, G., 2000. Autothermal fixed-bed reactor concept. Chem. Eng. Sci. 55, 5945–5967 [18].

Kolios, G., Glöckler, B., Gritsch, A., Morillo, A., Eigenberger, G., 2005a. Heat-integrated reactor concept for hydrogen production by methane steam reforming. Fuel Cells 5 (1), 52–65 [18].

Kolios, G., Gritsch, A., Glöckler, B., Eigenberger, G., 2005b. Chapter 1. Enhancing productivity and thermal efficiency of high-temperature endothermic processes in heat-integrated fixed-bed reactors. In: Sundmacher, K., Kienle, A., Seidel-Morgenstern, A. (Eds.), Integrated Chemical Processes: Synthesis, Operation, Analysis, and Control. Wiley-VCH Verlag GmbH., Weinheim, Germany [18].

Komiyama, H., Inoue, H., 1968. Selectivity of Gas-Solid Catalytic Reaction. J. Chem. Eng. Jpn. 1 (2), 142–148 [12].

Konno, H., 1986. Pulsed anodic reactions. In: Puippe, J.-C., Leaman, F. (Eds.), Theory and Practice of Pulse Plating, American Electroplaters & Surface Finishers Society Washington, D.C. 1986 [9].

Konopnicki, D., Kuester, J.L., 1974. Forced oscillations in a non isothermal continuous polymerization reactor. J. Macromol. Sci. -Chem. A8, 887–908 [7].

Konstandopoulos, A.G., Kostoglou, M., 2000. Reciprocating flow regeneration of soot filters. Combust. Flame 121 (3), 488–500 [18].

Kotter, M., Lintz, H.-G., 1991. Selective catalytic reduction of nitrogen oxides. An original concept. Int. Chem. Eng. 31 (4), 685–692 [18].

Kotter, M., Lintz, H.-G., Turek, T., 1992. Selective catalytic reduction of nitrogen oxide by use of the Ljungstroem air heater as reactor: a case study. Chem. Eng. Sci. 47, 2763–2768 [18].

Koubek, J., Pasek, J., Ruzicka, V., 1980a. Exploitation of a non-stationary kinetic phenomenon for elucidation of surface processes in a catalytic reaction. Proc. 7th International Congress on Catalysis, Tokyo. vol. VIIb. 853–865 [11] [13].

Koubek, J., Pasek, J., Ruzicka, V., 1980b. Stationary and nonstationary deactivation of alumina and zeolites in elimination reactions. In: Delmon, B., Froment, G.F. (Eds.), Catalyst Deactivation. Elsevier Science Publ. Amsterdam, The Netherlands, pp. 251–260 [11] [13] [15].

Koubek, J., Thullie, J., Renken, A., Kurpas, M., 2000. Determination of the mechanism and models of stop-effect in catalysis. Polska Akad. Nauk. Instytut Inzyn. Chem. 2 (1), 3–22 [11].

Kouris, Ch, Neophytides, St., Vayenas, C.G., Tsamopoulos, J., 1998. Unsteady state operation of catalytic particles with constant and periodically changing degree of external wetting. Chem. Eng. Sci. 53, 3129–3142 [17].

Kowler, D.E., Kadlec, R.H., 1972. The optimal control of a periodic adsorber. I. Experiment. AIChE J. 18 (6), 1207–1219 [22].

Koznacheev, I.A., Dobrego, K.V., Shmelev, E.S., 2011. Water purification of organic inclusions in a reverse flow filtration combustion reactor. Int. J. Heat Mass Tran. 54, 932–937 [18].

Krefting, D., Kaira, P., Rotermund, H.H., 2009. Period doubling and spatiotemporal chaos in periodically forced CO oxidation on Pt(110). Phys. Rev. Lett. 102, 178301 [3].

Krgulikov, S.S., Becker, P., Jankowski, O., 2000. Combination of pulse & constant current plating for the creation of desirable surface microgeometry. In: Proc. 2000 SUR/FIN Conf.. American Electroplaters & Surface Finishers Society, Washington, D.C. 672–679 [9].

Krgulikov, S.S., Tsupak, T.E., Peganova, N.V., 2008. Nickel pulse plating from acetate-chloride bath, Proc.2008 SUR/FIN Conf., American Electroplaters & Surface Finishers Society, Washington, D.C. [9].

Krischer, K., Lübke, M., Wolf, W., Eiswirth, M., Ertl, G., 1991. Chaos and interior crisis in an electrochemical reaction. Ber. Bunsenges. Phys. Chem. 95, 821–823 [3].

Kristof, P., Pritzker, M., 1998. Improved copper plating through the use of current pulsing & ultrasonic agitation. Plat. Surf. Finish. 85 (11), 237–252 [9].

Kruglov, A.V., 1994. Methanol synthesis in a simulated countercurrent moving-bed adsorptive catalytic reactor. Chem. Eng. Sci. 49 (24A), 4699–4716 [21].

Kruglov, A.V., Bjorklund, M.C., Carr, R.W., 1996. Optimization of the simulated countercurrent moving-bed chromatographic reactor for the oxidative coupling of methane. Chem. Eng. Sci. 51, 2945–2950 [21].

Krylova, A.V., 1990. Specific features of ammonia synthesis under non-stationary conditions. In: Matros, Yu.Sh. (Ed.), Unsteady State Processes in Catalysis. VSP, Publishers, Zeist, The Netherlands [2].

Krylova, A.V., Abdeligani, A.Kh., 1987. Catalytic synthesis of ammonia under pulse conditions. Kinet. Katal. 28 (5), 1126–1131 [2].

Kuczynski, M., Oyeavaar, R., Pieters, R., Westerterp, K., 1987b. Methanol synthesis in a countercurrent gas-solid-solid trickle flow reactor, an experimental study. Chem. Eng. Sci. 42, 1887–1896 [21].

Kulkarni, M.S., Duduković, M.P., 1996. Bidirectional fixed-bed reactor for coupling of exothermic and endothermic reactions. AIChE J. 42, 2897–2910 [18].

Kulkarni, M.S., Duduković, M.P., 1997. Periodic operation of asymmetric bidirectional fixed-bed reactors: energy efficiency. Chem. Eng. Sci. 52 (1), 1777–1788 [18].

Kulkarni, M.S., Duduković, M.P., 1998. Periodic operation of asymmetric bidirectional fixed-bed reactors with temperature limitations. Ind. Eng. Chem. Res. 37, 770–781 [18].

Kumar, G.P., Sastry, I.V.K.S., Chidambaram, M., 1993. Periodic operation of a bioreactor with input multiplicities. Can. J. Chem. Eng. 71, 766–770 [10].

Kundu, P.K., Zhang, Y., Ray, A.K., 2009. Modeling and simulation of simulated countercurrent moving bed chromatographic reactor for oxidative coupling of methane. Chem. Eng. Sci. 64, 5143–5152 [21].

Kunitski, V.Ya., Boreskov, G.K., Matros, Yu.Sh, Charny, V.Z., Bunimovich, G.A., Eremin, O.G., Alentov, P.N., Klyachko, L.I., 1984. Zvetnye Metally (Non-ferrous Metals) No. 2, 28-31 (in Russian) [18].

Kushwaha, A., Poirier, M., Hayes, R.E., Sapoundjiev, H., 2005. Heat extraction from a flow reversal reactor in lean methane combustion. Trans. Inst. Chem. Eng., Part A, Chem. Eng. Res. Des. 83 (A2), 205–213 [18].

Kushwaha, A., Poirier, M., Sapoundjiev, H., Hayes, R.E., 2004. Effect of reactor internal properties on the performance of a flow reversal catalytic reactor for methane combustion. Chem. Eng. Sci. 59, 4081–4093 [18].

Labastida-Bardales, J.R., 1991. Oxidation of propylene to acrolein over a copper-molybdena catalyst. Ph.D. Thesis. Univ. Waterloo, Ontario, Canada [4].

Labastida-Bardales, J.R., Hudgins, R.R., Silveston, P.L., 1989. Multiplicity in the partial oxidation of propylene over a copper-molybdena catalyst. Can. J. Chem. Eng. 67, 418–422 [4] [8] [13].

Lai, H.-M., 1999. Modeling and performance analysis for a novel absorption cycle operated by periodic reversal forced convection (PRFC). Proc. Intern. Sorpt. Heat Pump Conf. 561–572 [8].

Lai, H.-M., Li, C.-Y., 1996. Application of periodic flow reversal reactors to chemical heat pump systems based on solid/vapour non-catalytic reaction. Chem. Eng. Sci. 51 (11), 2951–2957 [8].

Lai, H.-M., Li, C.-Y., Zheng, D., Fu, J., 1993. Modeling of cycling process of $CaCl_2/CH_3OH$ solid-vapor chemical heat pump, Proc. Intern. Absorption Heat Pump Conf., Amer. Soc. Mech. Eng. 31, 419–423 [8].

Lai, H.-M., Zheng, D., Li, C.-Y., Fu, J., 1992. Chemical heat pump based on $CaCl_2/CH_3OH$: theoretical and practical performance, Proc. 2nd Joint Meeting on Appl. Sci. Techn. Chemical Industry Press, Beijing, China [8].

Lang, X.-S., Hudgins, R.R., Silveston, P.L., 1989a. Maleic anhydride formation from 1,3-butadiene oxidation over a vanadia molybdate catalyst. Can. J. Chem. Eng. 67, 423–431 [4].

Lang, X.-S., Hudgins, R.R., Silveston, P.L., 1989b. Application of periodic operation to maleic anhydride production. Can. J. Chem. Eng. 67, 635–645 [1] [4] [8].

Lang, X.-S., Hudgins, R.R., Silveston, P.L., 1991. Selectivity improvement using alternate flushing and reactant cycle steps. Can. J. Chem. Eng. 69, 1121–1125 [1] [4] [8].

Lange, R., Gutsche, R., Hanika, J., 1999. Forced periodic operation of a trickle-bed reactor. Chem. Eng. Sci. 54 (13–14), 2569–2573 [17].

Lange, R., Hanika, J., Hudgins, R.R., Silveston, P.L., 1994b. Experimentelle untersuchungen zur periodischen prozessführung eines trickle-bed reactors. Chem. Ing. Tech. 66 (3), 365–369 [17].

Lange, R., Hanika, J., Klein, H., 1992. Bessere Ausnutzung von Katalysatoren in Trickle-Bed Reaktoren durch eine Periodische Prozessführung. Preprint, DECHEMA-GVC Meeting, Würzburg Germany [17].

Lange, R., Hanika, J., Stradiotto, D., Hudgins, R.R., Silveston, P.L., 1994a. Investigations of periodically operated trickle-bed reactors. Chem. Eng. Sci. 49, 5615–5621 [17].

Lange, R., Schubert, M., Dietrich, W., Grünewald, M., 2004. Unsteady-state operation of trickle-bed reactors. Chem. Eng. Sci. 59, 5355–5361 [17].

Langer, S.H., Yurchak, J.E., Patton, P.D., 1969. The gas chromatographic column as a chemical reactor. Ind. Eng. Chem. 61, 10–21 [20].

Langner, B., Bandermann, F., 1978. Optimising of reaction conditions for living polymerization. Preparation of living polymers with narrow molecular weight distribution in a periodically stirred tank reactor. Angew. Makromolec. Chem. 71, 101–115 [7].

Larruy, B., Ayude, A., Font, J., Fortuny, A., Begoa, C., Fabregat, A., Stüber, F., 2007. Gas feed composition modulation in phenol CWAO over active carbon. Chem. Eng. Sci. 62, 5564–5566 [17].

Lashmore, D.S., Weisshaus, I., Pratt, K., 1986. Electrodeposition of nickel-chromium alloys. Plat. Surf. Finish. 73 (3), 48–55 [9].

Laurence, R.L., Vasudevan, G., 1968. Performance of a polymerization reactor in periodic operation. Ind. Eng. Chem. Proc. DD. 7, 427–432 [7].

Lavelle, K., McMonagle, J.B., 2001. Mass transfer effects in the oxidation of aqueous organic compounds over a hydrophobic solid catalyst. Chem. Eng. Sci. 56 (17), 5091–5102 [17].

Lawrence, P.S., Grünewald, M., Agar, D.W., 2005. Spatial distribution of functionalities in an adsorptive reactor. Catal. Today 105 (3–4), 582–588 [1].

Lazar, J.G., Ross, J., 1990. Changes in mean concentration, phase shifts, and dissipation in a forced oscillatory reaction. Science 247, 189–192 [11].

Lee, C.K., Bailey, J.E., 1974. Influence of mixing on the performance of periodic chemical reactors. AIChE J. 20, 74–81 [7].

Lee, C.K., Bailey, J.E., 1980. Modification of consecutive-competitive reaction selectivity by periodic operation. Ind. Eng. Chem. Proc. Des. Dev. 19, 160–166 [12] [16].

Lee, J.K., Ferrero, S., Hudgins, R.R., Silveston, P.L., 1996b. Catalytic SO_2 oxidation in a periodically operated trickle-bed reactor. Can. J. Chem. Eng. 74, 706–712 [17] [20].

Lee, J.K., Hudgins, R.R., Silveston, P.L., 1995. A cycled trickle-bed reactor for SO_2 oxidation. Chem. Eng. Sci. 50, 2523–2530 [17].

Lee, J.K., Hudgins, R.R., Silveston, P.L., 1996a. SO_2 oxidation in a periodically operated trickle bed: comparison of activated carbon catalysts. Environ. Prog. 15 (4), 239–244 [17].

Lee, I.D., Kadlec, R.H., 1989. Effects of adsorbent and catalyst distributions in pressure swing reactors. AIChE. Symp. Ser. 84 (264), 167–176 [22].

Lee, C.-Y., Lee, S.-J., Jung, K.H., Katoh, S., Lee, E.K., 2003. High dissolved oxygen tension enhances heterologous protein expression by recombinant Pichia pastoris. Proc. Biochem. 38, 1147–1154 [10] [16].

Lee, J.H., Nararajan, S., Lee, K.S., 2001. A model-based predictive control approach to repetitive control of continuous processes with periodic operation. J. Process Contr. 11, 195–207 [19].

Lee, J.-W., Parulekar, S.J., 1996. Periodic operation of continuous recombinant cultures improves antibiotic selection. Chem. Eng. Sci. 51 (2), 717–731 [1] [10] [23].

Lee, C.K., Yeung, S.Y.S., Bailey, J.E., 1980. Experimental studies of a consecutive-competitive reaction in steady-state and forced periodic CSTRs. Can. J. Chem. Eng. 58, 212–218 [12] [16].

Lefort, L., Amariglio, A., Amariglio, H., 1994. Oligomerization of ethylene on platinum by a two-step reaction sequence. Catal. Lett. 29, 125–131 [12] [23].

Leidheiser Jr., H., Ghuman, A.R.P., 1973. Pulse electroplating of silver-tin alloys and the formation of Ag_3Sn. J. Electrochem. Soc. 120 (4), 484–487 [9].

Leion, H., Larring, Y., Bakker, E., Bredesen, R., Mattisson, T., Lyngfelt, A., 2009a. Use of $CaMn_{0.875}Ti_{0.125}O_3$ as oxygen carrier in chemical-looping with oxygen uncoupling. Energy Fuels 23 (10), 5276–5283 [5].

Leion, H., Lyngfelt, A., Mattisson, T., 2009b. Solid fuels in chemical-looping combustion using a NiO based oxygen carrier. Chem. Eng. Res. Design 87, 1543–1540 [8].

Leion, H., Mattisson, T., Lyngfelt, A., 2008. Solid fuels in chemical-looping combustion. Int. J. Greenhouse Gas Control 2 (2), 180–193 [8].

Leisner, P., Nielsen, C.B., Tang, P.T., Dörge, T.C., Moeller, P., 1996. Methods for electrodepositing composition-modulated alloys. J. Mat. Proc. Technol. 58, 39–44 [9].

Lemaux, P.G., Herendeen, S.L., Bloch, P.L., Neidhardt, P.C., 1978. Transient rate of synthesis of individual polypeptides in *E. Coli* following temperature shifts. Cell 13, 427–434 [23].

Le Naour, C., Moisy, Ph., Arpigny, S., Madic, C., 1999. Electro-oxidation of dihydroxymalonic acid on polycrysatlline platinum electrode. Electrochim. Acta 44, 3505–3512 [9].

Lenas, P., Baltzis, B.C., Lewandowski, G.A., Ko, Y.F., 1994. Biodegradation of wastes in a cyclically operated reactor: theory, experimental verification and optimization studies. Chem. Eng. Sci. 49 (24A), 4547–4561 [10].

Lendvay, J., Raub, Ch.J., 1975. Mechanical properties of electrolytically deposited gold plating. Metalloberfläche 29 (4), 165–167 [9].

Leucker, J.-J., Greco, A., Richarz, W., 1978. Effect of pressure oscillations on mass transfer from a single porous sphere. Ger. Chem. Eng. 1, 161–165 [15].

Leupold, E.I., Arpe, H.-J., Renken, A., Schlosser, E.G., 1979. Verfahren zur Herstellung von Essigsaeureaethylester Ger. Patent No. DE 2545845C3 [11] [12].

Leupold, E.I., Renken, A., 1977. A new non-stationary process for the production of ethyl acetate. Chem. Ing. Tech. 49, 667 [11] [12].

Leupold, E.I., Renken, A., 1978. A new ethyl acetate process using periodic operation. Ger. Chem. Eng. 1, 218–222 [11] [12].

Levenspiel, O., 1988. Chemical Engineering's great adventure. Chem. Eng. Sci. 43, 1427–1435 [18].

Levenspiel, O., 2005. What will come after petroleum? Ind. Eng. Chem. Res. 44, 5073–5078 [18].

Lewandowski, D., 2000. Design of thermal oxidation systems for volatile organic compounds. CRC Press LLC. [18].

Lewis, W.K., Gilliland, E.R., Reed, W.A., 1949. Reaction of methane with copper oxide in a fluidised bed. Ind. Eng. Chem. 41, 1227–1237 [1] [4].

Li, F., Fan, L.-S., 2008. Clean coal conversion processes—progress and challenges. Energy Environ. Sci. 1, 248–267 [8].

Li, C.-Y., Hudgins, R.R., Silveston, P.L., 1985a. Modelling of non-stationary ammonia synthesis kinetics over an iron catalyst. I. Deriving a model and estimating parameters from transient response data. Can. J. Chem. Eng. 63, 795–802 [2] [13].

Li, C.-Y., Hudgins, R.R., Silveston, P.L., 1985b. Modelling of non-stationary ammonia synthesis kinetics over an iron catalyst. II. Modelling results of forced feed composition in periodic operation and for step changes. Can. J. Chem. Eng. 63, 803–809 [2] [13].

Li, F., Kim, H.R., Shridhar, D., Wang, F., Zeng, L., Chen, J., Fan, L.-S., 2009a. Syngas chemical looping gasification process: oxygen carrier particle selection and performance. Energy Fuels 23, 4182–4189 [8].

Li, C.-Y., Su, D., Wang, J., Jiang, Q., Ai, D., 1992a. Dynamic studies of ethylene selective oxidation on silver catalyst: (I) structure and kinetics of the elementary process sequences. J. Chem. Ind. Eng. (China) 43 (3), 256–261 [4] [8].

Li, C.-Y., Su, D., Wang, J., Jiang, Q., Ai, D., 1992b. Dynamic studies of ethylene selective oxidation on silver catalyst: (II) the performance under forced composition cycling. J. Chem. Ind. Eng. (China) 43 (3), 262–267 [4] [8].

Li, C.-Y., Yu, C.-C., Shen, S.-K., 2000. The main origin of CO_2 in partial oxidation of methane to syngas over a new Ni/ Al_2O_3 catalyst. Acta Phys. -Chim. Sin. 16 (2), 97–100 [4].

Li, T.-X., Wang, R.-Z., Oliveira, R.G., Kiplagat, J.K., Wang, L.-W., 2009b. A combined double-way chemisorption refrigeration cycle based on adsorption and resorption processes. Int. J. Refrig. 32, 47–57 [8].

Li, Y.-E., Willcox, D., González, R.D., 1989. Determination of rate constants by the frequency response method: CO on Pt/SiO_2. AIChE J. 35, 423–428 [13].

Li, F., Zeng, L., Velazquez-Vargas, L.G., Yoskovitz, Z., Fan, L.-S., 2010. Syngas chemical looping gasification process: bench scale studies and reactor simulations. AIChE J. 56 (8), 2186–2199 [8].

Liang, Z.-Z., Li, Y.-X., Li, C.-Y., Chen, B.-H., 2004. In-situ transient DRIFTS study of reaction pathway of n-butane selective oxidation over (VO)2P2O7 catalyst. Spectros. Spect. Anal. (China) 24 (11), 1309–1314 [4].

Liang, Z.-Z., Li, Y.-X., Li C.-Y., Chen, B.-H., Hu, S.-Y., 2003a. On-line mass spectroscopic transient measurement of vanadium-phosphorous oxide reoxidation. Chin. J. Catal. 24 (11), 881–884 [4].

Liang, Z.-Z., Li, C.-Y., Li, Y.-X., Chen, B.-H., Huang, X.-F., 2003b. Roles of adsorbed and lattice oxygen on surface of VPO catalyst for n-butane selective oxidation. J. Chem. Ind. Eng. (China) 54 (6), 854–858 [4].

Liao., Y., Gabe, D.R., Wilcox, G.D., 1998. Microstructural morphology of Zn-Fe alloy electrodeposited coatings. Plat. Surf. Finish. 85 (8), 62–66 [9].

Liapis, A.I., Rippin, D.W.T., 1979. Simulation of binary adsorption in continuous counter-current operation in comparison with other operating modes. AIChE J. 25, 455–460 [21].

Liauw, M.A., Plath, P.J., Jaeger, N.I., 1995. Periodic perturbation of an oscillating reaction, Preprint, 2nd Intern Unsteady State Processes in Catalysis Conf. Missouri. St. Louis. September [3].

Liaw, J.S., Bailey, J.E., 1974. Regulation of lumped periodic processes. AIChE J. 20 (3), 924–932 [19].

Liden, G., Vamling, L., 1989. Periodic operation of a tubular reactor: simulation study of consecutive reactions in a chromatographic reactor. Chem. Eng. J. 40, 31–37 [12] [20].

Lie, A.B.K., Hoebink, J.H.B.J., Marin, G.B., 1993. The effects of oscillatory feeding of CO and O_2 on the performance of a monolithic catalytic converter of automobile exhaust gas: a modeling study. Chem. Eng. J. 53 (1), 47–54 [6].

Lim, H.C., Tayeb, Y.J., Modak, J.M., Bonte, P., 1986. Computational algorithms for optimal feed rates for a class of fed-batch fermentation: numerical results for penicillin and cell mass production. Biotechnol. Bioeng. 28, 1408–1420 [10].

Limb, S.J., Edell, D.J., Gleason, E.F., Gleason, K.K., 1998. Pulsed plasma-enhanced chemical vapor deposition from hexafluoropropylene oxide: film composition study. J. Appl. Polym. Sci. 67, 1489–1502 [7].

Lindblom, M., Hertz, H.M., Holmberg, A., 2006. Pulse reverse plating for uniform nickel height in zone plates. J. Vac. Sci. Technol. B24 (6), 2848–2851 [9].

Litto, R., Hayes, R.E., Sapoundjiev, H., Fuxman, A., Forbes, F., Liu, B., Bertrand, F., 2006. Optimization of a flow reversal reactor for the catalytic combustion of lean methane mixtures. Catal. Today 117, 536–542 [18].

Liu, G., Duan, Y., Wang, Y., Wang, L., Mi, Z., 2005. Periodically operated trickle-bed reactor for EAQs hydrogenation: experiments and modeling. Chem. Eng. Sci. 60 (22), 6270–6278 [17].

Liu, B., Hayes, R.E., Checkel, M.D., Zheng, M., Misrosh, E., 2001b. Reversing flow converter for a natural gas /diesel dual fuel engine. Chem. Eng. Sci. 56, 2641–2658 [18].

Liu, B., Hayes, R.E., Yi, Y., Mmbaga, J., Checkel, M.D., Zheng, M., 2007. Three dimensional modelling of methane ignition in a reverse flow catalytic converter. Comput. Chem. Eng. 31 (4), 292–306 [18].

Liu, H., Huang, X.-F., Chen, B.-H., Li, C.-Y., 2001a. Modeling of partial oxidation of n-butane to maleic anhydride in a CFB system. J. Chem. Ind. Eng. (China) 52, 247–251. 316-321 [4].

Liu, G., Mi, Z., 2005. Hydrogenation of 2-ethylanthraquinones in a periodically operated trickle-bed reactor. Chem. Eng. Technol. 28 (8), 857–862 [17].

Liu, T., Temur, H., Veser, G., 2009. Autothermal reforming of methane in a reverse-flow reactor. Chem. Eng. Technol. 32 (9), 1358–1366 [18].

Liu, G., Zhang, X., Wang, Li, Zhang, S., Mi, Z., 2008. Unsteady-state operation of trickle-bed reactor for dicyclopentadiene hydrogenation. Chem. Eng. Sci. 63 (20), 4991–5002 [17].

Lobera, M.P., Téllez, C., Herguido, J., Menéndez, M., 2009. Pt-Sn/$MgAl_2O_4$ as n-butane dehydrogenation catalyst in a two-zone fluidized-bed reactor. Ind. Eng. Chem. Res. 48, 6573–6578 [4].

Lode, F., Francesconi, G., Mazzotti, M., Morbidelli, M., 2003b. Synthesis of methylacetate in a simulated moving-bed reactor: experiment and modeling. AIChE J. 49 (6), 1516–1524 [21].

Lode, F., Houmard, M., Migliorini, C., Mazzotti, M., Morbidelli, M., 2001. Continuous reactive chromatography. Chem. Eng. Sci. 56, 269–291 [1] [21].

Lode, F., Mazzotti, M., Morbidelli, M., 2003a. Comparing true countercurrent and simulated moving-bed chromatographic reactors. AIChE J. 49 (4), 977–990 [20] [21].

Logist, F., Vandewouwer, A., Smets, I.Y., Van Impe, J.F., 2007. Optimal temperature profiles for tubular reactors implemented through a flow reversal strategy. Chem. Eng. Sci. 62, 4675–4688 [18].

Lööf, P., Kasemo, B., Bjömkvist, L., Andersson, S., Frestad, A., 1991. TPD and XPS studies of CO and NO on highly dispersed Pt + Rh automotive exhaust catalysts: evidence for noble metal-ceria interaction. In: Crucq, A. (Ed.), Catalysis and Pollution Control II. Elsevier Science Publishers, Amsterdam, pp. 253–273 [6].

Lopez Ortiz, A., Harrison, D.P., 2001. Hydrogen production using sorption-enhanced reaction. Ind. Eng. Chem. Res. 40 (23), 5102–5109 [22].

Lorkovic, I.M., Noy, M.L., Schenck, W.A., Belon, C., Weiss, M., Sun, S., Sherman, J.H., McFarland, E.W., Stucky, G.D., Ford, P.C., 2004. C_1 oxidative coupling via bromine activation and tandem catalytic condensation and neutralization over CaO/zeolite composites. II. Product distribution variation and full bromine confinement. Catal. Today 98, 589–594 [18].

Lou, J.-Chi, Huang, Sh.- W., 2008. Treating isopropyl alcohol by a regenerative catalytic oxidizer. Separ. Purif. Technol. 62, 71–78 [18].

Lugovskoy, V.I., . Yu Zudilina, L., Matros, Yu.Sh., 1982. Unsteady-state process for purification of off-gases. In: Matros, Yu.Sh. (Ed.), Unsteady State Processes in Chemical Reactors, Inst. Catal. Siberian Branch of Soviet Academy of Sciences. USSR, Novosibirsk, pp. 85–88 (in Russian) [18].

Luk'yanenko, V.P., Shapovalova, L.P., Kutnyaya, M.Yu., Solodkaya, V.S., 1987. Effects of the diluent on the activities of molybdenum-bearing oxide catalysts in the oxidative dehydrogenation of n-butane to butadiene. Zhurnal Prikladnoi Khimii 60 (5), 1169–1171 (Plenum Publ. Corp. Translation), [4].

Luther, M., 2008. Fast temperature cycling of the catalytic CO oxidation using microstructure reactors, Forschungszentrum Karlsruhe, École Polytechnique Fédérale de Lausanne. Ph.D. thesis Thesis No. 4050 [16].

Luther, M., Brandner, J.J., Kiwi-Minsker, L., Renken, A., Schubert, K., 2008a. Forced periodic cycling of chemical reactions in microstructure devices. Chem. Eng. Sci. 63, 4955–4961 [1] [16].

Luther, M., Brandner, J.J., Kiwi-Minsker, L., Renken, A., Schubert, K., 2008b. Novel design of a microstructured reactor allowing fast temperature oscillations. Chem. Eng. J. 135S, S254–S258 [1] [16].

Luu, H.D., Hudgins, R.R., Silveston, P.L., 1992. Anomalous excitation of CO oxidation over vanadia by periodic concentration forcing, Preprint. Can. Symp. Catal. Banff, Alberta, Canada [3].

Lynch, D.T., 1983. Modelling of resonant behaviour during forced cycling of catalytic reactors. Can. J. Chem. Eng. 61, 183–188 [3].

Lynch, D.T., 1984. On the use of adsorption/desorption models to describe the forced periodic operation of catalytic reactors. Chem. Eng. Sci. 39 (9), 1325–1328 [3].

Lynch, D.T., Emig, G., Wanke, S.E., 1986. Oscillations during CO oxidation over supported metal catalysts: III. Mathematical modelling of the observed phenomena. J. Catal. 97, 456–468 [3].

Lynch, D.R., Walters, N.P., 1990. Frequency response characterization of reaction systems: external recycle reactor with a solid adsorbent. Chem. Eng. Sci. 45, 1089–1096 [13].

Lysova, A.A., Koptyug, I.V., Kulikov, A.V., Kirillov, V.A., Sagdeev, R.Z., 2009. An NMR imaging study of steady-state and periodic operation modes of a trickle bed reactor. Top. Catal. 52, 1371–1380 [17].

MacGregor, J.F., Penlidis, A., Hamielec, A.E., 1984. Control of Polymerization Reactors: A Review. Polym. Process Eng. 2, 179–206 [7].

Machocki, A., 1996. Methane oxidative coupling in an undiluted reaction mixture in a reactor-adsorber system with gas recirculation. Appl. Catal. 146, 391–400 [22].

Maciejowski, J.M., 2001. Predictive Control with Constraints. Prentice-Hall, Inc., New York [19].

Magagula, Z., van Steen, E., 1999. Time on stream behaviour in the (amm)oxidation of propene/propane over iron antimony oxide: cyclic operation. Catal. Today 49, 155–160 [4].

Magee, E.M., 1963. The course of a reaction in a chromato graphic column. Ind. Eng. Chem. Fund. 2 (1), 32–36 [20].

Magee, J.S., Mitchell Jr., M.M., 1993. Fluid Catalytic Cracking, Stud.Surf.Sci.Technol., 76. Elsevier Sci. Publ. Amsterdam, The Netherlands [1].

Mahishi, M.R., Goswami, D.Y., 2007. An experimental study of hydrogen production by gasification of biomass in the presence of a CO_2 sorbent. Int. J. Hydrogen Energy 32, 2803–2808 [8].

Mallada, R., Menendez, M., Santamaria, J., 2002. On the favourable effect of CO_2 addition in the oxidation of butane to maleic anhydride using membrane reactors. Appl. Catal. A-Gen. 231 (1–2), 109–116 [4].

Mancusi, E., Russo, L., Altimari, P., Maffettone, P.L., Crescitelli, S., 2007. Effect of the switch strategy on the stability of reactor networks. Ind. Eng. Chem. Res. 46, 6510–6521 [5] [21].

Mandich, N.V., 2000. Electrodeposition of binary (Au-Sb) and Ternary (Au-Cu-Cd) 18 karat gold alloy under pulsed and DC current regimes, Proc. 5th Intern. Pulse Plating Symp., American Electroplaters & Surface Finishers Soc. Florida, Orlando. 87–102 [9].

Mandur, J.S., Al-Harbi, M., Epling, W.S., Budman, H., 2011. Modeling the transient cyclic operation of a commercial NSR catalyst. Chem. Eng. J. 166, 607–615 [6].

Manovic, V., Anthony, E.J., 2007. Steam reactivation of spent CaO-based sorbents for multiple CO_2 capture cycles. Environ. Sci. Technol. 41, 1420–1425 [22].

Marbach, A., Bettenbrock, K., 2011. Lac operon induction in *Escherichia coli*: systematic comparison of IPTG and TMG induction and influence of transacetylase LacA. J. Biotechnol (On-line Publication), Article 22079752 [23].

Marconi, P.F., Pagni, R., 1979. Polimerizzazione di stirene in reattori a regime periodico: influenza sulla distribuzione dei pesi moleculari. Ing. Chim. Ital. 15 (9–10), 93–99 [7].

Marek, M., 1985. Periodic and aperiodic regimes in forced chemical oscillations. In: Rensing, L., Jaeger, N.I. (Eds.), Temporal Order. Springer Verlag, Berlin, pp. 105–115 [3].

Marek, M., Schejbal, M., Koči, P., Nevoral, V., Kubiček, M., Hadač, O., Schreiber, I., 2006. Oscillations, period doubling and chaos in CO oxidation and catalytic mufflers. Chaos 16 (037107), 1–13 [3] [6].

Marín, P., Hevia, M.A.G., Ordóñez, S., Díez, F.V., 2005. Combustion of methane lean mixtures in reverse flow reactors: comparison between packed and structured catalyst beds. Catal. Today 105, 701–708 [18].

Marín, P., Ordóñez, S., Díez, F.V., 2008. Combustion of toluene-hexane binary mixtures in a reverse flow catalytic reactor. Chem. Eng. Sci. 63, 5003–5009 [1] [18].

Marín, P., Ordóñez, S., Díez, F.V., 2009a. Systematic study of the performance of a reverse flow reactor for the treatment of lean hydrocarbon emissions. J. Chem. Technol. Biotechnol. 84, 1292–1302 [18].

Marín, P., Ordóñez, S., Díez, F.V., 2009b. Procedures for heat recovery in the catalytic combustion of lean methane–air mixtures in a reverse flow reactor. Chem. Eng. J. 147, 356–365 [18].

Marín, P., Ordóñez, S., Díez, F.V., 2009c. Performance of reverse flow monolithic reactor for water–gas shift reaction. Catal. Today 147S, S185–S190 [18].

Marín, P., Ordóñez, S., Díez, F.V., 2009d. Simplified design methods of reverse flow catalytic combustors for the treatment of lean hydrocarbon–air mixtures. Chem. Eng. Proc. 48, 229–238 [18].

Marín, P., Ordóñez, S., Díez, F.V., 2010. Demonstration of a control system for combustion of lean hydrocarbon emissions in a reverse flow reactor. Chem. Eng. Sci. 65, 54–59 [18].

Marković, A., Seidel-Morgenstern, A., Petkovska, M., 2008. Evaluation of the potential of periodically operated reactors based on the second order frequency response functions. Chem. Eng. Res. Des. 86, 682–691 [14].

Markusse, A.P., Kuster, B.F.M., Schouten, J.C., 2001. Platinum catalysed aqueous methyl α-D-glucopyranoside oxidation in a multiphase redox-cycle reactor. Catal. Today 66, 191–197 [11].

Marts, B., Hagberg, A., Meron, E., et al., 2006. Resonant and non-resonant patterns in forced oscillators. Chaos 16(3), 037113 [3].

Marts, B., Simpson, D.J.W., Hagberg, A., et al., 2007. Period doubling in a periodically forced Belousov-Zhabotinsky reaction. Phys. Rev. E 76(2), 026213 Part 2 [3].

Marwood, M., Doepper, R., Renken, A., 1996. Modeling of surface intermediates under forced periodic conditions applied to CO_2 methanation. Can. J. Chem. Eng. 74 (5), 660–663 [13].

Marwood, M., Doepper, R., Renken, A., 1997. In-situ surface and gas phase analysis for kinetic studies under transient conditions. The catalytic hydrogenation of CO_2. Appl. Catal. A-Gen. 151, 223–246 [13].

Marwood, M., Van Vyve, F., Doepper, R., Renken, A., 1994. Periodic operation applied to the kinetic study of CO_2 methanation. Catal. Today 20, 437–448 [2] [13].

Mason, T.J. (Ed.), 1990. Advances in Sonochem., 1. JAI Press, London, U.K. [15].

Mason, T.J. (Ed.), 1991. Advances in Sonochem., 2. JAI Press, London, U.K. [15].

Mason, T.J. (Ed.), 1993. Advances in Sonochem., 3. JAI Press, London, U.K. [15].

Mason, T.J., 1997. Sonochemistry. Oxford Univ. Press, Oxford, U.K. [1].

Mason, T.J., Lorimer, J.P., 1988. Sonochemistry: Theory, Applications and Uses of Ultrasound in Chemistry. Ellis Horwood Ltd, Chichester, U.K. [15].

Mason, T.J., Lorimer, J.P., 2002. Applied Sonochemistry. Wiley-VCH, Weinheim, Germany [1].

Mason, T.J., Luche, J.L., 1996. Ultrasound as a New Tool for Synthetic Chemists. In: van Eldik, R., Hubbard, C.D. (Eds.), Chemistry under Extreme or Non-classical Conditions. John Wiley & Sons Inc., New York, pp. 317–380 [15].

Mason, T.J., Peters, D., 2002. Practical Sonochemistry. Horwood, Chichester, U.K. [1].

Massa, P., Ayude, M.A., Ivorra, F., Fenoglio, R., Haure, P., 2005. Phenol oxidation in a periodically operated trickle bed reactor. Catal. Today 30 (107–108), 630–636 [17].

Massa, P., Ivorra, F., Haure, P., Fenoglio, R., 2009. Optimized wet-proofed CuO/Al_2O_3 catalysts for the oxidation of phenol solutions: enhancing catalytic stability. Catal. Commun. 10, 1706–1710 [17].

Mata, A.R., Smith, J.M., 1981. Oxidation of sulfur dioxide in a trickle bed reactor. Chem. Eng. J. 22, 229–235 [20].

Matros, Yu.Sh., 1977. Prospects of applications of unsteady-state processes in catalytic reactors. Zhurn. Dm. I. Mendeleev Chemical Society 22, 576–580 [18].

Matros, Yu.Sh., 1985. Unsteady Processes in Catalytic Reactors. Elsevier Science Publ., Amsterdam, The Netherlands [4] [5] [21].

Matros, Yu.Sh., 1989. Catalytic Processes under Unsteady-State Conditions. Elsevier Science BV, Amsterdam, The Netherlands [5] [18] [19].

Matros, Yu.Sh., 1996. Forced unsteady-state processes in heterogeneous catalytic reactors. Can. J. Chem. Eng. 74, 566–579 [18].

Matros, Yu.Sh., Bunimovich, G.A., 1995. Control of volatile organic compounds by the catalytic reverse process. Ind. Eng. Chem. Res. 34, 1630–1640 [1] [18].

Matros, Yu.Sh., Bunimovich, G.A., 1996. Reverse-flow operation in fixed bed catalytic reactors. Catal. Rev. -Sci. Eng. 38 (1), 1–68 [18].

Matros, Yu.Sh., Bunimovich, G.A., Boreskov, G.K., 1984. Unsteady-state performance of sulphur dioxide oxidation in production of sulfuric acid. In: Doraiswami, L.K., Mashelkar, R.A. (Eds.), Frontiers in Chemical Reaction Engineering, vol. 2. John Wiley Eastern Publishers, New Delhi, India [18].

Matros, Yu.Sh., Bunimovich, G.A., Noskov, A.S., 1993a. The decontamination of gases by unsteady-state catalytic method. Catal. Today 17, 261–274 [18].

Matros, Yu.Sh., Bunimovich, G.A., Patterson, S.E., Meyer, S.F., 1996. Is it economically feasible to use heterogeneous catalysts for VOC control in regenerative oxidizers. Catal. Today 27, 307–313 [18].

Matros, Yu.Sh., Bunimovich, G.A., Strots, V.O., Mirosh, E.A., 1999a. Reversed flow converter for emission control after automotive engine. Chem. Eng. Sci. 54, 2889–2898 [18].

Matros, Yu.Sh., Ivanov, A.G., Gogin, L.L., 1988a. Generation of high-potential heat from low concentrated gases and fuels in unsteady-state reactor. Theor. Osnovy Khim. Technol. (Theor. Funda. Chem. Technol.) 22, 481–487 [18].

Matros, Yu.Sh., Lugovskoy, V.I., Bunimovich, G.A., Puzhilova, V.I., 1981. Unsteady-state techniques for decontamination of gases of organic compounds and CO. In: Popovski, V.I. (Ed.), Catalytic 3rd All-Union Conf. Purification of Gases – Proc. Institute of Catalysis, 1. Novosibirsk, Russia, pp. 72–79 (in Russian). Part [18].

Matros, Yu.Sh., Noskov, A.S., Chumachenko, V.A., 1991. Catalytic decontamination of industrial waste gases. In: Parmon, V.I. (Ed.), Catalytic Purification of Gases – Proc. 1st All-Union Conference, Novosibirsk, Russia (In Russian) [18].

Matros, Yu.Sh., Noskov, A.S., Chumachenko, V.A., 1993b. Progress in reverse-process application to catalytic incineration problems. Chem. Eng. Proc. 32, 89–98 [18].

Matros, Yu.Sh., Noskov, A.S., Chumachenko, V.A., Goldman, O.V., 1988b. Theory and application of unsteady-state catalytic detoxication of effluent gases from sulfur dioxide, nitrogen oxides and organic compounds. Chem. Eng. Sci. 43, 2061–2066 [18].

Matros, Yu.Sh., Strots, V.O., Bunimovich, G.A., Krasik, I., Spreutels, W., Berkel, D.A., Vavere, A., 1999b. Sulfuric acid production from viscose production plant effluents, in Proc. Sulfur 99 Conf., Calgary, Alberta, Canada. Preprint: British Sulfur Publishing Ltd., London, U.K., October pp. 201–208 [18].

Matsen, J.M., Harding, J.W., Magee, E.M., 1965. Chemical reactions in chromatographic columns. J. Phys. Chem. 69 (2), 522–527 [20].

Matsumoto, S., Ikeda, Y., Suzuki, H., Ogai, M., Miyoshi, N., 2000. NOx storage-reduction catalyst for automotive exhaust with improved tolerance against sulfur poisoning. Appl. Catal. B-Environ. 25, 115–124 [6].

Matsumoto, S., Shinjoh, H., 2007. Dynamic behavior and characterization of automobile catalysts. Adv. Chem. Eng. 33, 1–46 [6].

Matsunaga, S.-I., Yokota, K., Muraki, H., Fujitani, Y., 1987. Improvement of engine emissions over three-way catalysts by the periodic operations. SAE Techn. Paper Series, Paper # 872098, Soci. of Autom. Eng. Warrendale, PA. [6].

Mattisson, T., Leion, H., Lyngfelt, A., 2009b. Chemical-looping oxygen uncoupling using CuO. Fuel 88 (4), 683–690 [8]

Mattisson, T., Lyngfelt, A., Leion, H., 2009a. Chemical-looping with oxygen uncoupling for combustion of solid fuels. Int. J. Greenhouse Gas Contr. 3, 11–19 [8].

Maunula, T., Vakkulainen, A., R.Heikkinen, Härkönen, M., 2001. NOx storage and reduction on differentiated chemistry catalysts for lean gasoline vehicles. SAE Tech.Paper 2001-01-3665, SAE International, Warrendale, PA. [6].

Mazzotti, M., Kruglov, A., Neri, B., Gelosa, D., Morbidelli, M., 1996. A continuous chromatographic reactor, SMBR. Chem. Eng. Sci. 51, 1827–1836 [21].

Mazzotti, M., Neri, B., Gelosa, D., Morbidelli, M., 1997a. Dynamics of a chromatographic reactor: esterification catalyzed by acidic resins. Ind. Eng. Chem. Res. 36, 3163–3172 [20] [21].

Mazzotti, M., Storti, G., Morbidelli, M., 1997b. Robust design of countercurrent adsorption separation processes 4. Desorbent in the feed. AIChE J. 43 (1), 64–72 [21].

Mazzotti, M., Storti, G., Morbidelli, M., 1997c. Optimal operation of simulated moving bed units for nonlinear chromatographic separations. J. Chromatogr. A. 769, 3–24 [21].

Mbaye, M., Aidoun, Z., Valkov, V., Legault, A., 1998. Analysis of chemical heat pumps (CHPS): basic concepts and numerical model description. Appl. Thermal Eng. 18 (3–4), 131–146 [8].

McAuley, K.B., Macdonald, D.A., McLellan, P.J., 1995. Effect of operating conditions on the stability of gas phase polyethylene reactors. AIChE J. 41 (4), 868–879 [7].

McCarty, J.G., Wise, H., 1979. Hydrogenation of surface carbon on alumina-supported nickel. J. Catal. 57, 406–416 [12].

McKarnin, M.A., Schmidt, L.D., Aris, R., 1988. Forced oscillation of a self oscillating bimolecular surface reaction model. Proc. Roy. Soc., A 415, 363–387 [3].

McNeil, M.A., Rinker, R.G., 1994. An experimental study of concentration forcing applied to the methanol synthesis reaction. Chem. Eng. Commun. 127, 137–149 [2].

Meerkov, S.M., 1973. Vibrational control. Automat. Remote Control 34 (2), 201–209 [19].

Meerkov, S.M., 1980. Principle of vibrational control – theory and applications. IEEE Trans. Automat. Control 25 (4), 755–782 [19].

Meira, G.R., 1981a. Forced oscillations in continuous polymerization reactors and molecular weight distribution control. A survey. J. Macromol. Sci., Rev. Macromol. Chem. Phys. C20 (2), 207–241 [1] [7].

Meira, G., Johnson, A.F., 1981. Molecular weight distribution in continuous 'living' polymerizations through periodic operation of the monomer feed. Polym. Eng. Sci. 21, 415–423 [7].

Meira, G.R., Johnson, A.F., Ramsay, J., 1979. Molecular weight distribution control in continuous flow reactors. In: J.N. Henderson and T.C. Bouton. ACS Sympos. Series, No. 104, 253–266 [7].

Meira, G.R., Luciani, C.V., Estenoz, D.A., 2007. Continuous bulk process for the production of high-impact polystyrene: recent developments in modeling and control. Macromol. React. Eng. 1, 25–39 [7].

Mekarapiruk, W., Luus, R., 2001. On solving optimal control problems with free initial condition using iterative dynamic programming. Can. J. Chem. Eng. 79, 777–784 [10].

Melo, P.A., Biscaia Jr., E.C., Pinto, J.C., 2001a. Thermal effects in loop polymerization reactors. Chem. Eng. Sci. 56 (23), 6793–6800 [7].

Melo, P.A., Biscaia Jr., E.C., Pinto, J.C., 2003. The bifurcation behavior of continuous free-radical solution loop polymerization reactors. Chem. Eng. Sci. 58 (23), 2805–2821 [7].

Melo, P.A., Sampaio, J.G., Biscaia Jr., E.C., Pinto, J.C., 2001b. Periodic oscillations in continuous free-radical solution polymerization reactors - a general approach. Chem. Eng. Sci. 56 (11), 3469–3482 [7].

Menge, M., Elsner, M.P., Agar, D.W., (2003). Adsorption enhanced Claus reaction: an old process in a new guise. 4th Intern. Confer. Unsteady-State Processes in Catalysis, Montreal, Canada, Oct. 26-29 [22].

Mensah, P., Carta, G., 1999. Adsorptive control of water in esterification with immobilized enzymes. Continuous operation in a periodic counter-current reactor. Biotechnol. Bioeng. 66 (3), 137–146 [21].

Mensah, P., Gainer, J.L., Carta, G., 1998a. Adsorptive control of water in esterification with immobilized enzymes. I. Batch reactor behavior. Biotechnol. Bioeng. 60, 435–444 [21].

Mensah, P., Gainer, J.L., Carta, G., 1998b. Adsorptive control of water in esterification with immobilized enzymes. II. Fixed bed reactor behavior. Biotechnol. Bioeng. 60, 445–453 [21].

Metzinger, J., Hasokowati, W., Hudgins, R.R., Silveston, P.L., Gangwal, S., 1992. Application of a periodically operated trickle bed to sulfur removal from stack gas. Chem. Eng. Sci. 42 (13/14), 3723–3727 [17].

Metzinger, J., Kühter, A., Silveston, P.L., Gangwal, S.K., 1994. A novel periodic reactor for scrubbing SO_2 from industrial stack gases. Chem. Eng. Sci. 49 (24A), 4533–4546 [17] [20].

Meurer, M., Altenhöner, U., Strube, J., Untiedt, A., Schmidt-Traub, H., 1996. Dynamic simulation of a simulated moving bed chromatographic reactor for the inversion of sucrose. Starch/Stärke 48, 452–457 [21].

Meyer, T., Keurentjes, J. (Eds.), 2005. Handbook of Polymer Reaction Engineering. Wiley-VCH Verlag, Weinheim, Germany [7].

Mielczarski, E., Monteverdi, S., Amariglio, A., Amariglio, H., 1993. Direct conversion of methane to higher alkanes by platinum loaded zeolites. Appl. Catal. A 104, 215–228 [12].

Migliorini, C., Fillinger, M., Mazzotti, M., Morbidelli, M., 1999a. Analysis of simulated moving-bed reactors. Chem. Eng. Sci. 54 (13–14), 2475–2480 [21].

Migliorini, C., Mazzotti, M., Morbidelli, M., 1999b. Simulated moving-bed units with extra-column dead volume. AIChE J. 45 (7), 1411–1421 [21].

Migliorini, C., Meissner, J.P., Mazzotti, M., Carta, G., 2000. Regioselective enzymatic diol esterification in batch and fixed bed adsorptive reactors. Experiments and modeling. Biotechnol. Prog. 16, 600–609 [20].

Mikalsen, R., Roskilly, A.P., 2007. A review of free piston engine history and application. Appl. Therm. Eng. 27 (14–15), 2339–2352 [15].

Mills, P.L., Dudukovic, M.P., 1980a. Analysis of catalyst effectiveness in trickle-bed reactors processing volatile or non-volatile reactants. Chem. Eng. Sci. 35 (11), 2267–2279 [17].

Mills, P.L., Dudukovic, M.P., 1980b. On the effectiveness factor of partially wetted catalysts in trickle-bed reactors. Ind. Eng. Chem. Fund. 19 (2) 231–233 [17].

Minceva, M., Gomes, P.S., Meshko, V., Rodrigues, A.E., 2008. Simulated moving bed reactor for isomerization and separation of p-xylene. Chem. Eng. J 140, 305–323 [21].

Mitri, A., Neumann, D., Liu, T.-F., Veser, G., 2004. Reverse flow reactor operation and catalyst deactivation during high temperature partial oxidation. Chem. Eng. Sci. 59, 5527–5534 [18].

Monnerat, B., Kiwi-Minsker, L., Renken, A., 2001. Hydrogen production by catalytic cracking of methane over nickel gauze under periodic reactor operation. Chem. Eng. Sci. 56, 633–639 [4] [8].

Monroe, D.R., Krueger, M.H., Beck, D.D., D'Aniello Jr., M.J., 1991. The effects of sulfur on three-way catalysts. In: Crucq, A. (Ed.), Catalysis and Pollution Control II. Elsevier Science Publishers, Amsterdam, pp. 593–616 [6].

Morari, M., Zafiriou, E., 1989. Robust Process Control. Prentice Hall, Englewood Cliffs, New Jersey [19].

Morávek, V., 1992a. Steady state and transient kinetics of displacement adsorption and educt inhibition in dehydration of alcohols on alumina. J. Catal. 133, 170–178 [11] [16].

Morávek, V., 1992b. Steady state and transient kinetics of displacement adsorption and educt inhibition in dehydration of alcohols on alumina, In: Progress in Catalysis. Elsevier Science Publ. Amsterdam [13].

Morris, C.E., Wheelock, T.D., Smith, L.L., 1987. Processing waste gypsum in a two-zone fluidized bed reactor. In: Yang, W.-C. (Ed.), New Developments in Fluidization and Fluid-Particle Systems, AIChE Symp. Series, 83 (255), 94–104 [8].

Mortazavi, Y., Hudgins, R.R., Silveston, P.L., 1992. Segregated operation of the oxidative coupling of methane on a LiMgO catalysts, Proc. 12th Can. Symp. Catal. Banff, Alberta, Canada May, 1992 [4].

Mortazavi, Y., Hudgins, R.R., Silveston, P.L., 1996. Catalytic methane coupling under periodic operation. Can. J. Chem. Eng. 74, 683–694 [4].

Moulton, K.J., Koritala, S., Frankel, E.N., 1983. Ultrasonic Hydrogenation of Soybean Oil. J. Am. Oil Chem. Soc. 60, 1257–1258 [15].

Müller, E., Hofmann, H., 1987. Dynamic modelling of heterogeneous catalytic reactions - I. Theoretical considerations. Chem. Eng. Sci. 42 (7), 1695–1704 [13].

Mukesh, D., Kenney, C.N., Morton, W., 1983. Island models and the catalytic oxidation of carbon monoxide and carbon monoxide-olefin mixtures. Chem. Eng. Sci. 38, 69–77 [3].

Muncrief, R.L., Kabin, K.S., Harold, M.P., 2004. NO_X storage and reduction with propylene on Pt/BaO/alumina. AIChE J. 50, 2526–2540 [6].

Munder, B., Rihko-Struckmann, L., Sundmacher, K., 2007. Steady-state and forced-periodic operation of solid electrolyte membrane reactors for selective oxidation of n-butane to maleic anhydride. Chem. Eng. Sci. 62, 5663–5668 [4] [12].

Murakami, Y., Hattori, T., Hattori, T., 1968. Study on the pulse reaction technique. II. Dealkylation and disproportionation of cumene. J. Catal. 10, 123–127 [20].

Muraki, H., 1991. Performance of palladium automotive catalysts. Paper # 910842. SAE Tech. Paper Series. Soc. Autom. Eng. Warrendale, PA [6].

Muraki, H., Fujitani, Y., 1986. NO reduction over noble metal catalysts under cycled feed streams. Ind. Eng. Chem. Prod. Res. Dev. 25, 414–419 [3].

Muraki, H., Shinjoh, H., Fujitani, Y., 1985b. NO-CO-O_2 reaction on Pd/Al_2O_3 catalyst. Nippon Kagaku Kaishi 9, 1682–1688 [6].

Muraki, H., Shinjoh, H., Fujitani, Y., 1986a. Reduction of NO by CO over alumina-supported palladium catalyst. Ind. Eng. Chem. Prod. Res. Dev. 25, 419–424 [3] [6].

Muraki, H., Shinjoh, H., Sobukawa, H., Yokota, K., Fujitani, Y., 1985c. Behavior of automotive noble metal catalysts in cycled feed streams. Ind. Eng. Chem. Prod. Res. Dev. 24, 43–49 [6].

Muraki, H., Sobukawa, H., Fujitani, Y., 1985a. Periodic operation effects on carbon monoxide oxidation over noble metal catalysts. Nippon Kagaku Kaishi 2, 176–181 [3] [6].

Muraki, H., Sobukawa, H., Fujitani, Y., 1986b. Dynamic behavior of automotive noble metal catalysts in model gas systems. Nippon Kagaku Kaishi 4, 532–537 [6].

Muraki, H., Yokota, K., Fujitani, Y., 1989. Nitric oxide reduction performance of automotive palladium catalysts. Appl. Catal. 48, 93–105 [6].

Murali, K.R., Chitra, K., Elango, P., 2009. Characteristics of periodic pulse deposited ZnSe films from non aqueous bath. 215th Mtg. Electrochem. Soc. (Abstracts) [9].

Murchison, C.B., Vrieland, G.E., Khazai, B., Weihl, E., 1993. Anaerobic oxidation of butane to butadiene on magnesium molybdate catalyst. Preprint. 11th No. Amer. Mtg. Catal. Soc., Pittsburgh, Pennsylvania [4].

Muzen, A., Fraguío, M.S., Cassanello, M.C., Ayude, M.A., Haure, P.M., Martínez, O.M., 2005. Clean oxidation of alcohols in a trickle-bed reactor with liquid flow modulation. Ind. Eng. Chem. Res. 44 (14), 5275–5284 [17].

Nagai, Y., Hirabayashi, T., Dohmae, K., Takagi, N., Minami, T., Shinjoh, H., Matsumoto, S., 2006. Sintering inhibition mechanism of platinum supported on ceria-based oxide and Pt-oxide-support interaction. J. Catal. 242, 103–109 [6].

Nakanishi, T., Ozaki, M., Nam, H.-S., Yokoshima, T., Osaka, T., 2001. Pulsed electrodeposition of nanocrystalline CoNiFe soft magnetic thin films. J. Electrochem. Soc. 148 (9), C627–C631 [9].

Nakatsuji, T., Yasukawa, R., Tabata, K., Ueda, K., Niwa, M., 1999. A highly durable catalytic NOx reduction in the presence of SO_2 using periodic two steps, an operation in oxidizing conditions and a relatively short operation in reducing conditions. Appl. Catal. B-Environ. 21, 121–131 [6].

Nalpantidis, K., Platte, F., Agar, D.W., Turek, S., 2006. Elucidation of hybrid N_2O decomposition using axially structured catalyst in reverse flow reactor. Chem. Eng. Sci. 61, 3176–3185 [18].

Nam, Y.W., Hudgins, R.R., Silveston, P.L., 1990. Storage models for ammonia synthesis over iron catalyst under periodic operation. Chem. Eng. Sci. 45 (10), 3111–3121 [2] [13].

Nappi, A., Fabbricino, L., Hudgins, R.R., Silveston, P.L., 1985. Influence of forced feed composition cycling on catalytic methanol synthesis. Can. J. Chem. Eng. 63, 963–970 [2] [12].

Na-Ranong, D., Aida, T., Niiyama, H., 2002. Application of periodic operation to kinetic study for NO–CO reaction over Pt/Al_2O_3. Appl. Catal. A-Gen. 234, 103–110 [13].

Na-Ranong, D., Jaree, A., Hudgins, R.R., Silveston, P.L., Yakhnin, V., Menzinger, M., 2003. Amplification of inlet temperature disturbances in a packed-bed reactor for CO oxidation over CuO-Al_2O_3 catalyst. Can. J. Chem. Eng. 81, 1215–1221 [3] [16].

Na-Ranong, D., Yuangsawad, R., Kitchaiya, P., Aida, T., 2009. Application of periodic operation to kinetic study of NO–CO reaction over Rh/Al_2O_3. Chem. Eng. J. 146, 275–286 [13].

Natarajan, S., Lee, J.H., 2000. Repetitive model predictive control applied to a simulated moving bed chromatographic system. Comput. Chem. Eng. 24, 1127–1133 [19].

Natter, H., Hempelmann, R., 1996. Nanocrystalline copper by pulsed electrodeposition: effect of organic additives, bath temperature and pH. J. Phys. Chem. 100 (50), 19525–19532 [9].

Natter, H., Krajewski, T., Hempelmann, R., 1996. Nanocrystalline palladium by pulsed electrodeposition. Ber. Bunsen. Phys. Chem. 100 (1), 55–64 [9].

Nekhamkina, O., Sheintuch, M., 2008. Structure of operating domains of loop reactors. AIChE J. 54, 1292–1302 [21].

Nele, M., Pinto, J.C., 2000. Retrofitting of industrial olefin polymerization plants: producing broad MWD through multi-objective periodic operation. J. Appl. Polym. Sci. 77, 437–452 [7].

Neophytides, S.G., 1999. The reversed flow operation of a cross-flow solid oxide fuel cell monolith. Chem. Eng. Sci. 54, 4603–4613 [18].

Neophytides, S.G., Froment, G.F., 1992. A bench scale study of the reversed flow methanol synthesis. Ind. Eng. Chem. Res. 31, 1583–1589 [18].

Neophytides, S.G., Tripakis, A., 1996. The transient operation of a solid oxide fuel cell monolith under forced periodic flow reversal. Can. J. Chem. Eng. 74, 719–728 [18].

Neumann, D., Gepert, V., Veser, G., 2004a. Some considerations on the design and operation of high-temperature catalytic reverse-flow reactors. Ind. Eng. Chem. Res. 43, 4657–4667 [18].

Neumann, D., Kirchhoff, M., Veser, G., 2004b. Towards an efficient process for small-scale, decentralized conversion of methane to synthesis gas: combined reactor engineering and catalyst synthesis. Catal. Today 98, 565–574 [18].

Neumann, D., Veser, G., 2005. Catalytic partial oxidation of methane in a high temperature reverse-flow reactor. AIChE J. 51, 210–223 [18].

Newman, J., Thomas-Alyea, K.E., 2004. Electrochemical Systems, third ed. John Wiley and Sons, Somerset, New Jersey [9].

Nguang, S.K., Chen, X.D., 1998. High frequency multi-input periodic operation of continuous fermentation process. J. Chem. Technol. Biot. 73, 233–236 [10].

Ni, M., Leung, Y.C., Leung, M.K.H., Sumathy, K., 2006. An overview of hydrogen production from biomass. Fuel Process Technol. 87, 461–472 [8].

Nibbelke, R.H., Campman, M.A.J., Hoebink, J.H.B.J., Marin, G.B., 1997. Kinetic study of the CO oxidation over Pt/γ-Al_2O_3 and Pt/Rh/CeO_2/γ-Al_2O_3 in the presence of H_2O and CO_2. J. Catal. 171, 358–373 [16].

Nibbelke, R.H., Nievergeld, A.J.L., Hoebink, J.H.B.J., Marin, G.B., 1998. Development of a transient kinetic model for the CO oxidation by O_2 over a Pt/Rh/CeO_2/γ-Al_2O_3 three way catalyst. Appl. Catal. B-Environ. 19, 245–259 [3] [6].

Nicolis, G., Prigogine, I., 1977. Self-organization in Non-equilibrium Systems: From Dissipative Structures to Order through Fluctuations. John Wiley & Sons. New York [1].

Nieken, U., Kolios, G., Eigenberger, G., 1992. Regulung eines instationär betriebnen Festbettreactors mit Fuzzy-Kontrollregeln. Chem. Ing. Tech. 64, 826 [18].

Nieken, U., Kolios, G., Eigenberger, G., 1994a. Fixed-bed reactors with periodic flow reversal: experimental results for catalytic combustion. Catal. Today 20, 335–350 [1] [18] [19].

Nieken, U., Kolios, G., Eigenberger, G., 1994b. Control of the ignited steady state in autothermal fixed-bed reactors for catalytic combustion. Chem. Eng. Sci. 49 (29B), 5507–5518 [18] [19].

Nieken, U., Kolios, G., Eigenberger, G., 1995. Limiting cases and approximate solutions for fixed-bed reactors with periodic flow reversal. AIChE J. 41, 1915–1925 [18].

Nijdam, J.L., van der Geld, C.W.M., 1997. Experiments with a large scale reverse flow reactor. Chem. Eng. Sci. 52, 2729–2741 [18].

Nijdam, J.L., van der Geld, C.W.M., 1999. A comparison of hybrid and numerical techniques to model heat transfer in reverse flow reactors. Appl. Therm. Eng. 19, 1045–1070 [18].

Nitipanyawong, C., Wilcox, G.D., Gabe, D.R., 2000. The formation of compositionally modulated zinc alloy coatings by pulsed electrodeposition, Proc. 5th Intern. Pulse Plating Symp., American Electroplaters & Surface Finishers Soc. Orlando, Florida. 118–123 [9].

Niwa, M., Murakami, Y., 1972a. Study on olefin oxidation by periodic-pulse technique I. Propylene oxidation into acrolein using various oxide catalysts. J. Catal. 26, 359–369 [4] [13].

Niwa, M., Murakami, Y., 1972b. Study on olefin oxidation by periodic-pulse technique II. Oxidative dehydrogenation of 1-butene using various oxide catalysts. J. Catal. 27, 26–33 [4].

Nolan, R.P., Lee, K., 2011. Dynamic model of CHO metabolism. Metab. Eng. 12, 108–124 [23].

Noldus, E.J., 1977. Periodic optimization of a chemical reactor system using perturbation methods. J. Eng. Math. 11, 49–66 [14].

Noorman, S., Galluci, F., van Sint Annaland, M., Kuipers, J.A.M., 2010b. Experimental investigation of a CuO/Al_2O_3 oxygen carrier for chemical-looping combustion. Ind. Eng. Chem. Res. 49, 9720–9728 [5].

Noorman, S., Galluci, F., van Sint Annaland, M., Kuipers, J.A.M., 2011. Experimental investigation of chemical-looping combustion in packed beds: a parametric study. Ind. Eng. Chem. Res. 50, 1968–1980 [5].

Noorman, S., van Sint Annaland, M., Kuipers, J.A.M., 2007. Packed bed reactor technology for chemical-looping combustion. Ind. Eng. Chem. Res. 46, 4212–4220 [5].

Noorman, S., van Sint Annaland, M., Kuipers, J.A.M., 2010a. Experimental validation of packed bed chemical-looping combustion. Chem. Eng. Sci. 65, 92–97 [5].

Nor, Z.M., Tamer, M.I., Mehrvar, M., Scharer, J.M., Moo-Young, M., Jervis, E.J., 2001a. Automated fed-batch culture of Kluyveromyces fragilis based on novel method of on-line estimation of the cell specific growth rate. Biochem. Eng. J. 9, 221–231 [10].

Nor, Z.M., Tamer, M.I., Mehrvar, M., Scharer, J.M., Moo-Young, M., Jervis, E.J., 2001b. Improvement of intracellular beta-galactosidase production in fed-batch culture of Kluyveromyces fragilis. Biotechnol. Lett. 23, 845–849 [10].

Noskov, A.S., Bobrova, L.M., Matros, Yu.Sh., 1993. Reverse-process for NOx – off gases decontamination. Catal. Today 17, 293–300 [18].

Nova, I., Castoldi, L., Lietti, L., Tronconi, E., Forzatti, P., 2002. On the dynamic behavior of NOx – storage/reduction Pt-Ba/Al_2O_3 catalysts. Catal. Today 75 (1–4), 431–437 [6].

Nowak, F.M., 2010. Sonochemistry: Theory, Reactions, Syntheses and Applications. Nova Science Publ, New York [1].

Nowobilski, P.J., Takoudis, C.G., 1986. Periodic operation of chemical reactor systems: are global improvements attainable? Chem. Eng. Commun. 40, 249–264 [1] [3] [11] [13] [14].

Özgülsen, F., Kendra, S.J., Çinar, A., 1993. Nonlinear predictive control of periodically forced chemical reactors. AIChE J. 39 (4), 589–598 [19].

Occelli, M.L. (Ed.), 2004. Fluid Catalytic Cracking VI. Stud. Surf. Sci. Catal., 149. Elsevier Science Publ., Amsterdam, The Netherlands [1].

Occelli, M.L. (Ed.), 2007. Fluid Catalytic Cracking VII. Stud. Surf. Sci. Catal., 166. Elsevier Science Publ., Amsterdam, The Netherlands [1].

Occelli, M.L., O'Connor, P. (Eds.), 1998. Fluid Cracking Catalysts. CRC Press, Boca Raton, Florida [1].

Ochoa-Fernandez, E., Haugen, G., Zhao, T., Ronning, M., Aartun, I., Boresen, B., Rytter, E., Ronnekleiv, M., Chen, D., 2007. Process design simulation of H_2

production by sorption-enhanced steam methane reforming. Green Chem. 9, 654–662 [22].

Ogihara, H., Takenaka, S., Yamanaka, I., Tanabe, E., Genseki, A., Otsuka, K., 2006. Formation of highly concentrated hydrogen through methane decomposition over Pd based alloy catalysts. J. Catal. 238 (2), 353–360 [8].

Ogura, H., Yamamoto, T., Kage, H., 2003. Efficiencies of $CaO/H_2O/Ca(OH)_2$ chemical heat pump for heat storing and heating and cooling. Energy 28, 1479–1493 [8].

Oh, S.H., Cavendish, J.C., Hegedus, L.L., 1980. Mathematical modeling of catalytic-converter lightoff - Single-pellet studies. AIChE J. 26 (6), 935–943 [3].

Oh, S.H., Hegedus, L.L., 1982. Dynamics of high-temperature CO chemisorption on Pt/Al_2O_3 by fast response IR spectroscopy. In: Bell, A.T., Hegedus, L.L. (Eds.), Catalysis Under Transient Conditions, Am. Chem. Soc, 178. ACS Symp. Ser, Washington, D.C, pp. 79–103 [3].

Oh, S.H., Hegedus, L.L., Aris, R., 1978. Temperature gradients in porous catalyst pellets. Ind. Eng. Chem. Fund. 17 (4), 309–313 [3].

Oh, G., Moo-Young, M., Chisty, Y., 1998. Automated fed-batch culture of recombinant Saccharomyces cerevisiae based on on-line monitored maximum uptake rate. Biochem. Eng. J. 1, 211–217 [10].

Ohmori, T., Nakaiwa, M., Endo, A., Akiya, T., Amemiya, T., Yamaguchi, T., 2001. Improvement in efficiency of a chemical reaction system for converting methanol to light olefins by periodic operation. Kagaku Ronbun. 27, 812–818 [16].

Ohmori, T., Na-Ranong, D., Yamaguchi, T., 2003. Periodic and recycling operations in a chemical reaction process for production of light olefins. Kagaku Ronbun. 29 (3), 374–377 [16].

Ohmori, T., Satoshi, T., Sho, K., et al., 2011. Global reaction enhancement by periodic operation. Kagaku Ronbun. 37 (2), 125–127 (Special Issue) [16].

Ohmura, N., Kataoaka, K., Watanabe, S., Okuba, M., 1998. Controlling particle size by self-sustained oscillations in continuous emulsion polymerization of vinyl acetate. Chem. Eng. Sci. 53, 2129–2137 [7].

Ohmura, N., Kitamoto, K., Yano, T., Kataoaka, K., 2001. Novel operating method for controlling latex particle size distribution in emulsion polymerization of vinyl acetate. Ind. Eng. Chem. Res. 40, 5177–5183 [7].

Ohmura, N., Nagamitsu, M., Kondo, K., Yano, T., Usui, H., 2005. Emulsion polymerization of vinyl acetate in a system of two CSTR in series. J. Chem. Eng. Jpn. 38 (9), 722–729 [7].

Ohtsuka, Y., Tamai, Y., Tomita, A., 1987. Iron-catalyzed gasification of brown coal at low temperature. Energy Fuels 1, 32–35 [8].

Okumura, K., Motohiro, T., Sakamoto, Y., Shinjoh, H., 2009. Effect of combinations of noble metals and metal oxide supports on catalytic reduction of NO by H_2. Surf. Sci. 603, 2544–2550 [6].

Olaj, O.F., Bitai, I., Gleixner, G., 1985. The laser flash-initiated polymerization as a tool of evaluating (individual) kinetic constants of free-radical polymerization. 1. Outline of method and first results. Macromol. Chem. Physic. 186, 2569–2580 [13].

Olaj, O.F., Bitai, I., Hinkelmann, F., 1987. The laser-flash-initiated polymerization as a tool of evaluating (individual) kinetic constants of free-radical polymerization 2. The direct determination of the rate-constant of chain propagation. Macromol. Chem. Physic. 188, 1689–1702 [13].

Oliveira, R.G., Wang, R.Z., 2007. A consolidated calcium chloride-expanded graphite compound for use in sorption refrigeration systems. Carbon 45, 390–396 [8].

Oliveira, R.G., Wang, R.Z., 2008. Study of a non-isothermal, non-isobaric consolidated reactive bed for chemisorption icemakers. Chem. Eng. J. 138, 416–424 [8].

Olsson, L., Monroe, D., Blint, R.J., 2006. Global kinetic modelling of a supplier barium- and potassium-containing lean NOx trap. Ind. Eng. Chem. Res. 45, 8883–8890 [6].

Olsson, P., Schöön, N.-H., 1985. Transient formation and reduction of sulfur trioxide in the presence of an unsupported platinum catalyst. Chem. Eng. Sci. 40, 1123–1134 [3].

Olsson, P., Schöön, N.-H., 1986. Suppression of sulfur trioxide formation in a monolithic catalytic converter for cars due to oscillating reaction conditions. Ind. Eng. Chem. Proc. DD. 25, 528–531 [3].

Omota, F., Dimian, A.C., Bliek, A., 2005. Partially hydrophobized silica supported Pd catalyst for hydrogenation reactions in aqueous media. Appl. Catal. A-Gen. 294 (2), 121–130 [17].

Onken, H.U., Wicke, E., 1986. Statistical fluctuations of temperature and conversion at the catalytic CO oxidation in an adiabatic packed bed reactor. Ber. Bunsen Phys. Chem. 90, 976–981 [3].

Opoku-Gyamfi, K., Adesina, A.A., 1999. Forced composition cycling of a novel thermally self-sustaining fluidised-bed reactor for methane reforming. Chem. Eng. Sci. 54, 2575–2583 [4].

Opoku-Gyamfi, K., Vieira-Dias, J., Adesina, A.A., 2000. Influence of cycle parameters on periodically operated fluidised bed reactor for CH_4 autoreforming. Catal. Today 63, 507–515 [4].

Otsuka, K., Kobayashi, S., Takenaka, S., 2000. Decomposition and regeneration of methane in absence and the presence of a hydrogen-adsorbing alloy $CaNi_5$. Appl. Catal. 190, 261–268 [8].

Otsuka, K., Mito, A., Takenaka, S., Yamanaka, I., 2001. Production of hydrogen from methane without CO_2

emission mediated by indium oxide and iron oxide. Intern. J. Hydrog. Energy 26, 191–194 [8].

Otsuka, K., Wang, Y., Nakamura, M., 1999. Direct conversion of methane to synthesis gas through gas-solid reaction using CeO_2-ZrO_2 solid solution at moderate temperature. App. Catal. A 183, 317–324 [4].

Otsuka, K., Wang, Y., Sunada, E., Yamanaka, I., 1998. Direct partial oxidation of methane to synthesis gas by cerium oxide. J. Catal. 175, 152–160 [4].

Otsuka, K., Yamada, C., Kaburagi, T., Takenaka, S., 2003. Hydrogen storage and production by redox of iron oxide for polymer electrolyte fuel cell vehicles. Int. J. Hydrogen Energy 28, 335–342 [4].

Padberg, G., Wicke, E., 1967. Stabiles und instabiles Verhalten eines adiabatischen Rohrreaktors am Beispiel der katalytischen CO-Oxydation. Chem. Eng. Sci. 22, 1035–1051 [16] [18].

Padeste, L., Baiker, A., 1994. Three way catalysts in a hybrid drive system; 1. Experimental study of dynamic behavior. Ind. Eng. Chem. Res. 33, 1113–1119 [6].

Pagotto Jr., S.O., de Alvarenga Freire, C.M., Ballester, M., 1999. Zn-Ni alloy deposits obtained by continuous and pulsed electrodeposition processes. Surf. Coat. Tech. 122 (1), 10–13 [9].

Paloukis, F., Neophytides, S.G., 2007. Numerical simulation of methane fuelled cogenerative SOFCs for the production of synthesis gas and electrical energy. Chem. Eng. Sci. 62, 3868–3881 [18].

Pan, Y.-D., Lee, J.H., 2003. Identification and control of processes with periodic operations or disturbances. Ind. Eng. Chem. Res. 42, 1938–1947 [19].

Paquet, D.A., Ray, W.H., 1994. Tubular reactors for emulsion polymerization. 1. Experimental investigation. AIChE J. 40 (1), 73–87 [7].

Park, D.W., Gau, G., 1986. Simulation of ethylene epoxidation in a multi-tubular reactor. Chem. Eng. Sci. 41 (1), 143–150 [1] [4].

Park, D.W., Ghazali, S., Gau, G., 1983. A cyclic reactor study of ethylene epoxidation. Appl. Catal. 6, 175–193 [1] [4] [8] [13].

Park, J., Kinney, K.A., 2001. Evaluation of a slip feed system for vapor phase bioreactors. J. Environ. Eng. 127 (11), 979–985 [18].

Parulekar, S.J., 1998. Analysis of forced periodic operations of continuous bioprocesses-single input variations. Chem. Eng. Sci. 53 (14), 2481–2502 [10].

Parulekar, S.J., 2000. Analysis of forced periodic operations of continuous bioprocesses: multiple input variations. Chem. Eng. Sci. 55, 513–533 [10] [12].

Parulekar, S.J., 2001. Forced periodic operations of continuous recombinant cell cultures subject to antibiotic selection pressure. Chem. Eng. Sci. 56, 6463–6484 [10].

Parulekar, S.J., 2003. Systematic performance analysis of continuous processes subject to multiple input cycling. Chem. Eng. Sci. 58, 5173–5194 [1] [12] [23].

Parulekar, S.J., Lee, J., 1993. Structure-analysis of continuous cultures subject to periodic medium tuning. Chem. Eng. Sci. 48 (17), 3007–3035 [10].

Patnaik, P.R., 1994. Continuous fermentation: the case of variable plasmid loss probability. Can. J. Chem. Eng. 72, 929–933 [10].

Pedersen, H., Cho, G.H., Hamilton, R., Chin, C.K., 1987. Bioreactor operating strategies for plant cell cultures. Ann. N.Y. Acad. Sci. 506, 163–170 [10].

Pennington, R.J., 1984. Heat recovery from thermal oxidizers. Plant/Oper. Progr. 3 (2), 72–74 [18].

Peper, S., Johannsen, M., Brunner, G., 2007. Preparative chromatography with supercritical fluids. Comparison of simulated moving bed and batch processes. J. Chromat. A. 1176, 246–253 [23].

Peper, S., Lübbert, M., Johannsen, M., Brunner, G., 2002. Separation of ibuprofen enantiomers by supercritical fluid simulated moving bed chromatography. J. Biochem. Biophys. Methods 54 (1–3), 85–102 [23].

Pereira, C.S.M., Zabka, M., Silva, V.M.T.M., Rodrigues, A.E., 2009. A novel process for the ethyl lactate synthesis in a simulated moving bed reactor. Chem. Eng. Sci. 64, 3301–3310 [1] [21].

Petkovska, M., 2001. Nonlinear frequency response of nonisothermal adsorption systems. Nonlinear Dyn. 26, 351–370 [14].

Petkovska, M., 2006. Nonlinear frequency response method for investigation of equilibria and kinetics in adsorption systems. In: Spasic, A.M., Hsu, J.P. (Eds.), Adsorption Systems in Finely Dispersed Particles: Micro, Nano- and Atto-Engineering. CRC - Taylor & Francis, Boca Raton, pp. 283–327 [14].

Petkovska, M., Do, D.D., 1998. Nonlinear frequency response of adsorption systems: isothermal batch and continuous flow adsorber. Chem. Eng. Sci. 53, 3081–3097 [14].

Petkovska, M., Do, D.D., 2000. Use of higher order FRFs for identification of nonlinear adsorption kinetics: single mechanisms under isothermal conditions. Nonlinear Dynamics 21, 353–376 [14].

Petkovska, M., Marković, A., 2006. Fast estimation of quasi-steady states of cyclic nonlinear processes based on higher-order frequency response functions. Case study: cyclic operation of an adsorption column. Ind. Eng. Chem. Res. 45, 266–291 [14].

Petkovska, M., Nikolić, D., Marković, A., Seidel-Morgenstern, A., 2010. Fast evaluation of periodic operation of a heterogeneous reactor based on nonlinear frequency response analysis. Chem. Eng. Sci. 65, 3632–3637 [14].

Petkovska, M., Petkovska, Lj.T., 2003. Use of nonlinear frequency response for discriminating adsorption kinetics mechanisms resulting with bimodal characteristic functions. Adsorption 9, 133–142 [14].

Petkovska, M., Seidel-Morgenstern, A., 2005. Nonlinear frequency response of a chromatographic column. Part I: application to estimation of adsorption isotherms with inflection points. Chem. Eng. Commun. 192, 1–34 [14].

Petroulas, T., Aris, R., Carr, R.W., 1985a. Analysis of the counter-current moving-bed chromatographic reactor. Comput. Math. Appl. 11 (1–3), 5–34 [20].

Petroulas, T., Aris, R., Carr Jr., R.W., 1985b. Analysis and performance of a countercurrent moving-bed chromatographic reactor. Chem. Eng. Sci. 40 (12), 2233–2240 [20].

Pfeifer, C., Puchner, B., Hofbauer, H., 2007. Insitu CO_2 absorption in a dual fluidized bed biomass steam gasifier to produce a hydrogen-rich syngas. Int. J. Chem. React. Eng. 5, A9 [8].

Pfeifer, C., Puchner, B., Hofbauer, H., 2009. Comparison of dual fluidized bed steam gasification of biomass with and without selective transport of CO_2. Chem. Eng. Sci. 64, 5073-5083 [8].

Phan, N.H., Schwartz, M., Nobe, K., 1994. Pulsed electrodeposition of Fe-Ni-Co alloys. Electrochim. Acta 39 (3), 449–453 [9].

Pholjaroen, B., Laosiripojana, N., Praserthdam, P., Assabumrungrat, S., 2009. Reactivity of Ni/SiO_2-MgO toward carbon dioxide reforming of methane under steady state and periodic operations. J. Ind. Eng. Chem. 15, 488–497 [4].

Pickett, A.M., Bazin, M.J., Topiwala, H.H., 1979. Growth and composition variation of *Escherichia coli* subjected to square wave perturbations in nutrient supply: effect of varying frequencies. Biotechnol. Bioeng. 21, 1043–1055 [10].

Pickett, A.M., Bazin, M.J., Topiwala, H.H., 1980. Growth and composition variation of *Escherichia coli* subjected to square wave perturbations in nutrient supply: effect of varying amplitudes. Biotechnol. Bioeng. 21 1213–1122 [10].

Pigford, R.L., Baker, B.I., Blum, D.E., 1969. Cycling zone adsorption, a new separation process. Ind. Eng. Chem. Fund. 8, 848–851 [14].

Pilgrim, A., Kawase, M., Matsuda, F., Miura, K., 2006. Modelling of the simulated moving bed reactor for the enzyme-catalyzed production of lactosucrose. Chem. Eng. Sci. 61 (2), 353–362 [1].

Pinjala, V.Y., Chen, C., Luss, D., 1988. Wrong-way behavior of packed bed reactors: II. Impact of thermal dispersion. AIChE J. 41, 1663–1772 [18].

Pinkerton, B., Luss, D., 2007. Hot zone formation during hydrogenation of ethylene and acetylene mixtures in a shallow packed bed reactor. Ind. Eng. Chem. Res. 46, 1898–1903 [2].

Pinto, J.C., Ray, W.H., 1995a. The dynamic behavior of continuous solution polymerization reactors: 7. Experimental study of a copolymerization reactor. Chem. Eng. Sci. 50 (4), 715–736 [7].

Pinto, J.C., Ray, W.H., 1995b. The dynamic behavior of continuous solution polymerization reactors: 8. A full bifurcation analysis of a full-scale copolymerization reactor. Chem. Eng. Sci. 50 (21), 3455–3475 [7].

Pinto, J.C., Ray, W.H., 1996. The dynamic behavior of continuous solution polymerization reactors: 9. Effects of inhibition. Chem. Eng. Sci. 51 (1), 63–79 [7].

Pismen, L.M., 1976. Convective currents induced by chemical reactions in partially-filled porous media. Chem. Eng. Sci. 31, 693–699 [15].

Poirier, M., Sapundzhiev, C., 1997. Catalytic decomposition of natural gas to hydrogen for fuel cell applications. Int. J. Hydrog. Energy 22 (4), 429–433 [18] [23].

Popov, K.I., Pavlovic, M.G., 1993. Electrodeposition of metal powders with controlled particle grain size and morphology. In: Bockris, J.O.'M., Conway, B.E. (Eds.), Modern Aspects of Electrochemistry, vol. 24. p. 299 [9].

Posarac, D., Hudgins, R.R., Jaree, A., 2003. Dynamic analysis of thermal behaviour of a laboratory reactor using a process simulator. Can. J. Chem. Eng. 80, 110–117 [16].

Prairie, M.R., Bailey, J.E., 1986. Application of feedback-induced bifurcation for evaluating steady-state and transient heterogeneous catalysis kinetic models. Chem. Eng. Sci. 41, 937–944 [2].

Prairie, M.R., Bailey, J.E., 1987. Experimental and modelling investigations of steady-state and dynamic characteristics of ethylene hydrogenation on Pt/Al_2O_3. Chem. Eng. Sci. 42 (9), 2085–2102 [2].

Prairie, M.R., Cho, B.K., Oh, S.H., Shinouskis, E.J., Bailey, J.E., 1988. Steady-state and transient studies of CO oxidation on alumina supported rhodium via transmission IR spectroscopy. Ind. Eng. Chem. Res. 27, 1396–1407 [3].

Price, G.J. (Ed.), 1993. Current Trends in Sonochemistry. Royal Society of Chemistry, Oxford U.K. [15].

Prinetto, F., Ghiotti, G., Nova, I., Lietti, L., Tronconi, E., Forzatti, P., 2001. FT-IR and TPD investigation of the NOx storage properties of BaO/Al_2O_3 and $Pt-BaO/Al_2O_3$ Catalysts. J. Phys. Chem. B. 105, 12732–12745. [6].

Pritzker, M.D., Hudson, J.L., 2000. Modeling the current distribution over a wire electrode during copper pulse plating, Proc. 5th Intern. Pulse Plating Symp., American Electroplaters & Surface Finishers Soc. Orlando, Florida, pp. 79–86 [9].

Prokopowicz, R.A., Silveston, P.L., Baudais, F.L., Irish, D.E., Hudgins, R.R., 1987. Design of a transmission cell for the high temperature study of transient catalytic adsorption and reaction in a flow system. J. Appl. Spectros. 42, 385–389 [3].

Prokopowicz, R.A., Silveston, P.L., Hudgins, R.R., Irish, D.E., 1988. Oxidation of CO over a Cu(II), oxide/silica catalyst. React. Kinet. Catal. Lett. 37 (1), 63–70. [3].

Promaros, E., Assabumrungrat, S., Laosiripojana, N., Praserthdam, P., Tagawa, T., Goto, S., 2007. Carbon dioxide reforming of methane under periodic operation. Korean J. Chem. Eng. 24 (1), 44–50 [4].

Puippe, J.-C., 1986. Pulsed deposition of rhenium. In: Puippe, J.-C., Leaman, F. (Eds.), Theory and Practice of Pulse Plating, American Electroplaters & Surface Finishers Society [9].

Puippe, J.-C., Leaman, F., 1986. Theory and Practice of Pulse Plating, American Electroplaters & Surface Finishers Society [9].

Purwono, S., Budman, H., Silveston, P.L., Hudgins, R.R., Matros, Yu.Sh., 1994. Runaway in packed bed reactors operating with periodic flow reversal. Chem. Eng. Sci. 49, 5473–5487 [18] [19].

Putnam, A.A., Belles, F.E., Kentfield, J.A., 1986. Pulse Combustion. Prog. Energy Combustion Sci. 12, 42–79 [5].

Pyatnitsky, Y.I., Ilchenko, N.I., 1996. Some kinetic aspects of unsteady-state partial oxidation reactions: dynamic processes on metal oxide surfaces. Catal. Today 32, 21–28 [8].

Qiao, C.-Z., 1998. Performance of a fixed-bed reactor for n-butane oxidation to maleic anhydride under forced unsteady-state operation. M.S. thesis. Beijing Univ. Chem. Technol, Beijing, China [4].

Qin, F., Wolf, E.E., 1995. Vibrational control of chaotic self-contained oscillations during CO oxidation on a Rh-SiO_2 catalyst. Chem. Eng. Sci. 50 (1), 117–126 [3].

Quaade, U.J., Jensen, S., Hansen, O., 2004. Microsystem with integrated capillary leak to mass spectrometer for high sensitive temperature programmed desorption. Rev. Sci. Instrum. 75 (10), 3345–3347 [16].

Quaade, U.J., Jensen, S., Hansen, O., 2005. Fabrication and modeling of narrow capillaries for vacuum system gas inlets. 044906. J. Appl. Phys. 97, 1–4 [16].

Quinta Ferreira, R., Almeida Costa, C., Masetti, S., 1999. Reverse-flow reactor for a selective oxidation process. Chem. Eng. Sci. 54, 4615–4627 [18].

Quiram, D.J., Jensen, K.F., Schmidt, M.A., Mills, P.L., Ryley, J.F., Wetzel, M.D., Kraus, D.J., 2007a. Integrated microreactor system for gas-phase catalytic reactions. 1. Scale-up microreactor design and fabrication. Ind. Eng. Chem. Res. 46, 8292–8305 [16].

Quiram, D.J., Jensen, K.F., Schmidt, M.A., Mills, P.L., Ryley, J.F., Wetzel, M.D., Kraus, D.J., 2007b. Integrated microreactor system for gas-phase catalytic reactions. 2. Microreactor packaging and testing. Ind. Eng. Chem. Res. 46, 8306–8318 [16].

Quiram, D.J., Jensen, K.F., Schmidt, M.A., Mills, P.L., Ryley, J.F., Wetzel, M.D., Kraus, D.J., 2007c. Integrated microreactor system for gas-phase catalytic reactions. 3. microreactor system design and system automation. Ind. Eng. Chem. Res. 46, 8319–8335 [16].

Rabl, A., Doepper, R., Renken, A., 1986. Steady state and transient kinetic studies of an industrial vinyl acetate catalyst. In: Proc. 3rd World Congress of Chem. Eng. vol. IV, 387–390. Tokyo [12] [13].

Rabl, A., Renken, A., 1986. Untersuchungen zur produktinhibierung unter katalysator-desaktivierung bei der vinyl acetat-bildung. Chem. Ing. Tech. 58, 434–435 [12].

Rambeau, G., Amariglio, H., 1978a. Ammonia-synthesis on iron conceived as an auto-catalytic reaction. 1. Difficulties in kinetic determinations due to auto-catalytic behavior of reaction. J. Chim. Phys. 75 (1), 110–115 [2].

Rambeau, G., Amariglio, H., 1978b. Ammonia-synthesis on iron conceived as an auto-catalytic reaction. 2. Mechanism of hydrogenation of pre-adsorbed nitrogen. J. Chim. Phys. 75 (3), 333–340 [2].

Rambeau, G., Amariglio, H., 1978c. Ammonia-synthesis on iron conceived as an auto-catalytic reaction. 3. Reaction mechanism and quantitative desorption of self-activation. J. Chim. Phys. 75 (4), 397–405 [2].

Rambeau, G., Amariglio, H., 1981a. Ammonia synthesis on ruthenium powder from 100° to 500°C and hydrogenation of pre-adsorbed nitrogen down to -70°C. J. Catal. 72, 1–11 [2].

Rambeau, G., Amariglio, H., 1981b. Improvement of the catalytic performance of a ruthenium powder in ammonia synthesis by the use of a cyclic procedure. Appl. Catal. 1, 291–302 [1] [2].

Rambeau, G., Amariglio, H., 1984. Reflections upon conditions required for obtaining rate improvement by periodically changing the feed composition of a catalytic reactor. In: Kaliaguine, S., Mahay, A. (Eds.), Catalysis on the Energy Scene, Surf. Sci. Catal. Ser. Elsevier Science Publ., Amsterdam, The Netherlands. pp. 213–220 [2].

Rambeau, G., Jorti, A., Amariglio, H., 1982. Improvement in the catalytic performance of osmium powder in ammonia synthesis by use of a cyclic procedure. Appl. Catal. 3, 273–282 [2].

Ramdani, K., Pontier, R., Schweich, D., 2001. Reverse flow reactor at short switching periods for VOC combustion. Chem. Eng. Sci. 56, 5131–5139 [18].

Ramos, R., Herguido, J., Menendez, M., Santamaria, J., 1996. Oxidation of hydrocarbons in an in situ redox fluidized bed reactor. J. Catal. 163, 218–221 [4].

Randall, H., Doepper, R., Renken, A., 1996. Model discrimination by unsteady-state operation: application to the reduction of NO with CO on iron oxide. Can. J. Chem. Eng. 74, 586–593 [13].

Rao, A.S., Prasad, K.B.S., Rao, M.B., 1992. An experimental study of cyclic operation of a benzene hydrogenation reactor. J. Catal. 136, 242–245 [2] [12].

Rasmussen, J., 2000. Taguchi optimization of pulse plating. Proc. 5th Intern. Pulse Plating Symp., American Electroplaters & Surface Finishers Soc. Orlando, Florida. pp. 38–40 [9].

Rasmussen, J., 2001. Pulse anodizing of aluminum. Part 1: Basic facts. Plat. Surf. Finish. 88 (12), 42–43 [9].

Rasmussen, J., Kartchner, T., 2000. Proc. 5th Intern. Pulse Plating Symp. American Electroplaters & Surface Finishers Soc, Orlando, Florida (Title not available) [9].

Ravi Kumar, V., Kulkarni, B.D., Despande, P.D., 1991. On the robust control of nonlinear systems. Proc. Roy. Soc. Lond. A 433, 711–722 [19].

Rawlings, J.B., Ray, W.H., 1987. The stability of continuous emulsion polymerization reactors: a detailed model analysis. Chem. Eng. Sci. 42, 2767–2775 [7].

Ray, W.H., 1967. Modelling polymerization reactors with application to optimal design. Can. J. Chem. Eng. 45, 356–364 [7].

Ray, W.H., 1968. Periodic operation of polymerization reactors. Ind. Eng. Chem. Proc. Des. Dev. 7, 422–433 [7].

Ray, W.H., 1972. On the mathematical modelling of polymerization reactors. J. Macromol. Sci. -Revs. Macromol. Chem. C8, 1–34 [7].

Ray, W.H., 1981. Advanced Process Control. McGraw Hill Inc., New York, New York [19].

Ray, A.K., Carr, R.W., Aris, R., 1994. The simulated countercurrent moving bed chromatographic reactor: a novel reactor-separator. Chem. Eng. Sci. 49, 469–480 [21].

Ray, A.K., Carr, R.W., 1995a. Experimental study of a laboratory-scale simulated countercurrent moving-bed chromatographic reactor. Chem. Eng. Sci. 50 (14), 2195–2202 [21].

Ray, A.K., Carr, R.W., 1995b. Numerical simulation of a simulated countercurrent moving-bed chromatographic reactor. Chem. Eng. Sci. 50 (19), 3033–3041. [21].

Ray, A.K., Tonkovich, A.L., Aris, R., Carr, R.W., 1990. The simulated countercurrent moving bed chromatographic reactor. Chem. Eng. Sci. 45, 2431–2437 [1] [21].

Ray, W.H., Soares, J.B.P., Hutchinson, R.A., 2004. Polymerization reaction engineering: past, present and future. Macromol. Symp. 206 (1), 1–14 [7].

Řeháček, J., Kubiček, M., Marek, M., 1992. Modeling of a tubular reactor with flow reversal. Chem. Eng. Sci. 47 (9–11), 2897–2902 [18] [19].

Řeháček, J., Kubiček, M., Marek, M., 1998. Periodic, quasiperiodic and chaotic spatiotemporal patterns in a tubular catalytic reactor with periodic flow reversal. Comput. Chem. Eng. 22 (1–2), 283–297 [18].

Rehrig, D.L., 1974. Effect of deposition method on porosity in gold thin films. Plating 61 (1), 43–46 [9].

Reichert, K.H., Moritz, H.U., 1989a. Polymer Reaction Engineering (Chapter 24). In: Allen, G., Bevington, G.J.C. (Eds.), Comprehensive Polymer Science, vol. 3. Pergamon Press, Oxford, U.K. [7].

Reichert, K.H., Moritz, H.U., 1989b. Polymer reaction engineering. Polym. Sci. 3 (1), 327–361 [7]

Reijers, H.T.J., van Beurden, P., Elzinga, C.D., Kluiters, S.C.A., Dikstra, J.W., van den Brink, R.W., 2006. A new route for hydrogen production with simultaneous CO_2 capture. 16th World Hydrogen Energy Conf, Lyon, France. June, 2006 [22].

Renken, A., 1972. The use of periodic operation to improve the performance of continuous stirred tank reactors. Chem. Eng. Sci. 27, 1925–1932 [1] [12].

Renken, A., 1974a. Verbesserung von Selektivitaet und Ausbeute durch periodische Prozessfuehrung. Chem. Ing. Tech. 46 (3), 113 (Extended Manuscript) [1].

Renken, A., 1974b. Improving selectivity and yield by conducting processes in a periodic fashion. Chem. Ing. Tech. 46 (3), 113 [12].

Renken, A., 1982. Unsteady-state operation of continuous reactors. Chem. Ing. Tech. 54 (6), 571–580 [1].

Renken, A., 1984. Unsteady-state operation of continuous reactors. Int. Chem. Eng. J. 24 (2), 202–213 [1].

Renken, A., 1990a. Application of unsteady state processes in modelling heterogeneous catalytic kinetics. In: Matros, Yu.Sh. (Ed.), Unsteady State Processes in Catalysis VSP Publ. BV, Utrecht, The Netherlands [1] [13].

Renken, A., 1990b. Instationäre reaktionsführung zur modellierung heterogen katalytischer reaktionen. Chem-Ing-Tech. 62 (9), 724–733 [13].

Renken, A., Doepper, R., Schmid, M., 1989. Application of non-stationary and periodic processes in the study of heterogeneous catalytic kinetics. Dechema Monographs, 120, VCH Verlagsgesellschaft, 273–290 [4] [12] [16].

Renken, A., Helmrich, H., Lenthe, M., 1974. Experimental studies on the behaviour of instantaneous isothermal fixed-bed reactors, Paper A3-1. Proc.GVC/AIChE, Joint Meeting, Verlag Chemie GmbH, Weinheim, Germany [12].

Renken, A., Mueller, M., Wandrey, C., 1976. Experimental studies of fixed-bed reactors by periodic operation the catalytic oxidation of ethylene. In: Proc. 4th Int. Symp. Chem. React. Eng., vol. III. Dechema, Frankfurt, Germany pp. 107–116 [4] [8].

Renken, A., Truffer, M.-A., Dettmer, M., 1984. The effect of cyclic operation on heterogeneous catalytic reaction with educt inhibition. Inst. Chem. Eng. Symp. Series 87, 117–124 [11] [12].

Renz, R.P., Taylor, E.J., Inman, M.E., 2000. Charge-modulated chromium III plating to dispel hydrogen effects,

Proc. 5th Intern. Pulse Plating Symp. American Electroplaters & Surface Finishers Soc, Orlando, Florida. 124–130 [9].

Reshetnikov, S.I., Ivanov, E.A., 1999. Spatial regulation of the catalyst activity enhancement for catalytic reactions with stop-effect. React. Kinet. Catal. Lett. 68 (2), 393–398 [11].

Reshetnikov, S.I., Ivanov, E.A., Kiwi-Minsker, L., Renken, A., 2003. Performance enhancement by unsteady-state reactor operation: Theoretical analysis for two-sites kinetic model. Chem. Eng. Technol. 26, 751–758 [2] [11].

Reshetnikov, S.I., Ivanov, E.A., Startsev, A.N., 2007. Benzene hydrogenation in the thiophene presence over the sulfide Ni-Mo/γ-Al_2O_3 catalyst under periodic operation: Kinetics and process modeling. Chem. Eng. J. 134, 100–105 [2].

Reuter, D.M., Daniel, B.R., Jagoda, J.L., Zinn, B.T., 1986. Periodic mixing and combustion processes in gas fired pulsating combustors. Combust. Flame 65, 281–290 [5].

Reyes, S.C., Sinfelt, J.H., Martin, G.J., Ernst, R.H., 1997. Frequency modulation methods for diffusion and adsorption measurements in porous solids. J. Phys. Chem. B 101, 614–622 [15].

Rhee, H.-K., Lewis, R.P., Amundson, N.R., 1974. Creeping profiles in catalytic bed reactors, continuous models. Ind. Eng. Chem. Fund. 13, 317–323 [18].

Richards, J.R., Congalidis, J.P., 2006. Measurement and control of polymerization reactors. Comput. Chem. Eng. 30 (10-12), 1447–1463 [7].

Rinnemo, M., Kulginov, D., Johansson, S., Wong, K.L., Zhdanov, V.P., Kasemo, B., 1997. Catalytic ignition in the CO-O_2 reaction on platinum: experiment and simulations. Surf. Sci. 376, 297–309 [16].

Roestenberg, T., Glushenkov, M.J., Kronberg, A.E., van den Meer, Th.H., 2010. On the controllability and run-away possibility of a totally free piston pulsed compression reactor. Chem. Eng. Sci. 65, 4916–4922 [15].

Roginskii, S.Z., Rozental, A.L., 1962. Chemical reactions in a chromatographic regime. Dokl. Akad. Nauk SSSR 146, 152–155 [20].

Roginskii, S.Z., Rozental, A.L., 1964. Kinetics for reactions under chromatographic conditions on heterogeneous surfaces. Kinet. Katal. 5 (1), 104–112 (in Russian) [20].

Roginskii, S.Z., Yanovskii, M.I., Gaziev, G.A., 1961. Chemical reactions under chromatographic conditions. Dokl. Akad. Nauk SSSR 140, 1125–1127 [20].

Roginskii, S.Z., Yanovskii, M.I., Gaziev, G.A., 1962. Catalytic reactions and catalysts under chromatographic conditions. Kinet. Katal. 3, 529–540 (English translation (1961), 771-785) [20].

Ronen, M., Shabtai, Y., Guterman, H., 2002. Optimization of feeding profile for a fed-batch bioreactor by an evolutionary algorithm. J. Biotechhnol. 97, 253–261. [10].

Roopsingh, G., Chidambaram, M., 1999. Periodic operation of bioreactors for auto catalytic reactions with Michaelis-Menten kinetics. Bioprocess Eng. 20, 279–282 [10].

Rose, K.C., Battogtokh, D., Mikhailov, A.S., Imbihl, R., Engel, W., Bradshaw, A.M., 1996. Cellular structures in catalytic reactions with global coupling. Phys. Rev. Lett. 76 (19), 3582–3585 [3].

Rouge, A., Renken, A., 2001a. Forced periodic temperature oscillations in microchannel reactors. In: Proc. of the Fifth Intern. Conf. on Microreaction Technology. Strasbourg, France, pp. 27–30 [11] [16].

Rouge, A., Renken, A., 2001b. Performance enhancement of a microchannel reactor under periodic operation. In: Froment, G.F., Waugh, K.C. (Eds.), Reaction Kinetics and the Development and Operation of Catalytic Processes. Elsevier Science BV Amsterdam, The Netherlands, pp. 239–246 [11].

Rouge, A., Spoetzl, B., Gebauer, K., Schenk, R., Renken, A., 2001. Microchannel reactors for fast periodic operation: the catalytic dehydrogenation of isopropanol. Chem. Eng. Sci. 56, 1419–1427 [11] [16].

Roy, S., Baiker, A., 2009. NO(x) Storage-reduction catalysis: From mechanism and materials properties to storage-reduction performance. Chem. Rev. 109 (9), 4054–4091.

Roy, S., Bauer, T., Al-Dahhan, M., Lehner, P., Turek, T., 2004. Monoliths as multiphase reactors: a review. AIChE J. 50, 2918–2938 [17].

Roy, S., Carder, D.A., Stephenson, A., Meuleman, W.R.A., 2000. Effect of agitation and electrochemical parameters on the composition of compositionally modulated metal multilayers, Proc. 5th Intern. Pulse Plating Symp. American Electroplaters & Surface Finishers Soc. Orlando, Florida [9]103–108.

Rubio, O., Herguido, J., Menenedz, M., Grasa, G., Abanades, J.C., 2004. Oxidative dehydrogenation of butane in an interconnected fluidized-bed reactor. AIChE J. 50, 1510–1522 [4].

Rubio, O., Mallada, R., Herguido, J., 2002. Experimental study on the oxidation of butane to maleic anhydride in a two-zone fluidized bed reactor. Ind. Eng. Chem. Res. 41, 5181–5186 [4].

Ruettinger, W.F., Ilinich, O., Mikhailovich, M., Farrauto, R.J., 2004. International Patent, WO/2005/035116 Operating conditions for copper-based water-gas shift catalysts [1].

Russo, L., Altimari, P., Mancusi, E., Maffettone, P.L., Crescitelli, S., 2006. Complex dynamics and spatiotemporal patterns in a network of three distributed chemical reactors with periodical feed switching. Chaos Solitons Fractals 3, 682–706 [5].

Russo, L., Mancusi, E., Maffettone, P.L., Crescitelli, S., 2002. Symmetry properties and bifurcation analysis of a class of periodically forced reactors. Chem. Eng. Sci. 57, 5065–5082 [5].

Ruthven, D.M., 1984. Principles of Adsorption and Adsorption Processes. Wiley-Interscience, New York [1] [22].

Ruthven, D.M., Ching, C.B., 1989. Counter-current and simulated counter-current adsorption separation processes. Chem. Eng. Sci. 44 (5), 1011–1038 [21].

Ruthven, D.M., Farooq, S., Knaebel, K.S., 1994. Pressure Swing Adsorption. VCH Publishers, New York [14].

Sad, M.R., Querini, C.A., Comelli, R.A., Figoli, N.S., Parera, J.M., 1996. Dehydroisomerization of n-butane in a chromatographic reactor. Appl. Catal. A-Gen. 146, 131–143 [20].

Sadeghbeigi, R., 2000. Fluid Catalytic Cracking Handbook. Gulf Professional Publ., Houston, Texas [1].

Sadhankar, R.R., Lynch, D.T., 1994. N_2O reduction by CO over an alumina-supported Pt catalyst: forced composition cycling. J. Catal. 149, 278–291 [3] [13].

Sadhankar, R.R., Lynch, D.T., 1996. Slow convergence during forced oscillations of the NO+CO reaction over a platinum catalyst. Chem. Eng. Sci. 51 (10), 2061–2068 [3].

Sadhankar, R.R., Lynch, D.T., 1996a. Transient and resonant behaviour for NO reduction by CO over a Pt/Al_2O_3 catalyst during forced composition cycling. Can. J. Chem. Eng. 74, 674–682 [3].

Sadhankar, R.R., Lynch, D.T., 1997. NO reduction by CO over a Pt/Al_2O_3 catalyst: reaction kinetics and experimental bifurcation behavior. Ind. Eng. Chem. Res. 36, 4609–4619.

Saito, H., Murakami, Y., Hattori, T., 1965. Conversion and product distribution in pulse reactor. Kagaku Kogaku 29, 585–590 [20].

Sakurai, M., Honda, H., Kameyama, H., 2007. Fundamental study of a non-steady operation for 2-propanol dehydrogenation. Intern. J. Hydr. Energy 32, 1303–1308 [11].

Salden, A., Eigenberger, G., 2001. Multifunctional adsorber / reactor concept for waste-air purification. Chem. Eng. Sci. 56, 1605–1611 [18].

Saleh-Alhamed, Y.A., Hudgins, R.R., Silveston, P.L., 1992. Periodic operation studies on the partial oxidation of propylene to acrolein and acrylic acid. Chem. Eng. Sci. 47, 2885–2896 [4] [8].

Saleh-Alhamed, Y.A., Hudgins, R.R., Silveston, P.L., 1993. The role of water in the partial oxidation of propylene. In: Guczi, L (Ed.), Proc. 10th Int.Congr.Catal., Budapest, Hungary. Elsevier Science Publishers, Amsterdam, The Netherlands, [4] [8].

Sales, B.C., Turner, J.E., Maple, M.B., 1982. Oscillatory oxidation of CO over Pt, Pd and Ir catalysts: Theory. Surf. Sci. 114, 381–394 [16].

Salinger, A.G., Eigenberger, G., 1996a. The direct calculation of periodic states of the reverse flow reactor: I. Methodology and propane combustion results. Chem. Eng. Sci. 51 (21), 4903–4914 [18].

Salinger, A.G., Eigenberger, G., 1996b. The direct calculation of periodic states of the reverse flow reactor: II. Multiplicity and instability. Chem. Eng. Sci. 51 (21), 4915–4922 [18] [19].

Salomons, S., Hayes, R.E., Poirier, M., Sapoundjiev, H., 2003. Flow reversal reactor for the catalytic combustion of lean methane mixtures. Catal. Today 83, 59–69 [18].

Salomons, S., Hayes, R.E., Poirier, M., Sapoundjiev, H., 2004. Modeling a reverse flow reactor for the catalytic combustion of fugitive methane emissions. Comput. Chem. Eng. 28, 1599–1610 [18].

Sanayei, R.A., O'Driscoll, K.F., Klumperman, B., 1994. Pulsed laser copolymerization of styrene and maleic anhydride. Macromolecules 27, 5577–5582 [13].

Sapoundjiev, H., 2000. High potential energy from dilute methane emissions, Proc. 5th Intern. Conf. Greenhouse Gas Control Technol, Cairns, Australia [18].

Sapoundjiev, H., Aube, F., Hossate, S., Gauthier, J., Trotier, R., 1999. Economic heat recovery and greenhouse gas elimination form ventilation air of underground coal mines. In: Proc. International Coal-bed Methane Symp. Tuscaloosa, Alabama. 269–278 [18].

Sapundzhiev, H., Bunimovich, G.A., Drobishevich, V.I., Grozev, G., Yausheva, L.V., Matros, Yu.Sh., Elenkov, D., 1988. Effect of heat losses on the behavior of the reactor with periodic flow reversal. Teor. Osnovy Khim. Teknol. 22 (3), 349–355 (in Russian) [18].

Sapundzhiev, H., Bunimovich, G.A., Matros, Yu.Sh., 1984. Mathematical modeling of non-stationary method for sulfuric acid production from low-concentrated metallurgy gases. Commun. Dept. Chem. Bulg. Acad. Sci. 17, 298–304 [18].

Sapundzhiev, C., Chaouki, J., Guy, C., Klvana, D., 1993. Catalytic combustion of natural gas in a fixed bed reactor with periodic flow reversal. Chem. Eng. Commun. 125, 171–186 [18].

Sapundzhiev, C., Chaouki, J., Guy, C., Klvana, D., 1994. Catalytic incineration of VOC's in a cyclic reactor. Stud. Environ. Sci.: Characterization and Control of Odors and VOC in the Process Industries (Elsevier Science publishers) 61, 479–490 [18].

Sapundzhiev, C., Grozev, G., Elenkov, D., 1990. Influence of geometric and thermo-physical properties of reactor layer on sulphur dioxide oxidation in transient conditions. Chem. Eng. Technol. 13, 130–135 [18].

Sapundzhiev, C., Grozev, G., Elenkov, D., 1991. Non-steady state catalytic decontamination of waste gases. Chem. Eng. Technol. 14, 209–212 [18].

Saravanane, R., Murthy, D.V.S., Krishnaiah, K., 2001. Treatment of anti-osmotic drug based pharmaceutical effluent in an upflow anaerobic fluidized bed system. Waste Managem. 21, 563–568 [10].

Sardin, M., Villermaux, J., 1979. Estérification catalysée par une résine échangeuse de cation dans un réacteur chromatographique. Nouv. J. Chim. 3 (4), 255–261 [20].

Sarkar, D., Modak, J.M., 2003. Optimization of fed-batch bioreactors using genetic algorithms. Chem. Eng. Sci. 58, 2283–2296 [10].

Sarmidi, M.R., Barker, P.E., 1993a. Saccharification of modified starch to maltose in a continuous rotating annular chromatograph (CRAC). J. Chem. Technol. Biot. 57 (3), 229–235 [20] [21].

Sarmidi, M.R., Barker, P.E., 1993b. Simultaneous biochemical reaction and separation in a rotating annular chromatograph. Chem. Eng. Sci. 48 (14), 2615–2623 [20] [21].

Sastry, I.V.K.S., Kumar, G.P., Chidambaram, M., 1994. Effect of time delay on specific growth rate on the periodic operation of bioreactors with input multiplicities. Bioprocess Eng. 10, 127–130 [10].

Satrio, J.A., Shanks, B.H., Wheelock, T.D., 2005. Development of a novel combined catalyst and adsorbent for hydrocarbon reforming. Ind. Eng. Chem. Res. 44, 3901–3911 [22].

Sayles, G.D., Ollis, D.F., 1989. Periodic operation of immobilized cell biocatalyst performance. Biotechnol. Bioeng. 34 (2), 160–170 [10].

Schädlich, K., Hoffman, U., Hofmann, H., 1983. Periodical operation of chemical processes and evaluation of conversion improvements. Chem. Eng. Sci. 38, 1375–1384 [1] [14] [17].

Schejbal, M., Štěpánek, J., Kočí, P., Marek, M., Kubíček, M., 2010. Sequence of monolithic converters for lean exhaust gas detoxification: a simulation study. Chem. Eng. Process. 49, 943–952 [6].

Schell, M., Kundu, K., Ross, J., 1987. Dependence of thermodynamic efficiency of proton pumps on frequency of oscillatory concentration of ATP. Proc. Nat. Acad. Sci. (USA) 84, 424–428 [11].

Schinor, N., Hohmann, W., Kraus, M., Müller, J., Münster, A.F., Schneider, F.W., 1999. Enhancement of production rates in a bistable chemical reaction. Phys. Chem. Chem. Phys. 1, 827–832 [11].

Schlatter, J.C., Mitchell, P.J., 1980. Three way catalyst response to transients. Ind. Eng. Chem. Prod. Res. Dev. 19, 288–293 [6].

Schlatter, J.C., Sinkevitch, R.M., Mitchell, P.J., 1983. Laboratory reactor system for three way automotive catalyst evaluation. Ind. Eng. Chem. Prod. Res. Dev. 22, 51–56 [6].

Schlesinger, M., Paunovic, M. (Eds.), 2010. Modern Electroplating. 5th Edition, Wiley-Interscience, John Wiley & Sons, New York [9].

Schmeisser, V., de Riva Perez, J., Tuttlies, U., Eigenberger, G., 2007. Experimental results concerning the role of Pt, Rh, Ba, Ce and Al_2O_3 on NOX-storage catalyst behaviour. Top. Catal. 42-43, 15–19 [6].

Schmeisser, V., Tuttlies, U., Eigenberger, G., Nieken, U., 2009. An improved model for NOx-storage catalysts. Top. Catal. 52, 1934–1939 [6].

Schmidt, A.D., Clinch, A.B., Ray, W.H., 1984. The dynamic behavior of continuous polymerization reactors: 3. An experimental study of multiple steady-states in solution polymerization. Chem. Eng. Sci. 39 (3), 419–432 [7].

Scholz, C.M.L., Gangwal, V.R., de Croon, M.H.J.M., Schouten, J.C., 2007. Model for NOx storage/reduction in the presence of CO_2 on a Pt-Ba/γ-Al_2O_3 catalyst. J. Catal. 245, 215–227 [6].

Schoubye, P., 1999. The REGESOX™ process for upgrading single-absorption sulfuric acid plants, Proc. Sulfur 99 Conf., Calgary-Alberta. Preprints. British Sulfur Publishing Ltd., London, U.K. pp. 183–190 [18].

Schoubye, P., Nielsen, P.E., 1993. Topsoe process combines catalytic combustion with regenerative heat transfer. Proc. AIChE Annual Meeting, Princeton. New Jersey [18].

Schütz, E., Hartmann, N., Kevrekidis, Y., Imbihl, R., 1998. Microchemical engineering of catalytic reactions. Catal. Lett. 54, 181–186 [2].

Schulte, M., Strube, J., 2001. Preparative enantioseparation by moving bed chromatography. J. Chromatogr. A. 906, 399–416 [23].

Schwankner, R.J., Eiswirth, M., Mueller, P., Wetzl, K., Ertl, G., 1987. Kinetic oscillations in the catalytic CO oxidation on Pt100 – periodic perturbation. J. Chem. Phys. 87, 743–749 [3].

Schwarz, J.A., Madix, R.J., 1974. Modulated beam relaxation spectrometry – Its application to study of heterogeneous kinetics. Surf. Sci. 46, 316–341 [13].

Schwartz, S.B., Schmidt, L.D., 1987. Is there a Single Mechanism of Catalytic Rate Oscillations on Pt. Surf. Sci. 183, L269–L278 [3].

Schwartz, S.B., Schmidt, L.D., 1988. The NO + CO reaction on clean Pt(100): multiple steady states and oscillations. Surf. Sci. 206, 169–186 [3].

Schweich, D., Villermaux, J., 1978. The chromatographic reactor. A new theoretical approach. Ind. Eng. Chem. Fund. 17 (1), 1–7 [1] [20].

Schweich, D., Villermaux, J., 1982a. The preparative chromatographic reactor revisited. Chem. Eng. J. 24, 99–109 [1] [20].

Schweich, D., Villermaux, J., 1982b. Model for catalytic dehydrogenation of cyclohexane in a chromatographic reactor: comparison of theory and experience. Ind. Eng. Chem. Fund. 21, 47–51 [20].

Schweich, D., Villermaux, J., 1982c. Evidence for a transient promoting catalytic effect in a chromatographic reactor. Ind. Eng. Chem. Fund. 21 (1), 51–55 [20].

Schweich, D., Villermaux, J., Sardin, M., 1980. An introduction to the nonlinear theory of adsorptive reactors. AIChE J. 26 (3), 477–486. 2001 [20].

Schwendeman, M.N., McBride, R., Reuter, D., Isaacs, M., 1983. Porous metal filters prevent fluid bed catalyst loss July. Chem. Process. July, 100 101 [1] [4].

Scott, K., 1991. Electrochemical Reaction Engineering. Academic Press, New York. New York [9].

Scott, A.S., Dennis, J.S., Hayhurst, A.N., Brown, T., 2006. In situ gasification of a solid fuel and CO_2 separation using chemical looping. AIChEJ. 52 (9), 3325–3328 [1] [8].

Sczechowski, J.G., Koval, C.A., Noble, R.D., 1993. Evidence of critical illumination and dark recovery times for increasing the photoefficiency of aqueous heterogeneous photocatalysis. J. Photochem. Photobiol. A. Chem. 74, 273–278 [11].

Sedmak, G., Hočevar, S., Levec, J., 2004a. Transient kinetic model of CO oxidation over a nanostructured $Cu_{0.1}Ce_{0.9}O_{2-y}$ catalyst. J. Catal. 222, 87–99 [3].

Sedmak, G., Hočevar, S., Levec, J., 2004b. CO oxidation kinetics over a nanostructured $Cu_{0.1}Ce_{0.9}O_{2-y}$ catalyst: a CO: O_2 concentration cycling study. Top. Catal. 30-31, 445–450 [3].

Seidel-Morgenstern, A., Blümel, C., Kniep, H., 1998. Efficient design of the SMB process based on perturbation method to measure adsorption isotherms and on a rapid solution of the dispersion model. In: Meunier, F. (Ed.), Fundamentals of Adsorption. Elsevier, Paris, pp. 449–454 [14].

Seidel-Morgenstern, A., Guiochon, G., 1993. Thermodynamics of the adsorption of Tröger's base enantiomers from ethanol on cellulose triacetate. J. Chromat. 48, 631–637 [14].

Seidel-Morgenstern, A., Kessler, L.C., Kaspereit, M., 2008. New Developments in Simulated Moving Bed Chromatography. Chem. Eng. Technol. 31, 826–837. [23].

Seiler, H., Emig, G., 1997. Reduction-oxidation-cycling in a fixed bed reactor with periodic flow reversal. In: Froment, G.F., Waugh, K.C. (Eds.), Dynamics of Surfaces and Reaction Kinetics of Heterogeneous Catalysis. Elsevier Science BV, Amsterdam, The Netherlands, pp. 479–488 [18].

Seiler, H., Emig, G., 1998. Reduktions-Oxidations-Zyklen im Festbettreaktor mit periodischer Strömungsumkeher. Chem.-lng.-Tech. 70, 1397–1402 [18].

Seiler, H., Emig, G., 1999. Reduction-oxidation cycles in a fixed bed reactor with periodic flow reversal. Chem. Eng. Technol. 22 (6), 479–484 [18].

Semenenko, E.I., Roginskii, S.Z., Yanovskii, M.I., 1964. Catalyzed pulsed dehydrogenation of n-butene. Kinet. Katal. 5, 490–495 (Russian) [20].

Severyanin, V.S., 1971. Prospects for pulsating combustion in power engineering. In: Brown, D.J. (Ed.), Proc. 1st Int. Symp. Pulsating Combust., Sheffield, U.K. (Sept. 1971) [5].

Severyanin, V.S., 1982. Application of pulsating combustion on industrial applications Proc. Symp on Pulse Combustion Applications. Atlanta, Georgia, GRI Report PB82-240060 NTIS Washington, DC, March, 1982 [5].

Shacham, M., Brauner, N., Cutlip, M.B., 1994. Exothermic CSTRs - Just how stable are the multiple steady states? Chem. Eng. Educ. 28 (1), 30–35 [19].

Shamin, T., 2005. Dynamic response of automotive catalytic converters to modulation in engine exhaust compositions. Int. J. Engine Res. 6 (6), 557–567 [6].

Shamin, T., 2008. The effect of oxygen storage capacity on the dynamic characteristics of an automotive catalytic converter. Energy Convers. Managem. 49, 3292–3300 [6].

Shang, H., Forbes, J.F., Guay, M., 2004. Model predictive control for quasilinear hyperbolic distributed parameter systems. Ind. Eng. Chem. Res. 43 (9), 2140–2149 [19].

Shanks, B.H., Bailey, J.E., 1987. Modelling of slow dynamics in the oxidation of CO over supported silver. AIChE J. 33 (12), 1971–1976 [3].

Shanks, B.H., Bailey, J.E., 1988. Experimental investigations using feedback induced bifurcation: carbon monoxide oxidation over supported silver. Chem. Eng. Commun. 61, 127–149 [3].

Shapovalova, V.I., Voznyuk, T.V., Lysukho, V.P., Luk'yanenko, M.Yu., Kutnyaya, V.S., Solodkaya Dolya, L.P., 1987. Influence of the calcination conditions on the activity of a catalyst for oxidative dehydrogenation of n-butane. Zhurnal Prikladnoi Khimii 60 (2), 369–373 (Plenum Publ. Corp. Translation), [4].

Sharma, M., Clayton, R., Harold, M.P., Balakotaiah, V., 2007. Multiplicity in lean NOx traps. Chem. Eng. Sci. 62, 5176–5181 [6].

Sharma, M., Harold, M.P., Balakotaiah, V., 2005. Analysis of periodic storage and reduction of NOx in catalytic monoliths. Ind. Eng. Chem. Res. 44, 6264–6277 [6].

Sheikh, J., Kershenbaum, L.S., Alpay, E., 1998. Analytical basis for separation enhanced reaction in continuous flow processes. Chem. Eng. Sci. 53 (16), 2933–2939 [22].

Sheikh, J., Kershenbaum, L.S., Alpay, E., 2001. 1-butene dehydrogenation in rapid pressure swing reaction processes. Chem. Eng. Sci. 56, 1511–1516 [22].

Sheintuch, M., 2005. Analysis of sensitivity of flow-reversal reactors: simulations, approximations and oxidation experiments. Chem. Eng. Sci. 60, 2991–2998 [18].

Sheintuch, M., Nekhamkina, O., 2004. Comparison of flow-reversal, internal-recirculation and loop reactors. Chem. Eng. Sci. 59, 4065–4072 [5] [21].

Sheintuch, M., Nekhamkina, O., 2005. The asymptotes of loop reactors. AIChE J. 51 (1), 224–234 [5] [21].

Shen, S.-K., Li, R.-J., Zhou, J.-P., Yu, C.-C., 2003. Selective oxidation of light hydrocarbons using lattice oxygen instead of molecular oxygen. Chin. J. Chem. Eng. 11, 649–655 [4].

Shen, L.-H., Wu, J.-H., Gao, Z.-P., Xiao, J., 2009b. Reactivity deterioration of NiO/Al_2O_3 oxygen carrier for chemical looping combustion of coal in a 10 kWth reactor. J. Combust. Flames 156 (7), 1377–1385 [8].

Shen, L.-H., Wu, J.-H., Xiao, J., 2009a. Experiments on chemical looping combustion of coal with a NiO based oxygen carrier. J. Combust. Flames 156 (3), 721–728 [8].

Shen, S.-K., Zhou, J.-P., Zhang, F.-S., Zhou, L., Li, R.-J., 2002. Effect of Ce-Fe oxides additives on performance of VPO catalyst for n-butane oxidation to maleic anhydride in the absence of gas-phase oxygen. Catal. Today 74 (1-2), 37–43 [4].

Shieh, M.T., Barker, P.E., 1995. Saccharification of modified starch to maltose in a semi-continuous countercurrent chromatographic reactor-separator (SCCR-S). J. Chem. Technol. Biot., 125–134 [21].

Shieh, M.T., Barker, P.E., 1996. Combined bioreaction and separation in a simulated countercurrent chromatographic-separator for the hydrolysis of lactose. J. Chem. Technol. Biot. 66, 265–278 [21].

Shinbrot, T., Grebogi, C., Ott, E., Yorke, J., 1993. Using small perturbations to control chaos. Nature 363, 411–417 [19].

Shinjoh, H., Muraki, H., Fujitani, Y., 1987. Periodic operation effects on automotive noble metal catalysts: Reaction analysis of binary systems. In: Crucq, A., Frennet, A. (Eds.), Catalysis and Automotive Pollution Control. Elsevier Science Publishers, Amsterdam, The Netherlands [2] [5] [6].

Shinjoh, H., Muraki, H., Fujitani, Y., 1989. Periodic operation effects in propane and propylene oxidation over noble metal catalysts. Appl. Catal. 49, 195–2004 [5].

Shu, X., Rigopoulos, K., Çinar, A., 1989. Vibrational control of an exothermic CSTR: productivity improvement by multiple input oscillations. IEEE Trans. Autom. Control 34 (2), 193–198 [19].

Sicardi, S., Manna, L., Banchero, M., 2005. Application of quasi steady-state model to slow mode periodic pulsation of the liquid in trickle bed reactors. Chem. Eng. Commun. 192, 908–917 [17].

Siegwart, P., Cote, J., Male, K., Luong, J.H.T., Perrier, M., Kamen, A., 1999. Adaptive control of low glucose concentration of HEK-293 cell serum-free cultures. Biotechnol. Prog. 15, 608–616 [10].

Silaban, A., Harrison, D.P., 1995. High temperature capture of carbon dioxide: characteristics of the reversible reaction between CaO(s) and CO_2(g). Chem. Eng. Commun. 137, 177–190 [22].

Silaban, A., Narcida, M., Harrison, D.P., 1996. Characteristics of the reversible reaction between CO_2(g) and calcined dolomite. Chem. Eng. Commun. 146, 149–162 [22].

Silva, V.M.T.M., Pereira, C.S.M., Rodrigues, A.E., 2011. PermSMBR – a new hybrid technology: application on green solvent and biofuel production. AIChE J. 57, 1840–1851 [23].

Silva, V.M.T.M., Rodrigues, A.E., 2002. Dynamics of a fixed-bed adsorptive reactor for synthesis of diethylacetal. AIChE J. 48 (3), 625–634 [20].

Silva, V.M.T.M., Rodrigues, A.E., 2005. Novel process for diethylacetal synthesis. AIChE J. 51 (10), 2752–2768 [21].

Silveston, P.L., 1980. Influence of composition modulation on product yields and selectivity in the partial oxidation of propylene. Preprint, 73rd Ann. Mtg, AIChE Chicago. November [4].

Silveston, P.L., 1987. Periodic operation of chemical reactors – a review of the experimental literature. Sadhana 10, 220–246 [1] [2].

Silveston, P.L., 1995. Automotive exhaust catalysis under periodic operation. Catal. Today 25 (2), 175–195 [6] [13].

Silveston, P.L., 1998. Composition Modulation of Catalytic Reactors. Gordon-Breach Science Publishers, Amsterdam, The Netherlands [1] [2] [3] [4] [5] [6] [7] [8] [9] [10] [11] [12] [13] [15] [16] [19].

Silveston, P.L., Forissier, M., 1985. Influence of composition modulation on product yields and selectivity in the partial oxidation of propylene over an antimony-tin oxide catalyst. Ind. Eng. Chem. Prod. Des. Dev. 34, 320–325 [1] [4] [8].

Silveston, P.L., Hanika, J., 2004. Periodic operation of three-phase catalytic reactors. Can. J. Chem. Eng. 82, 1105–1142 [1].

Silveston, P.L., Hudgins, R.R., 1981. Reduction of SO_2 emissions from a H_2SO_4 plant by means of feed modulation. Environ. Sci. Technol. 15, 419–422 [3].

Silveston, P.L., Hudgins, R.R., 2004a. Periodic pressure forcing of catalytic reactions. Chem. Eng. Sci. 59, 4055–4064 [1] [15].

Silveston, P.L., Hudgins, R.R., 2004b. Periodic temperature forcing of catalytic reactions. Chem. Eng. Sci. 59, 4043–4053 [1] [16].

Silveston, P.L., Hudgins, R.R., Abdul-Kareem, H.K., Jain, A.K., Fabbricino, L., Roussel, J., Wagler, R., Huber, S.L., 1983. Forced composition modulation of CO oxidation over various catalysts, Preprint. AIChE National Mtg., Detroit, Michigan [3].

Silveston, P.L., Hudgins, R.R., Bogdashev, S.M., Vernijakovskaja, N., Matros, Yu.Sh., 1990. Modelling of a periodically operating packed bed SO_2 oxidation reactor at high conversion. In: Matros, Yu.Sh. (Ed.), Unsteady-State Processes in Catalysis, VSP BV, Utrecht, The Netherlands [3].

Sinčić, D., Bailey, J.E., 1977. Pathological dynamic behavior of forced periodic chemical processes. Chem. Eng. Sci. 32 (3), 281–286 [10] [16].

Sinha, S., Ramaswamy, R., 1990. Adaptive control in nonlinear dynamics. Physica D. 43, 118–128 [19].

Siriwardane, R., Poston, J., Chaudhari, K., Zinn, A., Simonyi, T., Robinson, C., 2007. Chemical looping combustion of simulated synthesis gas using nickel oxide oxygen carrier supported on bentonite. Energy Fuels 21 (3), 1582–1591 [8].

Skála, D., Hanika, J., 2008. Dicyclopentadiene hydrogenation in trickle-bed reactor under forced periodic control. Chem. Papers 62, 215–218 [17].

Skála, D., Houserová, P., Hanika, J., 2001. Periodically operated trickling liquid film on vertical catalytic wall with isothermal reaction of gaseous component. Chem. Ing. Tech. 73 (6), 748 [17].

Skarstrom, C.W., 1960. Method and apparatus for fractionating gaseous mixtures by adsorption U.S. Patent 2,944,627 [14].

Skolpap, W., Nuchprayoon, S., Scharer, J.M., Douglas, P.L., Moo-Young, M., Grisdanurak, N., 2008a. Fed-batch optimization of recombinant α-amylase production by Bacillus subtilis using a modified Markov Chain Monte Carlo Technique. Korean J. Chem. Eng. 25 (4), 646–655 [10].

Skolpap, W., Nuchprayoon, S., Scharer, J.M., Douglas, P.L., Moo-Young, M., Grisdanurak, N., 2008b. Fed-batch optimization of α-amylase and protease-producing Bacillus subtilis using genetic algorithms and particle swarm optimization. Chem. Eng. Sci. 63 (16), 4090–4099 [10].

Skolpap, W., Scharer, J.M., Douglas, P.L., Moo-Young, M., 2004. Fed-batch optimization of α-amylase and protease-producing Bacillus subtilis using Markov Chain methods. Biotechnol. Bioeng. 86 (6), 706–717 [10].

Smit, J., Bekink, G.J., van Sint Annaland, M., Kuipers, J.A.M., 2005c. A reverse flow catalytic membrane reactor for production of syngas: an experimental study. Int. J. Chem. Eng. 3, 1–13 [18].

Smit, J., van Sint Annaland, M., Kuipers, J.A.M., 2004. Development of novel reactor concept for the partial oxidation of methane to syngas. Chem. Eng. Res. Des. 82 (A2), 245–251 [18].

Smit, J., van Sint Annaland, M., Kuipers, J.A.M., 2005a. Feasibility study of a reverse flow catalytic membrane reactor with porous membranes for the production of syngas. Chem. Eng. Sci. 60, 6971–6982 [18].

Smit, J., van Sint Annaland, M., Kuipers, J.A.M., 2005b. Experimental demonstration of the reverse flow catalytic membrane reactor for energy efficient syngas production. Part 1: Influence of operating conditions. Chem. Eng. Sci. 62, 1239–1250 [18].

Smit, J., van Sint Annaland, M., Kuipers, J.A.M., 2007. Experimental demonstration of the reverse flow catalytic membrane reactor for energy efficient syngas production. Part 2: Model development. Chem. Eng. Sci. 62, 1251–1262 [18].

Smith, W.R., Bobrova, L.N., 2002. Mathematical modelling of a reverse flow reactor with catalytic surface dynamics. Chem. Eng. Sci. 57, 393–407 [18].

Snoeck, J.-W., Froment, G.F., 1997a. Filamentous carbon formation and gasification: Thermodynamics, driving force, nucleation and steady state growth. J. Catal. 169, 240–249 [8].

Snoeck, J.-W., Froment, G.F., 1997b. Kinetic study of the carbon filament formation by methane cracking on a Ni catalyst. J. Catal. 169, 250–262 [8].

Snyder, J.D., Subramaniam, B., 1993. Numerical simulation of a periodic flow reversal reactor for sulfur dioxide oxidation. Chem. Eng. Sci. 48, 4051–4064 [18].

Snyder, J.D., Subramaniam, B., 1994. A novel reverse flow strategy for ethylbenzene dehydrogenation in a packed bed reactor. Chem. Eng. Sci. 49, 5585–5601 [18].

Snyder, J.D., Subramaniam, B., 1998. Numerical simulation of a reverse-flow NOx-SCR reactor with side-stream ammonia addition. Chem. Eng. Sci. 53, 727–734 [18].

Snyder, J.D., Subramaniam, B., 1999. Energy trapping during reverse flow operation with discrete power introduction: experiments and theory. Can. J. Chem. Eng. 74, 743–750 [18].

Sobyanin, V.A., Belyaev, V.D., 2000. Gas phase electrocatalysis methane oxidation to syngas in a solid oxide fuel cell reactor. Solid State Ionics 136/137, 747–752 [4].

Sofranko, J.A., Leonard, J.J., Jones, C.A., 1987. The oxidative conversion of methane to higher hydrocarbons. J. Catal. 103, 302–310 [4].

Sohn, H.Y., Aboukheshem, M.B., 1992. Gas-solid reaction-rate enhancement by pressure cycling. Metall. Mat. Trans. B. 23 (3), 285–294 [15].

Sohn, H.Y., Chaubal, P.C., 1984. Rate Enhancement of the Gaseous Reaction of Iron-Oxide Pellets by Pressure Cycling. Iron Steel Inst. Jpn. 24, 387–395 [15].

Soler, J., Lopez Nieto, J.M., Herguido, J., Menendez, M., Santamaria, J., 1999. Oxidative dehydrogenation of n-butane in a two-zone fluidized- bed reactor. Ind. Eng. Chem. Res. 38, 90–97 [4].

Solymosi, F., Cserényi, J., 1994. Decomposition of CH_4 over supported Ir catalysts. Catal. Today 21, 561–569 [12].

Solymosi, F., Cserényi, J., 1995. Enhanced formation of ethane in the conversion of methane over Cu-Rh/SiO2. Catal. Lett. 34, 343–350 [12].

Solymosi, F., Erdöhelyi, A., Cserényi, J., 1992. A comparative study on the activation and reactions of CH_4 on supported metals. Catal. Lett. 16, 399–405 [12].

Solymosi, F., Erdöhelyi, A., Cserényi, J., Felvégi, A., 1994. Decomposition of CH_4 over supported Pd catalysts. J. Catal. 147, 272–278 [12].

Solymosi, F., Erdöhelyi, A., Szöke, A., 1995. Dehydrogenation of methane on supported molybdenum oxides. Formation of benzene from methane. Catal. Lett. 32, 43–53 [12].

Somani, M., Viswanath, S., Khinast, J., Luss, D., 1997. Maximum temperature in a reverse flow reactor with two independent reactions. Chem. Eng. Sci. 52 (15), 2483–2495 [18].

Somers, J.M., Schultz, H.L., 2008. Thermal oxidation of coal mine ventilation air methane. In: Proc. 12th U.S./North American Mine Ventilation Symp. 2008, Wallace pp. 301–306 [18].

Song, J., Kinney, K.A., 2000a. Effect of vapor-phase bioreactor operation on biomass accumulation, distribution and activity: linking biofilm properties to bioreactor performance. Biotechnol. Bioeng. 68 (5), 508–516 [18].

Song, J., Kinney, K.A., 2000b. Effect of directional switching frequency on toluene degradation in a vapor-phase bioreactor. Appl. Microbiol. Biotechnol. 56, 108–113. [18].

Song, N., Rhodes, C., Bartley, J.K., Taylor, S.H., Chadwick, E., Hutchings, G.J., 2005. Oxidation of isobutene to methacrolein using bismuth molybdate catalysts: comparison of operation in periodic and continuous feed modes. J. Catal. 236, 282–291 [4].

Song, Q.-L., Xiao, R., Deng, Z.-G., Zhang, H.-Y., Shen, L.-H., Xiao, J., Zhang, M.-Y., 2008. Chemical-looping combustion of methane with $CaSO_4$ oxygen carrier in a fixed bed reactor. Energy Conv. Manag. 49, 3178–3187 [8].

Song, N.-X., Xuan, Z.-Q., Bartley, J.K., Taylor, S.H., Chadwick, D., Hutchings, G.J., 2006. Oxidation of butane to maleic anhydride using vanadium phosphate catalysts: comparison of operation in aerobic and anaerobic conditions using a gas-gas periodic flow reactor. Catal. Lett. 106, 127–131 [4].

Sorin, M., Sapundzhiev, C., Paris, J., 1994. Application of a new exergy analysis method to a flow reversal SO_2 convertor. Chem. Eng. Sci. 49 (14), 5603–5613 [18].

Sotowa, K.-I., Shiraishi, N., Sugiyama, S., 2006. Propylene partial oxidation under fast temperature cycling, Proc. 5th Int. Conf. on Unsteady State Processes in Catalysis USPC5, Suita City, Osaka, Japan. pp. 215–216 [16].

Sotowa, K.-I., Shiraishi, N., Sugiyama, S., 2008. Forced temperature cycling of a catalyst layer and its application to propylene oxidation. Chem. Eng. Sci. 63, 2690–2696 [1] [16].

Spitz, J.J., Laurence, R.L., Chappelear, D.C., 1976. An experimental study of polymerization reactor in periodic operation. AIChE Symp. Ser. 160 (72), 86–101 [7].

Stamicarbon, B.V., 1974. Procédé pour la préparation de cétones saturées et d'un catalyseur pour la mise en oeuvre du procédé, French patent No. 2,187,747 [4].

Stanek, V., Hanika, J., Jiricny, V., Stavarek, P., Tukac, V., Lederer, J., 2009. Prediction of improved performance of catalytic hydrogenation reactor by periodic modulation of the feed rate. Chem. Biochem. Eng. Quart. 23 (3), 251–257 [17].

Stanitsas, G., 1986. M.A.Sc. Thesis, Dept. of Chem. Eng. Univ. of Waterloo, Waterloo, Ontario, Canada [3].

Stankiewicz, A., Kuczynski, M., 1995. An industrial view on the dynamic operation of chemical converters. Chem. Eng. Proc. 34, 367–377 [1].

Stegasov, A.N., Kirillov, V.A., Silveston, P.L., 1994. Modelling of catalytic SO_2 oxidation for continuous and periodic liquid flow through a trickle bed. Chem. Eng. Sci. 49 (22), 3699–3710 [17].

Stegenga, S., Dekker, N., Bijsterbosch, J., Kapteijn, F., Moulijn, J., Belot, G., Roche, R., 1991. Catalytic automotive pollution control without noble metals. In: Crucq, A. (Ed.), Catalysis and Pollution Control II. Elsevier Science Publishers, Amsterdam, The Netherlands, pp. 353–369 [6].

Stephanopoulos, G., 1984. Chemical Process Control. Prentice-Hall, Englewood Cliffs, New Jersey [14].

Stephanopoulos, G., Fredrickson, A.G., Aris, R., 1979. The growth of competing microbial populations in a CSTR with periodically varying inputs. AICHE J. 25, 863–872 [10].

Stephens, M.L., Christensen, C., Lyberatos, G., 1992. Plasmid stabilization of an *Escherichia coli* culture through cycling. Biotechnol. Prog. 8, 1–4 [10].

Stephens, M.L., Lyberatos, G., 1987. Effect of cycling on final mixed culture fate. Biotechnol. Bioeng. 29 (6), 672–678 [10].

Stephens, M.L., Lyberatos, G., 1988. Effects of cycling on the stability of plasmid-bearing microorganisms in continuous culture. Biotechnol. Bioeng. 31, 464–469 [10].

Sterman, L.E., Ydstie, B.E., 1990a. The steady state process with periodic perturbations. Chem. Eng. Sci. 45, 721–736 [1] [10] [12] [16] [19].

Sterman, L.E., Ydstie, B.E., 1990b. Unsteady state multivariable analysis of periodically perturbed systems. Chem. Eng. Sci. 45, 737–749 [10] [12].

Sterman, L.E., Ydstie, B.E., 1991. Periodic forcing of the CSTR: an application of the generalized π criterion. AICHE J. 37, 986–996 [10] [12] [16].

Stojek, Z., 2010. The electrical double layer and its structure. In: Scholz, F. (Ed.), Electroanalytical Methods, second ed. Springer-Verlag, Berlin, pp. 3–9 [9].

Storti, G., Masi, M., Paludetto, R., Morbidelli, M., Carrá, S., 1988. Adsorption separation processes: countercurrent and simulated countercurrent operations. Comp. Chem. Eng. 12, 475 482 [21].

Storti, G., Mazzotti, M., Morbidelli, M., Carra, S., 1993. Robust design of binary countercurrent adsorption separation processes. AIChE J. 39 (3), 471–492 [21].

Stradiotto, D.A., Hudgins, R.R., Silveston, P.L., 1999. Hydrogenation of crotonaldehyde under periodic flow interruption in a trickle bed. Chem. Eng. Sci. 54, 2561–2568 [16] [17].

Strohlein, G., Mazzotti, M., Morbidelli, M., 2005. Optimal operation of simulated-moving-bed reactors for nonlinear adsorption isotherms and equilibrium reactions. Chem. Eng. Sci. 60, 1525–1533 [21].

Strots, V.O., Bunimovich, G.A., Matros, Yu.Sh., 2003. Catalytic oxidation of nitrogen containing VOCs. Petrol. Technol. Quart. Summer 2003, 125–131 [18].

Strots, V.O., Bunimovich, G.A., Matros, Yu.Sh., Zheng, M., Mirosh, E.A., 1998. Novel catalytic converter for natural gas powered diesel engines. SAE paper 961976. Soc. Autom. Eng. Detroit. Michigan. [18].

Strots, V.O., Matros, Yu.Sh., Bunimovich, G.A., 1992. Periodically forced SO_2 oxidation in CSTR. Chem. Eng. Sci. 47 (9–11), 2701–2706 [3].

Stuchlý, V., Klusáček, K., 1990. Global activity improvement by feed composition cycling in CO/H_2-Ni/SiO_2 methanation reaction system. In: Matros, Yu.Sh. (Ed.), Unsteady State Processes in Catalysis. VSP Publ., Utrecht, Netherlands, pp. 422–428 [2].

Su, S., Prairie, M.R., Renken, A., 1992. Reaction mechanism of methanol dehydrogenation on a sodium carbonate catalyst. Appl. Catal. A-Gen. 91, 131–139 [11].

Su, E.C., Rothschild, W.G., 1986. Dynamic behaviour of three-way catalysts. J. Catal. 99, 506–510 [6].

Su, S., Zasa, P., Renken, A., 1994. Catalytic dehydrogenation of methanol to water-free formaldehyde. Chem. Eng. Technol. 17 (1), 34–40 [11].

Suman, B., 2004. Study of unsteady state process operation. Chem. Eng. J. 104, 63–71 [1].

Sun, J., Park, K.H., Lee, S.Y., Oh, S.J., Kim, C.K., Chou, B.T., Park, Y.C., Seo, J.H., 2007. Effect of temperature shift strategies on humen preproinsulin production in the fed-batch fermentation of recombinant. *Escherichia coli*, Biotechnol. Bioproc. Eng. 12, 556–561 [23].

Sun, Q., Young, B., Williams, D.F., Glasser, D., Hildebrandt, D., 1996. A periodic flow reversal reaction: an infinitely fast switching model and a practical proposal for its implementation. Can. J. Chem. Eng. 74, 760–765 [18].

Sunstrom, D.W., Klei, H.E., Brookman, G.T., 1976. Response of biological reactions to sinusoidal variations of substrate concentrations. Biotechnol. Bioeng. 18, 1–14 [10].

Suresh, B., Chidambaram, M., 1997. Periodic operation of well mixed enzyme reactors with steady state multiplicities using relay feedback. Bioprocess Eng. 16, 225–227 [10].

Suslick, K.S., 1988. Ultrasound, its Physical, Biological and Chemical Effects. VCH Science Publishers, Weinheim, Germany [15].

Suslick, K.S., Casadonte, D.J., 1987. Heterogeneous sonocatalysis with nickel powder. JACS 109, 3459–3461 [15].

Suzuki, M., 1990. Adsorption Engineering. Kodansha Ltd., Tokyo, Japan [1] [22].

Suzuki, T., Chouchi, H., Naito, K., Watanabe, Y., 1989. Cyclic feed CO_2 gasification of iron loaded coal char. Approach to activate iron catalyst in coal gasification. Energy Fuels 3, 535–536 [8].

Suzuki, T., Inoue, K., Watanabe, Y., 1988. Temperature-programmed desorption and CO_2-pulsed gasification of sodium or iron loaded Yallourn coal char. Energy Fuels 2, 673–679 [8].

Suzuki, T., Sakashita, Y., Ohme, H., Watanabe, Y., 1990. Reactivities of various coal chars to carbon dioxide and oxygen measured with pulse reaction technique. Energy Fuels 4, 593–598 [8].

Svensson, P., Jaeger, N.I., Plath, P.J., 1988. Phase and frequency relations with forced oscillations of the heterogeneous catalytic oxidation of CO on supported Pd. J. Chem. Phys. 92, 1882–1888 [3].

Swaminathan, V., Subramanian, V., Murali, K.R., 2000. Characteristics of CdSe films electrodeposited with microprocessor based pulse plating unit. Thin Solid Films 359 (1), 113–117 [9].

Swift, W.M., Wheelock, T.D., 1975. Decomposition of calcium sulfate in a two-zone reactor. Ind. Eng. Chem. Process Des. Dev. 14 (3), 323–327 [8].

Sze, M.C., Gelbein, A.P., 1975. Synthesis of aromatic nitriles, Preprint. 1st Chemical Congress of the North American Continent, Mexico City, Mexico. November [1] [4] [8].

Sze, M.C., Gelbein, A.P., 1976. Make aromatic nitriles this way. HydrocarbonProcess, 103–106 [1] [4] [8].

Takahashi, N., Yamazaki, K., Sobukawa, H., Shinjoh, H., 2009. The low-temperature performance of NOx storage and reduction catalyst. Appl. Catal. B-Environ. 70 (1-4), 198–204 [6].

Takeuchi, K., Miyauchi, T., Uraguchi, Y., 1978. Computational studies of a chromatographic moving-bed reactor for consecutive and reversible reactions. J. Chem. Eng. Jpn. 11 (3), 216–220 [20].

Takeuchi, K., Uraguchi, Y., 1976a. Separation conditions of the reactant and the product with a chromatographic moving bed reactor. J. Chem. Eng. Jpn. 9 (2), 164–166 [1] [20].

Takeuchi, K., Uraguchi, Y., 1976b. Basic design of chromatographic moving-bed reactors for product refining. J. Chem. Eng. Jpn. 9 (3), 246–248 [1] [20].

Takeuchi, K., Uraguchi, Y., 1977a. The effect of the exhausting section on the performance of a chromatographic moving-bed reactor. J. Chem. Eng. Jpn. 10 (1), 72–74 [20].

Takeuchi, K., Uraguchi, Y., 1977b. Experimental studies of a chromatographic moving-bed reactor – catalytic oxidation of carbon monoxide on activated alumina as a model reaction. J. Chem. Eng. Jpn. 10 (6), 455–460 [20].

Tan, H., Dai, Y.-C., Yuan, W.-K., 1994. A novel kinetic model discrimination method: eigenfunction distribution via forced concentration oscillation. Chem. Eng. Sci. 49 (24b), 5563–5569 [13].

Tang, P.T., 2001. Pulse reversal plating of nickel and nickel alloys for microgalvanics. Electrochim. Acta 47 (1–2), 61–66 [9].

Tang, P.T., Christensen, T.R., Jensen, M.F., 2004. Electroforming of inserts for injection molding of optical or micro-fluidic components, Proc. 2004 SUR/FIN Conf. Amer. Electropl. Surf. Finish. Soc. 763–774 [9].

Tantavichet, N., Pritzker, M.D., 2002. Low and high pulse current plating onto a rotating disk electrode. J. Electrochem. Soc. 149, C289–C299 [9].

Tantavichet, N., Pritzker, M.D., 2003. Low- and high-frequency pulse current and pulse reverse plating of copper. J. Electrochem. Soc. 150, C665–C677 [9].

Taylor, K.C., 1984. Automobile catalytic converters. In: Anderson, J.R., Boudart, M. (Eds.), Catalysis: Science and Technology. Springer Verlag, Berlin, pp. 119–170 [6].

Taylor, L.T., 2009. Supercritical fluid chromatography for the 21st century. J. Supercrit. Fluids 47, 566–573 [23].

Taylor, K.C., Sinkevitch, R.M., 1983. Behavior of automobile exhaust catalysts with cycled feed streams. Ind. Eng. Chem. Prod. Res. Dev. 22, 45–50 [6].

Taylor, E.J., Sun, J.J., Inman, M.E., Zhang, L., 2000. Metallization of electronic interconnects (HDI and ULSI) by charge modulation control of metal distribution, Proc. 5th Intern. Pulse Plating Symp. American Electroplaters & Surface Finishers Soc, Orlando, Florida, 65–73 [9].

Tellaeche, C., Glöckler, B., Kolius, G., Nieken, U., 2009. A heat-integrated reverse-flow reactor concept for endothermic high-temperature synthesis. Part II: Development of a reformer prototype for hydrogen production. Chem. Eng. Technol. 32, 1348–1357 [18].

Teymour, F., 1997. Dynamics of semibatch polymerization reactors: I. Theoretical analysis. AIChE J. 43 (1), 145–156 [7].

Teymour, F., Ray, W.H., 1989. The dynamic behavior of continuous solution polymerization reactors. IV. Stability and bifurcation analysis. Chem. Eng. Sci. 44, 1967–1976 [7].

Teymour, F., Ray, W.H., 1992a. The dynamic behavior of continuous solution polymerization reactors. V. Experimental investigation of limit cycle behavior for vinyl acetate polymerization. Chem. Eng. Sci. 47, 4121–4132 [7].

Teymour, F., Ray, W.H., 1992b. The dynamic behavior of continuous solution polymerization reactors. VI. Complex dynamics in full scale reactors. Chem. Eng. Sci. 47, 4133–4142 [7].

Thiele, R., 1984. Dynamic behaviour and stability of continuous reactors for bulk polymerization. Chemical Reaction Engineering. Inst. Chem. Eng. No. 87 Pergamon Press, London, U.K. [7].

Thomas, C.O., Smith, H.A., 1959. Gas chromatography with hydrogen and deuterium. J. Phys. Chem. 63, 427–432 [20].

Thullie, J., Burghardt, A., 1995. Simplified procedure for estimating maximum cycling time of flow-reversal reactors. Chem. Eng. Sci. 50 (14), 2299–2309 [18].

Thullie, J., Chiao, L., Rinker, R.G., 1986. Analysis of concentration forcing in heterogeneous catalysis. Chem. Eng. Commun. 48, 191–205 [1] [2] [14].

Thullie, J., Chiao, L., Rinker, R.G., 1987. Generalized treatment of concentration forcing in fixed-bed plug-flow reactors. Chem. Eng. Sci. 42 (5), 1095–1101 [2].

Thullie, J., Renken, A., 1990. Transient behaviour of catalytic reactions with stop-effect. Chem. Eng. Comm. 96, 193–204 [11] [13].

Thullie, J., Renken, A., 1991. Forced concentration oscillations for catalytic reactions with stop-effect. Chem. Eng. Sci. 46, 1083–1088 [1] [11].

Thullie, J., Renken, A., 1993. Model discrimination for reactions with stop-effect. Chem. Eng. Sci. 48, 3921–3925 [11] [13].

Tohyama, M., Patarinska, T., Qiang, Zi-W., Shimizu, K., 2002. Modeling of the mixed culture and periodic control for PHB production. Biochem. Eng. J. 10, 157–173 [10].

Tomilov, V.N., Zagoruiko, A.N., Kuznetsov, P.A., 1999. Kinetic investigation and mathematical modeling of oscillation regimes for oxidative dehydrogenation of butene-1. Chem. Eng. Sci. 54, 4359–4364 [13].

Tonkovich, A.L., Carr, R.W., 1994a. A simulated countercurrent moving-bed chromatographic reactor for the oxidative coupling of methane: experimental results. Chem. Eng. Sci. 49, 4647–4656 [21].

Tonkovich, A.L., Carr, R.W., 1994b. Modeling of the simulated countercurrent moving-bed chromatographic reactor used for the oxidative coupling of methane. Chem. Eng. Sci. 49, 4657–4665 [21].

Tonkovich, A.L., Carr, R.W., Aris, R., 1993. Enhanced C_2 yields from methane oxidative coupling by means of a separative chemical reactor. Science 262, 221–223 [21].

Toru, M., Takashi, T., Saturu, M., 2010. Optimized culture conditions for the efficient production of porcine adalynate kinase in recombinant Escherichia coli, Appl. Biochem. Biotechnol. 162, 823–829 [23].

Trimm, D.L., 1977. The formation and removal of coke from nickel catalyst. Catal. Rev.-Sci. Engng. 16 (2), 155–189 [8].

Trivizadakis, M.E., Giakoumakis, D., Karabelas, A.J., 2006a. A study of particle shape and size effects on hydrodynamic parameters of trickle beds. Chem. Eng. Sci. 61, 5234–5243 [17].

Trivizadakis, M.E., Giakoumakis, D., Karabelas, A.J., 2006b. Induced pulsing in trickle beds—particle shape and size effects on pulse characteristics. Chem. Eng. Sci. 61, 7448–7462 [17].

Trivizadakis, M.E., Karabelas, A.J., 2006. A study of local liquid/solid mass transfer in packed beds under trickling and induced pulsing flow. Chem. Eng. Sci. 61, 7684–7696 [17].

Truffer, M.A., Renken, A., 1986. Transient behavior of heterogeneous catalyst reactions with educt inhibition. AIChE J. 32 (10), 1612–1621 [11] [12] [13].

Tsikoyiannis, J.G., Stern, D.L., Grasselli, R.K., 1999. Metal oxides as selective hydrogen combustion (SHC) catalysts and their potential in light paraffin dehydrogenation. J. Catal. 184, 77–86 [4] [8].

Tukač, V., Handlova, M., Chyba, V., Lederer, J., Kolena, J., Kubicka, D., Hanika, J., Jiřičný, V., Stavárek, P., 2007b. Hydrogenation of pyrolysis gasoline in pilot trickle-bed reactor with periodic modulated feed rate. Proc. European Cong. Chem. Eng. (ECCE-6), Copenhagen [17].

Tukač, V., Hanika, J., Chyba, V., 2003. Periodic state of wet oxidation in trickle-bed reactor. Catal. Today 79-80, 427–431 [17].

Tukač, V., Šimíčková, M., Chyba, V., Lederer, J., Kolena, J., Hanika, J., Jiřičný, V., Staněk, V., Stavárek, P., 2007a. The behavior of pilot trickle-bed reactor under periodic operation. Chem. Eng. Sci. 62, 4891–4895 [17].

Turco, F., Hudgins, R.R., Silveston, P.L., Sicardi, S., Manna, L., Branchero, M., 2001. Investigation of periodic operation of a trickle-bed reactor. Chem. Eng. Sci. 56 (4), 1429–1434 [17].

Turnock, P.H., Kadlec, R.H., 1971. Separation of nitrogen and methane via periodic adsorption. AIChE J. 17, 335–342 [22].

Twigg, M.V., 2007. Progress and future challenges in controlling automotive exhaust gas emissions. Appl. Catal. B-Environ. 70, 2–15 [6].

Underhill, M.F., Smales, C.M., 2007. The cold-shock response in mammalian cells: investigating the HeLa cell cold-shock proteome. Cytotechnol. 53, 47–53 [23].

Unger, J., 1999. Druckwechseladsorption zur Gastrennung - Modellierung, Simulation und Prozeßdynamik. Ph.D. thesis. Stuttgart Universität, Institut für Chemische, Verfahrenstechnik. [14].

Unger, B.D., Rinker, R.G., 1976. Ammonia synthesis reaction in the chromatographic regime. Ind. Eng. Chem. Fund. 15 (3), 225–227 [2] [20].

Unni, M.P., Hudgins, R.R., Silveston, P.L., 1973. Influence of cycling on the rate of oxidation of SO_2 over a vanadium catalyst. Can. J. Chem. Eng. 31, 623–630 [1] [3] [13].

Urseanu, M.I., Boelhouwer, J.G., Bosman, H.J.M., Schroijen, J.C., 2004. Induced pulse operation of high-pressure trickle bed reactors with organic liquids: hydrodynamics and reaction study. Chem. Eng. Process. 43 (11), 1411–1416 [17].

Vairo, M.L.R., Borzani, W., Dos Anjos Magalhaes, M.M., 1977. Response of a continuous anaerobic culture to variations in the feeding rate. Biotechnol. Bioeng. 19, 595–598 [10].

Vamling, L., 1987. On the use of periodic operation for improving chemical processes. Ph.D. Thesis. Chalmers University of Technology, Gothenburg, Sweden [12].

van de Beld, B., Borman, R.A., Derkx, O.R., van Woezik, B.A.A., Westerterp, K.R., 1994. Removal of volatile organic compounds from polluted air in a reverse flow reactor: an experimental study. Ind. Eng. Chem. Res. 33, 2946–2956 [18].

van de Beld, L., Westerterp, K.R., 1994. Air purification by catalytic oxidation in a reactor with periodic flow reversal. Chem. Eng. Technol. 17, 217–226 [18].

van de Beld, L., Westerterp, K.R., 1996. Air purification in a reverse-flow reactor: model simulation versus experiment. AIChE J. 42, 1139–1148 [18].

Vanden Bussche, K.M., Froment, G.F., 1996. The STAR configuration for methanol synthesis in reverse flow reactors. Can. J. Chem. Eng. 74 (5), 729–734 [5] [18] [21].

Vanden Bussche, K.M., Neophytides, S.N., Zolotarskii, I.A., Froment, G.F., 1993. Modeling and simulation of the reverse-flow operation in a fixed bed reactor for methanol synthesis. Chem. Eng. Sci. 48 (12), 3335–3345 [18].

Van den Heuvel, J.C., Beuling, E.E., Van Dusschoten, D., Roosenschoon, G.L., Verschuren, P.G., 1997b. Convective flow in methanogenic granules. Water Sci. Technol. 36 (6–7), 311–316 [15].

Van den Heuvel, J.C., Portegies Zwart, I., Vredenbregt, L.H.J., 1996. Convective acceleration of mass transfer in three-phase systems by pressure oscillations. Chem. Eng. Sci. 51 (10), 2391–2398 [15].

Van den Heuvel, J.C., Verschuren, P.G., Ottengraf, S.P.P., 1997a. Acceleration of mass transfer in loop reactors. Water Sci. Technol. 36 (1), 311–319 [15].

van de Rotten, B.A., Verduyn Lunela, S.M., Bliek, A., 2006. Efficient simulation of periodically forced reactors with radial gradients. Chem. Eng. Sci. 61, 6981–6994 [18].

van der Wielen, L.A.M., van Nunen, C.A.P.M., Luyben, K.Ch.A.M., 1998. Pulsed adsorbent transport in a countercurrent adsorptive reactor. AIChE J. 44, 83–92 [20].

van der Wielen, L.A.M., Potters, J.J.M., Straathof, A.J.J., Luyben, K.Ch.A.M., 1990. Integration of bioconversion and continuous product separation by means of countercurrent adsorption. Chem. Eng. Sci. 45 (8), 2397–2404 [20].

Van Doesburg, H., De Jong, W.A., 1978. Dynamic behaviour of an adiabatic fixed-bed methanator. In: Weekman Jr., V.M., Luss, D. (Eds.), Chemical Reaction Engineering—Houston. ACS Symp. Ser. 65, pp. 489–503 [2].

van Eldik, R., Hubbard, C.D., 1997. Chemistry under Extreme or Non-classical Conditions. John Wiley & Sons Inc, New York [15].

Vanin, G.V., Noskov, A.S., Popova, G.Ya., Andrushkevich, T.V., Matros, Yu.Sh., 1993. The industrial plant for unsteady-state purification of flue gases from acrylonitrile and cyanic acid. Catal. Today 17, 251–260 [18].

van Neer, F.J.R., Bliek, A., 1999. The feedback mechanism in self-oscillations for CO oxidation over EUROPT-3. Chem. Eng. Sci. 54, 4483–4499 [16].

van Neer, F.J.R., Kodde, A.J., den Uil, H., Bliek, A., 1996. Understanding of resonance phenomena on a catalyst under forced concentration and temperature oscillations. Can. J. Chem. Eng. 74, 664–673 [16].

van Neer, F.J.R., van der Linden, B., Bliek, A., 1997. Forced concentration oscillations of CO and O_2 in CO oxidation over Cu/Al_2O_3. Catal. Today 38, 115–128 [13].

van Noorden, T.L., Verduyn Lunel, S.M., Bliek, A., 2002. Acceleration of the determination of periodic states of cyclically operated reactors and separators. Chem. Eng. Sci. 57, 1041–1055 [22].

van Noorden, T.L., Verduyn Lunel, S.M., Bliek, A., 2003a. Optimization of cyclically operated reactors and separators. Chem. Eng. Sci. 58, 4115–4127 [22].

van Noorden, T.L., Verduyn Lunel, S.M., Bliek, A., 2003b. The efficient computation of periodic states of cyclically operated chemical processes. IMA- J. Appl. Math. 68, 149–166 [22].

van Sint Annaland, M., Kuipers, J.A., van Swaaij, W.P.M., 2001. Safety analysis of switching between reductive and oxidative conditions in a reaction coupling reverse flow reactor. Chem. Eng. Sci. 56, 1517–1524 [18].

van Sint Annaland, M., Nijssen, R., 2002. A novel reverse-flow reactor coupling endothermic and exothermic reactions: Part II: an experimental study. Chem. Eng. Sci. 57, 4967–4985 [18].

van Sint Annaland, M., Scholts, H.A.R., Kuipers, J.A., van Swaaij, W.P.M., 2002. A novel reverse-flow reactor coupling endothermic and exothermic reactions. Part I: comparison of reactor configurations for irreversible endothermic reactions. Chem. Eng. Sci. 57, 833–854 [18].

Vaporciyan, G.G., Kadlec, R.H., 1987. Equilibrium-limited periodic separating reactors. AIChE J. 33 (8), 1334–1343 [1] [22].

Vaporciyan, G.G., Kadlec, R.H., 1989. Periodic separating reactors: experiments and theory. AIChE J. 35 (5), 831–844 [1] [22].

Vaporciyan, G.G., Annapragada, A., Gulari, E., 1988. Rate enhancement and quasi-periodic dynamics during forced concentration cycling of CO and O_2 over supported Pt-SnO_2. Chem. Eng. Sci. 43 (11), 2957–2966 [3].

Vega, J.R., Frontini, G.L., Meira, G.R., 1991. Optimal periodic control of a continuous living anionic polymerization. III. Presence of impurities in the monomer feed. J. Appl. Polym. Sci. 42, 3181–3193 [7].

Velardi, S.A., Barresi, A.A., 2002. Methanol synthesis in a forced unsteady state reactor network. Chem. Eng. Sci. 57, 2995–3004 [5] [21].

Velardi, S.A., Barresi, A.A., Manca, D., Fissore, D., 2004. Complex dynamic behavior of methanol synthesis in the ring reactor network. Chem. Eng. J. 99, 117–123 [21].

Vermeijlen, J.J.T.T., 2002. Application of pulse plating and pulse reverse plating techniques in nickel electrode position – a review. Proc. SUR/FIN 2002, Amer. Electroplating Surface Finishing Soc. pp. 995–1002 [9].

Vernikovskaya, N.V., Zagoruiko, A.N., Noskov, A.S., 1999. SO_2 oxidation method. Mathematical modeling taking into account dynamic properties of the catalyst. Chem. Eng. Sci. 54, 4475–4482 [18].

Veser, G., 2010. Multiscale process intensification for catalytic partial oxidation of methane: from nanostructured catalysts to integrated reactor concept. Catal. Today 157, 24–32 [18].

Vicente, M., Leiza, J.R., Asua, J.M., 2001. Simultaneous control of co-polymer composition and MWD in emulsion co-polmerization. AIChE J. 47 (7), 1594–1606 [7].

Viecco, G.A., Caram, H., 2006. Comparison between simulated moving bed and reverse flow chromatographic reactors for equilibrium limited reactions. Chem. Eng. Sci. 61, 6869–6879 [18].

Visser, J.B.M., Stankiewicz, A., van Dierendonck, L.L., Manna, L., Sicardi, S., Baldi, G., 1994. Dynamic operation of a three-phase upflow reactor for the hydrogenation of phenylacetylene. Catal. Today 20, 485–500 [2] [12].

Viswanathan, S., Aris, R., 1974. Countercurrent moving bed chromatographic reactors. ACS Symp. Ser. 133, 191–204 [1] [20].

Viswanathan, K., Farrell-Epstein, M.A., Cheh, H.Y., 1978. Application of pulsed current electrolysis to a rotating disk electrode system—1 mass transfer. J. Electrochem. Soc. 125 (11), 1772–1775 [9].

Volterra, V., 1959. Theory of Functionals and Integral and Integrodifferential Equations. Dover, New York [14].

von Oertzen, A., Rotermund, H.H., Mikhailov, A.S., Ertl, G., 1998. Subsurface oxygen in the CO oxidation reaction on Pt(110): Experiments and modeling of pattern formation. J. Phys. Chem. 102 (25), 4966–4981 [3].

von Oertzen, A., Rotermund, H.H., Mikhailov, A.S., Ertl, G., 2000. Standing wave patterns in the CO oxidation reaction on a Pt(110) surface: experiments and modeling. Phys. Chem. B 104, 3155–3178 [3].

Vortmeyer, D., Shaefer, R.J., 1974. Equivalence of one- and two- phase models for heat transfer processes in packed beds: one dimensional theory. Chem. Eng. Sci. 29, 485–491 [18].

Vortmeyer, D., Yahnel, W., 1972. Moving reaction zones in fixed bed reactors under the influence of various parameters. Chem. Eng. Sci. 27, 1489–1496 [18].

Vos, H.J., Groen, D.J., Potters, J.J.M., Luyben, K.Ch.A.M., 1990a. Countercurrent multistage fluidized bed reactor for immobilized biocatalysis: I. Modeling and simulation. Biotechnol. Bioeng. 36, 367–378 [20].

Vos, H.J., van Houwelingen, C., M. Zomerdijk, Luyben, K.Ch.A.M., 1990c. Countercurrent multistage fluidized bed reactor for immobilized biocatalysis: III. Hydrodynamic aspects. Biotechnol. Bioeng. 36, 387–396 [20].

Vos, H.J., Zomerdijk, M., Groen, D.J., Luyben, K.Ch.A.M., 1990b. Countercurrent multistage fluidized bed reactor for immobilized biocatalysis: II. Operation of a laboratory-scale reactor. Biotechnol. Bioeng. 36, 377–386 [20].

Vrieland, G.E., Khazai, B., Murchison, C.B., 1995. Anaerobic oxidation of butane to butadiene over magnesium molybdate catalysts. II. Magnesia alumina supported catalysts. Appl. Catal. 134, 123–145 [4].

Vrieland, G.E., Murchison, C.B., 1995. Anaerobic oxidation of butane to butadiene over magnesium molybdate catalysts. I. Magnesia supported catalysts. Appl. Catal. 134, 101–121 [4].

Vu, T.D., Seidel-Morgenstern, A., Grüner, S., Kienle, A., 2005. Analysis of ester hydrolysis reactions in a chromatographic reactor using equilibrium theory and a rate model. Ind. Eng. Chem. Res. 44, 9565–9574 [20].

Wainwright, M.S., Hoffman, T.W., 1974. The oxidation of o-xylene in a transported bed reactor. In: Hulburt, H.M. (Ed.), Chemical Reaction Engineering II, ACS Adv. Chem. Series. Am. Chem. Soc. Washington, D.C. 113 [4].

Waldron, W.E., Hufton, J.R., Sircar, S., 2001. Production of hydrogen by cyclic adsorption enhanced reaction process. AIChE J. 47, 1477–1489 [22].

Walker, A.P., Allansson, R., Blakeman, P.G., Lavenius, M., Erkfeld, S., Landaly, H., Ball, B., Harrod, P., Manning, D., Bernegger, L., 2003. SAE Technical Paper 2003-01-0778. Society of Automotive Engineers, Warrendale, PA [6].

Walker, B., Renken, A., 1983. Direct microgravimetric measurement of sorption kinetics under conditions of reaction. Chem. Ing. Tech. 55, 806 [11] [12].

Walker, B., Renken, A., 1984. Direct microgravimetric studies of sorption kinetics under reaction conditions. Ger. Chem. Eng. 7, 227–232 [11] [12].

Wallace, K., Viljoen, H.J., 1995. Modeling of a monolith catalyst with reciprocating flow. AIChE J. 41 (5), 1229–1234 [18].

Wandrey, C., Renken, A., 1973. Zur Beeinflussung der Produktverteilung durch periodische Konzentrationsschwankungen bei der Oxidation von Kohlenwasserstoffen. Chem. Ing. Tech. 45 (12), 854–859 [1] [12].

Wandrey, C., Renken, A., 1974. On improvement of autothermal operation by enforced concentration variations, Paper A3-3 Proc. GVC/AIChE Joint Meeting. Verlag Chemie GmbH, Weinheim, Germany [12].

Wandrey, C., Renken, A., 1977. Ignition and extinction of autothermal reactors in periodic operation. Chem. Eng. Sci. 32, 448–451 [5].

Wang, J.S., Anthony, E.J., 2008. Clean combustion of solid fuels. Appl. Energy 85, 73–79 [8].

Wang, W.-Y., Ku, Y., 2006. Photocatalytic degradation of Reactive Red 22 in aqueous solution by UV-LED radiation. Water Res. 40, 2249–2258 [1] [11] [23].

Wang, Y.N., Larachi, F., Roy, S., 2009. Simulating the dynamics of gas-solid flows in a multichannel microcirculating fluidized bed. Ind. Eng. Chem. Res. 48, 7928–7937 [8].

Wang, Y., Sward, B.K., LeVan, M.D., 2003. New frequency response method for measuring adsorption rates via pressure modulation: application to oxygen and nitrogen in a carbon molecular sieve. Ind. Eng. Chem. Res. 42, 4213–4222 [15].

Wang, L., Tao, L., Xie, M., Xu, G., Huang, J., Xu, Y., 1993. Dehydrogenation and aromatization of methane under non-oxidizing conditions. Catal. Lett. 21, 35–41 [12].

Wang, C., Zhang, P., Wang, R.Z., 2008. Investigation of solid-gas heat transformer system with the consideration of multistep reactions. AIChE. J. 54 (9), 2464–2478 [8].

Wankat, P.C., 1986. Large Scale Adsorption and Chromatography. CRC Press, Boca Raton, Florida [22].

Ward, M., Gabe, D.R., Crosby, J.N., 2000. Pulse plating for PCBs: effect of eductor agitation with pulsing. Proc. 5th Intern. Pulse Plating Symp. American Electroplaters & Surface Finishers Soc, Orlando, Florida, pp. 54–63 [9].

Wardwell, A.W., Carr Jr., R.W., Aris, R., 1982. Continuous reaction gas chromatography: The dehydrogenation of cyclohexane over Pt/γ-Al_2O_3. Wei, J., Georgakis, C. (Eds.), Proc. of Chemical Reaction Engineering-Boston, ACS Symp. Ser. 196, pp. 297–306 [20].

Watanabe, N., Onogi, K., Matsubara, M., 1981. Periodic control of continuous stirred tank reactors—I. The pi criterion and its application to isothermal cases. Chem. Eng. Sci. 36, 809–818 [1] [14] [23].

Weber, A.E., San, K.-Y., 1989. Dynamics of plasmid maintenance in a CSTR upon square wave perturbations of the dilution rate. Biotechnol. Bioeng. 34, 1104–1112 [10].

Wei, J., 1975. Catalysis for motor vehicle emissions. Advan. Catal. 24C, 57–129 [6].

Wei, F., Li, M., Chen, M., Jiang, H., Zhao, Y., 2011. A novel pseudo simulated moving bed with solvent gradient for ternary separations. J. Chromatogr. A. 1218, 2906–2911 [23].

Wei, C.Y., Wang, D.H., Chen, J., 2008. Overproduction of gluthathione by l-cysteine addition and a temperature shift strategy. Biotechnol. Bioproc. Eng. 12, 347–353. [23].

Weibel, B., Garin, F., Bernhardt, P., Maire, G., Prigent, M., 1991. Influence of water in the activity of catalytic coverters. In: Crucq, A. (Ed.), Catalysis and Automotive Pollution Control II. Elsevier Science Publishers, Amsterdam, The Netherlands [6].

Weiner, D.D., Spina, J.F., 1980. Sinusoidal Analysis and Modeling of Weakly Nonlinear Circuits. Van Nostrand Reinhold Co, New York [14].

Welles, J.B., Blanch, H.W., 1976. Effect of discontinuous feeding on ethanol production by Saccharomyces cerevisiae. Biotechnol. Bioeng. 18, 129–132 [10].

Westerterp, K.R., Kuczynski, M., 1987a. A model for a counter-current gas-solid-solid trickle flow reactor for equilibrium reactions. The methanol synthesis. Chem. Eng. Sci. 42, 1871–1887 [21].

Westerterp, K.R., Kuczynski, M., Kamphuis, C., 1989. Synthesis of methanol in a reactor system with interstage product removal. Ind. Eng. Chem. Res. 28, 763–771 [21].

Wetherold, R.G., Wissler, E.H., Bischoff, K.B., 1974. An experimental and computational study of the hydrolysis of methyl formate in a chromatographic reactor. Adv. Chem. Ser. 133, 181–190 [20].

Wheelock, T.D., Fan, C.W., Floy, K.R., 1988. Utilization of phosphogypsum for the production of sulfuric acid. Proc. 2nd Intern. Symp. Phosphogypsum, Fla. Inst. Phosphate Res. Bartow, Flaorida. pp. 3–24 [8].

Wheelock, T.D., Morris, C.E., 1986. Recovery of sulfur dioxide and lime from waste gypsum. TIZ-Fachberichte 110, 37–46 [8].

Wheelock, T.D., Riel, T., 1991. Cyclic operation of a fluidized bed reactor for decomposing calcium sulfate. Chem. Eng. Comm. 109, 155–168 [8].

Wicke, E., Padberg, G., 1968. Einfluß von Stoff- und Wärmetransport bei Reaktionen gasförmig/feststoffe am Beispiel katalytischer Brennzonen in adiabatischer Kontaktschicht. Chem. Ing. Tech. 40, 1033–1038 [18].

Wicke, E., Vortmeyer, D., 1959. Zündzonen heterogener Reactionen in gasdurchströmten Körnerschichten. Berichte Bunsengesellschaft 63, 145–152 [18].

Wilderer, P.A., Irvine, R.L., Goronszy, M.C. (Eds.), 2001. Sequencing Batch Reactor Technology, Scientific and Technical Report No. 10. IWA Publishing, London, U.K. [10].

Wilhelm, R.H., Rice, A.W., Bendelius, A.R., 1966. Parametric pumping: a dynamic principle for separation of fluid mixtures. Ind. Eng. Chem. Fund. 5, 141 [14].

Wilhite, B.A., Huang, X., McCready, M.J., Varma, A., 2003. Effects of induced pulsing flow on trickle-bed reactor performance. Ind. Chem. Eng. Res. 42, 2139–2145 [17].

Wilson, H.D., Rinker, R.G., 1982. Concentration forcing in ammonia synthesis—1. controlled cyclic operation. Chem. Eng. Sci. 37, 343–355 [2].

Wiltowski, T., Mondal, K., Campen, A., Dasgupta, D., Konieczny, A., 2008. Reaction swing approach for hydrogen production from carbonaceous fuels. Int. J. Hydrog. Energy 33, 293–302 [22].

Winslow, P., Bell, A.T., 1984. Application of transient response techniques for quantitative determination of adsorbed carbon monoxide and carbon present on the surface of a ruthenium catalyst during the Fischer-Tropsch synthesis. J. Catal. 86, 158–172 [12].

Winslow, P., Bell, A.T., 1985. Studies of the surface coverage of unsupported ruthenium by carbon- and hydrogen-containing adspecies during CO hydrogenation. J. Catal. 91, 142–154 [12].

Woertz, J.R., Kinney, K.A., McIntosh, N.D.P., Szaniszlo, P.J., 2001. Removal of toluene in a vapor-phase bioreactor containing a strain of the dimorphic black yeast exophiala lecanti-corni. Biotechnol. Bioeng. 75 (5), 550–558 [18].

Wojciechowki, J., Haber, J., 1982. Swingtherm—a new economic process for the catalytic burning of flue gas. Appl. Catal. 4, 275–280 [18].

Wong, K.P., Chan, K.C., Yue, T.M., 1999. A study of surface finishing in pulse current electroforming of nickel by different shaped wave forms. Surf. Coat. Tech. 115 (2–3), 132–139 [9].

Wong, K.P., Chan, K.C., Yue, T.M., 2000. Modelling the effect of complex wave form on surface finishing in pulse current electroforming of nickel. Surf. Coat. Tech. 135, 91–97 [9].

Wong, K.P., Chan, K.C., Yue, T.M., 2001. Influence of spiked current in different shaped waveforms on the surface finish of nickel electroforms. Surf. Coat. Tech. 140 (3), 284–292 [9].

Wongsuwan, W., Kumar, S., Neveu, P., Meunier, F., 2001. A review of chemical heat pump technology and applications. Appl. Thermal Eng. 21, 1489–1519 [8].

Wright, W.F., 2005. Transient response of vapor-phase biofilters. Chem. Eng. J. 113, 161–173 [18].

Wright, W.F., Schroeder, E.D., Chang, D.P.Y., 2005. Transient response of flow-direction switching vapor-phase biofilters. J. Environ. Eng. 131 (7), 999–1009 [18].

Wu, J.-Y., 1998. Quick analysis of enantioselectivity by a chromatographic reactor. AIChE J. 44, 474–476 [20].

Wu, J.-Y., Liu, S.-W., 1999. Quick analysis of enantioselectivity by a chromatographic reactor. II. Experimental verification. J. Chem. Technol. Biot. 74, 974–978 [20].

Xiao, Q., Cheng, Z.M., Jiang, Z.X., Anter, A.M., Yuan, W.-K., 2001. Hydrodynamic behavior of a trickle bed reactor under forced pulsing flow. Chem. Eng. Sci. 56, 1189–1195 [17].

Xiao, R., Song, Q.-L., Song, M., Lu, Z.-J., Zhang, S.-A., Shen, L.-H., 2010a. Pressurized chemical-looping combustion of coal with an iron ore-based oxygen carrier. Comb. Flame 157, 1140–1153 [8].

Xiao, R., Song, Q.-L., Zhang, S.-A., Zheng, W.-G., Yang, Y.-C., 2010b. Pressurized chemical-looping combustion of Chinese bituminous coal: cyclic performance and characterization of iron-ore based oxygen carrier. Ener. Fuels 24, 1449–1463 [8].

Xiao, W.-D., Yuan, W.-K., 1996. Modelling and experimental studies on unsteady-state SO_2 converters. Can. J. Chem. Eng. 74, 772–782 [18].

Xiao, W.-D., Wang, H., Yuan, W.-K., 1999a. An SO_2 converter with flow reversal and interstage heat removal: from laboratory to industry. Chem. Eng. Sci. 54, 1307–1311 [18].

Xiao, W.-D., Wang, H., Yuan, W.-K., 1999b. A novel unsteady-state SO_2 converter for relatively concentrated SO_2 gases. Chem. Eng. Sci. 54, 1333–1338 [18].

Xiao, W.-D., Wang, H., Yuan, W.-K., 1999c. Unsteady-state SO_2 converters with inter-stage heat removal. Chem. Eng. Sci. 54, 4629–4638 [18].

Xiao, W.-D., Wang, H., Yuan, W.-K., 1999d. Practical studies of the commercial flow-reversed SO_2 converter. Chem. Eng. Sci. 54, 4645–4652 [18].

Xie, L., Wang, D.I.C., 1994a. Stoichiometric analysis of animal cell growth and its application in medium design. Biotechnol. Bioeng. 43, 1164–1174 [10].

Xie, L., Wang, D.I.C., 1994b. Fed-batch culture of animal cells using different medium design concepts and feeding strategies. Biotechnol. Bioeng. 43, 1175–1189 [10].

Xie, L., Wang, D.I.C., 1994c. Applications of improved stoichiometric model in medium design and fed-batch cultivation of animal cells in bioreactor. Cytotechnol. 15, 17–29 [10].

Xiu, G.-H., Li, P., Rodrigues, A.E., 2002a. Sorption-enhanced reaction process with reactive regeneration. Chem. Eng. Sci. 57, 3893–3908 [22].

Xiu, G.-H., Li, P., Rodrigues, A.E., 2003a. Adsorption-enhanced steam methane reforming with intra-particle-diffusion limitations. Chem. Eng. J. 95, 83–93 [1] [22].

Xiu, G.-H., Li, P., Rodrigues, A.E., 2003b. New generalized strategy for improving sorption-enhanced reaction process. Chem. Eng. Sci. 58, 3425–3437 [22].

Xiu, G.-H., Soares, J.L., Li, P., Rodrigues, A.E., 2002b. Simulation of five-step one-bed sorption-enhanced reaction process. AIChE J. 48 (12), 2817–2832 [22].

Xu, Q., Chen, B.Z., He, X.R., 2001. Periodic simulation for naphtha pyrolysis in SRT-IV cracking furnace. Comp. Appl. Chem. 18, 303–306 [1] [23].

Xu, J.G., Froment, G.F., 1989a. Methane steam reforming, methanation and water-gas shift, 1. Intrinsic kinetics. AIChE J. 35, 89–96 [22].

Xu, J.G., Froment, G.F., 1989b. Methane Steam Reforming. 2. Diffusional Limitations and Reactor Simulation. AIChE J. 35 (1), 97–103 [22].

Yadav, R., Rinker, R.R., 1989. The efficacy of concentration forcing. Chem. Eng. Sci. 44 (10), 2191–2195 [3].

Yakhnin, V.Z., Menzinger, M., 1999. Deactivation avalanches through the interaction of locally deactivated catalyst with traveling hot spots. Stud. Surf. Sci. Catal. 126, 291–298 [3].

Yakhnin, V.Z., Menzinger, M., 2008. Estimating spectral properties of the thermal instability in packed-bed reactors. Chem. Eng. Sci. 63, 1480–1489 [16].

Yakhnin, V.Z., Rovinsky, A.B., Menzinger, M., 1995a. Absolute instability of a tubular packed-bed reactor with recycling. Chem. Eng. Sci. 50, 1591–1593 [16].

Yakhnin, V.Z., Rovinsky, A.B., Menzinger, M., 1995b. Convective instability induced by differential transport in a tubular packed-bed reactor. Chem. Eng. Sci. 50, 2853–2859 [3] [16].

Yamada, Y., Fujita, Y., 1983. Production of useful compounds in culture (Chapter 21). In: Evans, D.P., Sharp, W.R., Ammirato, P.V., Yamada, Y. (Eds.), Handbook of Plant Cell Culture. Macmillan Publishing Co., New York, New York, pp. 717–728 [10].

Yamada, H., Goto, S., 1997. Periodic operation of trickle bed reactor for hydrogenolysis in gas-liquid-liquid-solid four phases. J. Chem. Eng. Jpn. 30 (3), 478–483. [17] [20].

Yamane, T., Shimizu, S., 1984. Fed-batch techniques in microbial process. In: Feichter, A. (Ed.), Advances in Biochemical Engineering/Biotechnology, 30. Springer-Verlag., Berlin, Germany, pp. 147–194 [10].

Yamazaki, K., Takahashi, N., Shinjoh, H., Sugiura, M., 2004. The performance of NOx storage-reduction catalyst containing Fe-compound after thermal aging. Appl. Catal. B-Environ. 53, 1–12 [6].

Yang, R.T., 1987. Gas Separation by Adsorption Processes. Butterworth, Boston [14] [22].

Yankov, D., Sapundzhiev, H., Beshkov, V., 1995. Modelling of enzyme hydrolysis of maltose in a porous bed operates in transient conditions, Bulg. Acad. Sci. Comm. Dept. Chem. 27 (2), 43–49 [23].

Ye, Y.-M., Rihko-Struckmann, L., Munder, B., Sundmacher, K., 2005. Partial oxidation of n-butane in a solid electrolyte membrane reactor: periodic and steady-state operations. Appl. Catal. A-Gen. 285, 86–95 [4] [12].

Ye, Y.-M., Rihko-Struckmann, L., Munder, B., Sundmacher, K., 2006. Partial oxidation of n-butane in a solid electrolyte membrane reactor. Influence of electrochemical oxygen pumping. J. Electrochem. Soc. 153 (2), D21–D29 [4].

Ye, K.-M., Shijo, M., Jin, S., Shimizu, K., 1996. Efficient production of Vitamin B12 from propionic acid bacteria under periodic variation of dissolved oxygen concentration. J. Ferment. Bioeng. 82 (5), 484–491 [10].

Yeramian, A., Silveston, P.L., Hudgins, R.R., 1970. Influence of inert gases on catalytic oxidation of SO_2. Can. J. Chem. 48 (8), 1175–1181 [3].

Yi, K.B., Harrison, D.P., 2005. Lower pressure sorption-enhanced hydrogen production. Ind. Eng. Chem. Res. 44, 1665–1669 [22].

Yokota, Y., Muraki, H., Fujitani, Y., 1985. Rh-free three-way catalysts for automotive exhaust control, Tech. Paper Series No. 850129, Society of Automotive Engineers, Warrendale, PA. SAE [6].

Yokoyama, K., Konno, H., Takahashi, H., Nagayama, M., 1982. Advantages of pulse anodizing. Plat. Surf. Finish. 69 (7), 62–65 [9].

Yongsunthon, I., Alpay, E., 1998a. Conversion-temperature trajectories for well-mixed adsorptive reactors. Chem. Eng. Sci. 53 (4), 691–696 [22].

Yongsunthon, I., Alpay, E., 1998b. Design and optimization of temperature cycled adsorptive reactors. Comput. Chem. Eng. 22, S733–S736 [22].

Yongsunthon, I., Alpay, E., 1999. Design of periodic adsorptive reactors for the optimal integration of reaction, separation and heat exchange. Chem. Eng. Sci. 54, 2647–2657 [22].

Yongsunthon, I., Alpay, E., 2000. Total connectivity models for adsorptive reactor design. Chem. Eng. Sci. 55, 5643–5656 [22].

York, A.P.E., Hamakawa, S., Hayakawa, T., Sato, K., Tsunoda, T., Takehira, K., 1996. Partial oxidation of C_2-C_4 alkanes into oxygenates using an Au/yttria stabilized zirconia|Ag electrochemical cell. J. Chem. Soc., Faraday Trans. 92 (19), 3579–3586 [4].

Young, B., Hildebrandt, D., Glasser, D., 1992. Analysis of an exothermic reversible reaction in a catalytic reactor with periodic flow reversal. Chem. Eng. Sci. 47, 1825–1837 [18].

Yu, W., Hidajat, K., Ray, A.K., 2003. Modeling, simulation and experimental study of a simulated moving bed reactor for the synthesis of methyl acetate ether. Ind. Eng. Chem. Res. 42 (6), 6743–6754 [21].

Yu, W.F., Hidajat, K., Ray, A.K., 2004. Determination of adsorption and kinetic parameters for methyl acetate esterification and hydrolysis reactions catalyzed by Amberlyst 15. Appl. Catal. A-Gen. 260 (2), 191–205 [20].

Yuranov, I., Kiwi-Minsker, L., Slin'ko, M., Kurkina, E., Tolstunova, E.D.A., Renken, A., 2000. Oscillatory behavior during CO oxidation over Pd supported on glass and mathematical modelling. Chem. Eng. Sci. 55, 2827–2833 [16].

Zabransky, R., Anderson, R., 1977. Simulated moving bed alkylation process, US Patent 4008291 [1] [21].

Zacca, J.J., Ray, W.H., 1993. Modeling of the liquid-phase polymerization of olefins in loop reactors. Chem. Eng. Sci. 48 (22), 3743–3765 [7].

Zafar, I., Barker, P.E., 1988. An experimental and computational study of a biochemical polymerization reaction in a chromatographic reactor separator. Chem. Eng. Sci. 43 (9), 2369–2375 [20] [21].

Zagoruiko, A.N., 2007. Simulation of selective reactions' performance in transient regimes with periodical separate feeding of reagents. Case study: propane oxidative dehydrogenation in adiabatic V-Ti catalyst bed. Chem. Eng. J. 134, 117–122 [8].

Zagoruiko, A.N., 2008a. Anaerobic catalytic oxidation of hydrocarbons in moving heatwaves. Case simulation: propane oxidative dehydrogenation in a packed adiabatic V-Ti oxide catalyst bed. Chem. Eng. Sci. 63, 4962–4968 [4].

Zagoruiko, A.N., 2008b. Modeling of reverse-flow reactor for VOC incineration with account of reversible adsorption: the way to minimize the negative influence of desorption phenomena. Intern. J. Chem. Reactor Eng. 6 Article A110 [18].

Zagoruiko, A.N., Matros, Yu.Sh., 2002. Mathematical modelling of Claus reactors undergoing sulfur evaporation and condensation. Chem. Eng. J. 87, 73–88 [18].

Zagoruiko, A.N., Matros, Yu.Sh., Kumar, V.R., Kulkarni, B.D., 1992. Reactor performance with periodic flow reversal for a multistep complex reaction. Chem. Eng. Sci. 47, 4315–4321 [18].

Zahn, V.M., Mangold, M., Krasnyk, M., Seidel-Morgenstern, A., 2009. Theoretical analysis of heat integration in a periodically operated cascade of catalytic fixed-bed reactors. Chem. Eng. Technol. 32 (9), 1326–1338 [21].

Zahn, V.M., Mangold, M., Krasnyk, M., Seidel-Morgenstern, A., 2010. Autothermal operation of an adiabatic simulated counter current reactor. Chem. Eng. Sci. 65 (1), 458–465 [21].

Zaza, P., de la Torre, A., Renken, A., 1991. Direkte Dehydrierung von Methanol zu Formaldehyd in einer zirkulierenden Wirbelschicht. Chem. Ing. Tech. 63, 640–642 [11].

Zaza, P., Randall, H., Doepper, R., Renken, A., 1994. Dynamic kinetics of catalytic dehydrogenation of methanol to formaldehyde. Catal. Today 20 (3) 325–324 [11].

Zhabotinsky, A.M., 1964. Periodic liquid phase reactions. Proc. Acad. Sci. USSR 157, 392–395 [1].

Zhang, T.-J., Amiridis, M.D., 1998. Hydrogen production via the direct cracking of methane over silica-supported nickel catalysts. Appl. Catal. A-Gen. 167, 161–172 [4] [8].

Zhang, K., Hidajat, K., Ray, A.K., 2007a. Modified reactive SMB for production of high concentrated fructose syrup by isomerization of glucose to fructose. Biochem. Eng. J. 35, 341–351 [1].

Zhang, Y., Dai, J., Wang-Iverson, D.B., Tymiak, A.A., 2007c. Simulated moving columns technique for enantioselective supercritical fluid chromatography. Chirality 19, 683–692 [23].

Zhang, Z.Y., Hidajat, K., Ray, A.K., 2001. Application of simulated countercurrent moving bed chromatographic reactor for MTBE synthesis. Ind. Eng. Chem. Res. 40, 5305–5316 [21].

Zhang, Z.Y., Hidajat, K., Ray, A.K., 2002. Multiobjective optimization of simulated countercurrent moving bed chromatographic reactor (SCMCR) for MTBE synthesis. Ind. Eng. Chem. Res. 41, 3212–3232 [21].

Zhang, Y., Hidajat, K., Ray, A.K., 2007b. Modified reactive SMB for production of high concentrated fructose syrup by isomerization of glucose to fructose. Biochem. Eng. J. 35, 341–351 [21].

Zhang, Z.-G., Scott, D.S., Silveston, P.L., 1994a. Steady state gasification of an Alberta sub-bituminous coal in a micro fluidized bed. Energy Fuels 8, 637–642 [8].

Zhang, Z.-G., Scott, D.S., Silveston, P.L., 1994b. Successive pulsing of an iron loaded sub-bituminous coal char. Energy Fuels 8, 943–946 [8].

Zhang, Z.-G., Scott, D.S., Silveston, P.L., 1995. Iron-catalyzed gasification of coal char under composition modulation. Energy Fuels 9, 479–483 [8].

Zhdanov, V.P., 2004. Periodic perturbation of the kinetics of heterogeneous catalytic reactions. Surf. Sci. Rep. 55, 1–48 [1] [3].

Zheng, M., Banerjee, S., 2009. Diesel oxidation catalyst and particulate filter modeling in active – flow configurations. Appl. Therm. Eng. 29, 3021–3035 [18].

Zheng, M., Reader, G.T., 2004. Energy efficiency analyses of active flow aftertreatment systems for lean burn internal combustion engines. Energy Conv. Manag. 45, 2473–2493 [18].

Zheng, M., Mirosh, E., Klopp, W., Ulan, D., Pardell, M., Newman, P., Bunimovich, G., Matros, Yu., Nishimura, A., 1999. Development of compact reverse-flow catalytic converter for diesel dual fuel LEV. SAE Paper 1999-01-3558. Soc. Automotive Eng. Detroit, Michigan. [18].

Zheng, M., Mirosh, E.A., Matros, Yu.Sh., Bunimovich, G.A., Strots, V.O., Sallamie, N., Checkel, M.D., Windawi, H., 1998. Novel catalytic converter for natural gas powered diesel engines to meet stringent exhaust emission regulations. Conf. Proc. NGV-98. Cologne, Germany. pp. 276–284 [18].

Zheng, M., Shen, L.-H., Xiao, J., 2010. Reduction of $CaSO_4$ oxygen carrier with coal in chemical-looping combustion: effect of temperature and gasification intermediate. Int. J. Greenhouse Gas Contr. 4, 716–728 [8].

Zhou, X., Gulari, E., 1986a. CO oxidation on Pt/Al_2O_3 and Pd/Al_2O_3: transient response and rate enhancement through forced concentration cycling. Chem. Eng. Sci. 41 (5), 1277–1284 [3].

Zhou, X., Gulari, E., 1986b. CO adsorption and oxidation on Pd/ Al_2O_3 under transient conditions. Langmuir 2, 709–715 [3].

Zhou, X., Gulari, E., 1986c. CO oxidation on Pt/Al_2O_3 and Pd/Al_2O_3 transient response and concentration cycling studies. Chem. Eng. Sci. 41 (4), 883–890 [3] [13].

Zines, D.O., Rogers, P.L., 1971. A chemostat study of ethanol inhibition. Biotechnol. Bioeng. 13, 293–308 [10].

Zinn, B.T., 1992. Pulse combustion: recent applications and research issues. Proc. Combust. Inst. 24 (1), 1297–1305 [5].

Zirkwa, I., Grünewald, M., Agar, D.W., 2001. Intensification of mass transfer in porous catalysts using pressure oscillations. AIChE Annual Mtg, Reno (USA). November [15].

Znad, H.T., Katoh, K., Kawase, Y., 2007. High loading toluene treatment in a compost based biofilter using up-flow and down-flow swing operation. J. Hazard. Mater. 141, 745–752 [18].

Zoval, J.V., Lee, J., Gorer, S., Penner, R.M., 1998. Electrochemical preparation of platinum nanocrystallites with size selectivity on basal plane oriented graphite surfaces. J. Phys. Chem. B. 102 (7), 1166–1175 [9].

Zoval, J.V., Stiger, R.M., Biernaki, P.R., Penner, R.M., 1996. Electrochemical deposition of silver nanocrystallites on the atomically smooth graphite basal plane. J. Phys. Chem. 100 (2), 837–844 [9].

Züfle, H., Turek, T., 1997a. Catalytic combustion in a reactor with periodic flow reversal. Part 1. Experimental results. Chem. Eng. Proc. 36 (5), 327–340 [18].

Züfle, H., Turek, T., 1997b. Catalytic combustion in a reactor with periodic flow reversal. Part 2. Steady-state reactor model. Chem. Eng. Proc. 36 (5), 341–352 [18].

Index

Note: Page numbers followed by "f" and "t" indicate figures and tables respectively

A

Acetal, 627
Acetic acid (HAc), 317–322, 585
Acoustic generators, 424
Adaptive control, 310–311
Adiabatic packed bed reactors, 553
Adsorbent beds
  adsorption isotherm parameters, 619
  CMCR model, 617
  fourth-order Runge-Kutta algorithm, 619
  fructose, 617
  methane oxidative coupling, 617f, 618–619
  MX, 618
  OCM, 618
  OX, 618
  PX, 618
  reaction rate, 619
  SCMCR cascade with, 616–617, 616f
  SCMCR system modifications, 618
  three-section SCMCR operation, 617, 617f
Adsorption processes, 388
  cyclic, 388
  quasi-steady state calculation, 389
A/F. *See* Air fuel ratio
AIBN initiator. *See* Azobisisobutyronitrile initiator
Air fuel ratio (A/F), 142
  irregular A/F variation effects, 161–162
  modulation converter performance improvement, 169–170
  modulation effect on poisoning and sintering
    CeO_2 in TWC, 160–161
    reported observations, 161, 161f
    sulfur poisoning, 160
    TWC activity loss, 161
Aldehydes
  MA, 119–120
  phthalic anhydride from o-xylene, 120–121
  tellurium-promoted vanadium molybdate, 120
  toluene partial oxidation, 120
α-methyl styrene (AMS), 476
Ammonia, 286
Ammonia synthesis, 23–24
  application
    hydrogenation reactions, 33
    nitrogen transport to active sites, 33
    reactant recycle, 33
    reactant switching, 33
  experimental studies, 25t–27t
  interpretation, 30
    reactant inhibition, 31
    reactant storage, 31–32
    relaxed steady state, 30–31
    surface activation/restructuring, 33
  iron catalysts, 24
    mass transfer interference, 28–29
    near-adiabatic operation, 29
  motivation, 24
  osmium catalyst, 30
  under periodic operation, 24
  ruthenium catalyst, 30
    bang-bang switching, 30
    modulation, 30
  thermodynamics, 24
AMS. *See* α-Methyl styrene
Anaerobic oxidation, 205
Anderson-Schulz-Flory distribution, 364
Anhydrides
  MA, 119–120
  phthalic anhydride from o-xylene, 120–121
  tellurium-promoted vanadium molybdate, 120
  toluene partial oxidation, 120
Animal cell culture, 307
Anionic polymerization, 192–193
Aris number, 577
Aromatic nitriles, 121–122
Aromatics, hydrogenation of, 46t
  benzene hydrogenation, 45–47
  phenylacetylene, 47
  using step-change experiments, 47
  toluene hydrogenation, 45
Atactic polymers, 172
Attrition, 587
Au-Cu-Cd alloys, 257
Au-Sb alloys, 257
Automotive emissions, application to, 522t
Automotive engines, 141
  internal combustion engines, 142–146
Autonomous oscillations, 50
  CO feed, 57
  composition modulation interaction, 50–57
  forcing frequency, 58
  using highly asymmetric modulation cycle, 57
  Pd catalyst, 57
  Pt catalyst, 50–57
  quasi-periodic response, 57–58
Azo dyes
  photocatalytic degradation, 332
  photonic efficiency comparison, 332, 333f
Azobisisobutyronitrile initiator (AIBN initiator), 193

B

Backfill cycle, 640
Balzhinimaev mechanism, 69–70
Bang-bang cycling, 24–28, 71, 89
  NH_3 synthesis, 28f
  using osmium catalyst, 30

Bang-bang cycling (*Continued*)
of syngas reactants, 40t
Base-peak liquid flow rate modulation, 478–479
LFM parameter effect, 479
maldistribution in bed, 479
BASF catalyst, 43
Belousov-Zhabotinsky reaction (BZ reaction), 9–10
Benzene hydrogenation, 45–47
Bifunctional monomers, 174
Bifurcation analysis, 45
Biochemical applications, 681
pressure modulation
ultrasonic irradiation, 683
product-inhibited biochemical reactions, 681–682
temperature modulation, 682
combination of precursor addition, 682
inducer concentrations, 682–683
Biochemical systems
See also Gas–solid systems; Liquid–solid systems
continuous moving-bed model, 629f
dextran biosynthesis, 630–631
enzymatic conversion of sugars, 629–630
enzyme lactase, 631
fructose, 628
syrups of, 627–628
glucose, 628
and fructose, 628
isomerization, 629f
high fructose syrups, 627–628
isothermality, 628
lactosucrose
and glucose, 632
production, 631–632
maltose, 631
Mg^{2+} form, 629
PCR, 630
rotating adsorbent columns, 628f
stationary reactor columns, 628f
sucrose feed, 630f
12-column SCMCR, 632f
Biofilters, 537
for off-gas treatment, 537
reverse-flow biofilters, publications on use of, 537t
Biological process modulation
See also Dissolved oxygen modulation; Fed-batch operations; Recombinant cell culture stabilization
applications to immobilized cells, 298
immobilized cell reactor, 300
liquid fluidized bed bioreactor, 302f
multi-step model, 301–303
phosphate nutrient flow, 301
S. cattleya, 298–300
S. griseus cells, 301
SBR operation, 301
specific growth rate, 303
two phase dynamic model, 300–301, 300t
unit operations, 303f
bioreactors, 273
cell cycle, 273–274
culture medium tuning
animal cell cultures, 287
bang-bang tuning, 287
Floquet multipliers, 287–288
model trajectories, 288f
OAB, 288
OCD, 288
phase trajectories, 288f
stable operating states, 288–289
stationary cyclic state, 287f
transitions, 287–288
dissolved oxygen, 312
eukaryotic hosts, 313
fed-batch operations, 275
flow rate modulation, 280
bang-bang flow rate cycling, 283
cell culture simplified model, 276t
cell mass in chemostat, 283
CSTR model, 281–282
culture model, 281
using cycling, 280–281
glucose feed concentration, 282, 282f
metabolite production, 281
specific growth equation, 281
yeast culture and culture performance, 283f
lignocellulose, 312–313
mathematical models, 275
π criterion application, 277
quasi-steady-state analysis application, 280
Michaelis-Menten kinetics, 274
modulation strategies, 313
periodic harvesting and culture replacement, 312
PO, 274
prokaryotic hosts, 313
sequencing batch reactor, 274
substrate modulation, 280
cell culture simplified model, 276t
CSTR model, 281–282
culture model, 281
cycle periods, 283
using cycling, 280–281
glucose feed concentration, 282, 282f
metabolite production, 281
specific growth equation, 281
survival in mixed cultures, 289
biomass, 289–290
dilution rate and cycling frequency operating space, 292f
pH cycling, 291–292
phase plane plots of competing microorganism species, 290f
using simulation and unstructured model, 289
with single limiting substrate, 289t
stable coexistence, 292
stable cycle under in phase modulation, 290, 291f
Bioreactors, 273–274
Bisphenol A (BPA), 609
Boudouard reaction, 670
BPA. *See* Bisphenol A
Bulk dissolution model, 376–377
Butterfly effect, 547–548
BZ reaction. *See* Belousov-Zhabotinsky reaction

C

C-920 chromatographic separation unit, 626
Calcium oxide regeneration, 211–212
Candidate adsorbents, 573
Carbon deposition control, 686–687
Carlsson's interpretation, 129
$CaSO_4.2H_2O$. *See* Gypsum
Catalyst surface, 64
Catalytic combustion
See also Non-catalytic combustion
fuels combustion
chemical looping, 130
CuO catalyst, 132–134
CuO re-oxidation, 131, 131f
packed bed, 130–131

perovskite catalyst, 131–132
reactor effluents comparison, 133f
spatiotemporal temperature distribution, 132f
modulation studies of, 125t–126t
single component combustion
Carlsson's interpretation, 129
CH_4 combustion, 129–130
effluent composition comparison, 130f
platinum gauze fluctuates, 128
propane combustion, 128
propane oxidation, 128
propylene combustion, 127–128
propylene conversion, 128f
transient concentrations, 129, 129f
Catalytic dehydrogenation
advantages, 107–108
for autothermal operation, 107
bang-bang cycling comparison, 102–106, 106f
β-$NiMoO_4$ catalyst, 102
conversion evolution and product concentration, 106, 107f
diolefins, 102
MgO-$Mg_3(VO_4)_2$ catalyst, 102–106
olefins, 102
of paraffins, 102
Pt-Sn catalyst, 106
two-zone system, 106–107
Catalytic gas-solid reactions
See also Non-catalytic gas-solid reactions
anaerobic oxidation, 205
catalytic gasification under modulation, 217
carbonaceous materials, 218t–222t
using circulating fluidized bed, 223
CO vs. time in initial exposure, 217–223, 223f
Forestburg coal, 217
gasification, 216–217
iron oxide reaction, 217
metal-metal oxide couples, 217
Yallourn coal, 217
chemical looping, 206
circulating solid oxygen carrier, gasification employing
ash-free hot gas, 224
bubbling bed gasifier, 228
$CaSO_4$–CaS cycle, 225–226
circulating fluidized beds, 223, 224f
CO_2 content of gas leaving mixed fluidized bed, 225f
coal, 223–224
countercurrent moving-bed gasifiers, 226–227
Cu_2O and O_2, 224–225
double fluidized bed system for steam gasification, 229f
4-part cycle, 225
gasifier fluidizing gas, 226
nominal 10 kW Swedish system, 226, 227f
non-steady state gasification, 223
Shenhua bituminous coal, 226
Xuzhou, 227–228
coal gasification, 206
combustion, in circulating fluidized beds
CO_2 capturing, 230–231
methane and syngas, 228
hydrocarbons oxidative dehydrogenation, 206
autothermal dehydrogenation process, 207–208
bang-bang cycling performance, 208f
electrophilic oxygen, 206
isophthalonitrile, 207
mixed dehydrogenation-oxidation catalyst, 209f
novel oxidative dehydrogenation process, 207
propane oxidative dehydrogenation, 207
propene, 206–207
hydrocarbons partial oxidation, 206
bang-bang cycling performance, 208f
Bi_2O_3 catalyst, 207
catalyst lattice, 206–207
circulating fluidized beds, 207
commercial process, 207–208
using MgO/V_2O_5 catalyst, 207
mixed dehydrogenation-oxidation catalyst, 209f
toluene partial oxidation, 206
methane cracking, 209, 210t
carbon whisker formation, 209
hydrogen production from, 211f
Ni gauze, 209
off-gas composition, 209, 212f
periodic reaction switching
in chemical looping, 233
circulating adsorbent bed process, 232f
using data and fluidization models, 233
fluidizing gas, 233
locational modulation, 232f
temporal and modulation, 232f
Catalytic gasification under modulation, 217
carbonaceous materials, 218t–222t
using circulating fluidized bed, 223
CO vs. time in initial exposure, 217–223, 223f
Forestburg coal, 217
gasification, 216–217
iron oxide reaction, 217
metal-metal oxide couples, 217
Yallourn coal, 217
Catalytic oxidation
CO oxidation, 49–50
autonomous oscillations, 50
miscellaneous catalysts, 61
modeling, 63
noble metal catalysts, 50
single crystal catalysts, 58
temperatures, 61
nitrogen oxide reduction, 70
N_2O reduction, 76
NO reduction over supported catalyst, 71
Pt single crystals, 75–76
SO_3 reduction, 70
sulfur dioxide oxidation, 66–67
modeling, 69–70
travelling waves
axially located thermocouples, 76f
in packed bed reactors, 77
wrong-way behavior, 76
Yakhnin-Menzinger simulations, 76–77
Catalytic packing, 553–554
Catalytic reaction system
identification, 388–389
optimal steady-state condition, 389
theoretical problems, 389
kinetics and temperature effects, 389
Catalytic reactor, model for, 557t–558t
Cell
current, 236
voltage, 236

Cell cycle, 273–274
Cell model, 578–579, 641
Cells-in-series (CIS), 642
Cellular-Automata based model, 547–548
Centrifugal partition chromatographic reactor (CPCR), 593–595
CFB. *See* Circulating fluidized bed
CH_4. *See* Methane
Chain polymerization, 172
  linear polymer growth, 172
  living anionic polymerizations, 174, 175t
Chattering control
  relaxed steady state, 546
  steady-state operating point stability, 548
  steady-state value forces, 547
  temperature-residence time plane, 546–547, 546f
  two applications, 547–548
Chemical electrolytic production, 267–269
Chemical heat pumps, 212–215
  ammonia condensation and vaporization, 215
  cyclic ammoniate formation and decomposition, 215
  gas-solid reactions, 215
  methanolation-demethanolation cycle, 216
  2-propanol to acetone dehydrogenation, 215
  reactions, 215–216
  using reactor block ammoniate formation and decomposition, 217f
  using salt transformations, 216f
Chemical looping, 124, 206
Chinese hamster ovary (CHO), 292–293
  cell line, 682–683
CHISA. *See* International Congress of Chemical and Process Engineering
CHO. *See* Chinese hamster ovary
Chromatographic reactor (CR), 3, 562, 570–571, 580
  C_6H_{12} reactant concentrations, 571f
  candidate adsorbents, 573
  catalyst and adsorbent converts, 572
  CMCR, 571–572, 572f, 587
    attrition, 587
    catalyst and adsorbent, 587
    dispersion terms, 588–590
    equilibrium controlled reaction, 589f
    heterogeneous model, 587
    literature, 588t
    pseudo-homogeneous model, 587–588
  CPCR, 593–595
  CRAC reactor, 572, 590–591, 590f, 591t
  cyclic steady state, 579, 580f
  cyclohexane catalytic dehydrogenation, 570–571
  dimensionless groups, 578t
  exploratory efforts, 570
  extractive reactor systems, 593
  fixed bed chromatographic reactor
    with flow direction switching, 592, 592f
  fluidized bed chromatographic reactor, 591–592
  general models
    distributed systems, 573–578
    lumped models, 578–579
  heterogeneous model equations for, 576t
  mathematical models, 570
  periodic flow reversal, 572
  pressure swing systems, 573
  pseudo-homogeneous model equations for, 574t
  pulse chromatographic reactor, 580
    cationic resin catalyst, 585–586
    dehydrogenation of cyclohexane in, 583–584
    dilution ratio calculation, 584
    fixed bed adsorptive reactor, 584–585
    heterogeneous catalytic reactions, 580–583
    injection, regeneration and cycle time, 584f
    ion exchange resins, 586
    literature, 581t–582t
    using matrix-based adsorption modeling, 586–587
    mixed cell model, 584
    optimal repetitive pulsing, 583f
    reactants, 583
    reaction experiment, 585, 585f
    regeneration experiment, 586f
    triacetine, 586
  SCMCR, 572
Circulating fluidized bed (CFB), 102
  CO_2 sequestration expensive, 123
  combustor, 228–229
Circulating solid oxygen carrier, gasification employing
  ash-free hot gas, 224
  bubbling bed gasifier, 228
  $CaSO_4$–CaS cycle, 225–226
  circulating fluidized beds, 223, 224f
  CO_2 content of gas, 225f
  coal, 223–224
  countercurrent moving-bed gasifiers, 226–227
  Cu_2O and O_2, 224–225
  double fluidized bed system for steam gasification, 229f
  4-part cycle, 225
  gasifier fluidizing gas, 226
  nominal 10 kW Swedish system, 226, 227f
  non-steady state gasification, 223
  Shenhua bituminous coal, 226
  Xuzhou, 227–228
CIS. *See* Cells-in-series
Claus reaction, 324–325, 325f
Closed loop
  control system, 551f
  controller, 552f
CMA. *See* Composition-modulated alloy
CMCR. *See* Countercurrent moving bed chromatographic reactor
CMM. *See* Composition-modulated metal
CO oxidation, 49–50, 454–455
  average CO conversion, 62f
  inlet O_2 modulation, 155–157, 156f
  miscellaneous catalysts
    Al_2O_3-supported CuO catalyst, 63
    hopcalite, 63
    multiple resonance, 62–63, 62f
    Ni catalyst, 62
    O_2 concentration, 61
    twofold enhancement, 62–63
  modeling, 155
    catalyst surface, 64
    catalyst surface oxidation steps, 63
    composition modulation behavior, 65, 65t
    cycle frequency and amplitude function, 64f

dual catalyst, 66f
gas phase, 64
gas shrinkage, 64
global enhancement, 66
initial condition, 64
kinetic models, 65–66
material balance equations, 63
mathematical models, 66
oxygen storage, 66
stationary cycling condition, 65
support surface, 64
transport steps and boundary conditions, 65
modulation frequency effect, 459f
noble metal catalysts, 50
autonomous oscillations, 50
PDEs and ODEs integration system, 155–157
simulated amplification in packed bed reactors, 443
simulated reaction rate comparison, 440f
single crystal catalysts, 58
catalytic oxidation, 59
fast Fourier transforms, 60f
Krischer-Eiswirth-Ertl model, 59–60
O_2 partial pressure modulation, 58, 59f
photoemission electron microscopy, 60
reactant concentration, 60–61
standing waves, 61
sub-harmonic behavior, 58
turbulent surface oscillations, 58–59
temperature modulation of, 456, 458f
temperature modulation question, 439–440
using temperatures, 61
time-average rate under temperature modulation, 440–441
TWC performance, 157
Coal, 223–224
Coal gasification, 206
Collapsing bubble cavities, 429f
Combustion, 123
catalytic
fuels combustion, 130
modulation studies, 125t–126t
single component combustion, 127–128
chemical looping, 124
cycling, 123
looping, 134
non-catalytic
modulated combustion, 124–127
modulation studies of, 125t–126t
pulsed combustion, 124
simulated loop reactors, 134
applications, 139–140
bifurcation analyses, 139
inert packing, 137
loop combustion systems, 135t–136t
N-reactor loop, 138–139
slow cycling, 137–138
stable rotating temperature waves, 139
stable spatiotemporal patterns, 138f
three bed, 137f
two-bed loop, 134–137
VOC, 137
Combustion in circulating fluidized beds
CO_2 capturing, 230–232
dual circulating fluidized bed combustor, 228–229, 230f
methane
$CO/CO_2/H_2/H_2O/CH_4$ gas mixture, 229
CO_2, 228
coke oxidation, 230
Fe_2O_3/Al_2O_3, 231f
oxygen carrier, 229–230
syngas, 228
burning either syngas, 230f
CO_2-free waste gas, 230
Cu_2O and Mn_2O_3, 230
exit gas composition and bed temperature, 231f
small scale pilot units, 228–229
Commercial applications, 538
catalytic decomposition of N_2O, 541–542
reverse flow reactor technology for NO_x removal, 540
RFR control, 539
thermal reversed flow reactor, 540–541
VOC mitigation, 538–539
Competitive reactions
ethanol dehydration
adsorption rate, 360
EtOH flow periodic interruption, 360
products, 359
methanol synthesis, 359
parasitic reaction systems, 360
ethene acetoxidation, 360
ethyl acetate formation rates, 360, 361f
Composition-modulated alloy (CMA), 260–261
Composition-modulated metal (CMM), 260–261
Concave curve (SS), 474–475
Continuous rotating annular chromatographic reactor (CRAC reactor), 572, 590
literature, 590, 591t
operating principle of, 590f
simulation comparison, 591, 591f
used for liquid systems, 590–591
Continuous stirred tank reactor (CSTR), 393
See also Dispersed flow tubular reactor (DFTR)
analysis results, 400
conclusions, 395–399
consecutive reaction in, 437
dimensionless form of, 395t
expression
for asymmetrical second-order FRF, 402
for FRFs, 395
integration constants, 400–401
model equations for homogeneous reaction, 394t
numerical simulation of, 401f
periodic, 549
second order functions, 399f
temperature modulation of, 438
Conventional free-radical polymerizations, 173–174
Conventional laboratory equipment
amplification phenomenon explanation, 443
experimental studies with, 439
experimental system and reactor, 441f
gain in amplitude of temperature variations, 442f
periodic temperature disturbance amplification, 442–443
reactant inhibition effect, 443f

Conventional laboratory equipment (*Continued*)
simulated amplification in packed bed reactors, 443
temperature amplification investigation, 441–442
temperature modulation question, 439–440
time-average rate of CO oxidation, 440–441
Conventional time-invariant proportional control, 549–550
Coordination polymerization
conversion fluctuations, 186–187
using multi-site Ziegler-Natta catalyst, 187
MWD, 179–186
nonlinear model operation simulation, 188f
phase-shifted square-wave modulation, 187f
propene polymerization simulation, 186
UNIPOL process, 186
Ziegler-Natta ethylene polymerization, 186
Copper zinc catalysts (CuO/ZnO catalysts), 41
CO_2-rich half-cycle, 42
CuO/ZnO catalyst, 42–43
H_2-CO mixtures, 41
H_2:CO_2 modulation, 43
modulation and steady state rate comparison, 42, 42f
optimum steady-state ratio, 41–42
time-average, 41–42
Copper-zinc-alumina catalysts (CuO:ZnO:Al_2O_3 catalysts), 39
bang-bang experiment, 41
modulation variables influence on enhancement, 41t
results, 42f
steady-state synthesis rate, 39
Countercurrent moving bed chromatographic reactor (CMCR), 571–572, 587, 598–599
attrition, 587
catalyst and adsorbent, 587
dispersion terms, 588–590
equilibrium controlled reaction, 589f
heterogeneous model, 587
literature, 588t
pseudo-homogeneous model, 587–588
steady-state concentration profiles in, 572f
Countercurrent moving-bed gasifiers, 226–227
CPCR. *See* Centrifugal partition chromatographic reactor
CR. *See* Chromatographic reactor
CRAC reactor. *See* Continuous rotating annular chromatographic reactor
Creeping waves, 507
CSTR. *See* Continuous stirred tank reactor
CuO catalyst, 132–134
CuO/ZnO catalysts. *See* Copper zinc catalysts
CuO:ZnO:Al_2O_3 catalysts. *See* Copper-zinc-alumina catalysts
Cyclic steady state, 579, 580f
calculation of, 388
Cycling reactor stability
cycle period, 554–556, 555f
flow-reversal reactor, 554
open loop and closed loop control, 554–556
CYCLOP–Cordis Project. *See* European Consortium for Cycling Operation–Cordis Project

D

Damköhler number, 607, 641
DC. *See* Double conversion
DCPD. *See* Dicyclopentadiene
DEA. *See* Diethanolamine
Deamination
alcohol dehydration reactions, 328
ethanol dehydration, 328, 329f
isopropanol flow modulation, 328–330, 331f
miniature microchannel reactor, 328–330, 330f
on alumina-silica catalyst, 327
amines and alcohols stop-effect, 326–327, 327t
disproportionation reaction, 326
modeling, 330–331
Koubek I mechanism, 331–332
Koubek II mechanism, 331–332
two-site mechanism, 331, 331f
rates, 326, 326f
stop-effect mechanisms, 327–328, 328t
DEE. *See* Diethyl ether
Deposition polymerization
fluorocarbon coatings, 193–194
plasmas, 193
square-wave modulation, 194
Detoxification reaction modulation
asymmetric modulation, 154
catalyst composition, 154
modulation, 146
stoichiometric number, 155
symmetrical cycle, 154
Devising optimization techniques, 686
Dextran biosynthesis, 630–631
DFTR. *See* Dispersed flow tubular reactor
Dicyclopentadiene (DCPD), 481–482
Diesel engines, 142
Diethanolamine (DEA), 352
asymmetrical cycling, 353
reaction network, 352
symmetrical cycling, 353
Diethyl ether (DEE), 359
Diffuse reflectance, infrared Fourier transform spectrometer cell (DRIFTS cell), 371
Dilution rate modulation, 293
Dispersed flow tubular reactor (DFTR), 393
See also Plug flow reactor (PFR)
analysis results, 400
dimensionless form of, 395t
expressions for FRFs, 395
integration constants, 396t–398t
model equations for homogeneous reaction, 394t
second order functions, 400f, 401
theoretical plates, 394
Dispersity, 172–173
fluctuates, 186–187
periodic operation, 178
polydispersity, 177
polymerization reactors, 202
polymerization simulations, 181t–185t
quasi-steady-state limit, 202
Dispersive fluxes, 573
Dissolved oxygen (DO), 305f

Dissolved oxygen modulation
See also Flow rate modulation; Substrate modulation
cell density and product concentration, 284f, 285
dissolved oxygen concentrations, 286
near-optimal operation, 286
production and productivity, 285t
Propionibacterium, 283 285
Propionibacterium freudenreichii, 283–285
single fermentor, 285–286
substrate level phosphorylation, 285
time-dependent balances, 286
Distributed systems
Aris Number, 577
catalyst and adsorbent isotropic beds, 573–575
catalyst-adsorbent bed, 573f
Danckwerts boundary conditions, 577–578
dimensionless groups definition, 578t
fluid phase materials, 577
heat and material balances, 575–577
heterogeneous model, 575, 576t
pseudo-homogeneous models equations, 574t, 575
separating reactor, 573
solid phase materials, 577
total void fraction, 575
DO. *See* Dissolved oxygen
DOCFE. *See* Double collocation on finite elements
Dolomite, 668
Double collocation on finite elements (DOCFE), 642
Double conversion (DC), 540
DRIFTS cell. *See* Diffuse reflectance, infrared Fourier transform spectrometer cell
Dusty gas model, 421
Dynamic methods, 369
advantage, 369–370
qualitative applications, 370
absorption spectra variation, 370–371, 371f
4-part cycles, 370
N_2-H_2 modulation, 370
SO_2 oxidation, 370
3-part cycles
2-part cycles
quantitative applications
CO oxidation kinetics, 373–374
DRIFTS cell, 371
DRIFTS-recycle-reactor, 372
fast chopping, 374, 375f
Marwood experimental system, 371
sinusoidal modulation, 374
symmetrical cycle, 372, 373f
time-dependent IR signals, 372, 373f

E

Early-lumping, 560
Electrical power source, 236
Electrochemical process
machining, 264–265
polishing, 264–265
reactions, 236
systems, 244
technique adoption, 683
Electrochemical cells, 243
chloro-alkali membrane cells, 243, 243f
electrolysis, 243–244
three-dimensional electrodes, 244
Electrochemical processes, 235–236, 243
anodization, 263
AC methods, 264
aluminum, 263
DC methods, 264
electrolyte composition, 264
using mixed sulfuric-oxalic acid electrolyte, 264
porous Al_2O_3 films forming, 263f
porous film, 263–264
catalyzed gas and gas-liquid reactions, 235–236
chemicals electrolytic production, 267–269
electrochemical cells, 243
chloro-alkali membrane cells, 243, 243f
electrolysis, 243–244
three-dimensional electrodes, 244
electrochemical machining, 264–265
electrochemical polishing, 264–265
electrochemical systems, 244, 245f
electrochemistry, 236
applying potential effect, 240–241, 240f
cathode and anodic reactions, 238, 239f
cathodes and anodes, 237
charging current density deviation, 243
current efficiency reaction, 237
electrical double layer, 241, 242f
electrochemical cell, 236
electrochemical reactions mechanisms, 239–240
electrode potential reaction, 237–238
electrode-electrolyte interface, 238–239, 240f
electrolysis of water reaction, 237
electron transfer reactions, 242–243
galvanostatic operation, 237
half-cell reactions, 236, 238
non-electrochemical systems, 243
OHP, 241–242
opposite direction reaction, 237
overpotential deviation, 241
during potentiostatic operation, 237
electroforming, 262–263
electrolytic cells, dynamics in, 244–246, 245f
electroplating, 246
of alloys, 256–257
applications, 259–260
single metals, 255
electrorefining, 265–266
electrowinning, 265–266
galvanic cells, 266–267
non-electrochemical reactions, 269–271
Electrochemical technique adoption, 683
Electrochemistry, 236
applying potential effect, 240–241, 240f
cathode and anodic reactions, 238, 239f
cathodes and anodes, 237
charging current density deviation, 243
current efficiency reaction, 237
electrical double layer, 241, 242f
electrochemical cell, 236
electrochemical reactions mechanisms, 239–240

Electrochemistry (*Continued*)
electrode potential reaction, 237–238
electrode-electrolyte interface, 238–239, 240f
electrolysis of water reaction, 237
electron transfer reactions, 242–243
galvanostatic operation, 237
half-cell reactions, 236, 238
non-electrochemical systems, 243
OHP, 241–242
opposite direction reaction, 237
overpotential deviation, 241
during potentiostatic operation, 237
Electroforming, 262–263
Electrolysis, 243–244
Electrolyte composition, 264
Electrolytic cells, 244
dynamics in, 244–246
Electromachining, 264–265
Electronic conductors, 236
Electroplating, 246
of alloys, 256–257
Au-Cu-Cd alloys, 257
Au-Sb alloys, 257
deposit morphology, 259f
iron-group alloys, 258
pulsed plating, 259
silver-tin alloys, 257
using simple unipolar waveform, 257–258
single metal, 258
unipolar deposition, 257
zinc-nickel alloys, 258–259
cathodic reaction, 246–247
DC plating, 249–250
duty cycle definition, 247–249
electrocrystallization, 250–251
high pulsing frequency, 253
macroprofile, 252f
metal adatoms role, 247f
microprofile, 252f
novel electroplating applications, 259–260
Co-Ni-Cu alloy layers, 261
Cu alloy layers, 261
non-porous graphite, 260
using periodic operation, 260–261
potentiostatic or galvanostatic pulse plating, 261
pulsed electrodeposition, 260
using pulsed potentials, 260
unipolar pulse plating waveforms, 262f
zinc-containing CMA, 261–262
particular challenges in, 253–254
PCB, 254
PTH, 254
pulsating and stationary boundary layer, 249f, 250
using pulse plating, 247, 251–252
pulse reverse plating, 252–253
SEM micrographs, 251f
of single metals, 255–256
square-wave pulse waveforms, 247–249, 248f
zone plates, 254–255
Electropolishing, 264–265
Electrorefining, 265–266
Electrowinning, 265–266
Eley-Rideal mechanism, 30–31
Eluent, 573
Emulsifier concentration. *See* Initiator
Emulsion free-radical polymerization, 172
Emulsion polymerization, 188–192
oscillating chemical reactions, 192
PSD, 192
Emulsion styrene polymerization, 192
Endothermic processes, 528
See also Periodic flow reversal
combined endothermic reaction, 534f
flow reversal systems for, 528–529
inert fluid heating system, 528
introduction ways for heat, 528
modified Catofin Houdry process, 528
reverse-flow heat supply, publications on, 528
single bed periodic flow reversal reactor operation, 535f
temperature control problem, 534
temperature profiles for reverse-flow operations, 528–534, 533f
Epoxidation
Ag catalyst, 95
4-part cycles, 92–95
8-part cycles, 92–95
reactions, 92
redox mechanism, 95
Equilibrium measurement, application to, 432–433
Escherichia coli, 291–292
EtAc. *See* Ethyl acetate
EtBz. *See* Ethylbenzene
Ethanol (EtOH), 585
Ethene oligomerization
C_2H_4 hydrogenation, 365–366
selectivity, 365, 366f
Ethyl acetate (EtAc), 585
normalized ethyl acetate production comparison, 323, 323f
production, 322–323, 322f
stationary cycling state, 323
Truffer and Renken model, 324, 324f
Ethyl adipate saponification, 351
continuous stirred tank reactors, 351–352
in-phase modulation vs. diester/saponification, 351, 351f
Ethyl lactate, 627
Ethylbenzene (EtBz), 483
Ethylene hydrogenation, 43, 44t
bifurcation analysis, 45
using composition modulation, 45
cycling experiments, 45
laboratory scale tubular reactor, 43–45
Pt catalyzed hydrogenation, 45
time-average conversion, 43–45
EtOH. *See* Ethanol
Eukaryotic cells, 273–274
Eukaryotic platforms, 682
European Consortium for Cycling Operation–Cordis Project (CYCLOP–Cordis Project), 480
CHISA, 480
fast mode cycling, 481
investigations, 480–481
Evolutionary algorithm, 307–308
Experimental periodic flow-reversal reactor, 562f
Extractive reactor systems, 593

F

FAME. *See* Fatty acid methyl ester
Fast chopping, 374
Fast FTIR spectrometer, 452–453
Fast temperature cycling (FTC), 444–445
device, 450f
system, 444–445
Fatty acid methyl ester (FAME), 613
Fed-batch culture, 303–304

Fed-batch operations
adaptive control, 307, 310–311
animal cell culture, 307
biomass and α-amylase concentration, 310f
bioreactor configuration, 304
DO concentration, 304
dynamic models for, 307, 307t
fed-batch culture, 303–304
using feed-forward feeding strategy, 306–307
using genetic algorithm, 309
iterative dynamic programming, 307–308
key limiting substrate, 304–305
Kluyveromyces fragilis fed-batch culture, 306f
lactose and DO time profile, 305–306, 305f
MAb, 307
maximal productivity in, 308–309, 308f
MCMC methods, 309
using Metropolis-Hastings algorithm, 309
MRAC, 310
MSUR, 305
optimum feeding profile, 311f, 312
performance criterion for, 308, 308t
pollulan, 311
submerged fed-batch culture, 304
Feed step. *See* Pressurization
Feedback control
LQ controller, 560–561
optimal state, 545
PID, 556
reverse flow reactor, 556
Feedback-feed-forward control strategy, 556–558
FFA. *See* Free fatty acid
Fixed bed adsorptive reactor, 584–585
Fixed bed chromatographic reactor, 592, 592f
Floquet multipliers, 287–288
Flow direction modulation, 685
Flow rate modulation, 280
See also Substrate modulation, Dissolved oxygen modulation
bang-bang flow rate cycling, 283
cell culture simplified model, 276t
cell mass in chemostat, 283
CSTR model, 281–282
culture model, 281
using cycling, 280–281
glucose feed concentration, 282, 282f
metabolite production, 281
specific growth equation, 281
yeast culture and culture performance, 283f
Flow reversal reactor, 554
basic, 496, 496f
thermal, 498, 498f, 540–541
Flow-direction switching, packed bed reactor control with
catalytic combustion, 555f, 556
catalytic packing, 553–554
experimental periodic flow-reversal reactor, 562f
using feed-forward controller, 556–558
flow-reversal reactor, 554
using heat exchanger, 554
heat trapping and re-use, 554
inert packing, 553–554
LQ optimal control, 560–561
model for, 557t–558t
MPC, 558
NMPC, 559f
nonlinear reactor model, 559, 559f
performance index, 558–559
periodic flow-reversal reactor combusting, 561
PID controller, 556, 558f
using pseudo-homogeneous dynamic model, 554
repetitive model-predictive controller, 560
stability limitations, 556
stochastic step-disturbances, 560
3-section reactor, 561–562
Fluid desorbent, 573
Fluid phase material, 577
Fluidized bed chromatographic reactor, 591–592
Forestburg coal, 217
Free fatty acid (FFA), 613
Free radical polymerization, 188, 193
concentration modulation, 193
material and heat balances for, 177, 178t
methyl methacrylate, 193
moment equations from, 180t
vinyl acetate and observed bifurcation, 188
Frequency response, 388, 392
Frequency response function (FRF), 390
See also Higher order frequency response function (HFRF)
adsorption column function, 408f
nonlinear system with two inputs and one output, 391, 392f
n^{th} order, 390
relation with output
DC component, 391
Volterra series n^{th} elements, 390–392
weakly nonlinear system, 390f, 391
Frequency response methods, 685
Fresh isomer feed, 618
FRF. *See* Frequency response function
FTC. *See* Fast temperature cycling
Fuels combustion
See also Single component combustion
chemical looping, 130
CuO catalyst, 132–134
CuO re-oxidation, 131, 131f
packed bed, 130–131
perovskite catalyst, 131–132
reactor effluents comparison for anaerobic combustion, 133f
spatiotemporal temperature distribution, 132f

G

GA. *See* Genetic algorithm
Galvanic cells, 266–267
Gas flow modulation (GFM), 464
Gas mixing, 130
Gas phase, 64
Gas shrinkage, 64
Gas-liquid systems
external pressure modulation effect, 425, 426f
fibrous granules, 424
first-order decomposition reaction enhancement, 426–427, 427f
gas pockets, 427
H_2O_2 decomposition mass transport limited rate, 427f
particle breathing induction, 425, 425f
transient mixer model, 425–426
Gas-solid reactions, 656
Gaseous reactants, 416–420
acoustic generators, 424

Gaseous reactants (*Continued*)
compressible flow continuity equation, 420
using compressor, 422
cracking catalyst, 421
external pressure fluctuations, 422–423
fractional conversion, 424f
granular particle reduction, 421–422
modulating pressure, 421
modulation amplitude, 421
nickel oxide reduction, 423, 423f
oxide reduction, 422
porous iron oxide pellet, 421
sinusoidal pressure modulation effect, 420–421, 420f
Gasification, 216–217
of carbonaceous materials, 218t–222t
Gas–solid systems, 206
See also Biochemical systems; Liquid–solid systems
effluent work, 623–624
MeOH readily oxidizes, 622–623
MES and TMC, 620, 621f
mesitylene hydrogenation, 620f
methane oxidative coupling, 622, 622f
non-uniform make-up feed, 622
partial oxidation of methane to methanol, 623f
Sm_2O_3 catalyst, 622
solenoid valves, 620
General Motors (GM), 146
Genetic algorithm (GA), 309
GFM. *See* Gas flow modulation
Glc. *See* Glucose
Glcac. *See* Gluconic acid
Global enhancement, 9
Gluconic acid (Glcac), 336
Glucose (Glc), 336, 631–632
Glucose oxidation
in aqueous solution, 337, 338t
gas-liquid reactions, 336
measured rate changes, 336–337, 337f
rate, 337, 338f
Glycolysis, 310–311
Glycolytic reaction, 280
GM. *See* General Motors
Graham and Lynch model, 63
Gypsum ($CaSO_4.2H_2O$), 209–211

H

H_2:CO mixture, 41
HAc. *See* Acetic acid
Hamilton-Jacobi approach, 275
See also Second variational approach
Hamiltonian, 275–276
Hansen microreactor, 456–457
HB. *See* Hopf bifurcation
HE. *See* Heat exchange
Heat exchange (HE), 540
Heat transfer coefficient, 199–200
Heat-trapping
See also Periodic flow reversal
conversion profiles, 496, 497f
flow reversal reactor
basic, 496, 496f
thermal, 498, 498f
periodic flow reversal, 496–497
energy efficient operation, 497
oscillation effect, 498–500
with temperature profile, 497–498
temperature evolution in combustion chamber, 498–500, 499f
temperature profiles, 496, 497f, 498
evolution, 498, 499f
periodic flow reversal with, 497–498
Hermitian terms, 276
Heterogeneous model, 575
Heterogeneous reaction, 239–240, 380, 401
See also Homogeneous reaction
comparison with homogeneous reactions, 405
using dimensionless variables, 402
material balance for reactant, 401–402
nonlinear adsorption isotherm, 402
changing sign, 403
expression for asymmetrical second-order FRF, 402
functions for heterogeneous CSTR, 403, 404f
low-frequency asymptotic value, 403–405, 404f
periodic operation advantage, 402
for unfavorable isotherms, 403
HFRF. *See* Higher order frequency response function
High ash bituminous coal. *See* Xuzhou
High purity H_2production, 671
multi-step, non-isothermal, combined swing process, 672f
High switching frequency model (HSFM), 505
flow rate of exchanging streams, 506–507
material balance equation, 506
pseudo-homogeneous model, 506
slow heat transfer process, 506–507
temperature profile at high reversal frequencies, 505–506, 506f
Higher order frequency response function (HFRF), 389
See also Nonlinear frequency response (NLFR)
derivation of, 394
dimensionless form of reactor models, 395t
nonlinear systems, 392
for periodic processes, 392
time-average process performance, 393
Hoelder inequality, 400
Homogeneous reaction, 380, 393
See also Heterogeneous reaction
analysis results, 400
behavior under periodic forcing, 401
comparison with heterogeneous reactions, 405
contact time, 400
DC component estimation, 395
diethanolamine formation
asymmetrical cycling, 353
reaction network, 352
symmetrical cycling, 353
dimensionless variables, 393–394, 394t
ethyl adipate saponification, 351
continuous stirred tank reactors, 351–352
in-phase modulation vs. diester/saponification, 351, 351f
expressions for reactor model FRFs, 395
Hoelder inequality, 400
integration constants, 396t–398t
model equations for, 394t
model parameters for calculations, 400t
modulation and steady state comparison, 349

reaction kinetics, 349
reactor performance improvement, 401
reactor types, 393
steady state conditions, 349–350
steady-state concentrations , 394
yield and selectivity, 350, 350f
Hopcalite, 63
Hopf bifurcation (HB), 332
points, 197–198
HSFM. *See* High switching frequency model
Human cell lines. *See* Chinese hamster ovary (CHO)
Hydrocarbons
oxidative dehydrogenation
autothermal dehydrogenation process, 207–208
bang-bang cycling performance, 208f
electrophilic oxygen, 206
isophthalonitrile, 207
mixed dehydrogenation-oxidation catalyst, 209f
novel oxidative dehydrogenation process, 207
propane oxidative dehydrogenation, 207
propene, 206–207
partial oxidation
bang-bang cycling performance, 208f
Bi_2O_3 catalyst, 207
catalyst lattice, 206–207
circulating fluidized beds, 207
commercial process, 207–208
using MgO/V_2O_5 catalyst, 207
mixed dehydrogenation-oxidation catalyst, 209f
toluene partial oxidation, 206
Hydrodynamics
electrical capacitance tomography, 487
NMR imaging, 487
reactor properties, 486
relative permeability concept, 486
shape and descriptive parameter of slow-mode liquid pulses, 486, 487f
temperature and MRI measurements, 487
temperature variations with time during on-off LFM, 487, 488f
two-phase flow concept, 486
Hydrogenation processes, 23
ammonia synthesis, 23–24
comments, 33
interpretation, 30–31, 33
iron catalysts, 24, 28–29
osmium catalyst, 30
ruthenium catalyst, 30
aromatics hydrogenation, 46t
benzene hydrogenation, 45–47
phenylacetylene, 47
using step-change experiments, 47
toluene hydrogenation, 45
ethylene hydrogenation, 43, 44t
bifurcation analysis, 45
using composition modulation, 45
cycling experiments, 45
laboratory scale tubular reactor, 43–45
Pt catalyzed hydrogenation, 45
time-average conversion, 43–45
methanation, 34
adsorbed hydrogen, 34–36
carbon dioxide, 36–39
dynamic model of, 36
nickel faujasite, 36
normalized data points, 39, 39f
time-average rate, 36
methanol synthesis, 39, 40t
commercial methanol catalysts, 43
copper zinc catalyst, 41
copper-zinc-alumina catalysts, 39
NO_x reduction, 33–34, 35t
under modulation, 34
Rh surface, 34
symmetrical cycling, 34
under UHVac conditions, 34
oscillatory behavior, 47
Hysteresis loop, 200

I

IHP. *See* Inner Helmholtz plane
Immobilized cells, applications to, 298
immobilized cell reactor, 300
liquid fluidized bed bioreactor, 302f
multi-step model, 301–303
phosphate nutrient flow, 301
S. cattleya, 298–300
S. *griseus* cells, 301
SBR operation, 301
specific growth rate, 303
two phase dynamic model, 300–301, 300t
unit operations, 303f
Immobilized live cell bioreactor, two phase dynamic model for, 300t
Impulse-response function, 389–390
Inducers, 682–683
Industrial-scale equipment, 14
for chemicals application, 15–16
molecular oxygen, 15
riser-fluidized bed system, 15–16, 16f
solids reactors circulation, 16
two fluidized beds and catalyst transfer, 14–15, 15f
Inert packing, 553–554
Initiator
AIBN, 199
chain terminations, 201
free-radical polymerizations, 173–174
living anionic polymerizations, 174
low initiator concentrations, 199
polymerization
mechanisms and kinetics, 176
simulations, 181t–185t
Inlet ports, 562
Inner Helmholtz plane (IHP), 241–242
See also Outer Helmholtz plane (OHP)
Internal combustion engines
catalytic converter, 143–144
engine behavior, 142–143
exhaust condition, 142, 143f
experimental systems, 146, 146f
using IR spectrophotometry, 145, 145f
noble metals, 143
Otto engine, 142
TWC
function, 144, 144f
structure, 143f
vehicular emissions control, 145
Internal model control operation, 551f
International Congress of Chemical and Process Engineering (CHISA), 480
Interphases, 273–274
Interpretation in ammonia synthesis, 30
reactant inhibition, 31
reactant storage, 31–32

Interpretation in ammonia synthesis (*Continued*)
bulk dissolution model prediction, 32f
predicted response and experimental data comparison, 32f
surface storage model, 32–33
using Temkin Pyzhev model, 32
relaxed steady state
Eley-Rideal mechanism, 30–31
pseudo-homogeneous, 31
relaxed steady-state approach, 31
surface activation/restructuring, 33
IPTG. *See* Isopropyl-1-thio-β-G-galactopyranoside
Iron catalysts, 24
iron oxide catalyst, 28
mass transfer interference, 28–29
near-adiabatic operation, 29
steady-state operation, 29–30
temperature profiles and range of fluctuations, 29f
NH_3 synthesis normalized rates, 24–28, 28f
steady state and time-average cycling rates comparison, 28f
step-change experiments, 24
Iron-group alloys, 258
Isopropyl-1-thio-β-G-galactopyranoside (IPTG), 682–683
Isothermal simulation, 644
See also Non-isothermal simulation
adsorption equilibria, 644
1-butene dehydrogenation, 648–649
catalyst and adsorbent, 648
disproportionation reaction, 648
equilibrium-limited disproportionation, 649
ethene product purity and total recovery, 649f
four-step cycle, 649
methylcyclohexane dehydrogenation, 648
non-adsorbing reactant, 649
parallel reaction, 648
using three-step pressure swing cycle, 648
Isothermal systems
See also Non-isothermal systems
heterogeneous model, 599
non-isothermal operation, 599
pseudo-homogeneous and isothermal mass balance, 599
Isothermality, 628

J

Jensen reactor, 452f

K

K matrix, 549–550
Karhunen-Loeve expansion, 382–383
Kluyveromyces fragilis, 305
Kolbe electrolysis, 267
Korela resin (V07C). *See* Purolite ion exchange resin (PCR)
Krischer-Eiswirth-Ertl model, 59–60

L

Laboratory reactors
microreactors, 14
Tee-junction microreactor, 14, 14f
Lactose, 305–306
Lactosucrose, 613
Langmuir isotherm, 609–610
Langmuir-Hinshelwood mechanism, 384, 438–439
Langmuir-Hinshelwood model, 461, 627
Lean burn spark-ignited engines, 162, 165f
NO/NO_2 behavior comparison, 167, 167f
NO_x reduction, 167–168, 167f
NO_x storage and destruction, 166, 166f
NSR performance, 167
NSR system, 162–165, 166f
ppm levels, 162
sulfur poisoning, 165–166
LFM. *See* Liquid flow modulation
Lifting concept, 552–553
Lightoff temperature, 158
Lime recovery from waste gypsum
calcium sulfide persists, 211
using fuel injection, 211
non-catalytic fluid-solid reactions, 211
phosphogypsum, 209–211
Linear driving force model, 648
Linear homopolymers, 171–173
microstructure, 172
polymerization mechanisms, 173t
Linear polymers, 174
Linear quadratic regulator (LQR), 520–521
Linear-Quadratic optimal control (LQ optimal control), 560–561
Linearization method, 562–563
Liquid flow, 444
Liquid flow modulation (LFM), 464
See also Trickle-bed reactor (TBR)
cycle period and cycle split effect, 482, 482f
DCPD hydrogenation, 481–482
effect on activity or stability, 479–480
enhancement factors for pyrolysis gasoline components, 485t
ethanol oxidation, 481
hydrodynamics under
electrical capacitance tomography, 487
NMR imaging, 487
reactor properties, 486
relative permeability concept, 486
shape and descriptive parameter of slow-mode liquid pulses, 486, 487f
temperature and MRI measurements, 487
temperature variations with time during on-off LFM, 487, 488f
two-phase flow concept, 486
hydrogenation, 485–486
inlet liquid volumetric flow rate, 474–475
liquid outlet temperature, 485f
liquid superficial velocity, 484f
under liquid-reactant limiting condition, 479
phenol oxidation, 481
phenylacetylene hydrogenation reaction, 481
pyrolysis gasoline hydrogenation, 482–483, 485
relative concentration of ethylbenzene, 483, 483f
TBR pressure drop, 485f
transient axial temperature profiles, 483, 484f
transient pressure drop, 484f
Liquid-phase Belousov-Zhabotinsky reaction, 58–59

Liquid—solid systems
See also Gas—solid systems; Biochemical systems
acetal, 627
Amberlyst 15 ion-exchange resin, 626
C-920 chromatographic separation unit, 626
esterification
of acetic acid and β-phenetyl alcohol, 624, 624f
of EtOH, 625
ethyl lactate, 627
experimental stationary concentration profiles, 624f
immobilized enzyme and ion-exchange adsorbent, 626
predicted stationary concentration profiles, 624f
raffinate and extract streams, 625, 625f
rotary valve, 625—626
Living anionic polymerizations, 174
Loop reactor, 197f, 632—633
Looping combustion, 134
Looping system, 212—215
Low-frequency pressure modulation, 430, 434
LQ optimal control. *See* Linear-Quadratic optimal control
LQR. *See* Linear quadratic regulator
Lumped models
cell model, 578—579
chromatographic peak characteristic, 579
pseudo-homogeneous system, 579
separating reactor cascade representation, 579f
Lumped parameter model, 300—301

M

m-xylene (MX), 618
MA. *See* Maleic anhydride
MAb. *See* Monoclonal antibody
Magnesium vanadate catalyst (MgO/V_2O_5 catalyst), 207
Maleic anhydride (MA), 108
bang-bang modulation, 108
Chinese research
aldehydes, 113—114
anhydrides, 113—114
C_4 oxidation kinetics, 116
Ce-Fe complex oxides, 117
cyclic experiments, 116
using domestic VPO catalyst, 113
O_2 effect concentration in C_4, 116f
transient responses, 113—116, 117f
Du Pont research, 108—109
catalyst attrition, 113
polysilicate reinforced catalyst, 113
selectivity to maleic anhydride, 113
European research
novel cyclic system, 118, 118f
time variation simulation, 118—119, 119f
two-zone fluidized bed examination, 117
tellurium-promoted vanadium molybdate, 108
Manipulated variable (MV), 543—544
Markov Chain-Monte Carlo (MCMC), 307—308
Mars-Van Krevelen reaction scheme, 206
Mass transfer acceleration
with compressible fluids, 416
gas-liquid systems
external pressure modulation effect, 425, 426f
fibrous granules, 424
first-order decomposition reaction enhancement, 426—427, 427f
gas pockets, 427
H_2O_2 decomposition mass transport limited rate, 427f
particle breathing induction, 425, 425f
transient mixer model, 425—426
gaseous reactants, 416—420
acoustic generators, 424
compressible flow continuity equation, 420
using compressor, 422
cracking catalyst, 421
external pressure fluctuations, 422—423
fractional conversion, 424f
granular particle reduction, 421—422
modulating pressure, 421
modulation amplitude, 421
nickel oxide reduction, 423, 423f
oxide reduction, 422
porous iron oxide pellet, 421
sinusoidal pressure modulation effect, 420—421, 420f
porous catalyst particles, 416
Mass transfer interference, 28—29
Mass trapping reactors
See also Reverse flow reactor (RFR); Thermal reactors
periodic reverse flow, 534—536
reactive adsorber operation, 535f
reverse-flow adsorptive reactors, publications on, 536t
reverse flow system simulation, 536—537
Mathematical models, 275
cell culture simplified model, 276t
Hermitian term, 276
living organisms, 276
nomenclature, 276
π criterion application, 275—277, 279f
bioprocesses, 277—279
complex models, 277
dilution rate modulation, 278f
growth rate, 277
non-forced input, 277
second variational method, 280
specific growth models, 277t
substrate concentration limitation, 279—280
quasi-steady-state analysis application, 280
structural models, 276
Maximum substrate uptake rate (MSUR), 305
Mazzotti model, 625
MCMC. *See* Markov Chain-Monte Carlo
MEA. *See* Monoethanolamine
Medium tuning, 287
MES. *See* Mesitylene
Mesitylene (MES), 599—605
Methanation, 34
adsorbed hydrogen, 34—36
carbon dioxide, 36—39
dynamic model of, 36
nickel faujasite, 36
normalized data points, 39, 39f
time-average rate, 36
Methane (CH_4), 34—36, 42
$CO/CO_2/H_2/H_2O/CH_4$ gas mixture, 229
CO_2 target sequestration, 228
coke oxidation, 230

Methane (CH_4) (*Continued*)
Fe_2O_3/Al_2O_3, 231f
oxygen carrier, 229–230
Methane combustion with heat recovery, 515–517
See also Volatile organic compound oxidation (VOC oxidation)
axial temperature profile, 519f
catalytic combustion of dilute methane streams, 515t–517t
catalytic RFRs, application of, 518
flow arrangement comparison, 517–518
heterogeneous two-dimensional model, 518–519
RFR for, 517
temperature profile comparison in reactors, 517, 518f
Methane coupling, 618
Methane cracking, 209, 210t
carbon whisker formation, 209
hydrogen production from, 211f
Ni gauze, 209
off-gas composition, 209, 212f
Methane homologation, 361
at high temperature, 361–363, 362f
at low temperature, 363
advantage, 365
Anderson-Schulz-Flory distribution, 364
CH_4 dehydrogenation, 364–365
CH_4 space velocity, 363
Cu-Rh catalysts, 365
helium flush, 363–364
with Pt catalyst, 363
Methane oxidative coupling
bang-bang cycling, 89
C_2H_6, 89
catalyst's primary function, 88–89
CH_4 to MeOH oxidation, 92
experiments, 88–89
Li/MgO catalyst, 89
neural network modeling, 92
with 3-part cycle, 89–92
Methane reforming
CH_4 and CO_2 conversion, 87, 87f
on cracking/gasification cycle, 88, 88f
cycling between CH_4 and CO_2, 87
steam reforming, 86–87
Methanol catalysts
BASF catalyst, 43
$CuO/ZnO/Al_2O_3$ catalyst, 43
cycle split, 43
ICI catalyst, 43
Methanol dehydrogenation, 325–326
Methanol synthesis, 39, 40t
commercial methanol catalysts
BASF catalyst, 43
$CuO/ZnO/Al_2O_3$ catalyst, 43
cycle split, 43
ICI catalyst, 43
copper zinc catalyst, 41
CO_2-rich half-cycle, 42
CuO/ZnO catalyst, 42–43
H_2-CO mixtures, 41
H_2:CO_2 modulation, 43
modulation and steady state rates comparison, 42, 42f
optimum steady-state ratio, 41–42
time-average, 41–42
copper-zinc-alumina catalysts, 39
bang-bang experiment, 41
modulation variables, influence on enhancement, 41t
steady-state synthesis rate, 39
flow direction switching, 39
Methanolation-demethanolation cycle, 216
4-Methylbenzyl bromide, 428–429
Methyl tertiary butyl ether (MTBE), 610–611
Metropolis-Hasting algorithm, 307–308
MgO/V_2O_5 catalyst. *See* Magnesium vanadate catalyst
Michaelis-Menten kinetics, 274
Michaelis-Menten relation, 276
Microchannel foils, 445
Microcreactors, 449
catalysts use in, 451
chip-based, 451–452
cylindrical, 452
ether concentrations response, 454f
designs and performance characteristics, 450t
experimental procedure with, 452–453
experimental studies with, 444–445
FTC system, 444–445
isopropanol concentrations response, 454f
Jensen Reactor, 452f
microstructure device effects, 445
modified microreactor before assembly, 451f
non-homogeneous temperature distribution, 449–450
propene concentrations response, 454f
Sotowa Reactor, 452, 453f
temperature change of carrier, 450–451
Minimal bromate reaction
bistability, 333, 335f
Ce^{4+} concentration bifurcation, 332, 334f
Ce^{4+} formation rate enhancement, 332–333, 335f
Noyes-Field-Thompson mechanism, 332–333
thermokinetic system, 334
Miscellaneous catalysts
Al_2O_3-supported CuO catalyst, 63
hopcalite, 63
multiple resonance, 62–63, 62f
Ni catalyst, 62
O_2 concentration, 61
twofold enhancement, 62–63
Miscellaneous catalytic reactions, 318t–321t
Mixed cell model, 584
Mixed cultures, survival in, 289
biomass, 289–290
dilution rate and cycling frequency operating space, 292f
pH cycling, 291–292
phase plane plots of competing microorganism species, 290f
using simulation and unstructured model, 289
with single limiting substrate, 289t
stable coexistence, 292
stable cycle under in phase modulation, 290, 291f
MMD. *See* Molar mass distribution
Model predictive control (MPC), 520–521, 550
Model Reference Adaptive Controller (MRAC), 310
Model-based control
adaptive strategies, 564
for chemical systems, 550
MPC, 550
parameter adaptation, 551–552
repetitive and predictive, 563
Modulated combustion, 124–127
natural gas flow rate, 127

NO_x emission vs. combustion modulation frequency, 127f
NO_x generation, 127
oxygen flow rate, 127
Modulating automotive exhaust, 158
A/F modulation effect, 159–160
automotive exhausts composition, 158
electricity generation, 159
light-off temperature effect, 160
lightoff temperature, 158
TWC research, 159
used symmetrical modulation, 158
Modulation, 2, 369
of light intensity, 374–375
Modulation of monomer
asymmetrical, 181t–185t
polymerization
experiments on, 189t–191t
simulations, 181t–185t
Ziegler-Natta catalyzed, 201
Molar mass distribution (MMD), 172
Molar masses, 171–172
Molecular weight distribution (MWD), 172, 179–186
Moments of distributions
MMD, 172
molar mass and number chain length, 172–173
MWD, 172
RTD, 174–175
Monoclonal antibody (MAb), 307
Monoethanolamine (MEA), 352
Monomer, 172
MPC. *See* Model predictive control
MRAC. *See* Model Reference Adaptive Controller
MSUR. *See* Maximum substrate uptake rate
MTBE. *See* Methyl tertiary butyl ether
Multi-bed loop reactor, 633–635
Multicomponent mixture separation, 684
Multiple inputs modulation, 366–367
Multiple reactions modulation, 343t–348t
classes, 349
enhancement, 342
instantaneous selectivity, 342
selectivity, 342
Multiple reactor inputs modulation, 687–688
MV. *See* Manipulated variable
MWD. *See* Molecular weight distribution
MX. *See* m-xylene

N

N-reactor loop, 138–139
N_2O. *See* Nitrous oxide
N_2O reduction, 76
Naphtha process, 686
Near-adiabatic operation, 29
steady-state operation, 29–30
temperature profiles and range of fluctuations, 29f
Neural network modeling, 92
Nickel catalyst, 62
Nickel faujasite, 36
Nitric oxide (NO), 70
Nitrogen dioxide (NO_2), 70
Nitrogen oxide reduction (NO_x reduction), 33–34, 35t, 70
comparison, 75f
under modulation, 34
modulation studies, 72t–74t
N_2O reduction, 76
Pt single crystals
cycling frequency function, 75f
damped oscillations, 75–76
Rh surface, 34
over supported catalyst
bang-bang cycling and two-part cycling, 71
with CO, 71
CO adsorbtion, 71
Pt/γ-alumina catalyst, 71
SN, 71
using superimposed cycling frequencies, 75
symmetrical cycling, 34
TWC, 70
under UHVac conditions, 34
Nitrous oxide (N_2O), 70
NLFR. *See* Nonlinear frequency response
NMPC. *See* Nonlinear model-predictive control
NO. *See* Nitric oxide
NO_2. *See* Nitrogen dioxide
Noble metal catalysts, 50
autonomous oscillations, 50
CO feed, 57
composition modulation interaction, 50–57
forcing frequency, 58
using highly asymmetric modulation cycle, 57
Pd catalyst, 57
Pt catalyst, 50–57
quasi-periodic response, 57–58
bang-bang composition modulation, 57f
catalytic oxidation of carbon monoxide, 51t–56t
composition modulation, 50
periodic operation, 57f
Pt catalyst, 50
Noble metals, 129
Non-adiabatic packed bed reactors, 553
Non-catalytic combustion
modulated combustion, 124–127
natural gas flow rate and oxygen flow rate, 127
NO_x emission vs. combustion modulation frequency, 127f
NO_x generation, 127
modulation studies of, 125t–126t
pulsed combustion
commercial gas-fired furnaces and water heaters, 124
pulse combustors, 124
rockets and combustion engines, 124
Non-catalytic decomposition of waste gypsum, 206
Non-catalytic gas-solid reactions
See also Catalytic gas-solid reactions
calcium oxide regeneration, 211–212
chemical heat pumps, 212–215
ammonia condensation and vaporization, 215
cyclic ammoniate decomposition, 215
cyclic ammoniate formation, 215
gas-solid reactions, 215
methanolation-demethanolation cycle, 216
2-propanol to acetone dehydrogenation, 215
reactions, 215–216
reactor block ammoniate decomposition, 217f

Non-catalytic gas-solid reactions (*Continued*)
reactor block ammoniate formation, 217f
salt transformations, 216f
lime recovery from waste gypsum
calcium sulfide persists, 211
using fuel injection, 211
non-catalytic fluid-solid reactions, 211
phosphogypsum, 209–211
Non-condensable gases, 33
Non-electrochemical reactions, 269–271
Non-isothermal simulation
See also Isothermal simulation
adsorption
and desorption, 649
and particle effectiveness factors, 652f
CO_2, 652
comprehensive simulation, 650
endothermic reactions, 649–650
erratic variations, 651–652
five-step PSR system, 651f
H_2, 652
using pseudo-homogeneous one-dimensional model, 650
slope calculation, 650
thermal capacity definition, 650
well-mixed continuous flow reactor, 650
Xiu simulation, 650–651
Non-isothermal systems, 616
See also Isothermal systems
Nonlinear frequency response (NLFR), 389
See also Homogeneous reaction
approximate calculations
indefinite Fourier series approximation, 405
mathematical foundation, 405
periodic quasi-steady state procedure, 407–412
Volterra series, 406
impulse-response function, 389–390
output, 392
with two inputs and one output, 391, 392f
Volterra series, 389–390
weakly nonlinear system, 390f, 391
Nonlinear Kalman filter-based estimator, 562–563
Nonlinear model-predictive control (NMPC), 550
periodically operated catalytic combustor, 559f
Novel control technique, 552
Novel electroplating applications, 259–260
Co-Ni-Cu alloy layers, 261
Cu alloy layers, 261
non-porous graphite, 260
using periodic operation, 260–261
potentiostatic or galvanostatic pulse plating, 261
pulsed electrodeposition, 260
using pulsed potentials, 260
unipolar pulse plating waveforms, 262f
zinc-containing CMA, 261–262
NO_x reduction. *See* Nitrogen oxide reduction
Noyes-Field-Thompson mechanism, 332–333
NSR. *See* Selective periodic catalytic reduction

O

o-xylene (OX), 618
OC. *See* Orthogonal collocation
OCFE. *See* Orthogonal collocation on finite elements
OCM. *See* Oxidative coupling of methane
ODE. *See* Ordinary differential equation
OHP. *See* Outer Helmholtz plane
Open-loop technique, 546
Operated fixed bed systems, 657
Optimal control problem formulation, 545
constraints, 546
cost function calculation, 545
original nonlinear model, 545–546
position dependent vector, 545
predictive controllers, 545
time dependent vector, 545
Optimal linear control strategies, 549
Optimal pulsing parameters, 255
Optimization
actual optimization task, 615–616
eluent requirement, 614
general design model, 615–616
MTBE production optimization, 614–615, 615f
Pareto-optimal solutions, 614
SCMCR iterative operation, 614
solvent-to-reactant ratio, 614
Ordinary differential equation (ODE), 548, 642
Original square-wave modulation, 549–550
Orthogonal collocation (OC), 642
Orthogonal collocation on finite elements (OCFE), 642
OTB model. *See* Ozaki-Taylor-Boudart model
Otto engine, 142
catalytic converter, 143–144
engine behavior, 142–143
exhaust condition, 142, 143f
experimental systems, 146, 146f
using IR spectrophotometry, 145, 145f
noble metals, 143
TWC
function, 144, 144f
structure, 143f
vehicular emissions control, 145
Outer Helmholtz plane (OHP), 241–242
See also Inner Helmholtz plane (IHP)
OX. *See* o-xylene
Oxidative coupling, 670
Oxidative coupling of methane (OCM), 618
Oxidative dehydrogenation, 206
autothermal dehydrogenation process, 207–208
bang-bang cycling performance, 208f
electrophilic oxygen, 206
isophthalonitrile, 207
mixed dehydrogenation-oxidation catalyst, 209f
novel oxidative dehydrogenation process, 207
propane oxidative dehydrogenation, 207
propene, 206–207
Ozaki-Taylor-Boudart model (OTB model), 30–31

P

p-xylene (PX), 618
Packed bed, 130–131
Packed bed reactors, 69
controls for, 553

distributed parameter systems, 553
with flow-direction switching, 553–562
lumped parameter system, 553
model for, 557t–558t
moving temperature and concentration fronts, 553
non-adiabatic and adiabatic, 553
Parallel hybrid algorithm, 17, 686
Pareto-optimal solutions, 614
Partial differential equation (PDE), 476, 553
Partial oxidation, 206
additives effects, 80
bang-bang cycling performance, 208f
Bi_2O_3 catalyst, 207
catalyst lattice, 206–207
circulating fluidized beds, 207
commercial process, 207–208
methane partial oxidation and reforming experiments, 80
methane reforming
CH_4 and CO_2 conversion, 87, 87f
on cracking/gasification cycle, 88, 88f
cycling between CH_4 and CO_2, 87
steam reforming, 86–87
using MgO/V_2O_5 catalyst, 207
mixed dehydrogenation-oxidation catalyst, 209f
periodic composition modulation for, 80, 80f
perovskite oxide catalysts, 85
CO selectivity, 86, 86f
CH_4 conversion, 86, 86f
reaction temperature effect, 85–86, 86f
TPSR/MS measurements, 85
syngas production, 80
toluene partial oxidation, 206
Particle size distribution (PSD), 192
Particle swarm optimization (PSO), 309
Pasteur Effect, 304–305
PCB. *See* Printed circuit boards
PCR. *See* Purolite ion exchange resin
PDE. *See* Partial differential equation
PEM fuel cell. *See* Proton exchange membrane fuel cell
Performance index (PI), 296
Perfusion operation, 287
Periodic flow reversal, 495–496, 572
biofilters, 537
for off-gas treatment, 537
reverse-flow biofilters, publications on use of, 537t
commercial applications, 538
catalytic decomposition of N_2O, 541–542
reverse flow reactor technology for NO_x removal, 540
RFR control, 539
thermal reversed flow reactor, 540–541
VOC mitigation, 538–539
comprehensive empirical investigation, 561
dynamic behavior
aperiodic operation, 509
asymmetric operations, 509
complex for wall-cooled reactor, 509, 509f
dependence of maximum temperature, 508f
dependence on linear velocity, 507–508
high temperature operations, 508
hysteresis behavior, 507–508
presence of three steady states, 507–508
steady-state solutions, 508–509
temperature and concentration profile evolution, 510f
temperature and conversion profiles, 497f, 507
experimental, 562f
heat-integrated catalytic reverse flow reactors
catalytic process modeling effect, 505
complications, 504–505
heat and mass transfer parameters, 504
heat balance, 500–503
modeling, 500
simplifications, 504
theory and mathematical simulation contributions, 500t–503t
two-dimensional radial-symmetric model, 505
history of, 495–496
HSFM, 505
flow rate of exchanging streams, 506–507
material balance equation, 506
pseudo-homogeneous model, 506
slow heat transfer process, 506–507
temperature profile at high reversal frequencies, 505–506, 506f
miscellaneous applications, 537–538
reverse flow, publications on use of, 538t
numerical techniques, 509–510
periodically operated reactors, 543
reverse flow reactor design parameter estimation, 510–511
simulation, 509–510
stationary heat front movement, 507
sulfuric acid plants, 539
DC/DA system, 540f
RFR at Plovdiv lead-zinc smelter, 540
triple bed converter operation, 539
UNILAB design, 539
Periodic operation (PO), 1–2, 2f, 274
adaptive strategies, 564
adsorption, 2
applications to systems with multiple objectives, 686
carbon deposition control, 17–18, 686–687
in methane cracking, 687
during propane steam reforming, 687
catalyst regeneration, 2–3
Cellular-Automata based model, 547–548
chattering control, 544
cycle structure
asymmetric cycle, 6
long cycle periods, 7
short cycle periods, 7
SMBCR, 8
stop-flow and steady-state operation comparison, 8, 8f
symmetric modulation, 6
devising optimization techniques, 686
electrochemical technique adoption, 683
enhancing selectivity, 680–681
equipment
industrial-scale equipment, 14–16
laboratory reactor, 13–14, 13f

Periodic operation (PO) (*Continued*)
identifying attractive reactions, 680–681
industrial applications, 21–22, 688
inherently, 9–10
measuring improvement
global enhancement, 9
periodic and steady-state operation comparison, 8–9, 8f
relative improvement measure, 9
modulation, 2
with multiple objectives, 17
multiple reactor input modulation, 18, 687–688
recombinant and non-recombinant microorganisms, 687–688, 18
series-parallel reaction, 18, 688
naphtha process, 686
non-electro chemical reaction adoption, 683
novel control technique, 552
objectives, 10, 10t
open loop operation, 553
optimization of, 688
origins, 3
parallel hybrid algorithm, 17, 686
periodic process optimization, 688
periodically operated process engineering, 688–689
photocatalysis improvement, 688
photocatalytic process, 18, 688
reaction systems, 16–17
series-parallel reaction, 688
strategy, 10
amplitude, 12–13
multipart cycle, 12, 12f
three reactant cycles, 12
two reactant cycles, 11, 11f
variables, 10
UV-LED, 688
variables, 4, 4f
catalysts, 6
cycle period, 5
flow reversal, 6
in reactor input modulation, 5
total pressure modulation, 5–6
Periodic operation modeling, 487–488
catalyst pellet behavior, 488–489
dynamic model, 488
dynamic non-isothermal TBR model, 489–490
external efficiency enhancement, 489, 490f
LFM impact on transient concentration profiles, 493
mathematical model of catalytic hydrogenation, 490–491
modeling slow flow modulation, 488
on-off LFM raises conversion, 489
pellet-to-pellet variation, 489
predicted and experimental temperatures, 492, 492f
predicted and performance of reactor, 492f
predicted mean conversion, 490–491, 491f
simplified particle scale model, 489f
styrene hydrogenation in toluene solution, 491
Periodic pressure reduction
catalytic deamination, 431
deamination of propylamine over 13X zeolite, 430f
low-frequency pressure modulation, 430
vacant adjacent surface sites, 431
Periodic processes
classes, 388
inherent, 388
optimization, 688
time-average process performance, 393
favorable periodic reactor operation, 393f
heterogeneous reaction, 401–405
using HFRFs, 393
homogeneous reaction, 393–401
Periodic quasi-steady state
adsorption column functions, 408f, 409f
approximate periodic outlet concentration, 408
cyclic operation of adsorption column, 407
equilibrium-dispersive model, 407
FRF
calculation, 407–408
series, 407
inlet concentration modulation, 408–411
oscillatory behavior, 408
outlet concentrations, 408–412, 410f, 411f
procedure for approximate calculation of, 407
Periodic reaction switching
in chemical looping, 233
circulating adsorbent bed process, 232f
using data and fluidization models, 233
fluidizing gas, 233
temporal and locational modulation, 232f
Periodic temperature jump, 444
Periodically operated process engineering, 688–689
Periodically operated reactor, 543, 544t
chattering controls
relaxed steady state, 546
steady-state operating point stability, 548
steady-state value forces, 547
temperature-residence time plane, 546–547, 546f
two applications, 547–548
control schemes
adaptive strategies, 564
control strategy components, 566f
cyclic linear system controller, 565, 565f
using differential equations, 564–565
gas cooling, 565
limit cycle process, 564f
scheme parameter adaptation, 565
transfer function calculation, 565
model-based predictive controllers, 565–567
open loop control strategy, 544
optimal control problem formulation, 545
constraints, 546
cost function calculation, 545
original nonlinear model, 545–546
predictive controllers, 545
position dependent vector, 545
time dependent vector, 10019
packed bed reactor, 544
packed bed reactor control
adiabatic, 553
distributed parameter systems, 553

with flow-direction switching, 553–562
lumped parameter system, 553
model for, 557t–558t
moving temperature and concentration fronts, 553
non-adiabatic, 553
periodic reactor operation, 543–544
SMBCR control, 562
linear adsorption isotherms, 563–564
linearization method, 562–563
reactant and solvent, 562
repetitive and predictive control, 563–564
stirred slurry and fluidized bed reactors, controls for
adaptive control element, 551–552, 552f
closed loop control system, 551f
conventional time-invariant proportional control, 549–550, 550f
cycle periods, 551
internal model control operation, 551f
linear time-varying model, 552–553
MPC, 550
NARX polynomial model, 551
novel control technique, 552
ODE, 548
optimal PI controller for, 549, 549f
PI controllers, 548–549
simulated closed loop responses, 549
standard model-predictive control, 550–551
time-varying tuning parameters, 548–549
Perovskite catalyst, 131–132
Perovskite oxide catalysts, 85
CO selectivity, 86, 86f
CH_4 conversion, 86, 86f
reaction temperature effect, 85–86, 86f
TPSR/MS measurements, 85
Perturbed system, 388
PET. *See* Poly(ethylene terephtalate)
PFR. *See* Plug flow reactor
Phenomenological equation, 573–575
Phenylacetylene, 47
Phosphogypsum, 209–211
Photocatalysis improvement, 688
Photocatalytic processes, 688
Photoemission electron microscopy, 60
PI. *See* Performance index
Pilot scale reactor, 670
Plasmid-bearing cells, 293–294
Plasmid-containing cells, 293
Plasmid-free cells, 293
Plasmids, 292–293
Plated-through-holes (PTH), 254
Plug flow model, 31
Plug flow reactor (PFR), 393
See also Dispersed flow tubular reactor (DFTR)
analysis results, 400
conclusions, 395–399
dimensionless form of, 395t
expressions for FRFs, 395
integration constants, 396t–398t
model equations for homogeneous reaction, 394t
second order functions, 399f, 401
PO. *See* Periodic operation
Pollulan, 311
Poly(ethylene terephtalate) (PET), 172
Poly-β-hydroxybutyrate, 285–286
Polymerization, 171
assessment
narrow polymer MWD, 201
parameter space, 202–203
polymerization reactor periodic operation, 202
reactor jacket temperature, 201–202
Ziegler-Natta catalyzed polymerization, 201
atactic polymers, 172
bulk polymerization, 172
chain polymerization, 172
conventional free-radical polymerizations, 173–174
experiments on polymerization under input modulation, 188
anionic polymerization, 192–193
deposition polymerization, 193
emulsion polymerization, 188–192
free radical polymerization, 193
Ziegler-Natta catalysts, 193
feed-back control loops, 176f
ideal polymerization mechanisms and reactor types, 175t
living anionic polymerizations, 174
molar mass and number chain length distributions, 172–173
number- and weight-chain-length distributions, 175f
periodic operation, 177
polymerization mechanisms, 173, 173f, 173t
reactors, 176–177
polymers, 171–172
polymers saturation
Bussard-Dooley system, 201
experimental screw extruder for, 200–201, 200f
post-polymerization processing, 200
RTD, 174–175
simulation under input modulation, 177
coordination polymerization, 179–186
free radical polymerization, 188
spontaneous oscillations, 194
bifurcation diagram, 199
CSTR, 199
dynamic flow sheeting package, 194–197
heat transfer coefficient, 199–200
Hopf bifurcations points, 198–199
hysteresis loop, 200
and multiplicity analysis of, 195t–196t
parameter space in, 197–198, 198f, 199f
polymerization reactors, 194
schematic of loop reactor, 197, 197f
step-polymerization, 174
stages, 176
synthetic polymers, 172
Polymerization experiments under input modulation, 188
anionic polymerization, 192–193
deposition polymerization
fluorocarbon coatings, 193–194
plasmas, 193
square-wave modulation, 194
emulsion polymerization, 188–192
oscillating chemical reactions, 192
PSD, 192
styrene polymerization, 192

Polymerization experiments under input modulation (*Continued*)
experiments on polymerization under periodic operation, 189t–191t
free radical polymerization, 193
concentration modulation, 193
methyl methacrylate, 193
Ziegler-Natta catalysts, 193
Polymerization simulation under input modulation, 177
chain length and monomer conversion, 177–178
coordination polymerization
conversion fluctuations, 186–187
using multi-site Ziegler-Natta catalyst, 187
MWD, 179–186
nonlinear model operation simulation, 188f
phase-shifted square-wave modulation, 187f
propene polymerization simulation, 186
UNIPOL process, 186
Ziegler-Natta ethylene polymerization, 186
emulsion polymerization, 177
free radical polymerization, 188
material and heat balances for, 177, 178t
moment equations from, 180t
vinyl acetate and observed bifurcation, 188
polymerization reactors, 178–179
under periodic operation, 181t–185t
transformations converting polymer mass balances, 179t
z-transform equation, 177
Polymers, 171–172
Polymers saturation
Bussard-Dooley system, 201
experimental screw extruder for, 200–201, 200f
post-polymerization processing, 200
Polymethylmethacrylate, 188
Polystyrene, 200–201
Pontryagin's Maximum Principle, 436–437
Post-polymerization processing, 200
Predictive control
algorithm, 550–551
MPC, 550
nonlinear, 548
periodically operated reactors, 564
Pressure and temperature swing reactors
adsorbate concentration on hydrotalcite, 672f
high purity H_2 production, 671
mixed catalyst and adsorbent bed, 673f
product gas composition, 675f
regeneration of adsorbent, 671
simulation studies, 671–674
Pressure modulation, 415
assessment and research opportunities, 433–434
for biochemical reaction within bioflocs, 683
compression and reaction combination, 431–432
equilibrium measurement, application to, 432–433
heterogeneously catalyzed reactions, 415–416
literature, 417t–419t
mass transfer acceleration, 416
gas-liquid systems, 424
gaseous reactants, 416–420
periodic pressure reduction
catalytic deamination, 431
deamination of propylamine over 13X zeolite, 430f
low-frequency pressure modulation, 430
vacant adjacent surface sites, 431
rate measurement, application to, 432–433
sonocatalysis
aryl aldehyde irradiation, 428
bubble collapse consequences, 428f
collapsing bubble cavities, 429f
disproportionation, 428–429
frequencies, 427–428
homogeneous reaction systems, 428
liquids ultrasonic irradiation, 428f
soybean oil hydrogenation, 430
ultrasonic excitation, 427
at ultrasonic frequencies, 433–434
Pressure oscillation, 415
Pressure swing reactor (PSR), 562, 638
See also Temperature swing reactor (TSR)
application, 676
backfill cycle, 640
basic operational cycles, 638–639, 639f
boundary and initial conditions, 643t–644t
computational considerations
ODE, 642
periodic operation, 642
shooting methods, 642
connections and bed arrangements, 638f
delivery step, 639–640
equilibrium limited reactions, 674
experimental studies, 652, 661t
catalyst deactivation, 655
CO_2 production, 652, 655f
CrO_2/Al_2O_3 catalyst and KY zeolite adsorbent, 655
gas-solid reactions, 656
hydrogen pilot plant, 656–657, 656f
Ni catalyst and hydrotalcite adsorbent, 655
using repressurization, 657
six-step PSR cycle, 652–655
steady-state kinetic model, 655–656
experimental work, 675–676
five-step PSR system, 651f
inlet and outlet, 639
low-aspect-ratio beds, 640
pressure and TSR combination
adsorbate concentration on hydrotalcite, 672f
high purity H_2 production, 671
mixed catalyst and adsorbent bed, 673f
product gas composition, 675f
regeneration of adsorbent, 671
simulation studies, 671–674
pressure swing system simulations, 645t–647t
isothermal simulation, 644
non-isothermal simulations, 649
purge cycles, 640
reactor models
boundary and initial conditions, 642
cell models, 641
Damköhler number, 641
material, momentum and energy balances, 640

partial pressures, 640
PDE models, 641–642
phenomenological models, 642
pseudo-homogeneous, one-dimensional model equations, 641t
specific dimensionless variables, parameters and groups definition, 642t
scale-up of experimental units, 674–675
two design alternatives, 638
Pressure swing system simulations, 645t–647t
isothermal simulation, 644
adsorption equilibria, 644
1-butene dehydrogenation, 648–649
catalyst and adsorbent, 648
disproportionation reaction, 648
equilibrium-limited disproportionation, 649
ethene product purity and total recovery, 649f
four-step cycle, 649
methylcyclohexane dehydrogenation, 648
non-adsorbing reactant, 649
parallel reaction, 648
using three-step pressure swing cycle, 648
non-isothermal simulations
adsorption and desorption, 649
adsorption and particle effectiveness factors, 652f
comprehensive simulation, 650
endothermic reactions, 649–650
erratic variations, 651–652
five-step PSR system, 651f
H_2 and CO_2, 652
using pseudo-homogeneous one-dimensional model, 650
slope calculation, 650
thermal capacity definition, 650
well-mixed continuous flow reactor, 650
Xiu simulation, 650–651
Pressure swing systems, 573
Pressurization, 639–640
Printed circuit boards (PCB), 254
Process control, 544
Product-inhibited biochemical reactions, 681–682
Productivity, 614, 300
in different operating strategies, 285t
performance criterion for fed-batch operation, 308t
in styrene hydrogenation, 465t–473t
Prokaryotic platforms, 682
Propane combustion, 128
Propane oxidation, 128
Propanol dehydrogenation
Langmuir-Hinshelwood kinetics, 335–336
spray-pulse operation, 334–335, 336f
Propene and butene partial oxidation, 96–101
acrolein, 95–96
acrylic acid, 101
ammoxidation, 96
bismuth molybdate catalyst, 96
C_2H_4 partial oxidation, 96
CFB reactors, 102
Fe-Sb catalyst, 96
selectivity ratio, 101
Propionibacterium species, 283–285
Propionibacterium freudenreichii, 283–285
Propylene combustion, 127–128
Proton exchange membrane fuel cell (PEM fuel cell), 266
PSD. *See* Particle size distribution
Pseudo-homogeneous dynamic model, 554
Pseudo-homogeneous model, 31, 504, 575
PSO. *See* Particle swarm optimization
PSR. *See* Pressure swing reactor
Pt single crystals
cycling frequency function, 75f
damped oscillations, 75–76
Pt/SnO_2 catalyst, 57–58
PTH. *See* Plated-through-holes
Pulse chromatographic reactor, 580
cationic resin catalyst, 585–586
cyclohexane dehydrogenation in, 583–584
dilution ratio calculation, 584
fixed bed adsorptive reactor, 584–585
heterogeneous catalytic reactions, 580–583
injection, regeneration and cycle time illustration, 584f
ion exchange resins, 586
literature, 581t–582t
using matrix-based adsorption modeling, 586–587
mixed cell model, 584
optimal repetitive pulsing, 583f
reactants, 583
reaction experiment, 585, 585f
regeneration experiment, 586f
triacetine, 586
Pulse combustors, 124
Pulse mode, 8
Pulse reverse plating, 252
Pulsed combustion
gas-fired furnaces, 124
pulse combustors, 124
rockets and combustion engines, 124
water heaters, 124
Purolite ion exchange resin (PCR), 630
PX. *See* p-xylene

Q

Quasi-steady state (qss), 7, 504
analysis application, 280
operation, 436

R

R&D challenges, 680
See also Periodic operation (PO)
adoption of electrochemical techniques, 683
attractive reactions identification, 680–681
biochemical applications, 681
pressure modulation, 683
product-inhibited biochemical reactions, 681–682
temperature modulation, 682
circulating solids *vs.* fixed beds, 685
control catalyst activity, using modulation to, 684–685
control system development for periodic processes, 684
flow direction modulation, 685
frequency response methods, 685
multicomponent mixture separation, 684
new cycles, research on, 681
reaction-separation applications
alternate separating techniques, identification, 681

R&D challenges (*Continued*)
SMBCR
SCF-SMB, 683–684
solvents, 683
step-change experiments, 685
swing reactors, 686
Rate measurement, application to, 432–433
RCO. *See* Regenerative catalytic oxidizers
Reactant, 205, 562
Reactant inhibition, 31
Reactant storage, 31–32
bulk dissolution model prediction, 32f
predicted response and experimental data comparison, 32f
surface storage model, 32–33
using Temkin Pyzhev model, 32
Reaction identification, periodic operation, 680–681
Reaction-separation systems, 681
combined, 3
Reactive chromatography, 584–585
Recombinant cell culture stabilization
antibiotic resistance, 295–296
antibiotics, 296
β-lactamase enzyme, 294
dilution rate modulation, 293
flow rate cycling, 295
and host cell culture in chemostat, 293t
human protein pharmaceuticals, 292–293
in-phase modulation, 297
optimal phase shifts diagram, 299f
optimal SS, 297
PI, 296
using π criterion, 297
plasmid containing cells, 293
cell concentration, 296f
plasmid-bearing cell variation fraction, 294, 295f
plasmid-free cells, 293
Q/V square wave modulation, 294
regions of PI improvement, 297–298, 298f
using specific growth rate model, 293–294
using stability theory, 296
Redox reaction, 380
Regenerative catalytic oxidizers (RCO), 538–539
Regenerative thermal oxidizers (RTO), 538–539
Relaxed steady state (rss), 7, 546
approach, 31
Eley-Rideal mechanism, 30–31
operation, 436–437
pseudo-homogeneous, 31
relaxed steady-state approach, 31
Repetitive model-predictive controller, 560
Residence time distribution (RTD), 174–175
Residual reactant, 573
Resin, 606
Resonance peaks, 51t–56t
Reverse flow reactor (RFR)
See also Methane combustion with heat recovery; Thermal reactors
applications, 521–522
ammonia synthesis, 525t
to automotive emissions, 522t
converter-prototype design, 521–522
methanol synthesis, 525t
to oxidation of sulfur dioxide, 522–523
selective catalytic reduction application, 522, 523t
control, 520t
control strategies, 520–521
function of control system, 519–520
o.d. reactor, 520–521
open-loop MPC strategy, 521
repetitive MPC, 521
thermocouple-based control option, 519
LQ-feedback control problem for, 560–561
for methane combustion, 517
temperature profile comparison in reactors, 517, 518f
VOC oxidation, 511
Reverse-flow chromatographic reactors, 572
Rival model testing, modulation application to, 375–376
bulk dissolution model, 376–377
CO conversion comparison, 381–382, 382f
dehydrogenation rates, 379, 380f
discrimination procedure, 383
experimental observations comparison, 378f
gas phase and surface concentration comparison, 380, 381f
heterogeneous reaction, 380
homogeneous reaction, 380
Karhunen-Loeve expansion, 382–383
Langmuir-Hinshelwood model, 379, 384
oxidative dehydrogenation kinetics, 384–385
Pd/SiO_2 catalyzed acetoxidation, 377–378
redox reaction, 380, 384
square-wave cycling, 384
stop effect, 378–379
storage class model, 376, 377f
surface storage model, 376–377
two-site models, 375–376, 376f
rss. *See* Relaxed steady state
RTD. *See* Residence time distribution
RTO. *See* Regenerative thermal oxidizers
Runge-Kutta algorithm, 619
Russian synthesis catalyst, 28
Ruthenium catalyst, 30
bang-bang switching, 30
modulation, 30

S

Saccharomyces cerevisiae, 291–292
SAE. *See* Society of Automotive Engineers
SBR. *See* Sequencing batch reactor
SCF-SMB. *See* Supercritical simulated moving bed chromatography
SCMCR. *See* Simulated countercurrent moving bed chromatographic reactor
Screen oxygen carrying solids, 130–131
Second variational approach
See also Hamilton-Jacobi approach
Segregated catalyst-adsorbent systems, 658
Selective periodic catalytic reduction (NSR), 162
application, 168–169
concentric arrangement, 169
diesel exhaust detoxification, 169

catalysts, 162–165
conversion to NO_2, 166f
performance improvement, 167
Semi-crystalline polymers, 172
Separate catalyst
adsorbent beds, 616–619, 670
CMCR model, 617
fourth-order Runge-Kutta algorithm, 619
fructose, 617
methane oxidative coupling, 617f, 618–619
MX, 618
OCM, 618
OX, 618
PX, 618
reaction rate and adsorption isotherm parameters, 619
SCMCR cascade with, 616–617, 616f
SCMCR system modifications, 618
three-section SCMCR operation with separate beds, 617, 617f
in rotating reactor, 658
sulfur poisoning prevention, 167–168
Separating reactors, 569
catalyst and adsorbent isotropic beds, 573–575
extractive reactors, 593
Sequencing batch reactor (SBR), 274, 301
unit operations during, 303f
Series-parallel reaction, 688
in periodic operation, 18, 688
Shallow catalyst, 47
Shenhua bituminous coal, 226
Silver-tin alloys, 257
Simulated countercurrent moving bed chromatographic reactor (SCMCR), 572, 597
applications
exothermic solid catalyzed reactions, 632–633
to gas–solid systems, 600t
to liquid–solid systems, 601t–604t
loop reactors, 632–633
multi-bed loop reactor, 633–635
three-bed simulated moving bed reactor, 633f
biochemical systems
continuous moving-bed model, 629f
dextran biosynthesis, 630–631
enzyme lactase, 631
glucose and fructose, 628
glucose isomerization, 629f
high fructose syrups, 627–628
isothermality, 628
lactosucrose and glucose, 632
lactosucrose production, 631–632
maltose, 631
Mg^{2+} form, 629
PCR, 630
rotating adsorbent columns, 628f
stationary reactor columns, 628f
sucrose feed, 630f
sugars enzymatic conversion, 629–630
12-column SCMCR, 632f
CMCR, 598–599
disturbances, 563
fluid flow rate, 599
functions, 599
gas–solid systems
helium eluent, 623–624
MeOH readily oxidizes, 622–623
MES and TMC, 620, 621f
mesitylene hydrogenation, 620f
methane oxidative coupling, 622, 622f
non-uniform make-up feed, 622
partial oxidation of methane to methanol, 623f
Sm_2O_3 catalyst, 622
solenoid valves, 620
inlet ports, 562
linearization method, 562–563
liquid–solid systems
acetal, 627
acetic acid esterification, 624, 624f
Amberlyst 15 ion-exchange resin, 626
β-phenetyl alcohol esterification, 624, 624f
C-920 chromatographic separation unit, 626
ethyl lactate, 627
EtOH esterification, 625
immobilized enzyme and ion-exchange adsorbent, 626
predicted and experimental stationary concentration profiles, 624f
raffinate and extract streams, 625, 625f
rotary valve, 625–626
modeling and simulation
isothermal systems, 599
non-isothermal systems, 616
optimization, 614
SCMCRs modeling, 599–605
separate catalyst and adsorbent beds, 616–617
numerical simulation
BPA production, 609, 609f
FAME and FFA, 613
HAc and β-phenetyl, 610
Langmuir isotherm, 609–610
MTBE, 610–611
pseudo-homogeneous model, 609
solvent-to-feed ratios, 611–613
stationary concentration profiles, 610, 610f
sugar system multi-reaction simulation, 613
switching time on concentration profiles, 612f
two conventional processes, 613–614
principle, 598f
repetitive and predictive control, 563
single bed reactor operation, 598
Simulated loop reactors, 134
applications, 139–140
bifurcation analyses, 139
inert packing, 137
loop combustion systems, 135t–136t
N-reactor loop, 138–139
slow cycling, 137–138
stable rotating temperature waves, 139
stable spatiotemporal patterns, 138f
three bed, 137f
two-bed loop, 134–137
VOC, 137
Simulated moving beds
exothermic solid catalyzed reactions, 632–633
loop reactors, 632–633
multi-bed loop reactor, 633–635
three-bed simulated moving bed reactor, 633f
Simulated moving-bed chromatograph reactor (SMBCR), 8, 562, 684

Single component combustion
  Carlsson's interpretation, 129
  CH_4 combustion, 129–130
  platinum gauze fluctuates, 128
  propane
    oxidation, 128
    combustion, 128
  propylene
    combustion, 127–128
    conversion, 128f
  transient concentrations, 129, 129f
Single crystal catalysts, 58
  catalytic oxidation, 59
  fast Fourier transforms, 60f
  Krischer-Eiswirth-Ertl model, 59–60
  O_2 partial pressure modulation, 58, 59f
  photoemission electron microscopy, 60
  reactant concentration, 60–61
  standing waves, 61
  sub-harmonic behavior, 58
  turbulent surface oscillations, 58–59
Single fermentor, 285–286
Single input-single output system (SISO system), 543–544
Single metal electroplating
  carbon content in gold coatings, 255f
  chromium coatings, 256
  DC plating, 255
  gold, 255–256
  nickel, 256
  zinc electroplating, 255–256
SISO system. *See* Single input-single output system
Slime, 265–266
Slow cycling, 137–138
SLPC. *See* Supported liquid phase catalyst
SMBCR. *See* Simulated moving-bed chromatograph reactor
SN. *See* Stoichiometric number
SO_2 oxidation. *See* Sulfur dioxide oxidation
SO_3 reduction. *See* Sulfuric acid reduction
Society of Automotive Engineers (SAE), 141–142
Solenoid valves, 620
Solid phase material, 577
Solids catalyzed reactions, 353
  acetylene hydrogenation, 355
    Girdler catalyst, 355
    phenylacetylene, 357
    simulation, 355–357, 356f
    wrong-way behavior, 357
  of ammonia and ethylene oxide, 353
  butadiene hydrogenation, 353–354, 354f
  reactor performance variation, 354–355, 354f
  total oxidation
    cyclohexene conversion time-average, 358, 358f
    feed system, 357
    hydrocarbons adsorption, 359
    O_2 modulation, 358–359
Solvent, 562
Sonocatalysis
  aryl aldehyde irradiation, 428
  bubble collapse consequences, 428f
  collapsing bubble cavities, 429f
  disproportionation, 428–429
  frequencies, 427–428
  homogeneous reaction systems, 428
  liquids ultrasonic irradiation, 428f
  soybean oil hydrogenation, 430
  ultrasonic excitation, 427
Sonochemistry, 427
Sotowa Reactor, 452, 453f
Spontaneous oscillations, 194
  bifurcation diagram, 199
  CSTR, 199
  dynamic flow sheeting package, 194–197
  heat transfer coefficient, 199–200
  Hopf bifurcations points, 198–199
  hysteresis loop, 200
  loop reactor, schematic of, 197, 197f
  and multiplicity analysis of, 195t–196t
  parameter space in, 197–198, 198f, 199f
  polymerization reactors, 194
  loop reactor, schematic of, 197, 197f
SS. *See* Concave curve; Steady-state
Stable system, 388–390
Stainless steel micro-channel reactor, 445
Standing waves, 61
Starch acid hydrolysis, 681–682
Starve-regenerate periodic operation, 298–300
Steady-state kinetic model, 655–656
Step-change experiments, 685
Step-growth reactions, 174
Step-polymerization, 172
Stirred slurry and fluidized bed reactor controls
  adaptive control element, 551–552, 552f
  closed loop control system, 551f
  cycle periods, 551
  internal model control operation, 551f
  linear time-varying model, 552–553
  MPC, 550
  NARX polynomial model, 551
  novel control technique, 552
  ODE, 548
  optimal PI controller for, 549, 549f
  PI controllers, 548–549
  simulated closed loop responses, 549
  standard model-predictive control, 550–551
  time-invariant proportional control, 549–550, 550f
  time-varying tuning parameters, 548–549
Stoichiometric feeding, 303–304
Stoichiometric number (SN), 61, 71, 155
Storage class model, 376, 377f
Streptomyces cattleya, 298–300
Streptomyces griseus cells, 300–301
Submerged fed-batch culture, 304
Substrate modulation, 280
  See also Dissolved oxygen modulation; Flow rate modulation
  cell culture simplified model, 276t
  CSTR model, 281–282
  culture model, 281
  cycle periods, 283
  using cycling, 280–281
  glucose feed concentration, 282, 282f
  metabolite production, 281
  specific growth equation, 281
Sulfur dioxide oxidation (SO_2 oxidation), 66–67
  cycle amplitude with symmetrical forcing, 68
  half-cycle, 69
  modeling
    Balzhinimaev model, 70
    kinetic model, 70

Russian researchers, 69–70
modulation studies, 67t
packed bed reactors, 69
rate enhancements, 67–68, 68f
step-change observations, 69
Sulfuric acid mists, 70
Sulfuric acid plants, 539
DC/DA system, 540f
RFR at Plovdiv lead-zinc smelter, 540
triple bed converter operation, 539
UNILAB design, 539
Sulfuric acid reduction (SO_3 reduction), 70
Supercritical chromatography, 683–684
Supercritical simulated moving bed chromatography (SCF-SMB), 683–684
Supported liquid phase catalyst (SLPC), 317–322
Surface activation, 33
Surface reconstruction, 33
on adsorption/desorption, 61
LEED measurements, 51t–56t
Surface storage model, 32–33, 376–377
Surface-reaction model, 58
Swelling data, 625
Swing reactors, 3, 637, 686
concepts and types, 638–640
dehydrogenation, 638
development, 637–638
models for, 640–642
PSR, 638
TSR, 663
Switching time, 599, 607, 631
on concentration profiles, 612f
effects, 611
influences, 618
parameters, 599–605, 610
productivity, 614
TMC, 620
uses, 620
Symmetric modulation, 6
Syngas, 228
burning, 230f
CO_2-free waste gas, 230
Cu_2O and Mn_2O_3, 230
exit gas composition and bed temperature, 231f
small scale pilot units, 228–229
Synthetic polymers, 172

T

TBA. *See* Tertiary butyl alcohol
TBR. *See* Trickle-bed reactor
TEA. *See* Triethanolamine
Temkin Pyzhev model, 32
Temperature modulation, 435–436
bang-bang, 441
experimental observations
amplitude and cycling frequency effect, 455, 457f
CO conversion, 455–456
CO oxidation, 454–455
CO_2 concentration comparison, 455f
cycle split, 457–458, 460f
enhancement dependence of rate CO oxidation, 457, 460f
enhancement under temperature modulation, 454–455
ether concentrations response, 454f
using Hansen microreactor, 456–457
isopropanol concentrations response, 454f
isopropanol dehydration, 453–454
modulation frequency effect on fractional enhancement, 459f
off-gas CO_2 concentration comparison, 455, 456f
propene concentrations response, 454f
rate enhancement from temperature modulation, 457
surface of CO oxidation rate enhancement, 457, 459f
temperature modulation of CO oxidation, 456, 458f
experimental studies
amplification phenomenon explanation, 443
with conventional laboratory equipment, 439
experimental system and reactor, 441f
gain in amplitude of temperature variations, 442f
periodic temperature disturbance amplification, 442–443
reactant inhibition effect, 443f
simulated amplification in packed bed reactors, 443
temperature amplification investigation, 441–442
temperature modulation question, 439–440
time-average rate of CO oxidation, 440–441
measuring temperatures, 445–449
micro devices for temperature modulation, 445, 449f
microchannel foils, 445
microreactors, 449
catalysts use in, 451
chip-based, 451–452
cylindrical, 452
designs and performance characteristics, 450t
experimental procedure with, 452–453
experimental studies with, 444–445
FTC system, 444–445
Jensen Reactor, 452f
microstructure device effects, 445
modified microreactor before assembly, 451f
non-homogeneous temperature distribution, 449–450
Sotowa Reactor, 452, 453f
temperature change of carrier, 450–451
reaction rate increasing mechanism, 459–460
CO adsorption-oxidation sequence, 461
composition forcing, 459–460
Langmuir-Hinshelwood model, 461
simulation studies, 437–438
cyclic temperature variations, 438
Langmuir-Hinshelwood mechanism, 438–439
numerical values, 439
simulated reaction rate comparison, 439f, 440f
temperature modulation of CSTR, 438
stainless steel micro-channel reactor, 445
temperature distribution, 449
temperature modulation experiments, 446t–448t

Temperature modulation (*Continued*)
theoretical studies
concentration-temperature trajectory, 438f
consecutive reaction in CSTR, 437
dynamic behavior, 437
exit concentrations, 437f
quasi-steady state operation, 436
relaxed steady state operation, 436–437
of trickle beds, 444
Temperature modulation, 682
combination of precursor addition, 682
inducer concentrations, 682–683
Temperature programmed desorption (TPD), 369
Temperature programmed reaction (TPR), 369
Temperature swing adsorbers (TSA), 660
Temperature swing reactor (TSR), 562, 657
See also Pressure swing reactor (PSR)
application, 676
computational considerations, 642
ODE, 642
periodic operation, 642
shooting methods, 642
equilibrium limited reactions, 674
experimental studies, 663–664
Boudouard reaction, 670
carbonaceous materials, 667–668
carbonate formation, 668
CO content, 667
comparison, 670f
dolomite, 668
HCN synthesis, 670
hydrogen production, 667
oxidative coupling, 670
packed bed reactor, 664
pilot scale reactor, 670
sorbent, 669
two calcination-carbonation cycles, 668–669, 668f
vehicle-based H_2 production, 669–670
experimental work, 675–676
modeling, 658–659
and PSR combination
adsorbate concentration on hydrotalcite, 672f
high purity H_2 production, 671
mixed catalyst and adsorbent bed, 673f
product gas composition, 675f
regeneration of adsorbent, 671
simulation studies, 671–674
reactor models
boundary and initial conditions, 642
cell models, 641
Damköhler number, 641
dimensionless variables, parameters and groups, 642t
material, momentum and energy balances, 640
partial pressures, 640
PDE models, 641–642
phenomenological models, 642
pseudo-homogeneous, one-dimensional model equations, 641t
reactor networks, 663, 664f
reactor types
with catalyst and adsorbent isolation, 659f
circulating fluidized bed, 657, 658f
operated fixed bed systems, 657
rotating bed reactors and catalyst-adsorbent systems, 658
rotating bed TSR with catalyst and adsorbent, 660f
segregation, 657–658
two-section, rotating bed TSR, 659f
scale-up of experimental units, 674–675
system simulation, 660, 661t
Claus reaction, oxidiation, 660
equilibrium conversion, 662f
HCN synthesis, 663
TSA, 660
two-bed TSR, 662, 662f
Temporal synchronization, 61
Tertiary butyl alcohol (TBA), 610–611
Tetrahydrodicyclopentadiene (THDCPD), 481–482
Tetrahydrofuran (THF), 192–193
THDCPD. *See* Tetrahydrodicyclopentadiene
Thermal reactors, 523–525
See also Periodic flow reversal
application to non-catalytic combustion processes, 525
water side of reverse-flow combustor, 527f
THF. *See* Tetrahydrofuran
Thiele modulus
compressible flow continuity equation, 416–420
conversion enhancement, 422
sinusoidal pressure modulation amplitude, 420f
steady-state effectiveness factor, 421
value, 479
Three bed simulated loop reactor, 137f
Three-bed simulated moving bed reactor, 633f
Three-dimensional printed circuit boards, 254
Three-way catalyst (TWC), 70, 141–142
activity loss, 161
CeO_2 in, 160–161
muffler, 128
performance in CO oxidation, 157
structure, 142–143, 143f
Thyomethyl-β-galactoside (TMG), 682–683
Tiny short-lived bubbles, 427–428
Tissue plasminogen activator (TPA), 292–293
TMB. *See* True moving bed
TMG. *See* Thyomethyl-b-galactoside
Toluene hydrogenation, 45
TPA. *See* Tissue plasminogen activator
TPD. *See* Temperature programmed desorption
TPR. *See* Temperature programmed reaction
Transient mixer model, 425–426
Travelling waves
axially located thermocouples, 76f
in packed bed reactors, 77
wrong-way behavior, 76
Yakhnin-Menzinger simulations, 76–77
Trickle beds, 463–464
temperature modulation, 444
Trickle-bed reactor (TBR), 464
base-peak liquid flow rate modulation, 478–479
LFM parameter effect, 479
maldistribution in bed, 479
CYCLOP Project, 480

CHISA, 480
fast mode cycling, 481
investigations, 480–481
liquid flow modulation
effect on activity or stability, 479–480
under liquid-reactant limiting condition, 479
periodic operation, 474
inlet liquid volumetric flow rate, 474–475
liquid flow rate effect, 474–475, 475f
on-off operation, 474–475
PO modeling, 487–488
catalyst pellet behavior, 488–489
dynamic model, 488
dynamic non-isothermal TBR model, 489–490
external efficiency enhancement, 489, 490f
LFM impact on transient concentration profiles, 493
mathematical model of catalytic hydrogenation, 490–491
modeling slow flow modulation, 488
on-off LFM raises conversion, 489
pellet-to-pellet variation, 489
predicted and experimental temperatures, 492, 492f
predicted and performance of reactor, 492f
predicted mean conversion, 490–491, 491f
simplified particle scale model, 489f
styrene hydrogenation in toluene solution, 491
rate or conversion enhancement, 475
AMS hydrogenation rate, 476, 477f, 478
dynamic model for SO_2 oxidation, 476
scatter of data, 475–476
slow bang-bang cycling, 475, 476f
temperature profiles, 475–478, 477f, 478f
steady-state operation
limiting reactant, 464
reaction-transport-wetting phenomena, 464
spatial segregation, 464, 474
studies on periodic operation, 465t–473t
TBR performance enhancement, 464
Triethanolamine (TEA), 352
True moving bed (TMB), 562–563
Truffer and Renken model, 324
TSA. *See* Temperature swing adsorbers
TSR. *See* Temperature swing reactor
Turbulent surface oscillations, 58–59
TWC. *See* Three-way catalyst
Two-part cycling, 71
Two-site models, 375–376, 376f
Twofold enhancement, 62–63

U

Ultrasonic excitation, 415, 427
Ultrasonic irradiation, 427–428
collapsing bubble cavities, 429f
Ultrasonic waves. *See* Pressure oscillation
Unipolar cycle parameters, 255
USS1 (unsteady state), 474–475
UV-LED, 688

V

Vacuum-jacketed tubular reactor, 76–77
Vanadia catalysts, 69–70
Vibrational control. *See* Chattering control
VOC oxidation. *See* Volatile organic compound oxidation
Volatile organic compound mitigation (VOC mitigation)
incinerators, 538–539
RFR control, 539
Volatile organic compound oxidation (VOC oxidation), 511
See also Periodic flow reversal
applicability of one-dimensional RFR model, 515
autonomous control features, 511
commercial designs, 511
industry uses burners, 511–515
RFR design for, 511
Volterra series, 389–390
approximate calculations, 406
NLFR, 389–390
*n*th elements, 390–392

W

Wall-cooled RFR
aperiodic operation, 509
concentration profile evolution, 510f
dynamic behavior complexity, 509
map of regions with different dynamic behavior, 509f
temperature profile evolution, 510f
two-dimensional model, 509
Water purge, 629–630
Wheelock's gypsum, 211–212

X

Xiu simulation, 650–651
Xuzhou, 227–228

Y

Yallourn coal, 217

Z

Ziegler-Natta catalyst, 186, 193
Zone plates, 254–255

Edwards Brothers Malloy
Thorofare, NJ USA
December 26, 2012